T0133137

SYSTEM-BASED VISION FOR STRATEGIC AND CREATIVE DESIGN

PROCEEDINGS OF THE SECOND INTERNATIONAL CONFERENCE ON STRUCTURAL AND CONSTRUCTION ENGINEERING, 23–26 SEPTEMBER 2003, ROME, ITALY

System-based Vision for Strategic and Creative Design

Edited by

Franco Bontempi
University of Rome "La Sapienza", Rome, Italy

Volume 2

A.A. BALKEMA PUBLISHERS   LISSE / ABINGDON / EXTON (PA) / TOKYO

Published by: A.A. Balkema, a member of Swets & Zeitlinger Publishers
            www.balkema.nl and www.szp.swets.nl

For the complete set of three volumes: ISBN 90 5809 599 1
Volume 1: 90 5809 600 9
Volume 2: 90 5809 601 7
Volume 3: 90 5809 630 0

Printed in the Netherlands

System-based Vision for Strategic and Creative Design, Bontempi (ed.)
© 2003 Swets & Zeitlinger, Lisse, ISBN 90 5809 599 1

Table of Contents

4. Cost control and productivity management

5. Human factor, social–economic constraints on the design process

6. Structural optimization and evolutionary procedures

7. Shaping structures and form finding architectures

8. Issues in computational mechanics of structures

9. Advanced modelling and non-linear analysis of concrete structures

10. Concrete and masonry structures

11. Steel structures

Volume 2

12. Bridges and special structures

13. Precast structures

14. Earthquake and seismic engineering

15. Geotechnical engineering and tunnelling

19. Innovative methods for repair and strengthening structures

20. Artificial intelligence in civil engineering

21. Knowledge management

22. Quality and excellence in constructions

23. Sustainable engineering

24. Life cycles assessment

25. New didactical strategies and methods in higher technical education

26. Durability analysis and lifetime assessment

Volume 3

27. High performance concrete

28. Concrete properties and technology

29. Innovative tools for structural design

30. Issues in the theory of elasticity

31. Special session on "Structural monitoring and control"

32. Special session on "New didactical strategies"

33. Special session on "Information and decision systems in construction project management"

34. Special session on "Innovative materials and innovative use of materials in structures"

35. Invited session on "Recent advances in wind engineering"

36. Special session on "Geomechanics aspects of slopes excavations and constructions"

37. Special session on "Fastening technologies for structural connections"

38. Special session on "Advanced conceptual tools for analysis long suspended bridges"

Preface

The present Proceedings collect over 360 papers that will be presented during ISEC-02, the 2nd International Conference on Structural and Construction Engineering, hold in Rome, Italy, on September 23–26, 2003.

The Theme of the Conference, *System-based vision for strategic and creative design*, includes the following concepts:

– *systemic framework*: it includes the capacity to see all the aspects and the connections of any problem and its solution;
– *creative work*: the tension to develop something new;
– *design*: one of the methods to improve the world and make it better.

The current practice and research in structural and construction engineering is characterized by increasing levels of complexity and interaction.

This is due to several reasons. First, the transition to a global, high technology environment demands sound safety requirements in all aspects of human life and activities. With respect to what concerns buildings, structures and infrastructures, a century of continuous progress in knowledge, materials and technology should have reduced occurrences of damage, failure, and misconception by large magnitudes. In spite of this expectation, small and large structural and construction deficiencies are still common episodes. This is, probably, due to the more ambitious objectives our society intends to pursue. But, without doubt, there are still problems and mismatches along the engineering path from concept development to practical realization. Secondly, there is special attention being paid in the world today to the interaction, full of uncertainties and constraints, of the structure with the environment. In this sense, competitiveness and sustainability require a systems approach in which research activities support the development of coherent, interconnected and ecologically-efficient civil engineering structural systems, responding to both market and social needs. Finally, there is the necessity to answer to socioeconomic needs, by stimulating holistic approaches and heuristic techniques, by strengthening the innovative capacity, and by fostering the creation of businesses and services built on emerging technologies and market opportunities. Research will turn into environmentally and consumer friendly processes, products, and services and will contribute to improve the quality of life and working conditions.

With this point of view, the main objective of the Conference will be to define knowledge and technologies needed to design and develop project processes and produce high-quality, competitive, environment- and consumer-friendly structures and constructed facilities. This goal is clearly connected with the development and reuse of quality materials, excellence in construction management, and reliable measurement and testing methods.

Franco Bontempi

Professor of Structural Analysis and Design,
Faculty of Engineering, University of Rome "La Sapienza", Rome (ITALY)
Postgraduate School of Reinforced Concrete Structures "F.lli Pesenti",
Polytechnic of Milan, Milan (ITALY)

Acknowledgements

The Editor gratefully acknowledges the promotion, the sponsorship, and the endorsement of the following organizations:

Promoting Association:
– CTE, Italian Association for Building Industrialization

Co-sponsors:
– AICAP, Italian Association for Reinforced and Prestressed Concrete
– CIB, International Council for Research and Innovation in Building and Construction
– ACI, American Concrete Institute
– ASCE, American Society of Civil Engineers
– IABMAS, International Association for Bridge Maintenance and Safety

Endorsers:
– Faculty of Engineering of the University of Rome "La Sapienza"

From an institutional point of view, the following persons will be gratefully remembered: Giandomenico Toniolo, President of CTE, Emanuele Filiberto Radogna, President of AICAP, Giselda Barina, Executive Secretary – CTE, Vivetta Bianconi, Executive Secretary of AICAP, Tullio Bucciarelli, Chairman of the Faculty of Engineering of the University of Rome "La Sapienza", Fabrizio Vestroni, Head of the Department of Structural Engineering, Amarjit Singh, Chair of ISEC-01.

The specific financial support of University of Rome "La Sapienza", Ministry of Instruction, University and Research (MIUR), Societa' Stretto di Messina S.p.A. and HILTI Italia S.p.A. are gratefully recognized.

The scientific framework of the Conference finds its origin within interesting discussions with Remo Calzona, Fabio Casciati and Pier Giorgio Malerba, whose experience and advices are, for the Editor, of the utmost importance.

The conceptual organization and the operative development of such a complex and large international conference couldn't have been possible without the outstanding commitment of several people. Specifically, the Editor wants to recognize the contribute, always made with efficacy and positive tense, of the following persons, at the same time friends and colleagues: Fabio Biondini, Luciano Catallo, Pier Luigi Colombi, Ezio Dolara, Elsa Garavaglia, Cristina Jommi, Simone Loreti, Giuseppe Parlante, Flavio Petrilli, Paola Provenzano, Luca Sgambi, Maria Silvestri, Angelo Simone, Maria Laura Vergelli, Paola Tamburrini.

The Proceedings are dedicated to Professor Francesco Martinez y Cabrera.

Rome, September 2003

System-based Vision for Strategic and Creative Design, Bontempi (ed.)
© 2003 Swets & Zeitlinger, Lisse, ISBN 90 5809 599 1

Organisation

International Scientific and Technical Committee:

Franco Bontempi, Chair, University of Rome "La Sapienza", Rome, Italy

Amarjit Singh, Chair of past ISEC Conference, University of Hawaii, USA
Hojjat Adeli, the Ohio State University, USA
Chimay J. Anumba, Loughborough University, UK
Ghassan Aouad, the University of Salford, UK
Larry Bergman, University of Illinois at Urbana-Champaign, USA
Fabio Casciati, University of Pavia, Italy
John Christian, University of New Brunswick, Canada
Richard Fellows, the University of Hong Kong, Hong Kong
Dan Frangopol, University of Colorado at Boulder, USA
Ian Gilbert, New South Wales University, Australia
Paul Grundy, Monash University, Australia
Takashi Hara, Tokuyama University of Technology, Japan
Makarand Hastak, Purdue University, USA
Osama Ahmed Jannadi, King Fahd University of Petroleum & Minerals, Saudi Arabia
Vladimir Kristek, Czech Techinical University in Prague, Czeck Republic
In-Won Lee, Korea Advanced Institute of Science and Technology, Korea
Anita Liu, The University of Hong Kong, Hong Kong
Pier Giorgio Malerba, Technical University of Milan, Italy
Giuseppe Mancini, Technical University of Turin, Italy
Indu Patnaikuni, RMIT University, Australia
Takahiro Tamura, Tokuyama University of Technology, Japan
Francis Tin-Loi, New South Wales University, Australia
Ali Touran, Northeastern University, USA
Richard N. White, Cornell University, USA
Frank Yazdani, North Dakota State University, USA

Local Scientific and Technical Committee

Nicola Pio Belfiore, University of Rome "La Sapienza"
Remo Calzona, University of Rome "La Sapienza"
Claudio Ceccoli, University of Bologna
Carlo Cecere, University of Rome "La Sapienza"
Mario Como, University of Rome, "Tor Vergata"
Antonio D'Andrea, University of Rome "La Sapienza"
Bruno Della Bella, Precompressi Centro Nord S.p.A, Novara
Alessandro De Stefano, Technical University of Turin
Valter Esposti, ICITE Director, CIB Treasurer, S.Giuliano Milanese
Roberto Guercio, University of Rome "La Sapienza"
Luigi Annibale Materazzi, University of Perugia
Antonino Musso, University of Rome "Tor Vergata"
Emanuele Filiberto Radogna, University of Rome "La Sapienza", AICAP President
Alessandro Ranzo, University of Rome, "La Sapienza"

Luca Romano, Consultant Engineer, Albenga
Franco Rovelli, Gecofin Prefabbricati S.p.A, Verona
Marco Savoia, University of Bologna
Renato Sparacio, University of Napoli, "Federico II"
Paolo Spinelli, University of Florence
Franco Storelli, University of Rome "La Sapienza"
Giandomenico Toniolo, Technical University of Milan, CTE President

Local Organizing Committee

Giselda Barina, (Executive Secretary), CTE
Fabio Biondini, Technical University of Milan
Fabio Bongiorno, CTE
Luciano Catallo, University of Rome "La Sapienza"
Francesco Chillè, ENEL Hydro, Milan
Pier Luigi Colombi, Technical University of Milan
Ezio Dolara, University of Rome "La Sapienza"
Elsa Garavaglia, Technical University of Milan
Claudia Gomez, Consulting Engineer, Milan
Simone Loreti, University of Rome "La Sapienza"
Corrado Pecora, Consulting Engineer, Milan
Flavio Petrilli, Consulting Engineer, Rome
Paola Provenzano, University of Rome "Tor Vergata"
Luca Sgambi, University of Rome "La Sapienza"
Maria Silvestri, University of Rome "La Sapienza"
Angelo Simone, Delft University of Technology
Maria Laura Vergelli, University of Rome "La Sapienza"
Paola Tamburrini, University of Rome "La Sapienza"
Giuseppe Parlante, Content Manager and Web Master

12. Bridges and special structures

System-based Vision for Strategic and Creative Design, Bontempi (ed.)
© 2003 Swets & Zeitlinger, Lisse, ISBN 90 5809 599 1

Croatian experience in design of long span concrete bridges

J. Radic, G. Puz & I. Gukov
Faculty of Civil Engineering, Zagreb, Croatia

ABSTRACT: During recent years several hundreds kilometers of new highways have been constructed in Croatia, including some exceptional bridges. Experience gathered on the design and construction of some large structures can be summarized in some recommendations for future design.

1   ARCH BRIDGES

1.1   *Service performance*

Four arch bridges, the ˇibenik bridge, the Pag bridge and the Krk bridges (two arches) were built during the sixties and the seventies on the Adriatic coast in Croatia. The most famous is the Krk I bridge, which held the world record for twenty years. These bridges were designed and built in accordance with the regulations in force in late 1960's. Minimum statically admissible dimensions of all structural elements were utilised, in conjunction with a thin concrete cover. The performance in service of the four older arch bridges cannot be deemed satisfactory. Results of inspections and testing performed on these bridges confirmed degradation of structural parts. Structural defects were found at an early age and hence cannot be blamed primarily on the aggressive environment, but are attributed to conceptual design and errors and negligence on site.

The Maslenica highway bridge (Fig. 1), completed in 1997, belongs to the second generation of bridge design, with the severity of the aggressive maritime environment taken into account, based on the experience gained by in-situ investigations of previously mentioned Croatian arch bridges. Dimensions of structural elements were increased in order to avoid reinforcement congestion and increase durability. Structural details and cross sections were simplified, in order to minimize execution problems.

All these arches were erected utilizing the free cantilevering method with strong control of construction (Fig. 2).

Design errors can be summarized, as follows:

– Misinterpretation of the codes in designing very small concrete cover
– Extensive usage of precasting not supported by satisfactory joining details

Figure 1.   Maslenica bridge (Croatia, 1997, arch span 200 m).

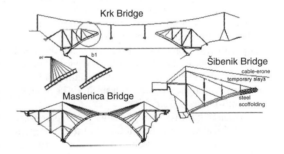

Figure 2.   Sketches of arch construction.

– Placing the precast prestressed girders of the superstructure directly on tops of piers with no fixed connection and no bearings
– Utilization of half-joints in the superstructure at middle supports
– Unsatisfactory drainage systems
– No waterproofing installed

Construction faults are the following:

– Permeable formwork
– Poor grouting of prestressed tendons
– Poor workmanship in slip-forming procedure
– Unsatisfactory concrete grade

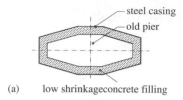

(a)   low shrinkageconcrete filling

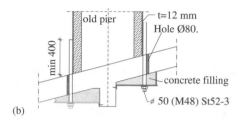

(b)

Figure 3.   Pag bridge piers repair. (a) Cross-section of the repaired pier (b) connection of the repaired pier to the arch.

Table 1.   Basic data of Croatian arch bridges.

Name	Year	Span (m)	Rise (m)	Rise/ span	Road width (m)	Max. stresses (N/mm^2)
Krk I	1979	390	60.0	0.15	11.1	19.7
Šibenik	1965	246	30.8	0.13	10.5	12.6
Krk II	1979	244	47.5	0.20	11.1	16.4
Maslenica	1997	200	65.0	0.33	20.4	13.0
Pag	1968	193	27.6	0.14	9.0	12.0

Figure 4.   (a) Bending moments in arch applicable traditional method. (b) Bending moments applicable Inverted load method.

All of these errors contributed to the quick deterioration of these bridges. The worst deterioration is the reinforcement corrosion caused by high chloride content, which penetrated into concrete.

All bridges have been investigated, and the Pag bridge was completely reconstructed. Repairs of piers and the superstructure started in 1978. Various techniques and procedures were tried but corrosion had not been stopped. In 1991 complete repair of the arch was performed, consisting of removing the damaged concrete surface by a hydro demolition device, of grouting all visible cracks, and of placing an additional reinforcement mesh on all outer arch surfaces, covered by a fine mortar layer 4 cm thick.

Various solutions for replacement or repair were studied. Finally, a radical solution had to be adopted by replacing the prestressed concrete superstructure with the new steel one, and by protecting all piers with a 12 mm steel outside shell (Fig. 3). The intermediate space between the steel shell and existing piers was filled with shrinkage-free concrete, resulting in composite action (Savor et al. 2001).

Performed numerical analyses of the arch stability revealed a very small margin of safety for eventual unexpected actions during service life (Table 1).

This is a lesson to be taken seriously, when other large Croatian arch bridges, notably the Krk bridge, designed on similar concepts, are due for thorough reconstruction.

1.2   The form (shape) of the arch

The arch is natural and appropriate structural solution, that is both aesthetically pleasing and clearly shows its function. According to some sources (Wooley 1963), first arches were built even before the great Flood, 4000 years B.C. A written record of the fact that the size of horizontal force in arch is dependent on the shape of the vault dates back to Leonardo da Vinci. With the appearance of reinforced concrete and the increase in bridge span, the arches have become more slender. Most dominant arches are those within range from 50 to 250 m. The dominant system for concrete arch bridges is fixed arch with the superstructure above.

Concrete is a material which compressive strength is by several times greater than the tensile strength. Arch concrete bridges are designed to transfer external loads primarily by compression. This is achieved by optimizing the shape of the arch axis. Bending moments are not desirable, and in case the arch axis is shaped in an optimum way, can be kept to a minimum. In case of concrete bridges where the proportion of traffic load is no more than 8 to 10 percent, the arch axis is usually determined for the dead weight loading.

The traffic load may cause tensile stresses in concrete, particularly at the point of force concentration at pier locations. In order to provide for an appropriate durability of a structures situated in an aggressive environment, it is desirable to make all cross sections in compression, or, at least, the tensile stress should exceed the allowed tensile stress.

Most analytical methods have replaced concentrated load by an appropriate uniform load along the arch and the solution of differential equation is given in form of a mathematical function (Faccio et al. 1995). Graphical methods provide more accurate solutions. However, loads in present day of arch bridges are not uniform. Spans of the superstructure become larger, and loads are transferred via piers to the arch in discrete points.

For that reason, it is no longer possible to determine the equation of the entire arch in closed form, as the curve is discontinuous at pier positions. On the other hand, availability of the complex numerical analysis asks for new solutions.

If we assume that only one force acts on the arch in the middle of the span and that the arch has no weight, the structural system is of triangular shape with no bending moment (Fig. 5a). If we wish to set a curve, it should be selected in such a way that it deviates as little as possible from this solution. (—). For the load of two forces, the solution is in a trapezoidal shape (Fig. 5b).

Both described examples could confirm they are in fact balanced positions of a catenary, hence we conclude: the problem of finding a thrust line of the arch is analogous to the problem of looking for the balanced position of the catenary subjected to similar forces, but of opposite direction. Therefore, we are looking for the structural system in which we have longitudinal forces only, without bending moments, what is in fact the catenary (Gukov et al. 2001). The problem of determining an optimum shape of axis for the known span and arch rise consists of determining coordinates for pier position points and determining set of equations for curves that link these points. Based on this idea, a computer program for finding an optimum bridge arch curve was created. For the selected bridge span and arch rise, the program finds a solution of selected accuracy by an iterative procedure. On several concrete arch bridges a comparison was performed, which proved the applicability of the method.

Results of parametric studies, utilizing both classical and new method confirmed significant differences in bending moments magnitude (up to four times) in favor of the new method. Moreover, there is a large difference in results under theory of 1st and 2nd order. Arch bridges should be calculated according to 2nd order theory, precisely seeking balance on a deformed system.

2 BEAM BRIDGES

One of today's most competitive procedures for building beam concrete lager bridges is free cantilever method. Segments are concreted on the site, or previously manufactured somewhere else, in order to be installed in appropriate place within construction. After manufacturing or assembling the sections, they bond with the previous ones by prestressing, becoming a part of the superstructure.

The main problem with the calculations of these bridges is not in the control of stresses and selection of the necessary number of tendons, but in assessment of the deflections necessary for construction camber (CEB 1997). Measured values of executed bridges in various parts of the world, show substantial increase compared with computed ones (Fig. 8). This increase is especially significant in the cases of monolithic construction, which, due to durability requirements and problems of working joints at the assembling construction, is most frequently applied.

Figure 7.   Bridge construction through the cantilever method. The Dubrovnik Bridge.

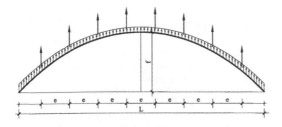

Figure 5.   Ideal shapes of "arches" for either one (a) or two concentrated forces (b).

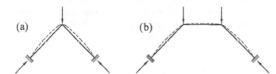

Figure 6.   Inverted load method.

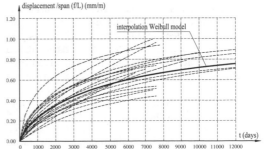

Figure 8.   Summary diagram of the change of the deflection in time.

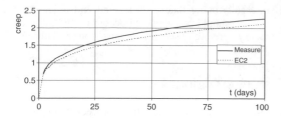

Figure 9. Comparison of creep coefficient according to EC2 regulation and measured values.

For the assessment of deflections it is necessary to know not only the final values of the creep and shrinkage coefficients of the concrete, but also their values in different time intervals. It is necessary to determine these coefficients as accurately as possible, because in the construction camber. Prior to the construction of Rijeka Dubrovacka values of creep and shrinkage coefficients were determined for the concrete mix to be applied.

Results differ from those ordained in EC2. Creep and shrinkage coefficients for the first 100 days were 11–23% bigger then EC2 values depending on the moment of load applying (Fig. 9).

Creep and shrinkage coefficient values and their time changes haven't been processed enough even in the latest regulations. The regulations are based on the old tests and are approximately adequate for the lower and middle class concrete, whereas at the high quality concrete >C45, the mix of which has changed in the last years, especially because of the plastificator and additive application, the influence of cement quality and water-cement factor should be included in their calculation.

Precast elements are normally prestressed to higher degree, utilize full. It is recommended to use prestressing even for the monolithically constructed bridges, with additional condition that the maximum stress (especially above the middle supports) for service stressing combinations (dead weight, prestressing, temperature and live load) does not exceed −2 MPa, i.e. materials should not be loaded to the limits of the allowed compression and tension capacity. It is necessary to prestress tendons in the process as late as possible when the concrete is at least three days old, and to reinforce places of anchoring area with larger reinforcement than prescribed by the codes for design concrete quality.

We normally design larger percentages of nonprestressed reinforcement then required by the codes. Joints between cross-section elements (i.e. between slab and web) should contain more reinforcement, (i.e. 6 bars φ25).

Consideration of bridge overhaul in the projecting phase is usually never done. Additional ducts are provided in construction stage to be utilized if some problems encountered with tendon installation. These ducts should be post grouted in any case. For future strengthening provisions for the installation of external prestressing are to be made at the design stage already.

If piers are not strong enough support non-symmetrical execution (one segment more on one side), which is a standard case, it is necessary to erect temporary supports. If double-wall piers are design temporary supports may be omitted. Hinges in span are to be avoided because of uncontrollable deflections, 80% larger then those of similar continuous bridges under the same conditions. In seismically very active region they should be definitely prohibited. The standard design of a bridge superstructure starts with the choice of an adequate cross section based on experience. The tendons layout and total quantity is determined in the second phase. We prefer another sequence. The assumption of the tendon layout and subsequently iteratively determining the adequate superstructure cross section, fulfilling all the required criteria.

3 CONCLUSIONS

During our work we came to the conclusion that the serviceability criteria always governs the final design, and not the ultimate limit state consideration. The performance in service of some older Croatian concrete bridges cannot be deemed satisfactory, because of degradation of structural parts found at an early age. Majority of bridges built in last two decades has been designed for durability, which means:

– Increased thickness and simplified shape of cross-sections
– Improved and simplified structural detailing
– Utilization of higher concrete grades (up to C 60), with water to cement ratio 0.35 to 0.38
– Increase in percentage of non-prestressed reinforcement.

Most of these bridges are designed according to DIN regulations and recommendations. Inspections performed revealed no significant damages on such bridges.

REFERENCES

ACI Committee 343. 1995. Analysis and Design of Reinforced Concrete Bridge Structures, *American Concrete Institute*, Detroit.
CEB, 1997, Serviceability models, Stuttgart.
Candrlić, V., Radic, J., Savor, Z. 1999. Design and Construction of the Maslenica Highway Bridge. Proc. Fib Symposium Structural Concrete – *The Bridge Between People*, Vol. II, Prague, 551–555.

E DIN 1045-1. 1998. Beton und Stahlbeton, *Bemessung und Ausführung*.

Gukov, I., Piculin, S., Puz, 2001. G. Selecting An Optimum Axis For The Concrete Bridge Arch. Proceedings of the 26. Conference on *Our World in Concrete and structures*, Singapore, pp. 255–260.

Radic, J., Djukan, P., Banjad, I. 1999. European Road Corridors in Croatia. *Österreichische Ingenieur- und Architekten-Zeitschrift* (OIAZ) 144, 3; pp.117–123

Radic, J., Savor, Z., Piculin, S., Puz, G. 2001. Large Concrete Arch Bridges in Croatia. Proc. ARCH'01, *Third international arch bridges conference*, Paris, pp. 49–58.

Savor, Z., Mujkanovic, N., Puž, G. 2001. The Pag bridge renovation – a case study. Proc. *Failures of concrete structures II* – International Conference, Bratislava, pp. 189–194.

System-based Vision for Strategic and Creative Design, Bontempi (ed.)
© 2003 Swets & Zeitlinger, Lisse, ISBN 90 5809 599 1

Hydrogen embrittlement of suspension bridge cable wires

K.M. Mahmoud
Director of Long Span Bridges, Hardesty & Hanover, New York City, USA

ABSTRACT: Deterioration of suspension bridge cables is demonstrated through stress corrosion cracking and hydrogen embrittlement. Stress corrosion cracking is usually associated with a cross-sectional area reduction of the wire. Hydrogen embrittlement, however, is associated with a considerable loss of ductility, but with an insignificant loss in the wire's cross-sectional area. Very little work has been focused on the assessment of deterioration of bridge wires due to hydrogen embrittlement. This paper outlines a newly developed model, quantifying the wire degradation due to hydrogen embrittlement.

1 INTRODUCTION

The safety factor evaluation of suspension bridges is usually evaluated without accounting for the effect of corrosive environments, sustained and cyclic loadings. Inspection reports and laboratory tests, however, show significant loss of ductility and reduced fatigue strength in old wires. In addition to the cross-sectional reduction associated with general corrosion, other forms of degradation compromise the strength and ductility of cable wires. Stress corrosion cracking, pitting, corrosion fatigue and hydrogen embrittlement are some of the major factors in reducing cable life. The presence of broken/cracked wires in cables and the loss of ductility associated with the corrosion phenomenon, are evident of a higher rate of deterioration of cable strength than that inferred from only the loss of wire cross-sectional area. Therefore, the conventional methods of calculating cable strength by summing the strength of all wires, without accounting for the loss of ductility of the wire material, obviously overestimate the true factor of safety for suspension bridge cables. Establishing the cause of bridge wire fracture, whether it is due to overload, metallurgical factors or environment-assisted cracking is a crucial step in the determination of the failure mechanism. The term environment-assisted cracking encompasses different forms of degradation of a bridge wire, such as stress corrosion, corrosion fatigue, and hydrogen embrittlement, which is also known as hydrogen-assisted cracking.

This paper discusses a newly developed model, quantifying the wire degradation due to hydrogen embrittlement is outlined. Crack growth rate data for suspension bridge wires is not currently available in terms of fracture parameters. However, the model validation is fulfilled through a comparison of the model's crack growth rate prediction, due to hydrogenation of the metal, versus the crack growth rate of different high-strength steel alloys, cited in the literature. Promising agreement is obtained when comparing the model's prediction with experimental data, by plotting the crack growth rate, (dl/dt) or (dl/dN), versus the stress intensity factor, (K_I), where t is the time variable and N is number of cycles.

2 FORMS OF WIRE DEGRADATION

2.1 *Stress corrosion cracking*

Stress corrosion cracking is associated with the effect of a static tensile stress coupled with corrosive environment. Failures under tensile stresses well below the yield strength of the material are characteristic in corrosive environments. The mechanics of stress corrosion cracking depends on metallurgical factors, environmental conditions and on the crack geometry, as well as stress state (Jones 1992). The corrosion severity of a bridge wire is reflected by the following criterion (Hopwood & Havens 1984):

Stage I: the zinc coating of wires is oxidized to form zinc hydroxide, known as "white rust".

Stage II: the wire cross-section is completely covered by white rust.

Stage III: appearance of a small amount of (20–30% of wire surface area) of ferrous corrosion due to broken zinc coating.

Stage IV: the wire cross-section is completely covered with ferrous corrosion.

Corrosion damage cannot be quantified using the above definition. These measures of corrosion are qualitative and don't provide a quantitative assessment for the corrosion damage of bridge wires. It should also be noted that these stages of corrosion do not model the hydrogen embrittlement process.

2.2 *Hydrogen embrittlement*

With a specific critical concentration, hydrogen can degrade the fracture resistance of the high strength wire material. With the development of pitting, the galvanic action at the pits generates free hydrogen, which causes embrittlement and eventual reduction in fracture toughness. Atomic hydrogen can diffuse into the interior of the wire weakens the inter-atomic bond of the wire high strength steel. Accurate modeling of hydrogen diffusion into the interior of the wire material is of paramount importance in the understanding of the mechanism of hydrogen-assisted cracking. Steps encountered by hydrogen during its diffusion through steel in an H_2S-CO_2-H_2O environment involves three important steps (Brickell et al. 1964):

1. Diffusible hydrogen generation on the surface $(xFe + H_2S \Rightarrow Fe_xS + 2 [H]_{\text{diffusible}})$.
2. Passage of hydrogen through the interior of the metal.
3. Emergence of hydrogen through the surface of the metal.

The earliest examples of a unique fracture mechanism, later to be known as *delayed failure* or hydrogen-stress cracking, were observed in cadmium-plated steel aircraft parts with yield strengths of 1100 to 1240 Mpa (McEowen & Elsea 1965). A characteristic of these failures was their occurrence at relatively low static load, even though the same part previously had withstood much higher dynamic loads. The fractures appeared to be brittle, but tensile specimens machined from failed parts exhibited normal ductility and strength in the conventional tensile test. It was also characteristic of these failures that the parts had absorbed hydrogen during some manufacturing process. High strength steel cable wires demonstrate a similar behavior.

3 ENVIRONMENT-MATERIAL SYSTEM

Hydrogen embrittles high strength steel, always decreasing the ductility and strength (the greater the concentration of hydrogen, the less the strength). The capability of hydrogen for dissolution and diffusion in metals is significantly retarded by the adsorption of oxygen and the presence of oxide films. The inhibiting effect of oxygen in hydrogen and water is explained by the much greater chemical activity of the oxygen-metal pair than the hydrogen-metal pair. Therefore, a thin oxide film is formed on the fresh surface of metal at the crack tip, protecting the metal from contact with hydrogen. When the oxygen feed stops, the process of reduction of oxygen by hydrogen or dissolution of the film by water begins to predominate (Johnson, H.H. & Paris, P.C. 1968).

The most favorable conditions for the occurrence of processes of absorption of hydrogen are found at the crack tip, in the small area of fresh metal surface not covered with a protective oxide film. The influence of moisture and hydrogen, therefore, is most significant in the process of sub-critical crack growth; where the incubation period is always highly dependent on the state of the surface of the smooth specimen and, when there is a notch, on its sharpness.

The rate of sub-critical crack growth dl/dt is a function of the stress intensity factor, K_I:

$$\frac{dl}{dt} = f(K_I) \tag{1}$$

The function $f(K_I)$ is equal to zero when $K_I \leqslant K_{Iscc}$, while when $K_I > K_{Iscc}$, it is increasing monotonically with increasing K_I. The ratio of K_{Iscc}/K_I for ductile metals is usually close to 1, while for high strength and brittle alloys it is much less than 1. The quantity K_{Iscc} is called the threshold stress intensity factor.

The validity of Equation 1 has been experimentally confirmed (Johnson & Willner 1965), where a center-cracked H-11 steel specimen was immersed in a water environment. In the process of crack growth, the load was changed so that the stress intensity factor remained constant. The results indicated that the rate of corrosion crack growth remained constant for the entire length of the crack, without correlation between crack growth rate and the stress in the specimen. In this experiment, the crack growth is explained, most likely, by the metal hydrogenation.

4 HYDROGEN DIFFUSION MODEL

The coefficient of diffusion of hydrogen in metals depends on the absolute temperature T as follows:

$$D = D_0 e^{-UR/T} \tag{2}$$

where U is the activation energy of the process, R is the Universal Gas Constant and D_0 is a constant of the metal.

The neighborhood of a crack tip in an elastic-plastic body is being considered, as shown in Figure 1.

The quantity $2v_0$ represents the crack opening displacement. Assume that the cavity of the crack is filled with a liquid or gaseous medium containing hydrogen. The atoms of hydrogen reach the material primarily through the surface of the fresh metal near point O. The plastic zone is not shown in Figure 1, its dimension being significantly greater than $2v_0$. At infinity, the body is subjected to uniform tension, which is fully described by the stress intensity factor K_I.

Let C represent the concentration of protons (hydrogen), varying from point to point in the body. The maximum value of concentration, equal to C_0 will be, obviously, at the crack tip (at point O). At infinity, the concentration of hydrogen will be assumed equal to 0. The diffusing hydrogen has a double effect. On one hand, it anchors dislocations, making the material brittle and decreasing its capability for deformation, thus decreasing the fracture toughness. On the other hand, hydrogen spreads the atomic lattice of the solid. The following mechanisms of local failure are possible in principle: (1) failure due to internal compressive stresses, (2) failure of the brittle plastic zone due to the external load, and (3) failure due to tensile stresses in the elastic area.

The assumption made here is that the crack growth takes place in sequential jumps, whose size is significantly greater than the crack tip opening displacement, $2v_0$. The crack is considered a mathematical slit (i.e. of zero thickness). To define concentration C, a boundary-value problem is formulated such that: at the tip of a stationary semi-infinite crack in a solid along $y = 0, x < 0$ at the initial time $t = 0$, a constantly acting point source of protons (from galvanic action) of intensity Q is switched on. It is assumed that the quantity Q is independent of time and the sides of the cracks are not loaded. The problem is to find the distribution of the proton concentration, C, in the wire (Mahmoud, in prep.) The proton diffusion is described by:

$$\frac{\partial C}{\partial t} = \left(\frac{1}{r}\right)\left(\frac{\partial}{\partial r}\right)[rD(\frac{\partial C}{\partial r})] \tag{3}$$

where $0 < r < \infty, 0 < t < \infty$, and $r^2 = x^2 + y^2$.

For the semi-infinite linear crack considered, the solution is self-similar and the concentration C is

given by:

$$C = \left(\frac{Q}{4\pi D}\right).E_1\left(\frac{r^2}{4Dt}\right) \tag{4}$$

where $E_1 = \int_x^\infty e^{-u}(du/u)$ is an exponential integral function.

5 MODEL VERIFICATION

Wire test data is not currently available in terms of fracture parameters. However, model validation is achieved through comparing the theoretical model prediction for crack growth with experimental data on high strength steels in corrosive environments. Type 350-grade maraging high strength steel, with a yield point of $210\,kg/mm^2$ after aging at 415 $C°$ for 8 hours, was tested in a 3.5% aqueous solution of NaCl (Carter 1970). Figure 2 illustrates the crack growth rate (micro-mm/cycle) versus the stress intensity factor range for the theoretical model prediction and Carter's test data. Figure 3 demonstrates the model prediction

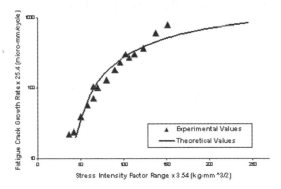

Figure 2. Crack growth rate versus stress intensity factor range for 350-maraging steel.

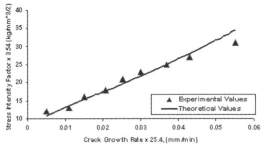

Figure 3. Crack growth rate versus stress intensity factor range for 13Cr-8Ni-2Mo steel.

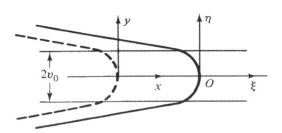

Figure 1. Crack tip in an elastic-plastic body.

for crack growth rate (mm/min) versus the stress intensity factor, compared with experimental data obtained for high strength steel by (Crooker & Lange 1969).

It is evident from both figures that the model prediction is in good agreement with the experimental data.

6 CONCLUSIONS

A new model for hydrogen embrittlement of high strength steel wire of bridge cables has been presented. The agreement between the theoretical model and experimental data is promising. It indicates that the model captures the mechanism of hydrogen embrittlement into high strength steel. It is clear that crack growth rate data for bridge cable wires is required for the proper application of the proposed model. The significance of having such data is to obtain the rate of crack extension with the specific environment-material-loading system at a given bridge. In other words, if a set of crack growth rate data is obtained, one would be able to obtain the fitting parameter for the bridge and determine the rate of crack growth due to the combined effect of the environment, bridge cable wrapping system and the loading. The latter is, of course, represented in the stress intensity factor.

REFERENCES

Brickell, W.F., Greco, E.C. & Sardisco, J.B. 1964. Corrosion of iron in an H2S-CO2-H2O system: Influence of (single iron) crystal orientation on the hydrogen penetration rate. *Corrosion*, 20 (7): 235–236.

Carter, C.S. 1970. The effect of heat treatment on the fracture toughness and subcritical crack growth characteristics of a 350-grade maraging steel. *Metallurgical Trans.*, 1: 1551.

Crooker, T.W. & Lange, E.A. 1969. Corrosion fatigue crack propagation studies of some new high strength structural steels. *Trans. ASME, J. Basic Engng., ser D*, 91: 570.

Hopwood, T. & Havens, J.H. 1984. Inspection prevention and remedy of suspension bridge cable corrosion problems. *Research Report UKTRP-84-15, Kentucky Transportation Research Program*. Lexington: University of Kentucky.

Johnson, H.H. & Paris, P.C. 1968. Sub-critical flaw Growth. *J. Engnr. Fracture Mech.*, 1: 33–45.

Johnson, H.H. & Willner, A.M. 1965. Moisture and stable crack growth in high strength steel. *Appl. Mater. Res.* 4: 34.

Jones, D. 1992. *Principles and prevention of corrosion*. New York: Macmillan.

Mahmoud, K.M. In preparation. Modeling hydrogen embrittlement in high strength steel wire of bridge cables.

McEowen, L.J. & Elsea, A.R. 1965. Behavior of high strength steels under cathodic protection. *Corrosion*, 21: 28–37.

System-based Vision for Strategic and Creative Design, Bontempi (ed.)
© 2003 Swets & Zeitlinger, Lisse, ISBN 90 5809 599 1

An universal approach to analysis and design of different cable structures

V. Kulbach
Tallinn Technical University, Tallinn, Estonia

ABSTRACT: Different prestressed cable systems for roof and bridge structures are under investigation in the report. Long-time research work at TTU has given reasons for preference of continuous models for analysis with use of relative deflection of the structure as the main unknown quantity in definitive system of equations. In spite of advantages of continuous analysis in some cases use of discrete calculation models is unavoidable, especially for suspension bridges and cable networks. Displacements of supporting structures under action of cable forces are taken into account immediately in definitive equations for both models. Determination of the stress–strain state of structures proceeds from non-linear conditions of equilibrium and equations of deformation compatibility of cables. In case of continuous analysis use of these equations brings us to a cubic equation (or system of equations) regarding the relative deflection (or deflection parameters). Displacements of cable supports are usually assumed as linear functions of cable forces. The main attention in our report is paid to prestressed cable trusses, girder-stiffened structures and cable-networks.

1 CALCULATION OF DISPLACEMENTS AND INNER FORCES OF AN ELASTIC CABLE

1.1 Cable as a spatial structure element

Determination of deflections and inner forces of a cable (Fig. 1) proceeds from nonlinear conditions of equilibrium and equations of deformation compatibility. Assuming axis x as the independent variable and y and z as dependent variables we may present the conditions of equilibrium of the cable in the form

$$\frac{dH}{dx} - (p_{ox} + p_x) = 0 \tag{1}$$

$$H\frac{d^2(y+v)}{dx^2} + \frac{dH}{dx}\frac{d(y+v)}{dx} - (p_{oy}+p_y) = 0 \tag{2}$$

$$H\frac{d^2(z+w)}{dx^2} + \frac{dH}{dx}\frac{d(z+w)}{dx} - (p_{oz}+p_z) = 0 \tag{3}$$

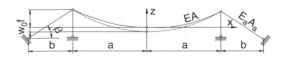

Figure 1. Single cable with inclined anchor cables.

where H = cable horizontal force after loading; w = the deflection function of the cable; p_{ox}, p_{oy}, p_{oz} = the load components applied to the cable in the state of prestressing; p_x, p_y, p_z = the additional load components.

The relative elongation of the cable element has the value

$$\varepsilon = \frac{ds}{ds_o} - 1 \tag{4}$$

where $ds_o = dx[1 + (\frac{dy}{dx})^2 + (\frac{dz}{dx})^2]^{1/2}$;

$$ds = dx[1 + 2\frac{du}{dx} + (\frac{du}{dx})^2 + (\frac{dy}{dx})^2 + 2\frac{dydv}{dxdx} + (\frac{dv}{dx})^2 + (\frac{dz}{dx})^2 + 2\frac{dzdw}{dxdx} + (\frac{dw}{dx})^2]^{1/2}$$

On the other hand according to the ratio of elastic deformation we have

$$\varepsilon = [(H-H_o) / (EA)] [1 + (\frac{dy}{dx})^2 + (\frac{dz}{dx})^2]^{1/2} \tag{5}$$

where EA = the stiffness of the cable in tension; H_o = cable initial force.

Equalizing Equations 4 and 5 we obtain the equation of deformation compatibility in the form

$$\frac{du}{dx} + \frac{dv}{dx}\left(\frac{dy}{dx} + \frac{1}{2}\frac{dv}{dx}\right) + \frac{dw}{dx}\left(\frac{dz}{dx} + \frac{1}{2}\frac{dw}{dx}\right)$$
$$= \frac{H - Ho}{EA}\left[1 + \left(\frac{dy}{dx}\right)^2 + \left(\frac{dz}{dx}\right)^2\right]^{1/2} \tag{6}$$

The system of non-linear Equations 1–3 and 6 with corresponding boundary conditions enables to determine the cable force H and the displacement components u, v and w.

1.2 Elastic plane cable under action of vertical loads

Determination of the stress–strain state of a plane cable proceeds from the system of Equations 3 and 6 with elimination of loads p_x and p_y and the variables y and v. In this case deflection w_o and the cable force H may be considered as the main unknowns; the initial value of the cable force is also to be calculated. The joined observation of Equations 3 and 6 brings us to the following expressions

$$H_o\frac{d^2z}{dx^2} = p_o \tag{7}$$

$$H\frac{d^2(z+w)}{dx^2} = p_o + p \tag{8}$$

$$\frac{du}{dx} + \frac{dw}{dx}\left(\frac{dz}{dx} + \frac{dw}{dx}\right) = \frac{H - Ho}{EA}\left[1 + \left(\frac{dz}{dx}\right)^2\right]^{3/2} \tag{9}$$

where p_o and p are initial and additional vertical loads of the cable.

For elimination of displacement u we have to integrate the Equation 9. Leaving aside the second power of du/dx as a very small value we may determine for the support with a guyed anchor cable (Fig. 3)

$$\int \frac{du}{dx}dx = (H - Ho)b/(E_aA_a\cos^3\beta) \tag{10}$$

where E_aA_a is the rigidity of the anchor cable in tension, β is the angle of its inclination.

After application of presented equations to the case of uniformly distributed load we get for determination of the relative deflection of the cable $\zeta_o = w_o/f$ a cubic equation

$$\zeta_o^3 + 3\,\zeta_o^2 + (2 + p_o*)\,\zeta = p*. \tag{11}$$

The cable force under action of initial and additional loads and the other factors may be calculated by expressions

$$H_o = p_o a^2/(2f); \quad H = H_o + \Phi\zeta_o(2 + \zeta_o);$$

$$p_o* = H_o/\Phi; \quad p* = P/\Phi; \quad P = pa^2/(2f);$$

$$\Phi = 2EAf^2/[3a^2(1 + \kappa + \theta)]; \quad \kappa = [2f^2/a^2 + 6f^4)]/$$

$$(5a^4); \quad \theta = Eab/(E_aA_a\,a\,\cos^3\beta), \tag{12}$$

where f = initial sag of the cable; p_o* = initial load parameter; $p*$ = additional load parameter; Φ = parameter of cable rigidity, κ = geometrical factor, θ = factor of rigidity of the anchor cable.

Given solution of Equations 7–11 corresponds to the exact values of deflections and cable forces with parabolic form of the cable before and after loading

$$z = f\left(\frac{x}{a}\right)^2 \text{ and } w = w_o\left[\left(\frac{x}{a}\right)^2 - 1\right]. \tag{13}$$

In case of approximation of the deflection function by

$$w = -w_0\cos\pi x/(2a) \tag{14}$$

with application of Galjorkin procedures we obtain the equation regarding the relative displacement in the form

$$\zeta_o^3 + 3\,\zeta_o^2 + [(2048/\pi^6 + 32/(3\pi^2)]\,\zeta_o$$
$$= (1024/\pi^5)\,p* \tag{15}$$

and regarding the cable force

$$H = H_o + (3/\pi)\,\Phi\,\zeta_o\,[2 + (\pi^3/32)\zeta_o]. \tag{16}$$

These values differ from the exact ones by unimportant quantities.

It is worth to mention, that the cubic Equation 11 may be used also for calculation of cable deflections in other load cases, using corresponding numeric factors given by Equation 12.

2 DOUBLE-CABLED PRESTRESSED STRUCTURE

In case of a structure consisting of the carrying and the stretching cables and vertical hangers (Fig. 2) the Equations 1–6 are to be applied to the both cables. For the initial load p_o in this case we have to consider the contact load between the cables. After loading the structure this contact load turns to a changed value p_c. Decisive system of equations may be presented after equalizing deflections of both cables in the

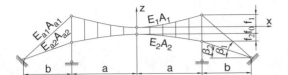

Figure 2. Double-cabled structure.

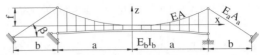

Figure 3. Girder – stiffened cable structure.

3 GIRDER-STIFFENED SUSPENSION STRUCTURE

In a girder-stiffened structure (Fig. 3) the outer loads are balanced both by the cable and the stiffening girder. For the cable we may write

following form

$$(H_1 + H_2)\, w_o + (H_1 f_1 - H_2 f_2) = p \tag{17}$$

$$(H_1 - H_{o1})\,(1 + \kappa_1 + \theta_1)\,/\,(E_1 A_1)$$
$$= \frac{4}{3} f_1\, w_o\, /\, a^2 + \frac{2}{3}\, w_o^2\, /\, a^2 \tag{18}$$

$$(H - H_{o2})\,(1 + \kappa_2 + \theta_2)\,/\,(E_2 A_2) = -\frac{4}{3} f_2\, w_o\, /\, a^2$$
$$+ \frac{2}{3}\, w_o^2\, /\, a^2 \tag{19}$$

where index 1 notes the carrying and index 2 the stretching cable.

If the initial form of cables is square parabola, the contact load is uniformly distributed one. After loading the structure by uniformly distributed outer load, the cables keep the form of parabola. For the state of pre-stressing we have

$$z_1 = f_1\, x^2\, /\, a^2\; ;\quad z_2 = -(f_o + f_2)\, x^2\, /\, a^2 \tag{20}$$

$$H_{o1} = p_o\, a^2\, /(2 f_1) \tag{21}$$

$$H_{o2} = p_o\, a^2\, /(2 f_2) \tag{22}$$

After application of Equations 17–22 we have the cubic equation for determination of the relative deflection of the structure as follows

$$(1+\psi)\,\zeta_o^3 + 3(1 - \alpha\psi)\,\zeta_o^2 + [\,2(1 + \alpha^2\psi)$$
$$+ p_o*(1+1/\alpha)]\,\zeta_o = p* \tag{23}$$

and expressions for the cable forces in the form

$$H_1 = H_{o1} + \Phi\zeta_o\,(2 + \zeta_o) \tag{24}$$

$$H_2 = H_{o2} - \psi\Phi\zeta_o\,(2\alpha - \zeta_o) \tag{25}$$

where $\alpha = f_2\, /\, f_1$; $p_o* = H_{o1}\, /\, \Phi$; $p* = P\, /\, \Phi$; $P =$

$$= pa^2\, /\, (2f_1);\; \Phi = 2\, E_1 A_1 f_1^2\, /\, [3a^2\, (1 + \kappa_1 + \theta_1)];$$

$$\psi = E_2 A_2\, (1 + k_1 + \theta_1)\, /\, [E_1 A_1\, (1 + \kappa_2 + \theta_2)]\; ;$$

$$H\, \frac{d^2(z + w)}{dx^2} = p_o + p' \tag{26}$$

Condition of equilibrium for the stiffening girder may be presented in the form

$$E_b I_b\, \frac{d^4 w}{dx^4} + p'' = 0 \tag{27}$$

The outer load p'' may be realized only after connecting the girder's mounting units. The load p_o consists of the girder's own weight and is balanced by the cable, the load p' is the additional load balanced by the cable.

Summation of (Equations 26 and 27) brings us to a differential equation

$$E_b I_b\, \frac{d^4 w}{dx^4} + H\, \frac{d^2(z + w)}{dx^2} + p = 0 \tag{28}$$

where $p = p_o + p' + p''$ is total load of the structure.

Using boundary conditions, after integrating Equation 15 we can obtain exact expression for the factor $c = (E_b I_b / H)^{1/2}$ in a complicated transcendental form [4]. On the other hand, after suitable approximation of the deflection function, we may obtain a proper result, very close to the exact one. Proceeding from the usual parabolic form of the cable and uniformly distributed outer load, we can approximate the deflection function in the form

$$w = -w_o \cos\,[\pi x/(2a)] \tag{29}$$

and obtain the solution of the Equation 28 with application of Galjorkin procedures as follows

$$\zeta_o^3 + (96/\pi^3)\,\zeta_o^2 + [2048/\pi^3 + 4\, E_b I_b\,(1 + \kappa + \theta)]\, /$$
$$(Eaf^2) + 32/\,(3\pi^2)\, p_o* = [1024/(3\pi^5)]\, p* \tag{30}$$

For the maximum bending moment of the stiffening girder we obtain

$$\max M = [\pi^2 E_b I_b / (4a^2)] \qquad (31)$$

Applying the analogy with the equations for the single cable and using more exact values of the corresponding numerical factors, we may write the equations for the relative deflection and inner forces as follows

$$\zeta_o{}^3 + 3\zeta_o{}^2 + (2 + \rho + p_o{}^*)\,\zeta_o = p^* \qquad (32)$$

The cable force

$$H = H_o + \Phi\zeta_o\,(2 + \zeta_o) \qquad (33)$$

is the same as for a single cable. For the maximum bending moment of the stiffening girder we obtain

$$\max M = \frac{4}{9}\,\rho\Phi f\zeta_o \qquad (34)$$

Equations 30 and 32 differ from corresponding equations for a single cable only by the factor ρ; the other designations are as in Equations 11 and 12.

4 PRESTRESSED CABLE NETWORKS

4.1 *Discrete analysis*

Determination of the stress–strain state of a cable network (Fig. 4) requires in general solution of a system of non-linear conditions of equilibrium of nodes with three unknown displacements components and equations of deformation compatibility for all cable sections (usually two sections per every node). For orthogonal networks loaded by vertical loads the simplified solution requires determination of one displacement component for every node and one equation of deformation compatibility for every cable. Initial form of an orthogonal network may be determined by solution of the system of linear equations with vertical co-ordinates of nodes.

The initial form of a prestressed orthogonal network in case of equidistant cables may be determined by the system of linear equations

$$z_{i,k} = [(z_{i,k-1} + z_{i,k+1}) + \lambda(z_{i-1,k} + z_{i+1,k})] / [2\,(1 + \lambda)] \qquad (35)$$

where $\lambda = H_{oy}a / (H_{ox}b)$, index i belongs to the carrying and index k to the stretching cables; a and b are distances between the stretching and the carrying cables respectively.

For the case of loading the network by vertical loads we have the condition of equilibrium for the node i, k in the form

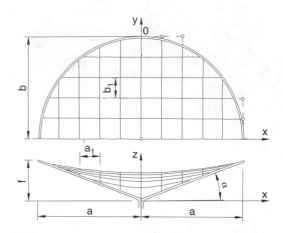

Figure 4.   Cable network with contour of two inclined arches.

$$w_{i,k} = \{(w_{i,k-1} + w_{i,k+1}) + (z_{i,k-1} - 2z_{i,k} + z_{i,k+1})$$
$$+ [H_{yk}a/(H_{xi}b)][(w_{i-1,k} + w_{i+1,k}) + (z_{i-1,k} - 2z_{i,k} + z_{i+1,k})]$$
$$- F_{i,k}a/H_{xi}\} / \{2[1 + H_{yk}a / (H_{xi}\,b)]\} \qquad (36)$$

and the equilibrium of deformation compatibility in the form

$$\frac{H_{xi} - H_{oxi}}{EA_i a^2} \sum \left[1 + \left(\frac{z_{i,k+1} - z_{i,k}}{a}\right)^2\right]^{3/2}$$

$$= \sum \{(w_{i,k+1} - w_{i,k})[(z_{i+1,k} - z_{i,k}) + \tfrac{1}{2}(w_{i+1,k} - w_{i,k})]\}$$

$$- \sum (u_{i,n} + u_{i,o}) \qquad (37)$$

$$\frac{H_{yk} - H_{oyk}}{EA_k a^2} \sum \left[1 + \left(\frac{z_{i+1,k} - z_{i,k}}{b}\right)^2\right]^{3/2}$$

$$= \sum \{(w_{i+1,k} - w_{i,k})[(z_{i+1,k} - z_{i,k}) + \tfrac{1}{2}(w_{i+1,k} - w_{i,k})]$$

$$- \sum (v_{k,m} + v_{k,o}) \qquad (38)$$

4.2 *Continuous analysis*

In case of a hypar-formed network (Fig. 5) the ordinates of its nodes is determined by the equation of the surface

$$z = f_x\, x^2/a^2 - f_y\, y^2/b^2 \qquad (39)$$

Both the carrying and the stretching cables have the form of square parabolas and the initial contact

954

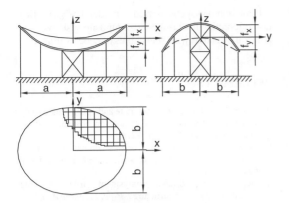

Figure 5. Hypar – formed cable network.

load between them is uniformly distributed. The pre-stressing forces of every family of cables considered to the width unit have values

$$H_{ox} = p_o a^2/(2f_x) \quad \text{and} \quad H_{oy} = p_o b^2/(2f_y) \qquad (40)$$

where a and b are the half-axes of the contour ellipse, f_x and f_y are the cable initial sags. The contact load p_o is reduced to a unit of the projection of the hypar.

Behaviour of the network loaded by vertical load p is characterized by the following equations with partial derivatives:

$$H_x \frac{\partial(z+w)}{\partial x} + H_y \frac{\partial(z+w)}{\partial y} = p \qquad (41)$$

$$\frac{\partial u}{\partial x} + \frac{\partial w}{\partial x}(\frac{\partial z}{\partial x} + \frac{1}{2}\frac{\partial w}{\partial x}) = [(H_x - H_{ox})/(Et_x)]$$
$$[1 + (\frac{\partial z}{\partial x})^2]^{3/2} \qquad (42)$$

$$\frac{\partial u}{\partial y} + \frac{\partial w}{\partial y}(\frac{\partial z}{\partial y} + \frac{1}{2}\frac{\partial w}{\partial y}) = [(H_y - H_{oy})/(Et_y)]$$
$$[1 + (\frac{\partial z}{\partial y})^2]^{3/2} \qquad (43)$$

where H_{ox} and H_x are forces of the family of carrying cables before and after loading the network, H_{oy} and H_y are corresponding values for stretching cables, u, v, w are displacements in direction x, y and z respectively, Et_x and Et_y are the tension rigidity of the carrying and stretching cables considered to the width unit.

By means of integrating Equations 38 and 39 we may express the members with $\partial u/\partial x$ and $\partial v/\partial y$ by corresponding displacements of the contour beam

under action of cable forces. For the case of uniformly distributed load, the deflection function may be approximated in the form

$$w(x,y) = w_o(x^2/a^2 + y^2/b^2 - 1). \qquad (44)$$

Going through the necessary procedures [3], we obtain for determination of relative deflection $\zeta_o = w_o/f_x$ and cable forces H_x and H_y the following equations

$$(1 + \psi + 4\xi)\zeta_o^3 + 3[(1 - \alpha\psi) + 2(1 - \alpha)\xi]\zeta_o^2$$
$$+\{2[(1+\alpha^2\psi)+(1-\alpha)^2\xi]+(1+\beta^2/\alpha)p_o^*\}\zeta_o=p^* \qquad (45)$$

$$H_x = H_{ox} + \Phi\zeta_o[(2+\zeta_o)+2(1-\alpha+\zeta_o)\xi] \qquad (46)$$

$$H_y = H_{oy} - \beta^2\Phi\zeta_o[(2\alpha-\zeta_o)\psi - 2(1-\alpha+\zeta_o)\xi] \qquad (47)$$

where $\alpha = f_y/f_x$; $\psi = [a^4 t_y(1+\kappa_x)]/[b^4 t_x(1+\kappa_y)]$;

$\kappa_x = 5f_x^2/(3a^2)$; $\kappa_y = 5f_y^2/(3b^2)$; $\beta = b/a$;

$\mu = 1 + 1/\psi$; $\Phi = 5 Et_x f_x^2/[9a^2(1+\kappa_x)(1+\mu\xi)]$;

$\xi = 5 Et_y f_x^2 a^3(a/b)^{1/2}/[72 E_c I_c a^2(1+\kappa_x)]$;

$p_o^* = H_{ox}/\Phi$; $p^* = P/\Phi$; $P = pa^2/(2f_x)$.

It is worth to be noticed the analogy between the equations for double-cabled plane structure and that for the hypar-network. The main difference consists in taking into account displacements of the cable supports (factors θ and ξ respectively). When the equations for the plane structure are exact ones that for the hypar-network are obtained by approximation of the deflection function. At the same time this approximation brings us to results, very near to exact ones.

5 CONCLUSIONS ABOUT ANALYSIS AND ERECTION OF CABLE STRUCTURES

5.1 *Remarks about continuous analysis*

Configuration and content of equations presented in the main paragraphs of the report demonstrate their radical analogy. Definitive equations for all structures under investigation are obtained in the form of cubic equations regarding to relative deflection of the structure. Understandably the equations for more complicated structures include more parameters as naturally as by additional factors in equations. It is to be mentioned that presented equations are exact ones for single cables and double-cabled structures and approximate for girder-stiffened structures and the cable networks. Collation of mentioned approximate equations with the exact ones obtained by means of discrete analysis demonstrate their very good accordance.

5.2 *Observation at the experience in design and erection of cable structures*

Experience of TTU upon suspension roof structures is connected with design, model testing and erection of different prestressed networks for acoustic screens in Tallinn and Tartu. The network in Tallinn is surrounded by two inclined plane arches; the latter are supported by two massive counterforts; the network in Tartu represents a hypar inside of a spatial inclined contour beam with elliptical layout. This contour beam may freely deform in directions of both axis of the ellipse; it is supported by vertical columns and does not need horizontal outer supports. The necessary prestressing forces of cables are comparatively small. Distribution of surface curvatures and cable forces is relatively uniform and collaboration between cables and the contour beam garantie a good damping capacity against possible vibrations of the network.

By our mind the best form of the cross section of the contour beam having a good torsion rigidity is the ring-shaped tubular one. Mentioned elliptical contour beam may be assembled of 16–24 straight tubular elements (including 5–7 different sections).

For the network's cables may be used as the wire ropes as the round steel bars; the latter may be preferred by simpler shaping of cable supporting nodes.

Experience on preliminary design of the bridge for the fixed link between Estonian mainland and the island Saaremaa enables to present the following observations. For the planned suspension bridge structure of the central navigable span may be recommended the self-anchored suspension bridge model. For simplification of erecting the bridge the composite system of suspension structure with stay cables may be examined.

REFERENCES

Aare, J. & Kulbach, V. 1984. Accurate and approximate analysis of statical behaviour of suspension bridges. *J. Struct. Mech.* (Helsinki), 17(3), 1–12.

Kulbach, V. 1991. Hanging roofs as acoustic screens for song festival tribunes in Estonia. *Trans. Tallinn Techn. Univ.*, No. 721, 12–20.

Kulbach, V. 1995. Statical analysis of girder- or cable-stiffened suspension structures. *Proc. Estonian Acad. Sci. Eng.*, 1, 2–19.

Kulbach, V. 1997. Plane and spatial suspension structures. In *Challenges to Civil and Mechanical Engineering in 2000 and Beyond*, Warszawa, 194–199.

Kulbach, V. 1999. Half-span loading of cable structures. *J. Constr. Steel Research*, 49, 167–180.

Kulbach, V., Idnurm, S. & Idnurm, J. 2002. Discrete and continuous modelling of suspension bridges. *Proc. Estonian Acad. Sci. Eng.* 8, 2, 68–83.

Kulbach, V. Idnurm, J. & Talvik, I. 2002. Analysis of saddle-shaped cable networks with different contour structures. *Proc. Estonian Acad. Sci. Eng.* 8, 2, 114–120.

Kulbach, V. & Talvik, I. 2001a. Analysis of self-anchored suspension bridge in Estonia. *Challenging Technical Limits. IABSE Reports*, 84, 170–171.

Kulbach, V. & Talvik, I. 2001b. Bridge structures for the fixed link Saaremaa. In J. Krokeborg (ed.) *Strait Crossings 2001.* Lisse, Balkema, 221–226.

Tärno, I. 1998. Effects of contour ellipticity upon structural behaviour of hyparform suspended roofs. *Licentiate Thesis.* Royal Inst. Of Technol., Stockholm.

System-based Vision for Strategic and Creative Design, Bontempi (ed.)
© 2003 Swets & Zeitlinger, Lisse, ISBN 90 5809 599 1

The Maya suspension bridge at Yaxchilan

J.A. O'Kon
O'Kon & Company, Atlanta, Georgia, USA

ABSTRACT: Seventh century Maya engineers achieved a unique feat of civil engineering technology that was not exceeded until the 14th century. These ingenious Maya engineers solved the transportation problems presented by the flood-ravaged Usumacinta River that surrounded the major city-state of Yaxchilan by constructing a 108 meter-long cable suspension bridge. The rediscovery of this lost landmark of civil engineering, located in the Yucatan Peninsula of Mexico, was recently documented using forensic engineering techniques, combined with archaeology and 3-dimensional computer simulation to recreate the geometry and civil engineering techniques that parallel present day suspension bridge design techniques.

1 MAYA TECHNOLOGY

Centuries before Columbus reached the Americas, the Maya, a unique culture that possessed advanced technological capabilities, created a widespread civilization ruled by city-states complete with high-rise temples towering over planned cities, a written language, a level of higher mathematics that understood the concept of the number zero 700 years before the Europeans, and created a calendar that is more accurate than our modern calendar.

This advanced culture, which encompassed over 100,000 square miles, was ruled by at least 50 independent states led by rulers that were believed to possess divine powers.

This civilization, which was at its peak for more than 1,000 years, from 200BC until 900AD, produced the Americas' first civil engineers. These creative engineers shaped the infrastructure of the expansive empire. They built multistory cast-in-place concrete temples and pyramids; an extensive all-weather highway system which was paved with concrete. All of these civil engineering feats were executed utilizing tools made of a jade that is harder than modern steel on the Mohs' scale.

These early civil engineers also produced the world's longest span bridge, which spanned 300 meters over the Usumacinta River. This river, the modern boundary between Mexico and Guatemala, presented unique transportation issues and a long-span bridge was required in order for the powerful city-state of Yaxchilan to operate and maintain its lifeline of commerce and military power on a year-round basis.

The Maya are termed a "stone age" culture by archaeologists. Applying the term "stone age" to a culture that was able to calculate numbers to the 20th power is an injustice. The construction of this 106 meter bridge, with a 63 meter center span, over the raging waters of the Usumacinta river (which has an increased water level of 15 meters during the annual flood stage) elevated the engineering capabilities of the Maya to a level not achieved by Europeans until 1377 when a 76 meter-long span was constructed over the Adda river in Italy.

2 THE POLITICAL AND ECONOMIC RATIONAL FOR THE BRIDGE

The city of Yaxchilan, along with Tikal, Copan, Piedras Negras, and Palenque, became one of the most powerful kingdoms during the Mayan classic period. The majority of these cities are located on the Yucatan peninsula of Mexico.

The ruler Yat-Balam founded the city of Yaxchilan in the fourth century. This powerful king and his descendants, including the legendary ruler Bird Jaguar, ruled Yaxchilan until its decline five centuries later. During their reign they build gleaming white temples atop pyramids and luxurious palaces clustered along the shore of the Usumacinta River. The magnificent structures were constructed on massive terraces built into hills, which provided a natural defensive barrier to the south of the city. The broad river formed a large U-shaped bend or "ox-bow" that encompassed the city on the remaining east, north and south sides of the city. These natural barriers protected the city from invasion from other city-states but in order to operate this city of 100,000 people, an all-weather passage across the treacherous river to the villages and farmlands was an absolute necessity.

3 ARCHAEOLOGY OF YAXCHILAN AND THE BRIDGE

The site of Yaxchilan is located in a lush rainforest environment, set against a backdrop of terraced hills. Today the ruined structures of its temples, palaces, and pyramids, stand as mute reminders of a once great culture. The brilliant plan of this city is centered around a grand plaza creating a vision of tranquility and an air of mystery envelopes this special place with an ambiance that is fraught with history and a tropical atmosphere. The air of mystery is heightened by the constant hum of insects and by the howler monkeys roaring their loud salutes to the past glory of the Maya.

Archaeologists have studied the ceremonial city of Yaxchilan, on the banks of a broad and swiftly moving river, for over one hundred and twenty years. The archaeological analysis consists of exhaustive studies that include site maps, renderings and detailed art and architecture drawings, but these academic studies totally overlooked the lost landmark of bridge engineering that was critical to the survival and success of this "island" city (Fig. 1).

4 CLUES TO THE EXISTENCE OF THE BRIDGE

Little remains today of this unique achievement in civil engineering; however, by analyzing clues scattered throughout early archaeological records, combined with contemporary evidence from numerous field surveys, aerial photography, remote sensing surveys, and computer simulation. Forensic engineering techniques have been used to conceptualize the geometry of the bridge structure, construction techniques used in the bridge as well as the critical role the bridge played in the political and economic activities of the kingdom of Yaxchilan.

This archaeo-engineering discovery presented a convincing argument for the reality of this unique bridge as well as the advanced levels of Maya engineering achieved during the classic period.

5 ARCHAEO-ENGINEERING INVESTIGATION

Forensic engineering investigation and computer-aided techniques indicated that the bridge was composed of a catenary rope/cable suspended superstructure supported by masonry/cast-in-place concrete bridge towers and abutments. The bridge spanned across the Usumacinta River between what is now Mexico and

Figure 1.

Guatemala linking the densely populated ceremonial center on the south side of the river with its rich agricultural territory, as well as allied Maya kingdoms to the North (Fig. 1).

The ceremonial city of Yaxchilan was the power base of a kingdom that included a vast empire located to the north of the river. The majority of the tall pyramids, temples and palaces of the city were founded on the terraced hillsides to the south as well as locations, which bordered the Grand Plaza. Maya city planners developed this great plaza along the curve of the river, which served as the focus of activity of this rich and powerful city-state.

The area around Yaxchilan receives one of the heaviest yearly rainfalls in the Yucatan peninsula. This heavy rainfall promoted an abundance of agricultural products that contributed to the wealth of Yaxchilan but raised the water of the Usumacinta to flood levels for as long as six months of the year.

A review of the archaeological studies shows that most written texts reveal a piece of the puzzle and indicate the need as well as the existence of an all-weather bridge. However, archaeologists did not believe that this "stone age" culture could create such an amazing engineering feat. More seriously, they overlooked the essential fact that this city could not exist without a bridge that could be used even during the seasonal floods.

Before archaeologists could be convinced that the bridge and its sophisticated technology had existed, they had to be convinced of this need for such a bridge. The archaeo-engineering study pointed out that in order to survive and operate efficiently, the seat of power in Yaxchilan required a dependable and safe passageway to cross over the river in all seasons. The swift water levels of the river rise more than 15 meters in height during the rainy season, therefore a river crossing was mandatory to maintain the political and economic power of Yaxchilan.

6 PROVING THE NEED FOR A BRIDGE

To survive and operate efficiently as a seat of power, the city required a safe and dependable passageway that would provide an uninterrupted flow of materials and human traffic across the river during all seasons. The solution to the problem was the construction of a bridge over the river. The basic requirements for the design of the bridge included an elevated span that would maintain the walkway deck above flood levels while spanning across the river from the center of the plaza to the causeway on the opposite northern shore (Fig. 1).

Investigations show that the design requirements for the bridge established by Maya engineers paralleled 21st century bridge design criteria. The bridge structure was sited where the opposed riverbanks were narrow and the bedrock for the tower foundations were near the surface. Design criteria included an elevated span that would maintain the bridge above the 15-meter flood levels while spanning the river from the richly endowed northern agricultural shore to the center of the Grand Plaza of the ceremonial city of Yaxchilan.

7 ARCHAEOLOGICAL INVESTIGATIONS

The art and architecture of the site has suffered minimal looting and a firm chronological and historical background has been established through studies of hieroglyphic inscriptions. The research indicated that recorded history of this city extended over a period of 500 years. Archaeologists have studied Yaxchilan for over 120 years and numerous photos, maps, renderings and drawings and narrative observations of the site, the buildings, and hieroglyphic art are included in their works.

An overview of their studies offers clues that point to the need and the existence of the bridge. However, archaeologists have never considered the fact that this unique structure had once existed or was necessary; possibly because of the "stone age" culture mind set that they retain, or a lack of knowledge of city planning and urban transportation needs.

Alfred Percival Maudsley (Maudsley 1989), who made the first archaeological visit to the site in 1892, references the ruins of the masonry and concrete bridge piers in the river in *Biologia Centrali-Americana*. A "pile of stones" in the river is noted as a landmark, appears in a photograph, and is located on his map of the site depicted in this publication.

Recently, Dr. Caroline Tate's (Tate 1993) book *Yaxchilan, The Design of a Mayan Ceremonial City*, describes each structure along the river side of the plaza and includes a reference to the ruins of the bridge pier which she considered to be the remains of a flood ravaged structure lost to the river. However, she actually identified the approach to the bridge in her description of Structure 5. This stone platform with a hieroglyphic stairway facing the Grand Plaza is described as "a long low platform paralleling the axis of the main plaza". She notes that no trace of masonry superstructure has been found on the top. To the rear of the building is a narrow esplanade and then a steep decent into the river. In my opinion, she was describing the configuration of a classic elevated bridge approach structure.

The façade of Structure 5 consists of a stairway made of six risers constructed of 188 individual stone blocks carved with hieroglyphics. Dr. Tate also describes a stairway down to the riverbank adjacent to Structure 5. She writes "in ancient times someone entering there would have passed between the structures as a gateway to the city". Dr. Tate provides a clear description of an elevated platform, which served as a grand entrance to the city by merging the bridge traffic from the northern shore with the traffic arriving by boat and gaining access to the plaza via the hieroglyphic stairway.

The site map indicates that the northeast end of Structure 5 lies in a direct line with the north bridge tower and the grand staircase from the river coincides with the northwest end of Structure 5. This geometrical alignment coupled with the description of Structure 5 as an elevated esplanade presents an obvious hypothesis: a bridge once spanned the river and terminated its approach on an elevated esplanade that also served as the terminus for a stairway extending up from the river landing. The confluence of the plaza, bridge, and stairway indicates that Structure 5 would have served as the grand entrance to the city, by merging river traffic and road traffic into a focal point on the plaza. However, field investigation, forensic engineering, and bridge design applications would be required to verify this hypothesis.

8 FIELD INVESTIGATION

Several field expeditions were carried out between 1989 and 1994 to survey the site and investigate the evidence remaining of this lost landmark. The surveys covered a wide range of advanced engineering methodologies including digital surface photography, aerial photography, remote sensing, videotaping and alignment studies. The collected data was synthesized through analysis of the multi media evidence, research and computer aided graphics.

During the visit to Yaxchilan in March 1989, the ruins of the pier structure in the river was first observed (Fig. 2). The position and configuration of the remains of the structure indicated that a bridge had connected the Ceremonial City of Yaxchilan with the north side of the river. However, detailed surveys were not possible at this time because the necessary equipment was not available on this expedition.

Figure 2.

To provide a comprehensive high angle overview of the plaza and the bridge structure, aerial photography was required. Aerial photographs were taken at low altitude directly along the centerline of the bridge. The photos (Fig. 4) clearly show the base of the bridge piers just below the water surface. The structure of the northern pier has suffered severely from the lateral forces generated by the action of the fast-flowing water on the broad outside curve of the river. Waterborne debris was thrust against the northern pier by the centrifugal flow along the outside river boundary.

The majority of the masonry and cast-in-place concrete structure of the north pier has been degraded and its stone block work is seen just below the water surface arranged in a spiral pattern extending downstream (Fig. 4). The geometry of the remaining south pier superstructure and the foundation of the submerged north pier is clearly visible. Its planar shape is similar in size and geometry to the south pier rising high above the water level. The grand plaza and its tall structure are clearly in view and the bridge piers appear to be in line with Structure 5.

The next visit to the site was in March 1993, working under the auspices of the Mexican Instituto National de Anthropologia y Historia (INAH) filming permit. A public broadcasting film crew under the direction of the Mayan archaeologist Dr. Nicholas Hellmuth accompanied the engineering team. The mission of the film crew was to photograph the engineering investigation of the concept of the Mayan bridge. The river sector adjacent to the bridge was surveyed, photographed, and videotaped. Field sketches were made of the alignment and geometrical configuration of various salient elements of the bridge structure. When the survey was completed and analyzed, the overall configuration of the bridge, its method of construction and its operating mechanisms became apparent.

The basic structural elements of the bridge structure identified during the survey included the following:

(a) *South Bridge Pier*: The south bridge pier is located 25 meters from the rivers edge. The remains of this pier extended 4 meters above the water level and have a base diameter of approximately 10 meters. The structure is constructed of a façade of worked stone masonry with and interior of cast-in-place concrete (Fig. 2).

(b) *North Bridge Pier*: The center of the north bridge pier is 62 meters from the center of the south bridge pier. The remains of the superstructure and foundations, located by aerial photography and verified by divers, was almost completely under water at the time of the survey. The foundation was formed by large flat stones (1.5 m × 0.3 m × 1.5 m), which were set in bedrock. The diameter of the pier was measured to be 10 meters, the same as the south structure.

(c) *Bridge Abutments*: The force of the strong water flow has displaced the bridge abutments. The abutments, which extended from the river edge to the approach structure (e.g. Structure 5), were founded on the soil of the riverbank and were undermined and dispersed downstream by the scouring of the swiftly flowing water. Large amounts of worked stone masonry facing were located downstream on both banks. These stones could have originated from the abutment structure.

(d) *Bridge Mechanisms*: Several unique carved stone mechanisms were observed downstream (Fig. 3). The devices consist of two parallel stone surfaces connected at midpoint by a rounded concave element. Observations indicated obvious smooth circular grooves in this stone guideway. The groove marks were apparently caused by friction of the

Figure 3.

Figure 4.

rope cable suspension system, which was supported by this mechanism. This device is similar to modern cable rope guideways used in modern bridge construction and could have been used as the suspension rope keeper for the bridge.

(e) *Grand Stairway*: Observations along the riverbank indicated the ruins of a 10-meter wide stairway that led from the riverbank to the western end of Structure 5.

9 FORENSIC ENGINEERING SYNTHESIS

The field data was assessed in the offices of O'Kon & Company in Atlanta and the research efforts were synthesized using computer simulation and forensic engineering analysis.

The advanced computer techniques were used to assess and verify the centerline location and geometry of the bridge structure. Archaeological maps from previous studies were computer digitized and integrated with data collected from the aerial photography and ground level surveys.

The geometry of the structures on the grand plaza was located exactly as surveyed on the site maps. The bridge oriented field data was interfaced with the historical site maps, which indicated the south pier, located within the river. Field survey measurements were used to establish and confirm the distance between the edge of the riverbanks and the centerlines of the bridge pier structures. The computer analysis indicated that the river was approximately 20% wider on the historical maps than was measured during the field surveys. The geometry of the river and pier structure locations was properly adjusted on the revised digitized map of the site (Fig. 5).

The centerline of the bridge structures was established by integrating data from the aerial photographs. The angle of incidence between the centerline of the bridge pier structures and the edge of the south riverbank was calculated from the aerial photographs.

Figure 5.

Using the known location of the south bridge pier as a fixed point, the angle of incidence between the bridge centerline and the river edge was introduced into the digitized site map.

The resulting computer graphic indicated that the centerline of the bridge extended southward over the riverbank and intersected with the eastern end of Structure 5 on the Grand Plaza. At this time, this result totally unexpected. It was quickly recognized that Structure 5 appeared to have the logical design configuration to serve as the terminus of the bridge on the Grand Plaza.

With the location of the bridge and its structures identified, it was vital to verify the capabilities of the bridge to span the river at a safe elevation above the high water level during flooding. The topographical maps indicated that the top surface of Structure 5 is approximately 22 meters above the riverbank at low water level. Historical records indicated that the flood levels of the river occurred at a height of 15 meters above the low water mark. The elevation of Structure 5 at 22 meters above the low water mark fulfilled another engineering requirement: Flood safety. The key elements of the bridge including the top of Structure 5, the causeway to the southern abutment, the surface of the bridge decking and the northern abutment were placed well above seasonal flood levels.

The unexpected discovery of the bridge centerline and the grand stairway coinciding with Structure 5 was an exciting development and the synthesis of the data had created a logical hypothesis for this necessary transportation conduit that was a lost landmark of engineering.

However, a further field survey was required to verify the alignment of the bridge structures through actual site observations. In April of 1994, the ceremonial city of Yaxchilan was visited to investigate and verify the surprising geometrical relationships uncovered during the forensic computer analysis of this important structure that was overlooked by archaeologists for over a century.

Working under the auspices of an INAH permit, the site was again visited with Dr. Nicholas Hellmuth. During this visit the alignment of the salient components (the bridge piers, the river stairway and the mysterious Structure 5), were observed, surveyed, photographed, videotaped, and mapped. The field evidence clearly indicated that the mystery of Structure 5 was solved.

This beautiful platform with a grand hieroglyphic stairway was in fact the terminus of the suspension bridge over the river as well as the Grand Stairway from the river. The bridge entered the hieroglyphic platform on the northeast side and the stairway on the northwest. Observations indicated vestiges of structures perpendicular to Structure 5, which extended northward from the platform. The causeways connecting to the bridge and the stairway verified the theory that Structure 5 was the grand entrance to the great ceremonial city of Yaxchilan.

Furthermore, a large number of commemorative steles are located on the plaza adjacent to the structures. These stele which are richly carved with historical figures and hieroglyphics commemorating the importance of the royal figures carved on their facades were placed adjacent to Structure 5 to be viewed by the many prominent personages that would pass that point as part of the traffic flowing from the grand entrance.

10   ENGINEERING HYPOTHESIS

Based on these revised data an engineering hypothesis for the bridge was developed a rope cable suspension bridge was constructed across the river Usumacinta (Fig. 1, 6). The rope-cable system was supported across the bridge towers and anchored by masonry supports at the north and south abutments. The rope-cable suspension system was connected to a series of vertical ropes that suspended the wooden bridge-deck system (Fig. 7).

The geometry of the base and centerline distances of the towers is based on the measured field dimensions. The walkway height was established by the

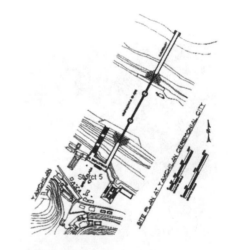

Figure 6.

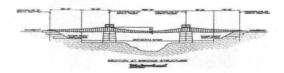

Figure 7.

962

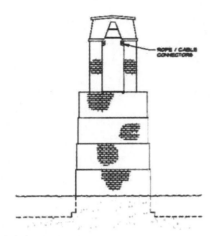

Figure 8.

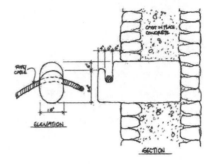

Figure 9.

elevation lf Structure 5. A vaulted arch at the top of the bridge piers, typical of other engineered structures at Yaxchilan, served as a three-dimensional structural support for the rope-cable connectors (Fig. 8). The bridge pier and the vaulted arch were constructed of a stone masonry exterior and a cast-in-place concrete interior. It was important that the rope-cable guideways be connected into the mass of the wall to optimize the stabilizing geometry of the vaulted arch. Large three-dimensional forces were generated by the rope-cable support system, and the supports required substantial resistance (Fig. 9).

The rope-cable suspension system for the bridge assumed a catenary shape typical of cable-supported bridge systems. The center-to-center span between the bridge piers is 63 meters and a vertical dimension of 5 meters was selected for the sag in the rope-cable system (Fig. 7). I calculated the required weight to be supported using the maximum number of people that could be assumed to be in a ceremonial procession crossing the bridge plus the dead load of the bridge deck. Beasts of burden did not exist in Mesoamerica, so rolling loads from carts were not included.

The bridge deck was assumed to be 3 meters wide and made of wooden planks spanning between cross beams suspended by the vertical ropes from the suspension system. Using the known strength of hemp rope, calculations indicated that the rope-cable system would require a bundle of six 5 cm diameter ropes on each side of the bridge walkway. Archaeologists have known for some time that large-diameter hemp rope was available at Yaxchilan. Finally, the stairway from the plaza to the riverbank, when plotted on the digitized site plan, connected to the western end of structure 5 and lead down to the riverfront esplanade (Fig. 6).

The concept of this lost landmark bridge structure provides a logical solution to the question of how this lively urban center could operate on a year-round basis. The engineering techniques were well within the technology of the Mayans. The size and configuration of the structural members were probably achieved by trial and error. This method of structural design was common in all societies until the last century.

The lessons learned from this exercise in archaeo-engineering indicate that a technologically advanced culture can achieve high levels of engineering design. The Maya achieved constructions that can be considered to be modern developments by their optimization of native materials combined with basic engineering mechanics. In the case of the suspension bridge at Yaxchilan the Maya paralleled the achievements of modern technology.

REFERENCES

Maudsley, A.P. 1889. *Biologia Centrali-Americana*. London: (Facsimile Edition) Dr. Francis Robicsek.
Tate, Carolyn. 1993. *Yaxchilan, The Design of a Mayan Ceremonial City*. Austin: University of Texas Press.

Roles of stiffening systems of steel plate girder bridges

E. Yamaguchi, K. Harada, S. Yamamoto & Y. Kubo
Kyushu Institute of Technology, Kitakyushu, Japan

ABSTRACT: Steel bridges usually have quite a few stiffening members. Nevertheless, the effects of stiffening systems on the structural behavior of a bridge are not clearly understood. Thanks to the advancement of computer technology and the finite element method, it has become possible to simulate the real structural behavior of steel bridges with good accuracy. In the present study, the three-dimensional finite element analyses of composite steel plate girder bridges are conducted, in which three levels of stiffening systems are studied. Under vertical loads, stiffening systems make little difference for the structural behaviors of the composite steel plate girder bridges, while under lateral loads, normal stress in the web becomes rather large when no stiffening members are used. It is also noted that normal stress in the vertical stiffeners becomes quite large under lateral loads, when cross beams are removed.

1 INTRODUCTION

For reducing the construction cost and enhancing the ease of inspection and maintenance, bridges with less stiffening members are preferable. However, since we have rather complicated loading conditions for bridges, steel bridges usually have quite a few stiffening members such as cross beams. Nevertheless, the roles of stiffening systems on the structural behavior of a bridge are not clearly understood. This is mainly because bridges are designed based on the simple beam theory in most cases. Thanks to the advancement of computer technology and the finite element method, it has become possible to simulate the real structural behavior of steel bridges with good accuracy. In recent years, therefore, some efforts have been made to help design a bridge with simple stiffening systems (Nagai et al. 1997, Natori et al. 1992). However, a good understanding of the roles of stiffening systems is yet to be achieved.

In the present study, the three-dimensional finite element analyses of composite steel plate girder bridges are conducted. Three levels of stiffening systems are considered, in which even a composite steel plate girder bridge with no stiffening members is included.

2 BRIDGE MODELS

The basic bridge model is shown in Figure 1. This composite steel bridge consists of a reinforced-concrete slab and four plate girders. The dimensions of plate girders, cross beams and vertical stiffeners are summarized in Table 1. The symbols in this table are illustrated in Figure 2. As for the slab, the thickness is 25 cm; Young's modulus and Poisson's ratio are 1/7 of the modulus of steel and 0.167, respectively. This basic bridge model is taken from the work of Nagai et al. (1997) and is called Bridge A in this study.

Two other bridge models, Bridges B and C, are also considered. Bridge B has no cross beams, otherwise it is the same as Bridge A. Bridge C is the model where vertical stiffeners are removed from Bridge B, thus it has no stiffening members.

For the analyses of these bridge models, the finite element method is employed. The slab is modeled by 8-node solid elements whereas 4-node shell elements

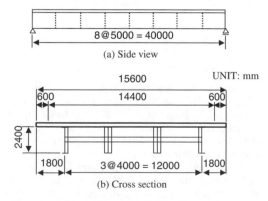

(a) Side view

8@5000 = 40000

15600   UNIT: mm

600   14400   600

2400

1800   3@4000 = 12000   1800

(b) Cross section

Figure 1. Basic bridge model (Bridge A).

Table 1. Dimensions of cross sections (Nagai et al. 1997).

	Main girder	Cross beam	Vertical stiffener
Bu (mm)	600	400	–
tu (mm)	20	20	–
Hw (mm)	2400	600 (1800)	300
tw (mm)	12	9	9
Bl (mm)	600	400	–
tl (mm)	50	20	–
A (cm^2)	708	214 (322)	27
Ix (cm^4)	6.94E + 06	1.70E + 06 (1.76E + 06)	1.823E + 00
Iy (cm^4)	1.26E + 05	2.13E + 04	2.025E + 03

( ): value of the end cross-beam when different from that of the intermediate cross-beam.

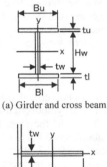

(a) Girder and cross beam

(b) Vertical stiffener

Figure 2. Notation in Table 1.

are applied to plate girders and vertical stiffeners. The cross beam at the mid-span is modelled by 4-node shell elements and the other cross beams are by 2-node beam elements so as to reduce computational time.

3 NUMERICAL RESULTS

3.1 *Vertical loads*

The symmetric vertical loads shown in Figure 3 are first considered. This loading condition is also employed in the work of Nagai et al. (1997). It is based on the L-load defined in the Japanese design specifications of highway bridges (Japan Road Association 1996), representing vehicle loads.

As a numerical result, the distributions of the normal stress in the longitudinal direction in the web of the outside girder at the distances of 0.1 m and 2.5 m from the midspan are presented in Figure 4.

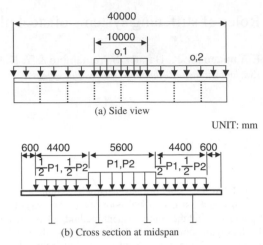

UNIT: mm

(b) Cross section at midspan

Figure 3. Vertical loads.

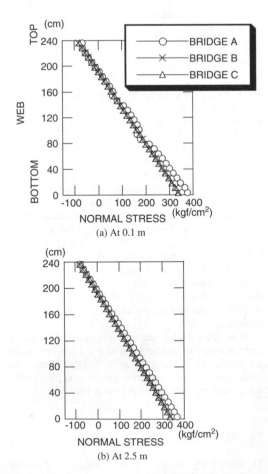

(a) At 0.1 m

(b) At 2.5 m

Figure 4. Normal stress in the longitudinal direction in the web of the outside girder.

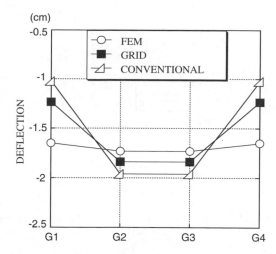

Figure 5.　Deflections of main girders.

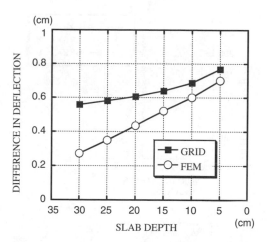

Figure 6.　Difference in deflection between G1 and G2 girders.

The differences in the stress distributions among the bridge models are insignificant. It is also noted that a similar stress state is obtained regardless of the distance from the midspan (stiffener). The eccentric loading of the L-load is also considered, and the same observation is made. From these results, it may be concluded that the cross beams and the vertical stiffeners are not necessary as far as the vertical loading condition is concerned.

Under symmetric loads, the deflections of the four main girders are computed also by two methods employed by Japanese steel-bridge design practices: the conventional method and the grid theory. The results are presented in Figure 5 together with the deflections of the three-dimensional finite element analysis. The discrepancies are significant, which are attributable to the way loads are distributed to main girders in those analyses.

The difference in terms of deflection between the end-girder (G1) and the central girder (G2) is smallest in the case of the three-dimensional finite element method. This appears to be due to the fact that the stiffness of the slab as a plate is taken into account. To confirm this observation, the thickness of the slab is reduced and further comparison is made with the grid theory. The results are shown in Figure 6. The difference in deflection between the end-girder (G1) and the central girder (G2) increases, as the slab becomes thinner in both analyses. The rate of the increase in the finite element analysis is faster, so that the discrepancy between the three-dimensional finite element analysis and the grid theory analysis becomes smaller, confirming that a major reason for causing the three-dimensional finite element analysis to yield the smallest deflection difference is the treatment of the slab stiffness.

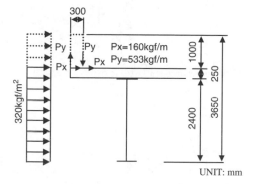

Figure 7.　Lateral loads.

3.2 Lateral loads

The lateral loads due to wind are taken into account. Referring to the Japanese design specifications of highway bridges (Japan Road Association 1996), 320 kgf/m^2 is applied to one side of the bridge. The loads on a handrail are transformed to the combination of the horizontal force and the torsional moment at the foot of the handrail, as has been done in the work of Nagai et al. (1997). The lateral loads thus applied are illustrated in Figure 7.

The numerical results in the form of the distributions of the normal stress (bending stress) in the vertical direction in the web and the normal stress in the vertical direction in the vertical stiffener are summarized in Figures 8 and 9, respectively. Very little bending stress is observed in the region close to the midspan in the case of Bridges A and B. This is because at the midspan not only the cross beam but also the vertical stiffener can suppress the bending deformation of the web. It is

967

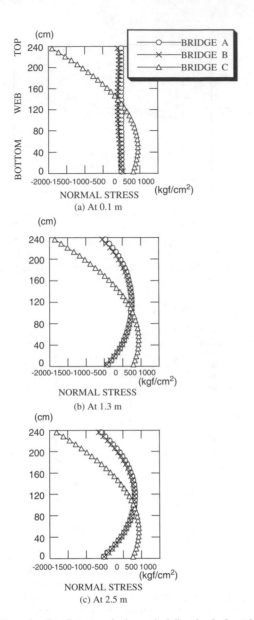

(a) At 0.1 m

(b) At 1.3 m

(c) At 2.5 m

Figure 8. Bending stress in the vertical direction in the web of the outside girder.

however noted that normal stress in the vertical stiffener is quite large in Bridge B, compared with that in Bridge A. The web of Bridge C that has neither vertical stiffeners nor cross beams undergoes bending deformation due to the wind, and noticeably large bending stress occurs in the web.

As the distance from the midspan (stiffener) is longer, the bending stresses taking place in the webs of Bridges A and B become larger. Yet the difference between the two bridges is very small, further implying

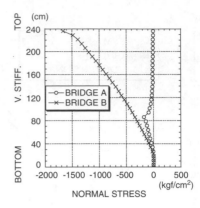

Figure 9. Normal stress in the vertical direction in the vertical stiffener of the outside girder.

that cross beams are not playing an important role for bending deformation. The bending stress in Bridge C is greater than those in Bridges A and B in all the cases of Figure 8.

4 CONCLUDING REMARKS

The composite steel plate girder bridges with simple stiffening systems have been studied under vertical and lateral loading conditions. As far as vertical loads are concerned, the removal of vertical stiffeners and cross beams does not make significant changes in the structural behavior of the steel bridges. It has been also demonstrated that the conventional analyses in design practices may yield a very conservative structure, which can be attributed to the treatment of the slab stiffness.

In the case of lateral loads, it has been found that not only cross beams but also vertical stiffeners help suppress the deformation of the web, reducing the bending stress induced in the web. Normal stress in the vertical stiffeners however becomes quite large when no cross beams are used. When all the vertical stiffeners and cross beams are removed, large bending stress takes place. Further investigation needs to be made into these large stresses, so as to make a composite steel plate girder bridge with simple stiffening systems feasible.

REFERNCES

Japan Road Association, 1996. *Design Specifications of Highway Bridges: Part I.* Tokyo: Maruzen.
Natori, T., Akehashi, K. & Oshita, S. 1992. Study on elimination of lateral bracings and intermediate cross frames. *Technical Report (Giho) of Yokogawa Bridge*, 21, 13–30.
Nagai, M., Yoshida, K. & Fujino, Y. 1997. Three dimensional structural characteristics of steel multi I-girder bridges with simplified stiffening systems. *Journal of Structural Engineering, JSCE*, 43A: 1141–1151.

System-based Vision for Strategic and Creative Design, Bontempi (ed.)
© 2003 Swets & Zeitlinger, Lisse, ISBN 90 5809 599 1

Continuity diaphragms in continuous span concrete bridges with AASHTO Type II girders

A. Saber, J. Toups, & A. Tayebi
Department of Civil Engineering, Louisiana Tech University, Ruston, LA, USA

ABSTRACT: The need for continuity diaphragms in continuous prestressed concrete bridge girders is investigated. The bridge parameters that are considered in this study are based on design of experiment techniques and the results of a survey sent to all states bridge engineers. The parameters include skew angle, length of the span, beam spacing, and the aspect ratio. The effects of diaphragms on stresses and deflections from truck and lane loading on continuous skewed bridges are evaluated. The results of this research provide a fundamental understanding of the load transfer mechanism in diaphragms of skewed bridges. The effectiveness of continuity diaphragms at skew angles larger than 30° in continuous bridges with AASHTO Type II girders is limited. The outcome of this research will reduce the construction and maintenance cost of bridges to the nation.

1 INTRODUCTION

The majority of highway bridges are built as cast-in-place reinforced concrete slabs and prestressed concrete girders. Composite action between the slabs and girders is assured by the shear connectors on the top of the girders. The design guidelines for bridges in AASHTO section 8.12 indicate that diaphragms should be installed for T-girder spans and may be omitted where structural analysis show adequate strength. Furthermore, the effects of diaphragms are not accounted for in the proportioning of the girders. Therefore, the use of diaphragms should be investigated.

Wong and Gamble (1973) studied the effects of diaphragms on load distribution of continuous, straight, right slab and girder highway bridges. The reported outcome indicate that diaphragms may improve the load distribution characteristics of some bridges that have a large ratio of beam spacing to span, the usefulness of the diaphragms is minimal and are harmful in most cases. Based on cost effectiveness, the authors recommended that diaphragms be omitted in highway bridges unless they are required for erection purposes. Sengupta and Breen (1973) reported that interior diaphragms were good only to distribute the load. The authors concluded that it was more economical to provide increased girder strength than to rely on improved distribution of load due to the provision of diaphragms. Saber (1998), investigated the lateral stability of prestressed girders. The analyses were for long span simply supported prestressed concrete bridge girders.

The results indicated that AASHTO 1996 recommendations for T-girder construction, of one intermediate diaphragm at the point of maximum positive moment of spans in excess of 40 feet, are conservative. Griffin, J.J. (1997) researched the influence of intermediate diaphragms on load distribution in prestressed concrete I-girder bridges. The study included two bridges, along the coal haul route system of Southeastern Kentucky, that were constructed with a 50 degree skew angle. One of the bridges has concrete intermediate diaphragms, while the other bridge has no intermediate diaphragms. Experimental field tests were conducted the researchers did not indicate a significant advantage in structural response due to the presence of intermediate diaphragms. Although large differences were noted percentage-wise between the responses of the two bridges, analyses suggested that the bridge without intermediate diaphragms would experience displacements and stresses well within AASHTO and ACI design requirements. However, the total elimination of intermediate diaphragms was not recommended since they were required during construction and in the event the deck was to be replaced.

1.1 Need for research

Continuity diaphragms used in prestressed girder bridges on skewed bents cause difficulties in detailing and construction. In general, details for small skewed bridges (<30° from perpendicular) have not been a problem, but as the skew angle increases or when the girder spacing decreases, the connection and the construction become more difficult. Even the effectiveness of the diaphragms is questionable at these high skew angles. Most of past research concentrated on simply supported straight bridges with limited number of parameters affecting diaphragm performance. The recommendations from these studies on the use of diaphragms in skew bridges are conflicting.

1.2 *Significance of research*

The results of this research provide a fundamental understanding of the load transfer mechanism in diaphragms of skewed bridges. The outcome of this research will reduce the construction and maintenance cost of bridges to the nation.

2 OBJECTIVE

The objectives of this research are (1) to determine the need of continuity diaphragms in continuous skewed bridges with AASHTO Type II girders. (2) To study the load transfer mechanism through diaphragms and their effects on stresses and deflections in bridge girders with AASHTO Type II girders.

3 SCOPE

The parameters that may affect the load distribution of a bridge can be divided into different categories. These include: the material properties of the slabs and girders, the relative dimensions of the slabs and girders, the geometry of the bridge, the type of loading on the bridge, and the location and stiffness of the diaphragms and the location of the supports.

4 METHODOLOGY

4.1 *Survey*

A survey of the bridge design agencies from the fifty states departments of transportation was conducted to collect information on the use of continuity diaphragms in prestressed concrete bridges. Twenty-two states replied to the questionnaires; of which seventeen states have owned, designed, or constructed skew continuous precast prestressed concrete girder bridges. The results of the survey were used in a design of experiment (DOE) to determine the combined effects of the various parameters on the behavior of bridges with continuity diaphragms.

4.2 *Design of experiment (DOE)*

The design of experiment is a series of tests in which changes are made purposefully to the input variables of a process or system, so that the reasons for changes in the output response are identified. Experimental design is a systematic approach to explore interrelationships to determine the main factors affecting a given system and optimize the design of the system for highest performance. In this research the DOE is used to isolate the contribution of continuity diaphragms to bridge system performance. The bridge system is assumed under controllable factors and the design variables in terms of material properties and geometry are the input parameters, while the output variables are the performance variables. Factorial experiments are an experimental strategy in which factors are varied together instead of one at a time. If there are k factors each at two levels then the factorial design will require 2^k experiments. The advantage of factorial design is capturing all possible combinations of the levels of the

Table 1. Bridge models using AASHTO Type II girders.

Bridge group	Model number	Skew angle diaphragm	Skew angle bridge	Span length (m)	Girder spacing (m)
A	D30B10S5	30°	10°	22.9	1.52
	D65B10S5	65°	10°	22.9	1.52
	D00B10S5	Not used	10°	22.9	1.52
B	D30B10S9	30°	10°	16.8	2.74
	D65B10S9	65°	10°	16.8	2.74
	D00B10S9	Not used	10°	16.8	2.74
C	D30B20S5	30°	20°	22.9	1.52
	D65B20S5	65°	20°	22.9	1.52
	D00B20S5	Not used	20°	22.9	1.52
D	D30B20S9	30°	20°	16.8	2.74
	D65B20S9	65°	20°	16.8	2.74
	D00B20S9	Not used	20°	16.8	2.74

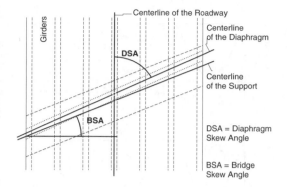

Figure 1. Skew angle for bridge and diaphragm.

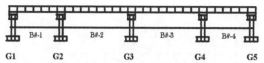

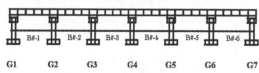

Figure 3. Typical bridge model girder spacing 2.74-m (top), girder spacing 1.52-m (bottom).

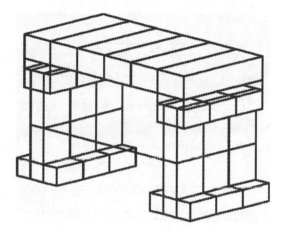

Figure 2. Typical girder, deck and diaphragm elements used in analysis.

factors. However, the larger the number of factors (k) included, the larger and more prohibitive the testing, (Toups et al. 2002). The different bridge designs presented in this paper are shown in Table 1.

In this study the skew angle of the bridge is the angle between the centerline of a support and a line normal to the roadway centerline, as illustrated in Figure 1. Also, shown the angle between the roadway centerline and the diaphragm centerline is the diaphragm skew angle, more details in (Saber 2002).

4.3 Bridge model and loads

The use of finite element modelling (FEM) is among the most popular methods of analysis. Significant advances in computer technology allow for detailed models to be constructed and analyzed. The finite element models used in this investigation simulate the behavior of skewed continuous span bridges. GTSTRUDL Version 25 software is used for this

investigation. The bridge girders presented are formulated using Type-IPSL, Tridimensional eight node elements. The bridge deck is formulated using Type-SBCR four node plate elements. Prismatic Space Truss Members are used to model the continuity diaphragms, as shown in Figure 2.

The bridges analyzed in this investigation are nine meter wide, three-span, continuous bridges. The slab thickness considered is 200-mm. The girder and diaphragm labels are illustrated in Figure 3.

The loading conditions specified in AASHTO LRFD (1998) are considered in this investigation, including lane and truck (HL-93) loads.

5 DISCUSSION OF RESULTS

The results of all bridge configurations listed in Table 1 are compared to determine the effects of continuity diaphragms on skewed continuous bridges.

Figure 4 presents the stress distribution in the critical girder (#3) of the bridge at 10° skew angle, Type II girders with 16.8-m span and spaced at 2.74-m (Group B Table 1).

The effects of the diaphragm skew angle on the maximum tensile stresses for bridges with 10° skew angle and AASHTO Type II girders spaced at 1.52-m with a span of 22.9-m (Group A Table 1) are presented in Figure 5. The results indicate that the tensile stresses for the continuous bridges with a diaphragm at 30° skew angle are lowest, and the change in the stresses for the cases of 65° skew angle and without a continuity diaphragm is insignificant. Therefore, the effectiveness of the continuity diaphragm with high skew angle is limited.

The effects of the skew angle of continuity diaphragms on the deflections of skewed continuous bridges are investigated in this study. The maximum deflections in the critical girder (#3) of Group D are presented in Figure 6. Similar graphs are presented in

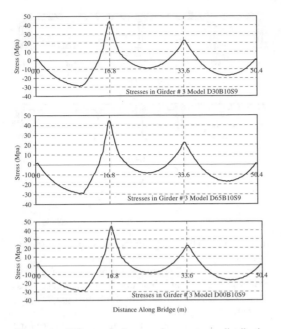

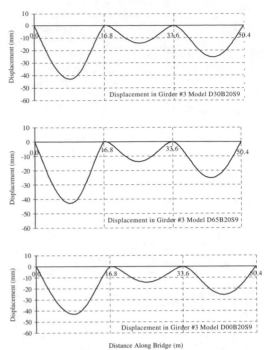

Figure 4. Effects of skew angle on stress distribution (Group B critical girder).

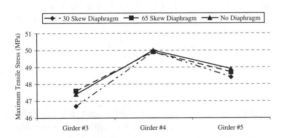

Figure 5. Effects of skew angle on maximum tensile stresses (Group A).

Figure 6. Displacements in bridges Group D.

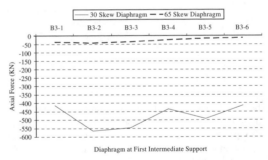

Figure 7. Effects of skew angle on axial load in diaphragm (Group A).

Saber (2002). The results indicate that the change in the deflections for the continuous bridges with a 65° continuity diaphragm and without a diaphragm is insignificant. Therefore, the effectiveness of the continuity diaphragm with high skew angle is limited.

The effects of the skew angle of the continuity diaphragm on their axial load are investigated in this study. The results of the bridge models Group A are shown in Figure 7. The axial force in the diaphragm decreased as the skew angle increased, indicating that the diaphragm is ineffective at high skew angles.

6 SUMMARY AND CONCLUSIONS

The effects of continuity diaphragm for skewed continuous concrete bridge girders, AASHTO Type II are

investigated in this study. Based on the results of the different finite element models that are analyzed, the conclusions of this study are: (1) as the skew angle of the diaphragm increases, the effects on the maximum stresses and deflections in the bridge girders are insignificant. (2) The axial load in the diaphragm decreases as the skew angle of the diaphragm increases. (3) The effectiveness of the continuity diaphragms diminishes as its skew angle increases. The second phase of this study which is still in progress is for bridges with AASHTO Type IV girders.

7 ACKNOWLEDGMENTS

The assistance provided by Professor Leslie Guice, Dean of College of Engineering and Science, Louisiana Tech University is gratefully acknowledged and appreciated.

Support of this work was provided by Louisiana Transportation Research Center under research project number [01-1ST] and state project number [736-99-0914].

The contents of this study reflect the views of the authors who are responsible for the facts and the accuracy of the data presented herein. The contents do not necessarily reflect the official views or policies of the Louisiana Department of Transportation or the Louisiana Transportation Research Center. This paper does not constitute a standard, specification, or regulation.

REFERENCES

AASHTO 1996, Standard Specifications for Highway Bridges. Sixteenth Edition. American Association of State Highway and Transportation Officials, Washington, D.C.

Wong, A. & Gamble, W. 1973. Effects of Diaphragms in Continuous Slab and Girder Highway Bridges. *Civil Engineering Studies Structural Research Series No. 391*, Department of Civil Engineering, University of Illinois, Urbana.

Sengupta, S. & Breen, J.E. 1973, The Effect of Diaphragms in Prestressed Concrete Girder and Slab Bridges. *Res. Report 1581F*, Texas University, Austin Center for Highway Research.

Saber, A. 1998. High Performance Concrete Behavior, Design, and Materials in Pretensioned AASHTO and NU Girders. *Ph.D. Dissertation*, Georgia Institute of Technology.

Griffin, J.J. 1997, Influence of Diaphragms on Load Distribution in P/C I-Girder Bridges. *Ph.D. Dissertation*, University of Kentucky.

Toups, J., Saber, A., Guice, L. & Tayebi, A. 2002. Continuity Diaphragm for Skewed Continuous Span Precast Prestressed Concrete Girder Bridges. *Report 1*, Louisiana Department of Transportation and Development, Louisiana Transportation Research Center Project No. 01-1ST.

Saber, A. 2002. Continuity Diaphragms for Skewed Concrete Girder Bridges in Louisiana. *ASCE* – Louisiana Civil Engineering Conference and Show, Kenner, Sept. 12–13.

GTSTRUDL Version 25. Georgia Institute of Technology, Atlanta, GA.

AASHTO 1998 LRFD Bridge Design Specifications. Second Edition. American Association of State Highway and Transportation Officials, Washington, D.C.

Modification of walking bridge into semi-heavy load bridge by new steel connection opener

A.A. Turk
Water division, KWPA, Khuzestan Water and Power Authority, Ahwaz, Kuzestan, IRAN

B.P. Samani
Manager of Fan Salar Eng. Co, Tehran, IRAN

ABSTRACT: A walking bridge was made by KWPA at 1992, on overflow structure, south of Iran. The traffic load forced to build a new bridge on main water channel that it should be carry semi-heavy traffic load. For each span; the one of two connections of the beam was jointed in each side and the other joint was without any connection with column. The main criterion was considered by designer into separate stiffness that the old bridge should be carry the weight of the dead load and the other stiffness should be computed to carry the live load. New steel bridge should be transferring the torsion into column that it is produced by motion of traffic live load. *New opener connection* could be solved the problem and it was invented by designer Afshin Turk 2002. The cost of modify bridge would be consider 7% of total cost of wide bridge (150 m).

1 INTRODUCTION

Design conception of bridge is varied by regional condition, population, traffic load and kinds of material. In bridges, main beam section was influenced by support connection. Steel or concrete walking bridge could be designed with live load traffic. Designed sections of bridge could be carry the load that it is allowed by requires load exerted upon the sections. Service load of pedestrian bridge should not be converted into semi-heavy load in structure engineering. A walking bridge was made on over flow structure that it belongs to MARED water channel (Irrigation and Drainage project, Sweden consulting engineering Co. 1975). In original design, the walking bridge has not been forecasted but after some years it was built by local population force. Firstly, the top of concrete cover should be damaged to tie the steel root bars (L-shape). Multi span bridge body was suggested with below condition:

1. 14 concrete column
2. 15 span concrete
3. column height = 1.4 m
4. span length = 6 m
5. total length of bridge = 120 m
6. initial width of pedestrian = 1.2 m

Roots of column steel bars was connected by L-shape with top mesh of overflow structure in desired points

that it could be carried the safety load. The column has a rigid connection in overflow structure and it is free at the top. Concrete slab with 6 m length was made by connection with columns. First a joint is in left span and the second a rolled joint in right span. In fact, there is only one connection between span and two columns.

2 MODIFICATION AND REINFORCMENT

After ten years that this walking bridge was built; the another semi-heavy load bridge should be made to pass new population traffic with span 150 m. The project must be spent a lot of money (600000 US$) to building a new heavy load bridge on Mared water channel. It is mentioned that the Mahab-Ghods consulting engineering of project had been declared to employer KWPA with a technical report about "making semi-heavy load bridge is impossible".

Afshin Turk who is the supervisor engineer and (p.samani) the mechanical contractor have been decided to build a new bridge on old walking bridge that it was approved by innovation design. If KWPA have decided for lengthy bridge 150 m on channel it should be paid more than 12 times of innovation design.

3 DESIGN PROBLEMS AND EXECUTIVE LIMITATION

3.1 Connection of old structure

The concrete connection between column and overflow structure has been acted on rigidity because the L-shape bend of bars has tied with top mesh of structure; therefore it could be carried torsion and moment into rigid support.

3.2 Slab and column concrete connection

The section of concrete member have been contained a slab ($120 \times 12\,cm^2$) and a beam $35 \times 35\,cm^2$ with 600 cm length in each span. In concrete members span, one of the connections was built by two bend steel bars that it has been acted the same simple joint connection with force vector R (Rx, Ry, Rz). Another connection could not be reacted any resistance that it is exerted on support. Inability is effected by non-reinforced root connection. One end simple joint beam was resulted by installation of two bar roots in end of each span. Therefore the torsion moment could not transferred in this case that it is produced by traffic live load. It is mentioned the initial design had been based on pedestrian traffic. Drought years and financial cost effected urgent starting of Mared main pump station. Below items should be recognized to make new bridge:

1. Making a new bridge on old pedestrian body.
2. Away from extra financial cost, building long bridge with 150 m.
3. Compulsory condition to built an accessory bridge behind walking bridge.
4. Project management could not be accepted more deletion against long bridge.

With above, innovation design could be solved the problems by using new connection or opener. It is mentioned that the cap modeling could be used in bridge structure to fasten joints.

4 MEMBERS IMPROVMENT STUDY TO CARRY THE LOAD

New connection opener specifications should be transferred the bending moment and forces that they are included the torsion, span directional moment, column directional moment and forces as same as the directions. Firstly, the connection weakness should be modified between beam and column. In ordinary condition, initial connection could not be carried the torsion and moment that it is exerted by beam into column. In each span, one connection has been fixed to carry the force and another useless joint was

installed by non-steel rod inside the concrete. Joint modification was improved by opener connection that it would be designed to resist the below items:

1. Weight of added steel members on bridge.
2. Live load traffic, semi-heavy load bridge.

5 DESIGN OF INNOVATION CONNECTION

Movement and rotation (Δx, Δy, Δz, θx, θy, θz) should be constrained by new connection on a belt acting. Torsion moment had been considered to avoid overturning rapture and it is a main target to satisfy the θ rotations in supports. Concrete connection is the only real anchorage that it could be resist on forces to exerted up on the steel new frame. If the concrete column has assumed on a glass bottle, the resistance force will appear to modeling on a bottle cap. This model of cap belongs the below items:

1. Cap strength material
2. Steel plate thickness and flexibility
3. Required cap root to acting the model

Corresponding above, this connection has been named the opener and it will act as same as the cap model. Notification, steel plate is used by plate flexibility characteristic to bend surfaces. Details of surface will explain in the text. Using the three items above could produce rigid and flexible connection. The plate thickness specially is used on surfaces it should be designed with thickness less or equal 6 mm to desire flexibility and workability installation. Also, the critical cap root length could be analyzed to determine minimum designed length (L_{cr} greater than L_{md}). Torsion and disturbing forces on beam is simulated by the opener cap characteristics to carry the load.

6 BASIC CONCEPTION AND COMPUTATION VISION

It is necessary to explain separate stiffness that the designer has determined to clear opener connection phenomena in this text. Traffic live loads should be exerted on steel stiffness and dead load must be carry by old concrete stiffnesss.

6.1 Cap modeling

Bridge connection was modeled by a bottle cap to determine the exerted forces up on the main connection. The simulation can be shown in the Figure 1 that the basic concept is used to analyze forces by opener phenomena. Stress distribution and scalar value are tried to show by next figures and reader will be noticing the cap modelling steps.

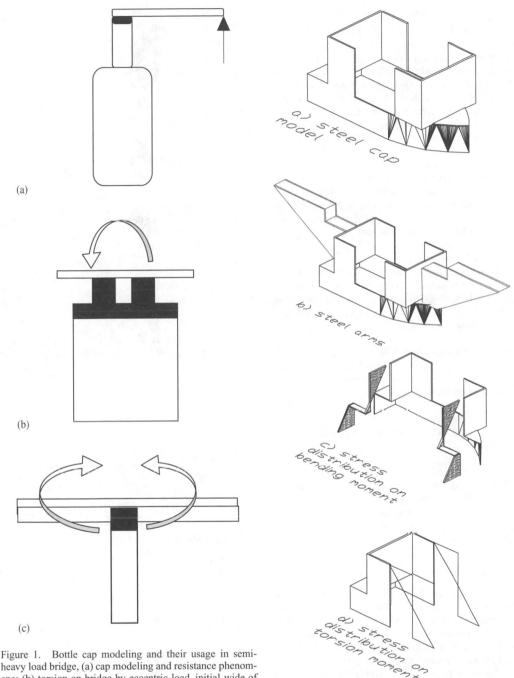

(a)

(b)

(c)

a) steel cap model

b) steel arms

c) stress distribution on bending moment

d) stress distribution on torsion moment

Figure 1. Bottle cap modeling and their usage in semi-heavy load bridge, (a) cap modeling and resistance phenomena; (b) torsion on bridge by eccentric load, initial wide of bridge (Li = 1.2 m) increase to desirable wide L = 2.2 m; (c) bending moment on new modified bridge by dead load, live load traffic and impact load. The black color is belonged the steel plate surfaces to show the cap action by flexible steel belt. In details, all surfaces should be welded correctly to insure the moving loads. It must be avoided to weld beam ends into supports.

Figure 2. Stress distribution, neutral axes is applied at mid point of vertical sections, (a) belt and support steel connected together. (b) Ox head and steel arms. (c) Vertical section of cap to show stress, Myy is exerted by live load and percentage of dead load of old bridge. (d) Vertical section on arm side, Mxx is applying.

977

6.2 Stress distribution value on steel cap sections

All forces should be determined on bridge to bending the steel plate in cap surface. Cap circumstance is used to cover flexible and rigid cap connection. More detail could be found in Figure 2. Main stresses are computed to draw by diagrams and sections at x and y directions (z direction is the column axial). Referring to the Figure 2, there are arranged four steps by new opener connection and the main stress could be evaluated by Mxx and Myy. Figure 2 is explained by below steps:

6.2.1 Steel cap model

Opener phenomena could be helped to know, how should the weak connection be modified and which part is needed to cover by steel plates. A long plate is replaced on ox head to help cap modeling and it could be tensed to keep tension on cap area.

6.2.2 Steel arms

Extra live load must be transfer to column and car's axial distance is determined by steel beam. This distance must be advised to avoid any extra un-desirable torsion. Steel arms have been supported to transfer bending moment into steel cap.

6.2.3 Stress distribution of bending moment

Critical section of cap is assumed to produce more possible stress by opener theorem that it is acted on un-bolted or un-welded cap by only root length belt's cap. The main assumption has been dictated to mid point of cap. It is mentioned that the cap model has not been connected to main structure with any physical operations.

6.2.4 Stress distribution on torsion moment

Steel cap could not be divided forces and moment symmetrically but the importance notice can be advised to welding steel arms into cap in high accurate. The mid point in torsion axes could be selected to take more stress.

6.3 Computation analyzing

In Figure 2, the maximum load on bridge could be considered by the local statistic traffic and it should be limited by maximum modified capacity of new bridge. Traffic live load 1 ton/m is carried by steel members that it is added to improve capacity and it should be strictly exerted on steel beams and cap by steel stiffener separated. To referring the Figure 1 and Figure 2, the maximum stress on x direction could be evaluated by using next formula. Myy is the allowable bending moment that it will be occurred on bridge and it is separated into below three parts:

1. Dead load, old body bridge and steel members.
2. Traffic live load.

$$\sigma_x = \frac{M_{yy}}{S_{yy}} \tag{1}$$

$$S_{yy} = \frac{I_{yy}}{C}, \quad C = 57/2 = 28.5cm$$

$$I_{yy} = 2\left[\frac{t.h^3}{12} + \frac{b.t^3}{12}\right], \quad h = 57cm, \quad b = 15cm$$

$$I_{yy} = 2\left[\frac{0.6 \times 57^3}{12} + \frac{15 \times 0.6^3}{12}\right] = 18520cm^4$$

$$\Rightarrow S_{yy} = \frac{18520}{28.5} = 650cm^3 \tag{2}$$

$$M_{yy} = q\frac{L^2}{12}$$

$$q = q_{LiveLoad} + q_{Im pact} + q_{DeadLoad} = 1.4q_L + q_D$$

$$q = 1\frac{ton}{m} + 40\% \times 1\frac{ton}{m} + 0.750\frac{ton}{m} = 2.15tons/m$$

$$\Rightarrow q = 2.15tons/m$$

$$M_{yy} = \frac{ql^2}{12} = 2.15\frac{6^2}{12} = 6.45 \qquad t.m$$

$$\sigma_x = \frac{645000}{650}\left(\frac{kg.cm}{cm^3}\right) = 992kg/cm^2 \cong 1000kg/cm^2$$

$$\sigma_x = 1000kg/cm^2$$

Torsion Moment:

$$\sigma_y = \frac{M_{xx}}{S_{xx}} = \frac{T}{S_{xx}} \tag{3}$$

$$S_{xx} = \frac{I_{xx}}{C}, \quad C = 28.5cm$$

$$I_{xx} = 2\frac{t.h^3}{12} = 2\frac{0.6 \times 57^3}{12} = 18520cm^4$$

$$\Rightarrow S_{xx} = \frac{18520}{28.5} = 650cm^3$$

$$M_{xx} = T_{torsion} = q_L.\Delta_{eccentric}.L \tag{4}$$

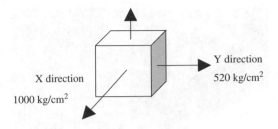

Figure 3.   3D stress elements and zero value on Z axe.

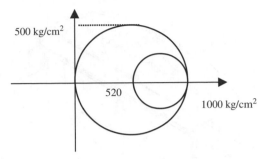

Figure 4.   Mohre circles to showing maximum shear stress 500 kg/cm2 on zero face or z direction.

$q_L = 1.4 tons/m$

$$\Delta_{eccentric} = \left(\frac{W_{Bridge} - W_{car}}{2}\right) = \left(\frac{2.20 - 1.40}{2}\right) = 0.40m$$

$$T_{torsion} = 1.4 \times 0.40 \times 6.0 = 3.36 t.m \qquad (5)$$

$$\sigma_y = \frac{T}{S_{xx}} = \frac{336000}{650}\left(\frac{kg.cm}{cm^3}\right) = 520\frac{kg}{cm^2}$$

$$\Rightarrow \sigma_y = 520 kg/cm^2$$

$$M_{MAX)Beam} = \left(1.4q_L\right)\frac{L^2}{8} = (1.4\tfrac{t}{m})\frac{6^2}{8} = 6.3t\,m$$

$$S_{req)Beam} = \frac{M_{MAX)Beam}}{\sigma_{Allowed}} = \frac{630000}{1400}\frac{kg.cm}{\frac{kg}{cm^2}} = 450 cm^3$$

$$S = 450 cm^3 \rightarrow \quad 2 IPE\,200mm$$

3) Impact forces by using coefficient 40% of live load.

Stress values should be substituted in 3D stress elements to studying maximum tension criterion. Each stress in tension or compressive could be presented in same direction simultaneously. Figure 3 is shown the 3D stress elements by directions.

7   STRESS CRITERION

By using Equation 1 and Equation 2, the maximum normal stresses could be applied to yield steel on cap sections that they are mentioned in Figure 3 by vectors and the Mohr circle will be used to control stress in Figure 4. Critical normal stresses are assumed by rigid and simple joint in each end of the beam. Therefore the maximum bending moment will be produced with ql²/8 at near mid point of beams. In connection joint, values of Equation 2 and zero in another end will appear the maximum bending moment. The beam has been connected with one rigid connection in span.

Thermal effects should be considered by joint connection of beam in each span. This system could be moved to expand by hot weather conditions that the new walking bridge has been passed traffic live load more than 16000 tons. Equation 6 will show the criterion values.

$$\tau_{MAX} = \left|\frac{\sigma_1 - \sigma_2}{2}\right| \le \tau_{critical} = \frac{\sigma_Y}{2} \qquad (6)$$

$$\tau_{MAX} = \left|\frac{1000 - 0}{2}\right| = 500\tfrac{kg}{cm^2}$$

$$\tau_{critical} = \frac{\sigma_Y}{2} = \frac{1400}{2} = 700\tfrac{kg}{cm^2}$$

$$\Rightarrow \tau_{MAX} = 500 \le 700 \quad \rightarrow \quad O.K$$

8   CONCLUSIONS

Based on Equation 6 and 7; opener connection could be used in modified bridge by referring to Figure 8 and 9. Steel connection action should be guaranteed to safe passing on bridge that it is mentioned; in duration 9 month, the total live load 16000 tons has been passed on bridge in the hottest area in south of Iran (Persian Gulf region). Day temperature difference will be appeared in range 40 degree in centigrade scale. Steel surface temperature was reported by laboratory more than +65 degree in hot days. Equation 7 shall be defined the safety factor and it should be added that the bridge was started at May 2002 and it is working continuously up to date (Feb–May = 9 month).

$$TotalWeight = m \times d \times n \times w \qquad (7)$$

$$TotalWeight = 9 \times 30 \times 30 \times 2 = 16200 tons$$

Where m = month; d = 30 days in month; n = Numbers of moving cars; w = weight of cars (tons).

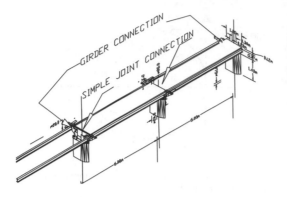

Figure 5. Bridge on overflow structure. It is contained to pass traffic by twenty equal spans in 120 m length of structure. Two steel beams is placed to transfer moment and torsion into opener connection.

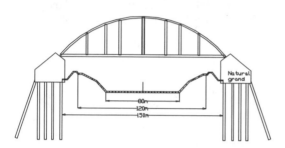

Figure 6. A bridge with 150 m length should be made to carry traffic load on Mared water channel. All discount is returned by new opener connection about 93% or 604500 US$.

Figure 7. Pedestrian: modification steps.

Figure 8. steel arms and new opener connection.

Figure 9. Bridge test by car.

REFERENCES

Ahwazy, R.T. 2000. Mahab-Ghods, 1992 consulting engineering, Notebook of overflow structure, Soeiko 1975, consulting Engineering.
Keynia, A.M. 1995. Analyses and Design of concrete structures. 10nd ed.
Mahab-Ghods, 1992. Consulting engineering, concrete laboratory report of pedestrian and overflow structure.
Popov, I.P. 1996. Engineering Mechanics of Solids 2nd Ed.
Salmon, C.G.1996. Steel Structures: Design and Behavior
Samani, P. 2000. Bouncy steel bridge on Karon River reports.
SES, 1992. Soil Mechanic Services, Soil Mechanic report of overflow Bahmanshir river, south of Iran, Persian Gulf.
Turk, A. 2002. A Report and Design of reinforcement of walking bridge. Submit into Karon Management Department.
Turk, R. & Delbakhsh, B. 1992. Photos and documents.
Turk, R. & Delbakhsh, B. 1992. Statistical report of traffic congestion in Mared area.

System-based Vision for Strategic and Creative Design, Bontempi (ed.)
© *2003 Swets & Zeitlinger, Lisse, ISBN 90 5809 599 1*

Bridge deck analysis through the use of grillage models

G. Battaglia
Structural Engineer, Milan, Italy

P.G. Malerba
Technical University of Milan, Milan, Italy

L. Sgambi
University of Rome "La Sapienza", Rome, Italy

ABSTRACT: The object of the paper is the study of the representativity of the grillage models with which different types of bridge decks are schematized. First, the theoretical principles on which this kind of modelling is based are recalled; the equivalent condition between bi-dimensional continuous elements and corresponding grillage models are imposed through the use of a kinematics and an energetic criterion. Secondly, the same technique is generalized to three-dimensional structures and specialized to the case of cellular decks. For this kind of deck, structural behaviours usually neglected by the current technical approaches, like shear lag, distortion and warping, are considered. The paper presents some methods introducing these effects in a grillage analysis; these methods provide a series of criteria with which it's possible to define the rigidities of the equivalent model. These criteria are applied and compared with finite element solutions. Finally, a series of applications are executed in order to verify the efficiency and the accuracy of this kind of approach.

1 INTRODUCTION

The analysis of the bridge decks through a grillage model is a technique diffused in the second half of the past century, after that some authors, Hrennikoff and Absì particularly, suggest the idea to study the elastic problems modelling the continuous systems through a finite number of elementary frameworks. This type of approach, first applied to the beam and slab decks, spreads widely and its application is extended to the case of much more complex structures, such as cellular decks, skew and curve bridges, and to the case of particular loading conditions, such us the temperature and pre-stress loads.

The fundamental principle lies on the bases of this modelling is clearly expressed in a Hrennikoff's note, referred to the case of bi-dimensional elastic continuous elements but generalizable to the one of three-dimensional structures:

"The basic idea of the method consists in replacing the continuous material of the elastic body under investigation by a framework of bars, arranged according to a definite pattern whose elements are endowed with *elastic properties suitable to the type of problem, in analyzing the framework and in spreading the bar stresses over the tributary areas in order to obtained stresses in the original body. The framework so formed is given the same external outline and the boundary restraints, and is subjected to the same loads as the solid body, the loads being all applied at the joints"* (Hrennikoff 1941).

Hrennikoff imposes the equivalence between continuous structure and grillage model through a kinematics principle, according to the two models are equivalent if, subjected to the same loading conditions, present equal strains. This technique is taken again by Absì after the spread of the Finite Element Method, and it is re-proposed in a different key; Absì, in fact, supposes continuous structure and grillage model are equivalent if, subjected to the same loading conditions, present equal total potential energy (Absì). The imposition of the equivalence let to the equations which define the axial, flectional, torsional rigidities of the grillage beams. The two approaches practically let to the same results.

2 DEFORMATIVE MODES OF BRIDGE DECK

2.1 The problem

In a bridge deck analysis through the use of a grillage model, the assignment of rigidities to the grillage members is certainly the main phase of this pattern. The expression of the rigidities must be assigned to the beams are given by various manuals for more common types of deck. These estimations of the equivalent rigidities derive from theoretical considerations and experimental observations referred to only "principal" deformation modes, or flections and torsion of deck in longitudinal and transverse directions. As for cellular decks, these principal modes are accompanied by "secondary" deformation modes usually negligent, such as shear lag, distortion and warping. For particular geometric and loading conditions, these effects can become significant and to neglect them can make inaccurate the grillage model. In this paper various techniques considering these effects are proposed.

2.2 Shear lag

From the basic assumptions of simple beam theory – where cross section remains plane – the distribution of stress across the top flange of a beam is constant. In broad flange "T" or "I" sections, this is true only for span sections; for end sections or for sections corresponding to points of contra flexure, normal stresses change with a maximum adjacent to the web and reducing to zero at the extremity of the flange.

This effect, usually called shear lag, occurs equally in a cellular deck, ideally composed of a series of adjacent "T" or "I" beams. Since shear lag reduces the effective stiffness of each beam, greater accuracy can be obtained from a grillage analysis if the "effective" section properties arising from shear lag are used in the grillage model. This phenomenon is influenced by loading-restraint conditions and by the type of section used; it's common use to consider an effective flange width established through theoretical valuations and experimental observations (Hambly & Pennels, 1975).

The dependence from the section properties is analysed referring to a unicellular deck of which span, width and height of cell, depth of webs and flanges are changed in turn in order to obtain a great number of cases. The deck is composed of only one span and it's subjected to a loading distribution so that a longitudinal flections arises (Fig. 1).

b = 600 cm s_1 = 25 cm
c = 150 cm s_2 = 35 cm
h = 150 cm   l = 3000 cm

The distributed load is worth 50 kg/cm

The deck is studied through a FE analysis and the value of the vertical displacement in mid-span (corresponding to the intersection of web and flange)

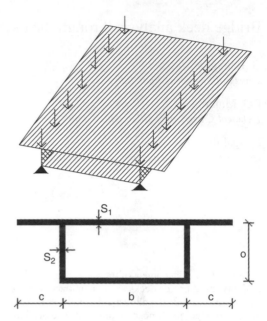

Figure 1. Loading condition and section properties of the reference deck.

is compared to the one deriving from the flections theory for a beam in the same loading-restraint conditions and with the same longitudinal inertia of the whole deck. Drawing the diagram of the ratio $f_{theoretical}/f_{numeric}$ as to the variation in turn of span, height, width and depth of cell, the curve obtained in each case can be considered about linear. It's believed to create a field containing the results of the different analyzed cases, in which the ratio $f_{theoretical}/f_{numerical}$ depends from the value of this ratio obtained in the conditions of l/l_o, h/h_o, w/w_o, t/t_o max and min. The cases corresponding to an increase and diminution of the 50% of these parameters as to the reference one are chosen as extremes of this field. These values are then multiplied by some functions which describe their linear curve in the field. The expression of these functions can be obtained with reference to the shape functions of a truss, of an ISOP4, of a three-dimensional 8 nodes element. The expression of the beam functions in a master field is

$$N_i (\xi) = (1/2)(1+\xi\xi_i) \tag{1D}$$

$$N_i (\xi,\eta) = (1/4)(1+\xi\xi_i)(1+\eta\eta_i) \tag{2D}$$

$$N_i (\xi,\eta,\zeta) = (1/8)(1+\xi\xi_i)(1+\eta\eta_i)(1+\zeta\zeta_i) \tag{3D}$$

For an isoparametric finite element ideally lies in a four-dimensional field, the expression is

$$N_i (\xi,\eta,\zeta,\mu) = (1/16)(1+\xi\xi_i)(1+\eta\eta_i)(1+\zeta\zeta_i)(1+\mu\mu_i)$$

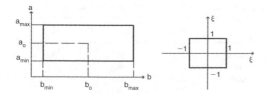

Figure 2. Generic and master field for a plane transformation.

where

ξ, η, ζ, μ represent coordinate system of a 4D field
$\xi_i, \eta_i, \zeta_i, \mu_i$ are the nodal coordinates or the extremes of the field

A coordinate transformation from a generic to a master field must be preventively effectuated in order to use these expressions. This transformation, consisting in a translation and a scale change of the field, is shown in Figure 2 for a plane field (offering much more clearly).

$$\xi = \frac{2(a_j - a_o)}{(a_{max} - a_{min})}$$

for $a = a_o\ \xi = 0$, $a = a_{max}\ \xi = 1$, $a = a_{min}\ \xi = -1$

$$\eta = \frac{2(b_j - b_o)}{(b_{max} - b_{min})}$$

for $b = b_o\ \eta = 0$, $b = b_{max}\ \eta = 1$, $b = b_{min}\ \eta = -1$

For a 4D field there are also the expressions

$$\zeta = \frac{2(c_j - c_o)}{(c_{max} - c_{min})} \qquad \mu = \frac{2(d_j - d_o)}{(d_{max} - d_{min})}$$

For the last, the same properties of the first are valid. The generic coordinates a, b, c and d represent now l/l_o, h/h_o, w/w_o, t/t_o. The values with the "pedice" zero are referred to those cases in which the ratios l_j/l_o, h_j/h_o, w_j/w_o, t_j/t_o, are max and min (where j is referred to the j-th analyzed case).

If $k = f_{theoretical}/f_{numeric}$, since the ratio of the 16 nodal coordinates is known, it's possible to obtain the coefficient for any point of the field using the relationship

$$k = k(\xi, \eta, \zeta, \mu) = \sum_{i=1}^{16} N_i(\xi, \eta, \zeta, \mu)k_i$$

If a grillage analysis is used to study the bridge deck behavior, when it's known the geometric inertia $I_{geometric}$ and the corresponding k coefficient, it's possible to

Table 1. Corrective coefficients for shear lag.

l/l_o	h/h_o	w/w_o	t/t_o	k
max	max	max	max	0,9164
max	max	max	min	0,6853
min	max	max	min	0,1553
min	max	max	max	0,5020
min	min	max	max	0,6158
min	min	max	min	0,3850
max	min	max	min	0,8557
max	min	max	max	0,9378
max	min	min	max	0,9894
max	min	min	min	0,9536
min	min	min	min	0,6836
min	min	min	min	0,9125
min	min	min	max	0,7615
min	min	min	min	0,3304
max	max	min	min	0,8571
max	max	min	max	0,9756

l – span; w – width of cell; h – height of cell; t – ratio between depth of web and flange: $t = s_{web}/s_{flange}$.

obtain an equivalent correct inertia $I_{equivalent}$ to assign to the longitudinal members of grillage mesh, considering so the shear lag effects. Under linear elastic hypothesis

$$k = f_{theoretical}/f_{numeric} = I_{geometric}/I_{equivalent}$$

and

$$I_{equivalent} = I_{geometric} * k$$

In Table 1 the k coefficients of the extremes of the corrective field are shown.

2.3 Distortion

Distortion of cells occurs when cells have few or no transverse diaphragms or internal bracing, so that a vertical shear force across a cell cause the slabs and webs to flex independently out of plane. The effects of distortion are usually considered in a grillage analysis by giving the transverse grillage members a low shear stiffness, chosen so that when the grillage members and cell are subjected to the same shear force, they experience similar distortion (Hambly, 1991). In this paper the distortional effects are considered assigning a correct inertia to the grillage transverse beams. Likewise shear lag case, the dependence to the section properties is analyzed referring to a unicellular deck of which the span, width and height of cell, depth of webs and flanges are changed in turn in order to obtain a great number of cases. The reference deck, having the same section properties used in shear lag study, is subjected now to a loading distribution so that distortion of cell is caused (Fig. 3).

983

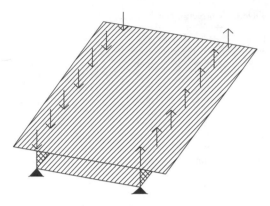

Figure 3.   Loading condition for the reference deck.

Table 2.   Corrective coefficients for distorsion.

h/h_o	w/w_o	t/t_o	d
max	max	max	0,8
max	max	min	1,6
max	min	max	0,95
max	min	max	1,6
min	max	min	0,6
min	max	max	1,2
min	min	min	0,7
min	min	max	1,3

w – width of cell; h – height of cell; t – ratio between depth of web and flange: $t = s_{web}/s_{flange}$.

The deck is studied through a FE analysis and the value of the vertical displacement in mid-span (corresponding to the intersection of web and flange) is compared to the one obtained from a grillage analysis in which distortion is first neglected. The distortional effects are considered by correcting the flectional rigidities of the transverse grillage members; particularly, a series of corrective coefficients, dividing the transverse inertia, are obtained to minimize the error of mid-span vertical displacement.

A corrective coefficient is considered valid when

$$\left| \frac{f - f'}{f} \right| * 100 \leq 1\text{‰}$$

where
f   vertical displacement of the FE analysis
f' vertical displacement of the grillage analysis

Drawing a diagram of the variations of the corrective coefficients as to the different parameters, the curve obtained can be considered about constant for the parameter l/l_o and about linear for the other parameters. It's so possible to trust distortion depend only from width, height, depth of cell.

Likewise shear lag case, it's believed to create a field containing the corrective coefficients of all the analyzed cases; these coefficients depend from the ones obtained in the conditions h/h_o, w/w_o, t/t_o max and min which are multiplied by some functions describing their linear curve in the field. As for a three-dimensional field, these functions are the beam functions of a 3D 8 nodes element; their expressions is

$$N_i \, (\xi, \eta, \zeta) = (1/8)(1+\xi\xi_i)(1+\eta\eta_i)(1+\zeta\zeta_i)$$

where
ξ, η, ζ represent the coordinate system of a 4D field
ξ_i, η_i, ζ_i are the nodal coordinates or the extremes of the field

A coordinate transformation from a generic to a master field must be preventively done in order to use this simple expression. The transformation is similar to the one used in shear lag case. Once the corrective coefficients of the 8 extremes of the field are known, it's possible to obtain the corrective coefficient for any point of the field by using the relationship

$$d = d(\xi, \eta, \zeta) = \sum_{i=1}^{8} N_i \, (\xi, \eta, \zeta) d_i$$

If a grillage analysis is used, it's so possible to consider the distortional effects correcting the inertia of the grillage transverse members through these coefficients. The equivalent inertia is

$$I_{equivalent} = I_{geometric}/d$$

In Table 2 the corrective coefficients d for the extremes of the field are given.

2.4   Warping

Warping is an out of plane displacement of point of cross-section. It's composed of two different components, torsional warping displacement, associated to a rigid twist of cross-section, and distortional warping displacement, associated to a distortion of cross section. Both these components give rise to the longitudinal normal stresses when warping is constrained (Maisel & Roll 1974).

In this paper only torsional warping is considered. It is not an immediate operation to introduce the effects of the no uniform torsion in a grillage analysis. A grillage model is avoid of the d.o.f. in warping direction and so is missing a parameter directly linked to warping displacement.

It's believed to introduce the phenomenon by operating on the terms of the torsion equation

$$EK_{xy}\beta^{IV} - GK_t\beta^{II} = ep$$

where
β torsion
p distributed load across span
e loading eccentricity
GK_t primary torsional rigidity
EK_{xy} secondary torsional rigidity

It's possible to obtain for this equation an approximate solution (Raithel, 1977). It's considered known, for a generic loading condition, the elastic line equation in the form

$$\eta(z) = \eta° f(z)$$

where $\eta°$ is the displacement of a particular section, $z = \zeta$, arbitrary, where $f(z) = 1$.

It's considered β in the form

$$\beta(z) = \beta° f(z)$$

It's supposed that approximately an analogy between $\beta(z)$ e $\eta(z)$ exists. The value $\beta°$ of the torsion is so the only one unknown of the problem. It can be obtained minimizing as to $\beta°$ the functional expression ξ composed of the elastic deformation energy and the loading work.

$$\xi = \int_0^1 \left(\frac{EI_x}{2}\eta^{II2} + \frac{EK_{xy}}{2}\beta^{II2} + \frac{GK_t}{2}\beta^{I2} - p\eta - ep\beta \right) dz$$

By making an integral as to z, neglecting the terms in $\eta°$ (constant as to $\beta°$) and imposing the following equalities

$$\int_0^1 f^{II2}dz = \Sigma^{II} \qquad \int_0^1 f^{I2}dz = \Sigma^{I} \qquad \int_0^1 pfdz = \Sigma^{P}$$

It's possible to obtain the value of $\beta°$

$$\xi = \frac{1}{2}\beta^{°2}(EK_{xy}\Sigma^{II} + GK_t\Sigma^{I}) - e\beta°\Sigma^{P}$$

$$\frac{\partial \xi}{\partial \beta°} = \beta°(EK_{xy}\Sigma^{II} + GK_t\Sigma^{I}) - e\Sigma^{P} = 0$$

$$\beta° = \frac{e}{EK_{xy}\Sigma^{II} + GK_t\Sigma^{I}}\Sigma^{P}$$

Being equal the external and the internal work

$$\Sigma^{P} = \int_0^1 pfdz = \frac{1}{\eta°}\int_0^1 \eta pdz = \frac{1}{\eta°}EI_x\int_0^1 \eta^{II2}dz = \eta°EI_x\Sigma^{II}$$

Substituting Σ^{P} expression in $\beta°$ one, is

$$\beta° = \eta° \frac{eEI_x}{EK_{xy} + GK_t\gamma} \quad \text{where } \gamma = \frac{\Sigma^{I}}{\Sigma^{II}}$$

In a grillage analysis is usually considered only the uniform torsion contribution. To neglect the secondary rigidity associated to the no uniform torsion, let to overestimate the value of $\beta°$.

If both uniform and non uniform torsional contributes are considered, the value of $\beta°$ is

$$\beta° = \eta° \frac{eEI_x}{EK_{xy} + GK_t\gamma}$$

Call this value β_1 and the corresponding primary rigidity $K_{t\ geometric}$.

If the non uniform rigidity is neglected ($K_{xy} = 0$), the value of $\beta°$ is

$$\beta° = \eta° \frac{eEI_x}{GK_t\gamma}$$

Call this value β_2 and the corresponding primary rigidity $K_{t\ equivalent}$.

It's clearly $\beta_2 > \beta_1$. If now $K_{t\ geometric}$ is considered known and $K_{t\ equivalent}$ unknown, from the equality $\beta_2 = \beta_1$ it's possible to obtain the value of primary torsional rigidity which includes no uniform torsional contribute. It's so possible to assign to the grillage longitudinal members the torsional constant

$$K_{t\ equivalent} = K_{t\ geometric} * \alpha$$

where α is a corrective coefficient subsequently shown.

$$\alpha = \frac{K_{t equivalent}}{K_{t geometric}} = \frac{1}{K_{t geometric}}\eta° \frac{eEI_x}{\beta_1 G\gamma}$$

$$\alpha = \frac{1}{K_{t geometric}}\eta° \frac{eEI_x}{G\gamma\left(\dfrac{\eta°eEI_x}{EK_{xy} + GK_{t geometric}\gamma} \right)}$$

and

$$\alpha = 1 + \frac{EK_{xy}}{\gamma GK_{t geometric}}$$

So therefore

$$\begin{array}{ll} \alpha = 1 & \text{if } K_{xy} = 0 \\ \alpha > 1 & \text{if } K_{xy} \neq 0 \end{array}$$

Once the section properties and the coefficient γ are known, it's possible to calculate the coefficient α correcting torsional rigidities of grillage longitudinal members. As for γ, a good approximation of this coefficient is

$$\gamma = \frac{l^2}{10}$$

where l is the span of deck.

3   AN APPLICATION

It's related the study of a multi-cellular deck in c.a. with a span of 30 metres. The deck is subjected to a vertical load distributed across the outside wall of 50 kg/cm and to a vertical load of 180 ton concentrated in mid-span for the same wall (Fig. 4).

Using the geometrical and loading symmetry, only one half of the deck is studied with a grillage mesh composed of 4 longitudinal members corresponding to the webs of deck and of 10 transverse members with a step of 150 cm (Fig. 5).

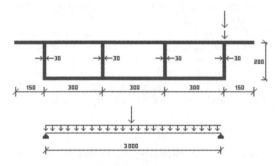

Figure 4.   Multi-cellular deck: cross-section and loading condition.

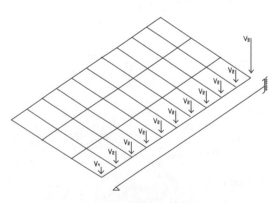

Figure 5.   Grillage mesh: geometry and equivalent nodal loads.

In the following tables (Tables 3–6), the equivalent nodal loads, the corrective coefficients, the elementary and correct rigidities are shown. It's also shown a comparison of the longitudinal deformation for the loaded wall obtained with the different analysis (Fig. 6).

It's possible to note how a grillage analysis which considered the effects of shear lag, distortion and warping, gives results which approximate accurately ones obtained from a FE analysis.

Particularly, the error of the vertical displacement in mid-span is reduced from a 13,91% for an "elementary" grillage analysis, to the 1,37% for a "correct" grillage analysis in which secondary deformation modes are included.

Note
The values are expressed in kg-cm. In Figure 6 one half of the deformation is shown; the section 1 corresponds to the bearing, the section 11 corresponds to mid-span.

Table 3.   Equivalent nodal loads.

Equivalent nodal loads	
V_1	3750
V_2	7500
V_3	93750

Table 4.   Elementary rigidities.

Beam	Longitudinal		Transverse	
	Internal	External	Internal	External
I	2*E8	1,5*E8	6,7*E5	3,35*E5
C	4,32*E8	2,16*E8	1,3*E6	6,75*E5
A	6*E3	6*E3	4,5*E3	2,25*E5

Table 5.   Corrective coefficients.

Corrective coefficients	
k	0,71496
d	1,33611
α	1,067

Table 6.   Correct rigidities.

Beam	Longitudinal		Transverse	
	Internal	External	Internal	External
I	2*E8	1,09*E8	5,05*E5	2,52*E5
C	4,34*E8	2,17*E8	1,3*E6	6,7*E5
A	6*E3	6*E3	4,5*E3	2,25*E5

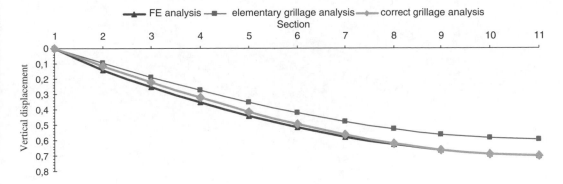

Figure 6. Longitudinal deformation of the loaded wall.

4 CONCLUSIONS

The object of this paper is the study of the representation of the grillage models with which different types of bridge decks can be represented. The purpose of the work is to contribute to this type of approach through the introduction of the effects of shear lag, distortion and warping, usually neglected. The introduction of these effects in a grillage analysis is obtained by applying a series of corrective coefficients to the elementary rigidities of the grillage members. These coefficients are obtained through theoretical considerations and comparisons with other techniques of modelling.

From the effected analysis, it's appeared how the use of these coefficients reinforce the physical equivalence between the real structure and the grillage model; particularly, a correct grillage analysis gives results in terms of stresses and displacements, comparable to the ones obtained from a FE analysis.

The accuracy of the results, the application of this kind of approach also to complex structures, the limited number of d.o.f. and so of dates, the direct use of the results, expressed in terms of generalized stresses (V, M, T), in design procedures, all these factors constitute the main advantages of this modelling.

REFERENCES

Absi, E. 1972. La Théorie des Equivalences et son application a l'etude des ouvrages d'art, *Annales de l'Istitut Technique du Batiment et des Travaux Publics*, Supplements au No. 298, October, 1972.

Absi, E.,Théorie des Equivalences. Application au genie civil, *Publication CEBTP*.

Cedolin, L. ed.1996. *Torsione e taglio di travi in parete sottile, una introduzione*, Edizioni Cusl, Milano.

CNR10024/86, 1986. Analisi di strutture mediante elaboratore: impostazione e relazioni di calcolo.

Curtiss, H. ed. 1997. *Fundamentals of Aircraft Structural Analysis*, WCB, McGraw-Hill.

Danusso, A. 1911. Contributo al calcolo pratico delle piastre appoggiate sul contorno. *Il Cemento*, No. 1–10, 1911.

Hambly, E.C. 1974. Discussion on the paper, "concrete box girder bridges", by Maisel, B.I., Rowe, R.E, and Swann, R.A., *The Structural Engineering*, Vol. 52, pp. 257–258.

Hambly, E.C. & Pennels, E. 1975. Grillage Analysis applied to cellular bridge deck, *The Structural Engineering*, July, No. 7, Vol. 53, pp. 267–274.

Hambly, E.C. ed. 1991. *Bridge Deck Behaviour*, Chapman and Hall, London.

Hrennikoff, A. 1941. Solution of problems of elasticity by the Framework Method, *Journal of Applied Mechanics*, December.

Kanok-Nukulchai, W. 1992. Mathematical Modelling of Cable-Stayed Bridges, *Structural Engineering International*, February.

Keogh, D.L. & O'Brien, E.J. 1996. Recommendations on the use of a 3-D grillage model for bridge deck analysis, *Structural Engineering Review*, Vol. 8, No. 4, pp. 357–366.

Maisel, B.I., Rowe, R.E. & Swann, R.A. 1973. *Concrete Box Girder Bridges, reprint from C&CA/CIRIA*, London.

Maisel, B.I. & Roll, F. 1974. *Techical Report*: Methods of analysis and design of concrete boxbeams with side cantivelers, November.

Malerba, P.G. & Toniolo, G. ed. 1991. *Metodi di discretizzazione dell'analisi strutturale*, Masson Italia Editore, Milano.

Martinez y Cabrera, F., Gentile, C. & Malerba, P.G. 1999. Ponti e Viadotti: concezione, progetto, analisi, gestio, *atti dei Corsi di Aggiornamento*, Pitagora Editrice, Bologna, 29 giugno-3 luglio 1998, 28 giugno-2 luglio.

Pietrangeli, M.P. & Zechini, A. Sul calcolo dei ponti a cassone unicellulare con pareti sottili.

Raithel, A. ed. 1977. *Costuzioni di ponti*, Liguori, Napoli.

Roark, R.J. & Young, C. ed. 1975. *Formulas for stress and strain*, McGraw-Hill, Kogakusha, Tokio.

Sawko, F. 1968. Recent developments in the analysis of steel bridges using electronic computers, *Proceedings of the Conference on Steel Bridges*, BCSA.

West, Recommendations on the use of grillage analysis for slab and pseudo-slab bridge decks.

System-based Vision for Strategic and Creative Design, Bontempi (ed.)
© 2003 Swets & Zeitlinger, Lisse, ISBN 90 5809 599 1

Hybrid truss and full web systems in composite bridges: the SS125 viaducts in Sardinia

M.E. Giuliani
Redesco srl, Milano, Italy

ABSTRACT: The idea of combining full web and truss elements in the composition of deck systems for composite bridges has been investigated in research and applications by the Author and Redesco design team in recent years; the basic issue is the conception of designs that proved to be competitive in global cost and quality with other standard realisations, combining elegance with efficiency to achieve remarkable aesthetical aspects.

This paper describes briefly the concepts and the perspectives, and some already built projects, that have been developed following this line of work.

1 BASIC PREMISES

The structure is composed of the following basic spatial elements.

A central "I" shaped beam and two lateral trusses laying on inclined planes and sharing a common lower chord, which coincides with the bottom flange of the central beam, are completed by an upper concrete slab. The slab is supported by the two external upper chords of the trusses and by the top flange of the central beam. The upper nodes of the trusses are connected to the central beam by transversal girders (Fig. 1).

This concept is based on the following fundamental premises

– constant depth full web beams represent a very efficient structural element and are easily manufactured by automatic welding; their efficiency is greater in sections where shear and bending moment are combined, while it declines in sections where shear decreases, due to the incidence of the web
– constant depth trusses are very effective in sections subjected to bending, while their shear stiffness is poor, unless they are given wide and expensive sectional areas and joint plates
– efficiency for resistance and stiffness for torsion is naturally obtained by using box or spatial arrangements
– a multiplicity of supporting lines for the concrete deck slab enhances its performance, for transversal

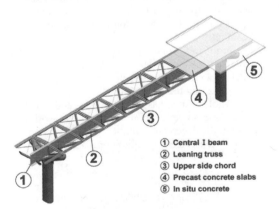

① Central I beam
② Leaning truss
③ Upper side chord
④ Precast concrete slabs
⑤ In situ concrete

Figure 1. Overall concept scheme.

Figure 2. Ground level view of a HFWT viaduct.

bending and for enlarging the effective slab width in service conditions
- spatial structures tend to produce benefical retroactions in terms of local bracing and stiffening, due to the interaction of different elements laying on converging lines and planes in space.

These premises, related to the structural behavior, are complemented by other, not secondary, issues:

- transportation and assembly of deck structures are main concerns both in congested and remote areas; elements to be transported are to be kept slender, in order to enter in standard road width allowance, and compact, to optimise load/volume ratio in transport
- a global cost reduction should not be attained by just reducing the complexity of the work or repeating well established industrial traditions, but instead by reducing raw material consumption while enhancing the human "added value", which means employment, labour qualification, culture
- in this perspective, we used industrial, low cost components as "raw material", and assembled them in the most convenient and efficient manner by design and construction skill
- structural concepts of new shapes and aesthetics foster research on formal and environmental aspects of the engineering works for road infra-structures and bridges.

The key points at this stage may then be summarised as follows:

- build a spatial truss system around a central monolithic element
- make it by simple members
- make it by simple joints
- minimise material consumption.

2 STRUCTURAL BEHAVIOUR

In the following, "hybrid full web and truss systems" will be briefly referred to as HFWT (Fig. 2).

HFWT are basically spatial structures, and thus their resisting scheme must be analysed in full tridimensional behaviour, resulting from the following concepts:

- composite section made by full web central beam and slab resists bending and shear in vertical plane
- inclined trusses with composite upper chords, slab and transverse beams resist torsion and horizontal loading as well as transversal bending and shear.

Although the contribution to vertical shear is somewhat a by-product of the truss orientation, this contribution is not negligible.

In well proportioned systems, i.e. with truss inclination around 30 degrees, the participation of the trusses in vertical shear has shown the following values:

Figure 3. Details of HFWT structure.

- construction (self-resisting, non propped steel structure during concrete casting): 40–47%
- service phase (full composite section under all loads): 16–21%.

The main effect of the truss systems in the longitudinal overall behaviour is that of enhancing the effective width of the concrete slab in composite action.

As confirmed by detailed finite element analysis, in service conditions effective widths as broad as the whole slab can be counted on.

Competitiveness of HFWT solutions can only be achieved by taking advantage of a full composite behaviour.

The upper chords and flange, and the transversal members that connect the upper nodes of the trusses to the central beam, are provided with full connection to the concrete slab.

The principle of composite action is fully applied in the upper chords of the trusses, whose steel section is reduced to a minimum.

The transfer of concentrated forces from the upper nodes of the trusses to the concrete slab requires a careful study of the interaction between steel struss and concrete.

The problem is solved by using flexible type connection, and placing a stronger connector in correspondence with the joints while distributing smaller connectors along the chord (Fig. 4).

The whole system has to be balanced and tuned in order to guarantee a good redistribution with-out excessive displacements and to provide a correct behaviour in service state condition.

2.1 Steel structural members

The central "I" beam is a standard automatically welded girder; thickness variation in the web and flanges is obtained by welding of different thickness panels.

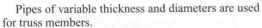

Figure 4.   Central lower joint.

Figure 5.   Side upper joint.

Pipes of variable thickness and diameters are used for truss members.

Weathering steel pipes are commonly available, provided a minimum quantity is required in order to limit production cost. Wide flange profiles are another possibility, but the unit cost reduction does not compensate for the efficiency decrease of the sections.

Standard "I" profiles are perfectly apt to be used as the upper transversal beams; a more efficient solution is provided by using folded sheet open sections, to be filled by reinforced concrete.

The upper chords of the trusses were developed as folded sheet open sections; this solution provides a fully composite member, since the hollow section is filled with concrete and the steel sheet is kept to a minimum area.

This configuration minimises the variation of position of the centre of gravity of the section between construction (only steel) and final phase (composite), reducing local bending effects.

2.2   Joints

Since constructability is one of the main concerns in the ideation of HFWT, it is very important to provide easy and flexible solutions for the joints. Maximum economy is obtained by welded joints.

The more complex joints, that are those of the tubular truss members, have been designed and realised by using simple plate connections. (Fig. 4, 5).

This type of connection is not rigid for bending in the direction perpendicular to the plates, without any influence on the overall efficiency, but the complexity and cost of rigid joints is not compensated by any significative structural enhancement.

Since the truss behaviour of the inclined members is basically efficient even with hinge joints, solutions with pins or bolts can perfectly be adopted.

As for the general organisation of the elements, the best way to optimise the system is to place vertical stiffeners of the central web in correspondence with the lower joints, thus concentrating the interventions on the web and minimising material consumption.

2.3   Concrete slab

Simplicity of realisation calls for a traditional reinforced slab, with a strict crack control obtained by adequate reinforcing, appropriate casting sequences, curing and mix design.

Anti-shrinkage additives for concrete are used to reduce stresses in the slab and steel structure.

The concrete slab may be cast over prefabricated plates for the whole section width, including cantilevers, or the prefabricated plates may be limited to the central part of the section, between outer chords, and the cantilevers be cast over a mobile formwork. In the latter case the prefabricated plates may be arranged longitudinally, that is supported by the transverse girders; this solution minimises the overall structural steelwork weight, because the outer chords are not loaded by the central part of the slab during construction, but on the other hand the slab execution results more complicated and time consuming.

The use of prefabricated elements to be completed by cast in situ concrete allows for any variation of the erection system, including the use of elements with adequate load carrying capacity fitted with route paths for the circulation of transport vehicles for placing the slabs in incremental progression starting from the abutments.

2.4   Bearings and piers

The natural support scheme for HFWT consists of a three point bearing arrangement: a lower support for the central beam and two higher lateral supports for the trusses.

991

Figure 6. Placing of steel segment by crane.

The central bearing is normally a "pot" PTFE and neoprene type, longitudinally sliding, while the two lateral bearings allow for multidirectional sliding.

In order to control the transversal distribution of the loads on the hyperstatic three-point scheme, and to avoid any uplift of the lateral bearings in eccentric loading conditions, the central bearing may be activated after the concreting of the slab; that means that a pre-defined amount of load is already transferred to the lateral bearings in a two-point isostatic arrangement.

The above said arrangement calls for a typical "Y" shaped pier head – or lintel –.

The "arms" of this kind of lintel may be pre-stressed in order to keep slender proportions for aesthetic and material consumption reasons. (Fig. 5)

Other pier shapes can be used, including solutions with steel or composite lintels or prefabricated elements.

2.5 Construction methods

The self-resisting, non propped construction method was applied to the SS125 viaducts.

Nevertheless, the concept of constant depth and full web steel girder takes into account the possibility of incremental launching, which represents the most efficient and often the only possible method in case of complicate site areas, river crossings or deep gorges.

The basic construction process insofar applied, provided abutments and piers are in place, may be summarised as follows:

– production at steel shop: complete central beam segments, including joint plates, stiffeners etc; complete outside chords; complete diagonal truss members and transverse elements
– transportation of components to the site; central full web steel girders segments may have typical length less than 14 m for standard trailer transport; other elements can be easily stacked on trucks

– on-site assembly of whole spans by welding; the geometry is easily controlled as the central beam acts as a template and all other elements are pre-cut in shop
– lifting on piers by cranes (Fig. 6)
– final welding of adjacent spans
– placing of prefabricated plates (and/or mobile formwork); placing of reinforcement (Fig. 8)
– segmental concreting; normally central parts of the spans are concreted first
– finishing.

2.6 Structural optimisation

As a result of the design and construction experiences acquired, some efforts have been made to determine the most important parameters in the optimisation of HFWT, and some relevant values

– optimal torsion behaviour requires the maximum of enclosed area of the section, which in HFTW solutions may be represented by the area contained between the trusses planes and the upper slab
– the planes on which truss elements are laying should have an adequate inclination, otherwise the effectiveness of the trusses participation in the global resisting scheme is remarkably reduced; this effect determines a necessary optimal balance between section width and depth, which can be expressed as the angle defining the truss planes
– platform width and limitation of external cantilevers determine the minimum transversal spacing of the trusses upper chords
– since the torsion resisting scheme develops in a spatial pattern, also longitudinal spacing of truss joints has an influence on the efficiency; this aspect is directly correlated to the optimisation of the trusses in their own planes, following standard criteria
– longitudinal spacing of truss nodes has a relation with the central web efficiency, since the optimum solution for the web is to place stiffeners corresponding with the common lower joints, and to avoid as far as possible any other stiffening of the panels
– allowing for automatic welding of "I" shaped beams requires keeping web depth under the limits of handling and acceptance of plant.

The multiplicity of parametrs which rule the optimisation problem does not easily allow to determine optimal solutions; through a trial and error process some results have been obtained, and some basic dimensioning rules for constant depth HFWT may be issued.:

– the optimal range of slenderness ratio (D/L) for continuous and constant depth girders, lies between 1/14 and 1/20; although these values are quite low, they do not affect the overall structural efficiency
– the angle (α) between horizontal and truss planes may range between 28 and 35 degrees

Figure 7.   A viaduct in landscape.

Figure 8.   Steel structure before completion.

– the longitudinal spacing (d) of the truss nodes should not exceed 2,5 times the depth of the central web, in order to limit the cost for stiffening of the latter.

In conclusion, the design process of HFWT in their basic arrangement is dominated by platform width: once the outside cantilever dimensions (b) have been fixed, the application of the abovesaid rules of thumb should give a fast estimate of the optimal section depth (D), and thus a quick verification versus the relevant slenderness of the girder (D/L).

2.7   Aesthetics

Aesthetics of HFWT derives from the natural interpretation of the resisting mechanism combined with the tuning of the rhythms, proportions, shapes and colours that constitute its essence.

Textures and colours of the materials are those typical of composite bridge structures; the use of non painted weathering steel brings the skeleton to its natural mat brown colour.

Some of the typical features of HFWT solutions, which are at the base of their formal definition are:

– resisting mechanism: the harmonic repetition of truss elements is a well perceived dynamic element, even with the background of the central web which eliminates the transparency typical of truss structures (Fig. 7)
– apparent slenderness: the central beam, which is the deepest element of the structure, lies on the centre line of the deck and is shadowed by the concrete slab, therefore appearing less intrusive, while the moderately enlighted spatial truss breaks its monolithical appearance introducing a subtle reverberation; this results in an enhanced apparent slenderness, and lightness (Fig. 7)
– sculptural piers and abutments: the "Y" shaped lintels call for interesting interpretations of the pier theme, and for natural, tree-like shapes.

3   THE SS125 SARDINIA VIADUCTS

The project involves a group of 9 viaducts, all based on steel spatial trusses, 6 of which are HFWT structures, for an overall length of about 2 km.

The main viaduct has a total length of 980 m as a continuous girder with no intermediate joints, based on 40 m span; pier heights range from 7 to 24 m, with "Y" shaped lintels; the platforms feature different gradients and curves. (Fig. 6, 7, 8)

Main statistics of HFWT SS125 viaducts (all are continuous girders):

– *Spans/overall length (m)*:
  $(25 + 30 + 25) = 80$, two viaducts
  $(30 + 2 \times 40 + 30) = 140$, one v.
  $(25 + 2 \times 30 + 25) = 110$, one v.
  $(30 + 4 \times 40 + 30) = 220$, one v.
  $(30 + 23 \times 40 + 30) = 980$, one v.
– *Deck width (m)*: 12.50
– *Section depth (m)*: 2.46
– *Construction method*: self resisting non propped steel structure, hoisted by crane
– *Slab*: cast on prefabricated plates for the central part of the section, cantilevers on mobile formwork
– *Construction year*: 2000/2001
– *Material consumption*: 92 kg/m^2 of structural steel, 0.24 m^3/m^2 of concrete
– *Contractor/steel manufacturer*:
  FABIANI Dalmine/CIMOLAI Pordenone
– *Price*: the final paid unit price (year 2000) for the completed steel structure was: 1.65 euro/kg

4   PERSPECTIVES

4.1   Range of application in span

As already summarised, the solution have been developed and tested in the range of small to medium spans. Increase of spans is strictly related to the described structural optimisation. The main parameters that determine the girder optimal dimensions without mention to piers and foundations cost, are the following:

Section depth:

- cost limitations in production, as the limit of about 3.5 m is typical for capability of standard machinery for automatic welding for the production of "I" beams
- cost limitations in material consumption: increased web height may lead to significant growth of stiffening requirements; this effect is not specific of HFWT structures only, but must nevertheless be taken into account
- cost limitations due to transport dimensions
- general optimisation of the section proportions, that is an height/width/truss-spacing relationships.

Constant or variable section height:
- effectiveness of the girders may be enhanced by introducing variable depth sections, but this means increase in production cost and may be unsuitable for incremental launching.

Deck width:
- as said before, the optimal section arrangement calls for a certain range of ratios between platform width and depth, and thus span length.

4.2 *Range of application in width*

All the described relations between basic dimensions are related to single-beam solutions.

In order to enhance the range of application to wider platforms, multiple sections have been studied and applied in some of our designs. The implementation of such solutions may lead to innovative combinations with interesting possibilities also correlated to the span range increment.

4.3 *Construction methods*

The possible developments in construction methods are based on the specific features of lightweight steel structures in composite bridges. The most promising paths may reside in the incremental launching method and in segmental construction using fast assembly joints, as already applied in another project.

The use of pinned or bolted connections for the joints, instead of welding, may be useful where fast non skill-needing solutions must be applied, taking into account the relevant cost increase due to greater steelwork weight implied by these solutions.

Slab casting, as said, may be performed according to anyone of the standard solutions applied to composite bridges, preferably by using semi-prefabricated systems.

4.4 *Piers*

The interaction between deck and pier presents an interesting field of investigation both from the morphological and the structural point of view.

The basic three point support arrangement, which naturally recalls the "Y" shaped lintel, can be developed in different ways according to specific conditions related to the project.

A basic list of hints may be the following:

- high piers: where the height of the pier calls for relevant broad sections, the accomodation of the three point support may be achieved within the width of the pier itself
- simple shape piers: solutions based on the above said concept may be applied to piers of any height, providing simple formwork needs
- steel lintels: the lintel may be designed as a steel or, even better, composite element
- deck diaphragm: the deck structure may include a rigid diaphragm over pier, and reduce the bearing arrangement to a two-point scheme.

5 CONCLUSIONS

As it was shown in the above described examples, the application of HFWT solutions may be extended to different bridge projects while maintaining the efficiency and competitiveness inherent to the concept. As any engineering and project problem is specific, and needs to be solved by ad-hoc ideas, it is not easy to make generalisations on the paths to be followed; nonetheless, the innovative and patented solution illustrated proved to be very efficient and suitable for further developments.

REFERENCES

Giuliani, M.E. 2001. Tough Competition. *Bridge , design & engineering* third quarter: 50–51.
Giuliani, M.E. 2001. Hybrid Truss and Full Web Systems in Composite Bridges: Concepts, Realizations, Perspectives. *Proceedings of the 3rd Intnl. Meeting on Composite Bridges – State of the Art in Technology and Analysis*. Madrid: Seinor.
Giuliani, G.C. & Giuliani, M.E. 2001. Strutture Innovative per alcuni viadotti della SS125 Orientale Sarda. *Proceedings of: Giornate Italiane della Costruzione in Acciaio*. Venezia: CTA.

Effectiveness of intermediate diaphragms in PC girder bridges subjected to impact loads

F.S. Fanous & R.E. Abendroth
Iowa State University, USA

B. Andrawes
Georgia Institute of Technology, USA

ABSTRACT: An analytical study was conducted to assess the effectiveness of different types of intermediate diaphragms in reducing the damage to the girders of a pre-stress girder (PC) bridge that is struck by an over-height object on a highway vehicle. Finite-element models were developed for non-skewed and skewed, PC girder bridges with three different diaphragm types. The bridge models were analyzed for a lateral-impact load that was applied to the bottom flange of the exterior girders at the intermediate diaphragm location and away from the diaphragm location. The induced strains and displacements in the girders were established for the different diaphragms. When a lateral-impact load was applied at the diaphragm location, the reinforced concrete, intermediate diaphragm provided more protection for the girders than that provided by the two types of structural steel, intermediate diaphragms, on the other hand, the different diaphragm types provided essentially the same degree of impact protection for the PC girders when the load was applied away from the diaphragm location.

1 INTRODUCTION

Bridge engineers are concerned about the response of prestressed concrete (PC) girder bridges which are hit by over-height-vehicle loads. According to (Shanafedt & Horn 1980), about 200 PC girder bridges in the United States are damaged each year, and about 162 of these bridges are damaged by over-height-vehicle loads. The actual number of impacts to bridges is probably significantly higher than these numbers, since many minor collisions are not reported to authorities.

Engineers with the Bridges and Structures Design Section of the Iowa Department of Transportation (Iowa DOT) require the use of reinforced concrete (RC), intermediate diaphragms for all overpasses, PC girder bridges that have traffic beneath them when they are constructed on the Federal and State Bridge Systems. To reduce the construction time and to simplify the construction process for PC girder bridges, bridge contractors in the State of Iowa have always expressed a desire to install steel intermediate diaphragms rather than to construct RC intermediate diaphragms for this type of a bridge.

This paper summarizes whether a steel intermediate diaphragm with simple connections to the PC girders can provide the same degree of damage protection to the PC girders as that provided by the RC intermediate diaphragm for both non-skewed and skewed bridges. This study involved the use of detailed finite-element models of several bridges using the ANSYS software (ANSYS User Manual 1998). Even though experimental work was not conducted during this investigation, published experimental test results for a similar structure was used to calibrate the finite-element models.

2 FINITE ELEMENT CALIBRATIONS

2.1 *Description of an experimental bridge model*

The single-span bridge that was constructed and tested by (Abendroth et al. 1991) at Iowa State University was used to guide the development of the finite element modelling of the prototype bridges presented herein. The bridge had three, Iowa DOT Type-A38, PC girders that were spaced at 6 ft – 0-in. (1830-mm) on center. The girders supported a 4-in. (100-mm) thick RC deck that was 40 ft – 4-in. (12300-mm) long and 18-ft (5490-mm) wide. At each end of the bridge model, a 42-in. (1070-mm) deep by 18-in. (460-mm) wide RC abutment supported the PC girders. The ends of the girders were embedded 8-in. (200-mm) into a full-depth, 14-in.

(360-mm) thick, RC, end diaphragm. The abutments were cast on the laboratory floor. Different types of intermediate diaphragms were considered in those tests.

2.2 *Finite-element model of the experimental bridge*

The deck and the PC girders of the experimental bridge were modeled using solid elements with eight nodes. Shell elements were used to model the end-diaphragms and the abutments. The different types of intermediate diaphragms were included in the finite-element models. Solid elements were used to idealize the RC intermediate diaphragms. For each diaphragm type, interface elements, which can model sliding and separation between the elements that are adjacent to interface surfaces were used in the finite-element-models. These models were analyzed for transverse horizontal loads that were applied to the bottom flange of one of the exterior girders at the mid-span of the bridge. These loads were applied as two equal concentrated forces that acted in a vertical plane. The complete details of the finite-element modelling techniques for the experimental bridge are summarized in (Andrawes 2001)

A 20% difference between the measured-maximum-horizontal and analytical displacements was observed. The calculated longitudinal strains in the PC girder were compared to the measured strains. The comparison showed that there was about a ±20% range between the measured and predicted strains. These differences were attributed to the presences of concrete cracks that existed in the experimental bridge model but were not included in the finite-element models. The relative closeness of the analytical predictions to the measured bridge responses revealed that the modeling techniques used were applicable to analyze PC girder bridges that are subjected to lateral forces.

3 BRIDGES SELECTED FOR THE ANALYSIS

3.1 *Non-skewed bridge*

The non-skewed bridge that was selected for this study has four spans, three frame-type piers, and two integral abutments. The length of each end span is 35 ft – 9-in. (10900-mm), while that for each inner span is 96 ft – 6-in. (29400-mm). An 8-in. (200-mm) thick bridge deck is supported by five, equally-spaced, Iowa Type-D, PC girders. A 3 ft (910-mm) thick, RC end diaphragm (abutment back wall) was cast at each abutment. At the ends of the PC girders and at the location of the RC intermediate diaphragm, two, ¾-in. (20-mm), diameter, coil rods pass through the bottom flange of each girder and extend into the end and intermediate diaphragms. Bent reinforcing bars connect the bridge deck to the end diaphragm.

3.2 *Skewed bridge*

A skewed, PC girder bridge was analytically investigated to determine the effect of a skew angle on the response of the bridge to lateral impacts. The skewed bridge that was selected for this study has a 20.4° skew angle, four spans, three frame-type piers, two integral abutments, and five PC girders. Each end span is 45 ft – 9-in. (13900-mm), long and each inner span is 96 ft – 6-in. (29400-mm) long. The bridge girders and deck have similar geometric and material properties and have the same connections between the diaphragms and the PC girders and RC deck as that for the non-skewed bridge.

4 INTERMEDIATE DIAPHRAGMS

Two steel and one RC intermediate diaphragm were considered in this study. The steel diaphragms were an X-braced diaphragm with a horizontal strut and a K-braced diaphragm with a horizontal strut. All three intermediate diaphragm types are standard diaphragms used by the Iowa DOT. Additional information regarding the description of these intermediate diaphragms is given in (Andrawes 2001).

5 FINITE ELEMENT MODELING

For all of the finite-element models, the PC girders and the 8-in. (200-mm) thick, RC slab were modeled using solid elements. An isometric view of the finite element model is shown in Figure 1. The 10-in. (250-mm) thick RC intermediate diaphragms, shown in Figure 2, were modeled by solid elements. A three-dimensional truss element was selected to idealize the coil rods that connected the RC intermediate diaphragm to the PC girders. For the two, steel, intermediate diaphragms shown in Figures 3 and 4, shell elements were used to model the horizontal struts, bent plates, and flat plates; and, beam elements were used to model the cross and K-bracing members.

Finite-element modeling of the connection between the members in the steel intermediate diaphragms and the bent plates that were used to attach the diaphragms to the webs of the PC girders included the use of rigid-link elements, contact elements, and pairs of counteracting compressive forces. The length of the rigid-link elements accounted for the eccentricity between the center of gravity for the diaphragm members and the outstanding leg of the bent plate. Contact elements with a coefficient of friction equal to 0.33 were placed between the interface surfaces of the connection to permit slippage between the parts. The counteracting compressive forces represented the clamping force that is developed in a connection when high-strength bolts are installed in a fully-tensioned condition.

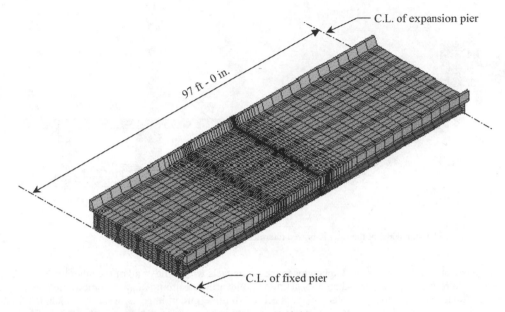

Figure 1. Finite-element model of the interior span of a non-skewed bridge.

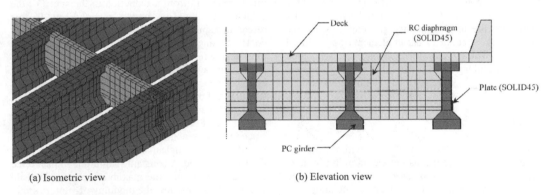

(a) Isometric view

(b) Elevation view

Figure 2. Finite-element model of the reinforced concrete intermediate diaphragm.

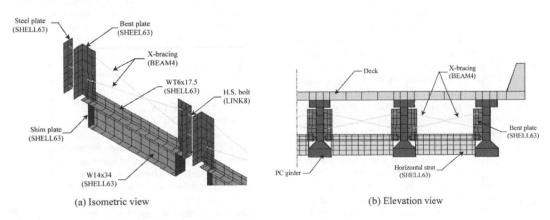

(a) Isometric view

(b) Elevation view

Figure 3. Finite-element model of the steel X-braced intermediate diaphragm.

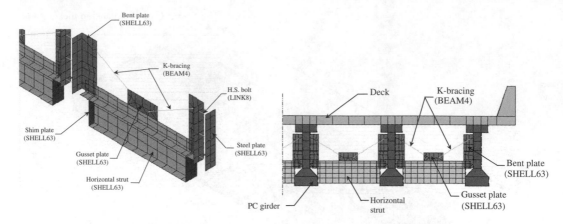

Figure 4. Finite-element model of the steel K-braced intermediate diaphragm.

The bolts that connected the steel diaphragms to the PC girders were modelled as three-dimensional truss elements. Since the high-strength bolts for the steel diaphragms and the coil rods for the RC diaphragm provided the connections between the diaphragms and the girders, common nodes were not used between the elements for the diaphragms and the elements for the bridge deck or girders. Contact elements were utilized on all interface surfaces where slippage and separation might occur along those surfaces. The concrete and steel material strengths were assumed to be linearly elastic.

6 LOADS

The major factors that are associated with the moving object and influence the magnitude of an impact load are its mass, speed, geometrical configuration, and hardness. A search of the published literature that addressed vehicle or object impacts did not reveal any information regarding over-height-vehicle loads striking bridges. Since the main objective of this research was to conduct a comparative study that evaluates the effectiveness of different types of intermediate diaphragms in minimizing structural damage to a bridge superstructure when a lateral-impact force was applied to the bottom flange of PC bridge girders, a precise forcing function did not need to be defined.

One scenario that may occur when an over-height-vehicle load strikes a bridge is as follows: First, the over-height object would impact the bottom flange or web of the first exterior girder (girder BM1). Then, because the vehicle would not suddenly stop, the object being transported could displace downward, as the vehicle-suspension system reacts to the impact, which would allow the object to pass beneath girder BM1. As the vehicle-suspension system rebounds, the object could displace upwards and cause additional

impacts with some or all of the other bridge girders at either their bottom flange or somewhere on their web. Multiple-girder impacts were not included in this study because the reduction in the impact-force magnitude resulting from a reduction in the speed of the vehicle after the first impact is unknown. In this research, a single-impact load was applied on BM1 or on BM5, since these loading conditions induce the most severe responses for an impacted girder.

To simulate the impact resulting from an over-height-vehicle load passing beneath the bridge and striking the bottom flange of an exterior bridge girder, an impact load was defined by its duration time and magnitude. Published articles that discussed car-crash tests were obtained during the literature search. Based on these articles, an 0.10 sec., impact-duration time was selected. A constant-magnitude impact force was selected to keep the maximum, principle-tensile strains near to or below the modulus-of-rupture strain for the concrete in the PC girders. Two different rectangular impact pulses were used to represent an impact force that is generated by an over-height-vehicle collision with a PC girder. A maximum load of 120 kips (530 kN) and 60 kips (270 kN) were selected when the impact load was applied at the mid-span and away from the mid-span, respectively.

7 ANALYSES OF ONE SPAN AND COMPLETE BRIDGE MODEL

To simplify the analytical work and to reduce computer computational requirements, an investigation was conducted to determine whether a finite-element model for only an interior span of a four-span bridge would predict, with sufficient accuracy, the PC girder strains and displacements that are obtained from an analysis of a four-span finite-element model. Even though the pier

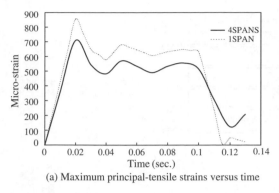
(a) Maximum principal-tensile strains versus time

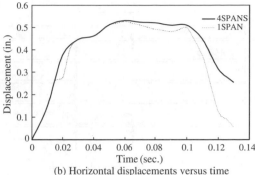
(b) Horizontal displacements versus time

Figure 5. Finite element results for the four-spans and one-span models when the load is applied at mid-span of girder BM1 (1-in. = 25.4-mm).

structures were not included in the finite-element models, their lateral stiffness was represented by horizontal springs at the pier locations. These two analytical models did not have any intermediate diaphragms.

The maximum, principal-tensile strains in the loaded girder and the maximum, horizontal displacement for the loaded girder of the four-span model and the one-span model are presented in Figures 5a and 5b, respectively. These strains and displacements were induced by a 120 kip (530 kN) lateral-impact load with a 0.10 sec. duration time that was applied at the mid-span of girder BM1. These strains occurred in the top fibers of the girder web at the girder cross-section containing the applied load, and these displacements occurred at the mid-span and at the bottom flange of the girder. Both the strain and displacement responses were the largest of those load effects for the loaded girder that were predicted by the finite-element models. The figure shows very similar strain and displacement behaviors. Only about a 15% difference occurred in the magnitudes of the maximum, principal-tensile strains. The displacement responses predicted by the two finite-element models were essentially identical for the 0.10 sec. period for the applied impact load.

The differences in the girder responses that were predicted by the two finite-element models and shown in Figure 4 were not significant enough to affect the objectives of the research. Therefore, the single-span, finite-element model was selected to adequately represent the girder responses to lateral-impact forces. A similar single-span model was developed for the skewed bridge. Additional information regarding the finite-element models is given in (Andrawes 2001).

8 ANALYSIS OF PROTOTYPE BRIDGES

For the non-skewed bridge, the impact load was applied at one of five locations, as shown in Figure 6a.

Load positions 1 and 2 were at the mid-span of girders BM1 and BM5, respectively. These locations matched the location for the intermediate diaphragms. Load positions 3 and 4 were applied at 16 ft (4880-mm) away from the intermediate diaphragm location. As the analytical study progressed, the researchers decided to apply an impact load at load position 5 that was 4 ft (1220-mm) away from the intermediate diaphragm location. This fifth load location was considered to account for an impact load that was applied close to but not at the intermediate diaphragm location. Figure 6b shows the impact-load locations for the skewed bridge.

When the impact load occurred at the intermediate diaphragm location, as shown in Figure 7a, the decrease in the maximum, principle-tensile strains from those strains for the bridge without intermediate diaphragms was significant and dependent on the type of intermediate diaphragm. While Figure 7b shows that when the impacted load occurred at 16 ft (4880-mm) from the midspan of girder BM1, the decrease in those strains was about 25% and basically independent on the type of intermediate diaphragm.

Figures 8a and 8b summarize the differences between the maximum principal strain in the impacted girder and the first interior girder. Figure 8a illustrates that the RC intermediate diaphragms provided the largest degree of impact protection to the impacted girder. The steel, X-brace and K-braced, intermediate diaphragms provided essentially the same amount of impact protection to the impacted girder as indicated by the essentially equivalent strain magnitudes. The presence of the RC intermediate diaphragms reduced the magnitude of the maximum, principal-tensile strain to about 28% and 35% of that strain for the no intermediate diaphragm condition for the non-skewed and skewed bridges, respectively. The use of either the steel, X-braced or K-braced, intermediate diaphragms reduced the maximum, principal-tensile strain in girder

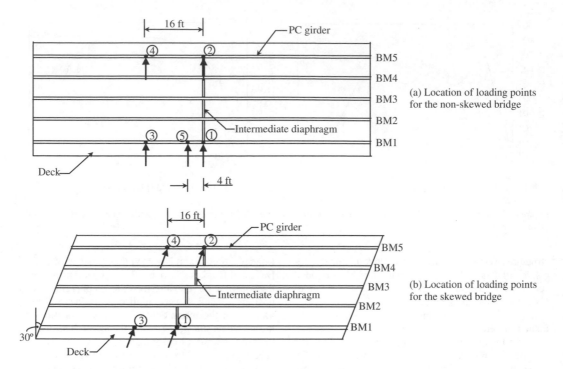

Figure 6.   Loading points (1-in. = 305-mm).

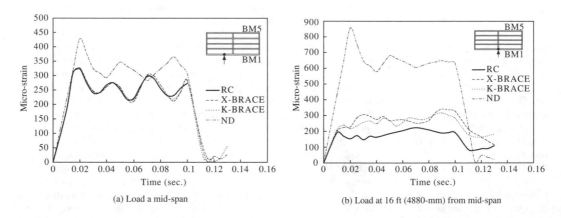

(a) Load a mid-span

(b) Load at 16 ft (4880-mm) from mid-span

Figure 7.   History of the maximum principal tensile-strain in BM1 for the non-skewed bridge.

BM1 to about 40% and 60% of that strain for the no intermediate diaphragm condition for the non-skewed and skewed bridges, respectively. A comparison of the strains that are induced in the first girder (girder BM2) for the non-skewed bridge (see Figure 8a) reveals that the RC intermediate diaphragms induced less strain in that girder than that for either configuration of a steel intermediate diaphragm. However, a similar comparison for the skewed bridge (see Figure 8b) reveals that

either configuration of a steel intermediate diaphragm induced less strain in girder BM2 than that for the RC intermediate diaphragms. This difference in the behavior between the skewed and non-skewed bridges was attributed to the staggered alignment of the intermediate diaphragms for a skewed bridge. When intermediate diaphragms are in alignment for a non-skewed bridge, a larger amount of the impact force, which is induced in the first intermediate diaphragm, is

1000

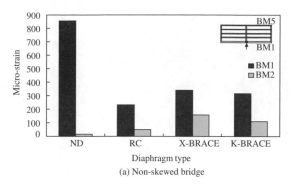 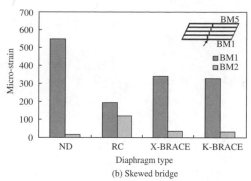

(a) Non-skewed bridge   (b) Skewed bridge

Figure 8.   Maximum principal-tensile strains in BM1 and BM2.

transferred to the rest of the diaphragms than that for the skewed bridge where the intermediate diaphragms are not in alignment.

9 SUMMARY

The results summarized in this manuscript illustrated that when a lateral-impact load was applied directly at the diaphragm locations in a bridge structure, the RC intermediate diaphragm provides the bridge girders with a higher degree of impact protection than that provided by either of the two steel types of intermediate diaphragms presented in this study. However, when an over-height-vehicle load strikes a bridge girder away from a diaphragm location, the degree of impact protection provided by each of the three types of intermediate diaphragms was almost the same.

ACKNOWLEDGEMENTS

The research presented in this paper was conducted by Iowa State University, administered by the Center for Transportation Research and Education, and sponsored by the Iowa Department of Transportation, Highway Division, through the Highway Research Board.

REFERENCES

Shanafelt G.O., & Horn W.B. 1980. *Damage evaluation and repair methods for prestressed concrete bridge members.* NCHRP Report 226.

De Salvo G.J., & Swanson, J.A. 1985. *ANSYS Engineering Analysis System User's Manual.* Volumes 1–4. Houston. Penn.: USA.

Abendroth, R.E., Klaiber, F.W. & Shafer, M.W. 1991. *Lateral load resistance of diaphragms in prestressed concrete girder bridges.* Iowa DOT Project HR-319, ISU-ERI-Ames-92076.

Andrawes, B.O. 2001. Lateral impact response for prestressed concrete girder bridges with intermediate diaphragms. *Thesis* submitted to Iowa State University.

Modelling aspects for the analysis of a box girder steel bridge

L. Catallo
University of Rome "La Sapienza", Rome, Italy

S. Loreti & S. Silvi
Structural Engineers, Rome, Italy

ABSTRACT: Bi-dimensional finite elements modelling allows to evaluate the tensional state with more precision, than one obtained by mono-dimensional elements. Anyway, the results achieved with these models are influenced by the level of discretization of the structure. Some fundamental indications (reported in CNR 10024/84, too) are: trying different models to obtain more results, repeating the analyses using different parameters and structural defects. Furthermore, in the structure two classes of situations can be found namely the B-regions, where the states of strain is produced by simple warp (with linear trend) and by the D-regions (diffusive stress place), where the absence of simple cinematic, allow state of strains however complex.

1 GENERAL CHARACTERISTICS

In this work the modelling of a steel girder box of a bridge of first category for a A type street (CNR 2000 for extra-urban streets), with a static model of continuous beam, with three 30 m long spans is presented (Fig. 1). The realization of two independent structures is expected for the two gear ways, with each deck 17 m wide.

The scope of this modelling is to evaluate the sensitivity of the model on changing some parameters (related both to numerical and mechanical aspects) as the level of discretization, the step of transverse diaphragms, the size of the steel plates, the thickness and the step of the ribs.

2 PREDIMENSIONING

The actions considered and the respective combinations have been evaluated by the Italian code, (D.M. 4 May 1990). Using a finite element commercial program, the structure has been modeled as a continuous beam on three spans using mono-dimensional elements, i.e. frame elements.

The section used in the code, is an hollowed rectangular shape different from the section chosen for the girder box (Fig. 2), therefore the geometric characteristics of the section has been calculated separately, and the introduction of multiplicative factors that have been assigned to the section used in the model (Fig. 3).

In particular, for the calculation of the torsional constant (Fig. 4) a hewn stone of deck one meter high has been modeled with bi-dimensional elements and in correspondence of the section at the quote of one meter it has been imposed tie of diaphragm and applied a torque while in the section at the zero level a fixed joint section was applied. The torsional costant it was foreed used in the calculation has been obtained using the eq. (1) and dividing the result by the tangential elasticity modul $G = 7.688\ 10^{10}\,\text{N/m}^2$.

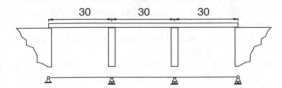

Figure 1. Static scheme bridge.

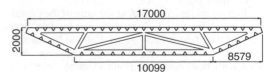

Figure 2. Cross-section of the bridge.

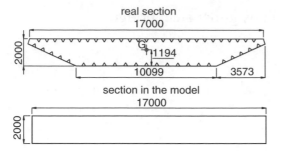

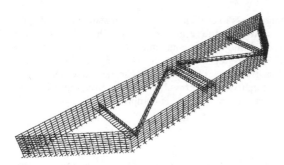

Figure 3. Schemes for calculated of the geometric property section.

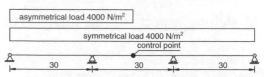

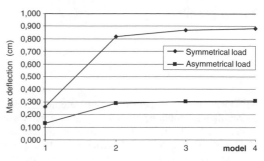

Figure 5. Load test.

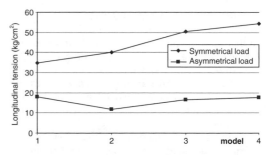

Figure 4. Modelling a hewn stone of deck one meter high to define torsional constant.

$$K = \frac{M}{\vartheta} \qquad (1)$$

The steel thickness of the plates has been taken by 12 mm (the Italian norms fixed minimum thickness 10 mm)

The dimension of the superior and inferior deck, have been fixed only changing the height of the girder. The height of the girder box (2 m), has been calculated by an iterative process.

3 RELIABILITY OF THE MODEL

With predimensioning, the girder box height has been fixed equal to 2 m.

To have an indication about the ratio of the modelling reliability used, four different models have created. The first called "model 1", used four mono-dimensional elements, while the others have been realized with bi-dimensional finite elements, obtaining respectively: "model 2" with 2250 elements, "model 3" with 9360, "model 4" with 26640 elements. These

Figure 6. A comparison about the deflection in the center line section.

Figure 7. A comparison about the state of strain in the center line section.

models have been loaded using two test loads located in a symmetrical and asymmetrical way with respect to the bridge longitudinal axis (Fig. 5).

A comparison about the deflection (Fig. 6) and the state of strain (Fig. 7) has been taken between the four models.

As can be noticed, the stress convergence is slower than the displacements one because the stress depends on differentially of the displacement.

A specific point to investigate is related to the fact that, with a girder box 17 m wide and a 30 m span long, the Bernoulli Theory can be applied only approximately as the section, doesn't remain plane. This assertion is convalitated by the shear-lag effect showed in Figure 8.

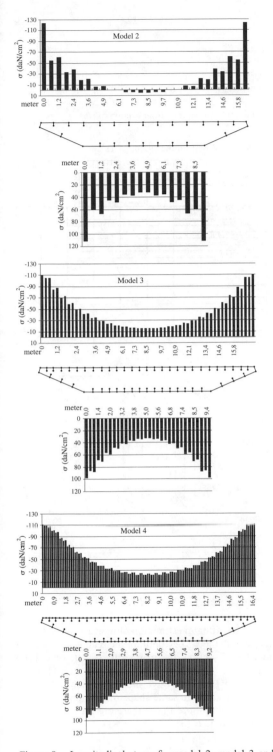

Table 1. Differences between the model 2-1, 3-1 & 4-1.

	Model 2-1	Model 3-1	Model 4-1
Step diaphragms (m)	5	5	4 and 3
Girder box plate (mm)	12	12	12
Upper ribs thickness (mm)	10	20	20
Lower ribs thickness (mm)	10	10	10

4 OPTIMIZATION

With "model 1" the influence lines for the bending moment on the first span's bearings section and center line section have been defined, as well as for the first span's torsional moment in the bearing section. The influence lines were used to load the model formed by bi-dimensional elements; therefore there are three identical models, in which the load disposition changes.

The discretization level of "model 3" has been considered acceptable: therefore, it has been used for the calculation of upper thickness ribs and the inter-axis between transverse diaphragms. Changing this two parameters the results in term of the ratio between ideal stress (Von-Mises) and steel volume used in the structure, have been compared.

Table 1 shows that the ribs thickness and the step of transverse diaphragms has been increased: this fact produce an increase of structural weight and a reduction of the ratio *stress/steel volume* (Fig. 9).

Transverse diaphragms were modelled with bidimensional elements and at this level of analysis they don't include openings for inspection and maintenance.

A new model is obtained with this optimization process called "model 4-1", in which the upper thickness ribs is 20 mm and the transverse diaphragms is 4 m.

5 SUBSTRUCTURING TECHNIQUE

Starting from the results obtained, a sub structuring process has followed in order to analyze the so called D-regions, or diffusions zones.

The zones where the concentrated loads and the restraints have been applied, have been analyzed: in both of them the discretization was improved. Furthermore, the thickness of bi-dimensional elements near the external constrain has been increased. With the "model 4-1" the optimization process of the structural weight was made, changing lower ribs step.

To these ribs has been assigned a different stretch modulus to consider the different shape between modelling ribs and those of the real structure. A different stiffness has been assigned to the upper ribs, too.

Figure 8. Longitudinal stress for model 2, model 3 and model 4 (shear-lag effect).

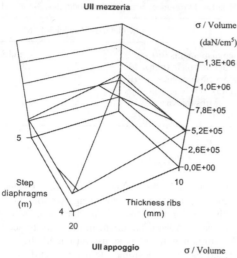

UII mezzeria

σ / Volume
(daN/cm⁵... rendering)

Figure 9 graph: σ / Volume (daN/cm^5), with axes Step diaphragms (m), Thickness ribs (mm). Values: 1,3E+06, 1,0E+06, 7,8E+05, 5,2E+05, 2,6E+05, 0,0E+00.

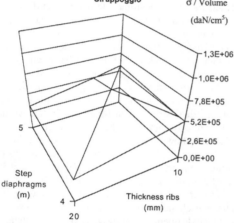

UII appoggio

σ / Volume (daN/cm^5)

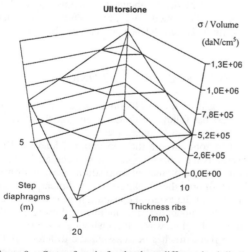

UII torsione

σ / Volume (daN/cm^5)

Figure 9. State of strain for the three different load dispositions, changing upper thickness ribs and the step of trasversation diaphragms.

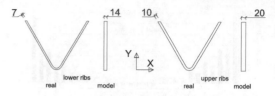

Figure 10. Upper and lower section of ribs.

Table 2. Weight for the "model 4-18".

DECK	Line n°	Thickness mm	Height mm	Long mm	Volume m³	Unit surface weight (kg/m²)	Incidence % of unit surface weight
UPPER RIBS	27	20	220	90000	10,7		
UPPER PLATE		16	17000	90000	24,5		
			Total volume m³		35,2		
			Steel weight kg/m³		7950		
			Total weight kg		279617	183	55%
CLOSED BOX	Line n°	Thickness mm	Height mm	Long mm	Volume m³		
RIGHT SIDE RIBS	3	12	220	90000	0,7		
LEFT SIDE RIBS	3	12	220	90000	0,7		
LOWER RIBS	8	20	220	90000	3,2		
LOWER PLATE		12	1000	90000	1,1		
RIGHT PLATE		12	4076	90000	4,4		
LEFT PLATE		12	4076	90000	4,4		
OBLIQUE PLATE		12	400	90000	0,4		
			Total volume m³		14,9		
			Steel weight kg/m³		7950		
			Total weight kg		118533	77	23%
STIFFENING	Line n°	Thickness mm	Surface mm²		Volume m³		
MIDDLE	22	30	18768262		12,4		
SUPPORT	3	30	21899896		2,0		
			Total volume m³		14,4		
			Steel weight kg/m³		7950		
			Total weight kg		114146	75	22%
			TOTAL WEIGHT kg		**512296**	**335**	

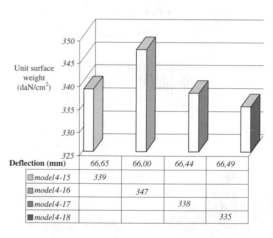

Deflection (mm)	66,65	66,00	66,44	66,49
model 4-15	339			
model 4-16		347		
model 4-17			338	
model 4-18				335

Unit surface weight (daN/cm²) axis values: 350, 345, 340, 335, 330, 325.

Figure 11. Maximum deflection for the model "campata".

In fact, after an estimation of inertia moment compared with the x-axes, both for the rectangular section and the V section, the new stiffness $E'(2)$ has been calculated multiplying the module E (1999×10^{11} [N/m²]) with the two inertia moment ratio. The lowers ribs has the same shape of the upper ones but dimensioning more thick: therefore they have a different module $E''(3)$.

$$E' = E \cdot \frac{I_{x,V}}{I_{x,R}} = 2{,}367E + 11 \ \left[N/m^2 \right] \tag{2}$$

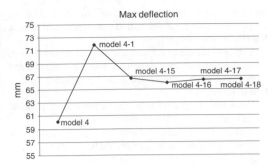

Figure 12. Structural weight for different lower ribs configuration.

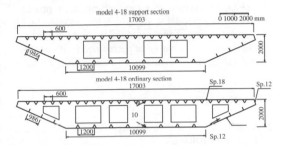

Figure 13. Section for the "model 4-18".

$$E'' = E \cdot \frac{I_{x,V}}{I_{x,R}} = 2{,}429E + 11 \left[N / m^2 \right] \qquad (3)$$

Results are different if one considers the inertia moment around the y-axes. They were considered those one respect the x-axes, considered the flexure for the moments (M_x).

In the "model 4-15" the lower ribs are smaller than the upper ribs, in the "4-16" both ribs are the same; in the "4-17" the lower ribs have double step in the current section, in respect of the upper ribs, while near the supports they have the same step, in the "model 4-18" they have double step for all the girder. In all the models, have been made openings for inspection and maintenance in the transversal diaphragms, as shown in Figures 14 and 15.

The steel thickness of the plate was fixed to 16 mm obtaining a state of stress in the plate itself smaller than 2500 daN/cm².

With this optimization process a model called "model 4-18" has been determined. In this model the structural weight has the lowest value.

6 FATIGUE SAFETY CHECKING

Steel structures are sensitive to the fatigue, in particular the bridges subjected to a high number of load

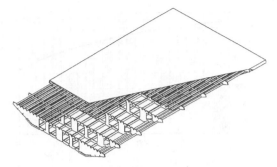

Figure 14. Three-dimensional schemes of box girder.

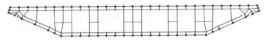

Figure 15. The stress value in the external knot were used for fatigue strength.

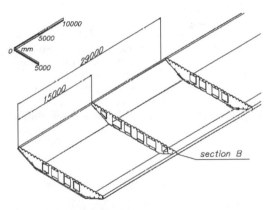

Figure 16. Section which was done the fatigue strength.

cycles. The fatigue checking has been carried out using the Wöhler method and supposing infinity life of the work ($n°$ cycle $> 2 \times 10^6$). With this hypothesis it is possible to design the Wöhler's curves on the $\sigma_m - \Delta\sigma$ plane neglecting the cycles number (for a cycles number $> 2 \times 10^6$, the projected curve is always the same). Fatigue checking will be verified if the representatives points of the fatigue-state in the model node (represented in this graphic the coordinates with $\sigma_m - \Delta\sigma$) are under the curve.

The inspections have been executed on the externals nodes (Fig. 15) of the section B (Fig. 16) with the load combination 1, including the own weight and accidental permanent weight and the load combination 2 in which were considered accidental weight besides load combination 1.

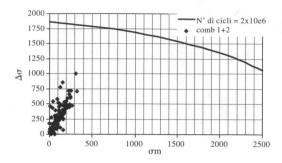

Figure 17. Wöhler's curve (daN/cm^2) the fatigue checking for the section B "model 4-18 camp".

The accidental loads are real, they aren't multiplied for a factor. The σ_m was calculated with load combination 1, the $\Delta\sigma$ was obtained as a difference between load combination 2 and 1 (Fig. 17).

7 CONCLUSIONS

A conscious modelling passes through the evaluation of the sensibility of the model to the change of some parameters like the discretization level, the transverse diaphragms step, thickness and the ribs step.

In the predimensioning phase, the height of the girder box has been fixed at 2 m. With the girder box's known dimension, the deflection and the state of strain has been analyzed changing the discretization level, to obtain a model called "model 3", in which 9360 shells were used.

Upper thickness ribs and the step of trasversation stiffening have been calculated, and the longitudinal stress have been comparated with the variation of this two parameters.

With this procedure it has been arrived at the new model called "model 4-1".

Starting from the results a sub structuration process of the problem examinated has followed thoroughly examining the D-regions analysis.

With this model it has been performed an optimization process of the structural weight changing the lower step ribs. There has been a convergence of the selected variable for the optimization process, obtaining the "model 4-18, which has provided with the lowest values of the structural weight.

On this model it has been performed (with positive result) the labor verification with the Wöhler method.

REFERENCES

Petrangeli, M.P. 1998. *Progettazione e costruzioni di ponti*. Masson (In italian).

Matildi, P. & Mele, M. 1971. *Impalcati a piastra artotropa ed in sistema misto acciaio calcestruzzo*. Italsider.

Hambley, E.C. 1976. *Bridge deck behaviour*. John Wiley & Son.

Zienkiewicz, O.C. & Taylor, R.L. vol.1. 1991, *The finite element method*. Mc Graw Hill.

System-based Vision for Strategic and Creative Design, Bontempi (ed.)
© 2003 Swets & Zeitlinger, Lisse, ISBN 90 5809 599 1

Structures for covering railways with real estate

Th.S. de Wilde

Holland Railconsult, Utrecht & Delft University of Technology, The Netherlands

ABSTRACT: Possibilities of realising real estate above railway infrastructure are studied because of a growing lack of space in inner city areas. The realisation of new buildings above existing railways involves, however, a complex process. New and innovative structural solutions can be used to enable constructing above tracks in such a way that it is financially and technically viable. For railways as well as for real estate, these (structural) solutions must comply with the demands that will be set for the flexibility to adapt to future changes. This paper discusses design considerations and basic structural solutions for such projects.

1 INTRODUCTION

Cities expand. The expansions are rapidly affecting the rural green spaces, and therefore they have an important environmental impact. In order to save these green spaces, alternative building sites are looked for. Within city limits, railway infrastructure takes up a lot of space on prime locations. By covering railway tracks, a new surface level is created on which real estate can be developed. Structures over the railway tracks are not only constructed to develop real estate on it, but they are also constructed to connect city districts sideways from the infrastructure in order to improve the urban fabric.

Building above railway tracks implies a complex process, which has organisational, technical, financial and safety challenges (Wilde 2002a). The fact that railway stations are located near historic city centres presents even more restraints to structural solutions as well as to applicable construction methods.

Structural solutions are guided by the possibilities to fit structural elements on station platforms and between railway tracks. The design is also subject to regulations set by railway companies exploiting the tracks. They demand flexibility, because future changes or extensions in railway infrastructure have to remain possible. Real estate developers, on the other hand, demand flexibility for the buildings they want to construct above the tracks, for they need the possibility to realise a large volume of different functions in order to make the project profitable. It stands to reason that the demands of these different parties are in conflict.

This paper discusses design considerations and different structural solutions, partly on the basis of international reference projects.

2 STACKING RAILWAYS AND REAL ESTATE

Railways and real estate have different functions, which are usually not combined vertically. The need for new floor area brings about a demand for solutions to realise these projects above railways, although they are not common. When considering the realisation of real estate above tracks, one should be conscious of the facts that their grid systems do not match and that there are different types of locations, where real estate can be positioned above the tracks. Furthermore, one should not forget that there is an important difference between new and existing situations.

2.1 Grid systems

Railways and real estate have different grid systems. Widthwise, the railway system differs from regular building structures. Longitudinally to the tracks, there are better possibilities to match the railway system with a standard structure for buildings.

2.2 Positioning the real estate

There are four different kinds of locations where real estate can be constructed above the tracks. Within the station, the position of platforms and tracks is mostly fixed. If there is enough space, structural elements can be placed on the platforms and between the tracks. The layout of the platforms mostly enables the use of a grid. In most cases, however, this grid does not comply with a standard building grid.

From the end of the platforms, the railway tracks are converging on the railway sidings. Building over railway sidings is difficult. There is hardly any space to place columns and a general grid system cannot be used. On these railway sidings, the flexibility of the railway tracks is important for transport capacity.

The third position to construct real estate above railway tracks is above the railway yards. Railway yards are used to put up trains and have a logistic function. In some cases train services such as maintenance work are also located there. These railway yards take up large sites close to the city centre with mostly parallel tracks.

The last of the four positions is found beyond the railway sidings where the tracks run more or less parallel. Building over parallel tracks is relatively easy compared to building over a series of platforms, because the stability of the structure can be found beside the tracks and the length of the span is limited.

Figure 1.   Covered platforms of Charing Cross Station.

2.3   Existing and new stations

Because of a lack of available building space, building over railway tracks is often considered in densely populated areas. As a consequence, building over tracks is considered predominantly for existing infrastructure. The design of buildings above new infrastructure can be geared to the demands of the real estate. Existing situations do not have this possibility, because in those cases the layout of platforms and tracks is an established fact.

Because of its complexity, this paper will focus on existing situations, where the layout of platforms and tracks determines the design of the real estate above it.

3   DESIGN CONSIDERATIONS

Before discussing basic structural solutions, the design considerations for real estate above railway tracks are dealt with. Important design considerations are quality, flexibility, structural design, construction, use of materials, pre-investments, and maintenance (Vákár & Snijder 2001).

3.1   Quality

When real estate is realised above the platform area, daylight is taken away. The space underneath the new building will be darker and the orientation of passengers towards their surroundings is disturbed. To ensure quality of space, the space must not be too low and daylight has to be present in such a way that it guides passengers. Adequate artificial lighting must be used to compensate for daylight that is taken away. Bringing in enough light at platform level is, however, not enough to guarantee sufficient quality. When platforms are covered, their exterior environment changes into an interior environment (Wilde 2002b). The use of materials of the platform area must match interior quality. The ceiling of the platform must therefore be more than the bottom of the building above it. A clear example can be found at Charing Cross Station in London.

If the platform area is used to place structural elements, it is important that social safety is guaranteed. If columns are too large, people can hide behind them and passengers may feel unsafe. Furthermore, columns must not hinder orientation and must not stand in the way of passengers getting on and off the trains.

3.2   Flexibility

Flexibility is a key word in designing real estate above railway tracks. Flexibility is needed on the one hand to enable changes and extensions of railway infrastructure, and on the other hand it is needed to be able to deal with the ever-changing plans of real estate developers. If one does not take flexibility into account, there will be long-lasting consequences of the chosen design, for the period of development and use of these projects is often a long one.

The structures should have a grid system that anticipates future changes. The buildings will have an economic life cycle of about forty years, while the infrastructure of railways is designed for more than a hundred years (Garvelink 2001). What is more, infrastructure is planned over a period of ten years, while real estate is planned over a period of five years or less.

3.2.1   Flexibility for railways
Flexibility for railway tracks is essential for their future capacity. In a project, choices have to be made on required changes and extensions, because these choices will be of importance to the use and capacity of the tracks. Changes in the platform areas are, however, less likely than changes on the railway sidings.

Where railway sidings are concerned, the railway companies' demands on flexibility will be more stringent.

3.2.2 Flexibility for real estate

The railway infrastructure in the station area is constructed for a use of more than a hundred years. The real estate above it will not have that life span and will be demolished at least twice during that period. Based on experience with many office buildings elsewhere the life span of real estate above rail tracks could be much shorter than that period; demolishing or restructuring is therefor very likely. Flexibility is required both to enable the construction of a new building, and to enable future adaptations of a building, which was already there. Especially the users of offices, who are paying high rents on station locations, have rapidly changing demands, which must be met by the real estate developers.

3.3 The paradox between flexibility and quality

Flexibility and quality are complementary. In order to create flexibility it seems logical to slightly overdimension the structure in order to build in redundancy. However, from the perspective of spatial quality it is desirable to have lighter structures to secure the entrance of daylight and to organise the orientation within the station. As a result, the structural engineer has the important task to weigh the demands between flexibility and quality into a balanced design.

3.4 Structural design

The design of the structure is subject to the length of the span that is imposed. When this length is less than about twenty meters the direction of the primal span can also be chosen longitudinal to the tracks. The length of the span over the tracks has also consequences for the stability.

3.4.1 Length of the span over the tracks

Because of the demands on flexibility for the tracks, the length of the span over the tracks will differ. If columns can be placed between all the tracks, the span can be limited to five meters. In situations where island platforms are used to place columns, this span will be about twelve to twenty meters. If all the tracks must be spanned in one go, the length can become more than eighty meters.

3.4.2 Direction of the primal span

When the span over the tracks is limited to about twenty meters, the direction of the primal span can also be chosen longitudinal to the tracks. In platform areas this direction can have its advantages, because the structure

can be placed between the trains instead of above them. In this way, the total height will be less than when a transverse structure with a girder above the tracks is used. The secondary span will be standardised for the use of a regular floor system within the building.

3.4.3 Stability

In longitudinal direction, the stability of the structure is realised between the tracks. Transverse to the tracks, the stability is much more difficult to realise. In most cases, there is no more space between the tracks than for a (single) column. The stability must be organised beside the tracks, which will be difficult when a structure is realised above the railway sidings. Other options are to take one of the tracks out of service, if that is possible, or to use a portal.

3.4.4 Connection with street level

A part of the structure, which is also important for the real estate is the connection of the building with its surroundings. This connection is also of importance to the arrangement of escape routes. A building above the railway tracks must be accessible and this may lead to bridge structures above the tracks. As this real estate is situated above street level, the route from street level to the real estate will often have a substantial difference in height.

3.5 Construction

The construction process of a building over railway tracks is an important factor for the design of the structure. Considerations on the building site, building time, and building noise are given below.

3.5.1 Building site

Building over railway infrastructure is mostly considered in inner city areas. When plans are made, one has to take into account that these building sites are not easily accessible, so that the use of large structural elements will be difficult. Usually, the dense station surroundings only allow for a very small building site. Therefore, structures will mostly be prefabricated and erected just-in-time on delivery. If the railway tracks are first covered with a basic layer, this layer could be organised as a building site.

3.5.2 Building time

Railway stations must be operational during the construction of the real estate and the logistics of construction must be co-ordinated with train schedules. As a consequence, building time must be reduced as much as possible and construction will have to take place mostly during the night (1 a.m.–5 a.m.) and during the weekends, when there is hardly any train traffic.

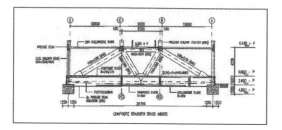

Figure 2. Transfer structure of the Malie-tower (Corsmit).

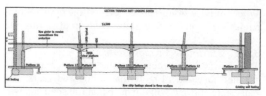

Figure 3. Columns of Bishopsgate 135–199 at platform level.

3.5.3 Building noise

As discussed above, the construction will mostly take place at night. Given the inner city environment, the regulations on building noise will limit the choice between different methods for foundations and on-site construction. Driving piles during night time will, for instance, not be accepted and it is likely that the structure will be prefabricated for the greater part.

3.6 Use of materials

Structures for real estate above railway tracks can be constructed from different materials, each with their own benefits.

3.6.1 Steel

Steel is often used in structures over railways. The most important reason to use a steel structure is that the construction takes a limited amount of time and there is only a limited weight to be hoisted. What is more, steel structures are relatively easy to change or extend, and they are more transparent, which has advantages for the required flexibility and for the quality of the platform area.

3.6.2 Concrete

Concrete can also be used to cover railways with real estate. Examples are found in Paris at Seine Rive Gauche and Gare Montparnasse. Concrete is used in these projects to limit the noise and vibrations of the trains. For concrete, pre-cast concrete is preferred over in-situ, because of a reduced building time.

3.6.3 Hybrid structures

Hybrid or mixed construction with pre-cast concrete means combined use of pre-cast concrete with other materials for the benefit of the project at large. An example is the structure of the Malie-tower above the motorway Utrechtsebaan that enters The Hague.

With the use of hybrid structures, the different materials are optimally used and 10 to 20 percent of construction time can be reduced (Vambersky 2002).

3.7 Pre-investments

Plans to build over railway infrastructure are often combined with changes on the railway infrastructure or the station. The connection of the station with the high-speed train network is an example.

When changes of the station are planned, there is often no clear plan for the real estate that will be realised above it. However, in order to enable the realisation of real estate above the station in a later stage, pre-investments will have to be made (Wilde & Suddle 2002). Because of the time gap between these investments and their actual use, they cannot be too high. It is best to minimise the pre-investments to a level in which building over the railways is not made impossible in the future.

3.8 Maintenance

Maintenance is an important aspect in the design, because trains must be disturbed as little as possible, while maintenance work is carried out. The railway area must therefore be designed with materials that need little maintenance.

4 BASIC STRUCTURAL SOLUTIONS

As discussed earlier, the layout of tracks and platforms is fixed in most situations. This means that the designer is not free to choose the grid of the structural elements at the level of the tracks. In order to construct real estate above the platforms, three structural solutions can be considered; the separating structure, the transfer structure, and the mega-structure.

4.1 Separating structures

The first structure to realise real estate above railway tracks is a separating structure. A separating structure is realised, by constructing a separating floor on columns that are put in the middle of the platforms or, where possible, between the railway tracks. This floor is over-dimensioned to take the construction loads of trucks and cranes. The structure of the real estate

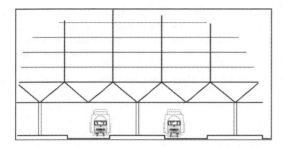

Figure 4. Basic scheme of a transfer structure.

Figure 5. Transfer structure of M7 in Seine Rive Gauche.

above the tracks adapts to the spans that are imposed by the columns below.

4.1.1 Reference project

An example of such a separating structure is found in London. The buildings of Bishopsgate 135–199 have been constructed above platforms 11 to 18 of Liverpool Street Station. The buildings are eight to ten storeys high. The columns are placed in the middle of the platforms and the primal structure spans 13.5 meters.

4.1.2 Considerations on application

When using a separating structure, the existing layout of platforms and tracks determines the structural grid of the real estate above. At the separating level, this structural solution is not very expensive. Extra costs will arise, because the floors of the real estate will have a deviant structural grid compared to normal real estate. An important advantage of this structure is that the new urban level is positioned almost directly above the tracks, which thus limits the difference in height with street level. The separating level will also be used in the construction process thus creating a larger building site. Furthermore the separating level will separate the trains from the construction

Figure 6. The Exchange House, London.

process with the consequence the construction process doesn't disturb the railway schedule.

4.2 Transfer structures

Transfer structures change an unfavourable grid of platforms and railway tracks into a feasible or even a standard grid for real estate. Transfer structures take loads where they can conveniently be collected and transfer them to where they can conveniently be resisted (Bird 1993). The transfer structure serves as an intermediate between the imposed grid of the tracks below and the desired grid of the real estate above. The use of a framework is logical, because bearing points can be created for the real estate directly above the trains on a convenient grid. Concrete beams can also be used. A transfer structure will take up two meters in height, or even more. If the height of the structure is made one storey high, the space can be used for parking facilities and installations. This extra storey can also serve as a buffer between the tracks and the real estate in case of calamities.

4.2.1 Reference

A good example of the use of such a transfer structure is found in the project Rive Gauche in Paris, which is still under construction. Parcel M7 is positioned directly above the newly constructed train station Gare Bibliothèque François Mitterand. During the construction, the trains were taken out of service. The transfer structure is constructed with in-situ concrete and the height of the transfer structure is used as a car park.

4.2.2 Considerations on application

The most important advantage of using a transfer structure as an intermediate is that the real estate above can have a convenient structural grid. This also has advantages for the flexibility of the real estate, for the position of the tracks can remain unchanged. Other advantages are that when the transfer structure is one storey high, the space can be used for parking facilities and installations. Finally, the standard grid that is created can be re-used when the building is demolished without disturbing the trains.

A problem with transfer structures is that they are relatively expensive. However, the extra costs of this expensive first level can be earned back by constructing sufficient building levels. All in all, one can say that the transfer structure enables a precise choice between the flexibility of the tracks and the flexibility of the real estate.

4.3 *Mega-structures*

If it is not possible to build on the platforms or between the railway tracks, a mega-structure can be used. A mega-structure is used to build over the infrastructure in one span. A mega-structure offers the flexibility to freely reposition the railway tracks after the building has been realised. An extension of the number of tracks, however, is difficult, because the large foundations for the mega-structure enclose the tracks.

4.3.1 *Reference project*
The Exchange House is constructed above the railway sidings of Liverpool Street Station. The structure spans almost eighty meters. The mega-structure was used, because no structural grid could be found between the tracks (Ridout 1989). The application of an arch is logical because it carries a permanent uniform load of the building.

4.3.2 *Considerations on application*
When a mega-structure is used, there will be maximal flexibility for the railway tracks below, but extension is difficult. Within the mega-structure a standard grid can be used for the building. A disadvantage of the mega-structure is that it is expensive and that changes on the building are difficult. It will also be difficult to demolish the building when it exceeds its economic life span.

4.4 *Building heights and functions*

Apart from the position above the tracks, the choice between the basic solutions that were presented above depends, to a large extend, on the height of the real estate and on its function.

4.4.1 *Building heights*
Transfer structures and mega-structures are expensive. If they are used, the costs must be earned back through the square meters of real estate that can be sold. For only one or two floors, a separating structure will therefore be a more logical choice.

If the real estate has more floors, the use of transfer- or mega-structures will become more attractive, because the buildings can be constructed in a standard grid for more floors. The choice between a transfer structure and a mega-structure depends on the possibilities to place large columns on the platforms or between the tracks. A mega-structure will be considered, when flexibility for the tracks is required. A mega-structure can also be considered when a building is high and the columns will take up too much of the space of the platforms.

4.4.2 *Functions*
The function of the real estate also has an influence on the choice between the structural solutions that were presented above. Office buildings depend on standard structures, mostly because an efficient structural grid enables the owner to optimise the lettable square meters. In this respect, transfer structures and mega-structures are to be preferred. Apartments are more flexible in their division. It is therefore possible to use a separating structure. In the light of flexibility, the use of a standard grid is desirable, so the application of a transfer structure is also preferred for apartments.

Shops and leisure functions are more flexible in their use of space. Large column-free floors are sometimes even desired. This can make the choice for a separating structure logical, especially when only two layers of facilities are constructed. Comparable considerations can be made for parking facilities.

5 CONCLUSION

Social developments demand an efficient use of scarce inner city locations for the realisation of new real estate. As a consequence, building over railway infrastructure is considered on many locations. In this paper, structural solutions to enable constructing above railway tracks have been considered.

When structures above tracks are designed, an important task of the structural engineer is to judge the demands both on flexibility and on quality, and these are, to a large extent, complementary. Moreover, choices must be made concerning construction in an inner city environment, which will often present important restrictions to the methods of construction that can be applied. Three basic structural solutions have been presented that can be used along with considerations on their application. The choice for each of these structures will depend on their position above the tracks, on the height of the buildings, and their function.

The realisation of structures above tracks includes more than adding square meters of floor area to the city. The buildings above railway tracks and the public space around them can substantially add to the spatial quality. The structures will form a connection between the city districts on both sides of the tracks and in that way they also improve the urban fabric.

REFERENCES

Bird, B. 1993. Transfer structures. In A. Blank (ed.), *Architecture and construction in steel*: 252–261. London: E&FN Spon.

Garvelink, J.J. 2001. Bouwen in stedelijke knooppunten, Een tour d'horizon, Utrecht: Holland Railconsult.

Ridout, G. 1989. Arch revival, *Building*, 17 March: 39–44.

Vákár, L.I. & Snijder, H.H. 2001. The design of stations in densely populated areas, *Structural Engineering International*, 11(2): 128–138.

Vambersky, J.N.J.A. 2002. Precast concrete in buildings: World trends, achievements and future. In ANO "NIZB-FORUM", Beton na rubezje tretjego tricjacjeletija, Moskow, Associacija Ijelezobeton: 376–387.

Wilde, Th.S. de & Suddle, S.I. 2002. Multiple use of land in The Netherlands, Concrete, 36(4): 34–35.

Wilde, Th.S. de. 2002a. Multiple use of space in railway station areas. In C.A. Brebbia et al. (eds.), *The sustainable city, Urban regeneration and sustainability*, Southampton, WIT Press: 529–538.

Wilde, Th.S. de. 2002b. Underground spaces at new heights, In ACUUS, *Urban underground space: a resource for cities*; Proc symp, Turin, November 2002.

Grain silos problem and solution

F. Shalouf

Faculty of Structural Engineering, University of Derna, Derna, Libya

ABSTRACT: During discharging the silo occurs high dynamic pressure, shear and pulsation. The silo must be withstand such things otherwise it will be damaged. The reinforced concrete silo can stand to shear but the steel silo must be taking in consideration the shear. For this reason the paper discusses the controlling of flow pattern to reduce of the shear by using Side wall discharge tubes.

1 INTRODUCTION

The most important tasks of a silo structure designer, especially as regards eccentric discharge from one or more outlets, still include the problem of determining the dynamic pressure and shear acting on the silo walls when the materials discharging and the possibilities of reducing it. It is generally agreed that a side discharge is less desirable than a center discharge from the viewpoint of wall loading. This is due to the fact that the side discharge tends to impose a non-symmetrical loading on the silo walls, thereby, introducing bending stresses onto the walls. For elimination of the above-mentioned disadvantageous effects during discharge needs to manage the bulk material flow pattern by anti-dynamic tubes.

The experience of the past forty years has shown that the use of discharge tube inserts is an effective and cheap way to reduce the horizontal pressure and shear evoked by bulk materials on silo walls and maintain low stresses inside the material. Many authors have recommended repairs of grain silos that developed vertical cracks by installing centrally located anti-dynamic tube (Kaminski et al., Ooms et al. 1985, Kota 1996). Moreover, according to American, Australian and Hungarian experiences (Kota 1996, Ooms 1985, Sargent 1979) it has been shown that tubes built into the center of the large silo were deteriorating after some time of operation, causing discharging without the benefit of such applied devices. These disadvantages do not have silos in which a discharge tubes is fixed to internal or external surface wall.

Tubes fixed to internal wall surface may have been proposed by Reimberts (1987) and subsequently, by Theimer (1958), recently by Kota (Kota 1996, Shalof 1998, Shalof 2000a, Shalof 2000b). In turn of tubes fixed to outside wall surface were proposed by Kaminski et al.

2 PROBLEMS OCCUR DURING DISCHARGING THE MATERIAL FROM SILO WITHOUT USING DISCHARGE TUBE

- some problems occur when the material discharging from silo such as an eccentric flow channel occurs in very big silo and leads to high horizontal and vertical bending moments;
- over pressures and shears occur during the discharging of materials from silos and they can exceed the static by many times;
- pressure pulsation occurs when the materials flow from the silo;
- high pressures and shear occur at connection of the cylinder with hopper in mass in flow;
- an eccentric flow channel may occur in several outlets malfunctioning simultaneously and leading to horizontal and vertical bending moments;
- bending moments occur when silo discharge by multiple outlets;
- mass flow occurring when silo design to funnel flow;
- bending moment of circular walls caused by eccentric outlet discharging. When the silo discharge by eccentric outlet, eccentric flow channel occurs and leads to non-uniform pressures around the circumference of the silo leading to horizontal and vertical bending moments.

3 MODEL DESCRIPTION

The test model was designed so that features such as the support structure were typical of existing structures. The model parameters are taken so that they will be similar to silo on a scale 1:10: a new concept of load reduction by designing silos with a deliberate increased flexibility (Kobielak et al. 1997). The structural system

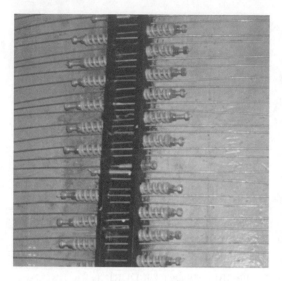

Figure 1. Test model, a steel auxiliary silo, support structure and a bulk material bucket conveyor and feeder.

Figure 2. Anchored strings by means of the springs in the steel plaster.

is analyzed as many isolated structure. The silo walls are reinforced concrete prefabricates and they are designed with assumed boundary conditions and interaction between individual fabricates is ignored. This concept of a load reduction-flexible silo design (Koster 1985, Feng 1996) led to the design of the stand to investigate phenomena in silos with flexible walls. In this paper the model is stiff. The model silo has 10 pressures cells mounted along two generatrix A and D, Each generatrix has 5 cells fixed at five levels. Distances to each level were measured from the bottom of the silo (level 1–11 cm, level 2–103 cm, level 3–165 cm, level 4–234 cm, level 5–300 cm). The presented results concern cell position placed at level 2 (103 cm from the bottom of the model).

3.1 The test model

The overall height of test model is 4.0 m, the diameter $D = 1.5$ m. The model is a flat steel bottom circular silo consisting of eight reinforced concrete (grade B25) prefabricates with wall thickness of 3.5 cm. On four wall prefabricates, openings are made to fix the pressure cells measuring bulk material pressures evoked on the model's wall, whereas on the other four elements, steel pillars are attached in which compressing strings are anchored by means of springs in bracing pilasters (Fig. 1). Figure 2 shows the steel pilaster with fixed string and springs. By changing the spacing of the working compressing strings, can get different values of silo wall flexibility. The diameter of the discharging hole is d = 216.7 cm.

3.2 The discharge tube

The discharge tubes are made of wood with square size 16×16 cm ($\pi_1 = 0.10667$) with vary vertical slot and horizontal slots, the discharge tube fixed in internal surface of silo walls (Fig. 3).

4 ANALYSIS OF EXPERIMENTAL RESULTS

4.1 Silo fitted with one discharge tube with fixed height and vary vertical slot widths

From Figures 4 and 5 which present coefficient curves for opposite and side tubes of different vertical slot width, it can be seen that, there is an optimal width equals 8 cm which has smallest average dynamic shears coefficient (1.061 in generatrix D and 1.121 in generatrix A). At level 2 however, using of discharge tube with width equal 8 cm causes reduction of dynamic shear equal 48.34% in generatrix A.

4.2 Silo fitted with one discharge tube with varied heights and fixed slot width

Analyzing the Figures 6 and 7 of the results of silo model discharging by a opposite and side tubes with

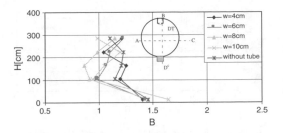

Figure 4. Effect the width of vertical slot width of one discharge tube in generatrix B with fixed height ($h_t = 3.9$ m) and barley height (H = 3.9 m) on dynamic horizontal pressures coefficient in generatrix D.

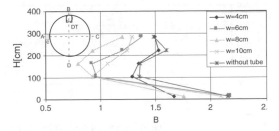

Figure 5. Effect the width of vertical slot width of one discharge tube along generatrix B with fixed height ($h_t = 3.9$ m) on the dynamic horizontal pressures coefficient in generatrix A.

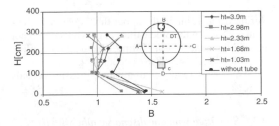

Figure 6. Effect the height of discharge tube with fixed vertical slot width (w = 6 cm) of one discharge tube in generatrix B with material height (H = 3.9 m) on the dynamic horizontal pressures coefficient in generatrix D.

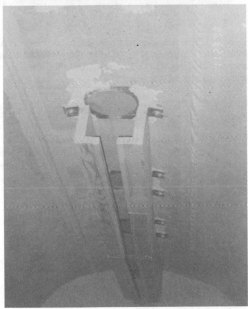

Figure 3. Side wall discharge tubes with vertical and horizontal tubes.

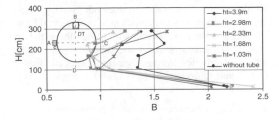

Figure 7. Effect the height of discharge tube with fixed vertical slot width (w = 6 cm) of one discharge tube in generatrix B with material height (H = 3.9 m) on the dynamic horizontal pressures coefficient in generatrix A.

different height and fixed vertical slot width, it can be seen that, there is an optimal height of discharge tube equals 2.98 m, which has smallest average dynamic shears coefficient (1.019 and 1.247 in generatrix D and A, respectively). By using this height, the shears

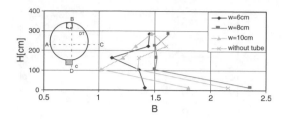

Figure 8. Effect the width of horizontal slots width of one discharge tube in generatrix B with fixed height ($h_t = 3.62$ m) and material height (H = 3.9 m) on dynamic horizontal pressures coefficient in generatrix D.

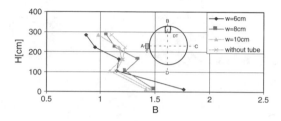

Figure 9. Effect the width of horizontal slots width of one discharge tube in generatrix B with fixed height ($h_t = 3.62$ m) and material height (H = 3.9 m) on dynamic horizontal pressures coefficient in generatrix A.

during discharging in the upper three quarter in generatrix D and in the middle third in generatrix A do not exceed the shears during filling. The maximum effective of tube ($h_t = 2.98$ m) occurs in generatix A at level 4 causing reduction of dynamic pressures equals to 40%.

4.3 Silo fitted with one discharge tube with fixed height and varied horizontal slots widths

Analyzing the Figures 8 and 9 of the results of silo model discharging by a opposite and side tubes with fixed height and vary horizontal slots width, it can be seen that, using one tube with different horizontal slot widths 10 cm, 8 cm and 6 cm does not give high flow shear reduction, it is more evident with slot equal 8 cm. The highest shears reduction with using tube with horizontal slot takes place in the tube of width 10 cm which has smallest average dynamic shears coefficient (1.151 and 1.366 in generatrix D and A, respectively).

4.4 Comparison between one tube with vertical and horizontal slots

Figures 10 and 11 show plots of dynamic shear coefficients for one discharge tube with vertical and horizontal slots. Clearly it can be seen that, discharging

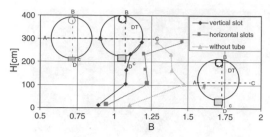

Figure 10. Effect the width of horizontal slots width of one discharge tube in generatrix B with fixed height ($h_t = 3.62$ m) and material height (H = 3.9 m) on dynamic horizontal pressures coefficient in generatrix D.

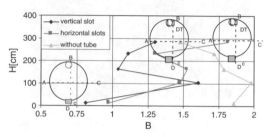

Figure 11. Effect the width of horizontal slots width of one discharge tube in generatrix B with fixed height ($h_t = 3.62$ m) and material height (H = 3.9 m) on dynamic horizontal pressures coefficient in generatrix A.

the silo by the use of discharge tube with vertical slot is more effective than silo fitted with tubes with horizontal slots. At level 5 the dynamic shear coefficient for silo discharging by a opposite and side horizontal slot are higher than discharging the silo without tube.

5 CONCLUSIONS

- Reduction of the flow shear depend on Geometric characteristic of discharge tube (i.e. with vertical or horizontal slots);
- There is an optimal width of one discharge tube with vertical slot along generatrix B equals to 8 cm ($\Pi_2 = 0.125$), which has smallest average horizontal dynamic pressures coefficient (1.061 in generatrix D and 1.121 in generatrix A, Figs 4 & 5), using of discharge tube with optimal width causes maximum reduction of dynamic pressures equal to 48.34% in generatrix A at level 2 (Fig. 5);
- Using tubes with horizontal slots widths give very small flow shear reduction (Fig. 8 and Fig. 9);
- Discharge tubes with vertical has more effective than discharge tube with horizontal slots (Fig. 10, Fig. 11).

6 NOTATION

β-dynamic coefficient equal (flow pressure/filling pressure); π_1 – hydraulic radius of tube/hydraulic radius of silo; π_2 – area of holes/surface area of discharge tube; C – pressure cell; d- diameter of discharging hole; D – diameter of the silo; H- height of stored barley; h_t – height of discharge tube, W – width of slot.

REFERENCES

Shalof, F. & Kobielak, S. 1998. Influence Of Geometric Characteristics Of Discharge Tubes Fixed To Internal Wall Surface On Reduction Pressures In Grain Silo, CD Rom Of Full Texts Of *13 International Congress Of Chemical And Process Engineering*, Symposium Powder Technology, Chisa.

Shalouf, F. & Kobielak, S. 2000(a). Low Cost Silo Design, Powder Handling And Processing, Vol. 2. No. 2. April/June 2000.

Shalouf, F. & Kobielak, S. 2000(b). Reduction The Dynamic Flow Pressures In Grain Silo By Using Discharge Tube, part one, CD Rom Of Full Texts Of *14 International Congress Of Chemical And Process Engineering*, Symposium Powder Technology, Chisa.

Shalof, F. & Kobielak, S. Reduction The Dynamic Flow Pressures In Grain Silo By Using Discharge Tube, part two, Submitted To Powder Handling And processing, Germany.

Feng, Y.T. & Hua, Y.L. 1996. Modified Janssen Theory for flexible circular bins, *Journal for Structural Engineering*, April.

Kamiński, M., Łagnowski, J., Suwalski, J. & Zubrzycki, M. The use of devices reducing apparent dynamic pressure in a large-diameter silo chamber. Przegląd Bud. No. 1/84.

Kobielak, S. & Klimek, A. 1997. Loads in the innovative type of silo with flexible walls, *Proceedings of the International Conference CCME*, Vol. III, p. 65–72, June, Woclaw – Poland.

Koster, K.H. 1985. Practical experiences with a 'flexilo', *Bulk Solids Handling*, Febr.

Kota, B. 1996. Operational tests of discharge tubes with models, *Powder Handling & Processing*, Vol. 8. No. 1. Jan–March.

Ooms, M. & Roberts, A.W. 1985. The reduction and control of flow pressures in cracked grain silos. *Bulk Solids Handling*, Vol. 5. No. 5, Oct.

Reimbert, M. & Silos, A. 1987. *Theory and Practice.* Lavoisier.

Sargent, L.N. 1979. General Layout & Structural Design. *The Complete Proceeding of the Original Elevator Design Conference*, Kansas City.

Theimer, O.F. 1958. On the storage of raw cocoa beans in silo compartments, *International Chocolate Review*, V Jahr. XIII, No. 3/4.

System-based Vision for Strategic and Creative Design, Bontempi (ed.)
© 2003 Swets & Zeitlinger, Lisse, ISBN 90 5809 599 1

Structural response characteristics of stack with hole

T. Hara
Tokuyama College of Technology, Tokuyama, Japan

ABSTRACT: There are many stacks which have been constructed for the industrial and power plants and have been composed of steel and/or R/C or bricks. These structure shows a cylindrical shell shape or a thin walled box shape. These should be designed to resist against a wind pressure and an earthquake load. Usually, these structure has a hole to intake fuel gas. The position of such cutout is different for each stacks due to the total arrangement of the plants. In this paper, the structural characteristics of the stack with a hole is studied numerically. In the numerical analysis, the quasi-static load deflection behavior is calculated and the responses of the structure are determined by the finite element method.

1 INTRODUCTION

There are many stacks which have been constructed for the industrial and power plants and have been composed of steel and/or R/C or bricks. These structure shows a cylindrical shell shape or a thin walled box shape. These should be designed to resist against a wind pressure and an earthquake load.

Usually, these structure has a hole to intake fuel gas. The position of such cutout is different for each stacks due to the total arrangement of the plants. In such structures, the shell wall around the hole is strengthened by improving the shell thickness or by adding additional rebars. Considering the stiffening design of such portion, not only the static analyses but also the dynamic evaluation should be taken into account.

In 1999, the stack of Tüpras plants at Kocaeli, Turkey, was collapsed under the strong earthquake ground motion. The stack has one hole around the lower portion the structure. Several failure causes are considered. The dynamic structural characteristics based on the geometric design of the structure, are one of the major causes of collapse. Most stack has a hole. Therefore, it is important to investigate the effects of the hole size and position on the dynamic characteristics of the structure.

To evaluate the dynamic load carrying capacity of the structure, the dynamic response analysis presents us the important information. Also, the capacity spectrum method enables us to design the structure against the earthquake (see Fig. 1). In such case, we must obtain the nonlinear analysis of the structure by use of the pushover analysis to evaluate the structural response capacity.

In this paper, nonlinear structural characteristics of the stack with a hole are studied numerically. In the

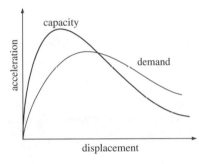

Figure 1. Acceleration displacement response spectra.

numerical analysis, the static load deflection behavior is obtained and the responses of the structure are calculated by the finite element method.

In FE analyses, the stack is modeled by use of shell elements and both geometric and material nonlinearities are taken into account.

From the numerical analyses, the static characteristics of the stack with hole are presented. Moreover, comparing the numerical results with the failure patterns of the real structure reported by the inquire report (Gould 2000), the failure mechanisms of such structure are clarified.

2 NUMERICAL MODEL

2.1 *Tüpras stack*

Figure 2 shows the dimensions of the stack. The height is 115 m and the shell thickness at the bottom

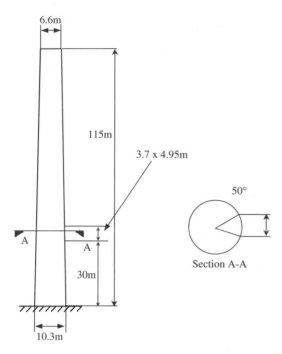

Figure 2. Tüpras stack.

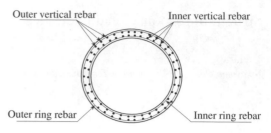

Figure 3. Rebars in the stack.

Table 1. Material properties.

Concrete	
Elasticity modulus	23.0 GPa
Density	23.5 kN/m³
Compressive strength	23.5 MPa
Rebars	
Elasticity modulus	205 Gpa
Ultimate strength	412 MPa

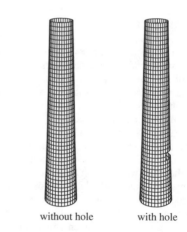

Figure 4. Numerical model.

and the top is 45 cm and 20 cm, respectively. The thickness changes from bottom to top linearly. The diameter also changes from 10.3 m to 6.6 m along with the stack height. Of particular interest, the stack has large rectangular hole of height 3.7 m and width 4.95 m at 30 m from the bottom to intake fuel gas into the stack.

Rebars are placed on both inner and outer surfaces of the shell. On each surface, rebars are placed in both meridional and hoop direction. The diameter of the hoop rebars changes from φ14 m to φ12 mm along with the stack height and are placed every 5 m spaces. The diameter of the longitudinal rebars changes from φ26 mm to φ10 mm along with the stack height. At the bottom, φ14 mm and φ26 mm rebars are implemented in hoop and meridional direction. The typical arrangement of rebars are shown in Figure 3.

The material properties are shown in Table 1.

2.2 Numerical modelling

Figure 4 shows the numerical model. The stack is divided into 46 and 32 elements in longitudinal and hoop direction, respectively. Therefore, the stack is modeled by 1472 elements. Each model is represented by the isoparametric degenerate shell element with layered approach. Each layer is composed of 8 concrete layers and 4 rebars layers because of representing the crack propagation through the thickness.

The left and the right figures show the model of a stack without hole and of a stack with hole, respectively. In this analysis, the additional arrangement of rebars around the hole are not considered because of clarifying the effect of a hole on the structural behavior. The stack is fixed at the base and is free on the top. Therefore, it behaves as the cantilevered beam.

2.3 Description of material nonlinearity

In this analysis material properties of concrete and rebars are defined individually.

It is assumed that the concrete behave as the isotropic material before cracking. To represent the material nonlinealities of the concrete, following assumptions are adopted.

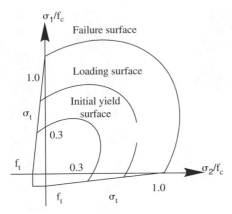

Figure 5.   Stress status in compression.

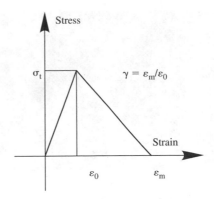

Figure 6.   Stress status in tension.

The Drucker Prager criterion is defined for the biaxial stress states (Hinton 1984).

$$f(I_1, J_2) = \sqrt{\beta(3J_2) + \alpha I_1} = \sigma_0 \qquad (1)$$

where I_1 and J_2 are, the mean normal stress and the shear stress invariant, respectively. α and β are constants obtained from experimental results. In this paper, α and β are taken as 0.355 σ_0 and 1.355, respectively, based on the Kupfer's experiment (1969). σ_0 is the equivalent stress. The Drucker Prager criterion in the two dimensional stress plane is shown in Figure 5.

For concrete in tension, the tension stiffening is introduced (Hara et al., 1996). In this model the concrete behaves as the linear elastic material up to the tensile strength and after the peak stress, the stress is assumed to degrade linearly as shown in Figure 6. The parameter of the tension stiffening is shown below.

$$\gamma = \frac{\varepsilon_m}{\varepsilon_0} \qquad (2)$$

where ε_m and ε_0 are the ultimate tensile strain and the cracking strain of the concrete, respectively. γ is the stiffening parameter and is defined as 20 in the numerical analysis. To define the crack criteria it is assumed that the cracks occur when the principal tensile stress at the integration point on each layer exceeds the tensile strength of the concrete. Therefore, the progress of the crack through the element thickness is considered in each loading stage.

The rebar is assumed to have a uniaxial bilinear constitutive relation in both compression and tension and only to have one dimensional stress strain relation.

2.4   Loading condition

To analyze the deformation characteristics of the stack under quasi-static loading, the applied load is calculated (Chopra 1995). If the x-axis coincides with the longitudinal direction of the stack, the applied load is calculated as

$$f = \overline{\Gamma} m(x) \psi(x) \qquad (3)$$

where $m(x)$ and $\psi(x)$ denote the mass distribution and the deformation mode, respectively. The factor $\overline{\Gamma}$ is calculated as

$$\overline{\Gamma} = \frac{\overline{L}}{\overline{m}} \qquad (4)$$

where

$$\overline{L} = \int_0^L m(x) \psi(x) dx \qquad (5)$$

and

$$\overline{m} = \int_0^L m(x) [\psi(x)]^2 dx \qquad (6)$$

The deformation patterns $\psi(x)$ is adopted from the eigen modes of the stack (see Fig. 7).

3   NUMERICAL RESULTS

3.1   Load deflection behavior under mode 1

Figure 8 shows the load deflection curves at the top of the stack on the meridian across the hole. In the case of the stack without hole, the target point is the same as that of the stack with hole. In the figure, α denotes the load factor defined based on the gravity g.

$$\alpha = \frac{\overline{L}}{g} \qquad (7)$$

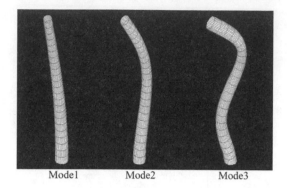

Figure 7. Deformation mode.

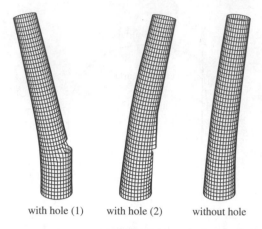

with hole (1)   with hole (2)   without hole

Figure 9. Deformation pattern.

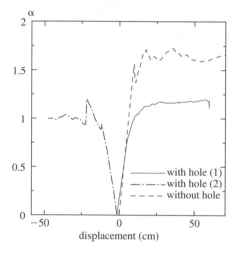

Figure 8. Load deflection behavior.

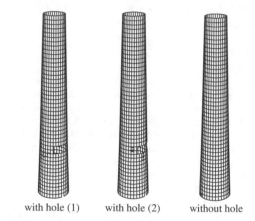

with hole (1)   with hole (2)   without hole

Figure 10. Crack pattern.

The dotted line shows the load deflection curve of the stack without hole. The ultimate strength of the stack shows the factor $\alpha = 1.7$.

In the case of the stack with hole, the solid line and the dot dash dot line denote the deformation when the earthquake motion acts from the hole side and the opposite side, respectively. The ultimate strength of the stack with hole shows $\alpha = 1.2$ and $\alpha = 1.0$, respectively. From these analyses, the strength of the stack is strongly influenced by the direction of the earthquake against the hole.

Figure 9 shows the deformation patterns of the stack. The deformations shown in the left and the center of the figure correspond to the solid line and the dot dash dot line shown in Figure 8. When the earthquake motion acts from right to left, the stack around the hole is subjected to the strong tensile force. Therefore, the deformations around the hole are quite large. Due to these deformations the cracks occur around these region (see Fig. 10, left). However, the stack shows the

ductile behavior after cracks occurred because of the ductility of reinforcement around the hole.

On the other hand, when the earthquake motion acts from left to right the concrete around the hole is subjected to the strong compressive force. Therefore, in the region below the hole, the concrete sprits vertically (see Fig. 10 center). The load carrying capacity of the stack drops suddenly. However, the hoop rebars carry the tensile force due to stress redistribution after concrete cracks under such a tensile force.

The failure patterns presented in the inquire report (Gould 2000) coincide with the failure patterns shown in Figure 9.

3.2 Load deflection behavior under mode 2

Figure 11 shows the load deformation curves of the stack under the load due to the deformation mode 2. It depends on the secondary bending vibration mode of the stack (see Fig. 7). The dotted line shows that the

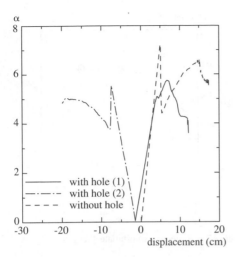

Figure 11. Load deflection behavior.

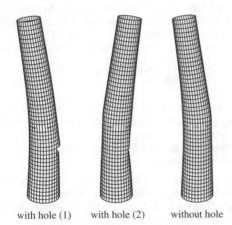

with hole (1)   with hole (2)   without hole

Figure 12. Deformation pattern.

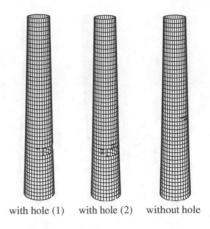

with hole (1)   with hole (2)   without hole

Figure 13. Crack pattern.

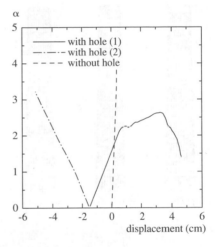

Figure 14. Load deflection behavior.

ultimate load of the stack is $\alpha = 6.5$. However, the deformation patterns under the load due to this mode is different from that due to mode 1. Cracks occurred in the middle of the total height and the load carrying capacity drops starkly (see Fig. 13 right). The solid line and the dot dash dot line show the load deformation curve of the stack with hole under the loading due to mode 2.

When the earthquake motion acts from the side of the hole to another side the load deformation curve behaves the same path as that of the stack without hole (see Fig. 12 center). On the other hand, the load deformation curve shows the ductile behavior shown in Figure 11 when the earthquake motion acts from the opposite side of the hole to the side of the hole. However, the ultimate load shows the same factor $\alpha = 5.7$.

Figure 12 shows the deformation patterns under the loading due to the deformation mode 2. From the figure, the same conclusions obtained by Figure 11. In the case of loading shown in Figure 12 center, the stack shows the same deformation as that of the stack without hole shown in Figure 12 right. In both cases, the load carrying capacities drops suddenly when the cracks occur on the concrete. On the other hand, the stack deforms with ductile characteristics under earthquake motion from opposite side (see Fig. 12, left).

3.3 Load deflection behavior under mode 3

Figure 14 shows the load deformation curve of the stack under quasi-static load under the assumption of the mode 3. In this loading condition, the local bending of the stack occurs and the load deformation curves are different from the previous two loading cases.

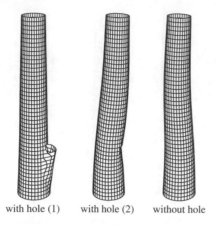

with hole (1)   with hole (2)   without hole

Figure 15.   Deformation pattern.

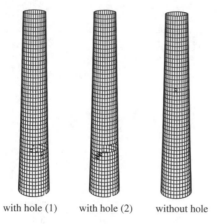

with hole (1)   with hole (2)   without hole

Figure 16.   Crack pattern.

In the case of the stack without hole shown as the dotted line, the ultimate strength of the stack shows the factor $\alpha = 4.0$. However, in this loading pattern, the deformation at failure is quite small comparing with other loading cases.

In the case of the stack with hole, the load deformation curve shows the different patterns depending on the direction of the earthquake motion.

The solid line denotes the load deflection curve under the earthquake motion from the side of the hole to another side. The dot dash dot line shows the curve under the earthquake motion from opposite direction. The former case shows the ductile behavior of the stack. On the other hand, the later case shows the brittle failure of the stack.

Figure 15 shows the deformation patterns of three cases correspond to the Figure 14. The deformation patterns shows the causes of the local failure under such loading conditions.

Figure 16 denotes the crack patterns.

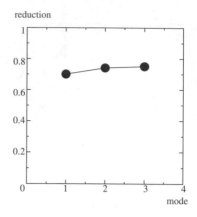

Figure 17.   Reduction of ultimate strength.

3.4   Reduction of the ultimate strength

Figure 17 shows the ratio of the ultimate strength of the stack with hole to that without hole. The reduction factor shows between 0.7 and 0.75 in spite of denoting the different ultimate strength for individual loading modes.

4   CONCLUSIONS

Nonlinear behavior of the R/C stack with and without hole are analyzed for the purpose of using the data for the earthquake resistant design.

From the numerical analyses following conclusions are obtained.

1. The ultimate strength of the stack strongly depends on the direction of the earthquake motion.
2. The behavior after cracking is controlled by the existence of rebars.
3. The reduction of the ultimate strength due to the hole is almost 70% and the effect of the reduction is large under the load due to mode1 in this model.
4. The relation between the dynamic deformation mode and the hole plays an important role for the failure mechanisms of the stack.

REFERENCES

Chopra, A.K. 1995. Dynamics of Structures. Prentice Hall
Gould, P.L. 2000. Report on visit to the industrial facilities in Turkey – September 20–22, 2000. NFS Project CMS-0084737.
Hara, T., Kato, S. and Gould P.L. 1996. Ultimate strength of R/C cooling tower shells with various reinforcing ratio. Journal of IASS 37(3): 153–163.
Hinton, E. and Owen, D.J.R. 1984. Finite element software for plates and shells, Prineridge Press, Swansea, UK.
Kupfer, H. and Hilsdorf, K.H. 1969. Behavior of concrete under biaxial stress. ACI Journal 66(8): 656–666.

Tendencies for solid structures in waterway engineering

C.U. Kunz

Bundesanstalt für Wasserbau, Karlsruhe, Germany
(Federal Waterways Engineering and Research Institute, Karlsruhe, Germany)

ABSTRACT: In former times solid concrete structures in waterway engineering such as locks and weirs had been designed with joints so as to avoid the consideration of constraint. Because of the restricted durability of joints on one hand and statical problems in case of large settlements on the other hand monolithic concrete structures have been developed and designed. Problems with constraint have to be solved such as early constraint rised by hydratation of young concrete and late constraint rised by structure-soil-interaction and settlements. The assessment of constraint by numerical methods, the optimization of suitable concrete mixtures and measurements in laboratory and on construction site are appropriate tools for the design of monolithic structures. Design methods, the construction of monolithic waterway structures for modern ship traffic and structural details are shown and discussed as well as future work is mentioned.

1 INTRODUCTION

Germany has a mesh of waterways of about 7.400 km, where the overall goods transportation with ships is about 60 Mrd. ton * kilometers per year representing nearly constantly 23% of all goods traffic. Ship traffic is not only a safe but also one of the most economic and ecological traffic system. Besides the waterways at themself the maintenance of the ship traffic system needs waterway structures for safe and easy use. About 350 locks clear the falls and about 300 weirs keep up the reaches of German waterways. These and other structures as culverts, bridges, dams and storm surge barriers are operated by the German waterway and shipping administration. The fixed assets amount to around 40 Mrd. €.

Bundesanstalt für Wasserbau gives expert opinions and consultancy to all offices of the waterway and shipping administration on the fields of structural engineering, geotechnics and hydraulic engineering.

2 PROBLEM

Solid waterway structures have to be impervious for water so as to prevent the exchange of basin water with groundwater and vice versa which could lead to soil transportation and suffosion. Soil transportation may cause settlements and other static problems.

So far most of the solid waterway structures have been built with joints, joint rubbers or joint fillers. German code DIN 1045 (1988) recommends joints as a constructional measure in concrete structures to restrict the cracking by strains out of temperature effects. Grundbautaschenbuch (1997) recommends settlement joints for slab foundations when a bigger difference in between different parts of the structure is expected. German code DIN 19703 (1995) wants to build locks with a block-length of 15 m normally.

Joints in solid structures are leading to higher costs for construction and maintenance and often to increasing damages within the lifetime, KORDINA (1996). Leaks may reduce stability and the load–bearing capacity for upper limits. Unforeseen and additional settlements may also limit the serviceability of the structure. So efforts have been made to avoid joints.

Experiences with monolithic slab foundations have been gathered in Germany with tower blocks and with electric power stations. From these waterway structures are in some kind different, for example by the influence of early and periodic temperature changes, by the massiveness of the structure and by changing of the water level, may be 20 times a day in locks. So ideas for monolithic waterway structures have to be developed carefully and step by step.

3 MONOLITHIC CONSTRUCTION METHOD FOR WATERWAY STRUCTURES

3.1 *First monolithic constructions*

Beginning with the 20th century locks had been constructed either without joints or with reduced numbers of joints. Ancient locks on Dortmund-Ems-Kanal, built around 1910, had been constructed without joints. A shaft lock near Minden with integrated recuperation basins built between 1911 and 1914 and founded on rock had got only three joints within a length of 140 m. Twin lock Eisenhüttenstadt near river Oder was built between 1924 and 1929 as a solid half-frame with a slab of 5 m and with continuous joints only between the lock chamber and the head structures. Above the tail-water level the upper parts of the lock walls got partially continuous joints. These and other existing waterway structures are examples of good experiences with structures where no joints or reduced numbers of joints had been built. Former construction methods as slow poured concrete and former materials as mortar with low parts of cement and bigger parts of aggregates favoured the monolithic principle.

3.2 Modern monolithic construction method

At the end of the 80's a new weir in Bremen in river Weser was built as a waterway structure firstly in modern times without joints. With five weir gates the dimensions amount to 180 m in the length and 37 m in the width, Frerichs & Grabau (1991), Figure 1. The structure has got a pile foundation and an underwater concrete sole plate which is 1.5 m thick. The base plate measures 2.3 m and a weir sill with nearly the same dimensions is on top. Base plate and weir sill have been built as a monolithic even bridging the gaps by the stages of construction. Because of the sensitiveness of the weir gates on top of the sill against deformations the structure had to be jointless. To provide extended stages of construction had been one of the advantages of this monolithic construction method. Over 10 years experience with the weir shows good results concerning cracking and the structure is still impervious.

The monolithic construction method launched a measurement campaign. The development of the temperature of young concrete by hydratation and relating strains had been measured within five sections M1–M5 and in different levels of the base plate and the sill, Ehmann (1991). The measurements aimed at an better understanding of constraints in young concrete and of cracking limitation by reinforcement. Indeed the stiffnesses of pile foundation and underwater concrete plate lead to a nearly complete impediment of strains and bending. When cracking started the cross sections had been subjected to a combination of tensile forces and bending moments. According to Hampfler (1991) the excentricity lead to a reduction of the cracking force and to a reduction of the anti-crack reinforcement especially at the bottom of the plate.

3.3 Monolithic lock chamber slab

The twin lock Hohenwarthe at the Mitteland-canal near by Magdeburg, built between 1999 and 2003, has been designed for modern barge traffic and as a recuperation lock with open basins besides the lock chambers, Figure 2. The useful length of the lock chamber measures 190 m, the useful width 12.5 m and the fall head is 19 m. The totally length of the structure is about 250 m, the width about 56 m and the height is about 30 m. The chamber base slab integrates the hydraulic filling and emptying system and is 5.5 m thick. The subsurface is coursed with sand, gravel, laminated and boulder clay. Coat thickness and pass are very different in longitudinal and cross direction. The subsurface is quiet heterogeneous. In most areas the foundation of the twin lock is on the clay layers which stiffnesses are to a limited extent. Under the head there are stiff sand and gravel layers. The total settlement is predicted with 40 cm which is mainly not caused by the lock structure at itself but by unavoidable fill towards the head so as to integrate the

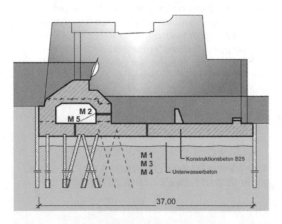

Figure 1.   Cross section of the new weir in Bremen with measurement sections M1–M5.

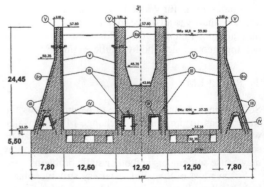

Figure 2.   Cross section of twin lock Hohenwarthe.

lock in the topography. Due to this different sur-charges the lock tilts in longitudinal direction.

The original design with solid blocks of lengths of about 15 m was abandoned because of the predicted settlements and resulting deflection and tilting of single blocks. The deformations within the joints in between the blocks did not seem to manage with commercial joint constructions. Some joint would be totally closed, some other opened by 5 cm where tensile and combined shearing forces would have been beared by the joint rubber. Even simulated hinges in the slab between each block did not improve the situation.

So feasibility studies had been made for a monolithic lock chamber plate whereas the lock walls are still divided by joints. The decisive factor for a monolithic plate is to realize constraint caused by hydratation in young concrete, by seasonal changes in temperature and by settlements later on when operating the lock. Constraint during the operating phase has to be super-posed with the load arrangement. The strains had been estimated and a crack limiting reinforcement had been designed, Kunz & Bödefeld (2000).

The settlement had been calculated realistically by considering the structure-soil-interaction. The struc-tural design used a finite-element-model which was bedded on elastic-isotrop halfspace. This theory con-siders also lateral loads for settlement calculations and is known to be realistic enough.

The results of the study gave effects of actions in longitudinal and cross direction according the hollow of settlement. Additional there is a torsional action. The maximum bending moment in longitudinal direc-tion resulted in about M = 47.000 kNm/m where a reinforcement of about As = 320 cm²/m is needed (8 levels of bars with diameters of 28 mm and bars spaced 15 cm), Figure 3. The deformations in the joint constructions of the lock walls had been calculated with a maximum of 16 mm for compression and with a maximum of 10 mm for gaping, Figure 4. The cross

displacement caused by torsion was restricted to less than 10 mm. Now each of the joints in the wall had been proved for commercial constructions.

The study proved the feasibility of a spread founded lock chamber where the monolithic base slab reduces the deformations in the remaining joints of the lock walls because of its stiffness. Comparing studies considering a pile foundation or a combined plate pile foundation which was constructed in the end proved the necessity of a monolithic plate without exceptence so as to improve the deformations in the joint constructions.

The estimated internal actions by constraint out of temperature and settlement has been verified during construction by extensive measurements within the concrete structure. The evaluation of these measure-ment and accompanying numeric simulations are going on up to now.

Monolithic slabs for spread founded lock chambers are now designed for nearly all new locks in Germany.

3.4 *Partly monolithic locks*

For the new lock Uelzen No. 2, at Elbe-Seiten-canal, present under construction not only the slab of the lock chamber but also the lower parts of the vertical lock walls are designed for monolithic construction. The useful length is 190 m, the width is 12.5 m and the fall is 23 m, Wachholz & Saathoff (1999), Figure 5. To improve the settlements because of alternating actions due to lifting and lowering the water level for lifting the ships the recuperation basins are integrated in the struc-ture. Therefore some block sections have walls which are 20 m thick and which are designed solid and mono-lithic for the lower 15 m. These walls exceed present experiences with mass concrete concerning early con-straint due to hydratation. Comprehensive studies have been undertaken by the Bundesanstalt für Wasserbau as well as by the construction company so as to estimate

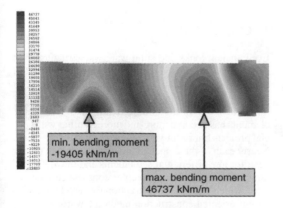

Figure 3.   Bending moments in monolithic lock chamber slab.

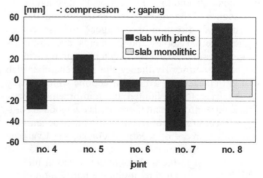

Figure 4.   Deformations in selected joints of the lock walls of lock Hohenwarthe.

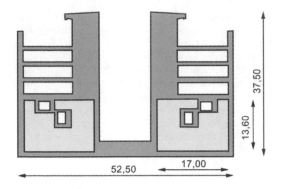

Figure 5.   Cross section of lock Uelzen No. 2.

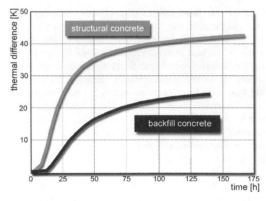

Figure 6.   Thermal differences between structural concrete and backfill concrete.

the constraint realistically and to get crack limiting reinforcement dimensioned economically.

Designing and placing concrete for these thick cross sections are adjusted to the hydratation and cracking problem of young concrete. The development of temperature due to hydratation may be influenced by the choice of the cementing material, the part of it and the mortar mix. In order to avoid separating cracks which impedes the durability of the structure the mix design has to ensure that the temperature within an element increases only a little. The hydratation temperature is controlled by a multi-columned structure, Westendarp (2001). Whereas on the edges of a cross section covering the reinforcement bars a structural concrete (i.e. about 270 kg/m³ cement) is used with specific durability features the inner area is poured with a backfill concrete whose content of cementing material is low (i.e. about 180 kg/m³ cement and up to 120 kg/m³ fly ash).

Within suitability tests the rise in temperature of structural concrete and backfill concrete shows differences of about 20 Kelvin within the first days under adiabatic circumstances, Figure 6. The required water penetration depth is limited to 30 mm for structural concrete and to 50 mm for backfill concrete.

Because of the monolithic construction method in longitudinal direction the bigger stiffness causes higher internal forces and moments due to settlements. These internal effects can be beared easily by the cross section. By theory the lock chamber slab will get cracked by settlement actions which is why cracks have to be limited by reinforcement bars. Cracking reinforcement therefore is designed according to recent results of research for water penetration and self-recovery of separated cracks, see Evardsen (1996). The crack width is limited to $w_{cal} = 0.25$ mm.

In contrast to the traditional construction method with blocks separated by joints the partly monolithic construction method will show a structural behaviour which is more robust.

3.5   Total monolithic lock

At present studies are prepared by the Bundesanstalt für Wasserbau concerning extended monolithic locks. Covered are further steps of

– a totally monolithic lock chamber from bottom to top but with joints to head and tail,
– a totally monolithic lock where no joints may be found any more.

The problems may arise from late constraint during the operation time when periodic changes in temperature and actions due to settlements have to be superposed with the changes of actions by lifting and lowering the water. Experiences with earlier steps of partly monolithic structures will be included. The feasibility of advanced monolithic structures will be influenced by the lock design itself, the stiffness of the structural parts, the transition between lock chamber and head or tail, the dimensions of the structural elements, and so on.

4   CONCLUSIONS

Even ancient locks and solid waterway structures had been built with reduced numbers of joints in contrast to others which was built with separated blocks so as to avoid constraint actions. In the last years the monolithic construction method i.e. with partly monolithic elements has become more and more familiar with special projects but is not yet a rule for constructing locks. Nearly each of these special projects showed an urgent need for low differential settlements and low deformations. With examples of recently constructed locks or those which are under construction the development of a joint-oriented construction method towards a monolithic method has been shown for waterway structures.

The bigger robustness of monolithic structures may lead to an advanced use of monolithic elements.

In Germany the monolithic construction method for locks will be developed at that moment and in the future. This development will be accompanied by on-site measurements, testing and numerical investigations. The monolithic construction method will use advances in materials design and manufactoring methods. Increases in the content of reinforcement which may be caused by the increased stiffness of a monolithic structure are of minor magnitude. Compared with this the improved robustness may lead to an improved durability of the structure. The client will decide the design method because of economical aspects and/or of safety aspects and needs.

REFERENCES

DIN 1045. 1988. Beton und Stahlbeton, Bemessung und Ausführung. Berlin: Beuth-Verlag.
DIN 19703. 1995. Schleusen der Binnenschiffahrtsstraßen, Grundsätze für Abmessungen und Ausrüstung. Berlin: Beuth-Verlag.
Ehmann, R. 1991. Bauwerksmessungen am Beispiel des Weserwehrs Bremen. In: Mitteilungsblatt der Bundesanstalt für Wasserbau Nr. 68. Karlsruhe: Eigen-Verlag.
Evardsen, C. 1996. Wasserundurchlässigkeit und Selbstheilung von Trennrissen in Beton. In: Deutscher Ausschuß für Stahlbeton, Heft 455. Berlin: Beuth-Verlag.
Frerichs, K. & Grabau, J. 1991. Bau des Weserwehrs in Bremen. In: Beton 41, H. 4.
Grundbautaschenbuch. 1997, Teil 3, Flachgründungen, 5. Auflage, Berlin: Ernst & Sohn.
Hampfler, H. 1991. Temperature- und Dehnungsmessungen während der Erhärtungsphase des Betons. In: Mitteilungsblatt der Bundesanstalt für Wasserbau Nr. 68. Karlsruhe: Eigen-Verlag.
Kordina, K. 1996. Behälter in Massivbauweise. In: Bautechnik 73, H. 5.Berlin: Ernst & Sohn.
Kunz, C. & Rödefeld, J. 2000. Bauwerk-Boden-Interaktion am Beispiel einer Schleusenanlage mit einer monolithischen Flachgründung. In: 5. FEM/CAD-Tagung der TU Darmstadt.
Wachholz, T. & Saathoff, J. 1999. Neubau der Sparschleuse Uelzen II. In: Binnenschiffahrt (ZfB) Nr. 1. Hamburg: Schiffahrts-Verlag Hansa.
Westendarp, A. 2001. Entwicklungen und Tendenzen bei Baustoffen und Bauausführungen im Schleusenbau. Beton-Informationen 1/2001, pp. 3 8.

Performance of a fibre-reinforced polymer bridge deck under dynamic wheel loading

A.F. Daly & J.R. Cuninghame
TRL Limited, UK

ABSTRACT: The paper describes the research carried out at TRL Limited on behalf of the UK Highways Agency to examine the performance of fibre-reinforced polymer (FRP) bridge decks under local wheel loading. The objective of the research was to produce a draft standard giving generic design requirements for technical approval of FRP deck systems. The project included the formulation of design guidelines for fatigue. A series of tests was carried out on a full-scale glass FRP bridge deck under static and dynamic wheel loading. The loads were imposed using the TRL Trafficking Test Facility, which replicates the effects of the wheel of a heavy goods vehicle. The deck was subjected to over 4.6 million cycles of a 4 tonne wheel load, equivalent to 30–40 years of service traffic. The paper includes a description of the FRP deck, the dynamic testing and a summary of the performance of the deck.

1 INTRODUCTION

1.1 Background

Most modern short span bridges have concrete decks. In general they are efficient and durable provided proper attention is paid to detailing and standard of workmanship. Concrete is likely to be the most common deck material for some considerable time. However, some concrete bridge decks have suffered corrosion, due in part to the increasing use of de-icing salts. As there is unlikely to be an economically viable replacement for rock salt for de-icing, increasing interest is being shown in materials that are corrosion resistant. In addition, rapid growth in the volume and weight of heavy goods vehicles (HGVs) has led to serious problems and many older bridges no longer meet current design standards. There is therefore, a need for methods of replacing bridge decks to deal with structural deterioration and to increase load capacity, without extensive and expensive bridge works.

1.2 Use of FRP in bridge decks

The use of fibre reinforced polymer (FRP) as a primary structural material is developing rapidly in the construction industry. FRP materials have considerable advantages in terms of weight, strength and corrosion resistance, and they have been used for several decades in the aerospace and automobile industries. The "product life cycle" is much longer in civil engineering so

uptake has been relatively slow, but FRP is in use in a number of bridges around the world. As production technology develops and design standards and guidelines become more generally available, these FRP materials will be used more widely to provide cost-effective alternatives to steel and concrete. Potential applications for FRP decks are new design, replacement of under-strength decks in existing bridges, and the provision of temporary running surfaces.

In spite of the advantages, however, there is still a lack of information on the behaviour of FRP components in bridge applications. This has impeded the development of generally accepted guidelines for the design and application of bridge schemes. In response to this, the UK Highways Agency commissioned the TRL Limited to carry out research into the performance of FRP bridge decks with the objective of developing design guidelines.

1.3 Outline of research project

The project was to examine the use of FRP in new and replacement bridge decks and produce draft design requirements. A desk study of current research and practice in bridge construction and other FRP applications was carried out to address issues such as design and detailing requirements, resistance to environmental effects, repair and maintenance.

Existing loading rules can be used for FRP decks and global static design is relatively straightforward provided material properties are known. Some design guides are available, but there are currently no fatigue design standards for FRP bridge decks. Local dynamic stresses at details potentially at risk of fatigue due to wheel loading may be difficult to calculate. There are likely to be high stress gradients and the stress cycle may be tensile, compressive or alternating. As these stresses are raised to the power of 10 to obtain fatigue life, it is important that they be calculated with sufficient accuracy. This is similar to the situation for steel orthotropic bridge decks, which are currently beyond the scope of the UK bridge design code (British Standards Institution 1980).

Because of the development costs, it is likely that most FRP decks will be produced as modular systems. The deck system tested was thought to be typical in that it consisted of top and bottom flanges supported by webs running transverse to the direction of traffic flow. The limited number of bridge deck systems that have been developed so far, notably in the US and Europe are of this form.

Therefore it was decided to carry out full scale tests to study the performance of FRP decks under local wheel loads and provide data on which to base generic fatigue requirements for technical approval of FRP deck systems.

2 TESTING PROGRAMME

2.1 Description of FRP bridge system

The design of the deck used in the dynamic testing was carried out by Maunsell Ltd, UK, who developed an FRP bridge deck system consisting of standard glass FRP cellular "planks" and connectors that can be assembled in a variety of configurations. The units have undercut grooves on the sides, which are used to connect them together using a rod with a "dog bone" cross-section to provide mechanical interlock; the joints are also glued. This Advanced Composite Construction System (ACCS) has been used in the Linksleader footbridge at Aberfeldy in Scotland and a lifting bridge at Bonds Mill in Gloucestershire, England.

The ACCS deck was not designed to resist local wheel loading, so a new Roadway panel was designed and manufactured. This has a similar geometry to the lighter ACCS plank but with thicker sections to resist direct wheel loads. It was designed to be laid transversely on a sub-structure built up from ACCS planks. The Roadway panel is shown in Figure 1, further details of the design are given elsewhere (Daly & Duckett 2002).

2.2 Test specimen

A full-scale section of bridge deck was built at TRL. It spanned 4.0 m, was 2.12 m wide and had a total construction depth of 0.8 m: see Figure 2. No surfacing

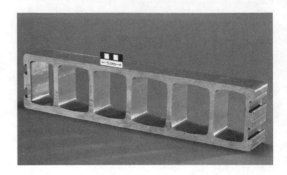

Figure 1. Roadway panel.

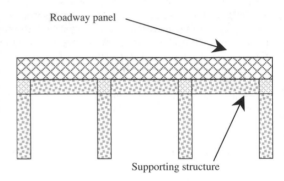

Figure 2. Section through FRP deck.

was used (in practice, a thin anti-skid dressing would be sufficient). The deck was instrumented with electrical resistance strain gauges to determine the strain influence lines as the wheel moved along the deck. Displacement transducers were also fitted to measure deflection of the deck under load.

2.3 Test rig

Rolling wheel tests were carried out in the TRL Trafficking Test Facility (TTF). The main features of the TTF are shown in the photograph in Figure 3. It is similar to a conventional pavement test machine, but is smaller and much cheaper to operate. A wheel, either single, twin or "super-single", rolls along a 3 m long track. The wheel load may be up to 5 tonnes and the load may be applied during both forward and reverse passes, or on the forward pass only. The wheel reaches a maximum speed of 20 kph and can complete one pass per second (bi-directional), equivalent to 80,000 per day. The wheel, carriage and drive mechanism are set up as a spring-mass system operating at its resonant frequency. This greatly reduces the energy required and hence the cost, of running the rig.

Figure 3. FRP deck in test rig.

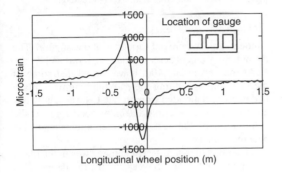

Figure 4. Influence line for strain at top of web.

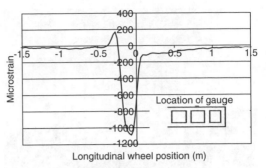

Figure 5. Influence line for strain in top flange.

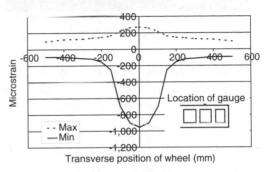

Figure 6. Transverse influence line for strain in top flange.

2.4 Test methodology

Initial static tests were carried out to determine the strain levels in the running surface and to select an appropriate load for the dynamic test. Each series of tests consisted of placing the wheel at various positions on the deck and recording the strains. Transverse and longitudinal influence lines for strain could then be plotted and used to determine the load to be used in the test.

Once the test load level was determined, the dynamic test was started, with the logging system set up to record maximum and minimum values of each gauge output every 10 or so cycles. At various rig cycle counts, a full sweep of strain and deflection was captured and retained to illustrate the behaviour of the deck panel through one loading cycle.

Figure 4 shows a typical influence line, giving the strain at the junction between the web and top flange of the roadway panel as the wheel traversed the deck (see insert in figure). This was one of the locations indicated by analysis to be subject to high stress range due to traffic loading. The strain shown is in the direction of the traffic movement (i.e. longitudinally in the deck, transversely along the Roadway panel). A strain range of 1000 microstrain (tension) to −1290 microstrain

(compression) was recorded. This was the highest range of strain recorded in the deck. The influence lines exhibited the expected strain reversal as the wheel rolled over the webs.

Another critical point for fatigue was thought to be in the top flange, mid-way between two webs. Figure 5 shows the influence line for longitudinal strain on the top surface of the second cell of the Roadway panel. Here, the peak strain recorded was −1090 microstrain (compression).

Figure 6 shows the influence line for the strain as the transverse position of the wheel is changed. Again, the strain shown is in the direction of the traffic movement. Two plots are shown, for the maximum and minimum strain as the wheel swept from one end of the deck to the other. The strains recorded for this gauge ranged from 270 (tension) to −950 microstrain (tension).

Influence lines were also recorded for deflection of the deck to indicate the global performance. The maximum deflection was just under 1 mm.

Once the test load level was determined, the dynamic test was started, with the logging system set up to record maximum and minimum values of each gauge output every 10 or so cycles. At various rig cycle counts, data were recorded continuously to provide influence lines

Figure 7   Cracking at support in FRP deck.

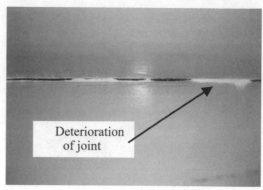

Figure 8   Deterioration of joint in sub-structure.

for strain and deflection to monitor any change in behaviour.

3   TEST RESULTS

3.1   *General performance of the deck*

The dynamic test was continued for as long as possible, within the constraints of the project time-scale. In all, the deck sustained about 4.6 million load cycles.

At various times, the test was stopped and close visual examination of the surface was carried out. Fibre-scope examination of the gauged area inside the Roadway panels did not indicate any damage. Subsequent examination of the surface (including sections and cores taken through the running surface) indicated that there was no visible deterioration and it was concluded that if this was a real structure there would be no cause for concern.

A 500 mm section of the Roadway panel, taken from along the line of the wheel loading, was subsequently tested statically to failure. Comparison with a similar test on a not-previously-loaded section indicated no different in strength. This confirmed the above conclusion, that the Roadway panel provided a robust structure capable of sustaining HGV traffic.

3.2   *Damage in the sub-structure*

The only visible damage in the deck was in the sub-structure. After 247,400 load cycles, damage was noticed adjacent to one of the bearings. This consisted of cracks in the first and second webs of the ACCS plank and debonding of the hard-wood insert (used to prevent local distortion and buckling of the section at the bearing): see Figure 7. At 256,000 load cycles, some debonding was noted in six of the eight supports,

although cracks were only apparent at one support. The dynamic movement across these cracks was as 1.1 mm.

At load cycle 247,400, it was noted that some of the FRP angles holding the end diaphragms in place had become debonded from the ACCS webs, with relative movement of about 0.07 mm occurring. At the completion of the test, two other angles had also become detached.

The only other significant damage was in the glued joint between the ACCS webs and the top flange. This was first noticed because the two surfaces were grinding together and the adhesive was being ground into a fine white powder. Figure 8 shows the joint (viewed from under the deck) at the conclusion of the test. The relative movement at this joint was about 0.5 mm.

4   FURTHER TESTING

The strain data generated by the rolling wheel tests will be used to design a test that can be carried out in a conventional servo-hydraulic testing machine. A short length (500 mm) of Roadway panel will be loaded via two steel backed rubber pads loaded alternately to simulate a wheel passing over the panel. Strain gauges installed at the same locations as in the rolling wheel test will be used to set up the loading to give the correct stress distribution. Tests can then be carried out at up to 5 Hz, to long endurances, and at several stress ranges to obtain points on an S-N curve, or to provide evidence that fatigue damage will not occur under design traffic loading. The options for testing are discussed in the next section.

5   FATIGUE DUE TO WHEEL LOADING

The rolling wheel tests were carried out to study the behaviour of the FRP deck, to develop a method by which a client could ensure that an FRP deck proposed for use on a UK highway bridge would have an adequate fatigue life. Fatigue damage may be divided into:

- "Global" damage: for example, the cracking that occurred in the test, adjacent to the bearings. The relevant loading is the total due to vehicles on the span.
- "Local" damage: where the cyclic stresses are due to individual wheel loads and cracking occurs under, or close to, the wheel tracks. "Local" cracking is the most common form in lightweight steel decks and it is necessary to ensure that similar cracking does not arise in FRP decks.

It is assumed that FRP decks will consist of components designed and developed by specialists. These components can then be used by non-FRP-specialist bridge designers and the cost of designing and developing the FRP components will be spread over a number of bridges. The use of FRP components in this way is analogous to the use of steel components, e.g. Universal Steel Beams, except that steel components are supported by product standards and are supplied with certified properties. In the absence of product standards for FRP components, it is proposed to base fatigue requirements on testing, at least as an interim measure.

It was recognised that fatigue is only one aspect of the development of an FRP deck and the fatigue strength of FRP materials in general is said to be high compared to welded joints in steel. Therefore, the fatigue requirements should not be so onerous as to form a disproportionate part of the total development costs. The rolling wheel tests have the advantage that all parts of the deck, including joints between components, are automatically included in the test, but the rate of testing is low (1 Hz) and the load is limited by the rating of the tyre and the capacity of the rig. It is proposed to base the fatigue requirements for the parts of the deck affected by local wheel loading on a test in which a standard fatigue test machine can be used to simulate the effect of a single wheel.

Traffic loading is the same for any type of deck, so the fatigue design loading in the UK bridge design code (British Standards Institution, 1980) can be used. Maximum axle loads for normal traffic are 60 kN for steering axles (30 kN per single wheel) and 100 kN for rear axles (50 kN per twin wheel – "super-singles" are not included in the code).

On an individual bridge, the number of HGVs and the mix of vehicle types may be different from those in the design code. For example, where an existing concrete deck is to be replaced by an FRP deck, the design loading may be based on a traffic survey. It is envisaged that the specialist FRP designer, or component supplier, should provide fatigue data so that the bridge designer can check that the deck is adequate for the traffic loading on a particular bridge.

Figures 4, 5 and 6 shows that the influence lines for potential crack locations in the Roadway panel are short, so a test that simulates the effect of an individual wheel is sufficient. There are two options for the testing regime.

5.1 Assessment without damage summation

All stress cycles due to wheel loads should be below the damage threshold (infinite life).

The most common existing method of fatigue design is simply to limit all dynamic stresses to a value at which damage will not occur. As there are other reasons to limit stress, e.g. to prevent excessive deflection, this approach may not be as over-conservative as it sounds.

To specify a laboratory test to demonstrate that all stresses are below the threshold for fatigue damage, it is necessary to know the number of cycles which corresponds to the constant amplitude fatigue limit. The test would consist of applying the heaviest wheel at the "worst case" location on the component for that number of cycles. It would be necessary to determine the worst case position of the load, but full influence lines would not be needed.

It seems to be accepted that FRP materials do exhibit a fatigue limit, but the corresponding endurance does not appear to be well defined (very few tests have been continued beyond 10^7 cycles). This problem could be avoided by applying the same number of cycles as the number of wheels passing over the component on a bridge, but this would require between 171 and 685 million cycles depending on the type of road for the normal UK design life of 120 years. Even for a reduced design life of 40 years, the cost of such a test would be too high. It is proposed that the required endurance should be 10^7 cycles.

The test load would be applied through a steel-backed rubber pad 200 mm^2, to represent a single wheel contact patch. The magnitude of the load would be 30 kN multiplied by a partial factor to be defined when the current project is completed. For most details a single loading pad would be sufficient, but where the influence line changes sign (as in the web to top flange detail), two pads loading alternately would be required to reproduce the correct stress ratio. Alternatively, a factor might be applied to allow for the stress ratio.

A further simplification is proposed, i.e. that the criterion of failure be the static strength of the component at the end of the fatigue test. Fatigue damage may be sub-surface, or in locations where visual inspection is difficult. More sophisticated NDE would add significantly to the cost of the test. Therefore, the component would be tested statically to failure at the end of the fatigue test and the result compared with a similar test on a component that had not been subjected to fatigue loading. The component would be acceptable provided its static load capacity after fatigue testing was within (say) 30% of its "as-new" capacity. This meets the bridge owner's requirement that the bridge should be capable of carrying the applied loads throughout its life.

5.2 Damage summation method

Some damage may occur but the estimated life is greater than the specified design life.

This option requires a method of estimating the total fatigue damage due to all the various wheel loads in the design traffic spectrum. Of the methods currently available, the Palmgren-Miner cumulative damage rule is proposed, as it is relatively simple and already familiar to bridge designers (but there is some doubt whether it is accurate for alternating stresses).

The procedure given in the UK Bridge Design Code, BS 5400 Part 10 can be used. The data required to estimate fatigue life is:

- The design traffic loading.
- The relationship between wheel load and stress, i.e. single wheel influence lines.
- A design S-N curve for the detail being assessed.

It is assumed that the assessment will be carried out in terms of stress (or strain) at the crack location. Single wheel influence lines would be calculated or measured for each detail, then the design traffic applied to derive stress spectra.

Design S-N curves exist for steel details which may be at risk of fatigue, following decades of research and testing. No such database of results exists for FRP components. In addition, stress ratio can be ignored when assessing welded steel details (and there is a simple method for non-welded details). For FRPs on the other hand, stress ratio is a significant variable, so has to be reproduced in the tests.

Around six fatigue tests would be required to define an S-N curve for each of the "at-risk" details (though it may be possible to cover more than one detail with each test, for example producing "lower bound" curves for details which did not fail). Joints between FRP components would need to be included in the assessment. Once the influence line and S-N data are available, it is a relatively simple matter to estimate the fatigue life for any given traffic loading.

It remains to be seen whether the amount of testing necessary to define a design S-N curve can be justified. A possible compromise option might be to carry out sufficient fatigue testing to define a fatigue limit, e.g. stress at 10^7 cycles, then to assume a design S-N curve based on this value.

5.3 Rolling wheel test equivalent life

It is of interest to know the equivalent service life simulated by the rolling wheel test in the TTF rig.

The traffic loading in BS 5400 Part 10 was used with two million HGVs per year (the value for a UK motorway). In the absence of a design S-N curve for the Roadway panel, the conservative assumption was made that the fatigue life of the most highly stressed detail on the panel (the top of the web) was about to fail at 4.6 million cycles. This provided one point on a notional S-N curve.

The S-N equation for FRP materials may be expressed by an empirical equation, for example:

$$\sigma_{max}/\sigma_u = 1 - k \cdot \log(N_L)$$

where σ_{max} is the maximum stress in a cycle, σ_u is the ultimate stress, k is an experimental factor (assumed 0.1), and N_L is the number of cycles to failure.

The load applied by the rolling wheel (37 kN) was around 10% of the ultimate capacity of the panel, so the applied stress was 10% of ultimate. Clearly the test was in the low stress – high cycle region of the S-N curve. It is not known whether the above equation is applicable at endurances greater than 10^7. Its form suggests that all cycles are damaging in proportion to the applied stress. This does not take account of the fact that stresses below the constant amplitude cut-off are not damaging until later in the life of the detail. To avoid this difficulty, it was decided to use an S-N equation of the form:

$$S^m \times N = k$$

where S is the maximum stress in a cycle, m is the inverse slope of the S-N line (assumed to be 10 for FRP), N is the number of cycles to failure, and k is a constant.

At high stress and low cycles the S-N curves produced by the two equations are similar, but the latter is less severe at high endurances. The constant in the equation which passes through the test point is $k = 4.66 \times 10^{40}$. Applying this equation, along with the influence lines, gives an estimated life for the Roadway panel of 39 years.

6 CONCLUSIONS

The dynamic testing of an FRP deck has shown that FRP components can provide a robust bridge solution complying with the general requirements of the UK design code and, in particular, are capable of resisting local wheel loads due to heavy vehicles for at least 30–40 years, without major damage. However, careful attention is needed to prevent local damage in highly stressed regions such as web to flange connections and close to bearing supports.

A procedure for fatigue design and a testing procedure are being devised for incorporation into a Highways Agency Standard for FRP decks. It is considered important to minimise the testing requirement so as to encourage development of FRP bridge decks. The results so far suggest that a simple test simulating a single wheel load may be sufficient.

A simple fatigue design method, not requiring a damage summation appears possible, based on demonstrating that fatigue damage will not occur under design traffic loading.

REFERENCES

British Standards Institution 1980. BS 5400: Part 10: 1980. Steel, concrete and composite bridges. *Code of practice for fatigue*, British Standards Institution, London, UK.

Daly, A.F. & Duckett, W.A. 2002. The design and testing of an FRP highway bridge deck. *Journal of Research*, Volume 5, No 3, Transport Research Laboratory, Crowthorne, UK.

Numerical simulation of steel and concrete composite beams subjected to moving loads

N. Gattesco
Department of Architectural and Urban Design, University of Trieste, Italy

I. Pitacco & A. Tracanelli
Department of Civil Engineering, University of Udine, Italy

ABSTRACT: The paper concerns a numerical procedure able to study the structural behavior of composite beams subjected to moving loads and considering the actual cyclic relationship between the shear load and the slip of the connectors. Such a procedure allows to follow, cycle after cycle, the slip history of each connector as well as to evaluate the distribution of the shear force along the connection. The procedure was used to study the behavior of a 40 m span simply supported bridge-type beam subjected to two different types of moving loads: fatigue model 1 (ENV 1991-3) and heavy abnormal 15 axles vehicle (3600 kN). The results show that large slip values are involved so that the stud collapse may occur after few thousands of cycles.

1 INTRODUCTION

Bridges are subjected to very complex loading conditions concerning mainly cyclic loads of different magnitude. Cyclic loads produce a progressive damage, more or less pronounced, in all parts of the structure. Actually the parts that are more sensitive to damage are bolted or welded joints and all parts where stress concentration occurs. Among these parts particular attention has to be paid to the connection between the concrete slab and the steel beam, because the presence of damage can not be surveyed with any inspection method; in fact studs are inaccessible because they are embedded in the concrete slab.

The loads acting on a bridge are moving commercial vehicles with different size and weight. For fatigue checks, most of codes of practice suggest to subdivide the variable cyclic stresses due to moving loads in a certain number of blocks of constant amplitude cyclic stresses according to an assumed cycle counting technique (e.g. rainflow, reservoir, etc.). Then the damage caused by a block of constant amplitude stresses is assumed proportional to the fraction of life used up by the event (ratio between the number of cycles and life to failure for the event); the total damage is obtained by summation of damage fractions due to each block of constant amplitude stresses (Miner rule). In the procedure the damage varies linearly with the number of cycles and the connection behavior is considered linearly elastic.

Actually, the various types of vehicles passing on the bridge produce significantly different damage levels, which are not varying proportionally with the number of cycles and depend on the sequence of loading events. Particularly, the most heavy vehicles, even though they pass over the bridge only few tenths of times per year, may cause the most part of the damage accumulated in the connection and moreover they can provoke the connection failure due to low-cycle fatigue (Gattesco et al. 1997).

These aspects are not yet a known cause of failures in service, probably because of some over-conservative points of the design, and because most composite bridges are still within the first third of their design life. However it is mandatory to develop numerical tools, able to simulate the actual behavior of composite bridge girders subjected to cyclic loads, which allow to preview the actual life to failure for new bridges and to assess the remaining life for existing bridges. It is, then, necessary to develop a numerical procedure able to study the structural behavior of composite beams subjected to moving loads and considering the actual cyclic relationship between the shear load and the slip of the connector. Thus, herein a numerical procedure to face these problems is presented. Such a procedure allows to follow, cycle after cycle, the slip history of each stud connector, as well as to evaluate the distribution of the shear force along the connection.

2 LOAD-SLIP MODEL FOR CONNECTORS

The constitutive relationships for concrete and steel are assumed linear elastic, whereas the load-slip relationship for the connection is assumed nonlinear. In fact, the stresses in the steel member are normally below yielding and in the concrete slab are rather limited. The load-slip curve of the connection is nonlinear even for very low values of the shear force and moreover the unloading curve is always different of the monotonic one (Gattesco 1997).

The cyclic load-slip relationship is derived from some experimental results (Gattesco & Rigo 1999, Gattesco 1999) and it is illustrated in Figure 1. The monotonic curve is described by the function

$$Q = \alpha \cdot (1 - e^{-\frac{\beta \cdot s}{\alpha}}) + \gamma \cdot s \qquad (1)$$

where Q is the shear force, s is the slip between the concrete slab and the steel beam and the coefficients α, β, γ are constants experimentally determined (Table 1). The unloading curves refer to a local axis with the origin in points (Q_i, s_i) where the unloading starts (point B in Fig. 1)

$$Q'_i = Q_i \cdot \lambda \cdot (1 - e^{-\frac{\eta \cdot z_i}{\lambda}} + \frac{\delta}{\lambda} \cdot z_i) \qquad (2)$$

$$z_i = \frac{s'_i}{s'_{li}}$$

where s'_{li} is the slip, referred to the local axis, corresponding to a zero value of the load (point C) and it depends to the shear load Q_i

$$s'_{li} = c_1 \cdot Q_i^3 + c_2 \cdot Q_i, \qquad (3)$$

η has the following expression

$$\eta = b_1 \cdot Q_i - b_2 \quad (\geq 0). \qquad (4)$$

The coefficients λ (Equation 2), c_1, c_2 (Equation 3) and b_1, b_2 (Equation 4) have to be experimentally determined (Table 1). The value of coefficient δ is obtained by Equation 2 imposing $Q'_i = Q_i$ for $z_i = 1$ ($s'_i = s'_{li}$). The value of the slip at the end of the unloading s'_{li} varies with the number of cycles according to the relationship (Gattesco & Giuriani 1996, Gattesco 1997)

$$s'^{j}_{li} = s'^{1}_{li} \cdot (1 + \rho \cdot \frac{(j-1)^\varepsilon}{25 + (j-1)^\varepsilon}), \qquad (5)$$

where ρ and ε are constants (Table 1) and j is the cycle number. At the first cycle the unloading curve ends at the intersection with the reverse monotonic curve (point E); a further increase of the negative load follows this curve (path E-F). From point F, the reloading curve has the same expression as Equation 2 (path F-G-H-I) and it ends at the intersection with the first branch of the unloading curve of the preceding cycle (point I); an increase of the load follows this last curve up to point J and then the straight line J-K. A further increase beyond point K follows the monotonic curve. The new unloading curve follows a path similar to the reloading one with changed sign. If the unloading path (B-C-D) stops before point E, the reloading curve coincides with the unloading one up to point J.

The s'_{Ji} which defines point J is derived from the equation

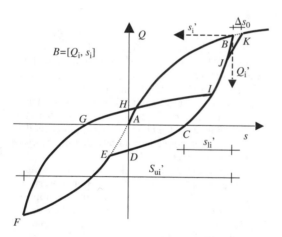

Figure 1. Cyclic load-slip relationship of the connectors.

$$-c_3 \cdot e^{-\frac{c_4 \cdot z_i}{c_3}} + c_5 \cdot z_i = 0, \qquad (6)$$

where c_3 is equal to

Table 1. Load-slip relationship coefficients for 19 mm stud connectors ($f_y = 350$ MPa, $f_t = 450$ MPa) embedded in concrete with compressive strength $f_c = 30$ MPa (Gattesco 1997).

$$c_3 = \frac{\Delta s_{oi}}{s'_{ui}}. \qquad (7)$$

$\alpha = 82$ kN	$\lambda = 0.90$	$c_4 = 2.3$	$\varepsilon = 0.85$
$\beta = 230$ kN/mm	$b_1 = 0.054$ kN^{-1}	$c_5 = 0.4$	$\rho = 0.31$
$\gamma = 5$ kN/mm	$b_2 = 1.34$		
$c_1 = 9.2 \cdot 10^{-7}$ mm/kN^3		$\nu = 1.16 \cdot 10^{-4}$ mm/kN	
$c_2 = 9.0 \cdot 10^{-4}$ mm/kN		$\mu = 1.7 \cdot 10^{-5}$ mm/kN	

and c_4, c_5 are constants. s'_{ui} is the slip corresponding to the beginning of the reloading curve, referred to the local axis.

The increment in slip Δs_{oi} at the end of the reloading curve (Fig. 1), which represents the damage at each cycle (crushing of concrete beneath the stud), may be determined in this way

$$\Delta s_{oi}^j = \frac{Q_i \cdot s_{ui}^{'j}}{s_{li}^{'j}} \cdot (\frac{\nu - \mu}{j} + \mu),$$ (8)

where ν and μ are constants. The coefficients c_3 and c_4 vary with the number of cycles as follows

$$c_3^j = c_3^1 - (\frac{\Delta s_{oi}^1}{s_{ui}^1} - \frac{\Delta s_{oi}^j}{s_{ui}^j}),$$ (9)

$$c_4^j = \frac{c_3^j}{c_3^1} \cdot c_4^1.$$ (10)

3 NUMERICAL MODEL

The model of the composite beam consists of two unidimensional elements, for steel beam and concrete slab, coupled by lumped springs, for shear connectors (Fig. 2). The main assumptions are: (1) linear behavior of the beams, both mechanically and geometrically; (2) the elements can slip along the connection without separation (i.e. uplifts neglected); (3) nonlincar behavior of lumped springs.

The symbol $\nu(z)$ is used for the transverse displacement of the elements, $\omega(z)$ for section rotations, $u(z)$ for the axial displacements and $\gamma(z) = \omega(z) + v'(z)$ for the shear strain, where $v'(z)$ means the first derivative of v with respect to z. A subscript c or s is placed on the above symbols to refer them to concrete or steel element. The kinematic of the problem is described by the six fields

$$v_s(z), \omega_s(z), u_s(z), v_c(z), \omega_c(z), u_c(z).$$

By assumption (2), the unknown fields are reduced to five by letting $v_s = v_c = v$ with the further consequence

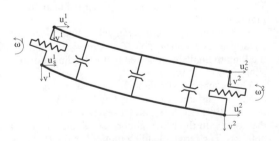

Figure 2. Finite element considered in the numerical model.

that $\omega_s - \omega_c = \gamma_s - \gamma_c$. As expected, and verified *a posteriori*, the shear strains γ are more than one order of magnitude smaller than rotations ω. Hence the approximation

$$\omega_s = \omega_c = \omega \Rightarrow \gamma_s = \gamma_c = \gamma$$ (11)

may be done, which allows to reduce the number of unknowns fields to four. By the above assumptions the shear slip of the i-th connector results

$$s_i = u_c(z_i) - u_s(z_i) + \omega(z_i)h,$$ (12)

where h is the distance between the centers of gravity of the two beams.

To develop a finite element formulation of the problem, it is temporarily assumed that the shear response Q_i is an external given quantity and that external forces are constant in time. Then the total mechanical energy is

$$J(v, \omega, u_s, u_c) := \frac{1}{2}\int_0^L E_c A_c u_c'^2 + E_c A_c u_s'^2 dz$$

$$+\frac{1}{2}\int_0^L (E_c J_c + E_s J_s)\omega'^2 + (G_c \overline{A}_c + G_s \overline{A}_s)\gamma^2 dz$$

$$-\int_0^L pvdz - \sum_{i=1}^{N_c} Q_i s_i$$ (13)

where $p(z)$ is the transverse load, E_a and G_a the elastic constants, A_a, J_a the area and the second moment of the area of the sections and $\overline{A}_a$ the shear area. Introducing a finite element mesh of N_{el} beam elements with N_{no} nodes of coordinates z_j, $j = 1, ..., N_{no}$, and four associated degrees of freedom

$$d^j = [u_s(z_j), u_c(z_j), v(z_j), \omega(z_j)]^T$$ (14)

the energy functional could be written as

$$J = \frac{1}{2}KD \cdot D - F \cdot D + Q \cdot D,$$ (15)

where $\mathbf{D} \in \Re^N$ is the generalized displacement vector, N the total number of degrees of freedom, $\mathbf{K}$ is the stiffness matrix formed by assembling the concrete and steel beam elements, $\mathbf{F}$ is the external load vector, made by the contribution of the elements and node loads, and $\mathbf{Q}$ is the load vector formed by the connector contributions. While $\mathbf{K}$ and $\mathbf{F}$ are standard quantities to form, $\mathbf{Q}$ requires some further remarks. The connectors are placed at the nodes of the mesh so that the slip s_i of the connector placed at node x_i depends only on the degrees of freedom of that node by the relation $s_i = -\mathbf{b} \cdot \mathbf{d}^i$ where $\mathbf{b} = [-1, 1, 0, -h]$

(see Equation 12). Called with $\boldsymbol{B}^i$ the vector of $\Re^N$ such that $\mathbf{B}^i \cdot \mathbf{D} = -\mathbf{b} \cdot \mathbf{d}^i$ for every $\mathbf{D} \in \Re^N$, ($\mathbf{B}^i$ is a vector with at most 3 components not zero) one gets

$$s_i = \mathbf{B}^i \cdot \mathbf{D}, \tag{16}$$

$$Q = \sum_{i=1}^{N_n} Q_i \mathbf{B}^i, \tag{17}$$

and, finally, from the minimization of the total energy (Equation 15), the equilibrium equation

$$KD + Q = F. \tag{18}$$

Now the static assumption can be removed allowing the external force $\mathbf{F}$ to vary with time $t \in [0, T]$ where T is the time a moving load needs to cross the beam. Furthermore the connector response $Q_i = Q_i(\mathbf{p}^i, t)$ depends on the local shear slip history $s_i(\tau)$, $\tau \in [0, T]$, through a set of local state parameters $\mathbf{p}^i$. Neglecting inertia and damping forces, equilibrium Equation 18 reads

$$\mathbf{KD}(t) + \mathbf{Q}(\mathbf{P}, \mathbf{D}(t)) = \mathbf{F}(t), \tag{19}$$

where $\mathbf{P}$ is the complete set of state parameters of the connectors.

The nonlinear set of Equations 19 are solved by the modified Newton Raphson algorithm. Starting from a uniform discretization of the time domain into N time steps of size $\Delta T = T/N$, the algorithm consists in the repetition of the following procedure, until $t_n \geqslant T$.

Let $t_n = n\Delta T$, $\mathbf{D}^n$, $\mathbf{P}^n$ be an estimate of $\mathbf{D}(t_n)$, $\mathbf{P}(t_n)$, and $^1\delta\mathbf{F} := \mathbf{F}(t_{n+1}) - \mathbf{F}(t_n)$ $^1\mathbf{D} := \mathbf{D}^n$, solve iteratively for $m = 1, 2, \ldots$, the problem:

1. $^1\delta\mathbf{F} := \mathbf{F}(t_{n+1}) - \mathbf{F}(t_n)$,
2. evaluate $^m\mathbf{S} = \partial\mathbf{Q}/\partial\mathbf{D} = (^m\mathbf{D})$,
3. solve $[^m\mathbf{S} + \mathbf{K}]d\mathbf{D} = {}^m\delta\mathbf{F}$,
4. evaluate $^{m+1}\mathbf{D} = {}^m\mathbf{D} + \delta\mathbf{D}$ and
   $^{m+1}\delta\mathbf{F} = \mathbf{F}(t_{n+1}) - \mathbf{K}^m\mathbf{D} - \mathbf{Q}(^m\mathbf{D})$,
5. evaluate $^{m+1}\varepsilon = \|{}^{m+1}\delta\mathbf{F} - {}^m\delta\mathbf{F}\| / \|^1\delta\mathbf{F}\|$ and return to step 1 if $^{m+1}\varepsilon > \varepsilon_0$,
6. update $\mathbf{D}^{n+1}$ with the last evaluated $^m\mathbf{D}$ and stop the procedure

In step 5 ε_0 is the required maximum equilibrium error. The rate of convergence of the above procedure is slowed-down, if not broken-down, by very frequent changes in the loading–unloading condition of some connectors, which produces a strong variation in a few components of $^m\mathbf{S}$. To overcome this difficulty, a variable time step strategy was introduced after step 4. Let $\mathbf{u}$ the subset of $\mathbf{P}$ made by that boolean parameters describing the loading-unloading state of the connectors. If passing from $^m\mathbf{D}$ to $^{m+1}\mathbf{D}$ it produces a change in $\mathbf{u}$, $^{m+1}\mathbf{D}$ is rejected, and the procedure is repeated, keeping the connector state frozen, with an

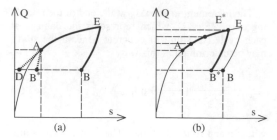

Figure 3. The variable time step size procedure.

increasingly smaller time step $\Delta T/\alpha$, $\alpha = 10 \div 100$. In Figure 3a, the thick solid line represents the load-slip path of a connector that passing from time t_n (point A) to t_{n+1} (point B) changes its state from a loading to an unloading condition.

Dotted lines A-D and A-B* represent the path the Newton Raphson procedure would follow without the variable step size strategy depending on the last updated tangent stiffness $^m\mathbf{S}$. In Figure 3b the thick line represents the actual path followed by the variable step size procedure. After the trial step of Figure 3a has been performed the procedure assumes a prediction of unloading, freezes the connector in the load state and proceeds with reduced time steps along the curve A-E until a new unloading condition is detected. Then, the system is reset to the last known load (point E*), the connector is unfrozen, the tangent stiffness matrix updated, and the original procedure restarted with the original time step size.

4 SIMULATION OF A BRIDGE GIRDER

The numerical procedure was used to simulate the behavior of a single span simply supported composite bridge (40 m) subjected to two models of moving loads, according to ENV 1991-3 (1998). The bridge deck carries a three-lane motorway and has an overall width of 11 meters. Two longitudinal steel girders placed at 5.5 m spacing support a concrete slab, 22 cm depth, connected to them through 19 mm headed stud. The cross section of the bridge girder is illustrated in Figure 4.

Stud connectors were arranged in nine blocks of equally spaced studs (Table 2) designed according to the longitudinal shear caused by the loading model 1 (ENV 1991-3 1998).

The numerical simulations consider two cases of moving loads: the fatigue loading model 1, defined in ENV 1991-3 (1998), and a severe combination of that load with the heavy abnormal vehicle weighting 3600 kN. The fatigue loading model 1 in the first lane consists of a conventional vehicle with two axles of

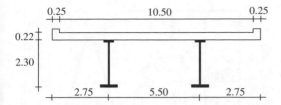

Figure 4. Cross section of the bridge girder.

Table 2. Distribution of stud connectors along half span.

Abscissa [m]	Studs per group	Spacing [mm]
0.00 ÷ 4.00	3	200
4.00 ÷ 10.00	3	250
10.00 ÷ 13.50	2	250
13.50 ÷ 16.50	2	300
16.50 ÷ 20.00	2	350

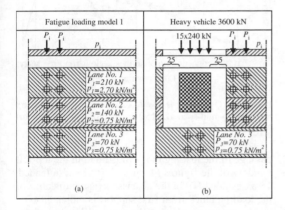

Figure 5. Moving loading patterns considered: (a) fatigue loading model 1, (b) heavy abnormal vehicle plus loading model 1.

210 kN each and a uniformly distributed load equal to 2.7 kN/m²; in the second lane the loads are 140 kN and 0.75 kN/m², respectively; in the third lane the loads are 70 kN and 0.75 kN/m², respectively. The loading scheme is illustrated in Figure 5a.

The other loading model consists of 15 axles equally spaced (1.50 m), weighting 240 kN each. In front of and behind such a vehicle an unloaded zone of 25 m is considered; outside this zone the fatigue loading model 1 is added, as shown in Figure 5b. The fatigue model 1 is normally used to check the unlimited endurance of the structure; the other loading condition was considered with the main purpose to assess the effect on the shear connection of the occurrence of several very severe loading events on the structure.

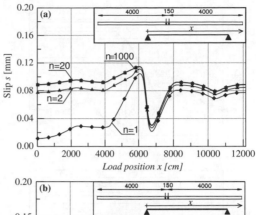

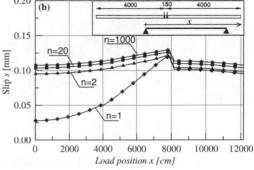

Figure 6. Oscillograms of the slip (load FM1): (a) near to mid-span (23.15 m); (b) close to the support (40.00 m).

The results obtained by the first numerical simulation (1000 cycles of the Fatigue Model 1) show small values of the slip at steel-concrete interface and low levels of load on shear connectors. In Figures 6a, b are plotted the oscillograms of the slip near to mid-span (23.15 m) and close to the second support (40.00 m), respectively, corresponding to the passage of the loading block illustrated in the top-left window of the figures (40 m p_i, 2 P_i, 40 m p_i). In the abscissa axis it is indicated the distance of the beginning of the loading block from the first support (x). The diagrams show that the maximum slip occurs when the two axles load is over the considered position in the beam. The corresponding oscillograms of the shear force are plotted in Figures 7a, b. As expected the maximum range of the shear force occur in stud connectors near the mid-span.

The numerical simulation concerning the application of the heavy abnormal vehicle (1000 cycles of the 15 axles of 240 kN) shows that significant slip values are obtained at the interface and the range of the shear force in all connectors is rather large. In Figures 8a, b are plotted the oscillograms of the slip near to mid-span (23.15 m) and close to the second support (40.00 m), respectively, corresponding to the

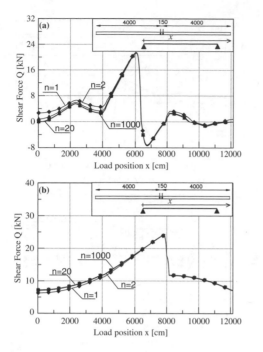

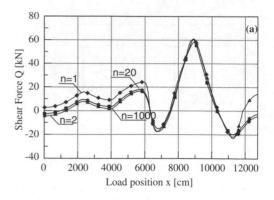

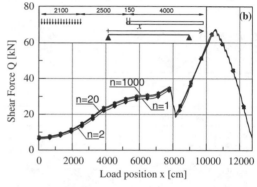

Figure 7. Oscillograms of the stud shear force (load FM1): (a) near to mid-span (23.15 m); (b) close to the support (40.00 m).

Figure 9. Oscillograms of stud shear force (heavy vehicle): (a) near to mid-span (23.15 m); (b) close to the support (40.00 m).

passage of the loading block illustrated in the top-left window of the figures (40 m p_i, 2 P_i, 25 m unloaded, 15 axles 240 kN). In the abscissa axis the distance of the loading block from first support is indicated (x).

The corresponding oscillograms of the shear force are plotted in Figures 9a, b. Two major cycles occur at each passage of the loading block in the connectors near the midspan (Figs 8a, 9a). Moreover the main cycle is large enough to cause the connectors to fail after a a few thousands of cycles (Gattesco & Rigo 1999).

The results show that, except for the first cycle, limited variations of the slip and the shear force at the increase of the number of cycles occur. Actually, a progressive increase of the slip has to be awaited due to the stud stiffness decrement with the damage of the connector shank. The cyclic load-slip relationship of the stud do not include such a damage.

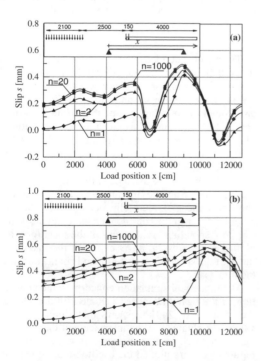

Figure 8. Oscillograms of the slip (heavy vehicle): (a) near to mid-span (23.15 m); (b) close to the support (40.00 m).

5 CONCLUDING REMARKS

In the paper a numerical procedure able to simulate the actual behavior of bridge steel-concrete composite girders subjected to moving loads (commercial vehicles) is presented. The procedure considers a special finite element made by two uni-dimensional elements connected at the extremities by two nonlinear springs (stud connectors).

The numerical simulations consider a simply supported bridge composite girder with a 40 m span. Two different types of loading were considered: fatigue model 1 of ENV 1991-3 (1998) and heavy abnormal 15 axles vehicle (3600 kN). The results show that in the first case very low values of the slip and of the shear force occur whereas in the second case both slip and shear force are large enough to cause a possible stud collapse after a few thousands of cycles.

The further step of the research work concerns the consideration of different abnormal vehicles passing on the bridge following a random sequence.

REFERENCES

Gattesco N., Giuriani E. & Gubana A. 1997. Low-Cycle Fatigue Tests on Stud Shear Connectors. *J. Struct. Engrg. ASCE*, 123(2), 145–150.

Gattesco N. & Rigo G. 1999. Stud Shear Connections in Steel and Concrete Composite Beams: Experimental Investigation of The Behavior Under Cyclic Loads. *Pers. comm.*

Gattesco N. & Giuriani E. 1996. Experimental Study on Stud Shear Connectors Subjected to Cyclic Loading. *J. of Constr. Steel Res.*, 38(1), 1–21.

Gattesco N. 1997. Fatigue in Stud Shear Connectors. *Int. Conf. Composite Construction – Conventional and Innovative*, Innsbruck, Austria, 139–144.

ENV 1991-3 (1998). Actions on Structures. – Traffic loads on Bridges. Part 3.

Ultimate and service load simulation of a masonry arch bridge scheduled for controlled demolition

P.J. Fanning & V. Salomoni
Department of Civil Engineering, University College Dublin, Ireland

T.E. Boothby
Department of Architectural Engineering, The Pennsylvania State University, USA

ABSTRACT: The new Ballincollig by-pass, to the south and west of Cork City in Ireland, requires the demolition of two existing masonry arch bridges to make way for the new road works. Agreement has been reached with the National Roads Authority in Ireland, Cork County Council and the Consultants and Contractors appointed to design and construct the by-pass, to load one of these bridges to failure during the summer of 2003. The ultimate load test will be preceded by a series of service load tests on the bridge.

The ultimate and service load responses are currently being simulated using three-dimensional finite element modelling techniques. The masonry of the single span bridge, including the arch barrel, spandrel walls and buttress supports are modelled at a macro level using the three dimensional solid elements with a smeared crack material model to facilitate initiation and propagation of any cracks formed in the masonry. The fill material is also modelled using three-dimensional solid elements but with a Drucker-Prager material model assigned. It is the purpose of this paper to describe the modelling, loading and solution simulation strategies used to predict both the ultimate and service load responses of the bridge in advance of the actual test.

1 INTRODUCTION

Stone and masonry arch bridges are systems whose structural behavior is determined by the composite masonry and mortar material, the contained fill material and the interaction between these and the surrounding soil medium. The simulation of their service and ultimate load responses is complex and requires consideration of possible cracking in the arch structure, plastic response of the fill material and the transfer of compressive and frictional stresses between the masonry arch and the fill material. This paper describes the numerical modelling strategies using to capture these effects, which will be validated against test data on a masonry arch bridge in Ballincollig, Co. Cork, Ireland, Figure 1.

Traditionally masonry arch bridges have been idealised in two dimensions and analysed and assessed under plane strain conditions. The stiffening and strengthening effects of the spandrel walls are generally ignored for what is accepted to be a conservative analysis. However service load tests on masonry arch bridges clearly indicate the three dimensional nature of the structural response, Fanning and Boothby (2001[1]), and to the authors' knowledge no masonry

Figure 1. Greenfields bridge.

arch bridge has ever failed in a classical four hinge mechanism under vehicle loading. On the other hand a significant number of arch bridge failures, such as spandrel wall blow-out, separation of the arch ring from the spandrel walls, or local punching failure of the arch ring have been known to occur. The capture of these modes of failure requires three dimensional bridge responses to be considered.

It is the intention of this paper to demonstrate that three-dimensional models provide a better basis for a more complete understanding of the structural response of masonry arch bridges.

2 GREENFIELDS BRIDGE AND TEST DETAILS

Greenfields bridge, Figure 1, is a masonry arch bridge spanning approximately 5 m with a rise above the springing points of 1.5 m. The barrel width is approximately 5.5 m with an arch ring thickness of 0.5 m.

At the time of writing the service and ultimate load tests scheduled for summer 2003 are being planned. For the purposes of service load tests a 4-axle vehicle, Figure 2, weighing approximately 31 tonnes, will be driven across the bridge. Measurements of strain and deflection at various points along the underside of the arch barrel will be captured and recorded as in previous tests undertaken by the authors (Fanning & Boothby 2001[2–4]). The bridge will be tested with the vehicle being full, empty and half-full to examine the degree of repeatability and linearity (if any) of bridge response.

For the ultimate load test a large reaction block, cast beneath the bridge to support a test reaction frame constructed over the bridge, will act as a counter balancing dead weight to the test loads. A series of hydraulic jacks, reacting against the reaction frame, will be used to apply a pressure at road surface level on the bridge. The distribution of the jack loads will be such as to replicate a relatively large axle load rather than a distributed load as has been usual in other ultimate load tests – this is to replicate more closely the type of loading that might occur in an overload situation. The applied pressure load will be increased progressively in a controlled manner until such time as the bridge fails, either globally or locally.

The analyses discussed in this paper have two primary objectives. Firstly a thorough understanding of the bridge response is required to properly plan and arrange both the service and ultimate load tests. Secondly it is hoped that the execution of the analyses in advance of the tests, and their subsequent comparisons with test data, will serve to demonstrate further the suitability of three dimensional finite elements models in reasonably predicting the response of such structures.

In addition a direct comparison of the finite element model results was made with the results obtained by a mechanism and equilibrium method, implemented in the computer program ARCHIE-M which examines the capacity of a bridge along the span length only.

3 MODELLING AND ANALYSIS OF GREENFIELDS BRIDGE

3.1 *Geometry and mesh*

For this study the geometry and profile of the bridge curvature was determined from a photographic survey of the bridge. The fill was assumed to be uniform through its depth and the barrel thickness was assumed to be equal to the thickness of the facing blocks.

Three-dimensional eight noded isoparametric elements, Solid65, were used for the masonry/mortar continuum, including buttress supports (ANSYS version 6.1). Overlaid on the arch barrel, and contained within the spandrel walls, three-dimensional solid elements are also used to model the fill material. To enable sliding or movement of the fill material relative to the masonry, without generating tensile stresses at the interface, three-dimensional frictional contact surfaces are included. The movement of the service load test truck on the bridge lane was simulated through appropriately distributed nodal forces applied to the roadway surface. The full model is shown in Figure 3.

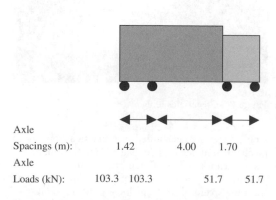

Axle				
Spacings (m):	1.42	4.00	1.70	
Axle				
Loads (kN):	103.3 103.3		51.7	51.7

Figure 2. Truck loads and loading arrangement.

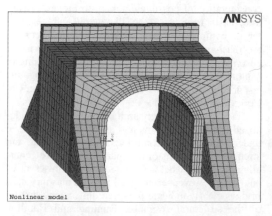

Nonlinear model

Figure 3. Finite element mesh.

3.2 Material properties

The Solid65 element allows takes into account cracking in tension zones through a smeared crack model and a plasticity algorithm to account for the possibility of crushing in compression zones based on a constitutive model for the triaxial behaviour of concrete after Williams and Warnke (1975). The fill, generally a soil material, was modelled using a Drucker-Prager material law. The material properties required and specified in the model, taken from Fanning and Boothby (2001[1]), are summarised in Table 1a and Table 1b.

3.3 Boundary conditions and loading

Vertical restraint to the arch and fill is provided at the base of the model. The outer edges of the modelled fill material are restrained in the span direction at the boundaries of the finite element model while the free edges of the spandrel walls are not restrained. In the transverse direction spandrel wall edges, at the base of the model on one side, are restrained to prevent rigid body motion in that direction.

The simulation of the moving vehicle was undertaken by a sequential series of static loadsteps with the equilibrium solution at the end of one loadstep resulting in a set of initial conditions for the subsequent load step.

Self-weight effects were accounted for by specifying acceleration due to gravity to the model in an initial load step. This load step also serves to simulate the in-situ stresses in the bridge system prior to test loading.

The ultimate condition was performed with load increments applied on a area, representative of that proposed for the in-situ tests, of the bridge width at road surface level (Fig. 4). The load was incremented until an unstable unconverged solution was reached – the state of the numerical model was then assessed to examine whether non-convergence was due to structural instability or numerical solution instability. In the event of the latter the magnitude of load increments was reduced until evidence of structural rather than numerical instability was confirmed.

3.4 Analysis of results

3.4.1 Service load analyses

The service load tests were analysed with both linear and non-linear constitutive laws for comparison purposes and deflections for two points on the arch barrel (Fig. 5) using linear (Table 1a) and nonlinear (Table 1b) material models are plotted in Figure 6 for passage of the test truck. Using nonlinear material data three successive crossings of the truck were considered.

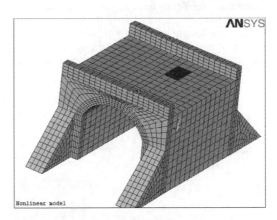

Figure 4. Applied pressure for ultimate load simulation.

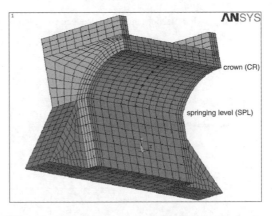

Figure 5. References nodes for service load analyses.

Table 1a. Material properties for FE model.

	Young's modulus (MPa)	Poisson's ratio	Density (kg/m^3)
Masonry	10.10^3	0.3	2200
Fill	15	0.23	1700

Table 1b. Material properties for FE model.

	Tensile strength (MPa)	Compressive strength (MPa)	
Masonry	0.5	10	
	Cohesion (MPa)	Angle of friction (degrees)	Angle of Dilatancy (degrees)
Fill	0.001	44	44

The maximum deflection at the crown is approximately 0.14 mm for the linear model, Figure 6. This maximum deflection is roughly 70% higher than deflections predicted using nonlinear material data. After initial passage of the truck in the nonlinear model there is a reduction in deflection for the second passage while the third passage result is similar to that of the second passage.

The variation in results for the linear and nonlinear material models are explained by examining the response of the fill material under loading in each model. The deflections at road surface level for both models are shown in Figure 7. In the linear model the fills deflections reach a maximum of approximately 5 mm compared to approximately 25 mm for the first passage in the nonlinear model. The greater deflection in the nonlinear model is due to plastic deformation of the fill material which has the effect of spreading the load, applied at road surface level, over a greater area of the arch barrel and hence reducing arch deflections. Furthermore in successive passages of the truck the fill deflections are reduced as the fill material has compacted. Clearly the behaviour, and hence representation, of the fill material plays an important role in the response of the bridge and tests on reconstituted samples of the fill material also form part of the testing program for the bridge.

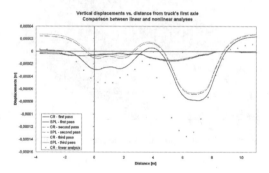

Figure 6. Arch deflections under service load condition.

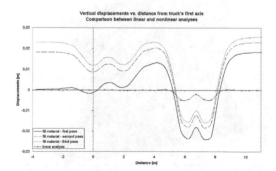

Figure 7. Fill compaction under service load condition.

3.4.2 Ultimate load analysis

The ultimate load analysis was undertaken to study the mode of failure of the bridge. The solid element used in the model allows cracks to form perpendicular to the direction of principal stresses that exceed the tensile strength of the masonry mortar continuum. In the numerical routines the formation of a crack is achieved by the modification of the stress–strain relationship of the element to introduce a plane of weakness in the requisite principal stress direction.

The analysis was undertaken with large deflection effects activated and nonlinear material properties using multiple successive loadsteps. Tensile strengths of 0.5 MPa and 0.25 MPa were considered for the masonry mortar continuum and a full Newton–Raphson iterative solution algorithm with force convergence was used. Automatic load incrementation was used with maximum and minimum load increments of 34.3 kN and 0.21 kN respectively applied as pressures on the area identified in Figure 4.

For each load increment a maximum of twenty five equilibrium iterations was specified. Initially a load increment of 34.3 kN was applied. On completion of an equilibrium iteration the stiffness matrix was updated for changes in geometry and stress stiffening effects and the out-of balance force was calculated. Successive equilibrium iterations were employed to seek an equilibrium solution. The solution was deemed to have converged once the out of balance force was less than 0.5% of the applied loading. In the event of non-convergence the load increment was reduced, to a minimum of 0.21 kN, and the solution re-started from the last converged solution. The ultimate load quoted is the maximum load for which a further 0.21 kN of applied load results in a non-converged solution.

Contours of deflection for the bridge, analysed with a tensile strength of 0.5 MPa, for the last converged solution are plotted in Figure 8. The applied

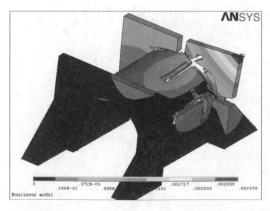

Figure 8. Deformed shape with $f_t = 0.5$ Mpa.

load at this point was about 700 kN. The elements whose tensile strain has exceeded the tensile strength divided by the Young's modulus, i.e. the cracked elements, have been removed.

The analysis predicts a hinge at the springing level spreading the width of the bridge. Additionally there is a partial width hinge near midspan. Significantly the arch barrel is seen to separate from the spandrel wall. Also two definite cracks are formed through the spandrel wall, one at the springing level and one at midspan resulting in overturning of the spandrel wall. The ultimate load analysis predicts a spandrel wall blow out as the primary mode of failure.

The deflections of points along a line indicated in Figure 9 are plotted in Figure 10 for increasing load. The first cracks to form are perpendicular to the span direction near midspan followed by those at the springing level. As the load increases these spread and at a load of approximately 375 kN the arch barrel separates from the spandrel wall resulting in the significant step in the load deflection response. Subsequent to this cracking and then overturning of the spandrel wall occurs.

The bridge response clearly demonstrates both longitudinal and transverse effects as the ultimate load is approached – the deflections at midspan below the spandrel wall furthest from the loaded area are significantly smaller than those on the opposite side of the bridge. This three-dimensional response requires a three dimensional analysis.

As in the service load tests the fill material has a significant effect on the ultimate strength of bridge. It distributes concentrated loads over greater lengths and widths of the arch barrel and provides longitudinal restraint to the arch by its interaction with the surrounding soil medium. This was reinforced by the results of another ultimate load in which the load was directly applied on masonry. The resulting maximum load was approximately 16% lower than when the load was applied through the fill material.

A second ultimate load analysis was performed, using a lower value for the tensile strength (0.25 MPa instead of 0.5 MPa). The maxim load reached was 267 kN. Contours of deflection with cracked elements removed are plotted in Figure 11.

As in the previous analysis the arch ring is seen to separate from the spandrel wall, however the spandrel walls remains stable in this instance. The deformed shape of Figure 11 shows that the ultimate capacity is not defined by a blow-out of the spandrel wall as was predicted previously.

Examination of the last converged solution indicates a localised punching failure of the arch barrel beneath the area of applied load.

In addition to the above analyses the strength of the bridge was assessed using the computer program ARCHIE-M (Obvis 2001). In this two-dimensional mechanism type analysis the load is assumed to be distributed across a unit width of the barrel. Transverse effects and the benefits of the spandrel wall strengths and stiffnesses are not considered. The program uses the zone of thrust to define the minimum depth of arch

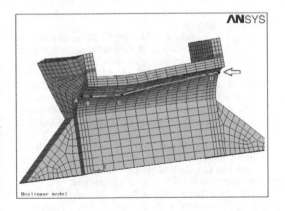

Figure 9. Deformed shape at collapse with indication of the nodes for reported deflections.

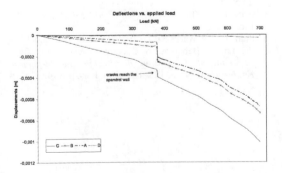

Figure 10. Arch deflections till collapse.

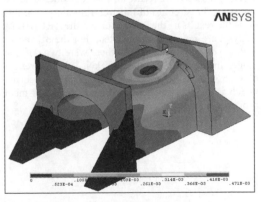

Figure 11. Deformed shape with $f_t = 0.25$ Mpa.

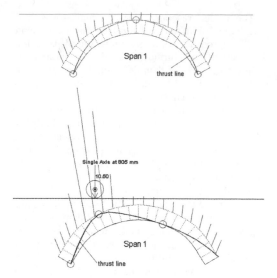

Figure 12. Formation of plastic hinges under self-weight (top) and thrust line for maximum safe load (bottom) – ARCHIE-M.

ring required to support a given set of loads. Provided a thrust line can be found with three hinges which is contained within the depth of the arch ring the bridge is assumed to be safe for the load applied.

Applying ARCHIE-M to Greenfields Bridge under selfweight conditions only produces three hinges and a line of thrust fully contained within the depth of the arch ring, Figure 12 (top). A safe load path, which is not a mechanism, has been found and hence the bridge is deemed to be safe under its own selfweight. Subsequently a single axle load was applied and incrementally increased until the thrust line begins to exit the depth of the arch ring, Figure 12 (bottom). In this instance a maximum load of 723 kN was attained.

The ARCHIE-M result may be considered to be assessing the strength of the arch ring alone. On this basis the results of the 3D model and the ARCHIE-M analysis are reasonably consistent. For the 3D model with 0.5 MPa tensile strength there were two definite hinges in the arch barrel at 700 kN. With a strength of 0.25 MPa a localized punching failure occurred which is beyond the scope of most 2D assessment procedures.

4 CONCLUSIONS

Service and ultimate load analyses on Greenfields Bridge, a 5 m span stone arch bridge in Ireland, have been undertaken in an attempt to predict the mode and load of failure in advance of service and ultimate load tests planned for the summer of 2003.

The analyses demonstrated the importance of considering transverse and three-dimensional effects in assessment programmes for stone and masonry arch bridges. The accepted important role of the fill material was also demonstrated by the analysis results.

The analyses predicted either a punching type failure of the arch barrel local to the applied pressure loading or a spandrel wall collapse, depending on the strength of the masonry mortar continuum.

The analysis results will be compared to test results, which will be available from mid summer 2003, during the conference presentation.

REFERENCES

ANSYS, 2000. *ANSYS Manual Set*, ANSYS Inc., Southpoint, 275 Technology Drive, Canonsburg, PA 15317, USA.

Fanning P.J. & Boothby T.E. 2001[1]. Three-Dimensional Modelling and Full-Scale Testing of Stone Arch Bridges, *Computers and Structures*, **79**, 2001, pp. 2645–2662.

Fanning P. & Boothby T. 2001[2]. Three Dimensional Modelling of Masonry Arch Bridges, *Proc of 2nd Int Conf on Current and Future Trends in Bridge Design*, Construction and Maintenance, Institution of Civil Engineers (UK), 25–26 April, Hong Kong, pp. 524–533, ISBN 0 7227 3091 6.

Fanning P. & Boothby T. 2001[3]. Nonlinear Three Dimensional Simulations of Service Load Tests on a 32 m Stone Arch Bridge in Ireland, ARCH '01, *Proc of 3rd Int Arch Bridges Conference*, 19–21 September, Paris, pp. 373–378, ISBN 2 85978 347 4.

Fanning P., Boothby T. & Gorman P. 2001[4]. An Experimental and Analytical Program for Assessment of Stone Bridges in Ireland, ARCH '01, *Proc of 3rd Int Arch Bridges Conference*, 19–21 September, Paris, pp. 219–228, ISBN 2 85978 347 4.

OBVIS, Ltd, 2001. *ArchieM Quick Guide*, Exeter, OBVIS, Ltd.

William K.J. & Warnke E.D. 1975. Constitutive Model for the Triaxial Behaviour of Concrete, *Proceedings of the International Association for Bridge and Structural Engineering*, **19**, ISMES, Bergamo, Italy, p. 174.

System-based Vision for Strategic and Creative Design, Bontempi (ed.)
© 2003 Swets & Zeitlinger, Lisse, ISBN 90 5809 599 1

Comparison of methods to determine strengths due to mobile loads, in simply supported bridge beams

H. Pankow
Universidade Federal do Paraná-UFPR, Curitiba-Brasil
Universidad Nacional de Asunción-UNA, Asunción-Paraguay

ABSTRACT: The determination of moments and shears strengths, using the traditional simplified methods, requires the use of tables, it has limitations and the process becomes burdened when the mobile load is not at midspan of the bridge. In this article these simplified methods of calculation will be used as reference and subsequently alternative methods by computer will be presented.

1 INTRODUCTION

The traditional methods used to figure strengths in main beams, referenced to, are:

(a) Simplified method of Engesser-Courbon.
(b) Simplified method of Leonhardt.

The suggested methods are:

(c) Finite elements without the contribution of the slab.
(d) Finite elements considering the contribution of the slab.

Methods (a) and (b) will be developed in Section 3 of this article and the suggested methods (c) and (d) will be developed in Section 4.

1.1 *The bridge used in the comparison has the following characteristics*

L = span length = 24 m; Showing in Figure 1: width of the bridge = 7.96 m; ε = beam spacing, center to center = 2.50 m; cast-in-place slab thickness = 0.19 m; asphalt thickness = 0.06 m; pre-cast false slab,

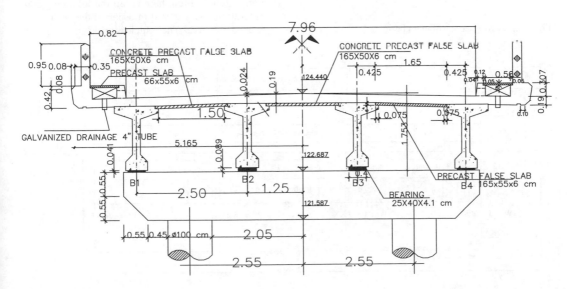

Figure 1. Measures, with unspecified units, are in meters.

thickness = 0.06 m; and length of false slab = 1.65 m. Four simply supported main beams with section properties as follows and shown in Figure 2: h = height = 1.45 m; bs = top flange = 1.00 m; bi = bottom flange = 0.50 m; t = web width = 0.18 m; as = 0.15 m; cs = 0.41 m; ds = 0.14 m; ci = 0.16 m; di = 0.16 m; ai = 0.20 m; A = area = 0.5310 m²; and JL = moment of inertia about centroidal axis xx = 0.1386 m⁴.

2 TYPE OF LOADS

2.1 *The permanent loads or dead loads*

As shown in Figure 3:

Cast-in-place slab load = 0.19 m × 25 KN/m³ = 4.8 KN/m; asphalt load = 0.06 m × 24.5 KN/m³ =

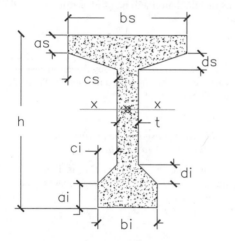

Figure 2. Cross section of main beam.

1.47 KN/m; curb loading = 0.22 m² × 25 KN/m³ = 5.5 KN/m; false slab load = 0.06 m × 25 KN/m³ = 1.5 KN/m; guard rail load = 1.6 KN; self-weight of the main beam = 0.5310 m³ × 25 KN/m³ = 13.28 KN; and self-weight of the diaphragm = 0.15 m × 1.19 m × 25 KN/m³ = 4.46 N/m.

The determination of strengths produced by the permanent loads will not be considered in this comparison but it can be found by any of the methods developed in this article, and then, added to the results reached here, due to the mobile loads.

2.2 *Mobile loads*

The norm used for the determination of strengths, produced by the mobile loads, is NBR 7188 (Brazilian Standards), ABNT (1984). It establishes that, the mobile loads are composed of a vehicle and a uniformly distributed load, in the following manner shown in Figures 4–5. Given that this is a comparative study, the impact factor due the mobile loads will not be taken into account.

2.2.1 *Considerations*

The following considerations are listed, due to the different transverse positions of the mobile load, the truck at midspan of the bridge, moving from left to right (beams are named B1, B2, B3 and B4, as they appear in Figure 1):

– Consideration 0: It corresponds to the uniformly distributed live loads, Figure 6.
– Consideration 1: The vehicle is near the left border, Figure 7. From here on all the following considerations are views at midspan of the bridge.
– Consideration 2: The vehicle is on beam 2, Figure 8.

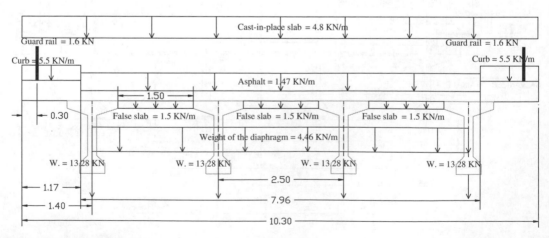

Figure 3. Dead loads in bridge.

- Consideration 3: The vehicle's wheels are on beam 2, Figure 9.
- Consideration 4: The vehicle is in the center of the bridge, Figure 10.

The next considerations that could be taken will be similar because they become symmetrical.

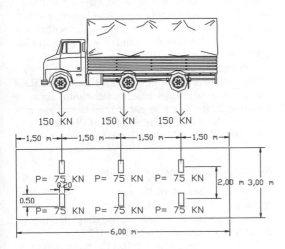

Figure 4. 450 KN truck load and spacing.

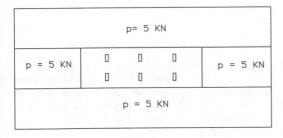

Figure 5. Uniformly distributed live loads, plan view.

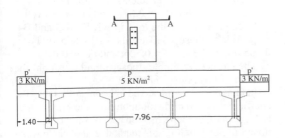

Figure 6. Live load out of midspan, view AA, consideration 0.

3 TRADITIONAL METHODS

(a) Simplified method of Engesser-Courbon

This method considers the case of the load applied on the diaphragm, giving it a supposed infinite stiffness and depreciating the torsion effect on the longitudinal beam, San Martin (1981).

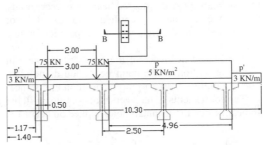

Figure 7. Live load at midspan, view BB, consideration 1.

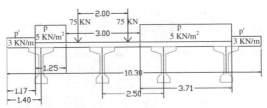

Figure 8. Live load at midspan, view BB, consideration 2.

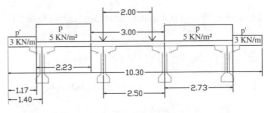

Figure 9. Live load at midspan, view BB, consideration 3.

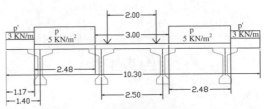

Figure 10. Live load at midspan, view BB, consideration 4.

The equation will be:

$$r_i = \frac{P}{N}\left[1 + \frac{6\lfloor 2i - (N+1)\rfloor \times e}{(N^2 - 1)\varepsilon}\right] \quad (1)$$

where r_i = coefficient in the considered beam; P = concentrated unitary load; N = number of longitudinal beams; i = considered beam; e = eccentricity where the P load is applied, $-$ to the left of center, and $+$ to the right and ε = separation among beams.

The line of influence of the reactions is found in each intersection of the main beam and diaphragm, due to a unitary load, varying its position in distance.

The coefficients of transversal distribution will be:
In beam 1 due to the unitary load applied on beam 1.

$$r_{11} = \frac{1}{4}\left[1 + \frac{6\lfloor 2\times1 - (4+1)\rfloor \times (-3.75)}{(4^2 - 1)2.5}\right] = 0.7$$

In beam 1 due to the unitary load applied on beam 2.

$$r_{12} = \frac{1}{4}\left[1 + \frac{6\lfloor 2\times1 - (4+1)\rfloor \times (-1.25)}{(4^2 - 1)2.5}\right] = 0.4$$

In beam 1 due to the unitary load applied on beam 3.

$$r_{13} = \frac{1}{4}\left[1 + \frac{6\lfloor 2\times1 - (4+1)\rfloor \times (1.25)}{(4^2 - 1)2.5}\right] = 0.1$$

In beam 1 due to the unitary load applied on beam 4.

$$r_{14} = \frac{1}{4}\left[1 + \frac{6\lfloor 2\times1 - (4+1)\rfloor \times (3.75)}{(4^2 - 1)2\times5}\right] = -0.2$$

In beam 2 due to the unitary load applied on beam 1.

$$r_{21} = \frac{1}{4}\left[1 + \frac{6\lfloor 2\times2 - (4+1)\rfloor \times (-3.75)}{(4^2 - 1)2.5}\right] = 0.4$$

In beam 2 due to the unitary load applied on beam 2.

$$r_{22} = \frac{1}{4}\left[1 + \frac{6\lfloor 2\times2 - (4+1)\rfloor \times (-1.25)}{(4^2 - 1)2.5}\right] = 0.3$$

In beam 2 due to the unitary load applied on beam 3.

$$r_{23} = \frac{1}{4}\left[1 + \frac{6\lfloor 2\times2 - (4+1)\rfloor \times (1.25)}{(4^2 - 1)2.5}\right] = 0.2$$

In beam 2 due to the unitary load applied on beam 4.

$$r_{24} = \frac{1}{4}\left[1 + \frac{6\lfloor 2\times2 - (4+1)\rfloor \times (3.75)}{(4^2 - 1)2.5}\right] = 0.1$$

The coefficients in beams 3 and 4 will be the same due to the symmetry of the bridge.

Considering that the longitudinal and transverse beams have the same E (modulus of elasticity). Then, the following relationship, in Equation 2, should be completed:

$$\lambda = \frac{l}{2L}\sqrt[4]{\frac{L}{l} \times \frac{N \times \rho L}{n \times \rho l}} \leq 0.30 \quad (2)$$

where l = 7.5 m = Bridge wide; L = 24 = Bridge long; N = 4 = number of longitudinal beams; n = 1 = number of diaphragm; ρL = main girder stiffness = $E \times JL = E \times 0.1386\,\text{m}^4$ = and ρl = diaphragm stiffness = $E \times Jl = E \times 0.0211\,\text{m}^4$.

Substituting values in Equation 2: $\lambda = 0.47$.

This method is not used because the bridge does not comply with the latter relationship done in Equation 2. It should be either thinner or longer; it either needs to have more number of diaphragms or the stiffness of the diaphragm must be increased.

(b) Simplified method of Leonhardt

The method considers the case of a unitary load applied on the diaphragm, accepting that its stiffness is no longer infinite and depreciating the torsion effect in main beams, San Martin (1981).

For main beams, with constant moments of inertia, simply supported and equally spaced out and the diaphragm beam acting at midspan, the stiffness degree of the structure is found:

$$\xi = \frac{Jl}{JL}\left[\frac{L}{2\varepsilon}\right]^3 \quad (3)$$

where Jl = moment of inertia of diaphragm = 0.021 m^4 and the following, similar to what are found in Section 1.1: JL = moment of inertia of the main beam = 0.1386 m^4; L = span of the bridge = 24 m; and = separation of main beams = 2.5 m.

Substituting values in Equation 3, $\xi = 16.76$ is obtained and Leonhardt's table (Table 1) was used to calculate the coefficients of transversal distribution.

In beam 1, due to the load in B1, B2, B3 and B4, respectively, these coefficients are obtained: r_{11} = 0.732; r_{12} = 0.366; r_{13} = 0.071; and r_{14} = −0.169

In beam 2, due to the load in B1, B2, B3 and B4, respectively, these coefficients are obtained: r_{21} = 0.365; r_{22} = 0.34; r_{23} = 0.225; and r_{24} = 0.070.

The values resemble those found through the method of Engesser-Courbon.

In fact, if $\xi = \infty$ is observed in Leonhardt's table, the coefficient values of distribution have the same values as Engesser-Courbon's method and as was earlier mentioned in Section 3(a), this method is applicable when diaphragm has infinite stiffness and a considerable length of span. This can be reached, by increasing the

Table 1. From Leonhardt's table to four beams, San Martin (1981).

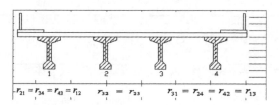

ζ	$r_{11}=r_{44}$	$r_{12}=r_{43}$	$r_{13}=r_{42}$	$r_{14}=r_{41}$	$r_{22}=r_{33}$	$r_{23}=r_{32}$	ζ
0.1	0.978	0.047	−0.028	0.003	0.878	0.103	0.1
0.2	0.962	0.079	−0.042	0.002	0.802	0.162	0.2
0.3	0.948	0.102	−0.048	−0.002	0.748	0.198	0.3
0.4	0.937	0.120	−0.051	−0.006	0.708	0.223	0.4
0.5	0.927	0.135	−0.052	−0.010	0.677	0.240	0.5
0.6	0.918	0.148	−0.052	−0.015	0.652	0.252	0.6
0.7	0.910	0.160	−0.050	−0.020	0.630	0.260	0.7
0.8	0.903	0.170	−0.049	−0.024	0.612	0.267	0.8
0.9	0.896	0.178	−0.046	−0.029	0.596	0.271	0.9
1.0	0.890	0.187	−0.044	−0.033	0.582	0.275	1.0
1.2	0.879	0.201	−0.039	−0.041	0.559	0.279	1.2
1.4	0.869	0.213	−0.034	−0.049	0.540	0.281	1.4
1.6	0.860	0.224	−0.029	−0.056	0.524	0.281	1.6
1.8	0.852	0.233	−0.024	0.062	0.510	0.281	1.8
2.0	0.845	0.242	−0.019	−0.068	0.498	0.280	2.0
2.2	0.839	0.249	−0.015	−0.073	0.487	0.279	2.2
2.4	0.833	0.256	−0.011	−0.078	0.478	0.278	2.4
2.6	0.828	0.262	−0.007	−0.083	0.469	0.276	2.6
2.8	0.823	0.268	−0.003	−0.087	0.462	0.274	2.8
3.0	0.818	0.273	0.000	−0.091	0.455	0.273	3.0
4.0	0.800	0.293	0.014	−0.107	0.428	0.265	4.0
5.0	0.786	0.308	0.025	−0.120	0.409	0.258	5.0
6.0	0.776	0.319	0.034	−0.129	0.395	0.252	6.0
7.0	0.768	0.328	0.040	−0.136	0.384	0.247	7.0
8.0	0.761	0.335	0.046	−0.142	0.376	0.243	8.0
9.0	0.756	0.341	0.051	−0.147	0.369	0.240	9.0
10.0	0.752	0.346	0.054	−0.152	0.363	0.237	10.0
12.0	0.744	0.353	0.061	−0.158	0.354	0.232	12.0
14.0	0.739	0.359	0.065	−0.163	0.347	0.229	14.0
16.0	0.735	0.364	0.069	−0.167	0.342	0.226	16.0
18.0	0.731	0.367	0.072	−0.170	0.338	0.223	18.0
20.0	0.729	0.370	0.074	−0.173	0.334	0.221	20.0
30.0	0.720	0.379	0.082	−0.181	0.324	0.215	30.0
40.0	0.715	0.384	0.086	−0.186	0.318	0.211	40.0
60.0	0.710	0.389	0.091	−0.190	0.312	0.208	60.0
100.0	0.706	0.393	0.094	−0.194	0.308	0.205	100.0
∞	0.700	0.400	0.100	−0.200	0.300	0.200	∞

moment of inertia of diaphragm or the length of span in Equation 3.

(b1) Maximum loads on beam 1
For beam 1, it will have the coefficients r_{11}, r_{12}, r_{13} and r_{14}, in the longitudinal beams B1, B2, B3 and B4 respectively, Figure 11.

To determine the strength in beam 1, due to the load consideration 0 and 1, the concentrated loads are multiplied by the ordinates of the line of influence, as well as the distributed loads by the area under the curve of the line of influence, Figure 12.

The longitudinal loads on beam 1 are calculated:

− 75 KN × 0.69 + 75 KN × 0.40 = 81.8 N
− 3 KN/m × 1 + 5 KN/m × 1.64 + 5 KN/m × 0.48 = 13.6 KN/m
− 3 KN/m × 1 + 5 KN/m × 0.48 = 5.4 KN/m

The diagrams of moment and shear are shown in Figures 13 and 14, and obtained with the program

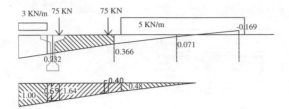

Figure 11.   Lines of influence of reactions in beam 1 – load consideration 1 – Leonhardt.

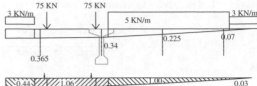

Figure 12.   Longitudinal loads distribution on beam 1, due to consideration 0 and 1.

Figure 13.   Bending moment (KN m) in beam 1, due to consideration 0 and 1.

Figure 14.   Shear strength (KN), in beam 1, due to consideration 0 and 1.

Amses2D, downloaded in its educational version from the website:

http://www.ainet-sp.si/AMSES/En/download.htm

(b2)   Maximum loads on beam 2
For beam 2 and the consideration of load 0 and 1, these coefficients r_{11}, r_{12}, r_{13} and r_{14}, are obtained, in the main longitudinal beams, B1, B2, B3 and B4, respectively.

Similarly to the beam 1:

- 75 KN × 0.36 + 75 KN × 0.34 = 52.5 KN
- 3 KN/m × 0.44 + 5 KN/m × 1.06 + 5 KN/m × 1 + 3 KN/m × 0.03 = 11.63 KN/m
- 3 KN/m × 0.44 + 5 KN/m × 1 + 3 KN/m × 0.03 = 6.41 KN/m

For beam 2 and the consideration of load 0 and 2:
The longitudinal loads on beam 2 are calculated:

- 75 KN × 0.35 + 75 KN × 0.29 = 48 KN

Figure 15.   Lines of influence of reactions in beam 2 – load consideration 1 – Leonhardt.

Figure 16.   Longitudinal loads distribution on beam 2, due to consideration 0 and 1.

Figure 17.   Bending moment (KN m) in beam 2, due to consideration 0 and 1.

Figure 18.   Shear strength (KN) in beam 2, due to consideration 0 and 1.

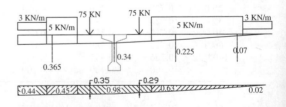

Figure 19.   Lines of influence of reactions in beam 2 – load consideration 2 – Leonhardt.

- 3 KN/m × 0.44 + 5 KN/m × 0.45 + 5 KN/m × 0.98 + 5 KN/m × 0.63 + 3 KN/m × 0.02 = 11.68 KN/m
- 3 KN/m × 0.44 + 5 KN/m × 0.45 + 5 KN/m × 0.63 + 3 KN/m × 0.02 = 6.78 KN/m

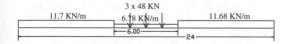

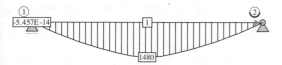

Figure 20. Longitudinal loads distribution in beam 2, due to consideration 0 and 2.

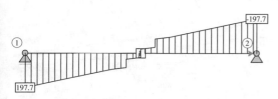

Figure 21. Bending moment (KN m) in beam 2, due to consideration 0 and 2.

Figure 23. Beam sections introduced in Ram Advanse.

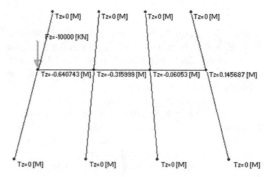

Figure 22. Shear strength (KN) in beam 2, due to consideration 0 and 2.

Figure 24. Deflection obtained.

As seen, moments and shears diminish for the combination of considerations 0 + 2, therefore considerations 0 + 3 and 0 + 4, are no longer verified, meaning that the maximum value will be with the combination of considerations 0 + 1.

The strengths produced in beam 1 (exterior left beam), due to load consideration 0 + 1 are greater than those produced in beam 2 due to the same consideration 0 + 1, these considerations are at the same time maximum in each considered beam (the load consideration 0 + 1 is what produces the greatest moment and shear in beam 1 and also in beam 2).

Note: shear strength is obtained only for comparison purposes, because they are higher when the loads act near the end of the bridge, and here it is considered that the loads are acting at midspan of the bridge, meaning the midspan of the main beam.

4 THE SUGGESTED METHODS:

(c) Finite elements without considering the contribution of the slab

Also those coefficients of distribution can be obtained through finite elements software (e.g. Ram Advanse, SAP2000, etc.).

By placing and moving a virtual load of 10,000 KN, on each intersection between the main

beams and diaphragm, this value is taken to see the deflection better.

The coefficient of distribution without dimension for each beam, will be equal to the deflection of the considered node, when the position of the load changes through the diaphragm, this number is divided by the sum of deflections of all the nodes.

If possible, a real beam section should be introduced in the finite element software, if not, an equivalent section with a similar area and inertia can be introduced (see next Section 4(d)).

As an example the calculation of the distribution coefficient r_{11} and r_{21} is shown:

$r_{11} = -0.64/(-0.64 - 0.316 - 0.06 + 0.1457) = 0.735$

$r_{21} = -0.316/(-0.64 - 0.316 - 0.06 + 0.1457) = 0.363$

$- r_{11} = 0.735;$ $r_{12} = 0.363;$ $r_{13} = 0.069$ and $r_{14} = -0.167$

$- r_{21} = 0.363;$ $r_{22} = 0.344;$ $r_{23} = 0.224$ and $r_{24} = 0.069$

It can be seen that the coefficients are practically identical to those found using the table in method (b), the methodology to find the longitudinal loads in beams, is also similar.

(*d*) *Finite elements considering the contribution of the slab*

As stated in Section 4(c), and as for the majority of cases, the software does not allow for input of the original beam section, therefore, an equivalent rectangular section is imputed, e.g.:

$$A = bh \qquad (4)$$

$$JL = \frac{bh^3}{12} \qquad (5)$$

Substituting *bh* in Equation 5 with A, found in (4).

$$JL = \frac{Ah^2}{12} \Rightarrow h = \sqrt{\frac{12JL}{A}} = 1.77m \qquad (6)$$

where A = area of the beam = $0.5310\,m^2$ (see Section 1.1); $JL = 0.1386\,m^4$ (see Section 1.1); and h = height of the equivalent rectangular beam = 1.77 m; b = width of the equivalent rectangular beam = 0.3 m found substituting h in Equation 4.

The bridge was divided transversely: in 6 segments of 0.47 m wide, in both sides of the bridge, in center, 15 segments of 0.5 m wide, they add up to 10.32 m.

Longitudinally: in 48 segments of 0.5 m long, in total 24.00 m, it is the span of the bridge.

The following loads shown in Figure 26 are applied:

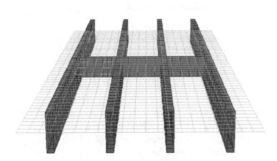

Figure 25. Finite element model of the bridge with slab passing through center of gravity of beams, Oñate (1992).

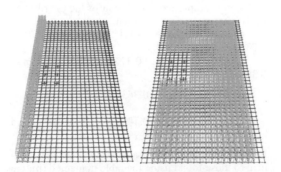

Figure 26. Distributed loads of 3 KN/m² left, and 5 KN/m² right.

Since the load of each wheel is 75 KN, it is distributed in an area of 0.25×0.5 m (the norms establish that 0.20×0.5 should be the contact area for the wheel but in order to simplify, the previous value is taken, because the elements were divided in 0.5 m long and wide, and 0.25 m is half).

That gives $600\,KN/m^2$ in 6 parts, distributed as shown in Figure 27.

From Figure 28: Moment in beam 1 due to condition 0 and 1 = 1988.20 KN m

Moment in beam 2 due to condition 0 and 1 = 1527.10 KN m

From Figure 29: Shear in beam 1 due to condition 0 and 1 = 237.33 KN

Shear in beam 2 due to condition 0 and 1 = 187.01 KN.

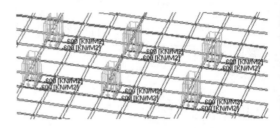

Figure 27. Truck load of 450 KN.

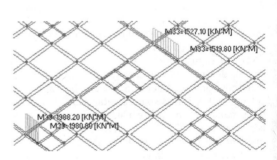

Figure 28. Bending moment (KN m).

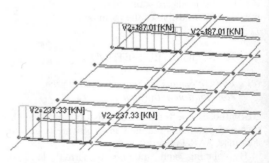

Figure 29. Shear (KN).

5 SUMMARY

The following Table shows the values of moments and shears founded for each method.

	Bending moment (KN m)	Shear (KN)
Traditional method (b)		
Simplified method of Leonhardt.		
Beam 1	2071.00	261.40
Beam 2	1542.00	203.30
Suggested method (c)		
Finite elements without the contribution of the slab.		
Beam 1	Values similar to the	
Beam 2	reference method (b)	
Suggested method (d)		
Finite Elements considering the contribution of the slab.		
Beam 1	1988.20	237.33
Beam 2	1527.10	187.01

6 CONCLUSION

The difference between the simplified method of Leonhardt (b) and the method (c) by finite elements, is that in the latter, tables are not needed to find the distribution coefficients and it is more general, since it can be used for "n" beams.

With the methodology (d) to determine strengths due to mobile loads, in simply supported bridge beams, through finite elements and considering the slab contribution, the following is possible:

1. A better distribution of the strength since the solidity beam-slab is considered and not only the solidity beam-diaphragm.
2. To calculate the strengths in slab, placing the mobile load in the most unfavorable position.
3. To calculate the maximum shears of the beams, when the truck load act near the end of the bridge.

REFERENCES

Associaçao Brasileira de Normas Técnicas. 1984. *Cargas móveis em Ponte Rodoviaria e Passarela de pedestres. Norma NBR 7188/84.* Rio de Janeiro: ABNT.

Oñate, Eugenio. 1992. *Cálculo de estructuras por el método de elementos finitos.* Barcelona: Centro Internacional de Métodos Numéricos en Ingeniería-Universidad Politécnica de Cataluña, 1ª Edición.

San Martin, F. J. 1981. *Cálculo de tableiros de pontes.* Sao Paulo: Livraria Ciencia e Tecnología.

System-based Vision for Strategic and Creative Design, Bontempi (ed.)
© 2003 Swets & Zeitlinger, Lisse, ISBN 90 5809 599 1

Design and construction aspects of a soil stabilized dome used for low-cost housing

M. Gohnert & S.J. Magaia
School of Civil and Environmental Engineering, University of the Witwatersrand, South Africa

ABSTRACT: A dearth of low cost housing in developing countries is a current and widespread problem. Although dome structures are not a common form of housing, the shape has many favorable characteristics which leads to a significant reduction in construction cost. The proposed structure is a composite, constructed out of stabilized soil bricks and plastered with a fibre impregnated structural plaster. However, the proposed structure has several challenging aspects – the analysis is complex and the construction difficult. This paper deals with both issues. The design adapts membrane equations from classical theory, but boundary conditions, which introduces complexities in the analysis, are dealt with in a simplified manner. A method of construction is also proposed to ensure quality control of materials and to ensure that the walls are constructed within acceptable tolerances. A prototype dome was constructed which is also described.

1 INTRODUCTION

In South Africa, the shortage of housing is in the order of 1.5 million homes. In addition, the number of new households increase by 200 000 per year which adds to the backlog. This shortage is primarily amongst the low-income group who cannot afford the cost of a home. The country, in response, has initiated several programs in an effort to alleviate this almost insurmountable problem. A current program, administered by the Department of Housing (Dept. of Housing 2002, Ensor & Radebe 2002), is a subsidy based program designed to reduce the cost of a home. The maximum subsidy is around $2000, which is granted to those who fall into an income group not exceeding $1350 per annum. Those who qualify for the subsidy generally cannot afford a home much greater than the value of the grant. For this reason, the cost of the house is of paramount importance.

It is evident, considering the income levels of subsidy recipients, that the total cost of the home should be reduced as much as possible (preferably, equivalent to the amount of the subsidy). This will require inventive schemes to make the house more affordable. An alternative housing design is presented here with the objective of reducing the cost of the house, but at the same time maintaining a reasonable amount of living space and ensuring a durable and sound structure.

The traditional western style of home in South Africa is made of bricks, a concrete ground slab, timber trusses and tiles or steel sheeting are used as roofing cladding. The roof comprises the majority of the cost. This type of house has been around for many years and optimized to such an extent that any further advancements would be of marginal benefit. In order for a significant savings to be achieved, the design principles of the traditional home must be scrapped and new types of housing developed. This paper describes a new type of home shaped in the form of a dome. The walls are constructed out of a composite material composed of stabilized soil bricks and plastered with a fibre impregnated mixture (Bolton 1998).

Although domes are extremely efficient, this structure has lost popularity due to two main reasons: firstly, the analysis is complex and requires specialized skills either to apply classical mathematical solutions or a finite element analysis. In rural areas, the level of skill is usually insufficient to perform complex tasks and therefore a simplified approach is desirable. Secondly, construction is difficult and tends to be expensive when creating double curvature structures. In addition, the walls of the domes must be constructed to close tolerances to prevent buckling and asymmetric gravity loading. Design and construction aspects are examined in this paper and various solutions are presented. In addition, a prototype dome is described which implements the design and construction techniques proposed.

2 ANALYSIS USING CLASSICAL METHODS OF DESIGN

2.1 Membrane forces and boundary effects

Classical mathematical methods of design are, by nature, complex. The difficulty lies in assessing the affects of boundary conditions. Boundary effects (i.e., the degree of fixity along edge of the dome) are the cause of bending moments and shears in the shell. If the shell is very stiff, the influence is greater. Dome shells, however, are usually thin and sufficiently flexible to minimize this effect. In addition, moments and shears are usually confined to the lower quarter of the shell and relatively small in magnitude. A typical moment and shear distribution is illustrated in Figure 1. As illustrated, peak moments and shears occur at the base and rapidly dissipate up the shell.

If the shell is subjected purely to membrane forces, the analysis is significantly simpler. Membrane equations are formulated by constructing equilibrium equations in the circumferential and meridian directions. Membrane stresses are solved by adapting these equations to various load configurations. The most common equations are those applicable to uniformly distributed load, given below:

$$N_\phi = \frac{-rq}{1+\cos\phi} \tag{1}$$

$$N_\theta = rq\left(\frac{1}{1+\cos\phi} - \cos\phi\right) \tag{2}$$

where, N_ϕ = meridian stress, N_θ = the circumferential stress, r = the radius of the dome, q = the uniformly distributed load applied over the surface of the dome and ϕ = an angle measured from the apex of the dome.

Equations 1 and 2 are easily solved. Similar equations for other types of loads are extensively published (Billington 1982, Gould 1999). Since the influence of boundary effects is confined to the lower extremities of the shell, designs based solely on membrane forces are safe if the structure makes an allowance for boundary forces. This is usually in the form of a localized thickening which gradually tapers out within the region of boundary influence (approximately one quarter of the height of the shell). Alternatively, the boundary effects could be resisted by strengthening the shell walls locally. Vertical starter bars projecting from the base into the shell walls is one way to strengthen this region.

Bending and shear effects at the base can be minimized by using form-finding techniques (Isler 1994). The most common method is to hang a round flexible material (e.g. fabric), allowing the shape of the dome to form naturally. In this state, the fabric is in tension and the boundary free of bending forces (fabric can only resist tensile forces). To maintain the shape, the fabric is moistened and frozen or a resin is applied. In a hardened state, the model is flipped over causing a reversal in stress – from tension to compression. In this hardened state, the coordinates (defining the shape of the shell) are recorded. By simply scaling the coordinates, the form of the shell is defined.

A grillage model could be formed out of pinned-ended bar elements arranged to resemble a circular flat "spider-web." Normal loads are applied to the nodal points and the magnitude of these loads should be in proportion to the tributary areas. Under loading, the plate will deform creating a natural shape in tension and free of bending at the base. These deformations can then be scaled to define the shape of a full size dome. This part of the analysis is simply a form-finding technique to determine a natural shape under gravity loads. To complete the analysis, the shape should be entered into a finite element analysis package and analyzed with shell elements under an assortment of load combinations (gravity, wind, earthquake, etc.). By doing so, the shape is confirmed for gravity loads and the magnitude of the boundary effects are determined for other loading configurations.

2.2 The analysis of point loads

Resistance against point loads is an inherent weakness of shells. If the load carrying capacity of point and uniformly distributed loads are compared, a vast difference in strengths exist. Distributed loads are carried primarily by membrane action and the structure is able to resist a significant amount of load. On the other hand, point loads are carried locally by

a. Moment distribution

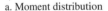

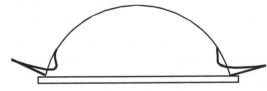

b. Shear distribution

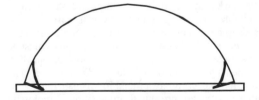

Figure 1. Moment and shear distribution in a shell due to boundary effects.

bending and shears; these forces are expensive forces and require a substantial amount of material to provide resistance. Designing for point loads is essential, simply because they exist and it is probable that the shell will be subjected to such loading during the life of the structure. Design procedures are identical to those of a slab. Bending is assessed by implementing Johansen's yield-line analysis (Johansen 1962), assuming a fan failure mechanism; this would imply that the material is sufficiently ductile. The shear strength of the shell is determined by a punching shear analysis, which is a common procedure in most codes of practice. It should be noted that since the surface of the dome is curved, a portion of the shell is diverted directly into the plane of the shell and therefore the load carrying capacity is increased.

2.3 Openings in the shell

Openings in shell walls introduce an assortment of mathematical complexities. Classical methods are restricted to standard shapes and simplistic loads. Closed formed solutions are not available to analyze shells with openings with the exception of a circular skylight located at the apex of the dome. Several methods, however, have been devised to assess the effects of openings.

Perhaps the most least common method is a photoelastic analysis. A scaled model is made of a plastic material referred to as CR-39 or alternatively Bakelite (Davis 1964). Polarized light is passed through the loaded model creating color patterns, which represent the state of stress in the shell. Concentrations of stress are identified and appropriately reinforced.

A finite element analysis is one of the most effective ways of determining the effects of openings in a shell. The only requirement is to ensure that a sufficient number of elements exist in regions of concentrated stress – typically at the corners. The resulting stress patterns not only illustrate the area of influence, but also the magnitude of stress.

A simple, yet effective technique is to consider the flow of hoop (circumferential) and meridian forces in the shell. The stress tends to flow around the opening as illustrated in Figure 2a. This causes a concentration of stress around the corners and sides of the opening.

The force due to a concentration of hoop stress (F_θ), located at the top and bottom of the opening, is equal to the stress times the height of the opening divided by two (Fig. 2b).

$$F_\theta = \frac{N_\theta h}{2} \tag{3}$$

where h = the height of the opening.

In addition to the hoop forces, the meridian forces will also cause stress concentrations around the top and bottom of the opening. This force can be estimated by applying deep beam theory. If a simple triangular truss is assumed, the force is approximated by the following expression:

$$F_\phi = -\frac{N_\phi b}{4} \tag{4}$$

The total force is therefore,

$$F_H = F_\phi + F_\theta = \frac{N_\theta h}{2} - \frac{N_\phi b}{4} \tag{5}$$

where F_H = the total concentrated horizontal force located at the top and bottom of the opening.

By inspection, equation 5 indicates that if both the hoop and meridian forces are compressive (which is desirable in concrete and brick domes), the meridian forces tend to reduce the effect of the hoop forces.

As a general rule, the degree of strengthening around an opening should not be less than the required strength of the materials displaced. For example, if the dome is constructed out of reinforced concrete, the amount of steel concentrated around the openings should not be less than the amount of steel displaced.

Along the height of the opening, the concentrated force to be resisted is determined using a similar equation.

$$F_v = \frac{N_\phi b}{2} - \frac{N_\theta h}{4} \tag{6}$$

where F_V = the total concentrated vertical force located along the vertical sides of the opening.

It is prudent to further strengthen the corners where the highest concentration of stress exists. If the dome is constructed out of reinforced concrete, additional diagonal bars are usually placed in the corners of the opening.

The proposed dome configuration is of composite construction; the walls are composed of compressed cement stabilized soil bricks and a structural plaster is applied to the exterior and interior walls (Magaia, in prep). The structural plaster is impregnated with fibres to resist any tension stresses in the wall. Strengthening

(a)   (b)

Figure 2. Flow of circumferential forces in a dome with openings.

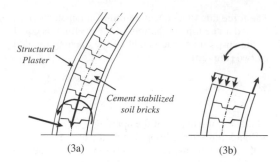

Figure 3. Boundary forces in the dome walls.

can be in the form of additional fibres, reinforcing steel placed in the plaster or stitching the brickwork with wire. The stitching technique is not a common method, but is used on occasion to repair brick walls. To repair a crack, the wall is stitched closed with wire. The application to domes is similar. A dome wall is similarly stitched to reinforce high stress areas.

2.4 Composite shell walls

As mentioned previously, bending moments and shear forces will occur near the base of the shell due to boundary effects. These forces are resisted by the shell walls. The proposed composite is illustrated in Figure 3a. Also included in this diagram are the forces which must be resisted by the shell wall. The shape of the bricks serve two purposes – they resist the compressive forces as well as the shears (Magaia, n prep). Each brick is moulded with a shear key on one face and an indentation on the opposite side. The indentation receives a shear key from an adjacent brick.

The bending moment, on the other hand, is resisted by the composite materials (i.e., brick and plaster). The mechanism is illustrated in Figure 3b. As drawn, the compressive stress is resisted by the brick and the tension stress is resisted by the fibre reinforced structural plaster. The properties of each of these materials are described.

2.5 Compressed stabilized soil bricks

The bricks used in constructing the shell walls are composed of a compressed cement stabilized soil (Magaia, in prep, Houben 1998, Carroll 1992). The soil is stabilized and bound by the addition of cement and the material is densified by applying a compacting pressure. The result is a high quality brick with 28 day compressive strengths ranging from 2 to 10 MPa – depending on the mix proportions and the compacting pressure. To economize, local soil is used to produce the bricks. A sieve analysis is performed to determine the gradation of the soil used in the tests. This

Table 1. Particle size distribution.

Material	Particle size (mm)	Passing %
Clay	0–0.002	22
Silt	0.002–0.06	24
Sand	0.06–2	44
Gravel	2–60	10

Table 2. Compressive strength of soil-cement bricks.

Test No.	Compressive strength (MPa)	Moisture content	Type of compression to form the bricks
1	3.82	>20%	Uncompressed*
2	3.14		(γ = 1795 kg/m^3)
3	6.86	10%	Compressed manually
4	5.26		(γ = 1800 kg/m^3)
5	6.38		
6	7.96		
7	9.40	10%	Compressed
8	9.13		mechanically
9	3.44		(γ = 1930 kg/m^3)
10	9.86		

* soil compressed to the given density

information is given in Table 1 for the material used to construct the prototype dome:

Not all soils are suitable for brick making; any soils containing organic material should be avoided. The correct proportions of clay for plasticity and sand for workability are needed to minimize the effects of shrinkage during curing and to optimize compressive strength. The amount of clay in the soil is critical to prevent excessive shrinkage and loss in durability. If the clay content is less than 5% or greater than 30%, more cement, clay or sand must be added to balance the mixture. The percentage of fines (clay + silt), the liquid limit, the plasticity index and the percent shrinkage are parameters considered in the design of the mix.

If fibres are added, the compressive strength may increase as much as 15%. Fibres also control shrinkage cracking; this is particularly useful with soil types which contain a high percentage of clay. Sisal fibres, between 40 and 60 mm in length, were used in the structural plaster of the prototype dome.

The amount of cement added to the mix will vary depending on the compressive strength required and the proportion of clay in the soil. Typically, the cement content will vary between 3 and 12%. However, 6 to 10% cement content is commonly used to obtain satisfactory compressive strengths. A cement content of 10% was used in the tests listed in Table 2. The amount of compressive force applied to form the bricks contribute significantly to the compressive strength

Table 3. Results of tensile tests of fibre reinforced structural plaster.

Specimen No.	Quantity of fibre (kg/m³)	Tensile strength (MPa)
1	20	1.10
2	20	1.20
3	20	1.25
4	20	1.68
5	20	1.45
6	20	1.58
7	20	1.55
8	20	1.70
9	20	1.45

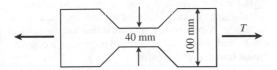

Figure 4. Shape of tensile specimens.

of the bricks. Three different categories of compression were applied – uncompressed, manually compressed and mechanically compressed.

2.6 Fibre reinforced structural plaster

The structural plaster used in the prototype dome is not a plaster in the conventional sense (i.e. sand, cement and water). The plaster is composed of the same materials as the bricks, but with a high percentage of fibres. The purpose of the plaster is to provide a durable membrane, bind the structure together and resist tension forces in the wall. The percentage of cement is usually between 10 and 12%. The plaster is applied in the same manner as other plasters (hand trowelled). The mix, however, is not compressed like the bricks and therefore has a lesser compressive strength. The role of the fibre is to provide tensile strength to resist bending moments. Table 3, contains the results of nine specimens tested in tension (Davis 1964).

The cross-sectional area of each specimen tested is 4000 mm². The shape of the specimens is given in Figure 4. A ramping tensile force was applied until failure occurred.

The tensile capacity is not only necessary to resist boundary effects, but also to provide ductility. The ductility requirement is essential when considering the effects of point loads and other safety issues.

3 METHOD OF CONSTRUCTION

The strength and stability of the dome is not only a function of the strength of the materials, but also

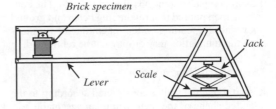

Figure 5. Compression testing apparatus.

directly related to the quality of construction. An "out-of-plumb" shell can result in buckling or over-stressing. It is imperative that the structure is constructed to the intended form with close tolerances. This is of particular importance considering the quality of labour that most likely will be involved in the construction of the dome. Therefore, inventive schemes are required to ensure a quality end product.

3.1 On-site testing of the compressive strength of the bricks

Luker (1999) identified problems determining the compressive strength of concrete at construction sites located in rural areas. More often than not, concrete mixed in rural areas is not checked simply because testing equipment and laboratories are not available. Luker devised a simple testing apparatus, which is cheap to construct and easy to transport. Although the intended purpose is to test concrete cubes, the same apparatus can be used to test the compressive strength of soil-cement bricks. A schematic of the apparatus is given in Figure 5.

As illustrated, a brick specimen is placed between two pressing plates. On the face of one plate, a bar is welded which is in contact with the specimen. A load is applied by gradually jacking a lever until splitting failure occurs. The splitting force is related to the compressive strength of the specimen. The apparatus works on a similar basis to a "nut cracker". The failure load is read directly from a scale located under the jack.

3.2 Ground slab and ring beam construction

The ring beam and ground slab are the only components of the dome that are constructed out of concrete. Depending on the soil condition, loads and size of dome, the ring beam may or may not be reinforced. Construction begins by clearing the site of vegetation. The first 300 mm is usually stripped to rid the site of organic matter. A spike is then driven into the ground at the location of the centre of the dome. A rope is attached to the spike and two rudimentary circles are drawn inscribing the extents of the ring beam foundation. A trench, usually about 500 mm deep, is dug and the soil compacted (and possibly stabilized). The concrete is then poured to form the ring beam and ground floor slab. Depending on the magnitude of ring tension, the foundation may or may not be reinforced.

3.3 Shell wall construction

It is vital that the wall is constructed according to the intended shape of the shell, within an acceptable tolerance. To achieve this, a guiding template is constructed as shown in Figure 6 below.

The base of the template is fixed and the template arm swivels horizontally 360 deg. The bricks are placed along the edge of the template arm to ensure that the shell has the correct curvature along the meridianal and circumferential directions.

3.4 Construction of the apex

As the brick wall is laid, the slope of the wall will reach a critical stage in which the brick will refuse to adhere to the top of the wall and tend to slide off while the mortar is wet. This is corrected by placing horizontal formwork at the apex of the shell (see Fig. 7). A hip is formed out of soil and the remaining

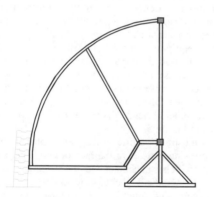

Figure 6. Guiding template to construct wall.

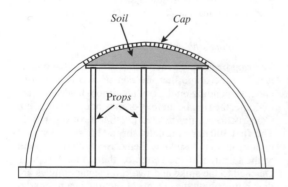

Figure 7. Construction of the dome cap.

bricks are placed into position. Once the mortar has cured sufficiently, the formwork is removed. The remaining job is plastering the interior and exterior wall with a fibre impregnated structural plaster.

6 CONCLUSIONS

The objective of this research is to devise a low cost home that is structurally sound and affordable. A dome structure, presented here, is a viable alternative. To economize, the walls of the shell are constructed out of local materials and the required skill level of the labour is minimal. The shape of the shell inherently eliminates the need for a conventional roof (the walls of the shell evolve into a roof using the same materials), which typically is the most expensive part of the structure. A cost survey indicated that the cost of the dome is approximately one-third the cost of a conventional home of similar size.

As indicated, analytical methods are complex and require a certain degree of understanding of shell theory. However, various methods of design have been devised to simplify the analysis and produce a structure that is robust and structurally acceptable. Boundary effects are dealt with by utilizing form-finding techniques to minimize moments and shears at the base. The walls of the dome are also strengthened locally to resist any residual stresses. Openings in the dome (such as doors and windows) also introduce analytical complexities. This problem is resolved by considering the flow of stress around the opening and calculating the resultant forces which act parallel to the edges of the opening. These areas of concentrated stress are similarly strengthened.

The structural integrity of shells is inept to poor construction. "Out of plumb" walls could lead to structural failures due to buckling or asymmetric loads. To ensure conformity to the original design, prefabricated guiding templates are utilized to ensure that the walls are constructed to the correct curvatures. At the apex, however, a soil hip is formed to construct a cap.

Other methods have been devised to ensure control over construction quality. For example, the brick compressive strength is monitored by using Luker's compression apparatus.

REFERENCES

Billington, D.P. (2nd ed.)1982. Thin shell concrete structures. NY: McGraw-Hill.
Bolton, M.C. 1998. Innovative use of site-based materials for homes. Report. Div. of Building Tech. CSIR. Pretoria.
Carrol, R.F. 1992. Bricks and blocks for low-cost housing. Report. Building Research Establishment, Watford, UK.
Davis, H.E. 1964. Testing of engineering materials. NY: McGraw-Hill.

Dept. of Housing 1997. A new housing policy and strategy for South Africa. *National Housing Code*. Pretoria.

Ensor, L. & Radebe, S. 2002. Government housing subsidy programme. *Business Day*. SA.

Gould, P.L. 1999. *Analysis of plates and shells*. NJ: Prentice Hall.

Houben, H. & Boubekeur, S. 1998. Compressed earth blocks: standards. *S.CRATerre-EAG*. Brussels: CDI.

Isler, H. 1994. Concrete shell derived from experimental shapes. *Structural Engineering International* (3): 142–147.

Johansen, K.W. 1962. *Yield-line theory*. Cement and Concrete Association.

Luker, I. 1999. A simple strength test for masonry units. *Concrete Benton* (93): 3–5

Magaia, S.J. *Domes used in low cost housing*. Dissertation in preparation, University of the Witwatersrand.

System-based Vision for Strategic and Creative Design, Bontempi (ed.)
© 2003 Swets & Zeitlinger, Lisse, ISBN 90 5809 599 1

The last mile – deployment of optical fibers through sewers

S. Gokhale
Department of Civil and Environmental Engineering, Vanderbilt University, Nashville, Tennessee, USA

M. Najafi
Department of Construction Management, School of Architecture, Michigan State University, East Lansing, Michigan, USA

E. Sener
Department of Construction Technology, Purdue University School of Engineering and Technology, Indianapolis, Indiana, USA

ABSTRACT: Over the past decade over 400,000 km of optical glass fiber network has been installed in the United States. However only 5% of all commercial buildings in the U.S. have access to fiber optics to their door, although most are within a mile of a fiber-optic connection. The "last mile" is a section of the network that connects the end user to the local area network in each major city. Using sewers as an avenue for the installation of fiber optic cabling is a relatively new concept. This paper will discuss the various technological advances achieved in the installation of fiber optic cables in sewers as well as the chief concerns faced.

1 INTRODUCTION

At a cost of over 10 Billion US$, the backbone of the long-haul fiber optic network has been put in place in the United States. With more than 110 million North Americans expected to telecommute to work by the year 2010, the focus has now shifted to establish the link between the urban area networks to the end user. The traditional method of utility installation, in other words the open trench method (also called as the dig & fill method) is not practical for the heavily congested, urban areas. Open-cut construction has several shortcomings, chief amongst which are: safety concerns of workers, surface disturbance, disruption to vehicular/pedestrian traffic and reduction of pavement life.

North America has already invested many trillions of dollars in the past century building over 2,000,000 km of sewers that service every business and the vast majority of the homes (Anderson 2002). Thus the concept of using these existing conduits to perform multiple functions, including the installation of fiber optic cable, seems a very obvious solution. Doing this would limit the open-cut construction that would need to be undertaken, and through leasing the public (sewer) space the telecommunication companies could infuse a much needed revenue to the local

municipalities for the maintenance and operation of sewer system. But as with any new idea, the believers were slow to come by. Over the past five years almost a dozen cities have participated in this experiment. While on the whole the experiment has been received well, their are its critiques chiefly due to some very legitimate concerns. This paper will describe the development of the fiber optic cable installation techniques in sewers and discuss some of the advantages and disadvantages there of.

Figure 1.   Long-Haul fiber optic network.

2 BRIEF HISTORY

The very first instance of using existing sewers for installing communication cables was achieved as a part of a research and development program by the Water Research Center (WRc), Swindon, United Kingdom (UK). WRc was issued a UK patent for their method in 1984 (UK Patent 1984). The first US patent was issued in 1987 to Cabletime Installations Ltd. (US Patent 1987). Japanese engineers assembled a robot (Fig. 2) in 1987, for the purposes of installing optical fibers in Tokyo sewers in order to remotely control the sewage treatment plants in Tokyo (Nakazato 1999). The Japanese applied for patents in Japan, Europe and the United States. In April 1989, US Patent No. 4,822,211 was issued to Nippon Hume, Tokyo Metro Government, and Tokyo Metro Sewer Service Corporation.

While the original robots were self-driven, the second generation of robots were assisted by winches. Most Japanese sewers were constructed of centrifugally cast concrete pipes and drilling into the pipe walls in order to affix the cables required considerable amount of power, which in addition to water & air had to be supplied through a single umbilical cable. In order to conserve the use of power the winch assisted robots became a preferred method for Nippon Hume.

The procedure for installing the cables used by Nippon Hume in the early stages involved the J-Hook method. In this process, the robot drills a 6 mm diameter hole, approximately 15 mm deep into the wall of the pipe. The holes are plugged using J-Hook anchor using a two-part resin system that hardens in the hole once the plunger pin is activated after the cable has been placed.

Tokyo Metro Government and Tokyo Metro Sewer Service Corporation promoted this concept in the early 1990s, and were instrumental in getting the Japanese public law changed in 1996 to permit a wider deployment of optical fibers in sewers (Saeki, Kingo, Saito & Satoshi, 1997). Subsequently Tokyo Metro installed over 850 km of fibers in sewers, more than any other city in the world (Fig. 4).

In 1997, Berlin Water (BerliKomm) acquired 3 robots from Nippon Hume and formed a joint venture with Nippon Hume called Robotic Cabling GmBH (RCC). In just over a year, RCC was instrumental in installing 1,500 m of optical fibers in the combined sewers in Berlin. The robots came in three sizes and were capable of installing cables in sewers 250–350 mm, 400–450 mm, and 600–1200 mm in size. The robots were steered by a control unit but not driven under their own power, rather pulled by a winch through the manhole.

The method used by Berlin Water was adapted from the earlier method used in Japan by Nippon Hume. Using this method robots were capable of achieving production of 150–200 m/day.

In 1999, a company called CityNet was formed in North America through a strategic alliance between CableRunner, Alcatel, and Carter & Burgess. CityNet

Figure 2.   First fiber optic cable robot.

Figure 3.   J-Hook anchors.

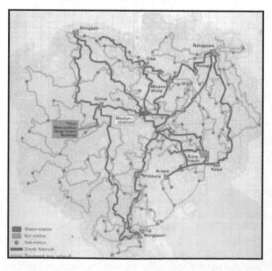

Figure 4.   Optical fiber network in Tokyo sewers.

turned towards the Swiss manufacturer KATE for a new method of deployment of fibers in sewers (Beyer 2000). One of the chief concerns with the *drill & dowel* technology was the damage inflicted (through drilling) to the host pipe. It was feared that this could result in a reduced design life of the pipe and also cause leaks. The concern was understandable since a 200 mm reinforced concrete pipe has a typical wall thickness of approximately 20 mm. Another concern was the blockage caused by the protruding J-Hook, especially in small diameter sewers.

The system developed in conjunction with KATE is called as the *Clamp & Ring* method and widely used today in North America. During this past decade roughly 40 sets of robots were manufactured by Nippon Hume while in the last three years alone KATE has produced and sold 50 robots.

3 CLAMP & RING METHOD

The Clamp & Ring method has been trade marked in the United States and goes by the trade name *CableRunner Method™*. This is a four pass system for installing cables in sewers 200 mm or larger in size. The method is illustrated in Figure 5.

3.1 *Cable installation*

The planning of the work commences with a CCTV inspection, mapping of sewer profile, and an analysis of the sewer line to evaluate its appropriateness for housing the cables. Any maintenance work required, such as sealing leaks, root removal, and cleaning, is carried out before the installation of the clip rings. Clip rings are spaced at 1.5 m intervals and are delivered to the specified location on board a robot. The rings are loaded on to a magazine that is attached to the robot, which travels through the sewer, using a laser guide to precisely place each ring at the prescribed location.

Each ring is fitted with between 3 and 9 clips that are used to fasten a steel conduit that houses a single mode optical fiber cable. Depending on the diameter of the sewer, up to 9 fiber optic cables in stainless steel conduits can be installed side by side in a sewer system (Fig. 7). The cables consist of 72, 144, or 216 fibers. The outer diameter of steel conduits varies between 1.5 to 15.5 mm. In order to install a clip ring, the spring box on the clip ring is unlocked, the ring is snugly pushed against the sewer pipe wall, and engaging four springs thus fixing the ring in place.

Once all the rings are in place the robot is withdrawn, fitted with a different head, in order to fasten the conduit to the clips, locking them securely in place. In the final step, the cable is threaded through the conduit using a push-pull method. The cable is terminated to a patch panel inside a manhole. Sewer manholes

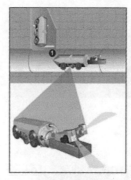

Step 1
Mapping & Analysis

Utilizing digital cameras, CityNet robots map and analyze the city's existing sewer infrastructure.

Click to View Animation

Step 2
Ring Install

Stainless steel alloy rings are positioned to fit flush against the sewer pipe walls.

Click to View Animation

Step 3
Conduit Install

Small, specially designed stainless steel alloy conduit/ tubes are secured to the rings.

Click to View Animation

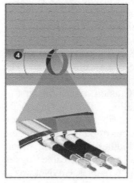

Step 4
Fiber Optic Install

Fiber optic cable is threaded into and through the conduit to build the network.

Click to View Animation

Figure 5. *CableRunner Method™*.

Figure 6. Sewer Access Module (SAM) robot used in *CableRunner™* Method.

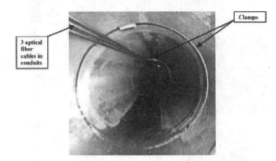

Figure 7. Stainless steel conduits with fiber optic cables in sewer.

form an important part of this technology. Manholes serve as access ports and store junction boxes and extra length of cables. The junction box is sealed with mechanical means against I/I and sewer gases.

4 CABLE INSTALLATION IN THE US

In October 2002, CityNet completed the installation of the first mini-ring in the downtown area of the City of Albuquerque, NM, with a total length of 6,700 m. Almost 60% of pipes used were storm sewers while the remainder were sanitary sewers with sizes ranging from 200 through 1800 mm. Currently work is underway or scheduled to be underway in a number of cities including Indianapolis, and Los Angeles (Fig. 8)

However the recent downturn in the telecommunication industry in the US has caused many of the projects to be put on hold or canceled entirely. The City of Houston is one such example.

In 2001, Houston City Council passed the first of three steps necessary to allow Silver Spring, Md.-based CityNet the right to put fiber optic lines through city sewers at a cost of millions of dollars. The process utilizing Sewer Access Modules, or SAMS, robots was to be used to that install fiber-optic cable rings in sewer pipes. After installing the cables, CityNet was to lease them to telecom companies and Internet providers that offer high-speed network services.

Figure 8. Cities with cable installation in sewers.

Bridging the last mile gap had been a major obstacle for the City of Houston as was the case with other large metropolitan areas in the US. But by laying these lines through the sewer system, CityNet proposed to unclogs the bottleneck by providing the required fiber optics, while avoiding the destruction and subsequent construction of city streets.

The contract agreement would allow CityNet to provide fiber connectivity to multi-tenant, high-rise office buildings within the city for a 27-year term. The value of the contract was estimated between US$17.4 million and US$19.1 million, according to city documents. The City of Houston was to have received a franchise fee (2.5% of the company's profits in Houston), a one-time fee of US$50,000 and an annual infrastructure fee of US$5.00 per meter of sewer line used for fiber optic installation. The city was also to receive connectivity to 12 municipal buildings and save maintenance monies by having CityNet perform some repair, cleaning and mapping of sewer lines.

But because the telecom sector experienced a severe downturn in 2002, CityNet was forced to building networks only after securing a guarantee of revenues through sufficient sales. Unfortunately Houston, and many of the original cities which had agreed to participate, have not reached that stage yet. Company officials say CityNet is well positioned to ride out the downturn in the sector and is looking toward developing enough potential customers to justify building the Houston network. However, at this point, the company has not committed to an entry date in Houston.

5 OTHER METHODS OF CABLE INSTALLATION

There are two other methods that are utilized, although neither have gained widespread acceptance.

In the Adhesive Bed Method, multiple conduits are attached to the pipe by use of a hose and a short liner and an injection line. The hose is inflated to fit closely to the pipe wall, trapping the conduits between the pipe wall and the lining. An ambient cure resin is used to secure the cable conduits to the pipe wall.

The Cured-In-Place (CIP) method also utilizes a liner material. Except in this method the fiber optic cables can be installed either inside or outside of the liner material. Using previous mapping results, the robot can be programmed with exact installation coordinates thus allowing placement of conduits at locations that are not in conflict with other sewer services. Typically the conduits are installed between the 10-O'clock and the 2-O'clock positions. The CIP liner can either be inverted or winched into place.

Each of the four methods discussed have advantages and disadvantages.

	Installation system			
	CIP	Adhesive bed	J-Hook	Clamp & Ring
Extends life of pipe	●			
Reduced maintenance cost	●			
Minimizes disruption	●	●	●	●
Cables fully protected	●	●		
Installation cost	H	M	L	L
Installation time	M	M	L	M

6 STANDARDIZATION

Currently no national standard exists in North America that regulates the deployment of fiber optic cables in sewers. Efforts are underway to develop such standards through the efforts of the American Society of Civil Engineers (ASCE) and the American Society for Testing and Materials (ASTM). ASCE and ASTM have formed technical committees, and are expected to formulate a State of Practice before the end of 2003, and technical standards by 2004.

The specific topics being investigated are:

1. Pipe Selection Criteria
   a. Access to sewer
   b. Sewer hydraulics
   c. Structural capacity
   d. Sewer maintenance issues
   e. Compatibility of sewer wall material
   f. Presence of excessive chemical reagents, calcium deposits, etc.
   g. Joint separation
   h. Condition & frequency of lateral connections

2. Test methods for material components
3. Installation Specifications
4. Standard Practice for Design, Operation, and Maintenance.

7 CONCLUSIONS

The deployment of fiber optics cables in existing underground pipelines offers a cost effective solution to overcoming the bottleneck in providing high-speed network capabilities to businesses and consumers. However prior to installing the cables a sound engineering analysis of the pipeline and its ability to serve as a host must be carefully evaluated. It is important to remember that the primary purpose of the sanitary sewer line is to carry wastewater, and to that end this function can never be compromised. Municipalities can succumb to the temptation of the additional revenue generated from leasing the pipeline right-of-way. However such an arrangement also creates additional liabilities. Should a sewer get over charged due to an exceptional rain event and damage the fiber optic cable thereby causing great loss to the business that depends on it, who is liable for the loss? These questions must be carefully reviewed and answered in proper contractual language in order to protect the cities and towns. Finally, a national standard for the design, testing, installation, and operation of fiber optic installations in sewers is very much needed.

REFERENCES

Anderson, G. 2002. Fiber Optic Deployment Methods. *Proc. of the Underground Infrastructure Advanced Technology Conference.* Nashville, Tennessee.

Beyer, K. 2000. Einbau Von Kommunikationskabeln in Abwasserkanalen. *Proc. of the 6th International Pipeline Congress:* 370–79. Hamburg, Germany.

Nakazato, T. 1999. Sewer Optic Fiber Network in Tokyo. *WQI:* 16–18.

Saeki, Kingo, Saito & Satoshi. 1997. Construction of the Optical Fiber Teleway Network Using Sewers. *Research Report* by Tokyo Metro Sewer Services Corporation.

UK Patent 2,129,627. Granted on 16 May, 1984, to Water Research Center for Installation of Communication Cables; Inventors: John Gale, Ian Swallow, Martin George, Henry Spooner, Steven Grosvenor, Timothy Henry, John Ceerny and Nicholas Paul.

US Patent No. 4,647,251. Granted on March 3, 1987, to Cabletime Installation for Installation of Communication Cables; Inventor: John Gale.

13. Precast structures

System-based Vision for Strategic and Creative Design, Bontempi (ed.)
© 2003 Swets & Zeitlinger, Lisse, ISBN 90 5809 599 1

Nonlinear analysis of prestressed hollow core slabs

B. Belletti, P. Bernardi, R. Cerioni & I. Iori
Department of Civil and Environmental Engineering and Architecture, University of Parma, Italy

ABSTRACT: In this work, a nonlinear, plane stress, finite element procedure capable of investigating the flexural and shear behaviour of prestressed hollow core slabs is presented. The material's mechanical nonlinear behaviour was simulated through the constitutive PARC model for uncracked and cracked reinforced concrete. The resulting secant stiffness matrix was then implemented into the FE code ABAQUS. The single panel was subdivided into a mesh of membrane elements, assigning to each of them a different thickness in order to follow the complex shape of the cross-section, characterised by hollows and very thin concrete webs. The longitudinal prestressing reinforcement was introduced on the basis of a smeared approach, proposing an effective method to model prestressing as cracking develops. The proposed procedure was applied in order to reproduce the experimental response observed in bending tests of factory-produced hollow core units. Comparisons with available code provisions (Eurocode 2) are also provided.

1 INTRODUCTION

Hollow core slabs represent nowadays one of the more common precast elements used in the building industry; nevertheless, their structural behaviour show some singular features which mean that a full understanding of their resistant mechanism is not yet available.

The cross-section is characterised by the presence of hollows, which are created through particular methods of production. For this reason, the introduction of anchorage and shear reinforcement is very difficult, and only the longitudinal prestressing strands are generally incorporated. These features undoubtedly make hollow core slabs advanced and economic structures, because, if compared to reinforced solid concrete floors, they can save nearly 50% concrete and 30% steel for the same performances. On the other hand, they present a series of problems, not only because of the total absence of transversal reinforcement, but also because of their geometry, which is characterised by very thin concrete webs.

European codes (Eurocode 2 1992), do not consider special rules for the analysis and design of single or assembled slabs, giving only general provisions for structural components (prestressed concrete structures without shear reinforcement), which aren't always satisfactory for hollow core units (Walraven & Mercx 1983), and only some special publications by FIP (1999) have tried to fill this gap. Moreover, it has to be remarked that the scientific literature about this subject is very restricted, both at experimental and at theoretical-numerical level, and a lot of analyses must still be conducted in order to clarify several aspects of their behaviour, in particular with respect to the way they manage to develop strength.

In order to improve the knowledge of the structural behaviour of hollow core slabs, flexural tests were carried out and in the following the more significant results will be reported. The experimental findings will be compared to those attained from numerical analyses developed through a FE code (ABAQUS) where the PARC model (Belletti et al. 2001) was implemented.

2 CONSTITUTIVE MODEL FOR REINFORCED CONCRETE

The PARC model (Belletti et al. 2001), which is able to describe the local behaviour of R/C both in the uncracked and in the cracked stage, was implemented into the ABAQUS code in terms of a constitutive stiffness matrix, in order to effectively carry on nonlinear analyses of R/C structures up to failure. The model is obtained for elements subjected to plane stress, basing on a smeared, fixed crack approach. The stiffness matrix is expressed in the local co-ordinate system through the sum of the concrete and steel contribution, in the form:

$$D' = D'_{c1,2} + D'_{s1,2}, \tag{1}$$

where 1, 2 (Fig. 1) represent the ortotropic material directions (assumed coincident with the principal

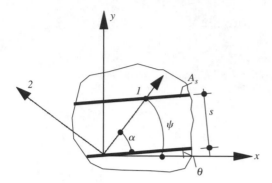

Figure 1. Adopted angles notation for the determination of the stiffness matrixes.

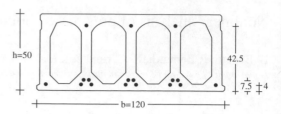

Figure 2. Slab H50 cross-section (all dimensions in cm).

stress directions) in the uncracked stage, while in the cracked stage they represent the principal stress directions correspondent to initial cracking.

Before cracking, concrete is assumed as a nonlinear, ortotropic, elastic material, whose stiffness matrix $D'_{c1,2}$ is expressed by:

$$D'_{c1,2} = \frac{1}{1-v^2} \begin{vmatrix} E_{c1} & v\sqrt{E_{c1}E_{c2}} & 0 \\ v\sqrt{E_{c1}E_{c2}} & E_{c2} & 0 \\ 0 & 0 & (1-v^2)G \end{vmatrix}, \quad (2)$$

E_{c1} and E_{c2} being the longitudinal secant moduli determined by duplicating the actual biaxial stress-strain curves with uniaxial relationships (Darwin & Pecknold 1977) for concrete in tension and in compression, modelled on the basis of the biaxial strength envelope proposed by Kupfer et al. 1969.

The matrix relative to steel contribution is:

$$D'_{s1,2} = T_\alpha' \begin{vmatrix} \rho E_s & 0 & 0 \\ 0 & 0 & 0 \\ 0 & 0 & 0 \end{vmatrix} T_\alpha, \quad (3)$$

where E_s is the steel secant modulus obtained from an exponential law proposed by Devalapura & Tadros 1992, effective for prestressing steel, and T_α is a transformation matrix, function of the angle α between strand direction and the 1 axis (Fig. 1).

Details of the procedure, particularly regarding the passage to equivalent uniaxial strains and the adopted stress-strain laws for concrete and steel, can be found in Bernardi 2003.

After cracking occurs, Expressions 2 and 3 are replaced by the following matrixes:

$$D'_{c1,2} = \begin{vmatrix} c_t a_m & 0 & -c_v a_m \\ 0 & \overline{E}_c & 0 \\ 0 & 0 & c_a a_m \end{vmatrix}, \quad (4)$$

$$D'_{s1,2} = \begin{vmatrix} \overline{E}_s^* \rho g c^4 + \frac{da_m}{st} s^2 c & \overline{E}_s \rho c^2 s^2 & \overline{E}_s^* g \rho c^3 s - \frac{da_m}{st} c^2 s \\ \overline{E}_s \rho s^2 c^2 & \overline{E}_s \rho s^4 & \overline{E}_s \rho s^3 c \\ \overline{E}_s^* g \rho c^3 s - \frac{da_m}{st} c^2 s & \overline{E}_s \rho s^3 c & \overline{E}_s^* g \rho s^2 c^2 + \frac{da_m}{st} c^3 \end{vmatrix}, \quad (5)$$

a_m being the fixed crack spacing and writing s, c for $\sin \alpha$ and $\cos \alpha$, respectively. In Matrixes 4 and 5 all the contributions affecting shear stiffness of the cracked section, such as aggregate interlock, (coefficients c_a, c_v), concrete in tension (c_t), dowel action (d), tension stiffening (g), are taken into account. A more detailed explanation of the single terms of the Matrixes 4, 5 can be found in Belletti et al. 2001.

From the formulation in the local axes, the stiffness matrix in the global (x, y) co-ordinate system is easily obtained from:

$$D = T_\varepsilon' D' T_\varepsilon, \quad (6)$$

being T_ε the transformation matrix expressed as a function of the angle ψ between the local direction 1 and the x axis (Fig. 1).

The so-determined stiffness matrix was introduced into the FE code ABAQUS through a user defined subroutine written in FORTRAN language; this subroutine was called by the FE procedure at each iteration of each load increment to determine the global stiffness matrix K at a given integration point.

3 TEST PROGRAM

The experimental program concerned two identical flexural tests on 50 cm height hollow core units (named in the following as "H50").

3.1 Test specimens

The slabs were manufactured by an extrusion machine using high-frequency vibrators and prestressed imposing an initial stress σ_{0p} (equal to 1350 MPa) on longitudinal reinforcement, constituted by 17 seven wire strands, whose nominal diameter ϕ_p was 12.5 mm (=1/2 inch) and nominal area 93 mm^2. The slabs were tested 30 days after casting. The cross-section shape is reported in Figure 2, with the indication of the

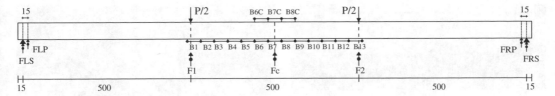

Figure 3. Loading arrangement and test setup for the tested slab specimens H50 (all dimensions in cm).

principal geometrical features characterising the slab section.

Figure 3 shows in details other geometrical properties of the tested hollow units, which were 15.30 m long and were designed to fail in flexure.

3.2 Materials

Concrete quality control is usually supplied by slab producers in terms of cubic compressive strength. However, materials properties differ from slabs to cubes, depending on the manner of casting, vibrating, curing, in other words on the manufacturing of the slab. For these reasons, additional tests on concrete extracted from the slab after the failure test were also performed.

The concrete compressive strength was first measured on cubes made of the same mix as the slab 28 days after casting. The average cube compression strength obtained from four 150 mm cubes was 49. 2 MPa. In order to determine the axial concrete tensile strength, which is fundamental for the determination of the cracking load, splitting tests were carried out on cylindrical cores, 100 mm in diameter and 150 mm high, extracted from cubes cast at the same time as the slab. The average splitting tensile strength measured on four cylinders was 4.0 MPa.

The axial tensile strength was then determined on the basis of Eurocode 2 prescription: $f_{ct,ax} = 0.9 f_{ct,sp}$.

From the tested hollow core specimens, six quadrangular prisms, 60 mm side and 200 mm high, were taken vertically from the central webs in order to determine the flexural tensile strength of the concrete in the slab. Bending tests were carried out on the prisms loaded at midspan up to failure. Flexural tensile strength was determined from the failure load P_u through the following relationship:

$$f_{ct,fl} = \frac{P_u}{4} l \left/ \frac{b^3}{6} \right., \tag{7}$$

where l is the span between supports and b is the cross-section side length of the prism. The obtained average strength value was 7.37 MPa. The axial tensile strength was also determined on the basis of Eurocode 2 prescription as: $f_{ct,ax} = 0.5 f_{ct,fl}$. On each of the two

Table 1. Average concrete mechanical properties.

$f_{c,cube}$ (MPa)	$f_{c,pr}$ (MPa)	f_c (MPa)	$f_{ct,sp}$ (MPa)	$f_{ct,fl}$ (MPa)	$f_{ct,ax}$ (MPa)
49.2	45.6	40.8	4.0	7.3	3.6

Figure 4. Experimental setup.

parts of the broken prism an uniaxial compression test was then performed, providing an average prismatic strength of 45.64 MPa.

Table 1 summarises the average concrete mechanical properties deduced from the performed tests and adopted in the finite element analysis.

Prestressing low relaxation grade seven wire strands with a tensile strength of 1860 MPa and a 0.01 per cent proof stress of 1670 MPa were used.

3.3 Testing arrangement and procedure

The specimens were loaded under two points placed at an equal distance from the middle of the slab (Figs. 3, 4), applied by means of two hydraulic jacks acting on rigid transverse steel beams. To stabilize the loading, the jacks were supported by a frame constituted by 45° ties (Fig. 4). The load points were designed to produce a shear span – to effective depth ratio a/d so that flexural failure was to be expected. A full width roller at one end and a pin support at the other end were used to permit end rotations and prevent axial constraints. The pin was constituted by a steel roller whose basis could shift on a 1 cm teflon layer.

Instrumentation, as shown in Figure 3, was constituted by three linear variable deflection transducers

Figure 5. First flexural cracking between the loading points.

Figure 6. Diagonal crack in correspondence of the last applied load $P = 321$ kN for H50-f1 slab.

Figure 7. Cracking pattern in correspondence of the last applied load $P = 321$ kN for H50-f2 slab.

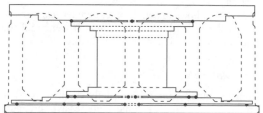

Figure 8. Superposition of the real slab cross-section with that adopted for the FE analysis.

(LVDTs), which measured slab deflections at midspan (Fc) and at load locations (F1 and F2); further on, thirteen 400 mm – strain gauges were placed on the bottom surface between the two load points, where bending moment is maximum, while three 400 mm strain gauges were bonded to the top compressed surface of the hollow-core unit. Dial gauges were installed at slab ends to monitor boundary displacements.

The load was applied incrementally until the first flexural crack occurred (Fig. 5), then the slab was unloaded and reloaded in order to determine the second cracking slab response. After that, all the instrumentation was taken off the slab and the panel was loaded to failure. For the first two loading cycles, at each loading stage the magnitude of the load, strain and deflection gauge readings were recorded.

Both in the case of H50-f1 and H50-f2 panel test, the ultimate load applied to the slab was not an effective failure load. However, it was not possible to increase any further the pressure to the jacks, because of the high deflections reached (about 30 cm). The ultimate applied load was, for both slabs, equal to 321 kN. The observed cracking pattern near the point load in correspondence of the ultimate applied load is reported in Figures 6 and 7 for slabs H50-f1 and H50-f2, respectively.

4 NONLINEAR FE MODELLING

The structural behaviour of the hollow core slabs was approached by simplifying the complex geometry of the slab cross-section, characterised by the presence of hollows, in the manner shown in Figure 8. The panel

was schematised as an equivalent double T beam, subdividing the cross section in a number of layers, each having a thickness equal to the sum of the solid parts of concrete. Moreover, in order to conduct a plane stress analysis, the slab was considered coincident with its middle plane, which was subdivided through quadratic, isoparametric 8 node membrane elements, with four Gauss integration points (reduced integration). To each membrane element a different thickness, according to Figure 8, was assigned.

At the prestressing strand level, a special layer was considered, constituted by elements characterised by a geometrical steel percentage equal to unity, or rather constituted only by steel. At the beginning of the FE analysis, as a given initial condition, a stress vector of the form:

$$\{\sigma_x \quad \sigma_y \quad \tau_{xy}\}^t = \{\sigma_{0p} \quad 0 \quad 0\}^t, \tag{8}$$

was applied to the elements which corresponded to the strand location. In Equation 8, σ_{0p} is the initial, not equilibrated tension given to strands, and x is the direction coincident with the slab longitudinal axis.

As the initial stress state is not in an exact equilibrium state for the finite element model, an initial step was included to allow ABAQUS to check for equilibrium and iterate, if necessary, to achieve equilibrium before applying the active external loads. In this way, the prestressing force will not remain a constant during all the analysis step, but it will vary following the development of the cracking pattern up to failure.

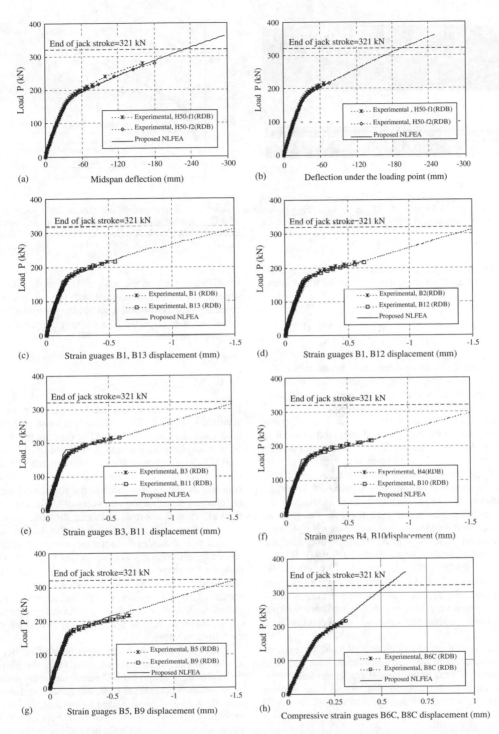

Figure 9. Comparisons between experimental and numerical results in terms of load applied vs (a) midspan deflection, (b) deflection under load points, for H50-f1 and H50-f2 slabs; in terms of load applied vs. displacement measured at the basis: (c) B1/B13; (d) B2/B12; (e) B3/ B11; (f) B4/B10; (g) B5/B9; (h) B6C/B8C, for H50-f2 slab.

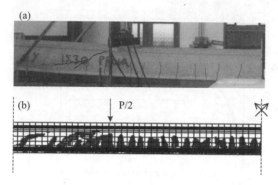

Figure 10. Comparison between (a) experimental cracking pattern observed for H50-f1 slab and (b) cracking pattern numerically obtained in correspondence of the ultimate load.

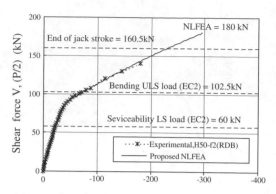

Figure 11. Experimental and theoretical response of H50-f2 slab compared with the serviceability and ultimate limit state loads (values without considering the slab dead load).

The initial strain ε_{0p} was added in the user's subroutine to the axial strain in the direction of the strands.

5 COMPARISONS BETWEEN NUMERICAL AND EXPERIMENTAL RESULTS

The nonlinear finite element analysis was subdivided into two steps: in the first, the prestressing force and the structure dead load were applied, in the second, the external load was incrementally applied.

In Figures 9–11, the numerically attained response is compared to the experimental one, first in terms of load applied vs. midspan deflection (Fig. 9a) and vs. deflection under the load point (Fig. 9b). As already explained, the experimental data were available only until the cracking load. Moreover, as in both the experimental tests (H50-f1 and H50-f2) the failure of the slabs didn't occur, the ultimate reached load must be intended as the end of the jack stroke. For H50-f2 slab, comparisons in terms of load applied vs. displacement, measured at strain gauges B1/B13 (Fig 9c), B2/B12 (Fig. 9d), B3/B11, (Fig. 9e), B4/B10, (Fig. 9f), B5/B9 (Fig. 9g), B6C/B8C (Fig 9h), are also performed. From the displayed graphs, can be observed as the numerical response was very close to experimental observations, particularly within the structure serviceability limit state.

The implemented model also allowed a visualization of the crack distribution at each load stage. In Figure 10a, b the experimental cracking pattern correspondent to the ultimate load stage (Fig. 10a) is reported above the numerical contour of the crack opening w (Fig. 10b), showing a good agreement with the physical reality.

Finally, in Figure 11, which reports the shear force V vs. midspan deflection, the values of the forces relative to the serviceability and the bending ultimate limit state (determined on the basis of EC2 provisions) are superposed to the theoretical and the experimental curves. As can be observed, within the prescribed limit states the slab response is accurately reproduced by the proposed model.

6 CONCLUSIONS

In this paper, a numerical method which is able to accurately simulate the structural response of hollow core slabs has been presented. The procedure is based on the implementation into a well-known FE code (ABAQUS) of a nonlinear constitutive matrix (PARC), which describes the nonlinear behaviour of reinforced concrete in plane stress, both in the uncracked and in the cracked stage, taking into account all the contributions affecting the normal and shear stiffness.

To perform a plane stress analysis, the slab was subdivided into a mesh of membrane elements, characterised by different thicknesses in order to accurately describe the shape of the cross-section. The numerical results were consistent with the corresponding experimental findings, providing a detailed description of the global and local behaviour, particularly regarding crack widths, as well as the stress and strain fields of the whole structure.

The reported analysis represents only a first step of a wider research programme, which has the aim of investigating, through refined numerical analyses and appropriate experimental tests, the general behaviour of hollow core slabs subjected to bending and shear, problem effectively approached by means of a plane stress schematization only for high a/d ratio (Bernardi 2003), but which needs a different schematization for structural cases that show a prevalent and fundamental three-dimensional response.

ACKNOWLEDGMENTS

The Authors gratefully acknowledge the support of ASSAP (Prestressed Hollow Core Slabs Manufacturers Association) and the RDB Group for the diffusion of the experimental data. Thanks are due to Engineers Bruno and Giorgio Della Bella and to Engineer Giuseppe Gazzola.

REFERENCES

Belletti, B., Cerioni, R. & Iori, I. 2001. A physical approach for reinforced concrete (PARC) membrane elements. *ASCE Journal of Structural Engineering* 127 (12): 1414–1426.

Bernardi, P. 2003. *Sulla capacità resistente di elementi in conglomerato armato e precompresso, senza armatura trasversale, (Ph.D. Thesis).* Department of Structural Engineering, University of Pisa.

Darwin, D. & Pecknold, D.A. 1977. Non-linear biaxial stress-strain law for concrete. *Proceedings of ASCE Journal of the Engineering Mechanics Division* 103 (EM2): 229–241.

Devalapura, R.K. & Tadros, M.K. 1992. Stress–strain modeling of 270 ksi low-relaxation prestressing strands. *PCI Journal* 37 (2): 100–106.

Eurocode 2 1992. *Design of concrete structures, part 1: general rules and rules for building.* London: Thomas Telford.

FIB 1999. *Special design considerations for prestressed hollow core floors, Bulletin 6.*

Kupfer, H., Hilsdorf, H.K. & Rusch, H. 1969. Behavior of Concrete Under Biaxial Stresses. *Proceedings of ACI Journal* 66(8): 656–666.

Walraven, J.K. & Mercx, W.P.M. 1983. The bearing capacity of prestressed hollow core slabs. *HERON* 28 (3).

R.C. shell revival: the Malaga airport new control tower

M.E. Giuliani
Redesco srl, Milano, Italy

ABSTRACT: Reinforced concrete shells have been widely used for the new Malaga Airport new Control Tower as main structural elements and for obtaining a strong architectural character of the construction.

A lower two story building is covered by two rings of thin hypar concrete shells; each ring lays on a different level and both are oriented along the radii.

The tower shaft is composed of six separate outer columns or giant ribs, which are bent along the radial planes thus enhancing the structure slenderness, which spring up from the inner shell ring. The ribs are composed of prefabricated concrete elements, which were match cast in a plant and in situ connected using epoxy and post-tensioned bars. Taking into account the limited thickness of the walls, the relevant structural behaviour is completely spatial so that the ribs too can be considered as shells laying in vertical planes.

1 IN GENERAL

The Malaga airport expansion plan includes the doubling of the runway and a new terminal (which will be built in the next future) and the construction of a new control tower that is located in a position allowing the direct eye view of both runways.

The tower functional scheme is described starting from the top (see Fig. 1):

– a platform for the short-range radar
– a glazed room for traffic control
– rest rooms and services for the controllers
– platform for equipment and air conditioning
– inner stair ramps between the aforesaid levels and the tower base, with a pair of elevators
– a floor with sleeping rooms and services
– a floor for equipment, offices, services
– a basement for parking and storage

The organization of the tower elements is based on a strict geometrical reference system, which allowed for the exact tri-dimensional generation of the shapes of all the parts, for the design, the prefabrication, the construction and the positioning within the prescribed tight tolerances.

The plane figure assumed as the base for the aforesaid generation is the hexagon in which the tower horizontal sections are inscribed.

By means of a similar hexagon, rotated at 90°, the plans and the roofs of the lower rooms are generated.

Also the inter-story heights are modular and directly related to the proportions of the base hexagon.

Therefore the dimensions of all the project result from geometrical, deterministic, quasi–Pythagorean relationships which yield an intrinsic formal and operative exactitude, without spoiling the freedom of the architectural composition.

The roofs of the base building are constituted by double curvature surfaces, cut from hyperbolic

Figure 1. The finished tower.

paraboloids, which are simply generated by the translation of straight segments along also straight directrices, which constitute the axes of the ridges and valleys and lay on diverging radii set on the basic hexagonal scheme.

2 STRUCTURES

The structures were conceived as an organism, which fulfills in a holistic way the functional, expressive and resisting requirements of the project.

The fundamental constraints on which the selected solution is based are the high seismic hazard of the area, the strict functional demands (limiting of the deflections and vibrations, high safety and durability and so on), the integration between shapes and structures and the optimization of the construction procedure both for the base and for the tower shaft.

The structures result from the composition of elements and techniques which are peculiar of reinforced concrete: thin double curvature shells mainly subjected to in plane stresses and huge precast elements, either planar or spatial, mainly connected by means of dry joints.

The leading theme of the project is the intensive use of precasting, which is meant, in its deepest sense, as a system used to generate the shapes and the textures, to assemble the resisting members, to develop the construction method.

The construction is located in a seismic area; in addition to the resistance requirements, the functional requirements imposed stringent limits to allowable accelerations and displacements (being these statical and dynamical bounds in reciprocal contrast) in order to ensure the operation of the tower even after a catastrophic event. The structural design had to search for a fine-tuning of the stiffness and of the dynamic response of the whole system.

The construction system, comprising a keen study of the production, precasting, erection and safety procedures and arrangements, was considered as a fundamental part of the design since its beginning.

2.1 Foundations

Down to around 5 m deep from ground level, the soil is characterized by surface layers of not settled very expansive clay and of loose sand which are not suitable for the direct foundation of the tower.

A layer of dense sandy gravel, suitable for a foundation raft, outcrops at the basement level.

A dodecagonal raft is then the adopted solution: the central part (corresponding to the shaft spring) is 1.20 m thick and the outer part (laying below the base building) 0.40 m thick and stiffened by radial ribs 0.70 m wide.

Two stiffening rings are constituted by the walls, which bound the annular parking and storing area; the walls, which bear the ribs of the tower structure, are anchored to the inner solid plate.

The aforesaid solution rationally solved the foundation problems either face to the allowable pressures on the soil and for the instant and delayed settlements, and at the same time complied with the requirements arising from the surge analysis which indicated the minimum thickness of 0.30–0.40 m.

As a protection against the possible effects of capillarity of aggressive salted water, a polyethylene membrane inside two layers of geotextile fabric was inserted between the plain concrete laid on the soil and the foundation; the relevant concrete was specified as sulphur resistant.

An alternate position of the sector casting was prescribed to allow for a reduction of the shrinkage effects.

2.2 Top of the inner basement ring

Because of the complex shape of this area, which is crossed by the walls supporting the tower and by several openings for stairs and electrical cables and air conditioning ducts, the structure is composed by a cast in situ slab.

A ring beam supported by cylindrical columns and fitted for cable passageways and for bearing of the outer part precast radial beams stiffens the outer perimeter.

2.3 Top of the outer basement ring

The floor, which constitutes the ground level of the base building offices, is composed of ad hoc precast prestressed beams (Fig. 2) supported by sliding bearings on the outer wall and pinned, inside the aforesaid steel boxes, to the inner cast in situ ring beam.

Aiming to a completely resistant section of these beams close to the supports, where the bonding of the

Figure 2.   Basement ring structure.

prestressing strands develops, a plate-head-clavette system was attached to the strands and the anchor effected during the stress releasing.

The beams bear precast slabs, which support the in situ casting.

The bearings of the supporting structure for the shell roof over the ground floor are connected to the outer ring wall.

2.4 Top of the inner ground and first floors

The structure is constituted by a cast in situ plate similar to the one of the floor below.

The perimetric beam is fitted with the supports of the shell roof.

2.5 Roof on the outer ground floor and on the inner first floor

The structural layout is based on an annular plate concentric to the tower shaft and houses the service and office rooms: the outer and inner bearings are constituted by façade precast elements and by the tower ribs respectively.

This roof constitutes one of the structural and architectural elements which imprint the construction: thin shells bent in a shallow negative double curvature (hypar parts) laying on opposite leaned directions. The mutual connection of the shells form a folded plate surface having an outer dodecagonal star perimeter (Fig. 3, 4, 5).

The structure, which is inspired by the great and long tradition of thin folded plate shells present in the contemporary architecture, is supported by aesthetical reasons as well as by structural optimization.

The shells can be built in concrete with a limited thickness and are subjected to reduced stresses because the resisting system, strongly tri-dimensional, is strictly related to the general form with the in plane stress flow mainly laying in the planes tangent to the resisting surface. Therefore a significant reduction of bending and shear stresses acting on the relevant of plane normal direction is achieved.

Figure 3.   Ground and first floor thin shell roof and façade.

Membrane stresses are of major importance in respect to the bending ones, which are typical of plate behavior.

The shells lay on ruled surfaces that allow for an easy form construction ad for simplifying the reinforcement laying.

The casting of the shells, which are reinforced by means of bars placed along the straight lines laying on the hypar, is performed by means of concrete shot against wooden formworks; this "shotcrete" technology allows for building thin elements of controlled thickness with the erection speed and the quality level prescribed for the project.

The characteristic prismatic strength of $3.5\,kN/cm^2$ for the shotcrete was prescribed.

The prefabricated X shaped façade elements, which fulfill both architectural and structural functions, constitute a perimetric close ring, the stability of it being given by the spatial behavior of the whole roof-façade system.

The aforesaid elements are restrained at the base by spherical steel bearings which constitute spatial hinge supports, while at the top a stiff connection

Figure 4.   Roof from inside.

Figure 5.   Shell roof during formwork erection.

with the shells is effected. The X struts lay in planes leaning outwards and are connected to each other by post-tensioned bars over the base supports.

2.6 Tower shaft

The shaft, which strongly imprints the construction also, is composed of six huge hollow core ribs featuring four not parallel surfaces bounded by vertical inner plane,s and by a wide radius outer torus shape. In this way the tower section is shrunk at mid height while having adequate base width and top dimensions suitable for supporting the control room (Fig. 8).

The shaft structure was designed for a complete use of precasting with elements that fulfill both structural and architectural functions.

The precast elements are constituted by the rib segments, the stair ramparts, the intermediate platforms.

The leading criterion lays in the warranty of quality, durability, safety and construction speed inherent to precasting procedure (Fig. 6, 7).

The shaft constitutes the stiff member which resists all the horizontal actions induced by the wind or originated by the inertial forces produced by the earthquake. It also provides the central supports of the base building and the relevant basement.

The "columns" are mutually connected to the platforms and stair ramparts by means of structural details which prevent the arising of bending moments.

The support of the upper rooms is constituted by inner cast in situ walls bearing on the inner edges of the ribs; the access to the control room is given by a couple of stair ramparts cast in situ also.

Figure 6.   Precast ribs during demoulding.

Figure 7.   Precast rib segment.

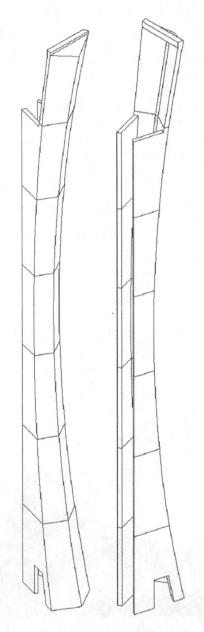

Figure 8.   Rib geometry and composition.

The aforesaid solution constitutes a very good balance for a structure which needs a stiffness high enough to resist the wind action without excessive deformations but, at the same time, not giving up the advantage of low frequencies, associated to a tuned flexibility, so as to place the structural response in a period range which is characterized by a lower energy content of the earthquake.

Each rib (Fig. 8), which is opened towards the tower central axis, is composed of six precast units that are superimposed with a dry joint using epoxy and post tensioned bars locally anchored and made continuous over the whole height. Casting of each rib was made against the lower one in order to obtain the perfect corresponding of the contact surfaces. This technique is normally used in segmental constructions and known as "matching concrete" (Fig. 6, 7).

The thickness of the rib webs is 0.30 m; the concrete was specified with white cement and characteristic prismatic strength of 4.0 kN/cm^2.

The rib structural system was designed as to ensure remaining compression stresses in the segment joints even in absence of prestressing but under the worst loading condition.

All the spaces resulting between adjacent ribs, but in the upper part, are not closed by façade elements.

2.7 Stairs and platforms

Plates 0.12 m thick, stiffened by edge beams 0.36 m and 0.54 m deep, which bear on four opposite ribs and effect the relevant connection also, constitute the platforms, which lay inside the shaft (Fig. 9); the two remaining ribs are connected to the platforms by means of precast beams.

The stair ramparts have a minimum 0.15 m thickness and bear on the platforms and are connected to the ribs which do not support the platforms (Fig. 9); the joints are reinforced by means of threaded

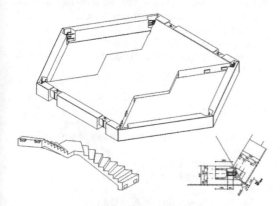

Figure 9.   Prefabricated slabs and stairs.

bars in addition to the in situ casting for filling and protection.

The three upper floors, where the bracing structures are positioned, feature a plan layout which is different from the typical one and are built by means of in situ cast plates connected to the ribs by means of shear keys and rebar splices. This solution allows for adapting the floor slabs to the functional requirements, mainly for inner inter-story stairs, while retaining the structural stiffening elements.

3   STRUCTURAL ANALYSIS AND DESIGN

Because of the complex shape of the tower, the acrodynamic pressures were obtained by means of wind tunnel tests performed on a scaled model.

The seismic action was evaluated through the response spectrum tuned to the site geophysical and geotectonic characteristics.

The analysis was effected by means of a finite element mathematical model for the general dynamic behavior, to detect the vibration modes and the effects of earthquake, vertical and wind loads, and time-dependent effects of concrete behavior.

In detail, extensive use was made of two-dimension elements "shell" type that can account for either the in plane actions (normal and shear forces) and the out of plane ones (bending and shearing) also implementing the model with "beam" type elements where appropriate.

Due to the characteristics of the dynamic response of the tower and the geometrical pattern of the solid and open surfaces of the shaft, no significant aeroelastic effects arise.

The interaction of the structure with the soil was implemented by means of Winkler's parameters, which were assigned as variable in the radial direction to account for the more important settlements of the central part of the foundation because of the influence of deeper layers.

Contributions of the aforesaid generalized loads were combined according to the Euro code 8 (seismic zone constructions) criteria and checks of the significant sections for the service and the ultimate states were effected.

The P, δ effects of the second order are very limited by the structural stiffness and by the low stress level; these design features allowed for implementing the analysis in the linear range only.

Construction details were studied to give the structure and the joints adequate ductility; the tying and the most important rebars were designed by using mechanical splices. The joints of the prefabricated parts were designed according with the same principles by using steel fittings and welding.

Figure 10. The tower during shaft erection.

Figure 11. The finished tower – view of the lower roofs.

. The post-tensioning force of the bars, which connect the tower shaft units, was selected within the criteria of assuring the overall efficiency at the limit state taking into account the partial opening of the joint.

4 CONSTRUCTION

The construction required the contributions of a general builder as well as of a firm specialized in precasting huge spatial elements.

For all the cast in situ structures composable forms made of plywood panels and stiffening frames were used.

The X shaped precast truss elements, which support the folded plate shells, were divided by the constructor into two parts in opposite V shape and connected by postensioned bars before the erection.

For the ruled double curvature surface of the shells, the casting was effected by using shotcrete on single wood planks.

The huge ribs of the tower shaft were cast inside double steel form-works (the inner part was retractable and the outer one fixed); the geometry of the structure allowed for a six time use of each form.

The erection of the units was performed by means of mobile cranes (Fig. 10), which temporarily laid these on steel boxes in order to effect the junction of the postensioning bars and the epoxy application before the final positioning.

Starting from the ground floor, or at the spring of the precast shaft structure, the erection for each of the precast vertical element was performed according to the following steps:

• positioning of the precast segments of the ribs: centering, connection of the postensioning bars by means of threaded sleeves which were accessible through boxes laying at the lower edge, lowering, tensioning and anchoring of the bars from the upper edge; every element corresponds to a two story height of the tower

• positioning of the intermediate story precast platform: lowering, horizontal positioning by using adjustment screws, provisional fixing of the joints

• positioning of the precast stair ramparts between the lower and the intermediate story: lowering, adjusting by screws, provisional fixing of the joints with the platforms and the ribs

• positioning of the precast edge beams at the intermediate story

• positioning of the upper story precast platform

• positioning of the precast stair ramparts between the intermediate and the upper story

• positioning of the precast edge beams at the upper story

• final fixing of all the joints by means of welding and in situ casting

• repetition of the cycle for the subsequent segments.

According to geometric, static and dimension reasons, the three upper floor slabs were cast in situ by using formworks attached to fittings created inside the ribs, thus avoiding the use of scaffoldings.

The structures of the control room, featuring tight limits imposed to the sections because of the stringent transparency requirements, were built in steel and were completely pre-assembled at ground level using welding and adding provisional bracings before the final hoisting as a complete unit by means of a mobile crane.

The construction time lasted from July 2000 to October 2001.

5 CREDITS

– *Client*: Aena (Aeropuertos Españoles y Navegaciòn Aérea)
– *Architect*: Bruce Fairbanks, GOP, Madrid
– *Structural Engineer*: M.E. Giuliani, Redesco, Milano
– *General Contractor*: FCC, Madrid
– *Precasting firm*: ALVISA, Madrid.

System-based Vision for Strategic and Creative Design, Bontempi (ed.)
© 2003 Swets & Zeitlinger, Lisse, ISBN 90 5809 599 1

Building system with joints of high-strength reinforced concrete

L.P. Hansen
Aalborg University, Denmark

ABSTRACT: A new building system has been developed during the last ten years. This new system consists of a column/slab system with 6 m × 6 m distance between the columns. The slabs are precast elements of size 2.9 m × 5.9 m and are connected through joints of very high strength fibre reinforced concrete, so-called Densit Joint Cast®. Also the connection between columns and slabs is made of this material. Using this material very short anchorage lengths for the reinforcement can be applied. This new building system has been used for some new buildings at Aalborg University in Denmark. The paper describes this new building system and the static tests carried out as well as some fire tests. In addition some fatigue tests of the reinforcing bars as well as fatigue tests of tensile specimens consisting of reinforcing bars embedded in Densit Joint Cast® are described.

1 INTRODUCTION

A new building system has been developed by the Danish architect firm Dall and Lindhartsen and the Danish consulting firm Carl Bro as. An essential part of this new system is the joints between prefabricated concrete slabs. These joints are made of a very high-strength fibre reinforced concrete developed at the Cement and Concrete Laboratory at Aalborg Portland, Denmark. Parts of this new building system have been tested at The Structural Laboratory, Department of Building Technology and Structural Engineering, Aalborg University, Denmark. The system has been applied for new buildings at Aalborg University. The principle of the new building system is shown in Figure 1.

1.1 Description of the building system

The load-bearing system of the building system is prefabricated concrete slabs at a thickness of 200 mm and circular concrete columns at a diameter of 350 mm. The columns are placed in a square net of 6 m × 6 m. The slabs are cast as 2.9 m ×5.9 m slab elements and are made of normal-strength concrete. At the building site the slabs are connected with the Densit Joint Cast® thus forming a continuous slab with columns as supports. No beams are used in this building system.

The building system gives the architects more freedom to create the plan for the different rooms in the building. It is also possible to allow different solutions at the boundaries. As no beams are present there are

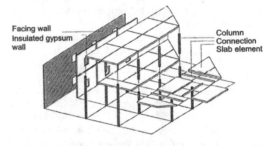

Figure 1. Sketch of the building system.

good possibilities to arrange the technical installations between the slab elements and the non-bearing ceilings.

The columns are spaced 6 m and on the top of the columns the slab elements are placed. As the slab elements are 2.9 m × 5.9 m there is an opening of 100 mm between the slabs. The slabs are reinforced in upper and lower sides with a net of 8 mm reinforcement – so-called Ks550 – and 80 mm of the reinforcement protruding from the slab. Ks550 are ribbed bars with a yield strength higher than 550 MPa.

After placing the slab elements on the columns, longitudinal reinforcement are placed in the joints and the joints are cast with Densit Joint Cast®. This means that slab elements now act as a continuous slab as the joints are able to transfer moments and forces. Very short anchorage length can be used, because Densit Joint Cast® gives this joint a sufficient load-bearing capacity.

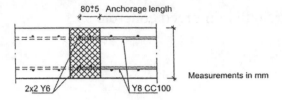

Figure 2. Typical joint between concrete slabs.

With the joints "welding" the slab elements together to a monolithic slab, it is possible to

- change position of the column
- change size of the slab element
- produce cantilevered parts of the slab

The vertical forces on the building are transferred through the slabs to the columns. The horizontal forces are also transferred through the columns and stabilizing walls are thus not needed.

1.2 *The joint*

Tests have been carried out on different types of joints and the final version of the joint between the slabs is shown in Figure 2.

In the following chapters a short description of the Densit Joint Cast® and some of the test results are given. More information can be found in Jensen (1995) and Hansen (2001).

A similar development has been reported by Harryson (2002). The objective in his work is to develop a joint that is highly applicable in industrial bridge concepts. Thus a connection, which is easy and fast to perform and makes the surrounding elements continuous, was a desirable feature. The basic material in this joint is also a steel fibre reinforced high-performance concrete. The width of the joint is also 100 mm. It is reported that the connection will be stronger than the surrounding concrete elements and fully moment-resisting. Different laboratory tests – static tests and fatigue tests – have been conducted and the results are compared with finite element calculations. It is concluded that a great deal of attention has to be paid to quality control at the construction site when using this type of connection.

2 DENSIT JOINT CAST®

In ultra high-strength concrete materials the binder is often composed of Portland cement powder and other fine particles. Often silica fume with ball shaped particles is added too and the binder is very dense and strong. Such concretes are brittle materials but the problem can be solved by adding steel fibres to the concrete. The material used in this investigation is based on the binder Densit Joint Cast® which is developed at Aalborg Portland, see Bache (1987).

In this steel fibre reinforced concrete the water/powder ratio is 0.15–0.18, the silica fume content is 20–25% and the compressive strength is in the interval 100–200 MPa. The steel fibre content is 6% by volume, the fibrelength is 12 mm and the diameter is 0.4 mm.

The failure criterion for this material is tested by Nielsen (1995). One of the results is that the theory of plasticity can be used in the ultimate limit state. Therefore, the design can be done according to modern theories of concrete design as for example described in the Eurocodes.

The price for the material is three to four times the price for ordinary concrete, which means that a general substitution of concrete with Densit Joint Cast® will probably not take place. However an evident use of the material is in joints carrying forces between structural elements. This is due to the eminent anchorage of reinforcing bars.

Typical properties of Densit Joint Cast® are:

Uniaxial compressive strength	100–200 MPa
Initial modulus in compression	50 GPa
Uniaxial tensile strength	10–15 MPa
Initial modulus in tension	50 GPa
Fracture energy	10–15 kN/m

3 STATIC TESTS

In this chapter, only a few of the results from the tests can be given. The static tests can be divided into 3 groups:

- Anchorage tests
- Slab tests
- Column/slab tests

3.1 *Anchorage tests*

A sketch of the test specimen used for the anchorage tests is shown in Figure 3 to the left. It is seen that two main reinforcement bars are embedded in a concrete block (500 mm × 300 mm × 100 mm) made of Densit Joint Cast®. Y in the figure denotes a ribbed bar. The concrete block is also reinforced to prevent failure of the concrete. The specimen is subjected to pure tension.

Before the tests at Aalborg University some tests were carried out at the Cement and Concrete Laboratory at Aalborg Portland with different diameters of the reinforcement. 16 mm reinforcement bars were used for all the tests at Aalborg University. The anchorage length for a 16 mm ribbed bar is only 170 mm corresponding to approximately 11 times the diameter.

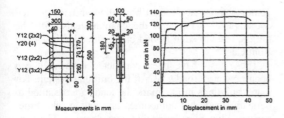

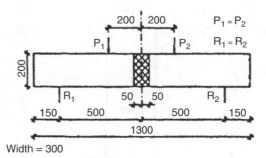

Figure 3. Sketch of pure tensile test specimen and load-displacement curve.

Figure 4. Sketch of slab in four-point bending. Measurements in mm.

The material properties for the reinforcement was an average yield strength of 600 MPa and an average ultimate strength of 685 MPa. A typical load-displacement curve for pure tension is shown in Figure 3 to the right. It can be noticed that there is a clear yield plateau. The average yield force was 113 kN corresponding to a stress in the reinforcement bar of 560 MPa and an average ultimate force of 133 kN corresponding to a stress in the bar of 660 MPa. These values are very close to the values obtained for the bars alone. All the failures were in the bars and no slip failures between the bars and the Densit Joint Cast® were observed. The same type of specimen was used for the fatigue tests described in chapter 5.

3.2 Slab tests

Several slabs elements were tested but only one type will be mentioned here. The loading was a four point load with two forces and two simple supports as shown in Figure 4. The slab width for this type of test was only 300 mm, so it is perhaps more like a beam structure. The reinforcement in the slab was of diameter 8 mm.

In Figure 4 the joint is the cross-hatched area made of Densit Joint Cast®. The two parts of the slab are made of normal strength concrete. Different types of joints were tested to develop a connection where the reinforcement is not pulled out of the Densit Joint Cast®. This means that the connection acts as if the slab is a "normal" concrete slab with respect to strength and stiffness. The tests also gave an estimate of the sensitivity for the anchorage length and the influence of horizontal reinforcement perpendicular to the principal reinforcement. The tests showed that for the 8 mm reinforcement an anchorage length of 60 mm was sufficient, but in the building system an 80 mm anchorage length was used. In these load tests forces, displacements, strains and crack widths were measured. All the specimens behaved in a plastic and ductile manner. A typical load-deflection curve is shown in Figure 5.

Other tests focused on the combination of bending and shear in the joint and with different surfaces

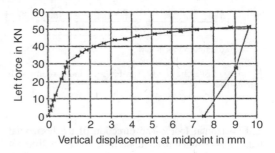

Figure 5. Load-deflection curve for a slab with the joint in the middle of the structure as shown in Figure 4.

(rugged and smooth) between the concrete slab parts and the connection made with Densit Joint Cast®.

Different upper and lower-bound solutions according to the plasticity theory were calculated and the measured values in the tests were found to lie in the interval between these bounds both for the yield state and for the ultimate state.

Also, some other types of slabs were tested. The slabs were constructed as parts of greater slabs and were loaded to simulate typical load cases and support conditions. All the slab tests were in close agreement with the expectations and the calculations.

The conclusion for all the slab tests was that the usual theory of plasticity can be used and that the joints are able to transfer the forces and moments as calculated.

3.3 Column/slab tests

As mentioned in the Introduction, the horizontal forces can be transferred through the columns and stabilizing walls are thus not needed. To be sure that the columns were able to transfer the forces to the columns, a series on both inner and outer columns were tested. The column/slab connection was subjected to both vertical

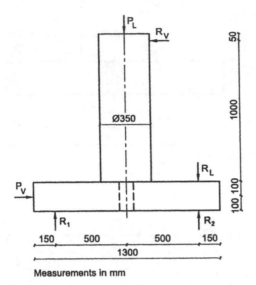

Figure 6. Sketch of test rig for an inner column/slab connection.

and horizontal loads. The direction of the horizontal loads was changed to simulate wind forces in different directions. A sketch of the test for an inner column/slab connection is shown in Figure 6. For practical reasons the connection was tested with the column above, and not below, the slab. Only the region of the slab near the column was examined in these tests.

All these tests showed a ductile rupture with crushing in the concrete and yielding in the reinforcement.

4 FIRE TESTS

Six beams with a Densit Joint Cast® joint were subjected to a combination of static load and fire exposure. In addition, two similar beams were tested only for static loads at room temperature. These beams acted as reference beams. The thermal exposure was according to a standard temperature curve corresponding to ISO 834 and the maximum temperature in the furnace was 900°C after one hour.

The conclusion of the fire testing was that the joints can be classified as fire resistant for at least 60 minutes. After such a fire exposure for 60 minutes the residual load-bearing capacity is minimum 75% of the original capacity.

5 FATIGUE TESTS

This chapter first deals with the fatigue tests of reinforcing ribbed bars – a type called New

Tentor – manufactured of hot-rolled and accelerated cooled steel bars. Next the chapter deals with the fatigue tests of such ribbed bars embedded in Densit Joint Cast®.

These tests have been conducted because it can open up for a broader use of the building system. For many buildings in Denmark the loads are assumed to be static loads but sometimes one has to take time-varying (dynamic) loads into consideration. The tests reported in this chapter are a little step towards a better knowledge of the fatigue properties of reinforcing bars and reinforcing bars embedded in Densit Joint Cast®.

5.1 Fatigue tests of ribbed reinforcing bars

The tests include 29 specimens with the diameter $\phi = 10$ mm and 33 specimens with the diameter $\phi = 16$ mm.

A 500 kN servohydraulic testing machine was applied for the testing. The testing machine was equipped with hydraulic grips. This is a very simple and quick way to fasten the test specimen to the testing machine. But a problem may arise during the fatigue testing, especially when ribbed bars are applied. If no precautions are taken, the failure will in many tests occur very close to the hydraulic grips. The hydraulic grips will destroy the surface material and the result can be a crack. These cracks will during the fatigue testing develop into larger and larger cracks and the result is that the fatigue properties will depend on these cracks.

To avoid this situation both ends of the reinforcing bar were equipped with aluminium tubes with a length of 85 mm. The pressure from the hydraulic grips was determined in accordance with the instructions from the manufacturer. In spite of the aluminium tubes some of the test specimens failed near the grips, see later. No considerable slip between the reinforcing bars and the aluminium tubes was observed during the testing.

The yield strength for both diameters of the reinforcement was determined to 630 MPa and the ultimate strength to 720 MPa.

All the fatigue tests were force controlled. The variation of the force was harmonic at different levels and the corresponding stresses were oscillating between constant values of σ_{max} and σ_{min}. The difference between these stresses is called $\Delta\sigma$. During the fatigue testing, values of the minimum and maximum force, the minimum and maximum grip stroke were measured for every 1000 cycles.

For the reinforcing bars with diameter $\phi = 10$ mm the following equation was found in the interval $4.5 \leq \log N \leq 6$, where N is the number of cycles:

$$\log \Delta\sigma = 3.82 - 0.24 \log N \qquad (1)$$

where $\Delta\sigma$ is in MPa.

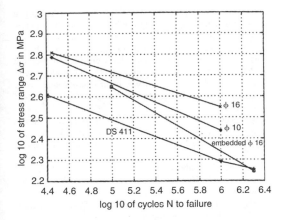

Figure 7. Fatigue strength curves (S-N – curves) for New Tentor reinforcing bars with diameters 10 mm and 16 mm, 16 mm New Tentor reinforcing bar embedded in Densit Joint Cast® and the curve from the Danish Code of Practice for the Structural Use of Concrete, DS411.

The equation given by the Danish Code of Practice for the Structural Use of Concrete DS411 is (independently of the diameter):

$$\log \Delta\sigma = 3.49 - 0.20 \log N \qquad (2)$$

For the reinforcing bars with diameter $\phi = 16$ mm the following equation was found in the interval $4.5 \leqslant \log N \leqslant 6$:

$$\log \Delta\sigma = 3.56 - 0.17 \log N \qquad (3)$$

These equations are shown in Figure 7, see later.

5.2 Fatigue tests of ribbed reinforcing bars embedded in Densit Joint Cast®.

The tests include 35 specimens. These specimens were of the same type as used for the anchorage tests described in section 3.1 and shown in figure 3. The testing machine and the way to fasten the specimens in the testing machine were the same as described in section 5.1.

The material properties for the reinforcement were as described in section 5.1 and the average compressive strength for the Densit Joint Cast® was 160 MPa. Seven different levels were used for the applied maximum forces, and the minimum force was approximately 10 kN in all the tests.

In some cases there was a failure in the reinforcing bar near the ends and this was an unwanted situation. The test was then automatically stopped by the testing machine. The test specimen was then equipped with a new aluminium tube and the testing machine was

restarted. For some of the tests there were thus some idle periods. These idle periods could have different length. If the failure was in the daytime, a new aluminium tube was installed quickly, but if the failure was for example in the night or in the week-ends, the idle periods could be much longer.

For 11 of the 35 specimens tested for fatigue it was not possible to obtain a failure in the concrete embedment zone for the reinforcing bar or in the reinforcement embedded in the Densit Joint Cast® because 2, 3 and 4 failures occurred in the reinforcing bar near the grips. It was thus impossible to continue the fatigue testing simply because it was not possible to fasten the test specimen to the testing machine. Such a specimen is here called a "run-out" specimen. For such a specimen the fatigue properties are better than given by this fatigue test as the number of cycles to failure is greater than the number of cycles when the test had to be stopped.

The equation for the 24 specimens without the "run-out" specimens in the interval $5 \leqslant \log N \leqslant 6.3$ was found to be:

$$\log \Delta\sigma = 4.19 - 0.31 \log N \qquad (4)$$

and for the 35 specimens including the "run-out" specimens:

$$\log \Delta\sigma = 4.02 - 0.28 \log N \qquad (5)$$

Equations 1, 2, 3 and 4 are shown in Figure 7.

6 CONCLUSIONS

From the test series described in this paper and the practical experience at the construction site the following can be concluded:

1. Casting with Densit Joint Cast® turns out to be easy to do in this type of joints.
2. The joint made of Densit Joint Cast® makes the slab act as a ductile structure with load-bearing capacities which can be calculated according to traditional methods.
3. For example the joint made of Densit Joint Cast® makes it possible to anchorage ribbed 8 mm reinforcing bars and a yield strength of 550 MPa at a length of only 60 mm. In practice an anchorage length of 80 mm was used.
4. It is possible to use the building system in Denmark without stabilizing walls. The horizontal forces on the building can be transferred through the slab/column joints to the foundations.
5. The joints made of Densit Joint Cast® can be classified as fire resistant for at least 60 minutes. After a fire exposure up to 900°C for 60 minutes, and

cooling the residual load bearing capacity is 75% of the original capacity.

6. The fatigue properties for both the reinforcing bars and for those embedded in Densit Joint Cast® seem to be better than given by the Danish Code of Practice for the Structural Use of Concrete, DS411.

REFERENCES

Bache, H.H. 1987. Compact Reinforced Composite, Basic Principles. *CBL Report No. 41, Aalborg Portland.* Aalborg.

Hansen, L.P. 2000. Udmattelsesforsøg med ribbestål indstøbt i Densit Joint Cast® (Fatigue tests of reinforcement embedded in Densit Joint Cast®). In Danish. *Department of Building Technology and Structural Engineering, Aalborg University, Denmark. Report No. 0052.* Aalborg.

Hansen, L.P. & Heshe, G. 2001. Static, Fire and Fatigue Tests of Ultra High-Strength Fibre Reinforced Concrete and Ribbed Bars. *Nordic Concrete Research, The Nordic Concrete Federation, Publication No. 26, 17–37.* Oslo.

Harryson, P. 2002. Industrial Bridge Construction–merging developments of process, productivity and products with technical solutions. *Department of Structural Engineering. Concrete Structures. Chalmers University of Technology.* Göteborg.

Jensen, B.C. et al.1995. Connections in precast buildings using ultra high-strength fibre reinforced concrete. *Proceedings of Nordic Symposium on Modern Design of Concrete Structures, Aalborg University, Denmark, May 3–5, 63–74.* Aalborg.

Nielsen, C.V. 1995. Ultra High-Strength Steel Fibre Reinforced Concrete. Part I & II. *Department of Structural Engineering. Technical University of Denmark. Report No.323 & 324.* Copenhagen.

Olesen, J.F. 1995. The Concrete Weld: Monolithic Slabs from Precast Elements. *Structural Engineering International, Vol. 4, 1995. 230.*

System-based Vision for Strategic and Creative Design, Bontempi (ed.)
© *2003 Swets & Zeitlinger, Lisse, ISBN 90 5809 599 1*

Moment-rotation relationships of hybrid connection under monotonic and cyclic loading

X. Sun & T.H. Tan
Nanyang Technological University, Singapore

ABSTRACT: Connection design has always been an important aspect in precast concrete construction. In this study, the ease of steelwork installation is adopted to replace the traditional reinforced concrete connection. This new hybrid connection consists of a single fin plate connection with HSFG bolts plus the insitu conventional reinforced concrete infill. For this connection, the semi-rigid behaviour of the steel component is complicated further by reinforced concrete infill. A series of experiments were carried out to study the behaviour of hybrid connection under monotonic as well as cyclic loading. A component-based mechanical model is proposed to predict the moment-rotation relationship of this hybrid connection under monotonic hogging and sagging moment. Using the hysteresis rule, the model can be extended for cyclic loadings.

1 INTRODUCTION

Precast technology offers the opportunity of compressing construction time schedules. Erection of precast members takes less time compared with cast-in-place concrete. Although the conventional precast technology has sharpened the competent edge over the normal reinforced concrete in many ways, however, it still posses some disadvantages and it is worthwhile for us to investigate and study them. The main shortcomings may be summarised as congestion of reinforcement in the connection, unstable structural system during construction period and labour-intensive job on site. Therefore, it is necessary to look for a more systematic alternative connection device that can simplify the complexity and congestion at the joint. Based on above considerations, a new hybrid beam-column connection method is proposed. The basic adoption of such hybrid connection to suit the construction industry is that the installation is simple, direct and fast, and structural system is stable during the construction period.

In the proposed hybrid connection, unlike those precast concrete beams that use reinforcement protruding out from the beam ends for connection to columns, the connection function is replaced by short steel I-beams partly embedded in a precast reinforced concrete beam and partly projecting out of the supporting ends of the beam. The precast reinforced concrete column is separated at the joint position by steel column partly embedded in the precast column. During the construction period, the precast beam is connected to the precast column through the single plate connection. The projecting structural steel I-beams are bolted to the steel column using High Strength Friction Grip (HSFG) bolts. Eventually, the beam-column joint is cast with matching concrete together with cast-in-place beam portion. The connection between the precast beam and cast-in-place beam are used by extending the shear links across the interface and anchored sufficiently in the cast-in-place portion. Continuity can be achieved for the beam across the joint through the use of top reinforcements as well as bottom reinforcements in the cast-in-place beam.

Using this kind of hybrid connection, it is much easier and simpler to bolt steel sections altogether. It eliminates the reinforcement congestion and possible honeycombed concrete inside the joints. In addition, it can eliminate or reduce the propping and re-propping during the construction. The most prominent value of this hybrid connection technology is its instant ability to undertake the self-weight of structural members right after installation during the construction period and able to ensure the stability of structure at construction stage. In order to use the proposed hybrid connection technology in practice, it is essential to study the moment capacity, rotational capacity and moment-rotation behaviour of the hybrid connection.

The primary objective of this research is to study the suitability and moment-rotation behaviour of proposed hybrid beam-column connection. This research

Table 1. Details of specimens.

| Test series | Specimen | Beam Reinforcement | | Loading |
		Top	Bottom	
Series A	BSC1			monotonic
	BSC2			monotonic
	BSC3			cyclic
Series B	HC1	2T16	2T13	monotonic
	HC2	2T16	2T13	monotonic
	HC3	2T16	2T13	cyclic
Series C	HCS1	2T16	2T13	monotonic
	HCS2	2T16	2T13	monotonic
	HCS3	2T25	2T16	monotonic
	HCS4	2T25	2T16	monotonic
	HCS5	2T16	2T13	cyclic
	HCS6	2T25	2T16	cyclic

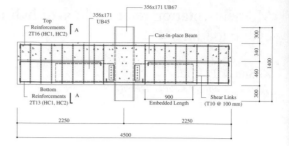

Figure 1. Test specimen with hybrid connection under monotonic loading.

is also intended to lay some bases for the design and analysis of the frames with hybrid connections under lateral loading.

2 EXPERIMENTAL PROGRAM

The experimental programme of this research is aimed at the study of the behaviour of hybrid connection under monotonic loading as well as cyclic loading. Based on the typical industrial building layout, the typical beam and column sizes are selected in this experimental programme. The test programme consists of three series. Test series A is for the bare single plate connection, which is aimed to study the moment-rotation relationship of single plate connection under monotonic and cyclic loading. Test series B is for the hybrid connections without topping concrete, which is aimed to study the moment capacity as well as the rotational capacity of hybrid connection under monotonic and cyclic loading. Test series C is for the hybrid connections with topping concrete and hollow core slabs, which is aimed to study the contribution of topping concrete as well as hollow core slab to the moment and rotational capacity. A total number of 12 specimens are tested and the details of each series are illustrated in Table 1. The configurations of specimens are shown in Figure 1 to Figure 5.

For the specimens with hybrid connection (series B & C), the fabrication procedures adopted is to simulate the actual construction sequence of hybrid connection structures. At first, the universal steel column was cast into the precast column. Then, the formwork of precast beam was built. The universal beam and all the reinforcing bars were placed in position. After the casting of precast beams, the precast beams were assembled to the steel column via single plate connections with HSFG bolts. The connection portion had a 200 mm distance away from the face of column flange to the end of

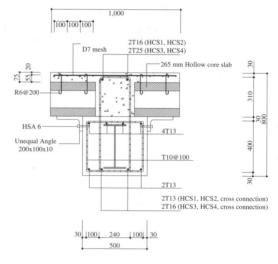

Figure 2. Section of test specimen (with hollow core slab and topping concrete).

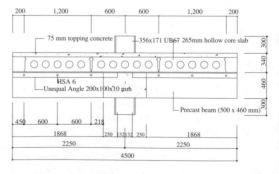

Figure 3. Test specimen with hollow core slab and topping concrete.

the precast beam. It will be left to cure for one week after the casting of precast beam. The connection portion was cast together with the cast-in-place beam as well as topping concrete. The cast-in-place beam was

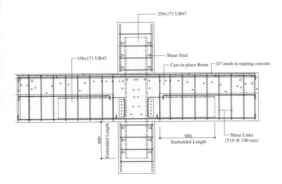

Figure 4. Test specimen with hybrid connection under cyclic loading.

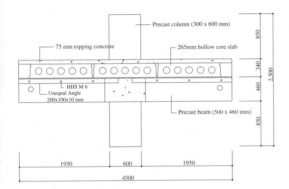

Figure 5. Test specimen with hybrid connection with hollow core slab and topping concrete under cyclic loading.

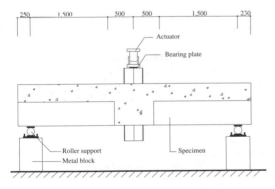

Figure 6. Test set-up for monotonic loading.

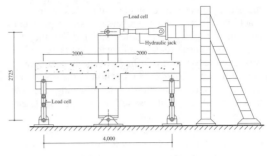

Figure 7. Test set-up for cyclic loading.

of 200T hydraulic jack and loading rig with 2000 kN capacity. Before testing, each specimen was white washed for easy identification of crack information and for marking crack pattern on the beam surface. The specimens were placed on the metal roller bearings at the support locations. Each test was started by applying the load in an increment of 5 kN at lower load level or before any crack appeared. After some flexure cracks had been generated, load increment was then increased to 10 kN until the beam failed. At each load increment, the displacement and load were recorded to a computer.

The test set-up for specimens under cyclic loading is shown in Figure 7. A reversible horizontal load was applied to the top of the column using a double acting 400 kN capacity hydraulic jack. The bottom of the column was pinned to the strong floor, and beam ends were connected to the strong floor by steel links that permit rotation and free horizontal movement of the beam but not vertical movement, thus providing the vertical reactions to the beams. The cyclic testing procedure on structural steel elements recommended by European Convention for Construction Steelwork (TC13 1986) was adopted. The short testing procedure C proposed in TC13 was used as a reference to derive the loading procedure. At the first and second loading levels, load increments based on force-control were used. Displacement control for the loading and reloading branches and force control for the unloading branches were used alternatively. In general, two or three cycles for each loading level were applied. A test was terminated if the failure of the specimen or the strike limitation of the actuator was reached.

3 EXPERIMENTAL RESULTS

The specimen HC1 and HC2 in series B had been tested so far. HC1 was tested to study the moment-rotation relationship of hybrid connection under monotonic hogging moment. The critical sections were the contact faces between the cast-in-place concrete and

cast over the precast beam and was continuous across the beam-column joint. Different casting sequence was used to achieve a cold joint between precast beam and cast-in-place beam.

The test set-up for specimens under monotonic loading is shown in Figure 6. The test set-up consisted the steel column face at the joint. Before the first cracking, the strains in the rebars and single plate connection were very small which indicted that they had very little contribution at this stage. After this, there was a reduction in the stiffness of the hybrid connection. This was shown by a reduction in the gradient of the moment rotation curve. After cracking, the steel bars were mainly responsible for carrying the tensile forces. This was reflected by the rapid increase in the strain readings of rebars under tension. At the same time, the single plate connection started to rotate which also contributed to the behaviour of the hybrid connection. With the progress of test, more flexural cracks occurred around the critical sections. The cracks along the critical section kept on developing and propagating towards the top of the connection. Some splitting cracks were also observed at the sections where the precast beam end met with the cast-in-place concrete. This is due to the weak bondage between these two parts. The rebars started to yield at a moment capacity of 248 kN.m. After this, the stiffness of the hybrid connection was further reduced until the specimen failed. The failure mode was due to the concrete crushing at the top of critical sections under compression.

The specimen HC2 was tested to study the moment-rotation relationship of hybrid connection under monotonic sagging moment. The critical sections consisted of two sections: the first was the section between the precast beam and the cast-in-place concrete; the second critical section was the contact face between the cast-in-place concrete and the steel column face at the joint. After cracking, the steel bars and single plate connection were responsible for carrying the tensile forces. There was also a reduction in the stiffness of the hybrid connection. This was shown by a reduction in the gradient of the moment rotation curve. With the progress of test, the cracks along the critical section kept on developing and propagating towards the top of the connection. The weakest critical section was the section between the precast beam and cast-in-place concrete. Because these two parts were cast in different stages, therefore, the bondage between these two parts was relatively weak. The cracks were progressively splitting the two parts under the increasing loading. These splitting cracks developed much faster than those normal flexural cracks and therefore resulted that the hybrid connection had less ductility and failed in a brittle manner under sagging moment.

4 COMPONENT-BASED MODEL

A component-based mechanical model is proposed to predict the moment-rotation relationship of hybrid connection under monotonic hogging and sagging moment. Since the end cross section of hybrid beam is within the cast-in-place beam, based on the plane

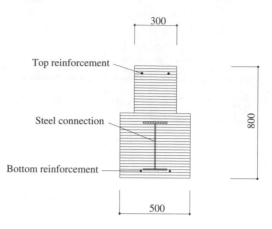

Figure 8. Subdivision of the hybrid beam section into monitoring areas.

section assumption, this section can be treated as a rigid body. Column is assumed to be rigid relative to the hybrid connection. The mechanical model consists of four components: steel connection, concrete beam section, the top reinforcements and the bottom reinforcements. The behaviour of reinforcements, steel connection and concrete can be represented by equivalent springs respectively. Because of the non-uniform deformations along the depth of the concrete beam section, therefore, it is subdivided into a finite number of layers n (Fig. 8). The deformational state of spring can be defined by two variables: the axial deformation and the rotation. If the position of neutral axis is $\bar{y}$, the axial deformation of a spring element i located at a distance y_i of the connection can be given as:

$$\Delta_i = (y_i - \bar{y}).\theta \qquad (1)$$

Once the deformation Δ_i is established, the corresponding force F_i to the connecting element can be derived based on corresponding load-deformation relationships.

$$F_i = K_i.\Delta_i \qquad (2)$$

The total axial force and moment transmitted by the connecting elements can be expressed as:

$$F = \sum_{i=1}^{n} F_i \qquad (3)$$

$$M = \sum_{i=1}^{n} F_i.y_i \qquad (4)$$

With the increase of rotation, the new neutral axis position is calculated. If the force-displacement relationship of connecting element is inelastic, iterations are needed until the force equilibrium condition is satisfied. Therefore, the numerical analysis procedure for the moment-rotation relationship of hybrid connection can be achieved. The force-displacement relationship of each type of the component will be described in the following sections.

4.1 Properties of concrete section

It is well known that the strength of concrete in structures can be well in excess of its uniaxial strength. This is due to the multi-axial or confinement effects arising from the existence of the transverse reinforcement of the reinforced concrete members. Different stress–strain relationships have been developed to describe the behaviour of confined concrete. In this study, the relationship proposed by Paulay et al. (1992) is adopted in describing the behaviour of hybrid beam section as well as the topping concrete. For a rectangular cross section, the stress–strain curve for confined concrete is expressed as:

$$f_c = f_{cc} \left[2(\frac{\varepsilon_c}{\varepsilon_{cc}}) - (\frac{\varepsilon_c}{\varepsilon_{cc}})^2 \right] \quad \text{for } \varepsilon_c \leq \varepsilon_{cc} \tag{5}$$

$$f_c = f_{cc} \left[1 - 0.3 \frac{\varepsilon_c - \varepsilon_{cc}}{\varepsilon_{ccu} - \varepsilon_{cc}} \right] \quad \text{for } \varepsilon_{cc} < \varepsilon_c \leq \varepsilon_{ccu} \tag{6}$$

where f_{cc} = confined concrete strength; ε_{cc} = confined strain at peak stress; ε_{ccu} = confined strain at ultimate compression point.

The tensile stress–strain relationship is assumed to be linear with the elastic modulus E_c:

$$f_{ct} = E_c.\varepsilon_{ct} \tag{7}$$

and the ultimate tensile strength is taken as 10% of its compressive strength.

4.2 Axial stiffness of reinforcements

The axial stiffness of top and bottom rebars in the cast-in-place concrete beam as well as the mesh reinforcements inside the topping concrete can be modelled as follows:

$$K_r = \frac{E_r.A_s}{L_s} \tag{8}$$

where E_r = elastic modulus of steel reinforcement; A_s = area of reinforcements; L_s = length of reinforcement bar experiencing the same extension.

4.3 Axial stiffness of single plate connection

The behaviour of single plate connection inside the hybrid connection can be represented by two springs

with different axial stiffness, which is located at the top and bottom of single plate connection respectively. The axial stiffness of spring at the top and bottom portion of single plate connection can be given as:

$$K_{spct} = \frac{[K_s.(\Delta_{st} - \Delta_{sb}) + K_{st}.(\Delta_{st} + \Delta_{sb})]}{2\Delta_{st}} \tag{9}$$

$$K_{spcb} = \frac{[K_s.(\Delta_{sb} - \Delta_{st}) + K_{st}.(\Delta_{st} + \Delta_{sb})]}{2\Delta_{sb}} \tag{10}$$

where K_s = axial stiffness of steel connection due to rotation; K_{st} = axial stiffness of steel connection due to stretching; Δ_{st} = displacement of steel connection at top portion; Δ_{sb} = displacement of steel connection at bottom portion.

4.4 Model prediction

The mechanical model of hybrid connection under monotonic hogging and sagging moment are shown in Figure 9 and Figure 10 respectively. Based on the equilibrium of forces,

$$F_{r,top} + F_{st} + F_{sb} + F_{r,bot} + \sum_{i=1}^{n} f_{ci} = 0 \tag{11}$$

The hogging moment applied on the hybrid connection can be given as:

$$M_{hog} = F_{r,top}.d + F_{st}.d_{st,hog} + F_{sb}.d_{sb,hog} + F_{r,bot}.d_{r,hog} + \sum_{i=1}^{n} f_{ci}.y_i$$

The sagging moment applied on the hybrid connection can be given as:

$$M_{sag} = F_{r,top}.d_{r,sag} + F_{st}.d_{st,sag} + F_{sb}.d_{sb,sag} + F_{r,bot}.d + \sum_{i=1}^{n} f_{ci}.y_i$$

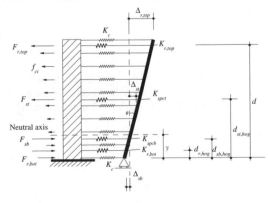

Figure 9. Proposed component-based mechanical model under hogging moment.

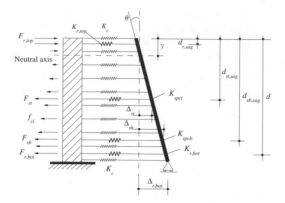

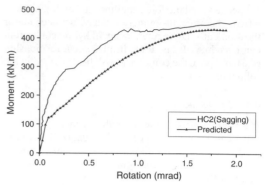

Figure 10. Proposed component-based mechanical model under sagging moment.

Figure 12. Comparison between test result and model prediction.

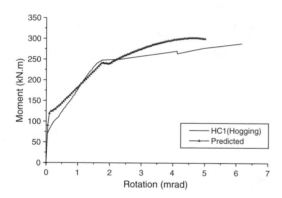

Figure 11. Comparison between test result and model prediction.

The comparison between tested results and model prediction are shown in Figure 11 and Figure 12. In general, the model can generate a satisfied prediction for hybrid connection.

5 CONCLUSIONS

Although only part of the specimens have been tested, valuable information has been obtained on the behaviour of proposed hybrid connection. The component-based mechanical model has been developed. The model can predict the moment-rotation relationships of hybrid connection under monotonic hogging and sagging moment. It can be extended to incorporate more components such as topping concrete and hollow core slabs. At next stage, the behaviour of hybrid connection under cyclic loading will be studied. The proposed model will be extended for cyclic loadings using the hysteresis rule. The design procedures for analysis of precast frames with hybrid connection will also be developed in the future.

REFERENCES

Chen, W.F. 1987b. *Joint flexibility in steel frames*, Elsevier Applied Science Publishers, Essex, U.K.
De Stefano, M., De Luca, A. & Astaneh-Asl, A. 1994. Modelling of cyclic moment-rotation response of double-angle connections, *Journal of Structural Engineering, ASCE*, Vol. 120(1), pp 212–229.
Guan, E.H. 1999. *Behaviour of precast hybrid steel/concrete joints*, M. Eng. Thesis, Nanyang Technological University, Singapore.
Nethercot, D.A. & Ahmed, B. 1997. Numerical modelling of composite connections and composite frames, *Proceedings of the 1996 Engineering Foundation Conference on Composite Constructions in Steel and Concrete III*, New York, pp.809–822.
Paulay, T. & Preiestly, M.J.N. 1992. *Seismic design of reinforced concrete and masonry building*, A Wiley Interscience Publication.

Mechanical response of pre-cast concrete wall panels to combined wind and thermal loads

A. Brencich & R. Morbiducci
DISEG Department of Structural and Geotechnical Engineering, Faculty of Engineering,
University of Genova, Genova, Italy

ABSTRACT: The interest for pre-cast concrete wall panels increased in Italy during the last years for aesthetic reasons and, in general, for a greater attention to quality and durability of pre-cast. Concrete wall panels provide thermal insulation to the structure, bear themselves loads of different origin and form the architectural elements of the facades. In Italy they are usually attached to precast concrete frames and may be typically found on industrial buildings, where they may span 10 m or more. These panels may present quite a diffuse cracking on the external surface, which does not usually affect their structural safety but reduces the expected life. In this work the structural behavior of two different kinds of pre-cast concrete wall panels are analyzed: a sandwich panel and a lightweight one. The mechanical response of the panels is analyzed considering the effects of wind, temperature distribution and the bending effects due to the accidental eccentricity in the building phase.

1 INTRODUCTION

Pre-cast wall panels have been given increasing attention in the last years, at least in Italy, due to the improvement of their aesthetic performance and to a greater attention to quality and durability performances of buildings. Pre-cast concrete panels typically consist of two wythes of concrete separated by a layer of insulation (continuous in the sandwich panel and discontinuous in the lightweight one). They provide insulation to the structure, bear different loads (i.e. mainly, wind, thermal loads) and design the facades in terms of different shape, typology and surface finishes.

Pre-cast panels can be used as exterior and interior walls for many types of structures, but in Italy they are usually attached to pre-cast concrete frames and are typically found on industrial buildings.

Regarding the static aspects in Italy, this kind of pre-cast panels are usually considered as elements of secondary importance, even if they have to support significant loads. Mainly in industrial buildings, they may span 10 m or more and they bear relevant wind and thermal loads; nevertheless the connection to the ground and to the column/beam structure is not always adequately analyzed.

These panels may present extensive cracking of the external surface, which does not usually affect their structural safety but reduces the expected life. Some of the cracks occur at reentrant corners, while transverse and longitudinal cracks may also be present, probably due to handling. In addition, some cracks have been observed at locations where the insulation is discontinuous, such as at the ends of the panels detailed with a solid end block or in the lightweight panel.

In this work, the structural response of two different kinds of pre-cast concrete wall panels are considered: a sandwich (Fig. 1a) and a lightweight one (Fig. 1b). They are two samples of the Italian production and employ different structural devices. The sandwich type is a stiffly joined sandwich panel, that is the two concrete wythes act together to resist applied loads by means of a continuous bent bar system. Even though it is not so frequently used, nevertheless it is an interesting example because it emphasizes some of the structural problems this kind of panel has to face. The lightweight type is a monolayer one with internal lightweight material blocks (expanded polystyrene) reinforced with steel zig-zag bars trespassing the insulation material from one wythe to the other. In particular, the internal reinforcement system is present in the longitudinal, transverse and intermediate ribs with different bar diameter in the different positions (14 mm for the longitudinal reinforcements, 10 mm for the internal concrete wythe, strands located at the centroid of external wythe using a diameter of 5 mm and a web of 250*200 mm.

The two types of panels are 2 m wide and 10 m high. In the sandwich panels, the overall thickness is 380 mm,

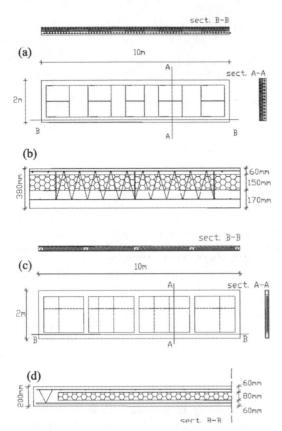

Figure 1. Vertical section of pre-cast wall panels: (a) sandwich panel; (b) particular of sandwich section; (c) lightweight panel; (d) particular of lightweight section.

Table 1. Thermal characteristics of the wall panels.

	Concrete wythe	Expanded polystyrene layer
Density: ρ (Kg/m³)	2000	15
Thermal conductivity: λ (W/mK)	1.160	0.054
Specific heat: c (J/kgK)	880	1220
Surface absorbitivity: α	0.6, 0.8, 0.9	
Thermal dilatation: d (°C⁻¹)	1×10^{-5}	7×10^{-5}
Heat-transfer coef. (W/m²K)	ext. h_e: 8.14	int. h_i: 17.0

the internal wythe being 170 mm thick, the external one 60 mm and insulation 150 mm. In the lightweight panel the concrete wythe thickness is equal to 60 mm, insulation thickness is 80 mm. Figure 1 shows the main characteristics of the two types wall panels.

The effects of wind, temperature distribution (considering the cycle of an entire day and the effect of different surface finishes) and the bending effects due to the accidental eccentricity in the building phase are considered and summed up to analyze the mechanical behavior of the two panels and to verify if the heaviest loading condition might be responsible for cracking of the surfaces. The structural response of the panels is analyzed by means of a non linear F.E. commercial code (Ansys 5.6).

2 THERMAL ANALYSIS

In this section the thermal behavior of the two pre-cast wall panels is analyzed to establish the heat flow through the panels and thus the thermal load condition to introduce in the mechanical analysis.

Temperature distribution in wall panels depends on many factors, including the external and internal temperatures, the orientation of the building, the season, the hour of the day, wind velocity, the different materials of the panel and surface finishes. Computer programs have been developed able of taking into account many if not all of these factors (i.e. Ho 1995, Xingde 1997, Chiu Liu & Denos 2001). In this work a standard procedure is used which, first, performs a thermal analysis of the panel, and then assign the temperatures to the structural model.

In the thermal analysis the summer (or winter) transient regime is considered since it has been found that almost no stationary thermal conditions are met. Many different situations have been taken into account: (1) the external temperature of a wall panel may be higher than the air temperature, the daily variation of the external surface temperature oscillates during daytime and gets an almost linear law during the night, (2) the amount of radiation on the a surface depends on the different components of solar radiation and material absorption. Table 1 summarizes the main thermal characteristics assumed for the materials of the two panels and the environmental data. The geographic location (Trieste) simulates extreme conditions: the UNI recommendations provide for this city high wind velocity (2.6 m/s) and high temperature differences between the day (≥31.1°C in summer) and the night (≤−5°C in winter) or between the external and internal wall surfaces (≤8°C). In particular, the UNI 10349 standards are used for the meteorological data. In summer the internal air temperature is assumed equal to 26°C for the day time, i.e. assuming conditioned air, while during the night it is assumed equal to the air temperature. In winter, the internal temperature is assumed as much as 18°C (internal temperature for industrial buildings) while the external air temperature varies from −5°C to +10°C.

On the basis of the standard assumption of a linear distribution of temperature inside the different panel layers, taking into account the global transmittance

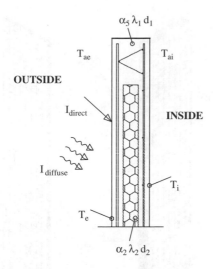

Figure 2. Scheme of heat flow process simulation by simplified procedure: lightened panel.

and solar absorption of the materials, the thermal analysis has been performed assuming a transient regime on the basis of a finite difference procedure. In this way, the temperature distribution inside the panels has been simulated throughout the day and night. Figure 2 shows a schematic representation of the parameters taken into account.

In particular the temperature transmission is studied through the Fourier equation considering *the equivalent temperature* (explicit method) (Patankar 1980). The heat flow from the external atmosphere to the internal air is assumed equal to $Q(\theta) = \Delta t_{eq}(\theta)$, where Q is the heat flow, θ is the time variable, K is the panel transmittance and Δt_{eq} is the equivalent temperature variation.

Figures 3, 4 show the results for the two wall panels in terms of temperature distribution in the case of summery transient regime, West exposition and $\alpha =$ 0.9, that is the heaviest thermal loading conditions. Unexpectedly, winter conditions are not the most severe ones since the highest thermal mismatch is attained in summertime, in the middle of the afternoon when the surface temperature gets the highest figure, being the internal air still conditioned at a 26°C temperature. This latter choice is not the most severe one, being possible, mainly for the administration offices, internal temperatures around 20°C. In both the wall panels, the curves of the temperature distribution have a positive concavity when the external solar radiation is high, otherwise the concavity is inverted when the internal concrete wythe is hotter than the external one.

The temperature distribution through the wall panels and the temperature gradient in the contact section between the external concrete wythe and the insulation

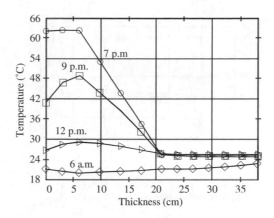

Figure 3. Temperature distribution for the sandwich wall panel in three different hours of a day.

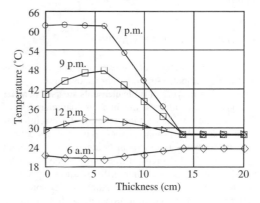

Figure 4. Temperature distribution for the lightweight wall panel in three different hours of a day.

layer, during the 24 hours, were used in the mechanical non linear model to verify if crack propagation may activate as consequence of the limited tensile strength of concrete.

In the lightweight wall panel, only the internal zones are constituted by three different layers, while near the boundaries a reinforced concrete beam stiffens the outwards parts of the panel. This concrete layer acts as a thermal bridge.

In the Figure 5 the temperature variations (over a 72 hours time interval) are shown considering the external surface, the internal one and their differences for the sandwich wall panel. The internal and external temperature diagrams are sinusoidal because they are dependent of the air-sun temperature daily oscillations. Their first pick is different to the successive because represents the transition phase of an oscillating phenomenon. It is worth noting that the diagram of the difference between the external and internal surfaces is different for the two panel: the maximum

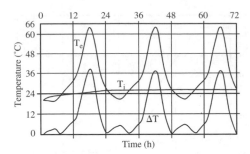

Figure 5. Example of temperature oscillations, case of sandwich wall panel: T_e = external surface temperature, T_i = internal surface temperature, $\Delta T = T_e - T_i$.

temperature mismatch is equal to 37.7°C for the sandwich panel, while it is equal to 36.5°C for the lightweight panel.

3 STRUCTURAL ANALYSIS

The numerical models of the two different wall panels for the F.E. models (Ansys 5.6) are shown in Figures 6. Three different types of finite elements are used in order to consider the different mechanical characteristics of the concrete wythes, the insulation and the reinforcement system. A 3-D solid element is used for modelling insulation layers, the element being defined by eight nodes with three degrees of freedom per node (translations); element loads include different surface pressure and temperature in each node. Another 3-D solid element is used for modelling concrete with reinforcing bars, which is allowed to undergo cracking in tension and crushing in compression. The material model is a non linear brittle one with the Willam-Warnke three parameter model as limit condition (Willam & Warnke 1975, Ansys 1999). The element is defined and has the same capabilities of the previous solid element. A membrane four nodes element is used for the continuous bent bar reinforcement system; this element has variable thickness, assumes translations as nodal degrees of freedom and a linear elastic material model.

The sandwich panel model is constituted by two layers of 30 mm of thickness for the external concrete wythe, six layers of 28 mm for the internal structural concrete wythe and two layers of 75 mm of thickness for the insulation. The reinforcements of the concrete wythes are located at the centroid of each wythe using a 5 mm diameter bar. The continuous bent bar system is constituted by membrane elements with equivalent thickness connected only to the concrete layers and not to the insulation ones.

The lightweight panel model is constituted by three layers of 20 mm of thickness for the external concrete wythe and three for the internal one. Four

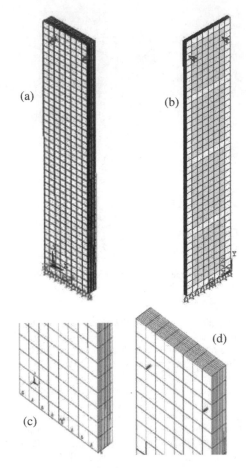

Figure 6. The geometrical models of the wall panels: (a) sandwich panel model; (b) lightweight panel model without the external concrete element layers; (c) a particular of the sandwich panel with the constrains on the basis; (d) a particular of the sandwich panel with the constrains on the top.

layers 20 mm thick have been used for the internal section of the panel in which there are both the discontinuous insulation layers and the reinforced concrete beams. The mechanical characteristics of the materials are summarized in Table 2, partially according to literature data (Brencich & Gambarotta 2001) and partially assumed from standard codes.

In the numerical models the constrains on the basis are supposed as simple vertical supports, connected to the structural wythe (the internal one) in the sandwich panel, located in the center of the lightweight panel, on the top only horizontal translation is prevented.

Several load conditions are analyzed to get the heaviest one:

1. wall panel weight + alignment error towards inside/outside,

Table 2. The mechanical characteristics of the materials.

	Concrete wythe	Bent bar system	Expanded polystyrene
Elastic modulus E [MPa]	31200	2103870	3300
Poisson ratio ν	0.15	0.3	0.4
Tensile strength σ_{rt} [MPa]	1.5	/	/
Compr. strength σ_{rc} [MPa]	30	/	/

2. wall panel weight + alignment error + wind effect,
3. wall panel weight + alignment error + thermal load in summer/winter,
4. the lifting phase (dead weight suspended in two intermediate points).

The numerical models showed that the wall panels generally have no structural problems both in the service and in the lifting phase. In details, some interesting considerations of the wall panels response can be pointed out, mainly from the technological point of view. In sandwich panel layers the stress are always lower than the concrete tensile and compressive strengths, Figure 7a. Nevertheless, rather significant tangential stresses (approximately 0.8 MPa) are developed at the concrete/insulation interface, showing that sliding at that interface can take place, Figure 7b. This fact, which is of marginal importance from the structural point of view, might give way to water absorption and consequent degradation of the concrete parts of the panel and of the bent bars.

The structural analyses of the lightweight panel show possible cracks for the wind load only (wind pressure = 1.2 kPa). Figure 8 shows an example of the possible cracking in the stiffened panel in which the tensile stress reach the tensile strength assumed for concrete (1.5 MPa) and the compressive ones are quite close to the compressive strength (30 MPa). The deformed configuration clearly shows a central zone in which the panel might crack. It has to be said that such extreme winds such as the assumed one would suggest to put some intermediate support to the panel; the entire height of 10 m is left without support intermediate horizontal supports only in areas where low winds are expected.

A relevant structural feature exhibited by the lightweight panel is due to the thermal expansion of the insulating material. In fact, in order to reduce the production costs, the central layer is given the cheapest material available on the market; the standard choice is that of expanded polystyrene. Poor attention is paid to choose a material with the same thermal expansion coefficient as concrete. When the internal insulation layer is warmed, i.e. at the end of the afternoon at

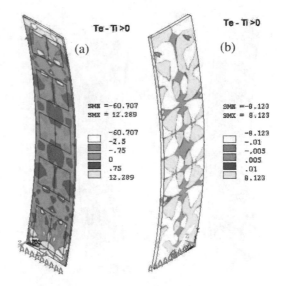

Figure 7. Vertical stress distribution in the load case of wall panel weight with alignment error and thermal load is summer (worst thermal condition): (a) vertical stress component distribution for the external concrete wythe; (b) tangential stress component distribution for the insulation [dN/cm²].

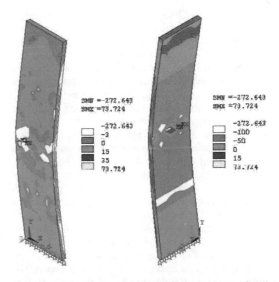

Figure 8. Vertical stress distribution in the load case of wall panel weight with alignment error and wind effect [dN/cm²].

7 p.m., Figures 3 and 4. In this case, the thermal expansion of the insulating layer induces a tensile stress state in the concrete whytes, Figure 9, which is not so severe (0.5 MPa), at least for the assumed materials, but is worthwhile noting since could be at the bases of some cracking in the case of bad workmanship.

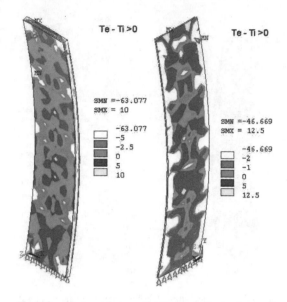

Figure 9. Horizontal stress distribution in the load case of wall panel weight with alignment error and thermal load in summer (worst thermal condition) [dN/cm²].

4 CONCLUSIONS

Pre-cast wall panels sometimes exhibit significant cracking on the external surface but the cause of this phenomenon is not yet clear.

Among the possible causes, the structural problem has been considered first, being wall panels low-cost components for industrialized buildings, light and designed with security coefficients very close to the code minimum requirements. Among the structural actions stressing a wall panel, errors in vertical positioning, wind and thermal loads have been considered. The latter were expected to produce important effects since a thermal analysis showed that the difference between internal and external temperatures might be as high as 40°.

The structural analyses performed for the different kinds of loads, and for all the possible combinations, show that, due to the large amount of reinforcement, seldom some structural problem is to be expected. The large amount of diffused reinforcement limits the tensile stresses in concrete to low values, usually below the material tensile strength. For this reason it can be concluded that, for the considered panels, the cracking phenomenon should not be caused by structural problems.

The main structural feature of the mechanical response of the panel is due to the different thermal expansion of concrete and of the insulating material (usually expanded polystyrene). Sliding seems to be possible at the concrete/insulation interface, which is not a structural damage but makes it easier for moisture to penetrate inside the panel.

Only extreme wind loads may produce some cracking in the panel, but this should not be considered as a problem since in such conditions it would be easy to provide the panel with intermediate horizontal supports.

The origin of the external cracking is not yet explained, but it seems to be not of structural origin. The performed analyses show that some optimization is possible for this kind of panels, i.e. choosing insulating material with thermal expansion very close to that of concrete, so relieving the stress state in extreme conditions. Nevertheless, the origin of external cracking should be sought in other causes, such as the wetting/drying cycles and the icing/de-icing cycles that may take place in wintertime.

ACKNOWLEDGMENTS

This research was partially carried out with the financial support of the University of Genoa for the 2001 year "*Comparative analysis of the mechanical behavior of the standardized typologies of pre-cast concrete wall panels*" research project.

REFERENCES

ANSYS. Revision 5.6, Swanson Analysis Systems, Inc., Houston, PA, 1999.

Brencich, A. & Gambarotta, L. 2001. Isotropic damage model with different tensile-compressive response for brittle materials, *International Journal of Solids and Structures*, 38: 5865–5892.

CENT 229/WG 1/TG8, 2002. Precast concrete products wall elements. Product properties and performances. 11th draft proposal.

Chiu Liu, P.E. & Denos C.G. 2001. Heat conduction waves in bilayer cement concrete structures, *Journal of Engineering Mechanics, ASCE*, 127(11), pp. 2044–2071.

Ho, D. 1995. Temperature distribution in walls and roofs. *Journal of Architectural Engineering* 1(3): 121–132.

Ozisik, M.N. 1993. Heat conduction, Wiley, *John Wiley*, New York, N.Y.

Patankar, S.V. 1980. Numerical heat transfer and fluid flow, *McGraw-Hill Hemisphere*, New York, N.Y.

Pratt, A.W. 1981. Heat transmission in buildings, *John Wiley*, New York, N.Y.

UNI 7357:1974. *Evaluation of the thermal needs for building heatings* (in Italian).

UNI 10349:1994. *Heating and cooling of buildings. Environmental data* (in Italian).

Willam, K. & Warnke, E.D. 1975. Constitutive model for the triaxial behaviour of concrete, *Proc. IASBE*, vol. 19, ISMES, Bergamo, 1975.

Xingde, L. 1997. Temperature calculation in the thickness direction for composite of arbitrary layers. *Computers & Structures*, 62(4): 763–770.

System-based Vision for Strategic and Creative Design, Bontempi (ed.)
© 2003 Swets & Zeitlinger, Lisse, ISBN 90 5809 599 1

Strength of precast slab subjected to torsion

S. Nakano

Department of Civil Engineering, Kure National College of Technology, Hiroshima, Japan

ABSTRACT: The precast slabs are mutually connected to each other with in-situ concrete. In the case of using a loop-shaped joint, there is a problem that occurs when the strength of the joint decreases because of the concentration of the reinforcing bars and the anchorage length of the reinforcing bars. There is a little research on the case that the joint receives torsion. So, for the purpose of checking the influence of torsion exerted on the joint of a precast slab, experiments were done on the loop-shaped reinforcing bar used. Two kinds of specimens, the unity and the composite slabs were produced. In the case of the composite slabs, the types of joints were a plain joint and a grooved joint.

By using the experimental results, and the theorem of Bredt and Coulomb's modified yield criterion, the carrying capacity is obtained. Theoretical values and experimental results are in good agreement.

1   INTRODUCTION

The opportunity for RC or PC precast slabs being used as composite slabs is increasing. As for the precast slabs, the fact is that the production and transport are easy and there is a little fieldwork and or maintenance is fine.

The way of joining precast slabs together becomes a big problem in the construction method that uses these RC, or PC precast slabs. In order for the precast slabs to be jointed with each other, there are two methods. One is infused with cement mortar and the other method is the use of prestressing bars.

In the case of the former, there is the problem that the joint reinforcement focuses on one place and the strength of the joining area between the precast slabs fall off where the reinforcing bar overlaps. Much research has been done in the case of the joining area subject to shear force and bending moment.

However, there is a little research with regard to the joining area as to the twisting range (Schiessl 1996). Therefore, this research is done for the purpose of studying the influence that is exerted on the joining area of the precast slabs caused by torsion.

In the failure model of the reinforced concrete members subjected to torsion, there is the space truss theory that was expanded the plane truss theory. And there is also the skew-bending model that analogizes from the bending destruction and predicts the ultimate strength of members.

On the other hand, the method that is able to estimate the load-deformation relation of the member corresponding to an optional load level has been tried. There is research on the plane stress method that does the torsion analysis by using the tension stiffening of concrete and the softening characteristic of cracked concrete (Collins & Mitchell 1980, Thomas 1988, 1991). However, this analytical method is fairly complicated.

Besides, the other method, the limit analysis method is used in the calculation of the carrying capacity of an elastic–plastic body (Jensen 1975). The solution in this method is physically deep and brief. The carrying capacity calculated with the limit analysis method is fairly easier than the stress calculation. This method has been used in shear force problems. This method is used where the place and the direction of the crack are evident. Especially, it is most suited when obtaining the strength of a joining face.

Thereupon, in this study, the ultimate torsion carrying capacities of a composite slab connected by two sheets of precast slabs subjected to torsion are obtained by using the limit analysis method. These calculations are based on experimental results. In the limit analysis method, the theorem of Bredt and Coulomb's modified yield criterion are included.

2   EXPERIMENT

2.1   *Specimens*

The specimen had side dimensions, a and b, of $50\,cm \times 50\,cm$, and a thickness h of about 8 cm shown

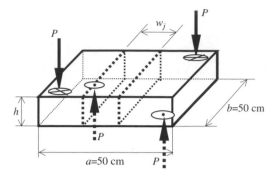

Figure 1. Loading method.

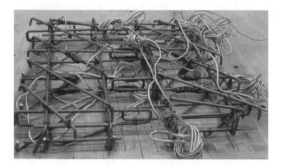

Figure 2. Arrangement of reinforcing bars.

Table 1. Dimensions of specimens.

Specimen	Joining face	h (cm)	w_i (cm)	f'_c (N/mm²)	Number of transverse steel bar
PR10	monolithic	8.0	–	45.0	3
PR11	grooved	8.0	10	45.5	3
PR12	plain	8.0	10	45.5	3
PR13	plain	8.0	14	51.8	3
PR20	monolithic	8.3	–	28.1	4
PR21	grooved	9.0	10	28.1	4
PR22	plain	8.2	10	34.0	4
PR23	plain	8.5	14	28.4	4
PR30	monolithic	8.0	–	42.2	5
PR31	grooved	8.0	10	26.0	5
PR32	plain	8.3	10	28.1	5
PR33	plain	8.3	14	30.9	5
PR40	monolithic	8.5	–	34.4	7
PR41	grooved	7.5	10	36.2	7
PR42	plain	9.0	10	33.5	7
PR43	plain	8.5	14	28.4	7

was produced in the slab, by supporting it at 2 points on one diagonal line and by applying an equal downward concentrated load at 2 points on the other diagonal line. Using this loading method, cracking loads, strains in the electrical resistance strain gages, sliding between the precast slab and in-situ concrete, and an ultimate load was measured.

2.3 Experimental result

The failure process of specimens was roughly divided into 2 types. One is a failure process that is almost the same as the monolithic specimens in Figure 3a. The other is the failure process that cracking causes along the joint at first, secondly, the specimen collapses by the diagonal cracking of the in-situ concrete area shown in Figure 3b.

The former is the composite slab with the grooved joint, specimen PR11, 21, 31 and 41. At first, a crack that connected the both supporting points caused and collapsed by the increase of crack width, after the first crack reached at the side. The latter is the composite slab with the plain joint. In this case, the specimens cracked along the joining face. After that, the specimens cracked diagonally in the in-situ concrete area. After cracks reached the side, the specimen collapsed with the increase of this crack width.

In Figure 4, 5 and 6, the relationship of load-transverse reinforcing bar strains of the in-situ concrete area of specimen PR40 (monolithic type), PR41 (grooved type) and PR42 (plain type) is shown respectively. The transverse reinforcing bar is vertical in the direction to the joining face. As seen from these figures, in either case, the transverse reinforcing bars

in Figure 1. First of all, producing two slabs, and two slabs were joined by in-situ concrete. The joining face was plain or rough (5 mm to 10 mm in height) to check the relation between the joint width and the strength of the composite slab. Widths of joining area w_j were varied (10 and 14 cm) to investigate its effect on the joint. And also investigating the influence of the joint, monolithic slabs were produced.

The reinforcing bars of all specimens were arranged toward the longitudinal, transverse, and diagonal direction of the top and bottom 2 layers. The diagonal reinforcing bars were arranged in the reverse direction in the upper and lower layer and the number of the transverse reinforcing bar were 3, 4, 5, and 7. Also, two precast slabs joined with loop-shaped reinforcing bars. Figure 2 shows an arrangement of reinforcing bars. Details of the specimens are shown in Table 1.

The reinforcements used in the specimens were round bars 6 mm in diameter. The overlap length of the loop-shaped reinforcing bar was limited to ten times of the diameter of the reinforcing bar.

2.2 Experimental method

The loading method by using a hydraulic jack test machine is shown in Figure 1. Constant torsion moment

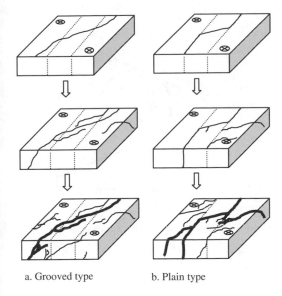

a. Grooved type   b. Plain type

Figure 3.   Failure process.

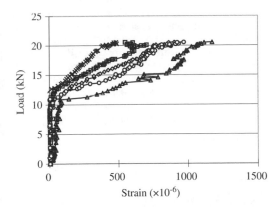

Figure 4.   Load–strain relationship (PR41).

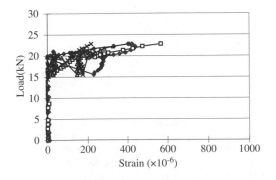

Figure 5.   Load–strain relationship (PR42).

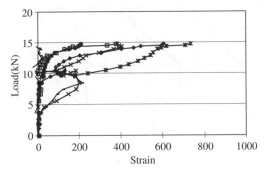

Figure 6.   Load–strain relationship (PR43).

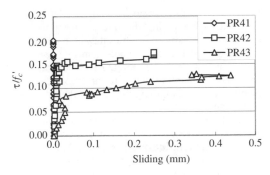

Figure 7.   Experimental relationship between τ/f_c' and sliding.

did not yield. And, the strains of transverse reinforcing bars of the monolithic type (PR40) and the grooved type (PR41) almost agreed, and the magnitude of strains was about a half of the yield strain. On the other hand, the magnitude of strains in the plain type (PR42) was about a quarter of the yield strain.

In the case of the monolithic and grooved type, after cracking, strains of all transverse reinforcing bars increased. On the other hand, in the plain type, there were transverse reinforcing bars that strains hardly increased. Such transverse reinforcing bars were seen to the specimens of wide joining width a lot even the same plain model.

Figure 7 describes the shear stress-sliding relationship of specimen PR41, 42 and 43. The sliding is the relative displacement of the horizontal direction between the precast slab and in-situ concrete. The sliding of specimen PR41 with a grooved jointing face was almost zero. However, in the case of the other specimens, a big amount of sliding was observed.

3 THEORY

The relation between torsion moment and shear stress can be obtained using the theorem of Bredt. Also, the

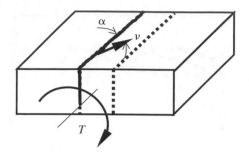

Figure 8. Relative displacement of composite slab subjected to torsion.

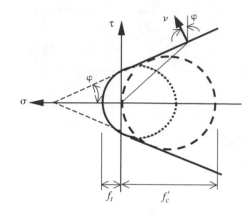

Figure 9. Coulomb's modified yield criterion.

shear stress of the composite slab at failure can be obtained by equalizing the internal work with the external work using the limit analysis method. The external work depends on torsion moment. The internal work depends on the reinforcing bar force and concrete force. Therefore, ultimate torsion moment can be obtained by combining the torsion theory and the limit analysis theory.

The internal work depending on the reinforcing bar force can be obtained by using a coefficient of reinforcement (Hashin & Sinan 1991). In monolithic slabs, cracks occur in a diagonal direction according to torsion. However, composite slabs connected with in-situ concrete collapsed along the joining face. The failure depends on the sliding along the joining face. At this time, transverse reinforcing bars connecting precast slabs and in-situ concrete did not yield. The degree of the magnitude of this reinforcing bar stress was expressed using a coefficient of reinforcement.

The torsion carrying capacity of composite slabs can be obtained by using the limit analysis method below. The composite slab that cracked along the joining face is shown in Figure 8. The failure mechanism is the yield line along the joining face between the precast slab and in-situ concrete. A relative displacement of in-situ concrete to the precast slab is v and the inclination of yield line is α. The yield line, in other words, a line of discontinuity is inserted into the plane of a paper.

For this displacement field, the work that is done by the shear stress τ caused by torsion moment T is given

$$W_E = \tau v \cos \alpha \cdot tb \qquad (1)$$

where $\tau = T/(2A_0 \cdot t)$ (theorem of Bredt); A_0 = area within the center-line of the shear flow zone; t = thickness of the shear flow zone.

The work done by the internal work consists of the contributory shares of reinforcing bars and concrete. Disregarding a dowel action, the internal work due to transverse reinforcing bars is given

$$W_{IR} = k_s A_s f_{ty} v \sin \alpha \qquad (2)$$

where A_s = cross sectional area of transverse reinforcing bars; f_{ty} = yield strength of transverse reinforcing bars; k_s = coefficient of reinforcement that shows the ratio of the reinforcing bar stress at failure to the yield stress.

The internal work by concrete depends on Coulomb's modified yield criterion in Figure 9; f_t = tensile strength of concrete, f'_c = compression strength of concrete, ϕ = internal frictional angle. The internal plastic work of concrete that is obtained from the limit analysis method is given

$$W_{IC} = v \left(\frac{1 - \sin \alpha}{2} f'_c + \frac{\sin \alpha - \sin \phi}{1 - \sin \phi} f_t \right) tb \qquad (3)$$

From $W_E = W_{IC} + W_{IR}$

$$\frac{\tau}{f'_c} = \frac{1 - \sin \alpha}{2 \cos \alpha} + \frac{\sin \alpha - \sin \phi}{(1 - \sin \phi) \cdot \cos \alpha} \frac{f_t}{f'_c} + \psi \cdot \tan \alpha \qquad (4)$$

where ψ is a degree of reinforcement.

$$\psi = \frac{k_s A_s f_{ty}}{btf'_c} \qquad (5)$$

The normal condition of the theory of plasticity requires that the displacement vector, given by displacement v in Figure 8, must be perpendicular to Coulomb's modified yield criterion in Figure 9. Then, $\alpha = \phi$ is valid along the straight line in the yield criterion.

4 DISCUSSION OF EXPERIMENTAL AND THEORETICAL RESULTS

The ratio τ/f'_c at failure for all specimens are shown in Figure 10. Ultimate shear stress τ is obtained from the theorem of Bredt.

From this figure, the following are obtained:

1. As for the ratio τ/f'_c the plain types are small in comparison with monolithic and grooved types.
2. The ratio τ/f'_c differs by the width of in-situ concrete area.
3. The number of transverse reinforcing bars exerts the influence on ultimate shear stress τ.

A relationship between the ratio τ/f'_c and the degree of reinforcement $\psi = A_s f_{ty}/(btf'_c)$ is described below. The coefficient of reinforcement k_s that shows the magnitude of the reinforcing bar stress is included in the degree of reinforcement ψ. From experimental results, strains of transverse reinforcing bars were about half of a yield strain in most of the monolithic and grooved type. Thereupon, as $k_s = 0.5$, the relation between τ/f'_c and ψ is obtained as shown in Figure 11. By using the method of least squares, the following relation is given

$$\frac{\tau}{f_c'} = 0.071 + 1.322 \frac{A_s f_{ty}}{btf'_c} \tag{6}$$

This relationship is shown in Figure 11.

In the case of the plain type, strains of transverse reinforcing bars were a quarter of a yield strain at failure. In the case of the plain type, as $k_s = 0.25$, the relation between τ/f'_c and ψ is obtained as shown in Figure 12. By using the method of least squares, the following relation is given

$$\frac{\tau}{f_c'} = 0.028 + 3.461 \frac{A_s f_{ty}}{btf'_c} \tag{7}$$

This relationship is shown with the straight line in Figure 12.

Furthermore, in obtaining the ratio τ/f'_c, from experimental results, an amount of transverse reinforcing bar was changed. In the case of the plain type, there were transverse reinforcing bars that strains hardly increased. It seems that transverse reinforcing bars along the joining face cracked in the in-situ concrete did not work effectively (Fig. 3b). From this reason, the transverse reinforcing bars of the only cracked joining area thought with effective. Thereupon, shear stress τ was obtained by using $(1 - w_j/b) A_s$ instead of A_s in Equation 6 and Equation 7.

Shear stress τ of all specimens at failure can be obtained by using Equation 6 or 7. Furthermore, the comparisons of the theoretical and experimental values of torsion carrying capacity P are shown in Figure 13.

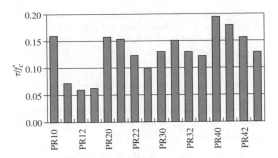

Figure 10. Experimental results of τ/f'_c.

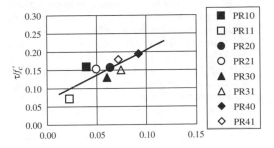

Figure 11. Relationship of $\psi - \tau/f'_c$ of monolithic and grooved type ($k_s = 0.5$).

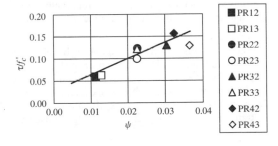

Figure 12. Relationship of $\psi - \tau/f'_c$ of plain type ($k_s = 0.25$).

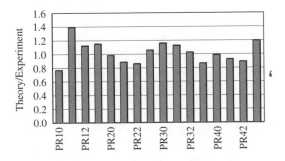

Figure 13. Ratio theory/experiment of carrying capacity.

From this figure, series PR1 that there is a little number of transverse reinforcing bars are not uniform. Also, theoretical values are 3% bigger than experimental values an average.

5 SUMMARY AND CONCLUSIONS

The following were obtained from the experimental results of composite slabs with the joining face of the grooved and plain type:

1. In the case of the grooved type (from 5 mm to 10 mm in height), cracks caused diagonally the same as the monolithic type.
2. In the case of the plain type, cracks caused along the joining faces. In this type, there were transverse reinforcing bars that the strain hardly results.
3. The values of the transverse reinforcing bar strain that crosses the joining face were about a half of the yield strain in the case of the grooved type. And in the case of the plain type, the values were about a quarter of the yield strain.
4. The torsion carrying capacities of the monolithic and grooved type almost became the same value. In the case of the plain type, the torsion carrying capacities of specimens with the wide joining width were smaller than that with the narrow joining width.

Using the limit analysis method, the torsion carrying capacity was obtained. The transverse reinforcing stress and the amount of the reinforcing bar were changed based on the experiment results:

1. The strain of the transverse reinforcing bar changed by the joining condition of the precast slab. Thereupon, the stress of the transverse reinforcing bar was obtained by using the coefficient of reinforcement.
2. In the case of the plain type, the amount of the transverse reinforcing bar was changed based on

the cracking and the values of transverse reinforcing bars.

The following were obtained from the comparison of experimental values and theoretical values of the torsion carrying capacities:

1. In the case of series PR1 that the amount of the transverse reinforcing bar is little, the differences between the theoretical values and experimental values of the torsion carrying capacity varied widely.
2. In the case of the monolithic and grooved type, the cracks did not cause along the joining face. This differs from a calculation assumption. However, the theoretical values and experimental values of the torsion carrying capacity agreed well.

REFERENCES

Collins, M.P. & Mitchell, D. 1980. Shear and Torsion Design of Prestressed and Non-prestressed Concrete Beams, *PCI Journal*, Vol.25, No.5, Sep.–Oct.
Hashin, M.S.A. & Sinan, Y.H.S. 1991. Prediction of Ultimate Shear Strength of Vertical Joints in Large Panel Structures, *ACI STRUCTURAL JOURNAL*, March–April, pp. 204–213.
Jensen, B.C. 1975. Lines of Discontinuity for Displacements in the Theory of Plasticity of Plain and Reinforced Concrete, *Magazine of Concrete Research*, Vol.27, No.92, September, pp. 143–150.
Schiessl, P. 1996. Drillsteigigkeit von Fertigplatten mit statisch mitwirkender Querschnicht, *Beton-und Stahlbetonbau*, Heft 3.
Thomas, T.C. & Hsu 1988. Softened Truss Model Theory for Shear and Torsion, *ACI Structural Journal*, Nov.–Dec., pp. 624–635.
Thomas T.C. & Hsu 1991. Nonlinear Analysis of Concrete Torsional Members, *ACI Structural Journal*, Nov.–Dec., pp. 674–682.

System-based Vision for Strategic and Creative Design, Bontempi (ed.)
© 2003 Swets & Zeitlinger, Lisse, ISBN 90 5809 599 1

Column-foundation connection for pre-cast elements

E. Dolara
University of Rome "La Sapienza", Italy

G. Di Niro
University of Strathclyde, Glasgow, UK

ABSTRACT: A large research work on an alternative system of making a column-foundation connection for pre-cast elements has been carried out and reported in this paper.
– Pad footing made with steel base plate has been the column-foundation connection investigated.
– Full scale tests on pre-cast concrete column-foundation connection have been performed.
– Moment capacity, required thickness of the base plate, anchorage characteristics have been studied.
– The failure mechanism of the connection components have been analysed.
– Recommendations and directions for structural design of the connection have been provided.

1 INTRODUCTION

Connections represent one of the most important aspect in prefabrication. Their correct design depends on many factors strictly related to the production, tolerances, delivery, erection and, of course, structural analysis, considering the whole system of forces applied to them during the service life.

In pre-cast concrete framed structures columns are commonly founded in *in-situ* concrete foundation (pocket pad footing).

The available depth of foundation may be influenced sometimes the choice of the column-foundation connection to be adopted.

For this reason sometimes could be useful to use pad footing with steel base plate directly bolted to the foundation. In fact base plate provide immediate stability when fixing the column on site, and the depth of the foundation is not excessive.

As regards the column-foundation connection with steel base plate two types are commonly adopted (Fig. 1):

• with larger base plate;
• with column-size base plate.

Normally base plates which are larger than the size of the columns are used where a moment connection is required.

The larger base plate increases effective bearing area, column corner reinforcing bars can be welded to

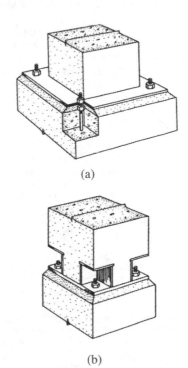

(a)

(b)

Figure 1. Column foundation details: a) lager base plate, b) column-size base plate (P.C.I Manual on Design of Connection for Precast, Prestressed Concrete 1986).

plate for anchorage, the oversize holes reduce tolerance problems, usually a thicker base plate is required.

The base plate with the same size in plan dimension of the columns allows continuous moulds to be used, corner recesses may be formed by steel angle, steel angle may be welded to base plate and column corner reinforcing bar.

It should be underlined that the attitudes towards the choice in using base plate rather than pockets tend to be based more on production than structural decisions.

2 MATERIALS

During this research projects the physico-chemical characteristics of the aggregate used in the pre-cast plant, the properties of both the wet and hardened concrete have been determined.

The mechanical behaviour of concrete used in a pre-cast plant for normal production has been investigated. Test on specimens made from this concrete have been performed.

In Table 1 the exact mix proportions for the mix MX 6S, used to cast the two columns tested, are reported. A superplasticizer has been added to this mix.

Slump tests have been carried out before casting each column. Specimens evaluation of the mechanical properties were casted together with the columns. In Table 2 are reported the results of these tests for the columns.

The properties of the steel used are reported in Table 3. They have been coded following the commercial production in Italy.

3 TEST SET-UP AND TESTING PROCEDURE

Pad footing made with larger steel base plate has been the column-foundation connection investigated and is reported in Figure 2. Only type A anchor bars are reported in Figure 2, no reinforcing bars needed in the current column cross section are shown.

All characteristics regarding the steel base plate, anchor bolts, welding, bolting, type A anchor bars and their welding to the steel base plate have been analysed and studied.

The collapse of one of these single elements involved the collapse of the whole connection. The experimentation aimed to determine the weak element of the connection.

Full scale tests on two pre-cast concrete column-foundation connection have been carried out.

Moment capacity, thickness of the base plate, anchorage characteristics have been studied.

The columns had a cross section of 45 cm × 55 cm and they were tested with the bending moment applied on both direction.

The columns were loaded with an increasing load with a fixed eccentricity and measurements were taken of load, deflection, and stresses in the reinforcement bars. The strain level in the top part of the column was also measured at discrete points with the use of strain gauges. The load was applied with an hydraulic jack set in an eccentric position compared to the column axis.

The column steel base plate was bolted to another steel plate welded to an HEA330 standard rolled-steel section.

A system of 8 tendons has been adopted to resist to the jack thrust (Figs 3 and 4).

Table 3. Steel properties.

Steel code	Use	f_t (N/mm^2)	f_y (N/mm^2)	ε_t (%)
Fe360	Base plate	$\geq$340	$\geq$235	$\geq$24
FeB44K	Column reinforcing bars	$\geq$450	$\geq$375	$\geq$12
FeB44K	Anchor bars type A	$\geq$540	$\geq$375	$\geq$12

Table 1. Mix proportions (1 m^3).

Mix code	W/C ratio	Cement (kg)	Water (l)	Fine aggregate 0–4 mm (kg)	Caorse aggregate 4–16 mm (kg)
MX 6S	0.45	450	203	1164	627

Table 2. Mechanical properties of the mix used to produce the two columns (air cured).

Mix code	W/C ratio	Slump (mm)	28-Day strength (Mpa)			Modulus of elasticity (MPa)
			Compressive	Tensile	Flexural	
MX 6S	0.45	210	66.80	3.45	4.05	42 750

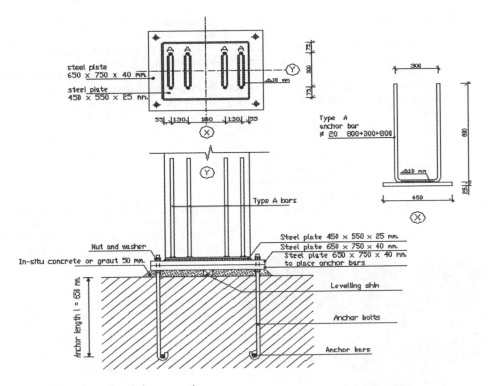

Figure 2.   Details of the column-foundation connection.

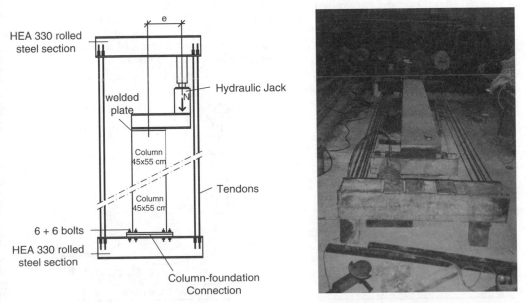

Figure 3(a).   Load scheme (axial load N with eccentricity e = 52 cm).

Figure 3(b).   Test equipment horizontally placed.

4 RESULTS AND DISCUSSION

Since the beginning of the study, it has been clear that the weak point of the connection could be the anchor bars and the base plate.

In fact the failure load was reached for an axial load N =38.5 tons with an eccentricity e =52.0 cm and a moment M =20.0 tons m.

Figure 5 shows the scheme for the anchorage bars that caused the failure.

The failure was due to the anchorage bars collapse, as it was expected, and is shown in Figures 6 and 7.

Strain-gauges were applied on the anchor bars and from the data recorded during the test it has been noticed that the anchorage bars failure occurred when the steel stress reached in the bars was lower than the steel yield strength.

This was probably due to a bending effect on the bars for their "U" shape.

But it should be underlined that all stresses acted on anchor bars that were already work-hardened by the cold-bending and by the welding with the base plate.

No cracks or separation has been noticed for the welding between the anchor bars and the base plate.

This means that the shape of the anchor bars and their system of direct connection to the steel base plate are the weak characteristics of the column-foundation connection investigated.

It seemed be useful to determine the maximum load that could be applied to the column-foundation connection due to the weak elements of the connection itself.

For this reason a design chart that could consider all the above aspects has been the following step of this study.

The load deflection curves for the top edge of the column have been plotted in Figure 8.

The maximum deflection reached was lower than the theoretical one. All column crack patterns have been followed and recorded during the test.

The full scale test results show that the column-foundation connection made with larger steel base plate performed satisfactorily.

As far as the failure occurred for anchorage bars colapse, probably due to "U" shape of the bars, an alternative method to connect the bars to the steel base plate is suggested.

As further study following the experimentation, M–N interaction diagrams for the cloumns have been processed and plotted as shown in Figure 9.

From the above interaction diagrams, design charts have been costructed and used to design a pre-cast underground parking in Rome.

Figure 4. Details of the column-foundation connection with external concrete strain gauges.

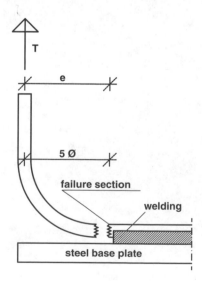

Figure 5(a). Failure scheme type 1.

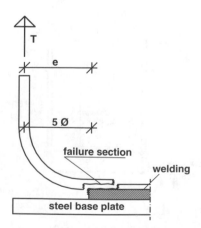

Figure 5(b). Failure scheme type 2.

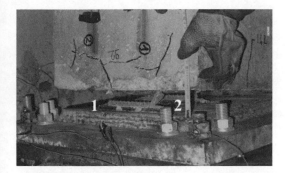

Figure 6. Anchorage bars collapse: type 1 and 2.

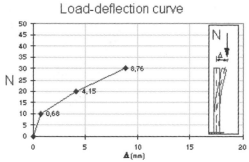

Figure 8. Load–deflection curve.

Figure 7. Anchorage bars collapse: type 1 and 2.

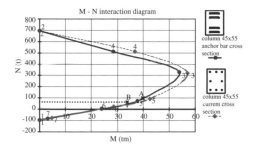

Figure 9. M–N interaction diagrams for the 450×550 mm column for the current cross section and anchor bar corss section.

Table 4. M–N diagram points related to modes of failure.

Point	Modes of failure	Current cross section			Anchor bars cross section		
		M (t m)	N (t)	e = M/N (cm)	M (t m)	N (t)	e = M/N (cm)
1	Tension failure	0	−96	0	0	−96	0
7	No reacting concrete	4.8	75.4	9.5	3.2	−82.3	5
6	Ductile failure ($\varepsilon_c = 2\text{‰}$)	28.4	23	60	24	4	60
5	Ductile failure ($\varepsilon_c = 3.5\text{‰}$)	42.7	93	36.5	37.5	74.5	50
3	Balanced failure	57	321	18.3	53.8	331.5	16
4	Fragile failure	36.5	511	7.5	28.3	512	5.5
2	Compression failure	0	696	0	0	696	0

REFERENCES

P.C.I. Manual on Design of Connection for Precast, Pre-stressed Concrete. *Second Edition. Prestressed Concrete Institute, Chicago, IL, 1986.*

Elliott, K.S. & Tovey A.K. 1992. Precast concrete frame buildings – A design guide. *British Cement Association, Wexham Springs, Slough.*

Elliott K.S. 1996. Multi-storey precast concrete framed structures. *BlackwellScience, 1996.*

14. Earthquake and seismic engineering

Modal damage of structures during earthquake

H. Kuwamura
The University of Tokyo, Tokyo, Japan

ABSTRACT: Since structures are, in general, multi-degree-of-freedom systems, damage caused by a hit of earthquake is distributed to all deformable portions from little to large extent. At present, the method of predicting the damage distribution is owing to dynamic inelastic analysis in time domain. This paper presents a new method of predicting the location and intensity of damage by means of modal analysis technique with an application of energy concept in frequency domain.

1 INTRODUCTION

One of the most critical concerns in seismic design is to predict the location and intensity of damage in a structure hit by a severe earthquake. Structures are, in general, multi-degree-of-freedom systems, and thus the damage is distributed to all deformable elements from little to large extent. At present, the most reliable method for predicting the damage distribution is regarded dynamic inelastic analysis in time domain, in which, however, designers may feel difficulties in clarifying the influential factors relevant to the damage distribution. In order to be released from such black-box analyses, some empirical formulae were proposed to evaluate the damage distribution as found in the articles by Akiyama (1985). However, it is criticized that the application of such formulae is limited to particular earthquake motions employed in the studies for deriving the formulae (Kuwamura & Suzui 1994). This paper presents a new method for predicting damage distribution by means of modal analysis technique with an application of energy concept, from which the influential factors involved in structures and input motions relevant to damage distribution will be noticed.

2 SCOPE OF INVESTIGATION

The structures investigated in this paper are shear-type multi-story building frames, which are often simplified to lumped mass systems as shown in Figure 1. Each story of the frame is assumed to behave elastic perfectly plastic as shown in Figure 2. The system, which is herein assumed undamped, is defined by the number of mass n, mass m_i, elastic stiffness k_i, and yield

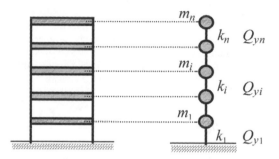

Figure 1. Shear-type multi-story building frames which can be simplified to lumped mass system.

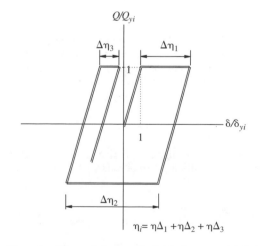

Figure 2. Hysteresis loops of individual story and definition of cumulative ductility.

1129

shear strength Q_{yi} of i-th story. It is well known about this type of structures that a slight change in yield shear strength of a particular story causes a significant change in damage distribution over the stories. Thus, this type of MDOF system is suitable for investigating the feasibility of a proposed method of damage prediction.

3 DEFINITION OF MODAL FACTORS

First, *modal shear force* τ_{ij} is defined as the i-th interstory shear force which causes j-th modal displacement of the MDOF system which remains elastic. The modal shear force is normalized as follows:

$$\frac{1}{2}\sum_{i=1}^{n}\frac{\tau_{ij}^{2}}{k_i}=e, \qquad j=1,2,\cdots,n \tag{1}$$

where $e = 1$. The modal shear force vector is represented by $\{\tau\}_j = \{\tau_{1j},\tau_{2j},\ldots,\tau_{nj}\}^{t}$.

Second, *modal energy* E_j is defined as the work done by j-th modal shear force until the end of earthquake input. Please note that damping is assumed zero in this study. It is known that E_j is related to the characteristics of the structural system and input earthquake with the following equation (Kuwamura, Kirino & Akiyama 1994):

$$E_j=\frac{1}{2}M\cdot\left(\beta_j\cdot\mathrm{smoothed}\left|F(T_j)\right|\right)^2 \tag{2}$$

in which M is the total mass of the structure, β_j is the modal participation factor normalized as follows:

$$\sum_{i=1}^{n}\beta_j^{2}=1$$

The smoothed$[F(T_j)]$ is the smoothed Fourier amplitude of the input accelerations at T_j, where T_j is the j-th natural period of the structure. E_j is the sum of E_{ij} of individual story as follows:

$$E_j=\sum_{i=1}^{n}E_{ij} \tag{3}$$

Third, *modal damage* η_{ij} is defined as the cumulative ductility of i-th story introduced by j-th modal energy as follows:

$$\eta_{ij}=\frac{E_{ij}}{Q_{yi}\delta_{yi}}=\frac{E_{ij}}{Q_{yi}^{2}/k_i} \tag{4}$$

Finally, damage of i-th story is given by the modal sum as follows:

$$\eta_i=\sum_{i=1}^{n}\eta_{ij} \tag{5}$$

The cumulative ductility η_i, which is used as damage index in this study, is the accumulation of drift during elasto-plastic excursions normalized by the yield drift as illustrated in Figure 2.

4 THEORY OF MODAL DAMAGE

The interstory shear force Q_i of i-th story at any instance during dynamic motion can be represented by a linear combination of the modal shear forces as schematically illustrated in Figure 3 and as follows:

$$Q_i=\sum_{j=1}^{n}a_j\sqrt{\frac{E_j}{e}}\tau_{ij}, \qquad j=1,2,\cdots,n \tag{6}$$

in which a_j is called *modal coefficient*, and then the elastic energy sustained by j-th mode at any instance is given by

$$E_{ej}=\sum_{i=1}^{n}\frac{1}{2}\left(a_j\sqrt{\frac{E_j}{e}}\tau_{ij}\right)^{2}/k_i, \quad \therefore E_{ej}=a_j^{2}E_j \tag{7}$$

It is known that a_j takes approximately uniform random numbers in the range of $-1\leqslant a_j\leqslant 1$ when the system remains elastic (Kuwamura & Tamura 1998). An example is demonstrated in Figure 4 for the case of a two-mass system excited by El Centro Earthquake. It is observed that the set of (a_1, a_2) moves over the area of a 2×2 square almost uniformly with few over-runs.

The yield condition of i-th story is given by $Q_i=\pm Q_{yi}$, which is represented by

$$\sum_{j=1}^{n}a_j\sqrt{\frac{E_j}{e}}\tau_{ij}=\pm Q_{yi}, \quad j=1,2,\cdots,n \tag{8}$$

This yield equation is drawn in the coordinate system of modal coefficients, a simple example of which is demonstrated in Figure 5(a) for the case of a two-mass

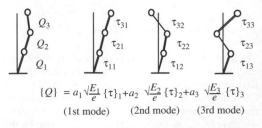

$$\{Q\}=a_1\sqrt{\frac{E_1}{e}}\{\tau\}_1+a_2\sqrt{\frac{E_2}{e}}\{\tau\}_2+a_3\sqrt{\frac{E_3}{e}}\{\tau\}_3$$

(1st mode)   (2nd mode)   (3rd mode)

Figure 3. Modal combination of interstory shear force.

system. The center white zone represents the elastic state of vibration, while the outer shaded zones beyond the yield limits. The shaded zones A_1^+ and A_1^- indicate the yielding of the first story in the plus and minus directions, respectively, while A_2^+ and A_2^- the second story in the same way.

Look at the route $O{\rightarrow}R{\rightarrow}S{\rightarrow}T{\rightarrow}S{\rightarrow}R{\rightarrow}O$, during which yielding occurs in a_1-direction or 1st mode. The energy is elevated as shown in Figure 5(b), which is in accordance with Equation 7. The energy accumulated in $O{\rightarrow}R{\rightarrow}S$ is released without damage in $S{\rightarrow}R{\rightarrow}O$ because of elastic loading and elastic unloading. The energy accumulated in $S{\rightarrow}T$ is not released in $T{\rightarrow}S$ and is stored in the structure as damage because of plastic loading and elastic unloading. Thus, the energy stored in the 1st story by the 1st mode vibration denoted by E_{p11} is given by integrating the energy over A_1^+ and A_1^-, which is the same each other, as follows:

$$E_{p11} = 2\int_{A_1^+} dE_{p11} = 2\int_{y_1}^{1}\left(1 - x^2\right)E_1 \cdot w_1 dy$$

$$\therefore E_{p11} = 2w_1 E_1 \int_{x_1}^{1}\left(1 - x^2\right)\left|\frac{dy}{dx}\right|_1 dx \qquad (9a)$$

where w_1 is the unit energy in a unit shift in y-direction, and $(dy/dx)_1$ is the slope in $a_1 - a_2$ coordinate system of the yield equation $Q_1 = Q_{y1}$ of the first story. The energy stored in 1st story by 2nd mode vibration, which is denoted by E_{p12}, is given by

$$E_{p12} = 2w_2 E_2 \int_{y_1}^{1}\left(1 - y^2\right)\left|\frac{dx}{dy}\right|_1 dy \qquad (9b)$$

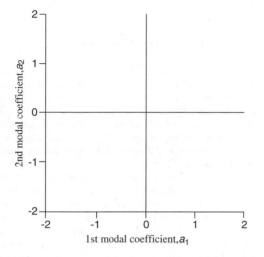

Figure 4. An example of the movement of modal coefficients.

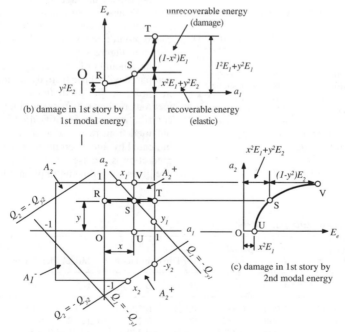

(b) damage in 1st story by 1st modal energy

(c) damage in 1st story by 2nd modal energy

(a) domain of modal coefficient and condition of yielding

Figure 5. Modal damage in terms of energy.

In the same way, the stored energy in 2nd story by 1st and 2nd modes is given by

$$E_{p21} = 2w_1 E_1 \int_{x_2}^{1} \left(1 - x^2\right) \left|\frac{dy}{dx}\right|_2 dx \qquad (9c)$$

$$E_{p22} = 2w_2 E_2 \int_{y_2}^{1} \left(1 - y^2\right) \left|\frac{dx}{dy}\right|_2 dy \qquad (9d)$$

Thus, the stored energy in i-th story by j-th modal vibration is generally represented by

$$E_{pij} = 2w_j E_j I_{ij} \qquad (10)$$

in which I_{ij} is the integral part of Equations 9a–9d and is called *modal moment*. Since the sum of E_{pij} over the all stories must be equal to the j-th modal energy E_j, then

$$\sum_{i=1}^{n} 2w_j E_j I_{ij} = E_j \qquad \therefore 2w_j = 1 / \sum_{i=1}^{n} I_{ij} \qquad (11)$$

Thus, Equation 10 is rewritten as

$$E_{pij} = E_j \bar{I}_{ij} \qquad (12)$$

Table 1. Structural model of 2-story building.

Structure	Stiffness distribution*	Strength distribution
St1	Soft 2nd story	Normal strength**
St2	Normal stiffness	Weak 2nd story***
St3	Normal stiffness	Normal 1 strength**
St4	Normal stiffness	Weak 1st story****
St5	Soft 1st story	Normal strength**

* See Table 2. $m_1 = m_2 = 1$, $M = m_1 + m_2 = 2$
** $Q_{y1} = 0.25\,Mg$, $Q_{y2} = 0.1825\,Mg$
*** $Q_{y1} = 0.25\,Mg$, $Q_{y2} = 0.1825\,Mg \times 0.8$
**** $Q_{y1} = 0.25\,Mg$, $Q_{y2} = 0.1825\,Mg/0.8$

where $\bar{I}_{ij}$ is normalized modal moment defined by

$$\bar{I}_{ij} = \frac{I_{ij}}{\sum_{i=1}^{n} I_{ij}} \qquad (13)$$

5 VERIFICATION OF THEORY

Damage distributions predicted by the proposed method of modal damage analysis are compared with those by dynamic response analysis in time domain. Five different types of two-story structures, denoted St1 to St5, are studied, where proportions of strength and stiffness in the 1st and 2nd stories are changed as listed in Table 1, and their modal properties are summarized in Table 2.

Three different types of earthquakes, denoted waveS, waveL, and waveE, are employed. First two are computer-simulated accelerograms with short predominant period (waveS) and long one (waveL). Third one is the El Centro earthquake accelerogram. Their Fourier amplitude spectra are shown in Figure 6, in which solid line is non-smoothed one and hollow and solid circles are smoothed ones with ductility factor 2.0 and 5.0, respectively. In the following calculation, the average of the two smoothed spectra is used.

The modal damage is calculated in accordance with the above-mentioned method. An example is shown in the process (1) to (4) in Table 3, and its associated coordinate system of modal coefficients is shown in Figure 7.

The amounts of predicted damage η_1 and η_2 are compared with those from dynamic response analysis in Figure 8. The predictions by the modal damage method are shown by bars, in which 1st and 2nd mode contributions are identified. The exact solutions obtained from response analysis in time domain are indicated by thick step lines. It is hard to say that the prediction is precise, but the method of modal damage gives the following important information:

Table 2. Model properties of structural model.

Structure	Stiffness k_1	k_2	Natural period T_1 sec	T_2 sec	Participation factor β_1	β_2	Modal shear force τ_{11}	τ_{21}	τ_{12}	τ_{22}
St1	206.7	51.67	1.000	0.382	0.851	0.526	10.69	8.65	17.30	−5.34
St2	103.4	103.4	1.000	0.382	0.973	0.230	12.23	7.56	7.56	−12.23
St3	103.4	103.4	1.000	0.382	0.973	0.230	12.23	7.56	7.56	−12.23
St4	103.4	103.4	1.000	0.382	0.973	0.230	12.23	7.56	7.56	−12.23
St5	84.2	336.8	1.000	0.234	0.998	0.062	12.54	6.66	3.33	−25.08

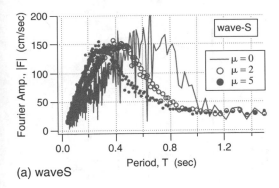

(a) waveS

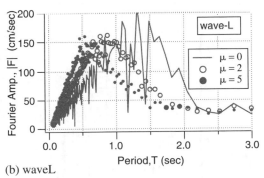

(b) waveL

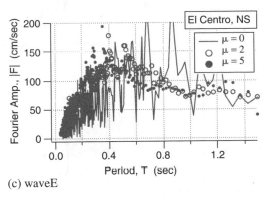

(c) waveE

Figure 6. Fourier amplitude spectra.

(2) Modal energy

j	T_j	β_j	$|F_j|$	$E_j = 1/2 \cdot M(\beta_j|F_j|)^2$
1	1.000	0.973	75	5325
2	0.382	0.230	125	827

(3) Yield condition and modal moment

i	$(E_1 \cdot \tau_{i1})a_1 + (E_2 \cdot \tau_{i2})a_2 = \pm Q_{yi}$	I_{i1}	I_{i2}	$\overline{I_{i1}}$	$\overline{I_{i2}}$
1	$892\,a_1 + 217\,a_2 = \pm 490$	0.890	0.901	0.530	0.649
2	$552\,a_1 - 352\,a_2 = \pm 358$	0.788	0.488	0.470	0.351
Sum		1.678	1.389	1.000	1.000

(4) Modal damage

i	E_{pi1}	E_{pi2}	E_{pi}	η_{i1}	η_{i2}	η_i
1	2822	537	3359	1.22	0.23	1.45
2	2503	290	2793	2.02	0.23	2.25
Sum	5325	827				

$$E_{pij} = E_j \times \overline{I_{ij}}, \quad \eta_{ij} = E_{pij}\backslash W_{yi}$$

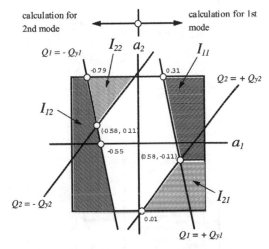

Figure 7. Modal moment in modal coordinate system.

1. Wave S with a short predominant period gives notable damage of 2nd mode, while wave L with a long predominant period gives slight damage of 2nd mode (e.g. (1) vs. (6) of Figure 8);
2. Different earthquakes give different damage distribution to the same structure (e.g. (3) vs. (8) vs. (13) of Figure 8);
3. Different stiffness proportion gives different damage distribution by the same earthquake (e.g. (1) vs. (3) vs. (5) of Figure 8); and
4. Damage is concentrated in weak story (e.g. (7) vs. (8) vs. (9) of Figure 8).

Table 3. Calculation procedure of modal damage for the combination case of St3 and waveE.

(1) Properties of structure

i	m_i	M_i	k_i	τ_{i1}	τ_{i2}	α_i^*	Q_{yi}^{**}	W_{yi}^{***}
1	1.0	2.0	103.4	12.23	7.56	0.25	490	2323
2	1.0	1.0	103.4	7.56	−12.23	0.365	358	1240

*yield shear coefficient
**$Q_{yi} = M_i g \alpha_i$, $g = 980$
***$W_{yi} = Q_{yi}^2/k_i$

Continued

1133

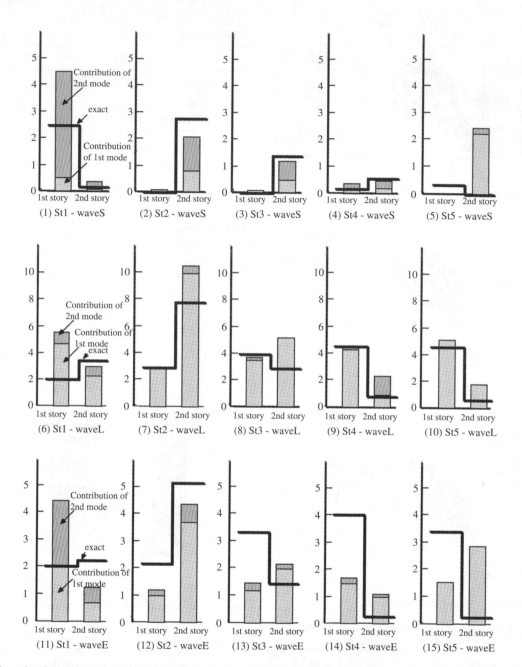

Figure 8. Comparison of damage estimation by modal analysis in frequency domain and response analysis in time domain.

6 CONCLUSIONS

A modal analysis method for predicting the distribution of damage in multi-story shear-type frames excited by earthquake is presented. The damage is given in terms of plastic ductility in individual stories. The method was compared with the numerical response analysis in time domain. It is hard to say that the method can precisely predict the damage distribution, but the method gives the useful information about the effects of higher modes of vibration and the spectral properties of input earthquake on the damage distribution.

REFERENCES

Akiyama, H. 1985. *Earthquake-Resistant Limit-State Design for Buildings*. Tokyo: Univ. of Tokyo Press.

Kuwamura, H. & Suzui, Y. 1994. Modal energy of elasto-plastic MDOF-systems excited by strong earthquake. *J. Struct. Constr. Eng. AIJ* 465: 71–79 (in Japanese).

Kuwamura, H., Kirino, Y. & Akiyama, H. 1994. Prediction of earthquake energy input from smoothed Fourier amplitude spectrum. *Earthq. Eng. Struc. Dyn.* 23: 1125–1137 (in Japanese).

Kuwamura, H. & Tamura, K. 1998. Modal damage of structures during earthquake. *J. Struct. Constr. Eng. AIJ* 507: 79–86 (in Japanese).

System-based Vision for Strategic and Creative Design, Bontempi (ed.)
© 2003 Swets & Zeitlinger, Lisse, ISBN 90 5809 599 1

Architectural design influence on seismic building behavior: evaluation elements

A. Gianni
Department of Structural Engineering, Politecnico di Milano, Milano, Italy

ABSTRACT: The seismic check of a multi-storey building, planned for a non seismic zone, is performed in the hypothesis of 1° level of seismicity. The choice of the building depends on its morphological and material characteristics of regularity, symmetry and compactness. In order to obtain a satisfying seismic behavior, some improvements are introduced to the structure; they result reduced owing to the shape characteristics of the building. The influence of architectural and structural design on the seismic behavior is evaluated on the basis of dynamic analysis criteria.

1 INTRODUCTION

The seismic strength of a building is usually checked on the exclusive basis of structural dynamics criteria; nevertheless it is well known that architectural design plays a major part in defining its seismic behavior (Naeim 1989); this is clearly shown by the history of damage, cracks and collapses of the buildings that must live together with the earthquake (Giuffrè & Carocci 1999); that is, the plan in seismic zone can not leave out of consideration either design criteria about architectural and structural shape, or the dynamic calculus of the structure; in other words architecture and structure must take shape side by side in the decision process of planning in order to reach a global organization of the building, the most effective possible to stand comparison with seismic forces.

The proposed work fits in the methodological context above defined; the aim is to evaluate the effects of morphological and material characteristics of a building on its seismic behavior, on the basis of classical dynamic analysis.

As an example a multi-storey building is considered, planned in non seismic zone, whose shape characteristics are fairly respectful of anti-seismic criteria provided by codes; the reading of the building is done trough shape parameters significative for its dynamic behavior; some improvements are carried out, in order to obtain a satisfying seismic response, that do not change in any way the morphology of the building; the influence of architectural design is evaluated on the seismic behavior trough dynamic analysis criteria.

This paper is derived from the degree thesis (Gianni 1999); supervisors were Prof. D. Benedetti and M.P. Limongelli, whose help was decisive.

2 SHAPE STRUCTURE RELATION IN SEISMIC ZONE

On the occasion of every seismic event, all the evidence makes it clear that most of damaged buildings are those suffering from construction faults before the earthquake, while the undamaged buildings are the *properly done* ones.

Properly done for a building means to be characterized by a correct and coherent organization of its structural elements and by the static effectiveness of each of them; the faults connected with the original establishment of the building, as well as its execution, initial or related to rehabilitation works, are the main causes of serious damage.

Properly done means to be constructed in conformity with rules of "good building" deep-rooted in the historical culture of the seismic site. The earthquake is in the memory of the site and teaches: constructive shapes, execution technologies and materials use evolved in time also referring to the advices that can be derived from recurring readings of damage, cracking and collapse mechanisms produced by seismic events. This explains why on the occasion of the last earthquakes in Italy, many buildings have been damaged, on which rehabilitation works had been carried out according to homologated and standard techniques of restoration, that is, ignoring the construction identity of the site.

Properly done means to be an organic system of components in which firmitas, utilitas, venustas are necessarily and harmonically fused in the resultant architecture (Quaroni 1995). That is, earthquake emphasizes strength values of architectural design in its whole meaning, as geometric form, material choice, construction features and execution quality.

As a conclusion, when earthquake occurs structural design and architectural design are requested to reach on the whole an effective building organization, respectful of the quality properly done. Therefore the question arises of defining optimal architectural shapes fitted to be resistant against uncommon, anomalous and exceptional stress, as the earthquake produces.

2.1 Architectural design and seismic behavior of buildings

On the basis of past experience and of structural dynamic criteria it's possible to define the main requirements that building design must satisfy in order to assure a good seismic behavior, and, over all, it's possible to individuate which characteristics must be avoided for constructions.

The set of design decisions that significantly are able to influence the seismic behavior of a building can be organized, as a rule, into three categories, that respectively regard:

– the shape of the building;
– the shape of structure elements;
– the characteristics of some non structural components that can cause loss of effectiveness and of functionality of the building.

It's clear that the distribution of strength inside a building depends in wide measure by its shape and the same can be said for the system of static ducts through which strengths are carried to ground; this aspect is emphasized in the seismic zone where the forces of inertia due to earthquake depend on mass distribution and therefore on: the ratio between the geometric dimensions, the size of the building, the internal system of floors and volumes. The question arises of defining shape parameters significative by this point of view, in particular, those that can induce deviations with respect to an optimum answer to horizontal forces. As an example three typical optimum constructive systems are showed in Figure 1, whose main characteristics are symmetry and regularity of plan and vertical sections.

As it regards the plan, symmetry and compactness are recommended; nevertheless some symmetric plans, in particular those with reentering angles, imply strength concentrations near angles and then dangerous behavior; the compactness, measured as ratio

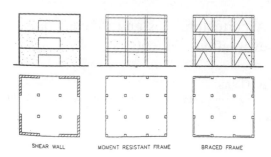

SHEAR WALL   MOMENT RESISTANT FRAME   BRACED FRAME

Figure 1.

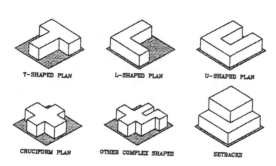

T-SHAPED PLAN   L-SHAPED PLAN   U-SHAPED PLAN

CRUCIFORM PLAN   OTHER COMPLEX SHAPES   SETBACKS

Figure 2.

between the dimensions of the plan, when inadequate, implies risky different dynamic behaviors along the two principal directions (Fig. 2).

As it regards vertical sections, irregularities as recessing or overhanging floors, produce strength concentrations and overturning moments. It must be observed that every kind of dangerous effects of shape irregularities is emphasized when building size increases.

Taking into consideration the shape of the structure significative parameters related to its seismic behavior are (Figs 3 and 5):

The structural density in plan, measured as the ratio between the area comprehensive of all the structural vertical elements in plan and the total area of the plan; constructions provided by considerable structural plan density can be considered seismically safe as ancient monuments show; as a classical example the Parthenon structural plan density is 20%.

The plan distribution of resistant elements; torsional effects are as weak as the distribution is regular and symmetric and this concerns above all shear walls and stiffening core; as a limit example, a circular layout characterized by lack of privileged inertia axes and shear walls outside, represents an optimum distribution (Figs 4 and 5). It must be observed that the stiffness of the perimeter also depends on its connection with non structural elements as enclosing walls,

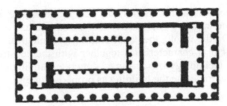

Figure 3.

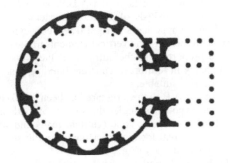

Figure 4.

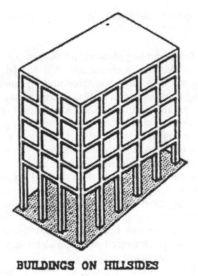

BUILDINGS ON HILLSIDES

Figure 6.

Structural plan density.	
Building, City, Date	Ground Level Structural Plan Density
1. St. Peter's, Rome, 1506-1626	25%
2. Temple of Khons, Karnak, 1198 B.C.	50
3. Parthenon, Athens, 447-432 B.C.	20
4. Santa Sophia, Istanbul, 532-537	20
5. Pantheon, Rome, 120-124	20
6. Sears Building, Chicago, 1974	2
7. typical contemporary steel high rise, 1975	.2
8. Monadnock Building, Chicago, 1889-1891	15
9. Chartres Cathedral, Chartres, 1194-1260	15
10. Taj Mahal, Agra, 1630-1653	50

Figure 5.

when they are stiff enough and connected with the structural elements trough sliding joints.

Dangerous irregularities concerning the vertical elements of the structure, are above all: differences of height of storeys, lack of enclosing walls, open storeys with pilotis structure, discontinuities in shear walls. Stiffness differences of columns belonging to the same storey, imply strength concentration in the more stiff ones and possible related brittle breakings (Fig. 6).

It must be observed that in general architecture and structure symmetry means that center of mass and center of stiffness coincides; asymmetry produces torsion and strength concentration.

2.2 Multi-storey buildings

In a multi-storey building horizontal and vertical elements can be organized and connected according to a number of possible framing systems. The dynamic behavior of the building mainly depends on the type of organization and connections in particular when, in presence of the earthquake, structural elements come out in stress–strain conditions of non linear nature, and therefore the building is required to work when damaged; that is, ductility characteristics of the structure assume a decisive role, as it is also provided by codes (Castellani et al. 1981). In this case the check of the structure behavior needs analyses of dynamic nature and then it can be of interest to study their influence on the architectural design. Nevertheless also in this case some general criteria are valid concerning those structural elements that, when earthquake occurs, are of great moment; among these:

– The stiffness characteristics of the floor, where the horizontal forces can be applied, the major part of the building mass being in them concentrated; floors in reinforced concrete, building cast, can be considered rigid, that is, working as stiff horizontal beam fitted to convey horizontal forces to the vertical elements provided for their transfer to ground.
– The bending characteristics of the framing system that mainly depends on the stiffness of the horizontal crosses; they, when quite stiff, induce a shear-type behavior of the structure, when characterized by negligible stiffness, imply a flectional bracket behavior.
– The shear walls location into the building that, when correctly placed, assure both considerable cross stiffness and torsional inertia to the framing system.

3 SEISMIC CHECK OF A MULTI-STOREY BUILDING

A residence building planned and recently built in the center of Milano is considered, that consist of six storeys out of ground and three under ground. The building is included in a wide intervention of rehabilitation, regarding an entire block of houses of nineteenth century, that provides the demolition of some decaying constructions and their replacement with the new building, with the respect of the morphology and the disposition of volumes demolished.

From the point of view of its shape characteristics (§ 2.1) the building is provided by adequate symmetry and regularity of the plan (Fig. 7) and presents some irregularities in height, due to the decrease of the areas of the last floors. The ratios between the dimensions are balanced and then the compactness level of the construction is good.

As it regards the structure, the building has been planned in order to assure the maximum flexibility of distribution of the inside spaces. Its structure consists in a beams-columns system in reinforced concrete building cast; beams are contained in the floor height; columns have different section depending on their location; brace walls coincide with the walls of the staircase and lift spaces.

The structural density in plan is sufficient (8%); the distribution of resistant elements in plan is not enough uniform depending on the brace walls location not quite symmetric. Significant irregularities are not found in the vertical elements of the structure.

On the whole, the constructive morphological and structural characteristics together with the constructive technology of reinforced concrete building cast, allow to define the building as properly done.

3.1 Seismic behavior of the building

The computing code used for the mechanical analysis of the structure belongs to the T.A.B.S. family (Three-dimensional Analysis of Building Systems) and is based on the hypothesis of rigid floors conveying the horizontal forces to the vertical elements of the structure in proportion to the stiffness of these last; the floors being in reinforced concrete building cast, the hypothesis is suitable.

With E.T.A.B.S. the structure has been checked in two load hypothesis: under the system of loads provided in non-seismic zone (vertical loads and wind forces in the real location in Milano) and in seismic zone of 1st category (Fig. 8).

In the first hypothesis the structure performs well within the admissible stresses.

When subjected to the static system of horizontal forces equivalent to the provided design spectrum of 1st category, the structure shows some intrinsic defectiveness and precisely: some beams (19, 20, 52, 53), that are joined on the staircase walls, show high values of bending moment and short values of ductility

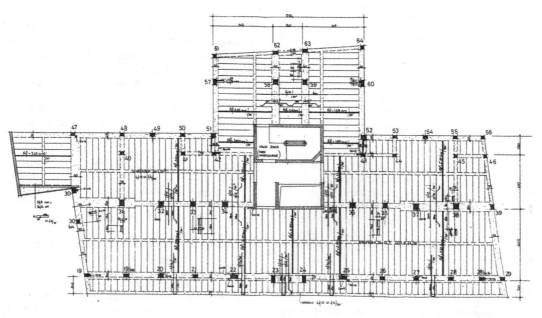

Figure 7.

(Table 1a); two columns (14, 16) show high values of axial stresses and short values of ductility; the building in its whole tends to torsion also when subjected to vertical loads only (Table 3a). The modal analysis points out relevant torsional modes, insufficiently

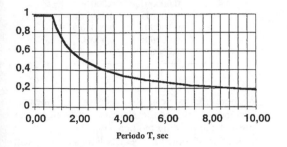

Figure 8.

Table 1a. Real case: beams ductility.

n beams	Section	Length	Ductility section	Ductility element
19	26 × 40	230	6.5689	3.3797
20	26 × 40	230	6.5689	3.3797
52	26 × 40	350	10.0392	4.0247
53	26 × 40	350	6.5689	2.0635

Table 1b. Modified case: beams ductility.

n beams	Section	Length	Ductility section	Ductility element
19	30 × 50	230	13.1713	9.2242
20	30 × 50	230	13.1713	9.2242
52	30 × 50	350	13.1713	7.1939
53	30 × 50	350	13.1713	7.1939

Table 2a. Real case: columns ductility.

n columns	Section	Highness	Ductility section	Ductility element
16	30 × 30	310	5.8600	2.9024
14	30 × 30	310	6.9842	3.3424

Table 2b. Modified case: columns ductility.

n columns	Section	Highness	Ductility section	Ductility element
16	30 × 50	310	13.0911	7.6692
14	30 × 50	310	15.2209	8.8439

decupled, as it can be deduced by the percentage of sharing masses.

When the spectrum of 1st seismic category is applied (in both the directions x and y), the earthquake exploits the intrinsic defectiveness of the building; in particular it is in evidence that the vertical elements of the structure far from the stiffness center are required to develop a relevant capability of following the torsional displacements of floors, with a consequent high request of ductility.

3.2 Anti-seismic changes of the structure

In order to improve the seismic behavior of the building, two ways are followed: to reduce the strain engagement required and the related required ductility, and to limit the torsional modes. With this aim:

– four shear walls, high as the building, are introduced (Fig. 7) in both the directions x and y, in order to increase torsional stiffness;
– with the purpose of improving the cooperation between beams and the new shear walls, the sections of the beams in them inserted are increased;

Table 3a. Real case: vibration modes.

Mode	T	M_x	M_y	M_z
1	0.79076	43.62%	0%	1.99%
2	0.64547	3.02%	21.06%	13.92%
3	0.57230	0.57%	28.69%	18.11%
4	0.28741	0.04%	6.4%	13.43%
5	0.16822	17.6%	0.05%	0.09%
6	0.14377	0.05%	17.31%	3.75%
7	0.10848	0.26%	5.99%	6.42%
8	0.08855	0%	1.51%	15.56%
9	0.7932	9.35%	1.88%	2.29%
10	0.07677	5.9%	6.98%	1.9%
Total		80.4	90.4	77.5

Table 3b. Modified case: vibration modes.

Mode	T	M_x	M_y	M_z
1	0.72624	36.56%	0.66%	8.11%
2	0.57543	1.21%	51.94%	0.12%
3	0.51078	10.56%	1.12%	31.47%
4	0.21288	2.25%	4.7%	10.05%
5	0.14605	13.15%	7.1%	0.26%
6	0.13637	5.2%	11.37%	4.1%
7	0.09598	0.64%	5.52%	15.57%
8	0.07528	1.3%	7.37%	6.61%
9	0.06999	15.53%	0%	3.98%
10	0.06413	0.04%	1.38%	2.75%
Total		86.4	91.2	83

Table 4a. Inter-floor displacements and internal forces in column 66.

Floor	Dspl. x	T_{max}	T_{min}	M_{max}	M_{min}
6 real	0.00099	0.99	10.07	5.02	16.41
6 mod.	0.00064	2.85	0.33	6.9	0.52
5 real	0.00091	0.70	7.58	3.09	11.18
5 mod.	0.00058	1.96	0.13	4.03	0.21
4 real	0.00079	0.69	7.62	3.52	11.49
4 mod.	0.00051	1.95	0.18	4.39	0.31
3 real	0.00044	1.63	7.07	3.73	10.34
3 mod.	0.00034	2.19	0.14	4.06	0.24
2 real	0.0004	1.26	6.73	3.29	10.02
2 mod.	0.00032	1.85	0.16	3.68	0.27
1 real	0.00034	1.37	6.18	3.1	9.16
1 mod.	0.00025	1.71	0.15	3.21	0.25
0 real	0.00024	1.31	5.54	2.18	8.22
0 mod.	0.00018	1.42	0.1	2.64	0.24

Table 4b. Real case. Displacement column 66.

Floor	Dspl. x	Dspl. y
6	0.00417	0.0002
5	0.00318	0.00012
4	0.00227	0.00006
3	0.00148	0.000025
2	0.00104	0.00001
1	0.00064	0.0000019
0	0.0003	0.000003

Table 4c. Modified case. Displacement column 66.

Floor	Dspl. x	Dspl. y
6	0.00287	0.00008
5	0.00223	0.00011
4	0.00165	0.00009
3	0.00114	0.00008
2	0.0008	0.00007
1	0.00048	0.00004
0	0.00023	0.00002

Table 5a. Real case. Torsion under vertical loads column 66.

Floor	Rotn. x	Rotn. y
6	0.00059	0.00028
5	0.00039	0.00018
4	0.00036	0.00018
3	0.00035	0.00012
2	0.00034	0.00009
1	0.00031	0.00009
0	0.00028	0.00009

Table 5b. Modified case. Torsion under vertical loads column 66.

Floor	Rotn. x	Rotn. y
6	0.00002	0.00030
5	0.00002	0.00015
4	0.000001	0.00016
3	0.00001	0.00012
2	0.000007	0.00010
1	0.000005	0.00010
0	0.000002	0.00008

– the more stressed beam (53) is lay out with the corresponding wall of the staircase, reducing shear stresses and bending moment (Fig. 7);
– the section of some peripheral columns is increased;
– in the modified elements the reinforced is thickened.

It must be observed that the changes introduced are of little account, their cost has been evaluated as the 3–5% of the total cost, and do not imply any shape alteration for the building. The shear walls are contained into perimetric walls; the increase of section of beams and columns is always contained in staircase and perimetric walls; the beam displaced is contained into the floor thickness.

The analysis of the seismic behavior of the improved building shows that: the local ductility increases in the modified beams and columns and consequently in the whole structure (Table 2b); from the modal analysis it is deduced that the sharing masses for every mode are consistent only in one direction at a time, that is, flectional and torsional modes are decupled (Table 3b); the more uniform distribution of stiffness brings near center of mass and center of stiffness and reduces the predisposition to torsion. After all a significative decrease of relative displacements among the floors is obtained (Tables 4a and 4b).

As a conclusion, the dynamic analysis points out that the criteria informing the design of the building, as it regards the architectural shape and the structural configuration, have a not negligible role in order to predispose the structure to a good seismic behavior; the effectiveness of the improvements introduced largely depends on this.

4 CONCLUSIONS

From the analysis of the structure displacement it can be observed that the improvements introduce a significant reduction of the relative displacements between a contiguous floors until decreases of 50%. Also horizontal relative displacements decrease until 30–50% and as a consequence also the required ductility will decrease and this in presence of disposable ductility unchanged when not increased (Tables 4b and 4c). The regularity introduced in the stiffness distribution with the improvements produce considerable effects also when vertical loads only are applied to the building; the predisposition to torsion in the real case decreases too in the real case when compared with the modified one (Tables 5a and 5b).

REFERENCES

Naeim, F. 1989. *The seismic design*, Van Nostrand Reinhold, New York.
Giuffrè, A. & Carocci, C., Ed. 1999. *Codice di pratica per la sicurezza e la conservazione del centro storico di Palermo*, Laterza.
Caffi, V., Pontina, D., Benedetti, D. & Gentile, C. 1991. Progetto di un edificio a torre in zona a diverso grado di sismicità, *Degree Thesis*, Politecnico di Milano.
Quaroni, L., Ed. 1995. *Progettare un edificio*, Gangemi.
Castellani, A. Benedetti, D., Castoldi, A., Faccioli, E., Grandori, G. & Nova R., Ed. 1981. *Costruzioni in zona sismica*, Masson.
Pizzetti, G. & Zorgno Triscinoglio A.M., Ed. 1979. *Principi statici e forme strutturali*, UTET.
Torroja, E. *Rason y ser de los tipos estructurales*.
Gianni, A., Benedetti, D. & Limongelli, M.P. 1999. L'influenza della verifica sismica sul progetto architettonico, *Degree Thesis*, Politecnico di Milano, 1999.

System-based Vision for Strategic and Creative Design, Bontempi (ed.)
© 2003 Swets & Zeitlinger, Lisse, ISBN 90 5809 599 1

Seismic retrofit of a historic arch bridge

S. Morcos & A. Hamza
HDR Engineering, Inc., USA

J. Koo
City of Los Angeles, USA

ABSTRACT: The seismic vulnerability of a historic multiple arch concrete bridge has been investigated under the Maximum Credible Earthquake (MCE) event with a magnitude of 7.0 and peak bedrock acceleration of 0.75 g. The bridge is located within 3 km from the Elysian Park Blind Reverse/Thrust fault. Built in 1928, Fourth Street Bridge located in Los Angeles, California, has been declared as a City of Los Angeles historic monument and is eligible for listing in the National Register of Historic Places. It consists of three main concrete arch spans with spandrel columns and three concrete T-beam end spans. The bridge carries two lanes of traffic in each direction, sidewalks, decorative railing, and decorative light posts. The bridge was retrofitted in 1994 for an acceleration of 0.6 g. In the current study, a seismic finite element analysis was performed to investigate the adequacy of the bridge structural members under the seismic loads induced by the higher acceleration of 0.75 g. The results of the seismic investigation indicated that the bridge structural members do not have adequate capacity to carry the seismic force demands induced by the higher acceleration and will collapse under the MCE event. A retrofit strategy was developed in which the main load carrying members were strengthened while other members were replaced in-kind. The retrofit strategy has minimum impact on the historic character defining features of the bridge.

1   INTRODUCTION

Bridge structures built in the early 1900s were not designed or detailed to resist large lateral seismic forces or accommodate large displacements induced by earthquake ground motion.

Stability of a bridge structure in seismically active zones is dependent on the ability of its structural elements to resist the seismic induced forces and accommodate the large displacements of the superstructure without sustaining severe damage or collapse. Recent earthquakes, particularly the 1994 Northridge earthquake in California have caused severe damage and collapse of several concrete bridges that were not designed or detailed to resist seismic forces. The seismic vulnerability of Fourth Street Bridge has been investigated and a retrofit strategy has been developed in this study.

Constructed in 1928, Fourth Street Bridge shown in Figure 1, has been declared a City of Los Angeles historic monument. It is also eligible for listing in the National Register of Historic Places. The bridge was retrofitted in 1994 to prevent its collapse during a moderate earthquake event. The bridge was reevaluated after recent earthquakes, as part of a seismic screening study of bridges conducted in 2000 and was determined

Figure 1.   Fourth Street Bridge over Lorena Street.

to be vulnerable to collapse under the Maximum Credible Earthquake (MCE) event.

The objective of this investigation is to conduct a seismic analysis of the bridge and develop a retrofit strategy to prevent the structure from collapsing or sustaining severe damage under the large seismic forces and displacements induced by the higher acceleration

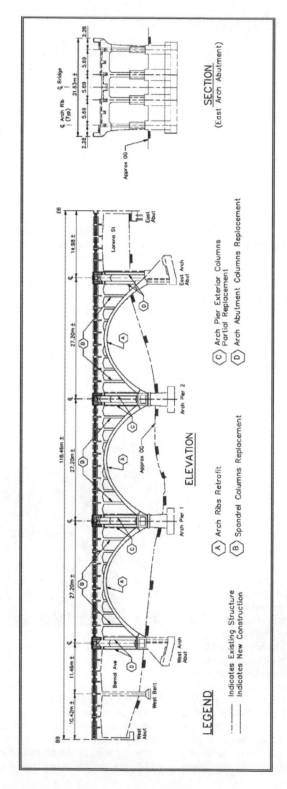

Figure 2. Fourth Street Bridge – retrofit srategy.

of 0.75 g, which is based on the MCE event. In order to preserve the historic structure, the selected retrofit strategy must have minimum impact on the historic character defining features of the bridge.

2 BRIDGE DESCRIPTION

Fourth Street Bridge consists of three main arch spans, two west approach spans, and one east approach span, as shown in Figure 2. The arch spans are supported on two middle Arch Piers and two end Arch Abutments. The bridge has straight alignment and is approximately 118.5 m long and 21.6 m wide. Each arch span consists of four arch ribs connected transversely by concrete struts. Each arch rib has six spandrel columns supporting the deck floor beams. The spandrel columns are spaced at 2.75 m. The arch ribs are fixed at their bases.

The west approach is a two span 21.87 m long frame supported on a four-column bent and spread footings. The height of the columns is approximately 9 m. The east approach span is 15 m long, simply supported at the East Abutment at one end and four columns at the East Arch Abutment, at the other end. The superstructure approach spans are cast in place concrete T-beams. The East and West Abutments are high cantilever type, approximately 7.3 m high with 0.45 m seat width. The West Abutment is supported on a spread footing, while the East Abutment is supported on concrete piles.

All bridge elements above the ground are considered historic especially the character defining features of the bridge, except the concrete deck. These features are: the open arch spandrels, arched window railing, pylons, decorative light posts, and the color and surface texture of the cast-in-place concrete. The surface texture, known as the board finish, resulted from using the old wood forms. Any proposed retrofit should have minimum impact on the above character defining features of the bridge and must confirm to the Secretary of Interior's Standards for Historic Preservation Projects.

3 PREVIOUS SEISMIC RETROFIT

In the early 1990s, the bridge was seismically investigated for a moderate earthquake event with an acceleration of 0.6 g. Based on that investigation, the bridge was retrofitted in 1994 by in-kind replacement of the eight interior columns at the Arch Piers and installing longitudinal cable restrainers at the East and West Arch Abutment expansion joints.

4 MATERIALS AND SECTION PROPERTIES

Concrete cores were taken from the columns, arch ribs, spandrel columns, abutment wall, and deck floor beams.

The average compressive strength of the cores was found to be 35 MPa. The yield strength of the reinforcing steel was assumed to be 227 MPa (Caltrans 2001). Cracked section properties were used for the columns, spandrel columns, and arch ribs, while gross section properties were used for the slab and the concrete T-beams.

5 SEISMIC PARAMETERS, MODEL, AND ANALYSIS

In this seismic analysis, a horizontal transverse base motion spectral input was used, corresponding to 0.75 g peak bedrock acceleration with 5% damping as per Caltrans' ARS curve (spectral acceleration versus period) for earthquake magnitude of 7.25 ± 0.25 and soil Type D. The spectral acceleration was increased by 20% since the bridge is located within 3 km from the Elysian Park Blind Reverse/Thrust fault. Modal combination was performed using the first 100 mode shapes.

Vertical acceleration must be considered in the seismic analysis of arch bridges since these are non-standard bridges (Caltrans 2001). Three load cases were considered in the analysis. Load Case 1 consists of 100% of the longitudinal ground acceleration coupled with 30% of the transverse acceleration and 30% of the vertical acceleration. Load Case 2 consists of 30% longitudinal ground acceleration coupled with 100% transverse acceleration and 30% vertical acceleration. Load Case 3 consists of 30% longitudinal ground acceleration coupled with 30% transverse acceleration and 100% vertical acceleration. The subsurface geotechnical investigation (Weeraratne 2002) determined that the bridge site has a low liquefaction potential.

Rigorous three-dimensional static and seismic analyses of the bridge, including the existing upgrades, were performed. The bridge deck slab was modelled by shell elements and the deck beams were modelled by beam elements. Each arch rib, column, and spandrel column was modelled by beam elements with cracked section properties. The bridge model is shown in Figure 3. The column to footing connection at the Arch Piers and Arch Abutments was modelled as a fixed connection as detailed on the as-built drawings. The bent column to footing connection was modelled as a pin, since the footing size is very small and cannot carry the plastic moment capacity of the column.

Soil springs representing the soil structure interaction were used in the analysis. The stiffnesses of the soil springs were determined based on the geotechnical investigation (Weeraratne 2002). The abutment stiffnesses in the longitudinal direction were calculated based on the effective soil pressure behind the back walls (Caltrans 2001). The soil spring stiffness in the transverse direction models the sliding resistance of the pile cap and lateral capacity of the piles at the East

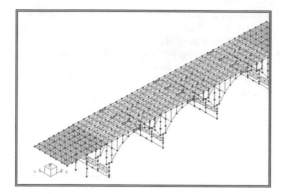

Figure 3.   Bridge model.

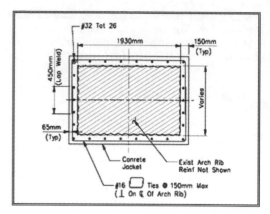

Figure 4.   Arch Rib cross-section.

Abutment. The soil spring stiffness in the transverse direction at the West Abutment models the sliding resistance of the abutment footing.

6   BRIDGE SEISMIC BEHAVIOR AND ANALYSIS RESULTS

In the existing arch span load path, the seismic forces due to the mass of the concrete deck slab transfer to the arch ribs through the spandrel columns, resulting in large seismic demands on the spandrel columns and arch ribs. Arch ribs are the main load carrying members of the span and therefore, their load carrying capacity must be maintained even if they sustain some local flexural damage. The arch ribs must remain essentially elastic (flexural demand/capacity ratio <2.0) during the maximum credible seismic event (Caltrans 1995). Spandrel columns and main columns may sustain significant damage without collapsing, i.e. behave inelastically, provided that they have sufficient ductility.

The analysis results indicated that the columns, spandrels, and arch ribs are susceptible to severe damage under the induced seismic forces. Longitudinal reinforcement is not provided in the sides of the arch ribs and therefore, the arch ribs have low flexural capacities. Lateral confining reinforcement ties are not provided in the arch ribs, and therefore the arch ribs have no ductility. The analysis indicated that the moment demand/capacity ratio for the arch ribs is 4.3, which is well above the elastic limit for non-ductile members, and therefore the arch ribs will fail. The main columns and spandrel columns moment demand/capacity ratios are found to be up to 7.4, which is well above the limit for unconfined non-ductile structural members, and therefore, the columns will fail. The shear demand/capacity ratios are up to 3.8 for the Arch Pier and Arch Abutment columns and 10.6 for the spandrel columns, which is well above the limit of 1.0. The lateral displacement demand/capacity ratio of the piles

at the East Abutment is 4.7, and therefore, the piles will fail.

The maximum displacements of the superstructure in the longitudinal and transverse directions were 142 mm and 292 mm, respectively. These displacements exceeded the ultimate displacement capacity of the columns and will cause failure. The first mode shape, having the lowest frequency, occurred in the transverse direction, while higher mode shapes were in the longitudinal direction, indicating that the structure is much stiffer in the longitudinal direction than the transverse. This behavior is expected in bridges with arch ribs, since the ribs are much more flexible in the transverse direction than the longitudinal direction. The structure's period was found to be 0.75 seconds.

It is concluded from the as-built seismic analysis that the main structural members of the bridge do not have adequate strength to resist the seismic forces and displacement demands and that the bridge will collapse under the MCE event. Further seismic retrofitting is required to prevent the collapse of the bridge.

7   RETROFIT STRATEGY

Retrofit strategies were evaluated (Morcos et al. 2002) based on preserving the historic character defining features of the bridge, aesthetics, construction cost, environmental impacts, and contractibility. The proposed retrofit strategy consists of measures to strengthen the main load carrying members, such as the arch ribs, and replace in-kind other structural members, such as the spandrels and columns. The retrofit strategy is shown in Figure 2 and detailed as follows:

(a) Arch Rib Strengthening: A 200 mm concrete jacket will be constructed around the entire length of all arch ribs. Longitudinal reinforcing steel and lateral ties will also be provided, as detailed in Figure 4.

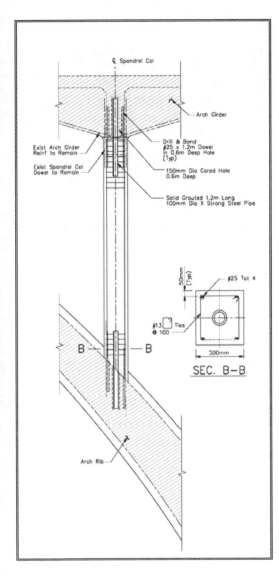

Figure 5.   Spandrel column replacement.

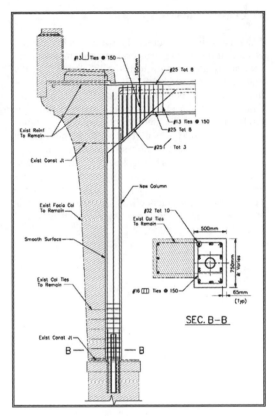

Figure 6.   Arch Pier exterior column partial replacement.

The longitudinal steel will be added by drilling holes in the Arch Piers and Arch Abutments and bonding the reinforcing steel. The provided U-shaped lateral ties around the arch ribs will be lap welded in order to provide the necessary cross-section confinement. The added longitudinal reinforcement will increase the flexural capacity of the arch ribs while the added confining lateral ties will increase the ductility of the ribs.

(b) Spandrel replacement: The spandrel columns will be replaced as detailed in Figure 5. The new spandrel columns will have identical dimensions as the existing ones, but will have much higher moment, shear, and displacement capacities.

(c) Arch Pier Exterior Column Partial Replacement: The exterior columns at the arch piers consist of two parts: A decorative facia part carrying the sidewalk and railing and a main rectangular part carrying the deck floor beams, as shown in Figure 6. The facia part is lightly reinforced and will not be replaced, while the rectangular part will be removed and replaced with a new column. The new column will have identical appearance as the old columns but will have higher moment, shear, and displacement capacities. A portion of the deck and floor beams at the new column locations will also be replaced.

(d) Arch Abutment Column Replacement: The East and West Arch Abutment columns will be replaced in-kind. The new columns will have identical appearance as the old columns but will have higher moment, shear, and displacement capacities. A portion of the deck and floor beams at the new column locations will also be replaced.

(e) Concrete finished surface: A board concrete surface finish will be applied to all the new concrete in order to match the historic concrete texture.

1149

8 CONCLUSIONS

The seismic investigation of Fourth Street Bridge indicated that the bridge structural main members do not have adequate strength to resist the forces and displacements induced by the MCE event, with a magnitude of 7.0 and peak bedrock acceleration of 0.75 g. Failure will occur in the arch ribs, spandrel columns, Arch Pier columns, and Arch Abutment columns. A retrofit strategy was developed, in which the main load carrying members were strengthened or replaced with ductile members having larger moment and shear capacities, while having the same exterior dimensions. The retrofit strategy will have very minimum impact on the historic character defining features of the bridge and also very minimum visual impact on the aesthetics of the bridge.

REFERENCES

Caltrans, 1995. Concrete Arch Bridge Seismic Retrofit Guidelines. Sacramento: California Department of Transportation.
Caltrans, 2001. Seismic Design Criteria. Sacramento: California Department of Transportation.
Caltrans, 2001. Bridge Design Specifications. Sacramento: California Department of Transportation.
Morcos, S., Hamza, A. & Koo, J. 2002. Seismic Retrofit of A Historic Concrete Arch Bridge. First Annual Concrete Bridge Conference Proceedings. Nashville: Tennessee.
Priestly, M.J.N. & Sieble, F. 1991. Seismic Assessment And Retrofit of Bridges. Report No. SSRP – 91/03, University of California, San Diego: California.
Weeraratne, S. 2002. Geotechnical Investigation – Seismic Analysis and Retrofit of Fourth Street Bridge Over Lorena Street. Diaz-Yourman & Associates. Santa Anna: California.

System-based Vision for Strategic and Creative Design, Bontempi (ed.)
© *2003 Swets & Zeitlinger, Lisse, ISBN 90 5809 599 1*

Effects of taking into consideration realistic force-velocity relationship of viscous dampers in the optimisation of damper systems in shear-type structures

T. Trombetti, S. Silvestri & C. Ceccoli
Department D.I.S.T.A.R.T., Faculty of Civil Engineering, University of Bologna, Italy

G. Greco
Civil Engineer in Cesena, Italy

ABSTRACT: In previous works, the authors have investigated the advantages offered by inserting added viscous dampers into shear-type structures and have identified a special damper scheme based upon the mass proportional damping component of Rayleigh damping matrices (referred to as MPD system) which is capable of providing good overall dissipative performances. All these analyses were performed using a linear force-velocity relationship to model the constitutive law of the damping devices. This paper investigates how the use of a more accurate non-linear modelling of the dampers affects those findings. The numerical simulations here illustrated (performed with reference to a 6-storey shear-type structure subjected to a series of 40 historical records of earthquake ground motions) confirm the effectiveness and the robustness of MPD systems with respect to other damping systems. The results also point out how special attention must be paid in the identification of the correct relationship existing between the linear modelling of viscous dampers and their non-linear actual behavior, as this may lead to large over- or under-estimations of their dissipative effectiveness.

1   INTRODUCTION

Dissipative systems have widely proven (Hart & Wong 2000, http://nisee.berkeley.edu/) to be able to effectively mitigate seismic effects on buildings. Nevertheless, as regards the introduction of dissipative elements into shear-type structures, published literature has yet to provide an exhaustive solution to the problem of optimising damping system efficiency.

In previous works by the authors (Trombetti et al. 2001, 2002a, b, c, d) and other researchers (Takewaki 1997 and 2000, Singh & Moreschi 2001 and 2002), different criteria and methodologies (based upon the system response to a stochastic base input and/or the modal damping ratios) were used to identify the systems of added viscous dampers which provide shear-type structures with overall good dissipative performances.

In general, these research works were carried out:

– under the "equal total cost" constraint. This constraint requires that the sum, c_{tot}, of the damping coefficients, c_j, of all n dampers introduced into the structure be equal to a set value $\bar{c}$:

$$c_{tot} = \sum_{j=1}^{n} c_j = \bar{c} \qquad (1)$$

– using a linear force-velocity relationship to model the constitutive law of the damper of the type:

$$F = c \cdot v \qquad (2a)$$

where F represents the damping force, c the damping coefficient and v the velocity developed between the two ends of the device.

The dissipative properties of commercial manufactured viscous damper are, however, more appropriately captured by the following non-linear force-velocity relationship:

$$F = c \cdot v^{\alpha} \qquad (2b)$$

with α generally <1.

This paper investigates how the numerically evaluated system (structure + dampers) responses are affected by the use of the two dampers modelling of above.

2   DAMPER PLACEMENT AND DAMPER SIZING

In previous works (Trombetti et al. 2002a, b, c, d, Silvestri et al. 2003), the authors have shown that the

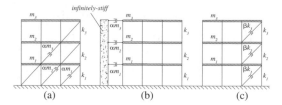

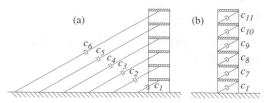

Figure 1.   (a,b) MPD system and (c) SPD system for a 3-storey shear-type structure.

Figure 2.   (a) FP and (b) IS placement for a 6-storey shear-type structure.

two components of the Rayleigh damping matrix, here referred to as the mass proportional damping (MPD) and the stiffness proportional damping (SPD) components, correspond to two damping systems which are physically separated and independently implementable: "*the MPD and the SPD systems*". Furthermore, it has also been shown that these two systems are characterised by specific damper placement and sizing:

– *MPD system:* the dampers are placed in such a way as to connect each storey to a fixed point (ground or infinitely-stiff vertical lateral-resisting element, as illustrated in Figures 1a, b for a 3-storey structure, respectively) and sized so that each damping coefficient c_j is proportional to the corresponding storey mass m_j;

– *SPD system:* the dampers are placed in such a way as to connect two adjacent storeys (Fig. 1c) and sized so that each damping coefficient c_j is proportional to the lateral stiffness k_j of the vertical elements connecting these two storeys.

Note that the damper placement of the SPD system is that typically adopted for dissipative bracing.

Given the above considerations, it is possible to define the following four categories of damper placement and damper sizing:

1. "*Fixed Point placement*" (FP placement): dampers connect each storey to the ground or to a fixed point as represented in Figure 2a. For this type of placement, dampers provide dissipative forces proportional to the storey velocities with respect to the ground. The MPD system is characterized by a FP placement.

2. "*Inter-Storey placement*" (IS placement): dampers are placed between consecutive storeys as represented in Figure 2b. For this type of placement, dampers provide dissipative forces proportional to the relative velocities between adjacent storeys. The SPD system is characterized by a IS placement.

3. "*Mass Proportional sizing*" (MP sizing): the damping coefficient c_j of each damper is proportional to the corresponding floor mass (mass of the floor at which the damper is connected to). The MPD system is characterized by a MP sizing.

Table 1.   Types of placement, sizing and damping.

Type of placement	Type of sizing	Type of damping
1. FP	MP	MPD
2. FP	Any type	Non classic
3. IS	SP	SPD
4. IS	Any type	Non classic

4. "*Stiffness Proportional sizing*" (SP sizing): the damping coefficient c_j of each damper is proportional to the stiffness of the vertical elements that connect the corresponding adjacent storeys. The SPD system is characterized by a SP sizing.

Table 1 summarizes the relationships existing between type of placement, type of sizing and type of damping.

Obviously, the four categories of damping systems described in Table 1 do not exhaust the possibilities of building a damping system; actually:

– as far as damper placement is concerned, it is possible to define also systems characterized by a mixed FP and IS placement and systems characterized by dampers connecting non-consecutive storeys;

– as far as damping sizing is concerned, obviously any sizing which is not proportional either to the floor masses or to the inter-storey stiffnesses does not fall within the previously defined categories.

3   SYSTEMS OF ADDED VISCOUS DAMPERS CONSIDERED

Without loss of generality, all numerical simulations here presented have been developed with reference to a 6-storey shear-type structure characterised by values of mass and lateral stiffness which do not vary along the building height ($m_j = m = 0.8 \cdot 10^5$ kg and $k_j = k = 4 \cdot 10^7$ N/m). This "reference" structure was selected for sake of comparison with results available in literature (Takewaki 1997).

In order to meaningfully compare the energy dissipation properties of various systems of added viscous dampers implementable in the "reference" structure, in all following analyses and for all damping systems considered (*i*) internal (intrinsic) damping is neglected, and (*ii*) the "equal total cost" constraint of the damping system is enforced. For sake of comparison with results available in literature (Takewaki 1997) $\bar{c}$ is set equal to $9 \cdot 10^6 \, \mathrm{N} \cdot \sec/\mathrm{m}$.

3.1 *Rayleigh damping systems*

As far as Rayleigh damping systems are concerned, in previous works the authors have numerically identified (Trombetti et al. 2002a, c, d) that the MPD system is capable of minimizes the storey mean square responses of a structure subjected to a white noise stochastic input. It has also been shown that the MPD system is capable of maximising the weighted average of modal damping ratios (Trombetti et al. 2002a, b, c). Table 2 gives the values of the damping coefficients c_j for the MPD system with reference to Figure 2. SPD system, on the other hand, leads to the mean square response (of a shear type structure subjected to a white noise stochastic input) being maximized (Trombetti et al. 2002a, c, d). Similarly, it has been shown (Trombetti et al. 2002a, b, c) that SPD systems minimize the the weighted average of modal damping ratios. Again, Table 2 gives the values of the damping coefficients c_j for the SPD system with reference to Figure 2.

3.2 *Dampers systems which optimise given performance indexes*

In previous works carried out by the authors (Trombetti et al. 2002a, Silvestri et al. 2003), use was made of genetic algorithms for the identification of the damping systems that minimise, under the "equal total cost" constraint, a number of performance indexes based upon the stochastic response of the system to a white noise excitation (top-storey mean square response, base shear standard deviation, average of the interstorey drift mean square responses, etc.). In these researches, the form of the damping matrix is left free, so that (*i*) classical and non-classical damping systems can be considered at once and (*ii*) no constraint upon the damper placement is imposed.

The genetically identified system (here referred to as "*GIO system*") which minimises the average of the

interstorey drift mean square responses has proven to globally provide optimal dissipative performances (Trombetti et al. 2003). Table 2 gives the values of the damping coefficients c_j for the GIO system with reference to Figure 2. Notice that the GIO system is characterised by FP placement only and may be seen as representative of the class of "optimal" damper sizing for given FP placement.

In recent research works (1997), Izuru Takewaki has identified, using an algorithm based upon an inverse problem approach proposed by the same author (1997), the damping system which minimises the sum of amplitudes of the transfer functions of interstorey drifts evaluated at the undamped fundamental frequency of the system. His investigations were carried out within the restricted class of dampers placed between adjacent storeys and therefore this system may be seen as representative of the class of "optimal" damper sizing for given IS placement. This system will be referred hereafter to as "*TAK system*". Table 2 gives the values of the damping coefficients c_j for the TAK system with reference to Figure 2.

3.3 *The selected systems of added viscous dampers*

Given the results summarized above, the following sections will present the seismic behaviour of the "reference" shear-type structure equipped with the following four damping systems:

– MPD system: MP sizing + FP placement (it represents the "optimal" damping system within the class of Rayleigh damping systems);
– SPD system: SP sizing + IS placement (it is characterised by the traditionally adopted interstorey damper placement);
– GIO system: optimised damper sizing for fixed FP placement;
– TAK system: optimised damper sizing for fixed IS placement.

4 THE MANUFACTURED VISCOUS DAMPERS

Viscous dampers are hydraulic devices that make use of the passage of viscous fluids through calibrated orifices to absorb energy (Infanti & Castellano 2001). Given the hydraulic nature of these devices, they are characterised by a highly non-linear behaviour.

Generally, the properties of these devices are characterised through the performance of a series of cyclic constant velocity tests.

Figures 3a, b provide an illustrative example of the results obtained with a single constant velocity test. Figure 4 plots the results of six constant velocity tests in terms of force-velocity points (sample points). The curve interpolating the points has, in most cases, the

Table 2. Damping coefficients of MPD, GIO, SPD and TAK, systems [$10^6 \, \mathrm{N} \cdot \sec/\mathrm{m}$].

	MPD	GIO		SPD	TAK
c1	1.50	0.45	c1	1.50	4.80
c2	1.50	1.35	c7	1.50	4.20
c3	1.50	1.80	c8	1.50	0
c4	1.50	1.80	c9	1.50	0
c5	1.50	1.80	c10	1.50	0
c6	1.50	1.80	c11	1.50	0

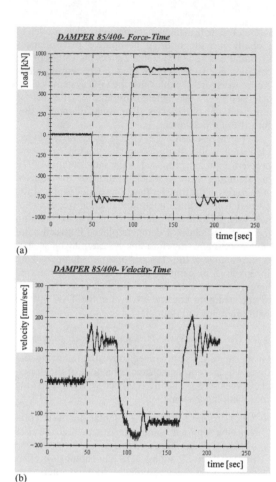

(a)

(b)

Figure 3. (a) Force-time and (b) velocity-time diagram for cyclic constant velocity tests.

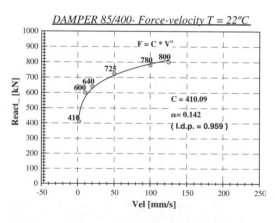

Figure 4. Force-velocity diagram for different test velocities.

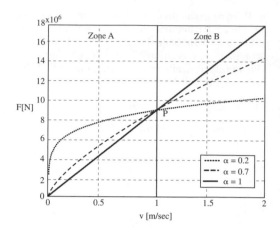

Figure 5. Force-velocity relationship for three α-values.

form $F = c \cdot v^{\alpha}$. Least square fitting the curve with sample points allows to experimentally estimate the values of c and α for any given damper. Notice that, in this case, c has dimensions:

$$\left[F(L/T)^{-\alpha} \right] \tag{3}$$

with F, L and T representing force, length and time, respectively.

In general, for manufactured viscous dampers, the experimentally determined values of the α-exponent are comprised between 0.15 and 1, such values depending on the specific damper type (i.e. on the application) and on the manufacturer.

As an example, among the products offered by Taylor Devices Company, dampers with α-values in the range between 0.4 and 0.5 are typically used for seismic applications, while dampers with α-values in the range between 0.5 and 1 are typically used for wind damping applications.

Figure 5 shows the force-velocity relationship for three values of the α-exponent. Notice that lower values of the α-exponent lead to:

– more rectangular hysteretic loops;
– the development of larger damping forces for small velocities (for this reason damping devices characterised by low α-values are often called "fast rise" dampers).

If $\alpha = 0$, the viscous damper degenerate into a device characterised by Coulomb damping (friction).

5 RESULTS OF TIME-HISTORY NUMERICAL SIMULATIONS

The dynamic behaviour of the reference shear-type building equipped with the selected damping systems

(SPD, TAK, MPD and GIO) has been extensively studied and this section presents selected results (in terms of maximum storey displacements and accelerations) as obtained for 40 historically recorded earthquake ground motions inputs.

Figures 6–9 present the results obtained by modelling dampers with $\alpha = 1$, $\alpha = 0.7$ and $\alpha = 0.2$, and keeping fixed the numerical value of $\bar{c}$ so that all force-velocity curves intersect in correspondence of $F = 9 \cdot 10^6 \,\mathrm{N}$ and $v = 1 \,\mathrm{m/sec}$ (point P in Fig. 5). Notice that an appropriate selection of the intersection point P is crucial for a correct relationship between the actual dissipative properties of manufactured dampers and the corresponding linear modelling. A wrong choice of the intersection point P (i.e. a not correct identification of the "effective" velocity range of the damper) may lead to over- or under-estimate the dissipative performances of the dampers, as clearly shown from analysis of Zone A and Zone B in Figure 5.

Figures 6a, b, c ,d show the average (over the 40 ground motions) values of maximum storey displacements for the SPD, TAK, MPD and GIO systems, respectively. Each graph shows the differences in the system displacement responses provided by numerical analyses carried out using the three different values of the α-exponent here considered. It can be seen that, in general (except for the upper storeys responses of the TAK system with $\alpha = 0.2$), small values of the α-exponent provide the structure with reduced storey displacements (this trend is more marked for MPD and GIO systems). In general, it can be observed that a non-linear analysis, developed with dampers characterised by $\alpha < 1$ and by an intersection (P) between the linear and the non-linear curve in correspondence of velocity values of about $1 \,\mathrm{m/sec}$, leads to drastically better results w.r.t. a linear analysis ($\alpha = 1$).

Figures 7a, b, c show the average values of maximum storey displacements for the SPD, TAK, MPD and GIO systems, as modelled with $\alpha = 1$, $\alpha = 0.7$ and $\alpha = 0.2$, respectively. Figure 7a briefly summarizes the results obtained for $\alpha = 1$ which indicates that, using a linear the linear modelling of the dampers, the MPD and the GIO systems provides the structure with the overall best dissipative performances. Figures 7b, c show that, using a non-linear modelling of the dampers ($\alpha < 1$), overall the relative dissipative properties the damping systems analysed remains the same: the MPD and the GIO systems (FP placement) are the damping systems which lead to the smallest storey displacements, whilst the SPD and TAK systems (IS placement) lead to the overall largest structural responses.

The main differences between the system responses obtained using the two different modelling are noticed for the TAK system. Using a linear modelling it always leads floor responses which are smaller to those of the SPD system. Using a non linear modelling it leads to floor displacements that are larger than those given by

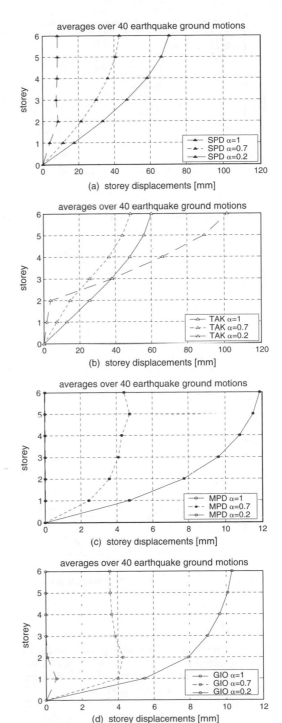

Figure 6. Averages of maximum storey displacements for $\alpha = 1$, 0.7 and 0.2 for (a) the SPD system, (b) the TAK system, (c) the MPD system and (d) the GIO system.

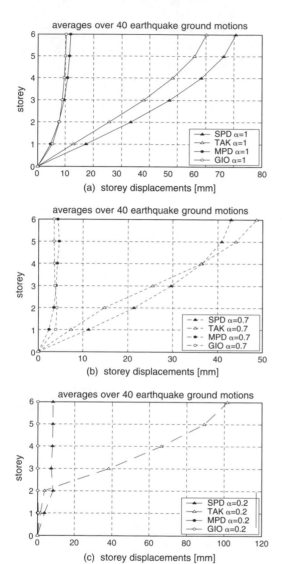

Figure 7. Averages of maximum storey displacements for the SPD, TAK, MPD and GIO systems for (a) $\alpha = 1$, (b) $\alpha = 0.7$ and (c) $\alpha = 0.2$.

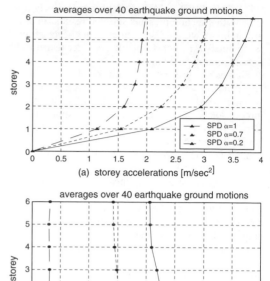

Figure 8. Averages of maximum storey accelerations for $\alpha = 1$, 0.7 and 0.2 for (a) the SPD system and (b) the MPD system.

those of the corresponding SPD system for both linear and non-linear modelling of the dampers.

6 CONCLUSIONS

This paper compares the dynamic behaviour of shear-type structures equipped with four different damping systems, as obtained using different (linear: $F = c \cdot v$ and non linear: $F = c \cdot v^{\alpha}$) modelling of the viscous dampers.

The results, obtained for a 6-storey shear-type structure subjected to 40 earthquake ground motion inputs, indicate that:

– by modelling dampers with $\alpha = 1$, $\alpha = 0.7$ and $\alpha = 0.2$, the best overall seismic performances are always offered by the damping systems characterised by a fixed-point damper placement (MPD and GIO systems), whilst interstorey placement leads to the largest structural responses (SPD and TAK systems);
– special care must be paid in the identification of the correct relationship existing between the linear modelling of viscous dampers and their non-linear

SPD systems from the 4th to the top and from the 2nd storey to the top, for $\alpha = 0.7$ and $\alpha = 0.2$, respectively.

Figures 8a, b show the average (over the 40 ground motions) values of maximum storey accelerations for the SPD and MPD systems, respectively (using the three α-values here considered). These figures confirm the general behaviour identified above: (*i*) smaller α-values leads to smaller system responses (acceleration) and (*ii*) the MPD system leads to system responses (accelerations) which are smaller than actual behavior, as this may lead to large over- or under-estimations of their dissipative effectiveness.

REFERENCES

Hart, G.C. & Wong, K. 2000. In John Wiley & Sons (eds), *Structural Dynamics for Structural Engineers*. New York. http://nisee.berkeley.edu/prosys/applications.html

Infanti, S. & Castellano, M.G. 2001. Viscous dampers: a testing investigation accordino to the HITEC protocol. *Fifth World Congress on Joints, Bearings and Seismic Systems for Concrete Structures, 7–11 October 2001 Rome*. Rome, Italy.

Singh, M.P. & Moreschi, L.M. 2001. Optimal seismic response control with dampers. *Earthquake Engineering and Structural Dynamics,* 30: 553–572.

Singh, M.P. & Moreschi, L.M. 2002. Optimal placement of dampers for passive response control. *Earthquake* Engineering and Structural Dynamics, 31: 955–976.

Takewaki, I. 1997. Optimal damper placement for minimum transfer functions. *Earthquake Engineering and Structural Dynamics*, 26: 1113–1124.

Takewaki, I. 2000. Optimal damper placement for critical excitation. *Probabilistic Engineering Mechanics*, 15: 317–325.

Trombetti T., Ceccoli C. & Silvestri, S. 2001. Mass proportional damping as a seismic design solution for an 18-storey concrete-core & steel-frame structure. *Proceedings of the Speciality Conference on "The Conceptual Approach to Structural Design", 29–30 August 2001, Singapore.*

Trombetti, T., Silvestri, S. & Ceccoli, C. 2002a. Added viscous dampers in shear-type structures: the effectiveness of mass proportional damping. *Technical Report N. T. n° 58. DISTART*. Università di Bologna, Italy.

Trombetti, T., Silvestri, S. & Ceccoli, C. 2002b. On the first model damping ratios of MPD and SPD systems. *Technical Report N. T. n° 64. DISTART*. Università di Bologna, Italy.

Trombetti, T., Silvestri, S. & Ceccoli, C. 2002c. Inserting Viscous Dampers in Shear-Type Structures: Analytical Formulation and Efficiency of MPD System. *Proceedings of the 2nd International Conference on Advances in Structural Engineering and Mechanics (ASEM'02), Pusan, Korea.*

Trombetti, T., Silvestri, S. & Ceccoli, C. 2002d. Inserting Viscous Dampers in Shear-Type Structures: Numerical Investigation of the MPD System Performances. *Proceedings of the 2nd International Conference on Advances in Structural Engineering and Mechanics (ASEM'02), Pusan, Korea.*

Silvestri, S., Trombetti, T., Ceccoli, C. & Greco, G. 2003. Seismic Effectiveness of Direct and Indirect Implementation of MPD Systems. *Proceedings of the 2nd International Structural Engineering and Construction Conference (ISEC-02), Rome, Italy.*

Rational estimation of the period of RC building frames having infill

K.M. Amanat
Department of Civil Engineering, BUET, Dhaka, Bangladesh

E. Hoque
Executive Engineer, Bangladesh Open University, Gazipur, Bangladesh

ABSTRACT: When the fundamental period of a building frame is determined by any rational analysis like modal eigenvalue analysis, design codes generally impose some limits on the calculated value of period if the period is longer than that predicted by empirical code equations. In this study, the fundamental periods of vibration of a series of regular RC framed buildings are studied using 3D FE modelling and modal eigenvalue analysis including the effects of infill. It has been found that the when the models do not include infill, the period given by the analysis is significantly longer than the period predicted by the code equations. However, when the effect of infill is included in the models, the time periods determined from eigenvalue analysis were remarkably close to those predicted by the code formulas. It is also observed that the randomness in the distribution of infill does not cause much variation the period if the total amount of infilled panels are same for all models. The findings of the study has shown us a practical way to determine fundamental period of RC frames using rational approaches like modal analysis and eliminates the necessity of imposing code limits.

1 INTRODUCTION

Building codes provide empirical formulas to estimate the fundamental period. These formulas are developed on the basis of observed periods of real buildings during ground motion and the period is generally expressed as a function of building height, type (frame or shear wall) etc. Building period predicted by these empirical equations are widely used in practice although it has been pointed out by many (Goel & Chopra 1997, 1998, Li et al. 1994, Smith & Crowe 1986) that there are scopes of further improvement in these equations. With the wide availability of high-speed personal computers it is now possible to develop a rigorous finite element (FE) model of a structure and determine its natural period by means of the *exact* eigenvalue analysis or by any rational method like Rayleigh's method. However, the period obtained by such rational method has been generally found to be significantly longer than the observed period of the buildings (Hossain 1997, Goel & Chopra 1997, 1998). For this reason, code specifications (BNBC 1993, UBC 1997, BSLJ 1987) generally put a limit on the period value if it is obtained by eigenvalue analysis of FE model. This, in fact, discourages the use of period obtained from computational modelling. Conventional FE modelling of reinforced concrete structures, which are widely used in strength

analysis and design, renders the structure too flexible than they actually are due to the fact that the effect of secondary components like the infills are not considered in the modelling. In reality, the additional stiffness contributed by these secondary components increases the overall stiffness of the building, which eventually leads to shorter time period as they are observed during earthquakes.

The objective of this paper is to investigate the natural period of vibrations of RC buildings by means of FE modelling under various conditions of geometric and other parameters including the effect of regular as well as randomly distributed infill. The diagonal strut model of infill proposed by Saneinejad & Hobbs (1995) and later enhanced by Madan et al. (1997) has been adopted in the present study. Three dimensional finite element modelling of a series of some idealized regular shaped building frames are analyzed and their time period are determined by means of modal (eigenvalue) analysis. These periods are then compared with the code formulas as well as periods obtained from analysis without considering infill. The study has been conducted under varying conditions of number of floors, floor heights, number of spans, amount of infilled panels etc. Infilled panels were distributed over the structure in regular as well as in random pattern. Comparisons of the periods obtained for different parametric conditions revealed the relative accuracy of the different methods as well as established the importance of incorporating the structural effect of infill in FE modelling to more accurately reflect the dynamic behavior of RC frames which is otherwise not possible to obtain in conventional frame modelling.

2 INFILL IN RC STRUCTURE

Infill of brick or stone masonry are frequently used in RC framed buildings. Although these are primarily intended to serve as partitions, their structural contribution in increasing the lateral stiffness of the frame is long recognized. There are several analytical models of infill available in the literature, which can be broadly categorized as (a) continuum models such as the models proposed by Lourenco et al. (1997) and Papia (1998) and (b) diagonal strut models such as the model proposed by Saneinejad & Hobbs (1995). For the type of work presented in this paper the diagonal strut model of Saneinejad & Hobbs (1995) has been found to be more suitable. This model has been successfully used by Madan et al. (1997) for static monotonic loading as well as quasi-static cyclic loading. They have also successfully verified the model by simulating experimental behavior of tested masonry infill frame subassemblage. A brief outline of the equivalent diagonal strut model is discussed below.

Considering the masonry frame of Figure 1, the maximum lateral force V_m and the corresponding displacement u_m in the infill masonry panel (Madan et al. 1997) are,

$$V_m^+(V_m^-) \leq A_d f_m' \cos\theta \leq \frac{vtl'}{(1-0.45\tan\theta)\cos\theta} \leq \frac{0.83tl'}{\cos\theta} \quad (1)$$

and $u_m^+(u_m^-) = \dfrac{\varepsilon_m' L_d}{\cos\theta}$ (2)

in which t = thickness of the infill panel; l' = lateral dimension of the infill panel; f_m' = masonry prism strength; ε' = corresponding strain; θ = inclination of the diagonal strut; v = basic shear strength of masonry; and A_d and L_d = area and length of the equivalent diagonal struts respectively. These quantities can be estimated using the formulations of the "equivalent strut model" proposed by Saneinejad and Hobbs (1995). The initial stiffness K_0 of the infill masonry panel may be estimated using the following formula (Madan et al. 1997),

$$K_0 = 2(V_m/u_m) \quad (3)$$

The parameters V_m, u_m, K_0, K_1 etc. are clearly shown in Figure 1. The degradation of strut stiffness from K_0 to K_1 was assumed to be a bilinear curve by Madan et al. (1997). A more rational degradation path would be a smooth curve shown by the heavy solid line in Figure 1. In this paper the form of the curve is suggested as given below,

$$V = \frac{(K_0 - K_1)u}{\left[1 + \left\{\dfrac{K_0 - K_1}{V_0}u\right\}^2\right]^{\frac{1}{2}}} + K_1 u \quad (4)$$

where,

$$V_0 = V_y \frac{K_0 - K_1}{K_0} \quad (5)$$

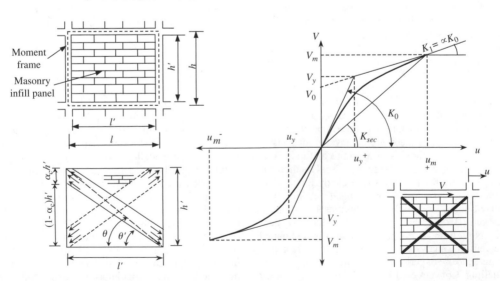

Figure 1. Strength envelope for masonry infill panel (Madan et al. 1997).

1160

3 COMPUTATIONAL MODELLING

In this study common two noded frame elements having six degrees of freedom per node has been used for the columns. For beams, similar elements with node offset capabilities has been used to model the web of T-beams (monolithic beam and slab). The floor slab has been modelled using common four noded plate elements. Point mass elements are used to represent the non-structural dead load like floor finish, partition walls etc. The infills are modelled as diagonal struts

Table 1. Properties of the reference RC frame.

Sl. No.	Parameters	Reference values
1	Modulus of elasticity of concrete	$2 \times 10^4 \, \text{N/mm}^2$
2	Density of concrete	$2.4 \times 10^{-9} \, \text{ton/mm}^3$
3	Size of column	$500 \, \text{mm} \times 500 \, \text{mm}$
4	Size of beam (width × depth)	$350 \, \text{mm} \times 500 \, \text{mm}$
5	Number of story	10
6	Height of each story	3500 mm
7	Number of spans (along the direction of motion and its transverse direction)	4×4
8	Size of each span	6000 mm
9	Thickness of slab	150 mm
10	Amount of infilled panels (percentage)	40%
11	Thickness of infill	250 mm

Table 2. Study parameters and their values.

Sl. No.	Parameters	Reference values	Studied values
1	Size of column	$500 \times 500 \, \text{mm}^2$	300×300, 400×400, 500×500, 600×600, $700 \times 700 \, \text{mm}^2$
2	Number of Story	10	4, 7, 10, 13, 16
3	Span	6000 mm	4000, 5000, 6000, 7000, 8000 mm
4	Story height	3500 mm	2500, 3000, 3500, 4000, 4500 mm
5	Size of beam	$350 \times 500 \, \text{mm}^2$	350×300, 350×400, 350×500, 350×600, $350 \times 700 \, \text{mm}^2$
6	Number of spans	4	2, 3, 4, 5, 6
7	Amount of infill (% of total panels)	40%	20%, 40%, 60%, 80%

using two noded truss elements having only three translational degrees of freedom at node. The reference RC frame has the properties given in Table 1.

Time period of a building is basically a function of its mass and stiffness. Thus any structural or building parameter that changes the stiffness or mass shall have influence on period. Typical parameters are building height, floor to floor height, span length of panels (span) in the direction of motion as well as in the transverse direction (bay), number of panels in the direction of motion and in the transverse direction, column and beam stiffness, infill effective as bracing etc. In order to asses the relative influence of these parameters, a systematic sensitivity analysis has been performed by varying such parameters one by one while keeping the other parameters unchanged. Table 2 lists the parameters, which were included in the investigation and their range of variation.

4 CODE EQUATIONS FOR TIME PERIOD

The empirical formula for the fundamental period of vibration of RC buildings specified in the Uniform Building Code (UBC, 1997) and Bangladesh National Building Code (BNBC, 1993) is of the form

$$T = C_t h^{3/4} \text{ sec.} \tag{6}$$

where h is the height of the building above the base. For framed RC buildings the numerical co-efficient $C_t = 0.03$ for UBC when h is in feet and $C_t = 0.073$ in BNBC when h is in meter. Both the codes permit an alternative way to calculate fundamental period T using the structural properties and deformational characteristics of the resisting elements in a properly substantiated analysis. The alternative formula for determining the period is

$$T = 2\pi \sqrt{\left(\sum_{i=1}^{n} w_i \delta_i^2 \right) \div \left(g \sum_{i=1}^{n} f_i \delta_i \right)} \tag{7}$$

The values of f_i represent any lateral force distributed approximately in accordance with rational distribution and $w_i's$ are the weights of individual floors. The elastic deflections, δ_i, shall be calculated using the applied lateral forces, f_i. However, Equation 7 is rarely used in practice. In Indian Standard (IS) Criteria for Earthquake Resistant design of Structures, the period T is calculated as,

$$T = 0.1n \tag{8}$$

where n is the number of floors above base. The Building Standard Law of Japan (BSLJ) uses the following formula for the fundamental period,

$$T = h \, (0.02 + .01\alpha) \tag{9}$$

where h is the building height and α the ratio of the total height of steel construction to the height of the building. In the National Building Code (NBC) of Canada the period is calculated by,

$$T=0.1\,N \qquad (10)$$

where N is the number of floor above exterior grade.

The fundamental period T, calculated using any of the Equations 6 through Equation 10 should be smaller than the *actual* period to obtain a conservative estimate for the base shear. Therefore, the code equations are generally calibrated to give a period lower that the *actual* by about 10–20%.

Code provisions generally permit the use of computational techniques like eigenvalue analysis using finite element modelling, but specify that the resulting period value should not be longer than that estimated from the empirical equations by a certain factor. The factor specified in UBC (1997) is 1.3 for high seismic zones and 1.4 for other zones. In BNBC (1993) the factor is 1.2 for all zones. These restrictions are imposed to safeguard against *unreasonable* assumptions in the rational analysis, which may lead to unreasonably long periods and hence unconservative values of base shear. When FE modelling is employed, a RC building is typically modeled using common frame elements to model the beams and columns and plate elements are used to model and slabs and shear walls. The structural contribution of the infill is generally not included. From designers' point of view, such modelling and analysis under static load results in a safer structure since only the primary structural system carries the load. However, such modelling without considering the structural contribution of infill also renders the structure too flexible, resulting in an unacceptably longer value of period.

Therefore, conventional modelling is not suitable for determining fundamental period necessary for calculating equivalent static earthquake load or doing dynamic analysis of RC frames having infill.

5 SENSITIVITY ANALYSIS

Earlier studies on sensitivity of different structural parameters (Amanat & Hoque 2002) reveals that when infill is considered in the analysis of building, the nature of the stiffness contribution of structural elements like beams, columns or the effect of building parameters like number of spans or floor panel size are significantly altered. Other parameters commonly established to have effect on period are number of floors and floor height or building height. In each case a complete 3D finite element model of the building is analyzed. Analysis is performed including the effect of infill by modelling them as diagonal struts (Madan et al. 1997) as well as without infill. The time period obtained from eigenvalue analysis of both the cases are then compared with the period obtained from code equations. These comparisons established the relative importance of incorporating infills in the FE model. Following subsections discuss some of the results in more details. It may be mentioned that although the results of Figure 2 correspond to the reference building (Tables 1 and 2), we can at least have some general idea of the characteristics of RC frames.

5.1 *Effect of column stiffness*

Figure 2a shows the determined period versus column stiffness (EI/L). For the frame modeled without infill, the period is highly sensitive to column stiffness, but for the frame with infill it can be observed that the

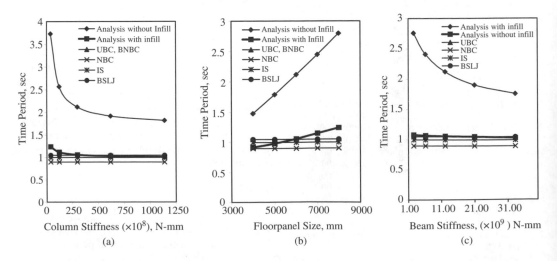

Figure 2.   Sensitivity analysis of (a) column stiffness, (b) floor panel size and (c) beam stiffness.

fundamental period does not depend much on the column stiffness. This is due to the fact that when infills are active as diagonal bracing, the stiffness induced by them against lateral deflection is much more pronounced than the effect of column stiffness. The periods obtained from code equations are also plotted for comparison. It can be observed that the analysis with infill closely agrees with code equations while the frame without infill grossly overestimates the period.

5.2 *Effect of floor panel size*

The code equations do not have any provision to incorporate the effect of floor panel size (span lengths) in determining time period. As such, the period predicted by these empirical equations are same for all values of floor panel size (span length) studied. The periods obtained from modal analysis of the infilled frame do show some variation with the changing values of panel size. It can be observed from Figure 2b that period increase with the increase in panel size. However, the periods obtained lies in close proximity with the code equations supporting the acceptability of the code equations. The modal analysis of the frames without infill produces periods significantly longer justifying the code provisions of limiting the period obtained from conventional rational analysis.

5.3 *Effect of beam stiffness*

When beam stiffness is varied, the period from the modal analysis without infill varied significantly. However, when infill is incorporated, the effect of beam stiffness becomes insignificant as can be seen from the nearly horizontal line of the corresponding data of Figure 2c Here again we observe that the code equation predictions are very close to the results of modal analysis with infill. It may be mentioned that the code equations do not have any provision to consider the effect of beam stiffness. As such all the equations produced horizontal lines. It can thus be inferred that in presence of infill the effect of beam stiffness does not have any significant effect and this parameter need not be considered in any attempt to improve code equations.

6 PERIOD WITH RANDOMLY DISTRIBUTED INFILL

It is already apparent from the results presented in the previous section that building period obtained by analytical method is close to the period predicted by codes when infill is present in the FE model. In all the results presented in previous section, the number of infilled frame panel was fixed at 40 percent. In practical cases, the amount of infill will vary from building to building. Also their arrangement or location will also be different from building to building. Therefore, the effect of different amount of infilled panels in the same FE model is studied. The amounts of infilled panels included in the investigation are 20%, 40%, 60% and 80%. For example, the reference building has ten stories and four bays in each direction. Thus the number of vertical frames in each direction is five. Considering this, the total number of frame panels in the building is 400 (4 span × 10 stories × 5 frames × 2 directions). Thus to provide 40% infill, we need 160 panels – 80 in each direction. These 80 panels are chosen randomly and diagonal strut is provided in those panels. In order to get an average result such random distribution was made five times for every percentage amount of infill.

Figure 3a shows the periods obtained when 20% of the panels were infilled. It is observed that the variation of period among different random distribution is not very significant. Average of the five periods corresponding to the five different random distributions are approximately same as that obtained when the infilled panels are distributed in a regular pattern. Figures 3b, c and d also reveal the same fact corresponding to 40%, 60% and 80% infilled panels respectively. It can thus be said that randomness in the distribution of infill in the structure does not influence the building period significantly. Figure 4 shows the change in building period with the increase of number

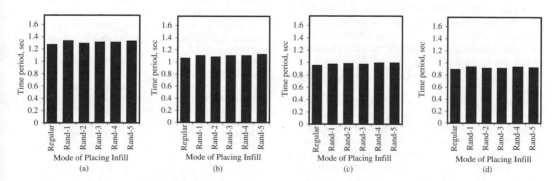

Figure 3. Effect of randomly distributed infill on period, a) 20% infill, b) 40% infill c) 60% infill and d) 80% infill.

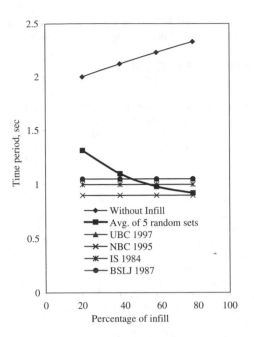

Figure 4. Period vs. amount of infilled panels.

of infilled panels. We observe that the code equations do not show any variation in period since these do not take into account the effect of infill in calculating building period. For conventional analysis, the period gets longer with increase of infill due to added mass from infill. When infill is incorporated in the FE analysis, the period becomes shortened with the increase in number of infilled panels due to added stiffness from equivalent diagonal struts.

7 CONCLUSION

Some sensitivity analysis has been performed in the present study which shows that in presence of infill, the beam and column stiffness have negligible effect on period. Size of floor panel and amount of infill in the structure are some important parameters in influencing the period which the code equations do not take into account. Study also shows that randomness in the distribution of infill does not have significant effect on the period, instead, it is the total amount of infill that matters.

While sophisticated computer packages are available to model and analyze building structures in a comprehensive way, code specifications still advocate the use of empirical equations for estimating the fundamental period of buildings. The approximations made in idealization of the structure during the generation of computer based FE model is basically the source of error in predicting the period though the same idealization is

generally thought to be comprehensive enough for strength design under static design loads. In this paper it has been shown by various sensitivity study that when the effect of infill is considered in the FE modelling, rational analysis do give reliable results and agree reasonably well with the code equations. Such modeling technique may be applied for dynamic time-history analysis or spectral response analysis of tall buildings where higher modes of vibration are also active and in which cases conventional analysis without infill may not accurately reflect the dynamic behavior.

REFERENCES

Amanat, K.M. & Hoque, E. 2002. A Reappraisal of Time Period Formulas of Design Codes for Framed Reinforced Concrete Buildings, *Proceedings Of The Sixth International Conference On Computational Structures Technology (CST)*, Prague, Czech Republic, 4–6 September, B.H.V. Topping and Z. Bittnar (eds), Civil-Comp Press, UK.

Goel, R.K. & Chopra, A.K. 1997. Period Formulas for Moment-Resisting Frame Buildings, *Journal of Structural Engineering*, ASCE, Vol. 123, No. 11.

Goel, R.K. & Chopra, A.K. 1998. Period Formulas for Concrete Shear Wall Buildings, *Journal of Structural Engineering*, ASCE, Vol. 124, No. 4.

Hossain, M.Z. 1997. *Influence of Structure Parameters on Period of Frame Structures for Earthquake Resistant Design*, MSc Thesis, BUET, Dhaka, Bangladesh.

Housing and Building Research Institute and Bangladesh Standards and Testing Institution. 1993. *Bangladesh National Building Code (BNBC)*, Dhaka, Bangladesh.

Indian Standards Institution (IS). 1984. *Criteria for Earthquake Resistant Design of Structures.* India.

International Conference of Building Officials. 1997. *Uniform Building Code*, Willier, California.

Li, Q. Cao, H. & Li, Gi. 1994. Analysis of Free Vibrations of Tall Buildings, *Journal of Engineering Mechanics*, Vol. 120, No. 9.

Lourenco, P.B., Brost, R.de & Rots, J.G. 1997. A Plane Stress Softening Plasticity Model For Orthotropic Materials, *International Journal For Numerical Methods In Engineering*, Vol. 40.

Madan, A., Reinhorn, A.M., Mander, J.B. & Valles, R.E., Modeling of Masonry Infill Panels For Structural Analysis, *Journal of Structural Engineering*, ASCE, Vol. 123, No. 10.

Ministry of Construction. 1987. *The Building Standard Law of Japan (BSLJ)*, Japan.

National Research Council. 1995. *The National Building Code (NBC)*, Canada.

Papia, M. 1998. Analysis of Infilled Frames Using a Coupled Finite Element And Boundary Element Solution Scheme, *International Journal For Numerical Methods In Engineering*, Vol. 26.

Saneinejad, A. & Hobbs, B. 1995. Inelastic Design of Infilled Frames, *Journal of Structural Engineering*, ASCE, Vol. 121, No. 4.

Smith, B.S. & Crowe, E. 1986. Estimating Periods of Vibrations of Tall Buildings, *Journal of Structural Engineering*, Vol. 112, No. 5.

System-based Vision for Strategic and Creative Design, Bontempi (ed.)
© *2003 Swets & Zeitlinger, Lisse, ISBN 90 5809 599 1*

Experimental results and numerical simulation of a concrete sandwich panel system proposed for earthquake resistant buildings

O. Bassotti
Emmedue srl, Fano (PU), Italy

E. Speranzini
Department of Civil and Environmental Engineering, Faculty of Engineering, Perugia, Italy

A. Vignoli
Department of Civil Engineering, Faculty of Engineering, Firenze, Italy

ABSTRACT: The aim of this work is to present the results of several laboratory tests conducted by the RITAM (University of Perugia's laboratory), on Emmedue System characterized of the use of sandwich panels for large-panel construction. It has been created a prototype of a two-storey house in real scale sized 3.45 × 4.2 m subjected to harmonic horizontal forces induced by a vibrodine jointed to the structure. The dynamic response of the most sensitive points has been measured by seven quartz accelerometers. These procedure has been carried out in order to obtain the principal parameters of dynamic behaviour of the prototype and to extend, after a process of dynamic identification, the numerical model to real structures. The high accelerating levels reached during the tests ($10 \, \text{m/s}^2$) have clearly shown the capacity of resistance against dynamic actions of this building system. The reduced width (0.370 mm) of the oscillations induced during the experimental tests of the System and the level of damping underline the high capacity of energy dissipation of this buildings.

1 INTRODUCTION

This research report the experimental results and the numerical simulations of a building System realized with special sandwich panels. Each panel is constructed with two reinforced concrete outside walls and interlaid polystyrene. These panels are produced by the Emmedue company in Fano (PU, Italy) and represent the main element of a building system that is proposed for the construction of civil and industrial buildings by on site assembling and completion.

The lightness of panels prior of the completion makes the transport and handling very easy; not only, the assembling does not require any special skills and permits to save time.

The joining of the several elements at the construction site realizes a non-traditional building system, lighter than other construction technologies and whose covered insulation has no thermal bridges. The advantage of this system is the use of light concrete panel, which is required to have higher quality characteristics easily obtainable by slightly increasing the concrete

batching and through the realization of a suitable mix-design both as a resistant element and finishing.

The tests were carried out at the RITAM Laboratory of Terni, at the University of Perugia (Italy). The test sequence aimed at examining the dynamic behaviour of the prototype subjecting to horizontal forces induced by a mechanical vibrodine and measuring the response by means of accelerometers and laser vibrometer.

Following the experiment results a well refined element model has been implemented and used, then, for the numerical analysis.

2 THE PROTOTYPE

The prototype is a two storey building, any of which is constituted by an unique room. It has been erected by Emmedue panels type PSM80 and has a rectangular shape, with dimensions respectively of 4.20 and 3.44 m. The two storeys are linked each other by an internal staircase realized by the elements PSSC EMMEDUE, parallel to its adjacent shorter side to it.

On the first floor, one window has been carried out in line with the ground floor's door in the centre line of the longer side. The roof is a couple-close type and its ridge line is parallel to the shorter side. The thickness of the panels is 15.5 cm for the walls and 17 cm for the horizontal slabs.

2.1 Panel specifications

The panels are made up of two electrowelded galvanized steel meshes, adjacent to the faces of an undulated foam polystyrene central slab. The panel is produced in automatic mode by numerical control panelling machines guaranteeing constant quality of the product. The steel meshes are realized in automatic and non-stop mode by suitable machines through which all the parameters affecting the welding are set. The density of the panel polystyrene slabs is equal to 20 kg/m³ and the thickness is of 8 cm The steel web reinforcement consist of φ 3.5 mm vertical bars with a spacing of about 7 cm, and of φ 2.5 mm horizontal bars at a spacing of 13 cm. Both the meshes are joined one another by means of double steel cross-ties placed in correspondence of the knots, in a number of about 72/m². The steel used is drawn by hot-galvanization treatment. The average tensile strengths turn out to be over 600 MPa.

The panels produced weigh about 50 N/m²; they have a standard width of 112.5 cm and the length changes according to the technical-designing requirements.

The Figure 1 shows the axonometric view of the described panel.

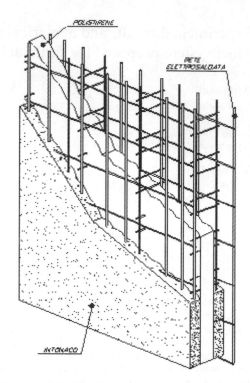

Figure 1. PSM Emmedue panel axonometry.

2.2 Prototype realization

The prototype was realized inside the RITAM Laboratory (ISRIM-University of Perugia) in Terni (Italy) for structural tests. The foundation was realized through a double casting of a thin concrete slab (5 cm approximately) and interlaid Emmedue PSM 80 panels. The vertical panels were connected to the foundation by a angular steel meshes, usually used for joining the elements at the crossings as shown in Figure 2.

Afterwards, the panels were laid vertically, paying special attention to alignment and verticality. The reciprocal joining of panels was realized by fastening made on the steel meshes having, on one side, an overlapping grid guaranteeing the continuity. The handling of the panels was extremely easy and did not require any construction site equipment for the examined length (about 6 m) either, as shown in the Figure 3.

The installation can be easily carried out by two-three persons who assemble remarkable portions of rough building in short time (Figure 4).

Doors and windows have been reinforced by means of flat meshes with 45° inclination at the lower

Figure 2. Detail of the panel anchorage to the foundation.

and upper corners of the openings. Both internally and externally, angular steel meshes have been positioned at all the angles, along the whole height of the corner. At assembling completion, the panels have been finished with sprayed concrete (Figure 5).

The concrete has been sprayed in two layers, the first of which till to cover the mesh and the second till to reach the thickness of about 2.5 cm over the steel mesh of the panel.

Figure 3.   Panel handling.

Figure 4.   Detail of panel laying.

Figure 5.   Detail of concrete finishing.

Figure 6.   Instrumentation arrangement.

3   METHOD OF TEST

In order to determine its dynamic behaviour, the prototype has been subjected to horizontal forces induced by the mechanical vibrator (vibrodine); the dynamic response of the most sensitive points has been measured by accelerometers and laser vibrometer.

For the processing of the results, an appropriate software has been used that enable, further to the input of the data, the visual control of the reply and the data storage.

The data stored in such a way are then processed using Windows programs Excel type or MatLab.

3.1   *Experimental layout*

The experiment has been executed by subjecting the prototype to horizontal forces according to two series of tests: the first one at the first floor on the axis of symmetry of the plan, the second on the roof level. Several tests have been executed for the two series by changing the order of applied forces, the frequency range and, at the same time, the maxima loads transmitted to the structure (Annunziata, Borri, Speranzini – part I and II – 1999). The Figures 6 and 7 show the layout of the used instruments.

3.2   *First series of tests*

Using the vibrodine placed at the first floor level in North–South direction, it was decided to execute four

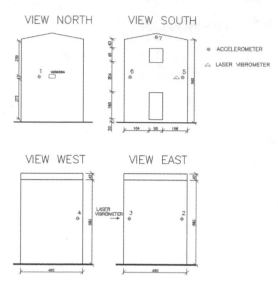

VIEW NORTH   VIEW SOUTH

VIEW WEST   VIEW EAST

Figure 7.   Instrumentation arrangement at the views.

series of tests during the first part of the experiment:

First test: the vibrodine gives to the structure an increasing excitation until reaching the frequency of 16.9 Hz; frequency that is maintained constant for approximately 2 min after which excitation to the structure is suddenly stopped by blocking the mechanical exciter. The maximum horizontal load applied to the structure is approximately 12.00 kN.

Second test: it has been carried out initially by giving to the structure a frequency of 8 Hz, maintaining it constant for approximately 30 s, then arriving to 16 Hz and maintaining excitation for approximately further 30 s and finally blocking it suddenly. The maximum given load is about 12.00 kN.

Third test: it must be reached a maximum load of 19.80 kN given in one unique ramp, that is progressively and continuously increasing the frequency from 0 to approx 20 Hz for a time history of 100 s.

Fourth test: it is carried out in the same way as the previous one reaching the maximum load of 20.00 kN for a duration of 80 s at the frequency of 21 Hz.

3.3   Second series of tests

The second series of tests has been carried out by loading the structure with horizontal forces, variable in time, placed at the covering level in two different directions (North–South and East–West).

The vibrodine was placed in correspondence of the covering slab respectively on the top of the roof for the tests from 1 to 6 and on the eaves line for the tests from 7 to 11.

First and second excitation (North–South direction): the mechanical exciter gives to the structure a growing excitation until reaching the frequency of 16.9 Hz. Such frequency is kept constant for about 30 s; and then excitation is suddenly stopped.

Third excitation (North–South direction): it has been carried out by initially giving to the structure a frequency of 8 Hz, keeping it constant for about 30 s, then bringing it to 16 Hz, maintaining such an excitation for further 20 s and finally stopping it dead.

Fourth and Fifth excitation (North–South direction): it is reached a peak load of about 18.00 kN given in a unique ramp, that is increasing progressively and with continuity the frequency from 0 to 20 Hz for a time history respectively of about 80 s for the fourth test and 30 s for the fifth test.

Sixth excitation (North–South direction): it is carried out in order to determine the first North–South resonance frequency; the mechanical exciter's revolutions are increased and decreased in correspondence of about 920 rpm (15.3 Hz).

Seventh and Eight excitation (East–West direction): the vibrodine gives to the structure a growing excitation until reaching the frequency of 16.9 Hz; such frequency is maintained constant for about 10 s, then excitation is stopped dead blocking the mechanical exciter. The peak horizontal load given to the structure with this excitation is about 12.00 kN.

Ninth excitation (East–West direction): it has been carried out by initially giving to the structure a frequency of 8 Hz, maintaining it constant for about 30 s, then reaching 16 Hz and keeping such excitation for about further 60 s and then suddenly stopping the excitation.

Tenth excitation (East–West direction): it is reached a peak load of about 18.00 kN given in a unique ramp, that is increasing progressively and with continuity the frequency from 0 to 20 Hz for a time history of about 100 s.

Eleventh excitation (East–West direction): it is carried out in order to determine the first East–West resonance frequency, varying the mechanical exciter's revolutions up to a maximum of 1003 rpm, equal to 16.7 Hz.

4   EXPERIMENTAL TESTS

With reference to the tests above-mentioned we can summarize the following results.

4.1   First series of tests

In the execution of these tests, whose results are given in the first part of this report, the load was applied on the North–South side direction at the level of the first floor, with a given maximum acceleration of $\pm 4 \text{ m/s}^2$

($\cong 0.4$ g) in the case of the first two excitations and $\pm$ 5 m/s^2 ($\cong 0.5$ g) for the other two excitations. Furthermore:

- the processing of the results demonstrated an asymmetrical behaviour of the structure, probably due to the asymmetry of the structure itself because of the presence of the staircase on the east side.
- it was noted a resonance frequency around 16–17 Hz.
- where the laser beam collimated, it has been noted that during the second test displacement was of 0.132 mm.

4.2 Second series of tests

Mechanical exciter applied on the roof ridge on the North–South direction: During this series of tests greater accelerations and displacements were obtained due to the application of stress with same direction but in correspondence of the roofing floor. During the first three excitations the accelerometer connected to the vibrodine measured a max. acceleration of $\pm$ 9 m/s^2 ($\cong 0.9$ g), whereas the accelerometer placed in correspondence of the roof ridge, on the South side, showed an acceleration of $\pm$ 7 m/s^2 ($\cong 0.7$ g). In the following three excitations the accelerometer no. 7, that was the most stressed, measured an acceleration of $\pm$ 8 m/s^2 ($\cong 0.8$ g).

During this series of tests it was reached a resonance frequency at various times; in correspondence of which it has been measured a displacement of the structure equal to 0.370 mm.

Mechanical exciter applied on the eaves line on the East–West direction: In this case the disposition of the vibrodine was such as to not produce torsion of notable value to the structure. The accelerometer connected to the mechanical exciter measured a maximum acceleration of about $\pm$ 10 m/s^2, whereas the accelerometers placed in correspondence of the roof ridge, on the West side, showed an acceleration of $\pm$ 6 m/s^2. It was also reached a resonance frequency of 16–17 Hz and a displacement measured by the vibrometer of 0.232 mm, against the displacement of 0.169 mm obtained at a frequency of 20 Hz.

4.3 Data processing and identification of the dynamic behaviour

So to evaluate the dynamic behaviour of the model we have analyzed the data obtained by the mechanical exciter oriented in the North–South direction and in the East–West direction.

The data obtained have been filtered by a digital low-pass filter with cut-off frequency at 50 Hz.

Afterwards the spectrum tensor of the recording – suitably detrended by means of a FFT algorithm

applied to consecutive time windows – has been calculated. The length of these time windows has been fixed in the minimal power of 2 necessary to obtain a resolution in frequency equal to rf = 0.05 Hz, therefore, since fc = 500 Hz is the frequency of signal sampling, the following number of data for each time window is obtained:

$$n = 2^{\lfloor \log_2 (f_c/r_f)\rfloor + 1} = 2^{14} = 16384$$

The number of windows averaged in order to obtain the signal spectrum changes according to the length of the signals themselves; anyway, we have tried to obtain a time superimposition between two consecutive windows equal to 10–15% of the length of each window. Finally, the spectra have been divided by f^4 so as to balance the rise in the force intensity which is proportional to f^2.

Once a datum system has been fixed – where x axis is in East–West direction, y axis in North–South direction, and z axis is equidirected with increasing elevations measured from the foundation pier layout – the accelerometer signals have been processed in order to obtain the three motions of rigid plane concerning the two levels into consideration.

Thanks to the spectrum tensor concerning the accelerometer measurements, through the pre-and postmultiplication by the pseudoinverse of the positioning matrix of the instruments, it has been possible to find out the spectrum tensor concerning the plane motions of the two instrumented levels.

4.3.1 Vibrodine arranged at floor elevation (level 1), North view, North–South direction

These are the results from the test No. 3: in the diagrams only the level 1 is shown.

In the Figures 8 and 9 are shown the spectra of the three plane motions and the norm of the phase and coherence tensors.

Observing the spectra of the plane motions, a peak close to 16.3 Hz is noticeable. The presence of a resonance near this frequency would be confirmed by the norm of the tensor of coherence close to 1 and by the norm of the tensor of phase close to zero.

4.3.2 Vibrodine applied at the level of the eaves line on the East view, East–West direction

These are the results from the test No. 10: in the diagrams the level 1 is in blue whereas the level 2 is in green.

In the Figures 10 and 11 are shown the spectra of the three plane motions and the norm of the phase and coherence tensors.

The norm of the tensor of coherence (Figure 11) is very close to 1 between 16 and 20 Hz. In the parallel way, the norm of the tensor of phase decreases around

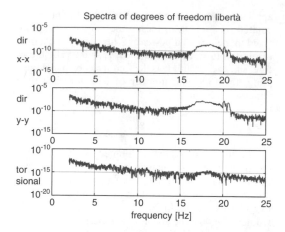

Figure 8. Spectra concerning the plane motions of both instrumented levels in E–W, N–S and torsional direction.

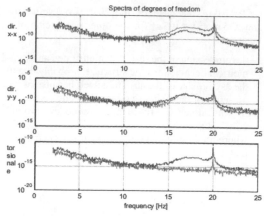

Figure 10. Spectra concerning the plane motions of both instrumented levels in E–W, N–S and torsional direction.

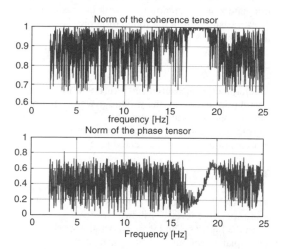

Figure 9. Norm of the tensor of coherence and phase of the signals as per frequency.

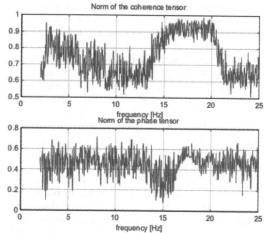

Figure 11. Norm of the tensor of coherence and phase.

16 Hz and still between 18 and 19 Hz. It seems, therefore, that on top of the frequency around 16.3 Hz in North–South direction there is something around 18–19 Hz in East–West direction.

It has to be noted, finally, the peak close to 20 Hz. This peak is supposed to show, together with the norms of the tensors of coherence and phase, that the structure gets sensitive to the excitations close to 20 Hz once again.

Anyway, it is very likely that the structure shows a resonance at the frequencies which are just above 20 Hz. It has to be noticed that, considering the limited range of utilization of the vibrodine used, it has not been possible to examine beyond the frequency of 20 Hz.

5 NUMERICAL MODEL

The numerical model that better schematises the behaviour of the real structure has been performed according to the previous knowledge of the examined structure and following the results of a series of static tests conducted on the panels during the same research sequence (Annunziata et al. 1999).

The structural analysis is carried out by means of a widely diffuse finite element code Ansys (Ansys, 1993) using a particular Shell 63 library element (Figure 12) whose peculiarity permitted to model the sandwich panel – having non-homogenous section – consisting of the two reinforced concrete thin layers with interposed polystyrene.

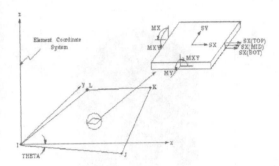

Figure 12.   Shell 63 element.

In particular the Shell 63 element is defined by four nodes having six degrees of freedom at each node (three translations and three rotations) and has both bending and membrane capabilities, so as to permit an elastic analysis.

The element requires mechanical and geometric characteristic such as: Young's moduli in the element plane (E_x, E_y), shear modulus (G_{xy}), Poisson's ratio (NU_{xy}), density (DENS) and thickness (TK), and definition of a special parameter (RMI) allowing the schematisation of the sandwich panel:

$$RMI = \frac{MIR}{MISI}$$

where:
MIR = moment of inertia of the non homogeneous section
MISI = moment of inertia of the section regarded as homogeneous

The calibration of the E_x, E_y and G_{xy} parameters was carried out by using the results of the static tests on the panels and has been made a numerical analysis of the single panels, more refined than this made on the prototype on its entirety. For this reason the panel has been modelled using the Solid65 element, which is three-dimensional with eight nodes having three translational degrees of freedom at each node, and permits the introduction of reinforcement bars according to three orthogonal directions.

As far as the mechanical features of the materials are concerned, for the steel has been given a bilinear elastic perfectly plastic law, and for the sprayed concrete has been given a non linear constitutive law (rectangular-parabola). This laws were obtained through the traction and compression laboratory tests carried out right on the same materials.

The model of the single panels provides for two layers of sprayed concrete with the steel web, each of them is 3 cm deep, at a distance of 9.5 cm. The modelled panel, perfectly bounded on the base, is 164 cm large, 273.5 cm high and 15.5 cm deep (Figure. 13).

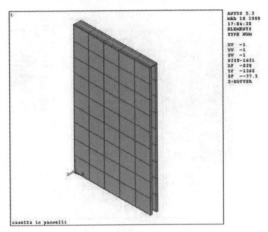

Figure 13.   Numerical model of the vertical partition.

5.1   Determination of the elastic modulus (E_x, E_y) at compression

The E_x and E_y elastic moduli are considered through the numerical analysis of the vertical panel loaded with a force distributed on the top surface. Once the "s" displacement on top has been measured, the modulus of elasticity is determined:

$$E = \frac{N \cdot l}{s \cdot A}$$

where:
E = modulus of elasticity of the panel
N = axial force applied
l = height of the panel
s = displacement in the load direction
A = area of the panel section

5.2   Determination of the shear modulus

The shear modulus has been obtained with the following equation:

$$K_0 = \frac{T}{\delta}$$

$$K_0 = \frac{G \cdot A}{1.2 \cdot H} \cdot \frac{1}{1 + \frac{1}{1.2} \cdot \frac{G}{E}\left(\frac{H}{B}\right)^2}$$

where:
K_0 = shear stiffness of the panel
T = horizontal force applied
δ = horizontal displacement on the panel head
G = shear modulus of the panel
E = Elastic modulus at compression (previously determined)

1171

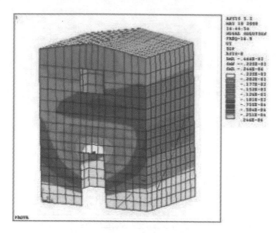

Figure 14. Displacements of the y direction.

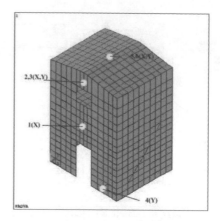

Figure 15. Tested areas, axonometric view from S–E.

5.3 Dynamic identification

Once obtained the results of the laboratory tests, we have operated through a step-by-step procedure improving the results as required by the procedure of parametric identification (Borri et al. 2000).

The choice of the numerical model that describes at best the behaviour of the real structure has been made according to the previous knowledge of the structure examined and the results of the experimental analysis.

In particular the dynamic identification has been carried out by mainly referring to the behaviour of the prototype in the North–South direction that turns out to be the weakest direction of the structure.

Once determined the mechanical parameters, the behaviour of the prototype has been numerically analyzed performing a dynamic analysis by means of harmonic forces.

As a comparison parameter between the reaction of the true structure and that of the numerical model, we have considered the displacement at the point where the laser ray of the vibrometer collimates.

The numerical model has been excited with actions similar to the ones caused by the mechanical exciter. Since we have observed that the characteristics of rigidity (the E_x, E_y and G_{xy} elastic moduli of the panel and consequently the elastic moduli E, G of the concrete) showed the higher indeterminacy, the E and G parameters have been assumed as unknown values of the identification process.

Their value has been calculated keeping to a minimum the error per cent of the results of the numerical analysis if compared to those of the experiment.

In the Figure 14 the displacements of the numerical model in y direction (direction parallel to the vibrometer laser ray) are shown.

As for example, from the data of the first test, we can see that the displacement measured at the point of

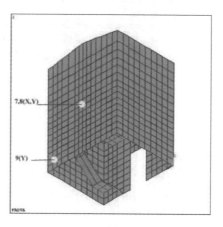

Figure 16. Tested areas, axonometric view from N–W.

collimation of the vibrometer is of 0.108 mm; if compared to the displacement of 0.132 mm, obtained through the experiment, it shows a difference of 18%.

These results show that the equivalent viscose damping value obtained from the experimental analysis is greater than the value of the numerical analysis, fixed at the conventional level of 5%.

The following step has been to investigate the stresses on the structure and check their compatibility with the structural characteristics of the materials.

5.4 Stresses state

The stresses state on the prototype has been fully determined and checks have be carried out in correspondence of the sections which are mostly stressed due to the stress produced (Figures 15, 16 and 17).

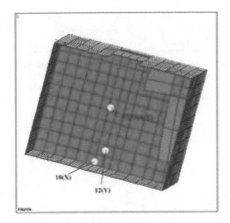

Figure 17. Tested areas, axonometric view from the bottom.

Table 1. More important shear and bending moment.

Sez	T_x N/m	T_y N/m	M_x N × m/m	M_y N × m/m
1	2076		429.39	
2	4186		429.39	
3		3280		538.99
4		−18902		−582.95
5	−5382		−487.68	
6		−10028		−582.95
7	−4253		429.39	
8		−5593		912.98
9		−5593		−208.97
10	−3198		−258.41	
11	1022		−946.21	
12		−10029		−208.97
13		−5593		−1331

In the Table 1 and 2 are summarized, for the sections taken into consideration, the values concerning the shear and bending moment and the results of the checks respectively.

The shear test of the section is left out, because the stresses turn out to be far under the allowable values for the maximum stress found out.

The results show that the tests have been largely satisfactory (as for stress obtained from the application of a force similar to the one concerning the experimental test) since they are below the assumed allowable values equal to 367 N/mm² as for steel and at least equal to 11 N/mm² as for concrete.

The values concerning the elastic modulus for the panels, calibrated according to the numerical simulation of experimental tests, turn out to be the following:

Vertical panel: $E_x, E_y = 8129$ N/mm²
$G_{xy} = 893$ N/ mm²

Table 2. Stress in the section mentioned at the table 1.

Sez	Y_N MM	σ_{fmax} N/mm²	σ_{fmin} N/mm²	σ_{cmax} N/mm²
1	8.7	123.72	5.95	−0.546
2	6.2	156.55	10.22	0.489
3	14.3	45.98	0.26	−0.349
4	All compressed	−2.29	−6.60	−0.475
5	144.1	41.28	−1.77	−0.521
6	106.3	90.3	−4.00	−0.388
7	130.4	29.81	−2.47	−0.423
8	30.1	33.2	−4.58	−0.608
9	All compressed.	−0.54	−2.09	−0.152
10	141.7	19.19	−1.19	−0.27
11	10.9	210.50	−5.18	−1.105
12	139.5	84.61	−6.67	−1.299
13	153.2	191.55	4.26	−1.469

Horizontal panel: $E_x, E_y = 6749$ N/mm²
$G_{xy} = 747$ N/ mm²
Stairs panel: $E - 3780$ N/mm²
$G = 415$ N/mm²

to which the elastic modulus of the sprayed concrete equal to 21 kN/mm² is associated.

6 STANDARD SEISMIC ANALYSIS

After subjecting the prototype to the actions of the vibrodine and realizing a relevant numerical model that has been suitably calibrated (by comparison of the displacements), the following step has been the analysis of the prototype after subjecting the same to "standard" stresses such as the seismic ones included in the Regulation.

6.1 Multi-modal response spectrum analysis

Once the elastic modulus has been fixed according to the test results and numerical analysis, a dynamic analysis is carried out by applying the accelerations response spectrum, in compliance with the provisions of the Italian Seismic Regulation.

The value of the spectral acceleration has been assumed as variable with the dynamic characteristics of the structure following the formula below:

$a\backslash g = C \cdot R \cdot \varepsilon \cdot \beta \cdot I$

as for the coefficients, we have the most unfavorable situation assuming the following:

$C = 0.1, R = 1, \varepsilon = 1.3, \beta = 1.4, I = 1.2$
$T_0 > 0.8$ seconds R $= 0.862/T_0^{2/3}$
$T_0 \leq 0.8$ seconds R $= 1.00$

1173

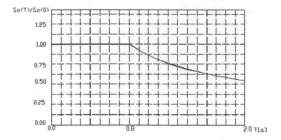

Figure 18. Spectrum of standardized design of the Italian norms.

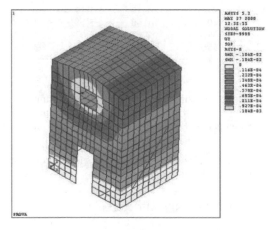

Figure 19. Dispalcement due to the spectrum in y direction.

Table 3. Most important axial forces and bending moments per unit of length for seismic action in y direction.

Section taken into consider ration and stress direction	Response spectrum analysis	
	T [N/m]	M [N · m/m]
1(x)	461	17.73
2(x)	1382	75.31
3(y)	687.29	92.97
4(y)	2729	157.92
5(x)	155.11	325.2
6(y)	687.29	19.35
7(x)	615.02	34.97
8(y)	687.3	112.63
9(y)	2049	112.63
10(x)	615.02	52.21
11(x)	308.41	34.97
12(y)	687.29	93.97
13(y)	687.29	38.01

Table 4. Most important axial forces and bending moments per unit of length for vertical loads.

Section taken into consider ration and stress direction	Static analysis with weight force only	
	T [N/m]	M [N × m/m]
1(x)	−1301	−325.2
2(x)	2685	315.1
3(y)	−412.22	394.79
4(y)	−12303	−112.63
5(x)	−4491	−17.74
6(y)	−7207	−424.28
7(x)	1090	155
8(y)	−3810	671.16
9(y)	−10005	−157.92
10(x)	1090	98.2
11(x)	−504.2	−645.48
12(y)	1287	118.43
13(y)	1287	−987.01

In Figure 18 this spectrum is reported.

Two analysis have been carried out by applying the spectrum in the two orthogonal directions (x and y) of the adopted reference system.

The main results referring to the spectrum in y direction are shown below being more significant than those in the x direction.

6.2 Spectrum in y direction

The results show a very rigid structure (Figure 19); the Displacement at the point of collimation of the vibrometer is less than the one concerning the simulation analyses of the experimental test, and more precisely, equal to 53 μm instead of 107 μm of the first simulation and 117 μm of the second one.

6.3 Stresses results

In the following tables the state of stress in term of axial forces, (T_x, T_y in horizontal and in vertical direction, respectively) and bending moments (M_x,

M_y on the horizontal and vertical shell face, respectively) due to the application of the seismic action in y direction is shown. The numerical results will be carried out in the sections which are mostly stressed. The stress due to the weight force have to be added to the stress coming from the seismic analysis.

In Tables 3 and 4 the values of the axial forces and bending moments per unit of length for the response spectrum and static analysis are shown. The resulting stress state in the structure and the results of the checks concerning the mostly stressed sections – obtained as the sum of the two contributions – are given in Table 5. The resulting stresses on the steel

Table 5. Stresses results [N/mm^2].

Section	Y_n [cm]	$\sigma_{s,max}$	$\sigma_{s,min}$	$\sigma_{c,max}$
1	13.86	46.83	−0.49	−0.411
2	0.59	146.41	9.87	−0.434
3	2.36	25.02	−1.85	−0.338
4	All compressed	−2.54	−4.54	−0.319
5	13.92	22.50	−1.82	−0.349
6	10.33	6.36	−3.11	−0.296
7	0.66	66.95	4.20	−0.222
8	3.00	29.64	−3.86	−0.525
9	All compressed	−1.84	−3.84	−0.272
10	0.62	66.27	0.60	0.190
11	15.61	117.02	4.88	−0.724
12	14.43	48.32	−2.00	−0.604
13	15.02	57.95	0.10	−0.534

and on the concrete obtained from this analysis are largely below the allowable ones.

7 CONCLUSIONS

The experimental tests and the numerical analysis conducted on the prototype of a new type of large-panel building obtained by the assemblage of sandwich panels have been presented.

The experimental tests has permitted to specify the mechanical and dynamic behaviour of the prototype and each element that composes it.

The test results show the high stiffness of the structure and the smallness of the displacements induced. The structural elements has not presented any cracks either/or limits to the working of the connections.

The seismic analysis, carried out after defining the typical parameters of the elements, has shown small stresses, perfectly compatible with the mechanical characteristics of the materials.

The high levels of acceleration obtained during the tests have pointed out the capacity of strength of the construction system to dynamic actions; we would like to remind that levels equal to 0.5–1 g have been reached, comparable to a strong earthquakes corresponding to following degree of the Modified Mercalli (MM) intensity scale (1956):

− XI–XII according to Cancani Sieberg correlation;
− IX–XI according to Richter correlation.

The small amplitude of the free oscillations of the system tested and the evident damping of the vibrations point out a good capacity of energy dissipation due to the viscous damping also in the elastic field.

In this way the structure has shown a good dynamic behaviour and has turned out to be able to take up great intensity seismic accelerations with no damage.

REFERENCES

Ansys 1993. *Swanson Analysis Systems* Inc., Houston Pennsylvania, USA.
Annunziata A., Borri A. & Speranzini E. 1999. Prove statiche sul sistema costruttivo EMMEDUE, Nota Tecnica Università di Perugia.
Annunziata A., Borri A. & Speranzini E. 1999. Prove dinamiche sul sistema costruttivo EMMEDUE, parte I, Nota Tecnica Università di Perugia.
Annunziata A., Borri A. & Speranzini E. 1999. Prove dinamiche sul sistema costruttivo EMMEDUE, parte II, Nota Tecnica Università di Perugia.
Chiostrini S., Facchini L. & Vignoli A. 2000. The effectiveness of light interventions of anti-seismic reinforcements on a masonry building, *Proceedings of the 12th World Conference on Earthquake Engineering*, cd-rom, Auckland, New Zealand, January–February 2000, Paper No. 2120.
Borri A., Speranzini E. & Vignoli A. 2000. Sistema costruttivo EMMEDUE: modello numerico di supporto all'attività sperimentale, Nota Tecnica Università di Perugia.
Bassotti O. & Ricci M. 2002, Pannelli sandwich in calcestruzzo alleggerito e loro applicazioni costruttive, CTE, 2002.

System-based Vision for Strategic and Creative Design, Bontempi (ed.)
© 2003 Swets & Zeitlinger, Lisse, ISBN 90 5809 599 1

Theoretical and experimental investigation on the seismic behaviour of concrete frames

F. Biondini & G. Toniolo

Department of Structural Engineering, Technical University of Milan, Italy

ABSTRACT: The paper investigates the seismic behaviour of one-storey reinforced concrete structures for industrial buildings. An incremental non-linear dynamic analysis is applied within a full probabilistic procedure based on the Monte Carlo method. The representative values of seismic capacity obtained from the statistical analysis show that precast structures have the same seismic capacity of the corresponding cast-in-situ structures. The experimental verification of this theoretical results is searched for by means of pseudodynamic tests on full-scale structures. Two structural prototypes have been designed, with monolithic joints for the cast-in-situ arrangement and with hinged joints for the precast one. On September 2002 the pseudodynamic test on the precast structure has been performed. The results of this test are presented and compared with the results of a numerical simulation. The good agreement between numerical and experimental results confirms the reliability of the theoretical model and, with it, the results of the statistical analysis.

1 INTRODUCTION

Eurocode 8 (EC8), like many other codes for seismic design of structures, refers to design seismic actions for an elastic analysis of the structure. These actions represent the forces corresponding to the peak acceleration of the structure subjected to the ground motion and are reduced to take into account the actual behaviour of the structure, which has some resources able to attenuate the seismic effects (ductility, overstrength, damping, redundancy). For the static force F equivalent to seismic action EC8 gives:

$$F = \frac{2.5\alpha_g}{q}\,\eta(T)W \qquad (1)$$

where W is the weight of the vibrating mass, $\alpha_g = a_g/g$ is the peak ground acceleration normalised with the gravity constant, $\eta(T)$ is the decreasing function of the elastic response spectrum computed for the natural vibration period T of the structure, and q is a proper reducing coefficient called behaviour factor.

EC8 gives a scale of q-factors related to the different types of structures, with their potential capacity of energy dissipation conventionally evaluated on the base of the ultimate failure mechanism. At present such q-values are defined more or less on the base of empirical choices, not supported by a rigorous investigation

of sufficient reliability. The experience of different seismic countries has contributed to this definition, through compromises reached during the works of the competent European committee. This empirical procedure may lead to unjustified inequalities between different materials and structures.

In order to verify the reliability of the design rules of EC8 for reinforced concrete frame systems, a large number of incremental non-linear dynamic analyses have been done based on a large set of accelerograms, artificially generated in order to obtain given frequency contents, and a large sample of material strengths so to apply a Monte Carlo process of probabilistic investigation of structural performance.

The results of the analyses are summarised in histograms and the "overstrength" ratio between the computed resistance and its design value is assumed as the principal random variable of the response. The distribution curves of the overstrength highlight a good reliability of the design rules of EC8 and confirm the value of the behaviour factor given to reinforced concrete frames. In particular the comparison of the responses coming from different arrangement of joint connections (rigid or hinged) shows that precast structures have the same seismic capacity of the corresponding cast-in-situ structures. This deny the discriminations set by precedent versions of the EC8 to some type of one-storey precast structures.

The experimental confirmation of these theoretical results is searched for by means of pseudodynamic full-scale tests carried out at ELSA Laboratory of Ispra (Italy). Two prototypes have been designed. The connections between columns and beams are made with monolithic joints for the cast-in-situ arrangement and with hinged joints for the precast one. On September 2002 the pseudodynamic test on the precast structure has been performed. The results of the pseudodynamic test lead to assess the analytical model used in the dynamic analyses. The very good correspondence of the computed and experimental responses investigated confirms the reliability of the analytical model and, with it, the results of the statistical analysis.

2 THEORETICAL INVESTIGATION

2.1 Cast-in-situ and precast concrete frames

The theoretical calibration of the behaviour factor q is shown with reference to the frame of Fig. 1, which refers to the two different solutions of a cast-in-situ construction and a precast construction. The structural details can be found in Biondini & Toniolo (2000).

As shown in Fig. 2, it is assumed that, under the same seismic force F, the arrangement of Fig. 2a, with four critical sections dimensioned for a moment $m = Fh/2$, may dissipate the same amount of energy which the arrangement of Fig. 2b dissipates in its two critical sections, dimensioned as they are for a moment $M = Fh$ double than the first one ($4u \cong 2U$). Thus, both the structures are intended to belong to the "prized"

type of frames, with the same force reduction factor. The confirmation of this assumption and the actual quantification of the reduction factor should come from the results of the analysis.

2.2 Dynamic non-linear analysis

The dynamic non-linear analysis for the systems under examination is based on the single degree of freedom motion equation:

$$m\,\ddot{d}(t) + c\,\dot{d}(t) + k(d)d(t) = -m\,a(t) \qquad (2)$$

where m is the vibrating mass, c is the viscous damping coefficient (assumed equal to 5% of the critical one), $k(d)$ is the degrading elastoplastic stiffness and $a(t)$ is the ground acceleration. The static term is directly given as:

$$k(d)d(t) = F(d) - \frac{N_{ad}}{h}d(t) \qquad (3)$$

where the last term represents the second order effect of the vertical load N_{ad} acting on the columns and $F(d)$ is read on the proper force-displacement model. Such model has been derived from Takeda model (Takeda et al., 1970), completed with a decreasing branch as introduced by Priestley et al. (1994) in order to

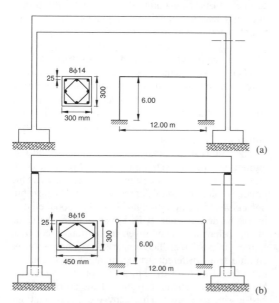

Figure 1. Scheme of the frame: (a) monolithic arrangement and (b) hinged arrangement.

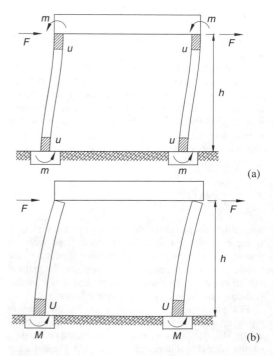

Figure 2. Energy dissipated by the frames: (a) monolithic and (b) hinged arrangement.

represent the ultimate phase of failure. Figure 3 shows the envelope curve of hysteretic cycles, where the limit points of the model correspond respectively to the first cracking of the critical section, to the full yielding of reinforcement and to the compression failure of the concrete core (for more details see Biondini & Toniolo, 2000).

Based on this model, the structural response to given ground motions is computed for increasing values of peak ground acceleration a_g starting from $\alpha_g = a_g/g = 0.30$ with increments $\Delta\alpha_g = 0.05$ up to collapse. The collapse itself is pointed out by the loss of the vibratory equilibrium, shown by unlimited increase of the displacement.

The last value of peak ground acceleration α_{gmax} before collapse is assumed to represent the "experimental" seismic capacity of the structure. This experimental capacity has been compared to the theoretical one given by the design rules of EC8:

$$\bar{\alpha}_g = \frac{q}{2.5\eta(T)} \frac{F_d}{W} \tag{4}$$

where $\eta(T)$ is chosen according to a subsoil type 1A (CEN-prENV 2000) and F_d is the ultimate limit value of the seismic force related to the resistant moments M_{rd} of the critical sections:

$$\begin{aligned} F_d &= 2\,M_{rd}/h \quad \text{for the cast - in - situ frame} \\ F_d &= M_{rd}/h \quad\ \text{for the precast frame} \end{aligned} \tag{5}$$

In this way, the following theoretical design capacities of the frames are obtained:

$$\begin{aligned} \bar{\alpha}_g &= 0.162q \quad \text{for the cast - in - situ frame} \\ \bar{\alpha}_g &= 0.158q \quad \text{for the precast frame} \end{aligned} \tag{6}$$

The seismic performance of the structures are then described in terms of the overstrength ratio between experimental and theoretical seismic capacities:

$$\kappa = \frac{\alpha_{g\,max}}{\bar{\alpha}_g} \tag{6}$$

of which a value equal 1 represents the perfect correspondence between design theory and "test" results.

2.3 Probabilistic model

In the statistical investigation, the randomness of the seismic action is considered by using a large set of artificial accelerograms generated so to match with good compatibility the response spectrum of the code. As an example, Figure 9 shows one of these accelerograms generated with the program SIMQKE (1976), together with its elastic response spectrum.

In order to account also for the randomness of the material properties, both concrete and steel strengths are assumed as random variables with lognormal distributions defined by their characteristic values (5% fractile) and standard deviations: $f_{ck} = 40\,\text{MPa}$ and $s = 5\,\text{MPa}$ for concrete; $f_{yk} = 500\,\text{MPa}$ and $s = 30\,\text{MPa}$ for steel. The corresponding random variation of the degrading stiffness model is shown in Fig. 4, where the envelope curves of the precast arrangement are drawn for about 50 couples of random values f_c, f_y.

2.4 Evaluation of the seismic performance

Figure 5 shows the density distributions of the overstrength ratio κ computed with $q = 5$ and obtained from a Monte Carlo simulation carried out for a sample of 1000 sets of random quantities. A direct comparison of these diagrams confirms the very good correspondence between the seismic performance of the two types of structures. A lognormal regression of the data leads to a 5% fractile which corresponds in both cases to the value $\kappa \cong 1.27$.

Finally, Fig. 6 shows the regression diagram with the couples of capacities computed for each of the 1000 simulations. A direct comparison of the theoretical

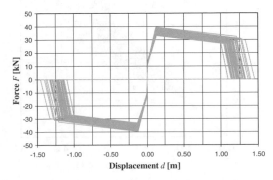

Figure 4. Sample of 50 envelope curves of the random model of degrading stiffness.

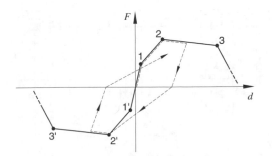

Figure 3. Degrading stiffness model.

continuous line of equal capacity with the dashed line given by the linear regression of the experimental data confirms the average coincidence of the seismic capacities of the two frames.

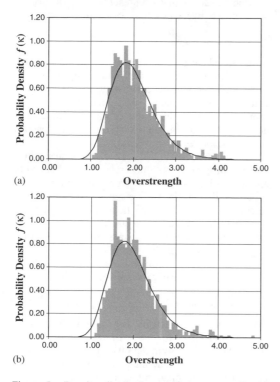

(a)

(b)

Figure 5. Density distribution of the overstrength: (a) monolithic and (b) hinged frames.

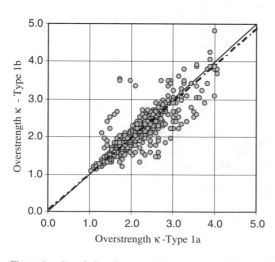

Figure 6. Correlation diagram of the overstrength for both frames (1000 simulations).

3 EXPERIMENTAL INVESTIGATION

3.1 *Precast prototype*

The pseudodynamic tests described in the following have been performed at ELSA European Laboratory for Structural Assessment of Ispra (Italy). Two structural prototypes have been designed, both consisting of six columns connected by two lines of beams and an interposed slab. The connections between columns and beams are made with monolithic joints for the cast-in-situ arrangement and with hinged joints for the precast one. The 5th and 6th September 2002 the pseudodynamic test on the precast structure has been performed.

Figure 7 shows the deck plan and the longitudinal section (parallel to the applied seismic action) of the

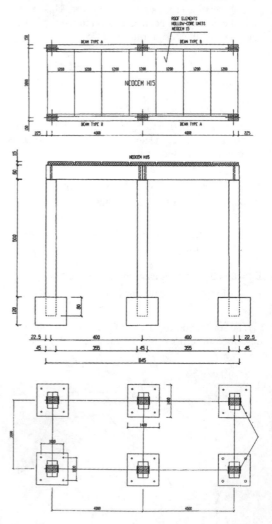

Figure 7. Plant and longitudinal view of the precast prototype.

prototype, as well as its overall dimensions. The columns have cross sizes 300 × 450 mm and are reinforced with 8 bars ɸ16 and stirrups ɸ6 spaced by 50 mm in the lower part (which includes the critical zone of the flexural strength) and by 150 mm in the upper part. Steel class B500 has been used. Concrete strength class is C40/50. For more detailed information see Biondini *et al.* 2002.

3.2 *Testing plant*

The prototype to be submitted to pseudodynamic test has been proportioned with the aim to reproduce

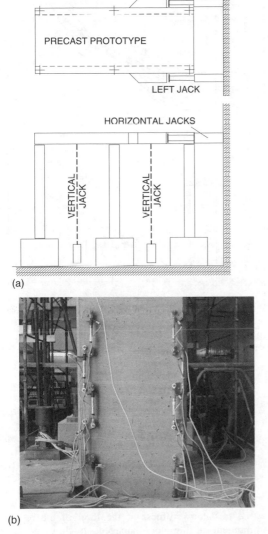

(a)

(b)

Figure 8. (a) Location of the actuators. (b) Detail of the instrumentation at the base of the columns.

possible real dimensions of a precast structure. For obvious economy reasons, all dimensions not affecting the structural behaviour under consideration have been reduced, replacing the missing weights by means of proper vertical jacks. The inertia forces are numerically simulated in the model governing the pseudodynamic procedure, together with the related second order effects. The two principal horizontal jacks have been connected with spherical joints to the deck, close to the midspan of the first bay. They were controlled in tandem with a master-slave system so to obtain an uniform translation of the deck. The displacements to be applied were controlled by means of digital transducers. Figure 8 shows the detailed scheme of the testing plant.

3.3 *Testing programme*

The seismic action for the pseudodynamic test has been simulated by an artificial accelerogram automatically generated so to be compatible with the response spectrum given by Eurocode 8 for a subsoil type 1B (very dense sand or gravel). Figure 9 shows this accelerogram together with the corresponding response spectrum superimposed to the Eurocode 8 spectrum. Through previous non-linear dynamic analyses, elaborated using the quoted model of degrading stiffness

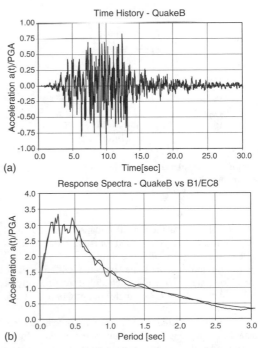

(a)

(b)

Figure 9. Time history and response spectrum of the used artificial accelerogram.

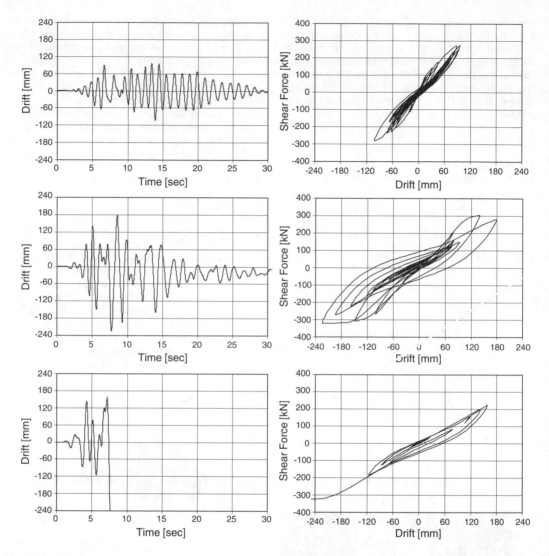

Figure 10. Displacement time-histories and load-displacement curves measured for the intensities $\alpha_g = 0.36$, $\alpha_g = 0.72$ and $\alpha_g = 1.08$.

of the columns, a theoretical collapse limit equal to $\alpha_g = 1.18$ has been quantified (versus a value of $\alpha_g = 1.08$ obtained from Eurocode 8 rules).

Taking into account the expected collapse limit, three load steps have been scheduled, respectively at 1/3, 2/3 and 3/3 of EC8 ultimate strength, or $\alpha_g = 0.36$, $\alpha_g = 0.72$, $\alpha_g = 1.08$.

For the interpretation of the results, some additional data have to be given. The ultimate resistant force, computed with the nominal properties of materials and taking into account the ultimate damaging of the structure, is $F_{rd} = 210$ kN. If the actual material properties

are referred to without damaging, the strength rises to $F_{rd} \approx 265$ kN. Finally, if the hardening of steel is accounted for, as measured with the tensile tests on the reinforcing bars, the strength attains $F_{rd} \approx 320$ kN. This could be the reaction force reached at the first cycles of the dynamic test, when the structural decay has just begun. Furtheron, the displacement at the first full yielding limit of the reinforcement, evaluated with the traslatory stiffness at the level of the resistant moment of the non damaged structure, is $d_y'' \approx 110$ mm, while the outer tensioned bars start yielding at $d_y' \approx 85$ mm.

1182

Figure 11.   View of the precast prototype at the end of the pseudodynamic test.

3.4   Experimental results

From the large set of data recorded during the test, this paper reproduce only the displacement vibration curves of the upper deck and the corresponding force-displacement diagrams. Such diagrams are shown in Fig. 10 for the three levels of seismic action investigated.

The diagrams of the first level at $\alpha_g = 0.36$ show maximum displacements between $+90$ and -100 mm, about the limit of the first yielding. They show again one only cycle with sensible hysteresis and a vibratory behaviour substantially elastic, stabilised along the reduced stiffness of the cracked columns and without any residual deformation. The cracks, observed at the maximum displacements and measured in terms of few tenths of millimeter, closed again at unloading. Then it can be assumed that the structure kept its full strength capacity.

The diagrams at the second level at $\alpha_g = 0.72$ show maximum displacements between $+180$ and -220 mm, far beyond the limit of full yielding of the reinforcement, with several cycles of large hysteresis, and forces raised to $+300$ and -320 kN corresponding to the ultimate strength of the sections with hardened reinforcement and with cracks up to some millimeter at the maximum displacements, only partially closed back at unloading. The residual displacement of about -25 mm indicates the irreversible effects of a certain structural damage. But on the whole, except for some spallings at the top edges of the columns, due to the concentrated pressure of the beams under the maximum deflection of the columns, the structure didn't show damages decisive with respect to its strength resources.

At the third level $\alpha_g = 1.08$, maximum displacements between 400 and 450 mm were expected at the limit of the maximum stroke of the jack pistons. The possibility of full completing of the test was not sure. Actually, after the first violent shocks, due to the residual displacement of the preceding loading step, the amplitude of the motion took the jacks to the end of stroke and the test had to be stopped. The only information which can be taken is that, for the maximum displacement of 400 mm, the cover of the critical zones of the columns was still intact and the reaction forces at the level of 320 kN without any incipient decay. The ultimate collapse limit was still far. This situation is reproduced in the picture of Fig. 11.

3.5   Calibration of the analytical model

The results of the pseudodynamic test allowed to assess the analytical model used in the dynamic analysis. Figure 12 reproduces the computed vibration curves superimposed to the experimental curves.

In the analytical model the strength parameters have been introduced with their actual values as tested in samples of the materials. For the two higher levels of seismic action, the point 1 of the model of Fig. 3 has been omitted, since the structure had already overcome the first cracking limit. A value $c = 0$ of the viscous damping has been assumed, consistently to the physical

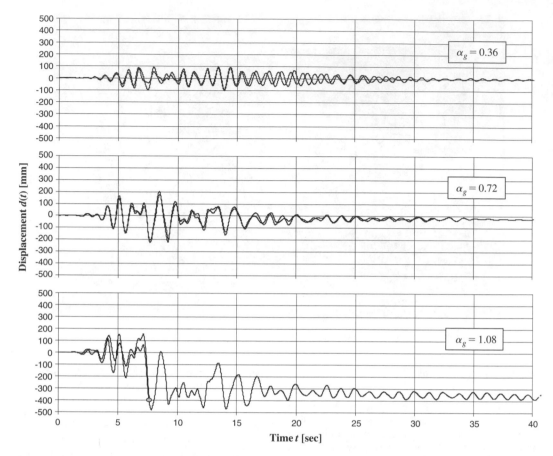

Figure 12. Displacement time-histories: numerical (thick lines) versus experimental results (thin lines).

behaviour of the prototype under the pseudodynamic test and to its control algorithm.

The very good correspondence of the computed and experimental responses confirms the reliability of the analytical model and, with it, the results of the preceding statistical analysis. At $\alpha_g = 0.36$ level some discrepancies occur around 20 sec and this could indicate the necessity to improve the model for postcracking small vibrations. At $\alpha_g = 0.72$ level the very good coincidence of the two curves indicates the good simulation of the pre- and post-yielding behaviour of the cracked structure. $\alpha_g = 1.08$ level the comparison can be made only up to 7.5 sec, time of the test termination. A plausible prediction of the subsequent behaviour seems to be given by the analytical model.

4 CONCLUSIONS

In this paper it has been shown by a theoretical probabilistic approach that cast-in-situ and precast concrete

frames have the same seismic capacity and that the value 4.5 given by Eurocode 8 to the behaviour factor for these frames seems to be correct. It is to be noted that the capacity of the frames for $q = 4.5$ is very high, more than double of what required in Italy for seismic zones of first category. This is due to the high values of their natural vibration periods (about 2 seconds), which lead to a reduced response to the ground motion. Therefore, these type of frames have a large margin of safety with respect to seismic collapse. They find their dimensioning from the non-seismic conditions (such as the wind pressure and the crane actions) and the seismic serviceability limit state referred to the storey drift.

In order to find an experimental confirmation of the theoretical results discussed above, real full scale pseudo-dynamic tests on precast and cast-in-situ reinforced concrete frames have been programmed at the Joint Research Centre of the European Commission at Ispra in Italy. This paper presents the first results on a precast prototype. In particular, the results of this test are compared with those obtained from a numerical simulation. The good agreement between numerical and experimental results confirms the reliability of the theoretical model and, with it, the results of the statistical analysis.

ACKNOWLEDGEMENTS

The tests have been performed within the Ecoleader Programme, which is reserved to the European Consortium of Laboratories for Earthquake and dynamic Experimental Research (JRC – Contract n° HPRI-CT-1999-00059). Particular thanks are given to Mr. Georges Magonette and Mr. Javier Molina who managed the setting up of the instrumentation plant and the execution of the loading tests, ensuring with their high professional ability the perfect accomplishment of the experimentation. Thanks also to Mr. Carlo Bonfanti, who menaged the design and the execution of the prototype, for the important contribution of his experience. The research has been led jointly with Prof. Matej Fishinger and his assistants of the Ljubliana University.

REFERENCES

Biondini F. and Toniolo G., Comparative Analysis of the Seismic Response of Precast and Cast-in-situ Frames. *Studies and Researches*, Graduate School for Concrete Structures, Politecnico di Milano, 21, 1–17, 2000.

Biondini F., Toniolo G. and Tsionis G., Design Reliability of Cast-in-Situ and Precast Concrete Frames under Recorded Earthquakes. *Studies and Researches*, Graduate School for Concrete Structures, Politecnico di Milano, 22, 2001.

Biondini F. and Toniolo G., Probabilistic Verification of the Seismic Performance of Precast and Cast-in-situ Reinforced Concrete Frames. *Studies and Researches*, Graduate School for Concrete Structures, Politecnico di Milano, 23, 1–17, 2002.

Biondini F., Ferrara L., Negro P. and Toniolo G., Results of Pseudodynamic test on the Prototype of a Precast R.C. Frame. *Proc. of 14° C.T.E. Congress*, Mantova, Italy, November 7-8-9, 2002 (in Italian).

Biondini F. and Toniolo G., Probabilistic parameters of the Seismic Performance of Reinforced Concrete Frames. *Proc. of the 1st fib Congress*, Paper E-228, Osaka, Japan, October 13–19, 2002.

Biondini F. and Toniolo G., Seismic Behaviour of Concrete Frames: Experimental and Analytical Verification of Eurocode 8 Design Rules. *Proc. fib Symposium Concrete Structures in Seismic Regions*, Athens, Greece, May 6–9, 2003.

CEN-prENV 2000, *Eurocode 8: Design of Structures for Earthquake Resistance*, European Committee for Standardization, Brussels.

Priestley M.J.N., Verma R. and Xiao Y., Seismic Shear Strength of R.C. Columns, *ASCE Journal of Structural Engineering*, 120(8), 2310–2329, 1994.

SIMQKE, *A Program for Artificial Ground Motion Generation. User's Manual and Documentation*, NISEE, Department of Civil Eng., Massachusetts Institute of Technology, 1976.

Takeda T., Sozen M.A. and Nielsen N.N., Reinforced Concrete Response to Simulated Earthquakes, *ASCE Journal of the Structural Division*, 96(12), 2557–2573, 1970.

System-based Vision for Strategic and Creative Design, Bontempi (ed.)
© 2003 Swets & Zeitlinger, Lisse, ISBN 90 5809 599 1

Seismic effectiveness of direct and indirect implementation of MPD systems

S. Silvestri, T. Trombetti & C. Ceccoli
Department D.I.S.T.A.R.T., Faculty of Civil Engineering, University of Bologna, Italy

G. Greco
Civil Engineer in Cesena, Italy

ABSTRACT: This paper illustrates the advantages offered by inserting added viscous dampers into shear-type structures in accordance with a special scheme based upon the mass proportional damping component of Rayleigh viscous damping matrices, which is herein defined as MPD system. A comparison between the dissipative performances offered by the MPD system and a number of other systems of added viscous dampers (comprising classical and non-classical damping systems) is provided herein. The dynamic behaviour of direct and indirect implementation of the MPD system is checked through numerical time-history simulations for a 6-storey shear-type structure subjected to seismic excitations. The results, here obtained with reference to a number (40) of earthquake ground motions, confirm what previously found by the authors by using indexes based on the stochastic response and on the modal damping ratios: systems characterized by most dampers placed so that they connect each storey to the ground (as it is for the MPD system) display larger efficiency in energy dissipation than systems characterised by interstorey damper placement (traditional placement). This indicates that the efficiency of damping systems depends much more on damper placement rather than on damper sizing.

1 INTRODUCTION

Dissipative systems have widely proven (Hart & Wong 2000) to be able to effectively mitigate seismic effects on buildings. However, still the issue is open of how to insert viscous dampers into shear-type structure in order to reach the best dissipative performances of the dynamic system (structure + dampers).

Actually, most of the research works available in literature regarding the problem of damper system optimisation deals with the search for optimal damper sizing for given traditional inter-storey damper placement (Hahn & Sathiavageeswaran 1992, Takewaki 1997 and 2000, Singh & Moreschi 2001 and 2002). Researches carried out by the authors (Trombetti et al. 2002a, b, c, d) have identified that the mass proportional damping (MPD) and stiffness proportional damping (SPD) components of the Rayleigh systems (characterised by completely different dissipative properties) correspond to two physically separated and effectively independently implementable systems.

In these works, the authors have also investigated the dissipative effectiveness offered by different damper placement/sizing (when applied to shear-type structures) using performance criteria based upon the system response to a stochastic base input and/or, for classical damping systems only, upon the modal damping ratios.

The aim of this paper is to compare the results (in terms of dynamic performances of selected damping systems) obtained in previous works using synthetic performance indexes with those obtained under seismic excitation (40 historically recorded earthquake ground motions are used as dynamic inputs).

2 BACKGROUND/PROBLEM FORMULATION

The aim of the body of the research work carried out in the last years by the authors (of which this paper would like to be a verification), is to identify the system of added viscous dampers which optimises the dissipative properties of the dynamic system composed of shear-type structure and viscous dampers.

The verification, without loss of generality, is here carried out using, as applicative example, the "reference" structure described below. To meaningfully compare different damping systems, a constraint about the "total" sizing of viscous dampers is here introduced as "equal total cost" constraint.

2.1 The "reference" structure

The "reference" structure is a 6-storey shear-type structure characterised by values of mass and lateral stiffness which do not vary along the building height. The lateral stiffness k_j of the vertical elements connecting each j-th storey to the one below is equal to $k = 4 \cdot 10^7 \text{N/m}$ and the floor mass m_j of each j-th sto-rey is equal to $m = 0.8 \cdot 10^5$ kg. Interstorey height is $h = 3$ m. Total height is $h_{tot} = 18$ m.

2.2 The "equal total cost" constraint

In order to meaningfully compare the energy dissipation properties of various systems of added viscous dampers implementable in the reference shear-type structure, in all following analyses and for all systems considered:

– internal (intrinsic) damping is neglected, and
– it is imposed that the sum, c_{tot}, of the damping coefficients, c_j, of all n dampers introduced into the structure, be equal to a set value, $\bar{c}$, as also imposed by Takewaki (1997 and 2000) and Singh & Moreschi (2001 and 2002).

The above constraint (in the following referred to as "equal total cost" constraint) mathematically translates in the following formula:

$$c_{tot} = \sum_{j=1}^{n} c_j = \bar{c} \qquad (1)$$

with $\bar{c}$ set equal to $9 \cdot 10^6 \text{ N} \cdot \sec/\text{m}$, for sake of comparison with results available in literature (Takewaki 1997).

3 CLASSICAL DAMPING SYSTEMS

For Rayleigh damped multi-degree-of-freedom systems, the damping matrix $[C]$ becomes:

$$[C]_R = \alpha[M] + \beta[K] \qquad (2)$$

where $[M]$ and $[K]$ are, respectively, the mass matrix and the stiffness matrix and α and β are two proportionality constants having units of $\sec^{-1}$ and $\sec$, respectively. Equation (2) allows to define the two following damping matrices:

– mass proportional damping (MPD) matrix:

$$[C]_{MPD} = \alpha[M] \qquad (3)$$

– stiffness proportional damping (SPD) matrix:

$$[C]_{SPD} = \beta[K] \qquad (4)$$

that correspond, respectively, to the MPD and SPD limiting cases of Rayleigh damping.

For the sake of clarity, the added-damper system that allows an MPD matrix to be obtained is defined herein as "MPD system" and, likewise, that which allows an SPD matrix to be obtained is referred to as "SPD system".

The reference 6-storey shear-type structure is represented in Figure 1a. Figures 1b, c and d provide a physical representation of the structure equipped with Rayleigh system, SPD system and MPD system, respectively.

From the physical representations of Figure 1, it can be seen that the dampers of the MPD system ($\alpha m_1 = \alpha m_2 = \cdots = \alpha m_6 = \alpha m$) link each storey to the ground, whilst the dampers of the SPD system ($\beta k_1 = \beta k_2 = \cdots = \beta k_6 = \beta k$) link each storey to the adjacent one.

As a direct result of this observation, we can give MPD and SPD systems the following alternative definitions in terms of damper sizing and placement:

– MPD system: the dampers are placed in such a way as to connect each storey to a fixed point (ground or infinitely-stiff vertical lateral-resisting element) and sized so that each damping coefficient c_j is proportional to the corresponding storey mass m_j;
– SPD system: the dampers are placed in such a way as to connect two adjacent storeys and sized so that each damping coefficient c_j is proportional to the lateral stiffness k_j of the vertical elements connecting these two storeys.

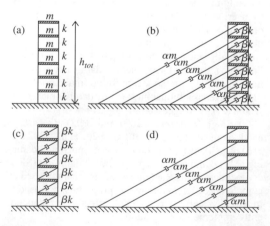

Figure 1. The reference 6-storey structure: (a) undamped, (b) damped with Rayleigh system, (c) with SPD system and (d) with MPD system.

For a generic N-storey Rayleigh system, the "equal total cost" constraint becomes:

$$\alpha \sum_{j=1}^{N} m_j + \beta \sum_{j=1}^{N} k_j = \bar{c} \tag{5}$$

Imposing the equal total cost condition to every possible type of Rayleigh system enables us to identify a system class characterised by the following α and β values:

$$\alpha = \bar{\alpha} \cdot (1 - \gamma) \text{ and } \beta = \bar{\beta} \cdot \gamma \tag{6}$$

where

$$\bar{\alpha} = \bar{c} \bigg/ \sum_{j=1}^{N} m_j \text{ and } \bar{\beta} = \bar{c} \bigg/ \sum_{j=1}^{N} k_j \tag{7}$$

and γ is a dimensionless parameter with values ranging between 0 and 1 that identifies each specific Rayleigh system within the class defined above. $\gamma = 0$ identifies the MPD system, whilst $\gamma = 1$ identifies the SPD system.

To compare the performances of MPD and SPD systems with other Rayleigh damping systems use is made of indexes based upon the system response to the following stochastic input:

- white noise, band-limited between 0 and $\bar{\omega} = 60\,rad$/sec, stationary, Gaussian with zero mean, and
- characterised by constant power spectral density of amplitude $A^2 = 0.144\,m^2/\text{sec}^3$.

These values have been chosen so that standard deviation of acceleration at the base of the structure supplied by this stochastic process is equal to 0.3 g.

Given a structure subjected at the base to the white noise acceleration above-defined, the mean square response (Crandall & Mark 1963), that coincides with variance for stochastic inputs with zero mean value, σ_j, of j-th storey displacement of the structure, is calculated as:

$$\sigma_j^2 = A^2 \int_0^{\bar{\omega}} |H_j(\omega)|^2 d\omega \tag{8}$$

where $H_j(\omega)$ is the j-th component (j corresponds to the coordinate of the j-th storey) of the system transfer function vector, $\{H(\omega)\}$, defined as:

$$\{H(\omega)\} = -\left(-\omega^2[M] + i\omega[C] + [K]\right)^{-1}[M]\{1\} \tag{9}$$

where ω represents the natural circular frequency, $i = \sqrt{-1}$ and $\{1\}$ is a vector whose elements are all unity for shear-type structures. According to the usual

notation of probabilistic theory, σ_j denotes the standard deviation of the j-th storey displacement.

Of all Rayleigh damping systems, the MPD system ($\gamma = 0$) represents the optimum solution as far as minimising σ_j is concerned, whilst the SPD system ($\gamma = 1$) provides the worst solution. This fundamental result is clear when analysing Figure 2 that shows σ_j as a function of parameter γ for each storey of the reference structure.

The curves $\sigma_j(\gamma)$ of Figure 2 are extremely smooth and feature an almost horizontal tangent at $\gamma = 0$, so that systems characterised by γ values close but not equal to zero still show very low values for σ_j. This clearly indicates the "robustness" of the dissipation efficiency of MPD systems. Figure 2 also illustrates the widening gap in the dissipative effectiveness of MPD and SPD systems following a corresponding increase with the number of storey j.

These first results are in accordance with those which can be obtained by taking into consideration an index based upon the modal damping ratios of the damped structure. A meaningful index of the system dissipative properties is the modal damping ratio weighted average, ξ_{av}^R (where subscript av stands for $average$), defined as follows:

$$\xi_{av}^R = \sum_{n=1} \bar{V}_{bn} \xi_n^R \tag{10}$$

where ξ_{av}^R is the n-th modal damping ratio of the generic Rayleigh system (Clough and Penzien 1993) and $\bar{V}_{bn}$ is the n-th base shear $modal\ contribution\ factor$, as defined by Chopra (1995):

$$\xi_n^R = \frac{\alpha}{2\omega_n} + \frac{\beta\omega_n}{2} \text{ and } \bar{V}_{bn} = \frac{M_n^*}{\sum_{n=1}^{N} M_n^*} \tag{11}$$

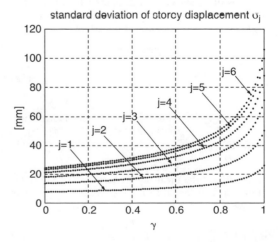

Figure 2. σ_j as a function of γ.

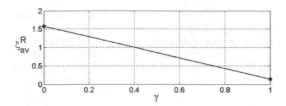

Figure 3. ξ_{av}^R as a function of γ.

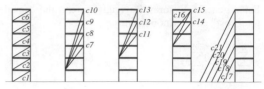

Figure 4. All possible damper placements for the reference 6-storey shear-type structure.

with M_n^* representing the *base shear effective modal mass* (Chopra 1995) of the structure's n-th mode of vibration.

For the reference shear-type structure considered, $\xi_{av}^{MPD} = 1.579$ and $\xi_{av}^{SPD} = 0.137$.

Figure 3 plots ξ_{av}^R as a function of γ. It is clear that ξ_{av}^{MPD} is always greater than any other ξ_{av}^R. Furthermore, notice that $\xi_{av}^{MPD}/\xi_{av}^{SPD} \cong 12$. Therefore, for typical Italian residential buildings (about 6-storeys high), it can be said that there is about one order of magnitude between ξ_{av}^{MPD} and ξ_{av}^{SPD}, and therefore between the respective damping efficiency of the MPD and SPD systems.

4 NON CLASSICAL DAMPING SYSTEMS

The investigations of the efficiency of damping systems are now extended to non-classically damped shear-type structures which are characterized by completely generic damper placement, thus including other non-conventional damper placements.

4.1 *Genetically identified optimal systems*

Genetic algorithms have been recently used in civil engineering applications for the identification of optimal systems (Singh & Moreschi 2002). Genetic algorithms are here used in order to numerically identify systems of added viscous dampers which minimise, for the reference structure and under the "equal total cost" constraint, the following two performance indexes:

– Mean Square Response Average:

$$\text{MSRA} = \frac{1}{6}\sum_{j=1}^{6}\sigma_j^2 \qquad (12)$$

– Interstorey Drift Mean Square Response Average:

$$\text{IDMSRA} = \frac{1}{6}\sum_{j=1}^{6}\overline{\sigma}_{IDj}^2 \qquad (13)$$

where $\overline{\sigma}_{IDj}$ is the mean square response of j-th interstorey drift of the structure and can be computed as follows:

$$\sigma_{IDj}^2 = A^2 \int_0^\omega \left| H_{IDj}(\omega)\right|^2 d\omega \qquad (14)$$

Table 1. Damping coefficients of SPD, TAK, MPD, GIOMSRA and GIOIDMSRA systems [$10^6\ N \cdot$ sec/m].

	SPD	TAK	MPD	GIOMSRA	GIOIDMSRA
c1	1.50	4.80	1.50	0	0.45
c2	1.50	4.20	0	0	0
c3	1.50	0	0	0	0
c4	1.50	0	0	0	0
c5	1.50	0	0	0	0
c6	1.50	0	0	0	0
c7	0	0	0	0	0
...	...	...	...	...	...
c16	0	0	0	0	0
c17	0	0	1.50	0	1.35
c18	0	0	1.50	1.50	1.80
c19	0	0	1.50	1.50	1.80
c20	0	0	1.50	2.25	1.80
c21	0	0	1.50	3.75	1.80

where $H_{IDj}(\omega)$ is the j-th component of the transfer function vector, $\{H_{ID}(\omega)\}$, of the interstorey drifts, defined as:

$$\{H_{ID}(\omega)\} = [T]\{H(\omega)\} \qquad (15)$$

with $[T]$ being a $N \times N$ constant matrix consisting of 1, -1, and 0.

In this search, the form of the damping matrix is left free, so that (*i*) classical and non-classical damping systems can be considered at once and (*ii*) no constraint upon the damper placement is imposed.

The genetically identified optimal (GIO) systems which minimise the performance indexes MSRA and IDMSRA are herein referred to as "*GIOMSRA and GIOIDMSRA systems*", respectively. With reference to Figure 4, the values of the damping coefficients of the GIOMSRA and GIOIDMSRA are given in Table 1. In both cases, these systems are characterised by dampers that connect storeys to the ground, as in the case of the MPD system.

4.2 *"Reference" damping system*

The damping scheme identified in the recent works by Izuru Takewaki (1997) as "optimal" (for the reference structure here considered) is also taken into account and will be referred hereafter to as *"TAK system"*. The specific values of the damping coefficients (given in Table 1) of the TAK system minimise the sum of amplitudes of the transfer functions of interstorey drifts evaluated at the undamped fundamental natural frequency ω_1:

$$\sum_{j=1}^{6}\left|H_{ID_j}(\omega_1)\right| \tag{17}$$

within the restricted class of dampers placed between adjacent storeys and satisfy the "equal total cost" constraint. This system was identified by Takewaki using an algorithm based upon an inverse problem approach, proposed by the same author (1997).

5 SEISMIC BEHAVIOUR OF THE SELECTED DAMPING SYSTEMS

In order to verify the indications regarding the dissipative properties of the selected damping systems (SPD, TAK, MPD, GIOMSRA and GIOIDMSRA) obtained by the authors from a number of performance indexes, a comprehensive study of the seismic behaviour of structures equipped with such devices has been carried out. This section presents selected results obtained for the reference structure excited at the base with 40 historically recorded ground motions including, among the others, Imperial Valley, 1940, El Centro record, NS component (270°), PGA = 0.215 g; Kern County, 1952, Taft Lincoln School record, EW component (21°), PGA = 0.156 g; and Kobe, 1995, Kobe University record, NS component (90°), PGA = 0.310 g.

Figures 5a, b show the average values (over the 40 inputs) of the maximum storey displacements and accelerations, respectively. In detail, from Figure 5a it is clear that the structure equipped with the SPD system develops the largest displacements, the structure equipped with the TAK system displays displacements that are reduced of about 15% w.r.t. those of the SPD equipped structure, and the structures equipped with the MPD, GIOMSRA and GIOIDMSRA systems provide the structure with the smallest displacements (reduction of roughly 85% at the top-storey w.r.t. the SPD equipped structure).

As far as the storey accelerations are concerned, Figure 5b indicates that, overall, the best performances (in terms of reduced floor accelerations) are offered by the MPD and GIOIDMSRA systems, while the worst performances are offered by the SPD system. The TAK and GIOMSRA systems provide "mixed" results: the TAK system is the worst performer at the top-storeys, but the best performer at the 1st storey, while the GIOMSRA system is the worst performer at the 1st and 2nd storeys, but the best performer at the top-storey.

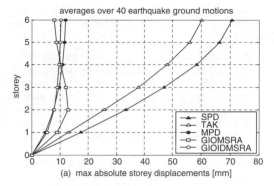

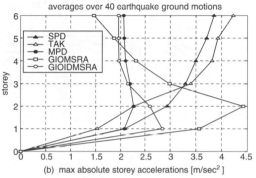

Figure 5. Average values (over 40 earthquake ground motions) of maximum storey displacements (a) and accelerations (b) of the reference structure equipped with SPD, TAK, MPD, GIOMSRA and GIOIDMSRA systems.

The above results are in accordance with those obtained through synthetic response indexes and, again, indicate that the best overall seismic performances are offered by damping systems characterised by fixed point placement (MPD, GIOMSRA and GIOIDMSRA systems), thus confirming the importance of damper placement.

6 DIRECT AND INDIRECT IMPLEMENTATION OF MPD SYSTEMS

With reference to the schematic representations of Figure 6, the MPD system can be implemented either (a) by placing damping devices on long buckling-resistant braces which connect the storeys to the ground (direct implementation: this leads to an "exact" MPD matrix, if damper sizing is chosen properly) or (b, c) by placing them so that they connect two structures characterised by different values of the lateral stiffness (indirect implementation: the resulting damping matrix is not equal to the "exact" MPD matrix).

In order to investigate the performances offered by the indirect implementation of MPD systems, a number of numerical investigations were carried out. This section presents the results obtained for the following two systems:

- the added viscous dampers connect the reference structure with a lateral-resisting element which deforms in a flexural way (Fig. 6b). The horizontal stiffness evaluated at the top storey of the vertical flexural concrete core is imposed to be equal to $3\ EJ/h_{tot}^3 = 5k = 2 \cdot 10^8$ N/m. This type of damper implementation is herein referred to as *"MPD-EJ-5k system"*.
- the added viscous dampers connect the reference structure with another shear frame characterised by different values of stiffness (Fig. 6c). The horizontal

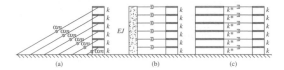

(a)   (b)   (c)

Figure 6. (a) Direct and (b, c) indirect implementation of MPD system.

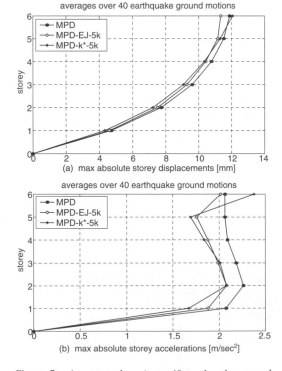

Figure 7. Average values (over 40 earthquake ground motions) of maximum storey (a) displacements and (b) accelerations of the reference structure equipped with MPD, MPD-EJ-5k and MPD-k*-5k systems.

stiffness of each storey of the shear frame is imposed to be again equal $k^* = 5k = 2 \cdot 10^8$ N/m. This type of damper implementation is herein referred to as *"MPD-k*-5k system"*.

Figures 7a, b compare the seismic response (in terms of average maximum storey displacements and accelerations) of the reference structure excited at the base with 40 historically recorded ground motions, when equipped with direct and indirect implementation of the MPD system. The results clearly show that direct and indirect implementation of MPD systems provide similar performances.

7 APPLICABILITY OF MPD SYSTEMS

Direct implementation requires the use of long dissipative braces for which, at the present time, the following technological solutions can be envisaged: (*i*) use of the "mega braces" of the Taylor Devices Company (which have been already employed even if not exactly following an MPD scheme, for the Chapultepec Tower best known as Torre Major and shown in Figure 8 in Mexico City), (*ii*) use of the "unbonded braces" of Nippon Steel Corporation (Clark et al. 1999) and (*iii*) use of the prestressed cables of SPIDER project (Chiarugi et al. 2001).

Indirect implementation of MPD system can be obtained with usual damping devices and appropriate positioning in the structural system.

Figure 8. The Chapultepec Tower in Mexico City with "mega-braces" (kindly provided by Pr. P. Duflot of Taylor Devices).

8 CONCLUSIONS

This paper, with specific reference to a 6-storey shear-type structure, presents selected results from a comprehensive numerical investigation, which assess the overall good dissipative performances offered by mass proportional damping (MPD) systems under seismic excitation.

The paper also introduces the "indirect" implementation of MPD systems, which is capable of providing dissipative effectiveness similar to the "direct" implementation.

REFERENCES

Chiarugi, A., Terenzi, G. & Sorace, S. 2001. Metodo di progetto di dispositivi siliconici per l'isolamento alla base: applicazione a due casi sperimentali. *Atti del 10° Convegno Nazionale ANIDIS*. Potenza-Matera, Italy.

Chopra, A.K. 1995. In Prentice Hall (ed.), *Dynamics of Structures, Theory and Applications to Earthquake Engineering*. Englewood Cliffs.

Clark, P.W., Aiken, I.D., Ko, E., Kasai, K. & Kimura, I. 1999. Design Procedures for Building Incorporating Hysteretic Damping Devices. *Proceedings, 68th Annual Convention, Structural Engineers Association of California*, Santa Barbara, California.

Clough, R.W. & Penzien, J. 1993. In McGraw-Hill (eds), *Dynamics of Structures, 2nd edition*, *Civil Engineering Series*. New York.

Crandall, S.H. & Mark, W.D. 1963. In Academic Press (ed.), *Random Vibrations in Mechanical Systems*. New York and London.

Hahn, G.D. & Sathiavageeswaran, K.R. 1992. Effects of added-damper distribution on the seismic response of buildings. *Computers & Structures*, 43(5): 941–950.

Hart, G.C. & Wong, K. 2000. In John Wiley & Sons (eds), *Structural Dynamics for Structural Engineers*. New York.

Singh, M.P. & Moreschi, L.M. 2001. Optimal seismic response control with dampers. *Earthquake Engineering and Structural Dynamics*, 30: 553–572.

Singh, M.P. & Moreschi, L.M. 2002. Optimal placement of dampers for passive response control. *Earthquake Engineering and Structural Dynamics*, 31: 955–976.

Takewaki, I. 1997. Optimal damper placement for minimum transfer functions. *Earthquake Engineering and Structural Dynamics*, 26: 1113–1124.

Takewaki, I. 2000. Optimal damper placement for critical excitation. *Probabilistic Engineering Mechanics*, 15: 317–325.

Trombetti, T., Silvestri, S. & Ceccoli, C. 2002a. Added viscous dampers in shear-type structures: the effectiveness of mass proportional damping. *Technical Report N. T. n° 58. DISTART*. Università di Bologna, Italy.

Trombetti, T., Silvestri, S. & Ceccoli, C. 2002b. On the first modal damping ratios of MPD and SPD systems. *Technical Report N. T. n° 64. DISTART*. Università di Bologna, Italy.

Trombetti T., Silvestri, S. & Ceccoli, C. 2002c. Inserting Viscous Dampers in Shear-Type Structures: Analytical Formulation and Efficiency of MPD System. *Proceedings of the 2nd International Conference on Advances in Structural Engineering and Mechanics (ASEM'02), Pusan, Korea*.

Trombetti T., Silvestri S. & Ceccoli C. 2002d. Inserting Viscous Dampers in Shear-Type Structures: Numerical Investigation of the MPD System Performances. *Proceedings of the 2nd International Conference on Advances in Structural Engineering and Mechanics (ASEM'02), Pusan, Korea*.

A practical code for seismic risk mitigation of historical centers: the case of Umbria (Central Italy)

F. Maroldi
Department of Structural Engineering, Politecnico di Milano, Milano, Italy

ABSTRACT: The earthquake occurred in Italian regions Umbria and Marche, called Assisi Earthquake, has proposed once more the problem of the definition of suitable strategies for seismic risk mitigation on a territorial scale. The Umbria landscape, characterized by a system of small urban centers only made up of historical construction, represent, in its whole, a monument. Therefore, the problem of its protection is faced on the basis of multidisciplinary research, as it is usually done for monuments. The data base used consists of the whole constructions set of Sellano, a urban center close to the epicentre of the earthquake of 1997. Damage suffered by the constructions has been studied in connection with the current characteristics of buildings (i.e., typology, structural system, component materials) often deriving by their localization and related with building histories. It is, here, shown how the memory of the transformation suffered by constructions during their life (qualitative data) influence their seismic behavior. Analysis performed allows to define: evaluation criteria of seismic vulnerability for urban agglomerates where great attention is spent in the identification of the best minimum unit of intervention (UMI); guide lines for the creation of a practical code for rehabilitation in urban centers of Umbria referring to technical-quantitative data well as to qualitative ones.

1 INTRODUCTION

The earthquake occurred in Italian regions Umbria and Marche in 1997, called as Assisi Earthquake, has proposed once more the problem of defining suitable strategies of seismic risk mitigation on a territorial scale; it cannot be happened by chance that existing buildings have fallen at every earthquake.

In the first inventory of the urban centers in Umbria after earthquake of 26 September 1997, it is written: "Entire old town centers collapse as houses of cards hit by an earthquake not particularly severe when compared with seismic events that over the centuries occurred in Italy center; these are episodes of an already written history".

The question concerns above all minor old town centers and minor buildings widespread on the region, the territory being characterized by the presence of "dwelling materials", building up a dwelling continuum, where the word "material" means a dwelling structure marked by the alternation between anthropic areas on different levels, but traceable back to agricultural use, and areas built up according to building techniques settled in traditions. A global revision of protection politics is needed for these minor buildings at their different levels and ways of aggregation. Experience shows that their integrity has vanished even

when rehabilitation works have been carried out in recent times and points out how strategies of risk mitigation, based only on static criteria of reinforcement of single residential units, are inadequate. A more complex approach to the problem seems necessary.

The preservation of the historical and architectural value of this building heritage requires rehabilitation and reinforcement strategies respectful of the identity of the sites; safety requirements add to preserving ones. The analysis of damage suffered by building on the occasion of the last earthquakes in Italy shows how safety cannot come off leaving out of consideration preservation criteria; risk mitigation design, apparently hovered between safety and preservation, finds in the culture of preservation effective principles of safety; that is a knowledge of construction is needed from different disciplinary angles and deep rooted in the site history. In this methodological context the paper takes its placed; the data base of reference consists in the urban centers of Sellano and Nocera Umbra seriously damaged by the Assisi Earthquake.

The predisposition to seismic damage of buildings is studied depending on their present state as well as on their typological evolution in time by different points of view, architectural, structural, material and constructive, and related to the evolution of the urban aggregation and the site to which they belong.

On the basis of the results obtained guide lines for the design of risk mitigation are defined related to different intervention scales; as a conclusion some hypotheses of rehabilitation design are studied for significant building of the considered sites.

2 TRASFORMATION-TIME-IDENTITY

Today, and not only in seismic zones, the urban identity is mentioned mainly for stating that it does not live anymore in the territories of the human settlement, that towns to day are abstract spaces of a last identity, probably never entirely owned: from this, the research of a definition of the term "urban identity" on the basis of different disciplines and disciplinary instruments.

The question of regaining and preserving identity must not reduce to a "compact and monotonic re-establishment", but conform to a "control of centrifuge and dissipative tensions" existing into the relation among subjects interacting in to the site (C. Levi Strauss).

The proposed paper tries to give a contribute in this sense starting from structural disciplines and analysis methodologies of seismic behavior of buildings.

From an analytical point of view it is necessary to indagate the conceptual binomial identity-difference, identity as outcome of processes of differentiation from what is something else; to research identity means to research differences; to think in terms of differentiation process implies to introduce the variable time. The concept of identity is defined by the trinomial transformation-time-identity.

The Clear effect of an earthquake is the physical annihilation of lives and sites, but the destroying phenomenon that it puts up is much deeper; earthquake introduces a discontinuity point into the site evolution, brings back at its origin the identification process through which the site just acquired its identity.

Building life in seismic zone, defined by destructions and rehabilitation interventions is the way in which hardly and slowly the site identity has been formed in time and along which is perhaps more difficult preserve it than elsewere; earthquake represents an hard testing as of structural strength as of the capability that a site identity has of remaining intact.

Urban centers in seismic zone mourn today the loss of their own identity there where reconstruction strategies have distorted building heritage from a typological, functional, technological and constructive point of view.

3 KNOWLEGE OF INTRUMENTS FOR THE DESIGN

The process of knowledge and research that supports the construction of the design of seismic risk mitigation must individuate its roots in the peculiar identity of the seismic site. The analysis of the present state of buildings is not enough; together with the regulations contained in building codes and in the dynamic analysis criteria an intermediate level of knowledge is needed that allows to consider the reasons of the site identity. This is what clearly comes out from the damage analyses of constructions hit by the last earthquakes in Italy when many damaged buildings were those that had been subjected to rehabilitation and reinforcement interventions before the earthquake, often standard interventions and homologated, that is ignoring the site identity. Together with the simulative models of structural analysis, qualitative models of a minute and detailed knowledge of the site constructive rules are needed.

With this aim guide lines are individuated to define the design process that have not the meaning of prescriptions and of homologation of the results. In them what is coercive is the approach methodology, not the rule. The guide lines are fitted into three scale levels respectively regarding:

– the intrinsic qualities of the site;
– the habitat;
– the constructions.

3.1 The intrinsic qualities of the site

The site identity is individuated by some elements characterized by strong potential, often unexpressed or in past time repressed. They are intrinsic qualities as the geographical position, the morphological and geological condition of ground, to be a focal point with respect to the landscape structure: qualities that have to be pointed out and exploited.

It is necessary to identify significant parameters suitable for reading and representing the signs of the contemporary territory from the point of view of the design. In seismic zones geomorphological and geological maps are introduced to represent the phenomena of earthquake amplification and the related distribution of hazard on the territory. A further consideration regards the question of attributing a significance to the landscape focal points when destroyed.

3.2 The habitat

By habitat it is meant the tangible shape of the human occupation of the ground; construction features represent a reflection of types of life, of social, political and economical organization. From the point of view of the architectural design it is impossible lo leave out of consideration the evolution analysis of architectural and constructive technologies and the study of the ways in which the relation shape-structure-construction combines in the site.

In the seismic zone the time history of constructions is defined by earthquake occurrence, the habitat transformations in time can be suitable read as a point process where points are the destroying seismic events. The earthquake teaches, and the evolution of the typologies is conditioned by seismic events occurence.

3.3 The constructions

Constructions have to be studied in their material becoming, methodologies for reading damage caused by the earthquake have to be defined related to tectonics of constructions and of their meaningful elements; that is, constructions have to be indagated in the time history of the material context to which they belong (initial conditions and boundary conditions of the mathematical structural model).

The question of the aggregation level of the constructions set that is under consideration is interpreted as a definition problem of seismic vulnerability at different scales; from the detailed methods of reading the predisposition to damage of monuments to the methodologies of vulnerability definition for wide sets of buildings through a vulnerability index.

4   THE CASE OF SELLANO

Some of the analysis carried out in the above mentioned sense for the urban center of Sellano are shown in the Figures.

The constructions in their whole are considered as a set whose sample element is the building with its time-space memory.

Cognitive maps of the ground characteristics (Fig 1) and of the related seismic local hazard are defined superimposable on buildings vulnerability maps. As focal points of landscape the historical layout of streets and the profile of hills are individuated (§3.1).

THE GREGORIAN CADASTRE OF SELLANO

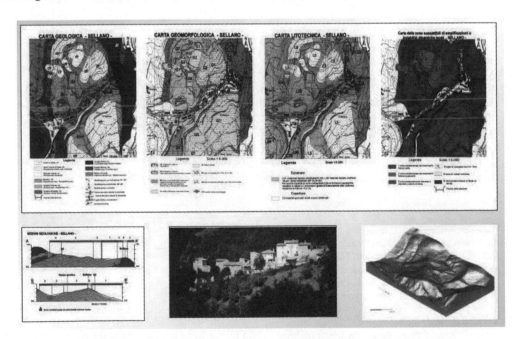

Figure 1.   Sellano. The intrinsic qualities of the site.

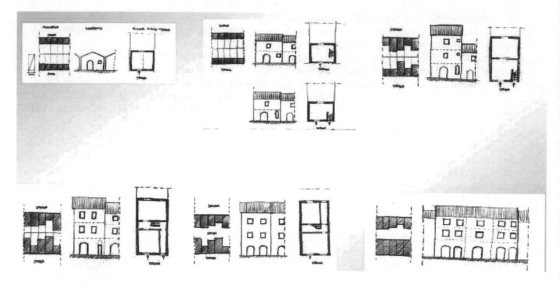

Figure 2a.   Sellano. Habitat evolution.

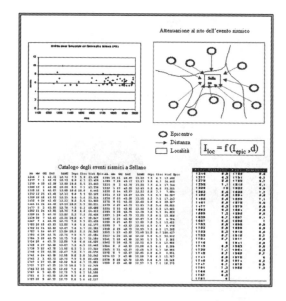

Figure 2b.   Sellano. Historical Seismic Catalogue.

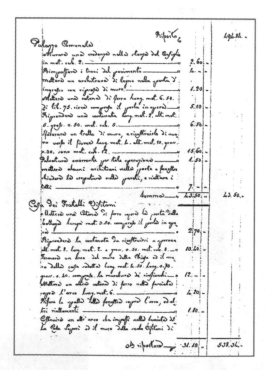

Figure 2c.   Sellano. Data base from the Historical Record Office.

Evolution analyses of architectural, structural and constructive typologies is performed. The constructions time history is related to the occurrences of the destroying earthquakes (§3.2) on the basis of the Historical Record Office data (Fig 2). The sequence of the transformations subjected by constructions in time is individuated also through the study of the present cracking frame (Fig 3) and of the materials in which buildings consist (Fig 4). The predisposition to damage (§3.3) for the building isolated or integrated into the aggregation to which belongs is studied (Fig 5). The consequences of this analysis are considered for the definition of the Unities Minimal of Intervention (UMI).

1198

Figure 3.   Sellano. Cracking examples.

Figure 4.   Sellano. Masonry degradation state.

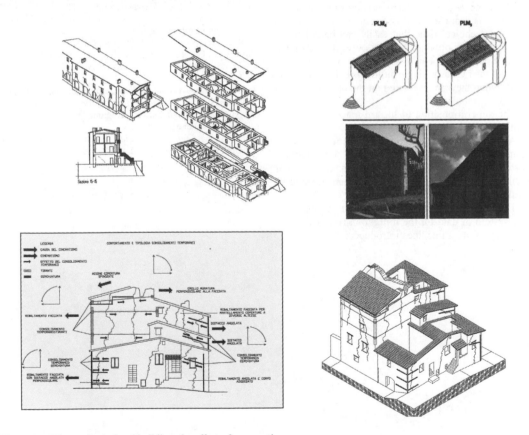

Figure 5.   Kinematics isolated building, the effect of aggregation.

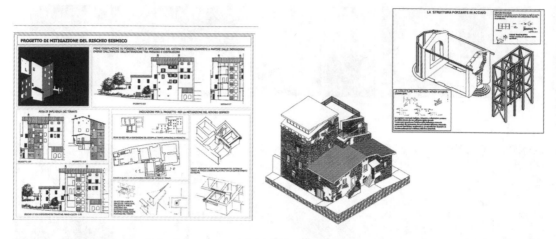

Figure 6.   Some hypotheses of seismic improvement for significant buildings.

Referring to a wider scale, the definition of a vulnerability index (i.v.) for the buildings on the basis of "poor data" (age, technical plants state, structural density in plane, regularity characteristics of the building shape…) is proposed, that is, on the basis of qualitative, not technical data, that easily can be collected. As a result, it is obtained that the i.v. evaluated on the basis of poor data for an UMI coincides with the mean value of the i.v. of the single buildings belonging to the UMI, when they are evaluated on the basis of technical, structural, data.

The superimposition of hazard map and vulnerability map gives the local risk map.

Analyses of damage are performed related to the characteristics of the rehabilitation interventions carried out before the Assisi Earthquake; different methodologies of structural reinforcement are studied as alternative hypothesis (Fig 6).

The study on its whole is framed into an informative system of data for the mitigation risk design, on the basis of the guide lines introduced.

REFERENCES

Tricart J., *Corso di geografia umana*, Ed. Unicopli.
G.N.D.T., *Le chiese e il terremoto* Ed. LINT.
AA.VV., *Primo repertorio dei centri storici in Umbria*, Gangemi Editore.
C.N.R., *La casa rurale nell'Umbria*, Ed. Olschki.
Giuffrè A., Carocci C., *Codice di pratica per la sicurezza e la conservazione del centro storico di Palermo* Ed. Laterza.
Giuffrè A., *La meccanica delle murature storiche* Ed. Kappa.
Guerrieri F., *Manuale per la riabilitazione e la ricostruzione postsimica degli edifici* Regione dell'Umbria, DEI tipografie del genio civile.
Doglioni F., *Codice di pratica (linee guida)*, Bollettino Ufficiale della Regione Marche.

Effect of column characteristics on its inelastic seismic behavior

A. Khairy Hassan
Department of Civil Engineering, Assiut University, Egypt

ABSTRACT: Nonlinear 3D FE is utilized to study the inelastic behavior of RC columns during seismic loads. Firstly, nonlinear FE was utilized to carry out inelastic response during earthquake motion for several RC columns with different natural periods and damping coefficients. From the numerical results, certain boundaries were established to determine the limits of natural periods at which the column is sensitive to displacement response considering the motion type. It was shown that motion characteristics play a significant role on the response of columns. Peak ground acceleration is proved to be not the main parameter to determine severity of an earthquake motion. Other parameters include the frequency, period and time interval of the motion. Secondly, the effect of input motion characteristics is clarified. The limits at which the inelastic response is sensitive to peak displacement, velocity or acceleration are obtained and summarized based on the interaction of the column and motion characteristics. The results are conservative in design of RC columns as both characteristics of the column and motion are considered simultaneously.

1   INTRODUCTION

Inelastic response of a structure during an earthquake motion depends greatly on two main parameters namely the characteristics of the structure and characteristics of input motion. The interaction of the two parameters usually determines the response and the level of severity. This interaction has not been qualitatively analyzed in spite of its importance. For example, the designers usually depend on the acceleration response and they usually scale the acceleration to have a stronger motion. This acceleration scaling leads to much larger scatter in the response than that when velocity spectral intensity scaling is used. Also, it is not clear – from previous studies – to which extent the response is dependent on acceleration, velocity and/or displacement. With relative to characteristics of input motion, previous studies (Reem 1987, Khairy 1999, Machida 1998, Elnashai 1996) showed that three groups of earthquake motions might be identified; normal motions having significant frequency, motions exhibiting large amplitude and high frequency, and motions containing a few severe, long duration acceleration pulses.

With relative to natural period of the structure, three groups are identified; structures of short period up to 0.5 second (sensitive to peak ground acceleration), medium period between 0.5 up to 3.0 seconds (sensitive to peak velocity), and long period more than 3.0 seconds (sensitive to displacement). However, such limits were established previously based on the characteristics of input motion only and based on linear analysis as shown in Figure 1 (Reem 1987, Khairy 1998). The target of the current study is to establish certain limits in such a way that the designer can judge when the response of a certain structure is sensitive to acceleration, velocity or displacement. This is the main purpose of the study.

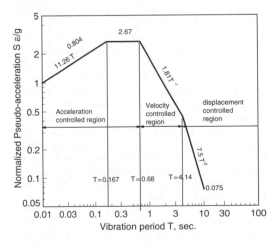

Figure 1.   Relation of natural period and maximum response (Elnashai 1996).

2 MODELLING

A nonlinear 3D finite element model was utilized in the study (Wu 1996, Wu 1989). The model was previously proposed and its accuracy was verified (Wu 1996, Wu 1989) Von Mises criterion with normality flow rule was adopted to consider the nonlinearity of steel box plate. Strain hardening was not considered. Nonlinearity of concrete was adopted through constitutive equations of nonlinear behavior of concrete in compression.

We utilized the model which was proposed for concrete under compressive loading and essentially based on classical work hardening plasticity theory (Wu 1996, Wu 1989), in which compression is either strain hardening or strain softening. In the implementation of the problem, the Drucker-Prager type yield surface relating to the well-known Mohr Coulomb criterion, in which mobilized friction angle and mobilized cohesion were introduced, was adopted with the associated flow rule. Nonlinear behavior of concrete was dealt with by an accumulated damage parameter and a subsequent loading surface in the stress surface which was defined depending on the plastic strain history through the introduced damage parameter. In order to simulate the formulation, the Druker-Prager type yield surface was adopted depending on the associated flow rule. The proposed model seems to rationally predict some essential features of concrete behavior under compressive loading of the strain softening tendency and microcracking.

The main characteristics of the model are; it includes strain hardening and strain softening behavior and it is easy because only four parameters are needed to determine the concrete behavior. The required parameters are effective stress, effective plastic strain, plastic work and initial strain. The method of determination of these parameters was given in references (Khairy 1999, Wu 1996, Wu 1989).

The model was modified to consider the effect of shear reinforcement on increasing ultimate strength and corresponding strain due to confinement. The modification was based on a model proposed by Mander et al (Mander 1988, Park 1990, Park 1994, Park 1975). MARC software was utilized to carry out the numerical calculations with the installed constitutive equations (MARC Manual 1995). In the second stage or cracked concrete, constitutive equations based on smeared crack model were used. Just the cracking occurs normal to the principle tensile stress, then concrete drops into orthotropic coordinates with axes parallel and normal to cracks. More details of the model were shown in previous work (Khairy 1999). Shear reinforcement ratio was defined as $A_{sh}/(e \cdot h)$ in which A_{sh} is the area of stirrup branches in the considered direction and e and h are spacing of stirrup and depth of the section respectively. Main reinforcement was defined as $A_s/(b \cdot d)$ where A_s is the area of main steel in tension side and d and b are effective depth and width of the section respectively. Main steel ratio was assumed to be constant and equal to 0.85%. MARC software was used to carry out the calculations with the installed constitutive equations. Axial stress ratio of the column was defined as $P/(A_c \cdot f'_c)$, where P is the axial load, A_c is the area of concrete section and f'_c is the compressive strength of concrete. The stress ratio was assumed to be constant and equal to 0.2.

3 ANALYTICAL MODEL

A FE program called MARC was used in the analysis (Marc Manual 1995). Figure 2 illustrates the analyzed RC column. Natural period is calculated according to the cantilever theory. The column is modeled by 3D 8-node brick element. The elastic assumption is adopted for axial and shear stiffness. The footing of the pier is assumed to be completely rigid. The superstructure is represented by a certain mass at the top of the pier and its value is assumed to give the initial natural period of the system.

Different cases of RC bridge columns were analyzed. Each column was subjected to earthquake motion. Different cases of damping coefficient were analyzed (0.0, 0.25, 0.45, and 0.75). For each damping coefficient, different cases of natural periods were analyzed (0.1, 0.2, 0.3, 0.4, 0.5, 0.6, 0.7, 0.8, 1.0, 1.2, 1.4, 1.6, 1.8, 2.0, 2.5, and 3.0 seconds). Inelastic displacement, velocity, and acceleration responses were obtained for each case of study. Recent motions were scaled to have 100 gal and they were utilized in the study. The motions are Loma prieta in Oakland, October 17, 1989 (EW) which has maximum acceleration of 270.361 gal, Miyagi-ken–Oki motion (transverse direction) in June 12, 1978 which has maximum acceleration of 413.5 gal, El Centro motion (NS direction) 1940, and Lima motion in Peru in 1970. Figure 3

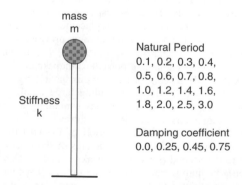

Figure 2. Analyzed column and parameters of the study.

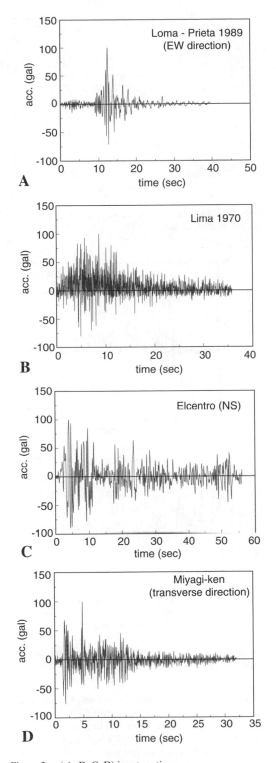

Figure 3.   (A, B, C, D) input motions.

(A, B, C, D) illustrates the mentioned motions which cover the known ranges of the motions.

4   EFFECT OF NATURAL PERIOD

In the previous studies, usually the designers concern with acceleration response. However, it was shown recently that peak acceleration is not a significant factor to determine the severity of an earthquake (Reem 1987, Khairy 1999, Khairy 1998, Elnashai 1996). Also, it was shown that RC bridge columns are not so sensitive to acceleration response. This depends greatly on the interaction of natural period and the motion type. For this reason, we are focusing on the displacement response in the current study. Figure 4 (A, B, C, D) illustrates inelastic displacement responses for RC piers with different natural periods during Loma-prieta earthquake when damping ratio equals 0.0, 0.25, 0.45 and 0.75 respectively. The response is very small for very short period columns and vice versa (Reem 19987). It is clear from the figure the dependence of the inelastic response on the period. It is obvious how much the difference of the response of pier with 0.1 seconds period as compared with another piers having period more than 1.6 seconds. This is because as the period is short, the pier is rigid and vice versa.

Figure 5 illustrates the effect of natural period on the maximum displacement response for different damping ratios. It is clear that the rate of change of peak displacement is very high for columns of natural period ranges between 0.5 to 1.6 seconds. This indicates that the designer should design the columns for displacement response if the column possesses natural period between 0.5 to 1.6 seconds considering the effect of input motion characteristics as it will be shown latter.

Also, it is clear that damping reduces the response and this agrees with the previously known fact. The new finding in the current analysis is that the effect of increasing the damping ratio on decreasing the peak displacement disappears for columns possessing long natural period. The reasons of this phenomenon require more analysis.

5   EFFECT OF MOTION CHARACTERISTICS

In previous studies (Khairy 1999, Machida 1998), we clarified the effect of the input motion characteristics on failure mode and ductility level of the bridge columns. In the current analysis, it is focused on the interaction of motion characteristics and natural period on the inelastic behavior.

Figure 6 (A & B) illustrates inelastic displacement responses of two columns having the same natural period during Loma-prieta and Miyagi motions

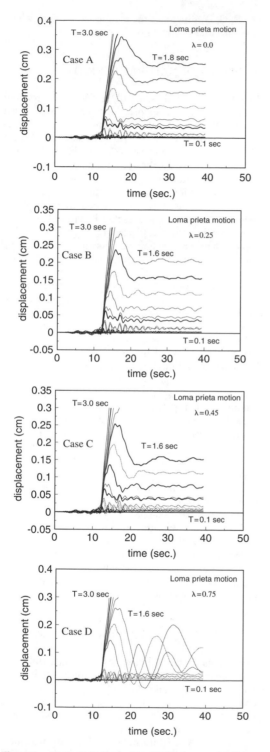

Figure 4. (A, B, C, D) Response of RC columns of different natural periods and damping ratios.

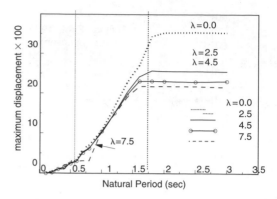

Figure 5. Effect of natural period and damping ratio on maximum displacement.

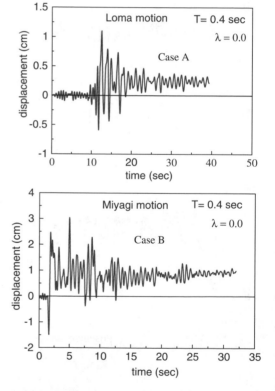

Figure 6. (A,B) Effect of input motion on inelastic response of RC column.

respectively. Even the columns possess the same natural period and the motions having the same peak acceleration, however there is big scatter in the response as well as in the peak displacement. This is due to the difference in the motion characteristics. The predominant frequency, frequency content, the

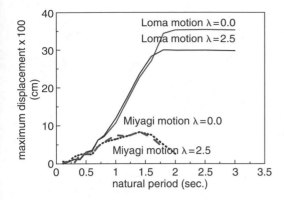

Figure 7. Effect of natural period and input motion on maximum displacement.

shape of the motion, the number of readings having high amplitude, number of cycles and the duration of motion are examples of the characteristics which have significant effect.

For this reason, it is not possible to fix a certain design motion due to the difference of motion characteristics. It is the purpose of future study to establish the fundamentals of a design motion, which can be considered in the codes. It may be an artificial motion such as sine or cosine wave.

Figure 7 illustrates the relation between maximum displacement and natural period during Loma-prieta and Miyagi motions for two cases of damping coefficient (0.0 and 0.25). It is clear that the response due to Loma motion is more than 3 times the response due to Miyagi motion. From the comprehensive analysis of the results, it was found that short period structures (T up to 0.5 second) should be designed based on peak acceleration for normal motions which have significant frequency and motions exhibiting large amplitude and high frequency. For medium period structures, (T ranges from 0.5 to 3.0 seconds) peak displacement should be considered in the design for the same motions. For long period structures, (t more than 3.0 seconds) peak velocity should be considered. Other cases of motions still require further study to analyze their interaction.

6 CONCLUSIONS AND REMARKS

Based on comprehensive nonlinear three dimensional finite element analysis, the following points are concluded:

- Both of column characteristics and input motion characteristics play significant role on the inelastic

response during an earthquake motion and this role is deeply interrelated.
- Inelastic response of a column may be sensitive to peak displacement, velocity or acceleration depending on natural period of the column and motion type.
- The limits at which the inelastic response is sensitive to peak displacement, velocity or acceleration are obtained and summarized based on the interaction of the column and motion characteristics. The obtained results are more conservative for seismic design of RC columns as they were established based on comprehensive analysis during earthquake motion.
- Further analysis is needed to illustrate the effect of other column characteristics such as column size and main reinforcement ratio. Besides, other recorded motions should be included in further study.

REFERENCES

Elnashai, A., & McClure, D. 1996. Effect of Modeling Assumptions and Input Motion Characteristics of RC Bridge Piers. *Earthquake Engineering & Structural Dynamics*, Vol. 25, 1996.

Khairy Hassan, A. 1999. *Analytical Study on Seismic Resistant Characteristics of RC Bridge Piers* D.Eng., dissertation submitted to Saitama University, March 1999.

Machida, A., & Khairy Hassan, A., Influence of input motion characteristics on inelastic behavior of RC piers of bridges. A three Dimensional finite element approach. *EASEC-6 International conference*, Taipei, Taiwan, Jan. 1998, pp. 597–602.

Mander, J.B., Priestely, M.N., & Park, R. 1988. Theoretical stress–strain Model for confined concrete. *J. of Structural Engineering*, ASCE Vol. 114, No. 8.

MARC Corporation "MARC Manual," Vol. A–F, 1995.

Park & Paulay 1990. *Bridge design and research seminar* Vol. 1, Transit New Zealand.

Park & Paulay 1975. *Reinforced concrete structures*. John Wiley, U.S.A, 1975.

Reem, H., & Anil, K. 1987. Earthquake Response of Torsionally Coupled Building Rep. No. UCB/EERC 87/20, Watson.

Zahn & Park 1994. Confining Reinforcement for Concrete Columns. *J. of Structural Engineering*, Vol. 120, No. 6, June 1994, Earthquake Engineering Research Center, University of California, Berkley.

Zhishen Wu 1989. Development of Computational Models for Reinforced Concrete Plate and Shell Element. *Ph.D. thesis*, Department of Civil Engineering, Nagoya University, 1989.

Zhishen Wu & Takada-aki 1990. A Hardening/Softening Model of Concrete Subjected to Compressive Loading. *J. of Structural Engineering*, Vol. 36B, 153–162.

System-based Vision for Strategic and Creative Design, Bontempi (ed.)
© 2003 Swets & Zeitlinger, Lisse, ISBN 90 5809 599 1

Effects of building vibration to low intensity ground motion: towards human perceptions

A. Adnan & T.C. Wei

Structural Earthquake Engineering Research (SEER), Faculty of Civil Engineering,
University of Technology, Malaysia

ABSTRACT: Because of no profound earthquake records for countries in low intensity earthquake regions such as Malaysia and Singapore in South East Asia, earthquake resistance design has never been seriously concerned. However, the questions arise from the issue are as follows; (i) Are local structures really affected or not at all by the earthquake in neighbouring regions? (ii) How far are we safe from earthquake hazards? Nobody has provided scientific evidences regarding this issue.

In this paper, the study would be focused on the response of the human toward building motion as a result of structural response under earthquake effect. The core of the study was based on the 20-storey frame and frame-wall buildings. The building of several heights were used for comparison purposes. This is to reflect the condition in Malaysia where only some people claimed had the feeling of the vibrations during strong earthquake in neighbouring countries. The earthquake acceleration applied in this study was only 0.045 g. With this scale, human beings cannot even perceive the motion at ground floor and structural defections is not expected to occur. However in the study it shows that the ground accelerations were amplified to almost 200 to 500 percent for buildings in a range of 4 to 40 storeys.

Besides structural safety, the building motions that have psychological or physiological effects on the occupant should be controlled to be in an acceptable range. In general, it is found that the acceleration is the agreed predominant parameter in determining the nature of human response to vibration.

This research illustrates the reactions or responses of the occupants in tall buildings in low intensity earthquake motions. Although the ground motion might not cause damages in structure, the vibration could result in public commotion. Fortunately, there is no report on structural destruction in Peninsular Malaysia after the Bengkulu Earthquake (2000) magnitude scale of 7.9, with the epicenter of about 800 km away from Kuala Lumpur. But, nobody could assure that we may be still that lucky if earthquakes take place in the Sumatra fault that is only 300 km to 450 km away from Kuala Lumpur with a large magnitude. Therefore, it is the responsibility of the engineers to comfort and convince the public by improving the current design standard for low intensity earthquake resistance.

1 INTRODUCTION

There are concerns of regions where seismic activities are low but the effect from other active regions located in the range of 400 km away provide quite a significant intensity to their regions. In this case the behavior of buildings would be much affected by the types of earthquake, local soil conditions, and types of buildings, followed by the building occupants comforts.

The study focused on the response of the human toward building motion rather than the structural response under earthquake effect. The core of the study was based on the 20-storey frame and frame-wall buildings. The building of several heights were used for comparison purposes. As to reflect the condition in Malaysia where only some people claimed had the feeling of the vibrations during strong earthquake in neighbouring countries, the earthquake acceleration applied in this study was only 0.0045 g or less than 0.05 m/s². With this scale, the occupants beings cannot even perceive the motion at ground floor and structural defections is not expected to occur.

2 HUMAN PERCEPTIONS

The building motions that have psychological or physiological effects on the occupant should be controlled to be in an acceptable range. The perception of building movement depends largely on the degree of simulation

Table 1. Human Perception Levels.

Range	Acceleration (m/s^2)	Effects
1	<0.05	Humans cannot perceive motion.
2	0.05–0.10	Sensitive people can perceive motion; hanging objects may move slightly.
3	0.10–0.25	Majority of people will perceive motion; level of motion may affect desk work; long-term exposure may produce motion sickness.
4	0.25–0.40	Desk work becomes difficult or almost impossible; ambulation still possible.
5	0.40–0.50	People strongly perceive motion; difficult to walk naturally; standing people may loss balance.
6	0.50–0.60	Most people cannot tolerate motion and are unable to walk naturally.
7	0.60–0.70	People cannot walk or tolerate motion.
8	>0.85	Objects begin to fall and people may be injured.

Table 2. Acceleration and human perception level for 20-storey buildings.

Frame building			Frame-wall building		
Level	Acc (m/s^2)	Range	Level	Acc (m/s^2)	Range
20 (Top)	0.8998E-1	2	20 (Top)	0.2126	3
18	0.5534E-1	2	18	0.1412	3
16	0.6097E-1	2	16	0.1149	3
14	0.6733E-1	2	14	0.1107	3
12	0.5839E-1	2	12	0.1333	3
10	0.7320E-1	2	10	0.1319	3
8	0.8123E-1	2	8	0.1445	3
6	0.8585E-1	2	6	0.1350	3
4	0.9513E-1	2	4	0.9435E-1	2
2	0.5962E-1	2	2	0.4242E-1	1
Base	0.4420E-1	1	Base	0.4420E-1	1

of body's central nervous system, the sensitive balance sensors within the inner ears playing a crucial role in allowing both linear and angular accelerations to be sensed. Human response to the building motion is influenced by many factors, such as the movement of suspended objects or fretting between building components. In general, it is the acceleration is the agreed predominant parameter in determining the nature of human response to vibration. The levels of human perception towards building motion that recommended by Yamada & Goto (1975) is presented in Table 1.

3 RESULTS AND ANALYSIS

Table 2 shows the maximum acceleration of every two floor in 20-storey building. As mentioned earlier, the ground acceleration of 0.0045 g which below human perception level was applied to the models. Therefore, the occupants at ground floor would not feel any movement resulted by the excitation. However, the structural interaction in the building had amplified the acceleration and resulted in greater vibration as the ground motion transferred to the upper levels that consequently sensed by the occupants.

For frame building, the movements induced by the earthquake were only be experienced by sensitive people from second floor and above; and the hanging objects might slightly moving. Therefore, the building was categorised as Range 2. The maximum accelerations along the height of the building did not differ much. The most crucial acceleration took place at level

four and the top of building, where the maximum acceleration closed to the acceleration limit in Range 3, hence, more people would perceive the motions.

If the frame-wall building of the same height is considered, the structural responses to counter the earthquake excitation resulted in different acceleration obtained in the building. As shown in table, the acceleration at all upper levels of frame-wall building were obviously greater than the frame system counterpart. This indicates that the motions experienced by the majority occupants in the frame-wall building might not be "felt" by the occupants in frame building, although both buildings are under the same earthquake excitation and they are about the same height.

Besides, the human perception level of different floors in the same building might not be the same as the 20-storey frame building discussed earlier. As demonstrated in table, while the upper floor in the frame-wall building suffered the vibration that could be sensed by most of the occupants and desk work were affected (Range 3), the situation at level 2 and below were still in Range I where nobody precept the motions. Moreover, it can be observed that the acceleration at second was even lower than the excitation acceleration. This explained the situation at Pekan Nanas Police Quarter, Pontian, during Bengkulu earthquake on 5 June 2000, where the occupants at upper level were rushing of escape while the occupants at lower level were not aware of the event.

As the building increases in height, the non-uniform in term of human perception level is even significant. Table 3 indicates that the human perception level of 40-storey frame-wall building and frame building ranged from Ranges 1 to 3. Moreover, range of human perception level was not increase proportionally with the height of building for frame-wall system. The analysis shows the acceleration at levels 4, 28 and 32 were below the perception of human and were categorised as

Table 3. Acceleration and human perception level for 40-storey buildings.

Frame building			Frame-wall building		
Level	Acc (m/s^2)	Range	Level	Acc (m/s^2)	Range
40 (Top)	0.8959E-1	2	40 (Top)	0.1001	3
36	0.8978E-1	2	36	0.6625E-1	2
32	0.1046	3	32	0.3773E-1	1
28	0.1551	3	28	0.4663E-1	1
24	0.8898E-1	2	24	0.6184E-1	2
20	0.7256E-1	2	20	0.6691E-1	2
16	0.7522E-1	2	16	0.8907E-1	2
12	0.8808E-1	2	12	0.9254E-1	2
8	0.9639E-1	2	8	0.7111E-1	2
4	0.7914E-1	2	4	0.3163E-1	1
Base	0.4420E-1	1	Base	0.4420E-1	1

Table 4. Acceleration and human perception level for 4 and 10-storey buildings.

Frame-wall building			Frame-wall building		
Level	Acc (m/s^2)	Range	Level	Acc (m/s^2)	Range
10 (Top)	0.2009	3			
9	0.1592	3			
8	0.1154	3			
7	0.1207	3			
6	0.1316	3			
5	0.1456	3			
4	0.1446	3	4 (Top)	0.1743	3
3	0.1228	3	3	0.1220	3
2	0.9402E-1	2	2	0.9410E-1	2
1	0.2816E-1	1	1	0.4796E-1	1
Base	0.4420E-1	1	Base	0.4420E-1	1

Range 1. In general, the structural behaviour that has been studied in previous sections plays the governing role in determining the human perception level in earthquake excitation. The rigidity and natural period of the structure has direct effect on the value of the floor acceleration which determining the human perception level. As can be notice, although the stiffer frame-wall undergone less deformation, the rigidity of the structure caused the inducing of greater acceleration, hence, higher range of human perception level.

If the low and medium-rise building is studied, the results show that it is not necessary the lower building suffer less critical response. As can be noticed in Table 4, the floor accelerations and the human perception level of 4 and 10-storey building is found higher than the 40-storey building. This explains that the human perception level to building also rely on the earthquake response spectrum.

4 CONCLUSIONS

This study illustrates the reactions or responses of the occupants in the taller buildings during strong earthquake. Although the ground motion might not cause damages in structure, the vibration could result in public commotion. Fortunately, there is no report on structural destruction after this Richter scale of 7.9 earthquake, with the epicenter of 650 km away from Kuala Lumpur. But, nobody could assure that we may be still that lucky if earthquakes take place in the Sumatra fault that is only 300 km to 450 km away from Kuala Lumpur with a large magnitude. Therefore, it is the responsiblity of engineer to comfort and convince the public by improving the current design standard for low intensity earthquake resistance.

REFERENCES

Aktan, A.E., Bertero, V.V. & Piazza, M. 1982. Prediction of Seismic Response of R/C Frame-coupled Wall Structures. *Report to National Science Foundation, Report no UBC/EERC-82/06, University of California, Berkeley, California.*

Chopra, A.K. 1995. *Dynamics of Structures: Theory and application to Earthquake Engineering.* New Jersey, Prentice-Hall, Inc.

Coull, A. & Smith, B.S. 1991. *Tall Building Structures: Analysis and Design,* New York, John Wiley & Sons, Inc.

Gosh, S.K. & Domel, A.W.,Jr 1992. *Design of Concrete Buildings for Earthquake and Wind Forces,* Illinois, Portland Cement Association and the International Conference of Building Officials.

Hung, V.V., Ramin S. Esfandiari 1997. Modelling and Analysis. *Dynamic Systems,* New York, Mc Graw-Hill, Inc.

MacLeod, I.A. 1990. *Analytical Modelling of Structural Systems: An Entirely New Approach with Emphasis in Behaviour of Building Structures,* New York, Ellis Horwood.

Rockey, K.C., Evans, H.R., Griffiths, D.W. & Nethercot, D.A. 1975. *The Finite Element Method: A Basic Introduction,* London, Crosby Lockwood Staples.

Shah, H.C. & Pan, T.C. 1993. Seismic Risk Management: Is this an Issue for Singapore? *Seismic Risk Management Course,* Nanyang Technological University, Singapore.

Staoian, V. & Olariu, I. 1995. Models, Simulations and Consideration in Design of a Steel-concrete Composite Structures in Seismic Areas, London, E & FN Spon.

Srivastav, S. & Abel, J.F. 1991. 3D Modelling of Building for Non-linear Analysis, *System Dynamics,* Rotterdam, A.A.Balkema.

Tamura, Y., Yamada, M. & Yokota, H. 1994. Estimation of structural damping of buildings, Proceedings of the Structures Congress, ASCE, Atlanta, pp. 1012–1017, Apr. 1994.

Kanda, J., Tamura, Y., Fujii, K., Ohtsuki, T., Shioya K. & Nakata, S. 1994. Probabilistic evaluation of human perception threshold of horizontal vibration of buildings (0.125 Hz to 6.0 Hz), *Proceedings of the Structures Congress,* ASCE, Atlanta, pp. 648–653, Apr. 1994.

System-based Vision for Strategic and Creative Design, Bontempi (ed.)
© *2003 Swets & Zeitlinger, Lisse, ISBN 90 5809 599 1*

Probabilistic hazard models: is it possible a statistical validation?

E. Guagenti & E. Garavaglia
Dept. of Structural Engineering, Politecnico di Milano, Milano, Italy

L. Petrini
Dept. of Structural Mechanics, Università degli Studi di Pavia, Pavia, Italy

ABSTRACT: Recently an index of credibility has been introduced in order to measure the variability of an estimated long-term hazard quantity. On its basis a probabilistic hazard model can be constructed reducing both epistemic and aleatory uncertainties. In this paper the definition of the credibility index is extended to estimate a short-term time varying hazard quantity.

1 INTRODUCTION

When a hazard quantity a concerning rare events has to be estimated, the estimated value $\hat{a}$ is affected by a very large uncertainty. The background hypotheses cannot be validated in an *absolute* sense: the physical knowledge of rare events is weak; the classic statistical analysis assigns equivalent degree of acceptance to different hypotheses; no direct quantitative evidence can be derived from the poor data set.

Here a *relative* criterion of validation is used based on an index called *credibility*; the effectiveness of the competing models is judged on the basis of the above index of credibility. The preferability of one model is decided in respect to a fan of conjectural realities.

The case of seismic hazard can be a paradigm for other decision problems under uncertainty conditions.

Three hazard problems are shown, different in time scale prediction. The hazard variable in time, before not analysed with the credibility index, needs an extension of the index that in this paper is proposed.

2 THE TWO KINDS OF UNCERTAINTY

Two kinds of uncertainties reside in a probabilistic model: epistemic and aleatory (or statistic). Epistemic uncertainty is concerning the formal model: it can be wrong. But, even if the model is correct, another uncertainty subsists because the model contains parameters which must be estimated from the available data. But the available sample of data is only one among all performable samples. So the set of estimated parameters is only one among all possible sets. This is the concern of statistic uncertainty.

Note that in seismic hazard analysis the unique available sample is the history catalogue of earthquakes. Moreover the interest of hazard analysis is focused on the large earthquakes that fortunately are rare events. So, our sample is unique and poor in data.

The questions: *"which is the true model?"*, *"which are the true parameters?"* cannot receive answer. The questions are wrong questions. Unfortunately even what is called "statistical validation" is not reachable. To be fully statistically substantiated, a proposed model "should be backed by a sufficient number of directly observed successes and failures to establish its performance at some agreed level" (Verc-Jones 1998).

About the term "validation" largely in use in the last years, it should be noted that never a probabilistic model can be validated in absolute sense.

A probabilistic model consists in a continuous way towards more robust degree of knowledge (Lakatos et al. 1974). In this sense has to be interpreted the famous statement of B. de Finetti "The probability does not exist" (De Finetti 1966). It is not a question of scepticism. The statement wants to stress the dynamic character of a probabilistic model. The Bayesian approach teaches how, starting from a basic guess, a basic prior probability, our confidence grows towards a posterior more credible probability when more information becomes available and can be incorporated in the probabilistic model. One needs to think an evolving field of probabilities, which can receive, in the standard statistical analysis, a certain degree of validation: a weak criterion of validation affected by the well known Type I and Type II errors.

Anyway, the criterion is a useful operating rule which recognises the inherent variability of the data and simplifies communication among scientists because the level of significance is declared. Moreover the classic statistical tests become a valid guide if other evaluations are done, typically: comparison between models and methods with relative likelihood functions, bayesian approach, Maximum Entropy Principle. They help to explore the reliability of one model (or hypothesis). Particular care deserves the underlying physics of the hazard: a probabilistic model is not a pure statistic consequence of a data set. Each probabilistic distribution contains some physical interpretations. So, in the case of seismic hazard for example the physical knowledge about the mechanism of earthquake generation can be incorporated: typically the heart crust deformation (sleep rate) (Wesnousky et al. 1984; Guagenti et al. 1988, ...).

In conclusion a probabilistic model can be supported by a wide patient data interpretation in such a way as to get a certain degree of confidence; even if it contains always a subjective choice, a careful analysis makes reasonable the choice. Obviously the most unsatisfactory residual field of uncertainty remains in problems concerning the rare events, i.e. the tail distribution. When different independent groups of experts are asked to evaluate the same quantity a the dispersion of the results is in general very large. Some years ago, an index Δ called credibility has been introduced (Grandori et al. 1998, 2002, 2003), in order to reduce, on its basis, this dispersion. The index Δ has the merit to focus the possible error in the quantity a estimation and, this way, to decide between two candidate models which one is the winner in the estimation of a.

Before introducing the index Δ let us outline three hazard problems significantly different in regard to time scale prediction.

3 THREE STUDY CASES

Let us see three cases significantly useful in programs aimed at risk reduction.

The peak ground acceleration (PGA) $a(500)$ at a given site, corresponding to 500 years return period, is typically assumed as a measure of the seismic hazard for engineering purposes: it is prescribed in the code as that acceleration which a normal building must resist to, without damage. *Which is a reliable estimate â of $a(500)$?* Note that such a PGA may have never occurred at the site in the period of observation. Nevertheless, the code requires an estimate of this value. The quantity $a(500)$ is a long term static estimate of average risk (here the term "risk" is used as synonymous of "hazard"). For this estimate a static model is reliable. In this first problem the hazard can be considered constant in time.

On the contrary a time varying risk model is needed if a short–medium term planning is required, which needs the knowledge of the hazard in the next years starting from today. A stationary memoryless model is not reliable. At least the model must feel how long time elapsed since the last earthquake. Renewal process (RP) is able to evaluate the probability of an earthquake in the next future, given information on the elapsed time. The hazard is variable in time continuously furnishing a background earthquake prediction.

Moreover, precursor phenomena seem exist and are studied. They are able to identify a sudden short time period with increased hazard. The RP coupled with this other information deriving from precursors can lead to very large hazard in next few days and then justify an alarm system, but this posterior earthquake prediction largely depends on the background one (Grandori et al. 1988).

Debatable questions obviously remain in all three subjects: the question of the tails and, for the third subject, the delicate question of ambiguous information because precursors may be false or missing.

Note that the three cases need *three different approaches* for the same site hazard assessment. A merit of RP is to be flexible, suitable in the three cases: it estimates the hazard as variable continuously in time; when coupled with a precursor process it is able to identify local peak of hazard; when analysed over a very long time period itself leads to stationary evaluation of hazard (Guagenti et al. 1988).

But RP defines a wide class of stochastic processes open to many choices for magnitude and time distribution. Our past research had led to results that got a certain degree of "validation" (always validation between inverted comma). Now we want improve it with a new statistical procedure, mainly aimed at magnitude distribution choice (first case) and at interoccurence time distribution (second and third case).

4 AN IMPROVEMENT IN VALIDATION

We give here an outline of the announced effort towards a validation of a model. Model here is essentially a *paradigm* to estimate a quantity a starting from a single sample. In fact the substantial difference with the standard statistical analysis consists in shifting the attention from the data fitting to the error in estimating the quantity a of interest.

Beyond discussion an absolute validation; more modestly we limit ourselves to the comparison between two models in competition to estimate a quantity a.

Which one of the two estimates coming from the two models is more reliable? This is the question. In this paragraph we recall the results already reached in evaluating the static hazard quantity $a(500)$; in the next paragraph we extend the method to evaluating the variable hazard quantity $\lambda(t_0)$.

Suppose that a region is the seat of a completely defined earthquake process. So our "truth" is completely defined and the quantity a assumes a precise value a^o. To be completely defined an earthquake process needs a complex knowledge. We focus our attention on a single component, supposed to be the main responsible on the estimation of the quantity a of interest. So, we limit ourselves to analyse a single distribution F which, coeteris paribus, the estimate $\hat{a}_r$ depends on. In the case of $a(500)$ evaluation (1st subject), such F is the magnitude distribution function. If $F_M(m; \vartheta)$ is given, both as form and parameters (ϑ can be a vector), then a known procedure (concerning distribution of events in space and time, and attenuation law), applied to F_M, gives the value a at the site.

From now on, a distribution with known parameters will be briefly indicated with an upper index; on the contrary lower index will indicate that estimable parameters are present in F.

The distribution F^o being known, it is possible (with Monte Carlo simulation) to draw many size ν samples from it. Starting from each sample our model (i.e. an F_r) leads to estimated values $\hat{a}_r$, one for each sample. All together they form the random variable $\hat{A}_r$, whose distribution is the sampling distribution of the PGA.

The index:

$$\Delta_r^o = \Pr\left\{ a^o - h < \hat{A}_r \leq a^o + h \right\} \tag{1}$$

is the probability that the estimated value $\hat{a}_r$ with the r-model falls in a given interval around the "true" value a^o. The index Δ_r^o takes into account both *epistemic* and *aleatory* uncertainty; it measures the *credibility* of r-model in respect to F^o when a is the estimable quantity (credibility not based on data fitting, but looking at the result).

Briefly: Δ_r^o is the probability that, on the basis of a random size ν sample drawn from F^o, the model F_r leads to estimate a with an error $\varepsilon_r^o \leq h$ (as absolute value)

$$\Delta_r^o = \Pr\left\{ \left| \varepsilon_r^o \right| \leq h \right\} \tag{2}$$

Analogously for the second model s – model (i.e. for an F_s):

$$\Delta_s^o = \Pr\left\{ a^o - h < \hat{A}_s \leq a^o + h \right\} \tag{3}$$

The difference

$$\Delta_{rs}^o = \Delta_r^o - \Delta_s^o \tag{4}$$

is a meaningful index of the *relative credibility* of the two models. We assume that r-model is more reliable

than s-model for the estimate of a if $\Delta_{rs}^o > 0$ and vice-versa. Then the sign of Δ is of critical importance: it decides which model is the winner.

What is the interest given that the true a^o is not known? We can start operating on the unique object available for us: the catalogue. We construct an empirical truth furnished by cumulative frequency polygon F^* and, proceeding as above described, construct the empirical index Δ_{rs}^*. The question now is: how much this empirical index is *representative* of the true unknown Δ_{rs}^o?

The degree of relationship is well represented by the two conditional probabilities:

$$^\downarrow H = \Pr\left\{ \Delta_{rs}^* > 0 \mid \Delta_{rs}^o = x > 0 \right\}. \tag{5}$$

$$^\uparrow \Omega = \Pr\left\{ \Delta_{rs}^o > 0 \mid \Delta_{rs}^* = x > 0 \right\}. \tag{6}$$

To make evident the importance and the difference between the two probabilities arrows are used.

The first is a probability "*towards down*": given the hypothesis that Δ^o is positive calculate the probability to observe Δ^* positive. So a first degree of agreement with the hypothesis is obtained.

More meaningful, but much more difficult to be evaluated, the posterior probability Ω which is a probability "*towards up*": given the observed data, evaluate the probability of hypothesis. In precise terms to calculate Ω we should know all the possible realities. So, in precise term Ω cannot be calculated. But with an enormous degree of patience we can explore a lot of plausible realities. For each of them we can repeat the simulations and the estimate above described.

The technical details are given in the quoted papers. We show here a graph (Fig. 1) that measures

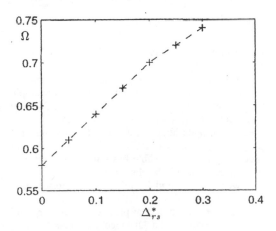

Figure 1. Posterior probability of hypothesis that F_r is better than F_s for evaluating $a(500)$, given the empirical value Δ_{rs}^*.

reasonably the *posterior probability* Ω *of the hypothesis*. So, the observed value Δ^* becomes much more than a symptom: it can measure the probability that one model is winner against the other one.

The panorama of conjectural realities explored is certainly not exhaustive, but this result encourages to go on. In fact the results date 1998; today more robust results are reached (Grandori et al. 2003). They show a general high credibility of a "polygon hybrid model" for magnitude, when PGA at site has to be estimated.

5 TIME VARYING RISK

Let us try to extend the definition of the credibility index Δ to the case of time varying risk. In this case, in the frame of the RP (renewal process), the model to be judged consists in the distribution of the inter-occurrence time F_τ in a given seismogenetic zone.

The prediction is governed by the hazard rate $\lambda(t_0)$

$$\lambda(t_0) = \frac{f_\tau(t_0)}{1 - F_\tau(t_0)} \qquad (7)$$

where t_0 is the elapsed time since the last earth quake, f_τ is the density of the random variable τ, the inter-occurrence time.

The immediate risk of an earthquake in the next very short time interval dt is

$$\lambda(t_0)\, dt = \Pr\{ t_0 < \tau \le t_0 + dt \mid \tau > t_0 \}. \qquad (8)$$

This quantity will be called *background earthquake prediction*.

In other words this is the posterior probability, given the information about the preceding gap, t_0.

It can be much greater than the prior probability

$$f_\tau(t_0)\, dt = \Pr\{ t_0 < \tau \le t_0 + dt \} \qquad (9)$$

in the same time interval, as qualitatively shown in Figure 2.

Particular care deserves, in assuming an F_τ model, the consequence on the hazard rate $\lambda(t_0)$. For example the apparently good two fittings of Figure 3 imply completely reversed posterior probabilities!

In spite of its large variability of behaviour, just this function $\lambda(t_0)$ has to be correctly estimated in order to prepare a credible basis in prediction, also when precursors will be observed.

If we want to extend the index Δ to the present case we face a larger difficulty: the quantity we want to estimate here is a function: precisely the hazard rate $\lambda(t_0)$. The "error" is the "distance" between the function $\hat{\lambda}_r(t_0)$ estimated with the r-model and the "true" function $\lambda^o(t_0)$. We propose the following index

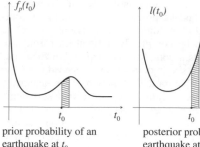

prior probability of an earthquake at t_0

posterior probability of an earthquake at t_0, given the t_0 gap.

Figure 2. The dashed areas are respectively the prior probability of an earthquake in a very short time interval at t_0 and its posterior probability, here called background earthquake prediction, given the information of the elapsed time t_0 from the preceding earthquake.

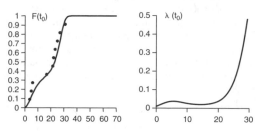

Generalized Exponential Distribution on direct observations (Umbria zone)

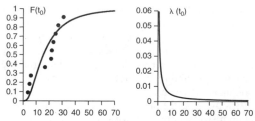

Lognormal Distribution on direct observations (Umbria zone)

Figure 3. The background earthquake prediction relative to one data set, interpreted according to two different models of F_τ: generalized exponential (maximum entropy) and lognormal.

$$\Delta_r^o = \Pr\left\{ \left| \int_{t_1}^{t_2}\left(\lambda^o - \hat{\lambda}_r\right) dt \right| < h_2 \right\} \qquad (10)$$

It measures the error in prediction during the time period $[t_2 - t_1]$, starting from a time t_1, after an earthquake. From conceptual point of view it can again be written as before:

$$\Delta_r^o = \Pr\left\{ \left| \varepsilon_r^o \right| < h \right\}. \qquad (11)$$

Table 1. W $1 - F = \exp[-(\rho x)^\alpha]$; $\alpha = 4$; $\rho = .03$; $t_2 = 50$ years.

Model	Δ_r^o	LTP^o	$\widehat{LTP}$	$\lambda^o(50)$	$\bar{\lambda}(50)$
$W(\hat{\alpha}, \hat{\rho})$	0.990	0.006	0.01	0.4	0.21
$W(\alpha^o, \hat{\rho})$	0.992	0.006	0.008	0.4	0.37
$W(2, \hat{\rho})$	0.0	0.006	0.12	0.4	0.08
$W(6, \hat{\rho})$	0.01	0.006	0.000006	0.4	1.40
$F*$	0.94	0.006	0.008	0.4	0.11

Table 2. Ex-W $1 - F = (1 - p)\exp[-bx] + p\exp[-(\rho x)^\alpha]$. $p = 0.0476$; $b = 0.162$; $\alpha = 8$; $\rho = 0.0189$; $t_2 = 56$ years.

Model	Δ_r^o	LTP^o	$\widehat{LTP}$	$\lambda^o(56)$	$\bar{\lambda}(56)$
Ex-W$(p^o, \hat{b}, \alpha^o, \hat{\rho})$	0.7920	0.00997	0.006408	0.2243	0.1382
Ex$(\hat{b})$	0.5790	0.00997	0.002445	0.2243	0.1074
$F*$	0.8960	0.00997	0.01551	0.2243	0.1081

It is worthwhile to note another meaning of the index. Because in a RP is

$$e^{-\int_0^t \lambda(x)\,dx} = 1 - F(t) = \Pr\{\tau > t\} \tag{12}$$

and then

$$\int_0^t \lambda(x)\,dx = -\ln \Pr\{\tau > t\} \tag{13}$$

we can state that the index measures the error in long term prediction (LTP). In other words, if the prediction period is starting just after an earthquake ($t_1 = 0$) the index measures the error in predicting a waiting time greater than t_2. Tables 1 and 2 show two examples. The first is a characteristic type earthquake process, governed by a Weibull distribution with a small coefficient of variation and increasing hazard rate; the second is governed by a mixture distribution Exponential-Weibull (Ex-W) with large coefficient of variation and bath-basin shaped hazard rate as qualitatively shown in Figure 2.

These are two cases paradigmatic of two seismogenetic behaviours: the first consists in a rather regular energy release; the second shows a kind of a double return period. Both of them remain debatable cases: the first because of the α value in interpreting Weibull (W) model, the second because of its apparently good fitting by a Poisson model (exponentially distributed interoccurrence time). The differences in the alternative estimates are not negligible when the t_0 gap reaches large values (two or more times the global return period) and the standard statistical analysis based on the data fitting is not sufficiently sensitive in choosing the better estimate.

The here proposed Δ index of credibility helps in the estimate of the hazard rate.

In Tables 1 and 2 both the processes are interpreted with true and wrong models operating on 1000 samples drawn from the two defined processes. The proposed index Δ is sensitive to recognise the true model and to denounce a decreasing effectiveness if the formal model is correct but the parameters are wrong. The second and third columns show the values of LTP ($\Pr\{\tau > t_2\}$) in the generator process and, as a mean value, in the estimator models; as usual the upper index zero means "true value" and the cap means estimated value (maximum likelihood method). The last two columns show the value of hazard rate at t_2. The differences between true and estimated values denounce the uncertainty affecting the background earthquake prediction, even when the "true" model is used. Moreover, compared with the relative good values of Δ, they indicate a not yet reached degree of severity of Δ, which, in the present definition, comes out to be only useful as a tool of comparison between models. The empirical polygon model $F*$ shows a good credibility and a good capability in long term prediction. The hope arises that on its basis a more credible model could be constructed, as in the case of long term prediction occurred.

6 CONCLUSIONS

The index Δ, successfully introduced as credibility of a probabilistic model in evaluating a static hazard quantity, can be extended to evaluate hazard variable in time.

REFERENCES

De Finetti B., 1966. L'adozione della concezione soggettivistica come condizione necessaria e sufficiente per dissipare secolari pseudo problemi, da I Fondamenti del Calcolo delle Probabilità, Pubb. Università degli Studi di Firenze. Firenze.

Grandori G., Guagenti E. & Tagliani A., 1998. A proposal for comparing the reliabilities of alternative seismic hazard models, Journal of Seismology, 2, 27–35, 1998.

Grandori G., Guagenti E. & Perotti F., 1988. Alarm system based on a pair of short term earthquake precursors, BSSA, Bull. of Seismological Society of America, 78, 4, 1538–1549.

Grandori G., Guagenti E. & Petrini L., 2003. Modelling magnitude distribution for local hazard evaluation: a new approach, Proceeding of PCEE 2003, 7th Pacific Conference on Earthquake Engineering, Christchurch, New Zealand, 2003 (to appear).

Grandori G., Guagenti E. & Tagliani A., 2002. Magnitude distribution versus local seismic hazard, BSSA, Bull. of Seismological Society of America (to appear).

Grandori Guagenti E., Garavaglia E. & Tagliani A., 1990. "Reccurance Time Distributions: A Discussion", *Proc. of the 9th European Conference on Earthquake Engineering*, EAEE, Moscow, Russia, Vol. I, pp. 33–41.

Guagenti E., Molina C. & Mulas G., 1988. Seismic risk analysis with predictable models, *EESD Earthquake Engineering and Structural Dynamic*, 16, 343–359.

Lakatos I. & Musgrave A., 1974. *Criticism and growth of knowledge*, Cambridge University Press.

Vere-Jones D., Harte D. & Kosuch M., 1998. Operational requirements for an earthquake forecasting programme in New Zealand, *Bull. of the New Zealand National Society for Earthquake Engineering*, 31, 194–205.

Wesnousky S.G. et al., 1984. Integration of geological and seismological data for the analysis of seismic hazard: a case study of Japan, BSSA, Bull. of Seismological Society of America, 74, 2, 687–708.

System-based Vision for Strategic and Creative Design, Bontempi (ed.)
© *2003 Swets & Zeitlinger, Lisse, ISBN 90 5809 599 1*

Fixing steel braced frames to concrete structures for earthquake strengthening

J. Kunz & P. Bianchi
Hilti Corp., Research, Schaan, Principality of Liechtenstein

ABSTRACT: Steel braced frames are widely used to upgrade concrete structures for earthquake loads. They are used both to strengthen the structure and to increase its ductility. In order to transfer the horizontal shear loads from the concrete structure to the steel braced frame, a reliable connection between the two is essential. It must be able to transfer the full capacity of the steel frame with only limited displacements. The behavior of two different fixing methods has been investigated in research programs in Japan and in Europe. A device to ensure an evenly distributed load transfer to a long row of anchors has been developed to meet the requirements of such applications.

1 INDIRECT CONNECTION: LOAD TRANSFER THROUGH MORTAR JOINT

1.1 Method

The first method investigated is to transfer the earthquake loads from the concrete to the steel frame through a mortar joint. In this case the steel frame is equipped with headed studs at its circumference and post-installed anchors with headed ends are fixed to the concrete structure. When the steel frame has been placed, a spiral reinforcement is put into the space between frame and concrete and a non shrink grout is injected into the space between the existing concrete member and the steel frame in order to connect headed studs and post-installed anchors.

1.2 Test series

This type of connection has been investigated in an extensive research program (Yamamoto 1999). Special attention was given to the fact that the concrete of the structure to be reinforced may be of very low quality. Since only very limited data on anchor behavior in low strength concrete are available, the first two series consisted of pullout and shear tests of single anchors in normal weight concrete with compressive strengths as low as 5, 10 and 15 MPa and in a lightweight concrete of 15 MPa. The conclusion drawn from the test results is that the standard adhesive anchors in low strength concrete can be designed according to standard methods developed for concrete strengths over 20 MPa (Cook et al. 1998). The results of these tests together with corresponding design recommendations have been published in (Kunz et al. 2001).

The test series to be presented hereafter assessed the strength, deformation and failure patterns under cyclic loads of indirect connections as shown in Figure 2. Twenty specimens were manufactured in real size and shape that was taken out from an actual indirect joint section. Rows of three adhesive anchors were cast into concrete blocks (Fig. 3). A spiral reinforcement was put between the anchors (Fig. 4), and

Figure 1. Frame prepared for indirect connection.

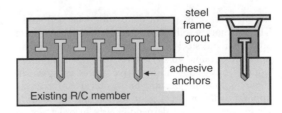

Figure 2. Indirect connection.

Figure 3.   Adhesive anchor with head.

Figure 4.   Group in formwork.

Figure 5.   Loading apparatus.

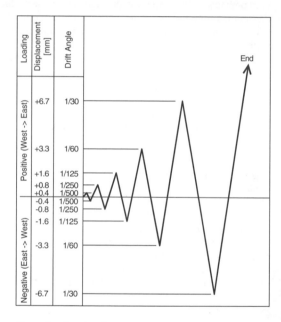

Figure 6.   Load cycles.

on top there was a steel plate with 4 headed studs. Then a mortar joint was injected between concrete specimen and steel plate. Figure 5 shows the loading apparatus. It applied shear force repetitively in directions of push and pull over five cycles as shown in Figure 6. In principle, only horizontal shear force was applied to the indirect joint section and the loading beam was confined perpendicularly in order not to apply tensile forces. The constraining force was approximately 0.1 MPa. The test parameters show up in Table 1.

Horizontal loads and displacement δ of the part filled with mortar were measured. The amount of slip was designed to be divided into relative slip at the interface between mortar and steel frame, δ_{SM}, and relative slip between mortar and base material, δ_{MC}.

1.3   Test program

Figures 7 and 8 show the load-displacement curves and final crack patterns for tests with 19 mm diameter bars, standard edge distances and a base block concrete strength of 5 MPa for figure 7 and 15 MPa for Figure 8. The maximum load was practically reached at the small

Table 1. Test parameters for indirect joint.

| Concrete strength [MPa] | Anchor diameter [mm] | Edge distances of anchor row in base plate [mm] | | Total |
		Standard 200/200	Eccentric 100/300	
5	D19	2	1	3
	D22	1	1	2
10	D19	2	1	3
	D22	1	1	2
15	D19	2	1	3
	D22	1	1	2
LC15	D19	2	1	3
	D22	1	1	2
				20 specimens

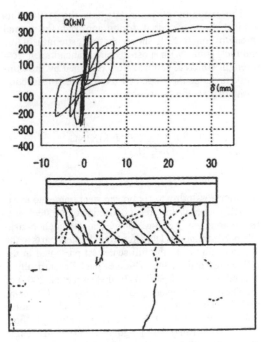

Figure 8. Cyclic shear test with f_c = 15 Mpa.

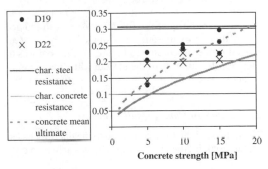

Figure 9. Comparison of normal concrete test results and EC4 design approach.

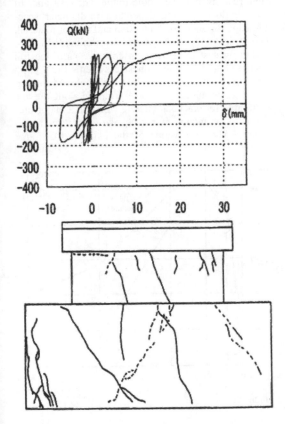

Figure 7. Cyclic shear test with f_c = 5 Mpa.

displacement of the first cycle in all tests. The same loads were reached again for the higher displacements of the subsequent cycles and the loading to failure at the end of the test showed a perfectly ductile behavior. Thus, the system performs in a stiff and elasto-plastic

manner as required. The higher the strength of the base material is, the less cracks will appear there and more cracks appear in the joint. For the eccentric tests, the obtained maximum loads were about identical to those in the standard edge distance tests and the tendencies in cracking were also the same. In the tests with light-weight concrete base material, the shear resistance was about 77% of that obtained in the corresponding tests with normal concrete.

The design of composite steel/concrete sections according to Eurocode 4 takes into account the bearing pressure on the concrete ahead of the connector and the steel shear strength by the formulae given in Figure 9.

The diagram of Figure 9 plots the test results against the Eurocode 4 design approach for normal strength concrete. It shows that the strength of the indirect joint with post-installed anchors in low strength concrete can be safely designed by the standard formulae developed for composite construction with normal strength concrete.

2 DIRECT CONNECTION: FIXING STEEL TO CONCRETE

2.1 Method

Another possibility to fix braced steel frames to existing RC frames is the direct connection of the steel parts to the concrete. Adhesive or mechanical anchors usually perform the load transfer from the steel parts to the concrete frame. In the full scale test presented in the following section, the "Ductile Steel Eccentrically – Braced System" (DSEBS) (Bouwkamp et al. 2001) of the Darmstadt University of Technology, in Germany, was fixed to a test frame designed by the National Laboratory for Civil Engineering, LNEC, Lisbon, Portugal, with an overall length of 12.50 m and a height of 10.80 m. The retrofitting procedure involved the introduction of a steel, one bay, ductile eccentrically braced frame with a vertical shear link as shown in Figure 10.

The overall aim of the full size test was to replace the infilled brick masonry of a single bay by the steel braced frame. According to Figure 11, the goal was to increase the ductility of the system while maintaining its actual strength.

2.2 Fixing of frame for full scale test

The enhanced ductility compared to the ductility based on the brick infilled walls (Fig. 10) was achieved only by the steel frame, in fact by the so called shear link. In the test, the earthquake shear forces were simulated by quasi static loads introduced by hydraulic jacks installed on the Reaction Wall of the ELSA Laboratories, Joint Research Center, Ispra, Italy. Special attention was given to the uniform load distribution along the steel beams, which requires an almost slip free connection between steel and metal. This can be achieved by avoiding clearance between anchor and drill hole in the concrete and between anchor and steel beam (see section 3.3).

The fastening elements were designed to resist shear loads of at least the maximum capacity of the shear link. Therefore, the resistance of the shear link HEA 120 and the geometry of the braced frame were decisive for the design of the anchors. The shear loads for the upper beam were expected to 260 kN for the maximum resistance of the whole frame. The resistance of each single anchor was depending on diameter, spacing, edge distance, load direction and anchor material. The choice of the anchor was in favor of the Hilti HVZ anchoring system. This is an adhesive cartridge foil system (Fig. 12). With adhesive systems there is no clearance between anchor and drill hole. Moreover, the HVZ, because of the cones arranged along the rod, is suitable for cracked concrete which is a major prerequisite for anchors in earthquake retrofitting applications. The dimensions of the existing RC frame

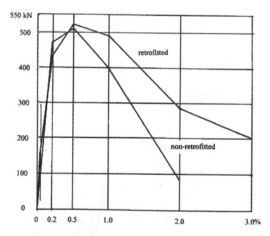

Figure 11. Design.

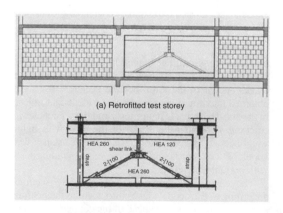

(a) Retrofitted test storey

Figure 10. Test setup.

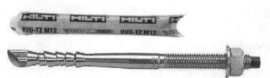

Figure 12. HVZ anchor system.

structure limited the size of the anchor significantly. Since the depth of the columns was only 20 cm, the maximum possible anchor diameter was 16 mm with a related drill hole depth of 125 mm.

The resistance of a single anchor was calculated by the concrete capacity design (CCD) method (Fuchs et al. 1995) which is available as a design guideline in ETAG Annex C (ETAG No 001). As an example, the shear design for the anchorage of the upper beam is shown in the following.

Basic shear resistance of a single anchor HVZ M16 in cracked concrete:

$$V_{Rd,c}^0 = 5.9kN$$

Shear design according to ETAG Annex C, taking into account spacing s, edge distance c, concrete quality, and load direction:

$$V_{Rd,c} = V_{Rd,c}^0 \cdot f_{B,V} \cdot f_{\beta,V} \cdot f_{AR,V}$$

Influence of the concrete quality $f_{B,V} = 1.0$ for concrete C20/25. Influence of the load direction $f_{\beta,V} = 2.0$ for load direction parallel to the concrete edge.

Influence of the spacing and edge distance $f_{AR,V}$:

$$f_{AR,V} = \frac{3 \cdot c + s_1 + s_2 + s_3 + \ldots\ldots s_{n-1}}{3 \cdot n \cdot c_{min}} \cdot \sqrt{\frac{c}{c_{min}}}$$

$$= \frac{3 \cdot 135\,mm + 20 \cdot 200\,mm}{3 \cdot 21 \cdot 85\,mm} \cdot \sqrt{\frac{135\,mm}{85\,mm}} = 1.0$$

Therefore, the design resistance in shear of one HVZ M16 × 105 anchor in this application is:

$$V_{Rd,c} = 5.9kN \cdot 1.0 \cdot 2.0 \cdot 1.0 = 11.8kN$$

Considering only the shear-transfer reaction, without axial anchor forces, the design example for the upper beam requires 22 anchors in order to limit the anchor loads to design level at the time when the full capacity of the shear link is reached. Moreover, the reaction force of the shear link is eccentric with respect to the upper beam, and therefore, a bending moment causes axial forces in the anchors close to the shear link connection. The magnitude of these tensile forces depends on the stiffness of the steel beam. For this reason, the number of anchors obtained by the calculation of the design shear resistance was increased to 38 anchors HVZ M16 × 105 in order to transfer shear and tension loads between the upper beam and the shear link without any damage to the anchorage in the full scale test. In the same sense, to transfer the shear and tensile loads, 20 anchors instead of 14 required by shear

design were installed in the bottom beam. 14 anchors were used for the column beams. This number was not increased because these straps were welded to both the top and bottom steel sections. Anchoring the straps to the column was considered necessary since the shear resistance of the concrete at the beam-column interface was judged to be inadequate.

Setting of the anchors was performed in the presetting mode. This was necessary because of the used HEA 260 profiles. A through setting of the anchors was not possible since the space between the upper and lower flanges of the steel beam is too small to introduce the anchors. In the pre-setting mode the conditions to install the upper steel beam over 38 set anchors was to be considered carefully. Since the clearance of the drilled hole in the steel beams had to be as small as possible in order to guarantee the correct position and alignment of the anchors, a strong focus on the precision of the setting process was required. To guarantee the accuracy of the positioned anchors a special drilling template for each steel beam was produced, which not only provides the exact position of the drill holes, but also the right direction of the protruding anchor thread. This template consisted two steel plates, separated by struts with the height of the threaded part of the protruding anchor (Fig. 13).

To transfer the shear loads uniformly over all anchors, the clearance holes were injected with a mortar (Fig. 14) using the Hilti Dynamic Set (Fig. 16,

Figure 13. Drilling with template.

Figure 14. Injecting clearance holes with Hilti HIT HY 150 Mortar.

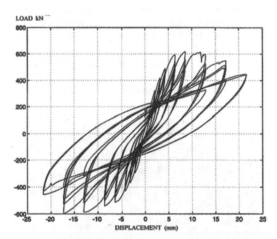

Figure 15. Total storey shear force versus storey displacement.

see section 3.3). The uniformly distributed shear transfer is necessary to prevent an overloading of a single anchor as well as a cone failure of the related concrete substrate area. In this application, the mortar did not only fill up the clearance holes between the steel beam and the anchor, but at the same time, it served as leveling grout, because it was also injected into the voids between the RC frame substrate and the steel beam.

According to the high safety factors for the anchoring of the braced frame, the results clearly showed that the anchor bolts acted fully satisfactory and slippage did not occur (Fig. 15).

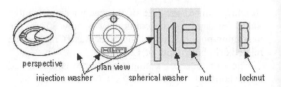

Figure 16. Injection washer "Dynamic Set".

2.3 *Distribution of load to anchors*

When fixing steel braced frames to the concrete directly, the anchors are set through clearance holes in the frame. Tightening the anchors creates a clamping force between the concrete and the steel frame so that small shear forces can be transmitted by friction from the concrete to the steel. But with higher shear forces introduced, the friction is overcome and there is a relative displacement between steel frame and concrete until the first anchors touch their clearance hole and take up more shear force. Since not all anchors are set exactly in the center of their clearance holes, only part of the anchors will first take the additional shear load. The load will be distributed to all anchors as the ones which first touched the frame are deformed. However, in case of cyclic earthquake loads, these deformations cannot be tolerated; all anchors must take up shear loads from the beginning. Therefore the so called "Dynamic Set" has been developed (Fig. 16).

The dynamic set contains a special washer which allows to inject the clearance holes between steel frame and the anchors with a mortar. Moreover, the set contains a spherical washer, which ensures that there is no bending in the anchor due to the tightening torque. This fact also improves the anchor behavior under cyclic loads.

Tests have demonstrated that the shear capacity of anchor rows can be significantly improved by injecting the clearance holes. Figure 17 shows the load displacement curves obtained on a plate with two anchors in void and injected clearance holes respectively, subjected to shear loading towards the concrete edge. With void clearance holes, the loading shows a stiff behavior until the capacity of the concrete edge ahead of the anchor closer to it is reached. Then a concrete cone breaks out in front of the first anchor and the plate slips towards the edge until the anchor behind can take up the load. After a further stiff increase of the load the concrete edge capacity in front of the second anchor is reached.

With injected clearance holes both anchors take up load from the start, the primary failure due to overloading of the edge by the first anchor is avoided and the maximum load is reached directly with very limited displacement.

These tests show that injection of the clearance holes is an efficient means of distributing the loads

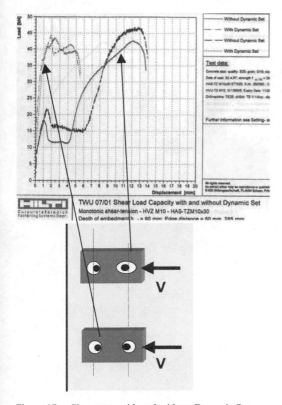

Figure 17. Shear tests with and without Dynamic Set.

to all fixing points in cases where only limited displacements can be allowed. When fixing steel braced frames to concrete, the clearance holes should always be filled in order to guarantee a good load displacement behavior. This method can also improve the efficiency of other applications such as fixing steel jackets to concrete columns.

3 CONCLUSIONS

Two methods to fix steel braced frames to concrete structures have been shown. The choice of the appropriate method will depend on the conditions of a specific project. For example, the direct connection will be faster in many cases, while the indirect fixing has the advantage that it does not require precision drilling.

Costs and availability of materials will also influence the choice.

The indirect connection consisting of adhesive anchors in the concrete, welded studs on the steel braced frame and a mortar joint can be designed according to the design rules for composite steel/concrete structures as given in the relevant structural design codes. Extensive testing has demonstrated that these rules can be applied even if the concrete strength in the structure to be rehabilitated is as low as 5 MPa.

The strength of direct connections will be determined according to the rules of anchor design. In order to avoid slip between the steel braced frame and the concrete structure under cyclic earthquake loads, the clearance holes should be injected. Tests have shown that the use of injection washers allows to efficiently distribute shear loads to all anchors. This method can be proposed not only for fixing steel braced frames to concrete, but in all applications of earthquake strengthening and rehabilitation, where displacements under shear loads should be avoided, for example also when steel jackets are fixed to concrete columns.

REFERENCES

Yamamoto, Y. 1999. *Experimental Research on Bonded Anchor Behavior in Low Strength Concrete and Lightweight Concrete*. Research for Hilti AG. Shibaura Institute of Technology, Tokyo.

Cook, R.A., Kunz, J. et al. 1998. Behavior and Design of Single Adhesive Anchors under Tensile Load in Uncracked Concrete. *ACI Structural Journal*, V. 95, No. 1, January–February.

Kunz, J., Yamamoto, Y. et al. 2001. Anchors in Low and High Strength Concrete. *International Symposium on Connections between Steel and Concrete*, Stuttgart, 10–12 September 2001. RILEM, 2001, ISBN: 2-912143-25-X.

Bouwkamp, J., Gomez, S., Pinto, A. et al. 2001. *Cyclic Tests on R/C Frame Retrofitted with K-Bracing and Shear Link Dissipator*. European Commission Research Centre. (ICONS Research Network, contract N. FMRX-CT96-0022).

Fuchs, W., Eligehausen, R. et al. 1995. Concrete Capacity Design (CCD) Approach for Fastenings to Concrete. *ACI Structural Journal*, Vol. 92, No. 6.

ETAG No 001: Guideline for European Technical Approval of Metal Anchors for Use in Concrete, Annex C: Design Methods for Anchorages. EOTA, rue du Trône 12, Troonstraat, B – 1000 Brussels, 1997.

15. Geotechnical engineering and tunelling

System-based Vision for Strategic and Creative Design, Bontempi (ed.)
© 2003 Swets & Zeitlinger, Lisse, ISBN 90 5809 599 1

Mechanical properties of steel-concrete composite segment for micro multi box shield tunnel

I. Yoshitake, M.I. Junica & K. Nakagawa
Department of Civil Engineering, Yamaguchi University, Japan

M. Ukegawa, M. Motoki & K. Tsuchida
Toda Corporations, Japan

ABSTRACT: The present study focused on the mechanical properties of the composite segment in order to obtain experimentally the fundamental data for Micro Multi Box Shield Tunneling (MMB) design. The test results indicated that the strength of the segment hardly decreased even if the initial load applied on. Furthermore, the results obtained from these experiments were compared with the strength values of a reinforced concrete (RC) member, calculated based on the allowable stress design method. The comparison showed the applicability of the design method, in which the experimental results were equal or higher than the calculated values.

1 INTRODUCTION

Micro Multi Box shield tunneling (MMB) is a large-scale rectangular tunnel, which consist of some small box shield segments as shown in Figure 1. MMB has

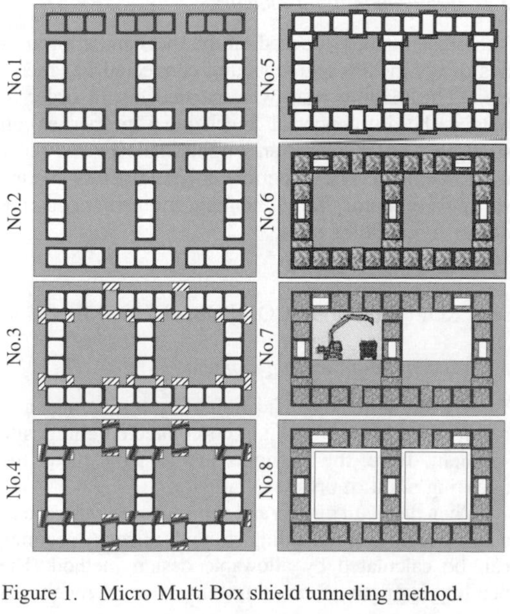

Figure 1. Micro Multi Box shield tunneling method.

an advantage in case of a need to excavate a tunnel in a limited public space, and an economically effect by reusing a small shield machine throughout the tunnel construction. However, fundamental data is insufficiently to be used yet for design or execution of MMB, and its design method is hardly established in the present situation.

A main segment in MMB is usually made from the steel-concrete composite structure, and the connection between the each segment is reinforced concrete (RC) structure for covering the construction error. The box segments are subjected by bending moment due to soil pressures during the construction, and this ultimate strength must be appropriately evaluated for its design.

The present study focused on the mechanical properties of the composite segment in order to obtain experimentally the fundamental data for MMB design. Furthermore, the results obtained from these experiments were compared with the strength values of a RC member calculated based on the allowable stress design method.

2 EXPERIMENTAL PROGRAM

2.1 *Specimens*

Figure 2 shows the basic structure of a beam specimen in the present study. These specimens were applied by a load of an oil-jack at 2 points of the center span

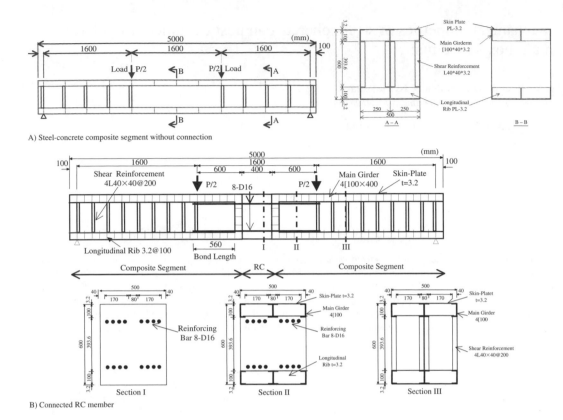

A) Steel-concrete composite segment without connection

B) Connected RC member

Figure 2. Detail of beam specimen and loading method.

Table 1. Mix proportion.

| | Unit Amount − kg/m³ | | | | | E_c |
Type	W	B	S	G	Ad	N/mm²
SCC	165~ 170	600~ 670	660~ 730	810~ 820	6~ 12	3.0 ~3.5 × 10⁴
Mortar	377	686	1017	—	—	1.8 × 10⁴

under a simple support condition. The specimens had 4 main-girders by employing channel steels, and longitudinal ribs with a certain interval. Skin-plates with 3.2 mm thickness were set on the outside of the main-girders. Width and height of specimen were 500 mm and 600 mm respectively. The sizes of the specimen were equivalent to 1/3–1/5 model of real structure.

In MMB method, the self-compactable concrete is usually used as filling materials. However, mortar concrete that has a compressive strength of 40 N/mm² was employed by reason for a scale-down of the model test. The mix proportion is given in Table 1.

2.2 Experimental parameters

The present study focused on the mechanical properties of composite segments and connected RC members. The loading tests were carried out in order to obtain the influence of mechanical properties on strengths, e.g. main girder, longitudinal ribs and shear reinforcement. The experimental parameters in this study were mainly initial bending moments, concrete defects and lifeline space.

3 PROPERTIES OF COMPOSITE SEGMENT

3.1 Strength of bending moment

Table 2 shows the specification of each specimen and results of bending tests, i.e. no-connected composite segment. Here, the strengths in this table mean the buckling point of upside skin-plate.

When the main girders and skin-plate are regarded as a reinforcing bar, a bending strength of each specimen can be calculated by allowable design method. The bending strength of experiments was as averagely 1.3

Table 2. Specification of no-connected composite segment specimen and strength.

Specimen	Main girder mm	Steel	Interval of L.R. (mm)	Shear reinforcement (mm)	Max Load (kN)	Calculation (kN) Bending	Shear	Experimental object
A-1	4[100	1.347	100	4L 40@200	928	747	1804	Base
A-2	4[100	1.347	100	4L 40@200	918	747	1804	Defect
A-3	4[100	1.347	400	4L 40@200	905	747	1144	L.R. Interval
A-4	4[40	1.013	100	4L 40@200	861	629	1850	Main-Girder
A-5	4[100	1.347	100	4L 40@200	962	747	1804	Lifeline space in
A-6	4[40	1.013	100	4L 40@200	860	629	1850	center
AI-7	4[100	1.347	400	4[100@800	916	747	1336	Initial stress
A-7	4[100	1.347	400	4[100@800	931	747	1336	vs. AI-7
AI-8	4[40	1.013	400	4[40@800	796	629	1046	Initial stress
A-8	4[40	1.013	400	4[40@800	816	629	1046	vs. AI-8
AS-1	4[100	1.347	100	–	838	747	482	Shear Re. Zero
AS-2	4[100	1.347	100	4L 40@800	941	747	812	Little Shear Re.
AS-3	4[100	1.347	100	4L 40@800	700	747	330	Lifeline space in
AS-4	4[40	1.013	100	4L 40@800	340	629	348	shear span

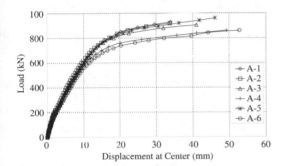

Figure 3. Load-displacement (A-1~A-6).

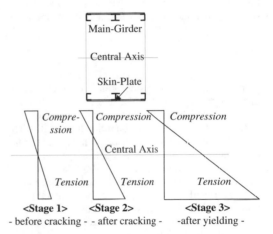

Figure 4. Change of strain distributions.

times as that of calculation. The strengths of A-5 and A-6 specimens, which had lifeline space in the center span, were almost equal to that of a normal specimen.

Figure 3 presents relationships between loads and displacements at the center span from A-1 to A-6 specimens. Every relationship is almost equal under a load of 600 kN, but that later is divided broadly into two groups over the load of 600 kN. This phenomenon is strongly related to the height of main girder. The volume ratio of steel between 100 mm and 40 mm height was approximately 1.3; the ultimate strength level of each group is different due to the yielding point of the steel member.

The failure process of the composite segment can be categorized into 3 stages as shown in Figure 4. At stage 1, all cross-sections of the member including tensile concrete resist the bending moment. When tensile stress reaches the tensile strength of concrete, cracks occur in lower side of concrete. A neutral axis moves up and the strain of each member increases as shown at stage 2. As the increase of the load after

yielding of tensile steels, the element in compression is increasingly affected by the applied load. Finally, the buckling of compression skin-plate occurs at stage 3.

3.2 Influence of shear reinforcement

Table 2 indicates that experimental values of max loads are higher than calculation of shear strengths of AS-1 or AS-2 specimen. The differences are due to the restraining effect of shear deformation by the skin-plate, which covers the top and bottom of concrete.

Here, the relationship between loads and displacements of AS-1 specimen is shown in Figure 5. This figure illustrates that the calculation strain of the tensile main girder at shear span becomes smaller than the experimental value as increasing the load. The

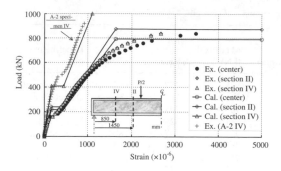

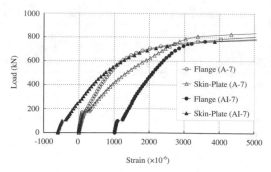

Figure 5.   Load-strain at shear span of AS-1 specimen.

Figure 6.   Strain of flange of main-girder & skin-plate.

calculation strain was obtained from an assumption that a shearing load was given by a shear-connector effect of longitudinal ribs. On the other hand, the shear load was given insufficiently to concrete member so that the concrete in shear span was separated from steel member due to the development of shear cracks.

3.3   Influence of initial stress

The MMB segment is normally affected by the initial moment due to soil pressures before concrete casting in the segment. A series of loading tests with specimens affected by initial stresses was conducted in order to obtain its influence on ultimate strengths. The specimen for this test was made by means of casting the concrete in an emptied segment with an initial stress. Here, the initial strain was 1000 micro resulted by applying a load of 50 kN in the emptied segment.

The strength of AI-7 specimen shown in Table 2 is lower than that of A-7, however, the difference of both is not so large. This tendency is similar to the case of AI-8 and A-8 specimens.

Figure 6 shows the comparison of tensile strains at the center span of AI-7 and A-7 specimens. In case of AI-7 specimen having an initial strain, the strain of skin-plate does not rapidly increase; in spite of the flange of main-girder yielded under relatively lower load. As shown in this figure, the ultimate behaviors of each specimen are almost similar. These results indicate that the initial stress with strain of 1000 micro has little effects on the strength and deformation of the composite segment.

3.4   Influence of lifeline space

A lifeline space is often provided in a segment when MMB tunnel is employed in an urban area. Accordingly, the present study employed two type specimens in order to survey the influence of the lifeline space, i.e. a cave in center of a shear span. Figure 7 shows the specimens with the lifeline space.

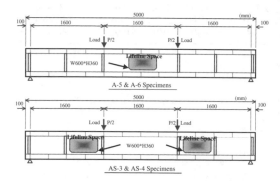

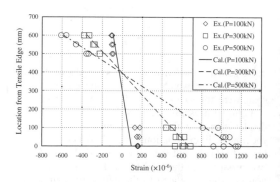

Figure 7.   Detail of specimen with lifeline space.

Figure 8.   Strain distribution in A-5 specimen.

As indicated in Table 2, the ultimate strengths of A-5 and A-6 specimens are almost equal or higher than that of A-1 and A-4 specimens. As shown in Figure 8, the strain distribution becomes likely a linear relation and the strength calculated based on allowable stress design method generally matches the experiments. These results represent no decline of the strength in case of an existence of the lifeline space in the bending span.

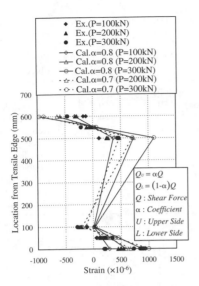

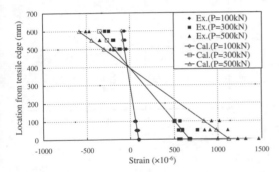

Figure 9. Strain distribution in AS-3 specimen.

Figure 10. Strain distribution in A-2 specimen.

On the other hand, strain distributions of AS-3 and AS-4 specimens having the lifeline space in the shear span exhibit non-liner relation as shown in Figure 9. These results were due to the influence of shear forces caused in the up-down beam separated by the lifeline space.

3.5 Influence of concrete defects

In order to obtain the influence of concrete defects, the loading test was conducted by employing A-2 specimen, which had styrene-foam on upper side of the main girder. The ultimate strength of A-2 specimen is almost equal to that of A-1 specimen, as indicated in Table 2. The strain distributions shown in Figure 10 are almost linear, and experimental values stay on the calculation line, which obtained based on the assumption of no-defects of concrete.

4 PROPERTIES OF CONNECTED MEMBER

4.1 Design of embedded reinforcing bar

In case of design on a connected member in MMB, the embedding length of reinforcement must be provided appropriately in order to obtain the sufficient strength. The minimum tensile strengths T were calculated from the yielding point of the reinforcing member, i.e. a main-girder of the segment (T_{sy}) or the reinforcing bar in the connection (T_{ry}). Here, the embedding length L_r were calculated from the bond strength f_{bod} between a reinforcing bar and concrete; embedding length L_s were estimated by considering the shear strength V_{scd} and the interval of longitudinal rib S. The designed length of embedding reinforcement L was selected from the longer length among the each ones.

$$\text{Minimum Tensile Strength: } T = \text{Min}[T_{ry}, T_{sy}] \quad (1)$$

$$\begin{aligned} \text{Designed length: } L &= \text{Max}[L_r, L_s] \\ &= \text{Max}[T/(f_{bod}*rl), T/(V_{scd}/S)] \quad (2) \end{aligned}$$

where, $f_{bod} = 2.5 \text{ N/mm}^2$ ($f_c = 40 \text{ N/mm}^2$) and rl means the round length of reinforcing bar.

4.2 Strength of bending moment

The specification of the specimens with the RC connection is given in Table 3. Furthermore, Table 3 gives the max-loads and calculation ones based on the allowable stress design method in the same way as no-connected composite structure test.

As given in Table 3, the strengths expressed by max-load are 8~32% higher than the calculation values in any case of the experiment. Those comparison results may represent the appropriateness of the design method and the employed embedding length.

4.3 Load-displacement relation

Figure 11 illustrates the relationship between loads and displacements at the center of the beam. Although the load-displacement curve of A-1 specimen is almost smooth, however, the curves of connected specimen have the definite two points of change.

First point means the occurrences of crack in concrete and second point represents the yielding point of the reinforcing bar. Furthermore, two groups can be categorized by the degree of yielding point, i.e. the ultimate strength of beam.

4.4 Strain in the connected member

Strains of the lower reinforcing bar in the connected member are shown in Figure 12, and strains of

Table 3. Specifications of connected composite segment specimen and strength.

Specimen	Main girder	Re. bar size & arrangement	Bond length (cm) L_r		L_s	Max load (kN)	Calculation (kN)	Experimental object
B-1	4[100 × 40	8-D16 Single	55	>	53	490	427	Base
B-2	4[40 × 40	8-D16 Single	55	>	45	579	440	Main Girder
B-3	4[100 × 50	8-D22 Single	76	>	74	820	714	Steel Ratio
B-4	4[100 × 40	8-D13 Single	44	>	42	480	397	Height
B-5	4[100 × 50	8-D13 Single	44	>	38	475	397	Metal Volume
B-6	4[100 × 40	16-D13 Double	44	<	84	782	715	Double
B-7	4[100 × 50	16-D13 Double	44	<	75	775	715	Double
B-8	4[100 × 40	8-D16 Single	55	<	106	520	427	Defect
BS-1	4[100 × 40	8-D16 Single	55	>	53	661	569	Shear Span
BS-2	4[100 × 40	8-D13 Single	44	>	42	675	529	Shear Span

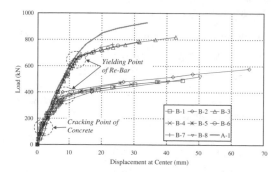

Figure 11.   Load-displacement (B-1~B-8).

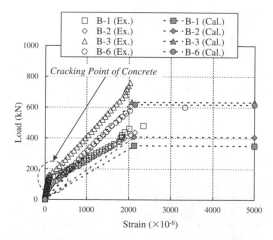

Figure 12.   Strain of reinforcing bar in tension.

the upper side concrete are presented in Figure 13 respectively.

It is observed from Figure 12 that the concrete has no crack under a load of 150 kN and the strains are almost equal to the calculation after the cracking.

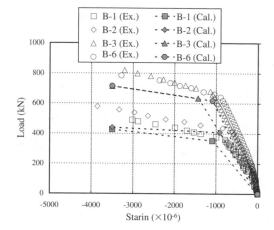

Figure 13.   Strain of concrete in compression.

On the other hand, concrete strains do not rapidly increase after the cracking of tensile concrete, and that increase gradually nevertheless the yielding of reinforcement as shown in Figure 13.

4.5   Behavior of connection in shear span

In MMB tunnel with a large space, the connected RC members are generally subjected to shear force in addition to bending moment. Thus, the loading tests were conducted by employing the specimen with connection in the shear span.

As mentioned above, the strengths by experiment were also higher than that of calculation in case of the specimen with connection in the shear span. Figure 14, which shows the load-strain relation of tensile main-girder of BS-1 specimen, represents that the calculations are nearly equal to or larger than the experimental strain.

These results may present that the RC member employed in this study can sufficiently connect the

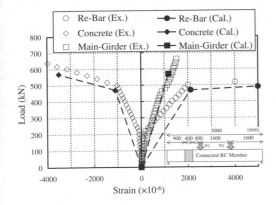

Figure 14. Load strain relation of BS-1 specimen.

each segment if shear force affects the connected RC member. Furthermore, the connected RC member seems to be able to be designed appropriately by using the normal design method in Japan, i.e. allowable stress design.

5 CONCLUSIONS

This paper presented the experimental data for MMB segment and its design method. The obtained conclusions of this study are summarized as follows:

1. The bending strength of experiments was as averagely 1.3 times as that of calculation by allowable stress design. The specimen strength with the lifeline space was almost equal to that of normal specimen.
2. As a result of loading test by means of an adoption of an initial stress on the specimen, the initial stress with strain of 1000 micro had shown little effects on the strength and deformation of the composite segment.
3. The strain distribution of the specimen with a lifeline space in the center span was likely a linear relation, on the contrary, that of specimens having the lifeline space in the shear span exhibited a non-liner relation. These results were due to the influence of shear forces occurred in the up-down beam separated by the lifeline space.
4. The connected RC member strengths were 8~32% higher than the calculations in any case of the experiment. In addition, the load-displacement curves of composite segment were almost smooth; however, the curve of the connected specimen indicated the development of cracks in concrete and the yield of reinforced bar on certain strains.
5. RC structure can sufficiently connect each segment if shear force can build its effect to the connected zone. Furthermore, the calculations were nearly equal to or larger than the experimental strains.

System-based Vision for Strategic and Creative Design, Bontempi (ed.)
© 2003 Swets & Zeitlinger, Lisse, ISBN 90 5809 599 1

To an estimation of dynamic stresses around of the immersed body with a rectangular form

L.V. Nuzhdin & A.O. Kolesnikov
Novosibirsk State University of Architecture and Civil Engineering, Novosibirsk, Russia

V.N. Popov
Institute of Theoretical and Applied Mechanics (SB RAS), Novosibirsk, Russia

ABSTRACT: On basis of a wave model the evaluation of the vibration process in a system the pile foundation – ground at vertical vibrations of an immersed body with the rectangular configuration in a plan will be carried out in view of a real foundation form.

1 INTRODUCTION

There are numerous data showing on essential influence of immersing at vibrations of the pile foundation. Majority of piles and pile-caps have the rectangular form in a plan. However definition of their dynamic characteristics will be usually carried out with the help of bodies having in a plan a circle of equivalent square. At the same time for a raise of an exactitude of dynamic accounts it is necessary to take into account real geometry of considered object.

In this paper the improvement of connection between movings and responses of the pile-cap lateral surface will be carried out at vertical vibrations of an immersed body with the rectangular configuration. For an initial mathematical model of a ground the elastic inert continuum is accepted. The foundation is considered as an absolute rigid body with a flat basis. The contact with a medium is carried out through a sole and lateral surface of a pile-cap. Ground lower than the sole of a pile-cap is considered as a half space, higher – as a layered structure. The connection between movings and reactions on a pile-cap lateral surface was defined from the decision of a task on vibrations of a plate with an undeformable rectangular cut.

2 METHOD OF ACCOUNT ON WAVE MODELS

Usually account method on wave models consists of two independent parts: definition of stiffness and damping parameters; account the basis amplitude with known dynamic parameters. The dynamic movings of the basis can be generally defined by a principle of a superposition as the sum of vibrations from force and kinematical influences.

$$a = \sum_{i=1}^{n} a_i \sin(\omega t + \varphi_i)$$

where a_i is the vertical or rotary-shearing components of vibrations amplitude from force ($i = 1$) and kinematical ($i > 1$) influences; n is the quantity of the taken into account foundations – sources of vibrations; φ_i is the casual phase shifts.

The amplitudes of forward vertical pile-cap vibrations at force excitation are calculated as

$$a_z = \frac{P_z}{K_{zz}} \cdot \frac{1}{\sqrt{\left(1 - \frac{m\omega^2}{K_{zz}}\right)^2 + \left(\frac{C_{zz}\omega}{K_{zz}}\right)^2}},$$

at kinematical excitation

$$a_z^{kin} = a_{sz} \sqrt{\frac{1 + \left(\frac{C_{zz}\omega}{K_{zz}}\right)^2}{\left(1 - \frac{m\omega^2}{K_{zz}}\right)^2 + \left(\frac{C_{zz}\omega}{K_{zz}}\right)^2}},$$

where P_z is the vertical dynamic loading on the foundation; a_{sz} is the amplitude of vertical vibrations of a ground surface; ω is the circular frequency of vibrations; m is the pile-cap mass; K_{zz} and C_{zz} are stiffness and damping parameters of the basis, respectively.

For a horizontal component of vibrations, neglecting rotation of pile-cap in vertical plane it is possible to use identical dependencies.

From consideration of the generalized rotary-shearing vibrations the following expressions are received

$$a_{X\varphi} = \sqrt{\frac{B_5^2 + B_6^2}{D_1^2 + D_2^2}}, \quad a_{X\varphi}^{kin} = \sqrt{\frac{B_1^2 + B_2^2}{D_1^2 + D_2^2}},$$

$$a_{Z\varphi} = \sqrt{\frac{B_7^2 + B_8^2}{D_1^2 + D_2^2}}, \quad a_{Z\varphi}^{kin} = \sqrt{\frac{B_3^2 + B_4^2}{D_1^2 + D_2^2}}.$$

The values B and D are defined through stiffness K_{ZZ}, K_{XX}, $K_{X\varphi}$, $K_{\varphi\varphi}$ and damping C_{ZZ}, C_{XX}, $C_{X\varphi}$, $C_{\varphi\varphi}$ parameters of the basis by summation of the appropriate values found for a pile-cap surface and for a pile basis (Nuzhdin 1994 & 1995). The definition of dynamic stiffness and damping characteristics for a surface of piles was considered in paper Nuzhdin et al. (2002). As is known from the literature, for the usual pile foundations the dynamic interaction happens only on a lateral surface. Let's note stiffness and damping parameters on a pile-cap lateral surface

$$K_{ZZ} = \rho_d V_{sd}^2 d_1 S_{w1}, \quad K_{XX} = \rho_d V_{sd}^2 d_1 S_{u1},$$

$$K_{X\varphi} = -\rho_d V_{sd}^2 d_1 \left(h - \frac{d_1}{2}\right) S_{u1},$$

$$K_{\varphi\varphi} = \rho_d V_{sd}^2 d_1 \left[r^2 S_{\varphi1} + \left(h^2 - hd_1 + \frac{d_1^2}{3}\right) S_{u1}\right],$$

$$C_{ZZ} = \frac{\rho_d V_{sd}^2}{\omega} d_1 S_{w2}, \quad C_{XX} = \frac{\rho_d V_{sd}^2}{\omega} d_1 S_{u2},$$

$$C_{X\varphi} = -\frac{\rho_d V_{sd}^2}{\omega} d_1 \left(h - \frac{d_1}{2}\right) S_{u2},$$

$$C_{\varphi\varphi} = \frac{\rho_d V_{sd}^2}{\omega} d_1 \left[r^2 S_{\varphi2} + \left(h^2 - hd_1 + \frac{d_1^2}{3}\right) S_{u2}\right],$$

Here S_w, S_u, S_φ are complex dimensionless parameters of environment resistance at vertical, horizontal and rotary-shearing movings for round body in a plan (Baranov 1967)

$$S_w = G(S_{w1} + iS_{w2}), \quad S_u = G(S_{u1} + iS_{u2}),$$

$$S_\varphi = G(S_{\varphi1} + iS_{\varphi2}).$$

For increase of calculation accuracy it is necessary to take into account the pile-cap form. For this purpose

a task about vibrations of an indefinitely thin plate with a rectangular cut $a \times b$ was considered.

3 DYNAMIC PARAMETERS FOR RECTANGULAR CUT

The equations of an elastic medium movements in cylindrical coordinates at a lack of volumetric forces in a case of a stratum axially symmetric vibrations has the aspect

$$\mu\left(\frac{1}{\rho}\frac{\partial}{\partial r}r\frac{\partial w}{\partial r} + \frac{1}{r^2}\frac{\partial^2 w}{\partial \theta^2}\right) = \rho\frac{\partial^2 w}{\partial t^2}, \quad (1)$$

where w is the moving along z; ρ is the density; μ is the stiffness of environment.

Let's consider originally round cut with the radius r_0, on which contour the boundary conditions are fulfilled

$$w(r_0, z, t) = w_0 e^{i\omega t}. \quad (2)$$

The solution of the Equation 1 is defined as

$$w = e^{i\omega t} \sum_{n=0}^{\infty} (A_n \cos n\theta + B_n \sin n\theta) \times$$

$$\times \left[C_n H_n^{(1)}(kr) + D_n H_n^{(2)}(kr)\right], \quad (3)$$

where $H_n^{(1)}$ and $H_n^{(2)}$ are Hankel functions, $k = \omega\sqrt{\rho/\mu}$.

From a condition of an axial symmetry follows, that $n = 0$. To satisfy with a principle of a radiation, the solution (Equation 3) should contain only separating waves and therefore $C_n = 0$. Further, assuming $D_n = 1$, we obtain

$$w = e^{i\omega t} \sum_{n=0}^{\infty} A_n H_n^{(2)}(kr).$$

From a boundary condition (Equation 2) the values A_n are defined

$$A_0 = w_0 / H_0^{(2)}(kr_0), \quad A_1 = A_2 = A_3 = \ldots = 0.$$

On a contour of a round cut the shear stresses work only

$$\tau_{rz} = \mu\frac{\partial w}{\partial r} = -\mu w_0 k e^{i\omega t} H_1^{(2)}(kr_0) / H_0^{(2)}(kr_0),$$

which are reduced to

$$q^* = -\int_0^{2\pi} \tau_{rz} \cdot 1 \cdot r_0 d\theta = 2\pi k r_0 \mu w_0 e^{i\omega t} \frac{H_1^{(2)}(kr_0)}{H_0^{(2)}(kr_0)}.$$

Thus we obtain, that the response of a stratum of a unit thickness affixed to a lateral surface of the pile-cap, is equal

$$q = S_w^c(kr_0)w_0 e^{i\omega t} = \mu w_0 e^{i\omega t}(S_{w1}^c + iS_{w2}^c),$$

where valid S_{w1}^c and imaginary S_{w2}^c dimensionless parts of S_w^c are defined

$$S_{w1}^c(kr_0) = 2\pi k r_0 \left(J_0 J_1 + Y_0 Y_1\right)/\left(J_0^2 + Y_0^2\right),$$

$$S_{w2}^c(kr_0) = 4/\left(J_0^2 + Y_0^2\right).$$

If $kr_0 \ll 1$, then

$$S_{w1}^c(kr_0) \approx -2\pi / \ln(kr_0), \quad S_{w2}^c(kr_0) \approx \left(\pi / \ln(kr_0)\right)^2.$$

Let's consider symmetric vibrations of a stratum in case of a rectangular cut with sides a and b (see Fig. 1).

Using relations $x = r\cos\theta$ and $y = r\sin\theta$, the boundary conditions on a contour of a rectangular cut in cylindrical coordinates can be presented as

$$w\left(\frac{a}{2\cos\theta}, \theta, z, t\right) = w_0 e^{i\omega t}, \quad -\theta_0 \le \theta \le \theta_0;$$

$$w\left(\frac{-a}{2\cos\theta}, \theta, z, t\right) = w_0 e^{i\omega t}, \quad \pi - \theta_0 \le \theta \le \pi + \theta_0;$$

$$w\left(\frac{b}{2\sin\theta}, \theta, z, t\right) = w_0 e^{i\omega t}, \quad \theta_0 \le \theta \le \pi - \theta_0;$$

$$w\left(\frac{-b}{2\sin\theta}, \theta, z, t\right) = w_0 e^{i\omega t}, \quad \pi + \theta_0 \le \theta \le 2\pi - \theta_0,$$

where $\theta_0 = \arccos\left(a/\sqrt{a^2 + b^2}\right)$.

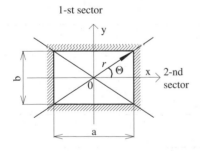

1-st sector

2-nd sector

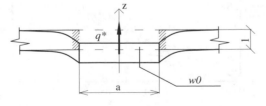

Figure 1.

The solution of a task, on one of rectangle sides, for example $x = a/2$, $-b/2 \le y \le b/2$, in a sector $-\theta_0 \le \theta \le \theta_0$ is defined

$$w = w_0 e^{i\omega t} H_0^{(2)}(kr)/H_0^{(2)}\left(k\frac{a}{2\cos\theta}\right).$$

Defining a shearing stress at symmetric vibrations on a considered side of a rectangular cut from

$$\tau_{xz} = \mu\frac{\partial w}{\partial x} = \mu\left(\frac{\partial w}{\partial r}\cos\theta - \frac{1}{r}\frac{\partial w}{\partial \theta}\sin\theta\right),$$

the following expression can be obtained

$$\tau_{xz} = -\mu w_0 e^{i\omega t}\frac{k}{\cos\theta}\frac{H_1^{(2)}\left(k\frac{a}{2\cos\theta}\right)}{H_0^{(2)}\left(k\frac{a}{2\cos\theta}\right)}, \quad -\theta_0 \le \theta \le \theta_0.$$

Thus, the size of a shearing stress for a considered side of a rectangle varies between values

$$\tau_{xz} = -\mu w_0 k\frac{H_1^{(2)}\left(k\frac{a}{2}\right)}{H_0^{(2)}\left(k\frac{a}{2}\right)}e^{i\omega t} = \tau_{xz}(a/2,0),$$

in a point $x = a/2$, $y = 0$ and

$$\tau_{xz} = -\mu w_0 k\frac{\sqrt{a^2 + b^2}}{a}\frac{H_1^{(2)}\left(k\frac{\sqrt{a^2 + b^2}}{2}\right)}{H_0^{(2)}\left(k\frac{\sqrt{a^2 + b^2}}{2}\right)}e^{i\omega t} =$$

$$= \tau_{xz}(a/2, b/2),$$

in a point $x = a/2$, $y = b/2$.

The size of a shearing stress on other side of a rectangle varies between values

$$\tau_{yz} = -\mu w_0 k\frac{H_1^{(2)}\left(k\frac{b}{2}\right)}{H_0^{(2)}\left(k\frac{b}{2}\right)}e^{i\omega t} = \tau_{yz}(0, b/2),$$

in a point $y = b/2$, $x = 0$ and

1237

$$\tau_{yz} = -\mu w_0 k \frac{\sqrt{a^2+b^2}}{b} \frac{H_1^{(2)}\left(k\frac{\sqrt{a^2+b^2}}{2}\right)}{H_0^{(2)}\left(k\frac{\sqrt{a^2+b^2}}{2}\right)} e^{i\omega t} =$$

$$= \tau_{yz}(a/2, b/2),$$

in a point $y = b/2$, $x = a/2$.

In the supposition, that the size of stress along a side of a rectangle varies linearly

$$\bar{\tau}_{xz} = \tau_{xz}(a/2,0) + 2[\tau_{xz}(a/2,b/2) - \tau_{xz}(a/2,0)]y/b,$$

$$\bar{\tau}_{yz} = \tau_{yz}(0,b/2) + 2[\tau_{yz}(a/2,b/2) - \tau_{yz}(0,b/2)]x/a,$$

from a condition of symmetry of a considered task on a contour the following expression is obtained

$$q^* = -4\left(\int_0^{b/2}\bar{\tau}_{xz} \cdot 1 dy + \int_0^{a/2}\bar{\tau}_{yz} \cdot 1 dx\right) =$$

$$= \mu w_0 e^{i\omega t} kb\left[\tau_{xz}\left(\frac{a}{2},0\right) + \tau_{xz}\left(\frac{a}{2},\frac{b}{2}\right)\right] +$$

$$+ \mu w_0 e^{i\omega t} ka\left[\tau_{yz}\left(0,\frac{b}{2}\right) + \tau_{yz}\left(\frac{a}{2},\frac{b}{2}\right)\right].$$

If to present sizes of a rectangle through a characteristic radius of a cut r_0,

$$a \cdot b = \pi \cdot r_0^2, \ b = \gamma \cdot a, \ \gamma \cdot a^2 = \pi \cdot r_0^2, \ c = \sqrt{a^2+b^2},$$

$$\frac{a}{2} = r_0\sqrt{\frac{\pi}{4\cdot\gamma}}, \ \frac{b}{2} = r_0\sqrt{\frac{\pi\cdot\gamma}{4}}, \ \frac{c}{2} = r_0\sqrt{\frac{\pi}{4\cdot\gamma}+\frac{\pi\cdot\gamma}{4}},$$

then the response of a unit thickness stratum affixed to a lateral surface of the pile-cap, is equal

$$q = \mu w_0 e^{i\omega t}\left(S_{w1} + iS_{w2}\right),$$

where valid S_{w1} and imaginary S_{w2} dimensionless parts are defined

$$S_{w1} = ka\frac{J_0(ka/2)J_1(ka/2) + Y_0(ka/2)Y_1(ka/2)}{J_0^2(ka/2) + Y_0^2(ka/2)} +$$

$$+ kb\frac{J_0(kb/2)J_1(kb/2) + Y_0(kb/2)Y_1(kb/2)}{J_0^2(kb/2) + Y_0^2(kb/2)} +$$

$$+ k(a+b)\frac{J_0(kc/2)J_1(kc/2) + Y_0(kc/2)Y_1(kc/2)}{J_0^2(kc/2) + Y_0^2(kc/2)},$$

$$S_{w2} = \frac{1}{\gamma}\frac{4/\pi}{J_0^2(ka/2) + Y_0^2(ka/2)} +$$

$$+ \gamma\frac{4/\pi}{J_0^2(kb/2) + Y_0^2(kb/2)} +$$

$$+ \left(\gamma + \frac{1}{\gamma}\right)\frac{4/\pi}{J_0^2(kc/2) + Y_0^2(kc/2)}.$$

4 CONCLUSIONS

The dynamic stiffness is complex function of vibration forces frequency ω, ground density ρ, stiffness of environment μ, sizes of a pile-cap sole and relation of sizes a and b. The graphs of valid S_{w1} and imaginary S_{w2} dimensionless parts at some of $\gamma = b/a$ are reduced on a Figure 2.

From the submitted results follows, that for frequency $\omega \to 0$, dynamic stiffness parameters aspire to zero at any geometry of a pile-cap.

For a comparison, the values of similar characteristics S_{w1}^c and S_{w2}^c are shown on a Figure 2. These characteristics were obtained for a round cut of the same square.

It is obvious, that during vibrations, there are shifts of phases between movings and revolting forces. The responses outstrip the appropriate movings for an angle φ, which is defined $\varphi = arctg(S_{w2}/S_{w1})$. The Figure 3 illustrates character of phase shift modification from parameter γ. Agrees of the received results follows that the maximal advancing in all considered cases occurs at a ratio of the sides $0.5 < b/a < 2$, and however it is less, than for a circle of the equivalent square.

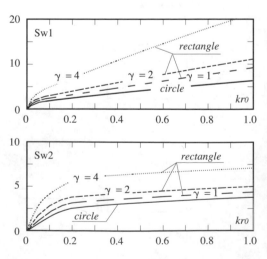

Figure 2.

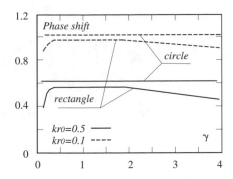

Figure 3.

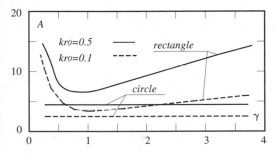

Figure 4.

The change of reaction amplitude A from γ, determined as $A = \sqrt{S_{w1}^2 + S_{w2}^2}$, is submitted in Figure 4. From the received results follows, that for various kr_0, the size A accepts the minimal meaning at the sides ratio $b/a = 1$. In this case distinction between A and $A_c = \sqrt{(S_{w1}^c)^2 + (S_{w2}^c)^2}$ is minimal also. It is obvious, that the changes of reaction amplitude parameter A very strongly depend from a ratio of the sides a and b.

In the conclusion it is necessary to note the following. With the help of wave model the estimation of dynamic characteristics in system the pile-foundation-ground was carried out at vertical vibrations of the immersed body with the rectangular form in a plan. Agrees of the received results are determined, that for increase of accuracy at definition of dynamic parameters it is necessary to take into account real geometry of considered object.

REFERENCES

Baranov, V.A., 1967. About account of the deep foundation forced vibrations. J. Dynamic questions. Riga, 195–209 [in russian].

Nuzhdin, L.V., 1994. Pile foundations designing and account of vibrations for the industrial equipment. *IV International conference. Pile foundation problems.* Perm, 167–172 [in russian].

Nuzhdin, L.V., 1995. The account of a pile-cap interaction with a ground at pile foundations vibrations. *Russian national conference with the international participation. Soil mechanics and foundation.* St. Petersburg, 505–510 [in russian].

Nuzhdin, L.V., Kolesnikov, A.O., Genze, P.A., 2002. Definition of dynamic characteristics of grounds for the account of pile foundations on wave loadings. *Proceedings of the international conference on coastal geotechnical engineering in practice.* Atyrau, 112–115.

System-based Vision for Strategic and Creative Design, Bontempi (ed.)
© 2003 Swets & Zeitlinger, Lisse, ISBN 90 5809 599 1

Design of a railway junction in Palermo: underground structures and building damage risks

D. Pelonero & A. Pigorini
ITALFERR S.p.A., Roma, Italy

E. Scattolini & E.M. Pizzarotti
S.IN.C. S.r.l., Milano, Italy

ABSTRACT: The open cut or underground tunnels construction with low overburden causes, in soft ground and urban areas, ground settlements closely connected to the potential damage of buildings and civil structures, located in the area involved with the disturbance. In the case of final design of a shallow tunnel, driven in soft ground, very close to or under-passing existing buildings, it is essential to perform an accurate assessment of the possible disturbance effects on the urban structures lying on the surface. In the final design of the railway junction between the "Palazzo di Giustizia" and the "Notarbartolo" stops, within the second railway track project of the "Palermo Centrale/Brancaccio-Palermo Notarbartolo" railway line, in the city of Palermo, due to the heavy urbanization of the area, evaluations of settlements and of possible damages to the buildings have been carried out. As regards the analyses, they are related to the structures influenced by the works in project and they are divided into effects induced on buildings by underground excavation and effects induced on buildings by open cut excavation. The methods adopted for the evaluations refer to the theory applied during the design of three main tunnels (such as the London Jubilee Line Extension), in a urban context as background and by mechanized excavation (Mair, Taylor, Burland, 1996), with the aim of settlement detection and connected damages evaluation in the case of underground excavation. In the stretches of the railway line carried by open cut excavations, the method proposed by Peck has been used. Also the interferences with existing road network have been taken into account, together with the disturbance caused to underground service lines, in order to detect the areas which may need special attention during the construction phases.

1 INTRODUCTION

The new line of the railway junction between the "Palazzo di Giustizia" and the "Notarbartolo" stops, within the second railway track project of the "Palermo Centrale/Brancaccio-Palermo Notarbartolo" railway line, in the city of Palermo, includes the following underground structures, placed adjacent to the existing ones:

- a new tunnel (even track), which doubles the existing line (odd track), between the railways area on the "Isola delle Femmine" side and the "d'Ossuna" street, near the "Palazzo di Giustizia" stop;
- two new stations, named "Lolli" stops, one pertaining to the even track and the other to the odd one;
- underground walking tunnel, connecting the two bodies of the even and the odd track stations;
- structures adjacent to the stations, containing machinery rooms and safety stairways, connections

between the ground level entrances and the station platform.

In the location of the existing track (odd track), the construction of the "Lolli" station is planned, including the connections with the bodies of the adjacent stations and the even track stop (Fig. 1); all the construction works in this part will be carried out by open cut excavations.

The doubling track (even track) alignment is developed lengthwise to about 1226 m of underground tunnel, that will be excavated by mechanised full section boring machines to limit the disturbance to the surface.

Grounds being excavated are calcareous sandstone and calcarenite-calcareous sands. More specifically the calcareous sandstone, with a stony consistency but to some extent fairly cemented and with loose sandy lenses, is generally located in the upper portion of the excavation and extends up to the ground surface; the

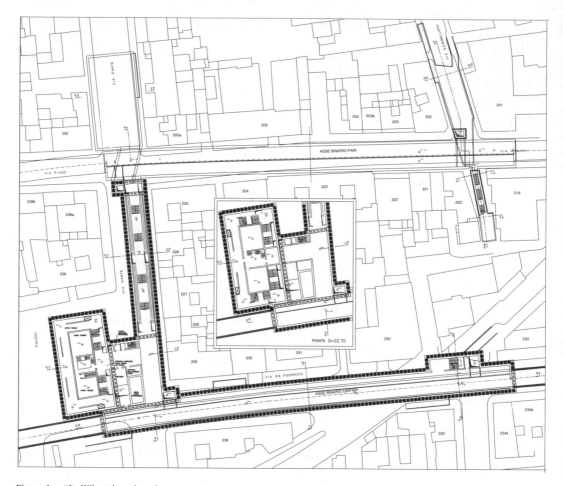

Figure 1. "Lolli" station plan view.

arenaceous calcareous sands is normally located in the lower portion of the tunnel.

The even track tunnel crosses an extensively built area with maximum overburden of about 10 m–11 m at the new "Lolli" stop location (Fig. 2). It's positioned under the centre of a street bordered on by buildings.

The water table has a depth of about 12 m from ground level and thus the excavation of the tunnel is submerged along the whole central portion of the line.

Interferences laying over the strip of the undergoing excavations are mainly apartment and office buildings, underground public facilities and road network.

The short distance between the buildings and the tunnel axis and the small overburden make it necessary to evaluate the possible static or functional damages which could be caused on buildings during the excavation phases of the railway line, of the station tunnels and of the open cut excavation bodies next to the existing underground structures.

Analyses of the subsidence deriving from the excavation and the effects theoretically induced on the adjacent structures have been carried out; the methods used for the above mentioned evaluations refer to the theory already applied for the design of big tunnels, in full urban context and with mechanized excavation (Mair et al. 1996), aimed to the subsidence and the damages assessment induced by the tunnel excavation. The method suggested by Mair, Taylor and Burland is based on the calculation of the building strains and angular distortions in the transverse direction with respect to the tunnel axis, which are then to be compared to the values reported by the technical literature (Burland et al. 1977, Boscardin & Cording 1989) to estimate a potential of damage level.

Further, for all the open cut tunnel branches, excavated between diaphragms walls, the method proposed by Peck has been applied which relates the

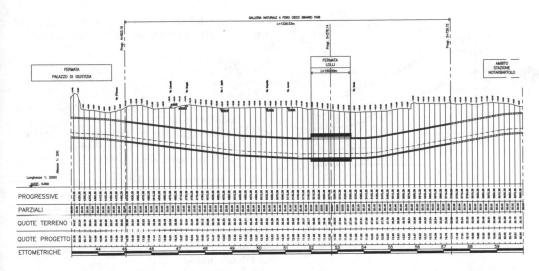

Figure 2. Even track longitudinal section.

settlement distribution to the distance from the wall and to the maximum settlement (that can be registered at the wall head location).

When prediction data about settlements obtained through one of the two methods above mentioned are available, it is possible to identify the potential damage class of each building. Then, according to the case, preventive measures during construction or remedial at construction completion will be examined.

2 GROUND MOVEMENT PREDICTION AND BUILDING DAMAGE RISK ASSESMENT METHODOLOGY ADOPTED

2.1 *The Mair, Taylor and Burland method*

Mair, Taylor and Burland proposed a method for the evaluation of subsidence and damages induced on buildings interfering within a tunnel excavation. It is based upon a theory empirically supported, developed in England for prediction of ground movements. It begins from the study of surface deformation without structures, caused by the excavation of a single tunnel (Fig. 3).

The surface deformation caused by the opening of a underground cavity, in absence of buildings, is well described (Peck 1969, O'Reilly & New 1982 /1991) by a Gaussian distribution curve, in which the settlement in a generic point, located at a distance y from the tunnel axis (Fig. 4), can be derived from the equation:

$$w = w_{max} \exp\left(\frac{-y^2}{2i^2}\right) = \frac{0.31 \cdot V_l \cdot D^2}{k \cdot z_0} \exp\left(\frac{-y^2}{2i^2}\right)$$

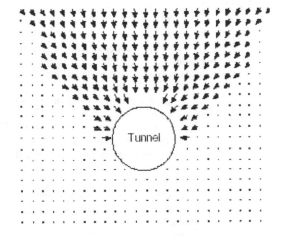

Figure 3. Deformations caused by a tunnel excavation.

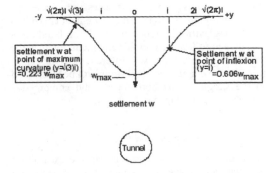

Figure 4. Settlement curve above a circular cavity.

1243

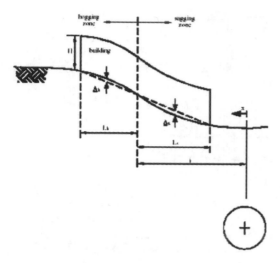

Figure 5. Hogging and sagging zones.

distance from the sheet wall) to the vertical settlement maximum value (by the bulkhead) through the relation:

$$w = w_{max} \cdot e^{-0,0545 \cdot x}$$

in which:

w is the surface settlement at a distance x from the sheet wall;

w_{max} is the maximum settlement of the surface, evaluated at the diaphragm location;

x is the horizontal distance from the diaphragm.

After the evaluation of the surface settlement profile as a function of the horizontal distance from the diaphragm, the risk of building damage has been estimated, applying again the method proposed by Burland. In this case, buildings lie always in the *hogging* zone, because of the convex bending of the subsidence profile.

The adopted value of w_{max} has been based on a calibration by a two-dimensional numerical modelling (finite difference method). Consistently, surface vertical settlements back of the sheet wall have been calculated. In particular, having adopted Peck's formulation with a value of w_{max} equal to 0,3% of the maximum excavation depth, results are consistent with the similar ones obtained by means of the numerical modelling.

3 WORK INTERFERENCE WITHIN EXISTING CIVIL BUILDINGS

3.1 Building survey

In order to achieve the above mentioned objectives, a selection of different types of building along the line has been identified, that are likely to be affected by the works.

During the preliminary design of the work subject of this debate, a survey of all civil structures interfering within the work has been carried out; the buildings catalogued are located into a strip of about 25 m width, the halfway line of which is the tunnel axis. The area affected by the disturbance is in the center of Palermo city and includes an overall amount of 194 buildings.

The census gathered information about different structural characteristics of each building, with particular attention to the typology and the depth of foundation and to the constitutive types of structural materials (masonry, reinforced concrete and combined structures).

3.1.1 Hazard index
Combining the ground settlement data with the buildings strains, with different modalities depending on the buildings stress conditions (bending strain or diagonal strain, related to the building position in comparison

in which:

w is the settlement;

w_{max} is the maximum settlement on the tunnel centre line;

y is the horizontal distance from the centre line;

$i = k \cdot z_0$ is the horizontal distance from the centre line to the changing point of inflexion of the settlement trough (z_0 is the depth of tunnel axis from the surface; k is a parameter depending on the geotechnical nature of the ground and on the depth of the water table);

V_l is the percentage of lost volume compared to the excavated volume (per metre length of tunnel);

D is the tunnel excavation diameter.

The calculation of the surface vertical settlement is the start point to evaluate the strain in the buildings above, in the transverse plane with respect to drive direction. To this purpose, it is necessary to note that the inflexion points divides the settlement curve into two zones, with different curvature; the building is assumed to follow the ground settlement trough at the foundation level, with two different stresses conditions (in the *hogging* zone all strain will be tensile and in the *sagging* zone both tensile and compressive strain may occur, Fig. 5).

2.2 The method proposed by Peck

To predict settlements caused by surface excavations between deformable diaphragms, Peck proposed to relate the settlement curve (function of horizontal with the settlement Gaussian curve), the total amount of settlement can be derived.

The maximum value of such a settlement must be compared to literature index values, relating the total settlement to the damage intensity and category (Boscardin & Cording 1989). In particular, the relationship between limiting tensile strain ε_{lim} and the potential damage identifies 5 damage classes (from 0 to 4). All the buildings marked with a value from 0 to 2 are considered not to be subjected to any risk, while the buildings with an index greater or equal to 3 need a detailed evaluation and an identification of preventive measures or an adequate rehabilitation plan. Defining as *hazard index* the probability that a building exhibits a damage related to an excavation (dealing with geometries, involved materials and mutual positions), it is possible in this case identify the category of damage with the previously mentioned *hazard* index.

3.1.2 Vulnerability index

By the study of every single building interfering with the designed switch line, a preliminary assessment on the buildings *vulnerability* has been given, considered as an intrinsic character apart from every subsequent action (such as, for example, the excavation at issue). The *vulnerability* index refers to a single structure and provides with information about the single building minor or major propension to be damaged, in relation to building typology, ground geotechnical characteristics and ground-structure interaction. Numerical values of the *vulnerability* index vary from 0 to 100 (*vulnerability* from "low" to "very high").

Dealing with this aspect, the mean value of *vulnerability* index of the buildings in the area V_m is 40; the most vulnerable buildings are the masonry ones ($V_m = 44$), followed by the combined structures ($V_m = 32$) and the reinforced concrete ones ($V_m = 28$). Buildings with the highest value of *vulnerability* index (raising from 60 to 69) are located near the "Palazzo di Giustizia" and "Lolli" stops zones.

3.1.3 Risk index

Having defined the *risk* linked to a specific event (in this case, the excavation of a shallow tunnel) as the product between the *hazard* index and the *vulnerability* index, the data collected, mentioned above, allowed to assess a risk rate related to the realization of the work in project, for each single building, through a simple multiplication.

3.2 Mair, Taylor and Burland method application

As mentioned, the value of maximum surface settlement is linked to the percentage of volume loss during the excavation: during the analysis carried out, two different values of this percentage were assumed to be representative for the effective situation and

consistent with data collected during real construction of others similar underground works.

In addition to this, closely related to the nature of the ground portion involved with the excavation (depth varying from 25 to 30 m), lithoid at intervals, the Gaussian curve assumed is wide spreaded, typical of cohesive materials. The registered buildings, regarded as interfering with the work in project, have been therefore analysed in four different conditions, related to the percentage of volume loss and parameter k judged as more meaningful.

3.2.1 Railway line natural tunnel, even track

The analyses carried out in this case have an equivalent diameter of excavation equal to 8,3 m. Among the registered buildings, a total amount of 80 results likely to be affected by the works of the even track underground line tunnel. None of them reaches a damage index of 3 (preponderance of 0 and 1) and just one exhibits a value of 2 (analysis with V% = 0,5% and k = 0,5). When the analysis has been carried out with V% = 1% and k = 0,5, the same "critical" building raises in the class of damage with index equal to 3, and the remainings reach the maximum value of 2.

3.2.2 "Lolli" stop underground tunnel, even track

In the portion of track by the "Lolli" stop, analyses have been carried out with an equivalent excavation diameter of 11,19 m. Buildings interfering with the work along the stop platform are 14 in their total amount. In the analysis with V% = 1% and k = 0,5, the damage index varies from 2 to 3, with a preponderance of 2; in the analysis carried out with V% = 2% and k = 0,5 all the detected buildings are in the category 3rd, with the exception of one, resulting in the 4th class of damage and thus requiring a particular attention.

3.3 Peck's method application

In all the portions expected to be excavated in a open cut way, between diaphragm walls, the potential rate of damage has been evaluated with the application of Peck's theory.

The analyses were carried out under the hypothesis of a settlement maximum value $w_{max} = 0,3\%$ of the excavation height.

3.3.1 "Lolli" stop artificial tunnel, odd track

All the buildings related to this part of work belong to the 3rd or 4th damage classes (Fig. 6).

3.3.2 Technical rooms of "Lolli" stop, even track

In all, buildings interacting with this stretch are 6, most of them marked by a 3 or 4 value of damage class. Three

Figure 6. Buildings interfering with the "Lolli" stop location.

of them are neighboring to excavation of low depth (around 6 m), but just one runs no risk (value 2), consistently with the analysis of "Lolli" stop natural tunnel, even track, classifying them in the 3rd range (Fig. 6).

3.3.3 "Lolli" stop underground passage between the even and the odd track

Along the pedestrians passage the buildings scheduled are 7: also in this case, all of them are countable in the 3rd or 4th damage classes.

3.4 Results interpretation

Examination of yielded results makes it clear that buildings in greater need of attention for possible repairs are those listed below.

- Railway line natural tunnel, even track: building n°241a.
- "Lolli" stop natural tunnel, even track: all the adjacent buildings.
- "Lolli" stop artificial tunnel, odd track: all the adjacent buildings.
- Technical rooms of "Lolli" stop, even track: all the adjacent buildings, except for n°250.
- Underground passage between the "Lolli" stop and the "Palazzo di Giustizia" stop: all the adjacent buildings, except for n°238.

4 INTERFERENCE WITH THE ROAD NETWORK AND UNDERGROUND FACILITIES

As done for studying the interference of the works in project with the existing buildings, particular attention has been paid to estimate their effects on road network: proper admittances and traffic diversions have been planned, in order to reduce discomforts and troubles caused on the city life routine by the sites and quickly restore the previous conditions.

Moreover, a research into the detected interactions between the works and the underground facilities has been carried out, with the aim to highlight the nodes in which canalizations, drainage systems, electric and telephone systems and telecommunication networks intersect the railway line in project.

5 CONCLUSIONS

In cases where unacceptable settlements were predicted beneath buildings, the utilisation of various forms of remedial works has been detected to protect structures wholeness and functionality, as well as people safety. These measures can be preventive, interactive (during settlement advancing) or recuperative, after the incidental instability.

Several strategies can be adopted depending on the ground conditions and the structure to protect. Where the buildings are low, of foundation located in granular soils, *permeation grouting* can be adopted. This involves grouting up the sands and gravels beneath the structure, prior to any tunnelling operation, to create a form of raft. Where unacceptable settlements have occurred, *compensation grouting* can be used to rectify such movements. This process involves the building raising by pumping grout through arrays of tubes with sleeved ports. Alternatively the same arrangement of sleeved ports can be used to carry out *concurrent grouting*, in which the development of settlement is controlled as the tunnelling or excavation work progresses.

During the realization phase, buildings need to be instrumented prior to tunnelling or excavation works so that any resulting movements could be closely monitored. In some cases sub-surface monitoring devices can also be installed in the ground below the building, in order to investigate soil-foundation interaction and to arrange in time the incidental measures needed to ensure safety conditions towards unforeseen limit states.

REFERENCES

Mair, R. J., Taylor, R. N., Burland, J. B., Prediction of ground movement and assessment of risk building damage due to bore tunnelling.
Celestino, T. B., 2000, Errors in ground distortion due to settlement trough adjustment, Tunnelling and Underground space technology, vol. 15, n°1, Pergamon.
O'Reilly, M. P. & New, B. M., 1982, Settlements above tunnels in the United Kingdom – their magnitude and prediction.
Bowles, J. E., 1991, Fondazioni – progetto e analisi, McGraw-Hill.

Computational methods for deep tunneling design

S. Francia
SPEA Autostrade S.p.A., Milano, Italy

E.M. Pizzarotti, M. Rivoltini & C. Pecora
S.IN.C. S.r.l., Milano, Italy

ABSTRACT: In the deep tunneling design for the Milano–Roma highway stretches known as "Variante di Valico", calculation procedures have been applied in which the analytical method of the Ground Reaction Curve (GRC) is coupled to continuum discretization with the Finite Differences Method (FDM).

Analyses procedure is aimed to evaluate the strain distribution on the cavity boundary surface as a function of the distance from the front face considering also strengthening operations of the front face. Furthermore the analyses provide indications on the axial strain (extrusion) of the excavation front and on the extension of plastic areas as well as absolute and relative radial displacements of the natural cavity at any distance from the front face and on the extensions of plastic zones.

Checks on the structural elements which are part of the first phase and final linings are carried out on the basis of the stress states produced by the succession of the excavation and strengthening phases, simulated through proper installation distances of these elements from the front face.

1 INTRODUCTION

This paper illustrates the computational procedures applied to deep tunneling design for the Milano–Roma highway stretches known as "Variante di Valico".

In this case calculation procedures have been applied in which the analytical method of the Ground Reaction Curve (GRC) is joined by continuum discretization with the Finite Differences Method (FDM) for the modelling of the supporting structures and stabilization means. The constitutive model of the considered rock mass follows the plasticity curve described by the Hoek & Brown formulation.

The verification analyses of the standard sections were carried out for tunnel stretches with homogeneous geomechanical conditions and sufficient length to allow the application of the hypothesis of plane strain state. Thus the analyses do not represent local areas of intense fracture and limited length, for which specific measures were disposed anyhow.

Procedure is aimed to evaluate the strain distribution on the cavity boundary as a function of the distance from the front face, considering also strengthening works of the front face, which are simulated by increasing the rock mass strength parameters or

introducing structural elements in the model. Furthermore the analyses provide indications on the axial strain (extrusion) of the excavation front and on the development of plastic areas as well as absolute and relative radial displacements of the natural cavity at any distance from the front face and on the extensions of plastic zones.

Checks of the structural elements which are part of the first phase and final lining are carried out following the stress states produced by the succession of the excavation and strengthening phases, simulated through proper installation distances of these elements in relation to the front face.

2 INTERACTION BETWEEN GEOLOGICAL AND GEOMECHANICAL ANALYSIS AND STRUCTURAL DESIGN

2.1 *Geologic characteristics*

The geologic-structural shape, in which a part of the layout of the new highway is designed, is the result of complex geologic evolution of the Apennines.

In application below shown, it is possible to characterize two geomorphological zones, coupled to geological soil formation.

The first one is characterized by hard morphology, frequently with steep slope, that are middle mountain areas with sandstone-marl rock, called "Macigno". In this zone the typical landslide are by collapse, by block movement and by detritus slide.

The second one, clearly crumbly, is characterized by allochthonous formation called "Argille Scagliose". In this zone it is possible to examine corrugated elevation and little steep slope, subjected to intense erosion, with flow slide and superficial gravitational distortion.

Below it is shown the analysis of a particular section of tunnel when the mass rock is characterized by Argille Scagliose.

2.2 Geomechanical characteristics

The rock has been classified referring to the Rock Mass Rating (RMR) of Bieniawski.

The geomechanical parameters, necessary for the evaluation of the RMR value and of the consequent rock qualitative class, are:

− R1: uniaxial compressive strength of the rock material;
− R2: Rock Quality Designation, RQD;
− R3: spacing of discontinuities;
− R4: condition of discontinuities;
− R5: groundwater conditions;
− R6: orientation of discontinuities.

By sum of absolute value of each parameter you have numerical evaluation of RMR, by which you assign one of the five rock mass classes of Bieniawski's method.

The behavior of the fractured rock has been characterized by the Hoek & Brown constitutive model:

$$\sigma_1 = \sigma_3 + \sigma_{ci} \cdot \left[\left(m_b \cdot \frac{\sigma_3}{\sigma_c} \right) + s \right]^a \tag{1}$$

where:
σ_1: maximum effective stress at failure;
σ_3: minimum effective stress at failure;
σ_{ci}: uniaxial compressive strength of the intact rock pieces;
m_b; s; a: parameters which depend upon the rock mass characteristics.

$$m/m_i = \exp[(GSI-100)/28] \tag{2}$$

$$s = \exp[(GSI-100)/9] \tag{3}$$

$$a=1/2 \tag{4}$$

In order to use the Hoek & Brown criterion for estimating the strength and the deformability of jointed rock masses, three properties of the rock mass have to be estimated:

− σ_{ci}: uniaxial compressive strength of intact rock pieces;
− m_i: Hoek & Brown constant for the rock mass;
− GSI: value of Geological Strength Index for rock mass.

The value of Geological Strength Index (GSI) is based on the value of Rock Mass Rating (RMR), by the equation:

$$GSI = URMR-5 \tag{5}$$

where:

$$URMR = RMR \tag{6}$$

evaluated by R5 = 15 and R6 = 0.

Moreover, to model the post peak behavior of the rock mass, it has been considered a residual value of GSI once reached the frontier of the plastic criteria.

About the Deformation Modulus (E) Serafim & Pereira proposed a relationship between the in situ modulus of deformation and Bieniawski's RMR classification. Hoek & Brown proposed the following modification:

$$E = \sqrt{\frac{\sigma_c}{100}} \cdot 10^{\frac{GSI-10}{40}} \tag{7}$$

Regarding the non associated flow rule, in the analytical solution there is a parameter f varying from 1.0 (no dilatancy) to f_a (associated flow rule):

$$f_a = 1 + \frac{m_b}{2 \cdot \left(m_b \frac{p_{cr}}{\sigma_c} + s \right)^{1/2}} \tag{8}$$

where p_{cr} critical radial stress to have plastic zone around the cavity.

2.3 Analysis tools and algorithm

The rock mass constitutive model is homogeneous, isotropic, elasto-plastic, following a flow rule non associated. Hoek & Brown model defines the criteria.

Instead the rock mass subject by excavation is non homogeneous and anisotropic one. This aspect simplifies the real behavior of the rock mass, which means a careful definition of the excavation and strengthening phases in order to contrast a potentially different behavior of the rock mass from results of analyses.

The analyses, using the method of the characteristic lines (Ground Reaction Curves, GRC, analytical solution to the problem of a hole subject to variable internal pressure in a continuous elasto-plastic, in junction with the numerical solution of Panet for the behavior to a certain distance from the front) and finite difference model, estimate the stress and the displacement of the rock mass around the cavity and therefore verify the dimensioning of the structures.

The finite difference method is a numerical technique to solve a set of differential equations, given initial values and/or boundary values. The rock mass is represented by a mesh, where each element (or zone) has a particular constitutive model.

The calculus starts from geostatic condition, where the model simulates the natural behavior. The next steps introduce stress changes.

Excavation could be simulated by deleting some zone. It is possible to apply pressures on the boundaries of the mesh. Also it is possible to simulate a strengthening phase by modifying the geomechanical parametres or introducing some structure: a beam, if the structures have flexural behaviour; cable, if it is necessary to simulate radial rockbolts.

The algorithm process is:

1. For each set of geomechanical parameters and overburden, it is carried out an analytical analysis by GRC, considering a cavity with radius 1.00 m.

   The results (absolute convergence and extension of plastic zone considering no pressure inside the cavity) allows the first evaluation about stability of cavity.

2. Carry out the finite difference analysis by axial-simmetric model of a cylindrical hole (circular cavity with radius 1.00 m) to evaluate the function convergence vs internal pressure. By that analyses it is possible to modify locally the mesh (extension and number of zones), because the results have to referee to analytical ones.

3. Axialsymmetrical analysis to simulate excavation progress. The results are:

   – to get displacements around cavity (without strengthening structures) correlated to the distance from front;
   – to evaluate strain and extension of plastic zones, particularly near front.

Regarding the rock mass where it is necessary to use strengthening structures on front, the analysis has been carried out with or without that structures. The strengthening phase is simulated by increasing geomechanical parameters just for the zones near by front correlated to pressure applied.

4. Elaborating the function displacements around the cavity vs distance from the front in junction with the function displacements around cavity vs internal pressure (results from A, B and C) you could get the function distance from front vs internal pressure. The internal pressure is equivalent to strength of the front and it is known as "excavation fictitious load" (in italian, "forza fittizia di scavo") (FFS below).

5. For each proper section of tunnel the analysis are carried out by finite difference method (plain strain hypothesis), considering the real geometry of excavation and strengthening phases (radial and/or umbrela rockbolts, steel frame support, spritz, reinforced concrete inverted arch). For each step of calculus (excavation or installation) it has been evaluated the value of FFS correlated to the proper distance of installation from front, to simulate the real behavior in progress of the rock mass.

6. The least step, i.e. when compatibility and equilibrium are obtained in all the rock mass zones and structures, the structural verify is carried out by Ultimate State Limit method.

The algorithm described is evaluated for each proper section of tunnel, i.e. for each length of tunnel where geomechanical conditions are near the same and approximate the plain strain hypothesis.

The analysis cannot evaluate behavior for local fractured rock, for which specific works must be designed.

3 APPLICATION

3.1 *"Variante di Valico"*

It is illustrated the same calculus results on a particular section of tunnel.

In table 1 are represented the geomechanical data used to carry out the analysis:

About Argille Scagliose's (AS, laminated clay) behavior, the swelling and viscoplasticity are

Table 1. Geomechanical data.

Tipo	GSI pic	R	GSI r res	G [kN/mc]	SIG Mpa	C Mi	Ed [Gpa]	Poisson	HOEK PICCO mb	s	HOEK RES. mr	sr
A	35	1,00	35	25	8	5	1,19	0,30	0.49	0,000730	0,49	0,000730
AS	45	0,78	35	25	8	5	2,12	0,30	0,70	0,002218	0,49	0,000730
MAC	40	0,36	14	26	30	10	3,08	0,25	1,17	0,001273	0,47	0,000074

Table 2. Structures and distance of installation.

Structure	d_{med} [m]	FFS
Front	0.0	18,4%
Radial rockbolts	1.0	17,2%
Spritz and steel	4.0	7,6%
frame + inverted arch		(extra-excavation
(thickness = 100 cm)		15 cm)
Lining	No boundary	0,0%
Medium thickness =		
85 cm		

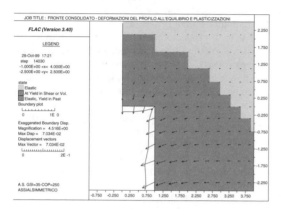

Figure 1. Axialsymmetrical analysis. Convergence at the front and along the cavity.

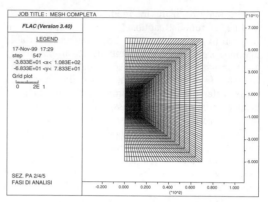

Figure 2. Complete mesh.

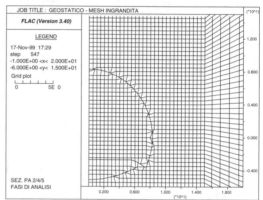

Figure 3. Mesh: particular of boundary of excavation.

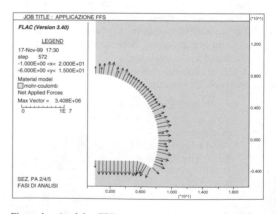

Figure 4. Applying FFS pressure along excavation boundary.

considered using a reduction of 10% for GSI value with reduction of the geomechanical parameters.

The section illustrated is PA-2/4/5, used in presence of Argille Scagliose and GSI value of 35. The overburden is 250 m.

In table 2 are illustrated: the strengthening structures (used in the analysis by FLAC), the proper distance from the front face and the value of FFS (evaluated at medium distance of installation). It has been considered an over-excavation of 15 cm.

The analysis has been carried out using the curve "distance from the front vs FFS", calculated by the axialsymmetrical numerical analysis including front consolidation (Fig. 1).

In the figures from 2 to 5 are illustrated the mesh and some steps of calculus.

In the figures from 6 to 9 are illustrated the results during the calculus process (extension of plastic zones, design loads for the structures).

In figure 10 are illustrated the uniaxial interaction failure surface of a reinforced concrete section and the design loads by Ultimate Limit Analysis for several steps of calculus.

It is important to note:

– the maximum absolute displacement cavity of the boundary is nearly 29.2 cm, after the installation of primary linings. Because the radial displacement at the front is 12.5 cm, the relative displacement is about 16.7 cm, i.e. the same engineering measure of over-excavation (15 cm);

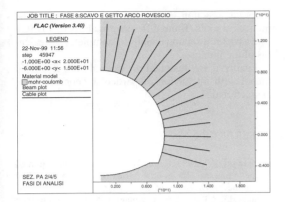

Figure 5.   Radial rockbolts.

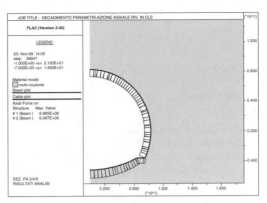

Figure 8.   Axial solicitation in final lining.

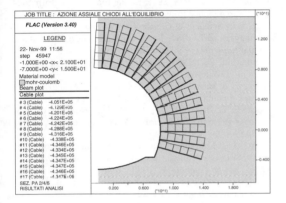

Figure 6.   Axial solicitation for rockbolts.

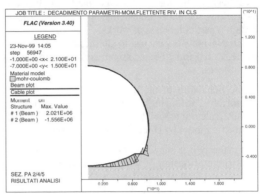

Figure 9.   Flexural solicitation in final lining.

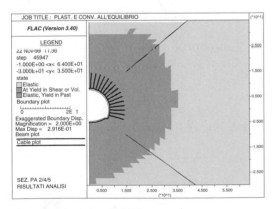

Figure 7.   Extension of plastic zones.

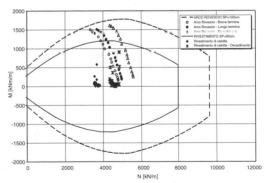

Figure 10.   Sectional verify.

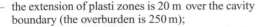
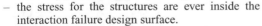

– the extension of plasti zones is 20 m over the cavity boundary (the overburden is 250 m);
– the stress for the structures are ever inside the interaction failure design surface.

At the final step, to simulate the swelling and the viscoplasticity of the rock mass it has been considered a reduction of 10% for GSI value, that it has involved an increment of the state of solicitation on the structures of definitive structures.

4 CONCLUSIONS

In deep tunneling design it is only the sinergy between:

- geolocical characterization;
- geomechanical characterization and estimation of geomechanical parameters of rock mass;
- analytical solutions and numerical models of excavation process, including installation of linings;
- knowledge of constructions to define the succession of the excavation and strengthening phases.

that allows the progress of geotecnichal and/or structural design. That means marriage (including disputes) of various technical knowledge and experiences to design a particular class of works.

This synthesis concurs to delineate a calculation process, that must be accompanied by the following hypotheses:

- the behaviour of the rock mass is homogeneous, isotropic, following an elasto-plastic criteria with flow rule not associated;
- plastic model by Hoek & Brown criteria.

Furtherly it is shown an algorithm to geomechanical/structural design of deep tunnel. It is evidenced the geomechanical parameters (RMR, URMR, GSI, Hoek & Brown constitutive model), analytical tools

(Ground Reaction Curve and Panet's solution) and numerical tools (finite difference analysis by FLAC). The structural sectional analysis completes the calculus of the linings.

Finally it is illustrated an example about a deep tunnel with interaction of particular geomechanical characteristics.

REFERENCES

Bieniawski, Z.T. 1993. Engineering classification of jointed rock masses. In South Africa Institution of Civil Engineers, *Transactions*.

Bieniawski, Z.T. 1974. Geomechanics classification of rock masses and its application to tunneling. In *Proc. 3rd Int. Congr. Rock Mechanics*, ISRM, Denver.

Bieniawski, Z.T. 1978. Determining rock mass deformability – experience from histories. In *Int. J. Rock Mechanics, Mining Sciences & Geomechanics Abstracts*.

Bieniawski, Z.T. 1984. *Rock mechanics design in mining and tunnelling*. Balkema, Rotterdam.

Bieniawski, Z.T. 1989. *Engineering rock mass classifications*. J. Wiley & Sons.

Hoek, E. & Brown, E.T. 1982. *Underground excavation in rock*. Institution of Mining and Metallurgy, London.

Hoek, E. & Brown, E.T. 1982. Pratical estimates of rock mass strength, *International Journal of Rock Mechanics and Mining Sciences*.

FLAC (Fast Lagrangian Analysis of Continua) version 3.4, 1988. *User's Manual*. Itasca Consulting Group, Inc.

Panet, M. & Guenot, A. 1983. *Analysis of Convergence behind the face of a tunnel*. Laboratoire Central des Ponts et Chaussées, Paris, France.

System-based Vision for Strategic and Creative Design, Bontempi (ed.)
© *2003 Swets & Zeitlinger, Lisse, ISBN 90 5809 599 1*

Life-cycle maintenance management of traffic tunnels – strategy assessment to develop new calcification reduction methods in tunnel drainage system

T. Gamisch & G. Girmscheid

Institute of Construction Engineering and Management, Swiss Federal Institute of Technology Zurich, Switzerland

ABSTRACT: The formation of lime deposits by seepage or ground water in tunnel drainages occur mainly because of the calcium carbonate in the water. Alkali compounds contribute considerably to the formation of lime deposits by the elution of calcium compounds and by raising the pH-value. Water hardness stabilization using polyaspartic acid has a positive influence on the lime deposits. Fewer and less harder deposits are formed. This renders it possible to lengthen the intervals between cleanings and the costs for the cleaning of the pipes can be reduced. The output per shift increases and the maintenance costs are reduced. The accompanying sequential inspection strategy leads to a significant reduction of the cleaning costs. As a consequence there is less strain on the pipes and the durability of drainage systems increases.

1   INTRODUCTION

Lime deposits in the drainage systems of tunnels frequently lead to considerable maintenance costs. Very often a persistent formation of hard to very hard deposits of varying strength can be observed in tunnels situated in the seepage area.

In order to maintain the functional efficiency of the drainage system it is essential to remove these deposits. If these deposits are allowed to grow, this can lead to a full-scale blocking of the piping which would result in the build-up of the water pressure on the tunnel. This pressure would mean an increase in the loading on the tunnel's arch which it is not designed for in drained tunnels and would thus jeopardize the stability of the construction as a whole.

The resulting costs for the maintenance are very high at present. To avoid an undue disturbance of the traffic the necessary maintenance often has to be carried out at night when the traffic is low. To illustrate this problem the railway tunnel of the high-speed connection Hannover-Würzburg can be mentioned: Between 1999 and 2002 the costs of the direct primary maintenance averaged 325,000 € per annum.

Not only does the cleaning cause high costs, but the maintenance also restricts the operability of the structures. The servicing of railway tunnels in particular often causes logistical problems such as blocking of certain routes or diversions and further company-internal costs. These include the secondary direct costs for the planning of and the assignment of personnel for the cleaning (planning of diversions, safety measures etc.) on the one hand and the indirect costs (e.g. the mechanical wear of the route because of additional acceleration and breaking manoeuvres, additional costs for energy, compensation for damages etc.) on the other hand. The Deutsche Bahn AG estimates that the additional company-internal costs depending on the utilization of the routes amount to 30 up to 60 per cent of the primary cleaning costs.

If these additional costs are taken into consideration the maintenance costs of the drainage system of the above-mentioned tunnel amount to 450,000 € per annum. After 100 years of operation the costs for the maintenance of the drainage system alone will amount to about 50 per cent of the construction price.

The removal of these hard deposits requires equipment which can exert pressure on the calc-sinter to detach it from the piping. Accordingly the piping itself is exposed to high mechanical strains which contribute to their wear and tear (Zimmermann 2000).

High-pressure water cleaning does not automatically lead to lasting damages in the medium term if the appropriate piping is used. Milling, however, puts bigger strains on the piping, especially on the arches and the jointing. The older the drainage system and the higher the number of cleanings the more likely are damages or even the destruction of the piping.

Deficiencies in the design of the drainage system increase the risk of damages considerably. The results of this are costly restoration measures which have already been carried out in many tunnels or will have to be carried out.

In another tunnel of the high-speed connection Hannover-Würzburg almost 2000 damages were found in a drainage system of 8500 meters. This tunnel has only been operating for 16 years but on average there is a damage every 2 meters. The drainage system was cleaned twice or three times a year in the past and milling was necessary almost everywhere.

If the maintenance cost per damage is estimated at 1000 €, the reconstruction of this drainage system will amount to about 2 Mio. €. These costs are a direct consequence of the cleaning of the drainage systems. This would correspond to an annual expenditure of 125,000 € to which other company-internal costs for the blocking of routes, diversions etc. will have to be added. The maintenance costs do not develop linearly over the age of the tunnel but they rise continuously with its increasing age.

In the following pages, the important mechanisms behind the formation of lime deposits in tunnel drainage systems will be discussed and basic measures with which to reduce the maintenance costs will be deduced. These include principles of design, some remarks on the choice of the correct material in reconstruction or building and the choice of a good maintenance strategy. In conclusion an outlook on the potential of hardness stabilization to reduce the costs of cleaning will be given and further steps in development will be discussed.

2 THE MECHANISMS BEHIND THE FORMATION OF LIME DEPOSITS

2.1 Chemical principles

The formation of lime deposits in tunnel drainage systems because of ground water can be ascribed to two main chemical reactions. On the one hand this is the property of water to dissolve salts. The most important mineral of calc-sinter is calcium carbonate. The solubility of this salt depends on the concentration of dissolved carbon dioxide in the water on the other hand. The interaction of both processes of solution is called calcium-carbon dioxide equilibrium.

Gases and volatile substances are absorbed at the gas phase – water limit. In accordance with the HENRY-DALTON law, the concentration of a gas in water is proportional to its partial pressure in the gas phase for constant temperatures and low pressures ($p < 10^5$ Pa). The factor of proportionality is specific to every gas and is called the HENRY-constant. The solubility of the gases decreases as the temperature increases.

The HENRY-DALTON law is only applicable when the concentration of ions in the solution is very low. Dissolved electrolytes (acids, bases, salts which dissociate ions in watery solution which renders them conductive) decrease the solubility of gases. The water molecules form hydrate envelopes around the dissolved ions. Thus the amount of water for a further solution process (gas solution) is reduced.

The solubility of most solids in water is limited, too. It is described by the solubility product (=product of anion and cation activities of a dissociated salt).

The solubility product increases with increasing temperatures and depends on the pH-value of the solution. This influences the solubility of the ions of most metals. There are only a few ions, e.g. potassium (K^+), sodium (Na^+), nitrate (NO_3^-) and chloride ions (Cl_l^-), which are readily soluble in almost the whole pH-range.

For the crystallization of salt deposits the two processes of nucleation and crystal growth are crucial. Crystal growth can occur at the slightest oversaturation of the solution while nucleation is only possible after a critical radius has been attained. Nuclei with a radius smaller than the critical one tend to dissolve before growing into bigger crystals. Those nuclei which reach the critical radius are more likely to grow into macroscopic crystals.

The Second Law of Thermodynamics states that every particle tends to assume the state of lowest energy. The accidental clustering of ions changes the relation between surface and conversion enthalpy. If the clusters reach a certain size their further growth is related to the reduction of free enthalpy. The clusters are thermodynamically stable. As the oversaturation of the solution increases the critical radius decreases. Thus there is an increase of the rate at which crystal nuclei are formed.

2.2 The mineralization of underground water and the formation of deposits

The formation of lime deposits because of seepage water is influenced by biological and chemical processes in the upper layers of the soil. The rain water seeps in the water-unsaturated soil whose pores are filled with subsurface air. This air contains less oxygen and considerably more carbon dioxide because of the soil respiration and the restricted adjustment with the atmosphere. The seepage water contains considerable amounts of carbon dioxide and mineral nutrients and absorbs numerous further substances such as humic acid, ammonium compounds, nitrite, nitrate etc. from the soil.

Calcitic deposits occur most frequently in the drainage systems. Thus the absorption of carbon dioxide and calcium carbonate explains the formation of lime deposits.

For the calcium-carbon dioxide equilibrium the following acids-bases-reactions are relevant:

- the solution of carbon dioxide and hydration to carbonic acid: $CO_{2(g)} + H_2O \leftrightarrow H_2CO_3^*$
- the dissociation in a first stage to hydrogen-carbonate-ions: $H_2CO_3^* + H_2O \leftrightarrow H_3O^+ + HCO_3^-$
- the dissociation in a second stage to carbonate-ions: $HCO_3^- + H_2O \leftrightarrow H_3O^+ + CO_3^{2-}$
- the dissolution of calcium carbonate in carbonated water: $CaCO_3 + CO_2 + H_2O \leftrightarrow Ca^{2+} + 2HCO_3^-$

Since all reactions are balance reactions which can go either way the solved species in the water are all in balance with each other, with the carbon-dioxide-potential of the gas phase and the supply of calcium carbonate.

As opposed to dissociation the hydration of dissolved carbon dioxide is a slow reaction. Therefore 99 per cent of the solution are in the form of molecularly dissolved gas ($CO_{2(aq)}$). Only 1 per cent of it has hydrated to real carbonic acid. Thus $H_2CO_3^*$ is written as compound carbonic acid.

If further seepage of the underground water occurs or in the ground water conduit, the water containing carbonic acid can dissolve rock containing lime or calcite. The solubility increases with the rising carbon-dioxide content of the water.

The solubility of the lime is determined by the pH-value of the water. The pH-value increases when there are changes of the content of carbonic acid and also when bases (OH^--ions) are added. This leads to a shift of the carbonic acid-dissociation-balance in the direction of carbonate ions. The solubility product of calcium carbonate is thus exceeded and the precipitation of macroscopic salt crystals follows.

Besides the rising of the pH-value because of the contact with concrete, the marked reduction of the partial pressure of carbon dioxide at the inflow of the water into the drainage piping is the main cause for the formation of lime deposits. In the soil the respiration of roots and micro-organisms and the restricted adjustment with the atmosphere allow the partial pressure of carbon dioxide to increase ten- or even hundredfold. In the atmosphere carbon dioxide is only found as a trace element (0.03 per cent by volume).

Because of the changes of the partial pressure the water has to yield dissolved carbon dioxide to the atmosphere over the boundary surface which decreases the solubility of calcium carbonate and leads to the formation of calc-sinters.

As the CO_2-yielding of the drainage water takes place on the boundary surface, all influences increasing this surface have an accelerating effect. These influences include over-dimensioned drainage piping and pipes with flat beds as even the smallest amount of water produces a large surface. All changes in the flow of the water, during which air is admixed, further the reduction of the content of carbon dioxide in the drainage water. Carelessly done pipe jointings, leaks, seepage slits in the flowing areas and drops are critical points for the formation of sinter.

3 MEASURES TO REDUCE THE COSTS OF MAINTENANCE

3.1 Measures for the design and building of tunnel drainage systems

Besides lime stone and rock containing lime concrete-bound building materials have been identified as the most important contributors of lime to the seepage water. In the construction of tunnels sprayed concrete, injection grout, anchor grout and one-size grain concrete are used for the dry packing. The degree of sintering cannot therefore be predicted on the basis of the occurring rock or the chemical composition of the underground water. It has to be assumed that the drainage water of tunnels which were built using concrete is saturated with lime. Accordingly measures have to be provided to counteract the changes of the chemical balance and the consequent precipitation of lime.

Elution depends on the amount of adhesive and the number of pores and on the size of the pores. Concretes with crushed aggregates have a higher lixiviation-potential than concretes with round aggregates (Maidl et al. 2001).

In the maintenance and the construction of tunnels one has to take care to use concretes with low elution in order to avoid the accumulation of lime in the drainage water. The amount of soluble calcium compounds has to be minimized. This can be achieved by replacing concrete with puzzolanic additives which do not contain any lime. One also has to be careful not to use additives containing lime or concrete respectively (e.g. made of recycling materials).

The calcium hydroxide ($Ca(OH)_2$) in the concrete, but also the sodium aluminate ($NaAl(OH)_4$) of alkali sprayed concrete catalysts raise the pH-value of the seepage water when it comes into contact with the building materials. A shift of the carbonic acid balance reduces the solubility of calcium-ions and results in the precipitation of the solved calcium-ions in the form of calcium carbonate. These will at first block all cavities in the drainage layer between the sprayed concrete and the inner tube and the dry packing until a low flow pressure is built up in which the precipitated amount of salt is equal to the solved amount. This pressure is reduced abruptly at the inflow into the drainage piping as the water's pressure is relieved after the seepage openings.

A particularly effective method to avoid the lixiviation of calcium-compounds from the concrete and to

1255

increase the pH-value of large amounts of water is the minimization of contact with underground water. Hence the already existing practice of diverting a local inflow of water straight from the ground into the drainage without its coming into contact with the sprayed concrete should be followed to an even greater extent in the future. Large-scale inflows should preferably be secured, collected and diverted.

Before the water reaches the drainage piping it flows through the dry packing. This has usually been made of one-size grain concrete. The open pore-structure of the filter concrete offers the water a very large surface on which it can absorb calcium compounds and hydroxide-ions (OH^-). Accordingly concrete-bound dry packings are one of the main origins of sintering. The new national norms in tunnel construction therefore require that new drainage systems are built with concrete- and lime-free materials. A new cavernous angle section made of PVC developed in Germany allows the use of unbound washed round gravel (Kirschke 2001).

The choice of the piping is the result of the interaction between a water drainage with little sintering and the guarantee that cleaning will be possible. The pipes must not be too big to avoid evaporation and to restrict the water-atmosphere-contact to a minimum. The hydraulic dimensions of the pipes should be based on the actually measured amounts in the driving stage and on the geological studies about the highest level of the ground water. A minimal circumference of 150 millimetres is necessary to ensure the cleaning of the pipes.

In order to prevent the admixing of the drainage water with air and to guarantee the access of hydro-mechanical and mechanical cleaning equipment a too narrow curving of the pipes should be avoided. The intakes of the pipes into the manholes should be built without drops because they favour a particularly high admixture of air and water and this can lead to spontaneous precipitations.

The gradient of the drainage pipes should be 1.5 to 1 per cent depending on the circumference of the pipes to drain the water efficiently. There also has to be a laminar flow, as a turbulent flow would increase the exchange with the air.

The interior surfaces of the pipes have to be smooth. On the one hand corrugated pipes generate areas of turbulent flows and on the other hand the large contact area leads to extreme adhesion forces of the sintering with the interior of the pipe which render the cleaning more difficult.

Only specific materials are suitable for an enduring drainage system. Due to the above-mentioned influences of building materials concrete and stoneware pipes should not be used because of the strains they impose on the hydro-mechanical cleaning equipment. The only suitable pipes are plastic ones with sufficient

wall strength. No materials containing plasticizer (e.g. PVC) should be used. This will start to leak in time and the pipes will become brittle resulting in more damages. Only pipes made of high-pressure polyethylene (HD-PE) – used e.g. in Germany for the construction of landfill sites – or made of PVC-U or PE respectively – provided that they guarantee sufficient form stability – should be used. Pipes made of recycling-PE are susceptible to impact sensitivity and could be damaged by the cleaning equipment.

Only round and partially perforated pipes with seepage apertures of 5 to 10 millimetres should be used. It takes longer for these wider apertures to be blocked with sintering. Moreover there is less sintering as the water drops are not held in the slits by the capillary force and can drip off more easily. The cleaning of sintered wide slits is possible because the sintering leads to a two-dimensional strain and can thus be broken up more easily.

In the construction but also in the reconstruction of the whole system possible changes in the flow regime should be allowed for in the planning stage. Cross-deflections of side wall drainages in regular distances constitute an unnecessary lengthening of the flowing distance resulting in more costly cleaning. The additional turns and the intake into the collecting pipe over drops contribute to the formation of sinter. The drainage of the water of every side wall leads to artificial evaporation ways with a very small gradient.

The drainage of the underground water should occur in the shortest possible distance. In order not to unnecessarily reduce the amount of water in the drainage pipes combined drainage transport pipes can be used. In the planning stage separate cross pipes leading into a middle collecting pipe should be provided so that the pipes will not have to be too large and that water with different chemical composition can be drained separately. The control of the water flow could be achieved by a manual opening and closing or by overflow brims in the manholes which would guarantee an optimal filling of the pipes. Overflow brims have the advantage that they do not cause more admixture with air. This makes it possible to reduce the tendency for sintering and the formation of hard deposits in the side wall drainage and to support the effect of hardness stabilization.

3.2 Measures and strategies for maintenance

The maintenance of drainage systems can be carried out following different strategies. Every operator determines the optimal strategy according to the value he assigns to the functioning of the tunnel drainage and by providing the necessary funds.

In the following only the strategy which is considered to be optimal for hardness stabilization will be discussed. Hardness stabilization is a pre-emptive maintenance practice which should reduce both the amount and the hardness of the deposits.

Originally this process derives from nature. Mussels and corals incorporate aspartic acid molecules in the formation of their shells which edge their way between the layers of the individual calcite crystals and thus influence the growth of these crystals. This is how the shape written down in the genome of crustaceans is formed. Without polyaspartic acid they would calcify uncontrollably. The agent of hardness stabilization – polyaspartic acid – is one of about 20 amino acids which constitute the elements of proteins. It is a natural material and is not toxic as it is completely degradable by micro-organisms.

Polyaspartic acid follows the threshold-effect. The formation of a solid stage from an oversaturated solution corresponds to a crystallization process. An oversaturated solution is a one-stage-system which is not thermodynamically stable. It tends to pass into a low-energy stage by precipitating the superfluous part in order to attain stability. This leads to the creation of a solid-liquid boundary surface. The overcoming of this boundary surface tension requires energy. Hence a system containing even more energy has to be created as a transitional state.

Small particles have a relatively large surface in relation to their volume. That is why they are particularly unstable und dissolve spontaneously. Only when the crystal nuclei reach the critical radius do they become stable and can go on growing. At this point threshold-inhibition sets in. Inhibitors are water-soluble compounds which are absorbed to a great extent by certain mineral particles on the surface.

This causes the structure of the surface to change which inhibits, disrupts or delays further growth. If inhibitors form an adsorption compound with crystal nuclei the probability of their reaching the critical radius is drastically reduced. This leads to a spontaneous dissolution of the nuclei.

As drainage pipes contain dirt and deposits of lime, primary heterogeneous and secondary formation of nuclei take place. The aim of hardness stabilization in drainage water is not to hinder the formation of nuclei below the critical radius, but to confine the crystal growth. Surface adsorption also applies here and can delay the further growth and disrupt it to a great extent so that no ordered crystallographic surfaces can arise. Thus irregular crystals are formed which are no longer able to accumulate hard deposits.

The effects of different inhibitors on certain slightly soluble salts vary. The substances for the inhibition of calcium or magnesium carbonates are not the same used for the inhibition of sodium, potassium or manganese compounds etc. Polyaspartic acid is particularly suitable for the prevention of lime deposits but is not very effective against other salt deposits as they occur e.g. at the inflow of deep ground waters.

The studies conducted so far use polysuccinimide-tablets for hardness stabilization. This has the advantage that the whole drainage system can be conditioned consistently and technical recesses are unnecessary unlike in liquid conditioning. The discharge of the agent is regulated independently depending on the flow of the water and on the pH-value.

Thanks to this new pre-emptive maintenance practice the dissolved lime stays in suspension. Fewer deposits can form and the intervals between cleaning can be longer. This results in lower maintenance costs and in a higher availability of the route. One has to take into account, however, the costs for the hardness stabilizator. The laying of the tablets can take place during the inspections or the remaining cleanings.

The determination of the cleaning intervals will have to be carried out sequentially according to the cleaning expenditure and the occurrence of lime deposits. After every maintenance measure the intervals are redetermined depending on the available time and funds.

Opportunistic strategies will have to be used since the maintenance depends on other tunnel infrastructure systems (catenary line, track superstructure).

The conditions of the underground water and in particular the biological activities in the drainage system are susceptible to fluctuations. The effects of hardness stabilization will have to be evaluated and amended in determined intervals. The above-mentioned reasons suggest that only a strategy combining sequential cleaning and inspections is feasible to ensure the success of the water conditioning and thus the functioning of the drainage system in the long term.

4   OUTLOOK

The use of polysuccinimide-tablets so far has reduced both the amount and the hardness of lime deposits. Increasing micro-biological infestation, however, has led to the increase of lime deposits which could be traced back to the reduction of dissolved carbon dioxide in the drainage water.

The micro-organisms themselves or the products of their metabolisms detract carbon dioxide from the water. The laying of the tablets provides nourishment for the bacteria thus allowing them to propagate excessively. As a consequence hard incrustations of lime are formed around the tablets and in their surroundings. The release of the agent is prevented and the tablets are worn down quickly under the incrustations.

To confine biological infestation small concentrations of a substance used for preservation in the food industry are added to the tablets. In this manner the effect of polyaspartic acid persists until the water is discharged into the runoff ditch.

Moreover the formation of lime soap (calcium stearate) on the tablets has been observed, but this has little influence on the release of the agent. High concentrations of calcium allow the agent to dissolve very quickly. The carrier is only slightly water-soluble and reacts with calcium. At the moment experiments are being carried out in which a reduction of the adhesive agent is attempted on the one hand to render it possible to include a higher portion of the agent in the tablets and on the other hand to minimize the saponification.

REFERENCES

Hildebrandt, E.E. 2001. *Allgemeine und organische Chemie für Forstleute – mit Querverbindungen zu (boden) ökologischen Fragstellungen.* Dritte, korrigierte und veränderte Fassung. Vorlesungsskript SS 01. Freiburg: Albert-Ludwigs-Universität, Institut für Bodenkunde und Waldernährungslehre.

Kirschke, D. 2001. Fortschritte und Fehlentwicklungen bei der Tunnelentwässerung. *Geotechnik* 24(1): 42–50.

Maidl, B. et al. 2001. Untersuchungen und Massnahmen zur Reduzierung des versinterungsbedingten Wartungsaufwandes am Saukopftunnel. *Bauingenieur* 76(9): 392–402.

Wood, J.T. & Greenwood, D.J. 1971. Distribution of carbon dioxide and oxygen in the gas phase of aerobic soils. *Journal of Soil Science* 22(3): 281–288.

Zimmermann, F. 2000. *Vergleichende Prüfungen zur Hochdruckspülfestigkeit verschiedener genormter Werkstoffe für Abwasserleitungen und – kanäle.* Prüfbericht. Zürich: Eidgenössische Technische Hochschule, Institut für Bauplanung und Baubetrieb.

System-based Vision for Strategic and Creative Design, Bontempi (ed.)
© 2003 Swets & Zeitlinger, Lisse, ISBN 90 5809 599 1

Foundation design formulation: a process approach

J.F. van den Adel, S.H. Al-Jibouri & U.F.A. Karim
University of Twente, Enschede, Netherlands

M. Mawdesley
Nottingham University, Nottingham, UK

ABSTRACT: This paper reports on the development of a system that uses a new approach in the decision on the selection of a foundation system. The new approach allows the geotechnical engineer to incorporate non-technical aspects in the decision-making process. The paper describes the structure of the decision system designed for this purpose to support with the selection procedure. A simple application example has been used to demonstrate the basis of the new approach and discussion and conclusion on this application are also included.

1   INTRODUCTION

In construction, the choice of a foundation system has not only technical consequences on integrity of the structures of the building but can also have substantial impacts on the programme and cost of the project as a whole.

Traditionally, in preliminary foundation design a foundation system is adopted based on available soil and site conditions as well as the project design requirements. As a result of this, foundation systems are often selected based solely on their technical feasibility and designers recent experiences. Whilst the choice of a particular foundation system based on such approach may be technically sound, it could however prove to be too impractical or costly later at the construction stage. At present, it is not common practice to take these aspects into consideration or to look for other design alternatives which may have more favourable effects on the construction process.

The treatment of foundation engineering in research and design practice as a technical subspecialty separated from the engineering system continues in practice despite a rising need for integration. No reference is made in the literature to any working system that integrates the foundation design process with the construction process in it's totality. A review of current practice in the Netherlands (Dooren et al. 1998, Adel 1999) showed that decisions on foundations are too often taken independently and without studying the possible effects of such a choice, in later stages, on the construction of a project. Systematic considerations of non technical aspects such as planning, costs, site layout of the foundation type and their integration with the project programme before embarking on detailed design can be very beneficial to the overall project. For example Fleming et al. (1992) state that it is more often than not the case that any one number of methods could be employed to yield a satisfactory end product. The real choice may rest on such items as site access for equipment, environmental impact, or on the important consideration of overall costs.

In this work a new foundation design approach, which allows the geotechnical engineer to incorporate the non-technical aspects mentioned above into the decision-making procedure, is proposed. The approach provides the designer with means to assess the effect of various foundation alternatives and ground conditions on the overall construction process. Optimisation of the construction process can then be expressed in terms of measures such as reduction in costs or duration, increased safety or quality or a combination of these. In this context, it is the whole process that is meant to be optimised and not the individual steps, which contribute to the process. In other words, it could be beneficial in this approach to have sub-optimal stages in order to allow other, usually later, stages to progress with greater benefit. Traditional design objectives providing piecemeal optimisation of a project are normally unsatisfactory. To be able to implement the new proposed approach, an integrated expert system is designed in order to provide the following facilities:

– Support to the geotechnical engineer during preliminary design of the foundations (and other

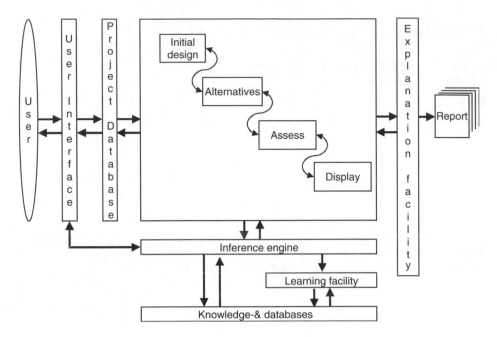

Figure 1. System architecture.

subsurface constructions) on the choice of foundation concept;
– Mechanisms to generate alternative foundation concepts based on initial design, the user knowledge and project specific information;
– Means to assess the effect of each foundation alternative on the whole project with respect to costs, time, construction method, site-layout and risk – the decision parameters;
– A display of the order of project alternatives with respect to the values of the decision parameters;
– Additional information to facilitate the decision making process on selecting suitable foundations;
– A report on the decisions and how they are derived;
– A basis for understanding how ground conditions will affect the whole construction process.

The following sections of the paper will describe the model on which the expert system is based. A simple application example to demonstrate how the new approach works is also presented. The last part of the paper will include summary and conclusions.

2 FOUNDATION DESIGN MODEL

2.1 Basic assumptions

The developed decision expert system is based on modelling the procedure used during preliminary foundation design stage. Minimum data will be available on

the project and project ground. At best, the following data will be available: client requirements, the site and building dimensions and location, structural loads and their distribution, a budget and a desired duration. The decision on the extent and type of site investigation will then be taken, and a first educated guess of the foundation type on basis of the minimum data available is usually made. Depending on the complexity of the construction and ground, a building is often designated to a geotechnical category, based on some national building regulations and geotechnical codes of practice. Following this, a foundation concept can become the final design if it satisfies the technical as well as the key data on constructability and project planning (site, cost, and duration).

The targeted system user is a foundation engineer during the initial design phase. The common foundation design methods are implemented in the system (strip foundation, bearing piles et cetera). The remaining components of the system have been developed to support the user in assessing the technical alternatives with respect to construction process aspects. The cases concern medium rise office buildings in an urban environment with or without a basement.

2.2 Model representation and techniques

Based on the provided information such as soil and site conditions and project design requirements, the system will generate several feasible foundation systems which satisfy the project technical requirements. The following decision process on the foundation concept in this model requires many iterations so that the consequences of each choice can be related to the project plans and costs, soil and site conditions, foundation construction method, and to the project environment. This new proposed formulation takes a more holistic view of the foundation design process than currently adopted in geotechnical practice. It also addresses the need, expressed repeatedly by clients and project participants and by researchers, for an integrated methodology in foundation design where ground conditions (soil and site), planning and cost factors are considered. A foundation concept will be acceptable and reasonable decisions can be made when at the end of iterations all these factors have been accounted for and their relationships tested and optimised. Flow of the logic applied in this methodology to reach these (so called reasonable) decisions is explained for the cases considered (the explanation mechanism). This flow, when established, is of further use to build up the knowledge required to test other cases and to make even better decisions on more complex problems (the learning facility).

The expert system incorporating this new approach is called GROUNDSS. The components of the system and their relationships are shown in the system diagram in figure 1.

The process to model is typically adaptable for programming in a Decision Support System (DSS) adapting techniques from artificial intelligence. Artificial intelligence, in particular knowledge-based, techniques in addition to traditional modelling techniques, is explored for use in the current research. A knowledge base basically consists of a set of rules combining both numerical and textual information on the field of knowledge. A rule would be for example like: "IF water-table higher then foundation level and shallow foundation is true THEN pumping is required and pumping-costs is true and pumping-time is true". Combining a larger set of these rules and applying them is called reasoning. Required data, e.g. initial soil data has been stored in a database.

3 EXAMPLE APPLICATION

The application of the outline prototype is limited to the simple case of a strip foundation with varying dimensions and depth of the embedment. This example is kept deliberately simple to demonstrate the effect of these few selected constructional aspects on the final costs and time of the generated foundation alternatives.

For this application example, two tests were carried out: the first test is using fixed ground conditions and is based on the following project information:

1. Water table: 0.8 m below surface;
2. Homogeneous soil: sand, $\phi = 25°$;

Table 1. Alternatives details.

Alternative	Strip width (m)	Foundation level (m-surface)	Concrete Thickness (m)
1	1.5	0.5	0.3
2	1.0	1.0	0.2
3	0.5	1.5	0.2

3. Design bearing capacity: $F = 125 \, kN/m^1$;
4. Strip length: $L = 20$ meter.

The above information has been used to formulate an initial design, referred to as alternative 1, using the system "Initial design" module. The produced initial design is based on calculating the initial strip width using rules of thumb. In the "Alternatives" module a number of feasible alternatives are then generated automatically as shown above.

The bearing capacity calculations for a shallow foundation are based on the project input using the (Brinch Hansen) formula stated in the Dutch code of practice (NEN6744 1991).

The following step in the procedure used by the system is to calculate the construction costs and duration for each alternative. This is done using the "Assess" module. The construction of these three foundations will involve additional some soil removal (excavation and transport/dumping) and a construction (partly) above or under the water table (and thus possibly pumping). Also, quantities of the materials involved (soil, concrete, brickwork, soil refill, compaction, etc.) and their associated costs differ. For each case (1 to 3) there will therefore be expected constructional time and cost consequences depending on the different quantities and their respective unit process. The program calculates automatically these quantities and assigns local (Dutch) market unit material and unit work prices (Bouwkosten 1997).

The example case as well as the basic costs and duration differences are displayed in figure 2 below.

In the second test the ground conditions are changed as indicated in table 2 below. Changing the ground conditions will affect the alternatives generated as well as the related construction cost and duration. The outputs of this test are the different strip widths, relative construction cost and duration differences as shown in table 2.

4 DISCUSSION OF RESULTS

The results of the first test of the example application discussed above show that although all the three design alternatives are technically feasible, their construction

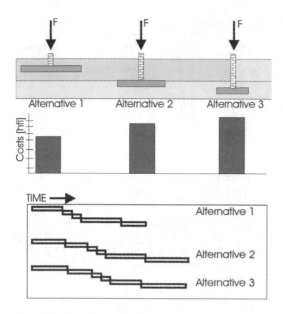

Figure 2. Example application.

Table 2. Test outputs.

Alternatives	1	2	3
Water table [m]	0.0	1.0	4.0
$\phi[°]$	25	25	25
Foundation level [m]	0.8	0.8	0.5
Strip width [m]	1.5	1.3	1.0
Relative costs	100	80	50
Relative time	100	90	80

costs and duration vary considerably. For example the need to lower the water table during construction will result in a considerable extra cost as in the case of alternative 3. If all the other parameters which may influence the cost and duration for all alternatives are considered to be fixed, alternative 1 would represent a better alternative design than the other two because of its lower cost and the fact that it takes less time to construct.

In the case of the second test the example application has shown the effect of changes in ground conditions on the associated construction cost and duration. A lower water table increases the bearing capacity of foundations and decreases the dimensions needed and hence the construction costs and duration.

In the case of a more complex foundation system within a large project, the differences in cost and duration of construction of various alternatives could have significant impact on the project programme especially if the activities related to foundation lie on the critical path.

5 SUMMARY AND CONCLUSIONS

It can be concluded that the choice of a foundation system has not only technical consequences on the structures of the building but can also have substantial impacts on the programme and cost of the project as a whole. The development and application of a new foundation design approach show that it is technically feasible and advantageous to integrate non-geotechnical aspects in the choice and design of a foundation system. These aspects include: costs, planning, site layout, environment and site.

REFERENCES

Adel, J.F. van den. 1999. Annual report 1999, modelling the effects of ground conditions on the construction process. Internal report. Twente university, Nl
Bouwkosten. 1997. (in Dutch), Elsevier Bedrijfsinformatie, Doetinchem, The Netherlands
Dooren, M. van, Mawdesley, M., Al-Jibouri, S. & Karim, U.F.A. 1999. A new approach to the process of selecting foundation type. *Concurrent engineering in construction, Proc. Of the 2nd. Int. conf. On Concurrent engineering in construction*, CIB-publication no. 236, Espoo, Finland, pp. 139–148
Fleming, W.G.K., Weltman, A.J., Randolph, M.F. & Elson, W.K. 1992. *Piling engineering.* 2nd ed., Blackie Academic & Professional, Glasgow, UK
NEN 6744 1991. (in Dutch) Berekeningsmethode funderingen op staal, NNI, Delft, The Netherlands

System-based Vision for Strategic and Creative Design, Bontempi (ed.)
© 2003 Swets & Zeitlinger, Lisse, ISBN 90 5809 599 1

Relationships between soil conditions and construction costs

J.F. van den Adel, S.H. Al-Jibouri & U.F.A. Karim
University of Twente, Enschede, the Netherlands

M. Mawdesley
University of Nottingham, Nottingham, UK

ABSTRACT: In construction, structures, including foundations, are designed in order to meet clients' demands. They should be designed to achieve the strength requirements with minimum costs. However in order to achieve a better balance of technology, quality, time and cost for the project as a whole, geotechnical engineers are required to develop awareness of the interactions between foundation design and construction project variables. This paper is based on ongoing research work that investigates the interrelations between geotechnical design variables and construction process aspects. Specifically this paper deals with the interactions between ground conditions and the costs of prefabricated concrete piles. Such interactions have been modelled and a number of experiments was carried out. These experiments for example involved changing bearing capacities of piles with differing lengths and diameters in order to show their effects on the foundation costs. Initial results have shown that the interactions between the ground conditions and foundation costs can be modelled and that the effects of changes in the variables can produce meaningful and useful relational trends.

1 INTRODUCTION

Geotechnical engineering has made significant advances over the recent years. More complex and different construction methods, design and engineering tools are applied than before. Also the process of exploring different designs is becoming fairly easier. The process will often lead to a number of possible technical feasible designs, which may vary in concept, material use, construction method and/ or dimensions. The majority of foundation design problems is routine and is solved by the combined application of experience, engineering judgment and local regulations (Whitlow 2001, Meyer 1992). Designs that are made according to national norms are likely to be considered to be technically feasible and hence meet the local regulations. Very often an engineer selects a proper design solution with which he/she is most familiar (Mohamed & Celik 2002), ensuring that the end product will meet the accepted quality standards The decision on whether or not to adopt a specific design however should not only be based on technical feasibility and experience but on an overall consideration of the project and the interrelations between its various aspects. Also although rarely involved in the choice of foundations, client do demand strong enough foundations and "optimum"

balance of technology, costs, time, quality, safety and environmental impact for their projects. A designer often assumes that "optimum" foundation will provide "optimum" project. In most cases however such assumption takes no account of the interrelations between the foundation and the rest of the project. It would therefore be surprising if an optimum project arose (Dooren et al. 1999).

Construction and its parts, including foundations, have to be designed in order to meet clients' demands. This means that geotechnical engineers, in addition to their technical abilities, are also supposed to develop an increasing awareness of the importance of subsequent relationships between geotechnical design and other construction project variables. This requires knowledge in the nature of the interrelations between ground conditions and hence the choice of foundation type and the construction process (including time, costs, environment, risk and quality). Such knowledge often is not available in the literature and experts in this field are unable to express it in a ready-to-use format. In the field of geotechnical engineering only a few systems have been found which integrate engineering design and other aspects such as construction cost, see (Cadogan et al. 1996, Fisher et al. 1995, and Mangini et al. 1994).

This paper uses a basic problem to show the importance of the inclusion of the cost element in the choice of foundation type. It also shows, based on the use of modelling, the nature of the relationships between ground conditions and the construction process. The application of this work is limited to medium size office buildings on soft soils.

2 BASIC MODEL

The modelling of the relationships between geotechnical design and construction project variables of foundations is carried out around an existing commercial software package for foundation design called MFoundation®, (MFoundation® 2002). This package enables the geotechnical engineer to design foundations based on Dutch codes of practice (limit state soil mechanics) (NNI 1991). In theory however any other geotechnical design package can be used without any expected major changes.

For any given project, the objective of the developed system in this research is to be able to produce a number of foundation design solutions and to estimate their associated construction costs and durations. The ability of any model to predict cost and duration of construction is dependent on a number of significant factors, which affect these two basic variables; see (Elhag & Boussabaine 1998). For example, in addition to the construction method used, some of the other important factors that could affect these variables are site conditions and layout.

The system described here has been developed in order to:

– Investigate the effect of changes in any variable on the project cost and duration;
– Investigate the effect of adopting different construction methods;
– Estimate the effects of such changes on the construction time;
– Estimate the effects of such changes on the associated costs.

The system, which is called GROUNDSS, is developed in a modular fashion. The calculation heart of the system is the MFoundation® program. Around this are a number of modules for assessment of time, costs, related-costs, risk, environment and site factors. These modules use knowledge-base like files containing knowledge on resources, costs, risk, environmental or site layout conditions. This knowledge is acquired from published data, rule-induction as well as experts.

The computer system is designed in such a way as to enable quick and parallel generation and comparison of multiple foundation types in one system run.

The relationships between the various elements of the computer system are shown in Figure 1. These elements are described in the following sections.

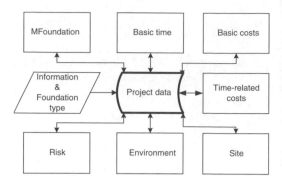

Figure 1. Links between software for integrated design.

2.1 The foundation design package

MFoundation® is the software program which has been integrated in the developed system in order to offer the possibility to calculate bearing piles according to the Dutch codes of practice NEN 6740, NEN6743 and shallow foundations in accordance with NEN 6740 and NEN6744 (NNI 1991). In addition it is also possible to perform design calculations for bearing piles and to use the software facility for optimising shallow foundations.

For the Bearing Piles model, two types of data are required. These involve data to determine the soil CPT (Cone Penetration Tests) with their corresponding soil profiles, including the ground water level and the expected ground level settlement, etc.. The vertical limit state bearing capacity of bearing piles is calculated using the principles of skin friction and tip resistance. The average cone resistance $\bar{q}_{c-1}$ over a depth up to four diameters below the pile toe, and the average $\bar{q}_{c-2}$ eight pile diameters above the toe as described by Meigh (1987). The ultimate base resistance is then given by Equation 1:

$$q_b = \frac{q_{c-1} + q_{c-2}}{2}$$

where q_b = Ultimate base resistance; q_{c-1} = Average cone resistance over a dept of up to four diameters below the pile toe; q_{c-2} = Average cone resistance over a depth of up to eight diameters above the pile toe.

The ultimate base resistance combined with the positive and negative skin friction over the pile length is used to calculate the overall bearing capacity of single piles.

Secondly, data is required to specify the construction (of the foundation), e.g. pile type, pile dimensions, pile plan, groundwater table et cetera. For the Shallow Foundations model two types of data are also required. The first type of data is provided to determine the soil profiles, including the ground water level and placement depth of foundation, etc. Unlike the bearing piles model, however, no CPTs are required here. The second type data is required to specify the construction (of the foundation), e.g. dimensions, foundation plan, et cetera.

2.2 Time assessment

Having calculated different foundation types based on the input data the system provides the option to calculate the associated foundation construction time based on the same input data as well as the outcomes of the geotechnical design. To calculate basic construction time it is necessary to have information related to:

- The construction activities to be performed
- The relations between these activities
- The expected resources to be used
- Productivities of these resources
- The project start date
- The amount of work required by the various activities

All of the above information is required to be provided to the system, either from direct user input or through automatic generation by the system.

Calculating the construction time for each foundation type is necessary because the choice of a foundation method might be based on construction time. The calculation of time-related costs is also dependent on the time to construct a foundation and some of the cost factors are directly related to the construction time of the foundation or elements of the foundation.

2.3 Costs assessment

Based on the information provided through input data, the engineered foundation type and the costs stored in the knowledge base, the cost-module calculates the construction costs for each foundation type separately. The costs are calculated based on a standard cost breakdown structure, using the quantities and the associated unit price for each cost element.

The system provides the opportunity to use standard published costs data, see for example (Archidat 2002) .The costs assessment module is limited to the basic costs. An example of the cost-breakdown structure and the costs of a prefabricated concrete pile are given in the Table 1.

The costs shown in the table are related to a prefabricated concrete pile with a length of 15 meter and a width of 0.35 meter. The basic costs in this case are the products of multiplying the unit costs by the calculated quantities.

The unit costs selected from the knowledge base are dependent on information such as the number of

Table 1.   Unit costs prefabricated concrete pile 15 m × 0.25 m.

Activity	Unit costs
Well	4000 €/pc
Excavate	1.94 €/m^3
Pile delivery	286.22 €/pc
Install and remove piler	2384.46 €/pc
Move and position piler	7.83 €/pc
Install pile	49.37 €/pc
Remove pile cap	17.06 €/pc
Beams	80.55 €/m^1
Backfill	1.97 €/m^3
Remove concrete waste	4.38 €/m^3
Remove soil surplus	6.81 €/m^3

piles, pile length, pile width, height of water table. An example rule to find the unit price for pile delivery is:

IF subtype = prefabricated concrete pile AND
IF pile length > 13 AND
IF pile length < 15 AND
IF pile width ⩾ 0.32 AND
IF pile width ⩽ 0.35 AND
IF numberofpiles < 60
THEN Ppiledelivery = 286.22'

The cost knowledge base at this moment contains 450 of those rules and is to be extended further in the future.

2.4 Time-related costs

The use of standard published costs data and construction time does not account for time-related costs. An example of time-related costs is the effect of difficult soil conditions on the costs of pile installation. Piling through hard intermediate soil layers takes more time than piling through soft soil and therefore both the associated time of pile installation and costs of piling are different. Therefore it is necessary to calculate these costs separately; this is applied in the time-related costs module of the GROUNDSS system.

The system uses a knowledge base to check which relations apply in the case under consideration. An example a relation to illustrate the set up of the system is: "IF upper soil = peat THEN Pmove and position piler = increase 20". The statement means that if the upper soil is peat then the piler should use plates to move, taking additional time and therefore costs.

2.5 Risk assessment

Risk is inherent in construction especially in relation to foundation and excavation works. For the kind of standard project type considered in this research, i.e. the medium size building in an urban environment, the system is designed to provide a list of potential risks associated with such projects. The system provides the ability to analyse these risks and determine their consequences on the project cost and duration.

3 BASIC PROBLEM APPLICATION OF THE SYSTEM

The application of the model to investigate the relationships between geotechnical design and basic construction project variables is presented. This is demonstrated on the basis of the design of a single pile in a foundation for a medium size office building. The costs of the beams are not included in the calculations of the foundation costs. Although the application to a complete foundation will make the analyses more complex, the final relationships will remain unchanged. The inputs related to the project site, soil, and environment conditions are based on an unrestricted site to obtain the basic relations. The soil and its associated parameters are based on two different CPTs. The experiments have been applied using both a Gibson soil with a constant strength and a Gibson soil with a strength increasing with depth. Soil details are described in the Table 2.

The CPT-value for the constant soil profile is set to 18 MPa, the CPT values for the soil with increasing strength vary from 1 MPa to 21 MPa on a depth of 20 meters. The water table and excavation level of the construction pit are both set to 1.00 meter below surface.

The additional soil parameters have been assumed as standard according to Dutch codes of practice.

Table 2. Soil data.

Dry unit weight	17 kN/m³
Wet unit weight	19 kN/m³
Angle of internal friction	30°
Cohesion	0
Median	0.30 mm

In the constant soil strength example, the soil friction parameter has a large effect on the pile length of soil replacement piles as well as on long prefabricated concrete piles. The analyses are applied to prefabricated concrete piles with a diameter of 0.22 to 0.45 m and a length of 4 to 20 m.

4 RESULTS

Firstly calculations have been carried out with different loads and therefore bearing capacities of prefabricated concrete piles with differing length and diameter. The results displayed in Figure 3 show that a single pile with higher bearing capacity costs more than a pile with a lower bearing capacity. This trend applies both to the case of increasing soil strength with depth as to the case with constant soil strength and is as would one expect.

In foundation engineering the option to choose between different foundation dimensions and therefore bearing capacity per pile exists. Except for the slopes of the trend lines, both soil profiles displayed in Figure 4 show the same trend for soil strength increasing with depth.

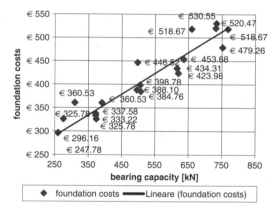

Figure 3. Trends in basic costs.

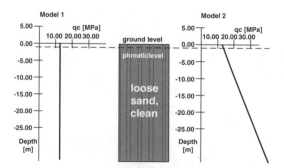

Figure 2. Soil graphs and parameters.

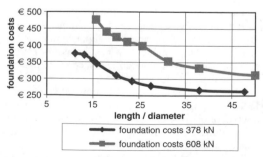

Figure 4. Trends in basic costs.

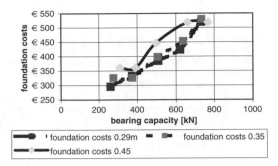

Figure 5. Trends in basic cost vs. pile type.

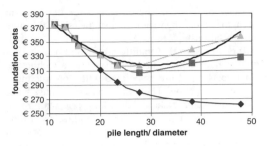

Figure 6. Trends in integral costs.

The lowest line shows the foundation costs for a pile with a bearing capacity of 378 kN, the upper line shows the costs for a bearing capacity of 608 kN. For length/diameter = 30 and a pile diameter of 0.25 a pile length of 7.5 m is required. The costs of a pile with a bearing capacity of 608 kN are found 33% higher. For an increase in costs of 33% there is an equivalent increase in bearing capacity of 61%. The results suggest that it can be beneficial to use less piles with a higher bearing capacity than a larger number of piles with a lower bearing capacity.

The results distinguished by pile type (length and width), have been displayed in the graph in Figure 5 to show an interesting trend.

In this figure it can be seen that in order to achieve a certain bearing capacity, the choice of pile dimensions doesn't really make much difference. The cost elements will differ since different pile dimensions have different unit costs due for example to the use of different type of piler, costs of pile cap removal, etc. The figure however shows that the basic costs of a pile 17.5 m long and 0.29 m wide are approximately equal to those a pile 12.5 m long and 0.35 m wide and to those of a pile 8.5 m long and 0.45 m wide. This trend can be explained by the distinction between basic costs, time-related costs, and other costs related to risk and site conditions. The costs shown in Figure 5 only include basic costs, and if other costs are considered then the relationship would look different as explained in the next section. The results indicate that an increase in bearing capacity of a single pile of 260% can be achieved for only 60% additional costs.

5 INTERPRETATION OF RESULTS

The results presented in this paper show some basic relations between the ground conditions, the choice of foundation type and the basic costs. Relations have been found for the increase in costs with bearing capacity of a single pile. This result is logical, but the curves also provide information about the increase of cost with the increase in bearing capacity. The inclusion of the beam in the analysis with more than one pile will change the outcomes, dependent on the load distribution, since the beam dimensions will change with the distance between the piles. The results suggest that it can be beneficial to use fewer piles with a higher bearing capacity instead of more piles with a lower bearing capacity. The results also show that it is possible to choose piles with different dimensions with equal bearing capacity for the same costs. This result is different from the conclusion of Fleming (1985) stating that is in general it is more "efficient" to have long slender piles than short stubby piles. But he also states that more reasons enter in the decision which pile might be most economic. Ultimately however the final choice between the various pile types is dependent on the effect on costs occur from other construction aspects such as duration, time-related costs, risk, environment and site conditions.

6 INCLUSION OF OTHER RELATIONS

The results of the experiments have shown some interesting points especially in terms of geotechnical engineering. For example although the results indicate that the costs of various pile dimensions are comparable on the basis of basic costs, the piling of a thin and long prefabricated concrete pile however has to be done with care and will therefore take a relative long time and therefore additional costs. The additional costs will be due to the risk of breaking piles during transport and the costs of delay associated with the piling process. It is therefore important that the time-related and risk related costs as well as the effects of environment and site conditions have to be included in the model. These are often the aspects which are poorly accounted for in current practice in the geotechnical design in the Netherlands.

The results of the interrelations between soil conditions and costs after including the various factors mentioned above are shown in the Figure 6. The outcomes for the soil with strength increasing with depth are shown in this figure for piles with a bearing capacity of 378 kN. The outcomes for the constant strength soil or other bearing capacities show the same basic relationships.

It can be seen that traditional method of cost estimating may neglect important factors affecting the final costs and time associated with constructing a project foundation. The results based on only basic costs for example show that a long pile with a thin shaft should be cheaper to apply. Integration of time and risk as well as the effects of site conditions and environment however has produced more realistic cost behaviour as shown in Figure 6. Generally the installation time and transport and handling risk of long prefabricated piles with a small shaft is relatively high compared to average or short length piles.

7 CONCLUSIONS

In current practice it has been stated that many subjective factors enter into the decision on which pile to select, making experience an important factor in such process (Elton & Brown 1991). The results presented in this work have shown that it is not an easy task to select an optimum foundation type and dimensions. The relationships between geotechnical design and basic construction variables are complex, but can be modelled using simulation.

The calculations presented in this paper are only related to prefabricated concrete piles. The model however can be applied to other pile types as well as to shallow foundations. The optimal solution in these cases may be different from the one currently presented. However the initial results presented in this paper are very useful in showing the interrelationships between the various factors affecting the choice of pile type. The paper has demonstrated that in order to achieve more informed decisions as to which foundation designs to adopt, it is important to take other aspects of the construction process into consideration.

REFERENCES

Archidat 2002. http:\\www.bouwkosten-online.nl
Cadogan J.V., Moore C.J., Miles J.C. 1996. Computational support for the design and costing of bridge foundations, *Information processing in civil and structural engineering*, B. Kumar (ed.), Civil-Comp Press, UK, pp. 113–119.
Dooren M. van, Mawdesley M., Al-Jibouri S., Karim U.F.A. 1999. Concurrent Engineering in Construction, Challenges for the future, Finland, 1999.
Elhag T.M.S., Boussabaine A.H. 1998. Factors affecting cost and duration of construction projects, *EPSRC Research report*, phase (1), Liverpool: UK.
Elton D.J., Brown D.A. 1991. Expert system for driven pile selection. *Geotechnical Engineering Congress*, ASCE, Boulder: Colorado USA.
Fisher D.J., O'Neill M.W.M- Contreras J.C. 1995. DS^2 drilled shaft decision support system. *Journal of construction engineering and management, ASCE*, v. 121 no. 1, march 1995.
Fleming W.G.K., Weltman A.J., Randolph M.F., Elson W.K. 1992. "Piling engineering", 2nd ed., Blackie Academic & Professional, Glasgow, UK.
Mangini M., Varosio G., Parker E. 1994. "EP4.4 COMBI KB-tool for foundation design", proc. ECPPM '94, *Product and process modelling in the building industry*, R.J. Scherer (ed.), Balkema, Rotterdam, the Netherlands.
Meigh A.C. 1987. *Cone Penetration Testing*, CIRIA-Butterworth.
Meyer S. 1992. Preliminary foundation design using EDESYN. *Optimization and artificial intelligence in civil and structural engineering*, ed. by B.H.V. Topping, Kluwer Academic publishers.
Mfoundation 2002. GeoDelft Software Systems, Delft: The Netherlands.
Mohamed A., Celik T. 2002. Knowledge-based systems for alternative design, cost estimating and scheduling, *Knowledge-Based Systems*, Elsevier Science, vol. 15, pp. 177–188.
NNI 1991. Dutch Normalisation Institute (in Dutch), NEN 6744 Geotechnics: Calculation method for shallow foundations, Delft, the Netherlands.
Whitlow R. 2001. *Basic soil mechanics*, Dorset: Prentice Hall.

System-based Vision for Strategic and Creative Design, Bontempi (ed.)
© 2003 Swets & Zeitlinger, Lisse, ISBN 90 5809 599 1

A mathematical model for the analysis of the rectangular machine foundations on a layered medium

M.Z. Aşık

Department of Engineering Sciences, METU

ABSTRACT: A mathematical model to predict the response of a rectangular machine foundation sitting on the surface of a layered medium is introduced. Model assumes that the displacements at base are zero; applied force and displacements are in the vertical direction and both harmonic; material and/or geometric properties of layers of foundations could be different.

1 INTRODUCTION

Rectangular machine foundations are often encountered in civil engineering practice. They may lie on the surface of naturally stratified medium and require the 3-dimensional dynamic analysis. Each stratum may have different thickness, modulus of elasticity, Poisson's ratio and other geometric and material properties.

So far, besides the numerical methods, a few analytical methods based on parameters that help to partly represent the behavior of machine foundations sitting on the layered medium are developed. The most famous and very old one is Winkler model known as one parameter model with many shortcomings in representing the behavior of a footing. The next development is on the two parameter models introduced to improve Winkler model (Pasternak 1954, Vlasov & Leon'tev 1966, Nogami 1987). Vlasov & Leon'tev (1966) developed a two parameter model by applying variational principles for the analysis of beams and slabs on elastic medium. Vallabhan & Aşık (2001) employed the Vlasov model for the dynamic analysis of strip and circular by assuming a one layer homogenous foundation. Aşik et al. (1996) and Aşik (1997, 2001) have developed a model for the dynamic analysis of circular and rectangular footings resting on a single layer. Aşik (1999a, 1999b, 2000, 2002) modified the model for the strip and circular footings on a layered medium for a static and dynamic loading. In this study, the model developed and verified by Aşık (1997) is improved for the dynamic analysis of the rectangular machine foundations resting on a layered medium.

2 FORMULATION

The governing equations to predict the response of a rectangular machine foundation subjected to harmonic concentrated force at center, shown in Figure 1, can be developed by using Hamilton's principle which states that

$$\delta \int_{t_1}^{t_2} (T - V)dt = 0, \qquad (1)$$

where, T is the kinetic energy of the footing and soil, and is the potential energy of the footing and soil. It is assumed that the footing and soil experience only vertical vibration (i.e. u(x, y, z, t) = 0 in the soil). Therefore, the vertical displacement in soil at any point could be defined as

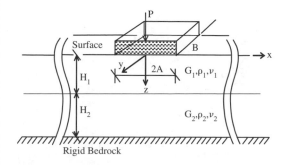

Figure 1. Cross-section of the rectangular footing and foundation.

$$\overline{W}(x,y,z,t) = \begin{cases} \phi_1(z)w(x,y,t) & 0 \le z \le H_1 \\ \phi_2(z)w(x,y,t) & H_1 \le z \le H_1 + H_2 \end{cases}.$$

Then, the following equilibrium equations and boundary conditions are obtained through variational principles by taking the variations in w, ϕ_1 and ϕ_2:

1. For $-A \le x \le A$ and $-B \le y \le B$ (soil surface under the footing)

$$(\rho_f h + \overline{m})\ddot{w} + kw = \overline{q} ; \qquad (2)$$

with boundary conditions (BC's) $2t\dfrac{\partial w}{\partial x}\delta w\Big|_{-B}^{B} = 0$;

2. For $x \le -A$ & $x \ge A$ and $y \le -B$ & $y \ge B$ (soil surface outside the footing)

$$\overline{m}\,\ddot{w} - 2t\nabla^2 w + kw = 0 \qquad (3)$$

with BC's

$$2t\frac{\partial w}{\partial x}\delta w\Big|_{-\infty}^{-A} = 0 \text{ and } 2t\frac{\partial w}{\partial x}\delta w\Big|_{A}^{\infty} = 0 ;$$

$$2t\frac{\partial w}{\partial y}\delta w\Big|_{-\infty}^{-B} = 0 \text{ and } 2t\frac{\partial w}{\partial y}\delta w\Big|_{B}^{\infty} = 0$$

3. For $0 \le z \le H_1$ (inside the soil)

$$\frac{d^2\phi_1}{dz^2} - \frac{\gamma_1}{H_1}\phi_1 = 0 \text{ and}$$

$$\left(\frac{\gamma_1}{H_1}\right)^2 = \frac{1-2v_1}{2(1-v_1)}\frac{\displaystyle\int_{-\infty}^{\infty}\int G_1(\overline{\nabla}w\bullet\overline{\nabla}w)dxdy - \int_{-\infty}^{\infty}\int\rho_1(\frac{\partial w}{\partial t})^2 dxdy}{G_1\displaystyle\int_{-\infty}^{\infty}\int w^2 dxdy} \qquad (4)$$

and for $H_1 < z < H_1 + H_2$

$$\frac{d^2\phi_2}{dz^2} - \frac{\gamma_2}{H_2}\phi_2 = 0 \text{ and}$$

$$\left(\frac{\gamma_2}{H_2}\right)^2 = \frac{1-2v_2}{2(1-v_2)}\frac{\displaystyle\int_{-\infty}^{\infty}\int G_2(\overline{\nabla}w\bullet\overline{\nabla}w)dxdy - \int_{-\infty}^{\infty}\int\rho_2(\frac{\partial w}{\partial t})^2 dxdy}{G_2\displaystyle\int_{-\infty}^{\infty}\int w^2 dxdy} \qquad (5)$$

with boundary conditions

$$\frac{d\phi_1}{dz}\delta\phi_1\Big|_0^{H_1} = 0 \text{ and } \frac{d\phi_2}{dz}\delta\phi_2\Big|_{H_1}^{H_1+H_2}$$

Where,

$$2t = \int_0^{H_1}G_1\phi_1^2 dz + \int_{H_1}^{H_1+H_2}G_2\phi_2^2 dz = G_1 H_1 c_t,$$

$$\overline{m} = \int_0^{H_1}\rho_1\phi_1^2 dz + \int_{H_1}^{H_1+H_2}\rho_2\phi_2^2 dz,$$

$$k = \int_0^{H_1}G_1\frac{2(1-v_1)}{(1-2v_1)}\left(\frac{d\phi_1}{dz}\right)^2 dz + \int_{H_1}^{H_1+H_2}G_2\frac{2(1-v_2)}{(1-2v_2)}\left(\frac{d\phi_2}{dz}\right)^2 dz = \frac{G_1}{H_1}c_k$$

and c_t and c_k are defined as:

$$c_t = \frac{1}{H_1}\left(\int_0^{H_1}\phi_1^2 dz + \frac{G_2}{G_1}\int_{H_1}^{H_1+H_2}\phi_2^2 dz\right)$$

$$c_k = H_1\left(\int_0^{H_1}\frac{2(1-v_1)}{(1-2v_1)}\left(\frac{d\phi_1}{dz}\right)^2 dz + \frac{G_2}{G_1}\int_{H_1}^{H_1+H_2}\frac{2(1-v_2)}{(1-2v_2)}\left(\frac{d\phi_2}{dz}\right)^2 dz\right)$$

Non-dimensional amplitude $\widetilde{W} = W_0 G_1 A/P_0$ is obtained from the Equation 2 and plotted with respect to non-dimensional frequency, $a_0 = \Omega A/V_1$. Governing differential equations are coupled through the parameters 2t, k and γ. Thus an iterative procedure is employed to solve them. W_0 is the amplitude of the displacement at center of a machine foundation; V_1 is the velocity of shear wave in the first layer; and P_0 is the amplitude of a concentrated load. $b_0 = M/\rho_1 A^3$ is a non dimensional mass ratio in which M is the mass of the rectangular footing.

3 DISCUSSION OF RESULTS

Three figures are plotted to understand the behavior of a rectangular footing resting on the stratified medium.

The effect of the layers with different material properties on the response of the rectangular footing is given in Figure 2. Layering effect is clearly noticed by comparing the curves 2 and 3 in Figure 2. The displacements are higher when the softer layer is at the top.

Figure 3 is plotted for the effect of E_1/E_2 ratio. The modulus of elasticity of the first layer is kept as constant for different ratios. Displacements are decreasing and the resonant frequencies are increasing as ratio is decreasing.

Figure 4 is plotted to see the effect of different layer thicknesses. Amplitudes and the resonant frequencies are affected by different H_1/H_2 ratios.

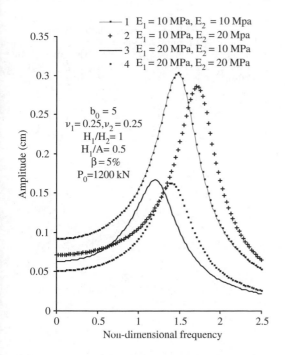

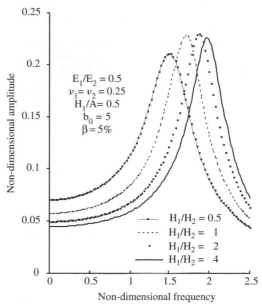

Figure 2. Comparison of vertical displacement amplitudes at center of footing.

Figure 4. Displacement amplitudes at center of footing for H_1/H_2 ratios.

REFERENCES

Aşık, M.Z. 2002. Dynamic Analysis of Flexible Circular Footings on a Layered Medium. *The Second International Conference on Advances in Structural Engineering and Mechanics*; Busan, Korea, 21–23 August, 2002.

Aşık, M.Z. 2001 & Vallabhan, C.V.G., Simplified Model for the Analysis of Machine Foundations on a Non Saturated, Elastic and Linear Soil Layer. *J. of Computers & Structures*, 79(31): 2717–2726.

Aşık, M.Z. 2000. A Mathematical Model for the Analysis of Circular Footings on a Layered Medium. *European Congress on Computational Methods in Applied Sciences and Engineering*; Barcelona, Spain, 11–14 September, 2000.

Aşık, M.Z. 1999a. Dynamic Response Analysis of the Machine Foundations on a Nonhomogeneous Soil Layer. *J. of Computers & Geotechnic*, 24(2): 141–153.

Aşık, M.Z. 1999b. Analysis of Strip Footings on a Layered Medium. *International Conference on Enhancement and Promotion of Computational Methods in Engineering Science*; Macao, China, 2–5 August, 1999.

Aşık, M.Z. 1997. 3-D Model for the Analysis of the Rectangular Machine Foundations on a Soil Layer. *6th International Symposium on Numerical Models in Geomechanics*; Montreal, Canada, NUMOG VI 1997.

Aşık, M.Z. & Vallabhan , C.V.G. & Das, Y.C. 1996. Vertical Vibration Analysis of Rigid Circular Footings on a Soil Layer with a Rigid Base. *Developments in Theoretical and Applied Mechanic*; Alabama, SECTAM, Volume XVIII 1996.

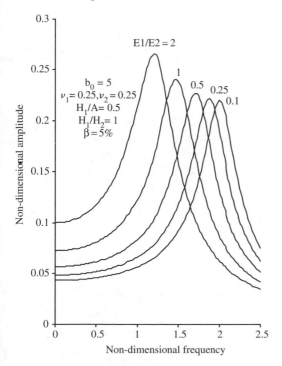

Figure 3. Displacement amplitudes at center of footing for different modulus ratios.

Nogami, T. 1987. Two-parameter Layer Model for Analysis of Slab on Elastic Foundation. *J. Engrg. Mech. ASCE*, 113: 1279–1291.

Pasternak, P.L. 1954. On a New Method of Analysis of an Elastic Foundation by Means of Two Foundation Constants. *Gosudarstvennogo Izatestvo Literaturi po Stroitelstvu i Architekutre*, Moskow.

Vlasov, V.Z. & Leont'ev, N.N. 1966. *Beams, Plates and Shells on Elastic Foundations*. Israel Program for Scientific Translations, Jerusalem, Israel.

Vallabhan, C.V.G. & Das, Y.C. 1991. The Analysis of Circular Tank Foundations. *J. Engng. Mech. ASCE*, 117(4): 789–797.

System-based Vision for Strategic and Creative Design, Bontempi (ed.)
© *2003 Swets & Zeitlinger, Lisse, ISBN 90 5809 599 1*

Numerical studies of seepage failure of sand within a cofferdam

N. Benmebarek, S. Benmebarek & L. Belounar
Civil Engineering Laboratory, Biskra University, Algeria

R. Kastner
URGC-Géotechnique, INSA Lyon, France

ABSTRACT: Urban development often requires the construction of deep retaining walls. These might be for deep basements, cut-and-cover tunnels, underground parking or underground transportation systems. The design of deep excavation is often dominated by the flow of water around the retaining wall. The seepage flow influences the stability of the wall where bulk heave, piping or failure by reduction of the earth pressure may occur. There are several methods of calculating the stability against seepage failure of the soil, but seepage failure sometimes occurs even in deep excavation designed by these methods. In this paper, the FLAC^{-2D} Codes in explicit finite difference method is used to analyze seepage failure of the sandy soil within a cofferdam subjected to an upward seepage flow. Based on this study, the conditions for seepage failure to occur by boiling or heaving of the soil behind retaining walls are clearly identified.

1 INTRODUCTION

The development of urban sites often requires the construction of deep excavations under retaining walls for deep basements, underground parking or tunnels. The design of the deep excavations is often dominated by the flow of water around the walls.

The seepage flow, induced by lowering of the groundwater table, influences the global stability of the wall and the stability of the excavation bottom where bulk heaving or boiling may occur. While the boiling takes place at the excavation level, the heaving is more spectacular and catastrophic.

There are several published methods for the assessment of bottom stability against seepage failure of soil, but failure sometimes occurs even in deep excavation designed by these methods (Tanaka 2002). More precise analysis is required to clarify the failure mechanisms.

To investigate the influence of seepage flow on stability excavation bottom, numerical analysis has been carried out using the code FLAC (Itasca 2000) for better understanding the seepage failure phenomena. From a series of numerical experiments, it is found that the failure mechanism shapes and the head loss at failure are significantly influenced by soil and interface characteristics.

2 OVERVIEW OF PREVIOUS WORK

Many methods have been proposed for examination stability against seepage failure of soil. Harza (1935) thought the maximum exit gradient is a problem at the downstream surface of seepage flow. He defined the safety factor against boiling by the ratio of the critical gradient $i_c = \gamma'/\gamma_w$ to the maximum exit gradient i.e. max, hence:

$$F_s = i_c / i_{e\ max} \tag{1}$$

The influence of seepage flow on the stability of retaining excavations was first addressed by Terzaghi (1943). From model tests, he found that within an excavation, the zone of danger of bottom heave is confined to a soil prism adjacent to the wall. He assumed, from experimental evidence, that the body of sand which is lifted by water has the shape of a rectangular prism (Fig. 1) with a width equal to the half of the wall penetration $D/2$ and horizontal base at some depth D_0 below the surface ($0 \leqslant D_0 \leqslant D$). It is assumed that at the instant of failure the effective horizontal stress on the prism vertical sides and the corresponding frictional resistance are zero.

Therefore the prism rises up and collapses as soon as the total excess water pressure U_e on the bottom of

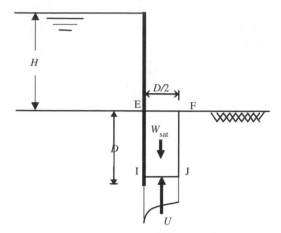

Figure 1. Failure by heaving Terzaghi.

the prism OA becomes equal to the submerged weight of the prism W'.

The safety factor against bulk heave is determined by the ratio between the submerged weight of the prism and the excess water force on the base of the prism:

$$F_s = W'/U_e = i_c / i_m \qquad (2)$$

where i_m is average gradient along EI.

The approach of Terzaghi gives a value of hydraulic head loss of $H/D = 2.82$ against seepage failure by heaving. Whereas the boilance, phenomenon which appears for a critical hydraulic gradient, occurs with a theoretical value of the hydraulic head loss equal to $H/D = 3.14 = \pi$.

McNamee (1949) identified two main types of failure:

- Local failure as failure by piping or boiling is most likely to begin at a point on the surface adjacent to the sheet pile as it lies within the shortest seepage path;
- General upheaval which involves a greater volume of soil.

McNamee states that piping occurs when a small prism of soil at the excavation level is not sufficiently heavy to resist uplift forces due to the upward flow. He define the safety factor against boiling as the ratio of the critical gradient to the exit gradient at the excavation level, hence

$$F = i_c / i_e \qquad (3)$$

The author presents charts enabling the determination of the corresponding factor of safety for a range of dimensions of the excavation.

Marsland (1953) undertook extensive model tests using both dense and loose homogeneous sands in an open water excavation. He concluded that in loose sand, failure occurs when the pressure at the pile tip is sufficient to lift the column of submerged sand near the wall of the cofferdam (the width of half the embedment is not mentioned); in dense sand, failure occurs when the exit gradient at the excavation surface reaches a critical value.

Bazant (1963) established a failure criterion with regard to the shear strength of the soil by plotting the excess hydrostatic pressure at the pile tip divided by the embedment depth of the pile against the internal friction angle of the soil.

Davidenkoff & Franke (1965) proposed a diagram based on model studies that can be used to determine the factor of safety against piping for excavations in open water with different thicknesses of the pervious stratum.

The comparison of these various approaches of safety factor in the case of a sheet pile wall driven in a semi infinite homogeneous and isotropic soil reveals significant variations being able to reach 75% (Kastner 1982).

From this overviews, it appears that seepage failure at bottom excavation has different mechanism of failure. Our aim in this paper is to propose numerical procedure to evaluate the stability against seepage failure at bottom excavation using the explicit finite difference method implemented in FLAC code.

3 NUMERICAL MODELLING PROCEDURE

3.1 Case study

This paper is concerned with the numerical study of stability against seepage failure at bottom excavation. We consider a sheet pile whose penetration depth is equal to D in homogeneous and isotropic semi-infinite soil and subject to hydraulic head H as shown in Figure 2.

The analysis was carried out using the computer code FLAC^{-2D}. FLAC (Fast Lagrangian Analysis of Continua) is a commercially available finite difference program. For the soil behaviour, a linear elastic-perfectly plastic Mohr-Coulomb model was adopted, requiring the specification of a shear modulus G, a bulk modulus K, a unit weight γ, a friction angle φ, and an angle of dilation ψ. The assessment of bottom stability against seepage failure are given for both associative and non-associative material.

In the case of a rough wall, modelling the interface between the soil and the wall is invariably an integral part of this analysis. The wall is connected to the soil grid via interface elements (Fig. 3) described by Coulomb law.

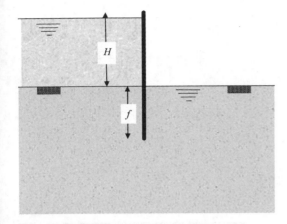

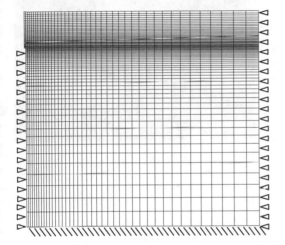

Figure 2. Case study.

Figure 3. Mesh used and limit conditions.

The interface has a friction angle, a null cohesion, a normal stiffness and a shear stiffness. The effective passive pressures in front of the wall diminish with the hydraulic head loss.

3.2 Modelling procedure

To identify the limiting cases corresponding to zero passive earth pressures and seepage failure by heaving or boiling, the following simulation procedure that includes three steps is adopted:

- First, the geostatic stresses were computed assuming the material to be elastic.
- Next, the field describing the distribution of the pore water pressure due to a hydraulic head loss was computed.

- In the third step, the mechanical response is investigated for the pore pressure distribution established in second step.

Steps 2 and 3 were repeated with hydraulic head loss progressively increased until failure.

Note that the simulation procedure does not impose a horizontal displacement of the wall as for the computation of the passive earth pressures. It should be mentioned that the evaluation of the precise value for which the passive pressures vanish is not easy to obtain. Therefore, only the state of the excavation floor stability is investigated for different ranges of the hydraulic head loss.

4 RESULTS AND DISCUSSION

The numerical results indicate that the excavation floor failure correspond almost to a bulk heave except in the case of a dense sand $\varphi \geq 40°$, a dilating material $\psi/\varphi \geq 1/2$ and a rough wall $\gamma/\varphi \geq 1/3$ where boiling would occur.

For $\psi/\varphi = 0$, a rectangular soil prism similar to that proposed by Terzaghi (1943) is observed. The prism has a width smaller than that Terzaghi's method. However, for dilating material $\psi/\varphi \geq 1/2$, a triangular soil prism is observed.

Figures 4–5 show respectively the displacement field and the corresponding distribution of maximum shear strain rates in two cases ($\psi/\varphi = 0, 1/2, \varphi = 30°$, $\delta/\varphi = 1/3$ and $H/D = 3.05$) where bulk heave is observed.

Figure 6 shows the displacement field and the corresponding distribution of maximum shear strain rates obtained in the case $\varphi = 40°$, $\gamma/\varphi = 2/3$, $\psi/\varphi = 1/2$ and $H/D = 3.2$, where the boiling phenomena is observed. In this case, the exit hydraulic gradient reaches the critical hydraulic gradient.

The critical hydraulic pressure loss H/D corresponding to the effective passive pressures zero depends on the soil friction angle and the interface soil-wall friction.

The boiling appears only for a dense and dilatant sand where $\varphi \geq 40°$, $\psi/\varphi \geq 1/2$ and roughness wall $\gamma/\varphi \geq 1/3$, whereas, rising occurs for the other cases. The boiling starting from a pressure loss of $H/D = 3.16$. This value is very close to the value of the theoretical critical pressure loss of $H/D = 3.14$.

For $\gamma/\varphi = 0$ and $\psi/\varphi = 1/2$, the critical H/D lies in the range $2.6 - 2.65$ for $\varphi = 20°$, $2.75–2.8$ for $\varphi = 30°$, and $2.9–3$ for $\varphi = 40°$. It is noted that for the various values of φ the solution of Terzaghi is $H/D = 2.82$ whereas that of Soubra et al. (1999) is $H/D = 2.78$.

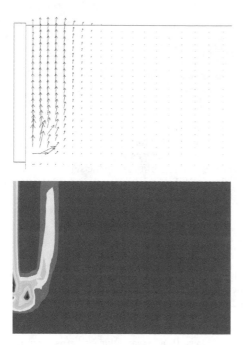

Figure 4. Displacement field and the corresponding distribution of maximum shear strain rates when $\varphi = 30°$, $\delta/\varphi = 1/3$, $\psi/\varphi = 0$, $H/D = 3.05$.

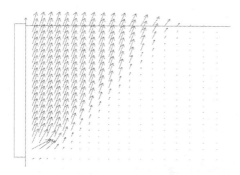

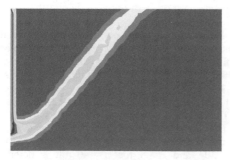

Figure 5. Displacement field and the corresponding distribution of maximum shear strain rates when $\varphi = 30°$, $\delta/\varphi = 1/3$, $\psi/\varphi = 1/2$, $H/D = 3.05$.

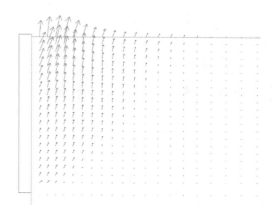

Figure 6. Displacement field and the corresponding distribution of maximum shear strain rates when $\varphi = 40°$, $\delta/\varphi = 2/3$, $\psi/\varphi = 1/2$, $H/D = 3.2$.

5 CONCLUSIONS

Many methods have been proposed for examination of seepage failure at the bottom excavation such heaving and boiling. Failure sometimes occurs even in deep excavation calculated by the conventional design methods. They may be broadly divided according to basic concepts. More precise analyses are required to clarify the cause of the failure.

The results of the numerical proposed procedure have shown that:

- The critical hydraulic loss H/D against failure by seepage flow depends on the soil friction angle, the soil dilatant angle and the interface soil/wall friction;
- The boulance appears only for a dense and dilatant sand where $\varphi \geqslant 40°$, $\psi/\varphi \geqslant 1/2$ and rough wall $\gamma/\varphi \geqslant 1/3$ which induces a gradient of exit equal to the critical hydraulic gradient, whereas, heaving occurs for the other cases;
- For a dilating material, a failure by heaving a triangular prism is obtained, whereas, a rectangular prism for the other cases;

- The rectangular prism has a width smaller than that Terzaghi's method.

REFERENCES

Bazant, Z. 1963. Ergebnisse der Berechnung der Stabilitat gegen Hydraulischen Grundbruch mit Hilfe der Elektronen-Rechenanlage. *Proc., Int. Conf. on Soil Mech. and Found, Engrg., Budapest*: 215–223.

Davidenkoff, R.N. & Franke, O.L. 1965. Untersuchung der raumlichen Sickerstromung in eine umspundete Baugrube in offenen Gewassern, *Die Bautechnik*, (9): 298–307.

Harza, L.F. 1935. Uplift and seepage under dams and sand, *Trans. of ASCE, N°100*: 1352–1406.

Itasca Consulting Group 2000. FLAC^{2D} – *Fast Lagrangian Analysis of Continua. User's Manuel*, Minneapolis: Itasca.

Kastner, R. 1982. *Excavations profondes en site urbain: Problèmes liés à la mise hors d'eau. Dimensionnement des soutènements butonnés*, Thesis of Docteur ès Sciences, INSA Lyon & University Claude Bernard, Lyon I.

Marsland, A. 1653. Model experiments to study the influence of seepage on the stability of a sheeted excavation in sand. *Géotechnique, The Institution of Civil Engineers, London, Vol. 7, N° 4*: 223–241.

McNamee, J. 1949. Seepage into a sheeted excavation. *Géotechnique, The Institution of Civil Engineers, London, Vol. 1, N° 4*: 229–241.

Soubra, A.-H, Kastner, R. & Benmansour A. 1999. Passive earth pressures in the presence of hydraulic gradients. *Géotechnique, The Institution of Civil Engineers, Vol. 49, N° 3*: 319–330.

Tanaka, T. 2002. Boiling occured within a braced cofferdam due to two-dimensionally concentrated seepage flow, *3rd International Symposium, Geotechnical Aspects of Underground Construction in Soft Ground, October, Toulouse, France*: 23–25.

Terzaghi, K. 1943. *Theoretical soil mechanics*. John Wiley & Sons, New York.

SYMBOLS

φ = angle of internal friction of the soil
ψ = dilation angle of the soil
γ' = submerged unit weight of the soil
δ = angle of friction at the soil-wall interface
γ_w = unit weight of water
c = soil cohesion
D = penetration depth of the sheet pile
G = shear modulus of the soil
H = total hydraulic head loss
i_c = Critical gradient
$i_{e\,max}$ = Maximum exit gradient
i_m = average hydraulic gradient along the prism
K = bulk modulus of the soil
K_n = interface normal stiffness
K_s = interface shear stiffness

Criterion of crack initiation and propagation in rock

Y.L. Chen & T. Liu
Department of Civil Engineering, Shanghai University, Shanghai, China

ABSTRACT: In this paper, the criterion and mechanism of crack initiation and propagation under creep condition were investigated using specimens collected from sandstone rock formations outcropping in Emei Mountain, the Sichuan Province of China. Cuboid specimens under three point bending were used in this investigation. All specimens were classified into four sorts and used for Mode-I fracture or creep fracture tests. The experimental result shows that due to creep deformation, rock crack will inevitably initiate and propagate under a load of K_I, which is less than fracture toughness K_{IC} but not less than a constant (marked as K_{IC2}). K_{IC2} indicates the ability of rock to resist crack initiation and propagation under creep conditions and is less than fracture toughness K_{IC}, it is defined as creep fracture toughness in this paper. K_{IC2} should be considered as an important parameter on design and computation of rock engineering.

1 INTRODUCTION

Rock fracture mechanics can give explanation of rock failure, crack occurrence and propagation for a wide range of geological and engineering situations. The principles of fracture mechanics have been employed widely and applied successfully to predicting rock fracture initiation and to solving rock engineering problems. (Whittaker et al. 1992). The creep behavior of rock is a very important phenomenon for many engineering geological problems, too. The previous investigation demonstrates that creep behavior has been found in some rocks such as rock salt, sandstone, and even some hard rocks. All these had been observed by site survey of Humen bridge project (Chen 1996) and so on (Cristescu & Hunsche 1998).

In some rock engineering problems, we had found that some cracks can not initiate and propagate at beginning, however, the cracks will initiate and propagate after a time-interval of sustained loading under creep condition. Up to now, research results about criterion of crack initiation and propagation for rocks under creep condition have never been found. In this paper, we will discuss this problem.

Here the cuboid specimens under three point bending were used, the specimen size is about $5\,cm \times 5\,cm \times 25\,cm$. All specimens were fabricated from three rock materials, namely

(i) Gray fine-grained sandstone, as shown in Figure 1(a).

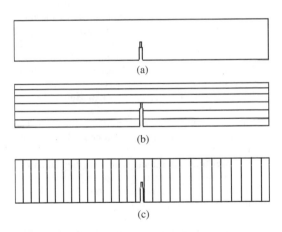

Figure 1. Illustration of the cuboid specimens for three point bending test: (a) fabricated from fine-grained sandstone without bedding planes; (b) fabricated from red fine-grained sandstone, with pre-existing cracks perpendicular to the bedding planes; (c) fabricated from red fine-grained sandstone, with pre-existing cracks parallel to the bedding planes.

(ii) Red fine-grained sandstone, as shown in Figure 1(a).

(iii) Red fine-grained sandstone with bedding planes, as shown in Figures 1(b),(c).

In the above-mentioned three materials, (a) and (b) have no bedding planes.

The specimens are classified into the following four sorts.

Specimen A – fabricated from gray fine-grained sandstone without bedding planes, as shown in Figure 1(a).

Specimen B – fabricated from red fine-grained sandstone without bedding planes, as shown in Figure 1(a).

Specimen C – fabricated from red fine-grained sandstone, with pre-existing crack perpendicular to the bedding planes, as shown in Figure 1(b).

Specimen D – fabricated from red fine-grained sandstone, with pre-existing crack parallel to the bedding planes, as shown in Figure 1(c).

The grain size seems to be a crucial factor in the measurement of fracture toughness of rocks since it not only affects the size of crack tip FPZ, but also influences the fracture parameters (namely fracture energy, fracture toughness, etc). Hence, the grain size cannot be ignored when evaluating fracture parameters of rocks. According to the obtained research results, the minimum specimen dimension must be at least 20–40 times the grain diameter in order to obtained a valid K_{IC} value for rock material. Since all specimens were fabricated from fine-grained sandstone rocks with a grain size of about $1/16$–$1/8$ mm, the grain size satisfies the requirements of the measurement of fracture toughness of rocks. In metallic materials, crack growth behavior under creep conditions may be classified as either creep-ductile or creep-brittle, the rock materials used in this paper are creep-brittle materials.

This paper firstly presents an experimental study into the behavior of crack development of fine-grained sandstone under three point bending test. The test was performed on a servo-controlled testing machine.

In order to study the criterion and mechanism of crack initiation and propagation for rocks under creep conditions, this paper secondly presents an experimental study into the behavior of crack development for fine-grained sandstone under creep condition by three point bending test. In order to avoid the drift instability of servo system under creep conditions, we have developed a rigid mechanical three point bending creep device. In this device, the displacements are measured by micrometers. It is loaded with weights. The experimental results show that due to creep deformation, rock crack will inevitably initiate and propagate under a load of K_I, which is less than K_{IC} but not less than K_{IC2}. K_{IC2} indicates the ability of rock materials to resist crack initiation and propagation under creep condition and is less than fracture toughness K_{IC}, it is defined as creep fracture toughness in this paper. And K_{IC2} should be taken as an important parameter on design and computation of rock engineering.

2 LABORATORY TEST

The rock specimens used in the tests are fabricated from Cambrian fine-grained sandstone rocks, which were collected from Emei, the Sichuan Province of China. And single-crack specimens were adopted and illustrated in Figure 1. As it is shown in Figure 1, specimen A and specimen B are fabricated from homogeneous and isotropic fine-grained sandstone materials, and specimen C and specimen D are very close to plane-isotropic materials. The crack was pre-fabricated and includes two parts: the wide part and the thin one .The wide part was pre-made by saw blade and saw web, and the thin part, whose width was about 0.2–0.3 mm, was made by very thin diamond saw web. In the present investigation, the labor tests were performed on a servo-controlled testing machine and a self-designed three-point bending creep test frame. The sketch map of three-point bending test is shown in Figure 2.

The test includes two parts, namely conventional fracture test and creep fracture test. And about 60 specimens are determined. The quantitative distribution of specimens is shown in Table 1.

2.1 Conventional fracture test

The conventional fracture test was performed on a servo-controlled testing machine. The applied load was measured by a pressductor. The Load Point Displacement (LPD) and the Crack Mouth Opening Displacement (CMOD) were monitored by linear transducers. Load versus LPD curves and Load versus CMOD curves were plotted automatically by

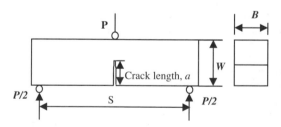

Figure 2. Geometries of cuboid specimen in three-point bending test.

Table 1. The quantity of various rock samples used in three point bending test.

Type of test	A	B	C	D
Conventional fracture test	2	0	3	3
Fracture test under creep condition	0	32	10	10

Table 2. The fracture toughness and corresponding dimension of the specimens used in the conventional fracture test.

Type	No.	W cm	B cm	S cm	a cm	P_{max} KN	K_{IC} MN·m$^{-3/2}$	Aver. K_{IC} MN·m$^{-3/2}$
Specimen A	A1	5.09	4.98	20.0	1.78	0.76	0.493	0.494
	A2	5.09	5.04	20.0	1.91	0.77	0.495	
Specimen C	C1	5.11	5.01	20.0	2.00	0.44	0.475	0.425
	C2	5.10	5.07	20.0	2.14	0.47	0.518	
	C3	4.98	5.09	20.0	1.95	0.41	0.281	
Specimen D	D1	4.98	5.19	20.0	1.97	0.17	0.113	0.243
	D2	5.06	5.12	20.0	2.05	0.23	0.250	
	D3	5.00	4.87	20.0	1.95	0.32	0.365	

Notes: W = Specimen width, B = Specimen thickness, S = The span between two supports under three point bending test, a = Crack length, P_{max} = Maximum load, and K_{IC} = Stress intensity factor for mode I.

testing machine. The sketch maps of the specimens are shown in Figures 1–2. The tested result is shown in Table 2.

2.2 Creep fracture test

2.2.1 Test device and samples cutting
The creep fracture test was carried out on a three-point bending creep test frame, which was self-designed and self- made. In the test we adopted three types of specimens, namely, specimen B, specimen C, specimen D. All specimens are cut from the red fine-grained sandstone materials. The quantitative distribution of specimens is listed in Table 1. The duration of sustained loading for a specimen under creep fracture test is from several hours to several months or even more than a year.

2.2.2 Test result
From the three point bending creep fracture test, the load-point deformation (LPD) versus time curves for various sorts of rock specimens under different levels of SIF K_{IC} are obtained. And some of the curves are shown in Figure 3.

From Figure 3(a) we can find

(i) When $K_I = 0.090\,\mathrm{MN \cdot m^{-3/2}}$, no obvious creep deformation occurs in rock material;
(ii) When $K_I = 0.103\,\mathrm{MN \cdot m^{-3/2}}$ or $K_I = 0.130\,\mathrm{MN \cdot m^{-3/2}}$, the creep deformation is apparent in rock material, after a time interval of sustained loading, however, it stops developing;
(iii) When $K_I = 0.179\,\mathrm{MN \cdot m^{-3/2}}$ or $K_I = 0.192\,\mathrm{MN \cdot m^{-3/2}}$ or $K_I = 0.195\,\mathrm{MN \cdot m^{-3/2}}$, the creep deformation is very obvious for rock material, and it will develop continuously until the crack initiates and propagates, and the specimen fractures rapidly after;

(iv) When $K_I = 0.250\,\mathrm{MN \cdot m^{-3/2}}$, the specimen fractures quickly.

Similar behavior can be observed from Figure 3(b) and Figure 3(c).

2.2.3 Test conclusion
Following criterion can be induced from the creep fracture test.

There exist three characteristic values in Mode I creep fracture test for red fine-grained sandstone materials under three point bending test, denoted as K_{IC1}, K_{IC2}, K_{IC3}, when

(i) $K_I < K_{IC}$, no obvious creep deformation occurs in red fine-grained sandstone materials;
(ii) $K_{IC1} \leqslant K_I < K_{IC2}$, the creep deformation is apparent in red fine-grained sandstone materials, after a time interval of sustained loading, however, it will stop developing;
(iii) $K_{IC2} \leqslant K_I < K_{IC3}$, the creep deformation is very obvious for red fine-grained sandstone materials, and it will develop continuously until the crack initiates and propagates, and the specimen fractures rapidly after;
(iv) $K_I \geqslant K_{IC3}$, the specimen fractures quickly.

The determined characteristic values for red fine-grained sandstone materials used in the creep fracture test are shown in Table 3. The conclusions obtained in the three point bending creep fracture test agree with the results of the pervious direct tensile creep fracture test.

3 CONCLUSIONS

In this paper, the following conclusions can be drawn.
(i) The parameter K_{IC2}, which is defined as creep fracture toughness of rock materials and smaller than

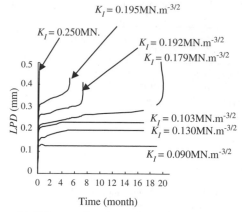

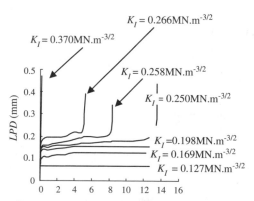

(a) *LPD* versus time curves for specimen B

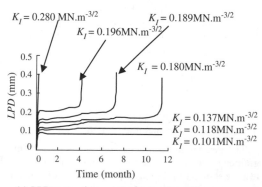

(b) *LPD* versus time curves for specimen C

(c) *LPD* versus time curves for specimen D

Figure 3. Variation of *LPD* with time for three types of sandstone specimens under three point bending creep test. *LPD* = Load-point displacement.

Table 3. The characteristic values of crack initiation and propagation for red fine-grained sandstone from Emei under creep conditions.

Type of specimen	K_{IC1} (MN·m$^{-3/2}$)	K_{IC2} (MN·m$^{-3/2}$)	K_{IC3} (MN·m$^{-3/2}$)
Specimen B	0.103	0.179	0.250
Specimen C	0.169	0.250	0.370
Specimen D	0.118	0.180	0.280

K_{IC}, is a very important parameter for rock engineering. Due to creep deformation, rock material can fracture under a condition of $K_{IC2} < K_I < K_{IC}$;

(ii) Based on this study, it is recommended that the experimental and theoretical studies of creep fracture behavior under mode II and mode III and mixed mode loadings in rock materials can be performed. Some other important problems should also be paid great attention, for instance, the creep fracture behavior under a dynamical load in rocks, the engineering application of above-mentioned theories and so on.

ACKNOWLEDGEMENTS

The financial support from the Natural Science Foundation of Shanghai (No. 02ZF14036) for this study is gratefully acknowledged.

REFERENCES

Cristescu, N.D. & Hunsche, U. 1998. Time effects in rock mechanics. *Chichester: John Wiley & Sons*,1–30.
Silberschmidt, V.G. & Silberschmidt, V.V. 2000. Analysis of cracking in rock salt. *Rock Mechanics and Rock Engineering*, 33(1): 53–70.
Chen, Y.L. 1996. Creep fracture of rock. *Chinese Journal of Rock Mechanics and Rock Engineering*, 15(4): 323–327 [in Chinese].
Whittaker, B.N., Singh, R.N. & Sun, G. 1992. *Rock Fracture Mechanics*, Elsevier Science Publishers B.V., Amsterdam, The Netherlands, 157–199.
Yuan, L.W. 1984. *Rheology*, Science Press, Beijing, China, 74–95 [in Chinese].

16. Structural problems and safety devices of road and railways

Load-carrying capability of PC beam during shear destruction by impact

N. Furuya, I. Kuroda & K. Shimoyama
National Defense Academy, Yokosuka, Kanagawa, Japan

S. Nakamura
Nihon Samicon Co. Ltd., Niigata, Japan

ABSTRACT: The authors conducted static-loading test, rapid-loading test and impact-loading test in order to make clear about shear failure behavior of PC beams by the force which was given near its support. In this report, such themes will be discussed as the maximum force to be needed for destruction of PC beams in shear mode, time-domain change of the force and displacement of the beam, energy absorption by the beam, comparison of influence between kinetic energy and momentum of falling steel weight onto the beam, and finally, influence of amount of stirrups in the beam on shear destruction.

1 PREFACE

Rock shed that protects roads from falling rocks or stones have been constructed with either prestressed concrete (PC), reinforced concrete (RC) or steel, among which PC rock shed has recently become popular in Japan because it can have compact cross section despite of its high strength; it can shorten construction time by utilizing precast method and from other reasons.

Researches regarding to concrete-made rock shed against impact have mainly focused upon flexural behavior (e.g., Enrin, Katsuki, Ishikawa & Ohta 2000), because the structure is usually designed so that flexural failure may go before shear destruction. However, it was reported that a PC rock shed experienced shear destruction when an impact load was given by drop weight falling near a support of the beam (Public Works Research Institute Structure Laboratory 1996), and we cannot thus deny risk for PC rock shed being designed in the above mentioned manner to have fore-running shear destruction under certain conditions.

As far as we know, only research done about shear destruction of PC beams by impact has been ours (Kuroda, Shimoyama, Furuya & Nakamura 2001) and (Shimoyama, Kuroda, Furuya & Nakamura 2002). In the former study, destruction pattern, maximum load, etc. were examined in both static-loading test and rapid-loading test (a certain static-load was given to a specimen during about 10 ms), and comparison was made between the measured load and calculated value by the currently used code in Japan.

In the latter study, impact loads were repeatedly given near a support point of a PC beam while the dropping height of steel weight was gradually increased until the beam was thoroughly destroyed in shear mode.

Here to an occasion of ISEC-02, both study results are summarized, and discussion is going to be made especially upon a question, which physical quantities (the momentum or the kinetic energy of a falling weight to a PC beam) govern shear destruction of PC beams. In this report, influence of arrangement of stirrups in the shear span on the shear destruction is also to be focused.

2 OUTLINE OF EXPERIMENT

Figure 1 shows shape, dimension and layout of inside steels of a pre-tensioned PC beams that were tested in our research. The total length of the beam is 2000 mm so as to secure sufficient anchoring length of PC wires, and anchoring plates with 10 mm thickness are provided to both ends of a beam for the same purpose. Two PC wires are arranged in 2 rows; the total number thus becomes four and each PC wire was given effective prestressing force of 66.6 kN. Condition between the wires and concrete was fully bonded. Table 1 shows property of the used material.

As for stirrups, D6 re-bars are placed at interval of 100 mm along the entire length of the beam. This amount of reinforcement against shear can be considered as weak level among actually designed real PC

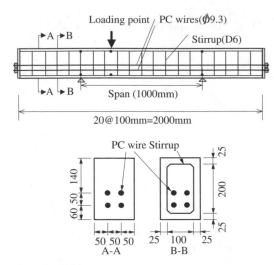

Figure 1. PC beam.

Table 1. Property of materials.

Material	Value
Concrete	W/C = 35%
	f'c = 59.7 N/mm²
PC Wire (SWPR7AN)	Strength at 0.2% of
	elongation: 1760 N/mm²
Stirrup and Erection Bar (D6)	f_y = 388 N/mm²

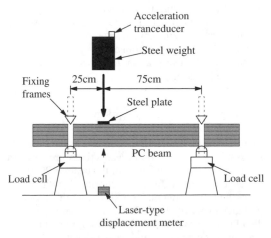

Figure 2. Impact loading test.

rock shed. However in NS series specimens, the stirrups are not placed in the shear span that is between the loading point and the nearer support.

Figure 2 shows how impact-loading test was carried out (see also Table 2), in which the load was

Table 2. Test cases.

Specimen	Mass of steel weight (kg)	Stirrups in shear span
S-type		
S-1000	1000	ctc: 100 mm
S-300	300	ctc: 100 mm
NS-type		
N-1000	1000	none
N-300	300	none

Table 3. Measured item.

Item	Equipment	Performance
Acceleration of steel weight	Acceleration tranceducer	Range: 500 G Frequency: 10 kHz
Reaction at shoe	Load cell	Capacity: 981 kN
Displacement	Laser-type displacement meter	Range: 300 mm Frequency: 915 Hz

repeatedly given to a PC beam while drooping height of the steel weight was gradually increased until the beam was thoroughly destroyed. Placement of the specimen, especially for the total span of the PC beam and relative position of the loading point and two supports, was kept same in all the experiments in order to guarantee the shear destruction, not flexural destruction. Table 3 displays measured items in the impact-loading test, which are basically same to the static-loading test and rapid-loading test.

3 RESULTS OF EXPERIMENTS AND INVESTIGATION

3.1 Time-domain response of impact force and displacement

Figure 3 shows the time-domain load change being measured on DF-1 specimen in the rapid-loading test, which has the stirrups with a pitch of 100 mm.

On the other hand, Figures 4a–4d show results by the first drop of the impact-loading test, in which a 1000 kg weight was dropped from 20 cm height onto a virgin PC beam. The impact force shown in Figure 4a was obtained by multiplying the measured acceleration at the steel weight and the mass of the weight. At least three independent peaks were found, among which the first peak was caused by the 1st collision of the falling steel weight. However, some degree of structural soundness was remained in the PC beam and the weight thus bounded off to make 2nd and 3rd collisions.

The reaction shown in Figure 4b is the sum of individually measured support points reaction.

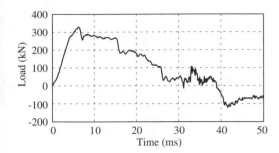

Figure 3.   Time-domain response of load.

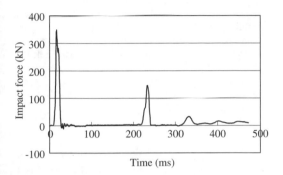

Figure 4a.   Time-domain response of impact force.

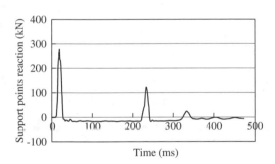

Figure 4b.   Time-domain response of reaction.

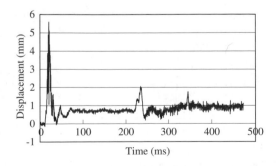

Figure 4c.   Time-domain response of displacement.

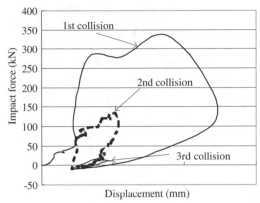

Figure 4d.   Relationship between impact force and displacement.

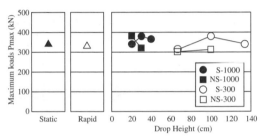

Figure 5.   Maximum loads.

Figure 4c tells that the 1st collision induced approximately 5.5 mm of deflection (measured at the lower surface of the PC beam beneath the loading point), but the deflection recovered to around 0.8 mm, which in turn became the residual displacement to be discussed later.

Based on Figures 4a and 4c, the relationship between the impact force and displacement is obtained as being shown in Figure 4d. It is clear in the figure that the 1st collision having most expansive hystereisis curve gave the largest damage to the PC beam and that the area of the curve of the rebound collisions was only 10% of the 1st one, from which we concluded that progress of the plastic deformation or destruction of the PC beam caused by the after-shocks could be neglected.

3.2   *Comparison of maximum load*

Figure 5 shows comparison of the maximum loads measured in the static-loading test, rapid-loading test and the 1st collision at impact-loading test. The maximum load can be concluded to be almost same. This point is especially noteworthy for the impact-loading test because some degree of structural damage might have been already accumulated in the specimen by repeatedly falling steel weight.

3.3 Relation between impact force and displacement

Because ignorance of the influence by the 2nd and 3rd collisions seems to be allowed as being mentioned in 3.1, the relationship between impact force and displacement for all the impact-loading tests is drawn as shown in Figure 6a–6d. A tendency that the curve turns toward the origin of coordinate after it passes the peak is observed unless the drop is final one, but the deflection observed in the final drop (3rd for S-series and 2nd for NS-series) cannot recover anymore, which fact tells that the PC beam is very near to the ultimate state at this drop.

3.4 Destruction pattern

Though minor shear cracks that directly connects the loading point and support appear (see Fig. 7a),

its opening is limited. However, they are fully developed as being shown in Figure 7b after the final drop of the steel weight was given. Every beam is concluded to reach the ultimate state in the shear destruction mode, not flexural mode, because flexural cracks seen on the lower portion of specimen are only 2 to 3 in the number and their width is less than 1 mm

The dropping height of steel weight in the impact-loading test (20–30–40 cm for 1000 kg weight, and 66–100–133 cm for 300 kg weight) was decided by an engineering judgement. We thus conducted one additional experiment by another sequence of the dropping height (10–20–30– ⋯ –140 cm for 300 kg weight), because the dropping height mentioned above could have possibility to influence on experimental results.

The result of this addition is flexural destruction as shown in Figure 8, and this fact seems to tell that final destruction pattern is effected by path of the loading and that appearance of the shear destruction mode might need quantities of energy from the beginning.

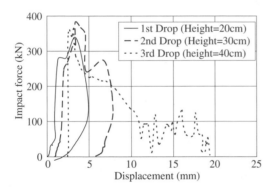

Figure 6a. Relation between impact force and displacement (S-1000).

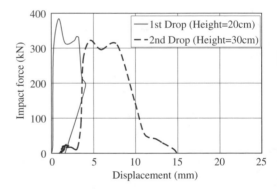

Figure 6c. Relation between impact force and displacement (NS-1000).

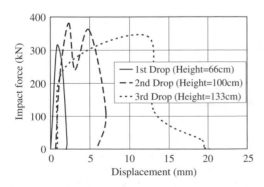

Figure 6b. Relation between impact force and displacement (S-300).

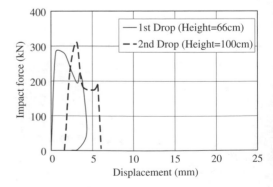

Figure 6d. Relation between impact force and displacement (NS-300).

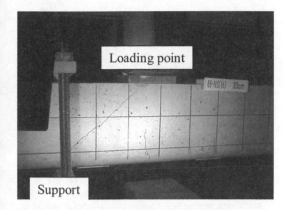

Figure 7a.   Minor shear crack after 1st drop.

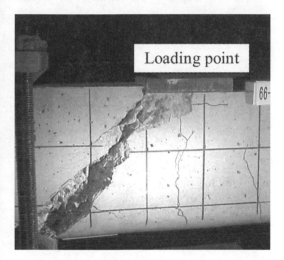

Figure 7b.   Destruction pattern.

Figure 8.   Final destruction pattern (flexural destruction).

3.5   *Comparison of influence between kinetic energy and momentum*

As far as we know, there has been no established conclusion about which quantity, the kinetic energy or momentum, of a falling object governs shear destruction of a PC beam.

As being shown in Figure 9a–9d, we draw the relationship between the kinetic energy or momentum and the residual displacement (see Fig. 6a–6d) at every drop. Same magnitude of the kinetic energy

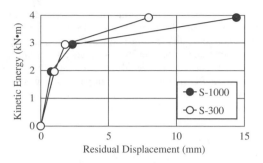

Figure 9a.   Relation between residual displacement and kinetic energy (S-type beams).

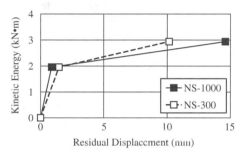

Figure 9b.   Relation between residual displacement and kinetic energy (NS-type beams).

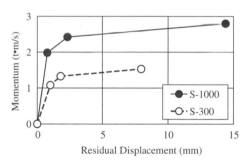

Figure 9c.   Relation between residual displacement and momentum (S-type beams).

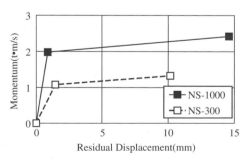

Figure 9d.   Relation between residual displacement and momentum (NS-type beams).

seems to generate the same value of residual displacement, but the same magnitude of momentum does not generate the same residual displacement. We thus concluded that the kinetic energy of a falling object might have larger influence on shear destruction of a PC beam receiving impact load near its support, though the number of experiments was limited.

3.6 Influence by stirrups

As being already shown in Figure 5, it is interesting that presence or absence of the stirrups in the shear span seemed to have little influence on the maximum load. But, existence of the stirrups raised deformation and thus energy absorption about twice (8.4 kN m vs 4.7 kN m) as shown in Figures 6a–6d and 9a–9d.

4 CONCLUSIONS

1) The maximum value of the impact load is almost same regardless the weight mass, existence of stirrups and structural damage accumulated by the foregoing drop in the impact-loading test, and the value is also approximately same to the results by static-loading test and the rapid-loading test.
2) In the relation between impact load and displacement, a tendency that the curve turned toward the origin of coordinate after it passed the peak is obvious, but the deflection observed in the final drop cannot recover anymore.
3) The ultimate shear failure of PC beams seems to depend on the kinetic energy of steel weight rather than the momentum.
4) Contradictory to the conclusion of 1), the presence of stirrups raises deformation and energy absorption of PC beams against impact about twice.

REFERENCES

Enrin H., Katsuki S., Ishikawa N. & Ohta T. 2000. The experimental study on dynamic flexural toughness of post-tensioned prestressed reinforced concrete beam under high speed loading, *Concrete Research and Technology, Vol. 11, No. 1: pp 19–27*. (in Japanese). Japan Concrete Institute: Tokyo.

Public Works Research Institute Structure Laboratory. 1996. Joint research report on design method of PRC rock shed, *Joint Research Report No. 148*. (in Japanese). Public Works Institute, Ministry of Construction: Tsukuba.

Kuroda I., Shimoyama K., Furuya N. & Nakamura S. 2001. Static and rapid shear loading experiment on PC beam members, *Journal of Structural Engineering, Vol. 47A: pp 1299–1308*. (in Japanese). Japan Society of Civil Engineers: Tokyo.

Shimoyama K., Kuroda I., Furuya N. & Nakamura S. 2002. Shear failure behavior of PC beams by drop weight impact experiment, *Concrete Research and Technology, Vol. 13, No. 1: pp 109–118*. (in Japanese). Japan Concrete Institute: Tokyo.

System-based Vision for Strategic and Creative Design, Bontempi (ed.)
© 2003 Swets & Zeitlinger, Lisse, ISBN 90 5809 599 1

Study on performance of aluminum alloy-concrete hybrid guard fence

T. Hida, R. Kusama & B. Liu
Department of Civil Engineering, Nagoya University, Japan

Y. Itoh
Department of Geotechnical and Environmental Engineering, Nagoya University, Japan

T. Kitagawa
Center for Integrated Research in Science and Engineering, Nagoya University, Japan

ABSTRACT: The tensile experiments were conducted to investigate the strain rate effect of aluminum alloy. Then, the finite element models of the aluminum alloy-concrete hybrid guard fence and a truck were developed to simulate the collision behaviors observed in the field full-scale collision experiment. The results of the numerical analyses were compared with that of the experiment and it was shown that the results of the numerical analyses agreed well with that of the experiment. The transfer of the collision energy and the adsorption of the energy by the hybrid fence were discussed.

1 INTRODUCTION

In recent years, guard fences are required to have more safety performance because the size of trucks becomes larger than ever and their gravity center becomes higher. A new code for the design of guard fences was implemented and issued in April 1999, in Japan (JRA 1999). The performances of guard fences were prescribed in the new code: (i) prevention of vehicle's deviation from the road, (ii) safety of persons in a vehicle, (iii) guiding a vehicle to a way, and (iv) prevention of spreading out the broken pieces. New structures and materials are able to be adopted in the design of guard fences as long as the performances mentioned above are satisfied. In the new code of guard fences, full-scale collision experiments for checking the performances of guard fences are also required in design of guard fences. However, it is difficult to investigate parametrically the performances of guard fences with the full-scale collision experiment because of enormous cost and time consumption. Numerical analyses to complement the full-scale experiment are then necessary.

The aluminum alloy-concrete hybrid guard fence shown in Figure 1 is one of the new guard fences designed in accordance with the new code. The hybrid guard fences were developed for the purposes of the improvement of a driver's view and a driver's feelings from oppression and surrounding by guard fences.

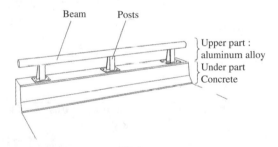

Figure 1. Aluminum alloy-concrete hybrid guard fence.

The upper part of the hybrid guard fence is made of aluminum alloy and the under part is made of concrete. The upper part plays a role to absorb the collision energy and the under part prevents the deviation of vehicles. The full-scale collision experiment for the hybrid fence was conducted in September 2001 at the Public Work Research of Japan (PWRC 2001). In this study, the numerical analyses on the truck-collision performances of the hybrid fence and on the behavior of the collision truck are carried out. Firstly, tensile experiments are conducted to investigate the strain rate effects of the aluminum alloy. Secondly, the finite element models of the guard fence and a truck are developed to simulate the collision behaviors observed in the field experiment. The FEM analyses are performed with a nonlinear dynamic analysis software (LS-DYNA). The results of numerical analyses are compared with the experimental results.

2 DYNAMIC TENSILE EXPERIMENT

2.1 Objectives and procedure of experiment

Dynamic tensile experiments were conducted to obtain the stress–strain relationship of aluminum alloy, in which the strain rate effects were reflected. Aluminum alloy A6061S-T6 and AC4CH-T6 are used in the posts and the beam of the aluminum alloy-concrete hybrid guard fence, respectively. A servo-valve-type material test machine (MTS) was used in this experiment. The strain rate in the dynamic tensile experiments was in the range of 10^{-4}–$10^{-0.5}$ (1/s).

2.2 Results of aluminum alloy coupon experiments

Figures 2 and 3 show the static stress-strain relationship for A6061S-T6 and that for AC4CH-T6, respectively.

In the case of both of the aluminum alloy, the stress changed smoothly as increase of the strain and upper yield point, lower yield point and yield plateau were not shown.

Figures 4 and 5 show the strain rate effects on yield stress of 0.2% offset. The vertical axes of Figure 4 and 5 are the dynamic response factor σ_d/σ_s where σ_d is the dynamic yield stress and σ_s is the static yield stress. In order to calculate σ_d/σ_s, it was assumed that σ_s was of the case under the strain rate of 10^{-4} (1/s). In both results, the strain rate effects on yield stress were at most 3%, and the strain rate effect on the aluminum alloy was very small. The small strain rate effect on the aluminum alloy hardly influenced the results of the numerical analyses (Itoh et al. 2001). Therefore, the strain rate effects does not need to be included in the numerical analyses on the aluminum alloy-concrete hybrid guard fence.

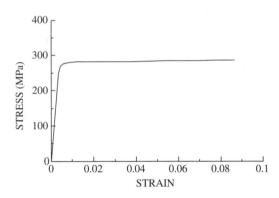

Figure 2.   Stress–strain relationship of A6061S-T6.

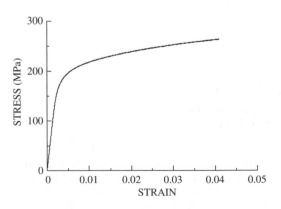

Figure 3.   Stress–strain relationship of AC4CH-T6.

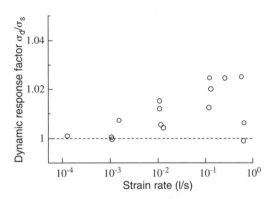

Figure 4.   Strain rate effect on yield stress of 0.2% offset of A6061S-T6.

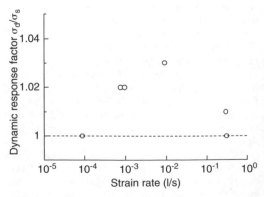

Figure 5.   Strain rate effect on yield stress of 0.2% offset of AC4CH-T6

3 FEM MODELS

3.1 Truck model

Figure 6 represents the FEM truck model that is used in this study, which was constructed by the authors (Itoh et al. 1998). The truck model was constructed based on the truck that was used in the field experiment. In this model, the number of nodes and elements were 8414 and 8498, respectively. The weight of the truck was 25 tf and the most parts such as the truck-frame, the driving room, the cargo and the tiers were mechanically and geometrically characterized in the model. The length, the width and the height of the model were 11890 mm, 2470 mm and 3300 mm, respectively. The height of gravity center was 1.4 m. The Young's modulus of steel was 206 GPa, and that of aluminum was 70 GPa. The Poisson's ratios of steel and aluminum were 0.30 and 0.34, respectively. The yield stress of steel was 235 MPa. In order to simulate the damage of the cabin in detail, many meshes were assigned to the cabin. The shapes of the bumper and the freight pallet of the model were the same as those of the truck used in the full-scale collision experiment.

3.2 Aluminum alloy-concrete hybrid guard fence model

Figure 7 shows the FEM model of the aluminum alloy-concrete hybrid guard fence. The height from the

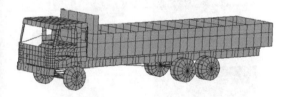

Figure 6.   Truck FEM model.

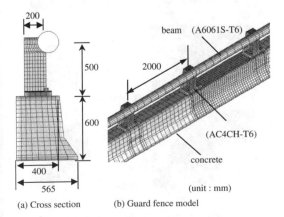

(a) Cross section   (b) Guard fence model

Figure 7.   FEM model of the hybrid guard fence.

center of the beam to the top face of concrete was 500 mm, the height of the concrete body was 600 mm. The interval between posts was 2000 mm and the total number of spans was 16. The beam model of the guard fence was made with shell elements. The beam and the posts of the guard fence were made of A6061S-T6 and AC4CH-T6, respectively. The Young's modulus of A6061S-T6 was 68 GPa and that of AC4CH-T6 was 72 GPa. The Poisson's ratios of A6061S-T6 and AC4CH-T6 were 0.3 and 0.33, respectively. The yield stress of 0.2% offset of A6061S-T6 and AC4CH-T6 were 277 MPa and 173 MPa, respectively. The Young's modulus and Poisson's ratios of concrete were 24.5 GPa and 1/6, respectively. The concrete volume modulus was 12.3 GPa. The concrete compressive and tensile strengths were 35.7 MPa and 2.90 MPa, respectively.

The aluminum alloy was assumed to be an isotropic elasto-plastic material following the von Mises yielding criterion. The strain rate effect on the stress-strain relationship of the concrete was not considered. It was assumed that the concrete corresponded to a general elasto-plastic material. It was also assumed that the stress of the concrete in the compressive side increased proportionally to the strain, and that in the tensile side maintained the cut-off stress when the strain became larger than the yield strain. All of the material properties mentioned above were obtained with the experiments. The total numbers of the nodes and elements of the hybrid guard fence model were 20469 and 19617, respectively. The boundary condition at the concrete bottom was fixed end.

3.3 Collision conditions

The collision condition of numerical analysis is shown in Figure 8. The collision conditions were set so as to be the same as those in the full-scale experiment, and the impact velocity was 84.6 km/h and the impact angle was 20.07 degree.

The performance of a guard fence is usually defined by its impact energy, which is a quantitative criterion to reflect the strength of a guard fence in the point of energy. The design impact energy can be defined as

$$Is = \frac{1}{2} \cdot m \cdot \left(\frac{V}{3.6} \cdot sin\theta \right)^2 \qquad (1)$$

where Is is the design impact energy, m is the quality of the design impact vehicle, V is the collision speed of the vehicle and θ is the impact angle. In the case of above-metioned condition, the impact energy was 653 kJ.

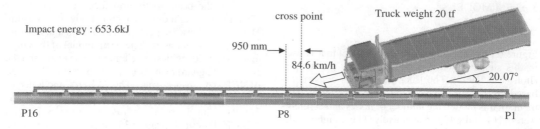

Impact energy : 653.6kJ

cross point

Truck weight 20 tf

950 mm

84.6 km/h

20.07°

P16   P8   P1

Figure 8.   Collision condition.

4   RESULTS OF NUMERICAL ANALYSIS

4.1   *Comparison of collision truck behavior*

Firstly, the result of the numerical analysis is compared with that of the full-scale experiment. As shown in Figure 8, the truck collided with the guard fence at the velocity of 84.6 km/h and the angle of 20.07 degree. The photographs taken at the full-scale experiment are shown in right-hand side in Figure 9, and the results of numerical analysis are shown in left-hand side in Figure 9. The 0 second corresponded to the time that the truck collided with the guard fence (Fig. 9(a)). The front left of the truck collided with the guard fence, firstly (primary collision). And then front wheel run on to the concrete portion. The loading platform of the truck leaned to the guard fence, and after about 0.4 seconds the loading platform collided (secondary collision). After the secondary collision, the truck ran along with the guard fence. Figure 10 shows the locus of the behavior of the truck from the top. Also, it can be seen that the truck ran along with the fence after the collision. Table 1 shows the breakaway velocity and the breakaway angles. In Figures 9, 10 and Table 1, it is found that the results obtained with the numerical analysis using the FEM models agree well with the experimental results.

There are two quantitative criterions to evaluate the performance for guiding a vehicle to a way in the new code for the design of guard fences. One is that breakaway velocity is over 80% of impact velocity. The other is that breakaway angle is under 80% of impact angle. The breakaway velocity of the experiment data and that of the numerical analysis were 64.3 km/h and 62.9 km/h, respectively. The breakaway angle of experimental data and that of the numerical analysis were zero and 3.5 degree, respectively. In this full-scale experiment, 80 % of impact velocity and impact angle were 67.8 km/h and 16.0 degree, respectively. The results of the experiment and that of the numerical simulation indicate that this aluminum alloy-concrete hybrid guard fence possesses the performance for guiding a vehicle to a way.

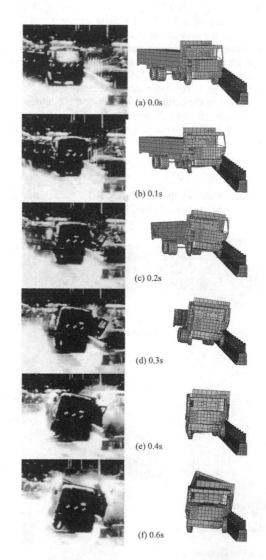

(a) 0.0s

(b) 0.1s

(c) 0.2s

(d) 0.3s

(e) 0.4s

(f) 0.6s

Figure 9.   The experiment results (right) and the numerical analysis results (left).

1294

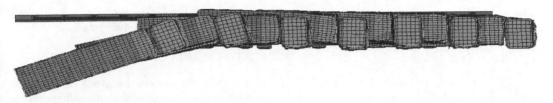

Figure 10. The locus of the behavior of the truck from the top.

Table 1. Comparison the results of experiment and the results of numerical analysis.

	Experiment	Numerical analysis
Impact velocity (km/h)	84.6	84.6
Impact angle (degree)	20.07	20.07
Impact energy (kJ)	653.6	653.6
Cross point	P7 + 1050 mm	P7 + 1050
Breakaway velocity (km/h)	64.3	62.9
Breakaway angle (degree)	0	3.5

Table 2. Response displacement of pillars.

	Maximum displacement (mm)		Residual displacement (mm)	
	Numerical analysis	Field experiment	Numerical analysis	Field experiment
Post 6	18.0	14.9	9.5	3.0
Post 7	26.3	–	9.5	3.0
Post 8	7.5	9.5	2.1	3.0
Post 9	7.5	–	1.0	2.0

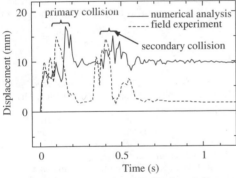

(a) Response displacement of the 6th pillar

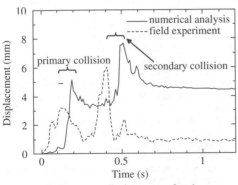

(b) Response displacement of under-part concrete of the 7th pillar

Figure 11. Comparison of calculated results and experimental results.

4.2 Response displacement of guard fence

Figure 11(a) shows the time histories of the horizontal displacement of the top part of the 6th post (see Fig. 8). The maximum displacement and the residual displacement of the posts around the collision point are shown in Table 2. As shown in Figure 11(a), the peak displacement of the experiment and that of simulation appeared twice at the time of the primary collision (0.2 s) and the secondary collision (0.4 ~ 0.5 s). The maximum displacement of the experiment and that of the simulation were almost in agreement. However, the simulated residual displacement was larger than the experimental result. In the vicinity of the collision part, the residual displacements of the simulation were much larger than the experimental results (Table 2).

Figure 11(b) shows the time histories of the horizontal displacement of under-part concrete near the 7th post. As shown in Figure 11(b), the peak displacement of the experiment and the simulation appeared twice and the displacement at the secondary collision was larger than that at the primary collision. At the secondary collision, the peak displacement of the simulation was 20% larger than that of the experiment. The time when the peak displacement occurred in the simulation delayed comparing with the experimental result. The residual displacement of the simulation was much larger than that of the experiment. It can be said that the numerical simulation in this research is useful to evaluate the maximum response displacement of the aluminum alloy-concrete hybrid guard fence.

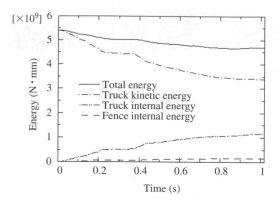

Figure 12. Energy shift of truck kinetic to truck and fence internal.

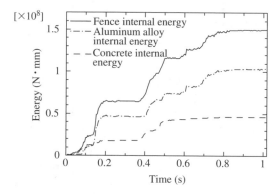

Figure 13. Energy absorption of the hybrid fence.

5 COLLISION ENERGY TRANSFER AND ADSORPTION

The transfer of the truck kinetic energy to the truck and the fence internal energy are studied and results are shown in Figure 12. The truck kinetic energy is the sum of the each element's kinetic energy that was calculated from the at the element center. The truck and the fence internal energy is the sum of the strain energy of each elements of the truck model and the fence model, respectively. The total energy including the truck kinetic energy, the truck internal energy and the fence internal energy decreased as the time increased. The internal energy of the truck model was obviously larger than the internal energy of the fence model. The most of the collision energy was absorbed by the truck.

The time histories of energy absorption of each guard fence components are shown in Figure 13. The aluminum alloy absorbed 68% of the hybrid fence internal energy approximately and the concrete absorbed 32%. This energy research can be useful to develop the

design methodology of the hybrid fences to involve the energy adsorption capacity in future.

6 CONCLUSION

In this study, the numerical analyses on the truck-collision performance of an aluminum alloy-concrete hybrid fence and on the behavior of the collision truck were carried out.

(1) From the dynamic tensile experiments, the stress–strain relationships of aluminum alloy were obtained, and it was not necessary to consider the strain rate effects in this simulation because the strain effects on aluminum alloy were very small.
(2) The FEM models were developed to simulate the full-scale field experiment. The truck behavior of the simulation agreed well with that of the experiment. The response displacement of the simulation generally agreed, but the improvements in the analysis accuracy are required.
(3) The transfer of the collision energy, which was difficult to measure in the field experiment, was obtained using the numerical analyses. And the performance to adsorb the collision energy of the hybrid fence was discussed.

REFERENCES

Hallquist, J. 1999. *LS-DYNA User's Manual (Version 950)*. Livermore Software Technology Corporation, The Japan Research Institute Ltd.
Itoh, Y. & Ohno, T. 1998. Numerical analysis on behavior of steel columns subjected to vehicle collision impact. *Journal of Structural Engineering, JSCE*, Vol. 40A: 1531–1542 (in Japanese).
Itoh, Y. & Usami, K. 2001. Numerical collision analysis of aluminum alloy guard fences. *Journal of Structural Engineering, JSCE*, Vol. 47A: 1707–1717 (in Japanese).
Itoh, Y. & Liu, C. 2001. Nonliner collision analysis of heavy trucks onto steel highway guard fences. Structural Engineering and Mechanics, Vol. 12, No. 5: 541–558.
JRA. 1999. *The specification of guard fence design*. Japan Road Association (in Japanese).
Ross, H.E. 1993. *NCHRP Report 350 Recommended Procedures for the Safety Performance Evaluation of Highway Features*. In National Cooperative Highway Research Program. Washington, D.C.: National Academy Press.
PWRC. 2001. A study on the aluminum alloy hybrid guard fences. *Research Report*. Japan: Public Works Research Center (in Japanese).
Reid, J. & Sicking, D. 1996. Design and Analysis of Approach Terminal Sections Using Simulation. Journal of Transportation Engineering, ASCE, 122(5): 399–405.
Sugiura, N. & Kobayashi, T. 1995. Strain rate dependency of impact tensile properties in an AC4CH-T6 aluminum casting alloy. *Journal of JILM*, Vol. 45, No. 11: 633–637 (in Japanese).
Wekezer, J. & Oskard, M. 1993. Vehicle impact simulation. *Journal Transp. Eng., ASCE*, 119(4): 598–617.

Dynamic actions on bridge slabs due to heavy vehicle impact on roadside barriers

G. Bonin & A. Ranzo
University of Rome "La Sapienza", Rome, Italy

ABSTRACT: In the past years the use of roadside safety barriers has changed: these devices were installed widely on the roads and increased in stiffness and resistance. This change was necessary because the early barrier design was found to be inadequate to contain and redirect the heavy vehicles. The change of barrier design led an increase of stiffness and resistance; consequently the action transferred by the device to the structure increased. The request of resistance on the bridge slabs can be too high, because the peculiar action of the roadside barriers was not taken in account in the oldest bridge design codes. Additionally, characterizing the actions transferred to the bridge slab is difficult due to the dynamic nature of the vehicle impact phenomenon on roadside barriers. The use of computational mechanics applied to dynamic impact/interaction problems is one of the best way to establish these actions. This approach is commonly used in automotive engineering to evaluate the impact behavior of the vehicle structure and can be used also to analyze the impact of a vehicle with a roadside barrier mounted on a bridge slab. The aim of this research is to use a 3D Finite Element Model of the bridge slab-barrier-vehicle system to perform a numerical simulation of the impact, following the procedure used for the roadside barrier homologation crash test, described in the European standard EN1317. The vehicle and barrier system has been validated with the results of a real crash test in similar conditions. The code used to perform all the analysis was LS-DYNA, a nonlinear explicit Finite Element Analysis code.

1 INTRODUCTION

1.1 *The adoption of concrete barriers*

In the late 80's in Italy there was a strong need of enhancing the road barriers: some accidents showed that the old steel barriers were really too low and structurally inadequate to contain the heaviest vehicles. Sometimes the truck hit the median barrier and, not contained, overran in the adjacent carriageway and collided with the vehicles running in the opposite direction. As a consequence a period of great changes for these safety devices started.

One of the first changes was the adoption of concrete barriers with New Jersey shape; these barriers were often mounted in pair in the median and sometimes they get stiffer by filling the space between them with soil.

These devices solved almost completely the median overrun problem and were quickly adopted in all the roads. Soon they began to be used as a protection for the lateral obstacles (piles, structures) and a new kind of concrete barrier was built explicitly for the bridge decks: it was higher than the median and had a circular steel beam on the top.

1.2 *The new test guidelines and the rise of steel barriers*

In the February of 1992 a new ministerial decree was published (DM LL PP 18/2/1992 n°223, *Regolamento recante istruzioni tecniche per la progettazione, l'omologazione e l'impiego delle barriere stradali di sicurezza*), defining the rules for road barriers design, test and homologation.

The decree specified the requirements for the devices to be tested with a full scale crash test, with conditions and vehicles assigned according to the required containment level.

In the 90's the steel barrier producers greatly improved their products and gained the major part of the market share. The new steel barriers are generally higher, more complex and stiffer than the old guard rail. Many of the new barriers are modular and have height and number of components growing with the containment level.

1.3 *Very high containment barriers*

The new guideline define the containment level categories, based on the lateral kinetic energy of the test vehicle; the top category is H4 (H4a and H4b), "very high containment level". The required containment level for H4a is 572 kJ and the test is done with a 30 ton 4 axles rigid vehicle hitting a barrier with a 20 degree angle and 68 km/h speed.

1.4 Barrier installation

The median and lateral concrete barrier are not restrained to the ground: their shape give them the redirection capabilities and some energy is dissipated through the heavy mass and high friction between concrete and road pavement.

The lateral and median steel barrier posts are partially embedded in soil, to give adequate restrainment: the post rotates (during elastic phase) and then a plastic hinge is created just below the pavement surface level, where the bending moment is maximum.

The bridge barrier, either the concrete and the steel ones, are fastened to the bridge deck with steel anchor bolts, that transfer to the structure shear and bending moment. It is necessary, on bridge decks, to reduce the lateral displacement of the barrier to prevent the vehicle from falling down the bridge; this requirement makes the barrier stiffer and this leads to higher actions transferred to the structure that can cause local structure faults. These faults can be dangerous for the vehicles on the bridge and for the eventual traffic or structures overpassed by and surrounding the bridge.

1.5 A new design tool: computational mechanics

The 1992 Italian decree and later the EN 1317 (European standard on Road Restraint Systems) gave to the roadside device designer and producer a guideline for the evaluation test and not a real design standard. The real device design, intended as structural project, is often missing.

This is due to the difficulty in performing an accurate structural calculation with the standard simplified rules, because these structures carry their loads during a crash event, so the large deformation and plastic behaviors have to be considered.

The requested performance for these devices is also one of the major causes of the calculation difficulties: the barrier has to contain a heavy vehicle (heavy mass vehicle test) but there is also a control on the impact severity (done with a light car test). The designer could be inclined to make the device stiffer to satisfy the heavy vehicle test, but that has to be balanced with the results on the light vehicle.

Therefore in the past only simplified calculations were performed, but the real behavior of the structure during the test remained unknown.

The situation changed when the growing speed of computers made possible the use of Finite Element

Figure 1. The two required tests for a high containment barrier.

codes specifically developed for impact simulation: these codes were born initially for military applications and ran on very expensive supercomputers.

These Finite Element codes made it possible to test design changes before the execution of a full scale crash test. This test will become in the future only the final confirmation of the computational mechanics simulations' results.

This is a valuable tool, because with the simulation is possible to test situations difficult to recreate in the reality (or simply too expensive or dangerous), getting from the analysis the complete stress–strain history of each part involved in the impact.

The main drawback to this application is the complexity of the code, that still needs highly trained engineers and, for some particular situation, additional research is needed.

All the calculations reported in this paper were performed using LS-DYNA v.950d, a nonlinear explicit finite element code. The current version of this code has a material model database, which counts more than 200 constitutive laws and is expanding to fulfill all the application needs, but often particular applications needs fine tuned user defined materials, valid only within limited conditions.

2 FINITE ELEMENT MODEL

This paper focuses on the evaluation of the actions transferred to a bridge deck when an heavy vehicle hits one of the newest very high containment steel barriers.

To do this it was built a model to reproduce the crash test for a H4a steel bridge barrier, mounted on a simulated bridge deck.

The complete model counts about 50,000 elements and is composed by three sub-models: the bridge deck, the barrier and the vehicle.

The typical run time to analyze a 1300 msec event, using the complete model, was 180 hours.

2.1 The vehicle model

The vehicle model was built at the University of Rome – Area Strade. In the past, because of the unavailability of the vehicle model and the long run

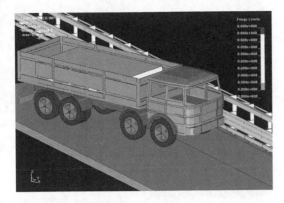

Figure 2. FE model of bridge barrier crash test.

Figure 3. FIAT IVECO F180NC vehicle model.

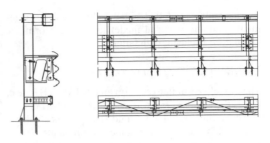

Figure 4. Side, front and top view of the bridge steel barrier used for the analysis.

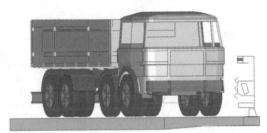

Figure 5. Bridge deck model (cross section view).

times, it was used a simplified vehicle model, with few elements, no suspension system and rolling tires and no correspondence to any real vehicle.

This choice was good for some kind of barriers, in particular the concrete barriers, but when the device got more complex, the analyses with simplified vehicle models showed that much of the real test behavior was lost.

The vehicle is a reproduction of a FIAT IVECO F180NC 4 axle vehicle with a total mass (ready for the test) of 29,600 kg. The model reproduces the rolling tires and the suspension system of the real vehicle. It was validated with the reproduction of crash tests on concrete and steel barriers, showing a good correspondence between the real test and the simulation.

The main part of the structure was modelled with 4 nodes shell elements; the material is elastic-plastic with failure (based on a maximum strain).

2.2 Barrier model

The barrier model. reproduces a three rail steel bridge barrier, with a containment level of 572 kJ (H4a). The barrier has a maximum height of 1.5 m and the post spacing is 1.333 m.

The barrier model is 78 m long.

The posts are mounted (welded in the real barrier) on steel plates fastened with steel anchor bolts to the bridge deck. In the model there is a central part (40 m) that reproduces the real installation of the device, with 4 beam elements for each plate connected to the bridge deck model. The remaining posts are connected to steel plates, which have 4 nodes fully restrained.

All the nodes on the barrier have the additional condition to avoid penetration of the bridge deck and pavement surfaces.

The barrier model is built with 4 nodes shell elements and the material model is elastic-plastic with failure.

2.3 Bridge deck model

The bridge deck was modelled to reproduce accurately the real restraining conditions of the barrier in the impact zone.

The test should be done in the same condition found on the road, so a bridge barrier should be tested on a simulated bridge deck, in order to give the device the correct restrainment and behavior.

The modelled deck is reinforced concrete slab, 42 m long and 6.5 m wide. The cross section is composed by a 4 m wide uniform height part and a cantilever slab, with thickness varying from 310 mm to 230 mm. The barrier is mounted on the thinnest extremity.

The deck is modelled with 8 node solid (brick) elements and the material model used to perform the simulation is elastic with failure.

2.4 EN 1317 TB71 test specifications

The model was built to perform an EN 1317 TB71 test, that is the standard test performed for H4a barriers.

The vehicle is a 4 axles rigid Heavy Goods Vehicle (lorry) that has to fulfill some requirements on the main dimensions, mass and Center of Gravity location.

The vehicle impacts the barrier with a constant speed of about 68 km/h with a 20 degrees angle that, with a mass of about 30,000 kg, have a lateral kinetic energy higher than 572 kJ.

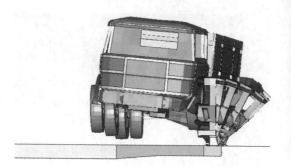

Figure 6. Computational mechanics model, side view (maximum dynamic load time step).

3 RESULTS

The actions transferred from the barrier to the bridge deck are basically shear (longitudinal and transversal) and bending moment (around longitudinal and transversal axis); due to the strength of the posts, the plates and the anchor beams, there is also a risk of local concrete crush. Our main objective was to evaluate shear and bending moment.

The anchor beams are the elements that connect and transfer actions from the barrier to the deck, so all the actions are calculated starting from shear and axial forces of the beam elements connected to the plate and to the bridge deck.

These forces vary dynamically during the test, so it is possible to represent the results with a diagram that shows the time history of each action.

The actions are calculated with an interval of 10 msec, that is a compromise value to cut out high frequency actions that have no practical significance.

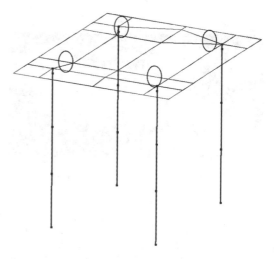

Figure 7. Beams used to evaluate actions.

3.1 Shear forces

Shear forces are calculated for each group (post) by summing the single beams shear forces (longitudinal and transversal component).

The following diagrams show the longitudinal and transversal forces for each time step of the simulation, in the most loaded post.

Maximum and minimum values of shear force computed for each group of anchor beams are reported in Table 1.

3.2 Bending moments

For every the two pairs of beams, bending moments are calculated around the transversal and longitudinal axis of the plate welded to the post.

Also the bending moments are variable during the test and are reported in the following diagrams for the most loaded post.

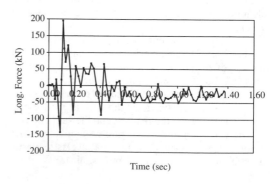

Figure 8. Longitudinal force time history (post 8).

It is to notice that (correctly) the bending moment is only positive in the transversal axis.

The results for the other posts are summarized in Table 2.

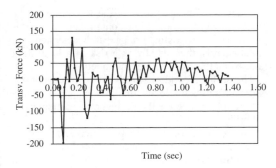

Figure 9. Transversal force time history (post 8).

Table 1. Maximum and minimum shear forces for each post.

Post	Transversal shear min (kN)	Transversal shear max (kN)	Longitudinal shear min (kN)	Longitudinal shear max (kN)
5	−137	118	−103	80
6	−119	113	−164	111
7	−138	121	−170	151
8	−198	129	−142	195
9	−110	117	−144	73
10	−101	83	−120	89
11	−104	137	−72	134
12	−95	118	−66	128
13	−74	114	−97	143
14	−82	117	−116	85
15	75	74	−101	49

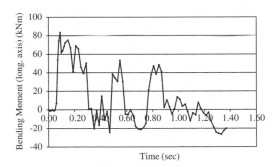

Figure 10. Bending moment around longitudinal axis (post 10).

3.3 Discussion of results and comparison with design standards

The analysis of the actions due to an heavy vehicle impact on a roadside barrier involves the bridge design and the barrier design.

The new stronger and stiffer barriers affect bridge decks design and calculation, as it is possible to see analyzing the bridge design standards.

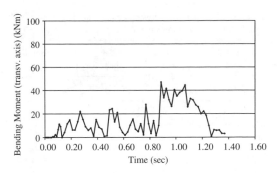

Figure 11. Bending moment around transversal axis (post 10).

Table 2. Maximum and minimum bending moment for each post.

Post	Bending moment (around long. axis) min (kNm)	Bending moment (around long. axis) max (kNm)	Bending moment (around transv. axis) min (kNm)	Bending moment (around transv. axis) max (kNm)
5	−24.80	43.06	−0.41	28.65
6	−33.84	44.87	−0.09	23.08
7	−23.17	50.17	−9.03	16.88
8	−31.26	38.73	−3.28	24.67
9	−36.93	77.90	−5.69	30.41
10	−26.57	83.32	−0.05	47.11
11	−63.62	74.30	−18.18	32.13
12	−46.92	72.74	−0.12	24.05
13	−30.49	63.03	−7.41	32.86
14	−24.04	23.30	0.00	27.35
15	−19.49	15.19	0.00	28.26

In Italy the current design standard was published in 1990 (DM 4/5/1990, *Aggiornamento delle norme tecniche per la progettazione, la esecuzione e il collaudo dei ponti stradali*) and specifies, on paragraph 3.11, that for the actions on parapets can be used a horizontal static force of 45 kN, at 0.60 m over the road surface, plus a static force of 30 kN applied on 4 posts maximum.

The currently published Eurocode 1 part 3 (ENV 1991-3:1995 – Basis of design and actions on structures, Part 3: *Traffic loads on bridges*) recommends a 100 kN static force, horizontal and transversal, applied 100 mm below the top of the barrier or 1 m above the road surface and acting on a line 0.50 m long. There is also a recommendation that the structure supporting the barrier should resist an accidental load effect 1.25 times the maximum characteristic local resistance of the vehicle parapet (e.g. the anchor bolt).

The new draft Eurocode 1 part 2 (draft prEN 1991-2 Eurocode 1 – Actions on structures Part 2: *General*

Table 3. Comparison between design standard values and computational mechanics results.

Standard	Shear force (kN)	Bending moment (kNm)
Italian DM 1990	52.5	31.5
ENV 1991-3:1995	100	140
prENV 1991-2 – Class A	100	140
prENV 1991-2 – Class B	200	280
prENV 1991-2 – Class C	400	560
prENV 1991-2 – Class D	600	840
Computational mechanics	198	83.3

actions – Traffic loads on bridges), that will supersede the current ENV 1991-3:1995, severely raise the actions due to collision with restraint systems. In paragraph 4.7.3.3 the draft Eurocode has the same recommendations, but there are now 4 classes of horizontal forces, ranging from 100 kN (that is now the minimum) to 600 kN. There is a clear reference to the European roadside hardware design standard, but it is no direct correlation between the higher classes of the Eurocode and the high containment classes for the EN 1317.

In Table 3 there is a summary of the values suggested by the design standards and the results of the computational mechanics model. The loads from design standards are calculated for a barrier like the one used in the computational mechanics model, 1.50 m tall.

The bending moment results of computational mechanics are lower than the values included in the Eurocodes, but the load is applied in a very different way. The Eurocode load has to be applied on a line 0.50 m long in a single position, that represents the worst possible action for that kind of load and has to be determined with the influence line method. The Eurocode says that the load "may be applied 100 mm below the top of the selected vehicle restraint system or 1.0 m above the level of the carriageway or footway,", but it is not clear if the load has to be applied on the barrier or on the deck itself. This is a key point, because if the force is applied on the barrier, the action will be transferred to the deck by many posts, with lower, distributed stresses on the deck.

The load resulting from computational mechanics is applied on a single post, and the other posts carry, in the same time, a similar load. These shear forces and bending moments are the real loads acting locally on the bridge deck.

4 CONCLUSIONS

There is no doubt that in the oldest design standards the loads were smaller than necessary to support the new devices. This is probably the key point of the problem: when a new high containment barrier is installed over an existing bridge, surely designed with old standards, there is always a risk of a local deck failure and therefore barrier malfunction.

The draft Eurocode seriously takes in account this issue and raises the loads, that in some cases can be very tough to face for the designer, in particular for old bridges.

The computational model presented and the draft Eurocode recommendations confirm that a barrier installation over a bridge is a complex operation that sometimes needs additional resistance from the deck, that has to be partially redesigned.

REFERENCES

EN 1317-1, Road Restraint Systems – Part 1: terminology and general criteria for test methods, April 1998.
EN 1317-2, Road Restraint Systems – Part 2: performance classes, impact test acceptance criteria and test methods for safety barriers, April 1998.
DM 4/5/1990, Aggiornamento delle norme tecniche per la progettazione, la esecuzione e il collaudo dei ponti stradali.
ENV 1991-3:1995 – Basis of design and actions on structures, part 3: traffic loads on bridges.
LS-DYNA Keyword user's manual – nonlinear dynamic analysis of structures, Livermore Software Technology Corporation, 1999.
Belytschko T., Liu W.K. & Moran B. 2000. *Nonlinear finite elements for continua and structures*, Wiley and Sons.
Hirsch T.J. & Arnold A.G., Bridge deck designs for railing impacts, TTI-Texas Highway department cooperative research, August 1985.
La Torre F. & Ranzo A. 1994. Simulazione per eventi discreti del comportamento dinamico di barriera da ponte in calcestruzzo ad ancoraggi pseudo-duttili, *Industria Italiana del Cemento*, n. 694.
Ranzo A. 1989. Barriere prefabbricate in calcestruzzo ad ancoraggi duttili per ponti e viadotti: calcolo e sperimentazione, *Industria Italiana del Cemento*, n. 630.
Ray M.H. & Wright A.E. 1996. Characterizing Guardrail Steel for LS-DYNA3D Simulations, In Current Research on Roadside Safety Features, Transportation Research Record No. 1528, *Transportation Research Board*, National Academy Press, Washington, D.C.
Ray M.H. 1996. Repeatability of Full-Scale Tests and Criteria for Validating Simulation Results, In Current Research on Roadside Safety Features, Transportation Research Record No. 1528, *Transportation Research Board*, National Academy Press, Washington, D.C.
Sicking D.L. 1995. Application of simulation in design and analysis of roadside safety features, *TRC* No. 435 p. 67, TRB.

The structural behavior of high strength aluminum alloy sections

Y.B. Kwon & K.H. Lee
Department of Civil Engineering, Yeungnam University, Gyongsan, Korea

H.C. Kim & D.H. Kim
Department of Aluminum Production, LG Cable Co. Ltd., Gumi, Korea

ABSTRACT: A research program has been carried out to study the structural behavior of the thin-walled high strength aluminum alloy sections for highway safety applications. Two series of static flexural tests on the aluminum alloy sections have been performed and the primary factors are the cross section shape, the plate thickness and the strength of material. The failure mechanism was investigated with particular attention given to the plastic folding of the loaded flange and the local buckling behavior of the webs. The effect of loading jig and support jig on the flexural strength of beam sections has been investigated. An advanced analysis using finite element method program has been executed and compared with the test results for the purpose of verification. In the numerical analysis, the nonlinear stress–strain relation of aluminum extrusions obtained by the tensile coupon test was used under the assumption of multi-linear curve rather than the smooth rounding one.

1 GENERAL INTRODUCTIONS

Aluminum has been widely used for the structural member for car body and building because of light weight, high durability and energy consumption capacity. Nowadays the concept of aluminum member has become attractive in the design of guard rails and barriers since the safety and beauty should be need for those structures. The static and dynamic behavior of the small size aluminum sections such as structural member for cars has been investigated by several researchers. Those aluminum alloy sections have shown excellent structural properties such as excellent ductility and high energy absorption capacity. On the contrary, due to low elastic modulus, the deformation was very large in comparison with steel sections. However, the size and material properties of the section which can be used for the guard rail are much different from those used for the car, the research results cannot be used directly.

In Korea, up till now, medium strength aluminum alloy A6061S T6 (nominal yield strength: 245 MPa) sections have scarcely been used as a substitute for steel or concrete bridge barrier.

In this paper, to apply the high strength aluminum alloy section for the guard rail, flexural tests for cross beam elements and cantilever tests for column members have been carried out. The tensile coupon test has been executed to investigate the material properties of high strength aluminum alloy A6024S T6 according to Korean Standard B0801, of which the nominal yield strength is 295 MPa(30 kgf/mm^2).

2 SECTION GEOMETRIES AND MATERIAL PROPERTIES

2.1 Section geometries

The test section was designed for practical use and full scale model. Types of the test section and geometries are shown Figure 1.

Two types of main and subsidiary beam sections and three types of rectangular hollow section(RHS) columns were selected for the flexural bending test and cantilever test respectively. The section properties of test sections are given in Table 1. The width of main beam section is 160 mm and 125 mm to 160 mm for subsidiary beams. The of depth of main beams is ranged from 200 mm to 250 mm and 65 mm to 90 mm for sub beams. The thickness of main beam is approximately 5 mm and that of sub beam is 5 mm–8 mm. The size of the rectangular sections for column member is 110 mm × 120 mm, 120 mm × 125 mm and 128 mm × 145 mm, and the thickness is 6 mm–9 mm.

2.2 Material properties

The tensile coupon test of aluminum alloy A6061S T6 and A6024S T6 was executed according to KS B0801 to investigate the material properties of test sections. The coupon was cut from the corner and flat area of the section to find out the changed properties from base material during extrusion and cooling procedures. The shape and dimension of tensile coupon are shown in Figure 2 and the coupon test setup is shown in Figure 3.

The test results are shown in Table 2 where coupon A and B stand for A6024S T6 and A6061S T6 respectively. The stress–strain curve shows gradual rounding after proportional limit up to ultimate strength. Therefore, 0.2% offset stress was assumed as a yield stress. The yield stress of A6024S T6 is higher than that of A6061S T6 by approximately 20% or so. Young's modulus of both aluminum alloys is about one third of that of steel. There is no difference in the yield stress between different locations of coupons for sub-beam and column. However, different yield strength has been observed between locations of

coupons. It was guessed that since the shape of main beam is complicated and the dimension is comparatively large, the material property has been changed during the extrusion and cooling procedures.

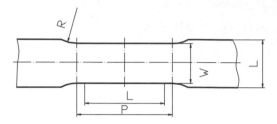

W	L	P	R	T
25 mm	50 mm	60 mm	15 mm	Original Thickness

Figure 2. Test specimen for tensile test.

Figure 3. Tensile coupon test setup.

Table 2. Measured material properties.

	Specimens		E (GPa)	F_y (MPa)	Avg. A/B
Main beams	PR-0210	A	65.343	347.84	1.20
			73.321	320.97	
		B	64.500	280.60	1.00
			65.372	276.20	
	PR-0231	A	67.063	342.00	–
			67.265	330.90	
Sub beams	PR-0075	A	66.826	331.75	1.21
			65.782	328.91	
		B	66.937	277.79	1.00
			66.883	267.53	
	PR-0226	A	65.386	340.88	–
			66.592	340.49	
Columns	PR-0212	A	67.427	313.54	–
			68.020	310.90	
	PR-0067	A	63.764	324.50	–
			66.766	318.90	
	PR-0225	A	60.399	347.30	1.25
			60.635	345.00	
		B	65.635	278.95	1.00
			64.166	272.71	

E: Modulus of elasticity.
F_y: Yield stress.

Figure 1. Section geometries and dimensions.

Table 1. Section properties.

Specimens	Depth (mm)	Width (mm)	A (mm^2)	I_x	I_y
				\multicolumn{2}{c}{I (cm^4)}
PR-0210	200	160	2902	1602	890
PR-0231	250	160	3620	2647	1284
PR-0075	65	125	2457	153	436
PR-0226	90	160	2464	323	749
PR-0212	110	120	2602	479	550
PR-0067	120	125	3022	671	677
PR-0225	128	145	4642	1140	1385

A: Cross sectional area.
I: Inertia moment of cross section.

3 FLEXURAL TESTS

3.1 Test method

The flexural beam test was executed to find out flexural and shear behavior of main and sub beams. The test beam length was decided as 2 m in consideration of the support distance of bridge barrier and the edge support condition was prepared for hinged boundary conditions. Both end supports were specially designed using supporting jig and rod. The concentrated load was applied at the center of beam through the specially designed loading jig. The loading and supporting jig was designed exactly same as the top and bottom flange shape of test sections respectively. The length of jig is 15 cm and the thickness is 2 cm. The test configuration is shown in Figure 4.

LVDTs were located on the top of loading jig and beneath the beam center to measure the central deflection. The vertical loading was applied using UTM (Shimadzu, capacity 1000 kN) by the displacement control method with loading rate of 1 mm/min. up to final failure of sections.

In this paper and Korean regulations, the collapse point was defined as the amount of displacement at the beam center reached 75 mm as shown in Figure 5.

The collapse moment M_0 was obtained using the equivalent energy method on the moment-deflection curve. If the moment-deflection relation was assumed as elastic-perfectly plastic, the collapse moment was defined as the moment which equated the areas A and B, and C and D in Figure 6 respectively.

3.2 Test results

The main beam was designed to provide large energy dissipation capacity and depth-thickness is approximate 25, but on the contrary, that of sub-beam is 4.5–9. Due to the large width-thickness, the local buckling initiated at the flange and web of main beam center.

The moment-deflection curve and deformed shape are given in Figure 7. The occurrence of local buckling decreased the flexural rigidity of the section. However,

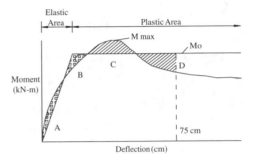

Figure 6. Computation of M_0 by equivalent energy method.

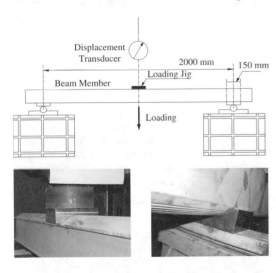

Figure 4. Flexural tests configuration.

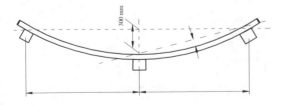

Figure 5. Assumption for deformation of barrier.

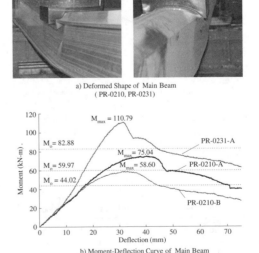

a) Deformed Shape of Main Beam
(PR-0210, PR-0231)

b) Moment-Deflection Curve of Main Beam

Figure 7. Moment-deflection curve and deformed shape of main beam.

the post-buckling behavior was observed as the load was increased up to ultimate load and after that the load decreased gradually. In the case that the loading jig was not used, the crippling deformation of top flange beneath the base plate initiated at the top flange under the loading point and then the local buckling occurred at the web at beam center. However, with the loading jig used, the crippling of the flange did not occurred and showed stable deterioration with the displacement increased.

The structural behavior of the sub-beam was different from that of main beams since the width-thickness of sub-beam is far small compared with that of main beam. Moment versus central deflection curves and deformed shapes of beams are given in Figure 8. It was found that the global flexural deformation was increased without local buckling as the load was incremented up to collapse load and after peak load the load was decreased gradually.

The flexural test results are summarized in Table 3. As shown in Table 3, PR-0210A and PR-0075A showed

a) Deformed Shape of The Sub Beam
(Respectively PR-0075, PR-0226)

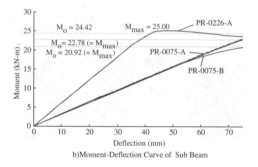

b)Moment-Deflection Curve of Sub Beam

Figure 8. Result curve and deformed shape of sub-beam.

Table 3. Flexural test results.

	Specimens		M_{max} (kN-m)	M_0 (kN-m)	A/B	
					M_{max}	M_0
Main beam	PR-0210	A	75.04	59.97	1.28	1.36
		B	58.60	44.02	1.00	1.00
	PR-0231	A	110.79	82.88	–	–
Sub beam	PR-0075	A	22.78	22.78	1.10	1.10
		B	20.92	20.92	1.00	1.00
	PR-0226	A	25.00	24.42	–	–

the increase in the maximum and collapse flexural strength by approximate 30% and 10% respectively in comparison with B sections. It is concluded that the increase of material yield strength affects the flexural strength of non-compact sections more effectively than compact sections.

4 CANTILEVER TESTS

4.1 *Test method*

A series of cantilevered column tests were carried out to investigate the structural behavior of aluminum alloy RHS. Cantilever test configuration is shown Figure 9.

The dynamic actuator (MTS, capacity 500 kN) was used to apply horizontal loads at the column top. Column height was 800 mm and embedded part was supported by the method of being anchored to massive concrete reaction block.

The collapse load is defined similarly to main beam flexural test as in Figure 5. It was assumed that the column was collapsed when the amount of displacement at the column top reached at 300 mm instead of 75 mm of central deflection. The collapse load P_0 was obtained in the same manner as flexural moment using the equivalent energy method on the load-displacement curve, where the load-displacement relation was assumed as elastic-perfectly plastic.

4.2 *Test results*

Since the width-thickness ratio of column section is small, inelastic local buckling occurred as the load was increased. Since rectangular hollow sections are very rigid in flexure and torsion, flexural or flexural-torsional buckling did not occur and the behavior of column was quite stable. Load-displacement curves

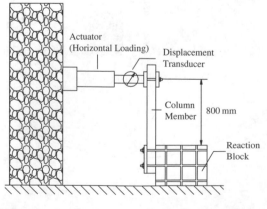

Figure 9. Cantilever test configuration.

obtained through test results are shown in Figure 10. As the load was increased, the local buckling occurred at compression side of column's bottom supported by the reaction block. After local buckling the flexural rigidity was decreased and the maximum load was reached gradually and then the load was decreased slowly without abrupt change of stiffness until failure.

Collapse and maximum load are summarized in Table 4. The test results of all the specimens showed enough strength for specified safety criteria. As shown in Table 4, A6024S T6 aluminum PR-0225A showed 13% and 15% increase for maximum and collapse load respectively in compared with mild aluminum alloy PR-0225B.

5 COMPARISON WITH NUMERICAL ANALYSIS

To verify the test results, the advanced analysis using finite element program LUSAS was carried out

a) Deformed Shape of Column Member

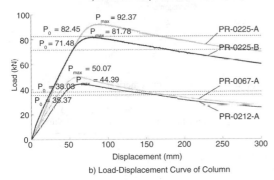

b) Load-Displacement Curve of Column

Figure 10. Load-displacement curve and deformed shape of column.

Table 4. Cantilever test results.

| | Specimens | | P_{max} (kN) | P_0 (kN) | A/B | |
					P_{max}	P_0
Columns	PR-0212	A	44.39	35.37	–	–
	PR-0067	A	50.07	38.03	–	–
	PR-0225	A	92.37	82.45	1.13	1.15
		B	81.78	71.48	1.00	1.00

which included material and geometrical nonlinearity. Modeling and meshes using Shell element and Stress Potential was adopted to consider nonlinear stress–strain properties of aluminum. The von Mises yield criteria was used for plastic flow. The deformed shape and moment(or load)-displacement relations obtained from tests were compared with numerical results in Figures 11 and 12.

The deformation shapes were simulated quite similarly and the maximum moment and collapse moment of sub-beams are quite well agreeable. The maximum moment of main beams obtained by numerical analysis is lower than tests by approximate 10% and the difference in collapse moment is about 20%. The reason is that since the deformation mechanism of main beam is more complex and the buckling, yielding and plastic deformation behavior is too complicated to simulate properly. However, the flexural test results seem to be reliable within engineering accuracy from comparison between test and numerical results.

The numerical results of columns are also given and compared with test results. As it is known by comparison, up to ultimate load, both results were in good agreement since the elastic and inelastic structural behavior was very stable. However, after ultimate load until the amount of displacement reached 300 mm, the plastic deformation at the buckled area was too large and collapse mechanism was too complicated to be clearly simulated numerically.

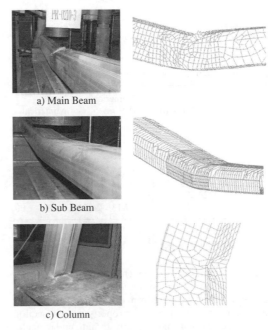

a) Main Beam

b) Sub Beam

c) Column

Figure 11. Comparison of deformed shapes.

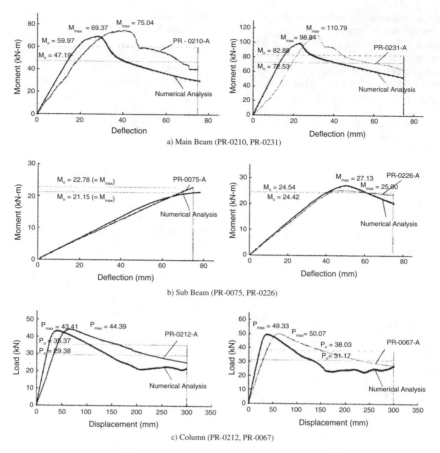

Figure 12. Comparison of result curve.

Table 5a. Comparison of numerical and test results.

Beam	Test (kN-m)		Numerical (kN-m)		Test/ Numerical	
	M_{max}	M_0	m_{max}	m_0	M_{max}/m_{max}	M_0/m_0
PR-0210A	75.04	59.97	69.37	47.19	1.08	1.27
PR-0231A	110.79	82.88	98.84	72.53	1.12	1.14
PR-0075A	22.78	22.78	21.15	21.15	1.08	1.08
PR-0226A	25.00	24.42	27.13	24.54	1.09	1.00

Table 5b. Comparison of numerical and test results.

Column	Test (kN-m)		Numerical (kN-m)		Test/ Numerical	
	P_{max}	P_0	P_{max}	P_0	P_{max}/P_{max}	P_0/P_0
PR-0212A	44.39	35.37	43.41	29.38	1.02	1.20
PR-0067A	50.07	38.03	49.33	31.17	1.01	1.22

6 CONCLUSIONS

Large size and complicated shape beams have shown considerable variance in the yield strength of tensile coupons which were cut from different locations. This non-homogeneous material property across section may affect estimation of design strength and expectation of structural behavior of sections. Therefore, an advanced extrusion process to produce homogeneous property across the whole section should be developed to improve the mechanical property.

The increase of flexural rigidity of aluminum alloy sections is resulted from the upgrading of material yield strength. The effect is more significant for non-compact sections than compact sections. Consequently, a thin section with large width-thickness ratio is more economic for high strength aluminum alloy sections.

1308

As mild strength aluminum alloy sections (A6061S T6) which are currently used for highway safety structures, so high strength aluminum alloy sections (A6024S T6) under flexural loads showed large extent of ductility and high energy absorption capacity which is critically necessary characteristics for road safety facilities. Therefore, even if the elastic modulus is smaller than that of steel and the deformation is comparatively large, the aluminum alloy section can be usefully and effectively applied to barrier. Since any complicated shapes of sections can be made through extrusion process and strong and safe barriers are needed more than ever due to the increase of traffic speed and number of heavy vehicles, the high strength aluminum alloy sections can be widely applied to highway safety structures.

REFERENCES

Bank, L.C. & Gentry, T.R. 2001. Development of a pultruded composite material highway guardrail. Composites. Part32 A. Applied science and manufacturing: 1329–1338.

Kecman, D. 1997. An engineering approach to crashworthiness of thin-walled beams and joints in vehicle structures. Thin-Walled Structures. vol. 28. nos. 3/4: 309–320.

Korea Standard Association. 1996. KS handbook – Nonferrous metal. Seoul: KSA.

Langseth, M., Hopperstad, O.S. & Hanssen, A.G. Crash behavior of thin-walled aluminum members. 1998. Thin-Walled Structures, vol. 32: 127–150.

Ministry of Construction and Transportation. 2001. Construction and maintenance of Road Safety Facilities – Barrier and Guard Rail.

Tabiei, A., Svenson, A. & Hargarve, M. 1998. Impact performance of pultruded beams for highway safety applications, Composite Structures. vol. 42: 231–237.

Yeungnam University. 2002. Development of high strength aluminum alloy sections for barrier and guard rail. Research report No. 2002-1, Department of Civil Engineering, Yeungnam University. .

System-based Vision for Strategic and Creative Design, Bontempi (ed.)
© *2003 Swets & Zeitlinger, Lisse, ISBN 90 5809 599 1*

Vehicle-structure interaction modelling

F. Giuliano
Structural Engineer, Master Mathematic Student, Rome, Italy

ABSTRACT: The paper focuses on the dynamic response and the interaction effects in bridge structures under moving loads. An algorithm able to model several kinds of moving loads is introduced.

A parametric analysis checks the main factors on the response of a beam, specifically the so-called amplification factor. Then an alternative design method capable of taking into account the main interaction effects is introduced.

1 INTRODUCTION

The design and the constructive development and the introduction of high performance materials change the structural solutions into slender and flexible ones, for which the structural element performances are optimized: so in bridge design the dynamic response and the vibration analysis for moving loads is more important than in the past.

The vibration shape and frequences computation is the first step to get an evaluation of resonance risks.

The amplification factor computation is the traditional way to take into account the dynamic amplification of deformation and stresses under the passage of vehicles; its value becomes more important for the increase of traffic volume, speed and frequence, and realistic load conditions have to be considered.

Most of national bridge design codes correlate a Factor of Impact (D) to the beam length or the First Frequence of the structure; this simplistic approach doesn't take into account the whole complex system of factors playing a role in the response, and the suggested values are often contradictory.

Besides the bridge dynamic response under moving loads is not ruled only by span or frequence, but these two values have the same importance, as shown by the expression of the dimensionless parameter β corresponding univocally with the Factor of Impact D in the simplest case of beam loaded by moving concentrated force.

The vehicle-structure interaction is a complex phenomenon, depending on structural typology, its dynamic attitude, the dynamic and geometric vehicle characteristics, their speed, number and reciprocal position, the structure and surface maintenance.

In the complex structure analysis it's not possible to check the response by only one physical coupled model: the computation of displacement, velocity and deck acceleration *time-histories* and Factors of Impact need global models of the structure, with realistic distribution of masses and stiffness along the girders, irregularities of surfaces modelled by power density spectra (PSD) and simple vehicle models capable of reliable evaluations about vertical box accelerations and comfort levels; on the contrary all the *serviceability* analyses need more sofisticated modelization of interfaces, and local computation of deformation and stresses in dynamic cases.

An algorithm for the computation of the dynamic response of simple structures, as Euler-beam with linear elastic behaviour has been realized.

The program has been tested on a simple supported beam and results have been compared with analytical results obtained by series developments published by C.E. Inglis for concentrated force case.

The discrete interaction modelization used and the results of a parametric analysis conducted by the code is presented. It gives evidence to the role of the main factors on the response.

Then an alternative, more rational and critical design methodology, is introduced: the interaction effects are taken into account even in static computations during the first phase of the design process.

2 STRUCTURAL MODELLING

The structure is modelled by a sequence of mono-dimensional Euler beam finite element: each node has three degrees of freedom: horizontal and vertical displacement and flexural rotation.

The distance between two adjacent nodes must be sufficiently small to get accurate results.

The mass distribution is modelled by consistent approach; for the structural damping the Rayleigh model has been used. The element matrixes are assembled in global structural ones. The motion equations of free vibrations in matrix form are given by the following expression:

$$[m]\ddot{q} + [c]\dot{q} + [k]q = 0$$

The code realized solves the dynamic problem of beam subjected to moving loads by Newmark's direct integration method, with constant average acceleration between two time-steps. By imposing adequate values to the parameters of the method, as $\alpha = 0.25$ and $\delta = 0.5$, the numerical scheme is unconditionally stable.

3 MOVING LOAD MODELLING AND RESPONSE OF THE STRUCTURE

To catch some design improvements by the analysis of the dynamic interaction between vehicle and structure, it has been necessary to develop a specific code, able to make dynamic analysis for different moving loads, simple or complex, capable or incapable of interaction effects.

The parametric analysis gets an evaluation of the role of several interaction effects in the structural response. By fixing a test beam, all the parameters have been changed and Factors of Impact calculated.

Concentrated force moving at constant speed

This load model is valid when the bridge span and mass are great in comparison with it.

This problem is governed by the equation

$$EI\frac{d^4y}{dz^4} + \mu A\frac{d^2y}{dt^2} = W\delta(z - vt)$$

This problem has been solved and studied by series envelopments by C.E. Inglis:

$$y = \frac{2WL^3}{\pi^4 EI}\left[\frac{sen\ \dfrac{\pi z}{L}}{1 - \dfrac{n^2}{f_1^2}}\left\{ sen\ 2\pi nt - \frac{n}{f_1} sen\ 2\pi f_1 t \right\} + \right.$$

$$\left. + \frac{sen\ \dfrac{2\pi z}{L}}{2^4 - \dfrac{2^2 n^2}{f_1^2}}\left\{ sen\ 4\pi nt - \frac{n}{2f_1} sen\ 2^2(2\pi f_1 t) \right\} + ... \right]$$

$f_1 = $ Fundamental frequency; $L = $ span length; $n = 2v/L$.

This load model allows to check the main factors for the response: the span length, the Fundamental frequency, the load speed. After fixing the boundary and the mechanical properties of the bridge, and so the dynamical ones, it is possible to determine the load speed that gives the greatest response of the structure.

The relationship between the length, the Fundamental frequency, the load speed and the Impact Factor D is summarized by a relation between D and a non dimensional parameter β

$$D = \frac{1}{1 - \beta} \sin \frac{2\pi\beta}{1 + \beta} \qquad con \quad \beta = \frac{vT_1}{2L}$$

$T_1 = $ Fundamental Period.

By the same beam models used for the code tests, it is possible to estimate the damping influence on the response; it has been demonstrated that the critical speed value is about inaltered by the damping effects.

The structural damping allows a fast restore of equilibrium state after the load passage and the reduction of dynamic effects.

Under a certain value of the damping the Impact Factor D grows monotonally with β if $\beta \in [0,2–0,6]$. For greater values of β, D decreases, and for very little value of β the dynamic deformation is close to the static one. For damping values greater than the critical, D decreases and becomes even smaller than 1.

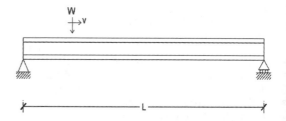

Figure 1. Concentrated force moving at constant speed.

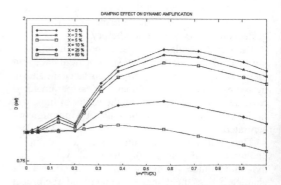

Figure 2. Damping effects on dynamic amplification.

Concentrated Mass moving at constant speed

The mass concentrated model is the simplest which is able to give interaction effects on the structure: during the passage the load increase the total mass and changes the dynamic characteristics, by reducing and then restoring the Fundamental frequency value.

The differential equation of the motion of a beam subjected to a noving mass, contains an inertial term: the force to which the beam is subjected is

$$Mg - M\ddot{y}_A(t)$$

$\ddot{y}_A(t)$ is the acceleration of the section A on which the mass at the time t runs.

$$EI\frac{d^4y}{dz^4} + \mu A\frac{d^2y}{dt^2} = M(g - \ddot{y})\delta(z - vt) \quad \delta = Dirac\_Function$$

So an equation with time-varying coefficients has to be solved by a numerical integration method.

After introducing the damping term, we obtain an equation system as follows:

$$[m]\cdot\ddot{q}(t) + [c]\cdot\dot{q}(t) + [k]\cdot q(t) = p(t)$$

The load mass introduces in the second member an unknown term that is an acceleration; it is the responsible of the dynamic interaction between load and structure in this simple modelling; it has been supposed that in the instantaneous contact point the beam loaded section and the load vertical displacements and acceleration coincide.

So the system keeps the number of degrees of freedom of the model with no interaction effects.

The load vector $p(t)$ is divided into two terms:

(1) $Mg\delta(z - vt)$, as the moving force case;
(2) $-M\ddot{y}_A(t)\delta(z - vt)$, inertial interaction term: at every time integration step the mass matrix [m] is

updated, and so the effective stiffness matrix and load vector:

$$[m]_{t+\Delta t} = [m] + [m^*]_{t+\Delta t} \; ;$$
$$[\hat{k}]_{t+\Delta t} = [k] + a_0[m]_{t+\Delta t} + a_1[c];$$
$$\hat{R}_{t+\Delta t} = R_{+\Delta t} + [m]_{t+\Delta t}(a_0 y_t + a_2\dot{v}_t + a_2\ddot{v}_t) + [c]_{t+\Delta t}(a_1 y_t + a_4\dot{v}_t + a_5\ddot{v}_t)$$

The instantaneous matrix $[m^*]_{t+\Delta t}$ has only one non zero element; its value is $-M$, corresponding to the loaded degree of freedom at the current time.

The same simply supported beam, with no damping, and an equivalent mass load instead of the previous force value have been used.

By comparing the *time-histories* of displacements in the middle of the span, it is possibile to say that the inertial effects are percentually greater after the passage of the load, under a free oscillation regime.

The inertial nature of the load causes a reduction of the Fundamental Frequency and changes the analytic diagram *D-β* obtained for the concentrated force case, especially if the vehicle mass becomes closer to the bridge one.

The inertial effects, at low speeds, or at low value of β, are not significant: the *D-β* curves are close even for high value of the vehicle mass *M*.

After fixing the rate *M/m*, there is a speed gap in which the dynamic response is much greater than the simple concentrate force case one; for no realistic β value, the latter gives a higher response.

If β < 0.2 and the rate of the masses is low, the differences between the responses is so small that the no interaction model can be accepted.

Single suspension vehicle at constant speed

A simple vehicle model but more realistic than a concentrated mass in motion is the single suspension: a mass M_c is connected to another mass M_e by a linear suspension system with an elastic spring and a linear damping C_v. M_e is in direct contact with the structure. This model allows to take into account the interaction effects in a quite effective way for structural design computation and for a valuation of comfort levels.

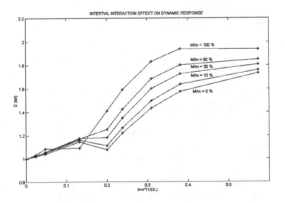

Figure 3. Inertial interaction effects on dynamic amplification.

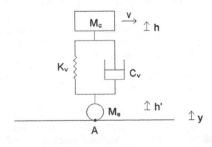

Figure 4. Single suspension vehicle.

The simple suspension vehicle has two degrees of freedom, for it's possible a no contact case; so the whole coupled structure-vehicle system has $3n + 2$ dof, ($n = number\ of\ nodes$).

In the present work a linearization of the phenomenon has been taken in account, because it is reasonable to foresee a continuous contact between wheel and surface and a repulsive interaction force system.

The vertical motion of the simple suspension in motion on an indeformable surface is governed by D'Alembert's law:

$$M_c \cdot \ddot{h}(t) + C_v \cdot \dot{h}(t) + K_v \cdot h(t) = 0$$

If the deck is deformable, the previous equation becomes:

$$M_c \cdot \ddot{h}(t) + C_v \cdot \left[\dot{h}(t) - \dot{y}_A(t)\right] + K_v \cdot \left[h(t) - y_A(t)\right] = 0$$

where

A = current vehicle position
$y_A(t)$ = current vertical displacement in A.

The equation couples the acceleration of the body of the vehicle (mass M_c) and the bridge one, in coincidence with the current transit section.

The consequence of the introduction of the new degrees of freedom is a re-definition of displacements, velocity and acceleration vectors.

With reference to the figure, it's possible to check the interaction forces between the body and the wheel and between the wheel and the structure surface:

$M_c \cdot g$ = vehicle body weight;
$M_e \cdot g$ = vehicle wheel weight;
$F_K(t) = K_v \left[y_A(t) - h(t)\right]$ = spring elastic force;
$F_C(t) = C_v \left[\dot{y}_A(t) - \dot{h}(t)\right]$ = suspension damping force;
$F_{Ie}(t) = -M_c \cdot \ddot{y}_A(t)$ = inertial force of the vehicle body.

In the linear hypothesis, the motion equation of the coupled system can be derived by suppressing one degree of freedom in correspondence with the loaded section at the generic time: as many cinematic conditions as the number of vehicle in motion are imposed at every time step.

Besides the introduction of this load model produces new dynamic phenomena, for the simple suspension has an own angular frequency:

$$\omega_v = \sqrt{\frac{k_v}{M_c}}$$

The dynamic characteristics of the vehicle, in combination with the structural ones, can give birth to another coupling effect, with a consequent

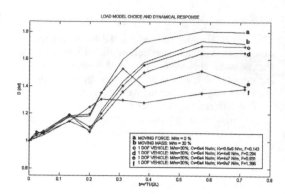

Figure 5. Load model choice and dynamical response.

amplification of the oscillations of the beam, just like happens for harmonic loads on a beam.

This phenomenon cannot be considered a resonance, because the vehicle in motion has no sufficient energy to offer to the system at its entrance, and the consequent oscillations are never so great.

In the present work the dynamic amplifications for an undamped system ($\xi = 0\%$) and for different vehicle models have been compared:

- concentrated force in motion: $W = 2.649\,KN$
- concentrated equivalent mass in motion at the same speed: $M = 270000\,Kg$;
- one degree of freedom vehicle in motion at the same speed and with the same total mass: $M_c = 266000\,kg$, $M_e = 4000\,kg$; $C_v = 60000\,Ns/m$, K_v variable.

The diagram above shows the interaction effects can be significant, and the force models unable to foresee with sufficient precision the structural response: beyond a speed limit, or a certain value of β, the force model in motion is prudent.

The vehicle damping C_v has a limited effect on the reduction of the structural response; on the contrary the effect of the suspension stiffness can be considerable.

4 SURFACE IRREGULARITY EFFECTS

The surface irregularities can depend on not accurate concrete casting, prefabricated element assemblage, on the progressive decay of bituminous and impermeable layers, or on the railway deformation.

A vehicle moving on an irregular surfaced deck exchanges a vertical interaction force system beyond the static load. The interaction forces depend mainly on the suspension characteristics, on the surface conditions and on vehicular speed; it can be source of disturb for the comfort and can accelerate the decay phenomena and the wheels and mechanical components damage.

Table 1. Surface irregularities effects.

Factors of impact D	V = 5 m/s	V = 10 m/s	V = 27.78 m/s	V = 42 m/s	V = 50 m/s	V = 65 m/s	V = 80 m/s	V = 120 m/s
Regular surface	1.0357	1.0688	1.1914	1.0672	1.1996	1.4039	1.5560	1.6988
Spectrum O.R.E. – level 1	1.0743	1.0948	1.2272	1.0897	1.2224	1.4407	1.5847	1.7358
Spectrum O.R.E. – level 2	1.0971	1.1355	1.2395	1.1244	1.2301	1.4488	1.6040	1.7593

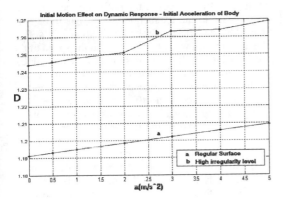

Figure 6. Initial motion effects on dynamic amplification.

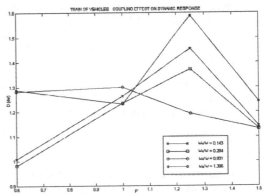

Figure 7. Coupling effects for train of vehicles.

A method for the generation of stochastic altimetric surface profiles uses Space Density Power Spectra.

The vehicles on the deck are excited by the spectrum and it is possible to get useful informations about comfort and road safety.

The computational algorithm uses the Power Density Spectrum introduced by O.R.E. (*Office de Recerches et Essais de l'Union Internazionale des Chemins de Fer*) for railways with two different levels of irregularity. The Spectrum for altimetric profile has an expression as follows:

$$S_v(\omega) = \frac{A_v * \Omega_C^2}{(\Omega^2 + \Omega_R^2) * (\Omega^2 + \Omega_C^2)}$$

The altimetric profile is set by the harmonic expression

$$y(z) = \sum_{n=1}^{N} A_n sen(\omega_n z + \phi_n)$$

$$A_n = \sqrt{2 * \Delta\omega * S(\omega_n)}$$

In the Table 1 is represented the comparison between the Factors of Impact for simple suspension vehicle $(K_v = 950000$ N/m; $C_v = 60000$ Ns/m$)$ at different speeds on irregular decks as results by applying Spectra O.R.E. for low and high levels of irregularity.

The surface irregularity can increase the dynamic amplification of the girder displacements, and the vertical body accelerations of the vehicles.

These effects generally affect the comfort levels and the serviceability of the structure in optimal quality standards.

For high span bridges these aspects can be significant and must be taken into account.

The relationship between the Factor of Impact and the vehicular body acceleration has been investigated by several computation with different initial accelerations and different levels of the surface regularity. $(\xi = 0\%;\ M_c = 266000$ kg; $M_e = 4000$ kg; $K_v = 950000$ N/m; $C_v = 60000$ Ns/m; $v = 27.78$ m/s$)$

Significant increases of the response show the importance of initial jumping of the vehicles for the irregularity at the entrance joint of the bridge.

In conclusion it is possible to say that the bridge response under the action of a simple suspension is controlled by four parameters:

– the speed parameter β;

- the rate between vehicular and bridge mass $x = M/m$;
- the natural frequency of the vehicle;
- the initial condition of the motion of the vehicle.

When a train of vehicles goes through a bridge, every section is subjected to a harmonic load, with a frequency that depends on the speed and the distance between the vehicles.

Besides the value of the force changes from a passage to the following one, because it depends on the interaction forces between structure and vehicle.

In the present work a uniform distribution of the vehicle has been supposed (a = distance between two subsequent vehicles); the frequency of the load on the generic section is

$$f' = \frac{1}{a/v} = \frac{v}{a}$$

In the present work has been introduced a Coupling Coefficient F' as the rate between the Fundamental Frequency of the girder and the load one:

$$F' = \frac{f_1}{f'}$$

The diagram above shows the relationship between F' and the Factor of Impact for several values of the rate ω_v/ω with reference to a 18 vehicles trains, different for their mutual distance, that has been supposed uniform: it is possible to check simil-resonant effects that can change the response, especially if the train length is comparable with the bridge span.

5 CONCLUSIONS

The first design codes for railway bridges used to increase the design loads to take into account the dynamic effects and keep the simple usual static load condition.

These codes, and most of the modern ones, correlate the span of the bridge and the Factor of Impact; other codes correlate the latter to the Fundamental Frequency.

By implementing the algorithm just presented, it has been shown also that other factors influence the response: the load speed, the mechanical characteristics, the initial condition at the entrance, the surface irregularities and the structural damping.

With regard to the choice of the structural model, the Euler-Bernoulli beam, supposed with mechanical characteristics equivalent to the real structure in every section, is considered suitable for the purpose of the analysis, for its simplicity and the contemporary ability to describe also in discrete formulation the global structural characteristics, the damping, the Fundamental Frequency and the modal frequencies and shapes.

The evaluation of the validity domain of each model is very complex: load models with no interaction can be suitable when the span of the bridge and consequently the rate between bridge and vehicle mass are high. The same model can be used when the speed parameter β is low; beyond a certain value, that depends also on the Fundamental Frequency and on the span, the no interaction models can be not adequate.

Taking into account all the parameters on the phenomenon is quite complex, because it is often impossible to foresee the characteristics of the vehicles that will pass through the bridge during its life.

Consequently in the design process the evaluation of the dynamic impact of the traffic on the structure, with regard to the interaction effects, has to be conducted by statistic and probabilistic approach: at first the designer knows only the span and the speed of the loads. After a preliminary design, in which the global characteristic of the structure are defined, the values of the Fundamental Frequency and the speed parameter are determined with sufficient accuracy.

By varying the vehicular speeds, the surface irregularities, the masses, the mechanical characteristics and the distribution of the moving loads, within their gaps evaluated by investigations on vehicular model in circulation, the Factors of Impact are determined. Consequently the envelop-diagram D-β is drawn; D increases with β at least at the ordinary traffic speeds.

For each β value, the greatest corresponding value of D is considered. This procedure has to be repeated after each structural change in the design.

This approach lets the designers to take into account the interaction effects by using the usual static load conditions for the analysis, as demanded for Performance Based Design philosophy.

This labourious approach, especially for important structure as suspension and cable-stayed bridge, allows a greater slenderness and lightness in agreement with the modern aesthetic and structural challenge.

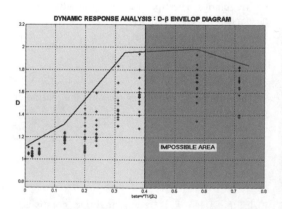

Figure 8.   Envelop diagram for the design process.

SPECIAL THANKS

The present analysis has been realized with the support of Prof. Franco Bontempi of the University "La Sapienza" of Rome, who suggested to the author the purposes of the whole work and indicated several useful researches. For this reason the author feels gratitude for the cultural and methodic growing occasion offered by the Professor during that experience.

REFERENCES

Inglis, C.E. 1934. *A mathematical treatise on vibrations in railways bridges*. Cambridge.

Wen, R.K. 1960. Dynamic response of beams traversed by two-axle loads. *Journal of Engineering Mechanics Division*. October 1960.

Biggs, J.M. 1964. *Introduction to structural dynamics*. New York: McGraw Hill.

Bathe, K.J. & Wilson, E.L. 1976. *Numerical methods in finite element analysis*, Elenwood Cliffs, N.J.: Prentice-Hall Inc.

Bathe, K.J. 1982. *Finite element procedures in engineering analysis*. Prentice Hall.

Clough P.W. & Penzien, J. 1993. *Dynamics of structures*. McGraw Hill Inc.

Fryba, L. 1996. *Dynamics of railways bridges*. Thomas Telford.

Panagin, R. 1997. *La dinamica del veicolo ferroviario*. Torino: Ed. Univ. Levrotto & Bella (*in Italian*).

Restat, F. & Prot, M. 1999. Ouvrages en interaction. Paris: Hermes Science (in French)

Petyt, M. 1990. *Introduction to finite element vibration analysis*. Cambridge University Press.

Coates, R.C., Coutue, M.G. & Kong, F.K. 1988. *Structural analysis*. Chapman & Hall.

Criteria for airport terminal roadway analysis post 911

H.W. Hessing
Professional Engineer, USA

ABSTRACT: This paper describes parameters and methodologies that may be utilized for planning terminal roadway and parking, post the September 11, 2001 tragedy.

1 INTRODUCTION

An airline carrier wishes to develop a new terminal. Typical and ordinary design, construction and construction support services are well defined by practice, an owner's request for a proposal and the Construction Management agreement. The technical approach presented herein will focus on the most important and critical issues involved in master planning the roadway portion of the project.

Based on an understanding of the project, its key issues the client's standards and requirements, we can develop an approach to master planning services by identifying the work program. Will an existing terminal be replaced and a new terminal constructed? Will the existing roadway system be adequate or need to be modified? What is the final configuration of the roadway cross section? Will separate arrivals and departures roadways be required? If elevated roadways are necessary, they will require footings and columns. What happens to the existing roadway traffic during utility relocation, and installation of footings and columns? Is a new parking structure required to accommodate the traveling public?

As program managers, we can develop an approach to determine curb configuration and operation, intersection capacity analysis, roadway lane requirements, and parking.

Of primary concern is determination of ground travel characteristics of air passengers and greeters in terms of average modal splits, vehicle occupancy rates, and vehicle curb dwell times for curbside operations.

Peak hour, air passenger forecasts for Departures and Arrivals are dependent upon flight schedules at the gate for both domestic and international sectors specifically on the arrival departure distribution profile that transitions passengers through the terminal. This will vary considerably based on changes to security and check-in procedures. An assumption for the load factor is required for each design year.

The paper will look at curb configuration and operation based on curb length demand estimates, intersection capacity analysis, roadway lane requirements, and parking. All are dependent upon the numbers derived from existing and future frontage requirements.

2 GROUND TRAVEL CHARACTERISTICS

Various ground transportation surveys and studies have been performed in the past for air passenger terminals at international airports. User data has been collected and developed which define travel characteristics of air passengers and related airport visitors. The attached tables summarize a sample set of ground travel characteristics of air passengers in terms of average modal splits, vehicle occupancy rates and vehicle curb dwell times for curbside operations.

By their very nature, these groundside characteristics are dynamic, and vary by flight sector, i.e., international and domestic, time of day, day of week, and season, weather conditions, level of traffic congestion on the regional highway system, and a host of other factors. As a consequence, the results of the empirical analyses produced by using the average values contained in these tables constitute best approximations suitable for facility planning purposes rather than rigid design parameters.

A rail rapid transit link is introduced as another parameter that will have an effect on the ground travel characteristics of air passengers using an airport. The figures shown in the tables are foot noted to account for the access modal split diversion that is expected to occur with the introduction of light rail service as well as the use of rail service for internal travel trips within the airport site, e.g., travel between remote parking and rental car lots and terminal buildings.

3 AIR PASSENGER FORECASTS

Peak hour air passenger forecasts for Departures and Arrivals are dependent upon flight schedules at the gate. For both Domestic and International sectors, the arrival departure distribution profile that transitions passengers through the terminal will vary considerably based on changes to security and check-in procedures. This information along with an assumption for load factor is required for each design year.

4 CURB CONFIGURATION AND OPERATION

Next, we determine an operation plan for Arrivals, then Departures.

5 CURB LENGTH DEMAND ESTIMATES

5.1 Methodology

1. Determine curb space required for arrivals curb
2. Determine curb length demands for each design year
– Peak hour
– Arrivals Curb 1
– Arrivals Curb 2
– Arrivals Curb 3
– Departures.

6 COMPARE CURB LENGTH DEMANDS VS CAPACITIES

6.1 Curb capacity

1. Total linear frontage (length) less
2. (–) Cross walks/signals
3. (–) Taxi stand
4. (–) Commercial (permittees) signed areas.

7 PEAK HOUR TRAFFIC VOLUMES

Required: Use the 24 hour passenger demand at the gate and convert it to peak hour traffic volumes at the curb.

7.1 Methodology

1. Enter the planning day schedules for analysis.
2. Calculate the peak 15 minutes demand ("at the gate") for the schedule.

3. Apply well-wisher/greeter ratios and arrival patterns to calculate demand at the curb of the terminal (e.g. entry to roads and parking).
4. Calculate the peak hour demand at the gate.
5. Calculate the peak hour demand at the curb by applying passenger arrival distributions.

8 EXAMPLE ASSUMPTIONS

	Current
Domestic Load Factor	80%
International LF	90%
Well wisher ratio for Domestic	3.0
Well wisher ratio for International	0.5
Greeter to passenger ratio	
For Domestic	0.4
For International	0.6

Tables 1, 2 and 3 follow. Two columns are shown. The column labeled Pre 911 is an example of data an airport planner would have had prior to September 11, 2001. The column labeled Post 911 is to be completed by the planner based on observations at the airport and terminals.

EXAMPLE PRE 911
Arrival Pattern for Passengers and Well Wishers for Domestic Departures

Table 1. Frontage/parking criteria example.

Key parameter	Pre 911	Post 911
Pax Modal Split-LOV		
Auto	37%	
Taxi	18%	
For-Hire	14%	
Rental Car	3%(1)	
Long Term Park	2%(1)	
Pax Modal Split-HOV		
Bus/Van	8%(1) Auto	
Courtesy Coach/Van	2%(1)	
Limo	12%	
Mass Transit/Off-Airport Bus	4%	
%Pax in Autos that Directly Access Parking Lot		
Domestic Arrivals-Terminating		
Auto	57%	
For-Hire	50%	
Domestic Departures-Originating		
Auto	24%	
For-Hire	10%	

(1) Shift to Light Rail (LRS).
(2) Did not shift to LRS.

Table 2. Frontage criteria example.		
Key parameter	Pre 911	Post 911
Vehicle Dwell Time (min) Departures-LOV	3.3	
Auto	1.4	
Taxi	2.9	
For-Hire	3.3 (1)	
Rental Car	3.3(1)	
Long Term Park		
Departures-HOV		
Bus	2.5 (1)	
Limo	1.8	
Courtesy Coach	1.8	
Mass Transit/Off-Airport Bus	2.5	
Arrivals-LOV		
Auto	3.3	
Taxi	1.4	
For-Hire	2.9	
Rental Car	3.3(1)	
Long Term Park	3.3(1)	
Arrivals-HOV	2.5(1)	
Bus	1.8	
Limo	1.8(1)	
Courtesy Coach	2.5	
Mass Transit/Off-Airport Bus	10	
Permittees/Van		

(1) Shift to LRS.
(2) Did not shift to LRS.
Note: Terminals have different peak times, therefore, parameters should vary.

Table 3. Frontage criteria example.		
Key parameter	Pre 911	Post 911
Vehicle Occupancy (Passengers) Departures-LOV		
Auto	1.5	
Taxi	1.5	
For-Hire	1.2	
Rental Car	2(1)	
Long Term Park	2(1)	
Departures-HOV		
Bus	10(1)	
Limo	4.5	
Courtesy Coach	2(2)	
Mass Transit/Off-Airport Bus	10	
Arrivals-LOV		
Auto	1.5	
Taxi	1.5	
For-Hire	1.2	
Rental Car	2(1)	
Long Term Park	2(1)	
Arrivals-HOV		
Bus	10(1)	
Limo	4.5	
Courtesy Coach	2(1)	
Mass Transit/Off-Airport Bus	10	
Permittees/Vans	3	

(1) Shift to LRS.
(2) Did not shift to LRS.
Note: Vehicle occupancy varies by terminal and time of day.

Min. before
Flight 150-135 135-120 120-105 105-190
%Arrival 10.0 10.0 10.0 10
90-75 75-60 60-45 45-30 15.0 15.0 20.0 10.0

EXAMPLE PRE 911
Arrival Pattern for Passengers and Well-Wishers
For International Departures
Min. before flight

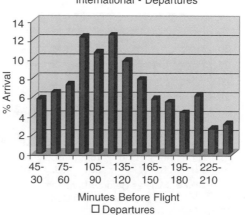

PAX Distribution
Domestic- Departures

PAX Distribution
International - Departures

1321

240-225-210-195
%Arrivals 180-165 150 225 210 195 180 165 150 135
3.2 2.7 6.2 4.4 5.5 5.8 7.9
135-120-105-90 75-60-45-120 105 90 75 60 45 30
9.8 12.5 10.7 12.3 7.3 6.5 5.8

EXAMPLE PRE 911
For Domestic Arrivals
Min. before flight
Min. after flight 45-30 30-15 15-0 0-15
%Arriving 25.0 25.0 30.0 20
Arrival Pattern (at the curb) for Domestic Arrivals
Min. after flight 0-15 15-30 30-45
45-60 %Arriving 15.0 50.0

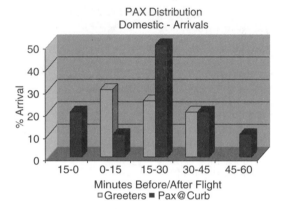

PAX Distribution
Domestic - Arrivals

Minutes Before/After Flight
□ Greeters ■ Pax@Curb

EXAMPLE PRE 911
Arrival Pattern For Greeters
For International Arrivals
Min. before flight 90-75 60-45-30-15-75-60 45-30 15 0
%Arriving 5.0 5.0 5.0 5.0 5.0 10.0

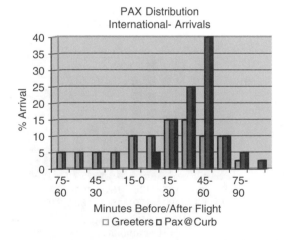

PAX Distribution
International- Arrivals

Minutes Before/After Flight
□ Greeters □ Pax@Curb

Min. after flight 0-15-30 45 60-75-90-15 30 45 6075
90 105
%Arriving 10.0 15.0 15.0 10.0 10.0 2.5 2.5

EXAMPLE PRE 911
Arrival Pattern (At The Curb) For Passengers For
International Flights
0-15-30-45-60-75-
Minutes after flight 15 30 45 60 75 90

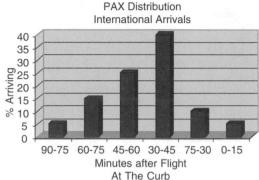

PAX Distribution
International Arrivals

Minutes after Flight
At The Curb

%Arriving 5.0 15.0 25.0 40.0 10.0 5.0
Vehicles entering the airport have four primary
movements

1. To terminals and then exit
2. To terminals, parking and then exit
3. To parking and then exit
4. To parking, terminals and then exit.

Determine the above four movements for each airline
based on "at the curb" passenger volumes, and compare to historical trends (%).
Use vehicle occupancy to convert the passengers
into vehicles.
Create a table to show peak hour vehicles for the
various LOVs and HOVs.

– Use actual survey data or
– Estimate percentage breakdown shown in planning
parameters. Analysis will require assigning, for
example, 45 MAAP, based on historical data and
passenger surveys. Use this to determine the various traffic movements. Make minor arithmetic
adjustments as necessary for roadway conditions.

Distribute Peak Hour Traffic Volumes On Roadway
Network.
Analyze Signalized Intersection.

9 CAPACITY ANALYSIS

– Do a capacity analysis at each intersection include
peak hour volumes, percentage of heavy trucks,
number of lanes, lane configuration, etc.
– Determine Level Of Service

This method is ok for cargo area but a Central Terminal Area (CTA) functions as a network of interconnected signals. Check the coordinated pattern of the CTA network.

9.1 *Roadway lane requirements*

Project volume during composite peak hour. As a base check look at the following:

- Current Standard vs. Proposed say 1,200 VPH/lane
- 800 VPH for roadways with signalized pedestrian crossings
- Compare projected volume to your current standard
- Check frontage lanes for sufficient capacity, blockage of access/egress, and triple parking then the interconnected roadways!

10 PARKING

The question is what are operational observations such as load factor?
Determine:

Parking Demand Estimate
Parking Entrance Requirements
Queue length
Parking Exit Requirements
Required number of lanes check queue length with one lane out of service.

17. Structural damage assessment

Assessment of fire damage in historic masonry structures

D.M. Lilley & A.V. March
University of Newcastle upon Tyne, UK

ABSTRACT: Extreme temperatures experienced in historic masonry structures as a result of severe fires result in varying levels of damage to the fabric of a building. In such circumstances, combustible elements are often completely destroyed or left in a condition where they are unusable and have to be replaced. Non-combustible material such as stone masonry can also be damaged by extreme heat. Large thermal deflections can render stonework unstable, and rapid local cooling from cold water used by those attempting to extinguish the fire risks cracking of hot masonry. Following a severe fire in a landmark church building at Brancepeth in County Durham in the North of England, the effects of extreme heat on sandstone masonry are currently being investigated at the University of Newcastle. A structural survey was made of the condition of the stonework, which showed unexpected localised variations in colour within individual stones. Damage in the form of peeling of some of the stones was observed and continued for several weeks after the fire was extinguished. Layers of stone became detached from the central core of some of the stonework columns in a similar fashion to the outer skins peeling from an onion. The paper reviews previous reports of fire damage to historic masonry, and discusses the investigations and assessments that took place following the fire. Brancepeth Church has been restored using as much of the original stonework as possible, and the temporary works required to support the upper sections of walls and columns were unusual in requiring ingenuity in concept and boldness of approach.

1 INTRODUCTION

Severe fires in structures of historic importance are relatively rare events and usually attract media attention outside the local area only if the building is of major national importance. Examples of such major events are the fires that occurred at Windsor Castle in November 1992, Hampton Court Palace in March 1986 and York Minster in July 1984.

Major damage involving partial or total collapse of main load-bearing elements of fire-affected structures is often obvious, but there may also be other effects which do not become evident for some considerable time after the fire has been extinguished.

There are two significant effects of a major fire on the structural stability of historic masonry. The first of these (and the most obvious) is the damage to combustible elements such as roof timbers, which can be completely destroyed. This removes any lateral restraint provided by these members to the masonry elements, thereby increasing their effective height and resulting in potential instability. A second and perhaps less obvious effect is that of the fire on the stonework itself. Although non-combustible, stone can be damaged by extreme heat, especially when followed by rapid local cooling from cold water used by those attempting to extinguish the fire. This creates large thermal gradients through the stone resulting in surface spalling and internal fracturing.

In historic masonry structures, the likely effects of fire on natural stone are to reduce its strength and to change its appearance, often by major changes from its original colour. There is clearly a need to identify where fire damage will enable stonework to be restored without replacement of stone and those situations where new stones will be required.

2 PREVIOUS STUDIES OF FIRE DAMAGE TO HERITAGE STRUCTURES

Chakrabarti et al. (1996) reported on a programme of work undertaken in the UK by the Building Research Establishment (BRE) to assess the effects of fire on natural stone in buildings. The programme also included work on possible guidance for restoration and conservation of stone in some historic buildings.

Aspects of the fire damage at Windsor Castle were presented by Dibb-Fuller et al. (1998), who

Figure 1.   Severe damage to Belford Church.

Figure 2.   Brancepeth Church after the fire.

concentrated on the behaviour of historic ironwork under fire conditions and its subsequent restoration.

The effects of the fire at Hampton Court and the subsequent restoration work were described by Dixon & Taylor (1993). Although the fire is thought to have burned slowly for some time before becoming severe, the brickwork masonry suffered relatively minor damage and required only localised repair. Stone balustrades, cornices and window surrounds (Portland limestone) experienced much more severe damage. It was reported, without further description, that individual stones were tested for internal flaws and any cracks were reinforced with resin and glass fibre. In some cases the damage was so severe that new stonework had to be introduced.

The fire in the South Transept of York Minster occurred through the night and is believed to have been caused by a lightening strike during a spectacular electrical storm. Damage to the stonework was clearly visible as there had been a change in the colour of the magnesian limestone from cream to pink in the areas around the southern gable wall and the tower arch of the South Transept. These areas are thought to have been in direct contact with flames and would have experienced extreme temperatures. Cracking of the stones was also apparent up to a depth of about 50 mm and large pieces of stone continued to break off and fall to the floor for about two days after the fire had been extinguished. The damage caused as a result of the fire was substantial and restoration work required the replacement of about 100 tonnes of limestone masonry.

Lightening strikes on masonry structures without effective protection systems can also generate extreme heat in stonework for very short periods of time. Such an event happened to the sandstone Tower of Belford Parish Church in Northumberland in May 2000 (see Lilley & March, 2001). The main damage was to the South West (SW) pinnacle, which was completely destroyed and obliterated down to roof level (see Figure 1). Sections of the masonry of the South and

West parapet walls immediately adjacent to the pinnacle were also missing. Fragments of stone fell over a large area of the church grounds and towards the surrounding buildings. Individual stones (some approximately 300 mm × 300 mm × 300 mm) were later found at distances up to 100 m from the base of the Tower.

The main focus of this paper is directed towards the events following an extreme fire which took place within a Grade 1 Listed church (see Figure 2) in the Parish of St Brandon, Brancepeth, County Durham, in the North East of England in the early hours of 16 September 1998.

3   BRANCEPETH CHURCH

The first church on this site was built in the 12th Century and gradually enlarged in stages up until the 15th Century. The internal appearance of the church building was much enhanced by the addition of decorative oak paneling and internal finishes at some time in the 17th Century. The church comprised a clerestoried Nave flanked by Aisles which clasped the West Tower, Transepts, Chancel with a Chapel and North and South Porches. The roof was formed from substantial lead sheeting supported by a structure made from oak timber. In view of the length of time over which the church buildings were developed and enhanced, it seems likely that stones used for construction came from several different quarries close to Brancepeth. The principal material used in the masonry construction is buff-coloured sandstone which occurs in abundance in the area around the church. This sandstone is from the carboniferous coal measure series and has a fine to medium grain with a density range of $2400 \, \text{kg/m}^3$ to $2600 \, \text{kg/m}^3$ and an expected average compressive strength of $55 \, \text{N/mm}^2$.

The church is in an isolated location and the fire was not discovered until 0400 hours on 16th September 1998. It was clear at the time of discovery that the fire had been in progress for some considerable

Figure 3. Damage and debris immediately after the fire.

time as the building was fully ablaze and the roof had collapsed. The Fire Service attended the fire immediately after the time of discovery, and pumped approximately 250,000 litres of water into the structure before finally extinguishing the fire later in the afternoon of the same day. Forensic investigation of the debris by experienced fire officers led to the conclusion that the fire had burned for several hours at a heat in the region of 800°C–1000°C. The main fuel for the fire had been provided by the extensive 17th Century oak paneling and furnishings for which the church was renowned. The fire had completely burned through the main oak timbers of the roof which were 300 mm × 300 mm in size. Only small fragments were left in the debris at ground level (see Figure 3).

4 EFFECT OF EXTREME HEAT ON SANDSTONE MASONRY

Chakrabarti et al. (1996) reported that earlier research at BRE and elsewhere had focused on changes to the colour of different types of stones and the effects of thermally induced stresses. In sandstones, extreme fire conditions can be expected to result in a colour change corresponding with the dehydration of iron compounds, which requires a temperature of 250–300°C. Brown- or buff-coloured sandstone changes colour to reddish brown but the change may not be apparent until the temperature of the stone has reached 400°C or more and can persist after heating to 1000°C. Internal cracking and shattering of sandstones often occurs if the temperatures exceed 573°C; this is a result of the quartz grains undergoing a phase transition from α-quartz to β-quartz at the normal transition temperature of 573°C. The stone is further weakened by the thermal shock which occurs on application of cooling water creating thermal induced stresses due to the low thermal diffusivity value of the stone.

Figure 4. Sandstone column with spalling.

Visual inspection of the stonework remaining in Brancepeth church shortly after the fire identified a clear and wide variation in colour change in individual stones, and clear evidence of spalling on the surface of many of the stones (see Figure 4). The spalling occurred in almost uniform layers with thicknesses of about 25–35 mm, and appeared to be similar to the effect of peeling an onion skin. General observation of the form of construction showed that, in general, the stonework had been laid so that the bedding plane of the sandstone was horizontal. However, the spalling followed the line of the surface contours of the stonework and up to three or sometimes four different layers of "onion skins" were evident.

Immediately after the fire, there was a period of assessment as to how to proceed with the severely damaged remains of the church. During that period decisions had to be made about the future of the structure, the structural implications for the stonework which had survived the fire, the likely costs of reconstruction or demolition, and the value of the church in terms of its contribution to and place within the heritage of its location. This period of time allowed the stonework to be monitored and simple non-destructive tests with a hammer were conducted on individual stones to try to establish their structural integrity. These simple tests revealed that stones which had been tested

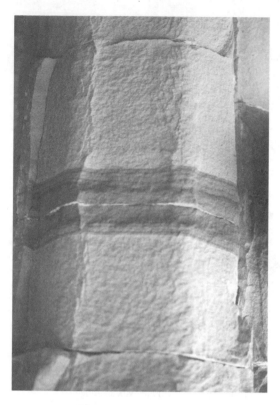

Figure 5. Stratified colour variation in a heat-affected sandstone voussoir.

Figure 6. Variation in change of colour with depth from exposed surface of sandstone.

after losing a first layer and which appeared to be intact were found to continue to spall. Long-term monitoring showed that spalling continued to occur for an extended period of 12–18 months. This was particularly noticeable in the voussoirs of the arches to the arcades.

There are three significant issues which were highlighted by the fire at Brancepeth. The most puzzling is the linear division in individual stones between sections where the sandstone has changed to its reddish colour and parts where the same stone has remained its original buff colour. This is clearly shown in Figure 5, where horizontal strips of two adjacent stones close to the mid-height of a voussoir are both discoloured red at the ends where they form a common connection. A possible explanation might be related to the material used for the lime mortar between the stones, but closer inspection of Figure 5 shows that there are similar joints with other stones above and below the affected stones which show no signs of change. In view of the ferocity and duration of the fire and the high temperatures believed to have been created, it seems unlikely that localised and precise variations in the temperature could have existed and produced the apparent stratification of the damage to the stonework.

An example of the apparent penetration of heat into an individual stone is shown in Figure 6, where the change to the reddish colour can be seen to emanate from a single location on the exposed face of a stone rather than simply being a function of the shortest distance from the exposed face. Gradations in the colour of the stone can clearly be seen, and suggest that thermal conductivity of the stone was not uniform. Figure 6 also indicates some impurities towards the centre and top of the stone. Such natural irregularities, if located close to the exposed face of a stone, might have produced deviations from linear change to the reddish colour.

The third issue and perhaps the most important from a safety viewpoint is the period of time over which the stonework continued to spall. Previously published work (see Chakrabarti et al. 1996) indicated that spalling was likely to continue for two or three days after the fire, but the fire at Brancepeth caused small sections of stone to fall to the ground for up to six months afterwards. This has serious implications for the safety of all those involved in investigation and emergency remedial works undertaken after a severe fire in a similar structure.

In view of the historic context and the importance of the church building, major funding was made available to support the decision to rebuild the church. It was intended to rebuild the church utilizing as much of the original masonry as possible. This made it essential that reliable information was found about the integrity of individual stones, particularly those forming the voussoirs and columns. In the light of the previous experience and misleading results given by non-destructive hammer tests which failed to reveal fracturing, it was decided to undertake a survey of the stonework in September 2000 using ground penetrating radar. This technique is an adaptation of geophysical method for remote sensing of sub-surface conditions, and has many uses ranging from archeological surveys to police investigations of possible sites of shallow graves.

5 RADAR SURVEY

Inspection using radar techniques has been successfully used by the authors (see Lilley & March 1998) in old masonry structures to make structural repairs with a minimum of disruption to the original fabric. One example is Dorothy Forster Court in Alnwick, Northumberland in North East England where the structure was located in a sensitive conservation area. The front wall of the building was a random rubble-filled sandstone wall, to which a brickwork facade had been added in the 18th Century. The facade was approximately 120 mm out of vertical alignment, contained localised bulges of up to 140 mm, areas of random cracking and generally appeared to be in poor condition. Radar inspection of the wall clearly indicated variations in the width of the cavity between the brickwork and stonework. Sections of wall with serious hidden problems were located and the dimensions of serious voids within the rubble core due to water penetration could also be determined.

At Brancepeth the equipment was adapted from a ground-penetrating radar system and used a GSSI SIR System radar with a 1.5 GHz antenna. Record length was 6 nanoseconds which equates with an approximate depth of penetration of 400 mm beneath the surface of the stonework. Data produced by the radar system was monitored and digitally recorded on site using a conventional modern Pentium-based computer with colour display. Data interpretation was made using Radmap software, which generates a colour diagram representing densities of material at different depths of penetration to the limit of the equipment and frequency of the wave-forms being used.

The basic elements of a ground-penetrating radar system consist of an antenna, processing console, a graphical display and a data recorder for further computer processing of data.

A radar survey proceeds by slowly moving the antenna, which is connected to the console by a cable, along a desired track-line. As the antenna is being moved it transmits radar impulses. These impulses are reflected from sub-surface horizons or interfaces, as well as discrete targets or individual objects.

Reflected pulses are sent via the cable to the processor, and then to the display, where a graphic presentation of sub-surface features is immediately produced. As the radar pulses are transmitted and received at a very high rate the data displayed on the screen represents a continuous sub-surface profile, or cross-sectional view of layering along the surveyed track-line.

The depth of penetration into the material, as well as the reflected energy from the various sub-surface features, depends on the electrical properties (conductivity and dielectric constant) of the materials and individual targets. The depth of penetration is also dependent of the central frequency of the antenna. In simple terms, the lower the frequency of the impulses, the greater is the depth of penetration but the resolution of the radar image is reduced. Conversely, higher radar frequencies penetrate to shallower depths penetration but give improved resolution. The 1.5 GHz antenna used for this survey can provide a maximum penetration of approximately 750 mm although this is very dependent on the material being tested.

Site conditions meant that the radar survey had to be performed using access provided by in-situ scaffolding and ladders. Prime interest lay in the stonework forming the arches and columns.

Each arch was made up of between 25 and 40 individual stones, and because of the shape in which the stones had been cut and shaped, each stone had five outer faces. Each face was surveyed to ensure good coverage of the stone and allow correlation between faces. The stones were surveyed in groups of between five and ten, depending on their size and the ease of access to a particular group of stones.

On either side of the arch stones were two faced cornice stones. Each of these was surveyed on one face only.

Four arches including the Tower Arch were surveyed in full but a further four arches were so badly damaged by the fire that only a small number of stones in seemingly random locations gave any quantifiable results. The difficulty in obtaining reliable results from the other stones was the result of their damaged uneven surfaces causing noise and distortion of the radar signals and data. The stones successfully surveyed on these arches were those which were visibly less severely damaged and which had a relatively smooth surface suitable for the antenna to be moved over. A typical result of a stone with a hidden delamination is shown in Figure 7.

The survey showed that even in the stones which appeared undamaged (and did not sound hollow when tapped with a hammer) there were varying degrees of internal delamination. The degree of the delamination found using the radar survey followed the 25–35 mm layering similar to the stones which had visibly failed.

6 COMPRESSION TESTS ON STONE SAMPLES

Samples stones within the Church were collected and accurately cut to cuboid shapes before being subjected to a uni-axial compressive test. Initial tests

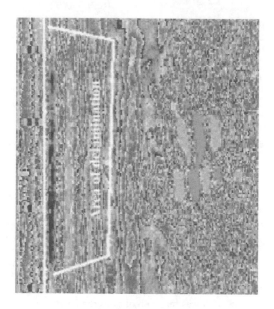

Figure 7. Results of a radar inspection of a stone with a hidden sub-surface delamination.

Figure 8. Sections of parent and delaminated sandstone taken from a fire-affected column.

on coarse grained stones indicated that the mean compressive strength of stone which appeared to be relatively unaffected by the fire was approximately 40 N/mm^2. Tests on sections of stone in which the colour had changed to red indicated that this had been reduced by about 20%. Further tests were undertaken on other stones which had a much finer grain size.

A stone was identified in which an outer section was at the point of delamination (see Figure 8), and separate cube samples were cut from the main part of the stone and the spalled section. These samples were tested to destruction and although the compressive strength was substantially higher than found with the coarse-grained stones, the strength of the outer leaf was approximately 10% less than that of the inner core.

Several of the stones which were removed from columns during the rebuilding works had suffered such a high degree of internal fragmentation that it was not possible to obtain a sample for testing.

7 CLOSING COMMENTS

Extreme fires in historic masonry buildings, although rare, can cause major damage to the load-bearing structure of walls, columns and arches. Very high temperatures possibly sustained for several hours can cause sandstone to change colour which is irreversible on cooling. The fire at Brancepeth demonstrated that this change of colour can be very localised and clearly defined in the local sandstone used for its construction. The compressive strength of individual stones is likely to be reduced by between 10% and 20% for the values for unaffected stones. In severe cases the internal fracturing can completely destroy the integrity of the stone.

Spalling of sandstone stonework can be expected where temperatures exceed approximately 570°C, and experience at Brancepeth has shown that this deterioration and delamination of stone can continue for several months after the fire.

Inspection of individual stones using a radar technique allowed previously hidden flaws to be identified and those with serious faults were replaced.

REFERENCES

Chakrabarti, B., Yates, T. & Lewry, A. 1996. Effect of fire damage on natural stonework in buildings, *Construction and Building Materials*, Elsevier Science Ltd, UK,10(7): 539–544.

Dibb-Fuller, D., Fewtrell, R. & Swift, R. 1998. Windsor Castle: fire behaviour and restoration of historic ironwork. *The Structural Engineer*, Institution of Structural Engineers, London, 76(19): 367–372.

Dixon, R. & Taylor, P. 1993. Hampton Court: restoration of the fire-damaged structure. *The Structural Engineer*, Institution of Structural Engineers, London, 71(18): 321–325.

Lilley, D.M. & March, A.V., 1998. "Methods of investigation and repair in old random rubble-filled masonry walls", *Proceedings of the Fifth International Masonry Conference*, ISSN 0950-9615, October, London, 337–341.

Lilley, D.M. & March, A.V., 2001. The effect of a lightning strike on a masonry church tower", *Proceedings of the Ninth International Conference on Structural Faults and Repair*, London, July, ISBN 0-947644-47-4, published in CD format only.

Detection of crack patterns on RAC pre-stressed beam using ultrasonic pulse velocity

E. Dolara
University of Rome "La Sapienza", Italy

G. Di Niro
University of Strathclyde, Glasgow, UK

ABSTRACT: The compressive strength of the original concrete in the demolished structure processed in the recycling site plant is almost in all cases unknown.

For the structural use of the Recycled Aggregate Concrete it would be useful sometimes to have a rough idea of the strength of the original concrete.

The possibility of a pre-cast concrete plant using RAC on a large scale means that there is a need to find a reliable method of testing the structures made from RAC.

Following the above observations a large part of this experimentation has been dedicated to the possibility of deducing the mechanical properties of RAC using non-destructive tests.

1 INTRODUCTION

The non-destructive tests adopted have been the Rebound Hammer and the Ultrasonic Pulse Velocity.

The experimentation has been carried out firstly on RAC specimens at different ages.

The measurements have been carried out using the direct transmission technique and indirect or surface transmission has been used as comparison.

The investigation described previously has been a preparatory study to the experimentation reported in this paper.

More detailed tests have been carried out on full scale pre-stressed beams made from RAC reported in this paper.

In fact during the last part of the study the ultrasonic method has been used to detect the depth of the cracks occurring during the pre-stressed beam test.

Comparing these results with the experimental results has checked the validity of this method.

2 NON-DESTRUCTIVE TESTING ON RAC PRESTRESSED BEAMS

The experimentation reported in this paragraph has been a forerunner to the later work in detecting the crack patterns during the test to failure of the pre-stressed beams made from RAC.

An investigation using rebound hammer and ultrasonic pulse velocity tests on different zones of the beams have been performed at different ages (3, 7, 28 days).

This part of the investigation has been carried out to deduce the compressive strength of concrete using non-destructive tests and to compare the results with the values coming from compressive strength cube tests cast with the same batch of the beams.

The mix-design named 525 MAR 2 has been the one chosen to cast the pre-stressed beams (with a section of 40 cm × 70 cm and a span of 1500 cm) because it had the best mechanical performances of the previous experimentation.

The first beam named with the code 40 AR 100 has been cast with a concrete made with 100% of recycled aggregate.

The second beam coded 40 AN 100 has been cast with concrete made with 100% of natural aggregate has for normal production in the Mabo s.p.a. site plant.

The third beam named with the code 40 RN 50 has been cast with a concrete made with 50% of recycled aggregate and 50% of natural aggregate. Cubes cast together with the beams had a compressive strength over 40 MPa.

Figure 1.   Beam casting phases.

Compressive, tensile, flexural strength and modulus of elasticity tests carried out on specimens cast together with the beams are reported in Chapter Five.

In Figure 1 a beam casting phase is shown. The reinforcement bars and the pre-stressing tendons together with the strain-gauges are also shown.

In Figure 2 the position of the points where the ultrasonic pulse velocity has been measured is reported together with a beam section that shows the path length of the ultrasonic wave inside the beams.

During test measurements the transducers have been placed in positions such that the reinforced bars and pre-stressing tendons would not influence the path length of the ultrasonic wave.

The measurement point grid is of 100 mm × 100 mm (Fig. 3).

The concrete surfaces on which the measurements were taken have been treated before the test to avoid any unevenness due to casting phases.

Full test results of non-destructive tests on specimens and full scale pre-stressed beams are reported in (Di Niro, PhD Thesis).

In particular compressive strength (Rc) (on 100 mm × 100 mm × 100 mm cubes), the rebound hammer number (I_R) measured on cubes and beams (on the previous shown grid), the pulse ultrasonic velocity (V) measured on cubes and beams (on the previous shown grid) are reported in Table 1.

The rebound hammer number (I_R) and the ultrasonic pulse velocity (V) are an average of a series of different measurements carried out on cubes and beams tested.

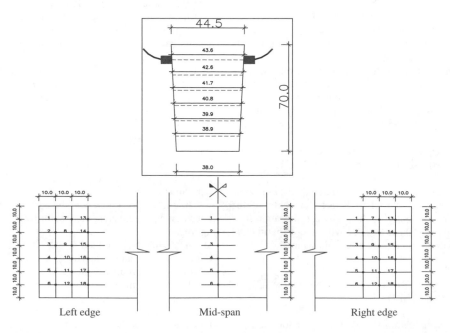

Figure 2.   Position for measurements for rebound hammer and pulse ultrasonic velocity tests.

2.1 Correlation rules between compressive strength (R_C), ultrasonic pulse velocity (V) and rebound number (I_R)

Compressive strength test results coming from crushing cubes (100 mm × 100 mm × 100 mm) have been used to check the reliability of results achieved with the non-destructive methods (rebound hammer and pulse ultrasonic velocity test) on the full scale pre-stressed beams and on cubes.

The correlation rules between the compressive strength (R_C), the rebound number (I_R), the pulse ultrasonic velocity (V) that have been used to evaluate the figures reported in Table 2 are given in the next page.

The Sonreb method is a way of evaluation of the compressive strength of concrete using results from both the Hammer test and the Pulse velocity test. Results from this method are also reported in Table 2.

Correlation rules that connect the compressive Strength R_C, ultrasonic pulse velocity V and the rebound Hammer number I_R are exponential and in particular:

– correlation rule R_C-I_R:

$$R_{C1} = k \cdot e^{c \cdot I_R} \tag{1}$$

where R_C is expressed in MPa, and I_R is the average rebound number.

– correlation rule R_C-V:

$$R_{C3} = a \cdot e^{b \cdot V} \tag{2}$$

where R_C is expressed in MPa, and the ultrasonic pulse velocity V is expressed in km/sec.

The Rules 1 and 2 are then combined into the Sonreb method with the use of others coefficients.

$$R_{C4} = A \cdot e^{(B \cdot V + C \cdot I_R)} \tag{3}$$

where $A = (a \cdot k)^{0.5}$, $B = b/2$, $C = c/2$.
Another correlation rule R_C-I_R commonly used is:

$$R_{C2} = s \cdot I_R{}^t \tag{4}$$

In the previous correlation rules there are coefficients a, b, c, e, K that are functions of the batch

Figure 3. Pre-stressed beam at support edge with the measurement point for rebound and pulse ultrasonic velocity tests.

Table 1. Comparison of the compressive strength R_C, the rebound number I_R, and the ultrasonic velocity V evaluated on specimens and on full scale beams.

Test		40AR100 (days)			40RN50 (days)			40AN100 (days)		
		3	7	28	3	7	28	3	7	28
R_C(Mpa)	cubes	33.9	36.6	44.4	34.0	42.4	45.8	43.3	47.8	59.4
I_R	cubes	32.4	35.3	36.8	34.3	37.5	38.1	36.5	38.5	39.7
I_R	beams	33.8	37.4	38.3	36.5	38.7	39.5	37.1	39.2	40.9
V (m/sec)	cubes	3680	3730	3845	3526	3825	3889	4065	4225	4372
V (m/sec)	beams	3702	3720	3751	3493	3955	4168	4108	4265	4427

Table 2. Comparison of the compressive strength R_C evaluated with destructive and non-desctructive tests on cube and full scale beams.

Test	R_C (Mpa)		40AR100 (days)			40RN50 (days)			40AN100 (days)		
			3	7	28	3	7	28	3	7	28
Crushing	R_C	cube	33.9	36.6	44.4	34.0	42.4	45.8	43.3	47.8	59.4
Schmidt Hammer	R_{C1}	cube	29.8	37.6	42.4	34.7	44.8	46.9	41.4	48.5	53.4
	R_{C2}	cube	26.9	32.6	35.9	33.6	41.2	42.7	42.2	47.7	51.1
	R_{C1}	beam	33.3	44.3	47.9	41.4	49.2	52.4	43.4	51.4	58.6
	R_{C2}	beam	29.6	37.1	39.3	38.7	44.2	46.3	36.5	41.4	45.4
Pulse Velocity	R_{C3}	cube	28.7	31.8	40.3	25.3	42.6	47.6	37.6	49.5	58.5
	R_{C3}	beam	31.2	30.1	33.2	23.9	53.4	77.4	40.1	50.18	63.3
Sonreb method	R_{C4}	cube	29.3	34.6	41.4	23.4	33.0	35.4	37.7	46.8	53.4
	R_{C4}	beam	32.3	36.7	39.9	24.5	38.0	45.7	39.8	48.5	58.2

Table 3. Coefficient values for the rules between, R_C, I_R, and V determined during this research.

Mix code	k	c	a	b	s	t	A	B	C
40AR100	2.25	0.08	0.015	2.060	0.010	2.27	0.184	1.0300	0.04
40RN50	2.25	0.08	0.110	1.430	0.012	2.27	0.475	0.7150	0.04
40AN100	2.25	0.08	0.055	1.745	0.011	2.27	0.352	0.8873	0.04

Table 4. Coefficient values for the rules between R_C, I_R, and V determined by Ravindrajah (Ravindrarajah & Tam 1985).

Types of aggregate	k	c	a	b	s	t	A	B	C
Natural	7.25	0.08	0.060	1.44	–	–	0.67	0.72	0.04
Recycled	7.25	0.08	0.068	2.06	–	–	0.24	1.03	0.04

characteristics, the curing environment, the type of cement, and the type of aggregate used.

These coefficients have been determined during the present investigation and they are reported in Table 3. In Table 4 the same coefficients coming from another experimentation (Ravindrarajah & Tam 1985) are reported. Comparing the two tables the differences between the coefficients coming from the two different experimentation can be noticed.

These differences are attributable to the different starting test conditions (batch characteristics, curing environment, etc.) and to the different recycled aggregate used. In particular the percentage of natural aggregate in the batch seems to be an important factor on the variation of the values of constant **a** and **b**. The test results coming from the indirect methods (indirect transmission) are shown in histograms in (Di Niro, PhD Thesis).

3 DETECTION OF CRACK PATTERNS ON RAC PRE-STRESSED BEAM USING ULTRASONIC PULSE VELOCITY

The investigation described in the previous paragraph has been a preparatory study to the experimentation reported in this paragraph.

In fact during this last part of the study the ultrasonic method has been used to detect the depth of the cracks occurring during the pre-stressed beam test.

Comparing these results with the experimental results has checked the validity of this method.

The indirect transmission has been the ultrasonic method used.

The beam zone investigated has been the mid-span one.

In Figure 4 the reinforcement bars and the pre-stressing tendons are shown. It can be noticed that

Figure 4. Reinforcement bars and the prestressing tendons inside the beam.

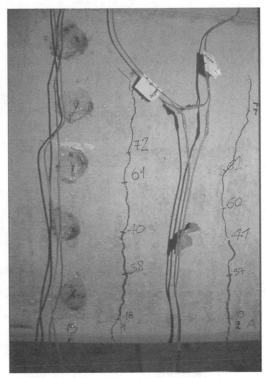

Figure 5. Crack detection in the beam mid-span.

reinforcement (bars and tendons) present in the mid part of the bottom of the beam are sufficiently far from the zone investigated with the ultrasonic method.

In this way the path length of the ultrasonic wave is not diverted by reinforcements.

Load rate for beam test was of 1000 kg (500 kg each jack). When cracks have been noticed they have been marked and coded with a capital letter (Fig. 5).

After that the ultrasonic transducers have been placed astride the crack layer with a distance of the presumed crack depth.

The distance between the two transducers was 20 cm.

To improve the ultrasonic wave transit, glycerine paste has been used.

The transducers have been placed in the mid part of the beam intrados where the ultrasonic wave are not influenced by the presence of reinforcement.

With the transducers in this position the ultrasonic wave is transmitted along a presumed path as reported in Figure 6.

The real path of the ultrasonic wave is very difficult to define for the refraction and reflection wave phenomena due to the shape of the cracks.

Moreover it is possible that there are contact points inside the crack layer that allow the undesired energy transmission of the sonorous wave.

As can be seen in the following Tables 5 to 7 (where a comparison between the optical and ultrasonic crack measurement is reported) that this inconvenience sometimes can give an erroneous measurement using the ultrasonic method.

The crack depth has been determined with the following relation, that is a partial modification of the relation in (Bocca 1983):

$$h_u = \alpha \cdot V_0 \cdot \sqrt{t_i^2 - t_0^2}$$

where:

α is a calibration coefficient for the evaluation of the cracks height (determined during this research);

$\alpha = 1.3$ is for concrete with 100% of RA;

$\alpha = 1.1$ is for concrete with 50% of RA and 50% of NA;

$\alpha = 0.7$ is for concrete with 100% of NA;

V_0 is the ultrasonic velocity inside the uncracked concrete;

t_i is the propagation time astride the cracks;

t_0 is the propagation time inside the uncracked concrete.

The measurements have been taken under two different load levels.

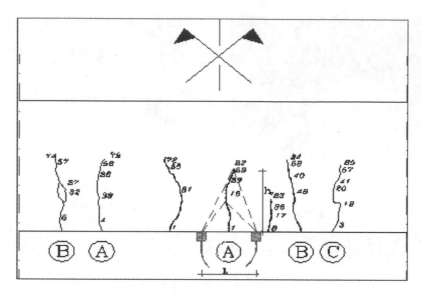

Figure 6. Path of the ultrasonic wave (—) on the crak layer.

Table 5. Comparison between the height of crack measures determined with the optical method and the ultrasonic one. Beam 40AR100.

	P = 157.5 N							
	Left edge				Right edge			
$\varepsilon = 315$ Crack	t_1 average (μsec)	Optical reading (h_O(mm))	Ultrasonic reading (h_v(mm))	Δ ($h_O - h_v$)	t_1 average (μsec)	Optical reading (h_O(mm))	Ultrasonic reading (h_v(mm))	Δ ($h_O - h_v$)
A	91.4	310	341	−31	75.0	350	237	113
B	87.1	270	316	−46	71.8	310	214	96
C	87.8	340	319	21	86.9	340	314	26
D	86.0	320	309	11	88.9	330	326	4
E	86.6	340	312	28	92.6	340	349	−9
F	80.8	330	276	54	87.4	330	317	13
G	77.5	280	255	25	85.7	360	307	53
H	91.8	340	343	−3	89.5	260	330	−70
I	92.4	280	347	−67	88.9	320	326	−4
L	90.7	220	337	−117	86.6	250	312	−62

The first measurement was taken with height of cracks of almost 20 cm.

The second one (reported in Tables in the Appendix A) when the crack height was over the half height of the beam.

4 CONCLUDING REMARKS

From the above results (Table 2) it can be concluded that the values of V and I_R are lower in RAC than NAC.

The use of RA in concrete seems not to affect the correlation Rule 1 but on the other hand it does influence the correlation Rule 2.

The Sonreb combined method gives results close to the values attained for crushing test on cubes for the mixes AR "All Recycled" and AN "All Natural".

The differences of values the compressive strength are almost 10 to 15%, better than the single method (rebound or ultrasonic).

Table 6. Comparison between the height of crack measures determined with the optical method and the ultrasonic one. Beam 40RN50.

| | P = 180 KN | | | | | | | | |
| | Left edge | | | | Right edge | | | | |
$\varepsilon = 360$ Crack	t_1 average (μsec)	Optical reading (h_O(mm))	Ultrasonic reading (h_v(mm))	Δ ($h_O - h_v$)		t_1 average (μsec)	Optical reading (h_O(mm))	Ultrasonic reading (h_v(mm))	Δ ($h_O - h_v$)
A	86.4	210	250	−40		83.2	190	235	−45
B	75.9	190	198	−8		85.7	150	247	−97
C	86.7	210	252	−42		82.8	220	233	−13
D	86.1	180	249	−69		87.9	250	258	−8
E	87.5	230	256	−26		96.4	40	297	−257
F	88.6	230	261	−31		89.3	140	264	−124
G	84.2	130	240	−110		89.9	210	267	−57
H	87.5	120	256	−136		90.6	190	270	−80
I	87.0	170	253	−83		87.8	210	257	−47
L	84.6	60	242	−182		86.1	180	249	−69

Table 7. Comparison between the height of crack measures determined with the optical method and the ultrasonic one. Beam 40AN100.

| | P = 185 KN | | | | | | | | |
| | Left edge | | | | Right edge | | | | |
$\varepsilon = 370$ Crack	t_1 average (μsec)	Optical reading (h_O(mm))	Ultrasonic reading (h_v(mm))	Δ ($h_O - h_v$)		t_1 average (μsec)	Optical reading (h_O(mm))	Ultrasonic reading (h_v(mm))	Δ ($h_O - h_v$)
A	113.7	200	283	−83		118.8	140	299	−159
B	81.5	250	178	72		111.7	100	277	−177
C	116.0	220	290	−70		84.8	90	190	−100
D	111.3	220	276	−56		104.8	140	255	−115
E	85.2	150	191	−41		108.6	150	267	−117
F	89.4	210	205	5		108.5	150	267	−117
G	84.6	220	189	31		110.2	190	272	−82
H	107.3	190	263	−73		81.8	90	179	−89
I	108.3	180	266	−86		81.5	40	174	−134
L	111.3	150	276	−126		67.2	40	124	−84

This is due to the fact that the Sonreb combined method attenuates the intrinsic imprecision of the single methods that are due to the curing conditions, the moisture content, and to the type of aggregate used.

More disperse are the values relative to the mix 40 RN 50 with differences of 30% to 40%.

From the previous study it can be concluded that this non-destructive investigation can be usefully used to check the real condition, for safety evaluation, of the in-situ concrete structures.

In this way if a concrete structure should be demolished the strength of the original concrete could be known for a recycling use of it.

In fact at the moment, except for pre-cast plant concrete waste material, the compressive strength of the original concrete in the demolished structure processed in the recycling site plant is almost in all case unknown.

Moreover this non-destructive method could be used to monitor the service life of a prototype structure made from RAC.

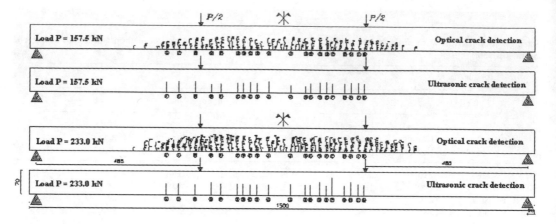

Figure 7. Comparison between the real crack patterns and the crack patterns evaluated with the ultrasonic method for the beam 40AR100.

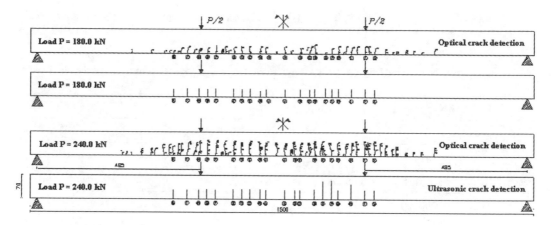

Figure 8. Comparison between the real crack patterns and the crack patterns evaluated with the ultrasonic method for the beam 40RN50.

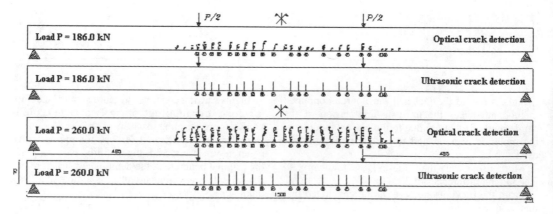

Figure 9. Comparison between the real crack patterns and the crack patterns evaluated with the ultrasonic method for the beam 40AN100.

5 ACKNOWLEDGEMENT

The authors are grateful for the support of Pescale S.p.a. in Castellarano (Reggio Emilia) and Mabo Prefabbricati S.p.a. in Bibbiena (Arezzo).

REFERENCES

Dolara, E., Ridgway, P. & Di Niro, G. 1995. L'aggregato riciclato. Caratteristiche chimico-fisiche. *Quarry and Construction. Edizioni Pei. Parma. Dicembre 1995.*

Dolara, E., Ridgway, P. & Di Niro, G. 1996. Proprietà e mix design del calcestruzzo fresco prodotto con aggregati riciclati. *Quarry and Construction. Edizioni Pei. Parma. Gennaio 1996.*

Dolara, E., Ridgway, P. & Di Niro, G. 1996. Analisi non lineare di una trave in conglomerato cementizio armato prodotto con aggregati riciclati *Quarry and Construction. Edizioni Pei. Parma. Marzo 1996.*

Dolara, E., Di Niro, G. & Ridgway, P. 1996. Il calcestruzzo prodotto con aggregati riciclati. *Mabogazine. Mabo Prefabbricati s.r.l.. Maggio 1996.*

Di Niro, G., Dolara, E. & Ridgway P. 1996. Recycled Aggregate Concrete (RAC). Properties of aggregate and RC beams made from RAC. *Proceedings of the International Conference "Concrete in Service of Mankind". Dundee, Scotland. 24–28 June 1996. Pag. 141–149.* E & F.N. Spon, London 1996.

Di Niro, G., Dolara, E. & Cairns, R. 1997. Recycled Aggregate Concrete (RAC). A mix-design procedure. *Proceedings of the International FIP 97 Synposium "The Concrete Way to Development". Johannesburg, South Africa. 9–12 March 1997.* Concrete Society of Southern Africa.

Di Niro, G., Dolara, E., Cairns, R. & Ridgway P. 1996. Recycled Aggregate Concrete (RAC). A mix-design procedure. *Proceedings of the International Conference "Our world in Concrete and Structures". August 1996. Orchard Plaza. Singapore.* M.

Di Niro, G., Dolara, E. & Cairns, R. 1998. The use of RAC (Recycled Aggregate Concrete) for structural purposes in prefabrication. *Proceedings of the XIIIth FIP – CEB 98 Congress "Challenges for Concrete in the next millenium" Amsterdam, The Netherlands. 23–29 May 1998.* A.A. Balkema Publishers.

Sciotti, M., Cairns, R., Di Niro, G. & Dolara, E. 1998. Bonding surface between recycled aggregate and new cement paste. *Proceedings of the International Symposium organised by the Concrete Technology Unit, University of Dundee and held at the Department of Trade and Industry Conference Centre, London, United Kingdom, 11–12 November 1998.* Thomas Telford Publishing.

Di Niro, G., Cairns, R. & Dolara, E. 1998. Properties of hardened RAC for structural purposes. *Proceedings of the International Symposium organised by the Concrete Technology Unit, University of Dundee and held at the Department of Trade and Industry Conference Centre, London, United Kingdom, 11–12 November 1998.* Thomas Telford Publishing.

Dolara, E., Di Niro, G. & Cairns, R. 1998. RAC prestressed beams. *Proceedings of the International Symposium organised by the Concrete Technology Unit, University of Dundee and held at the Department of Trade and Industry Conference Centre, London, United Kingdom, 11–12 November 1998.* Thomas Telford Publishing.

Cairns, R., Di Niro, G. & Dolara, E. 1998. The use of RAC in prefabrication. *Proceedings of the International Symposium organised by the Concrete Technology Unit, University of Dundee and held at the Department of Trade and Industry Conference Centre, London, United Kingdom, 11–12 November 1998.* Thomas Telford Publishing.

Di Niro, G. Recycled Aggregate Concrete for structural purposes. *PhD Thesis in Civil Engineering.* University of Strathclyde, Glasgow, UK.

Dolara, E. & Di Niro, G. 2002. Caratteristiche meccaniche del RAC: calibrazione dei coefficienti per NDT, *Quarry and Construction. Edizioni Pei. Parma. Giugno 2002.*

Ravindrarajah, R.S. & Tam, T.C. Properties of Concrete made with Crushed Concrete as Coarse Aggregate. Magazine of Concrete Research, 37, No. 130. 1985.

Bocca, P. I difetti delle strutture finite rilevati mediante ultrasuoni. Industria Italiana del Cemento 6/1983,

Structural failures: lessons for a brighter future

K.L. Carper
School of Architecture & Construction Management, Washington State University, Pullman, WA, USA

ABSTRACT: Investigating structural failures and providing the information gained from these investigations to practicing design professionals can bring about improvements in the performance of future designs. Today, despite the availability of advanced computational tools, failures continue to plague the construction industry. Many failures are caused by repeated errors. These could be avoided if designers were more diligent in heeding the lessons of the past. The community of forensic engineers is working internationally to address common concerns and to enhance the dissemination of failure-related information. If future designs are to be more beautiful, functional, economical and sustainable than those built in the past, then designers and builders must continue to learn lessons from the successes and the failures of their colleagues.

1 EXPERIENCE: A VALUABLE TEACHER

Failure is a costly way to learn, but lessons from structural failures have often contributed to improvements in design and construction practice. In 1856, while serving as President of the UK Institution of Civil Engineers (ICE), Robert Stevenson made these comments:

Nothing is so instructive to the younger members of the profession as the record of accidents in large works, and of the means employed in repairing the damage. A faithful account of those accidents, and of the means by which the consequences were met, is really more valuable than a description of the most successful works. Older engineers have derived their most useful store of experience from the observation of those casualties which have occurred to their own and to other works, and it is most important that they should be faithfully recorded in the archives of the Institution.

Forensic engineers have been called the pathologists of the engineering professions. The results of their investigations, when properly communicated to practitioners, can contribute to improvements in engineering design in the same way that medical pathologists have contributed to improved medical practices (Feld & Carper 1997). Currently, forensic engineers in the US are finding new ways to enhance the dissemination of failure-related information. They are actively publishing the results of their investigations and creating resources for use in undergraduate, graduate and continuing professional education. These activities, outside the litigation arena, will potentially contribute to a reduction in the frequency and severity of failures.

It is well-known that many professional engineering societies in Europe, the UK and elsewhere have actively engaged in failure-related information dissemination for many decades. In the US, the American Society of Civil Engineers (ASCE) has taken a leading role in this effort, especially since the establishment of its Technical Council on Forensic Engineering (TCFE) in 1985. This paper reviews TCFE activities, including the desire to initiate and strengthen liaisons with like-minded professional societies internationally.

Those who investigate engineering failures are in a unique position to identify emerging problems in the industry and to suggest strategies for mitigating against these trends. Several contemporary concerns are noted in this paper – topics that have been identified for focused mitigation strategies by the TCFE, the UK Standing Committee on Structural Safety (SCOSS), the Chamber of Forensic Engineers of the Czech Republic and other international organizations. These professional organizations are seeking ways to coordinate activities so that their combined efforts in addressing common concerns are more effective worldwide.

If future designs are to be more beautiful, functional, economical and sustainable than those built in the past, then designers and builders must continue to learn lessons from the successes and the failures of their colleagues.

2 ACTIVITIES OF THE ASCE TECHNICAL COUNCIL ON FORENSIC ENGINEERING

The creation of the TCFE came at the end of a period of intense public scrutiny within the US of the construction industry and its design professionals. A number of catastrophic structural failures occurred in the 1970s and early 1980s. These were widely published in the public media. The failures included collapse of completed buildings, as well as several fatal construction accidents. Prominent incidents included:

– Collapse of a 17-storey reinforced concrete building in Boston, Massachusetts, that killed four construction workers (January 1970).
– Progressive collapse of a 26-storey reinforced concrete residential tower at Bailey's Crossroads, Virginia, causing the deaths of 14 construction workers (March 1973).
– Total collapse of a longspan steel space truss roof over a completed coliseum in Hartford, Connecticut (January 1978).
– Failure of a completed longspan steel and aluminum dome in Greenvale, New York (January 1978).
– Collapse of a Cooling Tower scaffold and structure at Willow Island, West Virginia, accompanied by 51 deaths (April 1978).
– Failure of a completed longspan coliseum roof structure in Kansas City, Missouri (June 1979).
– Collapse of the Rosemont Horizon Arena near Chicago, Illinois, that killed five construction workers (August 1979).
– Collapse of a five-storey reinforced concrete condominium building in Cocoa Beach, Florida, causing the deaths of 11 construction workers (March 1981).
– Failure of walkways in a completed hotel in Kansas City, Missouri, resulting in the deaths of 114 occupants (July 1981).
– Collapse of the falsework supporting construction of a highway ramp in East Chicago, Indiana, accompanied by 13 deaths (April 1982).

These dramatic structural failures, occurring over such a short period, understandably caused a great deal of public alarm. The US Congress held a series of hearings regarding structural failures beginning in August 1982. A number of industry experts testified and several recommendations emerged from the hearings. Some of these recommendations have been addressed through innovative ASCE programs, including those undertaken by the Technical Council on Forensic Engineering.

The TCFE regularly sponsors symposia for practicing design and construction professionals, including three international Forensic Engineering Congresses held in Minneapolis, Minnesota; San Juan, Puerto Rico; and San Diego, California (Rens 1997, Rens et al. 2000, Bosela et al. 2003). The TCFE publishes guidelines for

investigations and for expert witnesses, and provides resources for civil engineering education. The ASCE *Journal of Performance of Constructed Facilities*, is published under the direction of the TCFE. This journal has become established internationally as a prominent resource for practitioners who wish to learn from failure case histories.

The TCFE maintains an Executive Committee, an Awards Committee and six standing committees:

– Committee on Dissemination of Failure Information
– Committee on Practices to Reduce Failures
– Forensic Practices Committee
– Committee on Education
– Committee on Technology Implementation
– Publications Committee.

In addition to the above committees, temporary task committees are regularly established to address specific topics. These have included such topics as project peer review, lift-slab construction, and avoiding failures caused by misuse of computers.

Further information on the history and current activities of the TCFE is given in Carper (2003).

3 ASCE JOURNAL OF PERFORMANCE OF CONSTRUCTED FACILITIES

An early survey undertaken by the TCFE determined the need for expanded discussion of failures in the professional engineering literature. Within the US, in particular, the engineering professions had been somewhat remiss in openly acknowledging failures and reporting on their underlying causes.

In response to these concerns, the *Journal of Performance of Constructed Facilities (JPCF)* began publication in February 1987. The journal aims to improve quality of the constructed project through interdisciplinary communication. Papers examine the causes and costs of failures and other performance problems. Catastrophic failures as well as serviceability problems are reviewed. Both procedural (human error) and technical causes of failures are included. Papers that discuss the interface between various professionals in the construction industry and management deficiencies are especially encouraged, as are papers submitted by international authors. The principal intended audience for the JPCF is the community of engineering practitioners. Many members of the editorial board are practicing design professionals, and the majority of published papers are written or co-authored by practitioners.

Guidelines, prepared by the journal's Committee on Publications, encourage ethical standards for publishing failure case histories. These guidelines emphasize the purpose of discussing failures. The goal is not to criticize the individuals involved in the unfortunate events, but rather to learn from their experiences in order to avoid repetition of errors. All papers are open to formal discussion, so that all relevant views can be expressed. In order to ensure interdisciplinary participation, an international editorial board comprised of members representing a wide variety of industry professionals is maintained.

The JPCF serves as a catalyst to encourage failure-related information dissemination. For example, in late 1988, incomplete forensic investigations of the complicated L'Ambiance Plaza lift-slab project construction failure were halted as the result of a remarkable and controversial mediated settlement. The JPCF Committee on Publications recognized an opportunity to bring the various investigators together to help assemble the missing pieces. A meeting was held in Arlington, Massachusetts, on March 14, 1989. While there may never be full agreement on the triggering cause of this complicated collapse, this well-attended meeting led to the publication of many useful papers, to a symposium, and to the formation of a task committee on lift-slab construction.

The *Journal of Performance of Constructed Facilities* is now enjoying its 17th year of publication. One measure of its influence is recognition received from the *Construction Index*. A paper by Frank Heger, "Public safety issues in the collapse of L'Ambiance Plaza," was listed as one of six outstanding papers published in 1991, after a review of more than 10,000 papers in 50 construction industry journals (Heger 1991).

4 IMPORTANCE OF INTERNATIONAL LIAISONS

At the outset, the TCFE adopted as one of its goals to establish and maintain critical liaisons with other likeminded organizations, both within the US and internationally.

Initially, it was useful to work with groups within the American Society of Civil Engineers that had already accomplished work with full-scale performance evaluation of structures, such as the Research Council on Performance of Structures (RCPS), the Technical Council on Research (TCOR) and the Structural Engineering Division.

As the TCFE has matured, a number of critical liaisons have been established with international professional organizations. The Institution of Civil Engineers (ICE) and Institution of Structural Engineers (IStructE) of the United Kingdom have been particularly helpful. Specifically, the UK Standing Committee on Structural Safety (SCOSS) has an active liaison relationship with the TCFE. The SCOSS sponsored two forensic engineering conferences in the UK, in 1998 and 2001, with participation by members of the TCFE (Neale 2001). The

ICE, IStructE and SCOSS are always represented at TCFE congresses.

The Chamber of Forensic Engineers of the Czech Republic is an influential organization in Europe that has sponsored a number of conferences on lessons from structural failures (Drdácký 1996). TCFE members have regularly participated in these conferences, and our Czech colleagues have actively supported TCFE events.

Further international liaisons are welcomed by the TCFE, through formal relationships, contributions to the *Journal of Performance of Constructed Facilities* or mutual participation in international forensic engineering conferences. It is clear that many of the underlying causes of failures exist throughout the world. We have much to learn through shared experience; the opportunities for coordination have never been more accessible.

5 CONTEMPORARY CONCERNS RELATED TO SAFETY AND RELIABILITY

Of the many failure-related issues dealt with by the TCFE and appearing in articles in the *Journal of Performance of Constructed Facilities*, several recurring topics are listed here. Many of these topics have also been noted as significant concerns by forensic engineering organizations outside the US. For additional discussion of these topics, including a comprehensive list of relevant references, see Carper (1998).

5.1 *Topics related to seismic performance*

The 1994 Northridge, California, earthquake caused 61 deaths and over 9000 injuries. Approximately 112,000 structures were damaged and the urban infrastructure was severely impacted. The Northridge earthquake proved to be the most costly natural disaster ever experienced in the United States, with estimated losses of $30 billion. Older non-ductile concrete bridges and buildings failed as expected. Some poorly-detailed precast concrete buildings failed dramatically, while other well-designed precast and cast-in-place concrete facilities performed admirably. The alarming cost associated with nonstructural component damage in buildings was especially disconcerting, as was the surprising number of brittle fractures in the connections of welded moment-resisting steel frames.

Seismic events in urban areas provide critical opportunities for learning directly from actual performance. Forensic engineering reports and related publications based on these events contribute to the development of improved design codes and construction practices.

5.2 Topics related to wind hazards

Wind hazards have been extremely costly the past two decades in the US and throughout the world. Hurricane Andrew, and several other subsequent extreme wind events, awakened the general population to the inherent vulnerability of coastal development. More research in this area is being funded, including research focused on enhancing the performance of low-rise residential construction. The ASCE published a new design standard for wind loading, and further improvements are expected as a result of ongoing research.

5.3 Construction safety

The 1987 collapse of L'Ambiance Plaza in Bridgeport, Connecticut, focused attention on the construction site safety issue in the US. The 16-storey post-tensioned concrete housing project was being constructed by the lift-slab technique when a temporary connection gave way, causing the deaths of 28 construction workers.

Construction is an inherently dangerous occupation for a number of reasons. The safety record of the construction industry in the US has improved due to efforts within the industry and to more aggressive enforcement of safety regulations. However, there are still far too many avoidable accidents on the construction site. These include preventable falls, construction equipment accidents, and excavation collapses. Likewise, the design of temporary structures intended to provide stability during construction is frequently deficient.

5.4 Redundancy and enhanced structural integrity

Improving structural integrity to reduce the potential for progressive collapse in an unusual event is a topic of intense technical and political debate in the US. This topic parallels discussion in the UK regarding "robustness" of structural systems. The aim of a robust design is that a local event should not have a disproportionate effect on the entire structure.

The advantage of structural redundancy and enhanced integrity is evident in extreme natural hazard events. A more immediate interest in this subject, however, is associated with concerns about terrorism. Forensic engineering investigations of the 1995 bombing of the Oklahoma City Federal Building, and the tragic events of September 11, 2001, at the World Trade Center and the Pentagon, led to enlightening published documents (FEMA/ASCE 1996, 2002). These reports show that design professionals can indeed make substantial contributions to improved performance, even in the event of attack by determined terrorists.

It is expected that much more research will be conducted on this issue. Americans are relative newcomers to the war on terrorism. We have much to learn from our colleagues in other parts of the world where terrorism has been more common.

5.5 Building envelope performance

A large proportion of construction litigation in the US is associated with building envelope problems: leaking roofs and water penetration through curtain walls. Many of these cases could be classified as "nonstructural" component failure, although the performance of the structural system is often a contributing factor when movements and deformations have not been properly considered or controlled.

A number of older buildings have experienced masonry cladding failures caused by long-term stress accumulation and missing or ineffective stress-relieving "soft" joints. In addition, there have been some costly facade replacements in modern buildings. Most of these were necessitated by the unanticipated response of thin natural stone sheets applied to the facades of tall buildings. Still other facade replacements were necessary when new, proprietary wall materials and systems were widely used without the benefit of experience.

The present interest in high-performance curtain walls may address some of the recurring problems. However, it is expected that new problems will arise, given the sophistication needed to properly maintain and operate some of the more complicated wall systems.

5.6 Infrastructure maintenance and repair

Each year, a larger portion of the US construction budget is spent on maintenance and rehabilitation of the aging infrastructure, including buildings, highway structures, water supply systems and other civil works. In addition, much research and experimentation is expended on retrofit of existing facilities to bring them up to current design standards. This is particularly true with respect to upgrading bridges and buildings in regions of high seismic risk.

Lives have been lost from failures of highway bridges due to deficient maintenance. Examples include the 1983 failure of the Mianus River bridge in Connecticut, and the 1987 Schoharie Creek bridge failure in New York. As a result of these failures, highway inspection standards have been upgraded, and new tools for nondestructive evaluation are available.

5.7 Management and procedural issues

There have been several structural failures in North America in which management and procedural deficiencies played a major contributing role. One prominent example is the 1988 Station Square Plaza shopping center parking deck collapse in Burnaby, British Columbia, Canada. Some projects are constructed under convoluted management schemes, with considerable opportunity for confusing roles and responsibilities. Several papers in the *Journal of Performance of Constructed Facilities* noted the contribution of failed management practices to the 1987 L'Ambiance Plaza lift-slab collapse discussed earlier.

Various innovative project delivery systems have been introduced to the construction industry and new professions have emerged, such as Construction Management. Design-Build contracts are increasingly popular, with successful and unsuccessful results. When non-traditional project delivery techniques are used, there is always the possibility for confusion unless the roles of the various parties are carefully defined. Professional responsibilities must not be compromised; the results can be catastrophic.

Understanding these procedural aspects that can contribute to structural failure is as important as understanding the technical characteristics of materials and structural systems. Research is needed to establish effective quality assurance and quality control techniques to reduce opportunities for human error, miscommunications and misunderstandings.

5.8 Integration of current research into practice

Recurring failures involving repeated design errors or construction deficiencies are common, suggesting the need for more effective feedback to the design professions. As the constructed project becomes more and more complex, involving a proliferation of new materials, systems and construction methods, practice often lags behind current research. The information overload puts intense pressure on designers as standards, codes and specifications multiply, producing a plethora of conflicting and redundant documents.

At a February 4–7, 1998 meeting of the Earthquake Engineering Research Institute in San Francisco, California, it was noted that there is a dangerous gap between researchers and practitioners in the field of seismic engineering, and the gap is widening. Practicing engineers simply cannot keep current with the confusing and redundant research coming out of the academic community, and the information often lacks the clarity and simplicity to be easily incorporated into practice. There is a great need for strategies that will help make the transfer from research into practice.

5.9 Misuse of computer software

The proliferation of engineering application software, while providing opportunities for designs never before possible, presents the potential for structural failures caused by misuse. There are already many examples of economic and physical disasters resulting from an undue reliance on computer solutions. The computer is an essential tool, one that can surely contribute to the resolution of current problems in the construction industry. Nevertheless, misapplication of this new tool has already caused new problems to surface.

Within the TCFE, a task committee identified more than 52 case histories of structural failures or near-failures caused by software misuse or over-reliance on the precision of computer solutions. Many more are expected in the future. It must always be remembered that the precision of computational results is never a guarantee of the correctness of the solution, nor is it a guarantee that the right questions are even being asked.

5.10 Failure-related information in professional education

A promising forum for disseminating failure-related information is the classroom, where the next generation of practitioners is being prepared for the industry. Unfortunately, not many universities include failure case studies in the professional engineering curriculum. Apparently, there is much opportunity for improvement in this area.

To help with the integration of failure-related information into education, the TCFE Committee on Education has produced a number of useful resources. A review of forensic engineering contributions to civil engineering education is given in Delatte & Rens (2002).

In addition to providing resources specifically for forensic engineering practice, TCFE projects will continue to support educational efforts. Currently, the Council is exploring innovative techniques to enhance the dissemination of failure-related information, including the development of Internet-based tools (Rens et al. 2000, Zickel 2000). Lessons from failures, when communicated effectively and efficiently through undergraduate, graduate and continuing professional education, have the potential to reduce the frequency and severity of failures.

The topics presented here should not be viewed as an all-inclusive list of important structural safety issues. These are, however, topics of intense current interest in the US. Many of these issues are receiving international scrutiny as well. At least six of these topics were addressed in a paper published in the UK (Menzies 1995). The eleventh report of the UK Standing Committee on Structural Safety (SCOSS 1997) also identified many of the topics under study by the TCFE.

The principal threats to structural safety are international in scope. It is hoped that combining our efforts through meaningful liaison activities might lead to progress in addressing these important topics.

6 SUMMARY

In 2002, the American Society of Civil Engineers celebrated its 150th anniversary. ASCE members and US civil engineers in general are justified in celebrating 150 years of successful achievements. However, we must also acknowledge that, from time to time, we experience failures along with our successes. These should be seen as opportunities to learn valuable lessons, and we should continue to be diligent in communicating about them.

The discipline of forensic engineering is firmly established. The term "forensic engineering" has become widely accepted, both nationally and internationally. It is understood that forensic engineering encompasses efforts aimed at reducing failures and improving the constructed project, as well as activities related to litigation (Campbell 2001, Carper 2001a, b, Neale 2001).

Design professionals and the society at large recognize that there is much to learn from experience, and that forensic engineers are skilled at finding the relevant lessons. For example, when the Alfred P. Murrah Federal Building in Oklahoma City was the target of terrorism on April 19, 1995, the Federal Emergency Management Agency (FEMA) turned to a former Chair of the Technical Council of Forensic Engineering Executive Committee, W. Gene Corley, to lead the technical Building Performance assessment Team (FEMA/ASCE 1996). The valuable insights provided by that report led FEMA to again call on Corley to direct the investigations at the World Trade Center and Pentagon sites following the tragic September 11, 2001 events (FEMA/ASCE 2002). Current TCFE members were appointed to provide review comments on the formal reports.

One critical TCFE goal is to strengthen established liaisons with like-minded professional international organizations. Today, there exists the unprecedented opportunity to coordinate the collection of failure-related information on an international and interdisciplinary scale. Forensic engineers throughout the world are discovering that modern societies share many common problems in the design, construction and operation of engineered facilities. Forensic engineers are working together in an effort to develop coordinated mitigating strategies, including the publication of resources that are universally useful.

Failure is a costly way to learn; few people enjoy learning from their own mistakes. None have the resources or the time to make all the errors themselves. A better approach is to learn lessons from the experiences of others. This is perhaps the least costly form of education. Design and construction professions must continue to record and disseminate information from failure case histories so that engineered projects of the future will be more successful.

REFERENCES

Bosela, P.A., Delatte, N.J. & Rens, K.L. (eds.) 2003. *Forensic engineering: proceedings of the third congress*. Reston, VA: American Society of Civil Engineers.

Campbell, P. (ed.) 2001. *Learning from construction failures: applied forensic engineering*. Caithness, Scotland, UK: Whittles Publishing.

Carper, K.L. 1998. Current structural safety topics in North America. *The Structural Engineer* 76(12): 233–239.

Carper, K.L. (ed.) 2001a. *Forensic engineering (2nd ed.)* Boca Raton, FL: CRC Press, LLC.

Carper, K.L. 2001b. *Why buildings fail*. Washington, DC: National Council of Architectural Registration Boards.

Carper, K.L. 2003. Technical Council on Forensic Engineering: twenty-year retrospective review. In P.A. Bosela, N.J. Delatte & K.L. Rens (eds.), *Forensic engineering: proceedings of the third congress*. Reston, VA: American Society of Civil Engineers.

Delatte, N.J. & Rens, K.L. 2002. Forensics and case studies in civil engineering education: state of the art. *Journal of Performance of Constructed Facilities* 16(3): 98–109.

Drdácký, M.F. 1996. *Lessons from structural failures 6: proceedings of the Sixth International Conference on Lessons from Structural Failures*. Prague, Czech Republic: SIA Praha.

Feld, J. & Carper, K.L. 1997. *Construction failure (2nd ed.)* New York, NY: John Wiley & Sons, Inc.

FEMA/ASCE 1996. *The Oklahoma City bombing: improving building performance through multi-hazard mitigation*. Washington, DC: Federal Emergency Management Agency, Mitigation Directorate.

FEMA/ASCE 2002. *World Trade Center building performance study: data collection, preliminary observations, and recommendations*. Washington, DC: Federal Emergency Management Agency, Federal Insurance and Mitigation Administration.

Heger, F. 1991. Public-safety issues in collapse of L'Ambiance Plaza. *Journal of Performance of Constructed Facilities* 5(2): 92–112.

Menzies, J.B. 1995. Hazards, risks and structural safety. *The Structural Engineer* 73(21): 357–363.

Neale, B.S. (ed.) 2001. *Forensic engineering: the investigation of failures: proceedings of the second international conference on forensic engineering*. London, UK: Thomas Telford Publishing.

Rens, K.L. (ed.) 1997. *Forensic engineering: proceedings of the first congress*. Reston, VA: American Society of Civil Engineers.

Rens, K.L., Rendon-Herrero, O. & Bosela, P.A. (eds.) 2000. *Forensic engineering: proceedings of the second congress*. Reston, VA: American Society of Civil Engineers.

Rens, K.L., Clark, M. & Knott, A.W. 2000. Development of an internet failure information disseminator for professors. In K.L. Rens, O. Rendon-Herrero & P.A. Bosela (eds.), *Forensic engineering: proceedings of the second congress*. Reston, VA: American Society of Civil Engineers: 441–453.

SCOSS 1997. *Structural safety 1994–96: review and recommendations: eleventh report of the Standing Committee on Structural Safety*. London, UK: SETO Ltd.

Zickel, L.L. 2000. Failure vignettes for teachers. In K.L. Rens, O. Rendon-Herrero & P.A. Bosela (eds.), *Forensic engineering: proceedings of the second congress*. Reston, VA: American Society of Civil Engineers: 421–429.

System-based Vision for Strategic and Creative Design, Bontempi (ed.)
© 2003 Swets & Zeitlinger, Lisse, ISBN 90 5809 599 1

Behavior of epoxy-coated reinforcement concrete beams with sodium chloride contamination

M.M.A. Elmetwally
Faculty of Eng. Zagazig University, Egypt

ABSTRACT: Premature corrosion of reinforcing steel has caused many concrete structures to deteriorate before their design life was attained. So, the effectiveness of reinforced concrete structural elements is how to protect the reinforcement from severe environment and weathering condition. This paper presents an experimental study on the strength and behavior of concrete beams with plain and epoxy-coated reinforcing steel. This study had two main objectives: 1) to develop and evaluate effective and economical methodologies for arresting or reducing the extent of steel corrosion due to chloride-contamination of concrete structures, thereby reducing maintenance costs and minimizing interruption to users. 2) to develop sound design and construction practices for preventing corrosion of reinforcement, hence minimizing future deterioration. To meet these objectives, 7-year research program in this study was carried out. In this study, 14-concrete beams were cast in concrete with 5% sodium chloride to simulate the severe environment, where the deterioration of concrete structures due to salt content can be considered as one of the major problems in the Middle East. The beams were tested after 28, 365, day and 7 years. The parameters of this study were the effectiveness of partial and full coating for long-term protection of steel against corrosion, effectiveness of type of steel grade in corroded environment and corrosion rates after long-term exposure as a function of time and environment. Effect of corrosion on mechanical properties of reinforcement is studied. A 7-year program indicated that, the rate of corrosion could be decreased significantly by using epoxy coating, which can give good long-term performance under severe conditions. Rate of corrosion in plain mild steel is double that of high-grade deformed steel. Pitting corrosion severely affected the load carrying capacity of beams. The beams were tested up to failure and the influence of variable factors on the structural behavior is reported.

1 INTRODUCTION

A large number of reinforced concrete structures are generally subjected to chloride ions that come from ingredients of concrete deicing salts used to melt snow and seawater. The chloride contamination may result in local disruption of the passive film. The source of chloride ions may be internal or external. Internal sources include contamination of the mix materials and the use of calcium chloride as a set accelerator in construction. Limitations are placed by current codes of practice on the acceptable levels of chloride contamination resulting from the use of contaminated mix materials, while the use of chloride-containing admixtures for reinforced concrete is generally not permitted. External sources of chlorides include de-icing salts and sea salt in marine environments. Corrosion of reinforcing bars is one of the main causes, which induces an early deterioration of concrete structures

and reducing their residual service life. The basic approaches that have so far been taken to prevent corrosion of reinforcing steel embedded in concrete are, improving the quality of concrete, protection at the concrete surface, implementing athodic protection, and protecting the reinforcement at the steel/concrete interface. Service life can considerably be increased by using high quality concrete with thicker concrete cover, which will reduce the effective chloride diffusion (Westers 1999).

(Okada 1988) carried out tests with sound, cracked, due to corrosion by spraying sodium chloride during 140–170 days, where cracks width ranging between 0.05 and 0.15 mm appeared at concrete surface. (Al-Sulaimani 1990) carried out an extensive research work to relate corrosion of reinforcement to bond deterioration. (Dunster 2000) studied relationship between the availability of moisture, concrete resistance and the rate of corrosion was studied where; the carbonation rate was the same for ordinary Portland cement and high alumna cement.

Corrosion of steel reinforcement was one of the main causes of concrete structure failures in many countries, (Emam 1989) carried out a study on the economic effects of corrosion. (El-sayed 1992) found that, the steel surface contamination significantly affects the corrosion activity. If the steel surface is contaminated with salt, it is likely to corrode sooner than the clean steel. (Hasson 1985) found that, higher rate of attack was measured from magnesium chloride than sodium chloride. The rate of corrosion could be decreased significantly by using epoxy-coating, where the results of experiments showed the existence of galvanic corrosion for epoxy-coated samples and that with no coating at all (Tavakkolizadeh 2001). Generally the steel is very intensive in cost and therefore must be protected against corrosion. A 2-year monitoring program on plain and epoxy – coated reinforcing bars in concrete slabs and exposed in the laboratory to synthetic seawater and 3% sodium solution, indicate that the corrosion current density was negligible for epoxy – coated bars with no damage to the coating regardless of the exposure conditions and that undamaged epoxy-coated bars provided excellent performance in preventing corrosion activity in reinforced concrete structures subjected to chloride (Erodgdu 2001). Pitting and crevice corrosion result from a failure of the passive film, where chloride is usually present as an essential ingredient to break down the passive film and initiate localized corrosion (Jones 1996).

Specified tests to prove the performance of coating for reinforcement is scarce. The suitability of tests to assess coating with respect to their performance in practice is questionable. The most valuable test is the actual performance in practice. Therefore a seven-year test program under natural conditions is presented in this study.

2 EXPERIMENTAL STUDY

2.1 Test program

Details of test program are given in Table 1. Seven groups of beams were manufactured for the experimental study. A total of 14 reinforced concrete beams were subjected to corrosion damage, by adding 5% sodium chloride from cement weight to the mixing water, and epoxy coating is used for coating some steel cages before casting and all beams are tested under flexure. All beams were cast in April 1995, and beams P1, …, …, P5 and D1, …, …, D5 were tested in April 2002.

The reinforced concrete beams were 1000 mm long and had a rectangular cross section of 200 mm depth and 100 mm width. Beams were divided into two main groups, one group is reinforced in bottom side with two steel bars of 10 mm diameter of high grade deformed steel and the second group is reinforced in bottom side with two bars of 10 mm diameter of normal mild steel. Two bars of 6 mm diameter used as top reinforcement and 6 mm diameter stirrups at 150 mm spacing were used in each beam. Top and bottom reinforcement were hooked in the form of U-shaped at the ends as shown in Figure 1. The loading configuration and testing machine is shown in Figure 2 where, hydraulic universal testing machine of 1000.0 kN capacity were used in testing.

Table 1. Test program details and results.

Beam	Steel type	Sodium chloride	Epoxy coating	Exposure period	Ultimate load (kN)	Corrosion cracks	Mode of failure
P0	plain	with	no	28-day	61.7	no	flexure
P	plain	with	no	365-day	62.5	no	flexure
P1	plain	without	no	7-year	63.1	no	composite
P2	plain	with	full	7-year	69.3	yes	shear
P3	plain	with	partial*	7-year	64	yes	shear
P4	plain	with	partial	7-year	36.4	yes	flexure
P5	plain	with	no	7-year	57.6	yes	flexure
D0	deformed	with	no	28-day	72.3	no	shear
D	deformed	with	no	365-day	73.1	no	shear
D1	deformed	without	no	7-year	75.4	no	shear
D2	deformed	with	full	7-year	77.5	yes	shear
D3	deformed	with	partial	7-year	68.8	yes	composite
D4	deformed	with	partial	7-year	74	yes	composite
D5	deformed	with	no	7-year	74.4	yes	flexure

* Partial coating means, some surface areas of bars were coated and others were left without coating, randomly.

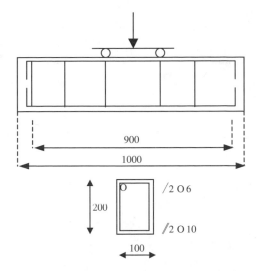

Figure 1. Beam specimen details.

Figure 2. Loading configuration and testing machine.

2.2 Materials

Ordinary Portland cement was used throughout. The fine aggregate was clean natural sand and the coarse aggregate was 20 mm maximum size clean gravel. The concrete mix used for the beams had proportions (by weight) 1: 2.23: 4.03

The concrete mix proportions are detailed in Table 2. For each concrete batch, compressive strength tests were performed on $15 \times 15 \times 15$ cm cubes. The average cube strength was $277 \, kg/cm^2$ after 28-day. Tension tests on three specimens for each diameter were performed on steel bar samples before casting to

Table 2. Concrete mix proportion.

Cement (kg/m^3)	Sand (kg/m^3)	Gravel (kg/m^3)	Water (l/m^3)
300	670	1210	180

determine yield, ultimate strength and total elongation. Mechanical properties as an average value for each diameter of reinforcing steel are given in Table 3. Epoxy was used in coating of reinforcement of some samples as shown in Table 1.

2.3 Corrosion in reinforcement

Corrosion in reinforcement was due to the sodium chloride salt, which was added to the mix as 5% of the cement weight in mixing water.

Two beams (P and D) with salt and without coating on steel cages were tested at 28 day after casting. Two beams (P0, D0) with salt and without coating on steel cages were tested at 365 day after casting. Ten beams (P1, P2, P3, P4 and P5) and (D1, D2, D3, D4 and D5) were tested after 7 year from casting. Details and environment conditions are given in Table 1.

After 7-year the beams were tested in flexure, after testing, the beams were completely destroyed carefully to obtain the steel cages. Parts of these steel cages were cut and cured to compute actual losses in steel weight and cross section area due to rusting, also samples of steel cages were cut to carry out tension tests to determine the mechanical properties of main reinforcement and reduction in cross section area due to corrosion as shown in Table 4.

2.3 Strain

Surface strains were measured using mechanical strain gauge of 200 mm length and 0.01 mm accuracy. The demic points were adhered carefully on the surface of the tested beams. The measured strain data is not presented here for the sake of conciseness.

2.4 Deflections

The deflections under the applied load were measured at the bottom side and at the middle of span of the tested beams using dial gauge of 10 mm capacity and 0.01 mm accuracy.

2.5 Weight loss of steel reinforcement

Specimens of main flexural reinforcement were cut from side of steel cages after testing and removing of concrete, to provide information on the total metal loss

Table 3. Mechanical properties of reinforcement before casting.

Steel type	Nominal diameter (mm)	Actual diameter (mm)	Yield strength (kg/cm^2)	Ultimate strength (kg/cm^2)	Elongation (%)
Mild	6	6	3003	4736	31
Mild	10	10	2640	4210	25
Deformed	10	10	3400	5380	21

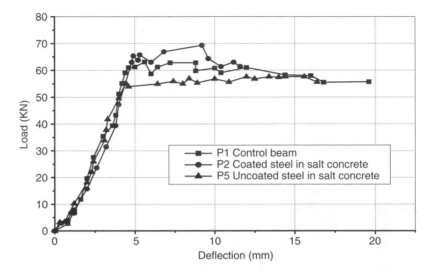

Figure 3. Load-deflection relationship after 7-year exposure for beams with plain mild steel.

as a result of corrosion at different periods, Each steel specimen was weighed after removal of the corrosion products with help of a cleaning solution to evaluate the total metal loss in steel weight, compared with the control specimens, and the reduction in cross section area is calculated as given in Table 4. Also, tension tests were carried out on specimens of steel cages (from side parts) after testing (after 7 year) to study the influence of corrosion and coating on mechanical properties of steel.

3 DISCUSION AND RESULTS

3.1 Flexural testing

At the age of 28-day, 365-day and 7-year after casting, beams were tested under four-point bending to determine their load-deflection relationship curves and the ultimate flexural strength.

3.2 Failure load

There is negligible difference between the corroded beams and the control ones in the ultimate load where

the ultimate load of severely corroded beam P5 reduces only by 8.7% of that of control beam P1. Also, the ultimate load of corroded beam D5 was reduced only by 1.3% of that of control beam D1. This demonstrates that usual damage will occur long before corrosion has resulted in a significant loss of steel section and load bearing capacity. It also, shows that the corrosion thus formed, in spite of being able to cause extensive cracking, did not affect the bond.

3.3 Load-deflection relationship

Figure 3 shows the load-deflection relationship for beams, P1 of clean concrete with uncoated cage, P2 of chloride ion contamination (CIC) concrete with coated steel cage, and P5 of CIC concrete with uncoated steel cage. It was found that beams P1 and P2 failed in mixed mode (shear and flexure) and shear mode respectively, but beam P5 failed in flexure mode. The ultimate load supported by beam P2 in spite of initiation of corrosion was greater by 7.8% than that supported by beam P1, and that supported by beam P5 was 91.3% of beam P1, where the loss by corrosion in flexural reinforcement of P5 was 15.4% in cross

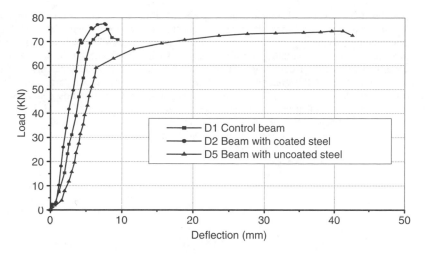

Figure 4. Load-deflection relationship after 7-year exposure for beams with deformed high-grade steel.

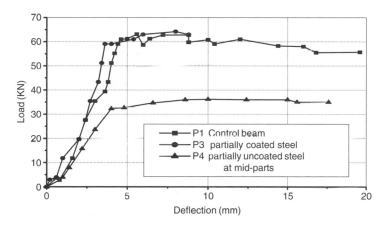

Figure 5. Load-deflection relationship after 7-year exposure for beams with plain mild steel.

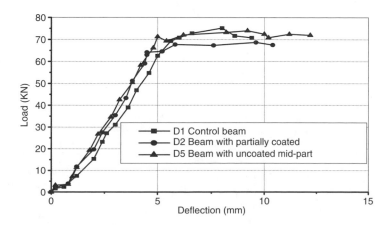

Figure 6. Load-deflection relationship after 7-year exposure for beams with deformed high-grade steel.

sectional area, also beam P5 was severely cracked and showed severe rebar corrosion which left thick corrosion layer in surrounding concrete.

Figure 4 shows the load-deflection relationship for beams, D1 of clean concrete with uncoated cage, D2 of chloride ion contamination (CIC) concrete with coated steel cage, and D5 of CIC concrete with uncoated steel cage. It was found that beams D1 and D2 failed in shear mode, but beam D5 failed in flexure mode. The ultimate load supported by beam D2 in spite of initiation of corrosion was greater by 3.0% than that supported by beam D1, and that supported by beam D5 was 98.7% of beam D1, where the loss by corrosion in flexural reinforcement of D5 was 7.8% in cross sectional area, also beam D5 was cracked and showed rebar corrosion which left thin corrosion layer in surrounding concrete.

Figures 5 and 6 show the load-deflection relationship for beams (P1, P3, and P4), and (D1, D3, and D4) respectively.

System-based Vision for Strategic and Creative Design, Bontempi (ed.)
© *2003 Swets & Zeitlinger, Lisse, ISBN 90 5809 599 1*

Damage analysis of masonry walls under foundation movements by distinct element method

Y. Zhuge
School of Geosciences, Minerals and Civil Engineering, University of South Australia, SA 5095, Australia

S. Hunt
School of Petroleum Engineering and Management, The University of Adelaide, SA 5005, Australia

ABSTRACT: Cracking failure in masonry houses founded on expansive soils has been widely reported throughout Australia and other countries. The cost associated with such damage is significant. In this paper a numerical model has been developed to study the behaviour of masonry walls under foundation movements as a result of expansive soils. The model is based on Distinct Element Method (DEM) which has been applied successfully by the authors to model the masonry walls under simulated (static) in-plane earthquake forces. The model is capable of predicting the crack initiation, propagation and failure modes of masonry walls under various footing movements (doming or dishing curvatures). The numerical solutions are validated by comparing the results with those obtained from the existing experiments.

1 INTRODUCTION

Cracking and damage in masonry structures founded on expansive soils has been widely reported throughout Australia and other countries. Buildings constructed on expansive soils are frequently subjected to severe movement arising from non-uniform soil moisture changes, with consequent cracking and damage due to distortion. The cost associated with such damage is significant. In Australia, approximately 30% of the total land area is covered by expansive soils.

The movement caused by expansive soil can be quite large. The extent of the movement depends mainly on the extent of soil moisture or suction change under the footing. These moisture changes are often induced by seasonal changes in rainfall and evaporation, watering of gardens, leakage from waterpipes, or extraction of water by trees and shrubs. If the soil is reactive, large relative movements could be expected in the soil producing either a "dishing" or "doming" of the soil profile under the building. The above effects can create angular distortions and therefore stresses in walls and can lead to problems such as jamming of doors and windows. This type of failure is particularly common for lightweight unreinforced masonry structures.

Unfortunately, the current codes of practice, AS 2870 (Residential slabs and footings 1996) and AS

3700 (SAA Masonry Code 2001) only provide broad guidance on the design principles for masonry wall/ footing systems due to a lack of research in the area. Therefore, there is a significant need for research aimed at developing a rational design procedure for footings and masonry structures on expansive soils.

In the literature, reference to masonry deformation due to foundation movements is very limited. A review of previous studies has revealed that only unreinforced masonry walls have been investigated at University of Newcastle for several years. Bryant (1993) performed a series of two-dimensional tests to study the response of masonry structures due to foundation movements. The major objective of the tests was to investigate the relationships between external deformation and structural cracking. An analytical model was developed based on their testing results using a linear finite element program (Strand 6). However, the stress redistribution effects and non-linear material behaviour could not be modelled, as isotropic elastic behaviour was assumed for both the masonry and concrete footing. The interaction of the foundation beam and soil was not considered in the model.

The problem was further studied by Masia et al. (2002) to consider the interaction of soil and structure and a probabilistic model was developed to predict the cracking in masonry walls. However, in order to simply the problem, all cracks in the masonry walls were pre-specified in their model. Automatic crack initiation and propagation were not included.

In order to study the behaviour of masonry structures and footings (concrete slabs) resting on expansive soils, a numerical model which is based on Distinct Element Method (DEM) is being developed to model the system as a whole, that is, wall and footing systems and it is expected that an improved design can then be engineered for an integrated system for structures on expansive soils. This paper will discuss the research work carried out at the stage one of the project, where DEM has been successfully applied to model unreinforced masonry walls under prescribed footing movements, where experimental results are available for comparison.

Masonry is not a simple material, the influence of mortar joints and bond as a plane of weakness is a significant feature which is not present in concrete and this makes the numerical modelling of masonry very difficult especially when the loading condition is complicated. Therefore, a simplified linear elastic one-phase (mortar joints were not modelled separately) model has been employed by many researchers to investigate the effect of foundation movements on masonry walls (Bryant 1993; Muniruzzaman 1997; Masia et al. 2002).

In order to model the discontinuous types of material, such as masonry, the investigators of this paper have carried out research for several years and found out that a distinct element method (DEM) could be used. Although DEM was primarily intended for analysis in rock engineering projects, it has been demonstrated by the investigators with their pioneering research that the non-linear behaviour of masonry walls may be simulated using DEM (Zhuge & Hunt 2002; Zhuge 2002). In their papers, the DEM has been applied to simulate the in-plane shear behaviour of unreinforced masonry walls where the testing results were available for comparison. The model was validated by comparing with the experiments of masonry shear walls. Two sets of results agreed very well and the comparison proved the abilities of the distinct element model developed by the investigators.

In this paper, the model has been further developed to study the structural behaviour of the masonry walls under foundation movement, where a progressively increased displacement boundary is applied at the bottom of the wall in the vertical direction. A brief outline of the Distinct Element Method and the model development are presented first. The material models for bricks and joints are then discussed as well as the selection of material properties. The analysis is performed and the results between the distinct element model and experiments are compared and discussed.

2 OUTLINE OF DISTINCT ELEMENT METHOD

Distinct Element Method has been progressively developed over the past two decades. Cundall (1971) first introduced the DEM to simulate progressive movements in blocky rock systems and the model has been implemented into a computer program UDEC since then. DEM simulates the response of discontinuous media subjected to either static or dynamic loading.

In the DEM method, a solid is represented as an assembly of discrete blocks. Joints are modelled as interface between distinct bodies. The contact forces and displacements at the interfaces of a stressed assembly of blocks are found through a series of calculations which trace the movements of the blocks (Itasca 2000). At all the contacts, either rigid or deformable blocks are connected by spring like joints with normal and shear stiffness k_n and k_s respectively (Fig. 1). Similar to Finite Element Method (FEM), the unknowns in DEM are also the nodal displacements and rotations of the blocks. However, unlike FEM, DEM is a dynamic process and the unknowns are solved by equations of motion. The speed of propagation depends on the physical properties of the discrete system. The solution scheme used by DEM is the explicit time marching and finite contact stiffness.

It should be noted that time has no real physical meaning if a static analysis was performed. Damping is utilised in the above equations. However, different methods were used for static and dynamic analysis. A detailed DEM formulation can be found in the literature (Zhuge 2002; Zhuge & Hunt 2002).

3 NUMERICAL SIMULATION OF MASONRY WALLS UNDER FOOTING MOVEMENTS

Numerical modelling of each unreinforced masonry wall resting on a concrete footing beam which was

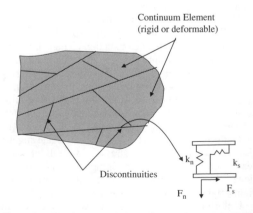

Figure 1. Continuum and discontinuum elements in DEM.

subjected to typical upward (doming) or downward (dishing) curvatures was carried out using the distinct element code UDEC (Universal Distinct Element Code) (Itasca 2000).

As it is introduced previously, DEM is fully dynamic and it deals with pseudo-static problems by allowing the dynamic behaviour to reach equilibrium with notional time. In general, a velocity-proportional damping (the magnitude of the damping forces is proportional to the velocity of the blocks) could be used for pseudo-static problems. However, it is proved from the current research that a local damping in which the damping force on a node is proportional to the magnitude of the unbalanced force, is more suitable for the type of problem where the progressive failure of the structure was the major interest of the research.

The dimensions of the wall are based on the experimental testing of Bryant (1993), where a total of 832 blocks were used. In order to calculate the internal deformation and stress distribution of blocks, the deformable blocks have to be discretised into finite difference triangular elements first. A typical discrete element mesh of the wall is shown in Figure 2 with more 50,000 elements.

3.1 Constitutive laws and failure criterion of joints

Micro modelling of masonry has to consider all basic types of failure mode including both mortar joints and the unit. The mortar joint cracks and the separation along the damp proof course (dpc) are the major failure modes being observed during the testing, therefore the major objective of the model is to simulate the crack initiation and propagation. In the model, the mortar joints are represented numerically as a contact surface formed between two block edges. The constitutive laws applied to the contacts are:

$$\Delta\sigma_n = k_n \Delta u_n \qquad (1)$$

$$\Delta\tau_s = k_s \Delta u_s \qquad (2)$$

where k_n and k_s are the normal and shear stiffness of the contact, $\Delta\sigma_n$ and $\Delta\tau_s$ are the effective normal and

shear stress increments, and Δu_n and Δu_s are the normal and shear displacement increments.

Stresses calculated at grid points located along contacts are submitted to the selected failure criterion. For the proposed model, the Coulomb friction is formulated:

$$|\tau_s| \leq C + \sigma_n \tan\phi = \tau_{max} \qquad (3)$$

where C is the cohesion and ϕ is the friction angle.

There is also a limiting tensile strength ft for the joint. If the tensile strength is exceeded, then $\sigma_n = 0$.

3.2 Constitutive laws and failure criterion of blocks

The selected constitutive law for the blocks is used to determine stresses at each grid point. For the present study, the relation of stress to strain in incremental form is expressed by Hooke's law in plane stress as:

$$\Delta\sigma_{11} = \beta_1 \Delta e_{11} + \beta_2 \Delta e_{22}$$
$$\Delta\sigma_{22} = \beta_2 \Delta e_{11} + \beta_1 \Delta e_{22} \qquad (4)$$

where $\beta_1 = E/(1 - v^2$ and $\beta_2 = vE/(1 - v^2)$

$$\Delta e_{ij} = \frac{1}{2}\left[\frac{\partial \dot{u}_i}{\partial x_j} + \frac{\partial \dot{u}_j}{\partial x_i}\right]\Delta t \qquad (5)$$

where Δe_{ij} is the incremental strain tensor, $\dot{u}_i$ is the displacement rate and Δt is time step.

As the tensile splitting or crushing of the brick is not a common type of failure for masonry walls under serviceability performance, in order to simplify the problem, the brick unit material is modelled with Mohr-Coulomb failure criterion with tension cut-off (Zhuge & Hunt 2002).

3.3 Material properties

The numerical model developed here will be compared with the existing experimental work (Bryant 1993) in the next section. However, the material properties of bricks and mortars were not provided from the Bryant's (1993) experiments, where only the material properties of masonry and concrete footing were available. From the constitutive relationship curves of brick, mortar and masonry shown in Figure 3 (Dhanasekar 1985), it can be seen that the masonry and brick curves are very similar and close to each other. Therefore the properties of brick are taken to have similar values as masonry. The material properties of the blocks are shown in Table 1 (Bryant 1993). k_n and k_s of the interfaces between the wall blocks are potentially important parameters in the numerical

dpc layer

concrete footing

Figure 2. Wall meshes modelled using UDEC.

analyses of masonry walls using UDEC. Unfortunately, there are very few testing data on stiffness properties for mortar joints are available. The only testing results the authors could find were the experiments conducted at the University of Delft, the Netherlands (Lourenco 1996). These testing results were used to validate the numerical model developed by the authors of masonry shear wall panels under in-plane lateral load (Zhuge & Hunt 2002). The values of kn and ks have been adopted again for the current model (Table 2). In table 2, the tensile strength of the bond was taken from Bryant's (1993) experiments.

3.4 Modelling of the damp-proof course (dpc)

The provision of dpc in domestic construction in Australia primarily is to provide a barrier to the upward movement of moisture from the ground.

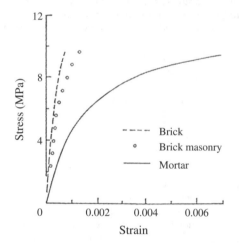

Figure 3. Average stress–strain curves for bricks, masonry and mortar (Dhanasekar 1985).

Table 1. Summary of blocks material properties.

Material	Elastic modulus MPa	Poisson's ratio	Density kg/m^3
Concrete footing	7000	0.2	2130
Clay brick masonry	9000	0.19	2000

Table 2. Summary of joint material properties.

Tension f_t N/mm^2	$\tan \phi$	Shear $\tan \psi$	C MPa	Normal stiffness k_n N/mm^3	Shear stiffness k_s N/mm^3
0.453	0.75	0.0	0.375	82	36

The experimental results of Bryant (1993) indicated that the dpc's have a secondary purpose as well that is acting as a horizontal plane of weakness in the wall panels, with vertical separation occurring along this plane under both dishing and doming curvatures.

During the testing, the dpc membrane was laid directly onto the brick course below, therefore a zero ft for the dpc layer could be assumed in the model. In order to model the shear sliding type failure along the dpc layer, a suitable value for the coefficient of friction along the dpc is required. Based on the experimental results carried out at the University of Newcastle (Page 1992), a constant value of 0.5 was suggested and this value has been adopted in this paper.

4 COMPARISON WITH EXPERIMENTAL INVESTIGATIONS

4.1 Experimental procedures for the wall tests

A series of full-scale tests on masonry walls supported on a foundation beam were carried out at the University of Newcastle, Australia (Bryant 1993). The tests have been limited to structural elements relevant to housing. In all cases the walls were supported by a foundation beam, the beam was subjected to either upward or downward curvature. The testing set-up is shown in Figure 4, the wall panel has a dimension of 6 m × 2.4 m and was supported on a standard 300 mm × 300 mm reinforced concrete strip footing. Vertical curvatures in the form of prescribed displacements at three points along the bottom of the foundation beam can be applied in either the upward or downward direction. A uniform vertical compressive load of 6 kN was applied via a simulated roof system consisting of timber joists at 600 mm centres resting on a timber top plate located along the top of the wall. A membrane type damp-proof course was located in the mortar joint near the base of the wall and acted as an isolation joint.

The testing results indicated that the wall suffered no distress when subjected to dishing or doming except for the separation and sometimes sliding that occurred

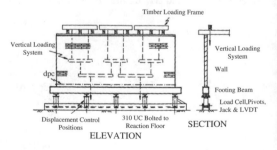

Figure 4. General arrangement of the testing rig (Bryant 1993).

at the damp proof course. In all cases the two courses of brick work below the damp proof course followed the profile of the foundation beam with all slip and separation occurring on the one weak plane. The experimental behaviour of the walls is shown in Figure 5.

4.2 Dishing curvatures

For the dishing case, the beam was deflected downwards until the load in the three jacking points approached zero, simulating the effects of soil expansion near the end of the beam. In this case separation occurred along the damp-proof course (dpc), in the central section of the wall, with some sliding along the damp-proof course. The brickwork was then spanning with

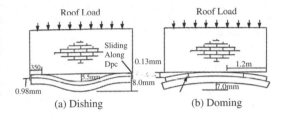

(a) Dishing   (b) Doming

Figure 5. Experimental behaviour of masonry walls under foundation movements (Bryant 1993).

some frictional restraint between the two ends of the beam (Bryant 1993).

The numerical and experimental results are compared in Figures 5a and 6. In the test at very small deflections vertical separation along the dpc was noticeable along the centre of the wall. At the maximum central deflection of 8 mm the separation along the dpc extended to within 350 mm of the ends of the wall (Fig. 5a). Figure 6 shows the computed crack initiation, progressive development and the wall movements for the dishing case modelled with DEM. The crack initiated at the centre of the joint containing the damp-proof course (dpc) (Fig. 6a). Vertical separation then taking place and extended as the downward displacement increased (Fig. 6b). Finally, when a central dishing deflection reached 8 mm, the separation along the dpc extended towards the ends of the wall with only a small connecting length on each side of the wall and also no crack was detected in the masonry wall above the dpc (Fig. 6c). It can be seen from the above figure, the behaviour of the wall is well captured by the proposed model.

The contours of horizontal stress distribution at a central displacement of 1 mm and 7 mm are shown in Figures 7a and 7b respectively. It can be seen from the figure, the increment of central displacement will only increase the stresses in the foundation beam; there are not much effects on the masonry wall above the dpc.

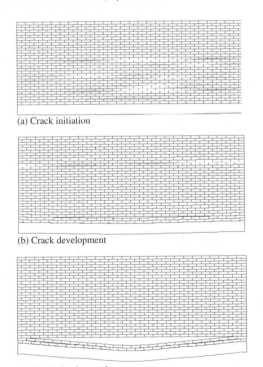

(a) Crack initiation

(b) Crack development

(c) Wall behaviour at the max curvature

Figure 6. Simulated behaviour of the wall under dishing curvature.

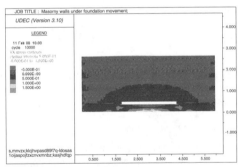

(a) At central deflection of 1 mm

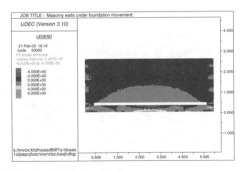

(b) At central deflection of 7 mm

Figure 7. Horizontal stress distribution – dishing.

1359

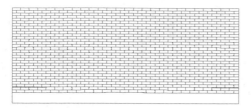

(a) crack initiation

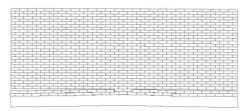

(b) Crack development

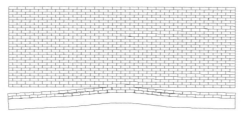

(c) Wall behaviour at the max curvature

Figure 8. Simulated behaviour of the wall under doming curvature.

It is again proved that the dpc acts as a horizontal plane of weakness for masonry walls under foundation movements.

4.3 Doming curvatures

For the doming case, the beam was deflected upwards until the load at the two support points approached zero, simulating the effects of soil shrinkage near the ends of the beam. Again separation occurred at the dpc level, but at the ends of the wall. At a central doming deflection of 7 mm the separation on both ends was about equal at 4.2 mm and 4.9 mm on either end of the wall (Fig. 5b). Again no distress occurred in the masonry wall above the dpc.

Figure 8 shows the computed crack initiation, progressive development and the wall movements for the doming case modelled with DEM. The crack initiated at the ends of the joint containing the damp-proof course (dpc) (Fig. 8a). Separation then taking place and extended as the upward displacement increased to around 3 mm (Fig. 8b). Finally, when a central doming deflection reached 7 mm, the wall supported only by the central section of the beam, with each end of the wall acting as a cantilever (Fig. 8c). Again the behaviour of the wall is well captured by the proposed model.

5 CONCLUSIONS

Cracking and damage in masonry structures founded on expansive soils has been a major concern for Australian structural engineers. A numerical model for predicting cracking and failure in masonry walls due to foundation movements has been described in this paper, which is based on the Distinct Element Method. The crack initiation, propagation as the footing curvature changes for both doming and dishing cases were successfully simulated in the model and the results compared well with those obtained from experiments.

The next step of the research will be aimed at including the soil moisture movements in the model and therefore to consider the soil/structure interaction. The results obtained in the current research have also shown the great potential of the method, especially for dynamic analysis.

REFERENCES

AS 2870 1996. Residential Slabs and Footings, *Standards Association of Australia*.

AS 3700 2001. SAA Masonry Code, *Standards Association of Australia*.

Bryant, I. 1993. *Serviceability of masonry walls subjected to foundation movements*, ME thesis, The University of Newcastle, NSW, Australia.

Cundall, P. A. 1971. A Computer model for simulating progressive large scale movements in blocky rock systems, *Proceedings of the Sym. of the International Society for Rock mechanics*, Nancy, Francs, Vol. 1, II-8: 11–18.

Dhanasekar, M. 1985. *The performance of brick masonry subjected to in-plane loading*, PhD Thesis, The University of Newcastle, NSW, Australia.

ITASCA Consulting Group 2000. *Universal Distinct Element Code*, ITASCA consulting Group, Inc., Minneapolis, Minnesota, USA.

Lourenco, P.B. 1996. *Computational strategies for masonry structures*, PhD Thesis, Delft University of Technology, The Netherlands.

Masia, M.J., Melchers, R.E. & Kleeman, P.W. 2002. Probabilistic crack prediction for masonry structures on expansive soils, *Journal of Structural Eng.*, ASCE, Vol. 128, No. 11, 1454–1461.

Muniruzzaman, A.R. 1997. *A study of the serviceability behaviour of masonry*, Ph.D Thesis, The University of Newcastle, NSW, Australia.

Page, A.W. 1992. *The design, detailing and construction of masonry – the lessons from the Newcastle earthquake*, Dept. of Civil Engineering and Surveying, The University of Newcastle, Research Report No. 073.031.1992.

Zhuge, Y. 2002. Micro-modelling of masonry shear panels with distinct element approach, *Proceedings of the 17th Australasian Conference on the Mechanics of Structures and Materials*, Gold Coast, Australia, 131–136. Rotterdam: Balkema.

Zhuge, Y. & Hunt, S. 2002. Numerical simulation of masonry shear panels with distinct element approach. *Structural Engineering and Mechanics*, An International Journal, under review.

18. Materials, composite materials and structures

Ultimate strength of CFT columns – an analytical method

R.V. Jarquio
New York City Transit, USA

ABSTRACT: This paper presents the analytical procedure for predicting the ultimate strength of circular concrete filled tube column section based on the concept that the concrete core develops its ultimate strength as a reference for determining the corresponding steel forces. The concrete forces for a circular section has already been presented at the ISEC-01 and SEMC2001 international conferences. Hence, this paper describes the method of determining the steel forces to be combined with the concrete forces to solve the ultimate strength of a circular CFT column section. Variables considered are concrete strain, ultimate concrete stress, steel yield strength, radius of the concrete core and wall thickness of steel tube. At ultimate condition of stress/strain of the CFT column section, the resulting steel stress volumes on the CFT column shell area are thus determined. Comparison is made to show the accuracy of the analytical method with the customary method of analysis.

1 INTRODUCTION

The current method of calculating the ultimate strength of CFT columns employs the concept of the column interaction formula for steel and reinforced concrete sections subjected to bi-axial bending. In contrast, the analytical method illustrated in this paper will eliminate the need to use the column interaction formula. This is done by using the diameter as the column capacity axis as the reference for equilibrium of internal and external forces and also to determine the capacities of the column section at every position of the concrete compressive depth "c".

This analysis involves the calculations of concrete and steel forces separately and then combining these to determine the ultimate strength of CFT columns. The concrete forces are determined using the true parabolic stress method of analysis presented in ISEC-01 and SEMC2001 structural engineering mechanics and computation. Hence, the calculation of the steel forces is the main focus of this paper. The same assumptions used in the ultimate strength of reinforced concrete columns are also applicable in the analysis for ultimate strength of CFT column section. As a matter of fact, the same equations for the determination of the centroid of the internal forces and the factors for the external loads are applicable in this case.

At ultimate condition of stress/strain, the concrete strain usually assumed equal to 0.003 by the ACI method and 0.0035 by Canadian practice is the reference point for determining the steel stress/strain since it is assumed that the concrete core develops it ultimate strength at this assumed strain value. The concrete and steel elements in a CFT column section undergo common deformation when resisting external loads.

This deformation is assumed linear with respect to the neutral axis, whose location is the concrete compressive depth "c". As the concrete compressive depth "c" is varied from the concrete edge it will generate a steel stress/strain diagram consisting of a triangular and rectangular shape. This steel stress/strain diagram will define the steel forces to be added to concrete forces to obtain the ultimate strength of a CFT column section. The compressive and tensile steel stress is limited to the yield strength "f_y" of the material.

The resulting stress volumes on the CFT column section are the measure of the steel forces for every position of "c". The integration of these volumes involves 6 limiting ranges of "c" defined by the shape of the steel strain diagram and the dimension of the CFT section. There are 79 equations to be written for the outer and inner circular section to obtain the total steel forces at ultimate strength of the CFT section. These equations are listed completely in the derivation.

The derived equations are programmed in a computer using Microsoft Excel'95 and a sample printout for a circular section is included in this presentation.

2 DERIVATION

In Figure 1, from the equilibrium conditions of external and internal forces acting on the column section,

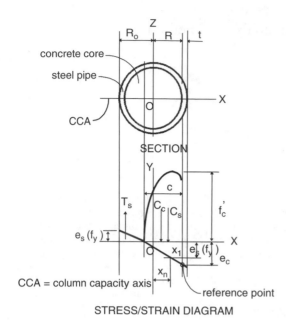

SECTION

CCA = column capacity axis

STRESS/STRAIN DIAGRAM

Figure 1. Circular CFT column section.

the following relationships become apparent i.e.,

$$\sum F = 0: \quad P = C_c + C_s - T_s \tag{1}$$

$$\sum M = 0: \quad M = C_c x_c + C_s x_s + T_s x_t \tag{2}$$

However, in this analysis only the steel forces C_s, T_s, $C_s x_s$ and $T_s x_t$ will be covered since the concrete forces C_c and $C_c x_c$ have already been determined in papers presented in ISEC-01 and SEMC2001 conference in structural engineering mechanics and computations. To solve the steel tensile and compressive forces, designate three main sections such as V_1, V_2 and V_3. V_1 is the compressive steel stress volume due to the triangular stress value, V_2 is the compressive steel stress volume due to the uniform stress value, f_y and V_3 is the tensile steel stress volume due to a varying stress value, f_y. These volumes will vary as a function of the value of "c".

Plot the stress/strain diagram such that the concrete reaches its assigned value of strain and the steel reaches the maximum yield stress, f_y. Write the governing equations for the tensile and compressive stress/strain lines to calculate the steel forces. This will require 6 limiting ranges for "c" to determine the values of the steel forces.

The equations for the outer circular section is first derived followed by the inner circular section. From analytic geometry, the equation of a circle is as follows:

$$x^2 + z^2 = R_o^2 \tag{3}$$

The equations of stress are:

$$y = k \{ x + (c - R) \} \quad \text{for triangular shape} \tag{4}$$

in which, $k = k_1$ or k_3 depending on the position of "c" from the compressive concrete edge of the CFT column section and the balanced condition (simultaneous development of ultimate concrete and steel stress).

$$k_1 = 29000 \, e_c / c \tag{5}$$

$$k_3 = f_y /(R + R_o - c) \tag{6}$$

$$y = k_2 = f_y \quad \text{for uniform stress} \tag{7}$$

The intersection of uniform with triangular shape is given by:

$$x_1 = (1/e_c) \, c \, (e_c - e_s) \tag{8}$$

$$x_n = R - x_1 \tag{9}$$

Evaluate the integrals of the stress volumes within their respective limits defined by 6 limiting ranges of the compressive depth "c" from the compressive concrete edge as it passes through the geometric limits of these 3 volumes and the strain diagram limits for the uniform and triangular shapes. This process should yield 35 equations for the outer circular section. Repeat procedure to obtain another 35 equations for the inner circular section. The difference between the outer and inner expressions is the value of the steel forces of the CFT column. Add to this the concrete forces to obtain the ultimate strength of the CFT column for every value of "c" on a particular column section. Plot the values of "P" and "M" to yield the ultimate strength capacity curve of the CFT column section.

The following are the derived equations for the outer circular section for every range of values of "c" on the CFT column section.

I. When $0 < c < R$

$$V_2 = k_2 \{1.5708 R_o^2 - [x_n (R_o^2 - x_n^2)^{1/2} + R_o^2 \arcsin (x_n/R_o)]\} \tag{10}$$

$$V_1 = (k_1/3) \{2 [R_o^2 - (R - c)^2]^{3/2} - 2(R_o^2 - x_n^2)^{3/2} + 3(c - R)[x_n (R_o^2 - x_n^2)^{1/2} + R_o^2 \arcsin (x_n/R_o)] - 3(c - R)[(R - c)[R_o^2 - (R - c)^2]^{1/2} + R_o^2 \arcsin[(R - c)/R_o]]\} \tag{11}$$

$$V_3 = - (k_3/3)\{-2[R_o^2 - (R - c)^2]^{1/2} + 3(c - R)[(R - c)[R_o^2 - (R - c)^2]^{1/2} + R_o^2 \arcsin [(R - c)/R_o] + 1.5708 R_o^2]\} \tag{12}$$

$$C_s = V_1 + V_2 \tag{13}$$

$$T_s = V_3 \tag{14}$$

$$V_1x_1 = (k_1/12)\{-6\ x_n[R_o^2 - x_n^2]^{3/2} + 3R_o^2\ [x_n\ (R_o^2 - x_n^2)^{1/2} + R_o^2\ arcsin\ (x_n/R_o)] + 6(R - c)\{[R_o^2 - (R - c)^2]^{3/2} - 3R_o^2[(R - c)[R_o^2 - (R - c)^2]^{1/2} + R_o^2\ arcsin\ [(R - c)/R_o]]\} - 8\ (c - R)[(R_o^2 - x_n^2)^{3/2} - [R_o^2 - (R - c)^2]^{3/2}]\} \tag{15}$$

$$V_2x_2 = (2k_2/3)(R_o^2 - x_n^2)^{3/2} \tag{16}$$

$$V_3x_3 = (k_3/24)\{3\pi R_o^4 - 12\ (R - c)[R_o^2 - (R - c)^2]^{3/2} + 6R_o^2\ [(R - c)[R_o^2 - (R - c)^2]^{1/2} + R_o^2\ arcsin[(R - c)/R_o]] - 16(c - R)[R_o^2 - (R - c)^2]^{3/2}\} \tag{17}$$

$$C_sx_s = V_1x_1 + V_2x_2 \tag{18}$$

$$T_sx_t = V_3x_3 \tag{19}$$

II. When $R < c < C_B$. $V_2 =$ Eq. (10), $V_2x_2 =$ Eq. (16), and

$$C_B = [e_c/(e_c + e_s)](R + R_o) \tag{20}$$

$$V_1 = (k_1/3)\ \{-2\ \{(R_o^2 - x_n^2)^{3/2} - [R_o^2 - (c - R)^2]^{3/2}\} + 3$$
$$(c - R)\{x_n(R_o^2 - x_n^2)^{1/2} + (c - R)[R_o^2 - (c - R)^2]^{1/2} +$$
$$R_o^2\ (arcsin(x_n/R_o) + arcsin[(c - R)/R_o])\}\} \tag{21}$$

$$V_3 = (-k_3/3)\{-2[R_o^2 - (c - R)^2]^{3/2} + 3(c - R)\{- (c - R)[R_o^2 - (c - R)^2]^{1/2} + R_o^2(1.5708 - arcsin[(c - R)/R_o]\}\} \tag{22}$$

$$C_s = V_1 + V_2 \tag{23}$$

$$T_s = V_3 \tag{24}$$

$$V_1x_1 = (k_1/12)\{3R_ox_n(R_o^2 - x_n^2)^{1/2} + 3R_o^2(c - R)[R_o^2 - (c - R)^2]^{1/2} + 3R_o^4\{arcsin(x_n/R_o) + arcsin[(c - R)/R_o]\}$$
$$- 6x_n(R_o^2 - x_n^2)^{3/2} - 6(c - R)[R_o^2 - (c - R)^2]^{3/2} - 8\ (c - R)(R_o^2 - x_n^2)^{3/2} + 8[R_o^2 - (c - R)^2]^{3/2}\} \tag{25}$$

$$V_3x_3 = (k_3/24)\{3\pi R_o^4 + 12\ (c - R)[R_o^2 - (c - R)^2]^{3/2} - 6R_o^2\ \{(c - R)[R_o^2 - (c - R)^2]^{1/2} + R_o^2\ arcsin[(c - R)/R_o]\} - 16\ (c - R)[R_o^2 - (c - R)^2]^{3/2}\} \tag{26}$$

$$C_sx_s = V_1x_1 + V_2x_2 \tag{27}$$

$$T_sx_t = V_3x_3 \tag{28}$$

III. When $C_B < c < (R + R_o)$. $V_1 =$ Eq. (21), $V_2 =$ Eq. (10), $V_2x_2 =$ Eq. (16), $V_1x_1 =$ Eq. (25) and

$$V_3 = (-k_1/3)\{-2[R_o^2 - (c - R)^2]^{3/2} + 3\ (c - R)\{- (c - R)[R_o^2 - (c - R)^2]^{1/2} + R_o^2\ (1.5708 - arcsin[(c - R)/R_o]\}\} \tag{29}$$

$$C_s = V_1 + V_2 \tag{30}$$

$$T_s = V_3 \tag{31}$$

$$V_3x_3 = (k_1/24)\{3\pi R_o^4 + 12\ (c - R)[R_o^2 - (c - R)^2]^{3/2} - 6R_o^2\ \{(c - R)[R_o^2 - (c - R)^2]^{1/2} + R_o^2\ arcsin[(c - R)/R_o]\} - 16\ (c - R)[R_o^2 - (c - R)^2]^{3/2}\} \tag{32}$$

$$C_sx_s = V_1x_1 + V_2x_2 \tag{33}$$

$$T_sx_t = V_3x_3 \tag{34}$$

IV. When $(R + R_o) < c < C_1$. $V_2 =$ Eq. (10), $V_3 = 0$, $V_3x_3 = 0$, $V_2x_2 =$ Eq. (16), $T_s = 0$, $T_sx_t = 0$ and

$$C_1 = Re_c/(e_c - e_s) \tag{35}$$

$$V_1 = (k_1/3)\ \{-2\ (R_o^2 - x_n^2)^{3/2} + 3(c - R)[x_n\ (R_o^2 - x_n^2)^{1/2} + R_o^2\ arcsin(x_n/R_o) + 1.5708R_o^2]\ \} \tag{36}$$

$$C_s = V_1 + V_2 \tag{37}$$

$$V_1x_1 = (k_1/24)\{-12x_n(R_o^2 - x_n^2)^{3/2} + 6R_o^2\ [x_n\ (R_o^2 - x_n^2)^{1/2} + R_o^2\ arcsin(x_n/R_o)] + 3\pi R_o^4 - 16(c - R)(R_o^2 - x_n^2)^{3/2}\ \} \tag{38}$$

$$C_sx_s = V_1x_1 + V_2x_2 \tag{39}$$

V. When $C_1 > c > C_2$. $V_2 =$ Eq. (10), $V_2x_2 =$ Eq. (16), $V_1 =$ Eq. (36), $V_1x_1 =$ Eq. (38), $V_3 = 0$, $V_3x_3 = 0$, $T_s = 0$, $T_sx_t = 0$ and

$$C_2 = [e_c/(e_c - e_s)](R + R_o) \tag{40}$$

$$C_s = V_1 + V_2 \tag{41}$$

$$C_sx_s = V_1x_1 + V_2x_2 \tag{42}$$

VI. When $c > C_2$. $V_1 = 0$, $V_1x_1 = 0$, $V_3 = 0$, $V_3x_3 = 0$, $T_s = 0$, $C_sx_s = 0$, $T_sx_t = 0$ and

$$V_2 = k_2\ \pi\ R_o^2 \tag{43}$$

$$C_s = V_2 \tag{44}$$

To write the set of equations for the inner circular section, replace R_o by R in the above list of equations. This process will yield another 35 equations.

The following are the derived equations for the inner circular section:

I. When $0 < c < R$

$$V_2 = k_2\ \{1.5708R^2 - [x_n\ (R^2 - x_n^2)^{1/2} + R^2\ arcsin\ (x_n/R)]\} \tag{45}$$

$$V_1 = (k_1/3) \{2 [R^2 - (R - c)^2]^{3/2} - 2 (R^2 - x_n^2)^{3/2} + 3 (c - R)[x_n (R^2 - x_n^2)^{1/2} + R^2 \arcsin (x_n/R)] - 3 (c - R)[(R - c)[R^2 - (R - c)^2]^{1/2} + R^2 \arcsin[(R - c)/R]]\} \tag{46}$$

$$V_3 = - (k_3/3)\{-2[R^2 - (R - c)^2]^{1/2} + 3(c - R)[(R - c)([R^2 - (R - c)^2]^{1/2} + R^2 \arcsin [(R - c)/R] + 1.5708R^2)\} \tag{47}$$

$$C_s = V_1 + V_2 \tag{48}$$

$$T_s = V_3 \tag{49}$$

$$V_1 x_1 = (k_1/12)\{-6 x_n (R^2 - x_n^2)^{3/2} + 3R^2 [x_n (R^2 - x_n^2)^{1/2} + R^2 \arcsin (x_n/R)] + 6(R - c)\{[R^2 - (R - c)^2]^{3/2} - 3R^2[(R - c)[R^2 - (R - c)^2]^{1/2} + R^2 \arcsin [(R - c)/R]]\} - 8 (c - R)[(R^2 - x_n^2)^{3/2} - [R^2 - (R - c)^2]^{3/2}]\} \tag{50}$$

$$V_2 x_2 = (2k_2/3)(R^2 - x_n^2)^{3/2} \tag{51}$$

$$V_3 x_3 = (k_3/24)\{3\pi R^4 - 12 (R - c)[R^2 - (R - c)^2]^{3/2} + 6R^2 [(R - c)[R^2 - (R - c)^2]^{1/2} + R^2 \arcsin[(R - c)/R]] - 16(c - R)[R^2 - (R - c)^2]^{3/2} \} \tag{52}$$

$$C_s x_s = V_1 x_1 + V_2 x_2 \tag{53}$$

$$T_s x_t = V_3 x_3 \tag{54}$$

II. When $R < c < C_B$. $V_2 =$ Eq. (45), $V_2 x_2 =$ Eq. (51), and

$$C_B = [e_c/(e_c + e_s)](2R) \tag{55}$$

$$V_1 = (k_1/3) \{-2 (R^2 - x_n^2)^{3/2} + 2 [R^2 - (c - R)^2]^{3/2} + 3(c - R)[x_n (R^2 - x_n^2)^{1/2} + (c - R)[R^2 - (c - R)^2]^{1/2} + R^2 (\arcsin(x_n/R_o + \arcsin[(c - R)/R])]\} \tag{56}$$

$$V_3 = (-k_3/3)\{-2[R^2 - (c - R)^2]^{3/2} + 3(c - R)[- (c - R)[R^2 - (c - R)^2]^{1/2} + R^2(1.5708 - \arcsin[(c - R)/R]]\} \tag{57}$$

$$C_s = V_1 + V_2 \tag{58}$$

$$T_s = V_3 \tag{59}$$

$$V_1 x_1 = (k_1/12)\{3Rx_n(R^2 - x_n^2)^{1/2} + 3R^2(c - R)[R^2 - (c - R)^2]^{1/2} + 3R^4\{\arcsin(x_n/R) + \arcsin[(c - R)/R]\} - 6x_n(R^2 - x_n^2)^{3/2} - 6(c - R)[R^2 - (c - R)^2]^{3/2} - 8(c - R)(R^2 - x_n^2)^{3/2} + 8[R^2 - (c - R)^2]^{3/2} \} \tag{60}$$

$$V_3 x_3 = (k_3/24)\{3\pi R^4 + 12 (c - R)[R^2 - (c - R)^2]^{3/2} - 6R^2 [(c - R)[R^2 - (c - R)^2]^{1/2} + R^2 \arcsin[(c - R)/R]] - 16 (c - R)[R^2 - (c - R)^2]^{3/2}\} \tag{61}$$

$$C_s x_s = V_1 x_1 + V_2 x_2 \tag{62}$$

$$T_s x_t = V_3 x_3 \tag{63}$$

III. When $C_B < c < (2R)$. $V_1 =$ Eq. (56), $V_2 =$ Eq. (45), $V_2 x_2 =$ Eq. (51), $V_1 x_1 =$ Eq. (60) and

$$V_3 = (-k_1/3)\{-2[R^2 - (c - R)^2]^{3/2} + 3 (c - R)[- (c - R)[R^2 - (c - R)^2]^{1/2} + R^2(1.5708 - \arcsin[(c - R)/R]]\} \tag{64}$$

$$C_s = V_1 + V_2 \tag{65}$$

$$T_s = V_3 \tag{66}$$

$$V_3 x_3 = (k_1/24)\{3\pi R^4 - 12 (R - c)[R^2 - (R - c)^2]^{3/2} + 6R^2 [(R - c)[R^2 - (R - c)^2]^{1/2} + R^2 \arcsin[(R - c)/R]] - 16[R^2 - (R - c)^2]^{3/2} \} \tag{67}$$

$$C_s x_s = V_1 x_1 + V_2 x_2 \tag{68}$$

$$T_s x_t = V_3 x_3 \tag{69}$$

IV. When $(2R) < c < C_1$. $V_2 =$ Eq. (45), $V_3 = 0$, $V_3 x_3 = 0$, $V_2 x_2 =$ Eq. (51), $T_s = 0$, $T_s x_t = 0$ and

$$C_1 = Re_c/(e_c - e_s) \tag{70}$$

$$V_1 = (k_1/3) \{-2(R^2 - x_n^2)^{3/2} + 3(c - R)[x_n(R^2 - x_n^2)^{1/2} + R^2 \arcsin(x_n/R) + 1.5708R^2] \} \tag{71}$$

$$C_s = V_1 + V_2 \tag{72}$$

$$V_1 x_1 = (k_1/24)\{-12x_n(R^2 - x_n^2)^{3/2} + 6R^2 [x_n (R^2 - x_n)^{1/2} + R^2 \arcsin(x_n/R)] + 3\pi R^4 - 16(c - R)(R^2 - x_n^2)^{3/2} \} \tag{73}$$

$$C_s x_s = V_1 x_1 + V_2 x_2 \tag{74}$$

V. When $C_1 < c < C_2$. $V_2 =$ Eq. (45), $V_2 x_2 =$ Eq. (51), $V_1 =$ Eq. (71), $V_1 x_1 =$ Eq. (73), $V_3 = 0$, $V_3 x_3 = 0$, $T_s = 0$, $T_s x_t = 0$ and

$$C_2 = [e_c/(e_c - e_s)](2R) \tag{75}$$

$$C_s = V_1 + V_2 \tag{76}$$

$$C_s x_s = V_1 x_1 + V_2 x_2 \tag{77}$$

VI. When $c > C_2$. $V_1 = 0$, $V_1 x_1 = 0$, $V_3 = 0$, $V_3 x_3 = 0$, $C_s x_s = 0$, $T_s = 0$, $T_s x_t = 0$ and

$$V_2 = k_2 \pi R^2 \tag{78}$$

$$C_s = V_2 \tag{79}$$

Equation 10 to 79 supersede equations for steel forces listed in the paper presented in the SEMC2001 structural engineering mechanics and computations because only a limited number of equations for the steel forces were listed in that paper.

3 COLUMN CAPACITY CURVES

Using Microsoft Excel, the above formulas will yield the ultimate strength curve for steel forces. When combined with concrete forces will generate the ultimate strength capacities of a CFT column section. the following examples illustrate the results of the following analysis.

Figure 2 is a printout of the Excel program showing the steel strength capacity curve for a steel pipe with a nominal diameter of 203.2 mm, t = 8.1 mm,

R_o = 109.5 mm, Rb = 101.4 mm, f'_c = 34.5 Mpa, and f_y = 344.8 Mpa. In which P_s = steel axial capacity and M_s = the steel moment capacity.

Figure 3 shows the column capacity curve of this example when the concrete forces are added to the steel forces in which P = axial capacity and M = moment capacity of the CFT column section. if desired the values of "P" and "M" at the customary key points maybe listed, i.e. at

C_B = Eq. (20) when ultimate concrete and steel stresses are developed.

c = 2R when whole concrete section in compression.

c = $2R_o$ when tension in steel is zero. The value of "P" is the minimum recommended axial capacity to account for minimum load eccentricity.

The theoretical maximum axial capacity when M = zero and the beam moment capacity when P = zero can be interpolated from negative and positive values while the maximum moment is obtained by interpolation from results shown in the Excel spreadsheets.

4 COMPARISON

A comparison of methods will be attempted by solving an example using the analytical method for a circular section presented in ISEC-01 by M. Chao and J. Q. Zhang.

Figure 4 shows the ultimate strength capacity curve of this column from the Excel program. Note that the shape of the capacity curve does not match the curves of the proposed methods shown by the authors especially at high eccentricity (M/P) values. One reason is the use of the column interaction formula for biaxial bending and the modification of the

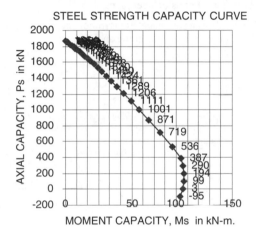

Figure 2.   Steel forces.

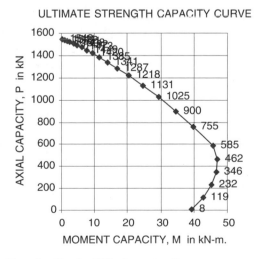

Figure 3.   Circular CFT column capacity curve.

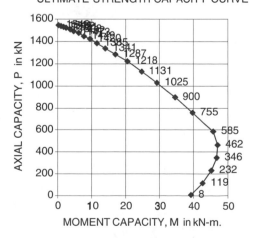

Figure 4.   Ultimate strength curve of example CFT.

internal capacity of the CFT column section inherent in their methods. Interestingly, the curve shown by the authors marked "Cross section exp. (5.9)" appears to show similar shape to Figure 4 albeit showing different numerical values.

The analytical method does not distinguish the length of the column since the strength of the section is invariant. The length of the column will magnify the external moment and will cause increase in eccentricities of the applied load. Multipliers including safety factors should be applied to the external load and not to the internal capacity. The resulting magnitude of external axial and moment loads should lie inside the envelope of the column capacity curve for safety in structural design.

5 CONCLUSIONS

1. This paper showed that an analytical solution is feasible for circular CFT column section, which is consistent with the analysis for reinforced concrete columns.
2. The analytical method yields the complete range of values of ultimate strength as defined by the compression depth "c" of the concrete core.
3. The strength of materials approach of this methodology predicts capacities of CFT column section accurately especially at higher eccentricities (M/P) of the internal forces.
4. This method precludes the use of equivalent rectangular stress block, interaction formula (sum of

stress ratios on orthogonal axes less than one) and the assumption of "forced equilibrium" for internal and external forces.

6 NOTATIONS

c = compressive depth of concrete core
C_B = concrete compressive depth when ultimate concrete and steel stresses are developed.
e_c = concrete strain = 0.003 (US)
    = 0.0035 (Canada)
e_s = steel strain
t = wall thickness of steel pipe
R = radius of concrete core or inner radius of steel pipe
R_o = outer radius of steel pipe.

Note: All other alphabets or symbols are defined in the context of their use in the analysis.

REFERENCES

Singh, A. 2001, *Creative Systems in Structural and Construction Engine*ering, 675–680, 777–781 and 783–787 AA. Balkema.
Zingoni, A. 2001, *Structural Engineering, Mechanics and Computation, Volume 1*, 319–326 and 343–350. ELSEVIER.
Beyer, W. H. *Handbook of Mathematical Sciences, 6th edition*: 31–32.

System-based Vision for Strategic and Creative Design, Bontempi (ed.)
© *2003 Swets & Zeitlinger, Lisse, ISBN 90 5809 599 1*

Shear transfer mechanism of Perfobond strip in steel-concrete composite members

H. Kitoh & S. Yamaoka
Department of Urban Engineering, Osaka City University, Osaka, Japan

K. Sonoda
Structural Research Center, Osaka Institute of Technology, Kyoto, Japan

ABSTRACT: Perfobond strip is a shear connection device between steel and concrete developed by Leonhardt et al. in 1987 as an alternative of a familiar headed stud connector. First, we carried out a series of its direct shear test. Second, three dimensional finite element analyses relevant to the tests were conducted considering not only contact condition on interface between the strip and concrete but also material nonlinearity. Last, its shear transfer mechanism is discussed based upon both the experimental and numerical results obtained.

1 INTRODUCTION

Perfobond strip is a shear connection device between steel and concrete developed by Leonhardt et al. (1987) as an alternative of a familiar headed stud connector. It is a steel strip with circular holes punched out to be filled with concrete, and directly welded upright to steel member, e.g. a top flange plate of steel girder to be joined with a R/C slab in a composite girder. Comparing with the stud connector, its main features are as follows: First, it can transfer shearing force due to concrete dowel action at the holes in the strip. Thus, it shows rigid-plastic deformation performance without carrying load degradation. Furthermore, its fatigue strength is enhanced owing to no localized stress concentration. Last, it can also lead an economical advantage because of a short construction period.

Its design method has been also proposed comprehensively by Leonhardt et al. (1987). However, continuous research works on the strip have been conducted even at present, e.g. Tategami et al. (2002), because their method was based on a limited number of push-out test results and its various applications are also now required for new type joints in steel-concrete composite construction such as a steel bridge girder rigidly connected to R/C pier reported by Hikosaka et al. (2001). In this paper, we examine its shear transfer mechanism through both experimental and numerical approaches, for its wide applications.

First, we carried out pull-out type direct shear loading tests of six element specimens, whose parameters were number of holes in the strip and existence of a transverse reinforcement penetrating the each hole. Second, we also conducted three dimensional finite element analyses relevant to the element specimens, in which both material nonlinearity and contact condition on the interface between the strip and surrounding concrete were considered. Last, we discuss the shear transfer mechanism of the strip based upon the stress distributions both in the strip and concrete as the numerical results obtained. It is noted additionally that there is little literature on such a numerical approach for the strip, except for Kraus & Wurzer (1997).

2 EXPERIMENT

2.1 *Perfobond strip specimens*

Figure 1 shows three types of the Perfobond strip specimens tested, whose thickness and height were constant. The first test parameter was number of circular holes of 40 mm in diameter to be filled with concrete, according to the number of the hole the strip length (L) was varied from 100 mm to 300 mm. To be controlled the specimens' failure by a dual surfaces' direct shear of filled concrete within the hole, the thickness and length between vertical edge and hole were defined not to cause a direct shear of strip itself nor a bearing of the filled concrete according to the design method of Leonhardt et al. (1987). Moreover, transmitting forces to the strip by the filled concrete solely, two styrene form block were arranged at both the vertical strip

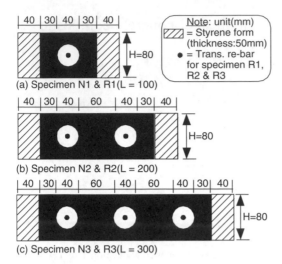

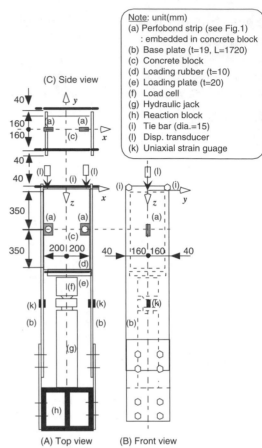

(a) Specimen N1 & R1(L = 100)

(b) Specimen N2 & R2(L = 200)

(c) Specimen N3 & R3(L = 300)

Note: unit(mm)
⬚ = Styrene form (thickness:50mm)
● = Trans. re-bar for specimen R1, R2 & R3

Figure 1.   Perfobond strips tested (12 mm thick).

Note: unit(mm)
(a) Perfobond strip (see Fig.1) : embedded in concrete block
(b) Base plate (t=19, L=1720)
(c) Concrete block
(d) Loading rubber (t=10)
(e) Loading plate (t=20)
(f) Load cell
(g) Hydraulic jack
(h) Reaction block
(i) Tie bar (dia.=15)
(l) Disp. transducer
(k) Uniaxial strain guage

(C) Side view

(A) Top view   (B) Front view

Figure 2.   Pull-out loading set-up.

Table 1.   Specimens.

| # | Tag | Test parameters | |
		Trans. re-bar	No. of holes
1	N1	non	1
2	R1	exist	1
3	N2	non	2
4	R2	exist	2
5	N3	non	3
6	R3	exist	3

edges to eliminate a bearing resistance of the edge against surrounding concrete.

The second test parameter was an existence of transverse reinforcement penetrating the center of the holes of the strips, which was a single deformed steel bar of 10 mm in nominal diameter. Combining these two test parameters, thus, all of the six specimens were prepared as shown in Table 1.

As to the material properties, cylinder strength and Young's modulus of concrete (f_c & E_c) and yield point and Young's modulus of steel (f_{sy} & E_s) were 30.4, 25.5×10^3, 283 and 202×10^3 in N/mm², respectively. Moreover, corresponding Poisson's ratios (ν_c & ν_s) were 0.183 and 0.333, respectively.

2.2 Pull-out loading test

Figure 2 shows the pull-out loading set-up used. The strip specimen was welded upright on the base steel plate coinciding the center of its bottom surface with the center of the area to be attached to a concrete block ($y = 0$ & $z = 350$ mm as shown in Fig. 2).

Subsequently, the concrete block was placed between two of the base plate with the strip apart from 400 mm, at which vinyl sheets were laid on inside surfaces of the both plates over the attached area to the block in order to cut natural bond between them off. The two base plates with the concrete block jointed each other only by the strips, then, was connected to a steel reaction block using some high tension bolts.

Direct shear loading on both the strips simultaneously was applied by a pull out load using a hydraulic jack, whose intensity was increased step by step until a relative slip between the plates and the block reached 20 mm. The pull out loading other than push out loading was necessary to prevent the plates from buckling.

2.3 Measurements

The intensity of the applied pull out load was measured through a load cell between the hydraulic jack and a loading steel plate as shown in Figure 2(A). In addition, longitudinal membrane forces of the plates were also measured to monitor the pull out load intensity of each plate, namely shearing force on the strip, by uniaxial strain gauges (k) in Figure 2. The relative slip, furthermore, was measured by twin displacement transducers attached to the free end surface of the block and anchored to each plate.

3 NUMERICAL ANALYSIS

3.1 Finite element modelling

Owing to symmetry in x and y axes in Figure 2, a quarter body of the plates and the block was idealized as an assemblage of three dimensional eight nodes solid finite elements as shown in Figure 3. Following boundary conditions were given, in which u, v and w were displacement components corresponding to x, y and z axes defined in Figure 2, respectively: (1) $u = 0$ on the x–z surface at $y = 0$ due to symmetry; (2) $v = 0$ on the y–z surface at $x = 0$ due to symmetry; (3) $u = v = w = 0$ for the plate on the x-y surface at $z = 1720$ mm due to its jointed end with the reaction block; (4) $u = 0$ on the free surface of the plate along the line at $z = 0$ due to two tie bars' restrict in Figure 2. Uniform incremental displacements in negative z-direction were imposed on the loaded surface of the concrete block at $z = 700$ mm, whose quantity was $1/200$ mm based on a convergency trial result previously examined. At the numerical execution, MSC Marc (2000) as a general purpose structural analysis code was used.

In case of existence of transverse reinforcement, the material properties of the concrete elements allocated at its arrangement were just replaced with the values for steel. Debonding of reinforcement, thus, could not account on.

3.2 Material nonlinearity

Concrete was assumed to be an elasto-plastic material according to associate flow rule with Drucker-Prager's criterion as follows:

$$f = \sqrt{J_2} + \alpha I_1 = k \tag{1}$$

where, I_1 = first invariant of stresses, J_2 = second invariant of deviatoric stresses and α & k = constants to be determined based on material properties. To determine the constants, it is necessary to previously define cohesion (c) and friction angle (ϕ) as Mohr-Coulomb's material. Various methods to define c & ϕ for concrete, e.g. Richart et al. (1928), have been proposed, because direct introduce of its compressive and tensile strengths (f_c & f_t) leads an over-estimation of shear strength in compressive region. The method by

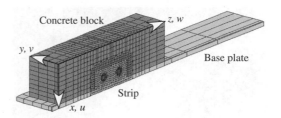

Figure 3. Finite element idealization for specimen N2 & R2 (see Fig. 2).

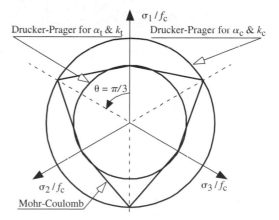

Figure 4. Yield surface on π-plane ($\phi > 0$; $c > 0$).

Table 2. Drucker-Prager's parameters ($f_c = 30.4$ N/mm^2).

	c/f_c	ϕ	α_c	k_c/f_c	α_t	k_t/f_c
Material test*	0.150	57.1	0.449	0.129	0.253	0.073
Richart et al. (1928)**	0.247	37.4	0.293	0.284	0.194	0.188
Ueda et al. (1984)	0.138	37.0	0.290	0.159	0.193	0.106

Note: * $f_t = 2.65$ N/mm^2; ** $m = f_c/f_t = 4.1$; Unit: ϕ in degree.

Ueda et al. (1984) was adopted herein, with which they could demonstrate a satisfactory numerical result for a direct shear test of R/C elements. Owing to the difference between radii of the compressive meridian ($\theta = 0$) and the tensile one ($\theta = \pi/3$) of Mohr-Coulomb's yield surface as shown in Figure 4, two combinations of α & k could be drawn as listed in Table 2.

Cracking of concrete was ignored because the observed failure mode was essentially controlled by not cracking but shear slip as shown in Figure 5, in which an obvious direct shear failure of the filled concrete within the hole of the strip could be confirmed.

Figure 5. Direct shear failure of filled concrete within the hole of specimen N1.

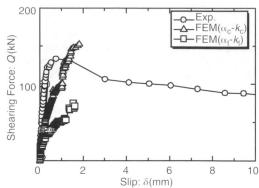

Figure 6. Shearing force vs. slip (Specimen N2).

Steel was to be elastic, because the dimensions of the strip and plate were too large to yield.

3.3 Contact condition

The finite element model shown in Figure 3 comprised the concrete block and the steel base plate including the strip to consider contact condition between them, so that double nodes were arranged on the interface between them. In aid of a special function for contact problem featured in MSC Marc (2000), the contact condition could be dealt without the usage of ordinary joint elements to connect the double nodes each other on the interface.

4 RESULTS

4.1 Shearing force – slip relation

Figure 6 shows the obtained relation between shearing force and slip of specimen N2, as a typical example of all the specimens. The numerical solution with α_c–k_c showed a fair agreement with the experimental curve from the points of view of the rigid stiffness and maximum shearing force, though another solution with α_t–k_t showed an obvious nonlinear behavior at an earlier loading stage. Because the former had a larger deviatoric plane area of yield surface than that of the latter as shown in Figure 4. However, both the numerical solutions terminated so suddenly, which suggested the occurrence of a brittle failure. Convergent solutions, thus, could not be obtained to follow the deformability after the peak shear strength observed experimentally.

4.2 Shear strength

Figure 7 shows the shear strength comparison of all the specimens from the experimental and numerical

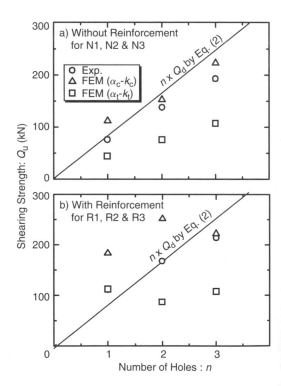

Figure 7. Shearing strength vs. number of holes.

results, in which the numerical strength at the ultimate state was defined as the maximum shearing force of convergent solutions. Furthermore, the design strength (Q_d) estimation by Leonhardt et al. (1987) was also plotted in the figure, which is given by the following empirical equation:

$$Q_d = \frac{\pi}{4} d^2 \times 1.08 f_c \times 2 \qquad (2)$$

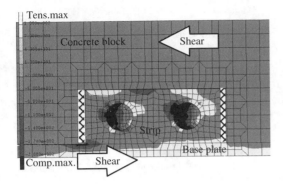

Figure 8. Horizontal normal stress σ_z distribution of N2 on x–z surface (see Fig. 3).

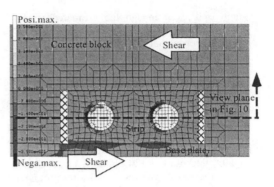

Figure 9. Shearing stress τ_{yz} distribution of N2 on x–z intersection including strip's surface (see Fig. 3).

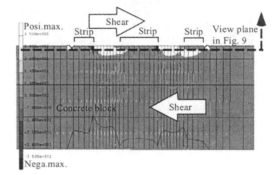

Figure 10. Shearing stress τ_{yz} distribution of N2 on y–z intersection through strip's hole center (see Fig. 3).

where, d = diameter of the circular hole of the strip (40 mm herein) and f_c = cylinder strength of concrete. It is noted that the coefficient for f_c in Eq. (2) is different from that of the original equation described in terms of cubic strength of concrete instead of f_c.

It can be found that all the experimental results ranged between the numerical results using the α_c & k_c for the compressive meridian and those using α_t & k_t for the tensile one. It could suggest the existence of shear meridian at $\theta = \pi/6$ (see Fig. 4) suitable for the direct shear problem (Chen 1982) similar to the object herein.

Furthermore, experimental strengths per a hole decreased slightly, as the number of the hole increased, from a contrast to the linear estimation by Eq. (2). The same tendency as the experimental strengths was observed in the numerical strengths without reinforcement, though those with reinforcement were somehow scattering.

It is, moreover, mentioned that both experimental and numerical results could show strength enhancement due to existence of reinforcement.

4.3 Horizontal normal stress distribution

As a typical example, Figure 8 shows the horizontal normal stress (σ_z) distribution on x–z intersection through the thickness-wise center of the strip at $y = 0$, as a numerical result with obtained using α_c & k_c at the ultimate state of specimen N2. Cross hatched regions adjacent to the strip in the figure were void corresponding to the styrene forms in Figure 1. The stress localization black-painted in both the right half regions of the circularly filled concrete in contact with the strip was remarkable, while the stress flow in another parts, i.e. the strip and base plate, was moderate.

4.4 Horizontal shearing stress distribution

Figure 9 shows a horizontal shearing stress (τ_{yz}) distribution on x–z intersection including the x–z surface

of the strip at $y = 6$ mm drawn as same manner as Figure 8, while Figure 10 shows the distribution on y–z intersection through the center of the holes of the strip at $x = 160$ mm. As shown in Figure 9, white-colored maximum stress occurred in overall regions of the filled concrete to suggest a direct shear failure occurrence illustrated previously with Figure 4. Furthermore, the maximum stress occurred in the hole portions extremely local-existed as outstanding white areas in Figure 10, whose shapes could correspond to the localized failure zone as shown in Figure 4.

5 CONCLUDING REMARKS

Shear transfer mechanism of Perfobond strips under direct shear due to pull-out loading, in which parameters were number of the hole in the strip and existence of reinforcement penetrating each hole, was examined experimentally and numerically. The following concluding remarks could be drawn:

(1) Experimental shear strength of the strip per a hole slightly decreased, as the number of the hole

increased, even though the estimated strength increases linearly to the number of the holes.

(2) Both experimental and numerical results could show shear strength enhancement of the strips due to existence of reinforcement.

(3) All the experimental strengths of the strips were located between the numerical strengths provided with the Drucker-Prager's constants based on compressive meridian of Mohr-Coloumb's yield surface and those on tensile meridian.

(4) Horizontal shearing stress distribution at the ultimate state numerically obtained could demonstrate the maximum stress localization around the filled concrete within the holes of the strip, which agreed with the direct shear failure mode of the concrete observed experimentally.

REFERENCES

Chen, W. F. 1982. *Plasticity in Reinforced Concrete*. New York, U.S.A.: McGraw-Hill International Book Company.

Hikosaka, H., Akehashi, K., Nakamura, K. & Imaizumi, Y. 2001. Construction of steel girder bridge rigidly connected to concrete pier with Perfobond plates. *Current and Future Trends in Bridge Design, Construction and Maintenance 2*: 166–174. London, U.K.: Thomas Telford.

Kitoh, H. & Sonoda, K. 1997. Bond characteristics of embossed steel elements. C. D. Buckner & B. M. Shahrooz (eds), *Composite Construction in Steel and Concrete III*: 909–918. New York, U.S.A.: ASCE.

Kraus, D. & Wurzer, O. 1997. Nonlinear finite-element analysis of concrete dowel, *Computers & Structures*, 64(5/6): 1271–1279, Pergamon, Elsevier Science Ltd, Great Britain.

Leonhardt, F., Andra, W., Andra, H. P. & Harra, W. 1987. Neues vorteilhaftes Verbundmittel fur Stahlnerbund-Tragwerke mit hoher Dauerfestigkeit, *Beton und Stahlbetonbau*, (12): 325–331.

MSC. Marc. 2000. Los Angeles, U.S.A.: MSC Software Corporation.

Richart, F. E., Bramdzaeg, A. & Brown, R. L. 1928. A study of the failure of concrete under combined compressive stresses, *Univ. Ill. Eng. Exp. Stn. Bull*. 185.

Tategami, H., Ebina, T., Uehira, K., Sonoda, K. & Kitoh, H. 2002. Shear connectors in PC box girder bridge with corrugated steel plate. J. H. Hajjar, M. Hosain, W. S. Easterling & B. M. Shahrooz (eds), *Composite Construction in Steel and Concrete IV*: 213–224. New York, U.S.A.: ASCE.

Ueda, M., Takeuchi, N., Higuchi, H. & Kawai, T. 1984. A discrete limit analysis of reinforced concrete structures. F. Damjanic, E. Hinton, D. R. J. Owen, N. Bicanic & V. Simovic (eds), *Proceedings of the International Conference on Computer Aided Analysis and Design of Concrete Structures*. 1369–1384. Swansea, U.K.: Pineridge Press.

Delamination of FRP plate/sheets used for strengthening of R/C elements

M. Savoia, B. Ferracuti & C. Mazzotti
DISTART – Structural Engineering, University of Bologna, Bologna, Italy

ABSTRACT: The problem of FRP-concrete delamination is studied. A non linear interface law is adopted, calibrated through results of experimental tests. A delamination model is developed, which is numerically solved via Finite Difference Method. Some numerical simulations are presented, concerning different delamination test setups and bond lengths. It is shown that the behavior of the typical pull-pull setup, adopted for experimental delamination tests, is characterized by a snap-back branch for FRP-concrete bond lengths greater than the minimum anchorage length.

1 INTRODUCTION

Several methods have been proposed so far for flexural and shear strengthening of R/C elements, based on the use of FRP-plates or sheets. These strengthening techniques allow for the improvement of structural behavior under ultimate loadings (Saadatmanesh & Ehsani 1991, Nanni 1993). Recent studies also showed that FRP-retrofit can be very useful to improve the behavior of R/C structures under both short term and long-term service loadings (Ferretti & Savoia 2003, Savoia et al. 2003a). They are competitive with respect to conventional retrofit criteria when low-weight increment and reduced application times are design requirements (Oehlers 2001).

The advent of this innovative reinforcement technique required a full body of experimental and theoretical data be available, in order to properly define the safety coefficients with respect to failure loads. For instance, it has been shown that unpredicted failure modes (FRP delamination, shear failure, etc.) may significantly reduce the theoretical bearing capacities of FRP-strengthened R/C beams when retrofit intervention is designed with respect to ultimate bending moment only (Arduini et al. 1997).

Delamination may start at the anchorage ends of the reinforcement as well as in the neighbourhood of concrete cracks (transverse cracks due to flexure or diagonal cracks due to shear). Delamination of external reinforcement from the anchorage ends may cause very brittle fracture mechanisms. Delamination originated by concrete cracks may be also critical, because it may arise even if a proper anchorage length at the extremities has been assured.

Current Codes suggest to perform the plate anchorage verification by calculating the maximum bond length and the maximum FRP force which can be transferred. Maximum transferable load cannot be predicted by linear elastic models, and the only attainment of maximum shear stress is not a criterion to predict delamination onset: in fact, maximum shear stress may be locally reached close to plate ends also under service loadings, i.e. for loads about 60/70 percent of failure load. The proposed relations for maximum load are usually based on fracture energy considerations (typically considering linear relations) or semi-empirical formulas (Taljsten 1996, Chen & Teng 2001, *fib* 2001). Since they do not consider the nonlinearity of interface law, their predictions are often very rough when compared with experimental results.

A comprehensive way of modelling FRP-concrete debonding phenomena is not yet completely available. The definition of a FRP-concrete interface law, for which some proposals can be found in the literature (see the next Section) is required. Moreover, due the presence of a softening branch in shear stress-slip law, the behavior of a specimen undergoing delamination may be highly non linear, and a snap-back branch can be observed in applied force – total slip curve.

In the present paper, delamination of FRP plates bonded to concrete is studied. A local bond-slip law for the FRP – concrete interface, recently proposed by the Authors (Savoia et al. 2003b), is adopted. The study confirms that stiffness of the interface requires compliances of both adhesive and a layer of concrete cover be considered (Brosens & van Gemert 1998). A kinematical model is then proposed to study the interface delamination phenomenon, and a solution scheme based on Finite Difference Method is developed.

In the last Section, numerical simulations are presented, where plate debonding under different setup conditions is studied. Results show that, for a pull-pull plate-concrete joint, the load – total slip curve exhibits a snap-back behavior when the length of plate bonded to concrete is greater than the minimum anchorage length. Snap-back may be even sharp if, as usual in experimental test setups, the external load is applied at the end of a portion of free plate. This circumstance explains the very brittle failure mechanism of plate debonding, so that the post failure branch cannot be experimentally followed even in the case of a displacement-controlled test.

A second setup is considered, where axial load is prescribed at the free end of the plate, and both plate and concrete are restrained at the other end. In this case, debonding is characterized by a stable post-delamination branch.

2 FRP-CONCRETE INTERFACE LAW

The definition of a constitutive interface FRP-concrete law is a difficult task for several reasons. From the experimental point of view, tests where shear force – maximum slip curves only are evaluated do not provide for sufficient data to define a local interface law. On the other hand, the transmission length of interface is very small, less than 80/100 mm from extremities or cracked sections, and several data on strains or displacements must be obtained within that length.

Moreover, the interface law is highly non linear (see for instance Fig. 1). At load levels typical of service loadings (about 50 per cent of maximum load), maximum slip may already exceed that corresponding to the peak shear stress, and softening branch of the

law is involved. As a consequence, maximum transmissible load also, related to fracture energy of interface law, strongly depends on softening branch.

Finally, the interface law also depends on the kinematic variables the law is referred to, i.e. on the features of the adopted structural model. For instance, in the proposed kinematical model, plane strain assumption is adopted for concrete and FRP plates (see Fig. 2), so that slip is referred to average displacement of concrete cross-section. Hence, in this case, interface compliance depends on shearing deformation of both adhesive and external cover of concrete, where the second term is the prevailing one due to the small thickness of the adhesive (about 0.1 mm in real applications). Usually, deformation of an effective external layer of concrete of about 30/50 mm thickness is considered in practical cases, allowing to predict a correct value of transmission length.

A FRP-concrete interface law must correctly reproduce elastic stiffness for low slip levels, maximum shear stress, and softening branch for higher slips. A bilinear law is suggested by *fib* (2001), with linearly elastic branch until local bond strength is reached, and linear descending softening branch until complete delamination. This law has been used for numerical simulations of delamination problem (Wu et al. 2002).

In Savoia et al. (2003b), an interface law has been obtained by post-processing experimental results by Chajes et al. (1996). The adopted interpolation law, based on a Popovics – like fractional law, reads:

$$\tau_p = \bar{\tau}\, \frac{s_p}{\bar{s}}\, \frac{n}{(n-1)+\left(s_p/\bar{s}\right)^n}, \qquad (1)$$

where $(\bar{\tau}, \bar{s})$ a reference point close to that corresponding to peak shear stress τ_{max}, and n is a free parameter mainly governing the softening branch. In the case considered, the values $\bar{\tau} = 6.93$ MPa, $\bar{s} = 0.051$ mm, $n = 2.860$ have been obtained using a best fitting procedure, and the corresponding interface law is reported in Figure 1, together with experimental results by Chajes et al. (1996).

3 THE PROPOSED MODEL FOR FRP-CONCRETE DELAMINATION

3.1 Governing equations

A kinematic model is developed to study the FRP-concrete delamination phenomenon. The notation adopted for displacements and stresses is reported in Figure 2. Plate and concrete are assumed to undergo axial deformations only. Axial displacements and forces for concrete and plates are denoted by u_c, u_p, N_c, N_p. Stresses due to bending can be neglected due to the very small bending stiffness of FRP plate with respect to concrete specimen.

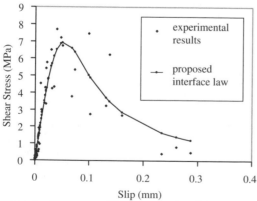

Figure 1. The proposed FRP-concrete interface law compared with data obtained from exp. tests by Chajes et al. (1996).

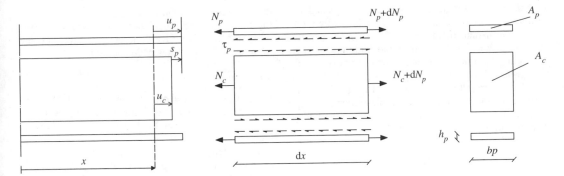

Figure 2.   The proposed model for FRP-concrete delamination: notation adopted for displacements and stresses.

The governing equations are equilibrium, constitutive and compatibility conditions, written as:

$$\frac{du_p}{dx} = \frac{1}{E_p A_p} N_p, \qquad \frac{dN_p}{dx} = b_p \tau_p \qquad (2)$$

$$\frac{du_c}{dx} = \frac{1}{E_c A_c} N_c, \qquad \frac{dN_c}{dx} = -2 b_p \tau_p, \qquad (3)$$

where A and E stand for area and Young modulus, τ_p is the concrete/plate interfacial shear stress, b_p is the plate width.

According to notation reported in Figure 2, interface law (1) is then written in the form:

$$\tau_p = k_p(s_p) \cdot s_p, \qquad (4)$$

with $s_p = u_p - u_c$ denoting FRP-concrete slip (see Fig. 2), and $k(s_p)$ is the non linear secant stiffness.

Hence, by substituting eqn (4) in (2) and (3), governing equations can be set in the form of a system of differential equations of the first order as follows:

$$\frac{d\mathbf{y}(x)}{dx} = \mathbf{A}(\mathbf{y}, x)\mathbf{y}(x), \qquad 0 \le x \le L, \qquad (5)$$

where L is the length of the bonded plate, vector $\mathbf{y}$ collects the unknown functions:

$$\mathbf{y}^T = \{N_p, N_c, u_p, u_c\} \qquad (6)$$

and the non linear matrix $\mathbf{A}$ is defined as:

$$\mathbf{A} = \begin{bmatrix} 0 & 0 & K_p & -K_p \\ 0 & 0 & -2K_p & 2K_p \\ 1/E_p A_p & 0 & 0 & 0 \\ 0 & 1/E_c A_c & 0 & 0 \end{bmatrix}. \qquad (7)$$

In (7), non linear coefficient $K_p = b_p k_p$ is the secant stiffness of FRP-concrete interface law. On the contrary, for the sake of simplicity, linearly elastic behavior is considered for both concrete and plates.

3.2   Numerical solution strategy

The non linear differential system (5) is numerically solved by the Finite Difference Method. The interval $[0, L]$ is divided using J nodes at uniform distance h, i.e., $0 = x_0 < x_1 < ... < x_J = L$. For the general j-th interior mesh point ($0 < j < J$), the derivative in eqn (5) is replaced by the central finite difference approximation $(\mathbf{y}_{j+1} - \mathbf{y}_j)/h$ centered at $x_{j+1/2}$.

Analogously, boundary conditions are written in terms of unknown vectors $\mathbf{y}_0$, $\mathbf{y}_J$ at the two end sections. The final system of non linear algebraic equations can be written as:

$$\mathbf{I} = \begin{vmatrix} \mathbf{S}_1 & \mathbf{R}_1 & & & \\ & ... & ... & & \\ & & \mathbf{S}_j & \mathbf{R}_j & \\ & & & ... & ... \\ & & & & \mathbf{S}_J & \mathbf{R}_J \\ \mathbf{B}_a & & & & \mathbf{B}_b \end{vmatrix} \begin{vmatrix} \mathbf{y}_0 \\ ... \\ \mathbf{y}_{j-1} \\ \mathbf{y}_j \\ ... \\ \mathbf{y}_J \end{vmatrix} = \begin{vmatrix} \mathbf{0} \\ \mathbf{0} \\ \mathbf{0} \\ \mathbf{0} \\ ... \\ \mathbf{a} \end{vmatrix} \qquad (8)$$

where $\mathbf{I}$ is the identity matrix and:

$$\mathbf{S}_j = -\frac{1}{h}\mathbf{I} - \frac{1}{2}\mathbf{A}(\mathbf{y}_j, x_j),$$

$$\mathbf{R}_j = +\frac{1}{h}\mathbf{I} - \frac{1}{2}\mathbf{A}(\mathbf{y}_{j+1}, x_{j+1}), \quad j = 1, 2, ..., J \qquad (9)$$

See Ferretti & Savoia (2003) for details. A quasi-Newton procedure has been used for the solution of non linear problem, by adopting secant values of stiffness coefficients appearing in governing matrix reported in eqn (8).

Numerical simulations have been performed by adopting different control parameters, basically force control or displacement control. If the control parameter must be changed during the simulation, only boundary conditions reported in the last row of matrix in eqn (8) must be modified, since both displacements and forces are unknown variables of the problem. As will be shown in numerical examples reported in the next Section, this versatility of the solution method is very useful when very highly non linear problems must be solved.

4 NUMERICAL SIMULATIONS OF DELAMINATION PROBLEM

The proposed model has been used to simulate some experimental tests concerning FRP-concrete delamination problems.

In Savoia et al. (2003b), the experimental results reported in Chajes et al. (1996) have been simulated. It has been shown that, adopting the interface law reported in Figure 1, strain distributions along the FRP plate can be well predicted, both in linear and non linear range. Maximum transmissible loads and minimum bond lengths obtained from the present method agreed well with experimental results.

In this Section, numerical results concerning FRP-concrete delamination are presented, considering two different test setups. The setups are depicted in Figure 3, together with mechanical and geometrical properties of specimens. For FRP-concrete interface, the law reported in Figure 1 has been adopted.

4.1 Setup N. 1

Setup N. 1 is a typical pull-pull test (Fig. 3a). Different bond lengths have been considered ($L = 50$ mm, 100 mm, 600 mm). The load-displacement curves are reported in Figures 4a–4c.

Displacement u_1 refers to the FRP section at the beginning of anchorage. First of all, by comparing the results corresponding to different bond lengths, it can be noted that maximum transmissible load increases with bond length. Moreover, the behavior after the attainment of maximum load strongly depends on the length of anchorage. For $L = 50$ mm (Fig. 4a), the

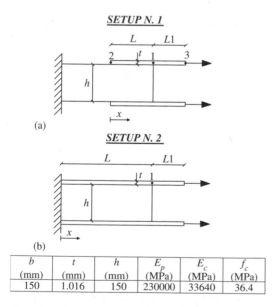

Figure 3. The two experimental setups considered in numerical simulations and mechanical/geometrical properties of specimens; L =bond length, $L1$ =400 mm.

b (mm)	t (mm)	h (mm)	E_p (MPa)	E_c (MPa)	f_c (MPa)
150	1.016	150	230000	33640	36.4

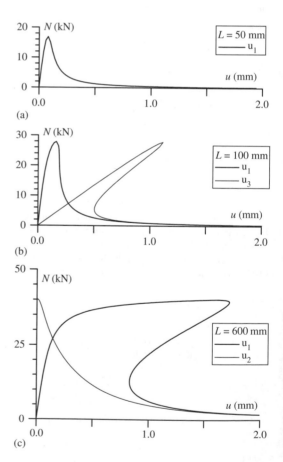

Figure 4. Pull-pull delamination test (Setup N. 1): load-displacement curves for different bond lengths. Displacements u_1, u_2, u_3 refer to points 1, 2, 3 in Figure 3a.

load displacement curve is almost linear until maximum load is reached, then the curve exhibits a softening behavior until complete delamination. For $L = 600$ mm (Fig. 4c), the linear initial branch is followed by a nearly horizontal plateau, where delamination progresses along the bonded length with almost constant axial load. Then, a very sharp snap-back behavior occurs, where both load and displacement u_1 decrease. Of course, this equilibrium branch cannot be followed numerically by adopting a control procedure where the displacement parameter is u_1. Hence, during the solution procedure, a shift of control parameter has been performed: at the beginning of the softening branch, the displacement of the end of anchorage u_2 has been used as control parameter, representing a monotonically increasing parameter during the whole delamination process (see Fig. 4c). Finally, results for the intermediate case $L = 100$ mm are reported in Figure 4b. The curve related to displacement u_1 shows a vertical softening branch, whereas the analogous curve related to displacement u_3 at the free end of FRP plate shows a very strong snap-back behavior. This is due to the elastic elongation of the portion of plate of length $L1$, which is maximum when the maximum load is reached, but rapidly decreases together with axial load during delamination. This result suggests that, from the experimental point of view, axial load or displacement must be prescribed as close as possible to the beginning of anchorage length.

The load-displacement curve for bond length $L = 200$ mm is reported in Figure 5; the corresponding distributions along the bonded length of main stress and displacement variables are given in Figure 6. These results are reported for three different equilibrium points (A, B, C), indicated in Figure 5: point A

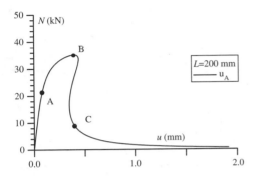

Figure 5. Pull-pull delamination test (Setup N. 1): load-displacement curve for $L = 200$ mm bond length.

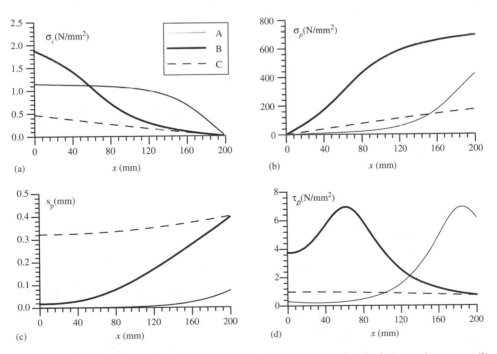

(a)

(b)

(c)

(d)

Figure 6. Pull-pull delamination test (Setup N. 1): Distributions along the bonded length of (a) stress in concrete, (b) stress in FRP plate, (c) FRP – concrete slip, (d) interfacial shear stress.

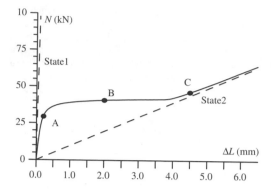

Figure 7. Delamination test according to Setup N. 2: load-displacement curve.

corresponds to about 60 percent of maximum load (then comparable with service loading condition), point B to maximum load and point C is the equilibrium configuration in the snap-back branch having the same displacement u_1 as point B. Figures 6c, d show that, for service loadings (point A) the slip may be greater than the value corresponding to maximum shear stress (0.051 mm in the present case), so that the softening branch of shear stress – slip law is involved. This circumstance has been verified also by postprocessing experimental data (Savoia et al. 2003b).

This phenomenon is even more evident when the maximum transmissible load is attained (point B). The corresponding FRP stress distribution (Fig. 6b) exhibits a change of curvature in the section corresponding to the maximum shear stress.

4.2 Setup N. 2

In Setup N. 2, both concrete and FRP plate are restrained at the end section (Fig. 3b). Bond length is $L = 1200$ mm. The load-displacement curve is reported in Figure 7.

The behavior of the specimen is a transition between *State 1* condition for low level loads (both stiffness contributions of concrete and FRP plates) and *State 2* condition after FRP delamination (FRP plates only contribute to specimen stiffness). Hence, the post-delamination branch is now stable and could be followed until complete delamination in an experimental delamination test.

Finally, Figure 8 shows the distributions along the bonded length of main stress and displacement variables, corresponding to three different situations A, B, C underlined in Figure 7. In particular, Figure 8d shows the translation along the specimen of the portion of plate contributing to bonding during delamination.

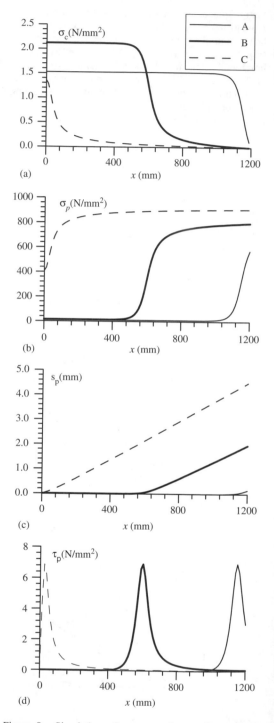

Figure 8. Simulation of test according to Setup N. 2: Distributions along the bonded length of (a) stress in concrete, (b) stress in FRP plate, (c) FRP – concrete slip, (d) interfacial shear stress.

ACKNOWLEDGEMENT

The financial supports of (italian) MIUR, PRIN 2001 Grant on *Theoretical and Experimental Analysis of Composite-Concrete Bonding for RC Members Reinforced by Composite Plates*" and C.N.R., PAAS Grant on "*Structural Effects of Rheological Properties of Composites*" are gratefully acknowledged.

REFERENCES

Arduini, M., Di Tommaso, A. & Nanni, A. 1997. Brittle failure in FRP plate and sheet bonded beams. *ACI Structural J.* 94: 363 370.

Brosens, K. & Van Gemert, D. 1998. Plate end shear design for external CFRP laminates. In H. Mihashi & R. Keitetsu (eds), *FRAMCOS-3 Proceedings*, Gifu, Japan: 1793–1804.

Chajes, M.J., Finch, W.W. jr, Januska, T.F. & Thomson, T.A. jr. 1996. Bond and force transfer of composite material plates bonded to concrete. *ACI Structural J.* 93: 208–217.

Chen, J.F. & Teng, J.G. 2001. Anchorage strength models for FRP and steel plates bonded to concrete. *J. Struct. Eng. ASCE* 127: 784–791.

Ferretti, D. & Savoia, M. 2003. Cracking evolution in R/C tensile members strengthened by FRP-plates. *Eng. Fract. Mech.* 70: 1069–1083.

fib 2001. *Externally bonded FRP reinforcement for RC structures*, FIB, technical report, Bullettin n° 14.

Nanni, A. 1993. Flexural behavior and design of RC members using FRP reinforcement. *J. Struct. Eng. ASCE* 119: 3344–3359.

Oehlers, D.J. 2001. Development of design rules for retrofitting by adhesive bonding or bolting either FRP or steel plates to RC beams or slabs in bridge and buildings. *Composites: Part A* 32: 1345–1355.

Saadatmanesh, H. & Ehsani, M.R. 1991. RC beams strengthened with GFRP plates. I: Experimental study; II: Analysis and parametrical study. *J. Struct. Eng. ASCE* 117: 3417–3433 and 3434–3455.

Savoia, M., Ferracuti, B. & Mazzotti, C. 2003a. Creep deformation of FRP-plated R/C tensile elements using solidification theory. In Bicanic N. et al. (eds.), *Comp. Modelling of Concrete Structures EURO-C*, St. J. im Pongau (Austria): 501–511.

Savoia, M., Ferracuti, B. & Mazzotti, C. 2003b. Non linear bond-slip law for FRP-concrete interface. In K.H. Tan (ed.), *FRPRCS-6 Conf. Proc.*, Singapore, 8–10 July 2003, 1–10.

Taljsten, B. 1996. Strengthening of concrete prisms using the plate-bonding technique. *Eng. Fract. Mech.* 82: 253–266.

Wu, Z., Yuan, H. & Niu, H. 2002. Stress transfer and fracture propagation in different kinds of adhesive joints. *J. Eng. Mech. ASCE* 128(5): 562–573.

System-based Vision for Strategic and Creative Design, Bontempi (ed.)
© 2003 Swets & Zeitlinger, Lisse, ISBN 90 5809 599 1

The abrasion resistance of fiber concrete surface

Y.W. Liu

Department of Civil and Water Resource Eng., National ChiaYi University, Taiwan

T. Yen & T.H. Hsu

Department of Civil Engineering, National Chung-Hsiung University, Taiwan New Institute, Gouda, Netherlands

ABSTRACT: In this research, we attempted to increase the abrasion resistance of concrete by adding fibers. Four different types of concrete were cast using plain concrete, carbon fibers concrete, and Polypropylene fibers concrete. And the abrasion resistance of concrete is compared with the ASTM C 944 testing method (rotating-cutter method). Test results show that, the abrasion resistance was also improved by the addition of steel fibers and carbon fibers. The abrasion resistance was better for concrete with carbon fibers than with Polypropylene fibers and plain concrete, but was worse for concrete with carbon fibers than with steel fibers.

1 INSTRUCTIONS

The percentage of concrete pavement used as highway transportation is higher recently. Scant attention has been paid to concrete abrasion resistance, despite the fact that poor abrasion resistance in highway concrete can accelerate pavement deterioration. Therefore, maintenance has to be carried out oftentimes. This is not only a waste of money and manpower, but also lowers the service quality of the pavement and consequently decreases its durability. The abrasion resistance of concrete is relevant to application, as rubbing, scraping, skidding or sliding of objects on the concrete surface commonly occur. It has been well established that a particular concrete's abrasion resistance depend on its compressive strength as well as coarse aggregate volume and hardness (Liu 1991, Laplamte 1991 & Dhir 1991). But the testing procedure is also a factor.

The recent technological breakthroughs in concrete technology should have a better abrasion resistance. For instance, fly ash, slag and silica fume as admixtures has been found to enhance compressive strength while decreasing cement content, pore size, and permeability that result in higher compressive strength and abrasion resistance (Manmonan et al. 1981). The development and application of fiber has made concrete has higher flexure, shear, tensile and impact strength, decrease shrinkage, and prevent crack, lead to increase the abrasion resistance of pavement concrete (Ench 1985 & Zeng et al. 1997). In this paper, a series of concrete specimens were made to investigate the affect of steel fibers, carbon fibers and Polypropylene fibers on the abrasion resistance of concrete.

2 TEST PROGRAM

2.1 *Material and proportion*

The cement used was Portland cement Type I, and complied with the CNS 61 requirements for Portland Cement Type I. River sand was used as the fine aggregate, with a fine Modulus = 2.98, specific gravity = 2.66 and absorption = 2.5%. The coarse aggregate used was a crushed basalt, with a size up to 20 mm, specific gravity = 2.64, absorption = 1.2%, and dry-rodded unit weight = 1682 kg/m³. Class F fly ash was obtained from Taichung Power Plant of Taiwan Power Company, with a specific gravity of 2.31, and Ground granular blast furnace slag with a specific gravity of 2.89 was used for binder that in the amount of 20% of total binder weight. The water/binder ratio was 0.36. The super-plasticizer, complying with ASTM C494 type-G range, and a specific gravity = 1.1 was used was used to improve the workability of the concrete. The mix proportions of the four types of concrete as given in Table 1. Samples of the mixtures were placed in cylinder forms (15 cm in diameter and 30 cm in height) for compressive strength test, and square molds of size 15 × 15 × 5 cm for abrasion resistance test. After 24 h, the samples were released and placed

Table 1. Mixture proportions of concrete (kg/m^3).

Mixture	Water	Cement	Slag 2.2	Fly ash 2.3	Coarse	Sand	Fibers	SP
PC	150	333	42	42	1015	800	–	6
SC	150	333	42	42	992	800	39	10
CC	150	333	42	42	998	800	9	8
PPC	150	333	42	42	999	800	4.5	7.5

Table 2. Properties of fibers.

	Steel fibers	Carbon fibers	Polypropylene fibers
Length	0.5 Φmm × 23 mm	7 μm × 12 mm	19 mm
Tensile strength (MPa)	1050	680	430
Specific gravity	7.8	1.8	0.9

under water for curing. After 28 days, all tests were measured of the concrete samples.

Four types of concrete were studied, namely (i) plain concrete (PC), (ii) concrete with steel fibers (SC), (iii) concrete with Polypropylene fibers (PPC), and (iv) concrete with silicon fume (SFC). Two types of fiber in the amount of 0.5 vol.% were used, and their properties are show in Table 2.

2.2 Test apparatus

The abrasion resistance was measured using ASTM C944-90a (Rotating–Cutter Method). The rotation cutting involved twenty-four No. 1 Desmond–Huntington grinding dressing wheels (diameter = 1.5 in, thickness = 3/32 in) and washer (diameter = 1 3/8 in, thickness = 1/32 in) between the dressing wheels, shown in Figure 1. The total cutting area was 71.12 cm^2. The load was 196 N during abrasion. The speed of the rotating cutting was 200 rpm. Each specimen was abraded once on each of its two opposite flat sides, such that time lasted 2 min. The weight after each time was measured using a balance to an accuracy of 0.1 g. Three specimens of each type of concrete were tested. The abrasion resistance as used in this paper is defined as the average abrasion depth that is the average of six values obtained for each type of concrete.

3 RESULTS AND DISCUSSIONS

The test results of each concrete are listed in Table 3. The steel fibers and silicon fume addition improved the abrasion resistance of concrete. Polypropylene fibers was not as effective as the above two concretes.

Figure 1. Ttest apparatus of rotating cutter.

Table 3. Mechanical properties of the concrete mixtures.

Parameter	Compressive strength (MPa)	Flexural strength (MPa)	Hardness of surface	The average weight loss (g)
PC	43.7	6.40	21.7	4.5
SC	41.9	8.36	46.3	1.5
CC	43.6	7.60	31.8	1.8
PPC	41.8	6.69	22.6	3.2

Figure 2 illustrates the influence of the compressive strength. It is found that the conventional compressive strength does not give quantitative information about the abrasion resistance of concrete. The plain concrete has the lowest abrasion resistance, even if it does not own the lowest compressive strength. The result illustrates that the effect of compressive strength of short-time abrasion resistance of concrete is little than that of long-time experiments. This point indicates that the mechanical properties of surface affect the abrasion resistance of concrete play a further important role.

Figure 3 shows a comparison between the four types of concrete and the flexural strength. It can clearly be distinguished; the effectiveness of steel fibers in improving the abrasion resistance is attributed to the

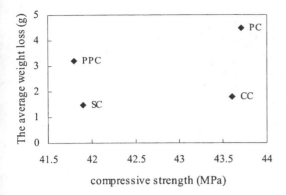

Figure 2. The influence of the compressive strength on abrasion loss.

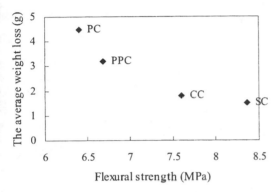

Figure 3. The influence of the flexure strength on abrasion loss.

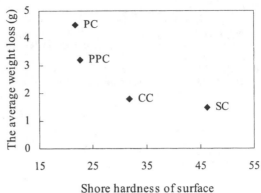

Figure 4. The influence of the hardness of concrete surface on abrasion loss.

the ASTM C 944 testing method. The effect relate to the increase in flexure strength and hardness result in resistance to fracture. The average abrasion depth of steel fibers and silicon fume concretes were only approach 1/3 of that of plain concrete. Obviously, steel fibers and silicon fume are superior materials resistance to abrade.

ACKNOWLEDGMENT

The supports from the National Science Council of Taiwan, R.O.C. under contract NSC 91-2211-E-415-001, and the National ChiaYi University of Taiwan are acknowledged.

increase in flexural strength of concrete. The feature points to a significant change in the abrasion behavior if fiber inclusions are added to the concrete. Because of the fibers increase the bond stress between cement matrix and aggregate, the cement matrix and aggregate be removed uneasily by abrade.

Figure 4 shows the influence of the hardness of surface on concrete abrasion resistance, which increases as the hardness of surface increases. In this case, the reason is the hardness of steel fibers is further great than concretes induce the cement matrix and aggregate around the steel fiber been protecting against attack. Under the action of Rotating–Cutter load (198N), the steel fibers can resistance against attack, results in decrease the cement matrix and aggregate been abraded.

4 CONCLUSION

The concrete abrasion resistance was found to be improved by adding steel fibers and silicon fume with

REFERENCES

Antoine E. Naaman. 1985/March. Fiber einforcement for concrete. Concrete International: 21–25. USA.
Ench fitzer. 1985. Carbon fibers and their composites. Springer-Verlag Borlin Heidelberg.
Zeng-Qiang Shi & D. D. L. Chung. 1997. Improving the abrasion resistance of mortar by adding latex and carbon fibers. Cement and concrete research, vol. 27, No.8: 1149–1153.
T. Yen. 1998. Mixture of low cement high performance concrete. Symposium on high performance concrete. Taiwan. 111–134.
A. M. Neville. 1997. Properties of Concrete. Putman Publishing Inc.
T. Yen. 1998. The produce and property of low cement high performance concrete. Symposium of National Science Council R.O.C.
Yoshihiko Ohama, Mikio Amano & Mitsuhiro Endo.1985/March. Properties of Carbon Fiber Reinforced Cement with Silica Fume. Concrete International: 58–62.
Y. W. Liu, J. C. Liou & Y. S. Chen. 1999. Influences of the Properties and the Contents of Mortar on the Abrasion Resistance of High Performance Concrete. 5th International Symposium on High Strength/High Performance Concrete Norway.

T. C. Liu 1991. Abrasion Resistance of Concrete. ACI Journal, 78(5): 341–350.

P. Laplante, P. C. Aitcin, & D. Vexina. 1991. Abrasion Resistance of Concrete. Journal of Material in Civil Engineering. Vol. 3, No.1.

R. K. Dhir, P. C. Hewleet, & Y. N. Chan. 1991. Near Surface Characteristics of Concrete Abrasion Resistance. Materials and Structure/Muteriaux et Constructions, Vol. 24, No.140: 122–128.

D. Manmonan & P. Mehata. 1981. Influence of Pozzolanic. Slag, and Chemical Admixtures on Pore Size Distribution and Permeability of Hardenc\ed Cement Pastes. Cement, Concrete and Aggregates, Vol. 3, No. 1.

Calculation and check of steel-concrete composite beams subject to the compression associated with bending

G.P. Gamberini
Faculty of Engineering, Departement of Structural Engineering, University of Cagliari, Cagliari, Italy

G.F. Giaccu
Freelancer in an engineer's office in Cagliari, Italy

ABSTRACT: In studying the question of compression associated with bending extended to mixed structures with steel and concrete, since the latter is not dealt with in terms of european regulations, it would be best to elaborate a complex finite element model allowing the study of stress on such a structure. From the practical standpoint, for example in the design and check of small slab bridge structures, this often entails complicated and laborious tasks. In the work that follows the behaviour of a mixed slab structure is examined through comparison of two models, obtaining simple relations which, by means of the calculation of an ideal area and a moment of inertia, allows easy check of the cross-section considered.

1 INTRODUCTION

1.1 Initial considerations

Analysis of an asymmetric bridge structure with backstays characterized by a mixed steel and concrete slab structure led to the detection of certain deficiencies in the Eurocode as concerns compression associated with bending stress transmitted to the slab by dead loads and backstays. In calculating the ideal area, the Eurocode contemplates exclusively the case of the simply bent-section and then a effective width of slab and therefore a moment of inertia relating only to bending stress in the slab structure. The calculation of internal stress, both in the concrete and in the steel, relates to bending only: this implies that the use of values supplied by the Eurocode in the case of compression associated with bending would lead to values that are completely different from real ones. From this perspective, the spirit of this work becomes that of studying, by means of a finite element model, the behaviour of the bridge floor system of the structure under examination in order to find a method allowing easy check of the section without recourse to a study of the floor system by means of finite elements with the perspective of future extension of such a method to other kinds of mixed structure floor systems.

2 DESCRIPTION OF THE STRUCTURE AND MODELS USED

2.1 Description of the structure

The structure under examination is, as stated previously, a bridge with backstays presenting a slab with a mixed steel and concrete structure with two side beams 13 m apart and 2 m in height and cross beams of the same kind 5 m on centre and 1.2 m in height; all these elements are connected by a reinforced concrete floor system 25 cm thick (Fig. 1); for the purposes of this description we shall examine the static behaviour of the first part of the slab described, having a length of 60 m and the configuration of a continuous beam on three rockers.

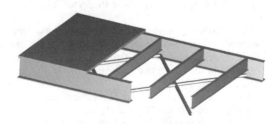

Figure 1. Representation of the slab.

2.2 *Comparison methods and study of the models*

At the base of the work performed is the study and comparison of two different methods of calculation. The first is composed of beam (*frame*) elements coplanar with the slab having moments of inertia the same as those supplied by the Eurocode whose stress in the linear elastic field are calculated by means of an appropriate software. The model described is illustrated in Figure 2 below and is composed of frame elements which schematize beams and crossbeams connected one to the other by plate elements simulating the behaviour of the concrete slab.

The comparison model (Fig. 3) is prevalently composed of shell elements which simulate both the steel elements in the beams and the concrete slab completing the floor system.

The web and the flange of the jack rafters are made up of shell elements composed of linear elastic material having the characteristics of the steel employed, while for the concrete slab a mesh of shell elements that takes into account the loads given by the code and the concentration of strains at particular points was employed; this makes it possible to find the stress of the floor system at any given point, thus allowing comparison with results obtained with the first model.

To simulate the distance of the axis of the slab from the steel beams, infinitely rigid frame elements equal in length to its half-thickness were used so as to be able to simulate a perfect beam-slab joint (Fig. 4).

2.3 *Definition of the cross section and effective width of the slab*

Determination of the geometric characteristics of the composed section is conducted by considering, on the basis of directions given in the Eurocode, an appropriate effective width of the slab and of the relative longitudinal reinforcement. The width supplied by the Eurocode takes into account the so-called shear lag effect and, therefore, the deformability of the concrete slab on its own plain. Operating in the linear elastic field, the section is considered homogeneous with reference to the values of the instantaneous moduli of elasticity of the concrete and the steel ($n = E_s/E_c$). In this case the position of the neutral axis is evaluated by applying the conventional static theory of reinforced concrete.

In calculating strains (bending moments), the Eurocode allows the use of a constant value for the effective width of slab which corresponds to the one evaluated in the centre line in the case of simply supported beams.

The effective width of slab B_{eff} is obtained as the sum of the magnitudes B_{e1} and B_{e2}: $B_{eff} = B_{e1} + B_{e2}$ (Fig. 5). Such magnitudes are given as a function of magnitude l_0 which represents the distance along the

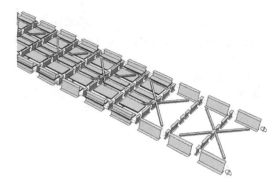

Figure 2. Representation of the first model.

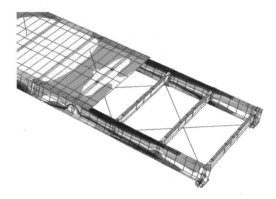

Figure 3. Representation of the second model.

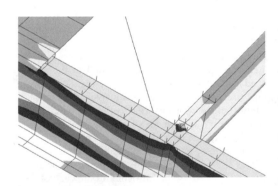

Figure 4. Details of the second model.

beam under examination of two null points on the bending diagram.

On the other hand, as concerns check of the cross-sections, the Eurocode recommends values of the differentiated effective widths for zones subject to positive and negative moment. All this, as we shall see, relates to bending only: indeed, the effective width of slab concerns exclusively the case of simply bent sections; in

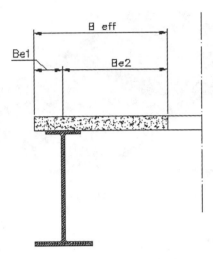

Figure 5. Representation of the effective width of slab.

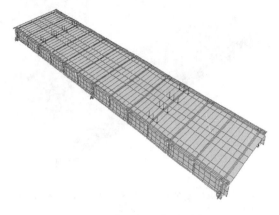

Figure 6. Concentrated loads on the model.

the case in which there is a compression force not contemplated by the Eurocode, the stress that would be created, considering the section supplied by the codes also in the case of compression, is totally different from the effective state of the structure.

2.4 Description of applied loads and strains on the structure

In the structure under examination several suitably combined loads were placed in an attempt to reproduce a situation as close as possible to the real one. The single load conditions were given by the dead load of the structure, by concentrated loads (Fig. 6) and the compression actions applied at the extremities of the floor system, which simulate the compression forces transmitted by the backstays (Fig. 7). To simplify the discussion, we took a symmetrical part of a continuous floor system restrained in the centre by two hinges and at supports by two rocker so that compression strains could be transmitted to the floor system.

2.5 Description of the stress of the first model

As previously mentioned, the first model is composed of frame elements which schematize beams and cross beams connected by plate elements simulating the behaviour of a concrete slab. From the study of the frame elements it is possible to obtain the strains and thus the bending moments and compressive actions acting on the beams; the distribution of the bending moments will be the following (Fig. 8) and the compressive actions will be equally distributed over the entire beam up to the hinge (Fig. 9). Once strains have been found, stress on the parts to be studied can be calculated by means of appropriate software.

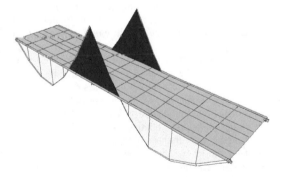

Figure 7. Compression loads on the model.

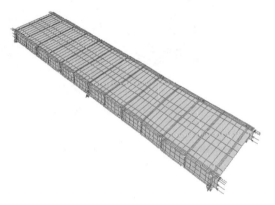

Figure 8. Representation of bending moments on the beams.

2.6 Description of the stress of the second model

From a first examination of the model contemplated, subjected to flexural loads exclusively, the shear lag effect, which is to say the concentration of compression or traction stress, is immediately noted on the

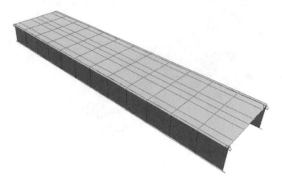

Figure 9. Representation of the compressive actions on the beams.

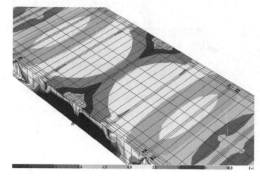

Figure 11. Enlarged detail of the model.

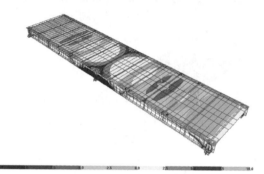

Figure 10. Diagram of normal longitudinal stress caused by bending.

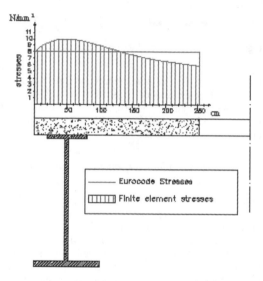

Figure 12. Comparison between slab traction stress.

concrete parts above the steel beams (Fig. 10). Such stress decrease with the increase in distance of the point under examination from the steel beam.

In particular, this effect is visible in greater detail in the enlargement of the structure under examination (Fig. 11).

To take into account the shear lag effect, the Eurocode proposes the use of an efficacious width of the effective width of slab for the purposes of static check evaluated as a function of the distances between two roots of the bending diagram. From the comparison of the values of the stress obtained by means of the model illustrated here with those obtained from the results of the first model, which take into account the section made homogeneous, we obtain the values represented in Figure 12. From the graph it is possible to find the stress on the upper part of the concrete wing as a function of the distance from the axis of the steel beam, compared to those obtained by using the Eurocode.

2.7 Comparison of the two models

Hypothesizing the use, should there be a compression component, of the ideal steel area made homogeneous as supplied by the code as well as the moment of inertia, all the force of compression would be absorbed by the previously defined effective width of slab and the stress on the cross-section would be calculated as follows in the case of the steel section:

$$\sigma_s = N/A_i + M\,y/J_i$$

and in the case of concrete:

$$\sigma_s = N/nA_i + M\,y/nJ_i$$

where N = force of compression; M = bending moment present at the point of the structure under examination; n = coefficient of instantaneous homogenisation; A_i = ideal area; J_i = ideal moment of inertia, all as defined by the Eurocode.

The stress state of the concrete in the comparison model is instead described in Figure 13 below, from

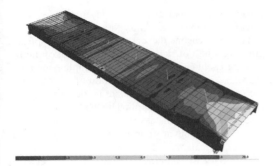

Figure 13. Diagram of normal longitudinal stress caused by compression associated with bending.

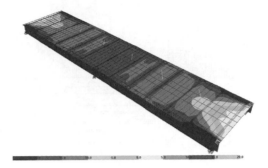

Figure 15. Diagram of normal longitudinal tensions caused by compression only.

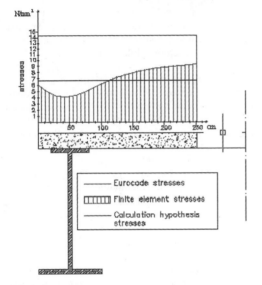

Figure 14. Diagram comparing normal longitudinal tensions caused by compression associated with bending.

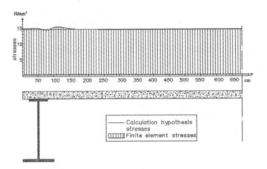

Figure 16. Normal longitudinal stress caused by compression.

the hypothesis of being able to spread the force of compression over the entire slab and not only on the so-called effective width of slab defined by the code, while the bending moment goes to solicit the effective width of slab only. Figure 16 below illustrates the trend of stress along the section, highlighting that in practice they are equally distributed over the latter.

2.8 Calculation hypothesis

Starting from the hypothesis formulated above, we can thus hypothesize calculation of on the section by means of the simple compression associated with bending formula analogous to the previous one, of the kind:

$$\sigma_s = N/A_{pi} + M \, y/J_i$$

in the case of the steel section;

$$\sigma_c = N/n \, A_{pi} + M \, y/nJ_i$$

in the case of the concrete, where J_i = moment of inertia defined by the code; A_{pi} = ideal floor system area, including the part of the slab outside the effective width one. For the purposes of static check it is necessary to take into consideration the fact that the most stressed part of the slab, which is to say the part in which the

which we can see the presence of lighter patches in correspondence to the central rocker and explained by the presence of bending.

On comparing the values obtained with the two models by means of the graph in Figure 14, we immediately note a large discrepancy between the compressive stress both of the concrete and the steel.

It is therefore evident that the stress in the comparison model are far below those obtained by using the model defined by the Eurocode.

On examining the stress in the comparison model subjected to compression only (Fig. 15), it is seen that in practice it is spread, once a certain distance from the point of application of the concentrated loads (the extremities of the floor system) is passed, uniformly over the entire concrete slab. From this comes

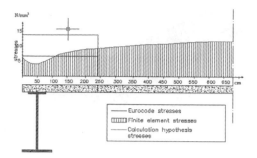

Figure 17. Compressive stress on the slab caused by compression associated with bending.

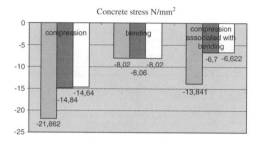

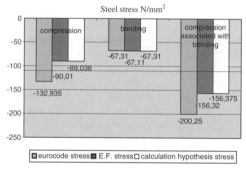

Figure 18. Diagram of stress on concrete and steel.

limit tension of the concrete on the floor system is reached, may not be the one adjacent to the steel beams but the one near the axis of the floor system, as in the case under examination: the contribution of bending is practically absent in this part of the slab (see Fig. 17).

In the graph below (Fig. 18) the mean compression on the concrete and the steel relative to bending only, compression only and a combination of the two, from which it can be observed that the value of the obtained by comparing the models, are practically coincident.

3 CONCLUSIONS

The study of compression associated with bending in mixed steel and concrete structures has led to the discovery of a deficiency in the Eurocode in the calculation and check of this type of structure.

For this reason, the code in question supplies a effectivè width of slab, and thus a moment of inertia, relating exclusively to the case of simple bending of the slab. Then, since the calculation of internal es in the concrete and the steel relate to flexure only, the use of the magnitudes supplied by the Eurocode in the case of compression associated with bending produces values that are completely falsified in comparison to real ones. Because of this deficiency, it is necessary to elaborate a complex finite element model capable of studying the stress on such structures. This procedure applied to the design and check of bridge floor systems is certainly laborious and complex from the practical standpoint.

Analysis of the behaviour of a mixed-structure floor system was addressed by means of the comparison between two models, one of which directly supplies the stress of the structure, while the other allows calculation of the bending moments and compressive actions. The study of stress by means of models led to the hypothesis of an equal distribution of the compressive action throughout the entire slab; then we obtained a relation which, starting from the results taken from the simplified model, allowed the finding of a method leading to easy check of the section, without recourse to a study of the floor system by means of finite elements.

In conclusion, we cannot exclude the possibility of future extension of this methodology to other types of mixed-structure floor systems.

REFERENCES

Naaman, A. & Breen, J. *External prestressing in bridges* Detroit: Patricia Kost.
Cadeddu, G. & Gamberini, G.P. 1989. Influenza dello shear lag sulla risposta di impacati da ponte con sezione scatolare. *L'industria italiana del cemento*, 635: 476–482.
De Miranda, F. 1980. *I ponti strallati di grande luce*. Roma: edizioni scientifiche A. Cremonese.
Gamberini, G.P. & Cultrera, L. 1995. Modello strutturale evolutivo per lo studio allo S.L.U. per taglio di una trave monodimensionale piana in calcestruzzo. *Atti della facoltà di ingegneria di Cagliari*, 36/95.
Gamberini, G.P., Cultrera, L. & Mancini, G. 1995. Modello di calcolo per la verifica allo S.L.U. per taglio delle travi in c.a. *Politecnico di Torino -Atti del dipartimento*, 53/95.
Křístek, V. 1979. *Theory of box girders*. Prague: John Wiley & Sons.
Messina, C. 1986. *Impalcato dei ponti* Firenze: Alinea editrice.
Rhodes, J. & Walker, A.C. 1979. *Thin-walled structures. Recent Technical advances and trends in design, research and contruction*. London: Granada.
Walter, R., Houriet, B., Isler, W. & Moïa, P. 1988. *Cable stayed bridges*. London: Thomas Telford.

Application of steel fibre reinforced concrete for the revaluation of timber floors

K. Holschemacher, S. Klotz, Y. Klug & D. Weiße
Leipzig University of Applied Sciences (HTWK Leipzig), Germany

ABSTRACT: For many years timber-concrete composite members are known in the context with the revaluation of timber ceiling. Thereby the existing timber beams are completed with special shear-connectors and strengthened by adding a concrete slab. The main benefits of such composite structures are the higher stiffness and load bearing capacity, the better sound insulation and the improvement of fire resistance. However, in most cases in the concrete slab a reinforcement is necessary, for obtaining a sufficient resistance against splitting forces around the shear-connectors and restrain caused by shrinkage. Because of the necessity to reach a sufficient concrete cover of the reinforcing bars, the thickness of the added concrete slab amounts in most cases more than 6 cm. This fact leads to a high dead load of the composite structure, combined with disadvantages for the supporting members. Therefore a new kind of timber-concrete members was developed, whereby the conventional reinforced concrete slab is replaced by steel fibre reinforced concrete, leading to smaller slab heights.

1 INTRODUCTION

Nowadays the protection of historic buildings and their reconstruction have increasing significance in modern town planning. In major German cities like Leipzig this fact concerns especially the dwelling houses erected in the late 19th and beginning 20th century, in which ones the ceilings were usually constructed as timber beam ceilings. Normally such ceilings fail in complying the present requirements of the building physics, particularly with regard to impact sound and fire protection. Furthermore these timber beam ceilings were not designed according to normative rules, but constructed only based on the experiences of the carpenters and architects at that time. As a consequence in most cases it is impossible to verify the timber beams by a calculation considering the existing design codes. This is an especially unfavourable circumstance, if there is a change in usage of the building, for instance from a residential house to an office or business building. Another aspect is, that for improvement of the impact sound protection normally a floating screed is placed onto an impact sound insulation layer. Therefore the dead load of the whole ceiling is increased and in many cases the load bearing capacity of the timber beams is exceeded or the deflection becomes too large.

Consequently, in case of rehabilitation of older buildings there is a high probability that also retrofitting and strengthening of the timber beam ceilings are necessary.

For reaching an improved load bearing capacity there are different possible solutions, reaching from complete replacement of the ceiling to strengthening the original structure. Also timber-concrete composite constructions are a well-known and approved strengthening technology in context with the reconstruction of existing timber beam ceilings, Natterer & Hoeft (1987), Blaß et al. (1995), Riedl & Seidler (2000), Holschemacher et al. (2001). Thereby a concrete slab is cast onto a timber beam ceiling and connected to the beams by special shear connectors.

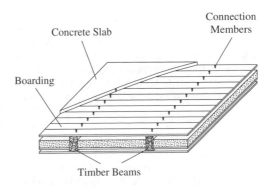

Figure 1. Principle structure of a timber-concrete composite ceiling.

The results of the strengthening procedure are an increase of stiffness and load bearing capacity, an improved sound insulation and a better fire resistance.

2 INITIAL SITUATION FOR THE OWN INVESTIGATIONS

In most cases reinforcement is necessary in the concrete slab of the composite construction for restrain caused by shrinkage of concrete and also to obtain a sufficient resistance against splitting forces around the shear connectors. Usually this reinforcement is placed in the middle of the slab. Due to the required concrete cover (min. 2 cm) and in consideration of possible lap splices of reinforcement the thickness of the slab results in a minimum of about 6 cm, a value that is often not essential regarding the bearing capacity. This fact leads to an unnecessary high dead load of the composite ceiling and will also increase the overall loading of supporting members.

Mostly steel mesh reinforcement is applied in the concrete slabs, leading to several disadvantages in construction practice. For example, the steel meshes must be transported into the rehabilitated buildings. So, often the transport way has to take difficulties like stairs, so that the dimensions of the meshes (width, length) need to get limited. Another fact concerns the amount and the position of the reinforcement in the slab. The necessary portion for shrinkage is often quite small, but available standard meshes exceed this requirement. Furthermore loads must be transferred by the concrete slab into the timber beams in transversal direction. For this purpose the mesh in the middle of the slab is not very efficient.

For the mentioned reasons a new kind of timber-concrete members was developed, whereby the conventional reinforced concrete is replaced by steel fibre reinforced concrete (SFRC). With this solution it is possible to reduce the slab thickness and the construction procedure can be more efficient. In the following first results of an experimental programme, carried out at the Leipzig University of Applied Sciences (HTWK Leipzig), are presented. In this study the mechanical behaviour of the timber-steel fibre reinforced concrete composite structures were investigated and the results were used in order to apply this strengthening method for a revaluation of an existing timber beam ceiling.

3 STEEL FIBRE REINFORCED CONCRETE

Steel fibre reinforced concrete (SFRC) is concrete according to DIN 1045-2 (2001), which is mixed with steel fibres in order to obtain certain properties, whereby for the steel fibres a national technical approval is necessary, according to the Deutscher Beton- und Bautechnik Verein guideline (2001). With normal contents of steel fibres (about 0,38–0,75 Vol.-% which is equivalent to 30–60 kg/m³), the influence on the compressive and flexural tensile strength as well as the modulus of elasticity is negligible. On the other hand due to the fibres it is possible to transfer forces over the crack plains after initial cracking, therefore a brittle failure can be prevented. So the tensile bearing behaviour of the SFRC is improved and the ductility increased.

In order to apply SFRC for structural components it is necessary to determine the influence of the fibres in the concrete matrix. The main reason for this fact is, that the presence of fibres modifies the post peak behaviour and this influence is depending on the fibre type, the fibre content and the fibre distribution besides the concrete properties itself.

For the determination of the properties of the SFRC there are several possibilities. In Deutscher Beton- und Bautechnik Verein guideline (2001) the test method is the 4-Point Bending Test (Fig. 2).

The RILEM Committee recommends the 3-Point Bending Test on notched beams as test method (Vandewalle et al. 2002). With these tests it is possible to receive specifications about the flexural tension strength and the post-crack behaviour.

The results of the above mentioned tests are load-deflection curves (Fig. 3). The amount of steel fibres,

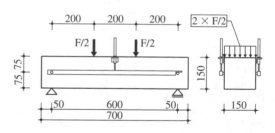

Figure 2. 4-Point bending test.

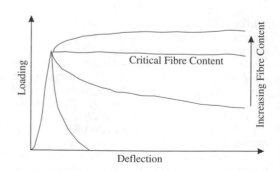

Figure 3. Critical fibre content.

whereby the forces at initial crack development are completely transferred, is called "critical fibre content". However, in most cases the used amount of steel fibres remains below this "critical content" and so the fibres can transfer only a part of the forces at initial cracking.

Out of these curves it is possible to calculate the residual tensile strength and wherewith to grade the SFRC into fibre classes, which are independent from the fibre type. So, the design of structural members reinforced with steel fibres can be performed by using the fibre classes.

4 TEST PROGRAMME

4.1 Building materials

The used building materials were timber, steel fibre reinforced concrete and ordinary wood screws as connection member. As timber coniferous wood with the quality grade S10 according to DIN 1052 (1988) was utilised. The used concrete with a strength class C20/25 was mixed with mill cut steel fibres. The fibre content was $50\,kg/m^3$, which corresponds to the required minimum reinforcement.

In a previous investigation the expedience of various connectors was implemented. A screw of 16 mm diameter and a length of 160 mm, arranged in an angle of 90° to the timber beam would be sensible under economical viewpoints.

4.2 Load-displacement behaviour of push-out specimens

In order to determine the stiffness and the shear bearing capacity of the connection system under application of steel fibre reinforced concrete two different arrangements of the push-out specimens (Fig. 3) were chosen, Köhler (2002). In series 1 and 2 the shear forces were carried by wood screws with single sided timber connector plates around them according to DIN 1052 (1988) and concrete keys. In order to prove the influence of the concrete keys on the stiffness, only wood screws were used in series 3. The first test series with plain concrete was implemented for comparing the stiffness and the load bearing capacity with series 2 and 3, where SFRC was applied.

The specimens were loaded in a servo-hydraulic testing machine, whereby the loading regime was implemented according to DIN EN 26891 (1991). At first the load was increased force-controlled with a loading rate of 20% of the estimated maximum load per minute until 70% of the maximum value (Fig. 5). Within this loading regime the force was kept constant for 30 secs at 40% and 10% of the estimated maximum load. After that the further loading was increased

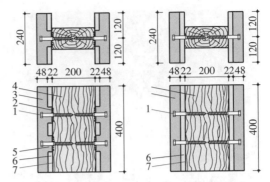

1 - Wood Screw 16/160 mm
2   Single Sided Timber Connector Plates Type D
3 - SFRC Slab
4 - Timber Beam
5 - Concrete Key
6 - Sheet of Plywood
7 - PE-Film

Figure 4.   Tested specimens.

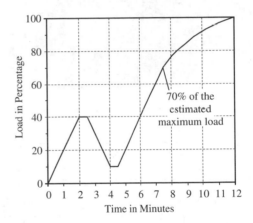

Figure 5.   Loading regime according to DIN EN 26891 (1991).

path-controlled until failure or a displacement of 15 mm.

In order to measure the displacement between the timber beam and the steel fibre reinforced concrete slab 4 LVDT's were applied (Fig. 6).

5 RESULTS

For the evaluation of the push-out tests the measured displacements of the 4 LVDT's were averaged. Figure 7 shows the load displacement curves of the selected tests in one diagram for comparison. The other specimens out of each test series showed a similar behaviour.

Figure 6. Testing machine with push-out specimen.

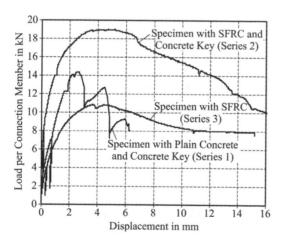

Figure 7. Load-displacement diagram.

The specimens of the first series failed through crater-shaped concrete breakouts around the wood screws, that means the concrete cracked around the screw heads because of high local stresses. After this initial crack development there was no further load

Table 1. Results of the push-out tests.

		Series 1	Series 2	Series 3
Ultimate load F_{max}	kN	15,2	19,3	11,0
Estimated maximum load F_{est}	kN	15,8	20,0	10,0
Initial slip v_i	mm	0,42	0,17	0,37
Modified initial slip $v_{i,mod}$	mm	0,51	0,22	0,44
Staying initial slip v_s	mm	−0,10	−0,05	−0,08
Elastic slip v_e	mm	0,16	0,10	0,21
Slip at $0,6 F_{est}$–$v_{0,6}$	mm	0,84	0,59	0,76
Modified slip at $0,6 F_{est}$–$v_{0,6,mod}$	mm	0,89	0,59	0,80
Initial sliding modulus k_i	kN/mm	18,4	51,7	10,9
Sliding modulus k_s	kN/mm	14,6	40,6	9,1
Sliding modulus $k_{0,6}$	kN/mm	12,5	20,8	7,6

increase. Nevertheless the load displacement diagram shows a sudden drop of the load, but an abrupt failure was avoided due to the 3 other connectors. When applying SFRC (series 2) the behaviour was more ductile, the crack propagation and opening were hindered by the steel fibres. Therefore it was still possible to transfer loads over the crack plains. Moreover there was a load redistribution to the other screws. Both effects led finally to higher ultimate loads. The same ductile failure behaviour was also observed in test series 3, but the stiffness and the ultimate loads were smaller due to the arrangement without concrete keys.

Table 1 contains some measured and computed results of all test series, which are imported for designing Timber-SFRC Composite Constructions. The complete evaluation of the test programme is represented by Köhler (2002).

6 FIRST APPLICATION OF SFRC ON AN EXISTING TIMBER BEAM CEILING

The second part of the investigation was applying SFRC for a rehabilitation of a timber beam ceiling in a building in Altenburg/Thuringia, Köhler (2002). The main tasks to investigate were:

– pumpability of the SFRC
– determination of the mid span deflections of the timber beam ceiling before and after reconstruction under a defined load
– short term monitoring of the deflection
– verification of the found results from the push-out specimens
– measurement of the impact sound behaviour

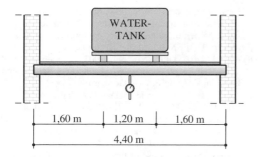

Figure 8. Loading arrangement on the timber floor.

Table 2. Measured deflections of the ceiling.

	Adjacent beam left side	Loaded beam	Adjacent beam rightside	Sum of deflec- tions
Deflection before rehabilitation [mm]	5,7	17,8	5,2	28,7
Deflection after rehabilitation [mm]	4,5	6,3	4,1	14,9
Calculated deflection after rehabilitation [mm]	–	–	–	15,8

Figure 9. Arranged ceiling before casting.

Before reconstruction the ceiling was reassessed according to EC 5 (1994). It was proved that the load bearing capacity was high enough, but in the serviceability limit states the deflection was too large. Therefore the ceiling had to be strengthened in order to reduce the deflection.

The first step was to measure the real deformations of the not yet strengthened timber beams under a certain load. For this a water tank was placed on one beam and the mid span deflections were measured on the loaded and the two adjacent beams (Fig. 8). So it was possible to evaluate the calculation and to assess the transversal load spearing.

Based on the push-out tests the configuration like in test series 2 was chosen for the rehabilitation (Fig. 9). The shear connectors (ordinary wood screws) were placed with a special tool that enabled to drill the hole in the boarding and the screw hole in one process.

Before casting the ceiling was supported and cambered in order to compensate deformations, which were

existing deflections and calculated additional ones due to the higher dead loads of the composite ceiling. 28 days after casting the deflection measurement was repeated. It was possible to show that the deflections could be reduced under the same load (water tank), because the stiffness of the reconstructed ceiling was increased due to composite action between the SFRC slab and the timber beams (Table 2). In this table the deflections of the three considered beams were added up in order to compare these values with a calculated one. In the computation it was only possible to derive the total deformations of the 3 beams, because the transversal load spearing was at that time unknown.

Besides the structural behaviour also the impact sound behaviour was investigated. The measurement resulted in the fact, that the rehabilitation did not induce an improvement of the impact sound protection according to DIN 4109 (1989). Nevertheless it was shown that the behaviour resembled rather a concrete ceiling than a timber beam ceiling. That means the isolation of mainly deep frequencies was improved and disturbing sounds appeared significantly quieter, Köhler (2000).

7 CONCLUSION AND OUTLOOK

The positive effects of timber-concrete composite constructions with reference to the bearing behaviour and the serviceability are well known. However, with the application of steel fibres as a replacement of conventional reinforcement further advantages are achievable. So it is possible to reduce the slab thickness to a static necessary value, because the steel fibres do not need any concrete cover. The crack development in the concrete is delayed and the stress distribution more uniform. Furthermore technological advantages can be realised in the construction process. With the performed rehabilitation of an existing timber beam ceiling it was shown that the application of SFRC is more economic than using mesh reinforced concrete. From the structural point of view the requested improvements could be reached, but some more short and long term investigations are necessary. One point will be the slab thickness, which could be reduced to about 3,5 cm in dependence of the steel fibre geometry. Another possible step would be a change from normal weight concrete to a lightweight aggregate concrete, but also reinforced with steel fibres.

REFERENCES

Natterer, J. & Hoeft, M. 1987. About the Bearing Behaviour of Timber-Concrete Composite Constructions. (in German) Research Report, CERS No. 1345, EPFL/IBOIS.

Blaß, H.-J., Ehlbeck, J. & v.d. Linden, M. 1995. Bearing and Deformation Behaviour of Timber-Concrete Composite Constructions (in German) Research Report T2710, IRB-Verlag.

Riedl, S. & Seidler, M. 2000. Tests on Timber-Concrete Composite Constructions: Investigation about the Influence of the Boarding (in German) Diploma Thesis, HTKW Leipzig.

Holschemacher, K., Rug, W., Pluntke, T., Sorg, J. & Fischer, F. 2001. Timber-Concrete-Compound. (in German) Holzbauforum 2001, Leipzig.

DIN 1045-2. 2001. Deutsche Anwendungsregeln zu DIN EN 206-1 Beton–Teil 1: Festlegung, Eigenschaften, Herstellung und Konformität. (Concrete, reinforced and prestressed concrete structures – Part 2: Concrete; Specification, properties, production and conformity; Application rules for DIN EN 206-1) (in German).

DBV-Merkblatt "Stahlfaserbeton". 2001. (Guideline "Steel Fibre Reinforced Concrete"), Deutscher Beton- und Bautechnik-Verein E.V. (in German).

Vandewalle, L., Nemegeer, D., Balazs, L. & di Prisco, M. et al. 2000. RILEM TC162-TDF: Test and design methods for steel fibre reinforced concrete-bending test, Materials and Structures, Vol. 33, January–February 2000, pp. 3–5.

DIN 1052. 1988. Holzbauwerke; Berechnung und Ausführung (Structural use of timber; design and construction) (in German).

Köhler, S. 2002. Composite Constructions made of Steel Fibre Reinforced Concrete and Timber (in German) Diploma Thesis, HTKW Leipzig.

DIN EN 26891. 1991. Holzbauwerke; Verbindungen mit mechanischen Verbindungsmitteln; Allgemeine Grundsätze für die Ermittlung der Tragfähigkeit und des Verformungsverhaltens (Timber structures; joints made with mechanical fasteners; general principles for the determination of strength and deformation characteristics) (in German).

Eurocode 5. 1994. Entwurf, Berechnung und Bemessung von Holzbauwerken. (Design of timber structures) (in German).

DIN 4109. 1989. Schallschutz im Hochbau; Anforderungen und Nachweise. (Sound insulation in buildings; requirements and testing). (in German).

System-based Vision for Strategic and Creative Design, Bontempi (ed.)
© 2003 Swets & Zeitlinger, Lisse, ISBN 90 5809 599 1

Application of fiber composite materials with polymer-matrix in building industry – new possibilities of structure durability rise

L. Bodnarova & R. Hela
Brno University of Technology, Faculty of Civil Engineering, Brno, Czech Republic

M. Filip & J. Prokes
Prefa Brno, Brno, Czech Republic

ABSTRACT: Composite profiles with polymer matrix belong to attractive building products especially for particular applications, where they successfully replace conventional materials and surpass them in some properties. Among significant priorities of these materials belong the low thermal conductivity, high dimensional stability, high wear resistance and plenty more. Involved are for instance applications of composite materials in surroundings with heavy corrosive load or structural members with high strength values and low mass which in shape variety and in possibility of being dyed through comply with particular esthetic points of view. The examined composite profiles and panels are prepared by pulltruding (drawing) method. The binder is a polyester or vinylester resin and as reinforcement oriented glass or sometimes carbon fibres and flat reinforcements are used. In spite of very good experience with this material and with particular products some physical-mechanical properties of polymer-composites are not yet verified in present time. In the paper the test results of these polymer-composites under normal temperature and under thermal load are presented.

1 PROCESS OF EXAMINED COMPOSITE PROFILES MANUFACTURE

The composite profiles were manufactured by the pulltrusion method. The pulltrusion is based on pulling a bundle of fibres, mats and fabrics through a resinous bath, where the saturation of the reinforcement takes place. The saturated reinforcement is afterwards shaped into the demanded cross section and at the same time the composite is hardened in a continuous hardening head. A very important factor is the temperature of the resin, which is at the input about 40°C and at the output in the range of 120°C till 160°C, and the pulltrusion velocity. Period how long the composite have to stay in the hardening head (1–5 min.) depends on the pulltrusion velocity. The hardening process of resin matrices is the critical point of the manufacture, for instance the gelling has to pass before the batch leaves the hardening part. The scheme of the processing you can see in Figure 1.

The matrix of these composite materials is a thermosetting one. Thermosetting resins are hardened in pulltrusion process by chemical reaction after having been mixed with an initiator. In comparison with thermoplastic resins the thermosetting resins form three-dimensional polymeric systems and it is perfect

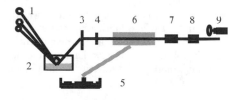

1- the stock of rovings, woven and unwoven mats
2 - tank for dipping the fibres with resin by means of rotating, "wringing-rollers"
3 - plate with holes, for drawing the fibres in the demanded direction
4 - plate which removes the excessive resin from fibres
5 - vessel for collecting the excessive resin
6 - heated hardening head
7,8 - drawing device
9 - saw for partition of the endless profile

Figure 1. The flow sheet of pulled composites manufacture – pulltrusion (Jancar, in press).

if the whole product is in the same time one macromolecule.

The correctly hardened thermosetting resin resists higher temperature (up to about 110°C) it doesn't soften, is creep-resisting and has many other good

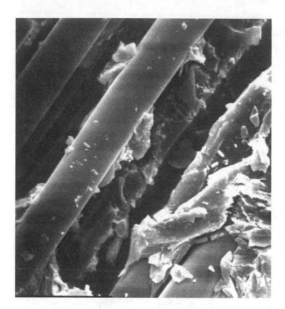

Figure 2. Glass fibres coated with a matrix, filler kaolin, magnification 1100 x.

properties. In comparison with thermoplastic resins it is however more brittle and it cannot be recycled in such a simple way. The widespread resins for the production of pulled building elements are the polyester resins. As reinforcement, glass fibres were used. These fibres with the diameter of $4\,\mu m$ till $20\,\mu m$ have been treated with a lubricating and a finishing layer and they were delivered by the manufacturer in rovings. The lubrication improves handling the fibres. The finishing improves the adhesion to the binder.

2 PROPERTIES OF PULLED COMPOSITE MATERIALS

The tested composite profiles are formed by two main components – the matrix (polyester binder) and glass reinforcement. The resin has low specific density ($\rho = 1500\,kg \cdot m^{-3}$) it has a relatively good resistance against chemical substances and aggressive medium. A disadvantage is a low elasticity modulus ($E = 6\,GPa$), strength only up to $100\,MPa$ and a small creep resistance. In comparison glass fibres, which are delivered as roving and as mats have high strength values, but they are fragile, susceptible to mechanical damage during handling and they have a smaller resistance to the effect of some chemical substances. By suitable composition and connection of fibres with the matrix to form the composite a material with very interesting properties is formed. The well hardened composite profile is lightweight ($\rho = 1800\,kg \cdot m^{-3}$), firm

($\sigma_t = 240$ till $700\,MPa$), tough, creep resistant, resistant to chemical agents, ultra-violet radiation and atmospheric conditions, non-combustible (class C, B, A). From the utilization point of view very interesting is the possibility of final application of composite elements to the manufacture of quite complicated profiles, the possibility to control the composite properties by addition of different additives (It is possible to increase the resistance to UV-radiation, to combustion, to static electricity etc.).

3 UTILIZATION OF COMPOSITE PROFILES

The PREFEN composite profiles have a wide utilization in special building applications. A great advantage of these profiles is the low volume mass and the specific strength (strength in relation to 1 kg of the material). From this follows the easy handling the readymade product. Very advantageous is the easy processing of profiles – the shaping of composite materials PREFEN is similar as the shaping of wood, aluminium and other comparable materials. We can use the same devices as for processing the metals. The profiles can be cut, sawed, bored, milled, turned, ground. The connecting can be made by sticking (epoxy resins or polyesters), screwing, riveting; very appropriate is the combination of a glued and mechanical connection. The profiles are manufactured in different shapes, as H, I, L, U – profiles, channel and tube profiles, circular and square bares and boards and plates and other specific shapes. Final products are structural profiles, whole load-bearing structures (load-bearing parts of walking surfaces, roof structures, load-bearing parts of halls, dividing walls in sewage treatment plants), steps, foot bridges, guard railings, ladders, grate floors, systems for cable lines etc.

4 THE AIM OF PERFORMED TESTS

The manufacture of composite materials undergoes in the Czech Republic a dynamic development. With the extension of the raw material base, with the offer of new additives, inhibitors, retarding agents and with new possibilities of functional surfaces for hardening moulds is it necessary to submit the resulting product – the pulled profile to a careful analysis. One not quite completely cleared area is the behavior of this material in a surrounding of a permanently higher temperature. Although the data concerning the degradation of used resins alone are exactly known, but the composite profile has to be evaluated as a whole, constituted by many factors (resin matrix, glass reinforcement, fillers, the technological process). For this reason we entered into cooperation with the manufacturer of these materials (the joint – stock company

PREFA Brno, Czech Republic) and with the Brno University of Technology, Faculty of Civil Engineering, Institute of Technology of Building Materials and Components, Czech Republic. The research work was oriented to the physical-mechanical properties examination of glass-fibre composites PREFEN, under normal temperatures and even after a thermal load (negative temperatures and even higher temperatures). We examined samples with different fillers (kaolin, kaolin with a wetting agent, talc, apyral, apyral with a wetting agent, combustion retarder). In the case of sample with combustion retarder (used instead of the filler) the behavior of the material under higher temperatures was tested and the grade of combustibility was determined.

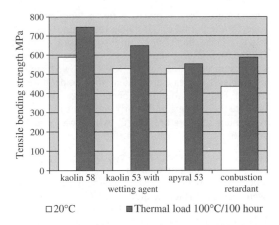

Figure 3. Influence of higher temperature effect on samples with different fillers, hardening temperature 140°C.

5 RESULTS OF EXPERIMENTAL WORKS

5.1 The heat load influence – long term influence of higher and of negative temperature effects

The test samples with different fillers (kaolin, kaolin with a wetting agent, talc, apyral, apyral with a wetting agent, combustion retarder) were prepared at two hardening temperatures – at 140°C and at 160°C. The samples were stored in a medium with elevated temperature. After a defined time period the deflections of test samples under different loads were measured and the tensile and bending strength values were determined. The test results you will find in following paragraphs.

5.1.1 The long term influence of higher temperatures on samples with hardening temperature 140°C

The prepared samples were for 100 hours stored in a chamber with the temperature of 100°C. After carrying out the bending strength tests, the original assumption, that the hardening temperature 140°C is insufficient to achieve a perfect hardening of the binder was confirmed. All samples showed that the effect of higher temperature (for instance 100°C for 100 hours) hardened the material i.e. the additional heat load had a positive influence and the strength values of all samples were higher. In the case of samples, where kaolin 58 was used as a filler, the increase was with 12% higher.

The strength increases after heat load is interesting, but it doesn't have any practical importance for now. This could easily lead to the idea to utilize this method for the application of difficult (indirect) elements as far as the shape is concerned. Actually internal tension would develop resulting from the glass reinforcement and with time cracks would emerge. For these purposes serve the composite profiles, manufactured by "thermoplastic pulltrusion".

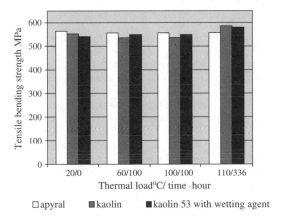

Figure 4. Influence of higher temperature effect on samples with different fillers, hardening temperature 160°C.

5.1.2 The long term influence of higher temperatures on samples with hardening temperature 160°C

Samples with different fillers hardened at the temperature of 160°C were placed into test boxes with conditioned medium and they were thermally loaded as follows: 60°C for the period of 100 hours, 100°C/100 hours, 110°C/336 hours.

We can state that when storing the samples at the laboratory temperature of 20°C the kind of usual fillers doesn't have any significant effect on the limiting strength of samples. The differences between achieved strength values in samples with different fillers are uninteresting. The achieved strength values of samples with different fillers didn't show significant differences, even under the effect of higher temperatures, see Figure 4. After a long-term heat load for the period of 336 hours the composites with usual fillers didn't show any significant material degradation. Surprising was the finding that even some samples hardened at 160°C showed similar effects as samples hardened at the temperature of 140°C. The value of limiting bending stress decreased admittedly but not significantly and in the case of the maximum temperature 110°C, which was effective for 336 hours this stress even slightly increased.

5.1.2.1 Composite-profiles containing combustion retarder

The samples with combustion retarder (replace filler) were tested at the temperature of 20°C, after the effect of 110°C for the period of 5 hours and under the effect of 250°C. The bending strength at the temperature of 20°C was 619 MPa, the bending strength at 110°C for the period of 5 hours was 605 MPa, which is not statistically significant difference. The shape and the size of samples were not changed, but the color became yellow and mildly stinking foam outlet from the sample.

After 30 minutes under the temperature effect 250°C a degradation of the sample took place, the resin turned black, no ignition occurred but pungent fume was given off. A significant decrease of strength took place as much as by 99%. This fact is very important, because it showed the suitability for using the profiles rather to applications where the temperature is permanently not higher than 100°C, but where fire hazard exists. During the direct effect of flame on the surface the retarder is the cause of foaming and it prevents the entering of high temperature into the material. As showed by test results, the composite profile with the retarder behaves worse in a surrounding where the material is for a longer period under the effect of a heat flow and its effect becomes evident positively only under the attack of high temperatures – of a fire.

5.1.3 *Frost resistance determination of composite profiles*

The composite samples were placed for 150 freeze/thaw cycles in the chamber for the period of 4 hours at the temperature of −18°C and in a water bath with the temperature 18°C for the period of 2 hours. In the Czech Republic there is no standard for the resistance of composite materials at low temperatures and therefore we have applied a method following the Czech Standard CSN 73 1322 for the frost resistance determination of concrete.

5.2 *Determination of the inflammability grade*

The test was carried out following the Czech Standard CSN 73 0862 "Determination of building

Figure 5. Sample of composite material with the combustion retarder at laboratory temperature.

Figure 6. Examination of high temperature effect on the sample of composite material with the combustion retarder – sample. after temperature effect of 250°C, evident "foaming" of the matrix.

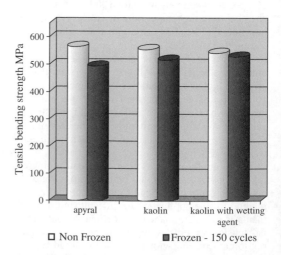

Figure 7. Tensile bending strength after 150 freeze/thaw cycles.

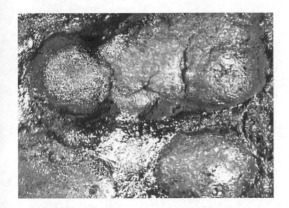

Figure 8. The sample after the inflammability test – the foamed sample crust after inflammability test.

materials inflammability grade". The principle of the test is the action of the gas burner flame for 10 minutes on the test sample of the tested building material and the determination of the mass loss of the sample in percents. The inflammability test was realized on samples containing combustion retarder. The test samples contained as matrix polyester resin and 20% of combustion retarder. The composite contained two mats – a woven one with the specific mass $\rho = 250\,g/m^2$ and a non woven mat with the specific mass ($\rho = 450\,g/m^2$).

Beginning with the moment of positioning the lighting burner under the test sample a foamed crust formation started on the sample surface and the sample near the crust turned yellow. During 30 seconds a continuous crust was formed. Four minutes after the burner ignition under the sample at the opposite side from the flame a white mildly stinking foam outlet from the sample. The sample started to glow slightly. After finishing the test and the removal of the burner the sample neither did burn nor glow and till it cooled down to the laboratory temperature it didn't change.

After completing the test the mass loss of tested samples was measured. The mean value of the mass loss was less than 2%, the time of spontaneous combustion was 0 seconds. Following the mass loss the tested composite profile PREFEN can be classified as far as the inflammability of building materials grade is concerned in the inflammability class A – inflammable.

6 RESULTS EVALUATION AND THE PERSPECTIVES OF FURTHER EXPERIMENTAL WORK

The subject of the paper is very stimulating for all manufacturers of composite materials and it is in a way unique. The Brno University of Technology and the manufacturer of composite materials the joint stock company Prefa Brno, makes plans of a further cooperation in the framework of proposed research projects and of mutual assignment and solving of dissertations and theses. However it is necessary to elongate the period of thermal load up to 10000 hours at temperatures between 80°C till 150°C. Further tests of direct flame-effect an the sample surface containing anti-flammable admixtures will be carried out, from the point of view of profile state description after the direct flame-effect and as far as the determination of mechanical properties is concerned (limit tension, E-modulus incl. the surface hardness after Barcol). The complex of tests will be extended by tests of samples with another binder – PES resin, where the deliverer gives a guaranty of resistance against permanent higher temperature (up to 160°C). To these conditions the temperature range of experiments will be adapted. The next phase of tests will pay particular attention to information concerning the chemical resistance of these materials.

ACKNOWLEDGEMENTS

This work was published with backing of GACR, within the research project No. 103/01/0814 "Microstructural features and related properties of cement based highly liquefied composites" and the GACR research project No. GACR 103/01/1144 and research project CEZ–MSM 26110008, name of the project is. "Research and Development of New Materials and Securing their Higher Durability in Building Structures."

REFERENCES

Bodnarova, L. & Filip, M. 2002. Properties of fiber glass reinforced composites with polymer matrix. In *The Concrete Days, Proc. international conference, Pardubice, Czech Republic, November, 2002.* Pardubice: CBZ.
Bodnarova, L. 2003. *Composite materials.* Brno, CERM.
Jancar, J. *The introduction of composite material engineering. Textbook.* In press. TU Brno, Faculty of Chemistry.

Flexural design of steel or FRP plated RC beams

P.M. Heathcote & M. Raoof
Loughborough University, Loughborough, UK

ABSTRACT: In recent publications by the second author and his associates, a semi-empirical model (the tooth theory) was developed for designing against the potentially dangerous (largely brittle) premature plate peeling failure of reinforced concrete (RC) beams strengthened in flexure with externally bonded steel or fibre-reinforced plastic (FRP) plates. Although previously backed by an extensive set of experimental data relating to 111 steel and 58 FRP plated beams, the original tooth model suffered from being computer-based, involving certain iterative procedures. The present paper reports the salient features of a simplified procedure, enjoying largely similar accuracy to the iterative method. The output of this simplified method is presently backed by test data relating to 317 steel and/or FRP plated RC beams with widely different beam design parameters. It is amenable to simple hand calculations, using a pocket calculator, hence of value to busy practicing engineers.

1 INTRODUCTION

With the development of strong epoxy adhesives back in the 1960's, there has been a considerable interest in the use of externally bonded steel and (more recently) fibre-reinforced plastic (FRP) plates for strengthening reinforced concrete (RC) beams in flexure. This method of strengthening has been shown to have considerable advantages when compared with other alternatives (Zhang et al. 1995), but, as documented extensively in the literature, it may suffer from a potentially dangerous (largely brittle) premature failure mode, commonly referred to as peeling failure: this involves the brittle separation of the externally bonded plate and concrete cover (as a unit) from the underside of the internal tensile reinforcement, Figure 1.

The second author and his associates (Zhang et al. 1995 and Raoof & Zhang 1997) have developed a semi-empirical model (called the tooth theory) which provides a reliable means of predicting such premature

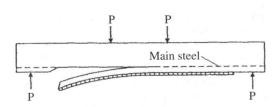

Figure 1.   Peeling failure.

peeling failures associated with either steel (Raoof & Zhang 1997 and Raoof et al. 2000) or FRP (Raoof & Hassanen 2000) plates, with the theoretical predictions backed by very extensive experimental results from other sources. Most importantly, it has been demonstrated that (contrary to the traditionally held view), due to the dependence of ultimate peeling moment on the spacings of stabilised cracks in the concrete cover zone (which may vary by a factor of, say, 2), a unique solution for the plate peeling ultimate load does not exist, and one has to resort to theoretical upper/lower bound solutions (with these not to be confused with the upper/lower bound solutions, in the sense used in, for example, plastic analysis of frames), and that the lower bound solutions are the suitable (i.e. safe) ones for design purposes.

The original tooth model, however, suffered from the potential drawback of being of an iterative (i.e. computer-based) nature. Hassanen & Raoof (2001) later reported a simplified version of the model which was amenable to hand calculations, using a pocket calculator. Although encouraging correlations were found between the numerical data based on the iterative and simplified approaches, in certain cases (to be discussed later) the simplified method has been found not to lead to practically acceptable predictions of the lower bound peeling moment. The purpose of the present paper is to report an improved version of Hassanen & Raoof's hand-based formulations which is applicable to a wider range of beam design details. The predictions of this alternative method are presently backed by a newly established set of test data, relating to 317 steel and/or FRP plated RC beams (including the 169 specimens from the database of Hassanen & Raoof 2001), covering a wide range of beam design parameters. Finally, the present paper includes a design flow-chart, enabling one to not only determine the lower bound peeling moment, but also easily predict the associated possible occurrence of concrete crushing and steel yield or FRP rupture. As a prerequisite to this, the salient features of the original tooth model are present in the next section for completeness- this will, then, enable the reader to better understand (and appreciate) the subsequent developments.

2 TOOTH MODEL

Zhang et al. (1995) and Raoof & Zhang (1997) suggested a mode of failure, Figure 2, which is controlled by the characteristics of the individual teeth in-between adjacent cracks within the concrete cover, with its associated final predictions of the plate peeling load very much depending on the size of stabilised crack spacings.

Very briefly, an individual plain concrete tooth formed between two adjacent cracks (within the concrete cover) is assumed to deform like a cantilever under the action of lateral shear stresses, τ, applied at the interface between the steel or FRP plate and concrete beam, Figure 2b. Ignoring the interactions between neighboring teeth, the tensile stress at point A in Figure 2b, σ_A, reaches the concrete tensile strength, f'_t, for the largely brittle plate peeling failure to initiate – i.e. at the critical state: $\sigma_A = f'_t$. Assuming elastic behaviour for the structural deformations of an isolated tooth up to failure, it can be shown that for simply supported plated RC beams subjected to symmetrical four-point loading (Zhang et al. 1995)

$$\sigma_{s(min)} = 0.154 \frac{L_p h_1 b^2 \sqrt{f_{cu}}}{h' b_1 t \left(\Sigma O_{bars} + b_1\right)} \qquad (1)$$

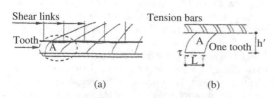

(a)   (b)

Figure 2. Assumed mode of failure due to premature plate peeling: (a) stabilised crack spacing; and (b) individual teeth within the concrete cover – Zhang et al. (1995).

where, $\sigma_{s(min)}$ = the lower bound plate tensile stress at the instance of peeling failure, h' = net height of the concrete cover, f_{cu} = concrete cube crushing strength, ΣO_{bars} = sum of main tensile reinforcing bar circumferences, $2h_1$ = the height of the assumed region of concrete in tension as defined by Zhang et al. (1995), b_1 = width of the plate, b = width of the beam, and t = plate thickness. For steel plated beams, the effective length of the plate, $L_{p,2}$, within the shear span over which the equivalent shear stresses at the plate/concrete interface may be assumed to remain uniform, may be calculated from the following Equations (Raoof & Zhang 1997),

$$L_{p,2} = l^p_{min}(21 - 0.25 l^p_{min}) \quad , \quad l^p_{min} \leq 72mm \qquad (2a)$$

$$L_{p,2} = 3 l^p_{min} \qquad\qquad , \quad l^p_{min} > 72mm \qquad (2b)$$

with L_p in Equation 1 being the lower value of the actual length of plate within the critical shear span (where plate peeling takes place) and $L_{p,2}$ as calculated from Equations 2.

For FRP plated beams, on the other hand (Raoof & Hassanen 2000),

$$L_{p,2} = l^p_{min}(24 - 0.5 l^p_{min}) \quad , \quad l^p_{min} \leq 40mm \qquad (3a)$$

$$L_{p,2} = 4 l^p_{min} \qquad\qquad , \quad l^p_{min} > 40mm \qquad (3b)$$

In Equations 2 and 3, the minimum stabilised crack spacing, l^p_{min}, is

$$l^p_{min} = \frac{A_e f'_t}{u(\Sigma O_{bars} + b_1)} \qquad (4)$$

where, u = steel or FRP/concrete average bond strength = $0.28\sqrt{f_{cu}}$, f'_t = concrete cylinder splitting tensile strength = $0.36\sqrt{f_{cu}}$, with the units of f_{cu} and l^p_{min} in N/mm^2 and mm, respectively. The concrete area in tension $A_e = 2bh_1$, with the other terms as defined previously.

For determining $\sigma_{s(min)}$ in FRP plated RC beams, Equations 3 (instead of Equations 2) should be used for calculating $L_{p,2}$ (and, hence, the corresponding L_p in Equation 1), with the procedure being, otherwise, exactly the same as that for steel plated beams.

With the magnitudes of lower and upper bound plate tensile stresses, $\sigma_{s(min)}$ and $\sigma_{s(max)} = 2\sigma_{s(min)}$, directly under the point load nearest to the support estimated, it is, then, possible to predict the corresponding lower and upper bounds to the peeling bending moment at this location: this was originally done by using an iterative procedure based on the traditional simple beam bending theory with the assumption of plane-section bending as fully explained elsewhere (Zhang et al. 1995), taking the sometimes important influence of concrete tensile stresses below the neutral axis into account (Raoof & Zhang 1997).

3 SIMPLIFIED PROCEDURE

3.1 Determining the depth of neutral axis

Unlike the original tooth model which suffered from the potential drawback of being of an iterative (hence, computer-based) nature, as previously mentioned, Hassanen & Raoof (2001) recently reported a simplified (hand-based) method for obtaining reasonable values of the lower bound peeling moment: a closed-form solution was reported for obtaining reasonable values for the depth of neutral axis, which was based on an assumed linear stress (and strain) distribution for concrete in compression while neglecting the presence of concrete tensile stresses below the neutral axis. With the depth of neutral axis determined, two alternative simplified methods were then developed for calculating the lower bound steel or FRP plate peeling moment. In both methods, the presence of tensile stresses in concrete below the neutral axis was ignored for simplicity.

In the first approach (method A), the stress–strain relationship for concrete as that recommended by the British code BS 8110 (1985) was adapted, and the bending moment was estimated using a parabolic stress distribution for concrete in compression. In the second approach (method B), the concrete compressive stress distribution was assumed to be given by a uniform stress block. In both methods, the maximum concrete compressive strain, ε_c, was taken to be less or equal to 0.0035, and plane-section bending was assumed. In method A, two cases (I and II) were considered, depending on the magnitude of the maximum strain in the concrete, ε_c: case I related to situations where $\varepsilon_c \leqslant \beta$, while in case II, $\beta < \varepsilon_c \leqslant 0.0035$, with $\beta = 2.44 \times 10^{-4}\sqrt{f_{cu}}$, where f_{cu} = concrete cube crushing strength in N/mm^2 (Hassanen & Raoof 2001).

Later investigations by the present authors, however, suggested that in certain cases (because of the neglect of the influence of concrete tensile stresses below the neutral axis), Hassanen & Raoof's simplified methods did not lead to practically acceptable predictions of the lower bound peeling moments: this happened for sufficiently low values of plate axial tensile stresses at the onset of peeling failure.

For the present purposes, the effect of concrete tensile stresses will be catered for in the estimation of the neutral axis depth, y, by assuming a triangular distribution of tensile stresses below the neutral axis: these are taken to linearly vary from zero at the neutral axis to a maximum value (equal to the cylinder splitting tensile strength) at the extreme concrete tensile fibre,

irrespective of the level of external loading on the plated RC beam, and assuming $f'_t = E_c\varepsilon_{ct}$, where E_c = Young's modulus for concrete = $5500\sqrt{f_{cu}}$ and ε_{ct} = maximum concrete tensile strain. Moreover, it is assumed that $f'_t = 0.36\sqrt{f_{cu}}$, where f_{cu} = concrete cube crushing strength in N/mm^2. Otherwise, similar to Hassanen & Raoof's (2001) approach, then, one can derive the following closed-form formula for determining y, by resorting to the equilibrium condition of tensile and compressive section forces:

$$y = \frac{d}{1-\frac{\varepsilon_{ct}}{\varepsilon_p}}\left(-\left(\alpha_{sc}(\rho'_s+\rho_s)+\rho_p\alpha_{pc}+\frac{\varepsilon_{ct}}{2\varepsilon_p d}(D_b+D)\right)\right)$$
$$+\frac{d}{1-\frac{\varepsilon_{ct}}{\varepsilon_p}}\left\{\begin{bmatrix}\left(\alpha_{sc}(\rho'_s+\rho_s)+\rho_p\alpha_{pc}+\frac{\varepsilon_{ct}}{2\varepsilon_p d}(D_b+D)\right)^2+\\ 2\left(\left(1-\frac{\varepsilon_{ct}}{\varepsilon_p}\right)\alpha_{sc}\left(\rho'_s\frac{d'}{d}+\rho_s\right)+\rho_p\alpha_{pc}\frac{D}{d}+\frac{\varepsilon_{ct}}{2\varepsilon_p}\frac{D_bD}{d^2}\right)\end{bmatrix}\right\}^{\frac{1}{2}} \quad (5)$$

where, $\alpha_{sc} = E_s/E_c$, $\alpha_{pc} = E_p/E_c$, $\rho_s = A_s/bd$, ($\rho'_s = A'_s/bd$, and $\rho_p = A_p/b_1t$, with E_c, E_s and E_p = concrete, steel and plate Young's moduli, respectively, A_s, A'_s and A_p = total areas of internal tensile steel, internal compressive steel, and the external plate, respectively, ε_p = plate axial strain = $\sigma_{s(min)}/E_p$, ε_{ct} = concrete maximum tensile strain, with the other parameters as defined in Figure 3, and $\sigma_{s(min)}$ must be less than or equal to the yield stress, σ_y (for steel) or the rupture strength, σ_{pu} (for FRP).

3.2 Determination of the peeling moment

3.2.1 Uniform concrete stress distribution

The lower bound plate peeling moment, M_{peel-R} (including the effect of concrete tensile stresses below the neutral axis) may be determined via a method similar to that developed by Hassanen & Raoof (2001), but assuming presence of a triangular concrete tensile

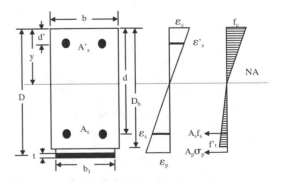

Figure 3. Assumed linear strain and stress distribution in concrete for calculating the depth of neutral axis.

stress distribution associated with a uniform stress block for concrete in compression:

$$M_{peel-R} = A_s f_s \left(d - \frac{a}{2}\right) + A_p \sigma_{s(min)}\left(D - \frac{a}{2}\right)$$

$$+ A_s' f_s'\left(\frac{a}{2} - d'\right)$$

$$+ 0.18b\sqrt{f_{cu}}(D_b - y)\left\{\frac{1}{3}(2D_b + y) - \frac{a}{2}\right\} \quad (6)$$

with a = the depth of the concrete compressive stress block, given by

$$a = \frac{A_s f_s + A_p \sigma_{s(min)} + 0.18b\sqrt{f_{cu}}(D_b - y) - A_s' f_s'}{bf_c} \quad (7)$$

and f_c = concrete compressive stress in the extreme fibre, where

$$f_c = \begin{cases} 0.67f_{cu} & \beta < \varepsilon_c \le 0.0035 \\ E_c\varepsilon_c + \dfrac{0.67f_{cu} - E_c\beta}{\beta^2}\varepsilon_c^2 & \varepsilon_c \le \beta \end{cases} \quad (8)$$

3.2.2 Parabolic stress distribution

The lower bound peeling moment, M_{peel-p}, based on the alternative concrete parabolic compressive stress distribution, is also derived in a manner similar to that of Hassanen & Raoof (2001), but with the effect of tri-angularly distributed concrete tensile stresses below the neutral axis catered for. The final outcome is

Case I, $\varepsilon_c \le \beta$

$$M_{peel-p} = y^2 b\left(\frac{E_c\varepsilon_c}{3} + \frac{0.67f_{cu} - E_c\beta}{4\beta^2}\varepsilon_c^2\right)$$

$$+ A_s' f_s'(y - d') + A_s f_s(d - y)$$

$$+ A_p \sigma_{s(min)}(D - y)$$

$$+ 0.12b\sqrt{f_{cu}}(D_b - y)^2 \quad (9a)$$

Case II, $\beta < \varepsilon_c \le 0.0035$

$$M_{peel-p} = 0.335f_{cu}b(y^2 - y_1^2)$$

$$+ y^2 b\left(\frac{E_c\beta^3}{3\varepsilon_c^2} + \frac{0.67f_{cu} - E_c\beta}{4\varepsilon_c^2}\beta^2\right)$$

$$+ A_s' f_s'(y - d') + A_s f_s(d - y)$$

$$+ A_p \sigma_{s(min)}(D - y)$$

$$+ 0.12b\sqrt{f_{cu}}(D_b - y)^2 \quad (9b)$$

where, $y_1 = (\beta/\varepsilon_c)y$, and the stresses in the tensile and compressive embedded steel bars, f_s and f_s', respectively, are

$$f_s = \varepsilon_p\frac{d - y}{D - y}E_s \qquad f_s \le f_y \quad (10a)$$

$$f_s' = \varepsilon_p\frac{y - d'}{D - y}E_s \qquad f_s' \le f_y \quad (10b)$$

with f_y = steel yield strength.

3.3 Results and discussion

Figure 4 presents a flow chart for the simplified lower bound design method which caters for the occurrence of all possible failure modes, involving concrete crushing and steel yield or FRP rupture.

In Figure 4, ε_{pu} = ultimate tensile strain of FRP, ε_{py} = yield strain of steel, σ_{pu} = ultimate tensile stress of FRP, and σ_y = yield stress of steel, and if in Figure 4 the plate yields or ruptures (i.e. $\sigma_p = \sigma_y$ or σ_{pu}), $\sigma_{s(min)}$ in Equations 6, 7 and 9a, 9b must be replaced by σ_y (for steel) or σ_{pu} (for FRP).

Figure 5 presents the very encouraging correlations between the predictions of neutral axis depth, y,

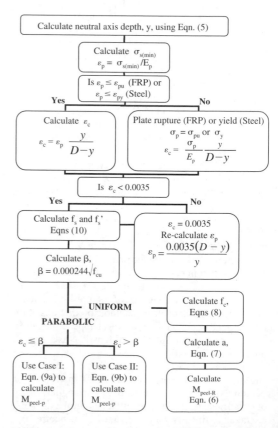

Figure 4. Flow chart of the simplified design method, allowing for all the possible modes of flexural failure.

based on the simplified and the more accurate iterative approaches, relating to 111 steel (99 as-cast, 12 pre-cracked) and 58 FRP (50 as-cast, 8 pre-cracked), from the existing database of Hassanen & Raoof (2001) and a further 64 steel (24 as-cast, 40 pre-cracked) and 84 FRP (63 as-cast, 21 pre-cracked) plated RC beams from Heathcote (2003), where the full details for various beam design parameters, experimental modes of failure and ultimate loads are given in considerable detail. For the present purposes,

Table 1 presents the very wide range of beam design parameters for the 317 beams in the database.

Figure 6 presents the very encouraging correlations between the predictions of lower bound peeling moments based on the simplified (uniform concrete compressive stress block) and the iterative methods relating to all the 317 steel and FRP plated beams. It is noteworthy that the database includes test results relating to both those beams which have been pre-cracked prior to external plating and also those beams

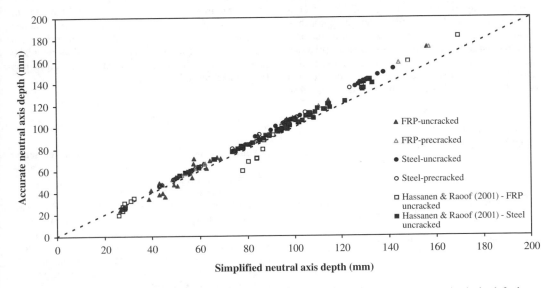

Figure 5.   Correlations between the presently proposed simplified and iterative predictions of neutral axis depth for beams plated with steel or FRP.

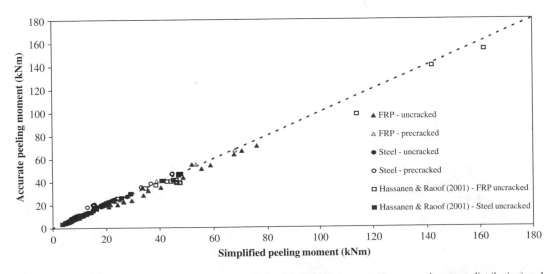

Figure 6.   Correlations between the presently proposed simplified (uniform concrete compressive stress distribution) and iterative predictions of plate peeling moment for beams plated with steel or FRP, including concrete tensile stresses.

which have been externally plated in an as-cast (i.e. uncracked) condition. Almost identical correlations to those presented in Figure 6 were found between the predictions based on the other type of simplified (i.e. parabolic concrete stress distribution) method and the iterative method, although, in view of the extreme space limitations, the corresponding results are not given in the present paper.

Finally, Figure 7 compares the lower bound theoretical predictions of the simplified (uniform concrete compressive stress block) approach with experimental ultimate peeling moments, relating to the 317 steel or FRP plated beams: the simplified design approach is found to predict conservative values of peeling moment for all the beams (except three with these being of a suspect nature). In the vast majority of cases, the predictions of failure modes, based on the lower bound design method, have been found to be the same as those observed in the experiments. It should be noted that the ultimate (as opposed to initial) peeling test data have been used in Figure 7, with some of the beams having end anchorage which have possibly been instrumental in significantly increasing their associated experimental ultimate loads. In the tooth theory, the teeth are all assumed to fail simultaneously, and the whole peeling failure occurs at once in a fully brittle manner. The theory is unable to predict the progression (as suggested in some tests) between the initial and ultimate peeling moments. It would have been desirable for the theory to have a progressive aspect, but this would have led to extreme complexities in the proposed formulations which (in view of the rather large differences between the theoretical lower and upper bounds) does not make practical sense: the model is probably best left as it is (i.e. simple but fully brittle).

4 CONCLUSIONS

A simple method based on a recently reported semi-empirical model (the tooth theory) for design against premature peeling failure of steel or FRP plated RC beams is presented. The proposed method is straightforward, hence, of value to busy practising engineers. The predictions based on this lower bound design approach have been supported by test data from other sources, relating to 317 steel and/or FRP plated beams, covering a wide range of beam design parameters, hence, providing ample evidence for its general suitability for use in practice.

Table 1. Range of RC beam design parameters in the compiled database.

Material	Elastic Modulus, E_p (Gpa)		Internal reinforcement ratio, ρ (%)		Shear span/ effective depth ratio, a/d		Beam span (m)	
	Min	Max	Min	Max	Min	Max	Min	Max
FRP	10.3	380	0.32	4.36	2.5	9.2	0.9	4.8
Steel	200	210	0.66	4.36	2.0	11.6	1.2	3.5

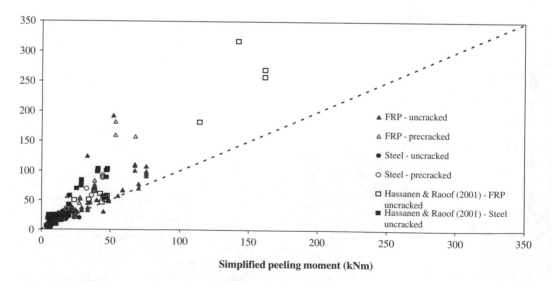

Figure 7. Correlations between the presently proposed simplified peeling moment (uniform concrete compressive stress distribution) predictions and experimental ultimate moments for beams plated with steel or FRP.

REFERENCES

British Standards Institution. 1985. *BS 8110, Structural Use of Concrete*. BSI.

Hassanen, M.A.H & Raoof, M. 2001. Design against premature peeling failure of RC beams with externally bonded steel or FRP plates. *Magazine of Concrete Research*. 53(4): 251–262.

Heathcote, P.M. 2003. Theoretical and experimental study of FRP or steel plated RC beams. *PhD thesis to be submitted to Loughborough University*.

Raoof, M., El-Rimawi, J.A. & Hassanen, M.A.H. 2000. Theoretical and experimental study on externally plated RC beams. *Engineering Structures*. 22: 85–101.

Raoof, M. & Hassanen, M.A.H. 2000. Peeling failure of reinforced concrete beams with fibre-reinforced plastic or steel plates glued to their soffits. *Proceedings of the Institution of Civil Engineers, Structures & Buildings*. 140: 291–305.

Raoof, M. & Zhang, S. 1997. An insight into the structural behaviour of reinforced concrete beams with externally bonded plates. *Proceedings of the Institution of Civil Engineers, Structures & Buildings*. 122: 477–492.

Zhang, S., Raoof, M. & Wood, L.A. 1995. Prediction of peeling failure of reinforced concrete beams with externally bonded steel plates. *Proceedings of the Institution of Civil Engineers, Structures & Buildings*. 110: 257–268.

Design methods for slabs on grade in fiber reinforced concrete

A. Meda
University of Brescia, Brescia, Italy

ABSTRACT: Slabs on grade are one of the more common applications of steel fiber reinforced concrete, especially in industrial pavement construction. This is one of the few applications where steel fibers can substitute conventional reinforcement. In this paper different methods for the design of slabs on grade are analysed. Due to the remarkable non-linear behaviour of these structures appropriate methods for the design should be used. By adopting methods based on non-linear fracture mechanics it is possible to closely predict the actual behaviour of slabs on grade and evaluate the bearing capacity of the structure. Also yield line methods give a good indication regarding the bearing capacity of steel fiber reinforced concrete slabs on grade and the results obtained can be applied in design practice.

1 INTRODUCTION

The use of Steel Fiber Reinforced Concrete (SFRC) is progressively growing in Civil Engineering constructions. The main applications are in shotcrete, pavements, precast, water retaining structures, and seismic structures (Vondran 1991, Li 1993, König et al. 2002). Slabs on grade are one of the more common applications of SFRC, especially in industrial pavement construction. This is one of the few applications where steel fibers can substitute conventional reinforcement allowing the saving of intensive labour and expensive reinforcing work. SFRC slabs on grade are often used for industrial pavements, roads, parking areas and airport runways. Pavement construction requires a great amount of concrete at considerable cost. For this reason pavements cannot be considered secondary structures and should be designed in a proper way.

At present there is a lack of design rules for SFRC slabs on grade. Engineers usually design SFRC slabs on grade by adopting the same rules for plain concrete slabs. However, the more extensive usage of SFRC and its toughness requires modified design rules and, most likely, SFRC would be used more often if reliable and simple design recommendations were available. The traditional methods, adopted for plain concrete, are often based on the assumption that the material has an elastic behaviour. The maximum stress, evaluated by means of Westergaard's theory (Westergaard 1926) or elastic Finite Element (FE) analysis, is compared with an equivalent strength determined according to the prescription of several standards (RILEM TC 162-TDF 2002, JCI 1984, Unicemento 2002). Since fibers activate after concrete cracking, elastic methods cannot take in account the overall fiber contribution. For this reason, SFRC slabs should be analysed by using non-linear methods: a linear elastic approach cannot take into account the overall beneficial effect of fiber reinforcement.

Design methods based on Non-Linear Fracture Mechanics (NLFM) (Hillerborg 1976) follow more closely the actual behaviour of SFRC, which exhibits a considerably non-linear response after cracking, when the fiber contribution becomes relevant. By using NLFM analysis, crack development can be more accurately predicted until a collapse mechanism occurs and the slab on grade fails.

In this paper a comparison of the results obtained from different design methods is presented. Traditional elastic methods are compared with different non-linear methods. In particular fracture mechanics analyses based on both smeared and discrete crack approaches are discussed. The results obtained are analysed herein and the suitability of the methods is discussed. In order to validate the methods, two full-scale slabs on grade have been tested and the experimental results confirm the applicability of the proposed method for pavement design.

The bearing capacity of SFRC slabs on grade can be also evaluated by adopting yield line method. These methods require the knowledge of the collapse mechanism and the crack pattern. The results obtained with this method are also discussed.

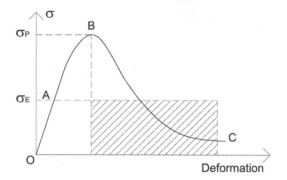

Figure 1. Equivalent post peak strength.

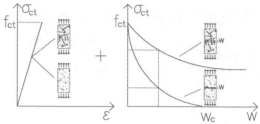

Figure 2. Pre-peak and post-peak behaviour of plain and steel fiber reinforced concrete, respectively.

2 TRADITIONAL DESIGN METHODS

Design methods usually adopted for SFRC slabs on grade are derived from the methods used for traditional slabs on grade (with or without reinforcement). These methods assume an elastic behaviour of the subgrade (Winkler soil).

A commonly used method is based on the comparison between the maximum stress in the slab and an equivalent flexural strength of the material. The maximum stress is determined by assuming an elastic behaviour of the slab and by adopting Westergaard's equations or FE analysis, while the equivalent strength is evaluated as the average value of the post peak strength (Fig. 1). The method leads to a significant underestimation of the bearing capacity of the slab: in fact, the equivalent flexural strength is lower than the crack flexural strength, the material is considered uncracked and the fiber effect cannot be taken in account. Moreover, after the onset of the first crack the load can still increase because of the post cracking concrete strength (Falkner et al. 1995).

3 NLFM DESIGN METHODS

Since fibers start working after cracking of the concrete matrix, where material response is no longer linear, SFRC slabs on grade can be better analysed by adopting methods based on Non Linear Fracture Mechanics (NLFM). The material is considered linear elastic until the peak strength is reached and, eventually, a post peak stress versus crack opening law is considered (Fig. 2). In this way the overall beneficial effect of the fiber reinforcement can be taken in account (Fig. 2).

In the NLFM analysis the crack can develop in the slab until a collapse mechanism occurs as in the actual behaviour. During the crack propagation the bearing capacity of the slab can increase: the load at ultimate is 4–5 times the first crack load (Meda et al. 2003).

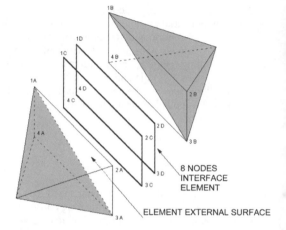

Figure 3. Crack interface elements in discrete crack model.

The numerical analysis of SFRC slabs on grade can be performed by adopting FE commercial programs that allow NLFM calculus.

Two different NLFM approaches are taken into consideration: a discrete crack approach in which all the crack process is concentrated in a single layer with zero thickness, and a smeared crack approach with cracks spread over the elements. The analyses were performed by means of MERLIN (Reich et al. 1994) and DIANA (DIANA 2002), for the discrete crack and smeared crack approach, respectively.

MERLIN considers the structure as a number of linear elastic subdomains linked by interface elements that simulate the cracks whose position must be known a priori. Interface elements initially connect the subdomains (as rigid links) and start activating (i.e. cracks start opening) when the normal tensile stress at the interface reaches the tensile strength of the material (Fig. 3). Afterwards, the crack propagates and cohesive stresses are transmitted between the crack faces according to a stress-crack opening (σ–w) law (which is given in the input for the interface elements).

With DIANA a conventional non-linear FE approach can be adopted, with a Galileo-Rankine criterion in

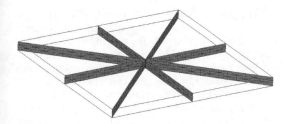

Figure 4. Position of the crack interface elements.

tension. Concrete tensile behaviour after cracking was modelled by means of a smeared crack approach based on a cohesive crack model (Hillerborg 1976), adopting a multi-linear σ–w curve.

A squared slab on grade with a point load in the centre was modelled by means of NLFM with both discrete and smeared crack approach. The slab rested on a Winkler soil and elastic springs were placed at the base element nodes.

Since in the discrete crack model the interface elements have to be placed where the crack is expected, these elements were localized along the diagonal and median lines (Fig. 4).

4 VALIDATION OF NLFM ANALYSES

Both discrete and smeared crack NLFM models were validated against two experimental tests on full-scale SFRC slabs on grade. Two square 3 × 3 m 0.15 m thick slabs were tested with a fiber content of 0.38% and 0.76% by volume, respectively. Hooked steel fibers with a length (l) of 50 mm and a diameter (Φ) of 1 mm (aspect ratio l/Φ = 50) were adopted.

The concrete mix adopted for the slabs had a cement content of 285 kg/m³ and a water–cement ratio of 0.6. The mechanical properties of the concrete at the time of the tests, measured on cores extracted from the two slabs, are shown in Table 1.

The slabs were placed on neoprene bricks (100 × 100 × 20 mm) located on a grid at 333 mm spacing and loaded by means of a hydraulic jack. A 100 kN load cell was placed between the jack and the slab while 6 LVDTs measured the slab deformation.

Load versus displacement measured in the slab centre curves for both the slabs are presented in Figures 5–6 with the crack pattern at the collapse.

The collapse is defined when the two main cracks reached the slab border. After this point the load can still increase because it is reacted by the elastic subgrade.

The experimental tests were numerically analysed. In both discrete and smeared crack approaches it is necessary to know the σ–w laws. These laws were obtained by performing an inverse analysis (Roelfstra & Wittmann 1986) from four-point bending tests on beam specimens according to the UNICEMENTO (2002).

Table 1. Mechanical properties of concrete at the time of the test on the slabs.

Compressive strength	30.6
Direct tensile strength	2.77
Young's modulus	38,000

Quantities in [MPa].

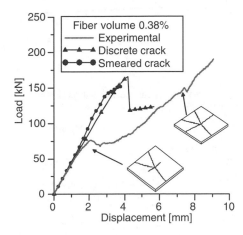

Figure 5. Load–displacement curves for 0.38% fiber volume.

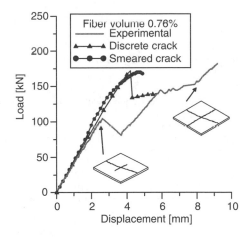

Figure 6. Load–displacement curves for 0.76% fiber volume.

The σ–w curves which provided the best fit of the experimental bending tests were determined: a bilinear σ–w curve was adopted in the discrete crack approach, whereas a multilinear curve was used in the smeared crack case (Figs. 7–8). In the latter case, the critical length (l_{cr}) transforming the crack opening in the smeared crack deformation, was taken as equal to 5 mm. Finally, the tensile strength in the smeared crack

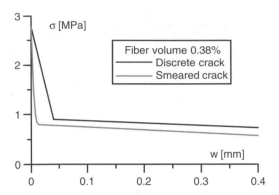

Figure 7. σ–w curves for concrete in tension (fiber volume equal to 0.38%).

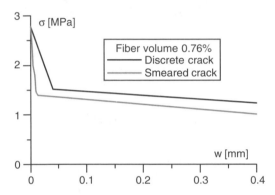

Figure 8. σ–w curves for concrete in tension (fiber volume equal to 0.76%).

Table 2. Ultimate load experimentally and numerically determined.

Slab	Fiber volume 0.38%	Fiber volume 0.76%
Experimental	147	154
Discrete crack	166	172
Smeared crack	152	170

Quantities in [kN].

s–w curve was reduced by 20% to increase the stability of the numerical solution and to account for size effects.

Figures 5–6 show a comparison between numerical and experimental results. Before the first crack development, both smeared and discrete crack models are close to the experimental evidence. Afterwards, due to the aforementioned asymmetric crack development, the experimental slab stiffness is lower than the numerical one, which assumes a symmetric crack development. In any case also the ultimate load for both slabs is closely predicted by the numerical analyses (Table 2).

The main advantages of using a discrete crack approach are related to a higher stability of the integration algorithm and to the possibility also of following highly unstable equilibrium paths. However, in order to use a traditionally discrete crack approach it is necessary to predict the crack pattern and so it can be used in a limited number of cases. A smeared crack approach can be used in all situations and might allow determination of the crack pattern for subsequent discrete crack analyses.

5 YIELD LINE METHOD

NLFM methods provide a good approximation of SFRC slabs on grade behaviour with good accuracy, as discussed before. However, programs based on non-linear fracture mechanics are characterized by remarkable computational complexities and they are not usually available in design offices.

For this reason simpler methods should be proposed for the designer. In any case the non-linear behaviour of SFRC slabs on grade have to be considered in order to take into account the fiber contribution.

SFRC slabs on grade can also be studied by adopting yield line theory (Johansen 1962), referring to that proposed from Bauman & Weisgerber (1983) for r.c. slabs on grade. The ultimate bearing capacity is estimated on the basis of a rigid-plastic slab resting on an elastic subgrade. By increasing the applied load to a SFRC slab on grade, a certain section reaches the plastic moment capacity. Eventually, the number of plastic sections increases with the onset of a number of yielding lines and a collapse mechanism forms. By imposing the collapse mechanism and by using the principle of virtual work it is possible to determine the ultimate load. To determine the plastic bending moment the equivalent strength is considered (Fig. 1) and a linear distribution of the stresses in the section is used. This is because the equivalent strength is determined by means of experimental tests on specimens, considering the failure section reacting entirely even when the crack is present.

Knowledge of the crack pattern and the collapse mechanism is necessary to apply the yield line method. The yield line method can be used for SFRC slabs on grade design if boundary conditions are imposed (Meda 2003). In the case of a square slab on grade with a point load in the centre, without links at the border and with a Winkler subgrade, the collapse mechanism is shown in Figure 9, where the crack pattern and the axes of rotation are marked. The chosen mechanism gives a crack pattern similar to the experimental evidence (Fig. 10) and only the part of the slab internal to the rotating axes is considered.

The internal work is given as:

$$W_{int} = 8 \cdot m_L \cdot \delta \qquad (1)$$

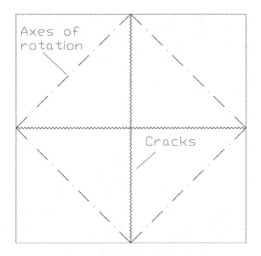

Figure 9. Collapse mechanism and crack pattern adopted in the yield line method.

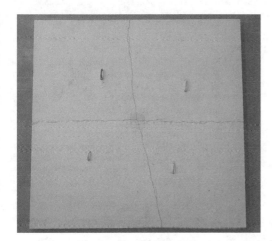

Figure 10. Experimental crack pattern.

with m_L moment capacity and δ virtual displacement at the point load application, while the external work due to the applied load P is:

$$W_{ext.load} = P \cdot \delta \qquad (2)$$

The external load due to the soil is:

$$W_{ext.soil} = \frac{k \cdot \delta^2 \cdot L^2}{12} \qquad (3)$$

with k Winkler soil constant and L slab side length.

By imposing the same collapse displacement $\delta = \Delta$ of a slab without support results (Bauman & Weisgerber 1983, Meda 2003):

$$\Delta = \frac{P_u^0}{K_S} \qquad (4)$$

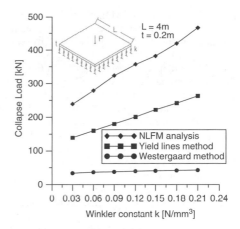

Figure 11. Comparison between collapse loads evaluated with different design method for different subgrade stiffness.

where P_u^0 is the collapse load without support, equal to $8\,m_L$, and K_S is the slab stiffness with simple support boundary conditions located along the axes of rotation.

By setting the internal work equal to the external one the collapse load P_u is:

$$P_u = 8 \cdot m_L \cdot \left(1 + \frac{1}{12} \frac{k}{K_S / L^2} \right). \qquad (5)$$

The moment capacity m_L can be evaluated, as discussed previously, by adopting a linear distribution of the stresses on the slab thickness t with a maximum value equal to the equivalent strength σ_E provided by the material:

$$m_L = \sigma_E \cdot \frac{t^2}{6}. \qquad (6)$$

Figure 11 shows the comparison between the obtained results:

• with a NLFM analysis by means of a discrete crack approach;
• by setting the maximum stress evaluated by the Westergaard's theory equal to the equivalent strength;
• with the Equation 5 (yield line method).

A slab with a side 4 m long and a thickness of 0.20 m is considered. The subgrade stiffness was varied, ranging within values significant for practical applications. The adopted material is a concrete with a compressive strength equal to 30 MPa reinforced with 0.38% by volume of hooked steel fibers. The equivalent strength was evaluated according to the Italian Standard (Unicemento 2002) as the average residual strength for crack tip opening displacement (CTOD) ranging from 0.6 mm to 3 mm ($f_{eq\,(0.6-3.0)}$).

It is seen that the collapse load obtained by adopting Westergaard's theory is very small and far less than the value that the NLFM method gives. The collapse load evaluated with the yield line method underestimates the NLFM collapse load by about 35%. The yield line solution gives a reasonable estimation of the collapse load, with a safe solution.

The underestimation of the collapse load with yield line method might be due to the definition of the equivalent nominal strength of the material. In fact from the NLFM analysis it appears that the crack opening at failure is still small, with residual stress greater than the equivalent strength. Another reason for the underestimation is due to the choice of a failure mechanism that does not consider the part of the slab external to the axes of rotation.

In any case the yield line method could be adopted for slab on grade design if the collapse mechanism can be predicted. Analyses performed with the NLFM smeared crack method can indicate the failure mechanism and the crack pattern for several load cases or different slab shapes (Meda et al. 2003).

6 CONCLUSION

SFRC slabs on grade should be considered as proper structures and designed with appropriate methods.

Traditional elastic analysis techniques such as Westergaard's or linear FE approaches, greatly underestimate the bearing capacity of SFRC slabs on grade and are excessively conservative. With these methods the fiber contribution cannot be taken into account.

The non-linear fracture mechanics approach gives reliable results and allows predicting the actual collapse load. In particular, the analyses with a discrete crack approach prove to give better results. The need for a knowledge of the crack pattern limits the use of the discrete crack approach. More complex loading cases may be more conveniently analysed with a smeared crack approach that might also allow determination of the crack pattern for subsequent analyses with a discrete approach.

Yield line analysis is suitable for practical design and gives a reasonable evaluation of the collapse load. As in the discrete crack case, the method needs the definition of the collapse mechanism and the crack pattern. Smeared crack analysis should give this information to the designer in the most common cases.

ACKNOWLEDGEMENTS

The author thanks Prof. P. Gambarova, Prof. G.A. Plizzari and Prof. P. Riva for their encouragement in writing the paper and their help in discussing the results.

The experimental part of the research project was partially financed by La Matassina s.r.l. (Castelnovo di Isola Vicentina, VI, Italy). The understanding and coope-ration of Mr. Giuseppe De Rossi are kindly acknowledged.

REFERENCES

Baumann, R.A. & Weisgerber, F.E. 1983. Yield line analysis of slab on grade. *ASCE Journal of Structural Engineering*, 109(7): 1553–1568.

DIANA – Finite Element Analysis. 2002. *User's Manual, V.8.1*, TNO Building and Construction Research: Delft.

Falkner, H., Huang, Z. & Teutsch, M. 1995. Comparative Study of Plain and Steel Fiber Reinforced Concrete Ground Slabs. *Concrete International* 17(1): 45–51.

Hillerborg, A., Modèer, M., & Petersson, P.E. 1976. Analysis of crack formation and crack growth in concrete by means of fracture mechanics and finite elements. *Cement and Concrete Research* 6: 773–782.

JCI. 1984. Method of Tests for Flexural Strength and Flexural Toughness of Fiber Reinforced Concrete. JCI Standard SF4 *Japan Concrete Institute, Standards for test methods of fiber reinforced concrete*: 45–51.

Johansen, K.W. 1962. *Yield line theory*. London: William Clowes and Sons Ltd.

König, G., Dehn, F. & Faust, T. 2002. *6th International Symposium on Utilization of High Strength/High Performance Concrete*. Leipzig: Institute for Structural Concrete and Building Materials, Leipzig University.

Li, V.C. 1993. From Micromechanics to Structural Engineering – The Design of Cementitious Composites for Civil Engineering Applications. *JSCE Journal of Structural Mechanics and Earthquake Engineering* 10: 37–48.

Meda, A., Plzzari, G.A. & Riva, P. 2003. Fracture behavior of SFRC slabs on grade. In A.E. Naaman and H.W. Reinhardt (eds) *High Performance Fiber Reinforced Cement Composites (HPFRCC4)*, 16–18 June 2003 Ann Arbor. Cahan: RILEM.

Meda, A. 2003. Yield line method for SFRC slab on grade design. *Studies and researches*. Politecnico di Milano (in press).

Reich, R., Cervenka, J. & Saouma., V.E. 1994. *MERLIN, a three-dimensional finite element program based on a mixed-iterative solution strategy for problems in elasticity, plasticity, and linear and nonlinear fracture mechanics*. Palo Alto: EPRI.

RILEM TC 162-TDF. 2002. Test and design Methods for steel fibre reinforced concrete – Bending test: final recommendation. *Materials and Structures* 35: 579–582.

Roelfstra, P.E. & Wittmann, F.H. 1986. Numerical Method to Link Strain Softening with Failure of Concrete. In Wittmann F.H. (eds) *Fracture Toughness and Fracture Energy*: 163–175. Amsterdam: Elsevier.

Unicemento. 2002. Steel fibre reinforced concrete – Part I: Definitions, classification specification and conformity – Part II: test method for measuring first crack strength and ductility indexes. Italian Board for Standardization (UNI), (in press).

Vondran, G.L. 1991. Applications of steel Fiber Reinforced Concrete. *Concrete international* 13(11): 44–49.

Westergaard, H.M. 1926. Stresses in concrete pavements computed by theoretical analysis. *Public Roads* 7(2): 25–35.

Economic aspects of steel composite beam stiffened with C-channel

M.Md. Tahir, A.Y.M. Yassin, N. Yahaya, S. Mohamed & S. Saad
Steel Technology Centre, Universiti Teknologi Malaysia, Skudai, Johor, Malaysia

ABSTRACT: Composite construction in buildings has known to increase the loading capacity and stiffness of the composite beam construction. The benefits of composite action result in significant savings in steel weight and in construction depth. However, the significant in weight savings can be further enhanced by stiffening the middle third of the composite beam with a C-channel. A new method of stiffening the composite beam is introduced in this paper by means of C-channel. This paper will discuss about experimental study on the flexural behavior of steel-concrete composite beams. Three full-scale composite beam specimens were carried out at UTM laboratory. The experimental work shown that the initial flexural stiffness, k_i, for stiffened beams were about 50% higher compared to the conventional composite beam. It was concluded that the composite beam stiffened with C-channel contributes to the strength and stiffness of the composite section, at both elastic and ultimate condition.

1 WHY COMPOSITE CONSTRUCTION?

Steel framing with in-situ casting of reinforced concrete lab construction was historically designed on the assumption that the concrete slab acts independently of the steel in resisting loads. No consideration was given to the composite effect of the steel and concrete acting together. However, with the introduction of mechanical shear connectors, the composite action became practical to resist the horizontal shear which result in the increase of bending strength. The typical advantages of the composite interaction are reduction in the weight of steel, shallower steel beam, increased floor stiffness, increased span length for a given member, and increased overload capacity (Lawson, 1989). The overall economy of using composite construction when considering total building costs appears to be good and is steadily improving (Tahir, 2000). One of the way to improve the composite construction is introduced by the authors in this paper by stiffened the composite beam using angle. Two of the specimens with C-channel sections where the opening part facing downward and upward act as a stiffener were welded to the lower flange of the steel beam. The stiffened length was provided at the center part of the beam, enough to the increase the bending capacity of the beam. Full shear connection was provided at the steel-concrete interface. The beams were simply supported and were loaded by two point loads. Measurements of ultimate load and maximum deflection were made in order to obtain the complete picture of the behavior of the beams.

2 DESCRIPTION OF TEST SPECIMEN

The work involved flexural testing on three full-scale composite beams. The cross-section of the beams is shown in Figure 1. The first specimen identified as specimen CB is a control specimen with no stiffener. The second specimen identified as specimen SB1 is a composite beam stiffened by C-channel with opening facing upward and the third specimen identified as specimen SB2 having C-channel as stiffener with opening facing downward. The steel beams and the C-channel used were locally produced by Perwaja Steel Sdn. Bhd. The concrete slabs were cast at Universiti Teknologi Malaysia with local supplier known as Industrial Hardware Supply Sdn. Bhd, delivered and installed the shear connectors. The details of

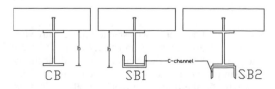

Figure 1.   Cross section of specimens to be tested.

Table 1. The details of the specimens.

Speci-men	Steel beam	C-channel	Yield strength, p_y (N/mm²)	f_{cu} (N/mm²)
CB	250 × 125 × 2	–	275	30.4
SB1	5.1	152 × 76 ×	(for both beam and channel)	30.5
SB2		17.9		32.1

the specimens are given in Table 1. The concrete slab was designed for C30 or 30 N/mm² of compressive strength. The width of the concrete slab was 1500 mm and the thickness was 125 mm. Transverse reinforcement of 0.9 per cent of the cross-section of the concrete was provided to prevent longitudinal splitting. Full shear connection was provided between the concrete slab and the steel beam. This has required an amount of sixty stud connectors, installed in each beam. The C-channel was welded to the lower flange of the steel beam through the whole length of the channel.

The stiffening length was determined based on "cut-off" length as in the design of cover-plated composite beam (Cook, 1977). Under that basis, the length of the C-channel must be long enough so that the unstiffened section can resist bending moment at the failure of the beam. SB2 is expected to have higher flexural capacity than SB1 due to the downward arrangement, which due to longer lever arm.

3 TEST PROGRAM

The beam's length was 6 m with an effective length of 5.7 m. The beams were simply supported at both ends and subjected to two point loads. Such loading was intended to produce a region of pure bending moment at mid-span. The beams were laterally restraint at several points as shown in Figure 2 to prevent any lateral torsional buckling. Stiffeners were welded at the steel web above the end supports to prevent crushing of the web. The loading was applied at an increment of 5 kN for 3 minutes interval until reaching the elastic limit. The applied loading was then control by the deflection of every 2 mm increment until failure.

4 TEST RESULTS

Each of the specimens was tested to failure. Three linear vertical displacement transducer or LVDT were placed at mid-span to measure the vertical deflections. For a simply supported beam, the most critical check is usually on the deflection (Johnson, R.P., 1994). A graph of moment versus mid-span deflection and load

Figure 2. Specimen in the test-rig.

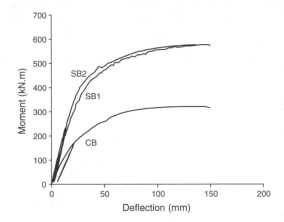

Figure 3. Plot of moment versus mid-span deflection for CB, SB1 and SB2.

versus mid-span deflection were plotted as shown in Figure 3, 4, 5 and 6 respectively. Significant increment in stiffness and moment strength of the stiffened beams compared to the conventional composite beam can be seen in Fig. 3 and in Table 2. It can be seen from Figure 3 to 6 and Table 2 that all beams exhibited good ductility where the deflections of up to 150 mm were achieved. Beam SB2, which was expected to have higher moment capacity, was shown to have similar moment capacity as SB2. Both beams of SB1 and SB2 failed in an equal ultimate load. However, the initial flexural stiffness specimen SB2 value was 15 per cent more than specimen SB1. There was no longitudinal splitting observed at any stage of loading thus the beams were failed solely due to flexural stresses. The 0.9% of steel reinforcement used was sufficient to prevent the longitudinal splitting.

The initial flexural stiffness, K_i, is defined as the secant value measured at 50% of the ultimate test load

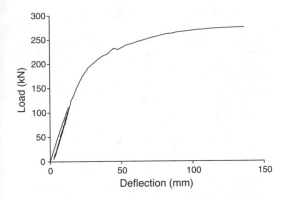

Figure 4.   Plot of load versus mid-span deflection for SB1.

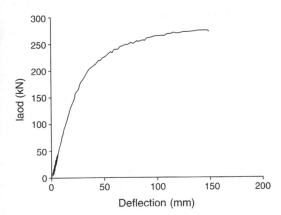

Figure 5.   Plot of load versus mid-span deflection for SB2.

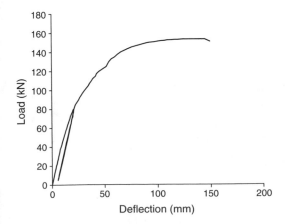

Figure 6.   Plot of load versus mid-span deflection for CB.

Table 2.   Test results for maximum load, moment, and deflection.

Spec-imen	Ulti-mate load, P (kN)	Max. test moment, M_R (kN/mm)	Mid-span deflec-tion at M_R (mm)	Initial flexural stiffness, K_i (kN/mm)	Total beam depth, h (mm)
CB	150.5	316.5	149.8	3.85	248.0
SB1	275.0	577.5	149.9	6.9	254.4
SB2	275.0	577.5	138.5	8.1	348.0

Table 3.   Test results of the ratio of unstiffened versus stiffened specimens for moment, stiffness, and total beam depth.

Specimen	Ratio of $M_R/M_{R(CB)}$	Ratio of $K_i/K_{i\,(CB)}$	Ratio of $h/h_{(CB)}$
CB	1.00	1.00	1.00
SB1	1.82	1.79	1.03
SB2	1.82	2.10	1.40

(Lam, D. et.al, 2000). For serviceability limit state the deflection depends on the flexural stiffness of the beam (EI), which is a product of the modulus of elasticity (E) and the moment of inertia (I) (Cain, J.A. and Hulse, R., 1990). The increase in stiffness of SB1 and SB2 compared to CB of about 44% and 52% is very significant. The increase in depth of about 2.5% and 29% in order to achieve the increment of the stiffness is very minimal especially for the SB1 specimen. This shows that the effectiveness of the upward arrangement of the C-channel can be proposed as an alternative to the stiffening of composite beam using cover plate. More works need to be done to this area of research so that standardized design can be developed.

5   CONCLUSION

From the test results conclusions can be drawn as follows:-

1. Full-scale testing has shown that by stiffening the composite beam stiffened with C-channel, the stiffness and flexural strength of the beam can be increased.

2. The increase in stiffness of the composite beam can be represented by the increase in initial flexural stiffness, k_i, in which the increment was about 50 per cent.

3. The increase in ultimate strength can be represented by the increase in maximum test moment, in which the increment was about 45 per cent.

1421

4. The arrangement of C-channel, either downward or upward, has been found to have little influence on the overall behaviour of the stiffened beams.
5. The increase in maximum moment of about 54 per cent is accompanied by increase in beam depth of 29 per cent.

ACKNOWLEDGMENT

Both authors like to thank Perwaja Steel Sdn. Bhd. and Industrial Hardware Supply Sdn. Bhd. for supplying the materials in the work. Thanks also to the technicians of Dept. of Structure and Material of UTM for their assistance in the laboratory work. The second author would like to thank his employer, UTM for allowing the study leave to complete his master degree. This work was part of IRPA Grant 72217 funded by the Ministry of Science and Environmental of Malaysia.

REFERENCES

Cain, J.A & Hulse, R. 1990. *Structural Mechanics,* The Macmillan Press Ltd. London. Cook, P. John, (1977). *"Composite Construction Method"* John Wiley and Sons Inc., New York.

Johnson, R.P. 1994. *Composite Structures of Steel and Concrete-Vol. 1*, Blackwell Scientific Publications, Great Britain.

Lam, D., Elliot, K.S. & Nethercot, D.A. 2000. Experiments on composite steel beams with precast concrete hollow core floor slabs, *Proc. Instn. Civ. Engrs Structs and Bldgs*, 2000, 140, May, 127–138.

Lawson, R.M. 1990. *Commentary on BS 5950: Part 3: Section 3.1 – Composite Beam*, The Steel Construction Institute, Berkshire.

Mahmood, Md. Tahir, Airil Yasreen Mohd Yassin, Nordin Yahaya, Shahrin Mohamed, 2000. Economic Aspects of Composite Beam Construction on Multi-Storey Braced Steel Frame, *4th Asia Pacific Structural Engineering and Construction Conference* (APSEC 2000), Palace of Golden Horses, Kuala Lumpur, Malaysia.

Deformation capacity of concrete columns reinforced with CFT

K. Maegawa
Kanazawa University, Kanazawa, Ishikawa, Japan

M. Tomida & A. Nakamura
Ishikawa National College of Technology, Tsubata, Ishikawa, Japan

K. Ohmori & M. Shiomi
Nippon Zenith Pipe, Chiyodaku, Tokyo, Japan

ABSTRACT: The aim of this research is to investigate the applicability of a concrete-filled tubular steel (CFT) as reinforcement for RC-structures, which are requested to have high deformation capacity. Both types of test specimens of the RC-column with CFTs and the RC-column with reinforcing bars have the same dimensions, and are designed so that they could have an equal strength due to the same quantity and/or strength of reinforcement. Two types of steel-tubes whose dimensions are ϕ 42.7 ×2.3 mm and ϕ 27.2 ×2.3 mm are also used for the CFTs. An axially constant load and a horizontally cyclic load are applied to the column top. It is found that the RC-column with CFTs is excellent at the deformation and energy absorption capacity compared with the RC-column with reinforcing bars, and that there is no effect of the ratio of diameter to thickness of steel-tubes on the behavior of the columns.

1 INTRODUCTION

It is well known that the stiffness, strength and ductility of a concrete-filled tubular steel, i.e., a CFT, become high, since the concrete in-filled prevents the steel pipe from buckling and is strengthened due to the confined effect (Maegawa 1995). The author has found that the deformation capacity of the CFT composite beam, in which a CFT is used as a compression reinforcement, is excellent as compared with that of the RC-beam with a deformed bar (Maegawa 1997, 2001).

Generally, relatively higher piers or bridge columns of reinforced concrete lack the bending ductility, i.e., the deformation capacity, against an earthquake load. Namely, compression rupture of reinforced concrete causes buckling of compression reinforcing bars and consequently decreases the bending strength of the RC-structure sharply. Therefore, in order to improve the deformation capacity of RC-structures it is effective to prevent the reinforcing bars from buckling.

CFT is hard to buckle as compared with a reinforcing bar. In this study, CFT is employed instead of a conventional reinforcing bar in order to increase the deformation capacity of the RC-column such as a pier subjected to the earthquake load. Therefore, the aim of this research is to investigate the applicability of the concrete-filled tubular steel as reinforcement for RC-structures. Both types of test specimens of the concrete columns with CFTs and ones with reinforcing bars have been tested under an axially constant load and a horizontally cyclic load. The effect of the diameter-thickness ratio of steel-tubes and the bond method of a tubular steel to concrete are also investigated.

2 TEST OUTLINE

2.1 *Test specimens*

Figure 1 shows the dimensions of three types of specimens. All specimens have the same cross-section of 200 × 250 mm and a height of 850 mm. First, the column part is manufactured, and then it is concreted in the base concrete of the 300 mm depth and the 800 mm square. The specimens are designed so that they could have same strength due to the same quantity and/or strength of reinforcements. However, the only difference is the type of reinforcements. Ten deformed bars of D13, four mild-steel-tubes of ϕ 42.7 × 2.3 or six mild-steel tubes of ϕ 27.2 × 2.3 are set as the reinforcement. The steel-tubes are filled with concrete. The advantage of filling concrete is not

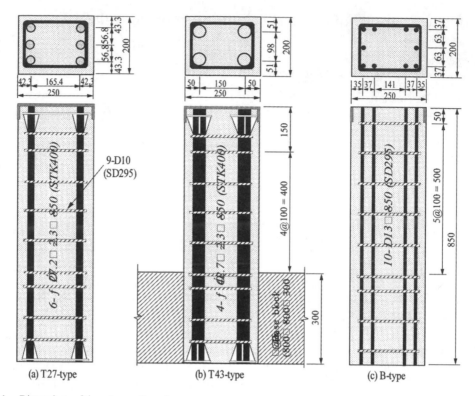

Figure 1. Dimensions of three types of specimens.

only to prevent the local buckling of the tubular steel but also to strengthen concrete by the confined effect. In order to prevent the steel-tubes from sliding out of the surrounding concrete, the anchor consisting of a steel disc and four steel blades is welded to all tube ends. The grit-blast treatment of SP-10 is also given on the tube surface for some specimens.

The centroid of five deformed bars is coincident with that of the tubular steel of ϕ 42.7 which is 50 mm from the edge of a cross-section. The outermost surface of the deformed bar and that of the CFT are at same distance from the edge of a cross section. The stirrups are attached at intervals of 10 cm from the column base. Owing to the test equipment, the shear-span ratio becomes 2.9, which is a little small as compared with the general values of 3 to 5 in bridge piers.

Eleven specimens are listed in Table 1 where the mechanical properties and the loading conditions are also summarized. The first one letter of the specimen name denotes the reinforcement types, i.e. a tubular-steel or a deformed bar. The second two letters are the diameter of the reinforcement. The next letters, N5 and N10 denote the applied axial load ratio of 0.05 and 0.1, respectively. The next one letter indicates the transverse loading condition, i.e. monotonic or cyclic.

The last letter G in some specimen names indicates that the tubular steel has been pretreated by the grit-blast named SP-10. The compressive strength of the concrete, f_c, is from 27 MPa to 35 MPa. The applied axial load ratio, σ_N/f_c, is 0.05 or 0.1, where σ_N = applied axial stress and f_c = compressive strength of concrete. The ratio of σ_N/f_c ranges about from 0.025 to 0.07 in general bridge piers.

2.2 Testing procedure

Figure 2 illustrates the outline of the loading set-up. Firstly, the axial load is applied and held constant by the actuator. Then the horizontal or transverse load by an oil jack is applied to the device, which is fixed on the column top. The length of moment arm becomes 720 mm from the column bottom, i.e. the top of the base concrete. The horizontal movement of the loading device to which an oil jack is connected is controlled in order to follow the loading procedure as shown in Figure 3. When the strain of the outermost surface of the CFT or the deformed bar becomes the yield strain at the position of 40 mm from the top of a base concrete, the horizontal movement of the loading device is called the yield displacement, δ_Y. The

Table 1. List of specimens.

Specimen type name	Reinforcements diameter × thickness n-φ × t (mm × mm)	φ/t	Tension steel ratio (%)	Concrete strength f_c (MPa)	Yield strength σ_Y (MPa)	Tensile strength σ_u (MPa)	Axial load σ_N/f_c	Transverse load condition	Grit-blast treatment for tube
T27 T27N5M	6–27.2 × 2.3	11.8	1.08	31.9	475	498	0.05	monotonic	non
T27 T27N5C	6–27.2 × 2.3	11.8	1.08	31.1	475	498	0.05	cyclic	non
T27 T27N10C	6–27.2 × 2.3	11.8	1.08	30.9	475	498	0.10	cyclic	non
T43 T43N5M	4–42.7 × 2.3	18.6	1.17	35.3	353	441	0.05	monotonic	non
T43 T43N5C	4–42.7 × 2.3	18.6	1.17	28.8	353	441	0.05	cyclic	non
T43 T43N5MG	4–42.7 × 2.3	18.6	1.17	26.9	423	511	0.05	monotonic	SP-10
T43 T43N5CG	4–42.7 × 2.3	18.6	1.17	31.0	423	511	0.05	cyclic	SP-10
T43 T43N10CG	4–42.7 × 2.3	18.6	1.17	29.4	423	511	0.10	cyclic	SP-10
B B13N5M	10–13 (deformed bar)		1.27	33.5	394	578	0.05	monotonic	–
B B13N5C	10–13 (deformed bar)		1.27	31.4	394	578	0.05	cyclic	–
B B13N10C	10–13 (deformed bar)		1.27	35.0	394	578	0.10	cyclic	–

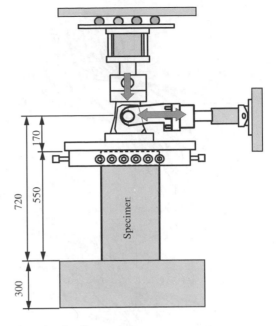

Figure 2. Loading set-up.

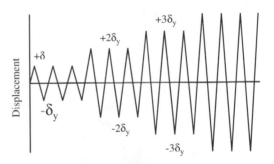

Figure 3. Cyclic procedure for transverse load or displacement.

yield load, P_Y is based on the similar definition as the yield displacement. In the first load step, the direction of the transverse load, for example the push in Figure 2, will be called positive.

3 TEST RESULTS

3.1 Load-displacement relationship

Figures 4a–4c show the load-displacement hysteretic curves of specimens T27N5C, T43N5C and B13N5C, respectively. The first crack load, the yield load and the maximum load are also indicated in the Figures.

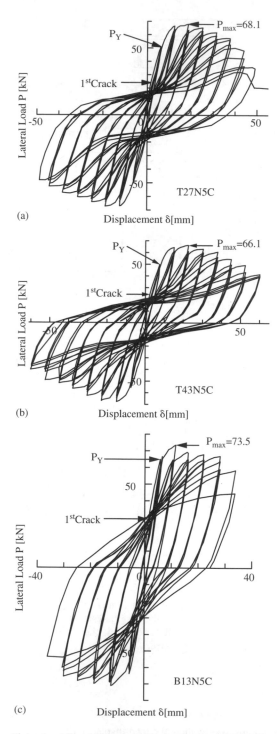

(a)

(b)

(c)

Figure 4. (a)Load-displacement hysteretic curves (T27N5C). (b) Load-displacement hysteretic curves (T43N5C). (c) Load-displacement hysteretic curves (B13N5C).

The peak load deteriorates significantly in the 2nd cycle in each displacement step of δ_Y. On the compression part of the specimen subjected to the positive load in the 2nd cycle, the compression reinforcement can carry higher stress, but the axial stress of concrete becomes low since the concrete has been cracked and opened on the previous negative load.

Therefore, the reduction of the shear stiffness of the concrete takes place and consequently brings the significant influence on the peak load or the restoring force in the 2nd cycle of each displacement-step. However, the deterioration in the 3rd cycle is mild since the stress re-distribution has already completed (Mishima 1992).

Due to the unloading, the recovered displacements of T27N5C and T43N5C are larger than that of B13N5C. For example within the unloading range from the positive displacement of $5\delta_Y$ to the non-loading state, the curve inclinations of specimens T27N5C and T43N5C are 0.24 mm/kN and 0.33 mm/kN, respectively. And that of B13N5C is 0.17 mm/kN. This indicates that the recovering ability of the specimen with CFT is better than that of the specimen with deformed bars.

The load and the displacement in Figures 5a–5c have been non-dimensionalized by the yield load P_Y and the corresponding displacement δ_Y, respectively. Figure 5a shows the load-displacement curves of the specimens subjected to a monotonic load. Figures 5b and 5c show the envelopes on the load-displacement hysteretic curves of the specimens subjected to a cyclic load, and the axial load ratios are different between them.

Figure 5a shows that there is little difference among the non-dimensionalized maximum loads of all specimens except B13N5M. However, the non-dimensionalized deformation capacities of them are different from each other. Especially the specimen with deformed bars, B13N5M has a great capacity of the deformation.

It is seen from Figures 5a and 5b that the non-dimensionalized load P/P_Y decreases more gently in the monotonic load as compared with the cyclic load. However, as for the specimens under the cyclic load, the maximum load of the specimens with CFT is larger than that of the specimens with deformed bars, and the deformation capacity of the specimens with CFT is excellent. The same result can be seen in Figure 5c.

It is also seen from Figures 5a–5c that the specimens of T43N5MG, T43N5CG and T43N10CG, for which the grit-blast treatment has been given on the tube surface, have the good capacity for the deformation as compared with the specimens of T43N5M, T43N5C and T43N10C.

3.2 Crack pattern

Figure 6 shows the final crack pattern on the side of the specimens shown in Figure 5b. The solid lines and the dotted lines represent the cracks caused by a

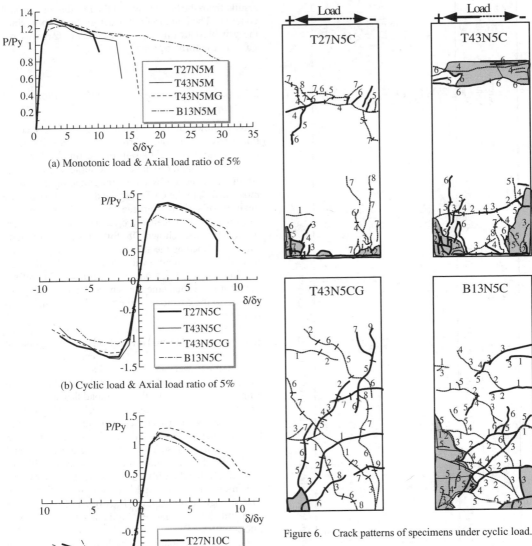

(a) Monotonic load & Axial load ratio of 5%

(b) Cyclic load & Axial load ratio of 5%

(c) Cyclic load & Axial load ratio of 10%

Figure 5.   Non-dimensionalized load-displacement relationship.

Figure 6.   Crack patterns of specimens under cyclic load.

positive load and a negative load, respectively. The solid arrow and the dotted arrow shown at the top of Figure 6 illustrate the load directions. The positive load direction corresponds with the push in the first cycle. The numeral near the crack line denotes the step number of the transverse load or displacement.

The hazy areas shown in Figure 6 represent the surface blocks of concrete, which separated and fell before the final state.

Firstly, in the specimens of T27N5C and T43N5C there is the flexure crack passing through the column bottom. Furthermore, the cracks are concentrating in two regions, namely, the bottom region including the relatively vertical cracks and the upper region including the horizontal cracks. Secondly, in the specimens of T43N5CG and B13N5C the cracks are generally forming X-shapes, though there are some vertical cracks in T43N5CG. Especially the cracks of B13N5C are widely formed. The difference between the crack patterns of the former group and the latter group

Table 2. Test results.

Specimen type name		Yield load moment displace			Maximum (positive) load moment displace			Maximum (negative) load & displace		Ultimate displace	Strength redundancy	Ductility index
		P_Y (kN)	*M_Y (kNm)	δ_Y (mm)	P_{max} (kN)	*M_{max} (kNm)	δ_{max} (mm)	P'_{max} (kN)	δ'_{max} (mm)	δ_U (mm)	M_{max}/M_Y	δ_U/δ_Y
T27	T27N5M	53.5	39.1	7.2	69.7	51.5	16.0	–	–	69.5	1.32	9.70
T27	T27N5C	50.8	37.1	6.3	68.1	50.5	18.9	−67.3	−19.0	42.3	1.36	6.71
T27	T27N10C	63.6	46.9	7.6	75.8	56.9	15.2	−74.9	−15.2	36.0	1.21	4.76
T43	T43N5M	54.0	39.5	6.8	69.0	51.1	15.4	–	–	92.0	1.29	13.46
T43	T43N5C	49.8	36.4	7.6	66.1	49.3	22.7	−67.7	−15.2	52.0	1.35	6.86
T43	T43N5MG	56.4	41.0	5.5	75.1	55.4	19.1	–	–	84.2	1.33	15.31
T43	T43N5CG	58.2	42.4	6.1	75.3	55.6	18.4	−72.9	−18.6	46.3	1.31	7.59
T43	T43N10CG	64.0	47.0	6.2	82.5	61.7	15.6	−77.3	−18.4	41.8	1.31	6.74
B	B13N5M	56.4	40.9	4.2	69.7	51.6	16.1	–	–	90.2	1.26	21.73
B	B13N5C	64.9	47.2	5.7	73.5	53.8	11.5	−71.1	−12.3	29.3	1.14	5.14
fB	B13N10C	72.1	53.2	7.2	80.1	60.2	14.5	−85.8	−14.6	29.0	1.13	4.03

*M_Y or *M_{max} = (Transverse load P_Y or P_{max}) × (Arm length of 0.72 m) + (Axial load) × (Lateral displacement δ_Y or δ_{max}).

results from the bond between the reinforcements and concrete. The bond in the former group is poor, since the grit-blast treatment for the tubular steel was not given to the former group.

3.3 Load carrying capacity and deformation capacity

Table 2 shows each experimental result of the specific transverse load and corresponding displacement under which the yield strain of the reinforcing steel and the maximum resistance were achieved. The yield strain was checked by the strain gauge on the outermost surface of the CFTs and the deformed bars at the position of 40 mm from the top of a base concrete. The ultimate displacement, δ_U and the ductility index, δ_U/δ_Y, are also shown. In this study the ultimate state is defined as the state in which the applied load level decreases again to the yield load, P_Y, after the maximum load, P_{max}, has been achieved. Therefore, the ultimate displacement has been determined by fitting P_Y to the envelope of the load-displacement hysteretic curve.

On the same loading conditions, there is little difference among the maximum loads and moments of T27-type, T43-type and B-type, because those types of specimens were designed to have the equal load carrying capacity. In the range of the axial load ratio employed in this study, the axial load has an effect on suppressing the tension yield of reinforcement. Consequently, both yield load and maximum load in the specimens under the axial load ratio of $\sigma_N/f_c = 0.1$ are higher than those in the specimens under the axial load ratio of $\sigma_N/f_c = 0.05$ under the same conditions. Conversely, the ductility index, δ_U/δ_Y, is higher in the specimens under the axial load ratio of $\sigma_N/f_c = 0.05$. The strength redundancy, M_{max}/M_Y, is from 1.13 to 1.36 and shows a relatively high tendency in the specimens with CFT.

On the cyclic load test, the buckling deformation and the shear deformation were detected in the deformed bars of the compression side and the tension side in the specimens with deformed bars, respectively. And the crack of CFT was detected in the specimens with CFT. They lead to the sudden deterioration in the resistance. From Table 2 the ductility index is comparatively high in the specimens subjected to monotonic load, and very high in especially B13N5M. However, that in the specimens under the cyclic load is higher in the specimens with CFT. The former indicates that the buckling of deformed bars is prevented in the case of the monotonic load until the severe rupture of the concrete occurs. The latter indicates that CFT contributes to the compression capacity even though the rupture of the surrounding concrete occurs due to the cyclic load.

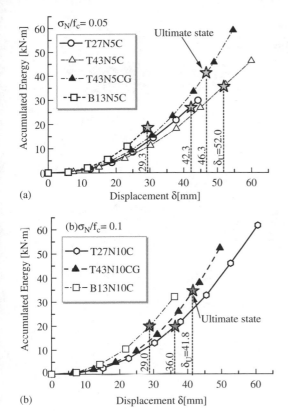

Figure 7. Accumulated energy absorption.

3.4 Energy absorption capacity

The energy absorbed into a specimen is obtained by integrating the load-displacement hysteretic curve such as Figure 4. The accumulation has been performed at the end of the 3rd cycle of every increment step of δ. Figures 7a and 7b show the relation between the accumulated energy and the displacement step. The end of a line indicates the unstable limit. The point of the ultimate state, which has been defined in section 3.3, is indicated by the mark of a star.

At the same level of displacement, the energy absorption is large in the specimen with deformed bars. Since the energy absorption becomes large as the plastic portion of a specimen is enlarged, it is explained that the damage of the specimens with CFT is relatively slight. Comparing the energy absorption at the ultimate state, the specimen with CFT is about two times as large as the specimen with deformed bars. Therefore, it is concluded that the specimens with CFT have an excellent energy absorption capacity.

4 CONCLUSIONS

The concrete columns reinforced with CFTs or with reinforcing bars have been tested under an axially constant load and a transversely cyclic load. The reinforcement type, the diameter-thickness ratio of steel-tubes and the bond method of a tubular steel to concrete have been investigated in relation to the load carrying capacity and the deformation capacity.

The following have been found.

- The cyclic load seriously reduces the deformation capacity, i.e. the ductility index, especially that of the concrete column with reinforcing bars.
- Both yield load and maximum load are high in the concrete columns under the axial load ratio of $\sigma_N/f_c = 0.1$ as compared with $\sigma_N/f_c = 0.05$, when the other conditions are same. Conversely, the ductility index, δ_U/δ_Y, is high in the specimens under the axial load ratio of $\sigma_N/fc = 0.05$.
- The concrete column with CFT is excellent in load-carrying capacity, deformation capacity, and energy absorption capacity as compared with the concrete column with reinforcing bars, when the concrete columns are subjected to an axially constant load and a transversely cyclic load.
- The diameter-thickness ratio of tubular steel does not influence the load-carrying capacity, and deformation capacity.
- The surface treatment on a steel-tube has a good effect on those capacities. It is important for us to improve the bond method in order to employ CFT as the reinforcement in a concrete column.

REFERENCES

Maegawa, K. & Yoshida, H. 1995. Impulsive Loading Tests on Concrete-filled Tubular Steel Beams Reinforced with Tendon. *JSCE Journal of Structural Mechanics and Earthquake Engineering*, No. 513/I-31, 117–127.

Maegawa, K. 1997. Ductility of Steel Tube Reinforced Concrete Composite Beams. *IABSE Conference Report on Composite Construction, Innsbruck, 16–18 September 1997*, 807–812.

Maegawa, K. et al. 2001. Static and Impact Tests on Cantilever Type Rock-shed of PC-beam with CFT-Compression-Reinforcements. *Proc. of EASEC-8., Singapore, 5–7 December 2001*, Paper No. 1136(CD-ROM).

Mishima, T. et al. 1992. The Analytical Approach to the Effect of Cyclic Loading on The Reduced Shear Stiffness of RC Crack Planes. *JSCE Journal of Materials, Concrete Structures and Pavements*, No. 442/V-16, 191–200.

System-based Vision for Strategic and Creative Design, Bontempi (ed.)
© 2003 Swets & Zeitlinger, Lisse, ISBN 90 5809 599 1

The deformability of hybrid (carbon and glass fibers) pultruded profiles

S. Russo
Venice Institute University of Architecture, Venice, Italy

G. Boscato
Vicenza, Italy

ABSTRACT: The experimental results on structural behaviour of "I" shape hybrid pultruded beams are presented in this study. The aim of this research was to provide more detailed informations on the structural behaviour – especially for a design structure view point – of hybrid pultruded beams, and also to give a comparison with the structural performance of GFRP pultruded beams.

The hybrid pultruded beam was made with carbon and glass fibers with thermosetting matrix vinylestere. The carbon fiber was applied in the flange, to improve the bending stiffness, beside the glass fiber was applied in the web.

However the approach to the analysis of the type of beam is not easy, in fact, for structural design we have not only the orthotropic behaviour of the material – with low value of longitudinal modulus E and the shear modulus G – but also the different mechanical characteristics of two kind of fibers.

The carried out test regards three points bending test with different slenderness parameters, the axial compression test, and also the evaluation of shear effect in evaluation of deformation in deflected beams.

Finally, the research proposes also the comparison – for a structural design point of view – with the pultruded shapes with the same geometry, the same type of matrix, but made only with glass fibers (GFRP).

1 INTRODUCTION

This research shows the experimental results on structural behaviour of hybrid pultruded profiles with "I" shapes in presence of carbon and glass fibers.

The aim regards the evaluation of effective structural behaviour of hybrid profile and the comparison with pultruded profiles reinforced only with glass fibers.

The hybrid profile has the carbon fibers in the flange, whereas the glass fibers are set in to the web; the matrix is type thermosetting vinylestere.

The presence of carbon fibers in the flange give an increment of flexural stiffness and the shear capacity of profiles is influenced of glass fibers, Figure 1.

Generally speaking the composite profile have the orthotropic behaviour with low values of E and G modulus.

The tests carried out regard the three point bending test – with different value of slenderness λ_g, ratio between length and height – to evaluate also the influence of shear deformation in flexural behaviour; the other type of test regards the compression behaviour of composite profiles.

Figure 1. "I" shape of HYBRID PROFILE.

2 MECHANICAL CHARACTERISTICS OF MATERIAL

As much known, the type of fibers with his percentage and direction influence the mechanical behaviour of HFRP (Hybrid FRP) material and consequently the behaviour of structure. In fact the theoretical approach on HFRP profile is strongly different than the profile made with traditional materials. Besides also the mechanical interaction matrix–fiber influence the performance of GFRP materials and the structural behaviour at ultimate state.

Table 1 shows the mechanical characteristics of matrix and fiber.

In our research we consider expecially the influence of E and G modulus in evaluation of displacement.

The tensile strength values has been determined with ASTM D 638 and UNI 5819, while the flexural modulus E with ASTM D 790 and UNI 790.

For a general view point the HFRP material have the linear – elastic behaviour until the collapse.

Table 1. Mechanical characteristics of materials.

Fibers	Diam. 10(µm)	Weight (g/cm³)	E (GPa)	Tensile strength (GPa)	Max Elongation (%)
E-Glass	10	2.54	72.4	4.35	4.8
Carbon HT	8	1.78	200–275	2.8	1

Resin	Tensile strength	E (GPa)	Max Elongation (%)	Flexural Strength (GPa)	E_{fl} (GPa)
Vinylestere	0.087	3.309	4.2	0.149	3.379

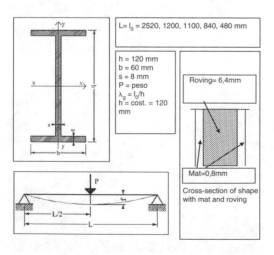

Figure 2. Static scheme and geometrical characteristics of shape.

The test has been carried out on three point bending test with different value of slenderness λ_g, ratio between length and height.

The static scheme and geometrical characteristics of "I" shape are showed in Figure 2.

The machine utilized for the tests is type Losenhausenwerk with 600 kN capacity and HP VL 4/50 system for automatical date acquisition.

Figure 3 shows the sample during the test, with transducers and strain gauge, to measure the local deformation of the material.

Figure 3. Test of beam with L = 1200 mm.

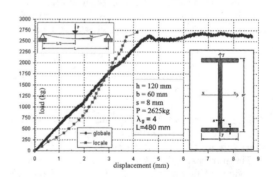

Figure 4. Load–deflection curve, L = 480 mm.

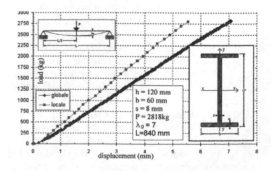

Figure 5. Load–deflection curve, L = 840 mm.

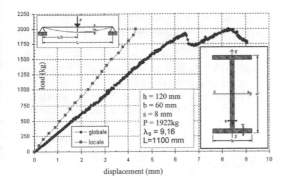

Figure 6. Load–deflection curve, L = 1100 mm.

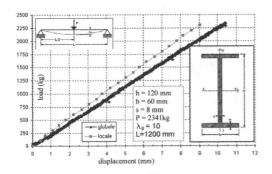

Figure 7. Load–deflection curve, L = 1200 mm.

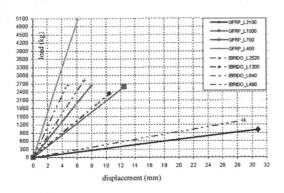

Figure 8. Comparison between experimental curves of GFRP and curve of hybrid profiles.

Table 2. Experimental results of load and displacement.

Length L(mm)	λ_g	Level of load P(kg)	Displacement η(mm)
2100	21	1010	30,8
1000	10	2580	12,5
700	7	2650	8,00
400	4	5100	6,2

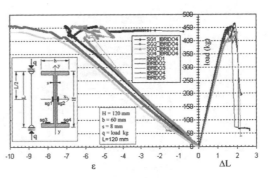

Figure 9. Load–displacement curve with deformation of materials, measured by strain-gauge.

Figure 10. Hybrid sample during the test compression.

Figures from 4 to 7 show the load–displacement curve for different value of λ.

Figure 8 shows the comparison between the structural behaviour of GFRP and HFRP beams. Same experimental values are presented in Table 2.

3 EVALUATION OF DEFORMABILITY IN HFRP BEAMS

For the evaluation of displacement in deflected beam we have employed the Timoshenko model that considers also the shear influence. The formula (1) has been employed for our static scheme.

$$\eta_{max} = [(P(l_o)^3)/48EJ][1+\chi12EJ(1/l_o^2GA)] \qquad (1)$$

The experimental value of modulus of elasticity, E_{exp} utilized in (1) has been calculated with formula (2), deduced directly from formula (3) in function of maximum displacement measured for beam with higher span, (2250 mm), in this way it is possible to reduce the shear influence in the evaluation of deformation.

$$E_{exp} = (P(l_o)^3)/48J\eta_{max} \qquad (2)$$

1433

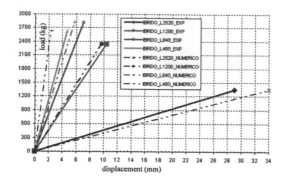

Figure 11. Comparison between numerical and experimental results.

$$\eta_{max} = (P(l_o)^3)/48EJ \qquad (3)$$

where:

η displacement at midspan
E modulus of elasticity
J minimum moment of inertia
l_o span between supports
χ shear factor calculated for our cross section
G shear modulus
A cross section

4 COMPARISON BETWEEN EXPERIMENTAL RESULTS AND NUMERICAL ANALYSIS

Figure 11 shows the comparison between numerical date obtained with formula (1) (whit E_{exp} value and G modulus equal to 4800 MPa) and experimental results.

As illustrated, this figure shows a good agreement between the experimental data and the numerical results obtained with Timoshenko's model.

In the detail, when l' influence of shear deformation is not so higher the experimental and numerical curves are really close; while for very low value of slenderness and a not negligible influence of shear the differences between two types of curves is more evident.

5 CONCLUSIONS

Considering all results we have pointed out the following aspects:

(a) Timoshenko's formula would give good results for evaluation of deformation of hibrid pultruded profile.
(b) A detailed evaluation of G modulus should be more appropriate to give a very good comparison between experimental and numerical results.
(c) For lower values of λ_g the maximum load of GFRP beams is higher than HFRP beams. In fact

the low value of slenderness $\lambda_g = 4$ give prominence to shear stress and the carbon fibers are not completely optimised.
(d) For λ_g value equal 7 and 24, the maximum value of HFRP beams is higher than GFRP beams (respectively 6% and 25%).
(e) The mechanical behaviour of hybrid profile subjected to axial load confirm a not good behaviour of carbon fiber in compression, in fact the risk of instability is easier than in the glass fiber.

ACKNOWLEDGMENTS

The authors are sincerely grateful to Carlo Tedeschi, Mario Celebrin, Italo Tofani, Lorenzo Massaria for their help during the tests.

REFERENCES

Ballinger, C. Structural FRP Composites, *Civ. Engrg., ASCE,* 60 (7), 63–65.
Chambers, E. 1997. ASCE design standard for pultruded FRP structures, *Journal of Composites for Construction,* Vol. 1 No. 1, pp. 26–38, Feb. 1997.
Bank, L.C. 1989. Flexural and shear modul of Full section fiber reinforced plastic (FRP) pultruded beams, J. Test. Eval. 17(1), 40–45, 1989.
Mottram, J.T., Brown, N.D. & Lane, A. The analysis of columns for the design of pultruded frames: Isolated centrally loaded columns, *ECCM, June '98,* Naples, Italy.
Bank, L.C. & Bednarczyk, P.J. A beam theory for thin-walled composite beam, *Composites Sci. Tech.,* 32, 265–277.
Bank, L.C. Shear coefficient for thin-walled composite beams, *Composite Structural,* 8(1), 47–61, 1987.
NORMATIVA ASTM D 638/UNI 5819.
NORMATIVA ASTM D 790/UNI 7219.
Mosallam, A.S. & Bank, L.C. 1992. Short-Term Behaviour of pultruded fiber reinforced plastic frame, *ASCE,* Vol. 118, No. 7, July 1992.
MMFG DESIGN MANUAL, Morrison Molded Fiber Glass Company, Bristol, Virginia, 1990.
FIBERLINE DESIGN MANUAL, for Structural Profiles in Composite Materials, 1995.
STRUCTURAL PLASTIC DESIGN MANUAL, ASCE Manual and reports on Engineering Practice, No. 63, NY, 1984.
Nagaray, V. & Gangarao, H.V.S. 1997. Static behaviour of Pultruded GFRP Beams, *ASCE Journal of Composites for Construction,* Aug. 1997, Vol. 1, No. 3.
Zureick, A., Hahn, L.F. & Bandy, B.J. 1994. Test on Deep I-Shape Pultruded beams, *49th, An. Conf., Composite Institute, The Soc. Of Plastics Industry,* Inc. Feb. 7–9, 1994.
Hollaway, L. 1984. Polymer Composites for Civil and Structural Engineering, *Published by Blackie Acad. & Professional,* Chapmann & Hall, 1984.
Timoshenko, S., Goodier, J.N. "Theory of Elasticity 2, *Mc Graw-Hill Book Company,* 1951.

System-based Vision for Strategic and Creative Design, Bontempi (ed.)
© 2003 Swets & Zeitlinger, Lisse, ISBN 90 5809 599 1

Comparison between BS5400 and EC4 for concrete-filled tubular columns

A.K. Alrodan
Department of Civil Engineering, Mutah Uni, Jordan

ABSTRACT: A wide range of experimental data and associated analysis results are used to examine the applicability of two widely used design codes (i.e. BS5400 and EC4) for calculating the capacity of concrete-filled steel tubular columns (CFSTs). The database consists of the measured results from past studies. The specimens include short and slender CFSTs made with normal- and high-strength steel tubes filled with normal- and high-strength concrete. To gauge the success of the code-based methods, the capacities are computed based on fiber analysis technique. The capacities are also computed through 3D nonlinear finite-element analysis. In addition to inherent differences between BS5400 and EC4, the reported study points to potentially large differences between the capacities as computed from these two methods. However, the capacities from EC4 and detail analytical results are generally closer. Neither design method appears to be appropriate for CFSTs made with high-strength steel tubes. A finite-element method based on full plasticity of the steel tube and material and geometrical nolinearity tends to produce significantly improved results for such cases.

1 INTRODUCTION

Investigations into the behavior of concrete filled tubular columns have been carried out since the beginning of this century. Theoretical and experimental studies were carried out in order to establish the behavior of such columns. Comprehensive lists of these publications are given elsewhere (McDevitt & Viest 1972, and Shakir-Khalil 1988). The interest generated by the work in the composite construction field, and also the economics of using this relatively new type of construction, lead to the development of a number of national standards in the UK, in the USA, in Europe and elsewhere for the design of composite structures.

The use of concrete-filled tubular columns (CFSTs) in high-rise buildings has become more popular in recent years as they provide several advantages over reinforced concrete or steel columns (e.g. elimination of formwork and reinforcement, high strength and stiffness by combining high strength tubes and high strength concrete, etc.). Previous studies have focused on a better understanding of the behavior of CFSTs and examined analytical methods with varying levels of complexity for analysis and design of CFSTs. Aho & Leon (1997) recently compiled a comprehensive database of past experimental studies. Analytical techniques with varying levels of complexity have also been proposed. Closed-form expressions are available

to compute capacity and cross section envelope values (Chen & Atsuta 1976; Grauers 1993; Bradford 1996). On the other side of the spectrum, nonlinear distributed plasticity finite-element models have been successfully used to examine the behavior of CFSTs (Gourley & Hajjar 1994).

The main goal of this paper is to gauge the success of the design method in British Standards BS5400, referred to sometimes as the Bridge Code, and the European Code EC4 methods for computing the capacity of CFSTs, and if necessary, to propose adjustments. For comparison purposes, the results from more detailed analytical techniques are also presented. The first analytical method is based on numerically generated moment–curvature relationships in conjunction with numerical integration methods [e.g., Newmark's method (1943)]. The second analytical method is based on the 3D finite-element model, where material and geometrical nonlinearity are taken into account. The modern commercial finite-element package ABAQUS (HKS 2002) was used to predict the experimental failure loads of all specimens. The reported study revolves around available experimental data from past studies. Short and long CFSTs made with normal- or high-strength steel and concrete are considered. The reported study is focused on square and rectangular CFSTs. Figure 1 shows some of the CFSTs used in practice, and which are covered by design methods in BS5400 and EC4.

2 SUMMARY OF EXPERIMENTAL PROGRAMS

The experimental data from a relatively large number of specimens were considered in this study. Important characteristics of the selected specimens are summarized in Table 1. The test specimens covered CFSTs made with normal-strength tubes (yield strength < 400 MPa) or high-strength tubes (yield strength above 400 MPa) filled with normal concrete (compressive strength < 46 MPa) or high-strength concrete (compressive strength above 46 MPa). The tubes were relatively stocky; the maximum tube width to thickness ratio B/t was 50. Most of the specimens meet the minimum tube thickness t as required by BS5400 and EC4 (i.e., B $\sqrt{(f_y/3E_s)}$) where B = width of tube taken as larger tube dimension; E_s = modulus of elasticity of steel tube; and f_y = steel tube yield strength. The concrete inside the tube could apparently restrain the inward buckles as suggested by Bridge et al (1995), Shakir-Khalil and Zeghiche (1989).

All tubes in the test specimens were cold formed. However, the tubes used by Tomii and Sakino (1979) had been heat treated before pouring the concrete to eliminate residual and cold form stresses. The length of the specimens tested by Knowles and Park (1969, 1970), Bridge (1976), and Shakir-Khalil and Zeghiche (1989) was significant (i.e., these columns are classified as slender). The remaining test specimens, on the other hand, provide data regarding the behavior and capacity of short CFSTs.

3 SUMMARY OF BS5400, EC4, ANALYTICAL, AND FE METHODS

To gauge the success of available design procedures in the Bridge Code BS5400 and the European Code EC4, the capacity of the specimens was computed and compared against the measured values. The capacities were also calculated from a detailed analytical method and FE method. This approach provided another measure for evaluating the design models. In the following sections, a summary of these methods is provided for completeness.

3.1 BS5400 method

The design method in BS5400 covers both axially and eccentrically loaded composite columns, where the latter may be subjected to either equal or unequal end eccentricities. The failure load of an axially loaded column is given as a function of both its squash load, N_u, and K_1, which is obtained from the column buckling curves and is only a function of the slenderness factor, λ, of the column. The carrying capacity of the uniaxially loaded column depends on both N_u and K_1,

on the ultimate moment of resistance M_u of the column cross-section, and also on two more coefficients K_2 and K_3 defined in the standard. These coefficients are in turn dependent on the ratio of the end moment, β, the concrete contribution factor, α_c, the slenderness factor, λ, and the type of steel section used which, in turn, dictates the column buckling curve to be used in the design.

3.2 EC4 method

Two design methods are given in EC4; a general design method, which may be applied to any type of composite column even to those with unsymmetrical cross-sections and simplified method which is valid only for columns with doubly symmetrical sections and with constant sections over the column length.

For the general method, the design has to include the influence of the deformations on the loads (second order theory) as well as the effects caused by the physical non-linearities of the materials structural steel and concrete. It is only possible to meet these requirements in verification by large numerical methods of analysis, which can generally only be performed with FE-computer program. As the simplified method is assumed to be the method which mostly will be taken for the design, the calculated capacity of CFSTs and given in Tables 1 and 2 based on this method.

For composite columns, structural steel according to EC3 (1993) where the elastic-perfectly plastic stress–strain relation is assumed. A uniform stress, as specified in EC2 (1992), is used for concrete. However, the ultimate concrete stress is taken as f_{cd} (f_{cd} = concrete compressive strength) to reflect that concrete experiences a better development of strength due to the complete shelter against the air, and split is prevented (Furlong 1967; Roik & Bergmann 1992).

3.3 Theoretical analysis

The capacity of the column specimens is computed from the moment-curvature response generated by fiber analysis. Compatibility and equilibrium requirements of the fibers are used to establish cross-sectional responses, provided the concrete and steel stress–strain relationships are known. A linear strain distribution through the section was assumed, and slip at the concrete-steel interface was ignored.

The behavior of confined concrete inside steel tubes was represented by a model proposed by Tomii (1991). As seen in Figure 2, the ascending part of the stress–strain curve up to a strain of 0.002 is parabolic, which was represented by an equation proposed by Hognestad (1951). The post-peak response, beyond concrete strain of 0.005, depending on B/t as indicated in Figure 2. A linear interpolation may be used for values of B/t other than those shown. The concrete is assumed to

1436

Table 1. Test variable and evaluation of computed capacities for short columns.

Specimen (1)	Specimen details				Measured/Computed			
	B (mm) (2)	T (mm) (3)	f'_c (MPa) (4)	f_y (MPa) (5)	BS5400 (6)	EC4 (7)	Fiber model (8)	FE model (9)
Furlong (1967)								
F-I-1	127	4.80	44.8	485	0.91	0.97	0.85	0.91
F-I-2	127	4.80	44.8	485	0.97	1.01	0.87	0.94
F-I-3	127	4.80	44.8	485	1.04	1.09	0.74	0.89
F-I-5	127	4.80	44.8	485	1.11	1.08	0.97	0.99
F-II-1	102	2.13	23.4	331	0.96	1.02	0.99	1.02
F-II-2	102	2.13	23.4	331	1.01	1.08	0.91	0.95
F-II-3	102	2.13	23.4	331	1.23	1.25	0.93	0.97
F-II-4	102	2.13	23.4	331	1.18	1.12	0.86	0.93
F-II-5	102	2.13	23.4	331	1.34	1.07	0.85	0.91
F-III-0	102	3.18	28.8	331	1.17	1.00	1.34	1.19
F-III-1	102	3.18	28.8	331	1.13	1.11	1.05	1.00
F-III-2	102	3.18	28.8	331	1.29	1.07	1.19	1.06
F-III-3	102	3.18	28.8	331	1.23	1.10	1.29	1.15
F-III-4	102	3.18	28.8	331	1.27	1.02	1.27	1.14
F-III-5	102	3.18	28.8	331	1.03	0.99	1.14	1.05
F-III-6	102	3.18	28.8	331	1.32	1.04	1.15	1.06
F-III-7	102	3.18	28.8	331	1.29	1.22	1.25	1.13
Average					1.146	1.073	1.038	1.017
σ					0.134	0.03	0.181	0.091
COV					0.117	0.068	0.174	0.089
Tomii & Sakino (1979)								
I-0	100	2.29	24.0	194	1.16	1.02	0.95	0.97
I-1	100	2.29	38.2	194	1.28	1.07	1.04	0.99
I-2	100	2.29	38.2	194	1.12	1.12	1.02	1.01
I-3	100	2.29	38.2	194	1.11	0.96	1.00	1.02
I-5	100	2.29	38.2	194	1.30	1.22	1.05	1.05
I-6	100	2.29	36.7	194	1.41	1.19	0.91	0.95
II-0	100	2.27	21.6	305	1.17	1.09	0.97	0.98
II-1	100	2.27	21.6	305	1.09	0.92	1.02	0.99
II-2	100	2.20	21.6	339	1.12	0.97	0.95	1.01
II-3	100	2.20	21.6	339	1.31	1.16	0.92	1.02
II-4	100	2.22	21.6	289	1.29	1.20	1.01	1.05
II-5	100	2.22	21.6	289	1.53	1.29	1.00	0.99
II-6	100	2.22	21.6	289	1.40	1.31	0.99	1.03
III-0	100	2.98	20.6	289	1.02	0.90	1.02	1.05
III-1	100	2.98	20.6	289	1.20	0.95	1.01	1.02
III-2	100	2.98	20.6	289	1.10	1.00	1.06	1.07
III-3	100	2.99	20.6	288	1.23	1.02	1.01	1.00
III-4	100	2.99	20.6	288	1.26	1.09	0.97	0.99
III-5	100	2.99	20.6	288	1.33	1.19	0.98	1.02
III-6	100	2.99	20.6	288	1.49	1.22	0.98	1.04
IV-0	100	2.25	18.6	284	1.14	0.92	1.11	1.09
IV-1	100	2.25	18.6	284	1.19	0.93	1.00	1.10
IV-2	100	2.25	18.6	284	1.32	1.11	1.00	1.11
IV-3	100	2.25	18.6	285	1.21	1.09	0.99	1.02
IV-4	100	2.25	19.8	285	1.18	1.08	1.01	1.00
IV-5	100	2.25	19.8	285	1.59	1.19	0.99	0.98
IV-6	100	2.26	19.8	288	1.39	1.15	1.02	1.09
Average					1.257	1.087	0.999	1.024

(Continued)

Table 1. *(continued)*

Specimen (1)	Specimen details				Measured/Computed			
	B (mm) (2)	T (mm) (3)	f'_c (MPa) (4)	f_y (MPa) (5)	BS5400 (6)	EC4 (7)	Fiber model (8)	FE model (9)
σ					0.140	0.116	0.040	0.041
COV					0.112	0.107	0.040	0.040
BRI (Fujimoto et al 1995)								
ER4-A-4-4.5	149	4.38	41.1	262	0.93	0.98	0.94	1.07
ER4-A-4-20	149	4.38	41.1	262	1.01	0.99	0.96	1.05
ER4-C-2-6	216	4.38	25.4	262	1.23	1.11	0.80	0.95
ER4-C-2-20	216	4.38	25.4	262	1.29	1.09	0.94	1.03
ER4-C-4-6	216	4.38	41.1	262	1.20	1.07	0.74	0.92
ER4-C-4-10	216	4.38	41.1	262	1.18	1.12	0.85	0.97
ER4-C-4-20	216	4.38	41.1	262	1.07	0.92	0.94	1.00
ER4-C-8-6	216	4.38	80.3	262	1.51	1.21	0.75	0.95
ER4-C-8-10	216	4.38	80.3	262	1.49	1.15	0.88	0.96
ER4-D-4-6	324	4.38	41.1	262	1.09	1.07	0.71	0.89
ER4-D-4-20	324	4.38	41.1	262	0.98	1.00	0.89	0.97
ER6-A-4-4.5	144	6.36	41.1	618	1.32	1.18	0.95	1.06
ER6-A-4-20	144	6.36	41.1	618	1.29	1.16	1.01	1.13
ER6-C-2-6	210	6.36	25.4	618	1.32	1.09	0.90	0.99
ER6-C-4-6	210	6.36	41.1	618	1.38	1.20	0.89	0.96
ER6-C-4-10	210	6.36	41.1	618	1.41	0.93	0.91	1.05
ER6-C-4-30	210	6.36	41.1	618	1.27	0.97	0.89	1.01
ER6-C-8-6	210	6.36	80.3	618	1.62	1.08	0.84	0.96
ER6-C-8-20	210	6.36	80.3	618	1.59	1.12	0.95	1.10
ER6-D-4-10	318	6.36	41.1	618	1.23	0.89	0.72	0.88
ER6-D-4-30	318	6.36	41.1	618	1.42	0.99	0.85	0.98
Average					1.278	1.063	0.872	0.994
σ					0.188	0.092	0.083	0.064
COV					0.147	0.087	0.095	0.064
BCI (Fujimoto et al 1995)								
ER8-A-4-01	120	6.47	40.5	835	1.43	1.13	0.99	1.07
ER8-C-2-04	174	6.47	25.4	835	1.32	1.09	0.95	1.02
ER8-C-2-06	174	6.47	25.4	835	1.51	1.16	0.98	1.06
ER8-C-4-025	174	6.47	40.5	835	1.29	1.04	0.96	1.04
ER8-C-4-04	174	6.47	40.5	835	1.39	1.06	0.96	1.03
ER8-C-4-06	174	6.47	40.5	835	1.36	1.05	0.91	0.98
ER8-C-8-04	174	6.47	77.0	835	1.66	1.12	0.94	1.00
ER8-C-8-06	174	6.47	77.0	835	1.71	1.14	0.94	1.00
ER8-D-4-04	264	6.47	40.5	835	1.32	1.02	0.90	0.98
ER8-D-4-06	264	6.47	40.5	835	1.41	1.10	0.95	1.02
ER8-C-2-025	210	6.36	25.4	618	1.19	0.98	0.94	0.99
Average					1.417	1.081	0.947	1.017
σ					0.149	0.053	0.025	0.029
COV					0.105	0.049	0.027	0.028

reach a constant level of stress after strain of 0.015. A cubic equation proposed by Saenz (1964) was used to model the post-yield behavior of steel. Zhang and Shahrooz (1997), show that the steel tube yield strength can influence the behavior of concrete confined by high-strength tubes. This factor was not included in these studies.

The maximum strength of a slender column can be obtained by well known deflection methods, in which the failure load may be defined as the peak of the load-deflection curve. Due to nonlinearity, the direct solutions to the differential equation of deflection are very complicated and impractical. The numerically generated moment-curvature relationship along the

1438

Table 2. Test variable and evaluation of computed capacities for slender columns.

| Specimen (1) | Specimen details | | | | | | | Measured/Computed | | | |
	B (mm) (2)	H (mm) (3)	t (mm) (4)	L (mm) (5)	e_o (mm) (6)	f'_c (MPa) (7)	f_y (MPa) (8)	BS5400 (9)	EC4 (10)	Analysis model (11)	FE model (12)
Knowles and Park (1967, 1970)											
K-1	76.2	76.2	3.33	813	7.6	41.4	324	1.19	1.07	0.92	1.05
K-2	76.2	76.2	3.33	1422	7.6	41.4	324	1.25	1.11	0.86	0.99
K-3	76.2	76.2	3.33	813	25.4	41.4	324	1.10	1.03	0.86	0.98
K-4	76.2	76.2	3.33	1422	25.4	41.4	324	1.16	1.06	0.74	0.93
Average								1.175	1.068	0.845	0.988
σ								0.054	0.029	0.065	0.042
COV								0.046	0.027	0.077	0.043
Bridge (1976)											
B-1	204	204	9.96	2130	38	30.2	291	1.03	0.97	0.95	1.01
B-2	203	203	10.0	2130	38	34.5	313	1.09	0.96	0.98	1.05
B-3	203	203	9.88	2130	38	33.1	317	1.05	1.00	0.98	1.04
B-4	203	203	10.0	3050	38	37.8	319	1.16	1.03	0.96	1.03
B-5	203	203	9.78	3050	64	32.1	317	1.19	1.05	1.02	1.08
B-6	153	152	6.48	3050	38	35.0	254	1.10	1.04	0.89	0.95
B-7	153	152	6.48	3050	64	35.0	254	1.21	1.11	0.86	0.93
Average								1.118	1.023	0.949	1.013
σ								0.064	0.048	0.051	0.050
COV								0.057	0.047	0.054	0.049
Shakir-Khalil and Zeghiche (1989)											
S-1	80	120	5.00	3210	24	34.0	386	1.19	1.23	1.03	1.07
S-2	80	120	5.00	3210	60	34.0	385	0.97	1.03	0.90	1.01
S-3	120	80	5.00	2940	16	37.4	385	0.97	1.04	0.82	0.97
S-4	120	80	5.00	2940	40	36.5	343	0.73	0.94	0.99	1.05
S-5	120	80	5.00	2939	8	35.7	357	1.02	1.12	1.08	1.13
Average								0.976	1.072	0.964	1.046
σ								0.147	0.097	0.093	0.054
COV								0.151	0.091	0.096	0.051

column in conjunction with Newmark's method of numerical integration (1943), have been found to be most convenient for this purpose. The second-order moments were taken into account by an iterative procedure. A computer program has been developed by the author based on this method.

3.4 Finite Element analysis

The Finite Element method (FE) is a useful tool for analyzing problems with complex geometry, material properties, and boundary conditions. This technique can produce comprehensive and reliable results if the method is used correctly. It is much cheaper than full-scale experimental testing. An accurate model to simulate the actual structures is necessary to accomplish a successful analysis.

In this study, ABAQUS version 6.2 (2002) was used. ABAQUS is a general-purpose nonlinear finite-element analysis program, which is used for stress, heat transfer and other types of analysis in mechanical and structural engineering. The pre- and post-processing work was done by ABAQUS/CAE version 6.2 which is additional module written to enhance ABAQUS. ABAQUS/CAE is a graphical user-interface program that allows a user to execute a FE analysis process from start to finish. The FE model can be viewed and checked interactively and the results (stresses, displacements, etc.) can be visualized graphically.

3.4.1 Finite Element model
In modeling a CFST columns, a properly graded mesh is essential. In particular, the sizes of the elements in the vicinity of the column mid-span. Moreover

by making use of symmetry in loading and geometry, only a half of the column needs to be modelled (Figs. 3 and 4). (no imperfection of boundary conditions or loading were investigated). In the FE model the concrete and steel were modeled using first order reduced integration brick-element (Element type C3D8R in ABAQUS).

The boundary conditions have to be applied correctly for the nodes, laying on the planes of symmetry to reflect the actual behavior. The nodal displacements perpendicular to the plane of symmetry are restrained while the two remaining transitional degrees of freedom are free; the nodal rotation perpendicular to the plane of symmetry is free while the two remaining rotational degrees of freedom are restrained. Furthermore, at the support, the nodal displacement in the Y-direction is restrained while the two remaining transitional degrees of freedom are free.

For the contact interaction between the outer surface of the core concrete and the inside surface of the hollow steel tube, deformable to deformable contact surfaces were used. This type of contact interaction has been developed by Hibbit et al. (2002) and implementation in ABAQUS. This can model contact between two deformable bodies in all dimensions and can model small or large sliding as well as separation. The two contacting surfaces can carry shear stress up to a certain magnitude across their interface before they start sliding relative to one another, this state known as sticking. The friction model adopted in this analysis assumes that the constant friction coefficient "μ" is the same in all directions (isotropic friction).

The average stress–strain curves obtained from the linear material tests were used to model the steel tubes. The steel is assumed to follow an elastic-perfectly plastic stress–strain relationship. For concrete, the compression response may be represented by a quadratic loading branch up to the ultimate strength and a linear softening branch down to the ultimate strain Aribert et al. (1995). The local energy release as the concrete begins to crack, are of a major importance in determining the response of such a structure between its initial recoverable deformation and collapse. The program offers an indirect technique for modelling these effects, known in literature as "tension stiffening".

The stress–strain curve of post-cracking concrete was determined through the sensitivity study conducted by Al-Rodan (2003). That is, various tension-stiffening curves were compared under full retension model. The tension stiffening was assumed to be linearly decreased and was reduced to zero at strain values of 10, 20, 30, 40 and 50 times of concrete crack strain, considering the simplicity of global analysis. The 10 times model was the best agreement with the laboratory tests data. However this 10 times model induced unstable behavior above failure load so that 1% crack stress was retained even after 10 times of the crack strain.

The reduction of the shear modulus after concrete cracking was determined as follows. The full shear retension model was compared with the model in which shear stiffness under cracking condition was reduced linearly to zero at 10 times the cracking strain. Both models were considered up to failure, where concrete cracking in tension extended for the whole region. The analysis was almost the same so that the full shear retention model was adopted because of better convergence of calculation. Figures 5 and 6 show the strain distributions in the concrete and steel tube respectively from the FE model.

4 DISCUSSION OF RESULTS

The capacities of the specimens calculated by the methods described in BS5400 and EC4 codes are summarized in Tables 1 and 2. For comparison purposes, the capacities are also computed based on relatively detailed analytical methods and FE method as described and verified previously. The results are presented and discussed separately for the short and slender columns.

4.1 Short columns

The ratio of the measured to computed capacity is summarized in Table 1. The EC4 method provides a very good estimate of the measured strength of the specimens tested by Furlong (1967). The average ratio of the measured to computed moment is 1.073 with a coefficient of variation of 0.068. For the same specimens, the BS5400 procedure is not as successful, and it tends to under estimate the capacity. The average ratio of the measured to computed moment is 1.146 with a coefficient of variation (COV) of 0.117. The capacities as computed by fiber analysis match the measured results very closely. The average value of the measured to computed moment is 1.038 with a COV of 0.174. The capacity of CFSTs is computed by a detailed three-dimensional finite element method as described previously. As seen from the Table 1 the finite element method (FE) gives very good estimate of the measured strength of the specimens tested by Furlong. The largest ratio of the measured to computed moment is 1.11, and the average value of the ratio is 1.012 with a COV of 0.077. The analytical values from the four methods are lower than the experimental data for specimens F-III-0 through F-III-7 (with B/t =32). Other researchers using nonlinear distributed plasticity finite-element models (Gourley & Hajjar 1994) have also observed a similar trend. The reported material properties for the steel tubes in specimens F-III-0 to F-III-7 were apparently nominal values and not the actual properties.

The measured to computed capacities for specimens tested by Tomii & Sakino are shown in Table 1. The table shows the EC4 methods is successful in calculating the capacities of CFSTs. On the other hand BS5400 over estimated the capacities, the average values of the ratio of the measured to computed moment are 1.087 and 1.257 for EC4 & BS5400 methods respectively, with corresponding COV of 0.092 & 0.112. Except for specimens I-0 to I-6, the capacities from the EC4 method to be within a reasonable range of the experimental data. The BS5400 calculated capacities are generally less than the measured ones.

The capacities as computed by fiber analysis match the measured results very closely. The largest ratio of the measured to computed moment is 1.11, and the average value of this ratio is 0.999 with a COV of 0.04. This observation should be expected as Tomii calibrated the concrete confined model used herein based on the test data obtained by Tomii & Sakino (1979). The FE model offers improved results, although the computed capacities tend to be larger than the measured values for some of the specimens. The largest ratio of the measured to computed moment is 1.09, and the average value is 1.022 with a COV of 0.037.

For the specimens tested at BRI, the results from the EC4 method are reasonable and, generally are similar to those obtained from fiber and FE analysis. The average ratio of the measured to computed capacities is 1.063 with a COV of 0.087. The BS5400 procedure, on the other hand, tends to consistently and appreciably under estimated the capacity as evident from the large ratios of the measured to computed capacity reported in Table 1. These trends are not impacted by the problems associated with the test set up in a few of the specimens as indicated in the table. The results reported by others support the tabulated values (Zhang and Shahrooz 1999). Note that for the specimens in the ER6 series, the measured capacities are generally lower than their count parts computed by the BS5400 method. For these specimens, high-strength steel tubes ($f_y = 618$ MPa) had been used, and the results suggest that the BS5400 method need to be adjusted. In an effort to explain this trend, the behavior of $120 \times 120 \times 5$RHS steel tube filled with a 30 MPa concrete was evaluated as an example. Both normal-strength (300 MPa) and high-strength (618 MPa) tubes were considered. The interaction curves computed by BS5400 method are shown in Figure 7. At a maximum compressive concrete strain of 0.0035 a large portion of normal-strength tube yields, whereas only the outmost fibers in a high-strength tube are expected to yield. This observation is clearly seen in Figure 7. It can be said that the BS5400 method can reliably be used for computing capacity of CFSTs as long as normal-strength steel tubes are used. However, BS5400 method tends to significantly underestimate the capacity of CFSTs made with high strength tubes and is not appropriate.

4.2 Slender columns

Table 2 shows the summary of the measured to computed capacity for the slender CFSTs. The table shows that BS5400 tends to underestimate the capacity, most notably those specimens tested by Knowls and Park, and Bridge. The results from the EC4 method appeared to be within a reasonable range of the experimental data, and generally, underestimate the capacity, which gives a safe design. The carrying capacity of the CFSTs was also computed from the details second order analysis and FE methods described earlier. The influence of the local buckling was ignored as the wall thickness of the steel tubes in the tested specimens were well above the minimum values required by BS5400 and EC4 design codes. The detail analysis method tends to overestimate the reported values despite the strain hardening in the steel stress–strain relationship was ignored. On the other hand, the FE method tends to reasonably estimate the carrying capacity of CFSTs. The computed capacities by the EC4 method are fairly closed to those computed from the FE method.

In view of the aforementioned results, it appears that the EC4 design procedure can reasonably provide a simple and reliable estimate of the capacity of slender CFSTs.

5 CONCLUSIONS

The success of BS5400 and EC4 standards techniques for computing the capacity of short and slender CFSTs was examined by using available experimental data. The capacity was also calculated by reasonably detailed analytical techniques and FE method. The experimental database included specimens made with normal- and high-strength concrete.

It can be concluded that, the capacities of CFSTs computed by BS5400 and EC4 standard methods can vary significantly. However, the results from the EC4 methods tend to match those from the FE method for short CFSTs using normal-strength steel tubes. Either method is reasonable for slender CFSTs made with normal strength steel tubes. Neither BS5400 nor EC4 method is applicable for cases in which-high-strength steel tubes are used.

The fiber analysis and numerical integration of moment-curvature relationship over the length in conjunction with iterative second-order analysis, can adequately estimate the capacity of short and slender CFSTs. On the other hand FE method proved to be highly successful of capturing the behavior and carrying capacity of such composite columns. Such analytical methods offer a viable alternative to expensive experimental work.

ACKNOWLEDGEMENTS

The author is grateful to Drs Michel Anderson, Russel Bridge, Richard Furlong, Jerry Hajjar and Shakir-Khalil for their inputs, suggestions, and interesting discussions. The author gratefully acknowledge the supercomputing resources supports provided by the Deanship of Engineering, Mutah Univeristy.

REFERENCES

Aho, M.F. & Leon, R.T. 1997. A database for encased and concrete-filled columns. *Rep. No. 97-01*, School of Civil & Envir. Engrg., Georgia Institute of Technology, Atlanta.

Al-Rodan, A. 2003. FE analysis of the flexural behavior of rectangular tubular sections filled with high-strength concrete, accepted for publication in the *Emirates Journal for Engineering Research, UAE University*.

Aribert, J., Hua, X. & Ragneau, E. 1995. Theoretical investigation of moment redistribution in composite continuous beams of different classes. *Engineering Foundation Conference, Composite Construction III*, Irsee, Germany, Jun 1995, pp. 125–140.

Bradford, M.A. 1996. Design strength of slender concrete-filled rectangular steel tubes. *ACI Struct. J.*, 93(2), 229–235.

Bridge, R.Q. 1976. "Concrete filled-steel tubular columns" Civil Engrg. Trans., Sydney, Australia, CE18, 127–133.

Bridge, P.Q., O'Shea, M.D., Gardner, A.P., Grigson, R. & Tyrell, J. 1995. Local buckling of square thin-walled steel tubes with concrete infill. *Proc., Int. Conf. On Strut. Stability and Design*, Sydney, Australia, 307–314.

BS5400, Part 5; 1979. Concrete and Composite Bridges; Code of Practice for Design of Composite Bridges. *British Standards Institution*.

Chen, W.F. & Atsuta, F. 1976. Theory of beam-columns, Vol. 1: In-plane behavior and design, McGraw-Hill, New York.

DD ENV 1992-1-1 Eurocode 2, Part 1, 1992. Design of concrete structures, General rule and rules for buildings. *British Standards Institution*, London.

DD ENV 1993-1-1 Eurocode 3, Part 1, 1993. Design of steel structures, General rule and rules for buildings. *British Standards Institution*, London.

DD ENV 1994-1-1 Eurocode 4, Part 1, 1994. Design of composite steel and concrete structures, General rule and rules for buildings. *British Standards Institution*, London.

Furlong, R.W. 1967. Strength of steel encased concrete beam columns. *Journal of the Structural Division*, Vol. 93, No. ST5, pp. 113–124.

Gourley, B.C. & Hajjar, J.F. 1994. Cyclic nonlinear analysis of three-dimensional concrete-filled steel tube beam-column and composite frames. *Rep. No. ST-94-3* University of Minnesota, Dept. of Civil Engineering, Minn.

Grauers, M. 1993. Composite columns of hollow steel sections filled with high strength concrete. Pupl. 93:2, Chalmers University of Technology, Goteborg, Sweden.

Hibbitt, Karlson & Sorenson. 2002. ABAQUS users manual: Version 6.21, Part 1 & 2.

Knowles, R.B. & Park, R. 1969. Strength of concrete filled tubular columns. *J. Structural Engineering*, ASCE, 95(12), 2565–2587.

Knowles, R.B. & Park, R. 1970. Axial load design for concrete filled steel tubes. *J. of Structural Engineering*, ASCE, 96(10), 2125–2153.

Newmark, N.M. 1943. Numerical procedure for computing deflection, moment & buckling loads. *Trans. ASCE*, 108, 1161–1188.

Saenz, L.P. 1964. Equation for the stress-strain curve of concrete. *ACI J. Proc.*, 61 (22), 1229–1235.

Shakir-Khalil, H. & Zeghiche, J. 1989. Experimental behavior of concrete filled rolled rectangular hollow-section columns. *The Structural Engineers*, London, 67(9), 346–353.

Tomii, M. 1991. Ductile and strong columns composed of steel tube, infilled concrete and longitudinal bars. *Proceeding 3rd Int. Conf. On Steel-concrete Composite Structures*, Fukuokda, Japan, 39–66.

Tomii, M. & Sakino, K. 1979. Experimental studies on ultimate moment of concrete filled square steel tubular beam-columns. *Trans Arch. Inst. Of Japan*, 275 (January), Tokyo, Japan, 55063.

Zhang, W. & Shahrooz, B.M. 1997. Analytical and experimental studies into behavior of concrete-filled tubular columns. *Report No. UC-CII 97/01*, Cincinnati Infrastructure Institute, Cincinnati.

System-based Vision for Strategic and Creative Design, Bontempi (ed.)
© 2003 Swets & Zeitlinger, Lisse, ISBN 90 5809 599 1

A cohesive model for fiber-reinforced composites

A.P. Fantilli & P. Vallini
Politecnico di Torino, Torino, Italy

ABSTRACT: A theoretical approach for the post cracking mechanism of fiber-reinforced concrete is proposed. In this model, considered as an extension and an improvement of the Hillerborg's one, the relationship between the stress and the crack opening displacement is defined from the cohesive law of the matrix, the bond-slip relationship between fiber and matrix and the stress–strain laws of both the materials. Moreover, the random inclination and the asymmetrical position of the fiber with respect to the crack surface, the aspect ratio and the percentage of the fibers are also included in the definition of the fictitious crack model for fiber-reinforced composites. In the case of steel fibers in a cementitious matrix, the proposed model furnishes acceptable results compared to the experimental measurements.

1 INTRODUCTION

Fibers are usually added to mortar and concrete in order to obtain a more ductile final product with a higher strength. For these reasons, fiber-reinforced cementitious composites are often called "pseudoductile" materials (Nanni 1991), in contrast to the quasi-brittle behaviour of ordinary concretes and mortars. The pseudo-ductility of fiber-reinforced cementitious composites (FRC) can be measured simple through experimental tensile tests. In particular, from tensile tests, it is possible to compare the structural response of the plain matrix with the one of composites obtained with different amount of fibers (Balaguru & Shah 1992). Sometimes, the so-called pseudo-plasticity and the strength of FRC elements are both evaluated with bending tests. Anyway, there is an evident similarity between the load P-midspan deflection η diagram of a beam in bending and a load P-elongation δ curve obtained from a tensile test (Fig.1a). As is shown in Figure 1b, after the linear branch of the first loading stage, when the material behaves in a linear elastic manner, different post-peak curves can be obtained in specimens with different amount of fibers. In particular, in the pre-peak phase, linear stress and strain profiles can be assumed in each cross-section (Stage I). When the maximum tensile strength is reached, a crack develops in the composite material. The external load, initially carried by the matrix, is progressively transferred to fibers by the increase of the crack width. There is not a great difference between the strength of beams made of plain concrete or FRC materials with a low

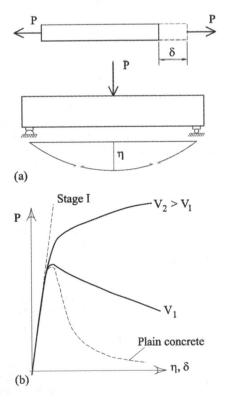

Figure 1. Behaviour of FRC structures: (a) tensile and bending tests; (b) load-displacement diagrams.

percentage of fiber (V_1 in Fig. 1b), even if an increase of ductility is clearly evident in the last case. Composites with a higher volume of added fibers ($V_2 > V_1$ in Fig. 1b) show an increase of the slope of the post-peak branch and sometimes even a hardening stage due to a widespread cracking. The analogy between the P–η and the P–δ diagrams, obtained from the experimental tests of Figure 1a in FRC elements, highlights the existence of a deep relationship between the pure tensile behaviour and the structural response with a strain gradient. As is well known, plain concrete beams in bending display an increase of tensile strength (Fantilli & Vallini 2001), which is usually called flexural tensile strength. In the case of fiber-reinforced cementitious elements, the flexural tensile strength can be five times higher than the direct tensile strength (Maalej & Li 1994).

The relationship between tensile strength and structural response of FRC elements is also evident in the post cracking shear resistance of the FRC beams tested without stirrups (Casanova et al. 1997). In fact, the fibers reduce the crack growth along the tensile stress lines and increase the ductility before complete failure, as is clearly shown by the torsion tests performed on steel fiber-reinforced concrete elements (Nanni 1990). Moreover, Swamy (2000) points out the capability of using both fibers and steel rebar in the concrete structures. Compared to ordinary reinforced concrete, bars and fibers develop their strength at different stages of external actions, so an increase of the cracking load and a higher tension stiffening can be obtained. In these situations, structures can improve their performances in the serviceability states, both under instantaneous and long term actions (Tan et al. 1994).

To reproduce theoretically the structural response of FRC elements subjected to different loads, it seems necessary to model adequately their behaviour in the case of simply uniaxial tensile or compressive stresses (Casanova et al. 1994). In case of uncracked fiber-reinforced concrete elements, the stress σ–strain ε relationship in compression can be obtained from the Sargin's diagram proposed in the CEB-FIP model code (1993) for normal strength concrete (NSC) and high strength concrete (HSC). Similarly, the linear σ–ε relationship of a FRC elements in tension is simply defined by the elastic modulus E_c and the tensile strength f_{ct}. On the contrary, it is more difficult to define the mechanical response of cracked FRC elements in tension. According to Hillerborg (1980), to apply the fictitious crack model (Hillerborg et al. 1976) to fiber-reinforced composites it is not so easy as for plain concrete, due to the complexity of the fiber-matrix interaction. As a matter of fact, in many of the cohesive models proposed (see Hu & Day 2000 for a review) several simplifications and hypotheses are introduced, so they can only give a qualitative description of physical evidences. In particular, in the

model proposed by Hillerborg (1980), fibres have the same length and diameter and are perpendicular to the crack surfaces. Moreover, in this approach a rigid-plastic relationship is assumed for the bond slip relationship, thus the bond stress remains constant independently of the slip values.

In order to obtain a more reliable stress σ–crack opening displacement w relationship for FRC materials, it seems necessary to improve and extend the Hillerborg's model by introducing, for example, a more appropriate bond slip (τ-s) relationship. In particular, in a huge number of experimental pullout tests (see Bartos 1981 for a review), a nonlinear relationship between the mean values of bond stresses and slips clearly appears. Generally, in a test conducted on smooth and straight steel fibres, the pullout force reaches the peak for small slips, before decreasing and approaching an asymptotic value (Balaguru & Shah 1992). Also, the random distribution of fibres in the matrix should be computed for defining the fictitious crack model for fiber-reinforced composites. By taking into account the real position of the fibers, it is possible to evaluate the number of fibers that cross the crack and their effective orientation with respect to crack surface.

2 A ONE-DIMENSIONAL IDEAL MODEL

Starting from the results of Hillerborg (1980), a new approach is here proposed to better define the ficti- tious crack model of fiber-reinforced materials. In particular, with reference to Figure 2a, a prism of cementitious matrix, whose length and cross-section are respectively L and A_s, is extracted from a FRC volume. The tensile stresses σ, applied uniformly over the end sections of the prism, produce a crack in this element. Both the bridging tensile stresses on crack surfaces $\sigma_c(w)$ and the fiber, positioned across the crack, prevent the growth of crack width w. The relationship between the stresses σ–w is the fictitious crack model of the FRC (Fig. 2b). In a first approach, the fiber length l is assumed to be perpendicular to the crack, which divides l into two asymmetric pieces of length $0.5\,l\,(2-\chi)$ and $0.5\,l\,\chi$ respectively (Fig. 2a). The σ–w diagram is a function of these lengths (defined by the coefficient χ): for a given value of crack width w, stresses decreases with decreasing in χ ($0 < \chi < 1$). When $\chi = 0$ the fiber bridging effect vanishes and the cohesive model for FRC corresponds to the σ–w relationship for plain concrete elements (Fig. 2b).

The proposed theoretical approach requires the constitutive relationship of fibers and matrix, the cohesive σ–w diagram of plain matrix and the bond–slip relationship between the materials (Fantilli et al. 1998). In particular, neglecting concrete strains and with the hypotheses of linear behaviour of fibers and

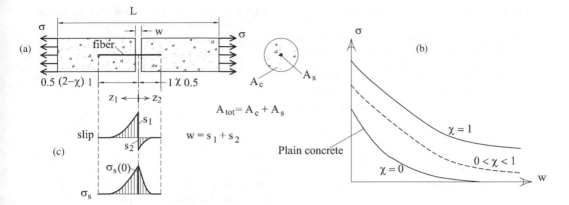

Figure 2. Proposed ideal model: (a) geometrical dimensions; (b) σ–w diagrams; (c) σ_s and s distributions along the fiber.

concrete and linear strain profile in each cross-section of the element, the proposed approach can be considered as an Interface Cohesive Model (Shah & Ouyang 1991). The problem consists in solving a system composed by an equilibrium equation of a portion of the bare fiber (Eq. 1) and a compatibility equation between matrix and fiber at the interface (Eq. 2):

$$\frac{d\varepsilon_s}{dz} = \frac{4 \cdot \tau}{E_s \cdot \Phi} \tag{1}$$

$$\frac{ds}{dz} = \varepsilon_s \tag{2}$$

where E_s = Young modulus of the fiber; ε_s = fiber strains; z = distance from the crack (z_1 or z_2 in Fig. 2c); ϕ = fiber diameter; τ = bond stresses; and s = slips between fiber and matrix. If the bond-slip function is known, Equations 1–2 can be rearranged in the following nonlinear second order differential equation:

$$\frac{d^2 s}{dz^2} = \frac{4 \cdot \tau(s)}{E_s \cdot \Phi} \tag{3}$$

In order to reproduce the behaviour of the fiber in Figure 2a, the adopted numerical procedure consists in evaluating of fiber stresses σ_s and the corresponding values of s (Fig. 2c) obtained for imposed values of crack width w and χ. In particular:

1. Select w and χ.
2. Assume a value for s_1.
3. Assume a value for $\sigma_s(0)$.
4. Integrate Equation 3 within the domain [0, 0.5 l (2-χ)].

5. If in any internal section of the domain [0 < z_1 < 0.5 l (2-χ)], the condition $\sigma_s = 0$ is satisfied, the slip in this section must be equal to zero. If this situation is verified, it is possible to obtain the distributions of slips s (z_1) and steel stresses σ_s (z_1) along the fiber (Fig. 2c). Otherwise, the procedure goes back to step (4) with a new trial value of $\sigma_s(0)$.
6. If in the end of the element ($z_1 = 0.5$ l (2-χ)), the condition $\sigma_s = 0$ is satisfied, it is possible to obtain the distributions of slips s (z_1) and steel stresses σ_s (z_1) along the fiber (Fig. 2c). Otherwise, the procedure goes back to step (4) with a new trial value of $\sigma_s(0)$.
7. Compute $s_2 = w - s_1$ ($\sigma_s(0)$ is known).
8. Integrate Equations 3 within the domain [0, 0.5 l χ].
9. If in any internal section of the domain [0 < z_2 < 0.5 l χ], the condition $\sigma_s = 0$ is satisfied, the slip in this section must be equal to zero. If this situation is verified, it is possible to obtain the distributions of slips s (z_2) and steel stresses σ_s (z_2) along the fiber (Fig. 2c). Otherwise, the procedure goes back to step (3) with a new trial value of s_1.
10. If in the end of the element ($z_2 = 0.5$ l χ), the condition $\sigma_s = 0$ is satisfied, it is possible to obtain the distributions of slips s (z_2) and steel stresses σ_s (z_2) along the fiber (Fig. 2c). Otherwise, the procedure goes back to step (3) with a new trial value of s_1.

For a couple of χ and w, it is possible to obtain the values of $\sigma_s(0)$ and $\sigma_c(w)$ from the numerical procedure and from the fictitious crack model of the plain matrix respectively, and consequently the value of σ:

$$\sigma = \frac{\sigma_s(0) \cdot A_s + \sigma_c(w) \cdot A_c}{A_{tot}} \tag{4}$$

where A_s = cross-section area of the fiber (Fig. 2a); and A_c = cross-section area of the matrix in the crack.

3 CONSTITUTIVE RELATIONSHIPS

Only if the materials and their interaction are correctly defined can the previous numerical procedure be activated. In this paper, both for the fiber and the uncracked matrix the linear elastic stress–strain relationship is assumed. When the first crack appears, the stress–crack opening displacement diagram, socalled fictitious crack model (Hillerbog et al. 1976), replaces the σ–ε relationship of the matrix. For cementitious composites, the Hordijk (1991) model can be assumed. In this model, called continuous function model (CFM), the envelope curve is defined by the following exponential formula (Fig. 3):

$$\frac{\sigma_c}{f_{ct}} = \left\{ 1 + \left(c_1 \frac{w}{w_c} \right)^3 \right\} exp\left(-\frac{c_2 w}{w_c} \right) -\frac{w}{w_c}(1 + c_1^3) exp-(c_2) \quad (5)$$

where f_{ct} = tensile strength of the matrix; and w_c = maximum crack width with non zero cohesive stresses. For normal strength concrete, the coefficients c_1 and c_2 in the Equation 5 are assumed to be equal to 3 and 6.93 respectively, while the value of wc is related to the values of f_{ct} and of the fracture energy G_F ($w_c = 5.14\ G_F/f_{ct}$).

The bond–slip mechanism between fibers and matrix cannot be clearly defined like the cohesive model for cracked matrix, because there is not a general relationship for all the materials. Anyway, for smooth steel fibers in cementitious matrix, the model proposed by Fantilli & Vallini (2003) can be adopted. It consists of an improvement and an extension of the classical model proposed by CEB-FIP model code (1993) for smooth steel reinforcing bars. In particular, both for bars and fibers, the post peak softening is introduced in conjunction with the size effect produced by the diameter on the model parameters. The ascending branch (Eq. 6)

and the post-peak stage (Eq. 7) of the proposed bond–slip relationship are depicted in Figure 4a:

$$\tau = \tau_{max} \cdot \left(\frac{s}{s_1} \right)^\alpha \qquad s \leq s_1 \quad (6)$$

$$\tau = \tau_{fin} + (\tau_{max} - \tau_{fin}) e^{k(s_1 - s)} \qquad s > s_1 \quad (7)$$

where the parameters s_1, α, τ_{min} and k of the Equations 6, 7 are defined in Figure 4b as a function of bond conditions, the type of smooth bar (hot rolled or cold drawn) and the concrete compressive strength f_c. The maximum value of bond stress must be considered as a function of the bar diameter through the Bazant's size effect law (Bazant et al. 1995) for hot rolled bars:

$$\tau_{max} = \frac{A}{\sqrt{B + \Phi}} \sqrt{f_c} \quad (8)$$

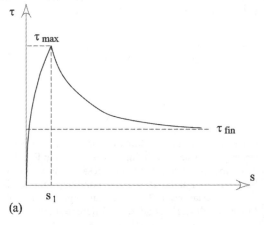

(a)

	Cold drawn wire	Hot rolled bars
	Good Bond conditions	Good Bond conditions
s_1	0.01 mm	0.1 mm
α	0.5	0.5
k	2 / mm	2 / mm
τ_{fin}	$0.067\ \sqrt{f_c}$	$0.1\ \sqrt{f_c}$

(b)

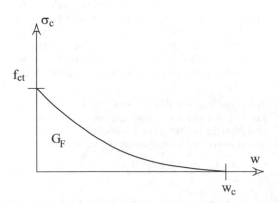

Figure 3. Fictitious crack model (Hordijk 1991).

and for cold drawn wires:

$$\tau_{max} = \left(\frac{A}{\sqrt{B+\Phi}} - 0.2\right)\sqrt{f_c} \qquad (9)$$

where $A = 1.572$ and $B = 12.50$ if τ_{max} and f_c are measured in MPa.

4 FROM THE IDEAL MODEL TO THE REAL MODEL

Due to the random distribution of the fibers within the cementitious composite, the values of A_s, A_c, and $\sigma_s(0)$ in the Equation 4 are still unknown. Therefore, to compute σ, it is necessary to define the value of χ (i.e. the position of the fiber with respect to the cracked cross-section) from which the value of $\sigma_s(0)$ depends. For a modest value of the crack width w, the transmission length is small as compared to the fiber length l and $\sigma_s(0)$ is not affected by χ. On the contrary, with the increase of crack width, there is also an increase of transmission length and a reduction of $\sigma_s(0)$ values. For these reasons, $\sigma_s(0)$ should be affected by both w and χ, and can be calculated through the following equation:

$$\sigma_s(0) = \sigma_s(\chi = 1) \cdot \beta(w) \qquad (10)$$

where $\sigma_s(\chi = 1)$ = steel stress in the cracked cross-section evaluated under the hypothesis of maximum effectiveness of the fiber; and β = reduction factor considered as a function of crack width w.

Equation 11 represents a simple way to evaluate β:

$$\beta(w) = \int_0^1 \frac{\sigma_s(\chi)}{\sigma_s(\chi = 1)} d\chi \qquad (11)$$

Only if the number N of the fibers that cross the cracked section of area A_{tot} (Fig. 2a) is known, can the values of A_s and A_c of Equation 4 be evaluated. Within the volume of a FRC element, a spherical volume of diameter l is taken into consideration, and it is supposed that all the N fibers included in this volume have the same length l. Under these hypotheses, the value of N can be evaluate by the following equation:

$$N = \frac{2 \cdot \gamma \cdot l}{3 \cdot p} \qquad (12)$$

where γ = fibers weight referred to a unit of FRC volume; and p = the weight of a single fiber of length l. If all the fibers cross the circular equatorial surface

of the considered volume, it is possible to evaluate the fibers area in the cracked cross-section:

$$A_s = \frac{\gamma \cdot A_{tot} \cdot \pi \cdot \Phi^2 \cdot l}{6 \cdot p} \qquad (13)$$

and consequently the matrix area of the same section

$$A_c = A_{tot} \cdot \left(1 - \frac{\gamma \cdot \pi \cdot \Phi^2 \cdot l}{6 \cdot p}\right) \qquad (14)$$

By substituting Equations 13-14 into Equation 4 it is possible to obtain:

$$\sigma = \sigma_s(0) \cdot \frac{\gamma \cdot \pi \cdot \Phi^2 \cdot l}{6 \cdot p} + \sigma_c(w) \cdot \left(1 - \frac{\gamma \cdot \pi \cdot \Phi^2 \cdot l}{6 \cdot p}\right) \qquad (15)$$

In the Equation 15 the crack surface A_{tot} does not compare, so the stresses σ are only functions of the values of $\sigma_s(0)$ obtained from the ideal model, the value of crack width w, the geometry and total amount of fibers.

Also the bond slip relationship defined in the previous paragraph regards fibers perpendicular to the crack surface. In case of a fiber that crosses the crack diagonally, the bond–slip relationship shows a constant value of the bond stress after the peak, independently from the slip (Naaman & Shah 1976). For these reasons, in absence of other experimental and numerical analyses aimed at a better definition of τ-s, Equation 7 can be substituted by the following expression:

$$\tau = \tau_{max} \qquad s \geq s_1 \qquad (16)$$

5 COMPARISON BETWEEN THE PROPOSED MODEL AND THE EXPERIMENTAL RESULTS

The fictitious crack model of a steel fiber-reinforced composite material, evaluated with the proposed numerical procedure, can be compared with the experimental measurements of Plizzari et al. (2000). The tests were performed on the FRC cylinders shown in Figure 5. The material properties and the experimental equipment are indicated in Table 1. Smooth cold drawn steel fibers with an aspect ratio of 60 have been used as fiber-reinforcement. Before testing the FRC specimens, the Authors tested the cementitious matrix in order to obtain its tensile strength f_{ct} and the fracture energy G_F. Both these parameters are necessary to define the cohesive model of Figure 3. In Figure 6 the good agreement between the numerical results obtained with the model and the experimental ones is clearly evident.

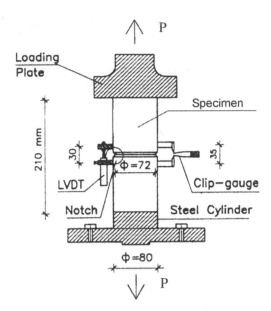

Figure 5. Geometrical characteristics of specimens and instrumentation adopted in the Plizzari et al. (2000) experimental tests.

Table 1. Mechanical properties of materials in the experimental tests of Plizzari et al. (2000).

Fibers	Concrete matrix
$\gamma = 30\,\text{kg/m}^3$	$f_c = 45.4\,\text{MPa}$
$\Phi = 0.5\,\text{mm}$	$f_{ct} = 4.01\,\text{MPa}$
$E_s = 210000\,\text{MPa}$	$E_c = 31600\,\text{MPa}$
$1 = 30\,\text{mm}$	$G_F = 151\,\text{J/m}^2$

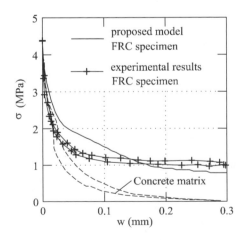

Figure 6. Comparison between numerical results obtained with the proposed model and the experimental results of Plizzari et al. (2000).

The same Figure also shows the higher fracture energy of FRC materials with respect to plain concrete matrix.

6 CONCLUSIONS

In this paper, starting from a suitable bond–slip relationship between fibers and matrix, a theoretical approach to extended the fictitious crack model to FRC materials has been proposed. With this model, it is possible to define the behaviour of composite cracked elements and their mechanical response under tensile stresses. The comparison with some experimental tests seems to validate the effectiveness of the proposed approach.

These results seem to have a relevant value, in particular in the light of the importance of tensile response for serviceability behaviour and for ultimate states. Moreover, the numerical procedure can immediately give information about the variation of mechanical properties due to different materials involved in FRC elements. This aspect is very important in designing fiber-reinforced composites and in optimising their mechanical properties. Therefore, the proposed model can be considered as an useful tool not only for practitioners but also for FRC producers.

REFERENCES

Balaguru, P.N. & Shah, S.P. 1992. *Fiber-Reinforced Cement Composites*, New York, USA: McGraw-Hill, Inc.
Bartos, P. 1981. Review paper: Bond in fibre reinforced cements and concretes, *The International Journal of Cement Composites*, 3(3):159–177.
Bazant, Z.P., Li, Z. & Thoma, M. 1995. Identification of Stress-Slip Law For Bar or Fiber Pullout by Size Effect Tests, *ASCE Journal of Engineering Mechanics*, 121(5): 620–625.
Casanova, P., Rossi, P. & Shaller, I. 1997. Can Steel Fibers Replace Transverse Reinforcement in Reinforced Concrete Beams?, *ACI Materials Journal*, 94(5): 341–354.
CEB, 1993. CEB-FIP *Model Code 1990. Bulletin d'information n°;203–205*, London, England: Thomas Telford.
Fantilli, A.P., Ferretti, D., Iori, I. & Vallini, P. 1998. Flexural deformability of reinforced concrete beams, *ASCE Journal of Structural Engineering*, 124(9): 1041–1049.
Fantilli, A.P. & Vallini, P. 2001. Softening behaviour of plain concrete beams. In K. Ravi-Chandar et al. (eds) *International Conference of Fracture ICF10, Honolulu-Hawaii, 1–6 December 2001*.
Fantilli, A.P. & Vallini, P. 2003. Bond–slip relationship for smooth steel reinforcement, *In EURO-C 2003, Computational Modelling of Concrete Structures, 17th–20th March 2003, St Johann im Pongau, Salzburger Land-Austria* (in press).
Hillerborg, A. 1980. Analysis of fracture by means of the fictitious crack model, particularly for fibre reinforced concrete, *The International Journal of Cement Composites*, 2(4):177–184.

Hillerborg, A., Modéer, M. & Petersson, P.E. 1976. Analysis of Crack Formation and Crack Growth in Concrete by Means of Fracture Mechanics and Finite Elements, *Cement and Concrete Research*, 6: 773–782.

Hordijk, D. A. 1991. Local approach to fatigue of concrete, Doctoral Thesis, TU-Delft.

Hu, X. & Day, R. 2000. Uniaxial stress-strain relationship of cementitious composites – a review. In P. Rossi & G. Chanvillard (eds), *Fifth RILEM Symposium on Fibre-Reinforced Concretes (FRC), Lyon, France, 13–15 September 2000* : 431–440.

Maalej, M. & Li, V.C. 1994. Flexural/Tensile-Strength Ratio in Engineered Cementitious Composites. *ASCE Journal of Materials in Civil Engineering*, 6(4): 513–528.

Naaman, A.E. & Shah, S.P. 1976. Pull-Out Mechanisms in Steel Fiber-Reinforced Concrete, *ASCE Journal of the Structural Division*, 121(8): 1537–1548.

Nanni, A. 1990. Design for Torsion Using Steel Fiber Reinforced Concrete. *ACI Materials Journal*, 87(6): 556–564.

Nanni, A. 1991. Pseudoductility of Fiber Reinforced Concrete. *ASCE Journal of Materials in Civil Engineering*, 3(1): 78–90.

Plizzari, G.A., Cangiano, S. & Cere, N. 2000. Postpeak Behavior of Fiber-Reinforced Concrete under Cyclic Tensile Loads, *ACI Structural Journal*, 97(2): 182–192.

Shah, S.P. & Ouyang, C. 1991. Mechanical Behavior of Fiber-Reinforced Cement-Based Composites, *Journal of American Ceramic Society*, 74(11): 2727–38, 2947–53.

Swamy, R.N. 2000. FRC for sustainable infrastructure regeneration and rehabilitation. In P. Rossi & G. Chanvillard (eds), *Fifth RILEM Symposium on Fibre-Reinforced Concretes (FRC), Lyon, France, 13–15 September 2000* : 3–18.

Tan, K.H., Paramasivam, P. & Tan, K.C. 1994. Istantaneous and Long-Term Deflections of Steel-Fiber Reinforced Concrete Beams, *ACI Structural Journal*, 91(4): 384–393.

19. Innovative methods for repair and strengthening structures

System-based Vision for Strategic and Creative Design, Bontempi (ed.)
© 2003 Swets & Zeitlinger, Lisse, ISBN 90 5809 599 1

Heat straightening repairs of damaged steel bridge members

K.K. Verma
Federal Highway Administration, Washington, D.C., USA

R.R. Avent
Louisiana State University, Baton Rouge, LA, USA

ABSTRACT: The use of heat straightening to repair damaged steel dates back to the 1930's. However, its use has tended to be more art than science with techniques passed from one practitioner to the next. Over the past 15 years, a significant effort has been devoted to providing a scientific and engineering basis for the application of heat-straightening repair. The purpose of this paper is to present the fundamental principles upon which heat straightening is based and to show how these principles can be implemented in practice. The latest research findings will be presented to illustrate the basic principles. Of key importance are: (1) heating temperature, (2) level of restraining force, (3) heating configuration, (4) heating pattern geometry, and (5) degree of damage. With proper consideration of these parameters, a wide range of damage configurations can be repaired without significantly changing the steel properties. The first portion of the paper will focus on basic principles of heat straightening. Topics will include basic heating patterns, effects on material properties, and behavior of plate elements during and after heat straightening. The second portion will focus on behavior of rolled and built-up shapes during and after heat straightening including effects on material properties, classification of damage and proper heating patterns. Suggested specifications for the conduct of heat-straightening repairs will also be presented. In terms of practical applications, the presentation will focus on how to heat straighten the typical types of damage found in practice. Included will be examples of how to analyze damage and design a repair scheme. Particular emphasis will be placed on: flexural damage about both the major and minor axes; localized bulges in plate elements; twisting damage; and specialty items such as stairs, tubular members and connections.

1 INTRODUCTION

The repair of damaged steel through the application of heat has long been considered an art rather than a science. As such both the methodology and results have varied considerably from one practitioner to another. Because steel tends to be a forgiving material, successful repairs out-number failures. Yet material properties may be compromised resulting in a weakened in-place structure. In addition, it is not unusual for a fracture to occur during the repair process. As a result the use of heat straightening (or flame straightening) has had limited application.

Many engineers are reluctant to allow heat straightening because of the lack of an established method of application as well as the lack of quantification of steel properties after the repair is complete. The purpose of this paper is to provide some engineering guidelines for use of the heat straightening process.

2 DEFINITION OF HEAT STRAIGHTENING

Heat straightening is a repair procedure in which a limited amount of heat is applied in specific patterns to the plastically deformed regions of damaged steel in repetitive heating and cooling cycles to produce a gradual straightening of the material. The process relies on internal and external restraints that produce thickening (or upsetting) during the heating phase and in-plane contraction during the cooling phase. Heat straightening is distinguished from other methods in that force is not used as the primary instrument of straightening. Rather, the thermal expansion/contraction is an unsymmetrical process in which each cycle leads to a gradual straightening trend. The process is characterized by the following conditions, which must be maintained:

1. The maximum heating temperature of the steel does not exceed either (a) the lower critical temperature

(the lowest temperature at which permanent molecular changes may occur), or (b) the temper limit for quenched and tempered steels.

2. The stresses produced by applied external forces do not exceed the yield stress of the steel in its heated condition. Only the regions in the vicinity of the plastically deformed zones are heated.

When these conditions are met, the material properties undergo relatively small changes and the performance of the steel remains essentially unchanged after heat straightening. Properly conducted, heat straightening is a safe and economical procedure for repairing damaged steel.

3 BASIC TYPES OF HEATING PATTERNS

There are several basic types of heats that are frequently used as illustrated by (Avent & Mukai 1999, Avent et al. 2000). For some applications only one type is required. However, in many cases a combination of these basic types are necessary. The three basic types most often used in heat straightening are describe as follows.

3.1 Vee heat

The vee heat is used to straighten strong axis bends in steel plate elements. As seen in Figure 1, a typical vee heat starts at the apex of the vee-shaped area using an oxy-fuel torch. When the desired temperature is reached (usually around 650°C or 1200°F for mild carbon steel), the torch is advanced progressively in a serpentine motion toward the base of the vee. This motion is efficient for progressively heating the vee from top to bottom. The plate will initially move upward (Fig. 1a) as a result of longitudinal expansion of material above the neutral axis producing negative bending. The cool material adjacent to the heated area resists the normal thermal expansion of the steel in the longitudinal direction. As a result, the heated material will tend to expand, or upset, to a greater extent through the thickness of the plate than longitudinally, resulting in plastic flow. At the completion of the heat, the entire heated area is at a high and relatively uniform temperature. At this point the plate has moved downward (Fig. 1b) due to longitudinal expansion of material below the neutral axis producing positive bending. As the steel cools, the material contracts longitudinally to a greater degree than the expansion during heating. Thus, a net contraction occurs. Because the net upsetting is proportional to the width across the vee, the amount of upsetting increases from top to bottom of the vee. This variation produces a closure of the vee. Bending is produced in an initially straight member, or straightening occurs (if the plate is bent in the opposite direction to that of the straightening

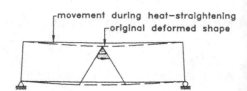

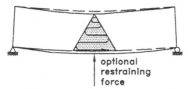

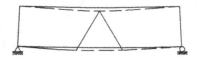

(a) **Plate movement during early heating phase (deflection decreases)**

(b) **Plate movement near the completion of heating (deflection increases)**

(c) **Final position after cooling (deflection decreases)**

Figure 1. Stages of movement during a vee heat.

movement, Fig. 1c). For many applications, it is most efficient to utilize a vee that extends over the full depth of the plate element but, partial depth vees may be applicable in certain situations. When using partial depth vees, the open end should extend to the edge of the element. The vee depth is varied by placing the apex at a partial depth location. For example, in deep plate elements the depth of the vee may result in too much cooling near the apex prior to completing the heat at the open end. A 3/4 depth vee may be applicable in reducing this problem. Other examples of partial depth vees will be shown in later sections.

3.2 Line heats

Line heats are employed to repair a bend in a plate about its weak axis. Such bends, severe enough to produce yielding of the material, often result in yield lines. A line heat consists of a single pass of the torch. The restraint in this case is often provided by an external force although movement will occur without external constraints. As the torch is applied and moved across the plate, the temperature distribution decreases through the thickness. The cool material ahead of the torch constrains thermal expansion, even if bending moment constraints are not present. Because of the thermal gradient through the thickness, more upsetting occurs on the torch (or hotter) side of the plate. During

1454

cooling this side consequently contracts more, creating a concave bend on the torch side of the plate. Thus, to straighten a plate bent about its weak axis, the heat should be applied to the convex side of the damaged plate. The movement can be magnified by the use of applied forces which produce bending moments about the yield line. In a manner similar to the vee heat mechanism, the material thus tends to expand through the thickness, or "upset". Upon cooling, the restraining moments tend to magnify transverse contraction. The speed of the travel of the torch is critical as it determines the temperature attained. With proper restraints and a uniform speed of the torch, a rotation will occur about the heated line.

3.3 Strip heats

Strip heats, also called rectangular heats, are used to complement a vee heat. Strip heats are similar to vee heats and are accomplished in a like manner. Beginning at the initiation point, the torch is moved back and forth in a serpentine fashion across a strip for a desired length. This pattern sequentially brings the entire strip to the desired temperature.

4 RELATIONSHIP BETWEEN FUNDAMENTAL DAMAGE PATTERNS AND BASIC TYPES OF HEATS

A convenient way to classify damage is to define four fundamental damage patterns (Avent 1995). For each damage pattern, there is sequence of specific heating types that will produces straightening. In most cases, actual damage to steel structures is a combination of two or more of these fundamental damage patterns. However, more complex damage can be repaired by sequentially using the heating patterns for these four fundamental damage patterns. For purposes of defining heating patterns, it is convenient to refer to the elements of a cross section as either primary or stiffening elements. The primary elements are plate elements damaged by bending about their major axes. The stiffening elements are those bent about their minor axes. Typically, vee heats are applied to primary elements while strip, line or no heat at all may be applied to stiffening elements. The combination of these heats, applied either simultaneously or consecutively, represent a single cycle of the heating pattern. Because the net change in curvature after one heating cycle is small, a number of cycles are required to straighten a damaged member. For each cycle the heating pattern should be shifted to a different location within the yield zone region, although the same location can be returned to after every third cycle. More heats should be placed in the region of largest curvature and fewer near extremities to reflect the difference in damage

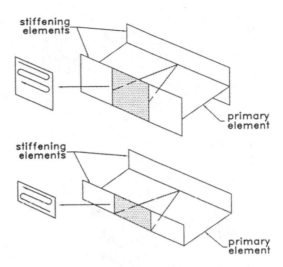

Figure 2. Heating patterns for wide flanges and channels bent about their major axes (category S).

curvature. The fundamental damage categories and their heating patterns are described in the following sections.

4.1 Category S damage

This type refers to bending about the "strong" or major axis. The heating pattern consists of a vee and strip heat as shown in Figure 2. This pattern is superimposed on a typical yield zone for a wide flange beam bent about the strong axis. The strip heat and open end of the vee are placed on the flange damaged in tension. The vee heat is first applied to the web. Upon completion, a strip heat is applied to the flange at the open end of the vee. The width of the strip heat always equals the vee width at the point of intersection. This procedure allows the vee to close during cooling without restraint from the stiffening element. No heat is applied to the flange at the apex of the vee. This vee/strip combination is repeated by shifting over the vicinity of the yield zone until the member is straight. The pattern is shown for a wide flange and channel in Figure 2.

4.2 Category W damage

This category refers to damage as a result of bending about the "weak" or minor axis. For rolled or built-up shapes the web is usually at, or near, the neutral axis. Consequently, it may not deform into the inelastic range. The heating pattern for this case is similar to the previous case but note the primary and stiffening elements are reversed. The vee heat is first applied to both flanges (either simultaneously or one at a time) as shown in Figure 3. After heating these primary

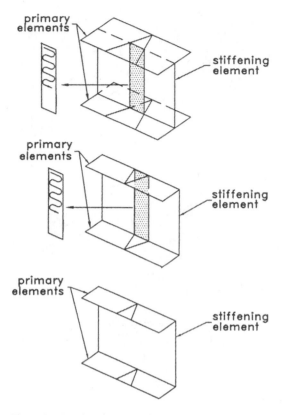

Figure 3. Heating patterns for wide flanges and channels bent about their minor axes (category W).

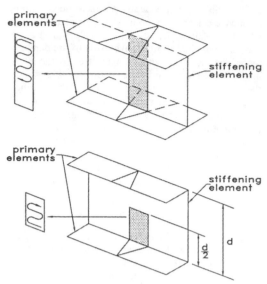

Figure 4. Heating patterns for wide flanges and channels with twisting (category T).

elements, a strip heat is applied to the web. The only exception is that no strip heat is applied to stiffening elements located adjacent to the apex of a vee heated element since this element offers little restraint to the closing of the vee during cooling. Note that the width of the strip heat is equal to the width of the vee heat at the point of intersection. For all cases the pattern is repeated by shifting over the vicinity of the yield zone until the member is straight.

4.3 Category T damage

This type refers to damage as a result of torsion or twisting about the longitudinal axis of a member. For rolled or built-up shapes, the flange elements tend to exhibit flexural plastic deformation in opposite directions. The web is often stressed at levels below yield. If one flange is constrained (such as the case of a composite beam), then only the unconstrained flange element is subjected to plastic deformation and yielding may also occur in the web. The heating pattern for this damage case is shown in Figure 4. The vees on the top and bottom flange are reversed to reflect the different directions of curvature of the opposite flanges. The vee heats are applied first and then the strip heat is applied. Note that for the channel, the strip heat need only be applied to half depth. This half depth strip allows the lower flange vee to close with minimal restraint from the web. If one flange is a composite connection, then that flange is not heated.

4.4 Category L damage

This category includes damage that is localized in nature. Local flange or web buckles, web crippling and small bends or crimps in plate elements of a cross section typify this behavior. A local buckle or bulge reflects an elongation of material. Restoration requires the bulging area to be shortened. A series of vee or line heats can be used for this purpose as shown in Figure 5. These vees are heated sequentially across the buckle or around the bulge. For web bulges either lines or vees may be used. If vees are used, they are spaced so that the open end of the vees touch. There is a tendency for practitioners to over-heat web bulges. For most cases, too much heat is counter-productive. The preferred pattern is the line heats in the spoke/wagon wheel pattern. For the flange buckle pattern (Fig. 5b) either lines or a combination of lines and vees may be used. For most cases, the line pattern with few or no vees tends to be most effective. Since the flange damage tends to be unsymmetrical, more heating cycles are required on the side with the most damage. When possible, the heating should be done on the convex side.

1456

Line heats   Vee heat star pattern

(a) Web bulge heating patterns

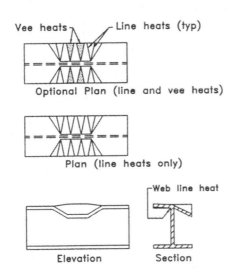

Vee heats — Line heats (typ)

Optional Plan (line and vee heats)

Plan (line heats only)

Web line heat

Elevation   Section

(b) Local flange damage heating pattern

Figure 5. Typical heating patterns for local damage (category L).

5 TEMPERATURE LIMITS DURING HEAT STRAIGHTENING

Temperatures greater than 727°C (1340°F) begin to produce a phase change in steel. This temperature is often called the lower critical (or lower phase transition) temperature. When steel cools below the lower critical temperature, it attempts to return to its former phase structure. Since this change requires a specified time frame, rapid cooling may not permit the complete molecular change to occur. Under these circumstances, a hard, strong and brittle phase called martensite occurs. The steel in this form may have reduced ductility and be more sensitive to brittle fracture under repeated loads. The maximum temperature recommended by most researchers such as (Avent 1995, Roeder 1986) is 650°C (1200°F) for all but quenched and tempered high-strength steels. The limiting temperature of 650°C (1200°F) allows for about 55°C (100°F) of temperature variation, which was found to be a common range among experienced

practitioners. To control the temperature, the speed of the torch movement and the size of the orifice must be adjusted for different thicknesses of material. However, as long as the temperature is rapidly achieved at the appropriate level, the contraction effect will be similar. Various methods can be used to monitor temperature during heating. Principal among these include: visual observation of color of the steel; use of special temperature sensing crayons or pyrometers; and infrared electronic temperature sensing devices.

6 LIMITS ON THE MAGNITUDE OF JACKING FORCES

On several occasions, the writers have observed sudden fractures during heat straightening. These types of fractures are not uncommon. The usual explanation is that small microcracks must have occurred as a result of initial damage. Thus, during heat straightening, one of these cracks propagates. However, no experimental evidence validating this assumption has been published. Research has shown (Avent & Fadous 1988) that high jacking forces can cause fractures during heat straightening. It is therefore recommended that stresses in the heated zone due to jacking not exceed 50% of the nominal yield stress at ambient temperature. Since the yield stress at 1200°F is reduced by approximately 50%, this criteria prevents the heated steel stress from becoming significantly larger than yield. In addition, because the entire cross-section does not reach the maximum heating temperature simultaneously, the 50% value includes an additional margin of safety. Thus, it is also recommended that jacks always be calibrated and the structure be analyzed to prevent over-stress.

7 SUMMARY AND CONCLUSIONS

Described in this paper are the fundamentals of applying heat straightening to repair distortion in damaged bridge girders. The principles discussed here can also be applied to heat curving members to produce camber, sweep or curved beams. Heat straightening is a safe and economical repair procedure when properly executed. It is a skill requiring practice and experience. The proper placement and sequencing of heats combined with control of the heating temperature and jacking forces distinguished the expert practitioner. In the past heat straightening has been more art than science. The basics presented here explain the behavior during repair.

Actual damage is usually a complex combination of the various damage categories described in this paper. Heat straightening of such damage is best conducted in a step-wise process where the damage associated with each of the fundamental categories is addressed individually, beginning with the most severe damage. A successful repair involves a clear understanding of the process, design of the heat straightening parameters (such as jacking level and heating patterns), and a continued monitoring of the progression of movement during repair. This combination of knowledge and skill requires experience.

REFERENCES

Avent, R. 1995. Engineered heat straightening comes of age. *Modern Steel Construction* 35(2): 32–39.

Avent, R. & Fadous, G. 1988. Heat-straightening prototype damaged bridge girders. *Journal of Structural Engineering* 15(7): 1631–1649.

Avent, R. & Mukai, D. 1999. *Heat-Straightening Repairs of Damaged Steel for Bridges*. U.S. Department of Transportation, Federal Highway Administration, Washington, D.C.

Avent, R., Mukai, D. & Robinson, P. 2000. Heat straightening rolled shapes. *Journal of Structural Engineering* 126(7): 755–763.

Roeder, C. 1986. Experimental study of heat induced deformation. *Journal of Structural Engineering* 112(10): 2247–2262.

Laboratory and field observations of composite-wrapped reinforced concrete structures

E. Berver, J.O. Jirsa, & D.W. Fowler
Civil Engineering Department, University of Texas at Austin, Austin, TX, USA

H.G. Wheat
Mechanical Engineering Department (Texas Materials Institute), University of Texas at Austin, Austin, TX, USA

ABSTRACT: Composite wrapping with fiber-reinforced plastics (FRPs) is being considered as a means of rehabilitating corrosion-damaged reinforced concrete structures, particularly bridges. However, little is known about the long-term effectiveness of wrapping. A research investigation consisting of a laboratory program and a field program was initiated to address some of the long-term issues. Phase I of the laboratory program consisted of 42 cylinders to model bridge columns and 18 rectangular blocks to model portions of bridge bents. The field program consisted of 12 corrosion-damaged bridges in the northern part of Texas, three of which were instrumented with corrosion rate probes prior to being wrapped. The laboratory specimens were cast, damaged, wrapped and then subjected to an alternate immersion in saltwater/drying in air sequence for a period of more than two years. The bridges were exposed to deicing salt applications as needed. The project is ongoing, but the results of autopsies on some of the cylindrical and rectangular specimens will be discussed. The paper will also include a discussion of the changes in appearance, corrosion conditions, and corrosion rates at four locations on one of the bridges.

1 INTRODUCTION

In order to increase the service life of existing bridges that have been damaged by corrosion, many remediation techniques such as cathodic protection and electrochemical chloride removal are being investigated. Another technique involves wrapping corrosion-damaged structures with fiber-reinforced plastics (FRPs). The successful use of this technique to repair structures that have suffered seismic damage has been demonstrated (Saadatmanesh & Eshani 1998). Fiber composites wrapped around columns in seismic zones are known to increase ductility and prevent collapse by providing confinement to the structural element. It is anticipated that such composites could minimize corrosion by acting as barriers against chlorides, water, oxygen, and other species that might be considered aggressive.

While there are many investigations of FRPs for the rehabilitation of infrastructure (Karbhari et al. 1996, Shiekh et al. 1997, Saadatmanesh & Eshani 1998, Debaiky et al. 2001, Chakrabarti et al. 2002, Soudki et al. 2002, Spainhour et al. 2002, Wooton et al. 2002), very little is known about the long-term effectiveness

of these composites. In addition, there are few means of determining the long-term effectiveness. One important concern is the long-term durability of the composites themselves. In many cases, the initial properties of these composites are outstanding, however environmental conditions may diminish them significantly. There are also some reservations about the use of this technique to rehabilitate corrosion-damaged structures since wrapping may hide additional damage. This was observed on a bridge in Florida where it was found that fiberglass jackets applied to bridge piles in a marine environment had actually accelerated the corrosion process. Capillary action allowed moisture to rise from the submerged portion of the pile, but the jacket prevented the concrete from drying out. The combination of moisture and the high levels of chlorides in the unrepaired portion of the pile resulted in severe corrosion damage (Sohangpurwala & Scannell 1994).

A field test program involving composite wrapping with FRPs has been used to rehabilitate 12 Texas bridges that have suffered corrosion damage as a result of repeated deicing salt applications for snow removal (Verhulst 1999). The major components of the bridges that have suffered damage are bridge columns and portions of bridge bents located at points where bridge deck runoff exposes the bent to water containing deicing salts.

A companion laboratory test program was developed to simulate components of the bridge systems, specifically bridge bents (rectangular specimens) and columns (cylindrical specimens). To incorporate many of the factors that contribute to field performance, some of the variables include cast-in chlorides, cracks, repair materials and concrete surface condition. Resin selection and extent/length of wrap are also being considered. (Fuentes 1999).

The work to date on the field and laboratory investigations of composite wrapping of the Texas bridges has been described (Fuentes 1999, Verhulst 1999, Berver 2001). The focus of this paper is the forensic analysis of the specimens in the second set of specimens that were autopsied after wrapping and re-exposure to a saltwater environment for approximately two years. These specimens belonged to Group B. Some comparisons will be made with the specimens in Group A, the first set of specimens to be autopsied. In addition, information will be presented based on monitoring four bridge bent locations on one of the bridges that was repaired, wrapped, and re-exposed to deicing salt application.

2 EXPERIMENTAL PROCEDURE

In an attempt to investigate the potential corrosion protection offered by composite wrapping in the laboratory, representative field conditions were chosen. To accelerate the corrosion process, many of the "worst case" field conditions were implemented in the laboratory tests. Brief summaries of the laboratory and field programs are presented below (Fuentes 1999, Verhulst 1999, Fuentes et al. 2000).

2.1 Laboratory research program (phase 1)

Beams and columns of the bridges were represented by rectangular and cylindrical specimens, respectively. In all, 18 rectangular and 42 cylindrical specimens were cast for this study. The rectangular specimens were 3 feet (914 mm) long with a 10 inch (254 mm) × 10 inch (254 mm) cross section and the cylindrical specimens were 3 feet (914 mm) long and 10 inches (254 mm) in diameter.

Specimens were constructed with steel cages formed of longitudinal and transverse reinforcement. Metallic tie wires were used to attach the rebar to the spiral. This also maintained electrical continuity necessary for monitoring the corrosion potentials. A small one inch (25 mm) cover was chosen to accelerate corrosion. The reinforcing bars were cut in 39 inch (991 mm) lengths so that 3 inches (76 mm) of reinforcement protruded out

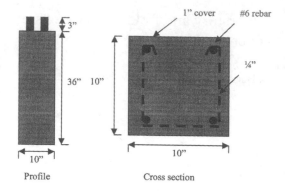

Figure 1. Geometry of rectangular specimens (Fuentes 1999).

of one end of the beams and columns. The geometry of the rectangular specimens is shown in Figure 1.

To insure timely results, a low quality, highly permeable mix with a water-cement ratio of 0.7 was used for the specimens.

In order to simulate chloride-contaminated concrete, half of the specimens had chlorides added to the mix. The chloride content was 1.5% by weight of cement. To investigate the effect of cracks, some specimens were cracked and flexural cracks approximately 20 mil (0.5 mm) in width were introduced. To investigate the effect of repair, some specimens were intentionally damaged and repaired with one of two different patching materials: latex modified concrete (LMC) and epoxy grout (EG). To study the influence of a corrosion inhibitor, a corrosion inhibitor was applied to some of the specimens following standard repair and just before application of the FRP wrap. To investigate the adhesive properties of a composite wrap in marine environments, some cylindrical specimens were wrapped while wet.

Two different composite systems were investigated, a commercial system manufactured by Delta Technologies (the Delta system, known as Fiberwrap) and a second system designed and fabricated at the Impact Laboratory at the University of Texas at Austin (the Generic system). The wrapping lengths were 24 inches (610 mm), just at the salt solution line; 30 inches (762 mm) below the salt solution line; or 6 inches (152 mm) below the salt solution line; or the full length of the cylinder, 36 inches (914 mm). The bottom ends were always unwrapped. This simulated field conditions where it is impossible to wrap a column to the foundation or below the waterline due to limited access. The rectangular specimens were placed vertically on timber supports, with the extruding reinforcement towards the floor. This stabilized the beams while allowing for proper encapsulation of the ends.

Table 1 gives a more complete description of the first and second set of specimens to be autopsied. The first letter denotes shape (R for rectangular, C for

Table 1. Specimen parameters (Fuentes 1999).

Specimen	Cast-in-chlorides	FRP Wrap Fabric type	L in	Resin system	Initial concrete condition	Concrete repair material	Corrosion inhibitor
Group A							
CC7	Chlorides	Fiber-wrap	24	TYFO S	Cracked		
CC18	Chlorides	None			Cracked		Ferrogard
CNC8		None			Cracked		Ferrogard
CNC13		Generic	24	Epoxy	Cracked		Ferrogard
CNC14		Generic	36	Epoxy	Cracked		
CNC19		Generic	24	Epoxy	Un-cracked		
RC4	Chlorides	None			Cracked		
RC7	Chlorides	Generic	30	Epoxy	Cracked		
Group B							
CC3	Chlorides	Fiber-wrap	24	TYFO S	Un-cracked	EG	
CC5	Chlorides	Generic	36	Epoxy	Cracked	EG	
CC6	Chlorides	Generic	36	Vinyl ester	Cracked	EG	Ferrogard
CNC10		Fiber-wrap	24	TYFO S	Cracked		
RNC6		Fiber-wrap	30	Epoxy	Cracked	LMC	
RNC7		None			Cracked		

Figure 2. Wrapped rectangular specimen (Fuentes 1999).

Figure 3. Test Setup for linear polarization (Berver 2001).

cylindrical) and the following letters differentiate chloride contaminated mixes from uncontaminated ones (C for chlorides, NC for no chlorides). All rectangular specimens were wrapped with dry surfaces.

For the specimens that were to be wrapped and monitored, a small hole was drilled to allow access to the steel in the concrete. This access hole was used for monitoring corrosion potentials. The wrapped specimens were fitted with removable plastic covers that allowed access to the steel in the concrete. The access hole for one of the wrapped rectangular specimens can be seen in Figure 2.

Specimens were exposed to a 3.5% salt solution on a cyclical basis that consist of a soaking period of 1 week in the solution followed by a 2 week drying period. In the case of the cylindrical specimens, the lower 1 foot (305 mm) of the cylinders was in the solution during the soaking period. Rectangular specimens, positioned at a slight incline, had solution irrigated over the top surface during their soaking period. Corrosion potentials were monitored every one to four cycles using a Cu/CuSO4 reference electrode. Corrosion potentials more negative than −350 mV vs the Cu /CuSO4 electrode have been associated with corrosion activity; corrosion potentials more negative than-500 mV vs the Cu/CuSO4 electrode have been associated with severe corrosion and the possibility of cracking (ASTM C876, 1991).

After approximately two years, Group B specimens were removed from exposure and allowed to dry for 6 weeks. Selected specimens were subjected to linear polarization tests from which corrosion rate were determined. Figure 3 shows a cylindrical specimen being subjected to linear polarization testing.

2.2 *Field research program*

The field program included 12 bridges and the major corroding elements were bridge bent endcaps and corresponding columns. Evaluation of the bridges prior to making the repairs and installing the wrap included visual inspection, measurement of corrosion potentials and corrosion rates at specific locations, determination of chloride contents at specific locations, and permeability testing based on selected cores removed from several of the bridges. The bridges were evaluated in 1998 and just prior to being wrapped in 1999 (Verhulst 1999). After traditional repairs were made on the bridges, embeddable corrosion rate probes (with dimensions $6.0 \times 6.0 \times 12.7$ cm) manufactured by Concorr, Inc. were installed on three of the bridges prior to application of the Delta wrapping system. This paper will specifically address the probes that were installed at four locations on bridge #8; locations 8.1, 8.2, 8.3 and 8.4 were all near bearing pads. The initial chloride contents by weight of concrete at

Figure 4. Corrosion damage on an endcap of bridge #8 (Verhulst 1999).

Figure 5. Bridge bent and column wrapped with FRP after repair of corrosion damage (Verhulst 1999).

the locations were between 0.12 and 0.17. Corrosion potentials and corrosion rates were measured at least once a year. The initial corrosion potentials at the four locations were -329, -362, -377, and -351 mV vs the Cu/CuSO4 electrode, respectively. The initial corrosion rates were 0.20, 0.62, 0.26 and 0.45 mpy respectively. Before and after composite wrapping photographs of one of the bridges are shown in Figures 4 and 5.

3 RESULTS AND DISCUSSION

3.1 *Laboratory research program (phase 1)*

Group B specimens were exposed and monitored for approximately two years. The final corrosion potentials prior to removal from exposure were more negative than -350 mV vs the Cu/CuSO4 electrode for all of the specimens except CC3. (The corrosion potential for CNC10 was -356 v vs the Cu/CuSO4 electrode). The specimens were allowed to dry for six weeks before being subjected to forensic analysis. Corrosion potentials just prior to forensic analysis were more negative than -350 mV vs the Cu/CuSO4 electrode for all specimens except CC3 and CNC10. Their values were both -275 v vs the Cu/CuSO4 reference electrode prior to forensic analysis. Corrosion rates taken for specimens CC3, CC5, CC6, and CNC10 were 0.34, 1.95, 2.39 and 0.58 mpy, respectively.

Since chloride levels are know to be a key factor in corrosion, chloride determinations were performed on the specimens to determine the chloride content in the concrete after exposure and prior to exposing the cages (Berver 2001). Chloride contents (by weight of concrete) at a depth of ¼ in. (32 mm) were 0.16 for CC5 and CC6, 0.09 for CC3 and 0.003 for CNC10. The chloride levels for RNC6 and RNC7 were 0.002 and 0.21 respectively. Note that CC3, CNC10 and RNC6, the specimens with the lowest chloride contents, were all wrapped with the commercial Delta system. RNC6, the specimen with the highest chloride content, was not wrapped. In addition, specimens CC3 and CNC10, which had the lowest corrosion rates, were wrapped with the commercial Delta system.

Specimens were opened systematically in order to examine the steel reinforcing cages while still preserving much of the concrete that had surrounded the cages. The results of the individual specimens are summarized below. All of the specimens were wrapped except RNC6.

For CC3, corrosion was found mainly in the splash zone (the level of the salt solution), which was also where a repair patch was located. This is not surprising since only the top two feet (508 mm) were wrapped.

For CC5, small areas of corrosion were found on the bars near cracks throughout the specimen, as well as near honeycombing, even though the entire surface was wrapped.

Similar findings as in CC5 were observed for CC6, which was also completely wrapped. However, in this case, moisture was trapped beneath the wrap in the splash zone. In addition, there was not a marked difference in the appearance of the cage (compared to CC5), even though a corrosion inhibitor was applied to CC6. A portion of the cage taken from CC6 is shown in Figure 6.

In CC10, corrosion was mainly associated with crack locations and there was an area of moderate corrosion just above the splash zone. Wrapping was confined to the top two feet (508 mm) as in CC3.

Perhaps the best comparisons can be made between RNC6, which was completely wrapped and RNC7, which was not wrapped. Moisture was trapped in the upstream end of the top face of specimen RNC6 and isolated areas of corrosion were found along bars in the patched areas. For RNC7, corrosion was mainly associated with crack locations.

The above findings are similar to those that were observed for the specimens in Group A that were exposed to a salt containing environment for approximately one year. For example, wrapping did restrict chloride ingress, however cracks, both pre-existing and subsequent, served as potential sites for corrosion.

The above findings are also in agreement with the results of other investigations that have been conducted over the last several years. The investigations involve various types of specimens; lollipop specimens (Wooton et al. 2002), small-scale square columns (Spainhour et al. 2002), as well as small to large-scale beams (Soudki et al. 2002). The investigators concluded that wrapping specimens led to prolonged test life, decreased reinforcement mass loss, and lower corrosion rates (Wooton et al. 2002) and that the use

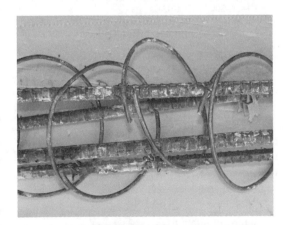

Figure 6. Steel reinforcing of CC3 with a corrosion rate of 0.34 mpy (Berver 2001).

of FRP sheets for strengthening corroded reinforced concrete beams is an efficient technique that can maintain structural integrity (Soudki et al. 2002). While some investigators indicated carbon fiber reinforced plastic (CFRP) wrapped samples had decreased corrosion probabilities, reduced chloride contents, and decreased reinforcement mass loss, they also noted that the results appeared to be highly dependent on the quality of the bond between the wrap and the concrete surface (Spainhour et al. 2002).

3.2 Field research program

Four locations on bridge #8; locations 8.1, 8.2, 8.3 and 8.4 were examined after one year. These locations were all near bearing pads. The previous year (right after repair and wrapping), the corrosion potentials for locations 8.1, 8.2, 8.3 and 8.4 were -329, -362, -377 and -351 V vs the Cu/CuSO4 electrode, respectively. One year later, corrosion potentials were almost the same; -344, -384, -395 and -359 V vs the Cu/CuSO4 electrode, respectively. In addition, the corrosion rates after repair and initial wrapping were 0.20, 0.62, 0.26 and 0.45 mpy respectively, compared to 0.20, 0.65, 0.56 and 0.60 mpy respectively, after one year of re-exposure. In addition, the wrapping seemed to have remained intact and the appearance of the wrapped areas was still quite good.

The excellent condition exhibited by the wrapping on bridge #8 was not always the case on all of the bents in the wrapped bridges. Some showed gaps, deterioration, and delamination. On one, there were even rust stains that could be seen through the wrapping. One of the columns showed vehicular damage as a result of being hit by an automobile.

While reported results of field use are not as numerous as laboratory results, other state Departments of Transportation (DOTs) are also using wrapping to repair and strengthen bridge columns and piers. The use of Aquawrap 22–77, which has been used successfully to repair and strengthen New York and Idaho DOT bridge columns, was recently described (Korff & Cercone 2002). Aquawrap 22–77 is a structural wrap that consists of a water-cured polyurethane resin impregnated into a variety of reinforcing fibers such as glass and carbon. The New York State DOT carried out repairs on a bridge that was suffering serious spalling on the columns due to weathering, freeze/thaw cycling, and the use of salt in winter conditions. In order to determine the long-term durability of the wrap system, durability samples were made on the structure. Enough material was put down to pull a sample each year and conduct mechanical tests. So far, the results are only visual and the investigators reported that the repair looked very good after one winter of freeze/thaw cycles. There was no evidence of delamination, voids, aging, or other defects.

3.3 Laboratory research program (phase 2)

Since some specimens in Group A appeared to benefit from the use of a corrosion inhibitor, a second phase of the laboratory program was initiated and involves 19 additional specimens and three inhibitors. The specimens have been exposed to cyclical salt water exposure for more than one year and corrosion potentials for all the specimens are close to $-350\,mV$ vs the Cu/CuSO4 electrode. However, to date, no specimens have been opened and autopsied.

4 CONCLUSIONS

A laboratory research program was developed to complement a field research program in which 12 bridges in northern Texas were wrapped with FRP. The objective of the programs was to evaluate the effectiveness of FRP wraps for the rehabilitation of corrosion-damaged reinforced concrete structures. The laboratory program was developed to simulate bridge bents and columns on actual bridges. Variables included specimen shape, flexural cracks, repair materials, length of wrap, resin selection and surface condition. Cylindrical and rectangular specimens, made using a highly permeable concrete with a water-cement ratio of 0.7, were exposed to salt water on a cyclical basis. Corrosion potentials were monitored at selected locations on the specimens for more than two years. Prior to forensic analysis, selected specimens were subjected to linear polarization test and corrosion rates were determined.

Based on the forensic analysis of the specimens in Group B, the wraps seemed to restrict chloride ingress; chloride contents after exposure were lower for the wrapped specimens. Moisture, however, was found trapped under the wrap in some of the fully wrapped specimens. Corrosion activity was especially severe near cracks, near defects such as honeycombing, and near repaired areas. In the wrapped portions of the specimens that were initially free from cracks and chlorides, the wraps helped to minimize corrosion activity.

For bridge #8, the portion of the field research program that was the focus of this paper, corrosion potentials and corrosion rates changed very little after one year. This bridge seemed to be wrapped especially well. It was free of visible gaps, deterioration, and delamination one year after repair and wrapping. This was not the case for all of the bridges, suggesting that performance after wrapping is very dependent on the quality of the wrap.

The results of the laboratory and field investigations seem to show that wrapping can retard chloride ingress into the wrapped concrete. Whether corrosion can be arrested is still not clear. If there is little initial damage and the initial chloride concentration is very low, FRP wraps may provide some initial benefit. However, long-term benefits are very dependent on the presence of cracks, the initial chloride concentration, the manner in which the structures are re-introduced to salt-containing environments, and especially the quality of the wrap procedure. Moisture retention in some of the wrapped laboratory specimens seems to be a problem.

More than 50 specimens from phases 1 and 2 of the laboratory programs are continuing to be exposed to salt water on a cyclical basis. In addition, the laboratory specimens and three of the 12 bridges are continuing to be monitored.

ACKNOWLEDGMENTS

The authors are very grateful for the support of the Texas Department of Transportation, especially Jon Kilgore, Randy Cox, and Allan Kowalik. The authors are also grateful for the assistance of Paul Guggenheim, Dr. Tess Moon, Dr. Peter Joyce, Lola Miret, David Whitney and Michael Rung.

REFERENCES

ASTM C876 1991. "Standard Test Method for Half-Cell Potentials of Uncoated Reinforcing Steel in Concrete," *American Society for Testing and Materials*, Philadelphia, PA.

Berver, E.W. 2001. "Effects of Wrapping Chloride Contaminated Concrete with Fiber Reinforced Plastics," M.S. Thesis, Civil Engineering Department, The University of Texas at Austin.

Chakrabarti, P.R., Millar, C. & Bandyopadhyay, S. 2002. "Application of Composites in Infrastructure-Part I and II," *Proceedings of the Third International Conference on Composites in Infrastructure*, San Francisco, CA, June 10–12, 2002 (ICCI-02).

Debaiky, A.S., Green, M.F. & Hope, B.B. 2001. "Corrosion Evaluation in CFRP Wrapped RC Cylinders," *Proceedings of the 9th International Conference & Exhibition on Structural Faults and Repair*, London, UK, July 2001.

Fuentes, L.A. 1999. "Implementation of Composite Wrapping Systems on Reinforced Concrete Structures Exposed to a Corrosive Laboratory Environment," M.S. Thesis, Civil Engineering Department, University of Texas at Austin.

Karbhari, V.M., Seible, F. & Hegemier, G.A. 1996. "On the Use of Fiber Reinforced Composites for Infrastructure Renewal – A System Approach," *Materials for the New Millenium*, American Society of Civil Engineers, NY, 1091–1100.

Korff, J.A. & Cercone, L. (2000). "Repair and Strengthening of DOT Columns and Piers with Water Activated Aquawrap 22–77" in ICCI_02.

Saadatmanesh, H. & Eshani, M.R. 1998. "Fiber Composites in Infrastructure," *Proceedings of the Second Intramural Conference on Composites in Infrastructure*, ICCI 1998 Tucson, AZ, Jan 5–7, 1998.

Sheikh, S., Pantazopoulou, S., Bonacci, J., Thomas, M. & Hearn, M. 1997. "Repair of Delaminated Circular Pier Columns by ACM," Ontario Joint Transporation Research Report, Final Report, MTO, Reference Number 31902.

Sohangpurwala, A. & Scannell, W.T. 1994. "Repair and Protection of Concrete Exposed to Seawater," *Concrete Repair Bulletin*, July–August, 8–11.

Soudki, K.A., Sherwood, T. & Masoud, S. (2002). "FRP Repair of Corrosion-Damaged Reinforced Concrete Beams," in ICCI_02.

Spainhour, L.K., Wootton, I.A. & Yazdani, N. (2002). "Effect of Composite Fiber Wraps on Corrosion of Reinforced Concrete Columns in a Simulated Splash Zone," in ICCI_02.

Verhulst, S.M. 1999. "Evaluation and Performance Monitoring of Corrosion Protection Provided by Fiber-Reinforced Wrapping," M.S. Thesis, Civil Engineering Department, University of Texas at Austin.

Improving interface bond in pile repair

G. Mullins, J. Fischer & R. Sen
University of South Florida, Tampa, FL, USA

M. Issa
TY Lin International, Chicago, IL, USA

ABSTRACT: The performance of repaired prestressed piles is determined by the efficiency of the interface bond between the substrate and the repair material. This paper summarizes findings from an experimental study that focused on improving bond using mechanical connectors. Powder actuated nails and epoxied dowel bars were installed as shear connectors and their performance evaluated from ultimate axial load tests on 1/3 scale models. Results indicated that ultimate capacity could reduce because of damage incurred during installation of mechanical connectors.

1 INTRODUCTION

Among the contiguous states in the United States, Florida has the longest coastline that is served by a large network of concrete pile-supported bridges. Unfortunately, the combination of a subtropical climate, tidal fluctuations and saltwater provides a lethal environment for reinforced and prestressed substructures. As a result, reinforcement located in the portion of the substructure subjected to periodic wet/dry cycles ("*the splash zone*") has been found to corrode rapidly.

Since at least 1940 "pile jackets" have been used to repair corrosion damage. The most commonly used pile jacket consists of a stay-in-place fiberglass form that is placed around a distressed pile and filled with grout. Such repairs are categorized as *non-structural*. In case where much of the steel has been lost, it is necessary to add new reinforcement. Such repairs are referred to as *structural*.

Pile size is generally dictated by foundation conditions, and structural capacity is rarely a consideration. However, there are situations where ultimate capacity of the repaired pile may be of critical importance, e.g. for resisting impact loads. This is particularly the case for piles where corrosion damage was severe with significant loss of cross-section. Moreover, as pile repairs are carried out in less than favorable conditions, the extent to which lost capacity is restored by the repair is unknown. For example, recommended practice for repairing corrosion damage, e.g. removing all the chloride-contaminated concrete, is not possible.

This paper presents a brief overview of an experimental study that addressed this problem. In the investigation, jacketed repairs were carried out on scale models of damaged prestressed piles and their capacity determined through ultimate load tests. Complete details may be found elsewhere (Fischer et al. 2000).

2 PROBLEM STATEMENT

Perhaps the most critical parameter influencing the efficiency of any pile repair is the integrity of interface bond between the deteriorated pile and the repair material. If this bond can transfer all the loads applied *after* the repair, then additional capacity may be realized. If on the other hand, only part of this load can be transferred, the increase in ultimate capacity would be limited. Thus, improving the performance of the interface bond is crucial if the repair is to be deemed successful.

Interface bond can be the result of chemical adhesion or due to mechanical action. Chemical adhesion is recognized to be limited, e.g. 1–2 MPa (Leet & Bernal 1997) though it may be augmented by the chemistry of the repair material or by the application of a bonding agent. Mechanical bond can arise from the geometry of the interface, i.e. by making the interface uneven by chipping. Alternately, dowel action of shear connectors widely used in bridge applications can be expected to enhance performance. Any study that attempts to improve the performance of repairs needs to consider both mechanisms.

3 OBJECTIVES

The objectives of this experimental study were two fold: (1) to evaluate the effectiveness of presently used pile jacketing schemes in restoring ultimate capacity, and (2) to investigate alternate repair methodologies that could enhance capacity by improving the interface bond.

4 EXPERIMENTAL PROGRAM

To meet the objectives of the study, two phases were required. In the first phase, ultimate load tests investigated the effectiveness of repairs under concentric and eccentric loading. The findings from this phase were used to develop the second phase.

This paper focuses primarily on the investigations conducted in the second phase. Essential background information on the first phase is included. More complete information may be found in the final report (Fischer et al. 2000).

4.1 *Specimen details*

The dimensions of the test specimens were selected on the basis of a field survey of observed damage. It was found that damage was typically observed in square pretressed piles with 45–50 cm side. A one third-scale model was selected taking into consideration fabrication and available testing facilities.

All specimens were 15 cm × 15 cm × 2.44 m long with a clear cover of 2.5 cm. They were prestressed using four, seven wire 7.9 mm low relaxation Grade 250 (1728 MPa) strands. Each of these strands was jacked to 51.5 kN to provide an effective prestress that exactly matched that in the prototype pile. Spirals ties were fabricated using No. 5 gage wire (5.28 mm) with an 11.5 cm maximum pitch.

4.2 *Fabrication*

The test specimens were cast at a commercial facility using the same concrete mix as prototype piles. Special plywood forms were used to simulate observed damage. These forms were fabricated with 19 mm plywood tapered at each end. This meant that the cross-section reduced nominally from 15 cm × 15 cm to 11.2 cm × 11.2 cm in this region. The length of this region was 55 cm to match the 165 cm average splash zone in distressed piles.

The average measured cross sectional area of this formed pile core (Fig. 1) was 130 cm² with the interface bonding area approximately 2700 cm². Most all of the repairs were carried out on this smooth, formed, damaged surface. Although this surface is unlikely to be encountered in an actual field repair, it is eminently

Figure 1. Formed damage (left) and chipped damage (right).

suited for the purposes of this study since it allowed improvements in the performance of the interface bond to be readily evaluated.

As damaged prototype piles continue to support sustained loads prior to repair, it was necessary to simulate this in the testing. For this purpose, a specially designed, self-straining frame was fabricated. This frame had to be compact, lightweight, easy to assemble or disassemble and inexpensive. More importantly, the applied load had to remain constant load until additional loading was superimposed during the ultimate load testing.

A self-straining frame relying on angles, threaded rods and bolts was developed that met all the required criteria. Threaded rods were tensioned using nuts to apply loads that were resisted by frictional resistance between two pairs of angles and the pile specimen (Fig. 2). The bolts were torqued to specification to provide an 89 kN load that corresponded to the ser vice loads in prototype piles. The torque was checked periodically using a calibrated mechanic's torque wrench. It may be seen from Figure 2 that the load was only applied to the repaired region that was expected to fail under the applied concentric loading.

4.3 *Jacketed repair*

Pile jackets were formed with plastic lined plywood that were sealed and maintained full of water. These forms were drained just prior to the placement of the cementitous filler material. Four filler material batches used for repairing prototype piles were used in the

Figure 2. Self-straining frame for simulating pre-existing pile loads. The frame was removed during the ultimate load test.

test. Complete details may be found elsewhere (Fischer et al. 2000).

4.4 Instrumentaion

Prior to testing 12 strain gages were mounted on each of the four faces of the piles. Two of these were located 30 cm from each end and were used to ensure the specimens were properly aligned inside the test frame. The third set was attached to the surface of the repair. Additional set of gages was mounted in this region in the second phase to provide information on debonding.

4.5 Test procedure

After instrumentation, piles were placed inside the loading frame. A nominal load was applied with a hydraulic cylinder and based on the strain measurements, the specimen was repositioned to ensure it was subjected to concentric loading. Once positioned properly, the 89 kN dead load simulation load was applied. Under this load, the self-straining frame was removed. The strain and deflection gages were reset and loading resumed until failure.

4.6 Phase I experimental program

Two types of repair – non-structural and structural, were performed on two types of interfaces – formed and chipped surfaces (Fig. 1). Chipped surfaces were prepared by removing the concrete cover and exposing the strands and spirals using a chipping gun. This resulted in approximately $1420 \, cm^2$ of interface bonding surface. This uneven surface was expected to provide good mechanical bond. The average cross sectional area of this chipped pile core is $81 \, cm^2$. This was much smaller than the $130 \, cm^2$ area for the formed surface.

4.7 Phase II experimental program

The principal objective of Phase II was to evaluate measures which could provide improvements to the interface bond. Powder actuated nails and epoxied dowel bars were selected to improve mechanical bond. Two proprietary products were also tested to evaluate their performance in enhancing chemical adhesion of the repair material.

Powder actuated nails are an unique type of fastener that is driven not by electricity or compressed air but by energy derived by a gun powder propellant. A special tool is required to fire the charge and release energy that accelerates a piston forcing the nail into the base material. Nails are made from specially hardened steel and usually have deformations to enhance anchorage performance. Construction is simplified since no pre-drilling of holes is necessary. Moreover, tools are simple and safe to operate, and do not require extensive training.

The idea of using powder actuated nails was first proposed by one of the Florida Department of Transport's district offices though it had not actually been tested. In their scheme it was proposed that powder actuated nails be placed in a 7.6 cm grid pattern above the water line. In the narrower model specimens such a grid was not possible and only a single line of nails was installed on each of the four faces. A modification of that scheme was tested in which powder actuated nails were also extended below the water line. Both structural and non-structural repairs were modelled.

As an alternative to powder actuated nails, epoxied dowel bars were evaluated as mechanical shear connectors. These were however installed in pre-drilled holes but their spacing was kept identical to the powder actuated nail system to facilitate comparison of their relative performance.

In addition to the mechanical connectors, proprietary products were also evaluated. In these repairs, the pile core was treated with a cement densifying and sealing compound and a mix water conditioner used in preparing the concrete filler.

5 TEST RESULTS

An assessment of any improvements to the interface bond required establishing baseline measurements. This was achieved in Phase I testing program considering the specimens with the formed surfaces. The other key result obtained in the initial round of testing was the *debond* load. This is the load at which loss of composite action within the repaired system is observed.

Figure 3 shows a load vs. axial strain plot. The debond event corresponds to the load at which the response becomes markedly non-linear. As shown in this figure, the debond load excludes the 89 kN preload that is included in the ultimate load.

5.1 Phase I results

The results of the ultimate load tests indicated that concentric rather than eccentric loads were more critical for evaluating interface bond performance. Results of tests containing information on the debonding and the ultimate load are summarized in Table 1. As expected, all failures were located in the repair region as its cross-section was smaller.

The debond load provides a measure of the efficiency of the repair. The higher debond load for chipped interfaces compared to formed surfaces is as expected. Lower ultimate loads for chipped interfaces are due to their reduced cross-sectional area indicated earlier. Structural repairs for formed surfaces showed higher capacities despite a smaller debond load. This is possibly because the load carried by the reinforcement was directly transferred by end bearing rather than shear. The confinement provided by the reinforcing ties used in such structural repairs had no apparent effect on bond.

In summary, the results from Phase I indicated that nonstructural type of repair essentially offered no improvement in ultimate axial capacity under concentric loading when compared to the damaged unrepaired controls. Structural repairs had no apparent beneficial bonding effect on the formed surface; however, they offered improvements for the chipped interface.

5.2 Phase II results

Three types of interface improvement where investigated in Phase II – powder actuated nails, epoxied dowel bars and chemical treatment. As before results for the debonding and ultimate loads are summarized in Table 2.

Compared to the debonding loads obtained in Phase I (Table 1), the introduction of powder actuated nails leads to increased debond load especially when the nails are positioned only at the top (543 kN vs 472 kN). Increases are more modest when nails are provided throughout (485 kN vs 472 kN). The results

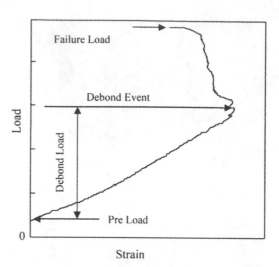

Figure 3. Load vs axial strain plot identifying debond load.

Table 1. Summary of Phase I axial test results.

Specimen		Capacity (kN)	Debond load (kN)
C	Undamaged control	1063	
D	Unrepaired control	658	
NF	Nonstructural formed	663	472
SF	Structural formed	859	463
NC	Nonstructural chipped	601	512
SC	Structural chipped	778	632

Table 2. Summary of Phase II axial test results.

Specimen		Capacity (kN)	Debond load (kN)
PAT	Nails top	676	543
PATB	Nails throughout	618	485
EDTB	Epoxied dowels throughout	743	463
CA	Chemical adhesion	712	423

were poorer for epoxied dowels and for the chemical treatment. However, for the latter two cases, ultimate capacity was greater (743 kN and 712 kN vs 663 kN in Phase I). Interestingly, the capacities were lower when nails were used throughout (618 kN vs 663 kN).

6 DISCUSSION OF RESULTS

It is instructive note that none of the repaired col umns were restored to their full ultimate capacity of 1063 kN (Table 1). The largest increase was for the

Figure 4. Damage due to installation of powder actuated nails.

Table 3. Debond load and average bond stress at interface.

Specimen		Debond load (kN)	Bond stress (MPa)
NF	Nonstructural formed	472	3.48
SF	Structural formed	463	3.41
NC	Nonstructural chipped	512	7.21
SC	Structural chipped	632	8.90
PAT	PA nails top	543	4.01
PATB	PA nails throughout	485	3.58
EDTB	Epoxied dowels throughout	463	3.41
CA	Chemical adhesion	423	3.12

structural repair that regained 81% of its original capacity. Among non-structural repairs, the highest strength gain was for the epoxied dowel system that regained nearly 70% of the strength.

Due to the loss of bond, ultimate capacity was largely dependent upon the condition of the pile core for all the repairs. Core damage imparted by the installation of powder actuated nails (Figure 4) decreased capacity from 658 kN (control) to 618 kN. Techniques which may be viewed as less invasive to the pile core such as epoxied dowels in pre-drilled holes and chemical admixtures fared better at 743 kN and 712 kN respectively even though debonding loads were lower.

7 AVERAGE BOND STRESSES

In order to evaluate improvement of the bonding interface it is instructive to compare the average bond stress corresponding to the debond load.

The average bond stress was determined by dividing the debond load by *half* the computed interface area. This assumes that the upper half of the pile takes the load from the core and the lower half transfers it back.

Table 3 summarizes these results. Inspection of this table indicates the average interface bond varied between 3.12 MPa (Chemical Adhesion) to 8.90 MPa (Structural Chipped). Interestingly, improvement in bond as a result of the introduction of shear connectors is marginal. Thus, it is reasonable to conclude that the debond load provides a measure of chemical adhesion between the substrate and the repair material.

The structural repairs carried out upon irregular chipped surfaces showed improvements to the composite behavior. The chipped surfaces reported bond stresses up to 8.90 MPa on the structural repair which is 1.69 MPa larger than the nonstructural repair upon that surface. This is probably because mechanisms other than shear, e.g. end bearing were responsible for the load transfer.

Although the mix water conditioner and the densifying agent did not lead to larger average debond stresses, it led to higher ultimate capacities. This is possibly because these agents appeared to minimize the effects of bleed water. A minimal amount of water rising to the top of the pile jacket was observed in comparison to other repairs that were carried out. Strain data (Fischer et al. 2000) showed that the upper part of the jacket tended to debond first. However, for these specimens this debonding occurred at a proportionately larger load (88% vs 64%).

8 SUMMARY AND CONCLUSIONS

This paper provides a brief summary of an experimental study conducted to evaluate measures that could be taken to improve the repair substrate interface bond in prestressed piles. Both mechanical connectors and proprietary chemical products were evaluated through ultimate load concentric tests carried out on one-third scale specimens. Two types of interfaces were examined – formed and chipped and both non-structural and structural repairs examined. All repairs were carried out using procedures specified for prototype piles using the same repair materials. Based on the results of the tests the following conclusions may be drawn:

1. The use of mechanical connectors such as powder actuated nails of epoxied dowels did not significantly improve performance of the bond interface. Moreover, damage caused by installation of powder actuated nails can potentially reduce ultimate capacity.
2. Ultimate capacities are largely dependant upon the core strength following debonding. Preservation of the minimum cross section is crucial in column repair strategy.
3. Structural repairs led to higher capacities despite compromises at the bond interface.
4. Interface bond is weakest at the top of the jacket. Measures to reduce the amount of bleed water rising to the top of the jacket can lead to better performance.

Although the scheme for installing powder actuated nails could not exactly model that in prototype piles, the results of the laboratory study suggest that considerable care must be exercised during their installation to limit damage to the core. Following debonding, this core must support the load. Rather than removing more material in the distressed portion of the pile it is preferable to prepare the ends of the damaged section so that they are square ensuring a near-constant repair depth. A constant depth facilitates direct load transfer by compression. This utilizes concrete's superior compressive strength rather than relying on its modest shear bond strength.

ACKNOWLEDGEMENTS

This investigation was carried out with the financial support of the Florida Departments of Transportation. However, the opinions, findings and conclusions expressed in this publication are those of the writer and not necessarily those of the Florida Department of Transportation.

REFERENCES

Fischer, J., Mullins, G. & Sen, R. 2000. Strength of Repaired Pile. *Final Report submitted to Florida Department of Transportation*, March, pp. 185.
Leet, K. & Bernal, D. 1997. *Reinforced Concrete Design*, Third Edition, McGraw-Hill, New York, p. 205.

System-based Vision for Strategic and Creative Design, Bontempi (ed.)
© 2003 Swets & Zeitlinger, Lisse, ISBN 90 5809 599 1

Effectiveness of FRP strengthening of masonry systems of arches and columns

U. Ianniruberto
University of Rome "Tor Vergata", Rome, Italy

M. Imbimbo & Z. Rinaldi
University of Cassino, Cassino (FR), Italy

ABSTRACT: The study investigates the collapse behavior of a simple masonry frame made by an arch and two columns, loaded by vertical and horizontal forces and strengthened by partial or complete FRP reinforcement of the inner surface of the structure. In the former case the limit analysis method is adopted while in the latter a new methodology, developed by the same authors, is used. The failure interaction domains in the plane of the loads are evaluated for both strengthening techniques. In the case of complete strengthening the collapse rarely occurs with a mechanism, but it is related to a failure condition at a local level. Therefore the material properties and the related failure criteria play a fundamental role in defining the ultimate loads of the structures. Herein the case of a collapse governed by the flexural crisis of the critical cross section is shown.

1 INTRODUCTION

Masonry constructions often require to be repaired, strengthened or reinforced in order to have their structural safe enhanced. Such constructions do often demand, as well, the maintenance of their architectural aspects. The use of composite sheets placed at the surface of the structures usually allows to satisfy both requirements.

Many experimental investigations (Briccoli Bati & Rovero 1999, Faccio et al. 1999, Modena et al. 1999) showed the efficacy of the technique in improving the strength of the structures and, in addition, the well known properties of the FRP sheets make the intervention fast, reversible and very well adaptable to different shapes with a strong concern to the architecture of the system.

The application of FRP sheets modifies the collapse behavior of the system without any reinforcement. In particular the effect is different if the composite is applied at the intrados or at the extrados and, as well, if the strengthening is partial or complete.

The problem of evaluating analytically the failure load of a strengthened arch has been addressed, with some simplifying assumptions, in many papers (Como et al. 2000, Faccio et al. 1999, Modena et al. 1999).

In Como et al. (2001) a simple method has been developed for calculating the failure load of a single arch completely reinforced at both intrados and extrados and subjected to vertical loading at the crown or at the haunch.

Starting from this paper in Ianniruberto et al. (2003) the more general case of simple masonry frame, made by two columns and a circular arch, loaded by lateral and vertical forces has been addressed. As a result the method to evaluate the interaction failure locus in the plane of the horizontal and vertical load is described.

In this paper a comparison between different techniques of FRP strengthening at the inner surface of the structure is shown for the reference frame of Figure 1.

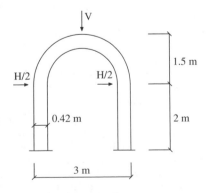

Figure 1.   Analysed masonry frame.

The thickness of the columns and the arch is chosen a bit larger than the minimum thickness of this structure (Heyman 1966 and 1982).

2 COLLAPSE ANALYSIS OF THE UNSTRENGTHENED MASONRY FRAME

It is well known that the collapse multiplier of a masonry frame without any reinforcement can be evaluated on the basis of the limit analysis.

In the framework of the rigid-in-compression no-tension model the collapse occurs when a pattern of "hinges", sufficient to give rise to a mechanism, develops (Heyman 1982). Hinges appear where the thrust line touches the extrados or the intrados and relative rotations between rigid blocks occur. In these cases, the mechanisms that do not require material penetrations are called cinematically admissible.

The evaluation of the collapse load can be obtained by using the cinematic or static theorem. In the first case for each of the cinematically admissible mechanisms, the upper bound of the collapse multiplier is computed on the basis of equilibrium conditions. In the second case for each of the statically admissible configurations the lower bound of the collapse multiplier is evaluated by means of compatibility equations that, in the case of the previously assessed constitutive assumptions, are equivalent to set the thrust line within the thickness of the structure.

The collapse behavior of the masonry frame reported in Figure 1, subjected to incremental vertical and horizontal forces, has been analyzed by considering symmetrical (Figure 2) and asymmetrical collapse mechanisms (Figure 3).

For each of the considered mechanisms the application of the principle of virtual works leads to:

$$\langle g, v(\alpha, \alpha_2, \ldots, \alpha_n) \rangle + H v_H(\alpha, \alpha_2, \ldots, \alpha_n)$$
$$+ V v_V(\alpha, \alpha_2, \ldots, \alpha_n) = 0$$

with:

- $(\alpha, \alpha_2, \ldots, \alpha_n)$ angles defining the position of hinges
- g dead loads
- $v(\alpha, \alpha_2, \ldots, \alpha_n)$ displacement field
- $H v_H(\alpha, \alpha_2, \ldots, \alpha_n)$ work made by the horizontal load;
- $V v_V(\alpha, \alpha_2, \ldots, \alpha_n)$ work made by the vertical load.

2.1 Collapse domains

For each of the analyzed mechanism, fixing the ratio $\eta = H/V$, the minimization of H or V with respect to $(\alpha, \alpha_2, \ldots, \alpha_n)$ leads to the definition of the collapse couple (H,V). Varying η the collapse relationships between H and V can be obtained. Then the failure domain of the structure is defined by the inmost envelop of the curves related to each mechanism.

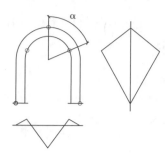

Figure 2.   Symmetrical collapse mechanism.

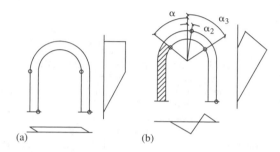

(a)   (b)

Figure 3.   Asymmetrical collapse mechanisms.

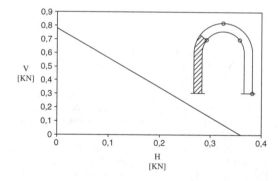

Figure 4.   Collapse domain of the analyzed masonry frame.

The collapse domain of the analyzed scheme is characterized by a linear pattern and it is related to the asymmetric mechanism reported in Figure 4. The other mechanisms give values of the multipliers always above the plotted line.

3 MASONRY FRAME PARTIALLY REINFORCED AT INTRADOS WITH FRP

In the following two strengthening methods are analyzed: the case of sheets applied at the crown of the arch and the case of sheets applied at the bottom of the columns.

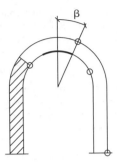

Figure 5.   FRP sheets applied at the crown of the arch.

Figure 7.   Hinge arrangement of the masonry frame strengthened at the bottom of the columns.

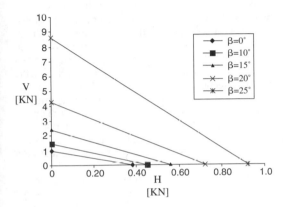

Figure 6.   Collapse domains H–V for different values of β.

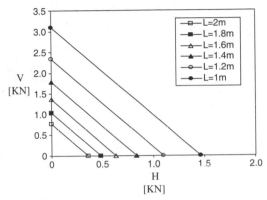

Figure 8.   Collapse domains H–V for different values of L.

3.1   FRP sheets applied at the crown of the arch

Let us consider the system reported in Figure 5 where the FRP sheets are placed on the central part of the intrados of the arch for a length defined by the angle β evaluated with respect to the keystone. The collapse mechanism is reported in the same figure. A hinge will form on the extrados in the cross-section at the end of the composite sheet portion and the other hinges will develop so as to still allow a collapse mechanism.

The ultimate domains of the analyzed masonry frame are plotted in Figure 6. The bottom line represents the collapse domain of the un-reinforced structure and the others are related to the same system strengthened with FRP sheets at the crown (β varying in the range 0°–25°). The Figure 6 shows the effectiveness of the intervention in improving the bearing capacity of the structure. The collapse domains enlarge with the FRP length, not in a proportional way. In particular a great increase of strength occurs when the β angle is higher than 20°. It is worth noting that the ultimate horizontal force can be increased of 3 times, when β varies from 0° to 25°, while the ultimate vertical load is increased of about 9 times in the same range of β.

3.2   FRP sheets applied at the base of the columns

The structure shown in Figure 7 is strengthened with FRP sheets placed on the bottom part of the columns for a length h.

In this case the composite is assumed to be anchored below the base section of both columns.

As in the previous case the FRP arrangement determines the position of hinges. In particular one hinge will form at the top of the reinforced part of the right column and the others will develop as shown in Figure 7. The ultimate domains of the analyzed masonry frame are plotted in Figure 8 where the first line represents the collapse domain of the un-reinforced system and the others are related to the strengthened structure. The legend shows the considered values of L (Fig. 7).

4   MASONRY FRAME COMPLETELY REINFORCED AT INTRADOS WITH FRP

4.1   General assumptions

The Figure 9 shows the masonry frame of Figure 1 with a composite sheet bonded over the entire inner surface.

It is assumed that the fibers are not bonded below both base sections of the piers. This arrangement of the strengthening sheet at the bottom is the most frequent because the foundation wall below the column is usually thicker than the pier, which lies on it.

The lack of anchorage length at the bottom implies that the thrust line must lie in the thickness of the pier at the base section. Moreover the frame can overturn around the right (left) corner of the bottom section of the right (left) column.

The method to evaluate the failure locus of the frame is shown in Ianniruberto et al. (2003). It is based on the following assumptions:

– the masonry is rigid in compression and has no tensile strength
– the self weight of the structure is negligible.

Both assumptions are widely discussed in the above-mentioned paper and are stated only with the aim of providing a simple method to evaluate the strength of this type of structure.

The classical mechanism of the un-reinforced structure can never form and the frame behaves as a statically determinate structure where the axial force, the shear and the bending moment can be easily evaluated in every cross section.

The collapse load is the minimum among the limit values obtained applying all the possible failure criteria, i.e. the crushing of the masonry or the rupture of the fibers in the critical cross section, the shear failure at the brick-mortar interface, the crisis of the adhesion close to the sheet-masonry interface or the overturning of the entire structure.

In the following the method is briefly summarized.

4.2 The case of a concentrated vertical load at the crown

If the structure is loaded by a concentrated vertical force V at the crown, as soon as V is greater than the collapse load of the un-reinforced frame, the thrust line moves outwards the keystone and the collapse occurs when four symmetrical hinges are already

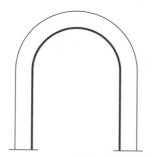

Figure 9. The strengthened masonry frame.

formed. Then the vertical failure load can be easily evaluated on the statically determinate structure shown in Figure 10.

4.3 The case of both vertical and horizontal loads

The safety of the structure, subjected to vertical and horizontal loads can be assessed only if the collapse domain in the plane (H, V) is known.

Following Ianniruberto et al. (2003) let's imagine to increase the ratio $\eta = H/V$ from 0 to ∞. As soon as η is greater than zero the line of thrust at the left base section of the pier moves inwards and then the collapse of the structure occurs with the hinge arrangement shown in Figure 11a.

While the frame without FRP collapses with the same hinge pattern for every value of η (Fig. 4), the behavior of the retrofitted frame is quite different. Indeed the system has the scheme of Figure 11a when the ratio η is lower than the value η_1, which produces the contact between the thrust line and the surface at the left base section. At this stage of loading a new hinge forms but it cannot convert the statically determinate structure into a cinematically admissible mechanism.

If the ratio η increases, the hinge at the left haunch must close and the frame returns to behave as statically determinate structure, which must be analyzed in order to obtain the system of internal and external reactions (Figure 11b). Increasing η the hinge at the right haunch of the arch moves downwards and stops at the top section of the right column. Let η_2 the value of η, which produces this event (Figure 11c). Further increases of η ($\eta > \eta_2$) never change this arrangement.

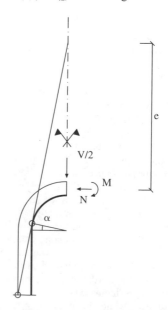

Figure 10. The line of thrust in presence of a vertical load.

The values of η_1 and η_2 for the reference frame are 0.105 and 0.26 respectively.

4.4 The application of the flexural failure criteria

In each phase shown in Figure 11 it is possible to evaluate the internal forces and then the internal stresses in every section of the structure. For every value of η the system of six equilibrium equations plus, if necessary, further compatibility equations, must be solved. Once known the internal forces as functions of V or H, it is possible to find the minimum V that satisfies the chosen failure criterion. As a consequence the collapse domain in the plane (H,V) for any failure criterion can be drawn.

Here we show the results obtained by applying the flexural failure criterion based on the crisis of masonry in compression or the rupture of the fibers in tension.

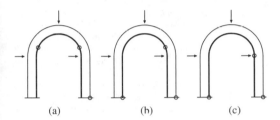

(a) (b) (c)

Figure 11. Evolution of hinge arrangement increasing η.

Table 1. Mechanical properties of the composite FRP sheets.

FRP mechanical properties	
Thickness [mm]	0.165
Tensile Young modulus [MPa]	230000
Tensile strength [MPa]	3430
Ultimate strain [%]	1.5

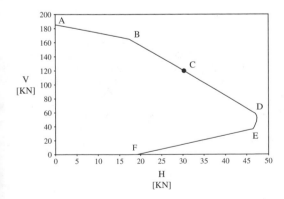

Figure 12. The collapse domain of the frame.

Indeed in the critical section the simplifying constitutive assumption for the masonry is removed and it is assumed the constitutive law reported in the EC6 with maximum strength equal to 6.8 MPa. The influence of the shear stresses on masonry strength is neglected. The sheets are linear elastic in tension and unable to sustain any compressive stresses and their mechanical properties are reported in Table 1. Furthermore the possibility of a global overturning is taken into account.

The results are shown in Figure 12. The segments AB and BC are obtained by the solution of the system of equations relative to the scheme of Figure 11a and 11b respectively. In both cases the most critical cross section is placed at the keystone and the crisis is always due to masonry crushing.

The curve CE is relative to the Figure 11c. In the branch CD the crisis occurs at the crown. At point D the critical section starts moving towards the left haunch. The point E corresponds to the simultaneous attainment of the failure of the section at about 50° from the keystone and of the global overturning of the frame. Along the whole curve AE the failure is governed by masonry crushing. The line EF is the locus of the couples (H,V), which produce the overturning of the frame around the right corner of the section at the bottom of the right column. This failure mode, which is never activated in the unstrengthened structure, must be always considered because of the great increase of the bearing capacity of the FRP reinforced structure.

5 COMPARISONS

The effectiveness of the FRP reinforcement is summarized in Table 2, where the pure vertical and horizontal ultimate capacities are reported for each considered case.

The partial and complete reinforcement of the inner surface provides a valuable increase of the ultimate capacity of the structure as shown in the previous paragraphs. Comparing the two types of interventions the following remarks can be addressed:

a) Behavior at collapse

The discontinuous strengthening modifies the position of the hinges with respect to the un-reinforced

Table 2. Vertical and horizontal ultimate capacity.

Ultimate capacity	H [KN]	V [KN]
Un-reinforced	0.38	0.8
Reinforced at the crown ($\beta = 25°$)	0.92	8.56
Reinforced at the column (L = 1 m)	1.46	3.1
Complete reinforcement	19.2	185

systems but still allows the activation of a collapse mechanism. This means that a limit analysis approach is still valid and no further information on the constitutive behavior of the materials are required.

When FRP sheets are applied on the entire surface of the system a collapse mechanism is not possible anymore.

The ultimate capacity is that of a statically determinate structure where the collapse load is due to local failures related to either the masonry crushing or slipping at some critical sections or the tensile failure of composite fibres or alternatively the bond failure of FRP/masonry interface. This means that the evaluation of the collapse load requires the knowledge of all the possible strength criteria of the materials which are not easy to define for existing masonry structures.

b) Ultimate capacity
The most valuable advantage of the complete FRP reinforcement with respect to the partial one relies on its efficacy in terms of ultimate capacity. Indeed the maximum vertical strength occurring for FRP at the crown is about ten times greater than the vertical collapse load of the un-reinforced structure.

The maximum horizontal strength, given by the reinforcement at the bottom of the columns, increases of about 284%, against the 137% provided by the FRP at the crown. The complete strengthening of the frame gives an ultimate vertical load of about 185 KN, which is about 232 times the capacity of the un-reinforced system. The horizontal ultimate strength is limited by the overturning of the frame; nevertheless it reaches the value of 19.2 KN, which is 50 times the lateral un-reinforced capacity.

6 CONCLUSIONS

The choice of FRP composite sheets in the strengthening of systems of masonry arches and columns represents a valuable technique for enhancing the structural capacity.

In this paper two strengthening techniques have been analyzed: the partial reinforcement of the inner surface and the complete strengthening of both the arch and the columns. In the former case two different sheet positions have been considered.

The results show the effectiveness of both strengthening technique. In particular the complete reinforcement of the inner surface of the structure gives a huge enlargement of the failure locus compared to the cases

of partial reinforcement. This means that both pure vertical and horizontal bearing capacity of the frame are greatly increased. It is worth noting that the obtained ultimate strength values are strongly influenced by the material properties, which, for this reason, need to be carefully evaluated.

ACKNOWLEDGMENTS

The authors thank the Italian Ministry of University and Scientific Research (MIUR) for the financial support.

REFERENCES

Briccoli Bati, S., & Rovero, L. 1999. Consolidamento di archi in muratura con composito in fibra di carbonio. In *Atti del II Convegno Materiali e Tecniche per il Restauro*, Cassino, Ottobre.

Como, M., Ianniruberto, U., & Imbimbo, M. 2000. La *resistenza* di archi murari rinforzati con fogli in FRP. In *Atti del Convegno Nazionale "La meccanica delle murature rinforzate con FRP-materials: modellazione, sperimentazione, progetto, controllo"*, Venezia, Dicembre 2000.

Como, M., Ianniruberto, U., & Imbimbo, M. 2001. On the structural capacity of masonry arches strengthened by FRP. In *Proceedings Third International Conference on Arch Bridges*, Paris, September 2001.

Como, M. 1992. Equilibrium and collapse Analysis of Masonry Structures. In *Meccanica*, Vol. 27, No. 3, Kluwer, Acad. Publ.

Eurocode 6 – Design of masonry structures, ENV 1996.

Faccio, P., Foraboschi, P., & Siviero, E. 1999. Volte in muratura con rinforzi in FRP. In *L'Edilizia, De Lettera Ed.*, Milano, n.7/8 pp. 44–50.

Foraboschi, P. 2000. FRP reinforcement used to upgrade masonry arch bridges to current live loads. In *Proc. of Advanced FRP Materials for Civil Structures*, Bologna, Italy, 19/10, pp. 109–119.

Heyman, J. 1966. The Stone Skeleton. In *Inter. Journ. of Solids and Structures*, 2.

Heyman, J. 1982. The Masonry Arch. Hellis Horwood Limited Publ., Chichester.

Kooharian, A. 1953. Limit Analysis of voussoir and concrete slabs. *Int. Journ.* American Concrete Institute, 24.

Ianniruberto, U., Imbimbo, M. & Rinaldi, Z. 2003. Strength of masonry frames reinforced at the intrados with FRP. Submitted to *ASCE Journal of Composite for Construction*.

Valluzzi, M.R., Valdemarca, M. & Modena, C. 2001. Behaviour of brick masonry vaults strengthened by FRP laminates. *Journal of Composite for Constructions*. Vol. V. No. 3, August.

System-based Vision for Strategic and Creative Design, Bontempi (ed.)
© 2003 Swets & Zeitlinger, Lisse, ISBN 90 5809 599 1

Uses of composite design for new and rehabilitated tall buildings

J.P. Colaco

College of Architecture, University of Houston, Houston, Texas, USA

ABSTRACT: Composite or Hybrid structures are increasingly used in construction of tall buildings in North America. In this keynote paper, the current practices in this field in North America are summarized. The typical structural systems are introduced in five categories: (i) Frames with composite columns and concrete spandrels; (ii) Frames with composite columns and steel spandrels, (iii) Frames with composite supercolumns and steel girders; (iv) Diagonally braced frames and (v) Composite shear walls with steel linking girders.

Another section will deal with Rehabilitation of a 47-storey tall building using Composite design.

1 FRAMES WITH COMPOSITE COLUMNS AND CONCRETE SPANDRELS

The modern era of Composite Systems using, both steel and concrete for columns, began with the work of the late Dr. Fazlur Khan (Khan 1969) in 1966. He thought that steel and concrete could be combined in the vertical plane just as efficiently as they had been in composite floor beams several decades earlier. His work led to construction of the first modern Composite building, the 20-storey Control Data Center in Houston in 1968 (see Fig. 1). From a systems standpoint, the structure was constructed as conventional steel frame except for the use of very small steel erection columns for the exterior columns. The steel frame was built approximately 8 floors ahead of the concrete placement in exterior columns and spandrel beams. The sequence of steel erection above and concrete columns and spandrels below, was continued till the structure was topped out. Temporary bracing had to be used in all the steel column portion of the structure till the concrete frame was completed. The resulting structure was then an exterior composite frame that carried gravity loads and all the lateral loads. Several advantages were quickly recognized.

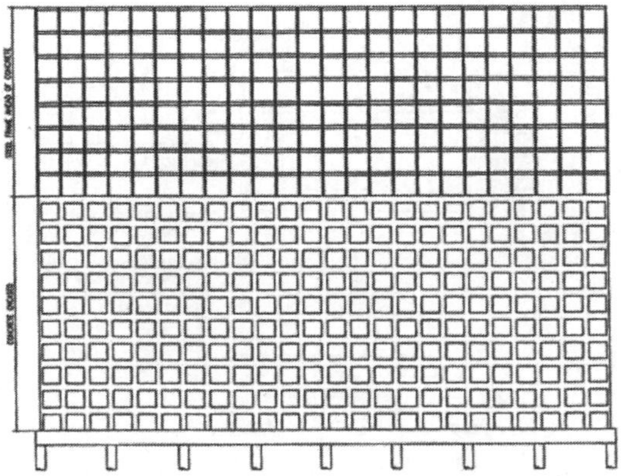

Figure 1.   Elevation of Control Data Center, Houston.

Figure 2.   75-Storey Chase Plaza, Houston.

(a) The steel structure could be built at its normal speed.
(b) The concrete encasement of the exterior columns provided structural rigidity and fireproofing.
(c) The composite structure was economical.

Currently, the tallest composite building in North America is the 75-storey J.P. Morgan Chase Plaza in Houston (Colaco 1985), which utilizes a composite column and concrete spandrel on the exterior as shown in Figure 2. Both materials were used to their optimum, thereby producing cost savings and an entire new structural genre' of building construction. Engineers are no longer restricted to discussing steel or concrete alternates, but added a composite system as viable third option.

2 FRAMES WITH COMPOSITE COLUMNS AND STEEL SPANDRELS

The next step in the evolution of the composite frame systems was the development of moment continuity between a composite column and a steel girder. Research done at the University of Texas developed the basic mechanism of the transmission of shears and moments from the steel girder to the concrete column. The key issue is the development of a diagonal compression strut within the composite column. In order to make this work; the steel girder had to have face bearing plates and adequate studs to develop this compression strut within the joint. The presence of steel girder within the composite columns requires special rebar details such as providing holes in the beam web to enable the column ties to be completed.

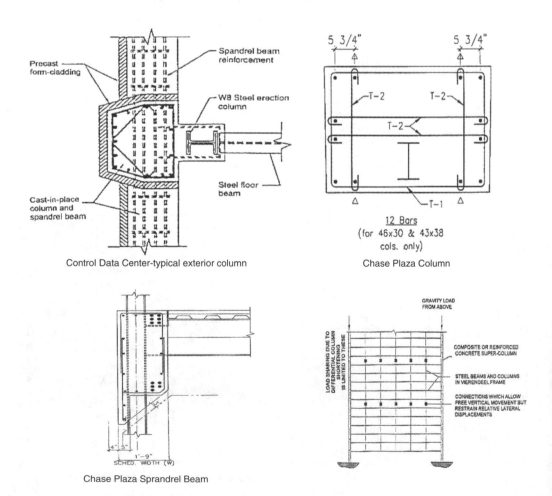

Figure 3.   Super-colums with steel girder.

1480

3 FRAMES WITH COMPOSITE SUPER-COLUMNS AND STEEL GIRDERS

It is generally recognized that concrete columns are substantially more economical than all steel columns in the United States. Further development has given rise to structural systems using only a few, but large columns (called super-columns) connected together by steel beams or braces as shown in Figure 3.

The composite super-columns are generally larger than four (1.2 m) feet in dimension and could be as large as twenty feet (6.0 m) in dimension. The detailing of the attachment of the steel girder is similar to that described in previous paragraph. The size of the composite super-column makes it easier to embed the steel girder. There are several ways in which the composite super-column can be built. One scheme uses a steel erection column with a formed and poured concrete section (see Fig. 4A). In this case the general details are similar to those for a regular composite column. For high-strength concrete and generally for larger sizes the heat of hydration is a consideration. The use of insulated forms and ice in the concrete can mitigate some of the thermal effects.

Another option for the column is a concrete filled large diameter pipe column (see Fig. 4B). Several buildings have been built on the West Coast of the United States using these composite super-columns. Generally, studs are placed on the inside face of the fabricated pipe and a reinforcing cage dropped in prior to concreting. The steel girder can run through the pipe with studs on both flanges. Heat of hydration problems and possible voids below the steel girder flanges need to be avoided. The use of superplasticizers and placement of concrete under hydrostatic head are techniques that have been used successfully.

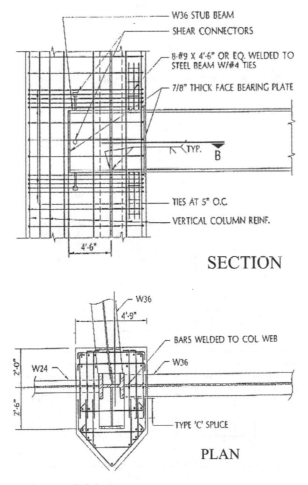

Figure 4A. Composite super-column steel girder.

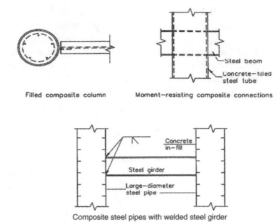

Filled composite column   Moment—resisting composite connections

Composite steel pipes with welded steel girder

Figure 4B.   Concrete-filled steel pipe column.

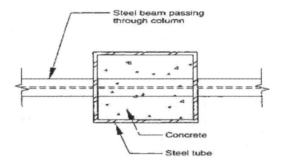

Figure 4C.   Concrete-filled steel tube with steel girder.

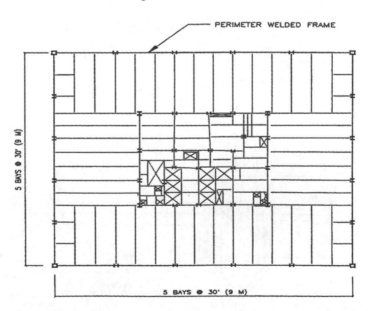

Figure 5.   Original building frame.

1482

A further option is a concrete filled large rectangular steel tube. The steel girder is generally placed in a I-shaped slot cut in the faces of the tube. The other problems and solution are similar to those of round steel pipes. The rectangular tube is not a efficient in the confinement of the concrete.

4 DIAGONALLY BRACED FRAMES

Composite columns have been used a part of diagonally braced system. The following alternates have been built.

4.1 Composite super-columns with diagonal bracing

This is an increasingly popular structural system. Special detailing problems are created due to the interference between the diagonals and the horizontal column ties. A description of the problems is detailed in Reference 4. It should be noted that one of the tallest composite buildings, the Bank of China in Hong Kong, uses this concept.

4.2 Concrete filled steel pipes or tubes with diagonal bracing

Several buildings on the West Coast of the United States use this system. On general, the diagonals are attached to the faces of steel pipe or tube. If the diagonal section goes into the column section, then care should be taken to preserve the integrity of the column ties.

5 COMPOSITE SHEAR WALLS WITH LINKING GIRDERS

Many structures have used composite shear walls as part of the lateral load resisting system. There are two construction methodologies that are used:

(a) A steel column is placed in the wall and the steel frame built ahead and the concrete wall follows approximately 8 to 12 stories behind.
(b) The concrete wall is built ahead of the rest of the steel frame.

The shear walls can be linked by steel girders and all the girder forces are transmitted to the wall. Research done by Shahrooz et al. (1993) has studied the problem and developed guidelines. When there is a steel column at the face of the wall and the girder is moment connected to it, most of the steel girder forces are transmitted directly to the steel column embedded in the composite wall. The detailing of the shear studs, rebar and ties are very important. Also, the differential axial shortening between the composite wall and the other steel columns requires consideration.

6 COMPOSITE RETROFIT OF A 47-STOREY STEEL BUILDING

The project is the retrofitting of an existing 47-storey all steel structure (with four basements) built in Houston in 1971. The original plan of the building is shown in Figure 5. The lateral resistance of the

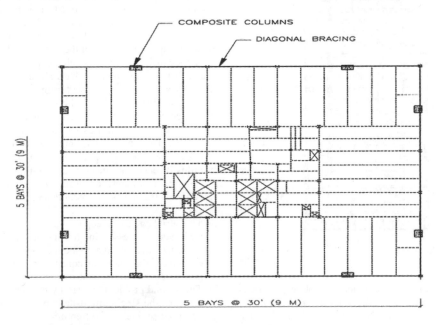

Figure 6. Composite retrofit.

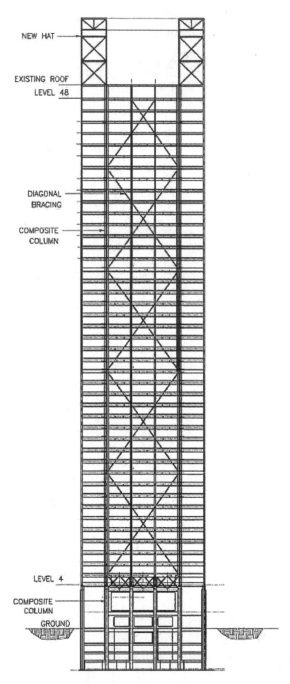

Figure 7. Elevation.

Labels on figure:
NEW HAT
EXISTING ROOF
LEVEL 48
DIAGONAL BRACING
COMPOSITE COLUMN
LEVEL 4
COMPOSITE COLUMN
GROUND

based on the wind loads of the existing code which were lower than the current code. It was apparent that the sway of the building and the P-Δ movements were not considered in the original design. The building was purchased in 1994 and it was decided that the building should be retrofitted for the requirements of the corporation, the 1994 Houston Building Code and the industry practice. The building tenants were relocated elsewhere prior to the start of the retrofit construction.

Six schemes were examined for the structural retrofit of the main wind-resisting frame of the building. The final selected system was to make a total of 8 of the existing exterior steel columns into composite "super-columns" with 9-story tall diagonal steel bracing these between these columns above floor 5 (see Figs. 6 & 7). Below floor 5 all exterior columns down to the mat foundation (at the fourth basement level) were made composite (see Fig. 8) and this composite frame carried the wind loads to the foundation. (A) It should be noted that the column bars were spliced throughout the height of the building with mechanical tension couplers, so that the length of bars was less than if they had been lapspliced. This allowed the bars to be transported in the existing building elevators. (B) A drawing of the construction of the diagonals is shown in Figure 9.

The original building was designed for strength only under the then existing Houston Building Code. This code prescribed a wind pressure of 20 psf (1.0 kPa) up to a height of 60 feet (18 m) and 30 psf (1.5 kPa) above that height. Under that code the base shear was 2,830 kips (12,578 kN) and the base overturning moment was 915,000 kip ft. (2,620,000 kN m). Hurricane Alicia went through downtown Houston in 1983 and this caused an upward revision of the code wind loads. The 1994 code loads were very much higher. Hence a wind tunnel test was run and the base shear is 3490 kips (15,511 kN) and the overturning moment is 1,59000,000 kip ft. (4,552,787 kN m).

The composite columns were built with concrete having a 56-day compressive design strength of 10,000 psi (68.9 mPa). This high strength concrete was used for three reasons, viz., economics, higher modulus of elasticity and reduced column size. The placement of concrete encasement in a diagonally braced steel frame has two technical problems:

1) The shrinkage and creep of the composite column will induce stresses in the diagonals. Hence to minimize the shrinkage induced stresses in the diagonals, the diagonal connections at the composite super-column used slotted holes.
2) The second problem created by the shrinkage and creep of the steel composite super-column is the potential problem of differential axial shortening between exterior composite super-column and

original building was provided by welding the exterior spandrel girders to the exterior W14 (357 mm) columns which are spaced at 30′–0″ (9.1 m) on centers. The original structure was designed for strength

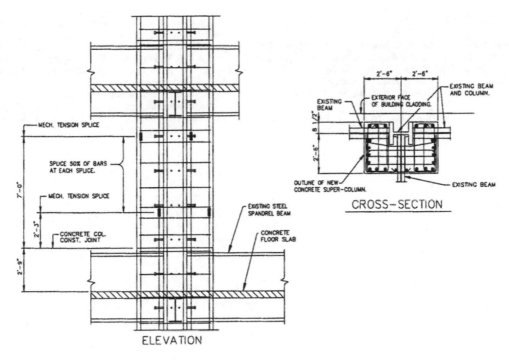

Figure 8.　Construction of a composite column.

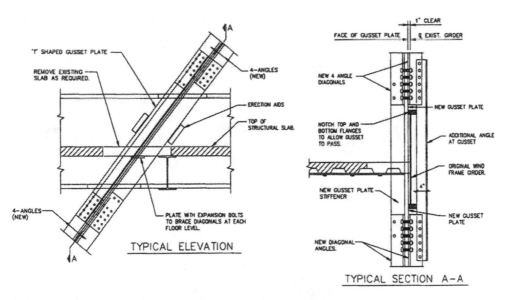

Figure 9.　Typical connection of diagonals at floor.

interior steel columns. Since the existing interior steel columns are set, any shortening of the exterior makes it go down relative to the core. Hence, it was decided to use expansive cement in the construction of the exterior composite columns.

7　CONCLUSION

Composite design is proving to be economical for the design of new tall buildings and the retrofit of older steel buildings.

REFERENCES

Khan, F.R. 1969. Recent Structural Systems in Steel for High- Rise Building, *Conference on Steel in Architecture*, Bethlehem Steel Corporation, Nov. 24–26

Colaco, J.P. 1985. 75-Storey Texas Commerce Plaza, Houston the use of High-Strength Concrete, *Special Publication SP- 87-1*, ACI, pp. 1–8

Sheikh, T.M., Deierlein, G.G., Yura, J. & Jirsa, J.O. 1989. Beam-Column Moment Connections for Composite Frames, Part I, *Journal of Structural Engineering, ASCE*, November, pp. 2858–2876

Viest, I.M., Colaco, J.P., Furlong, R.W., Grifis, L.G., Leon, R.T. & Wylie, L.A. 1997. *Composite Construction Design for Buildings*, Co-published by ASCE and McGraw-Hill, 1997

Shahrooz, B.M., Rememetter, M.E. & Quin, F. 1993. Seismic Design and Performance of Composite Coupled Walls, *Journal of Structural Engineerig, ASCE*, Vol. 119, November 1993, pp. 3291–3309

Structural repair of brick masonry chimneys

B. De Nicolo, Z. Odoni & L. Pani
Department of Structural Engineering, University of Cagliari, Italy

ABSTRACT: In this paper a method of repairing brick masonry chimneys is presented. Instead of utilizing external binding we propose inserting anchors inside the flu and passing a steel cable through the eye rings. The cable is closed in a ring form and prestressed. These rings should be placed at predetermined distances one from the other. This technology ensures the structural monolithicity and stabilizes the cracks in the chimney without damaging its architectural integrity. FEM was utilized to analyze the stress caused by the repair at the linear elastic field in the presence of the self-weight of the chimney and wind.

1 INTRODUCTION

Schincler defined brick masonry chimneys as "smoking obelisks" (Pevner 1986), the symbol and heritage of Industrial Revolution. Today these structures require conservation. They characterize the ski line of many cities and landscapes and bear testimony to the industrial characteristics of an area. The earliest chimneys were built on a square or rectangular bases utilizing ordinary bricks and the construction techniques of bell towers. As brick technology improved these earlier chimneys were replaced with chimneys that utilized *ad hoc* bricks. They have a circular base, truncated cone shape, continuous external tapering and internal stepping of the brickwork which corresponds to the progressive reduction in the thickness. This design proved to be successful because it was aesthetically acceptable (Breymann 1926), technically efficient (good draught without any convective upward turbulence) and stable (resistance to wind and temperature changes). Therefore all the chimneys that have survived have the above design even though they may differ in height or because some have a double flue which offers better wind resistance, functionality and durability (Pistone & Riva 2001, 2002).

Brick chimneys can be found throughout Italy and differ from each other in height and typology depending on geographic location, both for reasons of the windiness of the site and the skills of the brick layers at the time of construction. For example in Sardinia there are only single flue chimneys with a maximum height of 30 m while in Piemonte (Biella) chimneys with double flues and a height in excess of 35 m with a maximum height of 60 m are the most common and in the area of Veneto chimneys with both single and double flues with a height of 25–30 m are common.

Nowadays most of the chimneys show loss of verticality and considerable longitudinal cracks (Pistone et al. 1995, 1996) due to use. These cracks have deteriorated due to neglect and lack of maintenance.

There is no easy solution for repairing brick masonry chimneys because besides the static problems, the architectural integrity and their historical value must be respected (Riva & Zorgno 1995).

In this paper a method of repairing single flue chimneys is presented. It ensures the structural monolithicity of the chimneys and stabilizes the cracks without damaging its structural integrity. The method we propose consists of inserting a series of steel rings inside the flue for the full height of the chimneys at predetermined distances. These series of steel rings are transversally placed and joined to the wall by anchors and prestressed.

To evaluate the overall effects of this method on the stability of the chimney static linear elastic model, using the FEM method, was carried out.

2 MECHANICAL CHARACTERISTICS OF THE BRICKWORK

Numerous investigations both theoretical and experimental have been carried out (Hendry 1986) in order to analyze the average behavior of brickwork among the endless types of masonry bricks and mortar mixtures. The outcome of the above work is a series of tables which can be used to evaluate compression and shear strength of the brickwork in function of the main parameters and some correlations which describe the behavior of the brickwork.

Nowadays, however, we are not able yet to define unequivocally the mechanical characteristics of the brickwork especially in old structures. In fact, for example, compression tests on brickwork show up to 60% reduction in strength, from brickwork without defects to brickwork with defects (McDowall et al. 1966, Tursek & Cacovic 1971, James 1975).

Hendry & Sinha (1969, 1971), Chinwah (1972), Pieper & Traush (1971), Schneider (1976) proposed for shear strength the following correlation:

$$f_v = f_{vo} + \alpha \cdot \sigma_c \qquad (1)$$

where f_{vo} = shear strength of the brickwork without vertical load, σ_c = compression stress due to vertical load is considered in absolute value, α = number function of mortar and brick type; Italian standard (D.M. 20/11/879): $\alpha = 0.4$, $f_{vo} = 0.2$ N/mm^2 for bricks of compression strength $f_{bk} \leqslant 15$ N/mm^2 and $f_{vo} = 0.3$ N/mm^2 for $f_{bk} > 15$ N/mm^2, whatever the strength and composition of the mortar used.

Correlations of shearing are not exhaustive, in fact Pieper and Trausch (1971) demonstrated that shear strength reduces as the length and thickness of the brick masonry increases. In one study in to the influence of the degree of saturation of the bricks on shear strength showed that it is very low if the bricks were saturated or completely dried when laid.

Uncertainty is even greater regarding the calculation of tensile strength even if it is present (Laurence & Morgan 1975), the standards prudently exclude tensile strength because of its low value. It should be noted that the collapse of brickwork, under compression, induced by shear occurs for tensile stress along a diagonal. In the presence of low values of compression stress the fracture line is formed at an angle of around 45° between the bricks and mortar while, for high values, principal tensile stresses are inclined at an angle of less than 45° and crack lines cross both brick and mortar vertically. This behaviour was confirmed by experimental results (Samarasinghe 1980) and agrees with the following correlation:

$$\frac{f_v}{f_t} = \sqrt{1 + \frac{\sigma_c}{f_t}} \qquad (2)$$

where: f_v = maximum shear stress of the brick masonry, σ_c = absolute value of compression stress, f_t = principal tensile stress at collapse. The principal stress of tensile is not constant as its inclination diminishes as σ_c increases and the material is not isotropic, in addition it is strongly affected by the characteristics of the mortar in relation to the cement

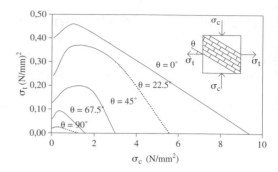

Figure 1. Tensile strength of the brickwork under biaxial tension-compression of relative inclination θ of the mortar joints.

content and water retention capacity. In Figure 1 (Samarasinghe 1980) also highlighted the variation of principal tensile stress at collapse as compression stress changes and the different orientation of the mortar joints with respect to the horizontal.

On the basis of the above considerations, experimental data from analogous structures and from the indication of the Italian standard, it is accepted that the compression strength $f_c = 1.2$ N/mm^2, shear strength $f_v = 0.2$ N/mm^2, tensile strength $f_t = 0.40$ N/mm^2, modulus of elasticity $E = 2000$ N/mm^2, Poisson's ratio $\nu = 0.20$, weight volume $\gamma = 16$ kN/m^3.

3 FAILURE CRITERIA

For the brickwork, valued principal stresses $\sigma_1 \geqslant \sigma_2 \geqslant \sigma_3$, should result:

$$|\sigma_1| \leq f_c \qquad |\sigma_3| \leq f_t \qquad (3)$$

$$\tau = \max\left(\frac{\sigma_1 - \sigma_2}{2}; \frac{\sigma_1 - \sigma_3}{2}; \frac{\sigma_2 - \sigma_3}{2}\right) \leq f_v \qquad (4)$$

4 THE PROPOSED STRUCTURAL REPAIR

The method of repair proposed strengthens the structure like external rings but utilizes an internal system which does not change the external aesthetics of the chimney.

Internal rings are placed at different levels and at predetermined distances one from the other and utilizing a steel cable for prestressing which is closed in a ring form and prestressed. The cable passes through the final eye ring point of a series of steel anchors fixed in the brickwork which generates the tensile force F on the brickwork (Fig. 2).

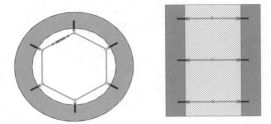

Figure 2. Method proposed using anchors and prestressed steel cable.

Table 1. Characteristics of the anchors (Hilti Corporation).

Symb.	Characteristics	Anchor type	
		A	B
l (mm)	anchor overall length	295	210
h_{min} (mm)	minimum thickness of base material	380	250
h_{eff} (mm)	dept of anchorage	190	125
d_f (mm)	hole in the brickwork	32	23
F^*_{max} (kN)	max tensile force recommended	29.8	13.9
V_{max} (kN)	max shear force recommended	35.9	17.1
$i_{s,min}$ (mm)	min center to center	h_{eff}	h_{eff}

*Center to center i_s between the anchors influences the recommended values of the tensile force F which is possible to apply to them on the basis of the following correlation: $F = f F_{max}$, where $f = 0.5 + i_s/(0.6 h_{eff})$, with $f \leqslant 1$ and $i_s \geqslant i_{s,min}$.

The intensity of F on a single anchor depends both on intensity of T and the number n of the anchors for each ring. For equilibrium, for n even and odd respectively, the result must be:

$$T = F_r/2 \cdot \sum_{i=1}^{n/2} \sin[\beta \cdot (2 \cdot i - 1)] \qquad (5)$$

$$T = F_r/(1 + \cos\beta) \cdot \sum_{i=1}^{(n-1)/2} \sin[\beta \cdot (2 \cdot i - 1)] \qquad (6)$$

where $\beta = 360°/(2n)$.

In Table 1 the geometric and mechanical characteristics of the two types of anchors used (from one producer) are shown. They can be utilised in brickwork that is cracked and in bad condition, in tension and compression areas.

5 CHIMNEY MODEL

On the basis of the geometrical data taken from various chimneys in Sardinia, we utilized FEM to model a chimney with a height of 30 m, external diameter of the base D = 3 m and diameter of the top d = 1.5 m, with a thickness of $s_{max} = 75$ cm at the base and $s_{min} = 25$ cm at the top. We assumed the fixed base hypothesizing optimum foundation base. For the modelling we utilized a solid 20 node brick element in all 13608 bricks and 68436 nodes.

We considered the brickwork as homogeneous and isotropic material in the static linear elastic field.

Besides the self-weight and the wind (equivalent static pressure) we considered the effects of the repair on the structure. We chose not to take into account the effect of heat because the repair is aimed at conservation of the structure and not at its reuse.

6 EFFECTS OF THE REPAIR

In the structural repair the following assumptions were made:

– for simplicity the same number of anchors were used for every ring,
– type A anchors were used up to the height of 20 m (thickness of the wall s $\geqslant$ 40 cm) and type B were used from 20 m to 30 m (25 $\leqslant$ s < 40 cm),
– the tensile force F, transmitted from anchors to the brickwork, was ideally applied at the mid point of the thickness,
– to verify the area of the brickwork in correspondence to the anchors it was felt sufficient to respect the characteristics supplied by the producer of the anchors (Tab. 1), so no plastic analysis was carried out,
– the stress induced is influenced by the longitudinal center to center of ring IL, by the transversal distance between two anchors IT and the intensity of F. By changing these variables an infinite number of combinations can be achieved and the best can be chosen to repair the chimney to ensure the best results from both a static and economic point of view.

On the basis of the above it is obvious that the repair of any chimney is dependent on a vast number of variables, the influence of which is not always clear.

Respecting the above we have also taken into account that the applied forces:

– reduce the circumference in correspondence to the reinforcement rings which must not produce, between the two successive rings, increments in the transversal sections compare to the original outline,
– generate, between the two successive rings, stress with only modest variations along the whole height of the chimney.

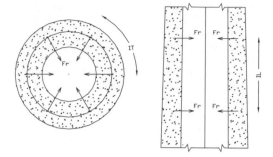

Figure 3. Longitudinal center to center IL and transversal IT of the applied forces.

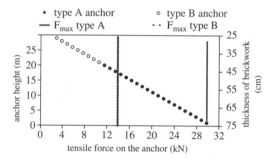

Figure 4. Types of anchors and tensile forces depending on the variation of height and thickness of the brickwork.

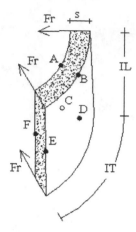

Figure 5. Significant points of brickwork portion longitudinally enclosed by two successive rings and transversally by adjacent anchors.

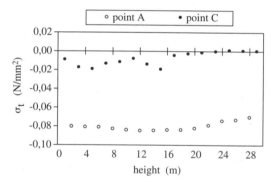

Figure 6. Stresses σ_t at point A and point C at varying heights.

While respecting all the demands of the design we were able, after many attempts, to define the following law of variation of force F for each single anchor at height z. It is valid for any longitudinal and transversal center to center of the anchors (Fig. 4).

$$F_r = -0,9509\, z + 30,736 \text{ (kN)} \qquad (11)$$

7 STRESS AND STRAIN

The compressive stresses caused by the self-weight of the chimney are almost linear with $\sigma_{z,max} \cong -0.35$ N/mm^2 at the base. The wind produces tensile stresses which are always less than the value limit ($\sigma_{z,max} = 0.35$ N/mm^2) also in areas of major stress.

7.1 Stress induced by the repair

Stress in a portion of brickwork which is longitudinally enclosed by the two successive rings and transversally by an angular portion equal to transversal center to center between two adjacent anchors (Fig. 5), represent the state of stress of the whole chimney, because it has been established that such stresses

repeat with little variations for the full height of the chimney (Fig. 6).

Figure 6 shows the σ_t relative to point A which is placed on the internal surface between two anchors of a ring and at point C, longitudinally aligned to point A, but between two rings. It can be seen that compression stresses are almost constant as the height changes: they are highest at point A (about four times). The σ_r values achieved at these points are obviously insignificant.

Table 2 shows the principal and shear stresses (positive tensile) of some points in the brickwork (Fig. 5) relative to repair using 6 anchors per ring placed at longitudinal center to center of 2 m, at a height of between 8 and 10 m for an angular portion of 60°. It showed that in general the state of stress produced by the anchors is very low. See Table 3 for the same points the components of stress σ_r, σ_t and σ_z, respectively in the radial, tangent and longitudinal direction, are reported.

Table 2. Principal and shear stresses (N/mm^2) at the same points reported in Figure 5.

Points	σ_1	σ_2	σ_3	τ
A	0.0280	−0.0020	−0.0800	0.0540
B	0.0001	−0.0100	−0.0160	0.0080
C	−0.0040	−0.0120	−0.0200	0.0100
D	0.0150	0.0008	0.0001	0.0074
E	0.0175	0.0015	−0.0110	0.0093
F	−0.0019	−0.0110	−0.0200	0.0090

Table 3. Stresses (N/mm^2) in cylinder coordinates system for the same points reported in Figure 5.

Points	σ_r	σ_t	σ_z
A	−0.0020	−0.0850	0.0300
B	−0.0007	−0.0098	−0.0160
C	−0.0005	−0.0120	−0.0220
D	0.0001	0.0009	0.0150
E	0.0014	−0.0009	0.0175
F	−0.0020	−0.0105	−0.0200

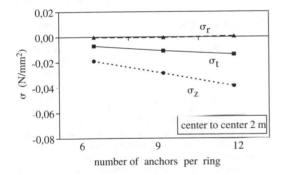

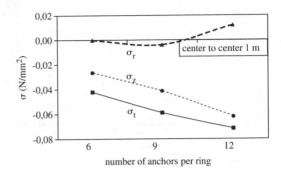

Figure 8. Stresses σ_r, σ_t and σ_z, the number of anchors per ring varying, at longitudinal center to center IL = 1 m relative to height of 14.5 m at point C.

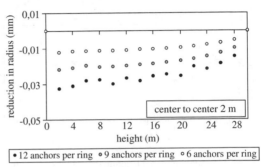

Figure 9. Variations of the radius in correspondence to the ring (longitudinal center to center IL = 2 m) at various heights.

Figure 7. Stresses σ_r, σ_t and σ_z, the number of anchors per ring varying, at longitudinal center to center IL = 2 m relative to height of 15 m at point C (Fig. 5).

Figure 7 shows the stresses σ_r, σ_t and σ_z relative to point C (Fig. 5) at a height of 15 m, located on the internal surface of the chimney, the number of anchors per ring varying (longitudinal center to center 2 m). The data show that doubling the number of the anchors the stresses σ_r and σ_t induced in the brickwork are almost constant, while σ_z stresses double.

If we assume a longitudinal center to center IL = 1 m, the stresses increase significantly from 6 to 12 anchors per ring and we have values of σ_r, even if modest, of tension (Fig. 8).

Generally the states of stress are low and even if they are added to the stress induced by the selfweight of the chimney and the wind they satisfy at failure criteria.

7.2 Strain caused by the repair

The rings increase the structural resistance of the chimney because they locally avoid transversal expansion and interrupt longitudinal discontinuities in the structure.

It must be taken into account however the strain induced by the repair. For this reason in the design we assumed that between two successive rings, the repair did not produce increments in the transversal sections in relation to the original configuration ensuring that, at the same time, the contraction in the ring sections was not excessive.

Figures 9 and 10 show the variations in the radius of the external circumference of the chimney for 6, 9 and 12 anchors per ring (longitudinal center to center IL = 2 m) in correspondence to the rings (Fig. 9) and the intermediate area between two rings (Fig. 10). In fact it can be seen that the variations of the radius of the chimney are not significant neither in the section of ring application nor between rings.

Figure 11, shows the maximum values in the reduction of the external circumference depending on

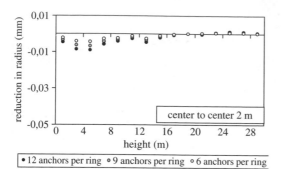

Figure 10. Variation of the radius between two successive rings (longitudinal center to center IL = 2 m) at various heights.

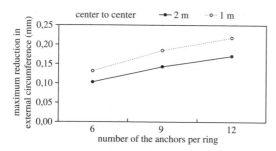

Figure 11. Maximum reduction in the external circumference depending on the number of anchors per ring at various center to center of the rings.

the number of the anchors per ring and the longitudinal center to center, confirms the results.

8 CONCLUSIONS

In this paper a method of repair, utilising internal rings, for single flue chimneys respecting the external aesthetics, is presented. The confinement effect is produced utilising a series of prestressed steel cables, which are closed in a ring form and prestressed, and fixed inside the chimney by steel anchors placed at a pre-determined longitudinal and transversal center to center.

This method must be completed utilising injections of mortar into the cracks to ensure closure. The mortar must be mechanically and chemically compatible with the brickwork.

The type, number of the anchors which join the single prestressed ring to the structure and the center to center between the rings, the geometrical characteristics and the condition of structure are the variable that have to be considered. In this study we established that between two successive rings, the rings did not produce increments in the transversal sections

compared to the original configuration and that the stresses produced by the repair continued for the full height of the chimney with only small variations.

The numerical analysis we carried out using FEM at elastic linear field allowed us, after many attempts, to define the pretension force which had to be applied to the rings as the height changed in respect to the demands of the design and valid for all longitudinal and transversal center to center of the anchors.

These repairs are relatively simple to carry out and their long term effectiveness is ensured because the anchor anchoring and the tension of the prestressed steel cable can be checked for any loss of stress that may occur due to temperature changes or to cable relaxation.

REFERENCES

Barbarito, B. 1994. *Collaudo e risanamento delle strutture.* Torino: Utet.

Breymann, G.A. 1926. *Trattato generale di costruzioni civili con cenni speciali alle costruzioni grandiose.* Vol. 1, Milano: Francesco Vallardi.

Chinwah, J.C.J. 1972. *Shear Resistence of Brick Walls.* Ph.D Thesis, University of London.

Hendry, A.W. 1986. *Statica delle strutture in muratura di mattoni.* Bologna: Patron.

Hendry, A.W. & Sinha, B.P. 1971. Shear tests on full scale single storey brickwork structures subjected to precompressing. *Civ. Engng publ. Wks Rev.* 66, pp. 1339–1344.

James, J.A. 1975. Investigation on the effect of workmanship and curing conditions on the strength of brickwork. *Proceeding of the third International brick Masonry conference* Bonn, pp. 191–201.

Laurence, S.J. & Morgan, T.W. 1975. Strength and stiffness of brickwork in lateral bending. *Proc. Br. Ceram. Sc.,* pp. 5–6.

Mastrodicasa, S. 1993. *Dissesti statici delle strutture edilizie.* Milano: Hoepli.

McDowall, I.C., McNeilly, T.N. & Ryan, W.G. 1966. *The strength of brick walls and Wallettes.* Special Report No. 1 Building Development Research Institute, Melbourne.

Pevner, N. 1986 *Storia e caratteri degli edifici.* Roma: Palombi.

Pieper, K. & Traush, W. 1971. Shear Tests on Walls. *Proceedings of the Second International Brick Masonry Conference* Stoke-on-Trent. pp. 140–143.

Pistone, G., Riva, G. & Zorgno, A.M. 1995. Structural behaviour of ancient chimneys. *5th Int. Conf. on Structural Studies, Repairs and Maintenance of Historical Buildings STREMA H'95.* Comp. Mech Publications, Southampton, Boston Vol. 3, Advances in Architectures Series, pp. 331–341.

Pistone, G., Riva, G. & Zorgno, A.M. 1996. Problems with restoration of old brickwork chimneys in northern Italy. *Proceeding of the 7th Int. Brick Masonry Conference.* University of Notre Dame, South Bend, Indiana, USA Vol. 1, pp. 408–417.

Pistone, G. & Riva, G. 2001. Le vecchie ciminiere in opera laterizia. *Costruire in laterizio.* N. 79, pp. 56–61.

Pistone, G. & Riva, G. 2002. Le ciminiere in laterizio tra conoscenza e conservazione. *Costruire in laterizio*. N. 85, pp. 56–63.

Riva, G. & Zorgno, A.M. 1995. Old brickwork chimneys: structural features and restoration problem. *4th Int. Conf.on Structural Studies, Repairs and Maintenance of Historical Buildings STREMA H'95*. Comp. Mech Publications, Southampton, Boston Vol. 2, Dynamics, Repairs & Restoration, pp. 317–327.

Samarasinghe, W. 1980. *In Plane strength of Brickwork*. Ph.D Thesis, University of Edinburgh.

Schneider, H. 1976. Tests on shear resistance of masonry. *Proceeding of the 4th Int. Brick Masonry Conference*. Brugge, Paper 4.b.12.

Sinha, B.P. & Hendry, A.W. 1969. Racking tests on storey-height shear wall structures with opening subjected to pre-compression. *Designing, Engineering and Constructing with Masonry Products*, Houston, pp. 192–199.

Tursek V. & Cacovic, F. 1971. Some experimental results on the strength of brick masonry walls. *Proceeding of the 2nd Int. Brick Masonry Conference*. British Ceramic Research Association, London, pp. 149–156.

System-based Vision for Strategic and Creative Design, Bontempi (ed.)
© 2003 Swets & Zeitlinger, Lisse, ISBN 90 5809 599 1

Replacement of Metro-North's Bronx River Bridge

Ted Henning
Parsons

Peter Pappas
Metro-North Railroad

ABSTRACT: Parsons inspected and designed the rehabilitation of Metro-North Railroad's Bronx River Bridge in Woodlawn, New York. The inspection and conceptual design phases determined that the existing steel trough deck had reached the end of its useful life and should be replaced by installing a new concrete deck. However, analysis of the remaining bridge components showed that the existing foundations could not support the significantly higher self-weight of the proposed concrete deck, and would thus have to be reinforced or replaced.

Conceptual design of the foundation rehabilitation, controlled by the many restrictions imposed by the surrounding site and by the need to minimize the impact of the construction on the operation of the railroad, was complex. Numerous conceptual design alternatives were thus carefully considered.

After much deliberation, it was decided to replace the existing foundations using the innovative technique of installing new mini-pile foundations through the existing substructure.

1 INTRODUCTION

The Bronx River Bridge is a two-span bridge carrying four railroad tracks approximately 120 feet across the Bronx River in Woodlawn, NY. Critically located immediately south of interlocking CP 112, the junction point for Metro-North's Harlem and New Haven lines, the bridge carries approximately 500 passenger trains per day.

The existing bridge comprises riveted steel plate deck girders with riveted steel plate trough deck, supported by a combination of masonry and concrete substructures. The concrete portions of the substructure are founded on piles. The nature of the foundations for the masonry portions is unknown.

Based on routine visual inspections by MNR in-house forces and by other consulting engineering firms, this bridge was selected for rehabilitation. Parsons Transportation Group (PTG) was contracted by Metro-North Commuter Railroad (MNR) in 1997 to provide inspection services, load rating analyses, conceptual and final design documents and construction support for the rehabilitation of five independent railroad bridges to establish a state of good repair for a 30-year duration, and to establish a load rating of Cooper E 80. Metro-North's Bronx River Bridge at Woodlawn was one of these five bridges.

The original two-span structure, built in 1898 by the New York Central and Harlem River Railroad, carried two ballasted tracks and comprised two riveted steel plate deck girder pairs per span with riveted steel plate trough deck supported on gravity-wall masonry abutments and a masonry pier. See Figure 1.

The bridge was modified in 1906 to carry four tracks. The substructure was widened by extending the existing masonry abutments and pier with concrete founded on timber piles. The existing bridge seats were also replaced with ones that were installed continuous across the full length of the widened hybrid abutments and pier. Typical to construction methods of the time, old running rail was used to reinforce the new pile caps and bridge seats. New riveted steel plate deck girder pairs and riveted steel plate trough decks were added to widen the superstructure.

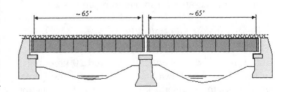

Figure 1. Elevation of existing bridge.

Inspection of the bridge during the first phase of the project determined that the existing steel trough deck was at the end of its useful life and should be replaced. It was determined during the conceptual design phase that the existing lightweight steel trough deck would be replaced with a concrete deck, significantly increasing dead loads. Thus, the capacity of the bridge's foundations would have to be improved.

To successfully address the issues related to the existing hybrid foundations, combined with site and schedule restrictions, described later in this paper, the unique substructure rehabilitation alternative of installing new mini-piles through the existing substructure was selected. This paper outlines the factors leading to this innovative design solution, and through its successful execution in the field.

2 PARSONS 1997 INSPECTION AND ANALYSIS RESULTS

The existing steel girders were generally found to be in good condition. However, significant corrosion and cracking of the steel trough deck. See Figure 2. Numerous missing rivets between the trough deck and the girders were also found.

The masonry portions of the substructure were generally in good condition while areas of minor deterioration were found in the concrete portions. However, settlement cracks at the centerline of both abutments and the pier were found, as well as downward bowing of the pier cap towards the centerline of the bridge. See Figure 3. This bowing runs the full length of the pier's bridge seat, indicating that settlement occurred after the 1906 bridge widening.

Load rating analysis showed that the existing superstructure, rated approximately Cooper E 55, was capable of carrying the current loads from passenger trains, rated approximately Cooper E 45 for a 60-foot span. However, significant reinforcement of the existing steel girders would be necessary in order to provide a rating of Cooper E 80.

3 CONSTRUCTION RESTRICTIONS

In addition to the structural needs to be considered during conceptual design, the following issues surrounding site access and the impact of the construction schedule on railroad operations also had to be addressed.

3.1 Site access and restrictions

The Bronx River Bridge at Woodlawn, which carries 500 trains per day, is critically located 200 feet north of the MNR Woodlawn Station platforms and immediately south of interlocking CP 112, the junction between the Harlem and New Haven lines. The bridge is flanked on the north and the east by a parkway,

Figure 2.   Crack in trough.

Figure 3.   Bowing of pier bridge seat.

located a top 30-feet of retained fill, and by parkland to the west. Thus, the tracks could not be relocated, temporarily or permanently. Nor could the track profile be raised to allow for the thicker concrete replacement deck. Additionally, the river beneath the bridge is a protected waterway, precluding any construction that would encroach on the river. See Figure 4.

Thus, contractor access to the site is restricted to a single, 15-foot wide path located off the southwest corner of the bridge. This path is located between the track right-of-way and a railroad substation and crosses underground duct banks, which had to be protected from the contractor's heavy equipment using layers of ballast and steel plates. An abandoned, 65-foot long steel truss signal tower that is founded on the side of this path also had to be removed prior to construction.

Figure 4. Overview of bridge.

3.2 Schedule restrictions

The railroad's primary goal at all times is to provide on-time service for its passengers. Thus, delays caused by construction are strongly discouraged, especially at this critically located bridge.

Consequently, only one of the four tracks crossing the bridge could be taken out of service long term at any time. A second, adjacent track out of service also could be taken for short durations during off-peak periods. Simultaneous outage of all four tracks, necessary to move materials across the track right-of-way, would be available only from 1 a.m. to 5 a.m. on weekends. Thus, the number of lifts required by the design had to be limited to as few as possible.

Also, work would not be allowed during the holiday/winter period from November 15 through March 15, and had to be complete by the end of 2003 in order to make tracks available for previously scheduled work farther north in the system. Thus, the design had to assure that the duration of the construction schedule's critical path fit within this window.

4 DESIGN ALTERNATIVES

4.1 Superstructure

Placement of reinforcement plates on the flanges and webs of the existing beams to increase their load rating

was first considered, but proved to be cost prohibitive because of the large number of field-drilled holes required for new bolts. Total replacement of the existing superstructure with precast beams was also considered. This design would eliminate the need for a separate deck, thereby reducing construction duration, and also would reduce future maintenance requirements. But it would increase the dead loads on the existing foundation too significantly. Design and construction of the precast beams would also be complicated by the need for several beam widths, necessitated by the varying widths of the four construction stages.

Total replacement of the existing superstructure with steel girders and reinforced concrete deck proved to be far more cost effective than reinforcing the existing. Compared to the use of precast beams, it allowed for more flexibility in placing them within construction staging lines and imposed less of an increase of loads on the substructure. For these reasons, this last alternative was selected.

4.2 Substructure

Conceptual design for the substructure was governed by the requirements of the American Railway Engineering and Maintenance-of-Way Association (A.R.E.M.A.) Manual for Railway Engineering, which prohibits the design of a combined spread and pile foundation. Therefore, the reinforcement of the existing hybrid foundations with additional piles was immediately eliminated. Thus, only alternatives calling for complete replacement of the existing foundations were considered, as follows.

The first alternative consisted of complete replacement of the substructure with reinforced concrete elements founded on piles. This required large amounts of deep excavation and the installation of temporary sheeting between tracks, which are very tightly spaced. This alternative proved to be expensive and time consuming, as well as quite risky given that the excavation, only inches from the third rail of the adjacent active track, would be approximately 20-feet deep.

Replacement of the existing foundations by underpinning the existing substructure with piles driven in front of the substructure faces was also considered. The substructure loads would be transferred onto these piles by jacking and connecting them to dead man beams installed through the existing substructure. The new piles in front would then be encased in concrete. While this alternative maintained the existing substructure, the new piles would intrude into the stream channel and permanently restrict it, raising issues pertaining to stream flow capacity and upstream flooding, and thereby requiring special permitting. In addition, uncertainties pertaining to the ability to develop full support of the dead man beams behind the existing abutments could not be resolved. See Figure 5.

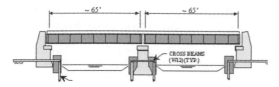

Figure 5.   Underpin with beams and piles.

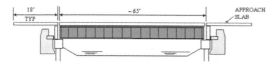

Figure 6.   Replacement with a single span bridge.

Another alternative considered consisted of underpinning the existing foundations with jet grouting. This process creates a bulb of grout with soil aggregate immediately beneath the existing foundations and thus effectively increases their bearing area. This method is very time efficient and requires no track outages. However, it could not be determined whether the jet grouting would improve the capacity of the existing pile footings. Nor could we adequately address concerns over differential settlement between the new jet grout spread footings under the inner tracks and the combined jet grout/pile footings under the outer tracks. In addition, this process requires a local mixing plant to supply the large volume of grout required. The only area large enough for this plant is in the adjacent parkland, which was not desirable.

Complete replacement of the existing two-span structure with a single span bridge was also considered. This alternative held promise for several reasons including reduction of the number of spans and thus the number of bearings, as well as elimination of the second channel in the river created by the center pier. This, however, raised concerns over of the extensive permitting process required to change the shape of the river channel under the bridge. See Figure 6.

The final alternative consisted of underpinning the existing substructure with mini-piles placed through the existing substructure and construction of a new bridge seat. The mini-piles would reach into the competent layer of sands and boulders approximately 50 feet below the existing footings. The new reinforced concrete bridge seat would also act as a pile cap, effectively removing all vertical loads from the existing foundations.

The existing abutments would be left in place to act as retaining walls, minimizing excavation and impact to the adjacent tracks and stream. The existing pier would be left in place to provide protection for the new piles against stream flow. In addition, removal of these existing elements would result in additional costs and

increased construction time. Concerns about shattering the existing substructure and the difficulties of cutting through running rail reinforcement in the existing bridge seats and foundations during coring operations were also addressed by discussing these operations with several contractors.

After considering the effectiveness of each alternative and its impact on the operation of the railroad during construction, as well as permitting requirements, it was decided that installation of mini-piles through the existing abutment and pier stems was the best solution.

5   DESIGN

All portions of the design were performed according to the requirements of A.R.E.M.A. The new mini-piles were designed to carry the entire foundation loads because A.R.E.M.A. does not allow the use of combined footings.

Placement and design of the piles was critical to the success of the project. In order to minimize the impact of construction on train service, and to keep construction equipment away from the protected waterway, the piles would be installed through the area of the existing bridge seat only after coring through the existing substructure. Thus, small diameter mini-piles were selected because of the need to install piles through these cored holes. But because the holes for the new piles were to be cored while the existing superstructure was still operational, the future piles had to be located around all existing members, and between staging lines.

6   CONSTRUCTION

Construction of the replacement of Metro-North's Bronx River Bridge in Woodlawn began in March 2001 and is being performed in four stages, one stage for each track, over the course of two years. Construction for each stage is limited to 4 months. To date, two of the four stages have been successfully completed.

Maintenance of this schedule is critical to the success of the project. Thus, as many construction items were removed from the project's critical path as possible. For instance, test piles were performed adjacent to the bridge on non-production piles, allowing this operation to be performed well in advance of actual production pile installation. The ability to core the holes for the mini-piles prior to taking the track out of service allowed this operation to proceed prior to deck removal. This was accomplished by using hand advanced, motorized drill presses set up beneath the bridge deck.

When the coring operation for each stage had advanced far enough ahead so that it would not interfere with pile installation, the track is removed from service. The track structure and ballast are then removed and

Figure 7. Drill press used to core holes through the substructure.

Figure 8. Narrow width pile drilling rig.

the mini-piles are installed from on-deck through holes cut into the existing trough. The mini-piles have a diameter of six inches and are constructed using 4,000 psi grout with a single number 20, 75 ksi reinforcing bar located in the center of the pile. The pile is protected by a 7" O.D. casing above the bond zone.

A duplex drill system is being used to install the mini-piles. The drilling rig had to be narrow in order to fit within the clearance envelope. See Figure 8.

Even with the narrow width, 7'-4", of the selected machine, mini-pile installation had to be restricted to those near the centerline of the stage while the tracks on both sides were in service. Installation of mini-piles at the outer edges of each stage is performed only when a temporary outage of an adjacent track can be scheduled, usually after 10 a.m. for the southbound tracks and before 3 p.m. for the northbound.

After completion of the pile installation, the remaining portion of the rehabilitation is performed using typical construction methods: remove the existing steel, saw cut and remove the existing bridge seat, form and cast new bridge seat around new piles, install new steel and concrete deck, waterproof, install new ballast and track structure and track utilities, and return the track to service.

Removal of the existing steel, and placement of the new, could only be performed from the southwest corner of the bridge because of site access restrictions cited earlier. Not only did this limit movement of these materials to weekend nights, but it also required a crane that could reach across the entire structure to the center of the northeast span, approximately 110 feet, and pick up a load of approximately 12 tons. The contractor selected the Liebherr LTM 1400, a 500-ton crane, for this purpose.

7 CONCLUSIONS

Construction ran smoothly during the first of two years of construction. As of January 2002, the project is ahead of schedule and under budget.

ACKNOWLEDGEMENTS

This project was performed in association with Mueser Rutledge Consulting Engineers, Geotechnical design subconsultant, and MATRIX Environmental and Geotechnical Services, environmental subconsultant. The involvement of the personnel of each company was essential for the successful completion of this project. Credit is also due to Ecco III Enterprises Inc, Contractor, and Nicholson Construction Co., pile subcontractor, for successful execution of this innovative project in a demanding environment.

The author would also like to thank those involved from Metro-North Railroad, especially Mr. Michael Feinberg, Project Manager and Mr. Paul Nietzschmann, Project manager at Parsons.

20. Artificial intelligence in civil engineering

System-based Vision for Strategic and Creative Design, Bontempi (ed.)
© 2003 Swets & Zeitlinger, Lisse, ISBN 90 5809 599 1

Joint stiffness estimation of thin-walled structure using neural network

A. Okabe, Y. Sato & N. Tomioka
Dept. of Mechanical Engineering, College of Science & Technology, Nihon Univ., Tokyo, Japan

ABSTRACT: The purpose of this paper is to study the method of quickly estimating the joint stiffness values at beginning stage of the automobile body structure design. The neural network, which is able to construct a non-linear mapping relation, was used to develop the method. The researched joint structure is a shell structure composed of three thin-walled box section beams. The proposed method by the neural network was made as follows. At first, the data of the pair of the design parameters and the joint stiffness values is prepared to train the neural network. Next, the relation between design parameters and joint stiffness values is constructed by using neural network. Once the trained neural network is given, the joint stiffness values can be estimated only by inputting the design parameters into the trained neural network.

1 INTRODUCTION

Recently, the design of the automobile body structure needs to develop in a short term. It is important to do a high-quality design in the beginning stage of the design, and not to bring the problem in to the ending stage of the body design as much as possible.

The jointed parts that two or more thin-walled box section beams mutually connected are not perfectly rigid. This flexible characteristic (called the joint stiffness after this) is one of the key factors dominating the basic characteristics of strength, vibration, and stiffness of an automobile body structure.

At a beginning stage of the design, the structure of the jointed part is simplified, and the structural sizes that satisfy the design specification of the joint stiffness tries to be decided. In general, the structural size of the jointed part like plate thickness and the section size, etc. is changed and a structural size which satisfies the target value of the joint stiffness is decided by the rule of trial and error. The stiffness of the jointed part cannot be easily estimated unlike the section rigidity of a general member, and it should depend on the numerical analysis method like finite element method (FEM). However, the analytical model is made again and calculated in FEM whenever the section sizes etc. are changed. It is difficult to correspond to the change in the dimension promptly because time is taken in the modelling. Moreover, the design division and the calculation division are dividing work in most motor vehicle manufacturing companies, and it is rare that the designer calculates with FEM. A convenient tool by

which the designer can calculate the joint stiffness values oneself might be useful for the design.

In this paper, the technique for evaluating the joint stiffness that the designer can use at a beginning stage of the design is proposed. This technique uses the neural network. The relation between the design parameters and the stiffness values of the jointed part is considered to be a mapping problem, and the relation between both is constructed with the hierarchical neural network. If the trained neural network is used, the joint stiffness value of high accuracy can be promptly estimated only by inputting the size of the jointed part.

2 JOINTED PART MODEL AND JOINT STIFFNESS DEFINITION

Figure 1 (a) shows the stress distribution in the vicinity of the jointed part of the L-shape thin-walled box section beams. The stress concentration has been generated in the ridgeline without transmitting the stress in the center part on the side. Figure 1(b) shows the deflection distribution of the L-shape joint structure. If the jointed part is assumed to be perfectly rigid, the calculation value by beam theory in elementary strength of material has the difference in the solution of FEM analysis. Therefore, it is necessary to evaluate the joint stiffness for designing the thin-walled box section beams with the jointed part.

In this study, the joint stiffness is represented by joint stiffness matrix that can represent a size and a property of the joint stiffness very well. Consider the jointed part composed of n beams. The jointed part was defined as follows. The number of beams is n. The longitudinal direction of beam, the shape of cross-section and the direction of the principal axis is assumed to be decided arbitrarily. It is assumed that the elastic deformation of the jointed part is allowed only against moments about three orthogonal axes.

On the above assumption, the jointed part may be considered – the elastic body consisted of the n nodes with same coordinates. The stiffness matrix can represent relationships between moments [M] applied to n nodes and rotation [Θ] and it defines the joint stiffness. It is here called the joint stiffness matrix [1]. In case of the jointed part that consists of 3 beams, the joint stiffness matrix can be represented by Equation 1 (Shimomaki et al. 1990).

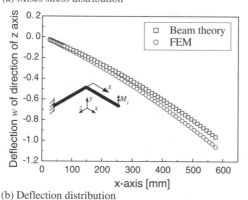

(a) Mises stress distribution

$$\begin{bmatrix} M_1 \\ M_2 \\ M_3 \end{bmatrix} = \begin{bmatrix} K_{11} & K_{12} & K_{13} \\ K_{21} & K_{22} & K_{23} \\ K_{31} & K_{32} & K_{33} \end{bmatrix} \begin{bmatrix} \Theta_1 \\ \Theta_2 \\ \Theta_3 \end{bmatrix} \quad (1)$$

where, $[M_i] = [m_{ix}\ m_{iy}\ m_{iz}]\ (i = 1,2,3)$
$[\Theta_i] = [\theta_{ix}\ \theta_{iy}\ \theta_{iz}]\ (i = 1,2,3)$
$[K_{ij}]\ (i,j = 1,2,3) =$ square matrix

The jointed part can be rotated in the rigid body without the action of the moment. Thus the following relation exists.

$$K_{i1} + K_{i2} + K_{i3} = 0 \quad (i = 1,2,3) \quad (2)$$

The joint stiffness matrix represented by Equation 1 is symmetric by reciprocal theorem.

$$[K_{ij}] = [K_{ji}]^T\ (i,j = 1,2,3\ \ T : \text{transpose of matrix}) \quad (3)$$

(b) Deflection distribution

Figure 1. Distributions of stress and deflection of the L-shape thin-walled box section beams.

Therefore the number of independent components is 6 in all the components of the matrix of Equation 1.

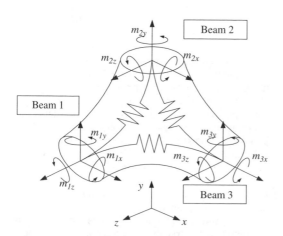

Figure 2. Jointed part consists of the three nodes.

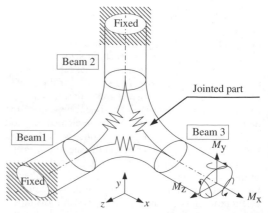

Figure 3. Joint structure consists of three beams.

The diagonal matrix $[K_{33}]$ of the joint stiffness matrix is represented as the following. The matrix $[K_{33}]$ means a structural rigidity of the structure that the edge in beam 1 and beam 2 is fixed. The eigenvalues are the positive real number. The eigenvectors are orthogonal. Thus the matrix $[K_{33}]$ is represented by Equation 4. The eigenvalues are called the principal stiffness of $[K_{33}]$. The eigenvectors are called the principal axis.

Using the principal stiffness, the stiffness matrix $[K_{33}]$ is represented by Equation 4. And the matrix $[A]$ in Equation 4 can be represented in terms of the Eulerian angles as shown in Equation 5. Similarly, $[K_{11}]$ and $[K_{22}]$ can be represented like $[K_{33}]$.

$$[K_{33}]=\begin{bmatrix} k_{77} & k_{78} & k_{79} \\ k_{87} & k_{88} & k_{89} \\ k_{97} & k_{98} & k_{99} \end{bmatrix}=[A]^T\begin{bmatrix} k_{p1} & 0 & 0 \\ 0 & k_{p2} & 0 \\ 0 & 0 & k_{p3} \end{bmatrix}[A] \quad (4)$$

where, k_{p1}, k_{p2}, k_{p3} = the principal stiffness; $[A]$ = the rotation matrix.

$$[A]=\begin{bmatrix} cos\,\gamma & sin\,\gamma & 0 \\ -sin\,\gamma & cos\,\gamma & 0 \\ 0 & 0 & 1 \end{bmatrix}\begin{bmatrix} cos\,\beta & 0 & -sin\,\beta \\ 0 & 1 & 0 \\ sin\,\beta & 0 & cos\,\beta \end{bmatrix}\begin{bmatrix} cos\,\alpha & sin\,\alpha & 0 \\ -sin\,\alpha & cos\,\alpha & 0 \\ 0 & 0 & 1 \end{bmatrix} \quad (5)$$

3 JOINT STIFFNESS ESTIMATION USING NEURAL NETWORK

The estimation tool with which the values of joint stiffness can be obtained by inputting the design parameters (thickness, dimensions of cross-section etc.) to the trained neural network is proposed. Figure 4 is flow chart of a basic idea to make the estimation tool.

Phase 1: The data of pair of the design parameters and the joint stiffness values is prepared as the learning data. The learning data are used to construct the relation between the design parameters and the joint

stiffness values by neural network. The learning data can be prepared by the experiment or the numerical calculation with FEM etc.

Phase 2: The relation between the design parameters and the joint stiffness values is constructed by neural network. The input data of neural network are the design parameters, and the output data of neural network are the joint stiffness values. This constructing is called "Training of neural network". And neural network that constructed the relation is called "the trained neural network".

Phase 3: Once the relation between the design parameters and the joint stiffness values was constructed, the joint stiffness values can be obtained quickly by inputting the design parameters to the trained neural network.

4 EXAMPLES OF CALCULATION

4.1 Learning data

Consider the joint structure as shown in Figure 5. The relation between the design parameters and the joint stiffness values is constructed by neural network. This structure consists of the L-shape structure with the square section and the beam with the rectangular section, and both beams are perpendicularly connected each other. The thickness t and the dimensions a_1, a_2 of cross-section were chosen as the design parameters. The stiffness of the joint structures with all combinations of the discrete values of the design parameter shown in Table 1 was calculated with FEM and the learning data was prepared. The number of the learning data is 255. The half of the learning data was used to construct the relation between the design parameters

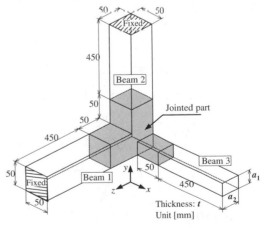

Figure 5. Joint structure that consists of the L-shape structure with the square section and the beam with the rectangular section.

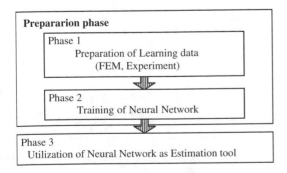

Figure 4. Flow chart of basic idea to make estimation tool by using neural network.

and the joint stiffness values. The other data was used to verify the generality of the trained neural network.

4.2 Joint stiffness estimation

Figure 6 shows the model of the neural network used in this study. The back-propagation algorithm was applied to the neural network used. The input data were the plate thickness t and dimensions a_1, a_2 of cross-section in the design parameter. The output data were the joint stiffness values. The joint stiffness values of output data were the elements $k_{77} \sim k_{99}$ of the partial matrix $[K_{33}]$ or the principal stiffness k_{p1}, k_{p2}, k_{p3} and the Eulerian angles α, β, γ. The number of units in input, hidden and output layers were 6:14:3.

Changing the number of units of hidden layer and conducting a numerical experiment decided the number. Figure 7 shows the relation between the time to the end of the training and the number of units of the hidden layer. If the number of units of the hidden layer is 14 or more, the training time of the neural network does not change so much. Thus the number of units in hidden layer sets 14 of the minimum number. Figures 8 (a), (b) and (c) show the values estimated by trained neural network. The vertical axis shows the estimated values output by the trained neural network. The horizontal axis shows the exact values that are the data prepared as learning data described in paragraph 4.1. As shown in Figure 8, the estimated values of not only the learning data but also unlearning data were in good agreement with the exact values. It was found that the relation between the design parameters and the joint stiffness values could be constructed well.

The neural network was able to construct the relation between the design parameters and the elements of the partial stiffness matrices other than $[K_{33}]$, too (Fig. 9).

The value of the joint stiffness used for the learning data was normalized between 0.2 and 0.8 by the linear transformation. Normalizing by the linear transformation usually converts minimum value and the maximum value of data into 0.2 and 0.8 respectively. There is a tendency that the error grows when the normalized small values are reversely converted because the joint stiffness has the value within the large range. The values normalized by the linear transformation, which are output values of the trained neural network, are reversely converted and the error of the obtained joint stiffness are shown in Figure 5(a). The errors included in small values of the joint stiffness have the tendency to grow. Therefore, the data of the joint stiffness was replaced with the logarithm, the logarithm value was converted between 0.2 and 0.8, and neural network was trained in this research. Figure 5(b) shows the error of the joint stiffness obtained by returning the output value of the neural network to the value before it is normalized. The errors included in small values of the joint stiffness decrease compared with Figure 5(a).

In the same way, about the joint structure as shown in Figure 11, the relation between the design parameters and the joint stiffness values was able to be constructed by using the neural network.

5 DESIGN PARAMETER ESTIMATION USING NEURAL NETWORK

The joint stiffness estimation method has been described above. The proposed tool is very useful for

Table 1. Design parameters of the joint structure model shown in Figure 5.

Design parameter (mm)	Dimension	Step
t	0.6 ~ 1.4	0.1
a_1	25 ~ 45	5
a_2	25 ~ 45	5

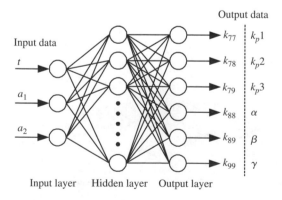

Figure 6. Neural network model to construct the relation between the thickness t and the dimensions a_1, a_2 of cross-section.

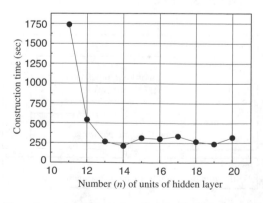

Figure 7. Relation between the time of the training end and the number of units of the hidden layer.

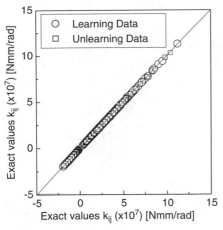

(a) Elements of the joint stiffiness matrices

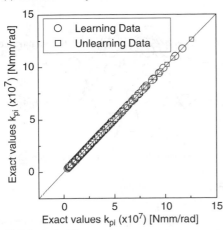

(b) Principal stiffness

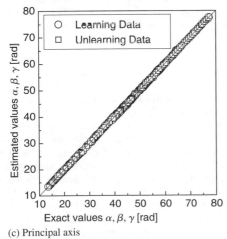

(c) Principal axis

Figure 8. Comparison of the estimated values and the exact values of the partial matrix [K_{33}].

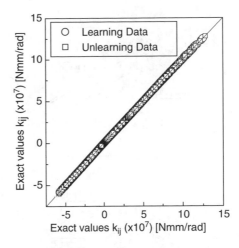

Figure 9. Comparison of the estimated values and the exact values (the partial matrices [K_{ij}] (i,j = 1,2,3)).

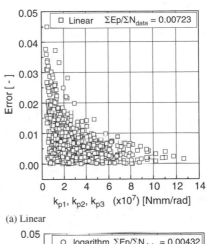

(a) Linear

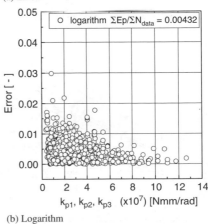

(b) Logarithm

Figure 10. Distribution of error.

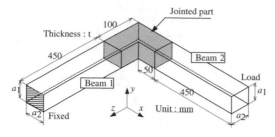

(a) L-shape joint structure with the same cross-section

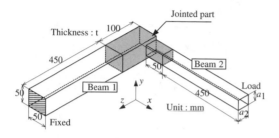

(b) L-shape joint structure with the different cross-section

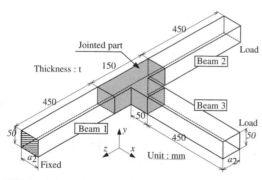

(c) T-shape joint structure with the same cross-section

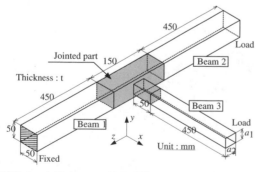

(d) T-shape joint structure with the different cross-section

Figure 11.   The other joint structure.

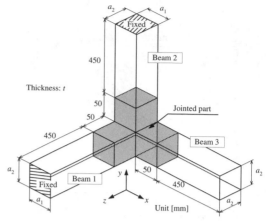

Figure 12.   Joint structure composed of three beams.

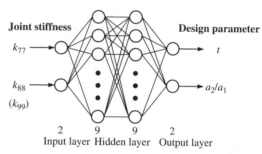

Figure 13.   Neutral network.

Table 2.   Design parameters of the joint structure model shown in Figure 12.

Design parameter (mm)	Dimension	Step
t	$0.6 \sim 1.4$	0.1
a_1	50	—
a_2	$25 \sim 45$	5

estimating the stiffness of the jointed part with the size given at the design. Moreover, the tool that instantaneously obtains a structural size of the joint structure where the design specification of the joint stiffness is satisfied will be very useful for design. Replacing the I/O of the neural network described before as shown in Figure 13 and learning it can develop the design parameter estimation tool.

The relation between the joint stiffness values and the design parameters of the structure as shown in Figure 12 was constructed by neural network. As shown in Figure 14, the estimated values of not only the learning data but also unlearning data were in good agreement with the exact values.

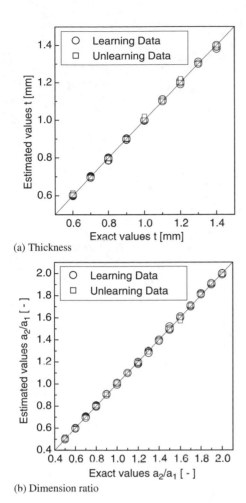

(a) Thickness

(b) Dimension ratio

Figure 14.   Comparison of estimated values and exact values.

6   CONCLUSION

The new method to estimate the joint stiffness values by using neural network was proposed. The main conclusion is as follows.

1. The relation between the design parameters and the joint stiffness values was constructed by neural network.
2. The trained neural network is able to estimate the joint stiffness values accurately by only inputting the design parameters.
3. The data of the joint stiffness was replaced with the logarithm, and the logarithm value was converted between 0.2 and 0.8. The error of the joint stiffness obtained by returning the output value of the neural network to the value before it is normalized was able to decrease.

REFERENCES

Shimomaki, K. et al. 1990, Joint Stiffness of Body Structure – Part 1: Its Evaluation Method, *transactions of JSAE*, No.43, pp.138–142

Okabe, A. & Tomioka, N. 1998, Joint Stiffness Identification of Body Structure Using Neural Network – Jointed part composed of 3 beams, *Proceeding of the 1998 annual meeting of JSME/MMD*, vol.1, No.410, pp.241–242

Okabe, A. & Tomioka, N. 1998, Joint Stiffness Estimation of Body Structure Using Neural Network – Jointed part composed of 2 beams, *Modeling and simulation based engineering*, vol.1, pp.588–593

System-based Vision for Strategic and Creative Design, Bontempi (ed.)
© 2003 Swets & Zeitlinger, Lisse, ISBN 90 5809 599 1

Integrating neural networks, databases and numerical software for managing early age concrete crack prediction

M. Lazzari,* R. Pellegrini & P. Dalmagioni[†]
Enel.Hydro, Seriate, BG, Italy
*University of Bergamo, Bergamo, Italy
†Qualitalia Controllo Tecnico, Milano, Italy

M. Emborg
Betongindustri, Stockholm, Sweden and Luleå University of Technology, Luleå, Sweden

ABSTRACT: This paper presents an expert system that supports early age concrete crack prediction. The system embodies knowledge gathered from experts both in procedural form (programs and spreadsheets) and declarative form (graphical assistants, users' guides, documentation). Moreover, processing tasks are performed by neural networks trained on data gathered from experimental and virtual models. The system comprises a large database for storing data about concrete mixes. The procedures of the expert system can access the database and exploit its data for processing purposes.

1 INTRODUCTION

This paper presents the achievements of a task of the IPACS project, a research granted by the European Communities to evaluate, integrate and extend the existing knowledge about early age concrete crack prediction in engineering practice[1].

The task was devoted to the design and development of software tools useful for contractors and designers for managing early crack prediction (Dalmagioni et al. 2002). These tools consist of a knowledge based system simulating basic early age transient concrete properties, and data bases for storing the same properties, standardized laboratory and field tests, and recommendations and specifications (Salvaneschi & Lazzari 1997). They have been collected and organized into an expert system (called IPACS) able to:

– support the evaluation of the cracking risk in concrete structures;
– indicate possible actions for optimization of technical quality and economy.

The system is focused on assisting the engineer in the decision-making process, offering guidance in assessing critical conditions leading to cracking. It is an integrated and easy to use tool hosting various pieces of codes, databases and logical rules taken by experts in this field, which cannot be all represented by formal mathematical algorithms.

The software may simulate the concrete hardening and building processes taking into account all important factors of influence, such as climatic conditions, non-uniform maturity development, restraints imposed by adjoining structures etc.

Some of the reasoning agents within the system have been developed through neural networks, which were trained on data gathered from experiments, site observations, mathematical modeling coming from other tasks of the IPACS project.

2 EXPERT SYSTEMS AND CIVIL ENGINEERING

Artificial intelligence (A.I.) technology was presented to the international structural community in the early 1980s and experienced an enormous growth of applications in the latter part of the same decade.

From the broad spectrum of the A.I. sub-fields, civil engineering applications belong essentially to the so-called expert systems, also known as knowledge-based systems.

They are software systems that include a *corpus* of expert knowledge and processing abilities to deal

[1]The research was funded by E.C. under the program Brite EuRam, Contract no. BRPR-CT97-0437 (IPACS; 1997–2001).

with data in a way that can be compared with that of experts of the domain.

Therefore, their development and success is strictly related to the availability of theoretical and practical experience on the application field, and to its formalization toward its embodiment into a software system – this process emerges from a joint effort of field experts and A.I. people to identify and correctly exploit the knowledge necessary for solving the problem to be faced by the expert system.

From the point of view of the applications, they belong to two main threads:

1 design;
2 diagnosis.

Generally, those interested in the applications to design, that is support systems which help designers in (part) of their tasks, do not deal with decision support in the field of structural assessment, and vice versa.

An intriguing aspect of the IPACS project is that the expert system to be develop had to be concerned both with supporting design of new concrete mixes and with diagnosis of existing mixes.

With reference to the concrete industry, the number of real application of A.I. technologies is still not large. Some of them deal with the support of site personnel when they choose the type of fresh concrete (BETVAL, Technical Research Center, Finland; COMIX, Central Laboratories, New Zealand); some others support the diagnosis and the definition of repair strategies of deteriorated concrete structures and pavements (ContecES, Darmstadt University of Technology, Germany; EXPEAR, Federal Highway Administration, U.S.A.; PAVE, French National Research Council, France).

The National Institute of Standards and Technology of the U.S. Bureau of Reclamation has also undertaken the development of several interesting A.I. based tools and databases to face several problems related to high-performance concrete production and management, and to develop and implement computational and experimental materials science-based techniques in order to enable the prediction and optimization of the initial cost and service life performance and minimize the environmental impact of concrete in the built infrastructure. They have also used cellular automata to model microstructural development of cement paste during hydration.

Eventually, the group of Knowledge-based systems of the Civil Engineering Department of the University of the Arabic Emirates has developed a system to support the production of concrete for hot weather conditions, that is the other side of the problem to be dealt by IPACS partners, who mainly come from northern Europe, were low temperatures affect concrete hardening.

3 THE SOFTWARE ARCHITECTURE

IPACS has been developed exploiting the Internet technology, so that users may easily share knowledge and data via Internet tools[2]. This solution is based on the development of a web site that embodies a database of information about concrete and intelligent tools to deal with these data, as well as with inputs provided by the users.

As a result, the system can be regarded as a client-server architecture over the Internet, where a server hosts the site and the partners may access it via a common web browser (the client). Each partner can get data from the system, process data and store data into the database. The resulting site, which may run on Windows 9X/NT personal computers, comprises the following components:

– an HTTP server (that is a Web server);
– a set of HTML pages;
– a database developed in Access;
– a layer of Perl programs to access the database (via an ODBC driver) and build on-the-fly HTML pages for interfacing database management functions;
– a set of software tools delivered by partners as executable programs and a set of Perl programs to interface them; Perl routines run these modules and, if necessary, feed them with data extracted from the material database;
– a set of JavaScript programs, to run some of the procedures of the expert system (neural networks, knowledge based and numerical modules);
– a download area, where users can find: tools implemented by partners as DOS executables or spreadsheets; documents which explain how to run the

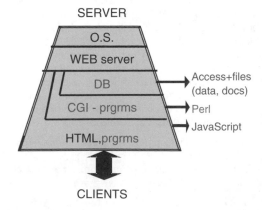

Figure 1.   The layers of IPACS server.

[2]Note that in 1997, when we started the project, the choice of developing an expert system coupled with a large database over the Internet was not as trendy as it is now.

tools and the modules of IPACS; documents which introduce users to the theoretical background of the IPACS tools.

4 THE FUNCTIONAL ARCHITECTURE

The detailed design of the system comprises the following modules:

- *culvert N.N.*: this module receives from the users data that describe a culvert section (a wuall on a slab), the ruling environmental conditions and the kind of concrete, and *evaluates the cracking risk*; it exploits *neural networks* to predict the cracking index; different networks may be used, on the ground of available data: users are driven by a graphical interface to select the most suitable net for their purposes;
- *plate N.N.*: it gets from the users data that describe a plate, the ruling environmental constraints and the kind of concrete, and *evaluates the cracking risk*; this module exploits the same techniques of the previous one;
- *knowledge based & numerical modules*: these procedures implement algorithms or rules of thumb with the purpose of studying cases not represented in the NN modules or to allow for a more engineering guided formalism of the early age response of the culvert and the plate cases; they comprise a module, exploiting both symbolic and neural processing, to *evaluate the restraint factor* for five main types of structure with several sub-cases; a graphical assistant to *simplify three-dimensional structures* for applying the proper case of the restraint evaluator; a *thermal solver*, to calculate the temperature evolution of a semi-infinite concrete wall during the hydration process; a *mono-dimensional solver for the viscoelastic problem*;
- *kind of structure*: it implements a choice point, where the users are given support to choose the processing module that is most suitable for their problem;

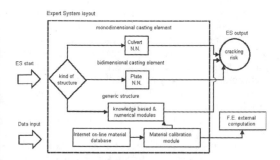

Figure 2. The functional architecture.

- *material database*: a relational database that collects data about concrete; the database comprises a main archive, containing records which describe concrete-mixes, eleven archives containing data related to concrete components (water, cement, …), and two archives for test data and processed data;
- *material calibration module*: an interface module that allows for the calibration of thermal and mechanical material models, by using data collected in the database; it can export data suitable to be managed by other processing modules. It may also export calibrated advanced material models to an external finite elements code, to solve the thermo-visco-elastic problem.

5 THE NEURAL NETWORKS

Several processing modules within IPACS have been implemented via neural networks. Neural networks were chosen for IPACS to provide a tool able to safely provide the cracking risk evaluation, which requires experience and use of advanced mathematical tools to manage material behaviour; and able to incorporate also data coming from actual structural response: starting from examples, neural networks were expected to be able both to *learn* how, for example, concrete evolves and generates cracks; and also to deal with data different from those used for training them, and therefore to generalize what they had learned.

The neural nets used by IPACS have been trained on the ground of data generated by other tasks of the project or derived from literature (ACI curves, by the American Concrete Institute).

For the training, a commercial tool has been used (Neural Works Professional); after the training the nets have been exported from the off-line environment as C programs, then translated to JavaScript and embodied as JavaScript code in the web pages of the system; some modules exploit a combination of different nets.

Each net is a *multiple-layer feed-forward net*, with 3 or 4 fully connected node layers. The *back-propagation* learning scheme has been used for training the nets.

The input layers are made of nodes representing the input variables of the problem, whilst the output layer is made of one node, representing the result of the net: the modules for culverts and plates produce an estimation of the *cracking risk*; the modules for the restraint factor evaluation generate values for the *restraint factor* or *slip factor*.

One or more hidden layers are present with nodes that can be varied in number to reach the desired accuracy of the solution: the absolute average error ranges, according to checks performed on a set of test data, from 1.2% to 6.5%, and has been considered as an excellent result by the partners. These figures come from the comparison of the expected result and that forecasted by the network when processing cases never used for the training phase.

6 THE MATERIAL DATABASE

The so-called material database is a relational database that collects data about concrete under a consistent format.

The database comprises a *main archive*, containing records which describe *concrete-mixes*, *eleven archives* containing data related to *concrete components* (water, cement, aggregates, …), and *two archives* for *test data* and *processed data*, that is data derived from test data by processing them (Fig. 3).

The *Concrete mix* archive can be regarded as the *master* archive: it contains the description of the different concrete mixes, as well as links to the other archives, which describe the results of tests (*Test data* and *Processed data*) or those of analyses of cement, silica, fly ash, and so on. The archives may contain data of different types:

- numbers (e.g. Cement content)
- strings (e.g. Responsible person)
- dates (e.g. Casting date) ·
- text files (descriptions of test methods)
- data files (e.g. temperature histories).

Several fields accept values coming from predefined dictionaries: these fields can be filled through graphical interactors, which increase system's user-friendliness.

Common database functions are available: for each archive users may run the functions *Load*, for inserting new data, *Search*, for querying the archive and getting records which satisfy users' search criteria, *Modify*, for modifying data already recorded, *Delete*,

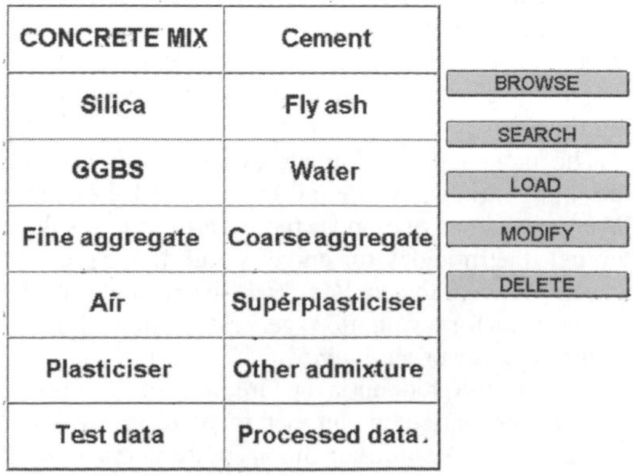

for removing records from the archive, *Browse*, to look at the whole set of data; export functions allow to export both data and files.

Data files can be presented to the users in a simple graphic format to give an immediate idea of the time series and allow users to export these files when they need them for more sophisticated processing or plotting.

The access to the database is protected via usernames and passwords; passwords are encrypted and stored into a metadata base; a super-user of the system can add new users, delete existing users, and modify passwords; normal users are allowed to modify their own password and requested to provide it to access the system.

Automatic users can access the database: some procedures of IPACS can access the database to use data related to the concrete that users wants to deal with. In this case users do not need to input data related to the concrete, provided that some of the partners has already loaded them into the database.

7 OTHER PROCESSING TOOLS

Graphical tools and written explanations support users throughout the system: whilst advanced users may directly use the most suitable path through IPACS, less experienced users will find different levels of explanations to select the tools they need, to feed the procedures with the right data, to load the database with properly formatted files and so on. The explanations can range from short in-line notes, to graphics, to hypertexts.

A graphical assistant supports users to simplify a three-dimensional structure to a two-dimensional one, when needed for applying 2D processing tools.

Other processing tools are integrated in IPACS and managed by the expert system: users are driven by the knowledge based module to use the proper tool. These tools, implemented via imperative programming and integrated into the expert system, are shortly described in the following.

A thermal solver implements a simplified formulation to calculate the temperature evolution of a semi-infinite concrete wall of a given thickness during the hydration process. It evaluates the average temperature increase in the wall (uniform) respect to the environment, through the use of a loss coefficient that incorporates the ability of the wall to dissipate heat.

A mono-dimensional solver for viscous-elastic problem calculates the stress evolution of a mono-dimensional concrete specimen due to restrained thermal loading.

A spreadsheet separates Thermal Dilation and Autogenous Deformation for a given free deformation test performed under a variable temperature history

CONCRETE MIX	Cement
Silica	Fly ash
GGBS	Water
Fine aggregate	Coarse aggregate
Air	Superplasticiser
Plasticiser	Other admixture
Test data	Processed data.

BROWSE
SEARCH
LOAD
MODIFY
DELETE

Figure 3.   The control panel of the database.

according to a measured/chosen/assumed development of the Thermal Dilation Coefficient.

A procedure to set up data to be run via the DIANA F.E.M. code computes age-dependent relaxation spectra associated to the Maxwell chain model.

Eventually, a program transfers semi-adiabatic calorimeter measurements (temperatures) to adiabatic temperature – and isothermal heat development, respectively. The program can also be used to provide a table with *maturity-heat release* data and to express the isothermal heat development in terms of continuous functions (the "Danish" or "Swedish" model).

8 HARDWARE AND SOFTWARE

The development of IPACS has been carried out on personal computers running the Windows operating system; several releases of the operating system have been exploited during the project's lifespan, from Windows 3.X to Windows 95/98 and NT.

The final release of IPACS has been installed on Windows NT computers and the final installation kit includes software for both NT and 95/98. Nevertheless, large parts of IPACS could be easily installed on Unix/Linux machines, provided that they run an HTTP server.

According to the initial choice of the Internet technology, IPACS requires an HTTP server to be run: the installation kit includes the free-of-charge Apache server and the instructions to install and run it on Windows PCs. Users are requested to run a common Internet browsers; the system has been tested on several releases of the most common browsers (Internet Explorer, Netscape Communicator, Opera).

IPACS includes HTML pages, which contain JavaScript programs, and a library of Perl programs. The Perl environment is included in the installation kit.

The neural networks have been trained within the software environment NeuralWare Professional Plus II and then saved as C routines and translated to JavaScript.

Some of the programs enclosed in IPACS have been written in FORTRAN and Visual Basic, while several modules available in the download area are Excel spreadsheets.

The material database is an Access archive, interfaced via the ODBC driver. Both the database and the HTML pages link documents in MS-Word, PDF, and zip format. Time histories within the *Test data* and

Processed data archives are ASCII files formatted according to rules that are well explained within the pages of the database which are used for loading or browsing data.

9 CONCLUSIONS

IPACS has been conceived and developed as a system where experience on the hardening of concrete of different kind is available to support users of different background: material science, constitutive modeling as well as structural modeling and construction technology are all embodied in the system.

The software architecture has been developed to allow for a wide exchange of data and to ease further incorporation of models/data.

The expert system has been put on-line on the Internet since its first prototypical version. This enabled partners involved in IPACS to use and test it throughout the lifespan of the project.

In such way they have shared data, knowledge and comments from the early stages of the development and provided developers with fruitful feedback and comments for improving the system. Moreover, the access to a unique version of the system on the Internet enabled the partners to fill the database with data coming from tests performed by all of them.

The final release of the system is currently used by the partners both for the support to cracking risk evaluation and for storing and managing data about concrete.

REFERENCES

Dalmagioni, P., Lazzari, M., Pellegrini, R., Salvaneschi, P., Emborg, M. 2002. An expert system for managing early age concrete crack prediction. In M. Schnellenbach Held & H. Denk (Eds.) *Advances in Intelligent Computing in Engineering – Proceedings of the 9th International Workshop of the European Group for Intelligent Computing in Engineering*, Darmstadt, Germany, 1–3 August 2002. Düsseldorf: VDI Verlag.

Salvaneschi, P., Lazzari, M. 1997. Weak information systems for technical data management. *Proceedings of the Worldwide ECCE Symposium on Computers in the Practice of Building and Civil Engineering (European Council of Civil Engineers)*, Lahti, Finland, 1 September 1997. Helsinki: Suomen Rakennusisinöörien Liitto RIL.

System-based Vision for Strategic and Creative Design, Bontempi (ed.)
© 2003 Swets & Zeitlinger, Lisse, ISBN 90 5809 599 1

An ANN model for biaxial bending of reinforced concrete column

M.E. Haque
Texas A&M University, Texas, USA

ABSTRACT: The reinforced concrete column cross-section and the area of reinforcing steel required to support a specific combination of axial load and moment can be established by using the column design interaction curves, where an interaction curve represents all possible combinations of axial load and moment that produce failure of the cross-section. The bending resistance of a column subjected to biaxial bending can be determined through iterations and lengthy calculations. These extensive calculations are multiplied when optimization of the reinforcing steel or column cross-section is required. This paper investigated the suitability of an Artificial Neural Network (ANN) for modelling a preliminary design of reinforced concrete columns. An ANN back-propagation multi-layered model was developed to design the column with biaxial bending, which predicted column cross-section for a given set of inputs, which were concrete compressive strength, column types (Tied and Spiral), reinforcing steel ratio (0.01–0.08), factored axial load, P_u, and moments, M_{ux} and M_{uy}. Several different ANN back-propagation trial models with different layers/slabs connections, weights and activation functions (including linear, Sine, Symmetric Logistic, Gaussian, Gaussian Complement, etc.) were trained. In addition, pattern selections including "Rotation" and "Random" were used with weight updates using Vanilla, Momentum and TurboProp. The training data was obtained from a series of column design Interaction Diagrams with $\gamma = 0.45$, 0.60, 0.75 and 0.9 for the column sizes in inches as determined by the equation: $h = 5/(1 - \gamma)$. The presented ANN back-propagation Multi-layer Perceptron (MLP) model with logistic activation function, "Rotation" for pattern selection, and "TurboProp" for weight updates was the best one among all other trials, which converges very rapidly to reach the excellent statistical performance. A set of data was randomly separated from the training set, and was used to evaluate the trained model. In addition, the trained ANN model has been tested with several actual design data, and a comparative evaluation between the ANN model predictions and the actual design has been presented.

1 INTRODUCTION

Applications of artificial intelligence (AI) have gained a broad interest in civil/construction/architectural engineering problems. They are used as an alternative to statistical and optimization methods as well as in combination with numeric simulation systems. Artificial neural network (ANN) is one of the AI algorithms that relates to the class of machine learning. ANNs are powerful computing devices. They can process information more readily than traditional computer systems. This is due to their highly parallel architecture inspired by the structure of the brain. Applications and research into the use of neural networks have evolved from their ability to understand complex relationships and hidden patterns within large data sets. These ANNs are modelling techniques that are especially useful to address problems where solutions are not clearly formulated (Chester 1993) or where the relationships between inputs and outputs are not

sufficiently known. ANNs have the ability to learn by example. Patterns in a series of input and output values of example cases are recognized. This acquired "knowledge" can then be used by the ANN to predict unknown output values for a given set of input values.

ANNs are composed of simple interconnected elements called processing elements (PEs) or artificial neurons that act as microprocessors. Each PE has an input and an output side. The connections are on the input side correspond to the dendrites of the biological original and provide the input from other PEs while the connections on the output side correspond to the axon and transmit the output. Figure 1 illustrates a simple PE of an ANN with the analogy of the human brain (Haque & Mund 2002). The activation of the PE results from the sum of the weighted inputs and can be negative, zero, or positive. This is due to the synaptic weights, which represent excitatory synapses when positive ($w_i > 0$) or inhibitory ones when negative ($w_i < 0$). The PEs output is computed by applying the transfer

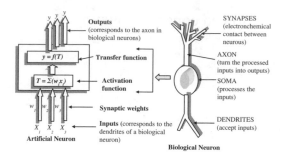

Figure 1. ANNs: The analogy to the brain.

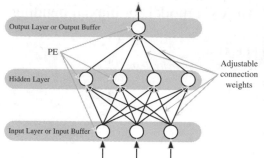

Figure 2. Schematic representation of a MLP.

function to the activation. The type of transfer function to be used depends on the type of ANN to be designed.

Currently, back-propagation is the most popular, effective, and easy to learn model for complex networks (Haque & Sudhakar 2001a, b). For the last few years, the author has been using various ANN back-propagation Multi-layer Perceptron (MLP) modelling techniques in materials science (Haque & Sudhakar 2001a, b, Sudhakar & Haque 2001), structural/construction engineering (Haque & Mund 2002, Haque & Karandikar 2002), and construction management (Choudhury & Haque 2001). To develop a back-propagation neural network, a developer inputs known information, assigns weight to the connections within the network architecture, and runs in the networks repeatedly until the output is satisfactorily accurate. The weighted matrix of interconnections allows the neural networks to learn and remember (Obermeier & Barron 1989). In essence, back propagation training adapts a gradient-descent approach of adjusting the ANN weights. During training, an ANN is presented with the data thousands of times (called cycles). After each cycle, the error between the ANN outputs and the actual outputs are propagated backward to adjust the weights in a manner that is mathematically guaranteed to converge (Rumelhart et al. 1986).

This paper describes an ANN back-propagation Multi-layer Perceptron (MLP) model to design a beam-column, which predicts column cross-section for a given set of inputs: concrete compressive strength, column types (square tied and circular spiral), reinforcing steel ratio (0.01–0.08), factored axial load, P_u, and moment, M_u. The trained ANN back-propagation model has been tested with several actual design data, and a comparative evaluation between the ANN model predictions and the actual design has been presented.

2 ANN-BP MLP MODEL

The neural network used for the proposed models was developed with NeuroShell 2 software by Ward Systems Group, Inc., using a back-propagation architecture with multi-layers jump connection, where every layer (slab) is linked to every previous layer. Figure 2 depicts a schematic representation of an ANN with multiple layers or slabs, i.e. a MLP (Haque & Mund 2002).

2.1 Network models

Two ANN models were developed. The model-1 network was trained for column size. The inputs were type of column (type = 1 for square tied column, and type = 2 for circular spiral column), factored load, P_u, Moment, M_u, and steel ratio, ρ_g. In this model concrete ultimate compressive strength, f'_c was 28 MPa (4,000 psi), and reinforcing steel yield strength, f_y was 414 MPa (60,000 psi). The model-2 network was trained for determining the values of P_o, P_{nx}, and P_{ny}, which determined the axial load capacity of a biaxial bending column using Bresler equation (Bresler 1960):

$$1/P_n = 1/P_{nx} + 1/P_{ny} - 1/P_o; for\ P_n \geqslant 0.1P_o \qquad (1)$$

where
P_n = nominal axial load capacity of a column under biaxial bending.
P_{nx} = nominal axial load capacity of a column when the load is placed at an eccentricity e_x.
P_{ny} = nominal axial load capacity of a column when the load is placed at an eccentricity e_y.
P_o = nominal axial load capacity of a column when the load is placed with a zero eccentricity.

The input for model-2 was h, ρ_g, e/h, and γ. The value of γ was estimated using the equation:

$$\gamma = 1 - (5/h) \qquad (2)$$

where the column size, h was in inches. Figure 3 shows the column dimensions.

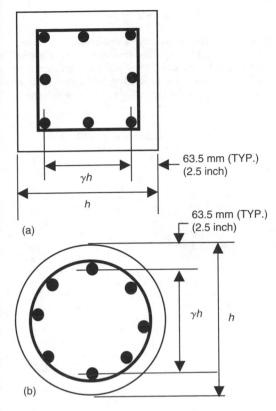

63.5 mm (TYP.)
(2.5 inch)

γh

h

(a)

63.5 mm (TYP.)
(2.5 inch)

γh h

(b)

Figure 3. (a) Square tied column dimensions; (b) Circular spiral column dimensions.

The number of hidden neurons was determined according to the following formula (NeuroShell 2 User's Manual 1996):

$$Number\ of\ hidden\ neurons = 0.5(Inputs + Outputs)$$
$$+ \sqrt{(Number\ of\ training\ patterns)} \qquad (3)$$

Given the properties of the training data used: 4 inputs, 1 output, and 340 training patterns – the number of processing elements was determined to be 21. The training data was obtained from a series of column design Interaction Diagrams with $\gamma = 0.45$, 0.60, 0.75 and 0.9 for the column sizes in inches as determined by the equation 2. A set of data was randomly separated from the training set, and was not used in the training model. These data were used to evaluate the trained model.

2.2 Network training and evaluation

Network training is an act of continuously adjusting their connection weights until they reach unique values

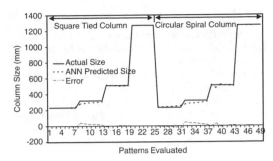

Figure 4. Column size prediction during ANN model evaluation.

that allow the network to produce outputs that are close enough to the desired outputs. This can be compared with the human brain, which basically learns from experience. The strength of connection between the neurons is stored as a weight-value for the specific connection. The system learns new knowledge by adjusting these connection weights. The learning ability of a neural network is determined by its architecture and by the algorithmic method chosen for training.

Back propagation method is proven highly successful in training of multi-layered neural nets. The network is not just given reinforcement for how it is doing on a task. Information about errors is also filtered back through the system and is used to adjust the connection weights between the layers, thus improving performance.

The accuracy of the developed model, therefore, depends on these weights. Once optimum weights are reached, the weights and biased values encode the network's state of knowledge. Thereafter, using the network on new cases is merely a matter of simple mathematical manipulation of these values.

Several different ANN back-propagation trial models with different layers/slabs connections, weights and activation functions (including linear, Logistic, Gaussian, etc.) were trained with pattern selections including "Rotation" and "Random", and weight updates using Vanilla, Momentum and TurboProp. The presented ANN back-propagation Multi-layer Perceptron (MLP) model with logistic activation function, "Rotation" for pattern selection, and "TurboProp" for weight updates was the best one among all other trials, which converges very rapidly to reach the excellent statistical performance as illustrated in Network Performance. Figure 4 demonstrate the graphical comparisons between the actual column sizes and the network predicted column sizes during the ANN model evaluation phase. They clearly demonstrate very good agreement between the actual and predicted performance. The evaluation was done using the data set, which was not used during the training phase.

1519

2.3 Network performance

The neural networks demonstrated an excellent statistical performance (Table 1) as indicated by the R^2 and r values. During network training, R^2 was obtained as 0.9980 and 0.9984 during network evaluation, which were very close to 1.0 indicating a very good fit between the actual and the network prediction. R^2 is a statistical indicator usually applied to multiple regression analysis, and can be calculated using the following formulae (NeuroShell 2 User's Manual 1996):

$$R^2 = 1 - (SSE/SS_{yy}) \qquad (4)$$

where $SSE = \Sigma\ (y - \acute{y})^2$, $SS_{yy} = \Sigma(y - \bar{y})^2$, y is the actual value, $\acute{y}$ is the predicted value of y, and $\bar{y}$ is the mean of the y values.

The correlation coefficient r, is a statistical measure of the strength of the relationship between the actual vs. predicted outputs. During network training, r-values were obtained as 0.9991, and 0.9993 during network evaluation, which were very close to +1.0 indicating an excellent fit between the actual and the network prediction. The formula for r:

$$r = SS_{xy}/\sqrt{(SS_{xx}\ SS_{yy})} \qquad (5)$$

where
$SS_{xy} = \Sigma xy\ -\ (1/n)\{(\Sigma x)(\Sigma y)\}$;
$SS_{xx} = \Sigma x^2\ -\ (1/n)(\Sigma x)^2$;
$SS_{yy} = \Sigma y^2\ -\ (1/n)(\Sigma y)^2$;
n = number of patterns,
x refers to the set of actual outputs, and y refers to the predicted outputs.

Table 1.   Statistical performance of the ANN trained model.

Items	ANN – training phase	ANN – evaluation phase
Pattern Processed	340	48
R^2	0.9980	0.9984
r	0.9991	0.9993
Mean Sq. Error	13.4 mm (0.529 in)	10.8 mm (0.425 in)
Mean Abs. Error	11.8 mm (0.466 in)	10.6 mm (0.418 in)
Min. Abs. Error	0.0 mm (0.0 in)	0.0 (0.0 in)
Max. Abs. Error	113.8 mm (4.479 in)	47.0 mm (1.854 in)
Top 5%	70.882	72.917
5% to 10%	18.529	18.750
10% to 20%	10.294	8.333
20% to 30%	0.294	0
>30%	0	0

3 DESIGN EXAMPLES

Several reinforced concrete short columns with uniaxial and biaxial bending moments were designed using the ANN models; a few of the results are shown in Tables 2 and 3. For predicting the required column sizes, the inputs for the ANN model-1 were type of

Table 2.   Column size prediction using ANN Model-1.

Col. type	P_u kN (kips)	M_{ux} kN-m (k-ft)	M_{uy} kN-m (k-ft)	ρ_g	h mm (in)	Select h mm (in)
1	1512 (340)	95 (70)	0	0.04	300 (11.8)	305 (12)
1	2847 (640)	447 (330)	0	0.03	497 (19.6)	508 (20)
1	2669 (600)	190 (140)	0	0.015	467 (18.4)	508 (20)
1(a)	5338 (1200)	406 (300)	169 (125)	0.03	574 (22.6)	610 (24)
1(b)	2002 (450)	136 (100)	176 (130)	0.04	381 (15)	406 (16)
1(c)	1601 (360)	95 (70)	108 (80)	0.04	333 (13.1)	356 (14)
1(d)	1681 (378)	256 (189)	214 (158)	0.02	483 (19.0)	560 (22.0)
1(e)	1922 (432)	146 (108)	195 (144)	0.03	414 (16.3)	457 (18)
2	1815 (408)	277 (204)	0	0.015	466 (18.4)	500 (20)
2	1886 (424)	172 (127)	0	0.041	397 (15.6)	406 (16)
2	1681 (378)	256 (189)	214 (158)	0.02	536 (21.0)	560 (22.0)
2	1922 (432)	146 (108)	195 (144)	0.03	447 (17.6)	457 (18.0)

Note: (a),(b),(c),(d), and (e) – See Table 3 for column load capacity using ANN Model-2.

Table 3.   Column load capacity using ANN Model-2.

Col type	P_o k-N (kips)	P_{nx} k-N (kips)	P_{ny} k-N (kips)	P_n k-N (kips)	ϕP_n k-N (kips)	Check $\phi P_n \geqslant P_u$
1(a)	11624 (2613)	9769 (2196)	10636 (2391)	9062 (2037)	6344 (1426)	OK
1(b)	5036 (1132)	4146 (932)	3719 (836)	3212 (722)	2247 (505)	OK
1(c)	3870 (870)	3176 (714)	3025 (680)	2580 (580)	1806 (406)	OK
1(d)	8083 (1817)	3666 (824)	4207 (946)	2585 (581)	1810 (407)	OK
1(e)	7266 (1634)	4752 (1068)	4158 (935)	3192 (718)	2235 (502)	OK

column (Type $= 1$ for square tied column, and type $= 2$ for circular spiral column), factored load, P_u, Moment, M_u, and steel ratio, ρ_g (selected within 0.01 to 0.08). For a circular spiral column with biaxial moments, the resultant moment, M_u was calculated by taking the square root of the sum of the squares of the moments from two perpendicular directions, M_{ux} and M_{uy}. For a square tied column with biaxial moments, M_u for the ANN model-1 was calculated using the equation (American Concrete Institute 1978):

$$\phi M_{nx}' = M_{ux} + M_{uy} \{(1- \beta) /\beta\} \qquad (6)$$

where β varies from 0.55 to 0.65.

In the example problems, the load-contour curvature with the parameter, $\beta = 0.65$ was used. Once the size of the squire tied column with biaxial bending was predicted through the ANN model-1, a reasonable size was selected, and the capacity of the column, ϕP_n was calculated using the Bresler equation as stated earlier from the values of P_{nx}, P_{ny}, and P_o. The values of P_{nx}, P_{ny}, and P_o were obtained from the ANN model-2. In general, all the cases, the ANN models produced pretty good results in determining column sizes with varying steel ratio.

4 CONCLUSIONS

This paper investigated the suitability of an Artificial Neural Network (ANN) for modelling a preliminary design of reinforced concrete beam-column. It was established in this paper that an ANN back-propagation model with Multi-layer Perceptron (MLP) network could be applied to design reinforced concrete beam-columns. The neural networks demonstrated an excellent statistical performance in the network training as well as in the evaluation of the trained networks. Several reinforced concrete short columns with uniaxial and biaxial bending moments were studied and found that the ANN models produced pretty good results in determining column sizes with varying steel ratio. The application of an ANN model certainly minimized the extensive calculations, especially required for column with biaxial bending during the optimization of the reinforcing steel or the column cross-section.

REFERENCES

American Concrete Institute 1978. Design Handbook, vol. 2, Columns, *ACI Publ. SP-17a (78)*.

Bresler, B. 1960. Design Criteria for Reinforced Concrete Column Under Axial and Biaxial Bending. *ACI Proc.*, Vol. 57, pp. 481–490.

Choudhury, I. & Haque, M.E. 2001. A Study of Cross-cultural Training in International Construction Using General Linear Model Procedure and Artificial Neural Network Approach. *Proceedings of the 3rd International Conference on Construction Project Management (3ICCPM)*, pp. 444–453, Singapore.

Chester, M. 1993. Neural Networks – A Tutorial, Prentice Hall: Englewood Cliffs, NJ, USA.

Haque, M.E. & Sudhakar, K.V. 2001a. ANN based Prediction Model for Fatigue Crack Growth in DP Steel. *International Journal of Fatigue & Fracture of Engineering Materials and Structures*, Vol. 24 (1), pp. 63–68.

Haque, M.E. & Sudhakar, K.V. 2001b. Prediction of Corrosion-Fatigue behavior of DP Steel through Artificial Neural Network. *International Journal of Fatigue*, Vol. 23 (1), pp. 1–4.

Haque, M.E. & Mund, A. 2002. Shoring Loads in Multistory Structure: An Artificial Neural Network Model. *Proceedings of the First International Conference on Construction in the 21st Century – Challenges and Opportunities in Management and Technology*, pp. 679–685, Florida, USA.

Haque, M.E. & Karandikar, V. 2002. A Study on Comfort and Safety in a Residential Housing Complex: A Neuro-Genetic Knowledge Model. *Proceedings of the Second International Conference on Information Systems in Engineering and Construction (ISEC 2002)*, Florida, USA.

NeuroShell 2 User's Manual 1996. Ward Systems Group, Inc., Frederick, MD, USA.

Obermeier, K. & Barron, J. 1989. Time to Get Fried Up, *BYTE*, 14 (8) pp. 227–233.

Rumelhart, D., Hinton, G. & Williams, R. 1986. Parallel distributed processing. MIT Press, Cambridge, MA, USA.

Sudhakar, K.V. & Haque, M.E. 2001. Mechanical behavior of Powder Metallurgy steel – Experimental Investigation and Artificial Neural Network-Based Prediction Model, *Journal of Materials Engineering and Performance*, 10(1), pp. 31–36.

Development of a high performance fully automated application system for shotcrete

S. Moser & G. Girmscheid

Institute of Construction Engineering and Management, Swiss Federal Institute of Technology, Zurich, Switzerland

ABSTRACT: Optimized shotcrete application techniques are required to guarantee the life-cycle-behavior of shotcrete shells. Today any spraying is done by hand or by manipulator causing heterogeneities while building up the shotcrete layers. With the automation of the shotcrete application, an important contribution to improve performance and quality may be achieved, while the over-all-costs of rock support by shotcrete may be reduced at the same time. With the development of the fully automated application system for shotcrete, the user will have a very effective tool at his disposal to spray concrete shells with a defined constant layer thickness or to provide the designed tunnel profile. With the new robot the user may choose from three different modes: manual spraying, semi automated and fully automated spraying. Especially the fully automated mode facilitates higher performance with less danger to the workman's health. The quality control is inherent in the application process in regard to layer thickness, compaction, homogeneity, evenness and rebound.

1 INTRODUCTION

Developments in the material technologies have enlarged the range of possible operation of shotcrete. Manipulators and shotcreting machines make high theoretical conveying capacities possible. To improve the over-all-economic-performance, the shotcrete application has now to be improved to enhance steady quality, to reduce rebound, and to improve work hygiene. To achieve these goals a fully automated shotcrete robot was developed at the Swiss Federal Institute of Technology, Zurich. The computer controlled automation system consists of the mechanical process control, the application process control and the application systematic.

2 STATE OF THE ART

2.1 *Spraying by hand*

Shotcrete is in many cases still applied by nozzle operators wielding tube and nozzle. The strain on the workmen limits the quantity of concrete that can be handled. The technique of application has to be trained and needs a lot of experience. The work demands high concentration even from experienced nozzle operators. To get an optimum of quality and a minimum of rebound the nozzle operator has to keep the right distance from, and angle, to the rock surface. In large tunneling sections the nozzle operator must apply the shotcrete from a lifting platform to maintain the optimum position of spraying. The experience on the sites shows that, due to human influence, it is not possible to keep all important application parameter in the best possible combination, especially not in large tunnel sections.

2.2 *Spraying by manipulator*

Performance can be improved by using a manipulator (Fig. 1). Because the strain on the workmen does not

Figure 1. MEYCO Robojet manipulator.

limit the spraying capacity, it may be much improved by using a shotcreting machine with larger capacity and a larger conveyor hose which additionally reduces the pulsation effects and thus improves the surface uniformity of applied shotcrete.

The workman steers the different joints with several joysticks to let the nozzle do the movements. The operation of the joints makes it difficult to keep the nozzle perpendicular to the surface and in the recommended distance. Even with a remote control it is still difficult to hold the quality on a steady level due to the poor visibility caused by the dust of spraying, the too large distance of the nozzle operators to the spray jet as well as the unfavorable angle of sight (EFNARC 1997).

3 AUTOMATED APPLICATION SYSTEM

3.1 Robot system

To improve the shotcrete quality and to simplify the application technique a robot was developed on the mechanical basic concept of the MEYCO Robojet manipulator (Meierhofer 1993).

The spraying robot is mounted on a vehicle that is not moving during the spraying process. The location of the nozzle is therefore always described with reference to the vehicle. The robot consists of the nozzle system (joints 7 and 8), the spraying-arm with boom (joints 1, 2, 3, 4 and 5) and lance (joint 6), a laser scanner, the remote control and the sensors and actors that are connected to the computer.

3.2 Mechanical process control

The electro-hydraulic spraying-arm owns eight degrees of freedom (Fig. 2). All joints are fitted with robust sensors, whereof six are working on angular and two on linear (joints 2 and 5) measuring principle, that detect the position of each joint to the next one simultaneously. The movements are controlled with standard control valves that are equipped with emergency manual control in case of break down.

In addition to the eight joints one joint is used for a rotational motion of the nozzle tip (opening angle $\varphi_{Rot} = 4°$) for a better distribution of the sprayed concrete. It has no effect on the kinematical model of the boom.

The vector φ is the joint vector that defines the workspace of the robot, the transversal and rotational vectors at the joints involved in the calculation are:

$$\varphi^T = [\varphi_1, \varphi_2, \varphi_3, \varphi_4, \varphi_5, \varphi_6, \varphi_7, \varphi_8]^T$$

The task requires the control of five degrees of freedom of the spraying robot, i.e. the position of the nozzle center point NCP (x, y, z) and 2 angles for the orientation of the nozzle (φ_7, φ_8). To solve the problem of the redundancy 3 constraints are required to link the supernumerary degrees of freedom:

$$\varphi_5 = -\frac{\varphi_1}{3}, \qquad \varphi_4 = \frac{1 + \varphi_2}{6}, \qquad \varphi_3 = \frac{1.85}{3} \cdot \varphi_6$$

The first condition limits the angle at the nozzle so that the perpendicularity of the nozzle to the rock surface is possible at any time. The second condition rules out the possibility of a collision of part 3 and part 6 (Fig. 2) and the third condition optimizes the workspace of the robot.

The computing is based on the inverse kinematic principle which means that for a given movement of the nozzle, a pattern of motion for each individual joint is computed by the mechanical process control. Due to the complicated kinematical structure no closed-form solution for the inverse kinematical model exists. The joint angles are thus calculated numerically with the Newton-Raphson method.

The nozzle operator uses a remote control with a 6D-joystick (Fig. 3), to steer directly the movement of the nozzle, due to the mechanical process control. The 6D-joystick is a large handle with integrated "dead man switch" and guarantees the water- and dust-resistance. The heart of the 6D-joystick is a modified piece of

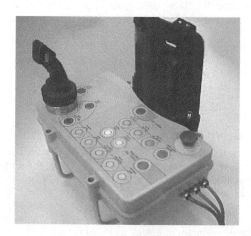

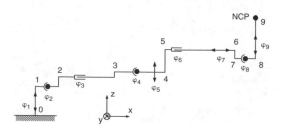

Figure 2. Definition of the angles of calculation.

Figure 3. Remote control with 6D-joystick.

equipment that is used as a standard in industrial robotics (Tschumi 1998, Tschumi 1999).

3.3 Application modes

3.3.1 Manual spraying

The nozzle operator uses the robot as a manipulator to apply shotcrete manually. The application is not supported by the application process control but the movements of the manipulator (boom, lance and nozzle) are controlled by the mechanical control system. After the machine having been positioned, the user is operating the application with the 6D-joystick. He does not have to take care of the individual boom joints but guides only the movement of the nozzle (Fig. 4). With the 6D-joystick are steered:

– Angle of the nozzle to the rock surface
– Path line and velocity of the nozzle v_n
– Distance d_{vp} from the nozzle tip to the tunnel wall.

This mode is thought for irregular conditions where a description of every movement is too difficult to be implemented into an operational program due to its complexity or for economical reasons. Such conditions could be as typical: Irregular local over profile, local covering of drainage half shells and anchor plates or filling of holes caused by rock fall.

3.3.2 Semi automated spraying

The user has the freedom to choose the path line; all other process functions of the application are controlled by the application process control and the internal mechanical process control. Contrary to the manual application mode the application process control generates out of the laser controlled measurements a virtual congruent plane to the scanned wall surface. On this plane, the nozzle movement is computer controlled in regard to the wall distance d_{vp} as well as to the perpendicularity of the nozzle to the scanned wall surface. The path of motion of the nozzle has to be manually controlled via 6D-joystick

and the spraying distance d_{vp} (distance virtual plane to the rock surface) has to be specified by the nozzle operator (Fig. 5). With the 6D-joystick are steered:

– Path line of the nozzle on the virtual plane
– Velocity of the nozzle v_n.

The semi automated mode avoids increasing rebound particularly in ranges which are badly visible or over head far away from the user due to optimized nozzle control in respect to the wall surface. The semi automated mode is an optional mode which can be used in areas where neither the manual application nor the fully automated modes are economically or technically useful.

3.3.3 Fully automated spraying

In comparison with the other two modes, the system has to take over the nozzle operator's experience and supervision functions with the resulting actions. The robot assumes the full control of the shotcrete application process. The measurement is effected in the same way as in the semi automated mode by defining the points from where the automated spraying starts and where it ends. Depending on the input given by the user (spraying distance d_{vp}, layer thickness, conveying capacity) the application process control program does the path planning for the nozzle and drives the nozzle automatically along these path lines, with the required velocity and in the according path line distance d to get the required layers, keeping the nozzle always perpendicular to the surface (Fig. 6). The 6D-joystick is locked, but for safety reasons the user still has to press the "dead man switch".

3.4 Profile measurement

Except when using the manual mode, the rock section under consideration has to be measured, but due to the spraying dust any measurement has to be executed before starting the spraying process. The measuring device is located at the head of the lance of the manipulator (in-between joint 6 and joint 7). Thus the range of

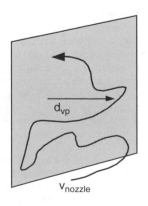

Figure 4. Manual spraying mode.

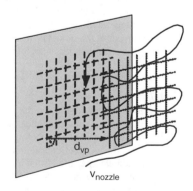

Figure 5. Semi automated mode.

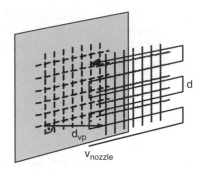

Figure 6. Fully automated spraying.

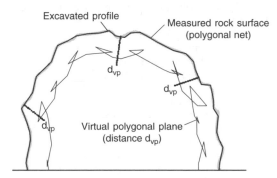

Figure 7. Geometrical place of the nozzle movement.

measurement along the tunnel axis is limited to 3.00 m due to the mechanical structure of boom and lance. The measurement of the tunnel profile is done by a laser measurement system, scanning the tunnel profile in a section of 180 degrees. The measuring principle is a reflector-less transit-time measurement in the infrared range. The user just has to position the lance that the distance laser device – rock surface is approximately the same over the whole tunnel profile (elimination of systematic measuring deviation) and to mark the required spraying range with 4 points by the laser device. The measurement of the tunnel profile is subsequently effected automatically in less than 30 seconds.

3.5 Application process control

3.5.1 Principle

The basic principle of the automation is the computation of a virtual polygonal 3-D plane parallel, in the distance d_{vp}, to the rock surface measured before (Fig. 7). The nozzle tip is in the consequence automatically guided in and perpendicularly to the virtual plane.

Depending on the chosen mode for the spraying process some parameters have to be given to the system as manual input by touch screen. The specified

functions are taken over by the application process control while reducing the nozzle operator's freedom of action: full robot control in the fully automated mode, no application process control in the manual mode (Honegger 1996). The unspecified functions are steered manually by the 6D-joystick that enables a user friendly handling.

3.5.2 Design of the system intelligence

If the knowledge of the independent relations of some spraying parameters was sufficient for spraying by hand or manipulator, it is not for spraying fully automated by the robot. The different influences have to be quantified and valued to be put into relation. All these dependent and independent factors have to be linked in the application process program (application process control). Important for the research is the adaptability of any experiments to site conditions because only this demand makes sure that the results are useful to be implemented into the robot application system. Because so many facts can't be influenced on site the data of the experiments shall not be taken to the last two digits after the decimal point. Much more important is the consideration of the interaction of the material and the application parameters.

The investigation of the meander-wise path planning of the nozzle moving turned out that the orientation (horizontal, vertical or any other orientation in between) of the path lines did not influence the concrete quality in regard to homogeneity and compressive strength. The rebound of the applied shotcrete did not vary significantly as well (Guthoff 1991). But to prevent the sagging of the layer-wise applied shotcrete the application has to be carried out from the bottom to the top (EFNARC 1997).

The path planning (vertical or horizontal) is thereby technically determined by the kinematics of the boom and the lance. Because of the orientation of the lance the application in horizontal meander-wise path lines is predominant.

The first step of the development of the application process control was to quantify the distribution of the shotcrete by the spray-jet in one layer along a stretched path line, considering spraying distance, nozzle rotation and nozzle moving velocity, concrete conveying capacity, air pressure and accelerator dosage. The quantified distribution curves, measured rectangular to the path of the spray-strips, are the basis for the full range application. An example is given in Figure 8 where the measured graphs of repeated executed spray-strips with the same application parameter combination are arranged.

The statistical evaluation showed that the distribution curves, summarized to a design profile (mean of graphs of distribution), correspond to the same entirety what means that the design profiles, gained through repeated execution of the experiments, may be compared. Further are the standard deviations

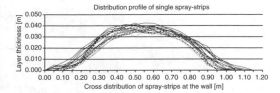

Figure 8. Shotcrete cross-distribution, sprayed in a single strip.

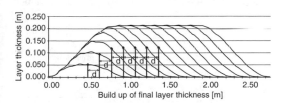

Figure 9. Structure of final layer thickness.

coincidentally. The path of nozzle movement may therefore be developed out of the design profiles.

The maximal design profile thickness that may be achieved vary between 0.02 and 0.07 m depending on the application parameter combination (spaying distance 1.50 or 2.00 m, nozzle moving velocity of 10, 15 and 20 cm/s and nozzle rotation velocity of 1 or 2 RPS) and the effective concrete conveying capacity of 9.5, 12.5 or 15.0 m³/h (Wijnhoff et al. 1999). Different to the spraying by hand or by manipulator only some application system adjustments can be done while spraying automatically. The reason is that once having left the finished part of the spraying range, which can not be reached by the spray jet in the same passage of spraying any more, the evenness of the sprayed surface has to be final. Therefore the evenness of the shotcrete strips along the stretched path line has to be high; otherwise the requirements of automated shotcrete application are not fulfilled.

The application model, the automated shotcrete application is based on, is the stretched horizontal meander wise overlapping of single spray-strips. Dependent on the required layer thickness and the sprayed layers (design profiles) the distance between the path lines of the nozzle movement has to be calculated by an algorithm as shown in Figure 9.

The basic design profile of Figure 9 i.e. has an extension of 0.95 m and a maximum thickness of 0.05 m (nozzle moving velocity: 10 cm/s, spraying distance: 2.00 m, nozzle rotation velocity: 1 RPS, effective concrete conveying capacity: 12.5 m³/h). With a spraying path line distance of d = 0.15 m, a theoretical total lining thickness of 0.22 m is achieved.

Figure 10. Input mask of the application process control.

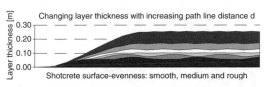

Figure 11. Evenness of shotcrete layers.

Table 1. Proposition of application parameter combination.

concrete conveying capacity			Shotcrete layer	Spraying distance	nozzle moving velocity	nozzle rotation velocity	path line distance
[m3/h]			[cm]	[m]	[cm/s]	[RPS]	[m]
15.0	12.5	9.5	l	d_{yp}	v_N	R_N	d
			5				0.26
		1b33	6	2.00	15	1	0.22
			7				0.19
			8				0.28
			9				0.25
			10				0.22
		1b13	11	2.00	10	1	0.20 / 0.21
			12				0.19 / 0.19
		1c12	13	1.50	10	2	0.17 / 0.18
			14				0.16 / 0.17
			15				0.15 / 0.15
			16				0.24 / 0.19
			17				0.22 / 0.18
3b13	2b13		18	2.00	10	1	0.21 / 0.17
			19				0.20 / 0.16
			20				0.19 / 0.15
			21				0.18 / 0.14

3.5.3 Simulation of shotcreting

The experimental data (design profiles, rebound, application parameters, water permeability, compressive strength, air pressure and concrete characteristics) are summarized in a data base. The path line distance is assigned according to the required layer thickness. Specifying the evenness of the surface to be sprayed: smooth, medium or rough, and classifying the relevance of rebound to time of spraying (Fig. 10) the best process control parameter set for the required layer thickness is evaluated automatically by an algorithm.

A simulation of layer thicknesses between 0.05 an 0.25 m showed that most parameter combination fit not the requirement of a smooth surface of the shotcrete shell. Further isn't it possible to spray any shotcrete layer thickness with the same parameter combination. This fact is represented in Figure 11.

Extracting the parameter combinations that fulfill a smooth evenness of the shotcrete layer the following recommendations for shotcrete application may be given (Tab. 1).

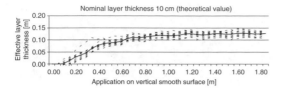

Figure 12. Experimental layer thickness.

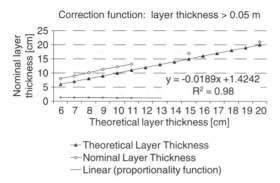

Figure 13. Correction function.

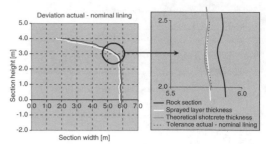

Figure 14. Shotcrete application in blasted tunnel section.

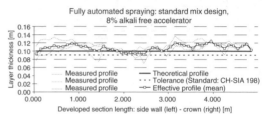

Figure 15. Layer control on site.

3.5.4 *Improvement of the theoretical model*

Due to the overlapping of the different spray-strips to achieve the theoretical layer thickness the rebound is decreasing and therefore the design profile enlarged. The value of increase is proportional to the gradient of the design profile. This effect was theoretically calculated to about 0.002 m for the design profiles. The quantified increase of 0.001 to 0.003 m fits well to the assumption and is reverse proportional to the thickness of the design profile.

Besides that the accuracy of the movements of the spraying-arm causes an additional uncertainty about the effective shotcrete layer (Fig. 12).

The difference between the theoretically calculated and the effective applied layer thickness depends therefore on rebound and accuracy. These two influences have to be considered and compensated by a correction function (Fig. 13):

- $y = -0.1357x + 1.7286$ $(R^2 = 0.93)$ for layer thickness up to 0.05 m and
- $y = 1.3298e^{-0.016x}$ $(R^2 = 0.98)$ for layer thickness above 0.05 m.

3.5.5 *Verification on site*

Considering the correction functions the transition from laboratory experimentation to application on site was achieved. The shotcrete application was carried out on pre-sprayed rock and on blasted rock, both washed with water before, with an accelerator dosage of 4, 6 and 8%.

That for any application the excavation line is evened out by the shotcreting (Fig. 14), that holes are partially filled while peaks are covered less (Teichert 1991) was confirmed by the fully automated shotcrete application.

In Figure 15 the developed length, from the side wall to the crown is shown for a required layer thickness (theoretical profile) of 0.10 m. According to (SIA 1993) the tolerances of excavation support for shotcrete shells less than or equal to 15 cm is minus 0.01 m. With the simulated and corrected parameter combination the fully automated shotcrete application on a blasted rock surface reaches very well the required shotcrete lining of 0.10 m and ensures that the minimal layer thickness is reached. The layer thickness is importantly more exact than applying shotcrete manually or by manipulator.

4 CONCLUSION

With the development of the "Fully Automated Application System for Shotcrete", companies will have a very effective tool to spray concrete shells as primary rock support or as final lining at their disposal. The application system offers three different modes: manual, semi automated and fully automated shotcrete application. Especially the fully automated mode facilitates higher performance with improved work hygiene. The quality control is inherent in the application process to guarantee the life-cycle-behavior of the tunnel shells in regard to layer thickness, compaction, and homogeneity. The time consuming profile and quality control, due to the optimized combination of the important application parameters with regard to performance, rebound, quality and time, the over-all-costs, may be reduced.

REFERENCES

EFNARC. 1997. Europäische Richtlinie für Spritzbeton. Aldershot, UK.

Guthoff, K. 1991. Einflüsse automatischer Düsenführung auf die Herstellung von Spritzbeton. *Technical Report 91-7*, Ruhr Universität Bochum, Institut für konstruktiven Ingenieurbau, Bochum, Germany.

Honegger, M. 1996. Steuerung für den Betonspritzroboter Robojet SC-30. *Interner Bericht*, Eidgenössische Technische Hochschule Zürich, Institut für Robotik, Zürich, Switzerland.

Honegger, M., Codourey, A. 1998. Redundancy Resolution of a Cartesian Space Operated, Heavy Industrial Manipulator. *Proceedings of ICRA 98, Int. Conference on Robotics and Automation, Leu-ven May 16–21 1998*.

Meierhofer, L. 1993. Marktstudie Programmgesteuerter Spritzroboter (Projekt SC-30). *Internes Papier* Meynadier AG, Zürich, Switzerland.

SIA. 1993. Untertagebau SIA 198. Zürich, Switzerland.

Teichert, P. 1991. *Spritzbeton*. Laich SA, Avegno, Schweiz.

Tschumi, O. 1998. MEYCO Robojet Logica. *CIM Conference 1998, Montreal, Canada, Internes Papier MEYCO Equipment MBT (Schweiz) AG, Winterthur, Switzerland*.

Tschumi, O. 1999. Fully Automated Shotcrete Robot – Machine Design. *Third International Symposium on Sprayed Concrete, September 26–29, Gol, Norway*.

Wijnhoff, M., Moser, S., Girmscheid, G. 1999. Forschungsbeitrag Spritzbeton. *Diplomarbeit*, Eidgenössische Technische Hochschule Zürich, Institut für Bauplanung und Baubetrieb, Zürich, Switzerland.

A learning-based design computer tool

H.M. Bártolo & P.J. Bártolo
School of Technology and Management, Leiria Polytechnic, Portugal

ABSTRACT: Designers need to explore huge amounts of knowledge to generate a set of good design concepts. Requirements for future computer supported design environments can be created through descriptive models of design, drawn from empirical research, integrating and adjusting evidence from real case studies. The solutions from previously solved problems are stored in a database of known design solutions, aiming to capture and process information in order to investigate successful solutions of similar problems. This work presents the initial stage of a research that intends to develop a learning-based computer to implement the methodological matrix that will allow the rapid generation of design solutions. The use of computational databases of innovative design solutions may assume a great importance in contexts of architectural design practice, as they can be particularly helpful in conceptual design, as well providing opportunities to gain insight into creativity itself.

1 INTRODUCTION

Architectural design is a complicated creative activity that requires the use of experience and knowledge reasoning. It involves analysis and synthesis, reasoning and judgment, numeric calculation and simulation. It can be described as a problem-solving activity comprising the establishment of goals, constraints and requirements. Hence, designers need to explore a great variety of knowledge to effectively generate a set of good design concepts (Gero 1990). Knowledge is generated and accumulated throughout the design process, so the use of databases of known design solutions assumes a great importance. These databases have the flexibility to store vast amounts of data relating many different aspects of the design domain relevant to design processes. Different types of solutions and knowledge require integration of different ways of representation and, consequently, of reasoning.

This paper presents the initial stage of a research that intends to develop a learning- or knowledge-based computer tool to allow the rapid generation of original design solutions. This computer tool is based on the premise that humans generally solve new problems by modifying solutions to previously solved problems. The model represents the design intent behind the design product, storing the how, why and what of a design, providing information on successful design solutions. These solutions are subsequently stored in a database of known design solutions. Whenever a new problem is addressed, this database can be investigated for solutions of similar problems as a basis for a solution of the current design project.

2 CASE-BASED DESIGN SYSTEMS

Several case-based reasoning (CBR) systems have been developed, stemming from psychological models of human memory structure. These systems are based on the idea that humans generally solve new problems by modifying solutions to previously solved problems (Schank 1982). A previously experienced situation can then be captured and learned in a way that enables its reuse to solve future problems or cases (Morisbak & Tessem 2001).

CBR is a cyclic and integrated process that consists of solving a problem, learning from this experience, solving a new problem, etc. Apparently, CBR can be considered a new research paradigm for exploring creativity, assuming that:

- Knowledge modelling is only performed by gathering past cases, not requiring domain understanding by the CBR developers
- Knowledge maintenance just requires the new cases to be properly added to the database of cases
- Applications aim at providing a similar past experience to help users to reason on the current situation they are facing.

General advantages of CBR can be found at Watson (1997) & Sycara (1992). Some issues are of particular interest for architectural design:

- Applications fit very well in complex domains
- Record of past experiences can play an important role in providing a corporate knowledge
- Applications can be put together and delivered through the Internet, helping the participants in the design process, sometimes geographically scattered, to share experiences.

3 CASE-BASED EXPERT SYSTEM

As previously described, in case-based reasoning systems, a set of cases represents the domain knowledge useful for problem solving. Each case and associated rules captures some heuristic of the problem, and each new case adds some new knowledge and thus makes the system smarter. The case-based reasoning system can easily be modified through changing, adding or subtracting cases.

The design expert system being developed under this research, based on the traditional structure of Design-Evaluation-Redesign process (Figure 1), is an hybrid of numeric calculation and logic reasoning system involving two main processes:

- Topological definition and optimisation
- The matching process

Numerical data information used within the expert system is categorised in three main types:

- Continuous (area, perimeter, cost, etc.)
- Discrete (number of floors, number of rooms, etc.)
- Categorical (location, gender and marital status of the client/user, etc.).

3.1 Topological definition and optimisation

Topological definition is used when there are no cases available in the database (initialisation), or if a solution cannot be found, and consequently a new design space should be created in the absence of any similar design memory cases. The topological definition for a building space begins with the definition of the starting space and its general dimensions. Next, other spaces are successively added to the initial one according to specific rules of desired adjacencies, forming different levels of possible topological sub-spaces of design solutions. From each level of building topology, the more appropriated design solution is defined according to the verification of specific goals.

Figures 2, 3 show the way the system operates. The initialisation of the topological definition process is made through the specification of the starting space and its dimensions, in this case a living room of dimensions $2L \times L$. The first level of topological definition is obtained through the addition of a second space (dining room, dimensions $L \times L$). Moreover, for the

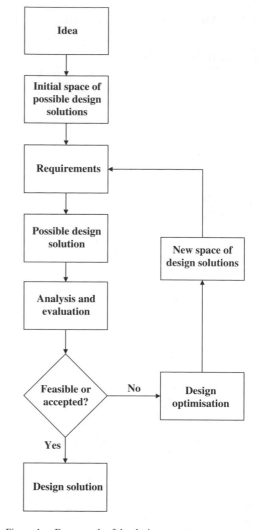

Figure 1. Framework of the design expert system.

definition of the available design spaces, it is assumed that the topological association of spaces is commutative, i.e. space_A + space_B = space_B + space_A.

Figure 3 represents part of the definition of the second level of topological definition. In this case, a novel space (kitchen, dimensions $L \times L$) is added to a first level design space. This design topological association is performed according to the following rule:

If a *space_kitchen* is to be added **then**
    If a *space_dining* room was previously added **then**
    Place it adjacently to the dining room
    End if
Else
Place it adjacently to any space
End if

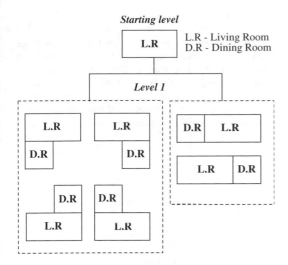

Figure 2. First level of a possible building topological definition.

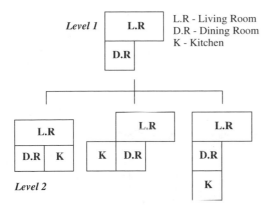

Figure 3. Second level of a possible building topological definition.

Assuming that one goal is the minimisation of the building design perimeter, the space A is selected as the space-solution for the second level. Afterwards, the system will sequentially generate different others topological levels.

This process, still in an early stage of development, can also be used to perform design optimisation of existing building designs (past cases), finding values for a set of building design parameters, fulfilling constraints. The adopted strategy follows a simplified variant of the propose & revise problem solving method (Van Eck et al. 2001) that involves a generator, which produces values for optimisation, and a tester

to check whether the current design violates any constraint.

3.2 Matching

Given a new problem, similar problems are selected from a database. The solution of the most similar case is transferred to the new problem and adapted. Finally, each new case is stored and the similarity relation is updated through a kind of computer learning.

In this expert system, the overall hierarchical structure of the knowledge is decided first. Classes and their attributes are identified, and hierarchical relationships among them are established. The architecture of this expert system not only provides a natural description of the problem, but also allows us to add new information, classes and attributes. The development of this expert system involves the following steps:

- Specify the problem and define the scope of the system
- Determine classes and their attributes
- Define instances
- Define displays
- Define rules
- Evaluate and expand the system.

Building descriptions and pictures should be stored separately, descriptions as text files (*.txt) and pictures as bit-map files (*.bmp). If we then set up a display that includes a text-box and a picture-box, we will be able to view a building design description and its two-dimensional and three-dimensional representations, 2Dbdesign*.bmp and 3Dbdesign*.bmp, in the display by reading the text file into the text-box and the bit-map files into the picture-box, respectively.

All existing building designs are introduced through appropriate databases (Table 1) that can be easily modified and accessed from the expert system, according to the structure indicated in Figure 4.

The next step is to list all the possible queries we might think of:

- What type of building do you want?
- Which suburb would you like to live in?
- How many bedrooms do you want?
- How many bathrooms do you want?
- How many living rooms do you want?
- Do you want to have a pool?
- What is the maximum amount you want to spend on?

Once these queries are answered, the expert system is expected to provide a list of suitable previous building design cases (Fig. 4). These cases should reflect and support the way in which architects design buildings, i.e., they should allow to tackle specific problems that are known to reoccur in the application domain.

Table 1. Building designs database.

Instance	Area (m²)	Suburb	Price (€)	Type
1	300	Central	500 000	House
2	200	Southern	300 000	House
3	160	Central	200 000	Townhouse
•	•	•	•	•
•	•	•	•	•
•	•	•	•	•

Bedrooms	Bathrooms	Living rooms	Picture files
5	4	2	2Dbdesign01.bmp
			3Dbdesign01.bmp
4	3	1	2Dbdesign02.bmp
			3Dbdesign02.bmp
3	2	1	2Dbdesign02.bmp
			3Dbdesign03.bmp
•	•	•	
•	•	•	

Text file

bdesign01.txt
bdesign02.txt
bdesign03.txt
•
•
•

4 CONCLUSIONS

Different types of design solutions and knowledge require integration of different ways of representation and, therefore, of reasoning. Knowledge based systems represent a merging of Artificial Intelligence Techniques with computer aided design technologies, and its ultimate goal is to capture the best design solutions into a corporate knowledge base. Whenever a new problem is addressed, this database will enable to investigate solutions of similar problems as a basis

CLASS: Building

[N] Area:
[Str] Suburb:
[N] Price:
[Str] Type:
[N] Bedrooms:
[N] Bathrooms:
[N] Living rooms:
[N] Dining rooms:
[Str] 2D picture:
[Str] 3D picture:
[Str] Textfile:
[N] Instance:

Figure 4. Class building and its instances.

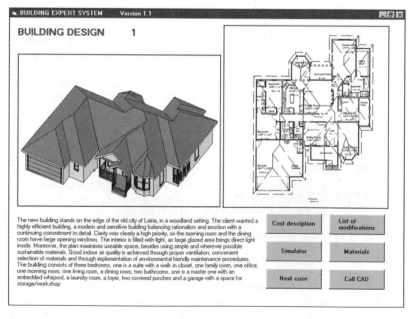

Figure 5. The building design information display.

for the possible solution of current design projects. Additionally, the ability to model and implement design processes as reasoning systems, making use of artificial intelligence, enables to support creative design.

A Knowledge Based system methodology can provide a friendly framework for formally capturing and defining the process of design creation within a system that can infer and then act on this information. The innovative use of computational databases of known design solutions may assume a principal importance in contexts of architectural design practice, particularly in the conceptual design stage, on top of providing opportunities to gain insight into creativity itself.

REFERENCES

Gero, J.S. 1990. Design prototypes: a knowledge representation schema for design, *AI Magazine*. 11: 26–36.

Morisbak, S.I. & Tessem, B. 2001. Agents for case-based software reuse, *Applied Artificial Intelligence*. 15: 297–332.
Schank, R. 1982. *Reminding and memory in dynamic memory: a theory of reminding and learning in computers and people*. Cambridge University Press.
Sycara, K. 1992. CADET: a case-based synthesis tool for engineering design, *International Journal for Expert Systems*. 2: 157–188.
Van Eck, P., Engelfriet, J., Fensel, D., Van Harmelen, F., Venema, Y. & Willems, M. 2001. A survey of languages for specifying dynamics: a knowledge engineering perspective, *IEEE Transactions on Knowledge and Data Engineering*. 13: 462–496.
Watson, I. 1997. *Applying case-based reasoning*. Morgan Kaufmann.

System-based Vision for Strategic and Creative Design, Bontempi (ed.)
© 2003 Swets & Zeitlinger, Lisse, ISBN 90 5809 599 1

Architectural reverse design through a biomimetic-based computer tool

N.M. Alves & P.J. Bártolo
School of Technology and Management, Polytechnic Institute, Leiria, Portugal

ABSTRACT: Architectural design is a problem-solving activity, where alternative solutions are generated and tested until an appropriate solution is found. It is a very complex and iterative process where each design solution opens new and potentially more complex, sub-spaces, implying a wide range of knowledge and expertise. Many experts from different fields generate spaces of design solutions in a collaborative process, starting from new ideas and concepts or existing design solutions through a redesign or reverse design process. This work describes a computational tool that will enable the reverse design of existing buildings through the replication of the human vision process. The subjacent principle of this system uses a commercial digital camera as the *retina* of the computer to take two-dimensional photos of existing buildings. These photos are then used by a robust computer routine, which works as the *visual cortex* of the computer, converting *observed* (input) photos into three-dimensional output computer models. This computational system combined with advanced rapid prototyping technologies may represent an important tool to enhance collaborative work with the building design process.

1 INTRODUCTION

Refurbishment and renovation of old and historic buildings has become an important aspect of the building construction industry. According to McMahon (1997) in Europe 50% of construction work is undertaken on existing buildings. In this field, the International Charter for Conservation and Restoration of monuments and sites, commonly known as the Venice Charter 1964, has been regarded as the key benchmark for principles governing architectural building re-use.

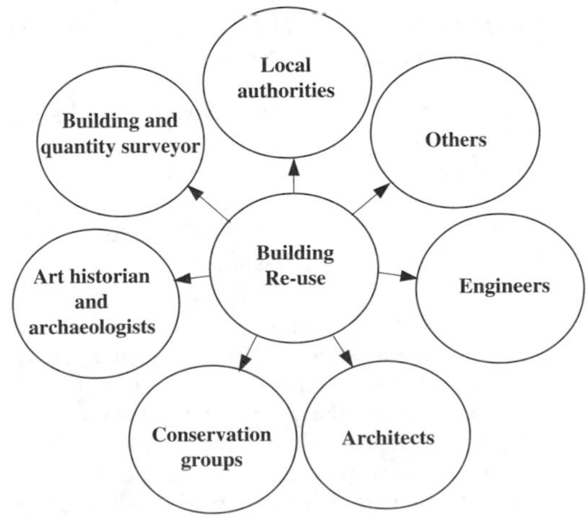

Most of the approaches regarding the unifying concept of building re-use include renovation, restoration, conservation, repair, rehabilitation, conversion, refurbishment and alteration. Building re-use requires a broad range of professional expertise as shown in Figure 1. Communication problems are particularly relevant so it is essential the implementation of a collaborative environment based on computational tools to support the activities of re-design or reverse design for re-use.

2 REVERSE DESIGN PROCESS

The design process of a new product in a computer aided design (CAD) environment starts with a set of ideas, progressing to the concept phase, where a computer model is created using CAD systems. Next, the design of a product is optimised by simulation and evaluation (Fig. 2). Conversely, in the reverse design process for building re-use, the shape of an existing building is captured and digitised and the resulting digitised data is transformed into three-dimensional models, which can then be manipulated and optimised by simulation and evaluation (Peng & Loftus 1999) (Fig. 3).

Figure 1. Collaborative environment for building re-use.

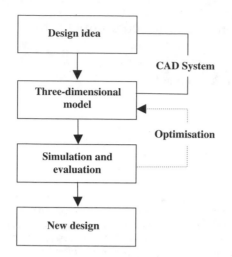

Figure 2. Optimisation of a design idea in a CAD environment.

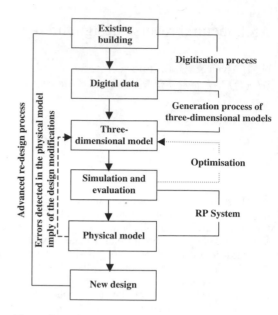

Figure 4. Advanced reverse design process in a CAD environment.

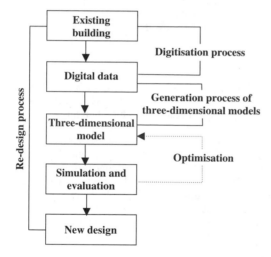

Figure 3. A reverse design process in a CAD environment.

are similar process to 2D printing and plotting technologies using both vector-based and raster-based imaging techniques.

The general process of building a physical model through RP technologies comprises:

– the creation of a solid model using a CAD system
– the slicing of the solid model into different layers using a RP appropriate software
– the creation layer by layer of the physical model using a RP machine.

According to Kai & Fai (1997), the computer solid model for RP processes can be represented by:

– a tessellated or STL model that approximates the solid with the set of triangles
– a boundary representation (B-rep) model representing the surfaces with a set of parametric curves and surfaces like the Bezier, B-spline and NURBS (Non Uniform Rotation B-splines).

RP systems enable all the distinct tasks of the design process to occur almost in parallel, through the generation of physical models quasi-instantaneously. Design integration can particularly benefit from these rapid prototyping techniques, as the prime purpose of physical prototyping is to promptly assist iterative design, allowing design participants to work with a three-dimensional hardcopy and use it for the validation of their design-ideas (Bártolo & Bártolo 2001a, b).

The reverse design process combined with rapid prototyping (RP) systems will allow the implementation of a new reverse design strategy, called advanced reverse design process, facilitating the re-design of an existing product (Fig. 4).

RP technologies are very sophisticated techiques, recently developed, for the generation of three-dimensional physical objects, comprising different techniques and different materials (Johnson 1994). These different techniques use the same principle, the transformation of a geometric CAD model into a physical model produced layer by layer. The RP techniques

3 BIOMIMETIC-BASED COMPUTER TOOL

3.1 Rapid generation of computer models from existing buildings

The rapid generation of computer models from existing buildings will enable to enhance creativity and innovation within the architectural design process. Over the past two decades, re-design has been extensively studied and several efforts to build computer models of architectural scenes have been made (Bártolo & Bártolo 2001a). However, traditional methods of constructing models from existing buildings are particularly labour-intensive, implying the survey of the site, the digitisation of existing architectural drawings, or the modification of existing computer-aided architectural design data. Therefore, a new approach based on the replication of the human vision sense is proposed for the rapid generation of computer models from existing building.

3.2 Human vision sense

The ability of the nervous systems to construct internal visual representations of the outside world represents one of the most important achievements in the evolution of human behaviour and cognition. "Seeing" allows humans to obtain information concerning the nature and location of objects in their environment without the need for direct or close physical contact, as required by other senses.

According to the theory of visual perception, the human vision is a complex information-processing system comprising different biological systems and regions: the physical world of objects (environment), the visual stimuli giving rise to visual events and the brain.

Seeing is initiated when the light, passing through the pupil of the eye, is projected by convex lens into the retina in an upside-down manner. These convex lenses are supported by ciliary muscles, which are capable of changing the shape of the lens, enabling the automatic focusing of the eye (Audesirk & Audesirk 1999).

The retina consists of millions of photoreceptors cells called rods and cones, acting as the screen of the image's projection. The cones provide chromatic images of high spatial resolution, and the rods required for achromatic vision with less spatial resolution, transform the projected image into electrical signals (Zrenner 2002). The visual information from the retina is then transmitted through the optic nerve to the primary visual cortex of the brain. The brain processes these electrical impulses and gives information about what we are seeing. In some visual areas the neurons are organized in an orderly fashion, called *topographic or retinotopic mapping*, forming a two-dimensional representation of the visual image formed on the retina in such a way that neighbouring regions of the image are represented by neighbouring regions of the visual area. Moreover, the corresponding points of the two slightly different images on the retinas (left and right eyes) are determined in a process called *binocular stereopsis* (Gordon 1989). Stereopsis relates to that extra sense of solidity and depth that is experienced by humans when using two eyes rather than one. Nevertheless, it is possible to obtain the relative location of objects using one eye (monocular cues). However, it is the lateral displacement of the two eyes that provides two slightly different views of the same object and allows acute stereoscopic depth discrimination. Each eye captures its own view and the two separate images are sent to the brain for processing. The two images arrive simultaneously in the back of the brain and are united into one picture through a process called *fusion*, that takes place when the observed objects are the same (Gordon 1989), allowing single binocular vision. The combined image is a three-dimensional *stereo* picture.

3.3 Three-dimensional reconstruction system

The biomimetic-based computer tool for reverse design, called Three Dimensional Reconstruction System (3DRS), has been developed in C++. This system uses a commercial digital camera as the retina of the computer to take two-dimensional (2D) photos of existing objects. These 2D photos are then used by a robust computer routine, which works as the visual cortex of the computer converting observed (input) photos into tri-dimensional (3D) output computer models. These 3D models can be used to produce very sophisticated photo-realistic renderings, generating solid models or three-dimensional meshes in triangular STL format for simulation purposes using finite element method codes and rapid prototyping applications for physical models production.

The code is organised into four main routines:

– the "ciliary" calibration
– the binocular stereopsis
– the images fusion and 3D reconstruction
– the STL mesh generation.

The "ciliariy" calibration process mimics the inter-relationship between the convex lens and the ciliary muscles in the human eye, establishing the projection from the 3D world co-ordinates to the 2D image co-ordinates, besides, correcting radial distortions of photographic images (Alves & Bártolo 2002b). Our system compensates the radial lens error through a proper mathematical model, similar to the model proposed by Pajdla (1997), establishing a relationship between the ideal (non observable distortion-free) pixel image co-ordinates and the corresponding real observed image co-ordinates.

The binocular stereopsis performed through the epipolar transformation will enable to identify the points in two or more images that are projections of the same point in the world. The epipolar geometry enables to describe the relations that exist between the two obtained images considering two cameras in different locations (Faugeras 1993).

The images fusion and 3D reconstruction routine performs the matching between points or edges from similar images. Through this operation all the different digital photos used to define a scene are blended into a single image corresponding to a wireframe representation of the object.

After this wireframe model is produced, it is necessary to generate an STL representation of the digital model and this information generated by the 3DRS is used into a rapid prototyping machine. Hence, it is possible to discus and optimise the initial design of the building making the necessary modifications for its re-use.

The generation of the STL representation follows two important rules:

- facet orientation rule: the facets define the surface of the three-dimensional object. The orientation of the facet involves the definition of the vertices of each triangle in a counterclockwise order
- vertex to vertex rule: each triangular facet must share two vertices with each of its adjacent triangles.

The flowchart of information through the use of the 3DRS system is described in Figure 5. The process starts by taking digital photos of an existing building. This 2D photos are then used to produce a 3D wireframe model, which can be used to generate photorealistic or STL models. Next, a RP machine can produce a physical model.

4 CONCLUSION

Reverse design capabilities may represent truly useful tools to create new and innovative design strategies. This paper describes a computational code to be used in reverse design activities. The code is based on a biomimetic approach replicating the human vision process into the computer. It comprises four main routines, that enable the rapid generation of digital models from existing buildings, and its subsequent physical representation through the generation of an appropriated tessellated model and the use of RP machines. This software can represent an important tool in a collaborative design environment, particularly in the case of building re-use involving a broad range of professional expertise.

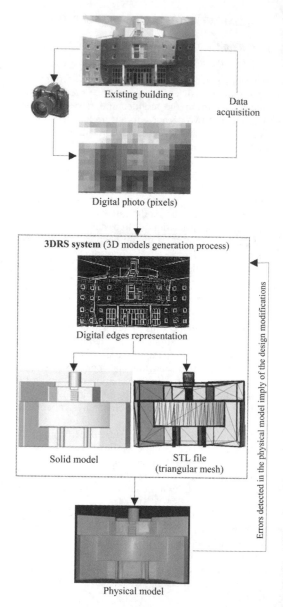

Figure 5. Three-dimensional models created by the 3DRS system from existing building and the correspondent physical model.

REFERENCES

Alves, N. & Bártolo, P. 2002a. A new image-based modelling system to support architectural redesign activities. In M. Sun, G. Aouad, C. Green, M. Omerod, L. Ruddock & K. Alexander (eds), *The Built and Human Environment; Proc. 2nd Intern. Postgraduate Research Conf.*: 322–333, *Salford University*. Blackwell.

Alves, N. & Bártolo, P. 2002b. Human vision principles supporting computer aided design. In C. Brebbia, L. Sucharov & P. Pascolo (eds), *Design & Nature 2002. Comparing Design in Nature with Science and Engineering; Proc. Inter. Conf.*: 411–420, *Udine*. London:Wit Press.

Audesirk, T. & Audesirk, G. 1999. *Biology: life on earth*. New Jersey: Prentice-Hall.

Bártolo, P & Bártolo, H. 2001a. Concurrence in design: a strategic approach through rapid prototyping. *Proc. CIB World Building Congress* 4: 54–64. Wellington.

Bártolo, P. & Bártolo, H. 2001b. The use of rapid prototyping techniques and internet tools to support a knowledge integrated design process. In J. Kelly & K. Hunter (eds). *Proc. COBRA* 2: 742–51. *Glasgow Caledonian University*. RICS.

Faugeras, O. 1993. *Three-Dimensional Computer Vision: A Geometric Viewpoint*. MIT Press.

Gordon, I. 1989. *Theories of visual perception*. Chichester: John Wiley.

Johnson, J. 1994. *Principles of computer automated fabrication*. Irvine: Palatino Press.

Kai, c. & Fai, L. 1997. *Rapid prototyping: principles and applications in manufacturing*. Chichester: John Wiley & Sons.

Pajdla, T., Werner, T. & Hlavváč, V. 1997. Correcting Radial Lens Distortion without Knowledge of 3-D Structure. *Technical report TR97-138*. Praha: FEL ČVUT.

Peng, J. & Loftus, M. 1999. Rapid prototyping based on image information in reverse design applications. *Proc. Instn. Mech. Engrs* 213: 317–322. London: IMechE.

Zrenner, E. 2002. Will retinal implants restore vision? *Science* 295: 1022–1025.

21. Knowledge management

System-based Vision for Strategic and Creative Design, Bontempi (ed.)
© *2003 Swets & Zeitlinger, Lisse, ISBN 90 5809 599 1*

Managing intellectual capital in construction firms

S. Kale & T. Çivici
Architecture Department, Balikesir University, 10145, Balikesir, Turkey

ABSTRACT: Intellectual capital is rapidly becoming a key strategic resource for creating competitive advantage in the construction industry. The emergence of intellectual capital as a key resource for creating competitive advantage poses enormous challenges to executives of construction firms. Construction firms can meet these challenges by revising their thinking about basis of competition and focusing on management and measurement of intellectual capital. This paper presents a working definition of intellectual capital and a framework for identifying and classifying components of intellectual capital. It also presents a three-stage process for effective management and measurement of intellectual capital in the construction industry.

1 INTRODUCTION

Construction business landscape is being transformed from production-based economy (p-economy) into a knowledge-based economy (k-economy). In p-economy, the basis of competition is organized around the control of financial capital (i.e., land, labor, money, machines) in the production of goods and services. The use knowledge as a production factor in was quite small. The main challenge facing construction executives in p-economy was managing the body of knowledge that produces and consumes financial capital (i.e., operations management). In k-economy, the basis of competition is organized around management and measurement of intellectual capital in the production of goods and services. Managing intellectual capital is the fundamental challenge facing construction executives in the k-economy. Yet research on managing intellectual capital in the construction industry is in its infancy stage of development. The paper presented herein focuses on this developing research area. It presents a simple framework for managing intellectual capital in construction firms. The main objectives of this framework are (1) to assist construction executives and (2) construction organizations to identify their intellectual capital resources and (3) to provide a foundation on which systems and processes for managing intellectual capital can be built.

The organization of the paper is as follows. Section 2 presents a working definition of intellectual capital. Section 3 traces the theoretical roots of intellectual capital. It also identifies three distinct generations of intellectual capital thinking and briefly discusses patterns to be recognized in each generation. Section 4 provides a three-stage process model for managing intellectual capital in construction firms. Section 5 offers some concluding remarks.

2 DEFINITION OF INTELLECTUAL CAPITAL

Intellectual capital is a term that has grown in popularity in the last decade. The term intellectual capital was first coined by economist John Kenneth Galbraith in 1960s (Stewart 1991). John Kenneth Galbraith, who in a letter to economist Michael Kalecki, wrote: "*I wonder if you realize how much those of us in the world around have owed to the intellectual capital you have provided over these past decades*" (Feiwal 1975). It is Stewart's (1991) influential article that recovers the term intellectual capital after more than three decades and brings it firmly on to the management and research agenda. The first use of the term is thus to describe the dynamic effects of individuals' intellect. Stewart (1991) makes intellectual capital the attribute of a firm. This attracted the attention of many business practitioners and researchers and created the main impetus for initiating research programs on intellectual capital. As a result, references to the term intellectual capital is common in books and journals in late 1990s (e.g., Edvinsson & Malone 1997; Roos & Ross 1997; Roos et al. 1997; Stewart 1997; Sveiby 1998; Nahapiet & Ghoshal 1998; Edvinsson et al. 2000; Pike & Roos 2000).

Intellectual capital is a complex concept. Different definitions have been set forth for exploring this complex concept. There is presently no universally acceptable definition of intellectual capital (Leon, 2002). Nahapiet & Ghloshal (1998) define intellectual capital as "the knowledge and knowing capability of a social collectivity such as organization, intellectual community, or professional practice". Stewart (1997) defines intellectual capital as "intellectual material knowledge, information, intellectual property, and experience – that can be put to use to create wealth". Klein & Prusak (1997) define intellectual capital as "intellectual material that has been formalized, captured, and leveraged to produce a higher-valued asset". Ulrich (1998) defines intellectual capital as "competence x commitment". Brooking (1996) argues that intellectual capital is the term given "to the combined intangible assets which enable the company to function". Williams & Bukowitz (2001) propose that intellectual capital embraces all forms of knowledge, ranging from the abstract (i.e., culture, norms, values, group dynamics, and individual members' knowledge and skills) to the concrete (i.e., presentations, documents, blueprints, process maps). Important underlying concepts in these definitions include the notion that; (1) intellectual capital is something invisible (2) it is closely related to knowledge and experiences of employees as well as customers/clients and technologies of an organization, (3) it offers better opportunities for an organization to succeed in the future.

These definitions provide a useful building block for understanding intellectual capital. Yet these definitions lack the specificity necessary to identify, classify and measure the components of intellectual capital. Several models have been set forth for identifying, classifying and measuring components of intellectual capital. The following section presents a review of these models.

3 INTELLECTUAL CAPITAL MODELS

The concept of intellectual capital management has dominated the managerial thinking and academic research studies in late 1990s. Its theoretical roots can be traced to the influential writings of Penrose (1959), Itami (1987), Wernerfelt (1984), Sveiby (1986) and Barney (1991). Penrose (1959) considers a firm as a collection of resources. Itami's (1987) groundbreaking work, which was originally published in Japanese in 1980 and not published in English until 1987, focuses on value of visible resources to the Japanese corporations. Wernerfelt (1984) proposes that firm resources are the primary determinant of creating value in product/services. Barney (1991) points out the importance of developing strategies for exploiting firm resources. Sveiby (1986) is considered to be the first person to recognize the need to manage and to propose a theory for managing and measuring intellectual resources of a firm. The work of Sveiby (1986), published originally

in Swedish, provided a rich and tantalizing view of the potential for valuing the enterprise based upon its intellectual resources. These views on firm resources, in particular intellectual resources, constitute the fundamental building blocks of the numerous intellectual capital models that have been set forth in the literature. The principal intellectual capital models include: the Balance Scorecard (Kaplan & Norton 1992), the Skandia Value Scheme (Edvinson & Malone 1997), the Intangible Asset Monitor (Sveiby 1997), the Intellectual Capital Index (Roos & Roos 1997) the digital IC-landscape (Edvinson et al. 2000) and the Holistic Value Approach (Pike & Roos 2000).

A succinct review of these intellectual capital models reveals that there are three distinct generations of intellectual capital thinking. The first generation of intellectual capital thinking (Edvinson & Malone 1997; Sveiby 1997; Kaplan & Norton 1992) is basically business based scorecard practices. The primary focus of first generation intellectual capital thinking is on identifying resources and measuring a firm's intellectual capital. *Skandia Value Scheme* is a good example of first generation of intellectual capital models. The second generation intellectual capital thinking focuses on transformations between and within resources (Roos and Ross 1997; Roos et al. 1997). The Intellectual Capital Index (*IC Index*) is a good example of second-generation intellectual capital thinking. It proposes that the presence of resources is a not sufficient to create value. The third generation intellectual thinking (e.g., Edvinson et al. 2000; Pike & Roos 2000) is about combining measures of different units into a totality measure that reflects the value perceived by a given observer. *The Holistic Value Approach* is a good example of third generation of intellectual capital thinking. The following sections present a brief discussion of the *Skadia Value Scheme* (Edvinson & Malone 1997), the *Intellectual Capital Index* (Roos & Roos 1997) and the *Holistic Value Approach* (Pike & Roos 2000).

3.1 Skandia Value Scheme

Skandia Value Scheme is developed by Skandia, a Swedish insurance and financial services firm (Edvinson & Malone 1997). The major driving force behind developing Skandia Value Scheme was that traditional financial statements represent only past financial information about a firm but additional information is needed to understand a firm's present and future capabilities. Skandia Value Scheme argues that a firm's total value is sum of its financial capital and intellectual capital. It also proposes that that intellectual capital of a firm takes three basic forms: human capital, structural capital and relational capital.

Human capital represents the knowledge, skills, and abilities of individual employees to meet the task

(Becker 1964). Human capital can also be defined as a combination of four factors: genetic inheritance, formal education, experience and social psychological attitudes about life and business. It is inherent in people and cannot be owned by firms. Therefore, human capital can leave a firm when people leave. Human capital also encompasses how effectively an organization uses its resources as measured by creativity and innovation.

Structural capital represents knowledge that stays within the firm at the end of the working day. It is the supportive infrastructure that enables employees (i.e., human capital) to function. *Structural capital* of a firm can be broken down into intellectual property and intellectual assets. Intellectual assets are knowledge that can be used exclusive of the creator – knowledge that has been articulated, codified, and often linked to the existing body of organizational knowledge. Intellectual assets include the management philosophy, corporate culture, management processes, information systems, and networking systems. It also includes the techniques, procedures, and programs that implement and enhance the delivery of products and services. Intellectual property is a subset of structural capital for which various forms of ownership are protected by law. It refers to the explicit, packaged result of innovation in the form of protected commercial rights, intellectual capital and other intangible resources and values. There are four major categories of intellectual property: patents, trade secrets, trademarks, and copyrights.

Relational capital represents knowledge embedded in organizational relationships with customers, suppliers, stakeholders, strategic alliance partners (Stewart 1997). It can be defined as "the actual and potential resources individuals obtain from knowing others, being part of a social network with them, or merely being known to them and having good reputation (Baron & Markman 2000). Therefore relational capital encompasses both actual and potential resources flowing through a relationship network established either individually or collectively.

3.2 Intellectual Capital Index

The Intellectual Capital Index (*IC Index*) is a tool for measuring a firm's performance of the value creation process. It was developed as an internal management tool as a means for managers to identify and articulate hidden value creating resources and processes and to create a structure for measuring the achievement of strategic goals. It is a generic measuring instrument rather than an assessment process. It consolidates different indicators of intellectual capital in single measure by using methods employed in experimental physics (Roos et al. 1997). First generation intellectual capital thinking (e.g., Skandia Value Scheme)

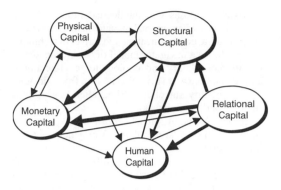

Figure 1. The Navigator for a construction firm.

focuses on identifying components of intellectual capital and measuring intangible resources. The first generation thinking was static in that transformations between categories of intellectual capital were not addressed. Intellectual Capital Index points out that merely identifying components of intellectual capital and in turn measuring the stocks of intellectual resources is not enough. It is also essential to measure, and thus, also manage, the flows of intellectual capital, that is the changes in the stocks of intellectual capital (Roos & Roos 1997). The term "stocks" herein refers a firm's human, structural and relational resources. The term "*flows*" herein refers to the transformations between and within stocks of human, structural, and relational and financial resources.

Intellectual Capital Index goes beyond merely identification of different forms of intellectual capital. It also attempts to identify the transformations (i.e., flows) between and within components of intellectual capital. Intellectual Capital Index proposes a model called "Navigator" to assess and visualize the impact of transformations on value creation. The Navigator reveals all value-creating stocks (tangible and intangible resources), their transformations (i.e., flows), and the relative importance of the resources and transformations (i.e., flows) in value creation. The circles represent stocks, the arrows the flows and their size their relative importance. The Navigator is a powerful visual tool reveals all value-creating stocks (tangible and intangible resources), their transformations (i.e., flows), and the relative importance of the resources and transformations (i.e., flows) in value creation (Figure 1).

3.3 Holistic Value Approach

Holistic Value Approach (Pike & Roos 2000) is a third generation intellectual capital model. It is a tool for integrating the separate issues of intellectual capital disclosure and management. Holistic Value

1547

Approach proposes that intellectual models should serve not only as an internal management support tool but also as a communication tool. It is a technique that combines a navigator model of the business with measurement theory and axiology to generate a non-dimensional view of firm value as seen from the point of view of any stakeholder (i.e., clients, suppliers, partners, and employees).

Holistic Value Approach considers a firm as a value generator. It argues that a firm generates value in some form or another. A firm generates value internally through: the values and quality of the corporate management, effectiveness of the deployed intellectual capital, effectiveness in activities and processes, quality of compliance with regulatory standards. The external environment awards value through: revenue from the sale of product/services, client value added after use, employee value added after receipt of salary and financial benefits, stakeholder value added, assessment of regulators and professional interests, assessment environmental and social impacts resulting from firm's activities and operation, reports from media and public opinion. Holistic Value Approach proposes that internal and external values represent a firm's inclusive value. Inclusive value of a firm is composed of two main categories: financial value (i.e., financial capital) and non-financial value (i.e., intellectual capital). The non-financial value of a firm is divided into the value performance of achieved with respect to operational objectives and the value resulting from assessments of external factors that evaluate the firm's achievement level with respect to their own objectives. Financial value of a firm is measured by using conventional methods such as net present value of cost or cash flows or option of costs or cash flows. On the other hand, the non-financial value of a firm is measured by using a standard set of combinatorial rules and practices consistent with measurement theory and axiology. Holistic Value Approach requires users (i.e., strategic leaders) to create hierarchies of categories of intellectual capital to which they assign weightings and value ratings (from 0 to 1) according to strategic priorities.

It visualizes the combination of financial and non-financial value of a firm by using a multi-dimensional contour map where normalized financial value of a firm defines the y-axis and non-financial (i.e., intellectual capital) value of a firm defines x-axis (Figure 2). The z-axis defines the overall combined value resulting from the combination of financial and non-financial values.

Thus far the concept of intellectual capital is defined, its roots are traced, and three distinct generations of intellectual capital thinking are reviewed and discussed. The following section explores the concept of intellectual capital and its management and measurement in the context of the construction industry.

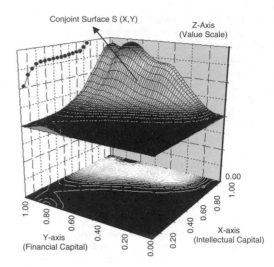

Figure 2.   The multi-dimensional contour map.

It integrates the concepts and frameworks that have been set forth by three distinct generations of intellectual capital thinking.

4   MANAGING INTELLECTUAL CAPITAL IN CONSTRUCTION FIRMS

Intellectual capital of construction firm is not conveniently warehoused and audited; it is invisible and does not appear in the balance sheet. It is scattered, difficult to detect and prone to leave. Therefore, managing intellectual capital in construction firm is fundamentally different from managing financial capital (i.e., physical and monetary capital), yet it requires similar detailed and painstaking attention.

A construction firm can manage its intellectual capital by using a three-stage process: (1) *Identification of stocks and flows* – construction firm identifies critical stocks and flows that are strongly related to its long-term strategy, (2) *Measurement of stocks and flows* – construction firm develops indicators to serve as proxy measure for each stocks and flows, (3) *Monitoring* – construction firm assesses and evaluates whether the firm has the intellectual capital it needs and is using its intellectual capital effectively.

Identification of stocks and flows. The first stage in managing intellectual capital in construction firms is identification of firm's stocks and flows between and within stocks. Intellectual capital of a construction firm includes its human, structural, and relational capital.

Human capital of a construction firm includes knowledge, capabilities, skills and experience of its employees. It is also a firm's combined capability for solving strategic, administrative and operational problems that prevail in the construction industry. Human capital is important for a construction firm because it is a source of innovation and strategic renewal.

Structural capital of a construction firm includes its strategy, structure, systems and processes that enable it to produce and deliver a product/service (i.e., contracting services, constructed facility) to its clients. Structural capital is an enabler of performance. It is the structural capital of a construction firm that promotes the continuous application of knowledge to construction business. Structural capital of construction firm is the critical link that allows intellectual capital to be measured at a firm level. Structural capital is also organizational culture and management philosophy that promote learning and sharing of those learning.

Relational capital of a construction firm resides in its relationships with clients, subcontractors, construction material vendors, sureties, design and engineering firms. These relationships enable a construction firm to receive resources (i.e., knowledge, information, labor, material, legitimacy) from its external environment. Maintaining good quality of relationships with these parties enables a construction firm to capture benefits of inter-organizational learning such as developing trust, overcoming communication and coordination problems.

Measurement of stocks and flows. The second stage in managing intellectual capital in construction firms is measurement of identified stocks and flows. Intellectual capital of a construction firm can be measured in turn managed by developing a set of indicators. A construction firm can follow a three-stage process for constructing indicators for measuring and in turn manage its intellectual capital: (1) developing indicators for stocks (i.e. human, structural and relational capital) (2) developing indicators that represent the flows between and within stocks and (3) developing a hierarchy of intellectual capital indices.

Human capital of a construction firm can be measured by such indicators as; (1) actual competence levels versus ideal levels, (2) success in planned ratios or completed development plans, (3) reputation of firm employees with headhunters, (4) years of experience in profession, (5) value added per employee, (6) proportion of employees making new idea suggestions.

Structural capital of construction firm can be assessed by such indicators as (1) the cost of transactions, (2) construction cost – year on year improvement in cost efficiency, (3) construction time – year on year improvement in time efficiency, (4) cost predictability – year on year improvement in cost predictability, (5) time predictability – comparing forecasted time of delivery to actual time of delivery, (6) defects – year on year reduction in defects,

(7) revenue per employee, and (8) product/service innovations – the number of new project delivery or construction methods introduced in each year.

Relational capital of construction firm can be measured by such indicators as (1) reduction in complaint resolution time, (2) product satisfaction – measuring clients' happiness with the end product, (3) contracting service satisfaction – measuring clients' happiness with the service, (4) relationship quality – measuring relationships with suppliers, sureties, subcontractors, design and engineering firms, construction material vendors (5) growth in contract awards (6) proportion of contract awards by repeat clients.

Managing intellectual capital in a construction firm implies not only identifying and measuring stocks (i.e., human, structural, and relational capital) that create value but also flows (i.e., transformations) between and within stocks. Therefore, it is also vital to develop indicators that represent the transformations between and within all resources, financial and intellectual capital (i.e., human, structural and relational capital) to create maximum value. Some examples of indicators for flows could include: (1) "the number of financially successful product/services developed in a given R&D" is an indicator for a flow from financial capital to structural capital, (2) "man-hours spent on a relationship" is an indicator for a flow from human capital to relational capital, (3) "percentage of available man-hours spent on developing and maintaining IT-based constructability library" as an indicator for a flow from human capital to structural capital, and (4) "cost savings due to use of an IT based constructability library in training new employees" as an indicator for a flow structural capital to human capital.

Monitoring. In this final stage the data for each indicator is collected. An intellectual capital report is then prepared that shows firm's existing levels of stocks and the flows between and within these stocks. It also includes conclusions about whether the firm has intellectual capital it needs and is using its intellectual capital effectively, and recommendations for action to achieve changes and improvements or to maintain existing levels of intellectual capital. Visual presentation tools (i.e., navigator models and multi-dimensional contour maps) can be used in preparing an intellectual capital report. These tools enables construction firms (1) to evaluate intellectual capital situation of firm holistically (2) to assess the impact of flows (i.e., transformations) on value creation, (3) to visualize growth or decline in capital and also the trade-offs to compare the same unit over time, or with other construction firms, (4) to analyze strategic impact of changes in intellectual capital, (4) to identify which categories of intellectual capital and flows are more important.

Managing intellectual capital in a construction firms is a continuous process. Therefore, a construction firm should collect data for the set of indicators for stocks and flows and produce a consolidated intellectual report at regular intervals.

5 CONCLUDING REMARKS

The traditional rules of competition in the construction industry are increasingly losing relevance in k-new economy. Intellectual capital, in contrast to financial capital (i.e., physical and monetary capital), is becoming a key strategic resource for gaining and sustaining competitive advantage in the construction industry. Managing financial capital (i.e., land, labor, money, machines) continues to be important factor in k-economy but its relative importance decreased through time as the importance intellectual capital has increased. The emergence of intellectual capital as a key strategic resource and a source competitive advantage poses enormous challenges to executives of construction firms. Construction firms can meet these challenging by revising their thinking about basis of competition and focusing on effective management and measurement of intellectual capital. Therefore, it is critically important that intellectual capital be well understood and properly managed if construction firms are to compete successfully in today's construction business environment.

REFERENCES

Barney, B. J. 1991. Firm resources and sustained competitive advantage. *Journal of Management* (17)1: 99–120.

Baron, R. A. & Markman, G. D. 2000. Beyond social capital: The role of social competence in entrepreneurs' success. *Academy of Management Executive* (14)1: 106–116.

Becker, G. 1964. *Human capital. A theoretical and Empirical Analysis*. New York: NBER.

Brooking, A. 1996. *Intellectual Capital*. London: International Thompson Business Press.

Chatzkel, J. (2002). A conservation with G. Ross. *Journal of Intellectual Capital* (3)2: 96–117.

Edvinsson, L., Kitts, B. & Beding, T. 2000. The next generation of IC measurement – the digital IC-landscape. *Journal of Intellectual Capital* (1)3:263–272.

Edvinsson, M. & Malone, M. 1997. *Intellectual capital: the proven way to establish your company's real value by measuring its brain power*. New York: Harper Collins.

Feiwal, G. R. 1975. The Intellectual Capital of Michael Kalecki: *A Study in Economic Theory and Policy*. Knoxville, TN: The University of Tennessee Press.

Itami, H. 1987. *Mobilizing Invisible Assets*. Cambridge: Massachusetts.

Kaplan, R. S. & Norton, D. P. 1992. The balanced scorecard – measures that drives performance. *Harvard Business Review* (70)1: 71–9.

Klein, D. A. & Prusak, L. 1994. Characterizing intellectual capital. *Ernst & Young Commercial Innovation Center*.

Leon, M. V. 2002. Managerial perceptions of organizational knowledge resources. *Journal of Intellectual Capital* (3)2: 149–166.

Nahapiet, J. & Ghoshal, S. 1998. Social capital, intellectual capital and the organizational advantage. *Academy of Management Review* (23)2: 242–266.

Penrose, E. T. 1959. *The Theory of the Growth of the Firm*. New York: Wiley.

Pike, S. & Roos, G. 2000. Intellectual capital measurement and holistic value approach. *Works Institute Journal* (42)6: 21–7.

Roos, G. & Roos, J. 1997. Measuring your company's intellectual performance. *Long Range Planning* (30)3: 413–426.

Roos, J., Roos, G., Dragonetti, N. C. & Edvinsson, L. 1997. *Intellectual Navigating the New Business Landscape*. Macmillan Press: London.

Stewart, T. 1997. *Intellectual Capital: The Wealth of Nations*. New York: Doubleday.

Stewart, T. 1991. Brainpower. *Fortune* (6): 44–57.

Sveiby, K. E. 1998. The intangible asset monitor. *Journal of Human Resource Costing and Accounting* (2)1: 73–97.

Sveiby, K. E. 1986. *Kunskapsföretaget, (The Know-how Company)* Liber.

Ulrich, D. 1998. Intellectual capital = competence X commitment. *Sloan Management Review* (39)2:15–26.

Wernerfelt, B. 1984. A resource based view of the firm. *Strategic Management Journal* (5)2: 171–180.

Williams, R. L. & Bukowitz, W. R. 2001. The yin and yang of intellectual capital management. *Journal of Intellectual Capital* (2)2: 96–108.

System-based Vision for Strategic and Creative Design, Bontempi (ed.)
© 2003 Swets & Zeitlinger, Lisse, ISBN 90 5809 599 1

Process- and success-oriented knowledge management for total service contractors

R. Borner
Institute for Construction Engineering and Management, Swiss Federal Institute of Technology, Zurich, Switzerland

ABSTRACT: A major obstacle to the wider acceptance of knowledge management in Swiss construction companies is that they find it difficult to recognize any concrete benefit for their day-to-day business to be gained from applying existing knowledge management concepts. It is important for construction companies to win orders in a competitive environment and to execute these orders profitably for the company and satisfactorily for the customer (project success). The identification and management of success factors in construction processes support the project management and help to achieve these goals. Based on these considerations, the author has developed a solution for knowledge management in his doctoral studies in association with leading Swiss construction companies. The solution pursues the approach that the knowledge that is valuable for construction companies is the knowledge needed to activate the success factors. This knowledge, which ensures success, is worth managing.

1 OBSTACLES FACING KNOWLEDGE MANAGEMENT AND APPROACHES TO IDENTIFYING A SOLUTION

1.1 *Difficulties facing total service contractors in respect of knowledge management*

The primary reasons why general and total service contractors do business are to successfully acquire contracts in their targeted area of business, to realize the customer's performance goals to the latter's satisfaction and in line with the latter's requirements, and to successfully complete the project from the perspective of their own companies (profits, image).

In the effort to achieve these goals, the attitude towards the element of knowledge plays an important role. For example, "rediscovering" tried and trusted solutions goes hand in hand with losses of efficiency in finalizing the project, which cause a deterioration in the competitive positioning of the supplier. Literature uses the "resource based view" to address the role played by knowledge within companies (Zack 1999). Various approaches and models have been developed in the field of knowledge management, especially during the 1990's, all of which aim to support companies in systematically managing their knowledge. Moreover, specialist literature offers a plethora of tools and measures for addressing the individual knowledge management processes of knowledge identification,

allocation, preservation, etc., such as intranet, experience discussion groups, databases, etc. (Girmscheid & Borner 2001).

Observing construction work in practice has shown that companies are frequently unable to identify any real visible benefit for their day-to-day business from adapting existing knowledge management concepts to reap advantages that would justify the investment of time and money needed to initiate the appropriate measures. Two major stumbling blocks preventing a broader level of acceptance and application of knowledge management in construction companies proved to be the unique character of each building project, and the lack of any practical tools for a company to use to pinpoint the knowledge that was of value to the said company and to identify the real benefit to be gained from systematically nurturing this knowledge.

1.2 *Approaches to identifying a solution*

In view of this situation, the Institute for Construction Engineering and Management (Swiss Federal Institute of Technology, Zurich) has developed the model approach of project-oriented and cluster-oriented knowledge management. The research work forms part of the overall construction system provider (SysBau) research approach developed by the Institute (Girmscheid 2000).

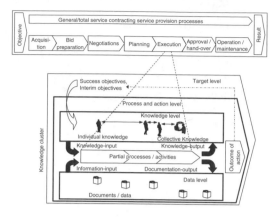

Figure 1. Knowledge cluster modelling.

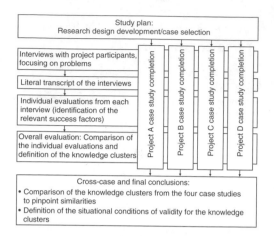

Figure 2. Flow pattern of the multiple case study.

When executing construction projects it has repeatedly been demonstrated that certain actions or decisions taken during the project have had a positive impact that played a major role in influencing the competitive success or success of the project (success factors). From a theoretical angle we know that knowledge only reveals its purpose or benefit when applied to actions, decisions or processes. However, whilst knowledge is based on data and information it is – unlike these two elements – also always linked to people (Nonaka & Takeuchi 1995).

As such, the knowledge that is particularly valuable for total service contractors is the knowledge that is needed to guarantee the best possible conversion of the individual success factors into actions that ensure success. This knowledge, which ensures success, is worth nurturing and promoting on the part of the company.

However, it is not a single piece of information or a single element of knowledge that is needed to activate success factors but rather a combination (cluster) of various elements of knowledge and data. The concept of "knowledge clusters" was derived as an aid to describing the success factors and the relevant requisite knowledge (Figure 1). Thus, these knowledge clusters can then be studied over the entire process of providing products and services for a construction project.

2 FUNDAMENTAL PRINCIPLES OF THE MULTIPLE CASE STUDY

2.1 *Objective and definition of the study*

The fundamental prerequisite to enable this approach to knowledge management to function at all is that there are success factors, respectively knowledge clusters, which occur similarly or are relevant in several projects of the same type. If all the knowledge clusters from one project were only valid for that one

specific case, these could not be reactivated for newly acquired projects and, as such, knowledge could not be managed by this means.

In order to verify this fundamental prerequisite, 4 major and complex structural engineering projects were examined to pinpoint their success factors, respectively knowledge clusters, and compared with each other to identify any similarities or differences (Borner 2003). Pre-determined selection criteria were used to first select 4 structural engineering projects, which were completed by different Swiss total service contractors using varying approaches to the project execution (project development, direct negotiations, total service competition).

When analyzing the success factors, the survey was restricted to the partial perspective of the interaction between the total service contractor and the other parties involved in the project in respect of the optimal achievement of all objectives within a pre-determined target system (competitive and project success).

In addition to verifying the suitability of the formulated knowledge management approach, the study also served to illustrate how knowledge clusters could be pinpointed, with the findings also being used to expand the project-oriented and cluster-oriented knowledge management model

2.2 *Research methodology*

An embedded multiple case design was chosen for the study, in line with the classification of research designs for case studies (Yin 1994). One elementary factor in designing multiple case studies, according to Yin, is that the research logic should limit itself to replication and not sampling logic.

The study in question was conducted in line with the layout illustrated in Figure 2, based on the flow pattern recommended by Yin for multiple case studies.

Table 1. Overview of the project characteristics of the selected cases.

Project	Relevant characteristics of the projects
Project A (PA)	– Project development by a promoter working with a famous architect – Direct negotiations on the part of the promoter with a single selected total service contractor – Tenant alone had responsibility for completing the interior – Sale of the construction project to an investor (public developer) during the completion phase – Total service contractor completed the basic works and the tenant finalized the interior
Project B (PB)	– Potential developer's plans to expand were well known – Total service contractor secured an ideally located plot of land early on – Total service contractor supported the developer during the initial stages (organization of ideas competition, negotiations with third parties, clarifications, first concepts) – Pre-project planning conducted by this total service contractor – Definitive award of the contract to the total service contractor on the basis of the cost estimate based on the pre-project planning – Realization as total service provider
Project C (PC)	– Preliminary study and pre-project planning drawn up by the developer and an architect, who was selected from a competition of architects – Organization of a total service competition with pre-qualification – Realization as total service provider
Project D (PD)	– Developer needed 1000 workplaces within a short space of time – Pre-selection of suitable total service contractors, specialist planners and architects – Organization of a total service competition with prior formation of teams on the basis of a list of participants drawn up by the developer – Organization of a workshop over several days incorporating all parties involved, following conclusion of the total service contract – Realization as total service provider

The four case studies were conducted using qualitative social research methods, whereby 10 interviews were conducted with the parties involved in the project in case study A, 6 each for case studies B and C

and 4 interviews for case study D. Not only were key personnel in the relevant total service contracting company questioned, but also representatives of the developers, users, planners and architects.

2.3 Case descriptions

The structural engineering projects that were examined featured the relevant project characteristics listed in Table 1.

3 FINDINGS FROM THE STUDY

3.1 Fundamental principles for comparing the knowledge clusters identified from the four case studies

15 knowledge clusters each were pinpointed in case studies A and C and 13 each in case studies B and D. Once these knowledge clusters had been identified, they were cross-compared using replication logic.

It was necessary to first define the relevant criteria on which to base a comparison of the knowledge clusters to reveal any similarities. The following criteria were defined on the basis of the knowledge cluster modeling (Figure 1), all of which had to be fulfilled by various knowledge clusters from the different case studies in order for these to be deemed sufficiently similar or even identical:

– Identical allocation of the knowledge cluster activity to the relevant process of providing products and services on the part of the service provider.
– Similarity of the knowledge cluster activities in the knowledge clusters being compared.
– Similarity of the outcome of actions respectively impacts achieved by actually executing the knowledge cluster activities. These outcomes serve to support the success targets of acquiring the contract, satisfying the customer and earning a profit for the company.

Once the identified knowledge clusters had been compared from every angle of each individual case study with the knowledge clusters from the other three case studies, these were then subdivided into case-specific knowledge clusters and cross-section knowledge clusters.

3.2 Case-specific knowledge clusters

Case-specific knowledge clusters – Case study A:

– Advice to the customer regarding possible solutions during the first contact
– Active incorporation of the customer's needs into the project
– Clear definition and demarcation of the individual work categories

- Minimization of possible losses of knowledge at personal interfaces
- Serious preparation, procurement of fundamentals and planning execution

Case-specific knowledge clusters – Case study B:

- Services to resolve the developer's initial problems were offered at an early stage/Securing an ideally located plot of land
- Close involvement of the developer in the project and definition of the customer's needs at all levels
- Careful preparation of the execution and discipline on the part of the subcontractors with regard to deadlines

Case-specific knowledge clusters – Case study C:

- Offering the customer supportive services at an early stage
- Actively responding to the needs and expertise of the developer/Channeling requirements
- Clear regulation of the planning permission process
- Particular attention to awkward aspects and minimization of errors thanks to the total service contractor's experience
- Careful preparation of the execution/execution planning
- Transfer of knowledge between project development (competition) and realization

Case-specific knowledge clusters – Case study D:

- Increasing the initial speed of the competitive teams by activating existing contacts
- Organizing a workshop to capture the customer's needs and integrate them into the project/involving the developer
- Minimization of knowledge losses thanks to personnel consistency throughout the project
- Careful preparation of the execution of construction elements with special quality requirements

3.3 Cross-section knowledge clusters

Cross-section knowledge clusters can be subdivided into qualified cross-section knowledge clusters, which occurred in two or three cases, and unqualified cross-section knowledge clusters, which were equally valid for all four case studies.

A cross-reference revealed that the following five unqualified cross-section knowledge clusters were equally relevant for all four case studies:

- Identification of potential areas of optimization on a competitive project and cost-effective inclusion of the same in the bid
- Risk minimization prior to signing the contract (cost-covering calculation and identification of qualitative problems before agreeing the contract)

Table 2. Summary of the qualified cross-section knowledge clusters.

Brief description of the knowledge cluster	Similar knowledge clusters from			
	PA	PB	PC	PD
Proof of experience and references from earlier and similar projects	X		X	X
Constructive collaboration with the architect to incorporate desired alterations (reaching a consensus)	X			X
Efficiency and transparency thanks to lean project organization and clearly structured planning process		X		X
Coordination of interfaces during the provision of the products and services		X	X	
Incorporation of changing needs of the customer/flexibility	X	X	X	
Identification and minimization of risks and potential mistakes at an early stage/quality control	X	X		X
Specific award criteria in case of difficult tasks	X	X	X	

Table 3. Number of case-specific knowledge clusters for each case study.

	No. of identified knowledge clusters	No. of case-specific knowledge clusters
Project A	15	5
Project B	13	3
Project C	15	6
Project D	13	4

- Transparency on the part of the total service contractor towards his customer (with regard to requests for alterations)
- Promotion of a constructive atmosphere of cooperation within the team
- Serious approach to providing warranties and eliminating defects

3.4 Conclusion from comparing knowledge clusters

After conducting the cross-case it was possible to take a closer look at the number of case-specific knowledge clusters and cross-section knowledge clusters.

A study of the cross-section knowledge clusters revealed that 7 of these occurred equally on two or three of the examined projects (qualified cross-section knowledge clusters) whilst 5 of them were equally relevant for all four of the cases studied.

As such, the conditions under which this study was conducted revealed that approximately 1/3 of the relevant knowledge clusters from any one project only related to the specific case of the project studied, whereas approximately 2/3 of the knowledge clusters were valid for several projects.

In consequence, the approach of using knowledge clusters to manage the knowledge of total service contractors revealed that the fundamental prerequisite was fulfilled in respect of the suitability of the approach, since some 2/3 of the knowledge clusters were equally valid for several structural engineering projects, and could therefore have been activated for specific use on these projects.

On the basis of replication logic, the situational conditions, under which the specific or cross-section knowledge clusters occurred in the context of the study, respectively the valid criteria were moreover defined by cross-case. These criteria are important subsequently for determining whether each knowledge cluster is even relevant or valid for the next specific case when applied once more to a new structural engineering project.

3.5 Conclusions from the contents of the knowledge cluster analysis

From the point of view of content, the five cross-section knowledge clusters are of particular relevance. When trying to win contracts the identification of potential areas of optimization on a competitive project and the use of intelligent solutions to achieve additional customer benefit in the bid would seem to ensure success. Moreover, the success of all projects was affected by the fact that the future development of market prices was realistically estimated when drawing up the bid, and that attention was paid to overcoming the contract risks.

An important factor in the realization of the projects seems to be that the customer is actively involved in the project, and that the contractor adopts a fair, open and transparent attitude towards the customer. The interpersonal relationships among the members of the team should also not be underestimated in order to ensure a constructive and cooperative atmosphere. According to the studies, it would also seem to be important that a lack of seriousness in handling the warranty obligations following the hand-over of the construction project can cause a lasting deterioration in any customer satisfaction that had been established up until that point, which, in turn, can impact the contractor's image.

The developer's feeling of security and his faith in the contractor's performance is therefore of major importance in ensuring both a trouble-free and successful completion of the project, and customer satisfaction.

4 IMPLICATIONS FOR A PROJECT- AND CLUSTER-ORIENTED KNOWLEDGE MANAGEMENT MODEL

4.1 Reactivation of knowledge clusters for new structural engineering projects

In order that the knowledge clusters identified from completed structural engineering projects can be applied respectively activated for new structural engineering projects, the results must be collated in such a way as to ensure that the suitability of the potentially applicable knowledge clusters can easily be verified.

In order to gain an overview over the ongoing identification of new knowledge clusters from completed construction projects it is necessary to collate the results in a "cluster pool". Whereby a cluster pool can have a variety of shapes and sizes, ranging from a collection of appropriate paper printouts to a collation of the information in databases or on the intranet.

If the contractor is now targeting a new structural engineering project, the following steps must be completed to reactivate any knowledge clusters:

– Right from the start the decision must be made in principle as to which acquisition strategy is to be applied: Should the contractor become involved in the project development at an early stage, or has the project development already been completed by the developer and other third parties, resulting, for example, in the decision to organize a total service contracting competition?

– On the basis of the specific project conditions stipulated for the new structural engineering project (e.g. type of project execution, customer's performance targets, general conditions, project participants, project organization) the individual conditions of validity of the knowledge clusters identified during earlier projects now need to be examined to verify that they comply with the new situation, resulting in the pinpointing of those knowledge clusters, which are applicable to and relevant for the new project.

– A resource schedule needs to be drawn up for the individual, applicable knowledge clusters, i.e. a determination of the appropriate ramifications for their use.

– And lastly, the individual knowledge clusters need to be activated at the right point in time (during the process of providing the products and services) and in the right places, respectively the project management team needs to control the activation of the same.

– Following completion of the project, the effectiveness of the steps taken needs to be assessed from the perspective of how successful they were, new knowledge clusters need to be identified and this information needs to be added to the cluster pool by means of a feedback process.

4.2 Expansion of the project- and cluster-oriented knowledge management model

The steps indicated so far for identifying and describing the knowledge clusters from completed projects and for activating valid knowledge clusters on new projects "only" serve to ensure that the success factors are consciously activated on new projects and, as a result, only potentially raise the chances of successfully winning the contract, achieving customer satisfaction and earning a profit for the company.

In order to achieve a real process of improvement or even development, a further step at the "strategic development level" needs to address the following points:

– Verify what knowledge is needed to activate the individual knowledge clusters
– This verification should reveal which knowledge has already been sufficiently processed internally or is already in the hands of the relevant colleagues, and which knowledge first needs to be developed or improved. The outcome reveals the potential areas of improvement and development with regard to the individual knowledge clusters.
– Prioritize the knowledge clusters
– Derive suitable measures for realizing the potential for improvement or development (e.g. strategic cooperation agreements, advanced training schemes, etc.)
– Approve and implement appropriate measures to improve or develop each knowledge cluster

An appropriate process model for project- and cluster-oriented knowledge management for total service contractors can now be sketched on the basis of the findings that have so far been discovered (Figure 3).

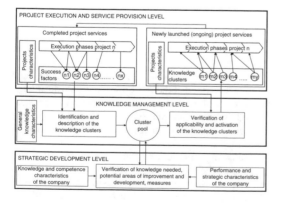

Figure 3. Process model for project- and cluster-oriented knowledge management for total service contractors.

4.3 Reflections on the organizational implementation

The knowledge management processes should be controlled by so-called "knowledge managers" from a central point within the company. If the company does not wish to set up a proprietary organizational unit to do this, the quality management unit could assume responsibility.

Consequently the knowledge manager would hold closing talks with the project and construction managers following the completion of a structural engineering project in order to determine the success factors. Moreover he would incorporate appropriate questions into the interview with the customer at the end of the project. Once the knowledge clusters have been described, he would compare the pinpointed knowledge clusters with the knowledge clusters identified during other structural engineering projects and look for any similarities. After determining the relevant conditions of validity for each knowledge cluster, the knowledge manager would then add these new findings to the cluster pool and manage the same.

At the strategic development level the knowledge manager would examine the knowledge needed to activate the individual knowledge clusters and derive the relevant areas of potential improvement or development on the basis of the same. After consultation with the management, the knowledge manager would prioritize the knowledge clusters and define suitable means of realizing the potential improvement and development. These measures would be submitted as a proposal to the management. As the knowledge clusters represent success factors, direct justification can be given as to which benefits would result for the company from implementing the relevant measures (cost vs. benefit).

In the case of newly acquired structural engineering projects, the knowledge manager would assume the role of coach. Once the conditions of validity had been verified, he would – at various stages in the project – draw the attention of those internal parties responsible for the project (calculation, competitive team, project manager) to the relevant knowledge clusters at the right point in time, and would advise them on using the knowledge clusters, whilst at the same time pointing out any tools, which might have been developed for the purpose. By doing so, an activation and consideration of the success factors, respectively knowledge clusters, can be guaranteed. The actual activation of the knowledge clusters during the execution of the project should then be controlled by the project management team.

5 OUTLOOK

The model approach to project- and cluster-oriented knowledge management, which has been outlined in this paper, will be refined in more detail in subsequent research work, whereby the processes in the knowledge management model presented here will be studied, developed and theoretically founded in more detail.

REFERENCES

Borner, R. 2003. *Win-Win-Erfolgsfaktoren bei Gesamtleistungen – Erfolgsorientiertes Wissensmanagement in GU- und TU-Leistungserstellungsprozessen.* Zürich: Institut für Bauplanung und Baubetrieb, ETH Zürich.

Girmscheid, G. 2000. Wettbewerbsvorteile durch kundenorientierte Lösungen – Das Konzept des Systemanbieters Bau (SysBau). *Bauingenieur* 75, 1/2000: 1–6.

Girmscheid, G. & Borner, R. 2001. Einsatz und Potenziale von Wissensmanagement in Unternehmen der Bauwirtschaft. *Bauingenieur* 76, Mai 2001: 256–260.

Nonaka, I. & Takeuchi, H. 1995. *The Knowledge-Creating Company.* Oxford: Oxford University Press Inc.

Yin, R.K. 1994. *Case study research: design and methods.* Thousand Oaks: Sage Publications.

Zack, M.H. (ed.) 1999. *Knowledge and Strategy.* Boston: Butterworth-Heinemann.

An intelligent cooperation environment for the building sector, why?

I.S. Sariyildiz, B. Tunçer, Ö. Ciftcioglu & R. Stouffs
Faculty of Architecture, Delft University of Technology, Delft, The Netherlands

ABSTRACT: Building design and construction process, from initiative until demolition, is becoming more complex not only in its infrastructure, communication and information, but also in the number of partners involved. Increasing requirements in terms of cost and time and high demands require better communication and co-ordination. In order to meet these requirements, there is a need for an Intelligent Modelling for Cooperative Engineering (IMCE) environment. In this environment, information must be selected, ordered and made available in an intelligent manner conforming to the needs of each partner, which may vary in character and type, and may conflict at times. Considering the complex as well as soft nature of the information concerned, the need for effective and efficient processing of this information leads us to make use of soft computing technologies. The resulting systems should actively support the user in the building process. This can be achieved by benefiting from existing emerging intelligent technologies. Referring to active research, this paper presents an overview of the IMCE project and provides insights into the state-of-the-art of this project.

1 INTRODUCTION

Building design is a multi-actor, multi-discipline, and multi-interest process. In this collaborative process, communication and information and knowledge exchange become increasingly important. Building design and construction is a teamwork among architects, consultants for various fields, contractors, and the government, each with their own specialized knowledge and information.

The design process is complex in the sense that many, often conflicting, interests and criteria are involved, and that many different types of expertise are required to find an optimal solution. The outcome of the design process has to fulfill different requirements of functional, formal, and technical nature. These requirements concern aspects such as usability, economics, quality of form and space, social aspects of architectural design, technical norms or laws, and technical and mechanical aspects of the design. Additionally, there is the uncertainty of the future use of the building, requiring the meeting of new criteria that are not defined explicitly at the moment of design. That means that a designer must have the ability to meet a certain range of criteria in a flexible way so that future demands are also met to a certain degree.

In relation to this, the need to bring together and assemble components that are manufactured elsewhere requires more and more integration and cooperation between the many partners in the construction process. Information required with respect to design, building technology, and logistics becomes ever more complex and extensive, both in terms of content and information logistics. The type of information and data varies intensely, being qualitative, quantitative, graphical, or numerical.

Increasing emphasis on cost and time constraints requires an improved communication. In order to fulfill on the increasing needs for communication and data and knowledge exchange, the information from suppliers and the information produced by the various partners and specialists in the building process has to be intelligently selected and presented according to the individual needs of each actor. At the same time, an Information, Communication and Knowledge Technologies (ICKT) environment or platform must be provided that stimulates and supports the collaboration and cooperation within a building team and between the various partners. Such an environment may also make it possible for the end user, and other possibly interested parties, to be involved in the process in a structural way.

2 HISTORICAL DEVELOPMENTS

2.1 Craftsmanship

When we look at the history of buildings and the building process, especially prestigious medieval edifices such as citadels and palaces, and temples, mosques and cathedrals, we see that it sometimes took ages to complete a building; some of the cathedrals even took 400 years to complete (e.g., Fig. 1). The building was a craftsmanship and, noteworthy, during the construction of these large buildings craftsmen from various cultures worked together (e.g., Fig. 2). Before the invention of book printing in the 9th century, communication, collaboration and knowledge exchange between craftsmen was by word of mouth. After book print technology became available, it became possible to extend this knowledge to a larger group of people.

Figure 1. Craftsmanship in Alhambra Palace, Granada-Spain.

Figure 2. Blue mosque, Istanbul, Turkey.

This craftsmanship from the past has now been complemented with a new kind of craftsmanship, in which the knowledge used in the building process is coupled with the machine. Increasing time constraints in relation to cost aspects and the complexity of data, information and knowledge require improved communication and co-ordination for cooperation between actors. As a result of globalization, these actors needs to work from a distance, independent of time and place. All these requirements necessitate a new approach in the building sector and the building process.

Looking back at these historical developments, we can conclude that technological developments in general always have had an impact on how people designed, built and lived. In the building sector, the discovery of new materials and techniques has always led to fresh challenges and changes. If construction iron and the human lift had not been invented, there would not be an Eiffel tower or skyscrapers; if the car hadn't been invented, we would still have the narrow streets of the middle ages. Developments in other fields of science, too, such as mathematics, materials science, mechanical and aerospace engineering have always influenced building design and construction. Numerous examples exist that demonstrate the impact of technological developments on society itself, in the form of changing habits and lifestyles, and consequently in the form of changes to the buildings and the built environment.

2.2 Computing technology

As a part of the technological advances, developments in information and communication technology also impact the building sector. A major area in which computing technology supports the building process is in exchanging and processing data, information and knowledge. For this, a *reliable model* for representing and integrating data, information and knowledge must be built. Such a model will not only bring efficiency and reduction of costs to the building design process, it is also of vital importance in the entire *lifecycle* of the building, from conceptual phase until re-use and demolition.

How are we to deploy computers when processing these data, information and knowledge? Could we go one step further and state that the computer can draw conclusions from these "the way human beings do"? Can the computer replace human intelligence?

Human intelligence is a unique phenomenon and the mystery of the way in which the human brain works is still not completely solved. If something cannot be fully defined and described, we will not be able to copy it. Therefore, we cannot compare human intelligence with the intelligence of a machine. The latter is merely a means in data processing, that is, data which humans feed into the computer with certain routines in order to be processed and annotated.

At best, the computer can imitate certain functions of human intelligence, but this does not mean that it can invent and develop things itself as humans can. The computer can only work on the basis of a program that it has been taught, whereas humans work on the basis of their education and upbringing, as well as from their imagination and creativity (Sariyildiz 1994). Nevertheless, intelligent computing techniques, such as neural networks, fuzzy logic, and agent technology can provide valuable support in the building process.

In order to make a contribution in solving such complex problems as mentioned before, we propose a web-based *Intelligent Modelling for Cooperative Engineering (IMCE)* environment that supports the modelling of all data, information and knowledge in the building process. This environment must be both flexible and dynamic, enabling each actor to specify his or her own requirements concerning the scope and organization of the information. In this environment, product and process information are integrated into a single Project Information Model (PIM).

Information must be selected, ordered and made available in an intelligent manner conforming to the demands of each user, which may vary in character and type, and may conflict at times. Considering the complex as well as soft nature of the information concerned, the need for effective and efficient processing of this information leads us to make use of soft computing technologies, including fuzzy logic, neural networks, and agent technology. Preferably, the resulting systems should actively support the user in the building process. This can be achieved by benefiting from emerging intelligent technologies, where intelligence refers especially to the ability to deal with complexity in a general sense with logic reasoning (Ciftcioglu & Durmisevic 2001).

3 INDUSTRIALIZATION NEEDS, COMMUNICATION AND COLLABORATION

Whereas in the Middle Ages stone was cut and tooled, hoisted up, fitted and often colored at the building site itself, our present-day practice has developed into the purchase of pre-fabricated, simple building materials and complex building components to be assembled only on site. Knowledge of measure coordination, tolerance and jointing techniques become predominant here. Building components offered by competing concerns have to be evaluated, such that a well-considered choice can be made from among the variants. Due to the ever-increasing complexity of these building components and the need, in spite of this, to include them

in detail in the project, designers are obliged to handle and process a large amount of components in three dimensions, for which they do not always have sufficient information available. Nowadays, manufacturers are providing information concerning building components on the web; however, it will have to be processable in three dimensions in order to be effectively used in the design process.

The advent of the Internet, and especially the Web, has strongly influenced the way data and information is processed and exchanged between building partners, and the way these partners communicate. The Internet reduces distance to a relative concept, and enables distance collaboration, independent of time and place. The Internet also allows communication to be intensified, thereby improving the quality of collaboration. For example, Computer Supported Collaborative Work (CSCW) allows designers, engineers, architects, and others all around the world to work synchronously on the same design using collaborative engineering methods and techniques.

Data warehousing and data mining techniques support information on building products to be digitalized and made available electronically. Access is provided through the Web and the user is assisted in accessing all this information and selecting the most appropriate product.

Next to the processing and exchanging of data, information and knowledge in the building preparation phase, the computer also plays a role in the building execution. Technologies such as CAD-CAM and CAE can be adopted as a replacement for heavy work done by humans under difficult circumstances (e.g., Figs. 3–4). The use of these techniques requires a number of considerations and components; these include the persons in charge of the realization as well as the materials used during the construction. In the construction process, information supply and demand can be immense and some information processing tools must be used for appropriate routing of the information flow to the prospective users, the level of information being within the user's capacity of involvement.

With the advancements in the building technology as well as with the related modern technologies, building design requires more comprehensive attention than that required earlier in order to meet the higher level of standards in demands. It increasingly involves multidimensional aspects that have to be considered with conflicting criteria, which themselves have to be reconciled for optimal design solutions. Also, it requires flexibility to accommodate the probable emerging demands in the course of the execution of the building design project. Such unforeseen demands may naturally occur due to various reasons like new technological possibilities, new limitations imposed. In such cases, the modifications and/or additions to the existing project should be reflected to the related bodies or

Figure 3. Industrialized building concept using CAD-CAM methods, Batibouw house, Brussels by Sariyildiz (1994).

Figure 4. Central cupola on the patio of Batibouw house.

individuals in a well-coordinated way so that concerted actions can be taken for the efficient executions in the course of the project. For instance, a certain change in design can interest mainly the architect rather than anybody else such as the contractor. Therefore such information on modification should be given with different depth to different people corresponding to their involvement capacity and/or relevance to the subject matter. In the terminology of architecture, this issue is identified as information ordering.

Using a four-dimensional, object-oriented, dynamic description of a building, robots can be instructed to assist in the realization of the building. There are already buildings in Japan constructed without any use of manpower on the building site. For example, the housing factory in Japan, Seki-Sui Heim, utilizes only robots during construction. This factory is one of the biggest pre-fabricated house producers in Japan. The client formulates the requirements, the design is made, and an expert system selects the right components. The assembly is done on the building site without any manpower.

During the whole process computer science plays a crucial role in data processing. Attempts have been made to fully deploy computers from the conceptual design phase to the completion of the final product. In the eighties "concurrent engineering" was developed in the various branches of industry in the United States, in which design and construction are simultaneous processes and in which teams of different specialists cooperate. The application of concurrent engineering in building is still in its infancy. In Europe building is still predominantly executed according to traditional methods, and is a monodisciplinary event.

In United States building is based on industrial production of standardized components. Consequently, buildings can be constructed quickly and efficiently. Conversely, the European way of building shows a large variety. In Japan, building is in the hands of large construction companies, and has been largely industrialized. The drawback of these Japanese and American building methods is the risk of creating a uniform and poorly varied architecture. In order to prevent this, additional research is required into the possibilities of varying industrialized building, notably in the field of form and design.

The current concern for the environment evokes fresh challenges for the building industry, such as the necessity of developing building materials that can be recycled and re-used. So our target need no longer be to build for eternity. We could develop such new building techniques, which would enable us to construct buildings cheaper, and more rapidly, or even make them portable and moveable, in that buildings could be dismantled.

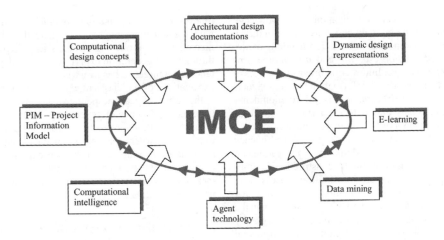

Figure 5. Components of the IMCE project environment.

4 IMCE – INTELLIGENT MODELLING FOR COOPERATIVE ENGINEERING

Our research group has committed itself for the next few years to the development of a Web-based IMCE environment for the building sector. Obviously, IMCE is a long-term research project that cannot be realized solely within the confines of one research group, as argued above. Instead, the group can contribute its expertise and research efforts towards the realization of this objective.

The design and development of an IMCE environment involves a large number of technologies (e.g., database, Web, representational, knowledge, and agent technologies, computational intelligence, data mining), takes on various forms (e.g., decision support system, electronic document management system, e-learning environment, Web tools), and applies to various areas (e.g., design analysis, design concepts, Industrial, Flexible and Demountable (IFD) design, and multicultural design) (Fig. 5). Below, we highlight and describe a number of aspects that are of importance in the design and development of an IMCE environment, and the research towards this goal.

Finding the right information and data with respect to a question or request within a large repository is becoming increasingly important (Tunçer et al. 2002). Supporting this requires a PIM, integrating a product and a process model with intelligent techniques for selecting and organizing information, and making this information available to the user. As a component in an IMCE environment, this will support cooperation and collaboration between the various partners and end users in the building process.

The realization of such a model will require a large amount of research and development. Creating a Web-based IMCE environment for the building sector is a complex task that will need to involve various participants, including the government, the building industry, and research institutes and universities. Practice has shown over and over again that it is impossible to develop this by one institute, one company, or the government alone. Instead, it concerns the communal interests of the entire building sector. On a national level, this development can be supported by a national ICKT institute for the building sector.

5 INTELLIGENT COMPUTING TECHNIQUES

The rich information environment of the building and construction industry makes the building and construction process a complex activity where many different individuals take part in. The advances in the information and communication technology provide possibility of distributing information in a wide range of the construction community in a cost-effective way with rapid dissemination. The advent of such desirable possibilities however urges the improvement of the existing means of information processing in order to be able to cope with the information in appropriate proportions. Due to the amount and complexity of information, the essential aid to deal with it may come from intelligent technologies since they are able to deal with complexity, but at the same time, they impose heavy demand on the computational power of the computers. Parallel to the advancements in the computer technology, there are considerable advancements in information acquisition and database management technologies. Due to this, data mining and knowledge discovery in databases is another emerging technology, which aims to maximize the gain of information utilization and knowledge acquisition. For the building and construction industry presumably two important information-handling bottlenecks are *information ordering* and *information processing in perspective*. Information ordering engages in optimal distribution of information to individuals involved in the construction, that is, information deliveries to individuals in a right proportion as much as appropriate without information redundancy. Such a policy in information distribution provides efficiency and clarity in communication while the redundant information-load imposed on the construction community is kept to a minimum. Information processing in perspective deals with information duly so that any piece of information is handled with due attention in the context of concern without any inappropriate gravity attachment to it. The essential gain from information processing in perspective is the ability of impartial planning and decision-making for cost effective and efficient building and construction process. *Soft computing* is a keyword in information technologies. It is a concept of computation similar to that involved in fundamental human activities. From soft computing viewpoint three basic features in a human activity become conspicuous. These are:

– parallel processing of information,
– ability of processing vague, imprecise even incomplete information, and,
– optimizing activities continuously based on some dynamic criteria which form the base for reasoning.

These three features can be simulated with soft computing methodologies to some extent with a remarkable success. Based on this success, soft computing is a potential information processing means to deal with exponentially increasing information-processing demand due to the advances in information and communication technologies. Three basic methodologies analogous to the features mentioned above are neural networks, fuzzy logic and evolutionary algorithms respectively. These methodologies have already established their associated "intelligent" technologies. Note that, these three basic technologies in any "intelligent" application ought to work synergistically for a maximum gain. Basically, fuzzy logic provides means to mimic the imprecise reasoning of human whereas neural network mimics massive parallel information processing capability of human brain by means of artificial neurons of an artificial neural network. In the fuzzy logic and neural network synergy, genetic search algorithms play the essential role of optimization process exactly mimicking the human behavior where optimization is almost a part of any act determined by a reasoning process in a human mind. The synergistic integration of "intelligent" technologies through the advanced capabilities of modern computer technology leads a new concept of machine "intelligence" which is

referred to as "*computational intelligence*" in contrast with the machine intelligence intended to be implemented by symbolic logic means as a hype in 80's, through first-generation expert systems. Today, computational intelligence has overwhelmingly surpassed the expectations of the classical expert systems, which are in the course of years desperately categorized and articulated as a sequence of generations. Today, expert systems should be conceived as a part of computational intelligence where several knowledge models may serve as mimicking several human experts in diverse domains while the knowledge models are formed by means of machine learning techniques in place of inputs provided by human experts, as was done traditionally.

6 SCIENTIFIC AND SOCIAL RELEVANCE

Many improvements and measures in order to increase efficiency are necessary in the building sector, such that ever-stricter quality requirements can be met, while taking into account a reduction in the workforce. The use of ICKT in company processes can lead to many positive effects in this respect, resulting in faster, cheaper, more flexible and more durable building and construction.

The development of an intelligent project model for building process management and an IMCE environment to support building project cooperation are essential for an intensive use of the Internet in the building sector. The resulting platform can support cooperation, coordination, collaboration and knowledge exchange between different partners in the building process, and enable the organization and presentation of project information during design and construction with respect to the needs of the user.

In order to stimulate the use of ICKT and the development thereof, a concerted effort of all parties involved in the building sector is needed. On a national level, the establishment of an ICKT institute can bring together the government, the industry, and the universities and research institutes. The law can set up, the financing of such innovations and developments, by directing a percentage of all building costs towards this purpose. As a similar example, in the Netherlands, 2% of all building costs are currently reserved for art projects.

Developments in the area of ICKT effectuate an improvement in information management in the building sector and boost:

– the efficiency and effectiveness of the building process,
– further industrialization in the building sector,
– the competitiveness of individual building companies as well as national building industries on the European or international market, and
– the ability of end users to participate structurally in the building process.

REFERENCES

Ciftcioglu, Ö. & Durmisevic, S. 2001. Knowledge Management by Information Mining. In B. de Vries, J. van Leeuwen & H. Achten (eds.), *Computer Aided Architectural Design Futures*. Dordrecht, Netherlands: Kluwer Academic.

Sariyildiz, S. 1994. X-dimension in the building, Inagurale rede TU Delft.

Tunçer, B., R. Stouffs & Sariyildiz, S. 2002. Document decomposition by content as a means for structuring building project information, *Construction Innovation* 2(4): 229–248.

System-based Vision for Strategic and Creative Design, Bontempi (ed.)
© *2003 Swets & Zeitlinger, Lisse, ISBN 90 5809 599 1*

Knowledge acquisition for demolition techniques selection model

A. Abdullah & C.J. Anumba
Department of Civil and Building Engineering, Loughborough University, UK

ABSTRACT: This research involves the development of an intelligent system for the selection of demolition techniques. The Analytic Hierarchy Process (AHP) was used to develop the intelligent system since it was widely used as a decision-aiding method to model subjective decision-making process. This paper discusses the Knowledge Acquisition (KA) process used to establish the relevant criteria for selecting the most appropriate demolition techniques. The criteria established were then developed as a hierarchic structure based on AHP model.

1 INTRODUCTION

This paper discusses the knowledge acquisition (KA) process, which is a part of research methodology to develop an intelligent system for the selection of demolition techniques. The intelligent system developed will act as a decision-making aid for demolition engineers as a decision maker in selecting the most appropriate demolition techniques. Analytic Hierarchy Process (AHP) is a decision-aiding method that was used to develop the intelligent system. Since the demolition engineers bases their judgments on knowledge and experience, then makes decision accordingly, the AHP approach agrees well with the behavior of the decision maker. AHP can provide the decision maker with the framework needed to model a complex decision scenario.

The KA process was an important aspect of developing the system because the decision making process by the demolition engineers need to be captured in order to develop a decision model for the system. The KA process will involves capturing and transforming appropriate knowledge from demolition engineers into some manageable form to develop the decision model. The expertise or human expert in this research is the demolition engineer who makes the decision on what technique to be used for a specified demolition project. The knowledge that's needed to be captured is the relevant criteria, which may affect the selection of demolition techniques.

The paper will review the concepts of KA, the methodology of the research, and findings on demolition criteria resulted from the KA process. The criteria captured from the experts then represented by a decision tree based on AHP approach to develop a decision model.

2 KNOWLEDGE ACQUISITION

2.1 Definition

Turban & Aronson (1998) defined KA as the process of extracting, structuring, and organizing knowledge from one or more source. It is referred also as the process of getting and transforming appropriate information from sources of expertise into some manageable form (McGraw & Harbison-Briggs (1989). In the process of KA the knowledge engineer carries out the activity of extracting the knowledge from an expert, checking it with the expert and then representing the knowledge in the knowledge base. This activity is known as the "elicitation of knowledge" (Turban & Aronson 1998).

2.2 Stages of knowledge acquisition

KA is made up of five stages, which are identification, conceptualization, formalization, implementation and testing (McGraw & Harbison-Briggs 1989).

Identification – The problem and the purpose of the Artificial Intelligence (AI) application to be built are identified. It also involves identifying the number of participants involved and the resources that are available for the system.

Conceptualization – In this stages the knowledge engineer made a decision on certain issues that would affect the overall structure of the system i.e. how the knowledge can be extracted, and how it will be represented.

Formalization – The knowledge engineer organizes the concepts, implementation and information into formal and clear representation.

Implementation – A prototype developed for testing out the design and the process by programming the knowledge into computer.

Testing – The prototype system is tested for its efficiency, accuracy and validates the rules used. The system and results are shown to the expert to see if they are satisfied and to see if the system working as required.

2.3 Methods of knowledge acquisition

Artificial Intelligence offers a range of knowledge acquisition methods that can be used under different situations (Ignizo 1991). Prerau (1987) & Surko (1989) in particular have provided some excellent guidelines for the knowledge acquisition process. Knowledge engineer can extract the knowledge from the expert either manually or with the aid of computers. Awad (1996) & Scott et al. (1991) identified that the acquisition methods can be classified in different ways and appear under different names. From the review the method can be classified into three categories: manual, semiautomatic, and automatic.

Manual methods – Most of the manual acquisition methods are well established that have been borrowed with several changes from system analysis or psychology. The three main methods are questionnaires or experts self report, interviewing, and tracking the reasoning process.

Semiautomatic methods – These methods can be classified into two categories. Firstly knowledge engineers have little or no help to the experts to build knowledge bases and secondly knowledge engineer carries out the necessary tasks with minimal participation by an expert.

Automatic methods – These methods enable a knowledge base to be built with minimal or no need for a knowledge engineer and an expert. Even though it is automatic there will always be the need for some sort of human builder such as a system analyst in automatic methods but without the help of the knowledge engineer and expert. An example of an automatic method would be *Induction*, which the expert would only be required for validation purposes and there would be no need for a knowledge engineer since the systems analyst could manage the whole process.

3 METHODOLOGY

3.1 Overview of approaches adopted

The AHP is a decision aiding method developed by Saaty in 1970s. It uses a multi-level hierarchical structure to represent the decision-making problem (Saaty & Vargas 2001). The hierarchical structure or decision tree consists of goal, criteria, sub-criteria, and alternatives. The knowledge acquisition process that involved capturing the expert knowledge plays a major role in developing the hierarchical structure. To construct the hierarchic structure, the knowledge that's need to be captured from the experts was the criteria and alternative used in the selection of demolition techniques.

The suitability of an acquisition method depends on the kind of knowledge being extracted and each method has it's own advantages and disadvantages. The elicitation of knowledge from the experts can be done manually or with the aid of computers. The process needs the aid of computer when there is limited time and fund existed in the research or the knowledge engineers lack of knowledge acquisition skills such as communication skills. Since there was ample time, sufficient fund and the researcher have the skills that a knowledge engineer may be called on to use during the knowledge acquisition process, therefore there are no needs of the aids of computers in the knowledge acquisition process and the manual method was selected for this research.

The research adopts three approaches of manual knowledge acquisition methods as the research methodologies, which are through industry survey, interviewing, and tracking method to develop a decision model based on AHP. Several researchers involved with AHP model also use either one, both or all of these three approaches for knowledge acquisition in their research (Al-Harbi 2001); (Alhazmi & McCaffer 2000) & (Amirkhanian & Baker 1992). The next sections will discuss each of the method in details.

3.2 Industry survey

In this research, an industry survey through postal questionnaires was used as an approach of obtaining preliminary knowledge from the demolition industry point of view in a broader perspective. The main objective of the survey is to identify a list of factors that may affect the selection of demolition techniques and a list of available demolition techniques to the industry. The researcher will then use these lists as a guide to develop a complete hierarchic structure, which will simplify the decision process of selecting the most appropriate demolition techniques.

A questionnaire can be defined as "a list or grouping of written questions which a respondent answers" (Adams & Schvaneveldt 1985). It also known as a "manual expert driven system" or "Expert's Self-report" (Turban & Aronson 1998). It is a data-gathering device that elicits from a respondent the answers or reactions to printed (pre-arranged) questions presented in a specific order. The method use to develop this questionnaire was based on Adams & Schvaneveldt (1985) & Chadwick et al. (1984).

The questionnaires consist of open-ended questions and closed-ended questions. It was designed in three parts: the introduction (cover letter), demographic or background questions (Section A) and the body of the study (Section B). In section A, the respondents have to answer 5 questions about his/her background information. In section B there are 13 questions have to be answered which carefully design to achieve the main objective of the survey. The content of the questionnaire was based on the literature review and closely referred to Code of Practice for Demolition BS 6187:2000. It has been designed in such a way that it can be answered in not more than 15 minutes.

Drafts of the questionnaire were pre-tested, with several lecturers in the construction management team at Loughborough University who have experience in questionnaire design, before being printed in final form to reveal confusing and other problematic questions that still exist in the question. The questionnaire distributed by mail, which include a cover letter, a copy of the questionnaire itself, and a self-addressed, stamped return envelope. To make sure only the expert will answer the questionnaire, it was directed to the person involved directly and technically with the demolition process, which as identified during the telephone inquiries.

The population for this research was demolition engineers in the United Kingdom (UK). National Federation of Demolition Contractors (NFDC) and the Institute of Demolition Engineers (IDE) provided the sampling frame for this survey. All the demolition engineers from the sampling frame were contacted by telephone to make sure of their willingness and confirmation of address before the questionnaires were sent by post. Finally, about 100 demolition engineers were included in the sample.

The response rate was 37 percent, out of 100 questionnaires delivered, of which 94.6 percent of the represented usable replies and 5.4 percent unusable replies. The experience and position of the respondents in the company stimulated confidence in the result of the survey. About 86 percent of the respondents has more than 10 years of experience and represents top-level management in the respondent's company. The experts assessed each criterion in the questionnaire by ranking their relative importance to each other. Statistical analysis was carried out to refine these criteria with the purpose of identifying the relative degree of importance of each criterion (Abdullah & Anumba 2002a). Table 1 shows the factors affecting the selection of demolition techniques.

One major advantage of industry survey through postal questionnaire is economy, as this type of research yields a maximum amount of data per research dollar (Chadwick, Bahr et al. 1984). However there are limitations to this approach. The major limitations of a questionnaire are the number of questions that can be asked and experts may forget to specify certain pieces of knowledge i.e. the important factors that may affect the selection of demolition techniques been left

Table 1. Factors affecting the selection of demolition techniques.

Criteria	Ranking
Health and safety	1
Stability of the structure	2
Location and accessibility	3
Presence of hazardous material	4
Environmental consideration	5
Shape and size of the structure	6
Client specification	7
Structural engineer approval	8
Time constraint	9
Extent of demolition	10
Financial constraint	11
Recycling consideration	12
Transportation consideration	13
Availability of plant or equipment	14

out from the list. It would be even harder for the expert to describe the process of actually getting to the decision made. The next two approaches adopted which were through interviews and tracking method was designed to overcome this limitations.

3.3 Interviews

Interviewing experts involves the knowledge engineer having a face-to-face interview with the expert. It is the most popular type of knowledge acquisition method and requires the knowledge engineer and expert to talk to each other about the actual problem that the expert system should solve. It involves collecting information via instruments such as tape recorders, video camera, questionnaires etc. It is also important that the knowledge engineer has good communication skills and the expert should be able to express his knowledge to the engineer (McGraw & Harbison-Briggs 1989). Two types of interviews are Unstructured and Structured.

Unstructured Interviews are informal and usually used as a starting point. It needs simple planning and is a brief way of understanding the structure of the problem domain and is usually followed by a more structured approach for understanding the attributes of the problem. It has the advantage of being a fast method to obtain the requirements but it has the limitation of being to vague (Wright & Ayton 1987).

Structured interviews are more meaningful and more specific method of knowledge acquisition. It is known as a "systematic goal-oriented process" as it uses a systematic approach and therefore being a well-organized approach (Wright & Ayton 1987). It reduces the problems found with unstructured interviews such as the vagueness and incorrect assumptions. It also enables more information to be obtained and has a planned answer to it.

In this research, structured interviews with six key experts were conducted after the entire postal questionnaire returned. The six experts were carefully selected so that they can provide the researcher with the required knowledge and cooperation. The experts selected must have several criteria such as have more than 10 years experience in the demolition practice, registered with the Institute of Demolition Engineers as a "Fellow", and play a role as a decision maker in the process for selection of demolition techniques in their company.

The refined criteria that resulted from the postal questionnaire were given to the key experts for reassessment and to ensure the relevance of the identified criteria. The result from the interviews was a set of criteria with each of the criterion has been justified why the experts selected it. The demolition techniques available were also identified and grouped into several types of main techniques.

The results from the interview must be subjected to through verification and validation methodologies before the prototype of a hierarchic structure can be developed. Therefore, the next step in the knowledge acquisition process will involved the use of tracking method to verify and validate the results from the interview.

3.4 Tracking method

Tracking method is a set of techniques that attempt to track the reasoning process of an expert (Turban & Aronson 1998). The most common tracking method used is protocol analysis. Protocol analysis is similar to interviewing but more formal and logical. It is a record of the expert's step-by-step decision-making behavior. The expert is asked to carry out a task but they have to think out aloud while working through the given scenario given by knowledge engineer. The experts will then talk about what they are doing to solve the problem, while the knowledge engineer is listening and recording what is being said.

Protocol analysis was adopted in this research because it can help the researcher to record, and later analyze with the expert the key decision point and finally develop a prototype of a hierarchic structure that represent the decision process for selecting the most appropriate demolition techniques. The approach was conducted in this research by asking the demolition engineer to think aloud while selecting a demolition technique when an example of a demolition project was given to them. The researcher used a video camera to record what is being said. The researcher then analyzes, interpret, and structure the protocol into a hierarchic structure for a review and validation by the experts. A series of prototype hierarchic structure models were developed for this purpose. The final hierarchy structure for the proposed model shown in Figure 1.

4 FINDINGS FROM KNOWLEDGE ACQUISITION

4.1 Demolition criteria

The identified criteria from the questionnaire survey were reassessed through the interview and protocol analysis approach, to ensure the relevance of the identified criteria. From these approaches of KA and based on the AHP principles, the criteria were grouped by the experts and a hierarchic structure then developed as a knowledge representation of the selection model.

In AHP, a complex decision problem is expressed as a hierarchy structure. A hierarchy structure has a top-down flow, moving from goal to general categories (criteria) to more specific ones (sub-criteria) and finally the alternatives. Located at level 0 of the model to serve, as a goal node is "the selection of the most appropriate demolition techniques". Levels 1 of the model serve as the main criteria and it include factors affecting the demolition technique selection, which had been classified into six categories. Level 2 of the model define sub-criteria nodes for categories in level 1. Level 1 and 2 of the hierarchy consisted of a total of 6 and 17 nodes, respectively. Finally the alternative solution (demolition techniques) occupied level 3 to serve as the choice available for the decision makers. (Fig. 1)

The first main criterion that may affect the selection of demolition techniques is "Structure Characteristics". Prior to demolition of any structure it is important to have in depth knowledge of the structure characteristics in order to predict and plan its demolition. There are five sub-criteria that been classified by the experts under this criterion which includes height, type, stability, degree of demolition and previous use of the structure.

The second main criterion is "Site Conditions". Four sub-criteria are group under this main criterion that includes, health and safety for the person on and off site, acceptable level of nuisance, proximity of the adjacent structure and site accessibility. In general, the demolition techniques selected by the demolition engineers should not at any time pose any threat to the health and safety of site personnel and members of the public, therefore site conditions with its sub-criteria were identified as one of the important factors that may affect the selection of demolition techniques.

The third main criterion is "Cost" and it was divided into two sub-criteria, which includes machinery cost, and manpower cost. It is important that the demolition engineers consider the financial implication of the demolition techniques selected for a particular demolition work since a client will probably give the contract on the basis of the cheapest cost option. However, the demolition engineers should not, under any circumstances compromise on safety while selecting a demolition technique despite the

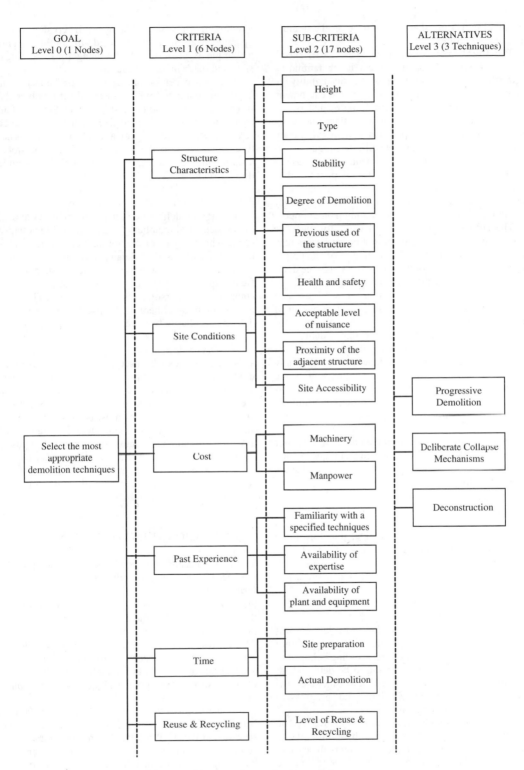

| GOAL
Level 0 (1 Nodes) | CRITERIA
Level 1 (6 Nodes) | SUB-CRITERIA
Level 2 (17 nodes) | ALTERNATIVES
Level 3 (3 Techniques) |

Figure 1.   Hierarchy structure for the demolition techniques selection model.

financial constraint that may impose on the demolition of a structure.

The fourth main criterion is "Past Experience". Three sub-criteria that may affect it are familiarity with the specified techniques, the availability of expertise and the availability of plant and equipment. The past experience of the demolition engineers will to some extent affect their choice of technique for dealing with a particular problem. A demolition engineer that is familiar with a specific technique or has the availability of plant, equipment, or expertise is more likely to apply that expertise if possible than search for another solution. If the problem falls outside the boundaries of their previous knowledge, they could only then be forced into examining other option.

The fifth main criterion is "Time". To select the demolition techniques the demolition engineers may have to consider the time factor in term of site preparation and actual demolition because each of the demolition techniques have its own time scale.

The sixth and final main criterion is "Reuse and Recycling". This criterion is an important aspect of the demolition project, which the demolition engineers need to consider when selecting a demolition technique. The level of Reuse and Recycling of the building materials and components will probably affect the choice of demolition technique to some extent. For example, if the level of reuse and recycling for a particular demolition project is minimal, the demolition engineer will possibly select the progressive type of demolition techniques which have a low level of reuse and recycling but much more faster compared to deconstruction type of techniques which have a higher level of reuse and recycling but have a slower finishing time.

4.2 Demolition techniques

The results from the KA show that, there are three main demolition techniques that can be selected as alternatives in the decision model. The demolition techniques are progressive demolition, deliberate collapse mechanisms, and deconstruction.

4.2.1 Progressive demolition

The progressive demolition is the controlled removal of sections of the structure, at the same time retaining the stability of the remainder and avoiding collapse of the whole or part of the structure to be demolished. The progressive demolition techniques include progressive demolition by machine and progressive demolition by wrecking ball.

In the progressive demolition by machine, the excavator attached with boom and hydraulic attachments such as crushers, impact hammer, shears etc. The crusher attachment breaks the concrete and the reinforcement by the hydraulic thrust through the long boom arm system. The hydraulic crusher can be operated from the ground outside the building. This technique is also suitable for dangerous buildings, silos and other industrial facilities.

Progressive demolition by wrecking ball consists of a crane equipped with a steel ball. It involves the progressive demolition of a structure by the use of an iron ball that is suspended from a lifting appliance (crawler crane) and then released to impact the structure, repeatedly, in the same or different locations. This technique is suitable for dilapidated or tumble down structure.

4.2.2 Deliberate collapse mechanisms

Demolition by deliberate collapse involves a removal of key structural members to cause complete collapse of the whole or part of the building or structure. This method usually used on detached, isolated, level sites where the whole structure is to be demolished. A sufficient space must be allocated to enable removal of equipment and personnel to a safe distance. The techniques include deliberate collapse by explosive and deliberate collapse by wire rope pulling.

4.2.3 Deconstruction

This technique usually used as part of renovation or modification work and prepares the way for deliberate collapse. The elements to be removed should be identified and marked on site. The effects of removal on the remaining structure must be fully understood and included in the method statement. If instability of any of the remainder structure could result in a possible risk to personnel on the site and to other people nearby, sections of the structure should not be removed. The deliberate removal of elements can be done manually or by machines.

5 DISCUSSION AND CONCLUSION

This paper presents the knowledge acquisition process in order to assess and identify the relevant criteria in formulating the AHP model to select the most appropriate demolition techniques. Knowledge acquisition is the most important element in the development of the proposed model because it involves extracting knowledge from sources of expertise and then presents it into a manageable form. Three approaches of knowledge acquisition methods were adopted, which include industry survey through postal questionnaire, interview, and protocol analysis.

The main aim of the knowledge acquisition adopted is to assist the researcher in developing the hierarchic structure in the AHP model. An industry survey was conducted as an approach of obtaining preliminary knowledge from the demolition industry since there are limited research or literature review in selecting the demolition techniques. To overcome the limitation in the questionnaire survey, such as a limited number of questions that can be asked, an interview approach was adopted. Interviewing also suffers from the drawback that the researcher can encourage the expert to speculate and not get valid knowledge. To overcome the limitation in the interview, a protocol analysis approach was adopted for verification and validation of the knowledge captured and finally developed a prototype hierarchic structure for the AHP model.

From the questionnaire survey, there are 14 factors identified by respondents that might effect the selection of demolition techniques. To ensure the significance of the identified criteria, it was reassessed through interview and protocol analysis approaches. Based on AHP method the criteria were group into 6 main criteria and 17 sub criteria and represented as a hierarchy structure. The hierarchy structure then can be used as one of the elements in the overall development of the decision support system using AHP.

The experts will compare the relative preference of all the three demolition techniques with respect to each of the sub-criterion in order to select the most appropriate techniques for a demolition project based on the AHP model. The next step in the model development process is the implementation of AHP model using Expert Choice and this step will not be discussed any further in this paper. The overall model development process has been discussed in (Abdullah & Anumba 2002b). In brief, Expert Choice is professional commercial software developed by Expert Choice, Inc. It help simplifies the implementation of the AHP's steps and automates many of its computations such as matrix calculation in pairwise comparisons.

As concluding remarks, two points can be learnt from the knowledge acquisition process on this research. First, knowledge acquisition is an important aspect of developing an AHP model, but it has rarely been discussed in the research literature. This paper is an attempt to filling the void in the literature by discussing the knowledge acquisition process in detail. Second, the suitability of an acquisition method depends on the type of knowledge extracted and each method has it's own limitation. For this reason a combination of several approaches is recommended to overcome any limitation on the approaches adopted.

REFERENCES

Abdullah, A. & Anumba, C.J. 2002a. Decision Criteria for the Selection of Demolition Techniques. *Proceeding of the Second International Postgraduate Research Conference In The Built and Human Environment, April 11–12, 2002*, University of Salford: Blackwell Publishers, 410–419.

Abdullah, A. & Anumba, C.J. 2002b. Decision Model for the Selection of Demolition Techniques. *Proceeding of the International Conference on Advances in Building Technology, December 4–6, 2002*, Sheraton Hong Kong Hotel: Elsevier Science Ltd, 1671–1679.

Adams, G.R. & Schvaneveldt, J.D. 1985. *Understanding research methods*. New York: Longman.

Al-Harbi, K.M.A.S. 2001. Application of the AHP in project management. *International Journal of Project Management*, 19(1): 19–27.

Alhazmi, T. & McCaffer, R. 2000. Project procurement system selection model. *Journal of Construction Engineering & Management*, 126(3): 176–184.

Amirkhanian, S.N. & Baker, N.J. 1992. Expert system for equipment selection for earth-moving operations. *Journal of Construction Engineering and Management*, 118(2): 318–331.

Awad, E.M. 1996. *Building Expert Systems*. Minneapolis: West Publishing Company.

Chadwick, B.A., Bahr, H.M. & Albrecht, S.L. 1984. *Social Science Research Method*. New Jersey: Prentice-Hall.

Ignizo, J.P. 1991. *Introduction to Expert Systems: The Development and Implementation of Rule Based Expert System*. Houston: McGraw Hill publication.

McGraw, K.L. & Harbison, B.K. 1989. *Knowledge Acquisition, Principles and Guidelines*. Englewood Cliffs, N.J.: Prentice Hall.

Muralidhar, K., Santhnam, R. & Wilson, R.L. 1990. Using the analytic hierarchy process for information system project selection. *Information and Management*, February: 87–95.

Prerau, D.S. 1987. Knowledge Acquisition in the Development of a Large Expert System. *AI Magazine* 8(2): 43–51.

Saaty, T.L. & Vargas, L.G. 2001. *Models, methods, concepts & applications of the analytic hierarchy process*. Boston: Kluwer Academic Publishers.

Scott, A.C., Clayton, J.E. & Gibson, E.L. 1991. *A Practical Guide to Knowledge Acquisition*. Reading, MA: Addison-Wesley.

Surko, P. 1989. Tips for Knowledge Acquisition – things they may not have taught you. *PC AI* (May–June): 14–18.

Turban, E. & Aronson, J.E. 1998. *Decision Support System and Intelligent Systems*. Jersey: Prentice-Hall.

Wright, G. & Ayton, P. 1987. Eliciting and modelling expert knowledge. *Decision Support Systems*, 3: 13–26.

System-based Vision for Strategic and Creative Design, Bontempi (ed.)
© *2003 Swets & Zeitlinger, Lisse, ISBN 90 5809 599 1*

Supporting verification of design intent

I. Faraj

Manchester Metropolitan University Business School, Manchester, UK

ABSTRACT: A major problem in the integration of CAD and construction applications lies in the difficulty of representing the component definition adequately for all applications. Feature-based design is considered as a main factor in the integration of such applications. Because various design, engineering, manufacturing, and construction data can be associated with a feature. Further, the use of features enables the development of high conceptual meaning to be associated with the component definition. However, tagging feature labels onto geometry does not guarantee the geometric correctness of the resultant feature and knowledge of the topology of the resulting feature and geometric analysis is necessary to correctly identify the validity of the resultant feature. Contradiction between the applied feature and resultant feature may occur due to feature interactions, wrong positioning or inadequate parameter supplied by the user.

This paper reports on an experimental environment which uses a product model that permits all geometrical and technological information associated with the design and construction stages to be represented. The design is represented as solid model. Individual features are extracted from the product model and analysed to determine their validity and accessibility. The knowledge and feature analysis parts of environment are structured around geometric modeller and the engineering information held in the product model. A user interfaces is developed to enable interaction.

1 INTRODUCTION

For many years, computers have been used to improve productivity in the design office and reduce the time element involved in the project development lifecycle such as planning, estimating, etc.

Different computer aided applications have been developed to handle various aspects of the product life cycle e.g., CAD, planning, site layout, etc. This led to the development of so-called "islands of information". Each system contains its own version of the product data. Changes made to the data in these applications may become inconsistent or obsolete. The concept of product model, where all the applications are interfaced to one central model capable of representing all the data related to a product, is used to overcome these problems.

The use of features enables the development of high conceptual meaning to be associated with product definitions by representing the geometry in terms of recognisable and meaningful forms. Features allow the grouping of low-level primitives found in a geometric modeller and other data for specific applications.

Features form a very important part in the development of the standard data models which are being developed by both STEP[1] (STEP, 1999) and IAI[2] IFC[3] (IAI, 1999). They enable geometric and other construction information to be associated with the various parts of a building.

In feature-based design systems, products are designed by adding or removing features. A feature may be translated and/or rotated in order to position it in the desired place. Contradiction between the *applied features* and *resulting features* may occur due to features interactions, wrong positioning, or inadequate parameters supplied by the user during the design definition, where applied features can be defined as *the features used by the designer to satisfy a particular function*, while the resulting features is *the effect of the applied features on the design*. Moreover, the application of other features may cause some features to degenerate to further features. Therefore, verification of the resulting features must be performed against the applied features to establish whether the resulting features conform to the underlying geometry. For example,

---

[1] Standard for the exchange of Product Model Data
[2] International Alliance for Interoperability
[3] Industry Foundation Classes

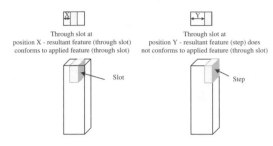

Figure 1.   Degeneration of features.

in steel-frame structure design, the subtraction of the volumetric feature "slot" from a blank (initial component) generates the surface feature "slot" and not a "step" (see Figure 1).

It is important to distinguish between different types of the same feature for a number of reasons:

– To improve the efficiency of features generation.
– To eliminate non-existing faces that can exist on the applied feature. Therefore,
– Improving the decision making process by provide better information to the down stream applications.

To improve the usefulness of the product model, features must be verified before they are used by other applications to ensure that there is no contradiction between the intent of the designer and the resultant feature or features. Any amended data should then be put back in the product model to enable other applications to use the correct features.

2   FEATURES REPRESENTATION IN A PRODUCT MODEL

To enable features to be represented in the framework of the product data model, a level of abstraction need to be introduced which should support the three stages of the life cycle: specification, definition, and actual.

Specification represents the list of requirements which the *actual* feature must satisfy. *Definition* represents how the specification will be realised. *Actual* contains the data of the actual feature.

The "definition" of the feature has the following attributes: feature geometry and construction information.

In this paper only the feature geometry will be discussed. The construction_information node contains construction information other than geometry to enable each construction application to perform its task.

2.1   *Feature geometry*

The feature geometry is represented as E^3 solid using the Constructive Solid Representation (CSG). The

E^3 *solid_objects* (e.g. features) is defined using a set of primitive halfspaces (planar, cylindrical, conical, spherical and toroidal) and rigid transforms. The CSG tree is represented with primitive halfspaces as its leaves and the Boolean operators and rigid transforms as its nodes.

For example, the volumetric feature hole can be defined by regularized intersection of two planar halfspaces at distance equals to the hole_depth with a cylindrical halfspace of diameter D. The result of the intersection is a bound feature, hole with diameter D and depth of hole_depth.

3   VOLUMETRIC AND SURFACE FEATURES

In order to define a product (e.g., building, steel component), and to determine the spatial clearance and, if needed, the machinability of the defined product; two types of features need to be considered. These are volumetric features and surface features. Volumetric features (V) are solid and can be defined as volumes of material that can be generated or removed using particular processes. Surface features are collections of faces, where faces can be defined as a 2-D region of the boundary of the solid.

A volumetric feature can be either a negative feature or a positive feature i.e. a negative feature is a result of a subtraction process (e.g., opening in a wall) while positive feature is an added volume (e.g., wall).

A surface feature is generated as a result of removing a volumetric feature from a design. For example, the surface feature door_opening is generated by removing the volumetric feature door_opening from a wall. Therefore, a surface feature can be defined as the contribution of the volumetric feature to the design boundaries (see Figure 2). Hence, the surface feature is the set of which the following is true:

$$(P \cap {}^* V = \phi \wedge P \cap V \neq \phi)$$

where P is the finished product; $\cap^*$ is regularised intersection; and $\cap$ is non-regularised intersection.

The above expression identifies that the applied feature must be part of the finished design. Notice that not all the faces of the volumetric feature can be found in the surface feature, these faces are called *virtual faces* and can be defined as *a face derived from a volumetric feature that contains no solid on either side.*

4   FEATURE VALIDATION

As mentioned earlier, when a volumetric feature is subtracted or added to a design the resulting collection of faces (surface feature) may not correspond to the definition of the intended feature due to inappropriate positioning, features interaction, or inadequate parameter. For instance, the width of the door_opening is less than the thickness of the wall (solid modellers may not support constraints to ensure the width of the opening is equal to the width of the wall therefore, the change of the wall thickness may render the door_opening to be invalid as shown in Figure 3).

Analysis on all or part of design model can be performed at any time during the definition stage, it can be used to determine:

– whether the resulting feature conforms to the definition of the applied feature,
– the spatial clearance of certain or whole parts of the product.

4.1 Feature transformation

As features are applied to the design the following will need to be determined for each feature: identification, dimensions, position, and orientation. The relative position of a feature to the building needs to be calculated. This is an important requirement for the feature verification and accessibility process where every face of a feature must be located.

When features are defined, the initial position and orientation of all the faces of a feature relative to the global coordinate system are known (default values). As features are applied to the design, they are likely to be rotated and translated. As a result, the position and orientation of each face of the feature is changed.

To verify the resultant features, the surface features have to be decomposed to a set of faces, edges and vertices. A method of transforming from volumetric feature data structure to that of surface feature is performed as follows using the door_opening as an example:

The form of the surface feature is determined from the pre-knowledge of specific form of the feature. The parameters of a specific surface feature are derived from those of its equivalent volumetric feature.

At this stage the local coordinate axes of the features are mapped to the global coordinate axes. Therefore, the orientation of every face of the features is known. In order to identify the vertices of the planar faces of the feature and the edges that link these vertices, a traverse from origin of the feature in the same direction of the face orientation by a distance that can be derived from the dimensional attributes of the feature that correspond to the traverse direction is carried out. A more detailed description is given in (Faraj 2002).

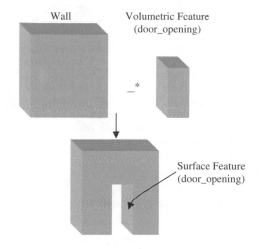

Figure 2.   Surface and Volumetric Features.

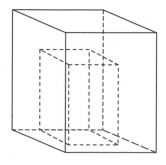

Figure 3.   Inaccessible feature.

5   SPECIFIC FEATURES AND THEIR VERIFICATION

The more computerisation of applications, the more is the need for verification applications to ensure that the integrity of every feature is maintained and the form of each feature is not violated by subsequent operations. For example, if a structural application places a number of columns to support a slab, then a subsequent operation modifies the design by introducing an opening in the slab, all the columns supporting the slab must still fulfil their function. Any column that falls within the boundary of the opening become redundant and may need to be removed. For a feature to be valid and accessible following conditions must be satisfied:

(1)   It must exist within the boundary of the product.
(2)   It must conform to its definition.
(3)   It must be accessible without interfering with the neighbouring features (spatial clearance).
(4)   It must conform to the tolerances (If tolerances are considered to be important).

The existence condition (1 above) ensures that the intended feature does in fact contribute to the boundary

1577

of the product (internal or external). This is done by ensuring that the coordinates of the feature are within the coordinates of the product.

Conditions (2) and (3) will be discussed next while condition (4) is outside the scope of this paper. Only one type of features, "door_opening" will be used to demonstrate the principles. Similar principles can be used on all 2.5D features.

5.1 Door_opening validity

The first validity of the surface feature "door_opening" is its existence within the building/component boundaries. It is required that either all or part of the feature must exist within the boundary in order to determine the other validity conditions.

The second condition of door_opening validity is the conformity of the door_opening to its definition. A door_opening must have 3 real faces and 3 virtual faces. A real face can be defined as *a face that contains a solid on one of its sides*. Figure 4 shows the representation of the surface feature door_opening where F1, F2 and F6 are real faces and F3, F4, and F5 are virtual faces. Also F1 is parallel to F2 and are connected by F3, which is at 90° to F1 and F2.

All the faces of the feature must be analysed to determine whether they are real or virtual, the results correspond to:

- IF F1, F2, and F6 are real faces and F4, F5, and F3 are virtual faces THEN the feature conforms to the definition.
- IF F3 is a real face THEN the feature does not conform to the definition – feature detected hollow rectangle.
- IF F1, F2, F3, F4, and F5 are real faces and F6 is a virtual face THEN the feature does not conform to the definition – feature detected is a pocket.

Depending on which face is real or virtual the feature may degenerate to different features. The problem of inadequate support to the structure which may be introduced to the design as a result of an error can also be addressed. This is referred to as "thin wall" phenomena. For example, depending on the position of the

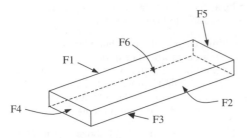

Figure 4. Surface feature door_opening.

door_opening, thin walls may exist if the feature is too close to the boundary of the building or neighbouring features. The existence of a thin wall can be verified by enlarging the feature by an amount equal to the value of the minimum permissible value of a thin wall (Stw), and then it is intersected with the boundary of the embedding wall. Thin walls exist if:

$$extended\_door\_opening \cap^* b(wall) \neq \phi,$$

where b is the boundary. Thin wall phenomena must be avoided in certain applications and therefore, the system must draw the user attention to their existence in those applications.

6 SPATIAL CLEARANCE OF FEATURE

A feasible approach direction is *a path along which an operator can approach and build a feature without being obstructed by any part of the building.*

For a feature to be constructable there must exist at least one such feasible direction. Approach directions and clearance of faces would enable other applications such as planning to determine the construction precedence i.e. whether a particular feature must be built before another due to accessibility. As a result, a more accurate data can generated at the early stage of the project and before any construction commences. This need is also evident in the design and production of steel components where some machining may be required and a tool path must exist to enable the machining of every feature on the part. If all the possible tool paths are obstructed in anyway then the part would be considered unmachinable and the design has to be revised.

6.1 Spatial clearance of planar faces

Volumetric features were defined as solids and are subtracted from, or added to, a design. The normals to the faces of the volumetric features point outwards (i.e. the material is present in the negative halfspace). Hence, all the possible approach directions to access these faces should lie only in the positive halfspace containing the face. Therefore, only the positive halfspace containing the face is checked for accessibility. In this research only those approach directions that are perpendicular to the global axes are tested for accessibility. To include approach directions that are not orthogonal to the major axis would not involve significant extension to the theory.

The position of a face, its type (e.g., virtual or real) and the feature type (e.g., positive or negative) are determined by the feature verification process. Depending on the type of feature, virtual faces (for negative features) or real faces (for positive features) can be considered as possible entities to be considered for accessibility. These faces are swept in the "normal" direction until one of the following is satisfied: the building boundary or a user-specified distance is reached, or solid is encountered. From this the clearance distance where clearance can be defined as *the perpendicular distance through which a face can be swept from its initial position before it comes into contact with solid*, can be calculated using:

Lets F_S be the volume swept by face F such that

$$\forall d \in [0, \alpha], F_s(\alpha) = (M(d) @ F) \cap W$$

where the clearance distance α is the maximum value of d such that

$$F_s(\alpha) = \phi, \alpha \in [0, \beta]$$

and @ denotes the application of a rigid motion (M), W is the design model, β is the building bound or specified distance.

Once the "clearance distance" (α) is determined and compared against the "clearance distance", the system can determine whether an operator can approach the feature from that particular direction.

6.2 Spatial clearance of cylindrical faces

A cylindrical face is approachable if all the points that lie on the part of the face can be approached by an operator without being obstructed. Thus a cylindrical face is tested for accessibility by enlarging it until the boundary of the product or a user specified distances is reached, or solid is encountered. If the result of the intersection of the enlarged face with the building or component is null then the face is approachable.

7 THE PROTOTYPE

The techniques of feature modelling and reasoning about applied features to determine feature accessibility are implemented in an experimental prototype system. The system provides an environment in which designs can represented using features.

Designs are represented in the product model by selecting features from the feature library. The library includes such features as holes, slots, pockets, walls, columns, door_openings, etc. Every feature in the library has an identifier. The parameters of the selected feature are supplied by the user and transformed to the required position by applying the geometric transformation data.

The product definition is then extracted from the product model. The application program requires the clearance distance from the user to verify the

Figure 5.   Simple case study.

accessibility of the feature. Calls are made to the geometric modeller by the application programs software to perform queries on the defined component in order to establish the conformation of the resultant features to those intended, and to verify the accessibility and approach directions of the features.

7.1 Simple case study

The component shown in Figure 5a consists of two features, through hole and through slot. The validity of the hole depends on its position. The result of the analysis is shown in below.

Modifying the component definition may affect the type of features used in the definition of the component. For example, if the position of the through slot is moved up or the depth is reduced, the feature through hole will become a blind hole as shown in Figure 5b. The accessibility of the hole is changed.

7.1.1 Results of case study

Defined Component

--- ---
SLOT
--- ---
Feature Verification

Feature conforms to definition

Checking Accessibility

Feature accessible in +x directions
Feature accessible in −x directions
Feature accessible in +y directions
Feature not accessible in −y directions
Feature not accessible in +z directions
Feature not accessible in −z directions
--- ---
HOLE
--- ---

Feature conforms to definition

Checking accessibility

Hole is accessible from end—1 direction
Hole is accessible from end—2 direction

Modifying the Slot

Only the results of the modifications are shown here

--- ---

SLOT

--- ---

Feature Verification

Feature conforms to definition

--- ---

HOLE

--- ---

Feature Verification

Feature does not conform to definition
Blind√Hole is detected

Checking accessibility

Hole is not accessible from end—1 direction
Hole is accessible from end—2 direction

8 CONCLUSIONS

This paper reports on the approach for constructing product models in terms of features. In particular, the work concentrated on features verification (i.e. whether the resulting features conform to the definition of the intended features), and features accessibility. The work was motivated by the inadequacy of the current systems to perform verification, and therefore accessibility, resulting in possible contradictions between the intended feature and the resultant feature. Verification would establish if the designer's intent is fulfilled by testing the resulting features against the applied features.

Designs are defined by applying volumetric features. In order to verify the resulting features of the defined design, the volumetric features are converted to surface features. Geometric tests are carried out on the surface features to verify the resulting features by determining the type of faces they have (i.e. real or virtual). Accessibility analyses are performed on features to determine whether the resulting features can be accessed by operators. The approaches reported in this paper can be applied in a number of construction-related applications where geometric elements can be represented as features, and verification, accessibility and machinability are issues of interest.

It is clear that features are being used in many applications to convey the intention of the user to satisfy a particular function. Therefore, it is vital that features must be verified to ensure that the integrity of each feature is maintained and the feature can still perform its function.

REFERENCES

IAI, 1999, International Alliance for Interoperability, *http://www.interoperability.com*
STEP Tools, Inc, 1999, The ISO STEP Standards, *http://www.steptools.com/library/standard/*
Faraj I. 2001, Verification and Spatial Clearance of Features in Product Modelling Environment, *Construction Innovation: Information Process Management*, Volume 1, Number 1, March, pp 55–74, ISSN 1471–4175.

Organization and decomposition of a complex structural project

S. Loreti & M. Salerno
Structural Engineers, Rome, Italy

ABSTRACT: This work aims at describing the different approaches that can be followed in the development of a complex project. In particular, are described the different procedures that are commonly used in a project. Starting from a view based on the Quality approach based on the Client satisfaction using an approach based on the process, the Authors have pointed the attention on the Knowledge Management and Information Technology as the key to involve all workers of a project in the way to create a modern and knowledge based organization.

1 INTRODUCTION

A project is an organized program of activities that must be carried out to achieve a fixed objective (O). To achieve such goal, it's necessary to operate according to a strict logic that expects the use of suitable resources (R), the best ones on the market compatibly with needs and means of the company, and according to times (T) and costs (C) (Fig. 1).

To design doesn't simply mean to develop a sequence of repetitive phases, but involves an analysis suitable to identify and develop the components of the management itself. In this optic, all the elements to manage need to be studied and reorganized constantly, in a way able to permit an integration between the different components to merge the work during the development phases.

Building an object doesn't mean the achievement of the best solution but only one of the possible ones. To solve this problem it's important to adapt the level of the product obtained in a single step with the one aimed

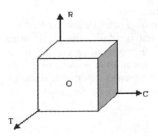

Figure 1. Key words for a project: (O) objective, (R) resources, (C) cost, (T) time.

to achieve, observing and modifying continuously the input data to obtain an excellent product and generally speaking, to find the correct solution by amalgamating the several solutions that had been proposed.

2 THE QUALITY SYSTEM IN A PROJECT

2.1 General features of the UNI EN ISO 9001:2000

The Quality system is a way of managing the organizations that extend the application of the Quality Control techniques to every field and must be exercised by all the workers.

The principal concepts of this way of managing correspond remarkably to the six literary values for the third millennium proposed by Italo Calvino in his *"Lezioni Americane"*, namely: *lightness, fastness, exactness, visibility, multiplicity, consistence.*

In the code *UNI EN ISO 9001*, Quality is defined as "the grade with which a whole of intrinsic characteristics satisfies the requisites". The definition is referred to a generic entity and not only to a product: in fact this concept can be associated to an activity, to a process, to an organization and to a person.

At a managerial level, Quality has an *all-embracing* meaning (it's an objective and a reference and so, it must incorporate all the aspects leading to obtain an excellent final result) or an *operative* meaning (the Quality is the satisfaction of the Customer).

The last definition leads to the concepts of *negative quality*, the measure of the variance between what we have obtained and what we expected, of *positive quality*, the measure of resetting of negative one, and of *latent quality*, that concerns about the required and

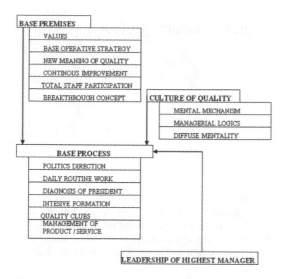

Figure 2. Global vision of the Quality system (Galgano A. 1990. *La Qualità Totale*. Il sole 24 ORE)

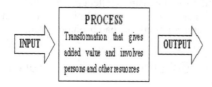

Figure 3. The process approach.

expected quality (we give to the customer something that he doesn't expect).

The most important principles of the Quality system are:

– the customer as absolute priority;
– the quality in all fields, applied by every component of the staff, to satisfy the customer;
– a continuous, constant and "endless" improvement (*kaizen*) and not only *control*;
– staff training to new mentality and leadership of direction of organization in this change.

A global vision of this system is offered in Figure 2. Nowadays, the quality system is applied all over the world by the most important companies. The referring code used is the *UNI EN ISO 9001:2000*, which represents the quality management system; it sets as a distinctive element the *process*, as a real mental disposition, based on activities "disassembly" to manage best and centre the objective.

The *process approach* is the base concept of quality management: it represents a new way of living the rules in the company because it gives great attention to the single activities and to the way in which they

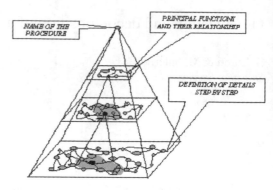

Figure 4. The pyramidal process composition.

interact, considering the weight of human component. The used model is a classical based on *input-output* (Fig. 3).

This is a easily manageable simple model based on four *macro-processes*:

– direction's responsibility;
– resources management;
– realization of the product;
– measure, analysis and improvement of the product.

2.2 *How to represent a process activity*

The basic tool to represent a process is the *flow charting*; this is a method to represent graphically a process using symbols, lines and words that describe the activities and the sequence. The principle of this methodology is the will to utilize the great power of the graphic and visual symbology: *"What we see influences the approach to the information and its understanding. The visual aspect influences the emotive behaviour on information: in fact it can motivate, involve, interest or distract."* (P. Forzano).

To reach this goal the "network's theory" is used: a network is a mathematical abstraction to model the information, a representation of objects (*nodes*) and their connections (*arches*). A flow chart must highlight clearly the logic of procedure and the way, underlining the entrance and the exit that give a quick comprehensibility.

Every process can be represented by a multilevel structure that is a hierarchy of elements: the top of pyramid is the name of procedure, the lower level is the description of base points and the other level represent each a greater descriptive detail (Fig. 4).

A procedure can be developed in three different ways:

– *Top-down*;
– *Bottom-up*;
– *Top-down + Bottom-up* (developed with the techniques of *Reverse Engineering*).

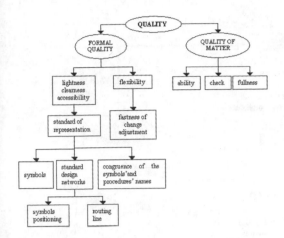

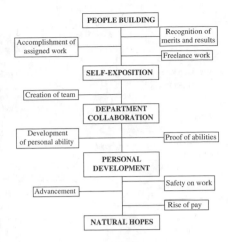

Figure 5. The quality of procedures (Forzano P. & Castagna P. 1998 *Dal disegno di grafi all'analisi strutturata dei problemi.* Franco Angeli).

Figure 6. People Building Process (Galgano A. 1990. *La Qualità Totale.* Il sole 24 ORE).

It's worth noting that nothing of the procedure described above represents the best choice to develop a project.

The most important aim to persecute is ordering: ordering is a fundamental human activity, necessitated by complexity. Ordering is very closely related with the idea of information and organization. Organization is strictly linked with quality and this, we perfectly know, is a typical procedure. So, it becomes important that quality of procedures can be evaluated through their suitability to suit customers needs that always change. Generally the quality of procedures is divided in two parts: *formal quality* and *quality of matter* as shown in Figure 5.

2.3 *The role of the human resources in quality system*

Once described the breakdown techniques of the project as a procedure, the quality system broach the problem of human resources. This theory is really innovative compared with the old Taylor's opinion about the clear split of the rules (*"Don't think, someone is paid to do it!"*).

Now, nothing is more important than the management of human resources because only these have no limits: *"The men can make miracles if:*

– *They are considered intelligent;*
– *their dignity isn't a compromise;*
– *they're always treated with respect;*
– *they're well trained;*
– *they can give a significant contribution to the success of the company"* (A. Galgano).

To simply understand this concept it is useful to know the sentence of Ishikawa: "A manager is really

formed when he's in a position to manage his own superior".

At the base of this new way of thinking there's a real PEOPLE BUILDING PROCESS, organized as shown in Figure 6.

The autonomy in work, the acknowledgement of efforts and the results obtained form a person that can independently react to background and develop natural psychic energy to give a bigger contribution to the company.

3 KNOWLEDGE MANAGEMENT

3.1 *General features*

Knowledge Management (KM) is the process of creating value from an organization's intangible assets (Liebowitz 1999). As such, knowledge management combines various concepts from numerous disciplines, including organizational behaviour, human resources management, artificial intelligence and so on. The focus is how best to share knowledge to create value-added benefits to the organization. It can be seen like a natural development of the total quality system and its ideas of improvement of productive cycles.

The process of creating a common, world-wide and free market represents one of the key features of our business system called *globalisation*. Globalisation was born and has grown up with the development of information and communication technologies allowing the fusion of local and national markets into a global one and it is one reason for the mergering of previously separated competitors.

Nowadays, markets rotate around the figure of the customer, his way of thinking and his needs, starting

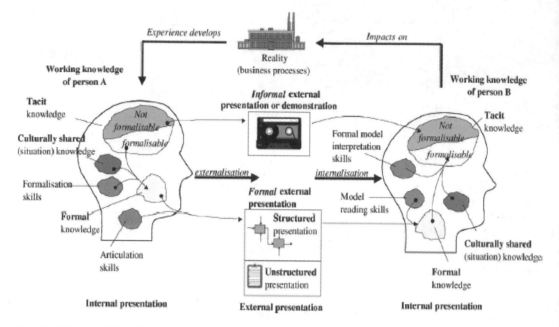

Figure 7.   The knowledge life-cycle model (Kalpic & Bernus 2001).

from an idea and until the construction of a product: these are elements of competitiveness.

At this point, quality, technical sophistication and price competitiveness of a product are for today on the market. The product must be suitable to fulfil the individual customer demands as reflected in the increasing individualisation of the production. With this vision, the product is only the base for the connection between a company and a customer.

Information and knowledge are becoming strategic resources in addition to the traditional ones and, in particular, knowledge must be considered as the key capital of companies.

Great attention must be paid to the fact that many people confuse information with knowledge. According to Hubert St. Onge of the Mutual Group, information is patterned data and knowledge is the capability to act. Knowledge includes the set of facts and rules of the thumb that experts may have acquired over many years of experiences; knowledge is what the master shares with the apprentice versus simply information.

KM aims at creating knowledge and in particular aims at transforming implicit and tacit knowledge into an explicit formal representation, distributing it throughout the organisation thus making possible the sharing and reuse of knowledge.

One of the most known model proposed has been given by Nonaka e Takeuchi. Here it is presented as an extension of this model of knowledge of life cycle, given by Kalpic and Bernus (Fig. 7).

In order to be able to execute business or decision processes, employees must possess some "working knowledge". This must be developed and updated receiving all the information about activities or phases of the processes that are going to be executed inside the entire organization or work group. This phase may seem to be simple, but it represents the main objective of KM: it is the *tacit* knowledge. In fact, very often knowledge holders do not use their own knowledge as explicit, formal and structured information. They understand and know what they are doing and how they have to carry out their tasks.

From the general point of view, working knowledge can be divided in two groups: formalisable and not-formalisable knowledge. Formalisable knowledge is connected with all the repetitive activities of a process. Not-formalisable knowledge is a difficult task to achieve; it is connected with creativity and innovative approaches to processes, such as some management, engineering and design activities. This part of knowledge is particularly important for KM, because it can be distributed and shared.

The set of constitutive processes is not predefined, nor is the exact nature of their combination well understood by those who have the knowledge. In contrast, the process of transformation of the formalisable part of working-tacit knowledge into formal knowledge represents one of the crucial processes in KM.

To perform the formalisation process it is necessary to have additional knowledge known as culturally shared or situation knowledge. This kind of knowledge plays an important role in the understanding of the process and its formalisation and structuring. In fact, the definition of an accounting process can only be done by an individual who understands accounting itself, but this formalisation will be interpreted by other individuals who must have an assumed prior culturally shared and situational knowledge that is not part of the formal representation.

As already said, the main objective of the KM is the externalisation of knowledge. To achieve this, different tools and approaches can be used, such as:

– *tacit and not formalisable* knowledge can be captured and presented as informal external presentations;
– *tacit and formalisable* knowledge can be captured and presented in formal external presentations.

This can be explained in a structured or in an unstructured form.

An example is given considering a textual description, like the quality procedure documents (ISO 9000), that can be seen like a presentation of process knowledge in unstructured form, while different company models are an example of structured form of knowledge.

The presented process of knowledge externalization has to be completed by a matching process of knowledge internalization necessary for the reuse of available knowledge. The internalization processes necessary are:

– *informal external presentation* of knowledge and its interpretation that can simply lead to build our working (tacit) knowledge;
– *formal external presentation* of knowledge avaible only after operating an interpretation of an existent business process model.

The passage from informal to formal knowledge is a natural process necessary for efficiency. The internalisation of externalised formal knowledge thereby closes the loop of the knowledge life-cycle.

The presented model proposes two criteria for knowledge explanation: the suitability of knowledge for abstraction (formalisable and not-formaalisable) and the state of appearance of knowledge (internalised and externalised).

3.2 *The role of KM in a project*

Today's market is taking to a modification of the companies structures. In fact, many of them are changing their classical hierarchical structure into a flat one, which enables them to (re-) act more flexibly.

As we know, globalization has taken to the formation of large companies which have to plan, manage and

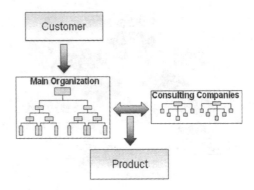

Figure 8. Project organization-flowchart.

design all the entire process of a project. To achieve their tasks, project teams have to be created. The project team members gain new experience and knowledge, which should become part of the organization's knowledge base. These knowledge-creating teams play key roles in the process of value creation of an organization.

The need of realizing a project in a short time has created the necessity of improving the skill of project workers. For a particular project, special teams must be created based on knowledge sharing.

Knowledge consists of truths and beliefs, perspectives and concepts, judgements and expectations, methodologies and know-how that is possessed by humans needed to determine what a specific situation means and how to handle it.

Today, it has grown in importance the use of Information and Communication technology (ITC) as an instrument to share knowledge. This is important due to the multidisciplinary of the projects that involve different teams working for the same goal.

For a long time, projects have been regarded as scheduling problems that could be managed simply using standard software packages available on the market according to quality standards. Now, technologies have modified this simplified way of managing making possible an approach based on knowledge.

It often happens that a project is broken-down in different teams not always of the same organization; special competence need special working teams and so, next to the main hierarchical organization, there are others hierarchically "flat" (Fig. 8).

This makes more difficult the information sharing and fusion making necessary for the creation of an instrument for the knowledge flow development. This can be built adopting all the internet-intranet technologies available, creating and adopting special software where to insert all the personal and project data.

These softwares must contain personal information of a team member starting from their individual history and knowledge acquired in their life, the

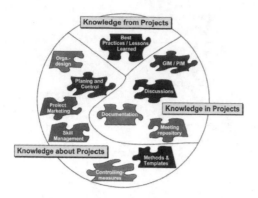

Figure 9. Modules of knowledge (Nonaka I. & Takeuchi H. 1995. The knowledge-creating company).

e-mail, calendaring and scheduling provided by the corresponding systems as well as groups information systems with the related group e-mail and teams correspondence systems.

It's necessary to implement a system of reportage and documentation in the way that in every moment all the items needed can be easily found and tracked. The documents must be written according to a prefixed quality standard. Next to the automatic system, it's important to organize regular meetings of the teams' members to report about status and problems; in this way everybody can share his personal or acquired knowledge, creating discussion from the doubts.

All the shown aspects refer directly to the working projects, but great attention must be given also to the knowledge which has been generated in projects that have already been completed.

In fact, during the development of a project, enormous efforts have been made by workers to solve problems. Problems that probably have already been encountered in other projects and that if available to consult may play an important rule for solving.

An organization is surely a sum of knowledge, but knowledge must be always improved through the personal will of learning continuously.

Finally, one can say that knowledge work is based on knowledge exchange, which requires a certain degree of trust as a pre-condition. As shown on Figure 9 new knowledge can be seen from three point of view:

- *from* projects;
- *about* projects;
- *in* projects.

Synergies is a primary goal of cooperation: the partners involved in inter-organizational projects need a high degree of collaboration and communication – projects are teamwork.

4 CONCLUSIONS

In this work the Authors wanted to fix the attention to the modern concepts related with an organization of a project. It's clear that, in order to develop a project, great attention must be given to the process breakdown. The quality system is not sufficient to describe and to guide a company because it doesn't guarantee the achievement of the optimum product, the *excellence*.

The *UNI EN ISO 9001:2000* represents a necessary but not sufficient condition to the improvement of a company structure. Necessary because it's not imaginable that any long time improvement of a company can be achieved unless starting from the fundaments and developing in a broad band, not sufficient because the survival and the success strictly depend of the strategic capacity of movements, of business, of *vision*.

Today, there is a huge tendency of abusing of the word quality; many companies having to respect quality codes, loose their entity, assuming a flat attitude towards on developing projects or problems, risking to relax or block their productive processes.

It's clear that the only right that can help to exceed this state of being is one of the most important human being: the human knowledge and so the knowledge management.

ACKNOWLEDGMENTS

The financial support of Stretto di Messina Spa is acknowledged. Anyway, the opinions and the results presented here are the responsibility of the Authors and cannot be assumed to reflect the ones of Stretto di Messina Spa.

REFERENCES

Galgano, A. 1990. *La Qualità Totale*. Il sole 24 ORE.
De Risi, P. 2001. *Dizionario della Qualità*. Il sole 24 ORE.
Garret, M. 1992. *Images: le metafore dell'organizzazione*. Franco Angeli.
Kaneklin, C. & Aretino G. 1997. *Pensiero organizzativo e azione manageriale*. Raffaello Cortina Editore.
Forzano, P. & Castagna, P. 1998. *Dal disegno di grafi all'analisi strutturata dei problemi*. Franco Angeli.
Sadler, P. 1991 *Progettare l'Organizzazione*. Franco Angeli.
Nonaka, I. & Takeuchi, H. 1995. *The knowledge-creating company*.
Nonaka, I., Reinmoeller, & Seno. 1998. *The "ART" of knowledge: Systems to capitalize on Market Knowledge*. European Management Journal Vol.16
Damm, D. & Schindler, M. 2002. *Security issues of a knowledge medium for distributed project work*. Int. journal of Project Management.
Kalpic, B. & Bernus, P. 2001. *Business process modelling in industry-the powerful tool in enterprise management*. Computer in Industry.

22. Quality and excellence in constructions

System-based Vision for Strategic and Creative Design, Bontempi (ed.)
© 2003 Swets & Zeitlinger, Lisse, ISBN 90 5809 599 1

Creativity and innovation

W.L. Kelly
Value Engineering Services Transworld (VEST), Walla Walla, Washington State, USA

ABSTRACT: This paper examines human potential, characteristics and mind. It defines creativity and innovation and qualities of creative people.

The paper describes two Value Engineering techniques, Function Analysis and FAST Diagramming and how they structure thinking to generate ideas.

Undirected creativity is of little worth. Purposeful creativity compounds benefits to both originator and recipients. Those points are demonstrated by an exercise. The paper describes typical scenarios encountered daily that offer opportunities for invention … They will be examined as time permits to practice visualization of creative ideas, problems ideas generate, cures for those problems and conversion to workable solutions.

1 HUMAN POTENTIAL

Endless efforts to determine how to achieve one's potential are fruitless. The answer is simple. Those who lack pride, curiosity and motivation have reached their potential.

Those who crave knowledge, seek challenges and find solutions will increase their talents. They will never achieve their potential since it multiplies and expands relative to personal achievement.

"A life unexamined is not worth living."[i]

I believe Supreme Intelligence has one question for humans, "What did you do with the mind you were given?" We can make it blob of sludge or wonder of enlightenment. Output that benefits civilization is a likely standard by which to judge.

Leonardo da Vinci more than any person in recorded history exemplified inventive powers of a mind. He not only immersed himself in most areas of science but also in music, physiology, the universe and became a proven master artist. He discovered root causes by experimenting and observing, not memorizing theories and dictates of others. His interests knew no bounds. Nothing was too small to attract his interest nor too large to intimidate his search. He assumed he could understand everything.

He has been called the most "gifted" man who ever lived. I disagree with "gifted". Minds are gifts to all. He was the most curious, observant, daring, motivated and productive human ever.

---

i. Socrates

2 HUMAN CHARACTERISTICS

People are the first computers. Our (mainframe) body contains all components necessary to support and protect the (hard drive) brain. Operating System Software manages involuntary functions such as breathing, digestion, temperature, immune system, senses and coordinating interactions.

It also manages subconscious storage, processing and retrieval of data. Subconscious software uses unique encoding to create links. It forms alternatives for pressing concerns. It stores default settings for all the above functions from body temperature to hard drive data storage locations.

Minds come loaded with hundreds of thousands of software programs for every subject imaginable. People activate, open and close software to reflect changing areas of interest throughout life. We tailor and develop software for special needs. We flavor information by coloring it with input from senses, emotions and linked events. This process helps define who we are.

It has been shown genetically identical twins turn on and off different genes (and software they represent) which accounts for their different personalities and characteristics. IBM is expending huge effort to produce a computer with 100 terraflops capability by 2004, the same power estimated for one human brain.

A brain can manage most level of challenge but often limits itself to the level in which it finds employment. This does not disparage types of work. Work ranges from low-demanding chores as weeding fields or hotel doorman to critical tasks such as open-heart surgeon or supreme court judge.

Work levels can equate to car gears. I often asked students which gear these occupations used: Air traffic controllers; consensus was fourth gear. Government regulators; consensus was neutral or reverse. Emergency Room (ER) personnel; consensus was fourth gear. I asked what air traffic control and ER had in common to tie at fourth gear. Comparisons began with volume and speed of decisions but evolved to variety of problems and skills and breadth of knowledge decisions required.

Analysis concluded making the same decisions at breakneck speed does not challenge the mind, only its endurance for monotony. Air traffic controllers always dropped one or two gears from fourth. An infinite variety of challenges spill into ER demanding instant and complicated major decisions and actions. It remained at fourth gear.

If one's employment does not support high gear thinking it must be found in avocation or hobbies. Like a rubber band, the mind needs occasional stressing to keep it elastic and prolong its life.

Never assume great minds are privy to any class of people. Great minds are everywhere. An isolated shepherd can generate grand visions and philosophy. Ideas care not who their parents are. Enlightened thinking can be found in common looking packages. Seek the mind, don't worry about the container.

Great minds examine problems from the highest viewpoint they can reach. Theirs are holistic views of problem origin, history and interface. They automate generating ideas plus evaluating cause and effect ideas spawn.

Average minds develop one solution for a problem and are content. They may produce one or even two alternatives but normally are happy following guidance and standards.

Little minds have little problems. They specialize pointing out imperfection such as undotted "i's" or uncrossed "t's". They are easily overwhelmed.

The larger breakdown is pragmatic vs. visionary minds. The world needs producers and dreamers for these reasons: "The common man tries to adapt himself to the world. The uncommon man tries to make the world adapt to him. Therefore all progress comes from uncommon men."[ii]

The world needs those capable of dreaming and those making dreams come true. Practical people make the world work. Must the two talents be separate? No. Is a Renaissance Mind still possible? Yes. For the fullest life one should develop both hemispheres to their maximum. Leonardo did. It only requires massive motivation and dogged persistence.

3   CREATIVITY AND INNOVATION

We cannot do something new unless we learn something new. We cannot build or perform something new unless we invent something new. Where those capabilities do not flourish people remain locked in the past tied to obsolescence. To compete globally requires rapid, unique, useful and marketable ideas and products whether a person, organization or country.

Opportunity surrounds us. "Chance favors only the prepared mind."[iii]

Some believe creativity only belongs to "gifted" people. Others believe innovation only comes from "clever" people. Both abilities lie within us. What are characteristics of creative people? Can they be absorbed by others?

Creative people have insatiable curiosity, unquenchable thirst for knowledge and regularly exercise abilities to both visualize and reason. These are talents we all possess but used in unison by a small minority. Truly creative people are self-confident and survive destructive criticism and ridicule. They are mostly upbeat, outgoing and seek company of like personalities. They are tenacious and share ideas readily. An idea is not a dangerous thing unless it is the only one you have. It then becomes an obsession and protecting it a duty. Creativity stops.

What are creativity and innovation?

Creativity develops new ideas to satisfy expressed or implied need of mankind.

Innovation adapts familiar items or actions to serve new uses unforeseen by others.

Of myriad human abilities three tower us over other life forms. They are curiosity, visualization and reason, a tool for analysis. They produce insight, discoveries, ideas and solutions. Their effectiveness is magnified by motivation and perseverance.

It is not how hard one tries but how long one keeps up effort that brings success.

"I'm a great believer in luck, and I find the harder I work the more I have of it."[iv]

Curiosity begins with life. I observed a baby in its mother's arms aboard a flight last year. Erratic movements of arms and legs were signs of its age. One hand flew by its face during gyrations and its head jerked back, eyes flew wide open and the reaction said, "Whoa! What is that thing?"

Nanoseconds later the other hand flew by and was greeted with the same reaction. The baby's head swiveled to compare both jerking wondrous appendages. His mother moved him closer to the seat in front and he touched it with one hand causing giggles and squeals. He hit it with both hands and stared

---

ii. George Bernard Shaw.

iii. Louis Pasteur
iv. Thomas Jefferson

at them in awe at feelings they transferred. His discoveries and reactions brought laughter, remembrances and renewal to aged viewers.

We are born curious. We eat bugs, burn fingers, build dams in gutters and endlessly ask two most important questions possible, "Why?" and "How?" We badger adults with hundreds of inquiries daily until we are banished outdoors to play. However, curiosity remains.

One must nourish curiosity forever; it makes us eager for tomorrow.

"Don't part with your illusions (dreams). When they are gone you may still exist but you have ceased to live."[v]

Unquenchable thirst for knowledge is imperative to become a complete person. We need to learn how to learn because all we know becomes obsolete over time and the pace accelerates yearly.

It is as important to unlearn as to learn. Minutiae may be important in trivia quizzes but not critical to productive lives. Let it go. You are not the same person today as you were yesterday. You will not be the same person tomorrow you are today. How you change is your choice. Curiosity and knowledge feed from each other and result in a tornado of growing, lifting one higher and broader with each revolution.

Visualization is magic. Our "inner eye" can convert every idea to reality even with eyes wide open. That lets us conduct detailed virtual experiments in our laboratory mind. It is the most underutilized resource we have. Concept, design, testing, redesign, etc., can largely be accomplished through visualization. It is critical to apply before making decisions.

"You can't depend on your judgment when your imagination is out of focus."[vi]

Unfocused creativity or innovation has less benefit than that with purpose. One must identify a need or problem before trying to solve it. It must be defined accurately to ensure creativity or innovation effort is most effective and efficient.

4 FUNCTIONS AND FAST DIAGRAMS

Function Analysis converts any item or problem to VE's common language, functions, a language readily understood by any person or discipline.

Function Analysis System Technique (FAST) Diagrams combine, expand and relate functions into a word picture of an item to be studied.

Together they structure any concept or existing problem on a single sheet of paper in terms that foster instant understanding of the big picture.

---

v. Mark Twain
vi. Mark Twain

Function Analysis defines intended use of an item or activity in two words, an active verb and measurable noun. The verb describes required action and the noun defines the recipient of action. Nouns are generic whenever possible. There is no single correct function description. Teams select preferred words from those suggested by members that reflect the team's understanding of intent.

Once function analysis is embedded in the mind and intended use of an item or activity is known, a person can determine functions of anything with precision. The way they view the world changes forever when clutter is clarified in two words.

Function Analysis System Technique (FAST) diagrams structure projects or concepts on one page. They are word pictures of the item being studied, formed by logic derived from three questions, How? Why? and When?

Why are functions being performed forces thinking to highest goals. How will functions be satisfied drives thinking to smallest details. When functions describe those which occur at the same time or are brought about by those lying on a critical path. When How and Why questions are logical in both directions the FAST diagram is complete.

FAST Diagramming Rules are on Figure 1 and simple FAST Diagram examples are on Figure 2.

Creativity and innovation by one person is not as efficient as several heads. Studies of teams indicate five members have highest synergism and produce more alternatives than other size team.

5 SCENARIOS TO SOLVE

To convert "Opportunity Surrounds Us" into patents the author developed scenarios seen daily to stimulate curiosity and inventiveness.

a. Dry earth does not conduct electric current, wet earth does. How can you control amount and frequency of irrigation automatically using these two observations? Function: Control Irrigation
b. Traffic light. Function: Alternate Movement. How can you satisfy the function without using wires?
c. Northern hemisphere cities experience snow storage problems after big blizzards. Snowplows and blades push snow into piles using most of street and sidewalk for snow storage. Function: Clear Pavement How else can you get rid of it?
d. How can aircraft push back from a gate without using wheeled prime movers? Function: Relocate Aircraft
e. How can one modify a two-wheeled scooter to make it more efficient, fun and marketable? Function: Propel Conveyance
f. We usually apply water on the surface of fields to soak ground for vegetation. Functions: Conserve

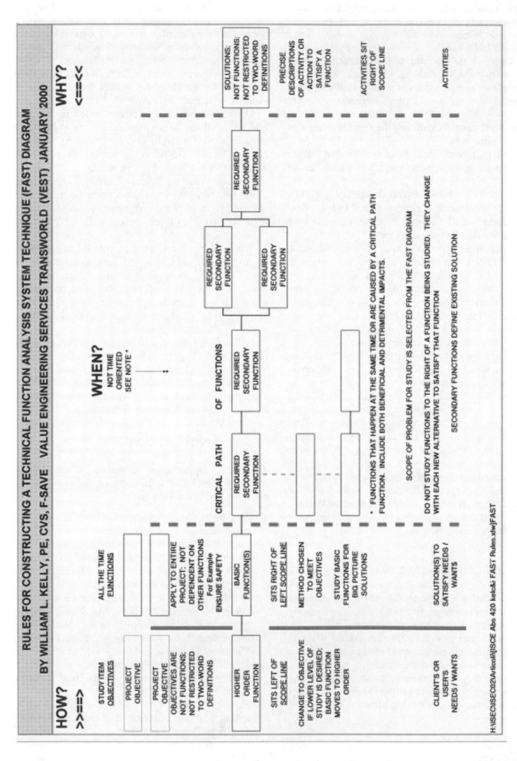

Figure 1. Rules for constructing a technical function analysis system technique (FAST) diagram by William. Kelly, PE, CVS, F-safe value engineering services transworld (VEST) January 2000.

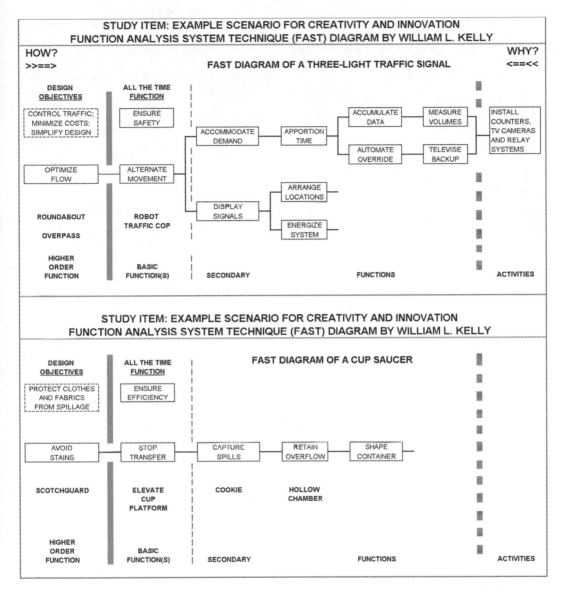

Figure 2. Study item: Example scenario for creativity and innovation function analysis system technique (FAST) diagram by William L. Kelly.

Water, Nourish Plants Water is scarce. Irrigation depth varies for type of soil, crop and evaporation. How can irrigation practices be changed to save 50 percent water?

g. How can you modify a diaper to outsell all competitors? Function: Improve Convenience

h. How can you modify commercial plane interiors to accommodate parents and babies more safely and comfortably? Functions: Protect Infants; Improve Convenience

i. How can you prevent a driver from falling asleep at the wheel? Function: Avoid Accident, Ensure Alertness

j. Is it possible to convert salt water to usable irrigation water other than reverse osmosis process? Function: Purify Water

k. How can you pressurize water to transport it without electricity or combustion engine power for conventional pumps? Function: Pressurize Water

l. Where sterile environments are necessary people wear thin rubber gloves which are discarded after one use. (Medical, Dental, Clean Lab technicians, etc.) How can one do that more effectively? Function: Sterilize Hands

m. How can you make a toy car using only three pieces of wire? Two are the same length and the third is six times longer. Function: Simplify Toy, Promote Creativity

6   CREATIVITY AND INNOVATION IN FUTURE GENERATIONS

Nurture children's curiosity to quicken creativity. Read stories two or three times then on occasion stop before the last page and ask them how the story should end. Endings differ and forming them stretches imagination. Have them draw what they wish. When finished say "Let's see what you drew." Search from every angle, then add an ear, eye, leaf, blossom, wing or fin and grow the picture together.

Search continually for unfulfilled needs, things that don't work, are inconvenient, unsafe or have flaws. Determine functions from intended use and create alternatives. Employ curiosity, reasoning and visualization powers to create. Make life worth living and enjoy the Leonardo life!

System-based Vision for Strategic and Creative Design, Bontempi (ed.)
© 2003 Swets & Zeitlinger, Lisse, ISBN 90 5809 599 1

Deploying and scoring the European Foundation for Quality Management Excellence Model (Part One)

P.A. Watson & N. Chileshe
School of Environment & Development, Sheffield Hallam University, Sheffield UK

ABSTRACT: The paper focuses on the key benefits to be obtained by Engineering and Construction organisations from the deployment of the EFQM Excellence Model. Empirical data collected from a sample of 100 engineering and construction firms and its analysis provides an overview of the extent to which the model has been implemented within the UK. In summary this paper provides a rationale for model deployment. A second paper presented by the same authors contains original work on how to present the scoring in the form of a Pentagonal Profile.

1 INTRODUCTION

The EFQM Model is a framework which recognises that there are many approaches to achieving a sustainable competitive advantage. Within this approach there are some fundamental concepts that underpin the model and these are identified below.

2 RESULTS ORIENTATION

Excellence is dependent upon balancing and satisfying the needs of all relevant stakeholders (this includes the people employed, customers, suppliers and society in general, as well as those with financial interests in the organisation).

3 CUSTOMER FOCUS

The customer is the final judge of product and service quality and customer loyalty, retention and market share gain are best optimised through a clear focus on the needs of current and potential stakeholders.

4 LEADERSHIP AND CONSTANCY OF PURPOSE

The behaviour of an organisation's leaders should create a clarity and unity of purpose within the organisation and an environment in which the organisation and its people can excel. A truly empowered organisation employs both a top down and bottom up approach to managing its activities within the context of aiming to fully satisfy all stakeholders.

5 MANAGEMENT BY PROCESSES AND FACTS

Organisations perform more effectively and efficiently when all interrelated activities are understood, systematically managed and decisions concerning current operations and planned improvements are made using reliable information that include stakeholder perceptions.

6 PEOPLE DEVELOPMENT AND INVOLVEMENT

The full potential of an organisation's people is best released through shared values and a culture of trust and empowerment, which encourages the involvement of everyone. This necessitates a holistic approach to people and their operational systems and organisational structure.

7 CONTINUOUS LEARNING, INNOVATION AND IMPROVEMENT

Organisational performance is maximised when it is based on the management and sharing of knowledge within a culture of continuous learning, innovation and improvement. Hillman (1994) suggested that *"… the EFQM Model provided a tried and tested framework, an accepted basis for valuation and a means to facilitate comparisons both internally and externally."*

8 PARTNERSHIP DEVELOPMENT

An organisation works more effectively when it has mutually beneficial relationships built on trust, the sharing of knowledge and integration with its partners. The introduction of the partnering concept in engineering and construction has been successful in addressing this critical issue.

9 PUBLIC RESPONSIBILITY

Adopting an ethical approach and exceeding the expectations and regulations of the community at large best serve the long-term interests of the organisation and its people.

10 THE EFQM EXCELLENCE MODEL

Corporate excellence is measured by an organisation's ability to both achieve and sustain outstanding results for its stakeholders, thus the enhanced version of the EFQM Excellence Model was developed. The fundamental advantages of the new Excellence Model included:

• increased cost effectiveness
• provision of a customer
• improved knowledge management
• a results orientation
• partnership development and monitoring focus
• enhanced performance and learning

The Model (European Foundation for quality Management 1999) was designed to be:

• simple (easy to understand and use)
• dynamic (in providing a live management tool which supports improvement and looks to the future)
• flexible (being readily applicable to different types of organisation and to units within those organisations)
• holistic (in covering all aspects of an organisation's activities and results, yet not being unduly prescriptive
• innovative (enabling new developments)

The EFQM Excellence Model consists of 9 criteria and 32 sub-criteria. The five criteria on the left-hand side of (Fig. 1) are called "Enablers" and are concerned with how the organisation performs various activities. According to Hillman (1994) *"The enablers are those processes and systems that need to be in place and managed to deliver total quality"*. The four criteria on the right of Figure 1 are concerned with the "Results" the

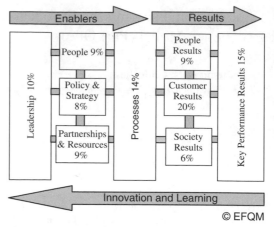

Figure 1.   The EFQM Model.

organisation is achieving with respect to different stakeholders. Hillman (1994) added that *"results provide the measure of actual achievement of improvement."*

Watson (2000) stated that *"the EFQM Model provided a truly service focused quality system which had an inbuilt mechanism for the attainment of continued organisational improvement."* Weile et al. (1997) identified that *"the criteria of the model helped managers to understand what TQM means in relation to managing a company."* However, managers also have to work at making the general descriptions of the criteria more specific to *"fit their particular situation and give them meaning within the context of their business activities."*

11 EMPIRICAL DATA

The theoretical advocated advantages for model deployment require testing in practice. This is necessary because engineering and construction firms expend valuable organisational energy in the application of the EFQM Excellence Model. Therefore the advantages have to be achievable or the implementational process will result in a waste of valuable corporate resources.

The next section provides a summary of empirical research conducted upon the implementation of the European Foundation for Quality Excellence Model within the UK. The results are based upon the following samples:

• for the advocated advantages of deployment a sample size of fifty companies;
• for the key aspects of implementation a sample size of 100 companies.

These sample sizes are representative of the population engaged in or having completed EFQM Excellence

Model application. Hence the conclusions do have validity.

The results of the research established that the majority of the sampled companies found the model simple, holistic, dynamic and flexible. They agreed that the model enhanced the understanding of TQM among senior management and enabled identification of the company's strengths and weaknesses via the self-assessment approach engendered within the model.

The research established the main problems faced by quality managers during implementation as:

- inexperience in the implementation of the Model;
- difficulties encountered during the documentation and data gathering stages;
- insufficient time and funds allocated to the project;
- resistance from employees within the host company.

Even though the research showed that senior management provided quality managers with full support and sufficient authority during the design and deployment phase, quality managers faced problems of insufficient funding and time allocation for the project. Full responsibility had been delegated without adequate authority.

The research confirmed that construction and engineering firms can obtain sustainable competitive advantages from the application of the EFQM Model.

Research in the field of construction and engineering, however, and in particular relating to TQM, is not utilised to its full potential. Research conducted within the manufacturing sector has identified the importance of adopting a post-modernistic paradigm when implementing quality systems. This fact seems not to have been embraced by both industries, thus the high failure rate in applying TQM within engineering and construction operational environments. Construction and engineering related enterprises must fully appreciate that the old style morphostatic change processes are not capable of sustaining an effective and efficient quality management system.

Empirical research was conducted to test some of the key elements of the deployment process. The first issue to be tested via the sampled companies' questionnaire related to an organisation's ability to respond to change. This encompasses criteria such as:

- empowerment;
- innovation;
- flexibility;
- cultural dynamics.

Since the previous studies on this topic (Watson 1996) there has been a great improvement for all firms in their ability to respond to changing environments. This has positive correlation with the number of firms now operating a post-modern philosophy.

Table 1. Distribution of sampled companies.

Size of firms	Sample size
Small	50
Medium	25
Large	25
Total Sampled	100

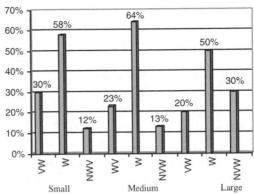

Key: VW-Very well; W-well; NVW-Not very well

Figure 2. Ability of organisations to respond to change as a percentage.

Table 2. Analysis of morphostatic and morphogenic process.

Organisations as a percentage having morphostatic process	Organisations allowing employees to be proactive at the following levels as a percentage		
	Senior management	Middle management	Operational level
Small 33%	60	30	10
Medium 56%	50	35	15
Large 11%	50	42	8

The above analysis in Table 2 indicates that morphostatic processes are common and the amount of proactive participation has not filtered down through the sampled firms. Therefore there are still changes required in cultural dynamics in order to fully obtain a sustainable corporate competitive advantage.

Construction and engineering firms require variety in their approach and hierarchical authoritarian organisations are poorly equipped to provide such variety. Only business enterprises based on a post-modernist model with vastly reduced bureaucratic control, a rich array of horizontal communication channels, and in which personnel are given a substantial share of

Table 3. A summary of EFQM Excellence Model deployment advantages for engineering and construction organisations. (Note: issues are not mutually exclusive).

Key deployment issues	Resulting benefits
• Process improvements	• A clear understanding of how to deliver value to clients and hence gain a sustainable competitive advantage via operations
• Attaining an organisation's objectives	• Enabling the mission and vision statements to be accomplished by building on the strengths of the company
• Benchmarking Key Performance Indicators (KPI's)	• Ability to gauge what the organisation is achieving in relation to its planned performance (Plan, Do, Check, Act)
• Development of clear, concise action plans resulting in a focused policy and strategy	• Clarity and unity of purpose so the organisation's people can excel and continuously improve
• Integration of improvement initiatives into normal operational activities	• Interrelated activities and systematically managed with a holistic approach to decision making
• Development of group/team dynamics	• People development and involvement. Shared values and a culture of trust, thus encouraging empowerment in line with a post-modernist company

authority to make choices and to develop new ideas, can survive under dynamic global market conditions.

12 CONCLUSIONS

Both IiP and BSEN ISO 9001:2000 are encapsulated within the EFQM Excellence Model. Therefore an incremental application to the full deployment of the EFQM Excellence Model could be via a strategy of implementing IiP, followed by BSEN ISO 9001:2000 leading ultimately to EFQM Excellence Model deployment.

This strategic incrementalist deployment would avoid the "big bang" approach with its associated cultural issues which are most noticeable and problematic in construction and engineering organisations.

Part Two of this paper explores how best to present model deployment results to senior management.

REFERENCES

Hillman, G.P., 1994. Making Self-assessment Success, *Total Quality Management*, Vol 6 (3), pp 29–31.

The European Foundation for Quality Management (EFQM, 1999) Excellence Model, accessed 17 January 2001, http:/www.efqm.org/award.htm.

Watson, P., 2000. Applying the European Foundation for Quality Management (EFQM) Model, *Journal of the Association of Building Engineers*, Vol 75 (4), pp 18–20.

Wiele van der, A., Dale, B.D. & Williams, A.R.T. 1997. ISO 9000 Series Registration to Total Quality Management: the Transformation Journey, *International Journal of Quality Science*, Vol 2 (4), pp 236–252.

Evaluation and improvement of public school facilities services using a TQM framework

K. Jayyousi & M. Usmen
Wayne State University, Detroit, MI, USA

ABSTRACT: A systematic management approach is needed to improve the delivery of facilities services in urban public school districts to ensure a clean, safe and healthy learning environment. A total quality management (TQM) framework was adopted to establish a continuous improvement system in the Facilities Department of the District of Columbia Public Schools (DCPS). A survey instrument was devised, called "Service-On-Wheels," and data was collected from 143 schools over 4 quarterly periods to evaluate 24 service categories, using statistical analysis tools such as mean, Pareto charts, histograms, and trend charts. The study showed that implementation of such a TQM framework improves the performance of the organization at all levels and provides avenues for problem solving.

1 INTRODUCTION

1.1 *Background*

Urban public schools are characterized by aging facilities and obsolete infrastructure. The average age of public school buildings in the U.S. is 45 years (US Department of Education 1999a), and many school districts are experiencing overcrowding while some are losing thousands of students every year. According to a U.S. Department of Education study (1990b), there are 16,783 public school districts in the US housing about 53 million students and the National Education Association (2000) has estimated that they need $322 billion for facilities repair, renovation and new construction. Seventy-five percent of the nation's schools are in need of repairs (ASCE 2001). Facilities management service departments in schools are struggling to respond to thousands of work order backlog in deferred maintenance to keep schools clean, safe and healthy. Students attend classes in buildings with serious deficiencies that include leaky roofs, damaged windows, obsolete electrical systems, aging heating systems, and deteriorated athletic facilities. As a result, the number of fire code and safety violations has increased, while scheduling renovations and repairs have become difficult during the school year. Additionally, due to the presence of asbestos, lead-based paint and poor air circulation, the environment of a school building has become vulnerable with increased risks of student exposure to contaminants.

In the face of such conditions, many believe that facilities departments in urban public schools are generally neglected by school district leadership, under-staffed, out-dated, lacking needed vehicles and equipment, and almost all are under-funded, resulting in poor service delivery and inefficiency.

1.2 *Problem statement and significance*

Aging buildings and shrinking facilities budgets have clearly contributed to the poor condition of urban school facilities nationwide. School facilities departments are finding it increasingly difficult to respond to customer needs in a timely fashion. Most importantly, many facilities departments lack an evaluation system that provides accountability and performance metrics.

A systematic management approach is needed to improve the delivery of services to ensure clean, safe and healthy learning environment. The study presented here examined functions of facilities service organizations as they relate to daily operation and maintenance (O&M) in urban public school districts, and developed a performance evaluation system through the application of Total Quality Management (TQM) principles, including the use of statistical analysis techniques. There is much at stake in bringing improvements to facilities and removing barriers to learning from all classrooms.

1.3 *Description of DC Public Schools*

DCPS is typical of urban school districts in the U.S. which have been faced with financial challenges.

Despite this fact, DCPS has successfully implemented key academic and operational reforms, including a re-organization of the facilities department and the completion of the first Facility Master Plan (DCPS 2000) in 30 years. DCPS is made up of 150 elementary, middle and high school buildings. More than 70,000 children attend schools in the Nation's Capital with an average building age of 65 years. There are over 4,000 classrooms, 15 million square feet of occupied space and 600 acres of real estate to manage. More than half the schools were built prior to World War II, some were built in the 1800s prior to the electricity age.

Towards the end of the 1960s, due to a continuing trend of increasing student enrollment, the District of Columbia planned and constructed a large number of schools to respond to the predicted increases. However, towards the late 1960s the enrollment started dropping off necessitating significant cuts in school funding and facilities budgets. As a result, facilities staff continued to shrink, and maintenance dollars became scarce. Capital improvement expenditures followed the same trend starting in the early eighties into the mid nineties. These conditions affected the facilities department's mission and function and culminated in a large backlog of deferred maintenance and inability to respond properly to customer work requests.

1.4 DCPS facilities management operations

The mission of a school facilities management organization is to ensure that every classroom is clean, safe and healthy. This can be a formidable task given the state of the buildings in many urban districts. In DCPS, a huge backlog of deferred maintenance was built up over the years as a result of budget cuts and service inefficiency. This resulted in a situation were almost every school service request is an emergency. Most of the building systems and equipment were observed to have surpassed their service lives. A facility condition assessment completed in 1998 by the US Army Corps of Engineers concluded that $2 billion was needed to renew public school buildings. It is interesting to note that an earlier condition assessment was conducted in 1990 by 3DI, which found that the need was about $500 million. This shows how over a period of eight years the condition of the buildings rapidly deteriorated, escalating the repair bill to four times the original estimate. Due to the lack of capital improvement money and reduced O&M budgets between 1990 and 1997, a 1998 condition assessment revealed that there were $400 million worth of failing or failed equipment (DCPS 2000c).

Large urban schools have a relatively large facilities staff, as compared to others. Table 1. shows a breakdown of the facilities staff at DCPS.

Table 1. DCPS facilities management staffing levels (2001).

Office unit	Staff	Description
Maintenance	100	Construction Trades
Building Operations	200	Operating Engineers and HVAC Mechanics
Housekeeping	600	Custodians
Design & Construction	15	Architects, Engineers
Grounds	40	Landscape, Hardscape
Administration	10	Management and Support
Others	35	Safety, Locksmith, Electronics, Service

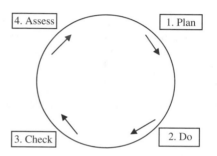

Figure 1. Shewhart's improvement cycle (Deming 1986).

2 TQM FRAMEWORK

Quality tools are used to improve processes, increase productivity, assist in problem solving, and bring change. A TQM framework, then, necessary to bring about the sought changes in the operations of a school facilities department. Shewhart's improvement Cycle (Deming 1986) presented in Figure 1 is useful in setting up the TQM framework.

Accordingly, in such a framework, improvement becomes a four-step, continuous and ongoing process. This is sometimes referred to as PDCA (Plan-Do-Check-Assess), consisting of the following steps:

Step 1: Plan: In the planning stage, stakeholders involved get together to study the current situation and key questions are asked. Alternatives may result, and the team finally decides which alternative to put to the test.

Step 2: Do: Once a course of action is determined, then it is put to the test. Process changes are exercised on incremental basis.

Step 3: Check: Collected data is checked and analyzed to note improvements or decline in measured parameters.

Step 4: Assess: The cycle is completed with an assessment of the situation and adjustment of plans and actions. This step represents a course correction in the improvement cycle.

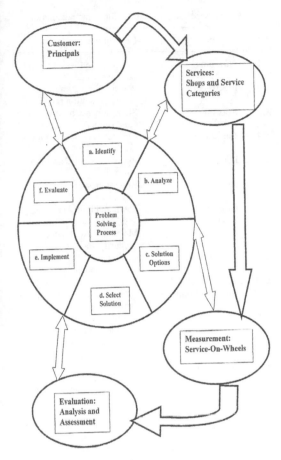

Figure 2. TQM framework (Adapted from Honeywell 1995).

Our research focused on a TQM framework based on Xerox's quality improvement process and practices (Honeywell 1995), which is based on, and is more detailed than PDCA. This framework is, again a, four-part structure: Customer, Services, Measurement, and Evaluation. The structure interacts with a unique problem solving process that continually improves the framework. Figure 2 illustrates the TQM framework representing our specific study.

In this figure, the "Customer" part is related to the "Planning" stage of identifying customers, customer expectations and requirements. The "Services" part relates to the "Doing" stage in which services are identified and carried out. This part of the framework represents the work processes and environment including workers, equipment, supplies and service delivery. This is the main part that requires process improvement. The "Measurement" part represents the "Checking" stage where performance and evaluation

measurement parameters are identified. This step includes the data collection instrument. A "Service on Wheels" survey was developed for this purpose, and will be elaborated on later in this paper.

The "Evaluation" part is the "Assessment" stage of Shewheart's cycle and represents the stage at which all collected data is analyzed, compared and processed to assess (based on customer feedback) any improvements on work performance. It is important to mention that the framework is linked with the problem solving process in all parts. This allows concurrent problem solving at all levels while the whole process of continuous improvement is taking shape.

3 FACILITIES IMPROVEMENT INITIATIVES AND EVALUATION

3.1 *Improvement initiatives*

There are three quality levels that can be identified in the facilities management environment: the leadership, the operations and the customer. Careful attention is required during the planning stage, in order to examine the impact of process improvement on the various levels.

A number of improvement initiatives were planned and executed in order to enhance department operations, ensure accountability and boost efficiency. These included a technical, safety and health management-training program for staff, the Service-On-Wheels customer service evaluation system, and an overall departmental re-organization to ensure accountability, support the employees, and focus on customer service.

Additional initiatives included a computerized maintenance management system to automate, track, monitor and manage customer work requests and work productivity, a maintenance plan for a comprehensive preventive maintenance program, establishing a service center to care for customer needs, and the development of an emergency management plan for foul weather procedures and disaster response. Further, a comprehensive database of facilities information was developed that included a Geographic Information System (GIS), building design guidelines, educational specifications, a facility master plan, and an asbestos management plan. Another initiative included a project control system for project management, a program management structure for the capital improvement programs; a deferred maintenance backlog management system, and a program to ensure that the schools are clean, safe and healthy. Table 2 summarizes the initiatives, showing the affected quality level by the symbol $\otimes$.

It should also be noted that the facilities employees were supported with a new inventory control system, new work tools and attire and a vehicle fleet management program.

Table 2. Facilities improvement initiatives.

Initiative	Customer	Operations	Leadership
Training		⊗	
Service-On-Wheels	⊗	⊗	⊗
Re-Organization		⊗	⊗
Computerized Maintenance Management System	⊗	⊗	⊗
Maintenance Plan		⊗	
Playground Personnel Certification		⊗	
Emergency Management Plan	⊗	⊗	
Service Center	⊗		
Service Managers	⊗		
Geographic Information System (GIS)			⊗
Master Plan		⊗	
Fleet Management		⊗	
Deferred Maintenance Backlog Management	⊗	⊗	⊗
Clean, Safe and Healthy Schools	⊗	⊗	
Project Control System		⊗	⊗
Safety Program	⊗	⊗	⊗
Design Guidelines			⊗
Work Tools		⊗	
Fire Code Violations		⊗	⊗
Inventory Control		⊗	
Personal Protection Gear		⊗	
Educational Specifications			⊗
Asbestos Management Program		⊗	⊗
Program Management			⊗

Table 3. Service categories of facilities management.

ID	Service category	Description
C1	Plumbing	Drainage and waste piping repairs
C2	Electrical	Power, lighting and communication repairs
C3	Roofing	Slate, shingle, built up, metal and rubber
C4	Carpentry	Cabinetry, doors, windows, walls
C5	Playground	Play sets, swings and safety mats
C6	Pest Control	Rodent, pest and animal control
C7	Grass/Field	Lawn and tree service, athletic fields, snow
C8	Fencing	Metal and iron fence, security gates
C9	Work Orders	Customer work request and service center
C10	Trash	Waste removal, dumpsters
C11	Building Supplies	Cleaning and toilet supplies, trash bags
C12	Service Managers	Facilities customer service representatives
C13	Custodians	Housekeeping, cleaning, dusting, mopping
C14	Restrooms	Disinfecting, sanitizing, scrubbing, upkeep
C15	Classrooms	Cleaning, dusting, mopping
C16	Floor Tile/Carpet	Vinyl / ceramic tile, wood floors, carpeting
C17	Heating	Boilers, unit ventilators, controls, radiators
C18	Operating Engineers	Heating plant, building maintenance
C19	Air Conditioning	Air conditioning repairs, controls
C20	Electronics	Fire alarm, security system, fire protection
C21	Locksmith	Locks and keying including security access
C22	Capital Projects	Planning, design, construction, renovation
C23	Safety Office	Environmental health and safety services
C24	Emergencies	Fire, flooding, vandalism, disaster recovery

3.2 Service categories

Twenty-four service categories were identified in the study. These services are provided by the various facilities units and represent most of the services requested by schools. The intent was to identify the most critical services that represent the facilities department mission and function, and those services that require performance measurement and evaluation. The selection of the 24 categories arose from meetings with service managers and supervisors. Additionally, several principals were consulted and they agreed with the categories. The goal was to ensure that the resulting measurement system would be useful for both the facilities employees and school principals. Table 3 shows the categories with the corresponding identification number.

3.3 Customer service evaluation system

In order to analyze and measure the reliability of services provided by the Facilities Management Department to its customers, a survey form was devised using the process analysis and brainstorming tools of TQM. The survey served as a data collection instrument for customer satisfaction, received from the principals for services rendered.

The rating scale was devised from 0 (poor) to 10 (excellent) as shown in Table 4. Simplicity and ease of use were the bases for this scale that enabled the principals to provide customer feedback.

Table 4. Service rating scale.

Interval	Rating designation	Description
0–1	Poor	Failure to deliver service
2–3	Needs Improvement	Service is insufficient
4–5	Fair	Service is marginal
6–8	Satisfactory	Service is suitable and sufficient
9–10	Excellent	Service is outstanding

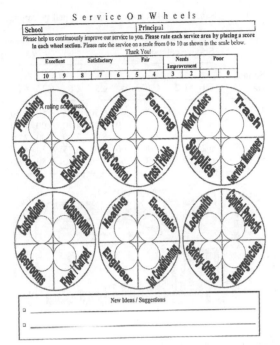

Figure 3. Service-on-wheels survey form.

Data was collected from 143 DC Public Schools over a period of one year. After the survey form was devised, data collection teams were organized to distribute the survey and collect them back. The data collection teams were made up of six customer service managers, each responsible for about 25 schools. All efforts were conducted under the supervision and direction of the Chief Facilities Officer.

The survey form, called "Service-On-Wheels" was designed with simplicity in mind. Six wheels each containing 4 service categories, were printed on a form, hence the namesake. The wheels were chosen to represent the concept of Shewhart's cycle, and also to represent service delivery to customers. The form intended to collect performance ratings, as well as recording customer ideas and suggestions as shown in Figure 3. The form presented the services provided by the facilities department in a simple manner to the customer and in a way to make him/her relate to his/her environment. The survey forms were explained to the customers in detail, and were collected back within two weeks. Survey results were organized using Microsoft Access® Software, and were then converted to Microsoft Excel® worksheets.

After the initial draft was completed in consultation with the supervisors of the various units responsible for providing services, it was tested by getting feedback from several principals and other supervisors, and was revised to improve clarity. The objective was to produce an evaluation instrument that is measurable, representative, objective and comprehensive. The instrument would serve the purposes of establishing a measuring tool for the quality of service provided to schools, and a process improvement tool for service quality. This was an attempt for building a foundation for a culture of continuous improvement in the organization.

3.4 Data collection

Initially, the plan was to collect one set of data for the whole year to measure the performance of facilities management service delivery, however the data made it clear that such a tool would be of greater benefit to supervisors on a quarterly basis. Thus, data collection encompassed 4 quarters, starting in June 2000 and running through September 2000, December 2000 and March 2001.

The data collected for a given school (S_i) for any given service category (C_j) were reduced to one data set by calculating the mean of the four quarters to produce one combined matrix containing a Mean Service Quality Rating, MR_{ij}, expressed as:

$$MR_{ij} = \frac{\sum_{k=1}^{4} R_{ij}}{N_m} \qquad (1)$$

where, R_{ij} =Mean rating at School i for Service Category j; and Nm =Number of sampling periods, which is 4.

3.5 Statistical analysis

The study employed a number of statistical tools to analyze the data including mean rating, Pareto chart, histograms, and trend charts.

3.5.1 Mean service quality rating

The mean MR_{ij} is a measure of service quality rating at a school for a service category. It covers the whole

Table 5. Sample result of mean service quality rating, MR_{ij}.

School ID	Service categories							
	C1	C2	C3	C4	C5	C6	C7	C8
1	7.75	6.00	7.50	7.67	6.50	5.75	6.50	6.50
2	6.25	5.50	5.67	5.50	5.00	4.50	6.00	6.00
3	5.75	6.00	2.00	2.00	5.00	7.00	5.00	2.00
4	6.25	6.00	6.00	6.00	6.00	6.75	4.50	4.50
5	6.75	9.00	6.50	7.67	6.50	9.00	5.50	9.00

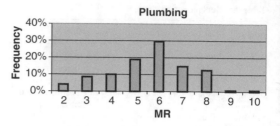

Figure 5. Plumbing (C1) histogram.

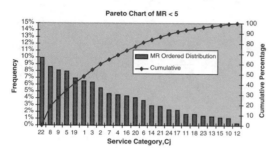

Figure 4. Pareto chart of MR_{ij}.

school district. The mean is simple to compute and easily understood by facilities employees. Table 5 shows a sample of the calculated mean service rating, MR_{ij}.

3.5.2 Pareto chart

Figure 4 shows the Pareto chart for MR_{ij} values less than 5. The cutoff point of 5 was chosen because it is the halfway point on the survey scale. The value of 5 represents the upper ceiling of the "Fair" rating interval and is usually considered by facilities employees and customers as the threshold under which focused attention to service quality is required.

The Pareto chart was used as a tool to classify and prioritize some 20,000-work orders in the deferred maintenance backlog and helped channel resources to the most needed areas. As a result, the deferred maintenance backlog was reduced to 3,500 work orders in about one year after the start of a focused maintenance management program.

It is interesting to note that the cumulative curve in Figure 4 does not follow the commonly observed 80–20 rule.

About half of the service categories accounted for 80% of the ratings less than 5. This is an indication of customer dissatisfaction across the board, and it calls on the facility manager to start an improvement program at a larger scale in the department. As expected, C22 (Capital Projects) is the category with the highest relative frequency of a rating less than 5. The plot also shows that 6 or 25% of the service categories

were responsible for about 50% of the low ratings (below 5). Those categories were C1 (Plumbing), C5 (Playground), C8 (Fencing), C9 (Work Orders), C19 (Air Conditioning) in addition to C22 (Capital Projects). The plot helps the facility manager prioritize the action plan to improve services; since those are in the order of importance. Consequently, it may be a good approach to take for improving those worst six service categories.

3.5.3 Histograms of service categories

Although studying the total picture of the school system service ratings is beneficial, it is also important to look at the ratings of the individual services. The histogram plot showing relative frequency is a suitable tool for this type of analysis. Figure 5 shows the histogram for C1 (Plumbing), which is one of the 24 services. Similar histograms were plotted for the others, but not presented in this paper in the interest of brevity.

The plumbing histogram indicates a mode of 6, and that about 25% of the ratings received are below 5. At the same time, it shows that less than 15% of the customers rated the service 8 or more.

3.5.4 Performance trend charts

Trend charts show the overall improvement across all services (across 4 quarters). There was variability for the improvement of the different series and a decline was noted in some quarters. A limited sampling of the 24 trend charts is given in Figure 6.

Continuous improvement or lack thereof, can be measured by monitoring performance. Trend charts are an excellent warning tool for progress or decline in a process. The charts can also show the rate or speed of improvement or decline. The trend charts in Figure 6 showed that some service categories improved faster than others. Naturally, we should not expect service improvement to occur at the same rate in all cases. Some service improvements require more time for better results and need more effort to mature.

Trend charts are an indication of the department's success in reaching its goals and performance targets. An example of a performance target for a particular shop is to increase customer service rating by 5% each quarter or by 10% each school year. Trend charts

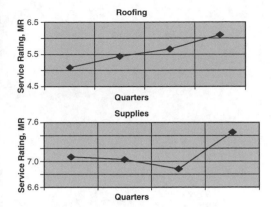

Figure 6. Sample trend charts.

represent the most powerful display and reporting method to employees. They can easily relate to progress and continuous improvement without using any statistical analysis jargon.

4 SUMMARY AND CONCLUSIONS

The primary objective of this research was to develop a customer service evaluation system of facilities services in urban public school districts. Total Quality Management techniques were applied to measure the performance of facilities management in the DC Public Schools District. The research followed a systematic approach in developing the evaluation system; first the survey instrument, then the evaluation system, and finally everything within a TQM framework.

Quantitative and qualitative techniques were employed to collect and analyze data obtained from a real survey. The application of this tool measured and monitored improvements in facilities management operations, increased efficiency, established accountability, and prepared a foundation for problem solving culture driven by continuous improvement. The survey showed that several improvement initiatives were introduced while applying this research and resulted in continuous performance improvement, as illustrated by the trend charts.

The following conclusions can be drawn from this study:

- A facilities management continuous improvement culture requires the establishment of a performance measurement tool or survey instrument, a customer service evaluation system and a department-wide TQM framework. The system can also be used as a facilities management decision-support tool.
- The use of Shewhart's cycle is a good TQM methodology in improving facilities management services, and requires careful planning and patient execution to show positive results.
- Trend charts showed continuous improvement in almost all service categories during the four quarterly periods of the study. All services showed improvements after one year. Trend charts showed a varying rate of improvement in service quality rating pointing to the fact that different services improve at different rates, and some services may need longer time for improvement to mature.
- The study recommended that the simple mean of all collected data be used to measure the overall service quality rating. On the other hand the simple concept of computing the mean value of the ratings to evaluate service quality can be effectively employed in measuring service performance at the service category or school level at any given time or for any given period.
- Statistical Analysis tools such as the mean, Pareto charts, histograms and trend charts are useful to facilities supervisors and managers in studying and monitoring the performance of their service shops and groups.

REFERENCES

ASCE. 2001. Schools. Fact Sheet, Issue Brief, Reston, VA.
Deming, W. Edwards. 1986. Out of the Crisis. Massachusetts Institute of Technology, Center for Advanced Educational Services, Cambridge, Massachusetts: 18–247.
District of Columbia Public Schools. 2000. Facility Master Plan, A New Generation of Schools.
Honeywell Corporation. 1995. Total Quality: User's Guide. 2nd Edition, Home and Building Control Division.
National Education Association. 2000. Modernizing Our Schools: What Will It Cost? Washington, DC: 15–18.
United States Department of Education. 1999a. How Old Are America's Public Schools? National Center for Education Statistics, Issue Brief, NCES 1999-048. Washington, DC.
United States Department of Education. 1999b. Back to School Special Report on Baby Boom Echo, No End in Sight. Washington, DC.

Warranty practices on DOT projects in the US

Q. Cui, M.E. Bayraktar & M. Hastak

School of Civil Engineering, Division of Construction Engineering and Management, Purdue University,
West Lafayette, Indiana, USA

I. Minkarah

Department of Civil & Environmental Engineering, University of Cincinnati, Cincinnati, Ohio, USA

ABSTRACT: The application of warranty provisions to state highway construction contracts in the US has expanded rapidly over the past ten years. The decision to replace conventional contracting by this innovative contracting method requires careful analysis of its impact on tangible and intangible components of a project such as cost, quality, construction duration, bonding, and disputes and litigation. However, because of limited industry experience, the impact of warranty provisions on highway projects is not clear. Therefore, there was a need for a comprehensive study to evaluate warranty provisions on highway construction projects in the US. This paper reports on a part of such a comprehensive survey, and provides general information about warranty use and its impact on two important project components, cost and quality. The analysis is based on responses from the surveyed agencies in the US including state Departments of Transportation (DOTs), contractors, and bonding companies.

1 INTRODUCTION

Since warranty contracting was first applied to state highway pavement projects in 1987 by the state of North Carolina, more and more state Departments of Transportation (DOTs) have considered its use to protect the initial state investment. Over time, however, the limited industry experience on warranted highway construction projects raised questions on the suitability of warranty contracting for the US highway construction industry. The most important concerns were related to cost, quality, construction duration, bonding, and disputes and litigation. Therefore, there was a need for a comprehensive study to evaluate the warranty provisions on highway construction projects in the US. This paper reports on such a survey and provides valuable information about warranty contracting in the US. The survey was conducted as part of a research project aimed at gathering information on the state-of practice and the important issues of warranty provisions. The surveyed agencies included state DOTs, contractors, and bonding companies. However, because of space limitations the results presented in this paper are limited to a general overview of warranty use in the US and the comparison of warranty projects to non-warranty projects with respect to project cost and quality.

2 CONSTRUCTION WARRANTIES

Construction warranties offer protection to owners with respect to the quality of a contractor's product in conformance with the contract requirements. It offers a legal avenue to the owner against parties, e.g., manufacturers who are indirectly related to the project through extended product warranties.

2.1 *Types of construction warranties*

There are two types of warranties used in the construction industry: (i) express and (ii) implied.

The most common type of warranty encountered in construction contracts is the express warranty. In an express warranty, the terms and conditions of warranty are clearly stated in writing. For construction contracts, these terms and conditions reflect statements about the product and services offered by the contractor and include commitment on the part of the contractor to remedy any defects or failures encountered by the owner (Blischke & Murthy 1996). The written promise that the project would be constructed according to the plans and specifications with the desired quality of workmanship and materials is an example of the express warranty for construction contracts.

Implied warranties, initiated by state laws, are unwritten promises made by the seller to the customer about the products or services sold. Implied warranties are effective as soon as an item is purchased, implying that the item is suitable for the intended purpose under the common law principle of fair value (Blischke & Murthy 1996).

2.2 Extended highway warranties

Additional warranty coverage over and above the express warranty, also known as extended warranty can be obtained for additional payment (Hancher 1994 & Russell 1999). One of the latest applications of extended warranties in the highway construction industry is warranty contracting. This innovative contracting method has two main features that differentiate warranty jobs from regular contract jobs: (i) the contractor is held responsible for any maintenance work that may occur over the warranty period and (ii) the contractor has the freedom to use the materials and techniques that he or she considers best for the job so long as the state standards are met.

3 WARRANTY CONTRACTING

3.1 The history of warranty contracting in the US

The Transportation Research Board (TRB) Task Force A2T51 was created in 1988 to study innovative contracting practices in the US and abroad. The methods studied under this project included among others: warranties, design-build, lane rental and cost-plus-time bidding. Upon the request of the Task Force for assistance, the Federal Highway Administration (FHWA) established Special Experimental Project No.14 (SEP-14-Innovative Contracting) to evaluate "project specific" innovative contracting practices undertaken by state highway agencies, effective February 13, 1990. Under the SEP-14 program eleven states, namely Arizona, California, Indiana, Michigan, Missouri, Montana, North Carolina, New Hampshire, Ohio, Washington, and Wisconsin were assigned to use warranty contracting on federal highway construction projects (Hancher 1994).

Warranty use on federal highway projects, however, was prohibited until the 1991 passage of the Intermodal Surface Transportation Efficiency act because it could indirectly result in federal-aid participation in maintenance costs, which, at the time, was a federal-aid non-participating item. Finally, on April 19, 1996 a warranty Interim Final Rule (IFR) was published to revise 23 CFR Part 635 to allow warranty clauses on federal highway projects to provide small contractors with the opportunity to enter the market with innovative practices (FHWA Briefing

2000). Currently, Michigan, Ohio, Indiana and Wisconsin are considered the frontrunners in warranty use in the US.

3.2 Description of the problem

Similar to the warranty of a manufactured product, a highway warranty is a guarantee which makes the contractor responsible for the repair and replacement of any pre-defined deficiency for a given period of time after project acceptance. Basically, the contractor is bound by warranty terms in the contract and forced to come back to maintain the highway whenever certain threshold values are met. Hence, warranty contracting simply shifts the risk of any early failure after project acceptance from owner to the contractor. In return for the shift in responsibility, the contractor is given the opportunity to select the construction materials and methods, within certain limits. However, the high degree of uncertainty included in the risk shift has raised several concerns in the industry. The effects of warranty provisions on project cost, quality, construction duration, bonding, and disputes and litigation are the most profound issues of concern regarding the warranty use in the US. The following paragraphs will give brief information about cost and quality issues associated with warranties.

The first major concern about warranty use in highway construction is the unknown cost impact, especially from a life cycle cost perspective. Contractors have generally increased bid prices to cover mandatory warranty bonds and future warranty scope costs.

Research efforts to date indicate that the impact of cost on a project is a function of type of project as well as the warranty period. Normally, warranty durations less than one year do not cost money to the state DOT. A 3% increase in the bid price was experienced on two hot bituminous pavement projects with 3-year warranties in Colorado (Aschenbrener & DeDios 2001). Similarly, Missouri DOT experienced no excessive cost overruns as a result of 3-year warranties on pavement preservation jobs of the agency (Webb 1994). Also, Michigan DOT observed no measurable impact of warranty provisions on cost of pavement, bridge deck and painting projects with short term warranties. However, Scheel (1996) found significant bid price increases on concrete pavement projects in California. As compared to the cost of similar types of projects, the increases were estimated to be 36% and 23% on concrete and rubberized concrete pavement projects with 3-year warranties respectively. With 5-year warranties, the increases were approximately 62% and 25% respectively. The Ohio Department of Transportation disclosed that bid prices on asphalt pavement projects with 5-year warranties increased by an average of 9% (ODOT 2000).

It is anticipated that warranties might reduce life cycle costs by spreading the initial investment over the entire warranty period (Russell 1999). However, assessing the life-cycle cost of warranty projects is very difficult because of insufficient data to study the long-term performance trends. Krebs et al (2001) compared the life cycle costs of 24 asphalt pavement projects with expired warranties to the life cycle costs of standard contracts. The comparison excluded the maintenance cost beyond the warranty period. Also, conflict resolution, distress survey and traffic count costs were found to be negligible. The study revealed that the standard contracts averaged $27.72/ton versus $24.34/ton for warranty contracts for the period from 1995 to 1999, excluding the state delivery costs.

The major anticipated advantage of warranty use is quality improvement. As a guarantee against the risk of early failure resulting from bad material and workmanship, contractors will provide better performance on warranted projects than on non-warranty projects to avoid potential repairs and maintenance. Contractors are expected to use efficient measures such as providing additional project supervision and using only qualified, experienced labor crews on warranted projects, in an effort to increase work-in-place quality. Also, given the opportunity to select materials and mix designs on warranty projects enables the contractors to use innovative processes and materials. However, contractors may implement a different maintenance strategy. They could hastily complete their contract work without regard to quality assuming that the maintenance and repair is seemingly imminent (Hancher 1994).

With respect to the quality, Indiana DOT was satisfied with the high performance of I 69 project with a 5-year workmanship and material warranty (FHWA Briefing 2000). Aschenbrener & DeDios (2001) found the performance of warranted projects and the traditional DOT controlled projects in Colorado to be very similar. Krebs et al. (2001) reported a measurable improvement in the quality and performance of pavements in Wisconsin after comparing 23 warranted asphalt concrete over flexible base pavements to historical pavement performance data for surface distress and ride quality. However, warranty provisions did not eliminate early failures. The state of North Carolina quit using warranty specifications because of an early failure on an epoxy pavement marking project. In Montana, similarly, significant failures occurred on a 4-year warranted pavement marking project.

4 SURVEY ON WARRANTY CONTRACTING IN THE US

4.1 Questionnaire survey

To determine the current state-of-practice of warranty usage in the US, three different types of questionnaires

were prepared for state DOTs, contractors and bonding companies respectively. The appropriate questionnaires were sent to 158 agencies in the U.S. A total of 63 responses were received including 40 state DOTs, 16 contractors and 7 bonding companies, giving a response rate of approximately 40%. Of the 63 responses, only 35 were tabulated, because 18 state DOTs had never used warranty provisions and 9 state DOTs had limited experience to participate in the survey. Similarly, a bonding company responded with a general information letter instead of filling out the questionnaire. Therefore, the questionnaire survey results are based on 13 state DOT, 16 contractor and 6 bonding company responses, unless stated otherwise (some respondents did not answer all the questions).

The questionnaires were divided into seven major sections: general information, cost issues, quality issues, construction duration issues, bonding issues, contract issues and unclassified issues. All responses were tabulated and analyzed.

The following sections of this paper summarize the results obtained from the general information, cost issues and quality issues sections. There were several questions to which respondents could provide as many answers as were applicable, for example contractor innovations used on warranty projects. Therefore, responses to these questions may not add up to the total number of the respondents or 100%, when reported in percentages. Also, it should be noted that the results of the survey only represent the responses received, and should not be construed as applying to the whole country.

4.2 Interview survey

As a follow-up to the questionnaire survey, an interview survey was conducted. The interviewed agencies included DOTs, contractors and bonding companies. A total of 20 interviews were conducted including 9 state DOTs, 6 contractors and 5 bonding companies respectively.

Because of space limitations, however, the opinions and viewpoints of the interviewees are not included in the subsequent sections and will be discussed in detail in a further publication.

5 FINDINGS OF THE QUESTIONNAIRE SURVEY

5.1 General information

The objective of this section was to gather information on the general use of warranty provisions by state DOTs.

5.1.1 State DOTs and warranty use

For a majority of the respondents (69%), the annual sale of warranty projects with respect to the percentage of total dollar project value is under 5%, whereas about 23% of the respondents indicated 10–20%. For very few (8%) the annual sale of warranty projects is over 30%.

A slight change was noticed when the annual sale of warranty projects was compared to the total number of the agency projects. A larger number (76%) of the respondents indicated that their annual sales were under 5%, whereas the remaining 24% was equally spread among 5–10%, 10–20%, and over 30%.

5.1.2 Warranty period requirements for different type of projects

The survey showed that the average warranty period required by state DOTs is 5 years for asphalt pavements, 7 years for concrete pavements and 2 years for most of the preventive maintenance applications. Figure 1 is a detailed summary of the results with the number of the respondents for each project shown in parenthesis.

5.1.3 The number of bidders on warranty projects

For a majority (84%) of 12 responding state DOTs, the average number of bidders on warranty and non-warranty projects did not change and was in the range of 3–7. Only one state DOT experienced a significant decrease in the number of bidders from more than 30 on non-warranty to a few on warranty projects, because local contractors were afraid of the related liabilities of warranty projects.

5.2 Cost issues

The objective of this section was to gather information on cost issues of warranty highway construction projects.

5.2.1 The average increase in the bid prices due to warranty provisions

Four (4) out of 10 responding state DOTs indicated 5–10% increase in bid prices due to warranty

provisions, whereas no respondent indicated over 50% increase. A detailed summary of the information obtained is given in Figure 2.

A majority (65%) of contractors indicated that the average increase in bid prices due to warranty provisions is 5–15%. About 20% of contractors indicated 0–5% increase, while the remaining 15% reported 15–20% increase in the bid prices.

According to the contractors, the future maintenance cost and the bond cost are the main sources of the increase in the bid prices. It was interesting to note that the contractors believed that despite the risk of fixing future warranty issues, the tight market tends to keep prices down, lower than what they would be in a less competitive market.

The questionnaire survey also revealed the competitive concerns of small contractors. According to the small contractors, large contractors have a competitive advantage over them, since they can spread out their cost of warranty among other jobs and reduce the bid price while carrying less risk than the small contractors. They also mentioned that the competition may be decreased due to small contractors' inability to obtain the required warranty bond.

5.2.2 Expected saving in the maintenance cost

Eight (8) out of 9 responding state DOTs indicated that the expected savings in maintenance costs as compared to non-warranty projects is less than 10%. Only one state DOT expected more than 50% saving in the maintenance cost of the agency's pavement marking projects.

The survey also revealed that longer warranty periods and tighter specifications are the major possible methods for state DOTs to reduce future maintenance costs.

5.2.3 Expected saving in the project life cycle cost

Very few (8%) of state DOTs indicated that they expect a substantial saving in the project life cycle cost of warranty projects as compared to non-warranty projects, whereas about 46% indicated little increase.

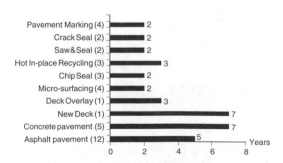

Figure 1.   Average warranty periods required by state DOTs.

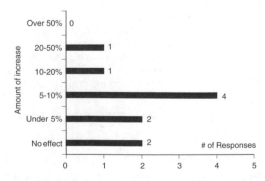

Figure 2.   Average increase in the bid prices after warranties.

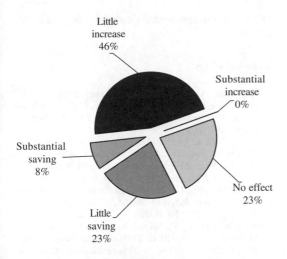

Figure 3.   Expected variation in the project life cycle cost.

A detailed summary of the information obtained is given in Figure 3.

5.3   Quality issues

The objective of this section was to gather information on quality issues of warranty highway construction projects.

5.3.1   The impact of warranty provisions on project quality

About 46% of state DOTs indicated that warranty provisions slightly improved the project quality, whereas 23% indicated great improvement. However, 31% of the respondents reported that the impact of warranty provisions on project quality is not yet clear.

About 62% of contractors indicated that warranty provisions encourage them to maintain higher quality. However, the remaining 38% of the respondents indicated that they have not observed any significant change in quality due to warranty requirements.

The survey also revealed that the contractors take a more conservative approach in case of warranty projects. Most of the contractors do not innovate on warranty projects, but are more quality conscious for fear of carrying additional risk. Figure 4 shows the most preferred innovations by contractors on warranty projects.

5.3.2   The impact of warranty provisions on site inspection

A majority of state DOTs (77%) indicated that warranty provisions have reduced the need for site inspection, whereas about 15% indicated no change, and the remaining 8% indicated more inspection.

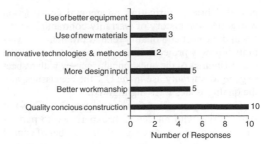

Figure 4.   Contractor innovations on warranty projects.

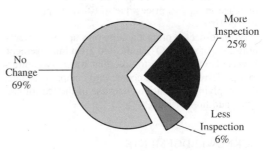

Figure 5.   Site inspection for contractors after warranties.

On the other hand, a majority of contractors (69%) reported no change in site inspection, while 25% indicated that warranty provisions increased the need for site inspection, as shown in Figure 5.

5.3.3   The impact of warranty provisions on record keeping

About 54% of state DOTs reported that warranty provisions resulted in less record keeping, whereas 31% indicated no change, and only 15% indicated more record keeping.

However, it was interesting to note that 94% of contractors indicated that warranty provisions have increased the need for record keeping, whereas only 6% indicated less record keeping. The contractors stated that their documentation and record retention is more important in case of a future dispute over the warranty.

6   CONCLUSIONS

A comprehensive survey was conducted as part of a research project to assess the state-of-practice and the important issues associated with warranty provisions on highway construction projects in the US. The surveyed agencies included state DOTs, contractors, and bonding companies. The first section of this paper provided brief information on general construction warranties, the history of warranty contracting in the US and the cost and quality related problems on warranty highway projects. In the subsequent sections the major findings of the questionnaire survey with respect to general warranty use in the US, the cost issues, and the quality issues were highlighted.

The survey revealed that for the majority (76%) of state DOTs, who had enough experience to participate in the questionnaire survey, the number of annual warranty projects was less than 5% of the number of total annual projects of the agency.

Another important result of the survey was that for 8 out of 10 responding state DOTs, the average increase in the bid prices due to warranty provisions was 0–10%.

With respect to project quality on warranty projects, only 23% of state DOTs indicated great improvement. It was also interesting to note that warranty contractors preferred to take a conservative approach rather than applying innovations because of the fear of carrying additional risks.

ACKNOWLEDGEMENTS

This study is supported by the Ohio Department of Transportation (ODOT). The opinions and findings expressed here, however, are those of the authors alone and not necessarily the views or positions of ODOT. ODOT has not approved the material presented in this paper.

REFERENCES

Aschenbrener, T. & DeDios, R. December, 2001. *Material and Workmanship Warranties for Hot Bituminous Pavement, Report No. CDOTS-DTD-2001-18.* Denver: Colorado DOT.

Blischke, W.R. & Murthy, D.N.P. 1996. *Product Warranty Handbook.* New York: Marcel Dekker, New York.

FHWA (The Federal Highway Administration) Briefing 2000. *Warranty Clauses in Federal-Aid Highway Contracts.*

Hancher, D.E. 1994. Use of Warranties in Road Construction. *Transportation Research Board NCHRP Synthesis 195.* Washington, D.C: NCHRP.

Krebs, S.W., Duckert, B., Schwandt, S., Volker, J., Brokaw, T., Shemwell, W. & Waelti, G. 2001. *Asphaltic Pavement Warranties, Five-Year Progress Report.* Madison: Wisconsin Department of Transportation.

ODOT (Ohio Department of Transportation) 2000. Implementation of Warranted Items on Construction Projects. *Ohio Department of Transportation Report.* Columbus: ODOT.

Russell, J.S., Hanna, A.S., Anderson, S.D., Wiseley, P.W., Smith, R.J. 1999. The Warranty Alternative. *Civil Engineering Magazine.* May, 1999: 60–63.

Scheel, D. 1996. *Test & Evaluation Project #014 (Warranty).* Sacramento: California Department of Transportation.

Webb, M. 1994. *Experimental Project No. M091-03 Final Report.* Jefferson City: Missouri Highway and Transportation Department.

Why are some designs more interesting than others? The quest for the unusual, the unconventional and the unique in design

L. Kiroff
School of Construction, Faculty of Architecture and Design, UNITEC Institute of Technology
Auckland, New Zealand

ABSTRACT: We live in a changing world. Buildings change with time. So do the notions of beauty because society, individuals, speed, duty, morals, and culture change. New and unconventional buildings as new ideas always make people ask questions: "What is new? What caused the change? New technology? New lifestyle requiring a new functional response by the building? New way to express the spirit of the times or simply a better and more creative way of solving old problems?" There are literally thousands of forces that shape buildings. When new ideas, resulting in a new form, appear upon the scene, the public is usually shocked. Buildings are like people – some are very physical, some are overemotional and some are deeply intellectual. Buildings that fascinate and catch the public's imagination and that people find worthy of reflection are usually the ones that are less predictable and more dynamic. Such buildings surprise you and at the same time stimulate your imagination. What is the feature then that makes them so unique and quite different? It should be an idea, readable by everyone or almost everyone. The search for an answer of the question "why are some designs more interesting than others?" represents the author's own interest in designs that fascinate and inspire. What is the place, if any, of "Sense and Sensibility" in design? It may be argued that an architectural design based on a sound sense will yield a reasonable but trivial result while a design based on sensibility will be fascinating and inspiring. Then why do some architects use a sound sense while for others sensibility is the guiding notion? Is it because for some this is simply an everyday routine and for others this is a creed and a way of life? The focus of this research paper will be on iconic buildings as they embody in a more explicit way the architect's quest for the unconventional, the poetic and the original in design. Identifying diverse sources of inspiration in the fuzzy front end of the design process deviating from common and standard practices (the actual site with its idiosyncrasies or similar existing buildings) proves to be the creative approach to design mastered by the great minds in architecture. The case study research method as described by Burns (1994) has been applied to this study. Three case studies representative of iconic buildings have been used in this discourse as they exemplify a unique approach to design and provide valuable insights – Renzo Piano's Tjibaou Cultural Centre in Noumea, New Caledonia, Toyo Ito's Mediatheque in Sendai, Japan and Jasmax's Te Papa, Museum of New Zealand in Wellington. Architecture inspired by nature and tradition that searches for poetry and lyricism in buildings is what unites the work of Piano, Ito and Jasmax. Tjibaou Cultural Centre – the "building that sings" (Architectural Record 2001), inspired by the traditional Kanak hut, the "lyrical grace of the Sendai Mediatheque" (Bognar 1997), inspired by the image of a floating seaweed and the Museum of New Zealand, Te Papa – "a statement of national identity" (Hunt 1998) illustrate the diverse and unconventional sources of inspiration that have informed these designs. Two techniques pertinent to the "case study" research method have been used – document analysis (articles about these three buildings) and non-participant observation (the author's own site visits).

1 INTRODUCTION

Modern life is mediated through the visual screen. Film and television and the Internet are not just the norm, they are life itself. Similarly, "architecture is so omnipresent in our lives that buildings have become our second nature – so much so that they affect us unconsciously" (Forster 1999). The new emerging globally shared visual culture becomes the underlying construct that explains and substantiates visual experience in everyday life. According to Walker & Chaplin (1997) the field of visual culture has four domains (fine arts, crafts/design, mass & electronic media and performing arts) and architecture belongs to the fine arts domain. The ability "to gauge the abundance of meanings that emanate from every building" (Forster 1999) and to decipher the message that it sends across needs a definition in a broader framework within the realm of visual culture. Barnard (2001) emphasizes the importance of studying visual culture, as one is more dependent on and subject to visual material. Generating "a more sophisticated, self-reflective and critical understanding of the visual world and one's place in it" through forming opinions and responses to visual culture encapsulates Barnard's (2001) viewpoint. In the context of architecture Forster (1999) asserts: "Any claims we make for buildings that fascinate us and that we find worthy of reflection prove hard to substantiate when our audience has little or no knowledge of the subject or does not incline to our point of view".

Each historical period has its best buildings with distinguishing qualities. Such buildings surprise the public and at the same time stimulate the imagination triggering further questions about the buildings' uniqueness. The most obvious question then is: "Why are some designs more interesting than others?" The author's personal interests in buildings around the world that are extraordinary and subsequent questions what makes them so unique have become the driving force behind this research. The focus of this discourse will be on iconic buildings as they embody in a more explicit way the architects' quest for the unconventional, the poetic and the original in design. The iconic buildings included in this study have been chosen in peer review as good examples of architecture – Renzo Piano's Tjibaou Cultural Centre in Noumea, New Caledonia, Toyo Ito's Mediatheque in Sendai, Japan and Jasmax's Te Papa, Museum of New Zealand in Wellington. These case studies become an integral part of a comparative analysis with everyday commercially driven architecture illustrating the flexibility with which the great masters of architecture explore unusual sources of inspiration to create unconventional designs – traditional huts with thatched roofs, seaweeds, paintings and sculptures. In contrast everyday commercially driven architecture remains surprisingly remote from such poetic notions.

The research methods adopted for the purposes of this study are: ethnography, case studies, and semi-structured interviewing involving ten architectural firms in New Zealand (five large, three medium and two small).

2 FROM THE STATUS QUO TO WHERE WE WANT TO BE

This section is structured around three themes starting with some of the great masters in architecture exploring their views on creativity in architectural design through the masterpieces they have created that embody their vision (analysis of the three case studies mentioned above) to the reality surrounding us where such ideas look remote and unpractical but definitely as a place where we want to be (analysis of the industry research in New Zealand).

2.1 *From the great Minds: Piano, Ito and Jasmax – inspirations, ideas, views and approaches*

"Architecture is about illusion and symbolism, semantics, and the art of telling stories" (Piano 2001). "Communicating architectural concepts is difficult because architecture inevitably has a dual character. It is both an abstract model of ideas and something that actually exists" (Ito 2001). "Architecture has the power to inspire the spirit" (Jasmax 2002).

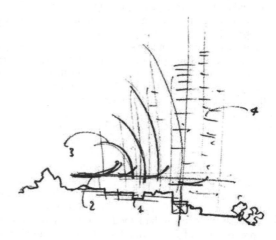

Figure 1. Piano's traditional Kanak hut (Piano's original sketch).

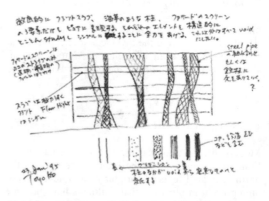

Figure 2. Ito's seaweeds (Ito's original sketch).

1614

Architecture inspired by nature and tradition that searches for poetry and lyricism in buildings is what unites the work of Piano, Ito and Jasmax.

For Renzo Piano "expression in architecture" is paramount. As a response to his desire to fight gravity, Piano makes the statement "Magic is essential in architecture" (Piano 2001). When he explains his approach to design Piano elaborates on his own evolution as an architect. "Developing objects" or "the piece-by-piece approach was essential to me". Later architecture was perceived as "more than putting things together. It's about the organic, about illusions, a sense of memory, and a textural approach" (Piano 2001). The architect states that the way he starts a new job is by visiting the site first and getting a basic understanding of the context. Architecture according to him not always needs to be integrated in this context, as sometimes it should contribute to that context. The way Renzo Piano carries out his site visits is fundamentally different from the conventional way that would put the emphasis on examining the site factors like sun traverse, north orientation, prevailing winds, noise, views, adjacent buildings and so forth. As the architect puts it in a poetic way he would "try to get a basic, fundamental emotion. Because that's what it's all about – building emotion" (Piano 2001). Tjibaou Cultural Centre – the "building that sings" (Piano 2001) in Noumea, New Caledonia with its poetic and at the same time dramatic forms is a concrete example of expressing that powerful emotion. Piano's "two-hour site visit" (Falkoner 2001) resulted in a concept inspired by tradition, a true celebration of the Melanesian culture of the Kanaks.

In contrast a competitor firm, one of the ten short-listed finalists "spent several weeks in Noumea, visiting the site each day in different conditions, clambering all over it, visiting other villages and interviewing many people" (Falkoner 2001). Unfortunately their strategy focused too much on the detail, failing to understand the whole or the emotion of the site, as Renzo Piano would approach it. Being self-critical, the team summarizes its experience in retrospect concluding that this ungrounded obsession with detail proved to be a wrong strategy.

The author's own site visit to Noumea in 2002 and 2003 shed more light on the issue. The models of five of the finalists were on display in one of the ten "cases" of the complex. The concepts were amazingly disparate. The author's perception was that Piano's design stood out as the most expressive of tradition and local culture. The author's personal opinion was of all the other four designs as universal schemes lacking specificity that would suit almost any other location around the world.

Inspired by new social and urban developments, and facilitated by new materials, structures, and technologies, Toyo Ito's unique approach to design promotes architecture with an almost immaterial lightness and transparency being metaphors of a new understanding of a world in flux. Ito's sleek, abstract and ephemeral concept-buildings are characterized by flowing spaces with no definition, and the use of light materials like aluminium and glass. Inspired by the ephemerality of the information society, and especially by its manifestations in the Japanese city, Toyo Ito progressively "dematerialized" his architectural designs. "The ephemeral, taken in a positive sense does not necessarily mean that the architecture is short-lived, but that new meanings are perpetually emerging" (Bognar 1997).

Ito's architecture of the ephemeral is replete with references to nature and environment. For Toyo Ito "all architecture is an extension of nature" (Ito 2001). The image of floating seaweed becomes his source of inspiration in the Sendai Mediatheque's design. Ito gives the following clarification about the use of the word "floating". "I often use the word floating not only to describe a lightness I want to achieve in architecture, but also to express a belief that our lives are losing touch with reality" (Bognar 1997). Deeply inspired by nature, he summarizes the main concept of the Mediatheque as a "technology forest (the mesh like columns) and the displays on the seventh floor as a "technology garden" (Ito 2001).

At a public lecture in Auckland, New Zealand in 2001 attended by the author, Ito explained passionately the philosophy of his approach to design by making the statement that "designs should resemble ripples in a water surface rather than resemble static rods in water". The metaphorical way of thinking and expression are present in all his work.

Jasmax is a leading New Zealand architectural and interior design practice with expertise spanning more than 35 years. The fact that the company possesses a diverse range of skills and specialist knowledge of many different building types and building sectors has contributed to making a reputation that places the firm in a high profile position on the New Zealand market as well as internationally. The firm's portfolio encompasses a wide range of building types like office buildings, recreational, hospitality and leisure developments, cultural and civic buildings, educational, retail and industrial facilities. Establishing a good working relationship with its clients is seen by the firm as an important premise to understand their culture and the particularities of each project. The company "believes that architecture has the power to inspire the spirit while meeting the practical requirements of human endeavour". The strong aspiration for beauty, innovation and professionalism pervades all aspects of the firm's creative approach to design.

Jasmax's design has distinctive qualities showing real appreciation and understanding of the bicultural nature of New Zealand making such issues an integral part of the building design rather than a simple addition as in Te Papa – the Museum of New Zealand in Wellington. Such innovative approach does not mean just displaying cultural artifacts in a museum setting; it is more about creating a whole new design language. This particular museum project designed by Jasmax is considered in many professional magazine articles as a huge success and as a building attempting "to express the total culture of this country" (Bossley 1998). Jasmax's innovative design approach has resulted in a winning scheme and a building, which tells in a poetic way the life story of one whole nation.

2.2 To the great buildings: Tjibaou Cultural Centre, Sendai Mediatheque, Te Papa

The analysis of the three iconic buildings presents the different sources of inspiration that have influenced the design – traditional huts, nature and the search for bicultural identity. Other aspects of the design remain beyond the scope of this study.

The "Building that sings"

"Real architecture, real painting, real poetry, real music is never detached from physicality. In architecture that's it" (Piano 2001). Piano proves what he believes in with the unconventional design of the

Figure 3. Piano's sketch of traditional Kanak huts.

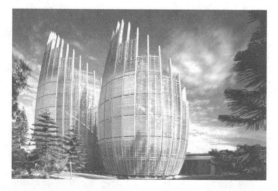

Figure 4. Inspired by tradition, formed by modern technology, this Centre celebrates and explains the Melanesian culture of the Kanaks.

Tjibaou Cultural Centre in Noumea, New Caledonia. It is an unambiguous statement, with its bold, modern style, both elegant and restrained, drawing inspiration from traditional Kanak architecture. The Centre is a unique embodiment of an age-old identity.

Learning from the Kanak culture, the vernacular architecture and the relatively small site with an iconic requirement was the starting point in the design process. "From the first, Piano was concerned to learn from local culture, buildings and nature, but he was determined not to end up with a kitsch replication of Kanak huts" (Mcinstry 1998). Piano himself says: "It was not feasible to offer a standard product of Western architecture, with a layer of camouflage over the top: It would have looked like an armoured car covered with palm fronds" (Findley 1998). Piano's innovative approach to the design has resulted in a unique complex inspired by tradition but at the same time consciously avoiding the accurate replication of the Kanak huts. The idea of the village cluster and the ribbed hut structure are elements of the vernacular architecture that were taken and interpreted by Piano. The architect's explanation of this design attitude is that he works "like a thief, reinterpreting what exists" realizing that the complex should be "reminiscent of a Kanak house not a copy" (Falkoner 2001).

Designing "a building that sings" can be considered as quite an eccentric and unusual approach to design but it has its cultural inspirations. Renzo Piano shares his experience: "In New Caledonia we learned from the local culture that buildings sing" (Piano 2001). That was successfully achieved in the design – the building actually makes a sound when the wind blows giving the impression that "the building sings". The abstracted tall huts catch the high winds and modify their effects through a system of horizontal slats carried by the curved ribs.

The "swelling wooden shapes, simultaneously alien and indigenous" as Findley (1998) calls them or "the gigantic buds, almost botanical in form" as Mcinstry (1998) describes them are probably the most prominent feature of the architectural design skillfully complemented by other design features – the main entry and the landscape design – both original and symbolically resonant with local tradition. The author's site visits in 2002 and 2003 resulted in a serendipitous find of a point of reference to the unusual "bud" shape. It was the equally unusual base of native palm trees growing in close proximity. That was the author's personal perception, the reason being Piano's fascination with traditional architecture and nature.

Falkoner (2001) analyzing Piano's design approach summarizes his view: "Design for him is a spiritual process, not a linear process but a circular reworking of an idea starting with tradition and a keen sense of place as points of departure". Findley (1998) qualifies the Centre as "a subtle and generous work, a model of diplomacy, and a compelling contribution to New Caledonia's native Kanak people.

The lyrical grace of the Mediatheque

Toyo Ito's Mediatheque is innovative at every level, fusing the physical and virtual worlds. The integration of diverse functions under one roof and the innovative steel-tube structural system have forged new approaches to spatial definition, building services, circulation and construction. "Inspired by new social and urban developments, and facilitated by new materials, structures, and technologies, the evolving new design paradigm embraced by Ito can evoke and construct architecture with an almost immaterial lightness and transparency" (Bognar 1997). The fascination that Ito has with layering semi-transparent skins and then drawing analogies between them and the natural phenomena of clouds or forests remains a genuine wish to be associated with the basic observable elements of nature. The "technology forest" (the mesh like columns) and the displays on the seventh floor – "the technology garden" embody and illustrate Ito's fascination with nature (Ito 2001).

The image of "the trees made of metal mesh", "the tubes" (Ito 2001) or the actual columns of the building was the dominant one in the initial stages of the design. The thirteen tubes, which are bent and twisted are all different in configuration and size and are as big in places as nine meters in diameter. The "trees made of metal mesh" were imagined as columns seemingly without mass, that is, as treelike structures without materiality.

Another unusual approach to the design is the perception of the entire building from outside when it is illuminated after the sun has set. The differences between the floors become more distinct not only because of their different height, ceiling finish and furniture but also because of the different method and colour of illumination. The unexpected effect is of an assemblage of floors as if from different buildings. At night the floors are more conspicuous than the tubes, and this results in a greater emphasis on the horizontality. This difference in perceptions during the day and at night reveals the creative application of lighting in design. It has the powerful ability to completely transform the entire building proving that ideas can evolve and are prone to metamorphosis.

Te Papa – "concepts in culture"

The design brief for Te Papa had specific requirements for the building to be a "bicultural" institution, a building that will "powerfully express the total culture of New Zealand" and will represent the "bicultural nature of the country, recognizing the significance of the two mainstreams of tradition and cultural heritage" (Bossley 1998). Jasmax's design was quite unique in that respect as it did not seek merely cultural presence through the display of indigenous artifacts. Overstating traditional forms incorporated in the design had not been seen as a possible alternative either. "At Jasmax, we were determined to ensure that any definition of the bicultural nature of the country would be integral to the building design, rather than a seemingly carved addition. We also intended that such reference would be within the larger language and idiom of the building, rather than a potentially patronizing enlargement of traditional-looking forms" (Bossley 1998).

The inclusion of a "living" marae in the design was also unique and a true expression of a nation's soul within the bicultural definition of the building. The Museum is a testimony to the validity of Maori and other Pacific cultures. The impression that the marae makes on the visitor is extremely powerful. Including references to both Maori and Pakeha had been considered by the design team as the first step that needed to be taken further by placing these references in an appropriate context where recognition, acceptance and celebration of cultural differences make biculturalism realistic and valued.

Looking at design as an opportunity and concurrently as a challenge for cultural issues to be aired is conducive to making the design process more creative and innovative. Such challenges spur the imagination to explore new exciting design philosophies that lack the ubiquity of conventional design, to test possibilities and debate controversial concepts. The concrete physical context and the site with its potential possibilities and limitations, overlaid by the cultural theme can be considered as a premise for uniqueness in design.

Figure 5. The "technology forest" and "the technology garden".

Figure 6. The marae in Te Papa.

2.3 And back to earth: analysis of the industry research

The purpose of the industry research involving ten architectural firms in Auckland was to establish everyday practices and analyze guiding notions, idea generators and metaphorical thinking in design. The next step was a comparison with the three case studies previously discussed and the way the great masters in architecture approach design. Ten architectural companies – five large (from 30 people upwards), three medium (10–15 people) and two small (3–5 people) were interviewed for the purposes of this research. The size of the practices was just one criterion for selection with the other one being the portfolio of the firm. The author's preference was to interview companies involved in commercial type of projects. A good cross section of people has been included – interviewees were Project Directors, Design Architects and Assistant Architects. The semi-structured interviews aided by a specifically designed questionnaire were complemented by observation and document analysis of projects presented at the interviews for discussion. The main purpose was to explore only certain aspects of the design process or the phase when sources of inspiration are explored and ideas generated. The research results have been analyzed by the author in that respect forming a representative study of a small segment of the market in New Zealand at a specific time. Visions, approaches to design, importance of creativity and innovation to the quality of design vary significantly from one firm to another reflecting a diverse range of values, interests and preferences.

The research results helped in forming an aggregate view. At the fuzzy front end of the design process when ideas are generated, for most architects looking at magazines with similar buildings or the work of others is a common practice, the reason for this being forming an opinion that will later inform the concrete design through the "crystallization of the idea on a subconscious level", as one of the interviewees put it. Pertinent images evoking feelings, emotions and associations and using references to nature, tradition, paintings and sculpture with connotations of remoteness rather than immediacy were not something experienced and experimented. Seeing seaweeds, traditional huts, and various forms of art as diverse sources of inspiration should not be perceived as a risky field. If we continue to strive to align our desires with the client's expectations we put ourselves at risk to simply add another conventional and uninspiring building to the heritage that already exists. Summarizing the research outcomes of the industry research it can be concluded, with the reservation that this study is a snapshot of a relatively small sample, that there is a profound difference between approaches to design, that ones of the great masters in architecture and the everyday routine practices of the other architectural professionals. The main reason that can be identified is the commercially driven architectural market imposing constraints in terms of time and budget as well as the strongly dominating figure of the client.

The logical question then is: "Where do we go from here?" People, who cannot see alternative viewpoints remain locked in the confined space of the stereotyped vision of reality; they merely help to make this world a practical but uninspiring place to live. As McKim (1980) puts it: "Creative seeing involves using imagination to recenter viewpoint; it is the ability to change from one imaginative filter to another".

3 CONCLUSION

Forster (1999) asserts: "Architecture has become our habitat and therefore one of the principal mediators between nature and human civilization". The author argues that there is a clear sign of change in the cultural significance of current architecture. Articulating symbols of utility or the mechanics of construction are not regarded any more as the only forces driving the art to create buildings. Other forces, mainly invisible ones, have begun to manifest themselves through the physical properties and the experiential effects of buildings. Forster continues his discourse on the meaning of modern architecture by expressing his opinion that where architecture merely aligns itself with its own conditions – exhibiting little more than economy, efficiency, and ambition – it fails to mediate between its own material existence and our need to locate ourselves in the world. "Only acts of imaginative transmission allow us to figure out how we came to fall into the place we occupy and what prospects lie before us" (Forster 1999). Utilitarian buildings employing high-tech construction technologies usually receive great acclaim but they often remain impersonal and somehow detached from our emotions, failing to touch our souls. Imaginative buildings speak to us and as Forster (1999) maintains, "they engage our senses by means of ingenious inscriptions of many-layered meanings no one can grasp, much less exhaust, at a glance". These imaginative acts manifested in a building are what make some designs more interesting than others. Diverse sources of ideas and unconventional approach to design are the essence of these imaginative acts. The research results revealed and emphasized the place of creativity in the complex and exciting process of idea generation. The three case studies illustrated creativity through the imaginative acts manifested in these buildings. Schwarzer (2000) poses the question about future architectural creativity, apart from matters of efficiency and comfort, suggesting that architects should rethink their identities and action within the commodified built landscape. In this view architectural sensibility and creativity emerge as the driving forces in the art of creating buildings.

REFERENCES

Barnard, M. 2001. *Approaches to understanding visual culture*. Hampshire: Palgrave

Bognar, B. 1997. Durability and Ephemerality. *Harvard Design Magazine* (No.3): 9–12

Bossley, P. 1998. Concepts in culture. *Architecture New Zealand* (Special Issue): 18–19

Burns, R. 1994. *Research Methods*. Melbourne: Longman Cheshire Pty Ltd.

Falkoner, G. 2001. The best of both worlds. *Landscape New Zealand* (July/August): 27–29

Findley, L. 1998. Piano Forte. *Architecture* (Oct.): 11–14

Forster, K. 1999. Conflicting Values. *Harvard Design Magazine* (No.7): 26–31

Hunt, J. 1998. Process of selection. *Architecture New Zealand* (Special Issue): 14–16

Ito, T. 2001. Interview with Ito. *Designboom* (Oct.): 10–13

Mcinstry, S. 1998. Sea and sky. *The Architectural Review* (Dec.): 5–8

McKim, R. 1980. *Experiences in visual thinking*. Boston: PWS Publishing Company

Piano, R. 2001. Interviews with Gehry, Calatrava and Piano. *Architectural Record* (Oct.): 3–6

Schwarzer, M. 2000. Sprawl and Spectacle. *Harvard Design Magazine* (Fall): 13–19

Walker, J. & Chaplin, S. 1997. *Visual Culture: an introduction*. Manchester: Manchester University Press

System-based Vision for Strategic and Creative Design, Bontempi (ed.)
© 2003 Swets & Zeitlinger, Lisse, ISBN 90 5809 599 1

Improved design by progressive awareness

S.P.G. Moonen

Eindhoven University of Technology, Eindhoven, The Netherlands

ABSTRACT: This paper refers to the effects of recognizing progressive awareness in a non-routine design. Progressive awareness figures largely in a non-routine design process, since specific knowledge of interacting conditions is mainly amassed by facing and reconsidering one's own previous design decisions. In this context it is essential to rethink preceding decisions when meeting new or renewed conditions. But in the hectic of a design process the causality of conditions is often forgotten. This is why incorrect decisions can be found in almost every non-routine design. The characteristic feature of these incorrect decisions is that they were correct when initially considered, but at the end work in a contrary way because of modified or lapsed conditions. When a decision is discovered that's no longer valid after a change by progressive awareness this is called a residue. This paper describes some residues that revealed in the author's own design practice. Also a design scheme is introduced to notice (and reduce) residues in a design process.

1 NON-ROUTINE DESIGN AND RESIDUES

Progressive awareness is of major importance in a non-routine design process. A non-routine design concerns the process of finding a solution to a novel problem (or generally to a problem that is novel to the designer). In a non-routine design the designer is not acquainted with an already found answer or emphatically looking for an original solution. Additional knowledge is often obtained by comparing and analyzing constituent solutions of similar problems developed by others. Figure 1 shows a schematic graphic of a non-routine design. This graphic demonstrates that initial design assumptions largely influence the final result, while validations of these assumptions just arrive when the design process is nearly finished. In this aspect there is a significant difference with a routine design-process, where as a rule implications of assumptions can be immediately assessed referring to common knowledge.

A mere routine design-process is a rarity in architectural designing. Designing a building is a complex interacting process governed by a plurality of multi-disciplinary conditions with incomplete information and inconsistent and conflicting requirements. Incomplete information in architectural design is inevitable because principal, and certainly a self-respecting designer himself, only too often expects a unique answer to a problem with a generally well-known answer. This explains among other things why

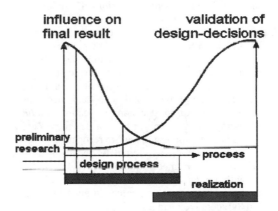

Figure 1. Schematic graphic indicating that initial decisions in a non-routine design have large effect on the final result. The possible influence on the final result rapidly decreases in early stages of a design process, while necessary information to decide on these design issues generally only just arrives at the realization phase (Moonen, 2001).

information obtained from analysis of known solutions is not exactly applicable in an architectural design. And that's why aspects of non-routine design as well as of routine design generally coincide in a "simple" design process.

Structural design is often a non-routine design with a complexity comparable to architectural design. When structural design amounts to structural calculations this can be considered as a routine design-process with significant other characteristics compared to a non-routine structural design.

2 A CORRECT FIRST STEP IS A CHANCE HIT

In a common design process specific knowledge is obtained by studying achievements of others as well as by analysis of consequences of previous design decisions. The latter means that a designer can amass knowledge by making a move in an unknown direction and study the outcome. In a non-routine design a first move is often more or less arbitrarily chosen. As a result a lucky first step in a non-routine design is more like a chance hit than a deliberate choice. But since the first steps strongly determine the direction of all possible final solutions, a designer must make it a habit to always reconsider first moves in a non-routine design. And also when introducing a significant additional, adapted or complementary condition, a designer would be well-advised to reconsider all preceding decisions carefully.

It is important to realize that repeated reconsiderations in a design process do not only regards preceding design decisions but also to specific conditions initially introduced by lapsed decisions. These improper conditions are often difficult to recognize in a design process, since in a little while effects of a condition are unlinked from the initial cause. The same is true for improper sequel deductions caused by a lapsed condition. In practice it is very difficult to recognize initial causality and discard all related conditions and sequel deductions. Improper conditions and deductions caused by a lapsed decision are called *residues*. Depending on the complexity of a design process a residue is easily made and hard to recognize.

Using a logbook or journal to document design decisions with related requirements and conditions *and* with a habit of a regular reconsidering of previous design decisions documented in this logbook can help to recognize residues.

2.1 *Intermezzo 1*

Running the risk of impressing myself unfavorably here an example of several years ago is given with a major residue that was unnoticed in the design-process. The example refers to a design of a plant for a radically changed production process made by a design team. We started the design process with initially strict requirements for air conditioning and dust control. That's why initially a two-story plant was designed (Fig. 2). The first floor was designed to accommodate the production process, the full second floor was considered necessary to locate all equipment for dust and air-control. Another strict requirement imposed by

the client was to minimize total building time. The plant was located abroad and we were told that a steel skeleton was not to be used because getting all required permits regarding fire-safety would lose time. Therefore we altered the initial designed structure with a steel skeleton in a prefabricated concrete structure. But as weeks went by requirements of the plant became clearer, resulting in an incessant reduction of requirements for air conditioning and dust control. In the final design only a small strip was left for equipment, as can be seen in the lower sketch in Figure 2.

Like this the plant is realized to complete satisfaction of the client. Gathered together on completion of the plant, the design team had its first moment's rest when we realized that the initial requirement of using a prefabricated concrete structure was lapsed during the process. A concrete structure was chosen to get a quick approval from the fire department, since we could not rule out maintenance engineers working on the second floor. But since the final design only left a second floor in a very small part of the plant a structure of partly prefabricated concrete leaning against a steel skeleton would have been also possible. Especially because the first two frames were already different from the following nine frames and because a steel workshop was also involved in the interior structure of the plant, this could have been easily realized. Afterwards a calculation is made indicating that changing 9 out of 11 frames from prefabricated concrete to steel could result in a substantial saving and a considerable reduction of construction period.

Although I am aware that it's easy to be wise after an event, the missed opportunity exemplifies the

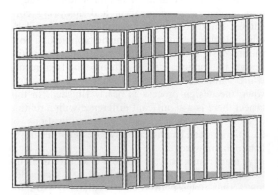

Figure 2. Simplified structure of a plant designed for a production process with clean rooms. Above: In the initial requirements a full second floor was considered necessary to accommodate equipment for air conditioning and dust control. Below: Realized structure where there is only one strip left having a second floor where all required equipment for air and dust is concentrated.

importance of paying attention to recognize residues in an early phase of the design process. Using a logbook or journal would certainly not slow down or hamper the hectic design process. I still wonder today whether this residue would have been discovered by a better documentation of the design progress in a logbook. ...

3 DESIGN SCHEME

In literature many possibilities to diagrammatize a design process can be found. Almost all found schemes represent a design process by a linear sequence, for one shown in Figure 3. Maybe a design process suggests linearity because there is always a clear beginning and a clear end. But every designer knows that all described steps are not necessary subsequently and sequel. In between beginning and end a design process looks more like a scheme with repetitive activities, called loops in a flow diagram. A real design process is also more complex because a design problem is often subdivided in smaller design problems and these, what's more, all have interacting correlations. A design scheme such as shown can be used more or less to see through the overall design process as well as a subdivided design process since the block scheme is meant an abstract representation. With this remarks kept at the back of one's mind the linear scheme of Figure 3–5 gives an appropriate structure because of the strong simplification.

Figure 4 sketches a loop fit in the scheme of Figure 3. As explained in the previous text an architectural design can encounter new insights in several phases of the design process. Applying new knowledge can be interpreted as a loop in the scheme. In literature this loop is often projected inside a block of the design scheme, shown in Figure 4. Besides the part of working out detailed plans, new insights will also appear in other phases, such as analyses or evaluation. And as already said, new insights resulting in progressive awareness is of more importance in non-routine designing than in routine designing.

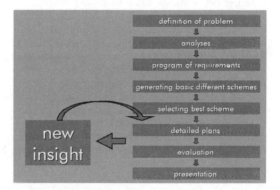

Figure 4. In a linear scheme of a design process the effects of progressive awareness is often ascribed to a certain design step. Especially when a detailed plain is worked out new insight is essential to improve a plan. This approach denies the design decisions with specific requirements and sequel deductions in earlier steps. Sometimes new insight will not only affect a considered plan but also call for a renewed program of requirements. This paper even gives examples of progressive awareness that was assessed in working out and evaluating a plan and that radically changed the first step: definition of problem.

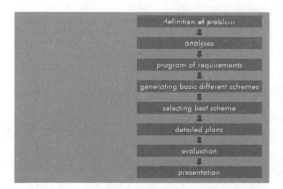

Figure 3. A possible scheme to be used in a design process. Like most design schemes this block diagram suggest linearity in activites. In practice reviewing and reconsideration characterize a design process so a flow diagram with repetitive activities is a more correct scheme. But considered as a sharp simplification representing a complex system this linear block diagram can well be used.

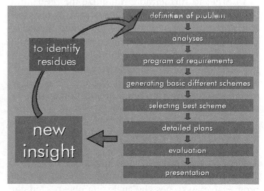

Figure 5. In the documentation of the authors design process examples are described of new insight (progressive awareness) that even has led to a redefinition of a design problem. To optimize design results the impact of a useful new insight has to be reconsidered from the very beginning of the design process. This can only be done if the design process is proper documented in a logbook or journal.

3.1 Intermezzo 2

Effects of progressive awareness causing a loop in the design process are illustrated in the analysis of the design process for a new foundation method for housing described in the doctoral thesis of the author (Moonen 2001). The result of this design process (an industrialized foundation) is also presented at this congress (Moonen 2003b).

As a result of progressive awareness the documented design-process knows three successive stages. The initially developed foundation (called accurate foundation) is especially designed to support industrially produced large building components as described in Moonen 2003a. Industrially produced components ask for a foundation with an extreme degree of accuracy (maximal deviation was aimed less than 2 mm). The workability is demonstrated in two trial projects. But these trial projects also indicated that this foundation method can offer important advantages in traditional building methods to simplify the building process of the foundation in addition to the appreciably exactness. This simplification (a progressive awareness) arises because all work below ground level (earthwork, foundation, and ground floor) can be completely contracted out to one specialized subcontractor. So the foundation principle again was worked out in detail (redefinition of problem) aiming a considerable simplification of the building organization for both building company and subcontractor. This version is called efficient foundation and was researched in a third trial project. And again a trial project confirmed the practical value and provided a progressive awareness. Now the need was found to develop a foundation method that can be executed on site immediately after getting all the permits.

In order to furthermore minimize the preparation period, a third version was developed using solely elements produced in advance (so-called industrialized foundation). In this improved foundation method it is possible to produce all required elements in a full industrial process without knowing specific data of a specific project and yet retaining the possibility to execute a predefined irregular foundation layout.

4 ADAPTED VERSION OF DESIGN SCHEME

This previous design process illustrates how verification of the design (here in trial projects) provided progressive awareness by analyzing unexpected consequences. With new knowledge the design is two times appreciably improved by redefining the design problem. But since all design decisions made before the problem was redefined, are made without awareness of the new knowledge it is likely that some of the previous made decisions are no longer correct and become residues. In the doctoral thesis several residues are described that were found later in the design. These examples of residues showed that it is very hard to recover all residues. Only by a logical documentation in a logbook *and* by a thorough reconsidering of all previous design decisions, conditions and sequel deductions it is possible to minimize residues.

5 CONCLUSION

Figure 5 shows a simplified design scheme indicating that after encountering a new insight all previous conclusions and deductions should be reconsidered from the very beginning irrespective of the actual design phase. In general design schemes new insight is regarded to influence preceding decisions. Residues, which ensued from this approach, are very difficult to recover. The documentation of a design process, entered in author's thesis, shows that a consequent reconsideration of all previous design phases by a backwards thumb through a logbook did benefit in this specific case.

REFERENCES

Moonen S.P.G. 1998. Simplified prefabricated foundation-concept. *In proc. intern. conf. XIII FIP on Challenges for concrete in the next millennium, Amsterdam, Netherlands, 23–29 may 1998*, Volume 1. ISBN 90 5410 945 9. Rotterdam: Balkema.

Moonen S.P.G. 2001. Ontwerp van een geïndustrialiseerde funderingswijze (Developing an industrialized foundation). *Thesis, in bouwstenen 59*. ISBN 90-6814-559-2. Eindhoven: Universiteitsdrukkerij TU/e.

Moonen S.P.G. 2002. Fundamentals of Industrialization: developing an accurate foundation. *In proc. intern. conf. Challenges of concrete construction, Dundee, Scotland 5–11 September 2002.*

Moonen S.P.G. 2003a. Wall panels for industrialized housing. *In proc. intern. conf. ISEC-02 Second International Structural Engineering and Construction Conference, Rome, Italy, 23–26 September 2003.* Rotterdam: Balkema.

Moonen S.P.G. 2003b. An industrialized foundation. *In proc. intern. conf. ISEC-02 Second International Structural Engineering and Construction Conference, Rome, Italy, 23–26 September 2003.* Rotterdam: Balkema.

System-based Vision for Strategic and Creative Design, Bontempi (ed.)
© *2003 Swets & Zeitlinger, Lisse, ISBN 90 5809 599 1*

Deploying the European Foundation in Quality Management (EFQM) Excellence Model (Part Two)

P.A. Watson & N. Chileshe
School of Environment & Development, Sheffield Hallam University, Sheffield, UK

ABSTRACT: Paper number one focussed on the deployment process of the EFQM Excellence Model. This paper builds upon the foundation established in part one and provides original work relating to the presentation of self-assessment results for organisational interpretation and action. This is a crucial activity because it enables engineering and construction firms to fully utilise their scarce resources in an efficient and effective manner directed at attaining both stakeholder satisfaction and a sustainable competitive advantage.

1 INTRODUCTION

The EFQM provides a scoring book approach to self-assessment and this should be consulted in order to obtain a full understanding of the benchmarking process.

However, this paper concentrates on the presentation of what can be confusing data for engineering and construction managers. A simplistic yet meaningful approach is adopted based upon the calculation of arithmetic means and their presentation on a "Pentagonal Profile". Senior managers can thus obtain a comprehensive picture of a host company without going through a full set of data analysis sheets which could number 60 pages.

Upon the completion of a full self-assessment process utilising the "assessment work book" a host company would produce Tables 1, 2 and 3.

Table 1 (see Appendix 1) incorporates the allocated scores for all five enablers. As can be seen from Table 1, these scores are totalled and divided by the appropriate number of sub-clauses. Thus, an arithmetical score for each enabler is obtained.

Table 2 (see Appendix 1) encapsulates the results obtained for performance criteria which in turn is multiplied by the appropriate weighting factor for each sub-clause.

The results of Tables 1 and 2 are incorporated into Table 3 (see Appendix 1) (calculation of total points). It is Table 3 that provides the summary total and point's allocation for enablers and performance criteria.

In order to put the score of 508.45 points in the context of best practice it should be noted that the EFQM will conduct a site visit on an organisation obtaining over 500 points. Also the EFQM award for excellence is usually awarded to organisations obtaining a score between 750 and 850 points. Therefore, a score of 508.45 points is a very respectable score. It is difficult to communicate the meaning of the scores to senior management. Thus this paper further develops the scoring system by demonstrating the usefulness of original work in the form of a "Pentagonal Scoring Profile". The pentagonal approach is an effective means of demonstrating the results of the self assessment process.

Based upon the RADAR approach (utilised in the self-assessment scoring book) Results, Approach, Deployment, Assessment and Review (Fig. 1 – see Appendix 2) the allocated points for each criteria and sub-clause are tabulated as indicated in Tables 4 and 5 (see Appendix 1). These figures are then used to calculate average scores (arithmetic means). These average scores are then plotted upon the Pentagonal Profile. The maximum score for each element is 100% (Fig. 1).

It is also a very quick and accurate method of benchmarking engineering and construction organisations. Senior managers must remember that the self evaluation process is designed to develop continuous improvement. Therefore, the benchmarking activity must be conducted on a regular basis so that corrective actions can be evaluated. From the example it is clear that the host company has some problems. However, they have a real issue in assessing and reviewing deployed actions.

2 CONCLUSIONS

The pentagonal profile can be utilised for benchmarking between corporate action periods. However, it is also of value in conducting generic benchmarking against other firms who operate more efficiently and effectively.

The EFQM Model does provide engineering and construction companies with a realistic and practical model for TQM application, hence attaining the associated advantages of deployment.

REFERENCES

The European Foundation for Quality Management, (EFQM, 1999), Excellence Model accessed 17 January 2001, http:/www.efqm.org/award/htm.

APPENDIX 1

Table 1. Scoring summary sheet (from completed self-assessment book) for enablers.

1. Enablers Criteria										
Criterion Number	1	%	2	%	3	%	4	%	5	%
Sub-criterion	1a	45	2a	50	3a	60	4a	50	5a	45
Sub-criterion	1b	40	2b	50	3b	35	4b	50	5b	60
Sub-criterion	1c	45	2c	40	3c	40	4c	55	5c	60
Sub-criterion	1d	50	2d	30	3d	40	4d	40	5d	50
Sub-criterion			2e	45	3e	50	4e	35	5e	50
Sum		180		215		225		230		265
		÷ 4		÷ 5		÷ 5		÷ 5		÷ 5
Score Awarded		45		43.2		45.2		46		53

Table 2. Results criteria.

Criterion Number	6				%	7				%	8				%	9				%
Sub-criterion	6a	50	x 0.75 =	37.5		7a	60	x 0.75 =	45		8a	50	x 0.75=	12.5		9a	60	x 0.50=	30.0	
Sub-criterion	6b	50	x 0.25 =	12.5		7b	50	x 0.25 =	12.5		8b	60	x 0.25 =	45		9b	55	x 0.50 =	27.5	
Score awarded		50					57.5					57.5					57.5			

Table 3. Calculation of total points.

Criterion	Score awarded	Factor	Points awarded
1. Leadership	45	× 1.0	45
2. Policy & Strategy	43.2	× 0.8	34.6
3. People	45.2	× 0.9	40.7
4. Partnerships and resources	46	× 0.9	41.4
5. Processes	53	× 1.4	74.2
6. Customer results	50	× 2.0	100.0
7. People results	57.5	× 0.9	51.8
8. Society results	57.5	× 0.6	34.5
9. Key performance results	57.5	× 1.5*	86.25
Total points awarded			508.45

* Note these are the factors from the model.

Table 4. Summary assessment sheet for "RADAR" (based on the EFQM approach) company profile.

Criteria (enablers)		Approach score	Deployment score	Assessment and review score
Leadership	1a	55	45	45
	1b	60	55	35
	1c	60	55	45
	1d	55	55	40
Policy &	2a	65	60	20
Strategy	2b	50	55	15
	2c	70	80	35
	2d	55	70	50
	2e	65	75	40
People	3a	55	65	50
	3b	65	70	60
	3c	50	35	25
	3d	50	65	25
	3e	30	40	30
Partnerships	4a	55	55	40
& Resources	4b	55	65	40
	4c	40	35	15
	4d	30	40	25
	4e	65	60	25
Processes	5a	70	85	40
	5b	50	60	30
	5c	45	50	25
	5d	60	60	40
	5e	65	75	50
Totals and average		1319 24 = 54.9 score	1415 24 = 58.9 score	845 24 = 35.2 score

Note: (based on the completion of a full corporate analysis).

Table 5. Summary assessment sheet for "RADAR" company profile.

Criteria (results)		Results	Scope
Customer	6a	55	65
	6b	40	30
People	7a	60	50
	7b	40	55
Society	8a	25	20
	8b	45	35
Key performance	9a	55	45
	9b	50	60
Totals and average		370 8 = 46.3 score	360 8 = 44.4 score

APPENDIX 2

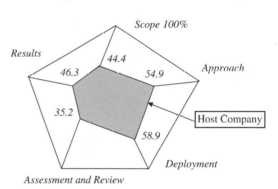

Figure 1. RADAR pentagonal profile.

1627

23. Sustainable engineering

Production of artificial fly ash lightweight aggregates for sustainable concrete construction

C. Videla
Department of Construction Engineering and Management, School of Engineering, Pontificia Universidad Católica de Chile

P. Martínez
School of Construction Engineering, Universidad de Valparaíso, Chile

ABSTRACT: Aggregates are the main volumetric component of concrete occupying between 65% to 75% of the total volume. The extraction of natural aggregates has caused world-wide erosion problems and generated a deficit of these materials. To face those problems different alternatives have been proposed: recycling technologies and production of artificial aggregates.

The use of fly ash allows the production of artificial aggregates for concrete. This solution also offer the possibility of using the industrial by-products of thermoelectric plants allowing to solve the disposal problems and the associated cost that these industries have had per decades.

Both problems, the shortage of natural aggregates and the use of industrial waste, are highly related to the necessity of Sustainable Development, one of the main concerns at the beginning of the XXI century.

In this research class F fly ash was used, agglomerated by an agitation process and hardened by "cold bonding". Major variables of the production process were the use of different types and contents of binders and cold bonding conditions. Results indicate that the best solution, considering aggregates physical and mechanical properties and economic and practical criterions, corresponds to aggregates manufactured with 5% wt of Portland Pozzolan cement subjected to out-door exposure hardening.

1 INTRODUCTION

With the beginning of the twenty-first century, the theme of the sustainable development has become an essential subject. Sustainability involves several social type aspects. Regarding the construction industry, the increased demand of resources and the associated environmental problems do not allow the sustainability of this economic sector for many generations under present conditions (Mehta 1998; Nixon 2000).

In this context an important feature of construction is that concrete is one of the most widely used construction material and its manufacturing process requires large amounts of raw materials. Therefore, concrete construction can have substantial negative impacts over the environment. Among these the extraction of natural aggregates produces erosion problems and loss of natural areas (CIB 2000), and the production of artificial aggregates appears as a promising alternative to mitigate these problems. It must be recalled that aggregates are the main concrete volumetric component occupying between 65% to 75% of the total volume of this material.

Synthetic lightweight aggregates can be manufactured from different materials and with different technologies, like pelletization of fly ash (Mehta 1986). This solution offers the possibility of using the industrial by-products of thermoelectric plants, and therefore allows meeting other goal of sustainable development.

2 RESEARCH OBJECTIVE

The objectives of the research was to evaluate innovations in the manufacture of Fly Ash Lightweight Aggregates (FALA) for lightweight structural concrete applications using a cold bonded production technique that avoids high energy requirements (Hwang et al. 1992; Shigetomi 1999; Baykal 1999). Particularly the effect of the use of different types and contents of binders and cold bonding conditions on aggregate properties were analysed.

3 EXPERIMENTAL PROGRAM

3.1 Fly ash and binders materials

The fly ash used in this research was obtained from the burning process of bituminous or anthracite coal and the chemical properties of the fly ash are shown in Table 1. According to the results shown in Table 1 the fly ash was characterised as Class F (ASTM C618-93).

The aggregate mixes were designed considering 3, 5 and 7% by weight of the following binders: lime (L) and two rapid-hardening cements. The cements were a Portland cement (P) and a Blended Portland cement with 19% of natural pozzolan (PP). Table 2 presents the measured physical properties of the fly ash and the binders.

3.2 Pelletizer disc

The production of Fly Ash Lightweight Aggregates (FALA) was made using a pan disc for fly ash agglomeration by the agitation method as shown in Figure 1. The diameter of the pelletizing disc was 1500 mm and with a depth of 250 mm. The applied revolution speed was constant and equal to 21 rpm and different slant angles were studied.

3.3 Curing and hardening system

The cold bonding technique was used for aggregates hardening. The following three different curing conditions were tried in order to evaluate the effect of the hardening system on the strength development of the aggregates:

- Out-doors exposure at ambient temperature and sprayed water at regular intervals, (OD).
- Moist-room at a temperature of $23 \pm 2°C$ and a relative humidity $\geqslant 90\%$, (MR).
- Sealed bags maintained at a temperature of $23 \pm 4°C$, (SB).

4 TEST PROCEDURE

4.1 Fly ash

The following tests were carried out to characterise the fly ash according to ASTM C618-93:

- Chemical: oxide composition and loss of ignition.
- Physical: specific gravity, specific surface and fineness.

The results of these tests are shown in Tables 1 and 2. Additionally, X-ray diffraction (XRD) studies were conducted to determine the mineralogical composition of the fly ash and the results are presented in Table 3.

Also the following two tests were developed and performed to analyse the potentiality and production viability of the fly ash to produce aggregates: Setting Time (ST) and Modified Pozzolanic Activity Index (MPAI).

The objective of the setting time test was to determine the minor setting time of the designed mixtures. This test was carried out in accordance to ASTM C191-99.

On the other hand the Pozzolanic Activity Index (PAI) test specified on ASTM C311-92, aims to determine the fly ash capacity to fix calcium hydroxide and therefore to harden and develop strength

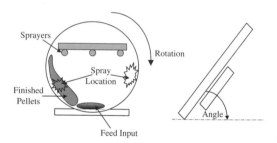

Figure 1. Diagram of the disc pelletizer.

Table 1. Chemical composition of fly ash.

Chemical analysis	SiO_2	Al_2O_3	Fe_2O_3	CaO	MgO	SO_3	K_2O	Na_2O	LOI
Fly Ash (%)	56.21	24.40	5.76	1.70	0.90	0.07	0.89	0.01	5.85

Table 2. Physical properties of fly ash and binders.

Material	Nomenclature	Specific gravity	Specific surface (cm^2/g)	Fineness (retained on N°325 sieve, %)
Fly Ash	FA	2.155	4380	23.00
Portland Cement	P	3.176	2860	
Portland Pozzolan Cement	PP	2.969	4500	
Lime	L	2.440		12.00

when mixed with lime. Because new binders were considered for the aggregate manufacture, a Modified Pozzolanic Activity Index (MPAI) test was proposed to represent the actual behaviour of the mixes to be used. The standard procedure was modified in the following aspects: materials constituents and proportions, and specimen shape and dimension. The sand was not used in the mixes and the compressive strength tests were carried out on 50 × 100 mm cylinder specimens. The storage and testing of the specimens was performed according to the specifications contained in the standard test. The fly ash – binder blend considered the following binders content: 3, 5 and 7% by weight of binder. The water requirement for each mix was determined according to ASTM C187-98.

4.2 Fly ash lightweight aggregates

In order to determine the mechanical and physical properties of the aggregates the following tests were carried out on manufactured fly ash aggregates:

- Aggregate Crushing Value (ACV) according to BS 812-75 (see Figure 3).
- Specific gravity, absorption and particle size according to ASTM V.4.02-98 (see Table 4).

Table 3. Mineralogical composition of fly ash (X-ray diffraction).

Composition	Nominal composition
Quartz	SiO_2
Mullite	$Al_6Si_2O_{13}$
Sillimanite	Al_2SiO_5
Tricalcium aluminate	$Ca_3Al_2O_6$
Hematite	Fe_2O_3

Scanning electron microscopy (SEM) analyses were also carried out to determine the degree of hydration of the aggregates as a function of binder type and age (see Figure 4).

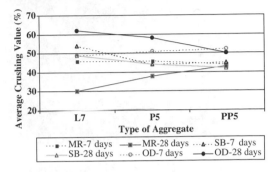

Figure 3. Relationship between fly ash aggregate type and aggregate crushing value for different curing conditions.

Table 4. Physical properties of 20 mm nominal maximum size fly ash aggregates.

Type aggregate	Hardening system	Dry specific gravity	Absorption (%)
L7	MR	1.150	35.0
	SB	1.230	31.6
	OD	1.222	32.5
P5	MR	1.119	35.2
	SB	1.236	33.1
	OD	1.253	29.6
PP5	MR	1.223	35.0
	SB	1.261	31.0
	OD	1.274	32.2

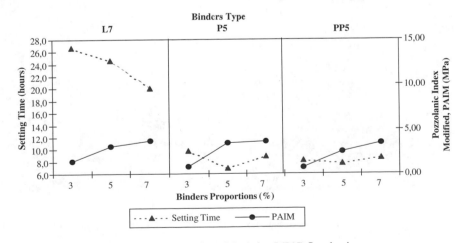

Figure 2. Setting Time (ST) and Modified pozzolanic activity index (MPAI) fly ash mixes.

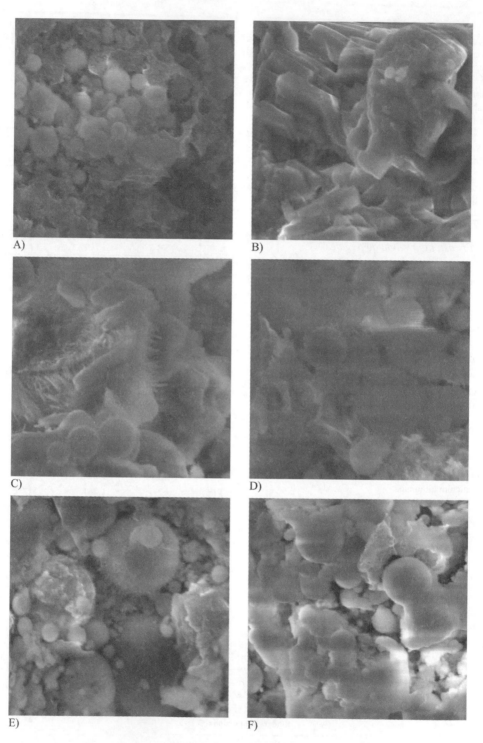

Figure 4.   SEM Micrographs × 5000 of hydrated aggregates. A) PP5 at 7 days; B) PP5 at 28 days; C) P5 at 7 days; D) P5 at 28 days; E) L7 at 7 days; F) L7 at 28 days.

5 RESULTS AND DISCUSSION

5.1 Fly ash characteristics

Tables 1 and 2 show that the used fly ash satisfies the chemical and physical requirements prescribed in ASTM C618-93, i.e.:

- $SiO_2 + Al_2O_3 + Fe_2O_3 = 86.37\% > 70\%$
- Loss on Ignition (LOI) = 5.8% < 6.0%
- Fineness (retained on N°325 sieve) = 24% < 34%

From the results the fly ash can be classify as class F and therefore has pozzolanic properties. Also the results of the XRD analyses (see Table 3) show that the main fly ash crystalline minerals are the ones required for the formation of the hydration products, such as: ettringite ($C_6AS_3H_{32}$) and calcium silicate hydrated (C-S-H).

5.2 Effect of binder type and content on fly ash mix behaviour

Figure 2 shows the effect of binder type and content on the pozzolanic reactivity of the fly ash. As expected, it can be concluded that independent of binder type, the larger the binders content the larger the compressive strength of the fly ash mixes. Also, for lime larger the lime content the smaller the setting time while for binders based on cements the results appears to indicate that there is an optimum dosage that minimise the setting time.

To select the best mix for aggregate manufacture the setting time was defined as the main variable because it showed to have a higher sensibility and a larger range of variation than the results determined by the MPAI test. It should be noted that changes of binder content between 3 to 7% wt produce a maximum change of MPAI of almost 3 MPa for any binder, while a minimum and a maximum of 1 and 6 hours setting time differences were measured. The larger setting time difference corresponds to lime binder while the minimum to Portland pozzolan cement. Also it should be noticed the large setting time difference between lime and cements; lime has almost 20 hours more setting time than cements.

Therefore, from Figure 2 it can be concluded that the best binder dosage according to binder type, taking into account the smallest ST, are:

- Fly ash + Lime $\rightarrow$ 7% wt of lime (L7)
- Fly ash + Portland cement $\rightarrow$ 5% wt of Portland cement (P5)
- Fly ash + Portland Pozzolan cement $\rightarrow$ 5% wt of Portland Pozzolan cement (PP5)

It must also be noticed that the L7 mix presents the simultaneous ideal conditions, i.e. the smallest ST and larger crushing strength of the mixture, but mixes manufactured with lime present excessively large setting time. However, mixes P5 and PP5 do not show an optimum binder content simultaneously satisfying both conditions. Finally, the results suggest that the optimum binder to be used with the available fly ash from a technical and economical point of view should be Portland Pozzolan cement.

5.3 Effect of curing and hardening system on lightweight aggregate mechanical properties

Figure 3 shows the relationship between aggregate type and aggregate strength (crushing value in per cent) as a function of hardening systems. From the figure it can be concluded that the effect of the hardening system on the strength gain of an aggregate depends on the aggregate age and type. In general at 28 days of curing the best results are obtained with the moist curing process and the worst with the outdoors exposure but at 7 days the hardening system appears to have a lesser effect on strength gain. On the other hand, when the effect of the hardening system is analysed with respect to a particular aggregate it can be seen that Portland Pozzolan cement and lime based aggregates are the least and most sensitive to curing conditions, respectively. It can also be observed that in all cases the larger strength obtained corresponds to PP5 aggregate type, except for the behaviour of L7 aggregate subjected to the MR hardening system during 28 days.

It is also interesting to compare the measured average crushing value of the manufactured fly ash lightweight aggregates with other lightweight aggregates: average crushing value of Chilean Pumice ranges between 79 and 97% and the corresponding value for expanded clay is 52%.

As previously mentioned PP5 aggregate presents the most stable behaviour with respect to the applied hardening system and the larger strength, independent of curing time. Also PP agglomerate is the cheapest binder and OD hardening system is the cheapest and feasibly to be used in practice. Therefore, taking into account economical and practical criterions it is concluded that PP5 aggregate maintained at outdoors exposure is the manufacture procedure presenting the greatest advantages.

5.4 Physical and microscopic characteristics of lightweight aggregates

Table 4 presents the physical properties of the manufactured fly ash lightweight aggregates. The dry specific gravity ranges between 1.15 and 1.274. Independent of the aggregate type the lower dry specific gravity values are for the MR hardening system.

However the observed differences between aggregates types and hardening systems are not significant and all the aggregate can be considered as lightweight aggregates. Corresponding values for pumice and expanded clay are 0.7 and 1.0, respectively.

Additionally, the results of water absorption range between 31 and 35%. This value is also very similar and characteristic of expanded clay lightweight aggregate.

Finally, SEM analysis were made on the selected aggregates types and subjected to the OD hardening system during 7 and 28 days. The hydration evolution processes are given in Figure 4. It is observed that at 7 days of curing the three aggregates presents similar hydration state development. However at 28 days of curing the PP5 aggregates present a notoriously more advanced hydration than the other two types of aggregates, and show a completely crystalline structures.

6 CONCLUSIONS

The following conclusions can be drawn from this experimental programme:

a) The fly ash used in this research complies with ASTM C618-93 and has been shown to have a potential use in different concrete applications.
b) Fly ash agglomeration by agitation methods by disc pelletizer is a simple and effective process for aggregate manufacture. The slant angle is a fundamental factor determining the particle size distribution; the best size distribution was obtained with a 57° angle for a 20 mm maximum size aggregate.
c) Paste mixtures agglomerated using cement binders show five times smaller setting times values than mixtures agglomerated with lime. This is a very important characteristic considering minimum early age strength requirements for large scale storage conditions.
d) The modified pozzolanic activity index test was shown no to have sufficient sensibility to discriminate between aggregates types.
e) The fly ash lightweight aggregate strength is affected in a different way by the applied hardening system, the effect being more noticeable for fly ash agglomerated with lime like the L7 type aggregate. The aggregates manufactured with Portland Pozzolan cement (PP5) are the least sensitive to curing conditions and do not show significant strength differences when are left to outdoors exposure. Also, this aggregate type presents the faster hydration speed. Therefore it is concluded that due to technical, economical and practical criterions, aggregates manufactured with 5% wt of Portland Pozzolan cement (PP5) are those possessing the greatest viability for use at full-scale production.

7 FUTURE WORKS

The following aspects has been considered for future work on this research area:

a) Study and analysis of physical and mechanical properties of fly ash lightweight aggregate structural concrete.

b) Full scale aggregates production and lightweight concrete application to prefabricated construction.

REFERENCES

ASTM C187-98 1998. Standard test method for normal consistency of hydraulic cement. American Society for Testing and Materials, U.S.A.
ASTM C191-99 1999. Standard test method for time of setting of hydraulic cement by Vicat needle. American Society for Testing and Materials, U.S.A.
ASTM C311-92 1992. Standard test methods for sampling and testing fly ash or natural pozzolans for use as a mineral admixture in portland-cement concrete. American Society for Testing and Materials, U.S.A.
ASTM C618-93 1993. Standard specification for fly ash and raw or calcined natural pozzolan for use as a mineral admixture in portland cement concrete. American Society for Testing and Materials, U.S.A.
Baykal, G. & Doven, A.G. 1999. Lightweight concrete production using unsintered fly ash pellet aggregate. *Proceedings: 13th International Symposium on Used and Management of Coal Combustion Products*. ACAA International, January 11–15, Orlando, Florida, USA, paper 3, pp. 1–14.
BS 812: Part 3 1975. Methods for sampling and testing of mineral aggregates, sand an fillers, Part 3, Mechanical properties. British Standard Institution, UK.
International Council for Research and Innovation in Building and Construction, CIB (2000). *CIB Publication 237*, AGENDA 21, Green Building Challenge Chile, Julio 2000.
Hwang, C., Lin, R., Hsu, K. & Chan, J. 1992. Granulation of fly ash lightweight aggregate and accelerated curing technology. *Fourth International Conference on Fly Ash, Silica Fume, Slag and Natural Pozzolans in Concrete*, May 1992, Istanbul, Turkey, SP132, pp. 419–438.
Metha, K. 1998. Role of the pozzolanic and cementitious material in sustainable development of the concrete industry. *Sixth CANMET/ACI International Conference on Fly Ash, Silica Fume, Slag and Natural Pozzolans in Concrete*, May 31–June 5, Bangkok, Thailand, SP178, Vol.1, pp. 1–20.
Metha, K. 1986. *Concrete: Structure, properties, and materials*. Ed. Prentice-Hall, USA.
Nixon, P. 2000. Concrete: Construction material for the next millennium?. *Concrete*, Vol. 34, N°1, pp. 20–23.
Shigetomi, M., Morishita, N. & Kato, M. 1999. Characteristics of high performance aggregate produced from fly ash using a rotary kiln and properties of concrete made using this aggregate. *Proceedings: 13th International Symposium on Used and Management of Coal Combustion Products*. ACAA International, January 11–15, Orlando, Florida, USA, paper 4, pp. 1–10.

Sustainable buildings with the infra$^+$ space floor

G.C.M. van der Zanden
PreFab Limburg B.V., Kelpen-Oler, the Netherlands

ABSTRACT: Within the scope of the conference "International Structural Engineering and Construction Conference" the development of the "flexible floor-construction", which is presented in this paper, will prove to create a breakthrough in the construction-industry.

The aim for cost-effective, efficient, low-weight and user-friendly buildings with a longer life-time and a good comfort for the users is in reach. Also the effect on the energy-consumption in buildings and (road) transport is of big importance.

The energy consumption for cooling for example, is one of the major reasons for the CO_2 production in North-America, and Europe is rapidly following this example.

As a result of the use of suspended ceilings and good insulation, the internal heat in buildings due to the people, use of office-equipment and lighting is becoming a problem. The need for cooling is required to maintain a good comfort level. The standard solution offered is "air-conditioning" with air as a carrier of energy, which is not an efficient medium and in addition this solution creates the risk for "sick-building" diseases. Circulation of water is a much more efficient method of transporting heat than air-based systems, offering many advantages as regards performance, capital costs and running costs.

1 LIFETIME OF BUILDINGS

The life-time of buildings becomes shorter, because of the changes in technology and therefore buildings

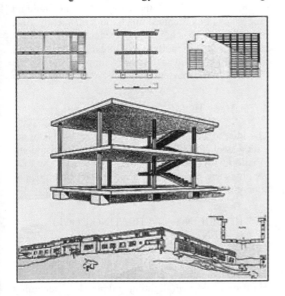

Figure 1. Domino-house of Le Corbusier.

should be adaptable. As John Habraken, Professor at the University of Harvard and chairman of the Open Building Association wrote already in 1963: "... *we should not try to forecast what will happen, but try to make provisions for what cannot be foreseen*".

This was already realised in the Dom-ino house of Le Corbusier in 1914 with maximum flexibility in for example the variation in floor plans.

The services are the major bottle-neck for flexibility in buildings as shown in figure 2.

2 SOLUTIONS FOR FLEXIBILITY AND ENERGY-REDUCTION

In 1997 the innovative consultancy group "A+" invented the INFRA$^+$ "space floor", which is further developed and marketed by PreFab Limburg B.V. This INFRA$^+$ "space floor" concept creates the flexibility in buildings by the integration of the services in the constructive floor.

Due to this basically simple solution it becomes possible to increase the life-time of buildings, because the interior can easily be changed and adapted to changing circumstances. But also the function of a building can be changed from office to residential use or other needs.

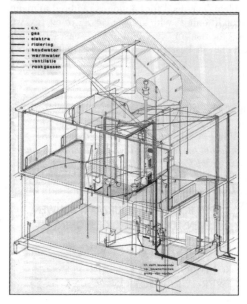

Figure 2.   Bottle-neck for flexibility.

Now often within 20–30 years buildings are already demolished and the building waste in The Netherlands for example increased from 10 million tons in 1990 up to more than 18 million tons in 2000.

On the other hand only the consumption of new materials in the construction industry is yearly another 15 million tons. This means in transport capacity daily more than 5,000 truck movements from building site to a manufacturer, raw material supplier, et cetera.

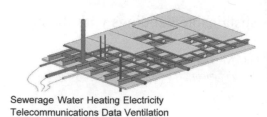

Sewerage Water Heating Electricity
Telecommunications Data Ventilation

Figure 3.   INFRA⁺ "space floor".

Figure 4.   Working and living – multifunctional buildings.

The transport energy and the road blockages will continue to increase considerably, if we are not able to change this trend.

Through the invention of the "space floor" as mentioned above, the buildings will become at least 50% lighter in weight, maintaining the constructive, acoustical and fire-resistance requirements.

By using the INFRA⁺ concept the construction height is reduced and the logistics of the building process can be organised much more efficiently. As a result the quality is higher and a reduction of expenses is achieved due to faults and failures. There is no need for scaffolding and assembly is simpler because the fitters no longer have to work "above their heads".

3   THE A+ HOME

The idea of Le Corbusier was used in the A+ home with a steel structure and the INFRA⁺ floor, in combination with prefabricated masonry walls. The total weight is less than 50% of a traditional brick/block or concrete structure.

4   THE CEILING AS A HEATING/COOLING ELEMENT

As a result of the use of suspended ceilings and good insulation, the internal heat in buildings due to the use of office-equipment, lighting and people is becoming a problem. The need for cooling is required to

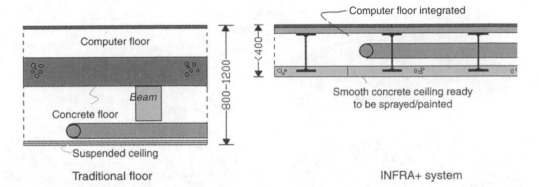

Figure 5. Comparison floor thickness with INFRA$^+$.

Figure 6. A+ Homes.

maintain a good comfort level. The standard solution offered is "air-conditioning".

The capacity of air as a carrier of energy is not high and increases the risk of "sick-building" diseases.

The solution of activating the concrete of the ceiling is a logic choice, which combined with the small mass of the 70 mm concrete of the "space floor", delivers a cooling/heating element by the constructive-elements. The material reduction and the lower (sustainable – heat pump technology) energy makes that this concept gives a considerable move forward in our society to use our resources more efficient.

5 HIGH RISE BUILDINGS

The INFRA$^+$ floor system combined with a steel structure meets all relevant requirements for high rise buildings.

The fire-resistance is more than 2 hours and the acoustical performance is excellent.

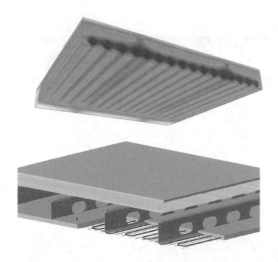

Figure 7. "Climate ceiling" INFRA$^+$.

In The Hague in 2003 the first apartment building will be realised with INFRA$^+$, which proved to be an economical solution and creates a new era for light-weight buildings.

6 CONCLUSION

In the future constructions will be based on "flexible floors". More prefabrication and a more efficient

Figure 8. "La Fenetre".

building process will be possible and as a result a reduction of costs and affordable housing for more people will become a normal situation.

Falling accident in the roofing work of residential houses

Y. Hino
National Institute of Industrial Safety, Tokyo, Japan

ABSTRACT: In Japan, there is about one hundred of falling accidents at the construction site of residential houses. However, the effective countermeasures against falling have not been established. In this study, the fundamental data for establishing countermeasure against falling from roof was especially obtained by review of some accident data reported by labour standard inspectors. From the results, suggestions on prevention of falling from roof was proposed, and also the fundamental data such as falling posture of the worker, falling velocity of worker, and etc were obtained.

1 INTRODUCTION

In Japan, falling accidents at the construction site of low rise houses have been recognized as the major cause of fatal accidents, since the occupational safety and health law (Industrial 2000) was instituted in 1972 (Editorial 1999). There is a decreasing trend of the accidents in the last two or three years. It is thought that the reasons are due to "popular use of precedent scaffold installation method" (Japan 1996) and "decrease in the number of construction works" (Construction 2001).

However, the number of accidents is still large. Especially, about 20 to 30 of falling accidents from roof periodically happen over the last decade, and the decreasing trend of the accidents has not been observed (Hino 2002). And in recent years, the construction work has changed from the construction of new house to extension or repair work. In some cases of extension or repair work, the construction period and the budget for installation of the countermeasure against falling may not be enough.

The purpose of this study is to get the fundamental data for establishing the countermeasure against falling from roof by using fatal accident data in Japan in 1997 (23 cases). And also, the requirements of prevention equipments against falling from roof were discussed.

2 CAUSES AND PROBLEMS OF THE FALLING ACCIDENTS

2.1 Inspection results from the disaster sites

Figure 1 shows the classification of basis cause of falling. Although about 50% of the accidents were not

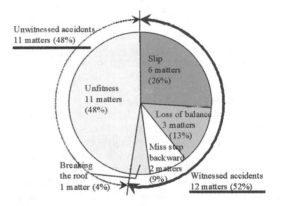

Figure 1. Basis cause of falling.

witnessed and valid data are a few, the falling accidents could be classified into 3 types which are "slip", "loss of balance", and "miss step backward". The type of "slip" has a largest rate of the accidents and more than half of this case (4 from 6 cases) happened under the wet roof condition caused by rain. However, the rain precipitation at all disaster sites of this kind of accidents was not larger than the maximum limit provided by the occupational safety and health regulation. Thus, it can be said that, not only the influence of heavy rain, but also the influence of light rain can not be neglected in the roofing work. Nevertheless, some roofing works under light rain condition have to continue without suspension. For example, some repair works of dwelling-house must be continued until it can assure that the dwellers can live there without problems. Hence, it is actually difficult to prevent basis

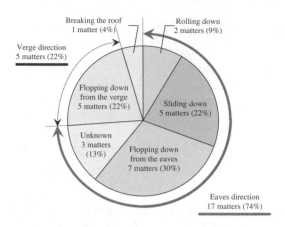

Figure 2.   Falling direction and falling posture.

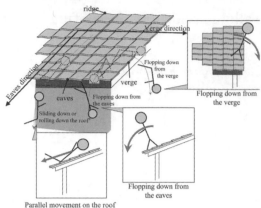

Figure 3.   Illustration of classified falling posture.

cause of falling such as "slip". Therefore, it is necessary to install such as guide rail as a protection against falling from roof.

Figure 2 shows the classification of the falling direction and falling posture of the worker. The falling direction was classified into 2 directions, eaves and verge directions by using the accident data. It can be seen that the number of falling from the "eaves direction" has the largest rate, 74% of the total. However, the cases of falling from the "verge direction" cannot be neglected because the number of this case of accident was also high. This result shows that not only a countermeasure for falling from "eaves direction" but also from every direction is need.

The falling posture of the worker could be classified into 3 types according to the witnesses, "sliding down the roof", "rolling down the roof", and "flopping down from the edge of the roof". From these classifications, combination from 2 types of falling direction and 3 types of falling posture of the worker gives totaling 6 cases of falling process from the roof. However, the falling direction of the type "sliding down the roof" and "rolling down the roof" is only the eaves direction and the movement of the worker on the roof is predominantly parallel to the roof surface. Thus, they were both reclassified as type of "parallel movement on the roof". For the falling type of "flopping down from the edge of the roof", it was also classified into 2 cases as they have 2 falling directions. Finally, the falling process from the roof could be classified into 3 types as shown in Figure 3. The "parallel movement on the roof" case shows the parallel motion on the roof surface due to sliding or rolling down the roof, and the "flopping down from the eaves" and "flopping down from the verge" cases show dominant motion in rotating around the eaves and verge of the roof, respectively. Therefore, it can be concluded that the appropriate falling countermeasure is to install

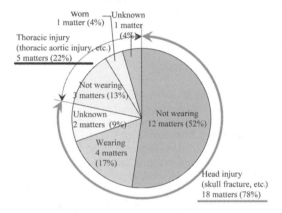

Figure 4.   Helmet wearing condition and cause of death.

some equipments that can protect the workers from above 3 kinds of falling.

Figure 4 shows the classification results of the helmet wearing condition and cause of death. About 20% of the accidents occurred when the victims were wearing the helmet. Concerning the cause of death, head injury is the main cause, about 80 (18 from 23 cases). Therefore, to prevent the fatal accident and heavy injury due to falling from roof, head protection is strongly required.

2.2   Countermeasures at the disaster sites

The countermeasures against labour accidents at the disaster sites were investigated corresponding to occupational safety and health regulation (Industrial 2000). Table 1 shows a list of the regulations and the amount of accidents which possibly violated those regulations. In general, the contents of these regulations can be categorized into 2 matters, installation of falling countermeasure and performance as a work

Table 1. Law violation at the disaster site in Japan.

Amount of accident[*]	Regulations	Contents
16	Article 519 Paragraph 1	To set some kinds of prevention equipment such as guide rail
6	Article 519 Paragraph 2	To use some safety equipments such as safety belt
6	Article 653 Paragraph 1	To set some kinds of prevention equipment such as guide rail around the winch
4	Article 517-12 Paragraph 1	Appointment of work chef
2	Article 517-13 Paragraph 1	Fulfillment of duty as the work chef
1	Article 526 Paragraph 1	To set some kinds of safety equipment for going up and down

* From total amount of 23 accidents.

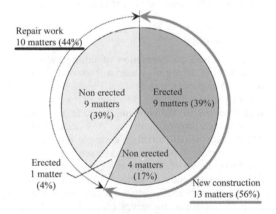

Figure 5. Availment condition of temporally scaffolds.

chef. Especially, about 70% of the accidents possibly violated the regulation on paragraph 1 of the article 519, which requires setting some kinds of prevention equipment such as guide rail, net, etc. Therefore, it can be concluded that the countermeasure for safety was not implemented at the disaster sites. However, there are some disaster sites that did not violate laws but the accident still happened. Somehow, it is presumed that such a construction site that has no law violation has effective prevention for safety and low possibility of labour accidents.

2.3 Problems in installation of guide rail

Generally, temporally scaffolds are used as an installation equipment for guide rail. Figure 5 shows the erected condition of the scaffolds at the disaster site. The scaffolds that were partially erected and the sites that had a plan to erect the scaffolds were also counted in the results in Figure 5. More than half of disaster sites did not use the scaffolds, especially for the site of repair work.

This is because of the differences in construction period, contract amount, site location, and living conditions of dwellers at the site between new construction and repair work. Therefore, it is necessary to develop an easy method for installation of the equipments and the new equipments that are easy to install and economical.

Figures 6(a) and (b) show the illustrations of falling accidents which happened in spite of the countermeasure for safety had been implemented. The accident in case of Figure 6(a) happened due to improper position of the guide rail installed on the scaffolds based on occupational safety and health regulation. The install position of guide rail is defined by the height to prevent falling from the scaffolds (over 750 mm from foot plate as shown in Figure 6(a)) without considering the position to prevent falling from roof. For Figure 6(b), the guide rail was installed on the scaffolds based on the guideline for "precedent scaffold installation method" as shown in Figure 6(c), an example of countermeasure for falling from roof proposed by the ministry of health, labour and welfare (Industrial 2000, Japan 1996).

The guideline states that the guide rail should be installed at the distance and height of less than 300 mm and over than 750 mm from the roof eaves, respectively. However, there is no any instruction for the proper position of the middle rail as the effective install position of guide rail should depend on the distance from the roof surface to the middle rail.

Moreover, based on the present guideline, there is still no protection to prevent falling from the verge. Therefore, it is necessary to consider the method that can prevent falling from all directions.

2.4 Requirements on the protection equipments

From the previous investigation, it was clarified that there are 3 types of falling accidents from roof, cause of the death is mainly due to head injury, and there is necessity to define the proper install position of the guide rail and middle rail. Moreover, it is necessary to use some equipments such as guide rail, net (Kinoshita 1971), etc. for protecting the worker from strong crashing at the ground surface with high impact load.

Therefore, requirements on the countermeasure against falling from roof can be concluded as follows:

1. There should be no space between roof surface and the guide rail so that the worker can pass through.
2. The equipment should not be destroyed by the impact load caused by crashing of the worker.
3. The equipment should not be heavy so that it can injure the head of worker by crashing.

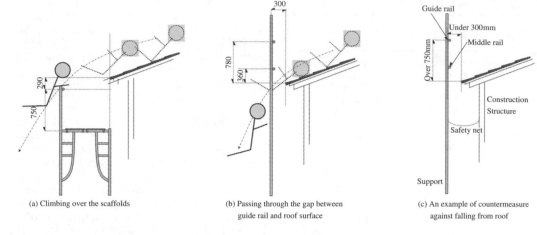

(a) Climbing over the scaffolds

(b) Passing through the gap between guide rail and roof surface

(c) An example of countermeasure against falling from roof

Figure 6.   Illustrations of falling accidents and an example of prevention method against falling accidents (unit: mm).

3 IMPACT LOAD ACTING ON PREVENTION EQUIPMENT

3.1 Basic equation of impact load

When the worker crashes to the point "A" of the equipment as shown in Figure 7(a), the impact load, F, can be expressed as the addition of the static load, F_s, and the dynamic load, F_D, as follows:

$$F = F_S + F_D \tag{1}$$

where: F = impact load; F_s = static load; and F_D = dynamic load.

F_s and F_D can be obtained from the variation of kinetic momentum on each part of worker's body and the equilibrium of forces as shown in Figures 7(b) and (c) as follows:

$$F_S = mg(\sin\theta - \mu\cos\theta) \tag{2}$$

$$F_D = \sum_{i=1}^{n} \frac{m_i v_i - m_i v_{0i}}{\Delta t} \tag{3}$$

where: v_i, v_{0i} = velocity on each part i of the body after Δt sec and at the moment when the worker crashes into the equipment, respectively; Δt = collision time; μ = coefficient of kinetic friction; m = mass of worker (assumed to be 70 kg in section 3.5); m_i = mass of each part i of worker's body; g = gravity acceleration; θ = pitch of roof; and n = the total number of body parts.

From the above equations, the magnitude of impact load depends on 6 parameters which are the velocities on each part of the body after Δt sec and at the moment when the worker crashes, collision time, pitch of roof, mass of worker, and coefficient kinetic friction.

3.2 Estimation on the falling velocity of worker

In this section, an estimation of falling velocity of worker at the eaves and ground surface will be carried out by using the accident data. Figure 8 shows the simplification of falling direction of worker on the roof surface. Only the accidents due to falling from the eaves, which happen more than 70% are considered. The worker was assumed as a simple model with a mass of mg at c.g. of the body as shown in Figure 7(b). A velocity vector of the mass at the eaves was assumed to be parallel to the roof surface. The falling path was assumed as a parabolic path to the diagonal projection from the origin at the eaves of the roof. Although the actual height of the c.g. at the eaves varies from about 120 mm (half of the maximum chest depth) to 1000 mm (navel height) (Research 1997), the height of the c.g. at the eaves was set to be 0 mm since they are not provided in the accident data. However, significant difference in the computation results by using value of 0 or 1200 mm as the maximum difference in the results is only about 20%.

The horizontal velocity and displacement along the falling path can be written as:

$$v_x = v_0 \cdot \cos\theta \tag{4}$$

$$x = v_0 \cdot \cos\theta \cdot t \tag{5}$$

As the worker falls vertically with a constant acceleration of gravity, thus,

$$v_y = v_0 \cdot \sin\theta + g \cdot t \tag{6}$$

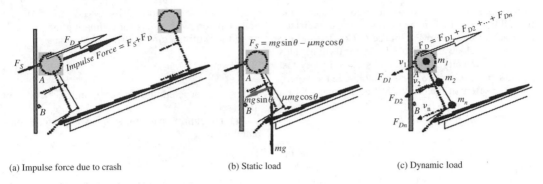

| (a) Impulse force due to crash | (b) Static load | (c) Dynamic load |

Figure 7.   Illustrations of impulse force due to crash.

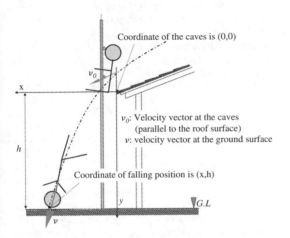

Figure 8.   Falling velocity estimation.

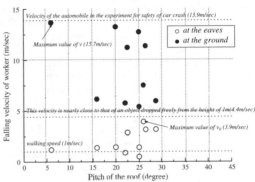

Figure 9.   Falling velocity of the worker.

$$y = v_0 \cdot \sin \theta \cdot t + \frac{1}{2} \cdot g \cdot t^2 \qquad (7)$$

where: x, y = the horizontal and vertical positions of the fall (the edge of the roof is set as the origin), respectively; v_0 = the velocity of the worker at the eaves; v_x, v_y = the velocity components of the worker; θ = the pitch of roof; and t = the time during falling from the eaves to the ground.

Then the velocity of the worker at the eaves, v_0, is

$$v_0 = \sqrt{\frac{g \cdot x^2}{2 \cdot \cos^2 \theta \cdot (y - \tan \theta \cdot x)}} \qquad (8)$$

The velocity until the impact at the ground can be described as the sum of the velocity components as follows:

$$v = \sqrt{v_x^2 + v_y^2} \qquad (9)$$

3.3   Estimation results of falling velocity

Figure 9 shows the relationship between pitch of the roof and estimated velocity obtained from the equations (8) and (9) (velocities at the eaves and ground surface). The results are based on the accident cases that have accurate information on the fall position of the worker, the height of the roof and the pitch of roof, without concerning cases that worker falls and crashes on the scaffolds.

It is found that the maximum velocity (indicated by white circle) at the eaves of the roof is about 4.2 m/sec (15 km/h). The estimated velocities for the pitch angle of 0° to 20° are nearly equal to the walking velocity of worker, while the pitch angle larger then 20° results in increasing velocity with the increasing pitch angle. Based on the accident data used in this study, only the accidents of "flopping down from the eaves" happened under the pitch condition from 0° to 20°. If the pitch of roof is small, the frictional force becomes large and slipping or rolling down the roof becomes difficult. Therefore, the accident type of "flopping down from the eaves" is a dominant case when the pitch angle is from 0° to 20°. Moreover, it is found that all the velocities at the ground surface (indicated by black circle) are over 5 m/sec (18 km/h), and the maximum velocity is about 14 m/sec (50 km/h). This maximum velocity at the ground surface is very fast as the setting velocity of the automobile in the experiment for safety of the car crash (Engineering 1998).

3.4 Simulation of impact load

In this study, the impact load can be estimated as previously described in equation (1). However, the magnitude of the dynamic load is much larger than that of the static load because of short collision time. Thus, only equation (3) was used in the simulation by assuming that (1) the head of worker crashes into the guild rail, (2) velocity on all parts of the body when the worker crashes into the guild rail are the same, and (3) the head's velocity becomes to 0 m/sec after Δt sec. Corresponding to the accident data and the estimated results from equation (9), the minimum and maximum falling velocity of worker (1 and 4.2 m/sec, respectively) are selected to use in the simulation as listed in Table 2. Cases 1 and 3 represent the cases of possibly smallest impact force acting on the head by assuming that other parts of body maintain the same velocity before and after crashing. Cases 2 and 4 represent the cases of possibly largest impact force acting on the head by assuming that the velocity after crash of all parts of body becomes 0 m/sec. Figure 10 shows the relationship between collision time and estimated impact load on each simulation case obtained from equation (3). In the figure, the simulation results are compared with the impact force obtained from the multiplication of assumed. human head mass (5 kg) with the experimental data (Prasad 1985) for skull fracture provided in form of the acceleration (indicated by white and black circle). White and black circle represent the experiments that had and had no skull fracture, respectively. Moreover, the approximate impact force based on the criteria of the acceleration that can cause head injury (HIC index) is also plotted in Figure 10 (Gadd 1966). The equation to compute that criteria

acceleration was proposed by GADD from Wayne State University in 1960s for estimating the possibility of the skull fracture by using cadavers as:

$$A = 155.3 \cdot T^{-0.4} \tag{10}$$

where: A = acceleration (m/sec²); and T = collision time (sec).

And the approximate impact load, F_A, can be computed from:

$$F_A = m_{head} \cdot A \tag{11}$$

where: m_{head} = mass of the head (assumed to be 5 kg).

From the results, it was shown that, firstly, the simulated impact load for case 3 is smaller than all data points of the experimentally based impact force. This means that case 3 has a low possibility to have the skull fracture. But for cases 2 and 4, the simulation results are larger than all data points of the experimentally based impact forces, which imply that both cases have possibility to have the skull fracture.

Secondly, it can be seen that the approximate impact load based on HIC index gives a good criteria on the skull fracture of the experimental data. The experimental cases with no skull fracture have more or less smaller impact force than the criteria curve and vise versa.

Considering the simulation results of case 4 that has possibly largest impact load for the case of 1 m/sec (3.6 km/h) falling velocity which corresponds to the accident type of "flopping down from the eaves", the results intersect the criteria curve at the collision time of 27 msec with 3.3 kN (340 kgf) impact load. This implies that a protection equipment which can introduce the reaction load of crashing less than about 3.4 kN (350 kgf) or prolong the collision time over than 30 msec would be able to assure safety for the accident type of "flopping down from the eaves".

Table 2. Parameters used in simulation.

	Velocities when the worker crashes into guild rail (v_0) (m/sec)	Velocities after Δt sec (v_i) (m/sec)	
	All parts of body	Head	Other parts of body
Case 1	4.2*	0.0	4.2
Case 2	4.2	0.0	0.0
Case 3	1.0*	0.0	1.0
Case 4	1.0	0.0	0.0

*4.2 m/sec = 15.0 km/h; 1.0 m/sec = 3.6 km/h.

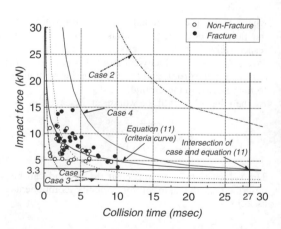

Figure 10. Impact force due to crash.

For the simulation results of case 2 that has possibly largest impact load for the case of 4.2 m/sec (15 km/h) falling velocity which corresponds to the accident types other than "flopping down from the eaves", they seem not to intersect the criteria curve or the intersection may take very long collision time. However, at very long collision time, the validity of the criteria curve is still in question. Thus, for the accident type of this case 2 simulation, it requires further study and other special equipments.

4 CONCLUSIONS

The major findings obtained in this study can be summarized as follows:

1. The falling process can be categorized into 3 types and there is a necessity to install some falling prevention equipments that can protect them all.
2. The maximum falling velocity is about 4.2 and 14 m/sec (15 and 50 km/h) at the eaves and the ground surface, respectively. The maximum velocity at the ground surface is much fast as the setting velocity of the automobile in the experiment for safety of the car crash.
3. It is necessary to apply the prevention equipment that can enlarge the collision area and extend the collision time as it can reduce the impact load.

REFERENCES

Construction Research Institute, 2001. *Statistical Yearbook on Construction (in Japanese).*

Editorial Committee for Pandect of Industrial Safety Technique, 1999. *Pandect of Industrial Safety techniquet (in japanese).*

Engineering and Safety Department, Road Transport Bureau, Ministry of Transport, 1998. *Manual for Screening Standard for New model car (in Japanese).*

Gadd, W. 1966. Use of Weighted-Impulse Criterion for Estimating Injury Hazard. *10th Car Crash Conference*: 164–174.

Hino, Y. & Nagata, H. 2002. Analysis of falling accident in the roofing work. *31st Symposium of Safety Engineering (in Japanese)*: 295 298.

Industrial Safety and Health Department, Ministry of Labour, 2000. *Manual of law of occupational safety and health (in Japanese).*

Japan Construction Safety and Health Association, 1996. *Instruction manual of guideline of precedent scaffold installation method (in Japanese).*

Kinoshita, K. & Ogawa, K. 1971. Efficiency Improvement of Safety Net. *Research Report of the Research Institute of Industrial Safety (in Japanese).*

Prasad, P. & Mertz, J. 1985. The Position of the United States Delegation to the ISO Working Group 6 on the Use of HIC in the Automotive Environment, *SAE Technical Paper Series.*

Research Institute of Human Engineering for Quality Life, 1997. *Japanese Body Size data 1992–1994 (in Japanese).*

Fly ashes thermal modification and their utilization in concrete

R. Hela, J. Marsalova & L. Bodnarova
Brno University of Technology, Faculty of Civil Engineering, Brno, Czech Republic

ABSTRACT: The aim of research work was the verification of fly ash activation in order to obtain an alternative material with better qualitative properties than common fly ashes for utilization in transport-concrete and in self-compacting concrete. The undemanding method of fly ashes thermal activation which should increase the puzzolana activity of common fly ashes was verified. This should increase the utilization of fly ashes in concrete manufacture and simultaneously achieve economical effects, especially cement savings. Two methods of thermal activation were tested experimentally and then a series of tests was made with the way treated activated fly ash in order to evaluate the differences of activity between common and activated fly ashes utilized as an active admixture in concrete.

1 FLY ASH FUNCTION IN CONCRETE INTRODUCTION

Power plant fly ashes have the power of puzzolana reaction. This reaction is defined as the reaction of silicon dioxide SiO_2 and aluminum oxide Al_2O_3 in fly ash with calcium hydroxide $Ca(OH)_2$ under the formation of calcium-silicate hydration products and calcium-aluminate hydration products. The evaluation of puzzolana reaction has to involve not only the power to bind calcium hydroxide but even the course of reaction between fly ash and the calcium hydroxide $Ca(OH)_2$. It was determined that the puzzolana reaction starts under normal hardening conditions after approximately 28 days and the positive influence on strength appears only after 90 days.

The fly ash is dosed in the quantity of up to 40% in relation to cement. It improves the mobility, the pumping properties and the workability of fresh concrete. In the same time it increases the necessary cement–water ratio to achieve the demanded consistency, because a part of water is used for the adsorption on the surface of fly ash grains. The hydraulically active fly ash increases the long-term strength values of concrete. The admixture of fly ash lowers the depth of concrete carbonation and the reversible shrinkage. More points of view can be considered for the determination of fly ash properties influencing the fly ash activity. As the most active are considered original or ground particles with the size 5 μm till 30 μm.

The fly ashes from the power plant Chvaletice (classic method of brown coal combustion) were chosen for the verification of fly ashes thermal activation possibilities and their effect on the activity of power plant fly ashes. Two processes of thermal activation were proposed and experimentally verified. The effectiveness of the way modified fly ashes was tested on cement mortars.

2 ACTIVATION OF POWER PLANT FLY ASHES

The following method was designed and experimentally tested for the activation of power plant fly ashes. Finely ground limestone was added to common fly ash from classical combustion of brown coal in the quantity of 15% till 20% and possibly even calcium sulphate in the form of building plaster up to 5%. The proper processing was as follows.

A homogenous mixture of fly ash, ground limestone and plaster was mixed with water to form a wet mixture from which pellets were formed. These pellets were treated in an autoclave in a hydrothermal process and afterwards the mixture was spontaneously cooled. The sample was dried to constant mass after the end of autoclaving. The way prepared sample is subsequent text marked as F1.

The second method of activation was: the samples were treated after drying in a short-term burning and afterwards spontaneously dried to the laboratory temperature. The way prepared sample is in the subsequent text marked as F2.

The fly ash Chvaletice was chosen for activation verification. It was mixed with lime powder and gypsum in the relation 78% : 17% : 5%.

After dosing the individual components the row mixture was homogenized thoroughly and by addition of a small quantity of water transformed into a moist mixture. This was in a sample carrier placed in an autoclave and subjected to hydrothermal treatment.

3 MINERALOGICAL COMPOSITION OF ACTIVATED FLY ASHES

Both activated samples were mineralogically tested by the diffraction X-ray method, by differential thermal analysis (DTA) and by electron microscopy. The X-ray pattern of activated fly ashes you can see in following figures.

As far as qualitative composition is concerned the X-ray diagrams of both phases are practically identical. They contain quartz and mullite from the fly ash and calcite from the original limestone. However the intensity of diffraction lines shows that the calcite content in sample F1 is significantly higher than in sample F2. The sample F1 shows besides a small content of gypsum ($d_{hkl} = 7.55$ Å). An accessorial quantity of portlandite ($d_{hkl} = 4.92; 2.627$ Å) cannot be excluded.

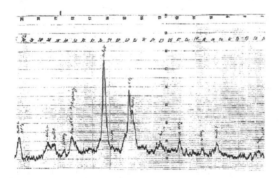

Figure 1.   X-ray diagram of fly ash F1.

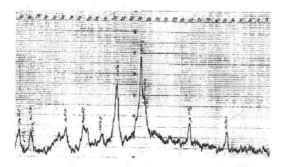

Figure 2.   X-ray diagram of fly ash F2.

Beside the above mentioned facts in both samples the diffraction line $d_{nkl} = 2.71$ Å was identified.

This diffraction line, which coincides with the hematite line $d_{hkl} = 2.68$ could belong to the hydrogarnet phases.

The DTA results of both samples show that up to the temperature 400°C an endothermic line of dehydration was observed (it is in accordance with the loss of hydrate water) with a double maximum, the first maximum is in the case of both samples at the temperature about 150°C, the second maximum is in the case of sample F1 at 200°C and with sample F2 at 300°C; after this first endothermic line follows a short indifferent zone which in the case of sample F1 turns in a slight exothermic waving as the result of organic substances burning from the fly ash. Beginning with the temperature of about 600°C follows an endothermic effect of calcium carbonate decomposition with a double maximum; the first at 700°C corresponds with the fine fraction, the second at 850°C with the coarse fraction of $CaCO_3$.

By the quantitative evaluation of the thermogrametric line it was found that the calcium carbonate content in sample F1 is in the region of 18% but in sample F2 it is about 9.3%. The ignition loss corresponding to the dehydration endothermic line up to 400°C is in the case of sample F1 – 5% and in sample F2 it is 3.6%. From the known gypsum dose used for the fly ash activation (5%) the conclusion was made that the mentioned difference in the ignition loss corresponds exactly with the ignition loss of gypsum, which was formed from plaster during the hydrothermal treatment.

The rest of the ignition loss is in the case of sample F1 in accordance with the dehydration till dehydroxilation of gelatinous silicic acid. In the case of sample F2 this loss of ignition corresponds by one half with dehydration of calcium-hydrate-silicates, by the second half it represents the phase corresponding with hydro-garnets.

The electron microscopy is illustrated in Figures 3, 4 and 5.

In Figure 3 you can see the round grains of original fly ash, mostly smooth without grains with a new shape.

In Figure 4 you can see, that during the hydrothermal treatment a significant change of the sample surface took place. This change was caused by the adsorption of activating admixture – calcium carbonate and even of the original gypsum on the surface of the sample.

Figure 5 provides documentary evidence for the morphology of activated F2 fly ash i.e. after the realization of burning. On the figure you can see the adsorption of activating substances at the surface of originally round, smooth fly ash grains. It seems that by the influence of partial reaction between fly ash and

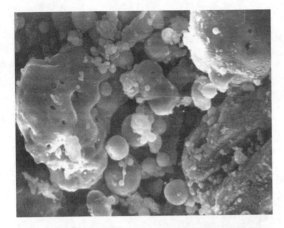

Figure 3. Fly ash grains: Sample F0 – not activated fly ash.

Figure 4. Fly ash grains: Sample F1 – autoclaved fly ash.

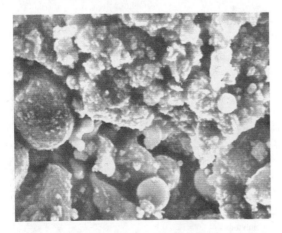

Figure 5. Fly ash grains: Sample F2 – autoclaved fly ash, subsequently thermally treated.

limestone a small portion of melt could occur which caused a certain "smoothing" of the grain surface.

During the hydrothermal treatment of sample F1 happens that the activating limestone and gypsum components practically don't react with the fly ash. On the base of thermo-gravimetric analysis evaluation it is obvious that the content of limestone after the hydro-thermal treatment didn't change and the plaster was transformed into gypsum. But it is obvious that the hydro-thermal treatment had an influence on the fly ash activity. Most likely the thermal process initiates the transformation of amorphous silicic oxide into silicic acid gel. These gels during the hydration with cement react more willingly with calcium hydroxide as a secondary product of reaction.

During the burning process a part of calcium carbonate reacts with the silicic acid gels and forms a calcium-hydro-silicate respectively a calcium-silicate-hydro-aluminate system. This can be documented by the lower calcium carbonate content of sample F2 in comparison with sample F1. At the same time on the base of X-ray analysis evaluation the mentioned calcium-silicate-hydro-aluminate phases were, regarding their composition, very near to the hydro-garnet phase. In this case during the further hydration with cement the mentioned CSH and CASH phases will act as crystalline nuclei for new hydration products and the rest of silicic acid gels will again relatively quite willingly react with portlandite under the formation of new calcium-hydro-silicate products.

4 VERIFICATION OF THERMAL ACTIVATION EFFECTS ON CEMENT MORTARS

The tests on cement mortars were performed in order to determine the activity of fly ashes and to compare their behaviour with unmodified fly ashes. The verified mixtures had always the same consistency and for this reason it was necessary to change the batch water dose.

The following tests were performed with hardened mortars on test samples with the dimensions 40 × 40 = 160 mm:

– compression strength after 3, 7 and 28 days of standard hardening
– tensile strength after 3, 7 and 28 days of standard hardening.

5 THE COMPOSITION OF USED CONCRETE MIXTURES

The composition you will find in Tables 1, 2 and 3. The not activated fly ash F0 and activated fly ashes F1 and F2 were dosed in the same way, therefore in the table

Table 1. Chemical composition of all three fly ash types.

	SiO_2 %	Al_2O_3 %	Fe_2O_3 %	CaO %	MgO %	Na_2O %
F0	56.82	28.93	6.18	1.79	1.31	0.32
F1	39.62	20.70	4.53	18.24	0.96	0.21
F2	42.62	21.70	4.64	22.34	0.98	0.24

Table 2. Chemical composition of all three fly ash types.

	K_2O %	SO_2 %	Overall %	Loss of ignition %
F0	1.79	0.50	97.64	2.36
F1	1.30	5.38	90.94	9.06
F2	1.34	6.14	100	0

Table 3. Chemical composition of all three fly ash types.

	Specific surface cm^3/g	Volume mass kg/m^3
F0	2426	2036
F1	2380	2420
F2	2401	2580

there is only the marking Fx. The chemical composition of all three fly ashes is in the following Tables 1, 2 and 3.

Standard quartz sands were used as fillers, the plasticizing admixture was from SIKA-company with the trade mark Viscocrete 5 – 800. For all batches the cement CEM I 42.5 R was used.

6 THE TEST RESULTS OF HARDENED MIXTURES

All tests on mortars were made with the aim to determine the activity of F1 and F2 fly ashes and to the compare it with the original unmodified fly ash F0 on the base of decrease evaluation of initial and final strength values. All mixtures were designed with the same consistency by means of water batch modification. The details of test results you will find in following tables and graphs, which provide documentary evidence for the whole work.

In following graphs examples of some compression and tensile strengths development are documented.

7 FINAL CONCLUSION

The described experimental work concerned the testing of activated fly ash properties in cement mortars

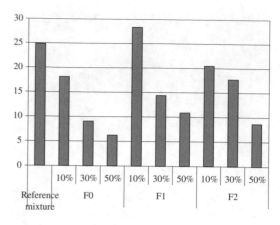

Figure 6.   3 day compression strengths in MPa.

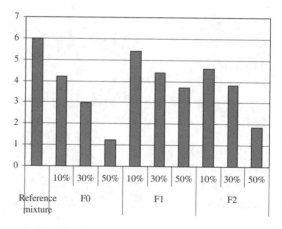

Figure 7.   3 day tensile strengths in MPa.

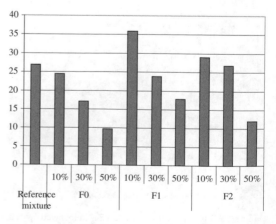

Figure 8.   7 day compression strengths in MPa.

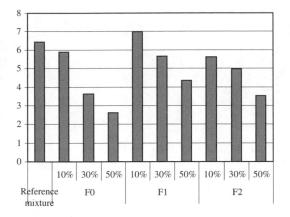

Figure 9. 7 day tensile strengths in MPa.

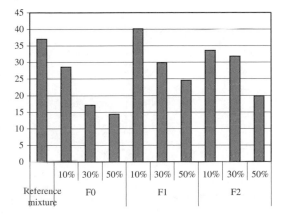

Figure 10. 28 day compression strengths in MPa.

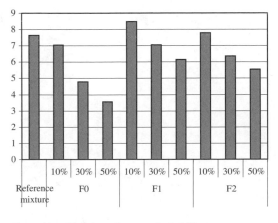

Figure 11. 28 day tensile strengths in MPa.

where the fly ash was used as a partial substitute of cement. The tests performed on cement mortars had to define the activated fly ash properties during their utilization as replacement of cement, further the determination of the initial strength acceleration and even the utilization of fly ash a material improving the rheological mortar properties for the possible application in self-compacting-concrete. The test results showed that the mortar containing the F1 fly ash has higher compression strength in comparison with mortars containing the fly ash F2. The latter has higher strength values than mortars with fly ash F0. The fly ash F1 in a certain percentage representation has even higher strength values than the mortar without fly ash with an overall higher dose of cement.

Mortars utilizing F1 fly ash have the highest tensile strength values from all tested variations.

Mortars with fly ash have a slower strengths increase. However this strength increase takes place even after the increase in mortars without fly ash was finished. This fact manifests itself in the age of 28 and more days when the difference of strength between the reference mixture without fly ash and with fly ash is in percentages slower than after 3 till 7 days of hardening.

ACKNOWLEDGEMENTS

This work was published with backing of GACR, within the research project No. 103/01/0814 "Microstructural features and related properties of cement based highly liquefied composites" and the GACR research project No. GACR 103/01/1144 and the research project CEZ – MSM 26110008, name of the project is: "Research and Development of New Materials and Securing their Higher Durability in Building Structures."

REFERENCES

Hela, R., Bodnárová, L. Experience with Self-Compacting Concrete Technology in Czech Republic, Performance in *Product and Practice, New Zealand, 2001-01-01*, CIB.

Hela, R., Sanhai, Zeng. Application of activated Fly Ash in Mortar and Concrete. In FIB Congress 2002. *Concrete Structures in the 21st Century, Osaka, 2002.*

Kratochvíl, A., Urban, J., Hela, R. Fine Filler and its Impact to a Cement Composite Life Cycle. In UVAR. *Non-Traditional Cement and Concrete, Czech Republic, Brno 2002.*

Sanhai, Zeng, Hela, R. *Analysis and Comparison of Portland Cement between Europe and Chine, Shanghai 2002.*

Necessity of study on countermeasure for climate change to avoid possible disasters

K. Baba
Kochi University of Technology, Japan

ABSTRACT: There is common threat of global climate change for human being in future. It is necessary to do something to prevent from this climate change by all means but also to prepare some countermeasure for the influence by this. By global warming, the sea water level will rise. This will be a potential of big disaster of shortage of food, especially in Asia where rice is major crop of food. If level of sea water rise, sea water will penetrate into paddy fields of rice production. Since rice plant cannot be grown in sea water, rice production will be reduced. It is necessary to do some countermeasure for this as one of important items in sustainable construction. The movement of rise of sea water will be quite gradual. Therefore, sometime these movements could be overseen. But, it is necessary to be considered as one of the big items in sustainable development especially in Southeast Asia.

1   INTRODUCTION

Since Japan has been developed so rapidly after the civilization of Japan – Maiji revolution in 1868 – and become one of the modern countries that were considered Western countries only could performed. And moreover this has been done through the difficulties brought by Second World War. In 1980's some well-known professor of US wrote the book "Japan as No. 1" And many countries followed the way of modernization of Japanese model.

Immediate after the Second World War GNP per capita of Japan was same as that of Philippines. After 50 years from that, GNP per capita of Japan becomes approximately 50 times as much as that of Philippines. How it become possible. Japan is, in principle, quite poor country since there is no resource except human resources. By industrialization Japan become rich by exporting many products all over the world.

And this historical fact established the interpretation that the development of the countries means the increase in GNP per capita by industrialization.

Now a day Japanese economic power has reached its peak and China becomes most rapid developing country. And Chinese way of development of the country is almost same as that of Japan. Industrialization and increase in GNP. Many other countries takes almost same steps therefore there are competition in industrialization and increase in GNP. This situation is quite worse from the global environment view point. It will bring shortage of resources and increase in pollution problem as well.

In order to prevent such tendency, it is necessary not only to adjust "Japanese model of development" in consideration of environment but also more active ways to prepare some countermeasure for the disasters caused by the sub-effects of development of the developing countries. One of the examples of this should be the countermeasure for the effect of climate change.

2   CONCEPT OF DEVELOPMENT

2.1   *Japanese model of development*

Japanese who worked hard in order to escape from poverty never realized that Japan can be one of the developed countries in such a short period of time as it was because Japan only has human resources, no oil, no mineral resources and even there is not enough food. Actually as far as income is concerned it has been remarkably increased.

However, in stead of their remarkable progress of development, many Japanese come to conclusion that to get out of poverty is not final object but something of means. Final object of the people should be welfare. Increasing GNP will bring some more choice for welfare. Therefore, the development that will bring something to destroy welfare in the future should be meaningless.

Actually, Japan has been experienced many cases such as pollution problem in cities of industry and some fatal deceased caused by poisonous materials discharged by industrial plants.

There are many discussions on both concepts of development and welfare of the people.

2.2 *Bhutan's unique policy based on development philosophy in their culture*

Higher authority of government of Bhutan, a small country in Himalayan mountainside, gives one of the most attractive opinions. Policy of Bhutan government is to attain welfare of the people not by developing the country in the way other developing countries took but to take the balance between their policy for sustainability and the philosophy in their national religion.

This was made certain by the author's meeting with the Minister of Planning of the Government of Bhutan and discussed on this matter. Their question was whether Japanese people did performed welfare of the people by rapid development.

Although they even evaluate very much the Japanese remarkable progress of the development, they do not want to follow Japanese model for welfare of people by development. Before Bhutan established this planning of development they were looking for the way of development in the same way of usual development as a least developed country. But they changed their policy not to place the priority in rapid development.

If we asses our experience and other opinion like the policy of Bhutan Japanese has reached to the conclusion that it is necessary to establish some philosophy of development for welfare, in consideration with not only environment but also culture of the people. Due to the trend of globalization human being may obtain big merit but also may loose something important and never can be obtained again.

Sustainability should be not only in the side of environment but also culture and heritages of the regional people. Especially, recent technology mainly based on the Western culture, it is necessary for the people of different culture to take the more consideration upon their own culture.

3 TRADE-OFF BETWEEN THE ENVIRONMENT AND ECONOMIC DEVELOPMENT

3.1 *Priority of environment*

As countries in Asia become richer and richer, the environment problems become a greater preoccupation. Then, many countries in Asia will be more unpromising about the trade-off between the environment and economic development.

There are not political parties of the same nature as "green party "in the European countries. However polls

shows that not only in Japan, but also such countries, Korea, Taiwan, Malaysia and the Philippines, a signi ficant number of people are prepared to have their earnings reduced in exchange for a less polluted environment. Many middle-class people in these countries are moving out of the cities to build settlements in the quiet country-side. Access provided so rapidly by increasing sophisticated telecommunication facilities, a small revise flow from cities to the countries has been observed.

Since, the social and environmental costs are incalculable, in rather rich country trade-off between development and environment are done in more consideration of environment. But, in less developed countries where GNP is considerably low put the priority more on development.

3.2 *Relation between GNP and air pollution*

Naturally, the trade-off between the environment and economic development should be done in the balance between priority of development in less developed countries and more consideration upon environment in relative developed countries.

This fact will be explained by the result of the study of the GATT on relation between GNP per capita and air pollution as shown in Fig. 1.

According to this report more consideration will be actually taken to environmental problem after their income per capita exceeding some around 4,500 in US $.

Of course, it is necessary to shift this curve, as a whole to the left but at the same time developed countries should do something more for environmental problems.

By doing so, South and North problems between the countries will be reduced and all country could do something for environmental problems.

In the Asia, many countries are suffered from environmental problems, especially pollution problem in the big city like Bangkok, Jakarta and Manila which are just located in drawing in Fig. 1 in the left side of the peak of the curve. Other countries like Japan,

Density of SO$_2$ in Atmosphere: in p.p.m.v.
Per GNP per Capita

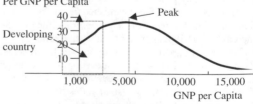

Figure 1.   Relation between GNP per capita and air pollution. *Source*: Grossman & Krueger 1991 in GATT (Trade & the Environment) Report.

Korea, Taiwan, Hong Kong and Singapore are located in the right side of the peak of the curve.

4 COMMON THREAT OF CLIMATE CHANGE FOR HUMAN BEING

According to the report of IPCC (International Panel of Climate Change: 1995) there will be 5 major influences on the life of human being by climate change as follows:

1. Spreading out of the epidemic decease of tropical region like malaria.
2. There will be more typhoon, cyclone & hurricane. Location of rainy and dry region will be changed.
3. Due to the rise of sea water level more erosion of land and penetration of sea water to river.
4. There will be changes in species of forests. Rapid shrinking of forest and expand of deserts will be.
5. Fatal shortage of food will cause famine especially in least developed countries.

From view point of sustainable construction (2), (3) and (5) will be field of studies.

5 CLIMATE CHANGE

5.1 *Predicting the near future*

There are natural questions to ask what we know about changes that will occur in the future. Is it possible to predict the characteristics of atmosphere in the future? Atmospheric scientists can, with reasonable confidence, forecast the patterns of the next few years, but their predictions become increasingly uncertain as they attempt to describe the more distant future.

5.2 *Models of the earth system*

Numerical atmospheric models differ greatly in complexity, especially in their degree of spatial and temporal resolution. Ideally, a model should be able to generate predictions specific to large and small geographical areas and for long and short the intervals. In practice, though, the information required to do that may not be available, or the capabilities of the computers may be insufficient. All computer modelling efforts at best represent tradeoffs: scientists must choose spatial and temporal resolution at the expense of physical, chemical, and meteorological detail, or vice versa.

Earth system models have several characteristic formats. The simplest model for atmospheric applications is the box models, envisioned as a box into which some things are added, from which some things are taken away, and within which changes occur.

Chemical species enter in two ways: they are emitted from sources within the box, or they enter by entrainment (the addition of air and its chemicals into the box

from the surroundings as a consequence of atmospheric motions). Conversely, detrainment (the loss of air and its chemicals into the surroundings) represents a loss of chemical species.

Computer models able to incorporate the properties of greenhouse gases along with specific emission scenarios can also be used to calculate likely temperatures for Earth's surface at different times in the near and intermediate future. In Fig. 2 several such calculations based on projections of future atmospheric emissions have been grouped to show ranges of possible low, intermediate, and high temperature increases throughout the next half-century.

The average global temperature over the next half-century will indeed increase to somewhere within the ranges indicated.

The source of greatest uncertainty in even the most sophisticated climate change predictions is thought to be the treatment of clouds in the models. Higher temperatures bring more water vapor into the atmosphere, which in turn results in the formation of more clouds, and clouds affect climate in two ways.

On the one hand, they reflect back to space part of the Sun's radiation before it reaches the ground, and thus they have a cooling effect. On the other hand, they trap upward-moving heat radiation and thus they also have a warming effect. The net result of these two processes depends on the clouds' altitudes. High (cirrus) clouds reduce outgoing heat radiation more than they reflect incoming solar radiation, thus adding energy to the Earth-atmosphere system; low clouds have the opposite effect.

The models must therefore predict the rates at which changes in temperature will produce clouds at different altitudes and seasons, all the while dealing with the complexities of cloud-droplet nucleation around particles, altitude and humidity changes over

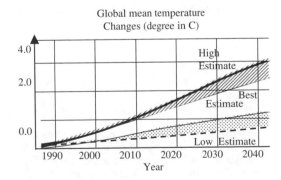

Figure 2. Predicted global temperature. *Note*: The annual mean surface air warming for computation that assume "business as usual" rates of greenhouse gas emissions but consider a variety of climate sensitivities, oceanic heat uptake rates, and other variables.

different continents and subcontinents, and the different possible air-parcel motions under different planetary temperature conditions.

6 RISE OF SEA WATER LEVEL

6.1 Global warming increase in global mean sea level

The possible global average temperature increase of one or two degrees centigrade over the next few decades may not sound very large until we compare it with past climate oscillations and consider what their effects have been.

Generally speaking, computer models estimate that among the results of such warming would be an increase in global mean sea level of perhaps 20 cm by the year 2030 and perhaps 45 cm by 2070. This will occur because the ocean water will expand slightly as it is heated and because a warmer climate will cause increased melting of mountain glaciers. Such sea level increases would produce major disruptions for the large fraction of the world's population living in coastal regions, especially the less developed nations in Southeast Asia. For example, in Bangladesh, most of the population lives at or near the water's edge. Finally, suppose that the higher-temperature warming scenarios occur, a prospect as likely as the lower ones. A change of 3 or more would be comparable to the temperature change that occurred between the last major ice age and the present, a transition that took place over several thousand years. In contrast, humankind would have produced, within a period of only a century, the warmest climate to exist in millions of years. We have only a vague idea of how our natural systems would respond.

In addition to predicting how changes in the concentrations of today's greenhouse gases might affect future climates, it is worth asking whether other gases might be involved in future climate change

6.2 Projections for future sea level rise

The National Research Council (NRC) Panel on Sea-Level Change projects a sea level rise of 50 cm ± 100 cm by the year 2100 (the assumed doubling time for CO_2). They assume a 3° to 6° global air temperature rise over the next hundred years based on results from general circulation models at NOAA's Geophysical Fluid Dynamics Laboratory, the National Center for Atmospheric Research, and NASA's Goddard Space Center. The thermal expansion (steric) part of this estimate is based on the work of Frei et al. (1988) who used two models for carrying heat down into and up out of the subsurface layers of the ocean, a pure diffusion model and an upwelling-diffusion

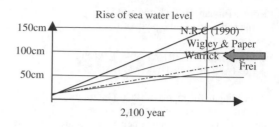

Figure 3. Estimated amount of rise of sea water level. *Note*: ◀━━━ shows target

model. They project a 10 to 50 cm rise over the next century.

For the contribution to sea level rise from ice wastage/melting the NRC panel used the results of the NRC Committee of Glaciology (NRC, 1985), whose "most likely" scenario was 55 cm ± 21 cm by the year 2100.

Warrick and Oerlemans (1990) (in the IPCC report on scientific assessment of climate change) project, for a "business as usual" scenario, a sea level rise of 21 to 71 cm by the year 2070, with a best estimate of 44 cm (66 cm for the year 2100). The thermal expansion part of this estimate is based on the upwelling-diffusion model of Wigley and Raper (1987), using projected global air temperature rises of 1.5°, 2.5° and 4.5°, with the middle value considered a best estimate. They used the glacial contribution calculated by Raper et al. (1990) using a simple global glacial melt model, and they assumed no contribution from the Greenland or Antarctica ice sheets. Both these review papers also list projections by various researchers with "best estimates" ranging from 20 to 100 cm.

All of the above projections for sea level rise in the next century are less than the earlier larger estimates that caused great concern, but they are not insignificant and, if they occurred, could have serious consequences. It is important for sea level research to continue, because there are so many unanswered questions, and the problem is so complicated.

There have been many estimation reports on the amount of rise of sea water level as shown in Fig. 3.

Since estimation of sea water level rise is not decisively fixed, it is better here to consider as some condition to consider the amount of the sea water rise by the end of this century might be some around 100 cm as a target.

7 EFFECT OF SEA LEVEL RISE ON RICE PRODUCTION

7.1 Rice as major crop in Southeast Asia

Major food crops in the world are wheat, rice and others. And distribution of the regions of production of these crops is depending on the geological and climate characteristics of lands. The production of rice is done mainly in Asian countries.

Therefore, the influence by rise of sea level on production of rice in Southeast Asia could be important to discuss the supply condition of rice in future.

7.2 Effect of rise of sea water level on rice production

Effects of rise of sea water level will be remarkably big on the agriculture. Especially effect on rice production is considered big, because rice is produced in the paddy field where water should be supplied from rivers or creeks near to sea. If sea water level will be rise, sea water will go to rivers or creeks to paddy field. Since rice cannot grow in sea water, production of rice will go down remarkably.

Since rice is major crop for food in the Southeast Asia where there are so many cases of damaged paddy field by penetration of sea water. Some of them became field for production of salt and other become pond for cultivation of shrimps.

8 EXECUTION OF SUSTAINABLE CONSTRUCTION

8.1 Sustainability in wide sense

Sustainability usually implies that the use of energy and materials in an urban area be in balance with what the region can supply continuously through natural processes such as photosynthesis, biological decomposition and the biochemical processes that support life.

Of course it is important and this concept of sustainability becomes common sense. But, this concept should be considered as narrow sense. It will be necessary to have some concept of wide sense such as global problem like climate changes.

By doing so, many global problems should be studied to prepare necessary countermeasures.

To take the balance between developed countries and developing countries for execution of sustainability, developed countries should do sustainability in wide sense and developing countries should do at least sustainability in narrow sense.

8.2 Role of civil engineer

The movement of rise of sea water will be quite gradual. Therefore, sometime these movements could be overseen. But, it is necessary to be considered as one of the big items in sustainable development especially in Southeast Asia.

Some countries like Bangladesh, Maldives and Ireland countries in the Pacific Ocean where level of land is so low that fatal damage will be expected occurred. Countermeasure for rise of sea water level should be studied as sustainable construction by Southeast Asian countries together with these low countries.

Since damages will become in reality in 22 century, therefore countermeasure should be prepared in this century as an important item of Agenda of 21 century.

Since issues concerning the countermeasure for global climate change are so complicated and subject will be studied in many different field of science and engineering, only one field cannot solve the problem. Civil engineering should be the field that handles the matter totally and generally.

The roles of civil engineer should be to take the leadership to prepare for common threat to human being of climate changes as many leading economist takes leadership for this subject.

8.3 Actual way of countermeasure

Actual way to prepare for rise of sea water level will be done by mainly to different systems. One is to avoid sea water to penetrate into river by some kind of wears and water gates. Another should be to make channel of water line along with shore lines to prevent penetration of sea water to paddy field near to coast with some protection of corrosion.

However, these countermeasures cannot be done by one country because many rivers are international rivers and in many cases there are borders of countries in major rivers.

International action and cooperation of the countries will be indispensable to prepare any countermeasure for these kinds of problem.

8.4 Required changes in nature of ODA (official aid of development)

ODA of Japan placed priority on the development of industrialization but more priority should be taken for sustainability, especially for prepare something for global climate change since it takes long period of time to make these effective.

9 THE TREATMENT OF RISK

9.1 Inaccuracy in estimation for prediction

In the case of environmental problem, it is not possible to determine with certainty what result some particular policy of countermeasure will be, because scientific predictions by estimating are quite imprecise.

Especially, it is significant that scientific uncertainty is afforded by the global-warming problem. Certain gases, when emitted into the atmosphere, are considered of causing the global temperature to rise. If this is correct, there could be very serious implication.

Global warming could be trigger a rise in the sea level and result in destruction of plants including rice plants that could not be sited for the sea water to which they would to be subjected.

The estimation that increased emission of carbon dioxide and other greenhouse gases are causing a rise in temperature is completely depending upon a computer model. This model itself has been partially validated. This is quite typical example showing the nature of problems.

Thus the subject concerned to global climate changes is depending on the prediction in future. And there are so many kind of estimate. Since any countermeasure for the influence by climate change cost tremendously. More accurate prediction should be needed to be depending upon. And also, risk is bigger, the more inaccuracy of estimation for prediction will be.

9.2 Risk treatment by policy

Treatment of this type of risk will have two dimensions as follows;

1. Identifying and quantifying risks
2. Determining the acceptable extend of risk.

In this case benefit-cost analysis may be applied. But, it could be more practical to set up several scenarios to study. It is not necessary to make a dominant policy. And it may be impossible to do so. The best and only way is to make several alternative policies to treat this kind of risk flexibly in accordance with the process.

10 CONCLUSION

For sustainable development usually many studies have been mainly done for environment problems and saving resources. However, one of the subjects to be studied before it will become too late should be study on countermeasures for disaster that might be caused by climate changes, especially global warming.

The biggest possible disaster will be the disaster caused by rise of sea water level. Because rice plant cannot grow in sea water, there will be fatal poor harvest of rice to bring famine especially in Southeast Asia.

The prediction of environmental problems is not clear. Therefore, more flexible approaches will be necessary.

This problem should be preoccupation for people of Asia. People in rather developed countries in Asia study this and prepare some countermeasures together with people in developing countries. Also, this will be studied and executed by the people of the world because the sustainable development of Asia, economically, politically, and culturally, is one of the important issues that will be taking place in the world of Today.

REFERENCES

IPCC 1990. Climate change: The IPCC scientific assessment. Cambridge University Press, 365 pp.

National Research Council, 1990. Sea-Level Change. Geophysics Study Committee of the National Research Council, National Academy Press, Washington, D.C., 234 pp.

Raper, S.C.B., Warrick, R.A. & Wigley, T.M.L. 1990. Global sea level rise: past and future. In: *Proceedings of the SCOPE workshop on rising sea level and subsiding coastal areas*, J. Milliman (Ed.), Bangkok, John Wiley and Sons, Chichester.

Roemmich, D. 1990. Sea level and thermal variability of the ocean. In: Sea-Level Change. Geophysics Study Committee of the National Research Council, National Academy Press, Washington, D.C., pp. 208–217.

Stewart, R.W., Kjerfve, B., Millimarl, J. & Dwivedi, S.N. 1990. Relative sea-level change: a critical evaluation. Unesco (COMAR) Working Group on Mean Sea-level Rise and Its Influence on the Coastal Zone, In collaboration with IOC. Unesco, reports in marine science No. 54.

Tietenberg, T. 1998. Environmental Economics and Policy, Addison-Wesley, N.Y.

CIB, 1995. cib Agenda 21 on sustainable construction, cib Report Publication.

Gradel, T.E. & Crutzen P.J. 1995. *Atmosphere, Climate, and Change*. W.H. Freeman and Company, N.Y.

Piirie, R.G. 1996. *Oceanography*, (Third Ed.) Oxford University Press, Oxford UK.

Sustainable bearing structures for (office) buildings

M.J.P. Arets & A.A.J.F. van den Dobbelsteen
Delft University of Technology, Faculty of Civil Engineering and Geosciences, Department of Building Technology and Building Process, The Netherlands

ABSTRACT: Taking a factor 20 environmental improvement in 2040 as a target, in the year 2000 the advance in the environmental performance of Dutch office buildings turned out to be behind schedule, and there is no reason to believe this is different in other countries. Hence a focus on the parts that contribute most in the environmental load is necessary. The energy consumption during the lifespan of a building causes the greater part (60%) of the environmental load; the use of materials cause 30% of the environmental load. The bearing structure of buildings is responsible for the greater part (approximately 60%) of the environmental load caused by building materials. Therefore, sustainability in building technology should be concentrated on improving the bearing structure. There are four ways to improve the bearing structure in a sustainable way, which are elucidated in this paper.

1   INTRODUCTION

1.1   *Factor 20 improvement*

In 1987 the "Brundtland Commission" released the report our Common future' in which the definition of sustainable development was stated. This definition made a connection between environmental and social goals in the world. In the year 1990 Ehrlich, Ehrlich and Speth reintroduced a formula, which makes it possible to quantify this connection between both goals:

$$EP = P * W * E \qquad (1)$$

In this formula EP stands for environmental pressure, which was said to be too high and had to be halved in 50 years from 1990. P stands for the world population, which is predicted to double within 50 years. W stands for the average welfare rate of the world population, which is predicted to improve with factor 5 in 50 years. Therefore E, the environmental impact by welfare per citizen, has to decrease a factor 20 in order to solve the equation.

This factor 20 is a guideline for the environmental performance of, for instance, office buildings.

1.2   *Environmental importance of the bearing structure*

Taking this factor 20 improvement in 2040 as a target, in the year 2000 the advance in the environmental performance of Dutch office buildings turned out to be behind schedule. And there is no reason to believe this is different in other countries. The arrears are partly due to attention to aspects and building parts in which just a small environmental improvement is possible. If we want to achieve sustainable development in buildings, a focus on the parts that contribute most to the environmental load is necessary.

When analyzing the environmental load of (office) buildings it turns out that the greater part (60%) of the environmental load is caused by energy consumption during the lifespan of the building (assuming the lifespan is 75 years). The use of materials cause 30% of the environmental load, in which the bearing structure is the most important part (60%) (van den Dobbelsteen, Arets & van der Linden, 2002).

Buildings often don't achieve a lifespan of 75 years; offices are functionally obsolete after 25–35 years. The shorter the lifespan the bigger the share of the bearing structure in the environmental load of buildings. Therefore, sustainability in building technology should be concentrated on improving the bearing structure. There are four ways to improve the bearing structure in a sustainable way:

1. Finding the best materials given a certain building structure span
2. Making optimal combinations of different building materials
3. Finding new materials for buildings structures
4. Developing innovative solutions for building structures.

In the next chapters these four principles are elaborated.

2 THE BEST MATERIALS GIVEN A CERTAIN BEARING STRUCTURE SPAN

2.1 Horizontal elements of the bearing structure

When improving the bearing structure, consisting of the foundation, columns, beams, floors and roof structure and sometimes walls and facades, the first question is which part of the bearing structure is determining the greater part of the environmental load caused by the bearing structure. It turns out that the horizontal elements of the bearing structure are responsible for 60%–80% of the environmental load, especially the floors (including roof, ground floor and storey floors) are an important part of the environmental load (Arets 2001). The keynote behind this is that horizontal elements of the bearing structure consist of a lot of material.

2.2 The functional unit

To draw conclusions about the most sustainable floor construction given a certain structural span, a proper comparison is needed. The functional unit forms the basis for this comparison. The functional unit consists of structural, safety and comfort requirements as shown in Figure 1. When comparing the floor variants, it is important that they are designed in such a way that they meet all requirements.

The comparison is carried out as follows. The cross-sections of the floor variants are calculated for different structural spans, meeting all requirements in the functional unit. Then all floors are equipped with floor finishing and a lowered ceiling. Based on the cross-section of the floor (including ceiling and finishing) the material use is determined; from that the environmental load is derived.

2.3 The environmental cost and integral cost of different floor variants

The environmental load is expressed in terms of cost, making it possible to compare the economic and the ecological cost. The environmental costs are the cost necessary to prevent environmental effects and to restore the damage caused by the environmental effects. Although the environmental costs are specific for each product, they are devolved upon society and therefore paid by public funds. The environmental costs are determined with the database of GreenCalc, a Dutch LCA (Life Cycle Analysis) based computer-program which calculates the environmental load of buildings (in terms of cost).

The realization costs are of importance when choosing between floor variants. These costs are influenced by the requirements mentioned in the functional unit,

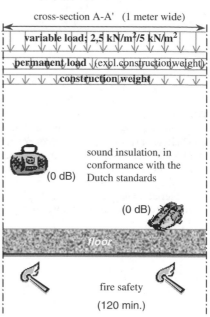
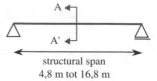

Figure 1. The functional unit, corresponding with the Dutch standards.

and also by aspects like the size of the project, the layout of the floor plan, etc. Adding the realization cost, the price of a floor readymade in the building, to the environmental costs leads to the integral costs.

The environmental costs, the realization costs and the integral costs are calculated for the several floor variants selected from the large amount of available floor alternatives. Selected are the floor variants which are often used in the Netherlands and which are different enough from each other. These are the hollow core slab, the composite floor-plate (reinforced/prestressed), the TT-slab, the steelplate-concrete floor, the BubbleDeck floor and a timber floor consisting of timber beams and timber floor elements.

An important difference between the BubbleDeck floor and the other floors is that the BubbleDeck floor has a load bearing capacity in two directions whereas the other floors need beams to lead the load to the columns.

For a proper comparison of the BubbleDeck floor with the other variants, the beams have to be taken into account for the other variants. This is achieved by subtracting the environmental costs and the realization costs of the beams from the costs of the BubbleDeck floor for the different structural spans.

2.4 Assessment of floor variants

For different structural spans the dimensions of the floors have been calculated; subsequently the environmental costs and the realization costs are determined.

The floor variants always meet the requirements as mentioned in the functional unit; in a few cases a lowered ceiling is necessary to meet the sound isolation requirements and for the timber floors plaster cardboard slabs are used to meet the fire safety requirement.

Figure 2 shows the environmental costs and Figure 3 shows the integral costs of the floor variants.

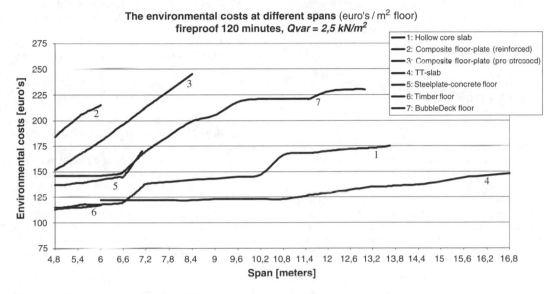

Figure 2.   The environmental costs at different spans.

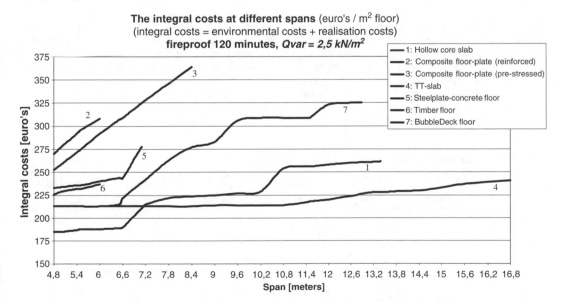

Figure 3.   The integral costs at different spans.

The first graphic shows that the timber floor and the hollow core slab have the lowest environmental costs for short spans; for large spans the TT-slab is environmentally preferable. Since the timber floor is more expensive per m² than the hollow core slab, the integral costs of the hollow core slab are the lowest for short spans, as is shown in Figure 2. For large spans the integral costs of the TT-slab are the lowest (Arets 2001).

Notice that these conclusions are only valid for the functional unit as mentioned earlier, that the realization costs are influenced by made assumptions and that the realization and environmental costs are calculated for the Netherlands.

3 THE RIGHT MATERIAL FOR THE RIGHT FUNCTION

3.1 Materials and the way of loading

Concrete, steel and timber are mostly used for the bearing structure. These materials have different qualities with respect to tensile strength, compression strength, bending strength, shear strength and so on.

Diminishing the environmental load is a combination of using less material and using material with low environmental impact. In order to use as less material as possible, it's logical to make use of a suitable shape and use a material of which the qualities fit the appearing load.

To find out which material should be used from an ecological point of view, the environmental costs of concrete (B35 and B65 precast), steel (Fe E 235) and timber (K24 and K24 glulam) have been calculated for different ways of loading. In Table 1 the results are shown; the environmental costs are indexed, the material with the lowest environmental costs gets index 100.

Looking at the table it's clear that steel has high environmental cost. Although, for example the compression-strength of steel is much better than the compression-strength of concrete and timber, the environmental cost are so big that the index is 5 times higher than the environmental cost of the "best choice".

Concrete without any reinforcement is almost never put into practise. With reinforcement the compression-strength and the tensile-strength of concrete get better, but due to the high environmental cost of steel the environmental cost of reinforced concrete rise.

If the concrete is pre-stressed, the material qualities will improve more than when it's reinforced, meaning that the environmental costs of pre-stressed concrete are lower then of reinforced concrete (Arets 2001).

Analysis of the relation between material qualities and environmental costs leads to these conclusions:

- High strength concrete (like B65) has the lowest environmental cost with regard to compression-strength.
- Timber (especially glulam) has the lowest environmental cost with regard to tensile-strength.

When applying the materials in a building construction, the shape of construction elements is also important. To get an impression of the environmental cost and the integral cost of beams, made of different materials and applying the most common shape for that material, the dimensions of the beams are calculated (Hogeslag 1995).

The span is 8 meter with one intermediate point of support; the load (without construction weight) is 15 kN/m. Knowing the dimensions, it's possible to calculate the environmental cost and the integral cost (in euro's). The results are shown in Table 2.

The results in Table 2 show that from economical and environmental point of view timber (glulam) is the best solution for beams and pre-stressed concrete is just a little bit worse. Notice that this only valid for this short span.

The results in Table 2 show that from economical and environmental point of view timber (glulam) is the best solution for beams and pre-stressed concrete is just a little bit worse. Notice that this only valid for this short span.

Also important for the environmental load of elements of the bearing structure is the shape of the element. As is well know an I-profile is much more

Table 1. Environmental cost (indexed) for different ways of loading.

	Weight	Volume	Compression-strength	Tensile-strength	Deformation by compression	Deformation by tension
Concrete						
B35	100	200	186	–	124	–
B65	100	200	100	–	100	–
Steel						
Fe E235	964	620	515	217	571	327
Timber						
K24	261	100	160	101	261	101
K24, glulam	324	124	158	100	259	100

effective (for bending, buckling, etc.) then a massive cross-section with the same amount of material.

3.2 *The right materials for the bearing structure*

There is much interest from construction engineers in the environmental comparison of bearing structures as a whole. The use of light weighted floors for instance lead to less load on the columns, which lead to less material use for columns. But what is the effect of less materials in the columns on the environmental load of the whole bearing structure? And what are from environmental point of view the most suitable bearing structure variants for often applied structural concepts for buildings?

The results of the study on the environmental and integral cost of floor variants are put together with the results mentioned above in a study on environmental improvement of bearing structures as a whole. This research is going on at this moment.

Before calculating the dimensions of the structural elements of the bearing structure, a model has to be made. Since there are many variables, the span in two directions, the height, the different materials and different shape that can be used for the structural elements, the main purpose of this model is to limit the amount of calculations that has to be carried out while maximizing the amount of bearing structure variants that can be assessed. Therefore the main constructional elements are calculated separately for different loads and then put together to a whole bearing structure. At the same time, together with construction engineers the most commonly used grids for bearing structures will be defined so that research can focus on today's building practice.

In conformance with the comparison of the floor variants a functional unit has to be formulated for the different constructional elements in order to come to a proper comparison. This functional unit will consist mostly of structural and safety requirements; all according to the international standards (Eurocode).

Table 2. Environmental and integral costs for different beams (indexed).

Material	Dimensions	Environmental cost indexed	Integral cost indexed
Reinforced concrete	250 × 490 mm²	251	114
Prestressed concrete	200 × 450 mm²	186	105
Steel	HE 260 A	543	229
Timber (glulam)	100 × 800 mm²	100	100
Aluminium	HE 240 A + 4 studs (15 × 30)	553	282

Of course not only the environmental cost are important but the realization costs and the integral cost as well. These are also calculated in this research.

The aim of the research is to draw up understandable graphs in which for different spans (in the 3 dimensions) the environmental costs, the realization costs and the integral costs of whole bearing structures can be read.

4 NEW MATERIALS FOR BEARING STRUCTURES

4.1 *Diminishing the environmental load of materials*

In order to achieve the factor 20, the environmental load of the bearing structure has to decrease. Two ways have been mentioned, but there are more possibilities:

- decrease material use with 95%
- use material with 95% less environmental load
- prolonging the lifespan by factor 20.

An example of decreasing the amount of material are the concrete floor slabs. Using hollow core slabs or BubbleDeck floors instead of massive floors, the amount of concrete is decreased from about 20% up to 45%. Decrease of material use can also be achieved by using the right material for the right function. Note that one has to be careful; although some materials are more suited for a function then others, the environmental load can be much higher.

Recycled materials and "natural" materials have less environmental load than "standard" materials. Note that when using these recycled or "natural" materials, they have to meet the same requirements as "standard" materials.

Prolonging the lifespan is possible on different levels. First of all the lifespan of the whole bearing structure (whole building) can be prolonged. Second the lifespan of elements of the bearing structure can be prolonged. This requires elements that are demountable and not polluted and that they are re-used in a sustainable way. Third the lifespan of the materials can be prolonged.

The greatest step towards the factor 20 is combining the three mentioned principles. Use less material and make sure that the used materials have a low environmental load and that they have a long lifespan (Arets 2001).

4.2 *"New" materials*

As said before, in the western world concrete, steel and timber are commonly used materials for bearing structures. There are also other materials with specific qualities like glass, synthetic materials, bamboo and all kinds of other natural materials, for instance round timber from young trees.

It will take time to find out what the exact qualities of these materials are and how to use them in a way their qualities are utilized. But there are certainly opportunities in research on "new" materials which meet the same requirements and have a less environmental impact.

5 EXAMPLES OF INNOVATIVE SOLUTIONS FOR BEARING STRUCTURES

Prefab demountable concrete constructions could be a solution for the shortening lifespan of buildings. The bearing structure consists of prefab pre-stressed concrete elements, put together at the construction site. The connections are mostly bolted and therefore demountable. Examples of these constructions are: the SMT system, the CD-20 system and the Bestcon system (Straatman 2000).

A similar kind of construction is developed by Prof. van Herwijnen. It consists of specially developed TT-slabs, steel-concrete beams and concrete columns. All the elements are prefab and all connections are bolted and demountable (van Herwijnen 2000).

One of the problems with this kind of construction is the question who is willing to take the used elements back, approve of them and use them in another building.

An example of an innovative solutions for floors is the IXO-floor. This floor consists of a three dimensional steel frame, with plates on both sides as ceiling and floor. The service-pipes can run through the frame. The floor construction can be demounted as far as cubic elements from about 600 mm and these elements can be demounted further (IXO Sustainable Technology 2001).

The last example comes from "historical" constructions, it's an "under-stressed construction". In these constructions the occurring tension and compression while bending is disengaged, and a large lever arm is created. Through this large lever arm the construction is very resistant for bending. And because of the disengaging different materials can be used; where there is tension, for instance timber, where there is compression, for instance concrete. Applying this principle in new constructions, floors could be under-stressed as well.

Just like the innovation in materials, the innovation in bearing structure needs to continue. More research of the whole bearing structure is needed. Also research on other building elements, on prolonging the lifespan and more efficient use of the building is necessary.

REFERENCES

Arets, M.J.P. 2001. Milieuvergelijkingen van draagconstructies ("Environmental comparison of bearing structures"), *Master thesis*, Delft University of Technology, Faculty of Civil Engineering and Geosciences, Department of Building Engineering, The Netherlands.

van der Dobbelsteen, A.A.J.F. & van der Linden, A.C. 2000. Finding the most effective sustainable measures, *Proceedings Sustainable Building* 2000 Maastricht, The Netherlands, 667–669.

Ehrlich, P. & Ehrlich A. 1990. *The Population Explosion*, Hutchinson: London, UK.

van Herwijnen, F. 2000. Development of a new adaptable and demountable structural system for utility buildings, *Proceedings Sustainable Building* 2000 Maastricht, The Netherlands: 378–380.

Hogeslag, A.J. 1995. Algemene constructieleer – deel III: Hout, Delft University of Technology, Faculty of Civil Engineering and Geosciences, The Netherlands.

IXO-Sustainable Technology, 2001. brochure IXO-floor, Amersfoort, The Netherlands.

Speth, J.G. 1990. Can the world be saved? *Ecological economics*, vol. 1, 289–302.

Straatman, R. 2000. Environmental related issues in precast demountable construction, *Master thesis*, Delft University of Technology, Faculty of Civil Engineering and Geosciences, Department of Building Engineering, The Netherlands.

World Commission on Environment and Development, 1987. Our common future, Oxford University Press, Oxford/New York.

System-based Vision for Strategic and Creative Design, Bontempi (ed.)
© 2003 Swets & Zeitlinger, Lisse, ISBN 90 5809 599 1

Recycled aggregate concrete for structural purposes: ten years of research 1993–2003

E. Dolara
University of Rome "La Sapienza"

G. Di Niro
University of Strathclyde, Glasgow, UK

ABSTRACT: An overview on a large research on recycled aggregate concrete (RAC) for structural purposes, started in 1993 and still in progress, is reported in this paper. The characteristics of the recycled aggregate (RA) have been determined including grading, particle shape and texture, fineness modulus, density, water absorption, sulphate soundness, analysis for contaminant materials, durability and frost sensivity. More than 20 different trial mixes were investigated using recycled aggregates, natural aggregates and a mixture of natural and recycled aggregates. The properties of the fresh concrete mixes were measured and the mechanical properties measured at various stages. Tests were carried out on six full size reinforced concrete beams and on three pre-stressed beams made using the same three types of aggregate i.e. all recycled aggregate, all natural aggregate and the blend of the two. The results indicate that it is possible to produce reinforced concrete beams and pre-stressed beams made from RAC with satisfactory mechanical performance. Finally a large part of this experimentation has been dedicated to the possibility of deducing the mechanical properties of RAC using non-destructive tests.

1 INTRODUCTION

Today the culture of recycling pervades each sector of our life. Yet in the world of civil engineering there is no great evidence of the use of demolition waste. Lots of international studies and research, have, however, reported an interest in recycling of demolition waste from concrete structures.

The increasing demand for natural aggregate and the quantity of demolition waste in industrialised countries mean that it must now be seriously considered. The term recycled aggregate concrete (commonly abbreviated to RAC) is used to describe concrete in which the natural aggregate is partially or totally replaced by the aggregate coming from demolition of concrete structures.

Since 1993 the aim of all the researches carried out by the authors on recycled aggregate and recycled aggregate concrete has been the use of this material for structural purposes. As opposed to the use for road foundations or as fill-in material the use of RA for concrete structures requires more tests and processing of results. In fact to be able to use a material for construction it is essential to asses more than just its compressive strength.

After the physico-chemical characteristics of the recycled aggregate and the properties of both the wet and hardened RAC have been determined, it is important to check if the mathematical models and numerical correlation normally used for design of ordinary concrete (such as mix-design procedure, design codes, non-linear analysis) are suitable for RAC. For this reason the main task of the previous investigations has been to ensure that RAC could have satisfactory mechanical performance for a structural use and later to guarantee a consistency of the results using methods checked for RAC.

A mix-design procedure suitable for RAC to attain the desired workability and the target strength has been the first step reached by the authors.

Three full scale long span prestressed beams made from RAC have been cast using the results of the previous investigations and critical discussions. In this paper general considerations and tests results of hardened RAC for structural purposes are reported.

2 PROPERTIES OF RECYCLED AGGREGATE

All recycled aggregate used throughout this research has come from the Pescale site-plant and therefore from a plant that usually produces and sells recycled aggregate. The RA used is named "R.O.S.E. calces-truzzo" and it contains only demolished concrete structures. This means that the RA came from unknown sources of demolished concrete structures and for this reason the study mirrors a real situation and a practical possible application of this material with all its disadvantages and advantages.

It is clear that the most important factor is to produce a constant quality of aggregate from the plant.

The first part of the experimentation was carried out in 1995 at the University of Strathclyde on recycled aggregate and on fresh and hardened concrete made from RAC. The results of this experimentation were used to direct the second part of the tests carried out in 1997. For this reason in this paper the tests performed to investigate the physical and chemical characteristics of recycled aggregate have been mainly divided into two parts: 1995 and 1997.

The aggregate sent to the Strathclyde Laboratories had a grading of 0–30 mm. In the 1997 tests the aggregate grading used was 0–16 mm, according to the grading used in the Mabo pre-cast plant for normal production.

2.1 Grading, particle shape and texture

Representative samples of the 1995 aggregate were sieved in accordance with BS 812 part 103:1985.

The samples of the aggregate and the mixes are coded as follows: XXNNYY. Where XX is the target strength of the mix design, NN is the type of aggregate used i.e. AR signifies all recycled, AN signifies all natural and RN signifies a blend of the two. The last digit YY is the percentage of natural aggregate used in that mix. Typical grading curves for the recycled aggregate, natural aggregate, and blends of the two (50% of RA and 50% of NA) are shown.

The grading curves for a significant number of samples of the 1997 recycled aggregate are reported in Figure 1. From the sieve analysis carried out, the recycled aggregate used seems to lack fine material.

This scarcity is compensated with the trend to produce fine material for abrasion and crushing during the mixing operation inside the concrete mixer.

The recycled aggregate examined comply with the BS 882 fine grading Zone 1, but does not comply with the Fuller-Bolomey grading zone that guarantees the best compromise between the density requirement and the non-segregation of the aggregate.

The specific weight of the concrete is another property that depends on the grading distribution of the aggregate. Usually increasing the weight increases the mechanical strength performance of concrete.

An ideal grading distribution, that means the correct quantity of fine and coarse aggregate, can produce also a better filling of the interstices and consequently a higher specific weight of concrete.

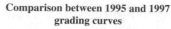

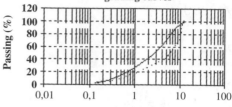

Figure 1. Comparison between 1995 and 1997 grading curves.

In order to check the consistency of production of the recycling plant the sample called CP 1, used in 1997, has been compared in Figure 1 with a sample of recycled aggregate coming from the same plant but used in 1995. There is a difference between the two grading curves. The reason for this non homogeneity of production during that time can be ascribed to the different original demolition concrete that arrives at the plant (in terms of quality of aggregate, cement and w/c ratio). It is clear in fact that a concrete made with a high performance cement will not have after crushing the same grading curve of a concrete made with a low performance cement. The same situation will pertain to different types of original natural aggregate.

It could be seen from the literature review that the particle size distribution is influenced also by the crusher characteristics (jaw crusher, cone crusher, impact crusher or swinghammer mill).

In general for concrete applications the choice of the maximum size of the aggregate is governed by the type of product produced.

Because one of the aims of this research is to cast and test prefabricated prestressed concrete beams a small diameter has been chosen as maximum size of the aggregate, in accordance with normal pre-cast production. With regard to this problem it should be said that increasing the maximum size of the aggregate decreases the total amount of water needed for consistent workability. This is because large diameter coarse aggregate has a lesser specific surface and requires less water to wet all particles. Using an aggregate with a bigger D_{max} results in a lower cement content with less shrinkage and less cost in order to achieve the same mechanical strength with the same w/c ratio. This encourages the use of mixes with large diameter aggregate but care has to be taken to avoid segregation.

It is usual to define the maximum diameter as the value equal to the sieve aperture of the total passing aggregate, starting the grading curve from this point.

Table 1. Grading characteristics of samples of recycled aggregate (1997).

Sample	Date	Starting weight	Final weight	Difference (%)	D_{max}	MF
CP1	18/02/97	1777.7	1781.9	0.236	9.024	4.361
CP2	18/02/97	1784.2	1786.2	0.112	9.145	4.390
CP3	18/02/97	1793.8	1794.1	0.017	9.323	4.582
CP4	19/02/97	1809.3	1805.1	0.232	9.122	4.544
CP mix	19/02/97	1840.8	1839.2	0.087	9.123	4.561
CP sand	03/03/97	1000.0	998.1	0.190	2.697	3.446

It should be said that this valuation is not representative of the real maximum diameter.

The maximum diameter D_{max} should satisfy the following formula: $D_{max} = d_1 + (d_1 - d_2) \cdot x/y$ where: d_1 is the first sieve on which the material is retained; d_2 is the directly next sieve; x is the percentage retained on d_1; y is the percentage retained on $d_1 + d_2$.

With this method it is possible to avoid going out of the grading zone especially where it is particularly delicate as is the case with the sand.
The value found in 1997 is reported in Table 1. The average value is around 9 mm.

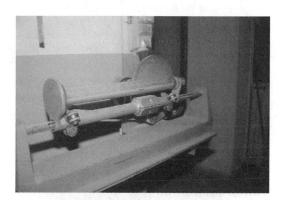

2.2 Fineness modulus

The MF of the sand is of fundamental importance in determining the workability of concretes. They should have MF between 2.3 and 3.3 according to the UNI 8520 parts 5°.

If a sand shows a fineness modulus MF not included in the previous values, a segregation of the coarse aggregate can occur during the mixing operation and consequently a decrease in workability.

The higher the value of MF the lower is the fineness in general. This parameter on its own can define the grading distribution of an aggregate. Table 1 reports the values found in 1997.

2.3 "Equivalent in sand" test

The "equivalent in sand" (ES) is defined as the volume percentage of the clean sand to the total sediment aggregate. Because the ES test does not distinguish between the slimy-clayey material (that is noxious for concrete) and the finest material coming from the crushing of rocks, (that is useful for concrete) this test has been carried out (when required) with the value of the indicator *blue of methylene*. This has been done with the aim of qualifying the fine material passing the 0.075 mm UNI 2332 sieve.

The ES test is carried out on fine aggregate passing the 4 mm UNI 2332 sieve. A reasonable value for ES is 85. The value of the ES has been determined also by optical measurement in accordance with the French standards.

Figure 2. Graduated cylinders with dosing and mechanical agitator for the determination of the "Equivalent in sand" ES.

Results determined with this method are around 70. This is lower than the value suggested by the Italian standard for an aggregate complying with class C. Further tests have been carried out to study the influence of the washing away solution used in this research. From the above results it can be seen that because there are no clayey or slimy fractions in the recycled aggregate, the low value of the ES can be attributed to the old cement paste attached to the aggregate. During the mechanical agitation of this test this fine fraction is dispersed into the washing solution and provokes a gel reaction that remains in suspension for a time longer compared with normal values for a natural aggregate. This phenomenon could negate the test result and it can be concluded that the recycled aggregate is not suitable for concrete production if clay-lime fractions are present.

Values that are reasonable for natural aggregate should be reviewed for recycled aggregate because of the different nature and behaviour of the aggregate.

A reasonable value for this test seems to be 70.

2.4 Relative densities and water absorption

One of most important physical characteristic of the recycled aggregate is the relative density of the aggregate. Great care should be paid in the determination of the SSD (Saturated Surface Dried condition) density to ensure that the material is in a SSD conditions.

The low density and the high water absorption of recycled aggregate are two important factors that should be considered influential in the procedure for the mix design of RAC. For this reason great care should be paid to the determination of the real water absorption value of this aggregate. In fact it is very important to determine the extra water to add to the mix for good workability.

In the pre-cast plant it is very important to measure the moisture content of the aggregate continuously and to change the total amount of water in the mix.

During this research work for the grading 0/5, 5/30 and 0/30 in 1995 and 0/4, 4/15 and 0/15 in 1997 the following characteristics have been determined: water absorption, relative density on an oven dried basis, relative density on a saturated surface dried basis, apparent relative density, moisture content (only in 1995), voids (only in 1995).

The tests have been performed on material dried in an oven at 105°C for 24 hours and eliminating the fine material passing to the ASTM 0.075 mm. Table 2 shows the results from 1995 investigation. Table 3 reports the test results from 1997 investigation.

As expected it can be concluded that the density of the recycled aggregate is lower than that of the natural aggregate (normally between 2600 and 2700 kg/m^3). This is due to the relatively low density of the old mortar which is attached to the original aggregate particles. The water absorption of the RA is much higher than the water absorption of the original aggregates. This is also due to the higher water absorption of the old mortar.

2.5 Contaminants

2.5.1 Chemical analysis by water extract of recycled aggregate

Test results from 1995 and 1997 experimentation are reported in Table 3.

Table 2. Physico-chemical characteristics of recycled aggregate determined during the 1995 and 1997 research.

	Grading	Relative density on an oven-dried basis (kg/m^3)	Relative density on a saturated surface dried basis (kg/m^3)	Apparent relative density (kg/m^3)	Water absorption (%)	Voids (%)	Moisture content (%)
1995	0/5	2.030	2.255	2.620	11.09	49	
	5/30	2.363	2.474	2.653	4.62	51	1.26
	0/30	2.153	2.314	2.567	7.50	43	
1997	0/4	1.972	2.218	2.616	12.49	–	–
	4/15	2.311	2.438	2.649	5.53	–	–
	0/15	2.109	2.280	2.550	8.21	–	–

Table 3. Chemical analysis by water extract of RA: 1995 and 1997 experimentation.

	Contaminants (by weight of both the fine and coarse aggregate fraction)						Water extract
	Organic substances (%)	Sulphate SO_4^- (%)	Chloride Cl^- (%)	Potassium K^+ (%)	Sodium Na^+ (%)	Calcium Ca^{++} (%)	PH of water extract l/log[H$^+$]
1995	1.97	0.001	0.005	0.0139	0.0155	0.0157	11.3
1997	0.02	<1 ppm	0.022	0.0214	0.0128	0.093	11.9

Sulphate (SO$_3$) by acid extract: 0.62%, Magnesium (Mg^{++}) < 1 ppm.

2.6 Durability tests on recycled aggregate

2.6.1 Los Angeles test

The tests have been performed following the C.N.R. anno VII, No. 34 of 1992. The class of the sample is determined comparing the results with given values reported in tables in the C.N.R. standards. Test result show a value of 26.7% for the Los Angeles coefficient.

It has been demonstrated that the L.A. abrasion lossed percentage for RA depends on the original concrete strength (low or high). According to ASTM Designation C-33-99ae1 aggregate may be used for production of concrete when L.A. abrasion loss percentage does not exceed 50%. According to BS 882, 1201, Part 2, aggregates may be used for production of concrete wearing surface when the aggregate crushing value does not exceed 30%, or 45% for other concrete. It may be concluded that RA produced from all but the poorest quality concrete can be expected to pass ASTM and BS requirements to L.A. abrasion loss percentage, BS crushing value even for production of concrete wearing surfaces, but probably not for granolithic floor finishes.

2.6.2 Frost sensitivity of recycled aggregate

Concretes could be sensitive to freeze and thaw cycles. This phenomenon depends on the porosity of both the cement paste and the aggregate. The water absorbed when freezing increases its volume. If this volume exceeds 91% (critical saturation) there is no more space to allow the ice dilatation and the zone surrounding the cement pores are subjected to compressive stresses that can produce cracks and disaggregation of the cement paste. To avoid this phenomenon the aggregate is normally tested for frost sensitivity following the *Raccomandazioni C.N.R. anno XIV n. 80, 15 Novembre 1980*. The aggregate is completely saturated and subjected to 20 cycles of freeze and thaw from $-20°C$ ($\pm 5°C$) to $+20°C$ ($\pm 5°C$). After the freeze and thaw cycles (it takes twenty days) the Los Angeles test is performed on the aggregate and the factor of frost sensitivity is determined from the following relation:

$$G = 100 \cdot \frac{(LA_2 - LA_1)}{LA_1}$$

where LA_1 is the Los Angeles test value before the freeze and thaw cycles and LA_2 is the Los Angeles test value after the freeze and thaw cycles.

Test results showed a value of 36.16% as Los Angeles coefficient after the freeze and thaw cycles and $G = 24.2$ as frost sensitivity. From the above figures it can be deduced that the frost L.A. abrasion loss percentage increased 20% compared to the corresponding normal L.A. Of course the frost sensitivity depends also on the pore system and the strength of the old cement paste. For this reason it is advisable to use recycled aggregate concrete designed with low water/cement ratio to reduce the pore system of the concrete.

3 MIX DESIGN PROCEDURE

Three different modifications of existing mix design procedures have been investigated. They are:

1. Concrete with pre-soaked recycled aggregate
2. Recycled aggregate concrete with different percentages of 20 mm of natural aggregate
3. Recycled aggregate concrete with different percentages of 0–20 mm of natural aggregate.

The third procedure has been suggested for structural uses of RAC. The first and the second procedure are briefly discussed in the next paragraphs.

Two important factors should be considered to influence this procedure for the mix design of RAC: the low density and the high water absorption of the recycled aggregate. Great care should be paid to the determination of the real water absorption value of this aggregate. In fact at this stage it is very important to determine the extra water to add to the mix for a good workability.

Referred to this mix design procedure the aggregate type for coarse and fine recycled aggregate has been assumed to be crushed.

The maximum aggregate size has been 20 mm. The relative density on a saturated surface-dried basis (SSD) of the recycled aggregate has been assumed to be: 2400 Kg/m^3. Concrete density: 2200 Kg/m^3. Proportion of fine aggregate: 44%.

The same procedure has been followed for the concrete mix design with all natural aggregate. The proportion of fine natural aggregate was 34%.

3.1 Concrete with pre-soaked recycled aggregate

Following this procedure the recycled aggregates have been left in the horizontal pan mixer to pre-soak for 10 minutes with the extra water for water absorption.

The effect of 10 minutes of pre-soaking of the recycled aggregate can be noticed comparing the V–B test figures of mix 40AR100PS (the digit PS signifies pre-soaking) and 40AR100 in Table 3.1. For the mix with the recycled aggregate pre-soaked the time (sec) for V–B test is shorter indicating a higher workability. More significantly this workability is maintained for longer periods which is important for real life concreting on site.

However 10% reduction in the compressive strength of RAC has been noted from the reports. This could be explained with the following considerations. The crushing process for production of recycled aggregate produces aggregate with diffuse microcracking especially in the old mortar attached to the original aggregate particles. This creates a sort of "sponge effect" when the water is added in the concrete mixer during the concrete production. In the first moment the recycled aggregate absorbs all the water and in a second time it tends to expel the part no strictly stechiometric with a consequent alteration of the w/c ratio.

3.2 Recycled aggregate concrete with different percentages of 20 mm of natural aggregate

Mixes with 100% of recycled aggregate (AR) have been produced and it has been found that even with a water/cement ratio of 0.35 which gave a strength of 60 MPa for the AN mixes, the AR mixes gave a strength of 35 MPa.

It is important to underline that the recycled aggregates used in all the investigations by the reports are aggregates that came from *unknown sources* of demolition concrete structures, with *different strength* of the original concrete. Furthermore the recycling process involves the possibility of contamination by weak aggregate like refractory brick, gypsum, clay balls, etc.

For these considerations during this investigation to improve the mechanical characteristics of the RAC different percentage (10–30%) of 20 mm natural aggregate has been added to the mix.

There was no significant improvement in cube strength with replacement by natural aggregate. The greatest improvement measured was less than 10%. This was probably due also to the fact that the 20 mm natural aggregate were not statistically well distributed in the concrete mix.

Anyway this investigation demonstrated that the workability characteristics can be improved by partially substituting 20 mm coarse aggregate. As little as 10–15% substitution has important effects on both the workability and the loss of workability with time. This observation is important since previous workers have usually found it necessary to include workability aids (admixtures) in the mix in order to obtain satisfactory behaviour for the fresh recycled aggregate concrete. It should be said that for RA is not so easy, like for the natural aggregates, the use of admixtures. In fact for the contaminants that the RA has in it, the action of admixtures could be entirely annulled. So the use of partial substitution of the aggregate by natural coarse aggregate seems to be an alternative method to achieve a good workability.

All these previous results have been processed and they were used to direct the third investigation referred to in the next paragraph.

3.3 Recycled aggregate concrete with different percentages of 0–20 mm of natural aggregate

It seems logical to the reports to attain a satisfactory strength for *structural uses* of RAC to blend different proportion of the two base mixes made by recycled aggregate concrete and natural aggregate concrete (e.g. the mix 40RN30 is a mix in which the 70% is recycled aggregate (all in) and the 30% is natural aggregate (sand, 10 mm, 20 mm).

This is also suggested by the improving of the workability characteristics as referred in the previous paragraph. On the other hand it seems also easier for the production of RAC to add natural aggregate (sand, 10 mm, 20 mm) to the mix instead of cutting the 0/5 grading of the recycled aggregate and to replace it with natural sand, it has been suggested from many authors to improve the mechanical characteristics of RAC. Extra water for water absorption must be added to the mix to attain the exact water demand. In this case a value of 5% for the water absorption for the grading 0/20 mm of RA has been adopted.

3.4 Properties of fresh RAC

Slump, V–B and Compacting Factor tests have been performed on different mixes of RAC. The results are reported in Table 4. It can be noticed by the effect of partial substitution of natural aggregates on the workability of the fresh concrete.

In particular it can be noticed that there is an optimum percentage of natural aggregate to replace in the mix to have a significant improving on workability. This value is close to 20–30%, as confirmed in previous investigations.

The mix design adopted has been found to be appropriate for the design of RAC, but the following considerations are suggested.

Instead of discarding the 0/5 mm grading of the recycled aggregate and replacing it with natural sand (as advised by many authors), it seems easier to produce mix design blending different percentage of natural aggregate and recycled aggregate.

For a good mix design is very important to determine the correct *water absorption* value of the recycled aggregate. This extra water can influence markedly the workability of the concrete. It is very difficult to accurately assess the surface dry condition in an aggregate containing such fine material to establish the correct water absorption. It is advised worthwhile to carry out a trial mix to obtain this figure.

It is important to obtain a correct value for the relative density of the aggregate since this has been shown to have a significant effect on the mix proportion and thus the compression strength. In fact comparing with a previous research, for a designed value of w/c the a/c ratio seems to influence the compressive strength of RAC.

It is apparent that the fine fraction absorbs appreciably more water than the coarse. Pre-soaking the recycled aggregate seems to have marked effect on workability during the time. A percentage of 20–30% of recycled aggregate can be substituted in a normal

Table 4. Workability of the fresh concrete with different percentage of recycled and natural aggregate: V–B test; slump and compacting factor tests (1995).

No.	Mix code	V–B consistometer readings (seconds)			Slump (mm)			Compacting factor
		Time 0	15 minutes	30 minutes	Time 0	15 minutes	30 minutes	
1	40AR100PS	3	4	5	62	47	45	0.98
2	40AR100	5	7	10	46	36	8	0.89
3	40RN30	4	6	6	54	39	36	0.95
4	40RN50	5	6	7	50	38	17	0.91
5	40RN70	6	7	11	28	11	0	0.89
6	40AN100	7	11	14	0	0	0	0.80

Table 5. Workability of the fresh concrete with different percentage of recycled and natural aggregate (1997).

Mix Code	w/c design	Slump design (mm)	Cement type	Slump (mm)		
				Time 0	5 minutes	15 minutes
CP1AR	0.45	10–30	325 Pozzolanic	100	0	0
CP2AR-F	0.45	10–30	325 Pozzolanic	105	60	40
CP325AR1	0.45	10–30	325 Pozzolanic	160	80	80
CP325AR2	0.45	10–30	325 Pozzolanic	110	100	90
325CAR1	0.45	60–180	325 IIb Portland	105	85	65
325CAR2	0.45	60–180	325 IIb Portland	105	85	65
CP525AR1	0.45	60–180	525 Portland	190	165	150
525MAR1	0.45	10–30	525 Portland	30	0	0
525MAR2	0.45	60–180	525 Portland	105	80	60
525MAR3	0.45	10–30	525 Portland	25	0	0
525MAR4	0.45	60–180	525 Portland	100	90	80
525AR50	0.45	60–180	525 Portland	70	60	40
40ANI	0.45	10–30	525 Portland	20	0	0
40AN2	0.45	60–180	525 Portland	150	130	90

concrete mix, without any significant differences in mechanical performances of the hard concrete. Table 5 shows test results for 1997 experimentation.

4 PROPERTIES OF HARDENED RAC

In the 1994 investigation mixes with 100% of recycled aggregate (AR) have been produced and it has been found that even with a water/cement ratio of 0.35 which gave a strength of 60 MPa for the AN mixes, the AR mixes gave a strength of 35 MPa.

During 1994 investigation to improve the mechanical characteristics of the RAC different percentages (10–30%) of 20 mm natural aggregate have been added to the mix.

There was no significant improvement in cube strength with replacement by natural aggregate. The greatest improvement measured was less than 10%.

All these previous results have been processed and used to direct the 1995 investigation.

It seemed logical that to attain a satisfactory strength for *structural uses* of RAC to blend different proportion of the two base mixes made by recycled aggregate concrete and natural aggregate. This is what has been done during the 1995 investigation.

For the 1995 investigation the mixes are coded as follows: XXNNYY. Where XX is the target strength of the mix design, NN is the type of aggregate used i.e. AR signifies all recycled, AN signifies all natural and RN signifies a blend of the two. The last digit YY is the percentage of natural aggregate used in that mix. e.g. the mix 40RN30 is a mix in which the 70% is recycled aggregate (all in) and the 30% is natural aggregate (sand, 10 mm, 20 mm). The digit PS signifies pre-soaking.

Ordinary Portland cement conforming to BS 12 was used for both recycled aggregate and natural concrete throughout 1995 research. By processing the tests results of the previous investigations in 1997 further mixes were investigated with different types of cement.

Those mixes are coded as follows: ZZZ KKK J. Where ZZZ is the type of cement used (e.g. 52.5,

42.5, 32.5 Portland cement); KKK is the type of aggregate used i.e. AR signifies all recycled, AN signifies all natural and RN signifies a blend of the two, CP/C signifies sample, M signifies mix and F signifies a type of admixture (superplasticizer). The last digit J is just the chronological order of casting of the samples.

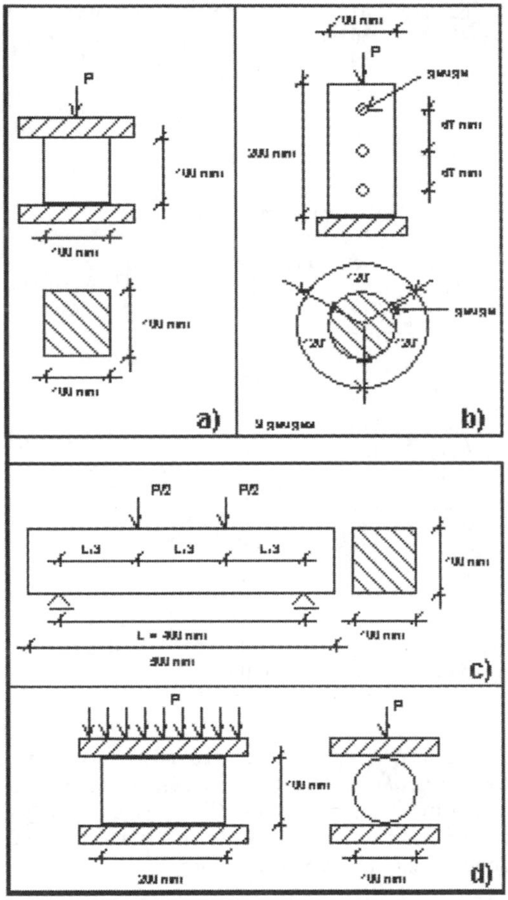

For samples cast together with the prestressed beams the same notation of the 1995 investigation has been adopted.

4.1 Compressive strength

Test geometry specimens are shown in Figure 3. For the 1995 experimentation test results are reported in Table 6.

Figure 4 is a plot of the compressive strength versus time for the above mixes with different percentage of natural and recycled aggregate.

It can been noticed that there is a 20% of difference of the compressive strength at 28 days between the 40AR 100 and 40AN100. This difference is reduced to 15% for the mix 40RN50 and to only 4% for the 40RN70.

The 40RN70 is the only mix that reaches the target strength of 40 Mpa. To attain the design target strength it is necessary to blend the natural and the recycled aggregates. In Table 7 test results for the 1997 experimentation are reported. The differences in the compressive strength between the different mixes are due to the different cement types used during this investigation.

Samples with Portland 32.5 pozzolanic cement, 32.5 composite type IIB cement and 52.5 ordinary Portland pre-blending cement have been cast and tested. The

Figure 3. Test geometry specimens. Compressive test (a), E-value (b), flexural test (c), splitting test (d).

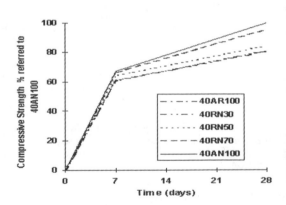

Figure 4. Compressive strength versus time for mixes with different percentage of natural and recycled aggregate.

Table 6. Compressive strength tests results for 1995 investigation on RAC.

Mix code	w/c ratio	Compressive strength at 7 days (MPa)	Compressive strength at 28 days (MPa)	R/R max at 28 days (%)
40AR100PS	0.45	27.8	32.1	71.3
40AR100	0.45	27.9	36.3	80.1
40RN30	0.45	27.3	36.6	81.3
40RN50	0.45	28.8	38.0	84.4
40RN70	0.45	29.9	43.0	95.6
10AN100	0.45	30.3	44.95 (R_{max})	100.0

last one showed the best mechanical performances. It should be noticed that the mix 525AR50 has been designed for a target strength of 50 MPa and it reached 53 MPa at 28 days.

The mixes 525MAR2/3/4 were designed for a target strength of 40 MPa and they reached this strength only with the use of 52.5 Portland cement.

4.2 Tensile strength

The tensile strength of RAC has been evaluated using the indirect tensile test (Brazilian test) or splitting test. Test results for the 1995 research are given in Table 8. In Table 9 test results for the 1997 experimentation are reported.

4.3 Flexural strength

For the 1997 investigation test results are reported in Table 10.

Table 7. Compressive strength tests results for 1997 investigation on RAC.

| Mix code | Compressive strength (MPa) | | | | |
	24 hours	3 days	7 days	28 days	90 days
CP1AR	16.63	29.38	28.50	36.38	–
CP2AR-F	19.38	33.80	33.25	40.50	–
CP325AR1	6.63	15.25	24.75	33.75	–
CP325AR2	3.13	9.75	18.15	26.25	–
325CAR1	3.63	12.05	14.03	19.25	26.90
325CAR2	4.63	15.00	21.13	26.55	36.20
CP525AR1	18.26	26.00	35.25	41.25	–
525MAR1	17.00	28.00	34.63	39.00	–
525MAR2	21.25	33.35	38.88	41.85	50.55
525MAR3	15.75	30.63	37.88	46.25	–
525MAR4	18.75	35.50	40.38	47.85	55.00
525AR50	22.88	37.75	47.00	53.00	–
40AN1	28.75	46.38	55.45	62.26	–
40AN2	–	45.50	55.25	62.75	–

Table 8. Tensile strength tests results for 1995 investigation on RAC.

| Mix code | w/c ratio | Tensile strength | | T/T_{max} at 28 days (%) |
		at 7 days (MPa)	at 28 days (MPa)	
40AR100PS	0.45	1.97	3.67	68.1
40AR100	0.45	2.20	3.71	69.1
40RN30	0.45	2.29	3.86	73.0
40RN50	0.45	2.58	3.67	68.1
40RN70	0.45	2.83	3.79	96.7
10AN100	0.45	3.00	3.92 (T_{max})	100.0

4.4 Modulus of elasticity

Test results are shown in Table 11.

4.5 Effects of freezing and thawing on RAC

In order to investigate the durability of RAC freezing and thawing tests have been carried out. This test has been performed at 256 days. During this time all cubes were water cured at 20°C. No cubes were tested before the freezing and thawing test. For this reason the compressive strength of concretes at this age have been determined by the maturity equation suggested by Plowman. The compressive strengths at 256 days are reported in Table 12.

It can be noticed from Figure 5 a reduction of 10% in the compressive strength of RAC after 300 cycles of freezing and thawing (−20°C/20°C in six hours).

The loss of weight is almost the same for all the different mixes and it is equal to 1% by weight of concrete. The RAC shows a good behaviour considering the losses of strength and the losses of weight after

Table 9. Tensile strength tests results for 1997 investigation on RAC.

| Mix code | Tensile strength | |
	at 7 days (MPa)	at 28 days (MPa)
525MAR3	2.21	2.79
525MAR4	1.94	2.60
40AN1	4.27	4.24
40AN2	3.85	3.68

Table 10. Flexural strength tests results for 1997 investigation on RAC.

| Mix code | Flexural strength at | | |
	3 days (MPa)	7 days (MPa)	28 days (MPa)
525MAR3	2.51	3.28	4.20
525MAR4	2.31	4.00	4.40
40AN1	8.80	6.44	8.40
40AN2	6.20	6.00	6.00

Table 11. Modulus of elasticity tests results for 1997 investigation on RAC.

Mix code	Modulus of elasticity at 28 days (MPa)
525MAR3	23 535
525MAR4	25 437
40AN1	26 765
40AN2	28 008

Table 12.   Compressive strength.

| Mix code | Compressive strength (MPa) | |
	at 28 days	at 256 days by [9]
40AR100	36.3	39.1
40RN30	36.6	39.4
40RN50	38.0	40.9
40RN70	43.0	46.3
40AN100	45.0	48.4

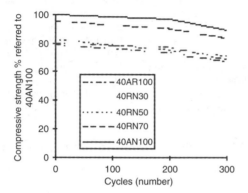

Figure 5.   Compressive strength (%) of RAC versus number of cycles of freezing and thawing.

300 cycles of freezing and thawing fatigue. The reduction in the freezing and thawing resistance is independent upon the proportion of replacement of natural aggregate. The freezing and thawing tests indicate that RAC is as durable as concrete made with natural aggregates.

4.6   Bonding surface between recycled aggregate and new cement paste

In recycled aggregate concrete (RAC) the aggregate is not a natural aggregate from the same rock fragment. It could be a mix of different rocks with different rheological properties bonding together with the old cement paste.

Normally the waste elements are stacked for long periods outside before being processed and recycled. It is therefore evident this material could have a progressive alteration that could influence its behaviour and each rock inside the recycled granule could have a different alteration depending on its mineral sources.

The aim of this part of the research is to understand what level of alteration is reached from the recycled aggregate and how it could influence the behaviour of the new concrete.

The failure mechanics is one of the aspects involved in this process. It is briefly discussed in this paragraph.

4.6.1   Petrographic analysis

A petrographic analysis has been carried out on samples of recycled aggregate concrete.

The microscopic analysis has been performed using a ZEISS optical polarizer microscope on six thin slides made with different orientation from different parts of the same sample of recycled aggregate concrete.

The objective being to study the petrographic and geological characteristics of the aggregate and the nature of the bonding between the aggregate and the new cement paste.

The petrographic analysis can be used to determine the composition of the aggregate. In this case it was found to be composed of: rock fragments, mono-mineral granules, concrete fragments of different shape and dimensions. In the concrete fragments, granulated elements were found which had the same petrographic and/or mineralogical nature of the natural aggregate that have been found isolated in the new cement paste.

In general, for all three, good adhesion to the new cement paste was observed and only in few points of the contour of the concrete fragments was there separation between the aggregate and the old cement paste. In some thin sections micro cracking was observed in the aggregate/cement paste interface. The pattern was short and irregular. This could be due the shrinkage of concrete. In Figure 6, section is reported, 25 magnifications, made with a microscope with interference with crossed ray and the same sections is reported but made with a microscope with single ray. It was noticed that the rock fragments in the sample have a diversified petrographic nature.

The different minerals appear mostly unbroken and not altered but the feldspars present in the sandstone in the old cement paste show evident signs of alteration (clayey phenomenon).

The micro photos performed on the thin section of RAC show the texture of the material and, in particular, they emphasised the bonding surface between the granule of the old cement paste and the new cement paste.

These granules appear on the whole compacted. They are distinguished by a more diffused carbonation of the cement paste.

It seems useful to remember that the carbonation phenomenon occurs due to the penetration of the CO_2 (contained in air) into the superficial surface of concrete. CO_2 reacts with the alkaline hydroxide making the corresponding carbonates and consequently with the calcium hydroxide dissolved into the water solution provokes a progressive degradation of concrete.

Anyway it should be underlined that during the preparation of the thin sections of RAC the abrasive process required to reduce the thickness of the material

(a)

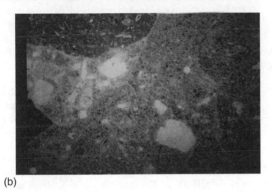

(b)

Figure 6. Thin section of a RAC sample using a polarizer microscope with interference with: a) crossed ray, b) single ray.

has provoked the disaggregation of some of these composite granules. It is probable that these were granules with a lower level of cementation.

The recycled aggregate comes from unknown sources of demolition waste and consequently from concretes with different w/c ratio.

The correlation between the pore system in hardened cement paste, the w/c ratio and the carbonation of concrete is well known. It is therefore logical that concretes with high w/c ratio could be less durable compared with concrete of low w/c ratio.

Anyway this consideration should not influence the recycling process in a site-plant because poorer concretes cannot be discarded.

For this reason particular attention should to be paid to the design of RAC to reduce the risk of diffused carbonation. A solution could be to have mix-design with a low w/c ratio to reduce the pore system of concrete or the use of particular admixtures so could prevent the carbonation phenomenon.

The use of admixtures is recommended so that the benefits of using RAC are not lost due to the presence of contaminants. Wet curing (high humidity period) is recommended for the recycled aggregate concrete so

that hydration of cement continues and reduces the depth of carbonation.

The type of cement used to make the recycled aggregate concrete is important (such as blended cement: silica and fly ash cement). The blended cement leads a lower $Ca(OH)_2$ content in the hardened cement paste and a smaller amount of CO_2 is required to remove $Ca(OH)_2$ by producing $CaCO_3$.

RAC with higher strength was more protected as regards the carbonation.

Prefabrication can encourage the use of RAC for its wet curing (e.g. steam curing) and for its high compressive strength in the production of prestressed elements. Normally the failure of concrete could occur through the cement paste, through the aggregate or along the bonding surface between the aggregate and cement paste. In recycled aggregate concrete the aggregate is its own concrete and so the failure mechanism could occur again with the same process. If the bonding surface is altered, sufficiently to be seen in the thin section of concrete, this phenomenon could be accelerated.

5 RAC REINFORCED BEAM

Two beams were cast for each of the three mixes. Cubes were taken from all of the mixes in order to check their strength. As expected the ANB mixes reached their expected strength with an average of 41.9 MPa at 28 days. Both the ARB and ARN mixes were of lower strength with averages of 26.4 MPa and 24.0 MPa respectively. These figures are comparable with those for the earlier mixes considering that the beams and the cubes were air cured rather than water cured. On the basis of preliminary tests it was decided to produce 6 full scale reinforced concrete beams (cross section: 100 mm × 200 mm; overall length: 2130 mm) based on the nominal 40 MPa mixes and made with the recycled aggregate 40ARB, the natural aggregate 40ANB and the mixed aggregate 40RNB which had 15% substitution of natural 20 mm aggregate.

Concretes made with the recycled aggregates or the blended aggregates show low strengths than concretes made with normal aggregates. There is also a maximum strength which can be achieved with these aggregates and this appears to be around 35 MPa.

The load–deflection responses for 40ARB2 and 40RNB2 beams are shown in Figure 7. Note that all of the beams failed at loads of 58 MPa to 62 MPa with very little difference between the three mixes although the deflection at maximum load was almost double that of the 40ANB for the 40ARB and 40RNB beams.

The finite element method was used for a non linear analysis of the behaviour of the beam made with recycled aggregate.

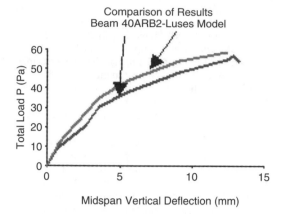

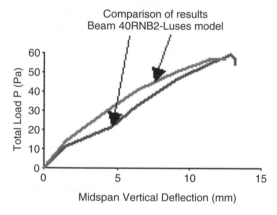

Figure 7. Load–deflection response for beam 40ARB2 and 40RNB2.

The load–deflection curves derived from the finite element model can be compared with those determined experimentally. The results can be seen in Figure 7.

In all cases, the model closely predicts the behaviour of the real beams. The tensile strengths and moduli of elasticity of the mixes were also measured and were found to be consistent with the cube crushing strengths. The full results may be found elsewhere (5). It would appear practicable to make reinforced concrete elements using concretes made with recycled aggregate and these elements would have satisfactory mechanical performance.

6 RAC PRESTRESSED BEAM

In 1994 the authors of the present paper cast and tested six reinforced concrete beams made from RAC. This has been their first attempt at a structural use of RAC. Satisfactory mechanical performances have been attained by the RAC beams. Full test results can be

found elsewhere . Since then the attention has been concentrated on mechanical performances of RAC and in particular on the use of RAC in prestressing. More than thirty different mixes have been studied and tested in order to choose a mix for production of RAC prestressed beams.

Compressive strength of RAC has not been the only criteria adopted for choosing the mix. Workability requirements have to be satisfied to be compatible with the plant production.

A mix-design procedure that could guarantee mechanical strength and acceptable slumps has been determined in this case. For this reason all the previous studies carried out before this practical application of RAC have been oriented to modify an existing mix-design procedure with all the information coming from tests on recycled aggregate.

6.1 Experimental details

Three prestressed concrete beams (cross section 43 cm × 55 cm; overall length 1500 cm) have been cast using 100% of recycled aggregate (RA) for the beam named 40AR100, 50% of RA and 50% of natural aggregate (NA) for the beam named 40RN50 and 100% of NA for the beam 40AN100. 16 mm maximum size recycled aggregate was used for the mixes with 100% of RA and 50% of RA. River sand and 16 mm maximum size gravel coarse natural aggregate was used for the mixes with 100% of NA and 50% of NA. 525 Portland Cement corresponding to the IA class of UNI-ENV 197/1 has been used for all the mixes. The mix-design adopted is reported in. The design target strength for the three mixes was 40 MPa. The exact mix proportions for the mix 40RN50 are reported in Table 13.

The prestressing reinforcement was of 10 tendons of 93 mm² of area each, with a prestressing stress of 142.5 kg/mm² and a total prestressing force of 132.5 tons for each beam. The design code used has been one for ordinary concrete prestressed beams. Accelerated curing was adopted for all beams as in the case in normal production in the plant and the thermal cycle inside the beam has been recorded.

The beam 40AR100 cast with a concrete made with 100% of RA performed very well at the release of the tendons and it had a precamber almost double compared with the beam 40AN100% casted with a concrete made with 100% of NA.

Slump tests have been carried out before casting each beam. The designed slump was 60–180 mm. Specimens for evaluation of the mechanical properties of RAC were casted together with the beams. In Table 14 are reported the results of these tests for the beams.

It should be underlined that the previous results of mechanical strength of concretes are referred to cubes air cured as has been for the beams. Together with the three beams, another series of specimens have been cast

Table 13. Mix proportions (1 m³).

Mix code	w/c ratio	Cement (kg)	Free water (l)	Recycled aggregate, all in kg	Natural fine sand 0–4 mm (kg)	Aggregate coarse 4.12 mm (kg)
40AR100	0.45	500	225	1415	–	–
40RN50	0.45	528	238	707	531	226
40AN100	0.45	556	259	–	757	757

Table 14. Mechanical properties of the three mixes used to produce the prestressed beams (air cured).

Mix code	w/c ratio	Stump (mm)	Compressive strength at 28 days (MPa)	Tensile strength at 28 days (MPa)	Flexural strength at 28 days (MPa)	Modulus of elasticity at 28 days (MPa)
40AR100	0.45	84	39.75	2.49	3.94	23 281
40RN50	0.45	72	43.00	2.62	4.56	26 516
40AN100	0.45	68	51.50	2.80	5.30	33 921

Figure 8. Test arrangements.

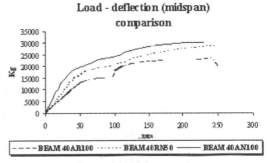

Figure 9. Load–deflection curves.

and they have been water cured. An increase more than 10% on the mechanical strength of concretes have been noticed. This is a confirmation that a RAC mix could reach the designed target strength only following particular conditions.

6.2 Test set-up and testing procedure

Before test, beams have been left for 28 days at 20°C. The test set-up adopted for all the three beams is shown in Figure 8. The beams were loaded at third point and measurements were taken of load, mid-deflection, setting at the support and stresses into the wires.

The strain level in the top part of the beam was also measured at discrete points with the use of strain gauges. Different load cycles have been performed for each beam. All crack patterns have been followed and recorded during the test. A supplementary investigation

on the beams made with the use of pulse ultrasonic velocity test has been carried out.

6.3 Results and discussion

The load–deflection curves for the three beams are reported in Figure 9. It is interesting to notice how the 40RN50 curve (mix with 50% of RA and 50% of NA) is almost in the middle of the curves with 100% of RA and 100% of NA.

An evaluation of the mechanical behaviour of beams with different percentages of RA and NA could be done by interpolating figures in this diagram.

7 DETECTION OF CRACK PATTERNS ON RAC PRE-STRESSED BEAM USING ULTRASONIC PULSE VELOCITY

The compressive strength of the original concrete in the demolished structure processed in the recycling site plant is almost unknown in all cases.

For the structural use of the recycled aggregate concrete it would be useful sometimes to have a rough idea of the strength of the original concrete.

The possibility of a pre-cast concrete plant using RAC on a large scale means that there is a need to find a reliable method of testing the structures made from RAC.

Following the above observations a large part of this experimentation has been dedicated to the possibility of deducing the mechanical properties of RAC using non-destructive tests.

The experimentation has been carried out firstly on RAC specimens at different ages.

The measurements have been carried out using the direct transmission technique and indirect or surface transmission has been used as comparison.

More detailed tests have been carried out on full scale pre-stressed beams made from RAC reported in. In fact during the last part of the study the ultrasonic method has been used to detect the depth of the cracks occurring during the pre-stressed beam test.

Comparing these results with the experimental results has checked the validity of this method.

7.1 Non-destructive testing on RAC prestressed beams

This part of the investigation has been carried out to deduce the compressive strength of concrete using non-destructive tests and to compare the results with the values coming from compressive strength cube tests cast with the same batch of the beams.

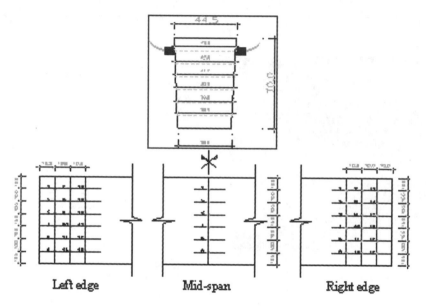

Figure 10. Position for measurements for rebound hammer and pulse ultrasonic velocity tests.

Table 15. Comparison of the compressive strength R_C evaluated with destructive and non-destructive tests on cube and full scale beams.

Test	R_C (MPa)		40AR100			40RN50			40AN100		
			3 days	7days	28 days	3 days	7 days	28 days	3 days	7 days	28 days
Crushing	R_C	Cube	33.9	36.6	44.4	34.0	42.4	45.8	43.3	47.8	59.4
	R_{C1}	Cube	29.8	37.6	42.4	34.7	44.8	46.9	41.4	48.5	53.4
Schmidt	R_{C2}	Cube	26.9	32.6	35.9	33.6	41.2	42.7	42.2	47.7	51.1
Hammer	R_{C1}	Beam	33.3	44.3	47.9	41.4	49.2	52.4	43.4	51.4	58.6
	R_{C2}	Beam	29.6	37.1	39.3	38.7	44.2	46.3	36.5	41.4	45.5
Pulse	R_{C3}	Cube	28.7	31.8	40.3	25.3	42.6	47.6	37.6	49.5	58.5
velocity	R_{C3}	Beam	31.2	30.1	33.2	23.9	53.4	77.4	40.1	50.18	63.3
Sonreb	R_{C4}	Cube	29.3	34.6	41.4	23.4	33.0	35.4	37.7	46.8	53.4
method	R_{C4}	Beam	32.3	36.7	39.9	24.5	38.0	45.7	39.8	48.5	58.2

In Figure 10 the position of the points where the ultrasonic pulse velocity has been measured is reported together with a beam section that shows the path length of the ultrasonic wave inside the beams.

During test measurements the transducers have been placed in positions such that the reinforced bars and pre-stressing tendons would not influence the path length of the ultrasonic wave. Full test results of non-destructive tests on specimens and full scale pre-stressed beams are reported.

Compressive strength test results coming from crushing cubes (100 mm × 100 mm × 100 mm) have been used to check the reliability of results achieved with the non-destructive methods (rebound hammer and pulse ultrasonic velocity test) on the full scale pre stressed beams and on cubes.

The correlation rules between the Compressive Strength R_c the Rebound Number (I_R), the Pulse Ultrasonic Velocity (V) that have been used to evaluate the figures reported in Table 7.1 are given. The Sonreb Method is a way of evaluation of the Compressive Strength of concrete using results from both the

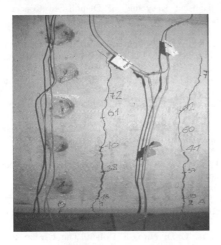

Figure 11. Crack detection in the beam mid-span.

Hammer test and the Pulse velocity test. Results from this method are also reported in Table 15.

2.1 Detection of crack patterns on RAC pre-stressed beam using ultrasonic pulse velocity

During this last part of the study the ultrasonic method has been used to detect the depth of the cracks occurring during the pre-stressed beam test.

The beam zone investigated has been the mid-span one. After that the ultrasonic transducers have been placed astride the crack layer with a distance of the presumed crack depth. The real path of the ultrasonic wave is very difficult to define for the refraction and reflection wave phenomena due to the shape of the cracks. Moreover it is possible that there are contact points inside the crack layer that allow the undesired energy transmission of the sonorous wave.

The measurements have been taken under two different load levels. The first measurement was taken with height of cracks of almost 20 cm. The second one when the crack height was over the half height of the beam.

In Figure 12 the comparison between the real crack patterns and the ultrasonic detection of the crack patterns is reported for the 40AR100 pre-stressed beams. The Sonreb combined method gives results close to the values attained for crushing test on cubes for the mixes AR "All Recycled" and AN "All Natural". The differences of values the compressive strength are almost 10–15%, better than the single method (rebound or ultrasonic). This is due to the fact that the Sonreb combined method attenuates the intrinsic imprecision of the single methods that are due to the curing conditions, the moisture content, and to the type of aggregate used.

More disperse are the values relative to the mix 40RN50 with differences of 30–40%. From the previous study it can be concluded that this non destructive investigation can be usefully used to check the real condition, for safety evaluation, of the in-situ concrete structures. In this way if a concrete structure should be demolished the strength of the original

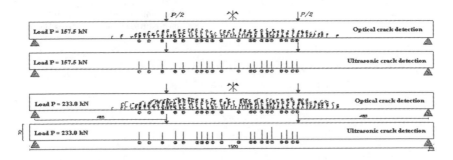

Figure 12. Real crack patterns versus the crack patterns evaluated with the ultrasonic method for the beam 40AR100.

concrete could be known for a recycling use of it. In fact at the moment, except for pre-cast plant concrete waste material, the compressive strength of the original concrete in the demolished structure processed in the recycling site plant is almost in all case unknown. Moreover this non-destructive method could be used to monitor the service life of a prototype structure made from RAC.

ACKNOWLEDGEMENTS

The authors are grateful for the support of Pescale S.p.a. in Castellarano (Reggio Emilia) and Mabo Prefabbricati S.p.a. in Bibbiena (Arezzo).

REFERENCES

E Dolara, G Di Niro, P Ridgway: Il calcestruzzo prodotto con aggregati riciclati. Mabogazine. Mabo Prefabbricati s.r.l.. Maggio 1996.

E Dolara, G Di Niro, R Cairns: RAC Prestressed Beams. Proceedings of the International Symposium organised by the Concrete Technology Unit, University of Dundee and held at the Department of Trade and Industry Conference Centre. London, United Kingdom. 11–12 November 1998. Thomas Telford Publishing.

E Dolara, G Di Niro: Caratteristiche meccaniche del RAC: calibrazione dei coefficienti per NDT. Quarry and Construction. Edizioni Pei. Parma. Giugno 2002.

E Dolara, P Ridgway, G Di Niro: Analisi non lineare di una trave in conglomerato cementizio armato prodotto con aggregati riciclati Quarry and Construction. Edizioni Pei. Parma. Marzo 1996.

E Dolara, P Ridgway, G Di Niro: L'aggregato riciclato. Caratteristiche chimico-fisiche. Quarry and Construction. Edizioni Pei. Parma. Dicembre 1995.

E Dolara, P Ridgway, G Di Niro: Proprietà e mix design del calcestruzzo fresco prodotto con aggregati riciclati. Quarry and Construction. Edizioni Pei. Parma. Gennaio 1996.

G Di Niro, E Dolara, P Ridgway: recycled aggregate concrete (RAC). Properties of Aggregate and RC beams made from RAC. Proceedings of the International Conference "Concrete in Service of Mankind". Dundee, Scotland. 24–28 June 1996. Pag. 141–149. E & F.N. Spon, London 1996.

G Di Niro, E Dolara, R Cairns: recycled aggregate concrete (RAC). A Mix-Design Procedure. Proceedings of the International FIP 97 Symposium "The Concrete Way to Development". Johannesburg, South Africa. 9–12 March 1997. Concrete Society of Southern Africa.

G Di Niro: recycled aggregate concrete for Structural Purposes. PhD Thesis in Civil Engineering. University of Strathclyde, Glasgow, UK.

G Di Niro, E Dolara, R Cairns, P Ridgway: recycled aggregate concrete (RAC). A Mix-Design Procedure. Proceedings of the International Conference "Our world in Concrete and Structures". August 1996. Orchard Plaza. Singapore. M.

G Di Niro, E Dolara, R Cairns: The Use of RAC (recycled aggregate concrete) for Structural Purposes in Prefabrication. Proceedings of the XIIIth FIP – CEB 98 Congress "Challenges for Concrete in the next millenium" Amsterdam, The Netherlands. 23–29 May 1998. A.A. Balkema Publishers.

G Di Niro, R Cairns, E Dolara: Properties of Hardened RAC for Structural Purposes. Proceedings of the International Symposium organised by the Concrete Technology Unit, University of Dundee and held at the Department of Trade and Industry Conference Centre. London, United Kingdom. 11–12 November 1998. Thomas Telford Publishing.

T C Hansen: Recycling of Demolished Concrete and Masonry. E & FN Spon/Chapman & Hall, 1992.

T C Hansen: recycled aggregates and recycled aggregate concrete. Second State-of-the-art Report Developments 1945–1985 RILEM TC-37-DRC. Materials and Structures (RILEM), Vol. 19, No. 111, May–June 1986, pp. 201–246.

M Sciotti, R Cairns, G Di Niro, E Dolara: Bonding Surface between recycled aggregate and New Cement Paste. Proceedings of the International Symposium organised by the Concrete Technology Unit, University of Dundee and held at the Department of Trade and Industry Conference Centre. London, United Kingdom. 11–12 November 1998. Thomas Telford Publishing.

A M Neville: Properties of Concrete. Fourth Edition. Longman, 1995.

P J Nixon: Recycled concrete as an Aggregate for Concrete – A Review. First state-of-the-art report RILEM TC-37-DRC. Materials and Structures (RILEM), Vol. 11, No. 65, Sept.–Oct. 1978, pp. 371–378.

R Cairns, G Di Niro, E Dolara: The Use of RAC in Prefabrication. Proceedings of the International Symposium organised by the Concrete Technology Unit, University of Dundee and held at the Department of Trade and Industry Conference Centre. London. United Kingdom. 11–12 November 1998. Thomas Telford Publishing. The Design of Normal Concrete Mixes BRE 1975.

24. Life cycles assessment

System-based Vision for Strategic and Creative Design, Bontempi (ed.)
© 2003 Swets & Zeitlinger, Lisse, ISBN 90 5809 599 1

Estimating the environmental aspects of an office building's life cycle

S. Junnila

Helsinki University of Technology, Construction Economics and Management, Finland

ABSTRACT: The environmental design of office buildings holds a particular interest for many stake holders in society, but at the moment it is still difficult to find comprehensive life cycle-based environmental data from offices. In this study, a life cycle assessment (LCA) was performed to determine the significant environmental aspects of a new office building. Those aspects of the office identified as the most significant were the heat lost in ventilation and conduction; the electricity used in outlets, lighting and HVAC systems; the manufacture and maintenance of steel, paints and non-ferrous metals; and the use of water and wastewater services. The most significant environmental aspects were quite dominant, since with 5% of all aspects they covered over 50% of the studied life cycle impacts.

1 INTRODUCTION

The significance of the environmental design of buildings has increased constantly ever since the importance of building-related environmental issues was recognized. At the moment, buildings are estimated to be one of the largest environmental impact producers in the society (UNEP 1999, Worldwatch 1995); consequently, many articles have expressed the opinion that the environmental dimension should already be included in the design phase of the building (Arena & de Rosa 2003, Ball 2002, Pilvang & Sutherland 1998).

The environmental design of office buildings holds a particular interest for many companies. A large proportion of all ISO14001 certified companies is operating in light and service industries (LSI) (ISO14001 2001), and the potential for reducing the environmental impacts of LSI companies has been found to be significant (Rosenblum et al. 2000). Building-related environmental impacts are often the ones with greatest potential for reductions in those companies. For example, one of the world's largest reinsuring companies, Swiss Re, has stated that around 60% of their environmental impacts are connected to the use of buildings (Swiss Re 2002). Other LSI companies have also presented similar figures (Kesko 2002, Royal & SunAllinace 2001).

The environmental design of buildings is essentially based on the knowledge about environmental issues of a building's life cycle (Gangemi et al. 2000). The environmental knowledge enables the control of environmental aspects and therefore helps to minimize the environmental impacts (Roberts & Robinson 1998). The term "environmental aspect" is used here as defined in the ISO 14001 (1996) environmental management standard: "Environmental aspect is an element of an organization's activity, product or service that can interact with the environment."

Several articles have already discussed the environmental aspects and life cycle impacts of buildings. However, most life cycle assessment (LCA) studies seem to have concerned residential buildings (Klunder 2002, Ochoa et al. 2002, Thormark 2002). Some of the LCA studies have also examined office buildings (Junnila & Rintala 2002, Treloar et al. 2001), but most of them have concentrated on either a limited set of life cycle phases or one or two environmental impact-indicators, or have presented the results at a relatively high, life cycle phase level and have not presented the result by life cycle elements. Generally, it is still very difficult to find comprehensive life cycle-based environmental data from offices.

The purpose of this paper is to examine the environmental aspects of an office building. The paper tries to determine what the significant environmental aspects of an office building are, and in which life cycle phase they occur. In order to get an extensive picture of the life cycle aspects, the paper emphasizes the wide range of life cycle elements in data collection.

2 METHOD

A life cycle assessment (LCA) method was used to analyze the environmental aspects of a new office building in Southern Finland. The performed study consisted of three main phases: (1) inventory analyses for collecting the data; (2) impact assessment for evaluating the environmental impacts and determining the aspects; and finally, (3) the interpretation of the results to determine the most significant aspects.

The identification and quantification of material and energy flows (inputs and outputs) of the studied system was performed in the inventory assessment. The inventory comprised the whole life cycle of the building. Fifty years was the estimated service life of the building, and it had five main life cycle phases: manufacture of the building materials, construction processes, use of the building, maintenance, and demolition. Transportation of materials was included in each life-cycle phase.

The emission data that was linked to identified energy and material flows were mainly collected from the actual building materials and energy producers in Finland. The age of emissions data was typically 2 to 5 years old, and it had been verified by an independent third party organization. The benefits (reduced emissions) gained by combined heat and power production, typical in Finland, were allocated equally to the heat and electricity.

The quality of data used in inventory was evaluated with a six-dimensional estimation framework recommended by the Nordic Guidelines on Life-Cycle Assessment (Lindfors et al. 1995). The quality of data was targeted to the level of "good", which corresponds to the indicator score 2 in the framework. In the life cycle impact assessment, the environmental impacts of the building's life cycle elements were evaluated. The impact categories used in the study were chosen according to the recommendation of the Finnish Ministry of the Environment (Rosenström & Palosaari 2000). The studied impacts were climate change, acidification, eutrophication and dispersal of harmful substances, which included summer smog and heavy metals. From the original list, ozone layer depletion and biodiversity loss were excluded due to lack of emission data.

The significance of different life cycle aspects was evaluated by the 80/20 management rule recommended for environmental studies (Vasara 1999). In practice, this means an explicit identification of life cycle elements that cause at least 80% of each examined environmental impact.

3 PRESENTING THE CASE BUILDING

The building used in the study is a new office building in southern Finland. The building has 4 400 m² of gross floor area, and a volume of 17 300 m³. An estimated two hundred people work there. The building has four floors, consisting mostly of office space. The first floor of the building also has some commercial space.

The total amount of building materials used in the building was 1 180 kg/gross-m², which is close to the amount (1 100–1 300 kg/gross-m²) presented to other office buildings in Finland (Kommonen & Svan 1998,

Table 1. Summary of material and energy flows used in the performed life cycle assessment. The estimated service life of the building used in the study was fifty years.

Office material and energy flows		Building materials	Construction	Use	Maintenance	Demolition
Energy	MWh	0	310	46 500	7	0
Fuels	Mg	0	39	22	9	23
Building mat.	Mg	9 138	84	4	1 254	0
Water	m³	0	847	138 700	0	0
Waste, Recycl.	Mg	0	55	752	338	2 468
Waste, Landfill	Mg	0	28	1 495	913	6 672

Raiko et al. 1998, Hara-Lindström 2001). The estimated heat energy consumption of the building is 36 kWh/m³/yr, which is 6% above the average heat consumption of the energy audited offices in Finland, and the estimated electricity consumption is 18 kWh/m³/yr, which is some 30% bellow the average in Finland (Motiva 2001).

The energy and materials needed during the life cycle of the building were primarily derived from the drawings and specifications of the building. The amount and type of building materials used were determined by building elements according to the Finnish building classification system. The categories included in the study were the elements on plot, substructure elements, foundations, structural frame elements, external envelope, roof elements, internal complementary elements, internal surfaces, elevators, mechanical services and electrical services (Kiiras & Tiula 1999). Sixty-nine different building elements consisting of forty-two different building materials were identified.

The energy use of the building was estimated using the WinEtana energy simulation program by the designer. The amounts in other services (use of water, wastewater, courtyard care and office waste management) were drawn from statistical data, and they presented a regional or Finnish average for offices. The materials and energy used in the construction and demolition phases were derived from other similar cases studied and were adjusted to this particular case (Junnila & Rintala 2002, Junnila & Saari 1998). A summary of energy and material flows used in the LCA is presented in Table 1.

4 RESULTS

4.1 Environmental impacts of the office building

The results of the impact assessment of the office building are presented in Table 2 and Figure 1. Almost

Table 2. The environmental aspects of an office building. The bolded figures indicate the significant environmental aspects, i.e. the aspects that account for 80% or more of the impacts in each category. (Number 0 in the table indicates that no emissions were reported in the used data set.)

Office building Environmental aspects	Climate change [ton CO_2 equiv.]	Acidification [kg SO_2 equiv.]	Summer smog [kg H_2C_4 equiv.]	Eutrophication [kg PO_4 equiv.]	Heavy metals [kg Pb equiv.]
Building materials					
Landscaping (gravel, etc.)	7	79	6	14	0,00
Asphalt paving (access roads, etc.)	2	11	2	2	0
Concrete	**510**	**1 600**	45	230	0,06
Steel reinforcing	56	160	110	17	**0,54**
Steel, cast iron	450	**920**	**1 300**	70	**0,87**
Nonferrous metals	83	560	37	40	**0,77**
Masonry	26	140	2	11	0,02
Timber	1	21	2	4	0,00
Plastic, rubber, etc.	31	310	27	23	0,08
Building boards, paper	56	600	44	45	0,00
Insulation	77	**620**	42	34	0,04
Waterproofing	1	9	5	1	0
Glass	58	370	17	54	0,00
Finishing (flooring, glues, etc.)	18	100	12	12	0,00
Paints	13	110	**360**	9	0,04
Others	**2**	39	0	3	0
Construction					
Materials in construction	46	440	39	40	0,07
Electricity	47	160	4	14	0,02
Heat	23	58	1	5	0,00
Machinery	100	**900**	100	150	0
Steam	4	7	0,1	1	0
Transp. of building materials	40	340	14	62	0
Constr. waste, transp. & landfill	1	8	1	1	0,00
Water	0,2	1	1	51	0
Use of building					
Electricity, cooling	**670**	**1 700**	130	160	0,07
Electricity, HVAC	**1 100**	**2 600**	210	250	0,11
Electricity, lighting	**1 400**	**3 400**	280	**330**	**0,14**
Electricity, outlet	**1 600**	**4 000**	330	390	**0,17**
Heat, loss through air leakage	220	530	55	49	0,01
Heat, conduction	**2 500**	**6 000**	610	**550**	0,10
Heat, hot water	410	**1 000**	100	92	0,02
Heat, ventilation	**3 200**	**7 600**	780	700	**0,12**
Courtyard care	64	600	30	99	0,00
Office waste, transp. & landfill	**790**	**1 400**	**500**	210	0,00
Water and wastewater	58	260	130	**8 400**	0
Maintenance					
Landscaping (gravel, etc.)	0,1	1	0,1	0,2	0,00
Asphalt paving (access roads, etc.)	5	28	4	4	0
Concrete	41	130	5	19	0,01
Steel reinforcing	0,0	0,1	0,1	0,0	0,00
Steel, cast iron	240	500	**680**	37	**0,50**
Nonferrous metals	53	290	41	15	0,11
Masonry	26	140	2	11	0,02
Timber	2	36	4	6	0,01
Plastic, rubber, etc.	9	87	10	7	0,02
Building boards, paper	56	600	44	45	0,00
Insulation	60	490	19	25	0,01
Waterproofing	1	9	5	1	0
Glass	58	370	17	54	0,00
Finishing (flooring, glues, etc.)	18	100	12	12	0,00
Paints	130	**1 100**	**3 600**	90	**0,40**
Constr. & transp. of materials	49	480	57	76	0,02
Demolition					
Demolition	41	350	43	56	0
Waste, landfill transp.	120	**1 300**	240	220	**0,15**
Waste transp. for recycling	14	120	1	22	0
Total	**14 600**	**42 800**	**10 100**	**12 800**	**4,47**

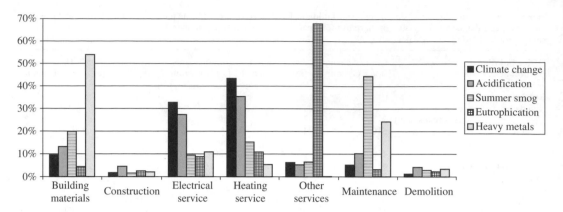

Figure 1. The environmental impacts of an office building presented by life cycle phases. The estimated service life used in the calculation was 50 years.

all life cycle phases seem to contribute significantly to the environmental impacts. However, it would appear that the heating service, manufacture of building materials, and electrical service cause most of the impacts. Maintenance and other services are also important, but in fewer impact categories. In contrast, the construction and demolition phases have little environmental impact. The results also show that in most of the impact categories, the two life cycle phases with the most impacts account for 75% of the overall impact in the category.

4.2 Significant environmental aspects of the office building

In this chapter, the environmental impacts of the office building are studied in more detail. Each life cycle phase (building materials, construction, use, maintenance and demolition) is further divided in life cycle elements, and the impacts caused by each element are analyzed. Altogether, 56 life cycle elements and 325 environmental aspects were defined. The life cycle elements with significant impacts, so-called "significant aspects" (elements that produce over 80% of the life cycle impact) are shown in the table in bold. The results are presented in Table 2.

As we can see in Table 2, all life cycle phases have significant environmental aspects, amounting to forty-four significant aspects in total. Most of the significant aspects (27 aspects) occurred in the use phase of the building. The second highest number of significant aspects (14 aspects) were found in the manufacture of building materials and the maintenance life cycle phases. The construction and the demolition life cycle phases had noticeably less – only three – significant aspects.

5 INTERPRETATION OF RESULTS

5.1 Most significant aspects

As we saw in the previous section, the identification of significant aspects (life cycle elements that cause 80% of the impacts) amounted to 44 different aspects. That is still too many to be considered in most of construction projects in practice. In this section, we further narrow down the number of significant aspects to bring forward only the so-called "most significant aspects." The most significant aspects are defined as those aspects that cause more than 50% of the environmental impacts during the life cycle of the building.

Figure 2 shows the most significant aspects. The list of identified significant aspects has now decreased from 44 to 13. The listed most significant aspects are quite dominant, since with 5% of all aspects they still cover over 50% of the life cycle impacts of the office. The most significant environmental aspects were caused by the following life cycle elements:

– heat lost in ventilation
– heat conduction through structures
– electricity used in outlets
– electricity used in lighting
– electricity used in HVAC systems
– manufacture and maintenance of steel products
– manufacture and maintenance of paint products
– manufacture and maintenance of non-ferrous metals
– use of water and wastewater.

5.2 Data quality

In this chapter, the data quality issues are discussed. The results of data quality assessment are presented in Table 3. The scores in the data quality table are rounded to the nearest whole number.

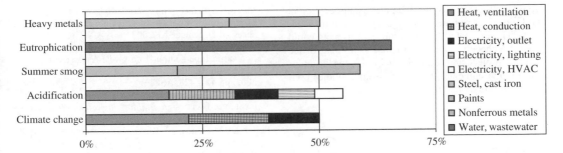

Figure 2. The most significant environmental aspects of an office building and their proportion of the studied environmental impacts.

Table 3. Summary of the data quality assessment. The detailed description of the method used in the data quality assessment has been presented by Lindfors et al (1995) and Weidema & Wesnæs (1996).

Data quality* table	Acquisition method	Independence of data supplier	Representa- tiveness	Data age	Geographical correlation	Technological correlation
Building materials	2	1	2	2	2	2
Construction	3	1	2	2	3	4
Heating service	2	2	1	1	1	1
Electrical service	2	2	1	1	1	1
Other services	2	2	1	1	3	4
Maintenance	2	1	2	2	2	2
Demolition	3	2	3	3	3	4

* Maximum quality = 1; minimum quality = 5.

The data quality indicators score two or better with all indicators in the building materials, electrical service, heating service and maintenance life cycle phases. Thus, the quality of data is as targeted in most of the major life cycle phases. This supports the findings presented in the result section.

The life cycle phase referred to as "other service" causes uncertainty in the eutrophication category. Other services scored only 3 in geographical and 4 in technical correlation, and since it is the major cause of the eutrophication impact (66%), the uncertainty of the result is higher than targeted in that category.

The data quality is also low in the construction and demolition phases, but since these have only a minor influence on the overall result, this does not cause significant uncertainty in the results.

6 DISCUSSIONS

The purpose of the study was to determine the significant environmental aspects of an office building.

The results showed that significant environmental aspects could be found in all life cycle phases of the building. However, most of the identified significant aspects were caused by the heating service. In addition, the manufacture of building materials, the electrical service and maintenance had several significant aspects. The identified life cycle elements causing the most significant aspects were the heat lost in ventilation and conduction; the electricity in lighting, HVAC and outlets; the manufacture and maintenance of steel, paints and non-ferrous metals; and the water and waste water services. The most significant aspects were quite dominant, since with the 5% of all aspects they covered over 50% of the life cycle impacts.

The findings of this study are consistent with the results presented in other studies. Typically, the life cycle assessment studies of buildings have concluded that the operational energy causes the majority of environmental impacts. In Finland, for example, the operational energy of buildings is estimated to cause roughly 80–90% of climate change and acidification impacts (Hara-Lindström 2001, Saari 2000, Kommonen & Svan 1998, Raiko et al. 1998). In this study, the operational energy accounted for 60–75% of those impacts. The difference is mainly due to the larger amount of life cycle elements included in this study.

Although the study aimed at extensiveness, it still has several limitations. Subjective choices had to be made in defining the scope of the study. For example, the outlet-electricity, the use of water, and the office waste management were all included as an element of office building's life cycle, but some other office-related life cycle elements such as furniture, computers and commuting were not included. Another limitation is the lack of sensitivity analysis, which could have improved the interpretation of the results. Finally, some important environmental impacts such as ozone depletion, resource consumption, biodiversity and indoor air quality could not be covered in the study because of a lack of data.

A single case such as this one is not yet enough to make a good generalization based on the results.

However, the results could find practical application in the field of environmental design and design management. The presented list of significant aspects could work as a checklist of environmental issues in an office building project, and could also help to focus the attention in environmental design or in other studies concernig environmental impacts of offices.

REFERENCES

Arena, A. P. & de Rosa, C. 2003. Life cycle assessment of energy and environmental implications of the implementation of conservation technologies in school buildings in Mendoza – Argentina. *Building and Environment* 38, 2003, pp 359–368. Pergamon.

Ball J. 2002. Can ISO 14000 and eco-labeling turn the construction industry green? *Building and Environment* 37, 2002, pp 421–428. Pergamon.

Gangemi, V., Manlanga, R. & Ranzo, P. 2000. Environmental management of the design process. Managing multidisciplinary design: the role of environmental consultancy. *Renewable Energy* 19, 2000, pp 277–284. Pergamon.

Hara-Lindström, E. 2001. *Life Cycle Assessment of Leija-Environment House*. Technical Report, Espoo – Vantaa Institute of Technology, Finland.

ISO 14001 1996. *Environmental management systems, Specification with guidance for use*. SFS-EN ISO 14001. Finnish Standards Association SFS. Finland.

ISO 14001 2001. *Types of US Companies Registered to ISO 14001*. The ISO 14001 Information Center, http://www.iso14000.com/Community/cert_us_analysis.html#USTypes, accessed October 14, 2002.

Junnila, S. & Rintala, T. 2002. Comprehensive LCA Reveals New Critical Aspects in Offices. *Proceedings of CIB/iiSBE International Conference on Sustainable Building 2002*. September 23–25, 2002. Oslo. Norway.

Junnila, S. & Saari, A. 1998. *Environmental Burdens of a Finnish Apartment Building by Building Elements in the Context of LCA*. Helsinki University of Technology. Construction Economics and Management. Report 167. Finland.

Kesko 2002. *Indicators of Environmental Responsibility. Report of Corporate Responsibility*, Kesko. http://www.kesko.com/users/26/227.cfm, accessed October 10, 2002.

Klunder, G. 2002. The Search for the most Eco-Efficient Strategies for Sustainable Housing Construction. *Proceedings of CIB/iiSBE International Conference on Sustainable Building 2002*, September 23–25, 2002. Oslo. Norway.

Kiiras, J. & Tiula, M. 1999. *Building 90 – The Finnish Building Classification System*. The Finnish Building Centre Ltd., Helsinki, Finland.

Kommonen, F. & Svan, T. 1998. *The Environmental Aspects of Ambiotica*. Senate Properties, Helsinki, Finland.

Lindfors, L.-G., Christiansen, K., Hoffman, L., Virtanen, Y., Juntilla, V., Hanssen, O.-J., Rønning, A., Ekvall, T. & Finnveden, G. 1995. *Nordic guidelines on life cycle assessment*. Nord 1995:20. Nordic council of ministers. Copenhagen. Denmark.

Motiva 2001. Energiankatselmustoiminnan tilannekatsaus. Motiva *Publication 1/2001, Information Centre for Energy Efficiency and Renewable Energy Sources*. http://www.motiva.fi/kirjasto/Energiakatselmukset/, accessed December 10, 2002.

Ochoa, L., Hendrickson, C. & Matthews, H. S. 2002. Economic Input–output Life Cycle Assessment of U.S. Residential Buildings. *Journal of Infrastructure Systems*, 2002, pp 421–428. ASCE.

Pilvang, C. & Sutherland, I. 1998. Environmental management in project design. *Building Research & Information* 26(2), 1998, pp 113–117. E&FN Spon.

Raiko, E., Miettinen, A., Kopra, J. & Kommonen, F. 1998. *The Environmental Aspects of Pöyry-House*. Technical Report, JP-Building Engineering Ltd., Espoo, Finland.

Roberts, H. & Robinson, G. 1998. *ISO 14001 EMS Implementation Handbook*. Butterworth-Heinemann Ltd., Oxford, UK.

Rosenblum, J., Horvath, A., & Hendrickson, C. 2000. Environmental Implications of Service Industries. *Environmental Science & Technology*, 34(4), American Chemical Society.

Rosenström, U. & Palosaari, M. 2000. *Indicators for sustainable development in Finland*. Finnish Environment 404. Ministry of Environment.

Royal & SunAllinace 2001. *Carbon Dioxide Emission Equivalents, Benchmarking*. Environmental Issue, http://www.royalsunalliance.com/pdfs/environment/environmentalissues.pdf, accessed October 10, 2002.

Saari, A. 2000. Management of life-cycle costs and environmental impacts in building construction. In: *Integrated life-cycle design of materials and structures ILDES 2000. Proceeding of the RILEM/CIB/ISO international symposium* PRO 14, May 22–24, 2000. pp 117–122. RILEM. Finland.

Swiss Re 2002. *Internal Environmental Performance Indicators 2001 for Swiss Re Group*. Internal Environmental Management, Swiss Re, http://www.swissre.com/, accessed October 13, 2002.

Thormark, C. 2002. A low energy building in a life cycle – its embodied energy, energy need for operation and recycling potential. *Building and Environment* 37, 2002, pp 429–435. Pergamon.

Treloar, G., Fay, R., Ilozor, B. & Love, P. 2001. Building Materials Selection: Greenhouse Strategies for Built Facilities. *Facilities*, 19(3–4), pp 139–149, MCB University Press.

UNEP 1999. *Energy and Cities: Sustainable Building and Construction*. United Nations Environmental Program, http://www.unep.or.jp/ietc/Focus/Sustainable_bldg1.asp, accessed October 14, 2002.

Vasara, P. 1999. Environmental Adaptive Benchmarking: A Framework for Environmental Assessment. *Acta Polytechnica Scandinavica, Chemical Technology Series* No. 268, Finnish Academy of Technology, Espoo, Finland.

Weidema, B. P. & Wesnæs, M. 1996. Data quality management for life cycle inventories – an example for using data quality indicators. *Journal of Cleaner Production*, 4(3–4), pp. 167–174. Elsevier Science Ltd. Great Britain.

Worldwatch. 1995. A Building Revolution: How Ecology and Health Concerns are Transforming Construction. *Worldwatch Paper 124*. http://www.cityofseattle.net/sustainable building/overview.htm, accessed August 10, 2002.

Quality and life cycle assessment

J. Christian & L. Newton
*Construction Engineering and Management Group, University of New Brunswick, Fredericton,
New Brunswick, Canada*

ABSTRACT: Difficulties arise in attempting to measure quality and making life cycle assessments. An owner must consider a trade-off between design and construction costs and operating and ownership costs. A case study of buildings regarding the cost-effectiveness of rehabilitating older structures in lieu of new construction is described. This particular study showed that, depending upon the design of the building and its structural soundness, it is equally cost effective to consider the option of rehabilitation over new construction. Another study of roofing assets at two public sector organizations showed that planned routine and preventative maintenance can have a significant impact on prolonging the life cycle of facilities. Without a quality maintenance programme, the best-designed building will fail to reach its service life.

1 INTRODUCTION

Decisions during the design and construction of facilities are based on short or long-term costs and the level of quality. Life cycle assessment is a method sometimes used by both the public and private sector to determine the most cost effective design for the duration of the life of a building. Life cycle costs are dependent on the facility thus a life cycle cost assessment cannot be undertaken without considering the level of quality in each phase of the service life of a facility. However, difficulties arise in attempting to measure quality and making life cycle assessments. The ability to influence life cycle costs is generally considered to be the greatest during the design phase. The reality is that a favourable influence is seldomly the case as owners do not have unlimited funds to spend on design and construction. Thus, the owner must consider a trade-off between design and construction costs and operating and owning costs.

A second consideration for an owner in life cycle assessment is sustainability and the minimization of the impact of construction on the environment. Can a well-built facility be rehabilitated for another use rather than being demolished and a new purpose-built structure erected in its place?

Quality can be defined in different ways. It can be defined as a degree of excellence, conformance to requirements, or as performance to meet the needs of the user. A dichotomy occurs in a construction project when quality at the end of construction is measured as conformance to requirements, whereas when the entire life cycle of the project is considered quality tends to be measured as the performance to meet the needs of the user. A large proportion of the total lifetime cost occurs after construction. Achieving quality therefore may include several stages where the different definitions are the most appropriate.

The scope of the paper covers research on the life cycle assessments of costs and maintenance which has been conducted on residential buildings on military bases and roofing assets at two public sector organizations, and any conclusions drawn are therefore extremely limited to these situations.

2 MEASURING QUALITY

Over the life of a facility the measure of quality can be considered as consisting of seven dimensions: performance, reliability, conformance, durability, serviceability, aesthetics and perceived quality. The relative importance of each dimension as a measure of quality depends upon the stage of the life cycle of a facility. (McGeorge & Palmer 1997).

In a recent study (Christian & Newton 2002a) a questionnaire was sent to designers, contractors, facility managers and users asking each to rate the importance of 25 "quality" factors. The results indicated that performance was the most important of the seven dimensions while aesthetics and perception were the least important.

A second questionnaire was developed to gain an understanding about the factors that users felt had the greatest impact on building quality and how they perceived the buildings, in which they worked, rated according to the factors. The results of the user study indicated that users were not overly concerned with building construction as an indicator of quality but facility managers were. For the user, the most important quality factors were those which affect an individual's ability to live or work in a particular building. There appears to be an inherent assumption that a building is structurally sound and will protect the occupant from the elements. Therefore, the occupants are only affected by the construction if the building has very obvious flaws.

However, a building occupant's "perception" of what constitutes building performance or reliability is not necessarily the same as that of a facility manager. The requirement to consider sustainability in construction is increasingly important and has become another "quality" factor that designers must take into consideration. Anecdotal evidence would suggest that sustainability, with both users and facility managers, is more of a consideration in Europe than North America.

3 DURABILITY OF CONSTRUCTION

The ability to influence life cycle costs is generally considered to be the greatest during the design phase. Theoretically, money well spent on design and construction will result in savings in the operation, maintenance and minor rehabilitation (O,M&R) of the facility and may ultimately determine if total rehabilitation of a building, as opposed to demolition and reconstruction,

is a cost effective option. The potential impact of this is illustrated in the cumulative cost of the five buildings shown in Figure 1. As can be seen, if the trend continues, the life cycle costs for Building 4, a newer building, will exceed those of the other buildings despite the fact that it has yet to undergo a mid-life retrofit as the older buildings have already experienced.

While this premise appears valid when individual buildings are considered, it is very difficult to prove in a large sample of buildings due to the many other variables that impact upon building life cycle costs. A statistical evaluation of historical cost including initial construction costs, maintenance, building services and rehabilitation costs was carried out using MINITAB® statistical analysis software. The data were compiled form historical facility cost records from the Canadian Department of National Defence. The initial findings of a study of 48 residential buildings indicated that there was evidence of a weak negative correlation (p = −0.32) between initial construction costs and historic maintenance and rehabilitation (M&R) costs when considering the first 20 years of a buildings life. That is, low initial construction costs resulted in higher M&R costs. In addition, high maintenance costs in the first 20 years of building life were moderately correlated to lower maintenance costs later on in life.

Durability is an important consideration in sustainability if the building is to be rehabilitated for continued use at the end of its initial service life. A steel frame, solid brick or reinforced concrete building is more likely to be considered for a major retrofit that will extend its service life as compared to a wood frame building as it is more economical than building

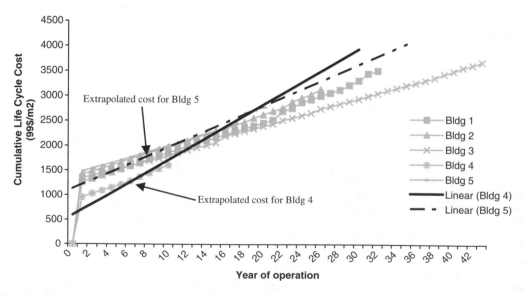

Figure 1. Cumulative life cycle costs showing impact of initial construction costs.

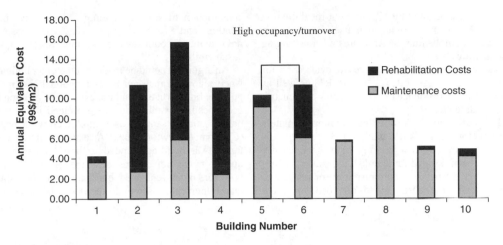

Figure 2.   Impact of occupancy on M&R costs for timber-framed residential structure.

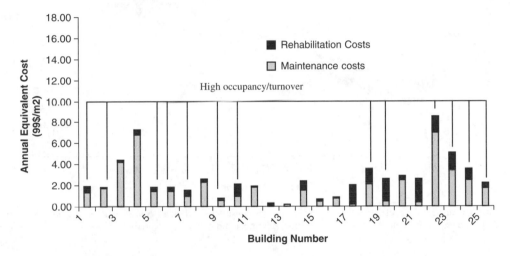

Figure 3.   Impact of occupancy on M&R costs for reinforced concrete residential structure.

a replacement structure. In addition, the impact that occupancy (level of use) and function have on the life and on the M&R costs of a building can be significant.

Buildings with a high profile or those which had undergone a change in profile/function were found to have higher combined M&R costs than buildings with an average or low profile occupant (Christian & Newton 2002b) regardless of the type of building construction. However, if the building had a high turnover of personnel, the impact of high occupancy on a timber-framed building was much more pronounced than that on a concrete structure. This is shown in Figures 2 and 3 where two types of residential buildings are compared. Note that Buildings 2, 3 and 4 in Figure 2 have high rehabilitation costs due to a design error and thus

only the maintenance costs should be considered for comparison purposes.

4   CASE STUDIES IN REHABILITATION

4.1   *Residential buildings*

Thirty-five older residential buildings were designed to have a secondary administration or institutional function. Ten of the residential buildings considered in this paper had been converted to the secondary use at some point during their life cycle while 14 had undergone some degree of rehabilitation. Four of these buildings were studied in greater depth to determine if there was a benefit to rehabilitation (subsequent change in use denoted by "C", straight rehabilitation denoted by "R") to extend their useful service life instead of demolishing the structure and constructing new facilities.

Figure 4 shows the cumulative life cycle costs for these four buildings (C1, C2, C3, & R1) as well as plots for a new residence (Res) and a new administration building (Admin).

The cost to retrofit the buildings was approximately $250/m^2$ for C2, C3 & R1 while C1 cost $400/m^2$ as a new roof was added at the same time as the retrofit. If it had been decided to demolish the existing buildings and construct new administration or residential infrastructure the costs, based on costs of recent

construction of similar buildings, would have been $880/m^2$ and $1075/m^2$ respectively. These costs do not take into account any costs/savings associated with sustainability.

An argument could be made that the older buildings when renovated may not be as functional as a newer purpose built structure and not perform in meeting the users needs as well. This was examined in a second quality survey sent to users and facility managers (Christian & Newton 2002a). As expected, it was found that newer buildings tended to have higher overall quality perception scores than the older buildings; however, older renovated buildings had comparable quality perception scores to the newer facilities. This

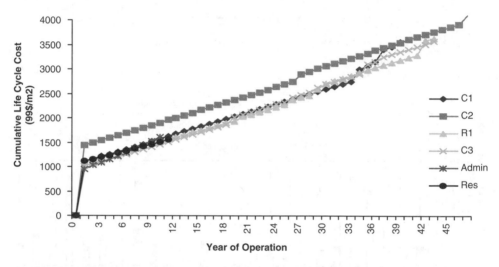

Figure 4.   Cumulative life cycle costs for renovated and newly constructed buildings. *Note:* C denotes rehabilitation accompanied by a change in function, R denotes rehabilitation, Admin(istration) and Res(idential).

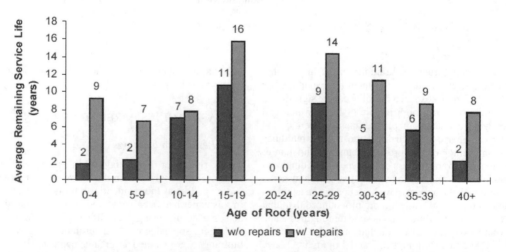

Figure 5.   Average remaining service life for UNB roofs (with and without repairs).

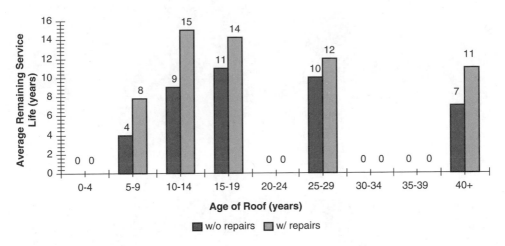

Figure 6. Average remaining service life for CFB Gagetown roofs (with and without repairs).

lends strong support to considering rehabilitation of a structurally sound older building as a viable option to new construction in this particular type of facility.

4.2 Roofing assets

Two organizations were investigated in a research project (Gamblin 2002) to determine the influence of maintenance management and quality management on the life cycle costs of roofing systems. The University of New Brunswick is an educational institution located in Fredericton, New Brunswick, Canada. UNB has 35 buildings, totalling 150 flat or low-slope roof sections on its Fredericton campus encompassing over 39 000 m² of roof area. Canadian Forces Base (CFB) Gagetown is a military combat training centre located in Oromocto, New Brunswick, Canada.

CFB has a total of 156 buildings but only 101 flat or low-slope sections were investigated in this research encompassing 47 000 m² of roof area.

It can be seen from Figures 5 and 6 that if repairs were made to the roofs, the remaining service life would increase.

The average increases in service lives for CFB Gagetown roofs appear to be smaller than the increases at UNB. This is because UNB roofs are presently in worse condition than CFB Gagetown roofs.

5 CONCLUSIONS

From the very limited scope of the research data discussed in the paper there is evidence of a slight correlation between low initial construction costs resulting in higher M&R costs and also that high maintenance costs in the first 20 years of the life of facility result in lower maintenance costs later on in life. Some buildings very obviously prove the premise some do not. For roofs, the research more clearly showed that a good maintenance management system and early repairs to roofs resulted in an extension to service lives.

ACKNOWLEDGEMENTS

Financial support for the research mentioned in this paper is gratefully acknowledged from the Natural Sciences in Engineering Research Council of Canada, the Canadian Department of National Defence, and the M. Patrick Gillin Chair in Construction Engineering and Management at the University of New Brunswick.

REFERENCES

Christian, J. & Newton, L. 2002a. Consideration of the User's Perspective in Design and Construction. *Proceedings of the Canadian Society for Civil Engineering Conference 2002*. CD Rom. Montreal: Canada.

Christian, J. & Newton, L. 2002b. A Prediction Model for Life Cycle Costs Based on Design Quality. *Proceedings of the 9th International on the Durability of Building Materials and Components*. Paper 149. Brisbane.

Gamblin, T. 2002. *Influence of Maintenance Management and Quality Management on Roofing System*. M.Sc.E. thesis, UNB. New Brunswick: Canada.

McGeorge, D. & Palmer, A. 1997. *Construction Management New Directions*. Oxford, UK: Blackwell Science Ltd.

Integrated life cycle design of structures

A. Sarja
Technical Research Centre of Finland, VTT Building and Transport

ABSTRACT: Traditionally structural design has been focused mainly on the construction phase, optimising construction costs, static and dynamic performance and safety. The future core competence of structural engineers should be the design of structures for sustainability over their life cycles. This field will here be referred to as integrated life cycle design. This means that the structural design process must be renewed. Furthermore, new methodologies and calculation methods must be adopted, e.g., from mathematics, physics, systems engineering and other natural and engineering sciences. A new model of integrated life cycle design includes a framework for integrated structural life cycle design, a description of the design process and its phases, special life cycle design methods regarding different aspects like life cycle economy, life cycle environmental impact, design for reuse and recycling, service life design, durability design, multi-attribute optimisation and multi-attribute decision-making.

1   INTRODUCTION

During last decade there have been a lot of discussion on principles and goals of sustainable construction and sustainable build environment. Now it is time to start creation of real technical content for filling the gap between general discussions, strategies, goals and building and civil engineering practice.

Lifetime engineering is an innovative idea and a concretisation of this idea for solving the dilemma between a very long term product and a short term design, management and maintenance planning. The integrated lifetime engineering methodology is aiming at regulating, optimising and guaranteeing the life cycle human conditions, economy, cultural compatibility and ecology of buildings and civil infrastructures with parameters of technical performance and economy, in harmony with cultural and social requirements.

For the *life cycle design*, the analysis and design are expanded into two economical levels: monetary economy and ecology, which means the economy of nature. The life cycle expenses are calculated into the present value or into yearly costs by discounting the expenses from manufacture, construction, maintenance, repair, changes, modernisation, rehabilitation, reuse, recycling and disposal. The monetary expenses are treated as usual in current value calculations. The expenses of nature are the use of non-renewable natural resources: materials and energy, the production of air, water or soil pollution. Consequences of air pollution are health problems, inconvenience for people, ozone depletion and the climatic global warming. These impacts dictate the environmental profiles of the structural and building service systems. The goal is to limit the natural expenses under the allowed values and to minimise them. Integrated lifetime design is an important link in construction: interpreting the requirements of owners, users and society into performance requirements of the technical systems; creating and optimising technical solutions, which fulfil those requirements; and proving through analysing and dimensioning calculations that the performance requirements will be fulfilled over the entire design service life.

2   INTEGRATED LIFETIME DESIGN

The integrated lifetime design includes the framework, process, methods and calculation models. These will be described in short in the following chapters.

2.1   *Design framework*

The objective of the integrated life cycle design is to make concrete the design methods and methodologies for structural design in order to meet the requirements of sustainable development during the entire life cycle of the structures (resources, transports,

Table 1. Generic classified requirements of the structures.

1. Human requirements	• business culture
• functionality in use	• aesthetics
• safety	• architectural styles and
• health	trends
• comfort	• image
2. Economic requirements	4. Ecological requirements
• investment economy	• raw materials economy
• construction economy	• energy economy
• lifetime economy	• environmental burdens
3. Cultural requirements	economy
• building traditions	• waste economy
• life style	• biodiversity

manufacture, use, recycling and reuse, demolition, wasting). The objective is to design for life cycle quality in relation to the generic requirements listed in Table 1. The lifetime quality means the capability of the structures to fulfil the requirements of the users, owners and society (Table 1).

The integrated life cycle design includes the mechanical, physical, economical, energy, health and environment aspects. The integrated life cycle design shall manage the multiple requirements in a systematic way.

2.2 Design process

The main phases in the model of integrated life cycle design process are:

1. analysis of the actual requirements
2. interpretation of the requirements into technical performance specifications of structures
3. creation of alternative structural solutions
4. life cycle analysis and preliminary optimisation of the alternatives
5. selection of the optimal solution between the alternatives and
6. the detailed design of the selected structural system and its modules and components.

The conceptual, creative design phase is very decisive in order to utilise the potential benefits of integrated design process effectively. Controlled and rational decision making when optimising multiple requirements with different metrics is possible through the application of systematics of multiple attribute optimisation and decision making. In detailed design phase, life cycle aspects rise needs for total performance over the life cycle, including durability design and design for mechanical and hygrothermal long term performance.

At the life cycle planning, a modular systematic is preferred. This allows the systematic allocation and optimisation of the target service life as well as life cycle economy and ecology of different parts of the building. Generic hierarchical classification is as follows:

1. Object (building or an infrastructural facility: bridge, tunnel, harbour, road, etc.)
2. Module (a partly autonomous part of an object: bearing structures, foundations, etc.)
3. Component (beam, column, etc.)
4. Detail (bearing support, connection, etc).
5. Material (metal, concrete, etc.).

A suited modularisation at the highest level of hierarchy is the following: Bearing frame, envelop, foundations, partitions, heating and ventilating services, information, water and sewage system, control, data processing and communication services and waste management system. All of these assemblies are specified during the development or design process on continuously increasing precision starting from general performance specifications and ending into detailed designs.

2.3 Design methods

2.3.1 General

In order to be able to design for entire planning and design period, which usually is 50 to 100 years, the designers have to apply several specific design methods, in addition to the traditional planning and design methods. These methods are typically connected each to a certain phase of the design process. A summary of the integrated life cycle design phases and the specific design methods are presented in Table 2.

The design methods have been described in the reference (Sarja 2002).

2.3.2 Life cycle economy and ecology

The life cycle costs are calculated using the ordinary current value discounting method, which is applied to both the monetary economy and the natural economy, using Equations [1, 4]

$$E_{tot}(t_d) = E(0) + \Sigma[N(t) \bullet E(t)] - E_r(t), \quad (1)$$

where
E_{tot} is the design life cycle monetary cost as a present value
$E(0)$ construction cost
$N(t)$ coefficient for calculation of the current value of the cost at the time t after construction
$E(t)$ cost to be borne at time t after construction
$E_r(t)$ residual value at time t

$$E_{tot}(t_d) = E(0) + \Sigma[N(t) \bullet (1-k_r) \bullet E(t)] - E_r \quad (2)$$
$$(t_d)$$

Table 2. Specific lifetime design methods of design phases.

Design phase	Life cycle design methods
1. Investment planning	Multiple criteria analysis, optimisation and decision making Life cycle (monetary and natural) economy
2. Analysis of client's and user's needs	Modular design methodology Quality Function Deployment Method (QFD)
3. Functional specifications of the object	Modular design methodology Quality Function Deployment Method (QFD)
4. Technical performance specifications	Modular design methodology Quality Function Deployment Method (QFD)
5. Creation and sketching of alternatives	Modular design methodology
6. Modular lifetime planning and service life optimisation	Modular design methodology Modular service life planning Life cycle (monetary and natural) economy calculations
7. Selection between alternative solutions and products	Modular design methodology Quality Function Deployment Method (QFD) Multiple criteria analysis, optimisation and decision making
8. Detailed design of the selected solution	Design for future changes Design for durability Design for health Design for safety Design for hygro-thermal performance User's manual Design for re-use and recycling

where
$E_{tot}(t_d)$ is total natural life cycle cost in relevant terms
$E(0)$ natural cost at the construction phase
$E(t)$ natural cost to be borne at time t after construction
K_r efficiency factor of recycling at renewal of each product

The time coefficients $N(t)$ and N_S can be calculated using the known Equations [1, 4]

$$N(t) = 1/(1+i)^n$$
$$N_S = ((1+i)^n-1) / i \bullet (1+i)^n$$
(3)

where
i is the rate (interest)
n the time (y) from the date of discounting

For the ecological calculations the expenses are, as presented above, the environmental burdens, e.g. consumption of non-renewable raw materials and energy, and the production of pollutants into the air, soil and water, including CO_2, CO, SO_2, N_{ox}, dust and solid wastes. It is recommended to also use a virtual rate when calculating the current value of future expenses of the nature. This is needed, because the technology in the future is assumed to be more environmentally effective than the current technology (Sarja 2002).

2.3.3 Ranking and selection between design alternatives, materials and products

Ranking between design alternatives ends the draft design phase. The system solution and some key products and materials which are connected to the solution of the structural system are selected even if the selection of products is mainly done in later phases of the design. All classes of requirements are systematically taken into account at the ranking.

As a ranking method Multiple Criteria Analysis is applied. The core properties are mainly calculated quantitatively with numerical equations, but some of the added properties are evaluated qualitatively only. The properties are normalised through comparison with reference alternatives. The selected alternatives can fulfil some of the following criteria (Sarja 2002):

- Best in all requirements
- Best weighted properties with reasonable cost level
- Best in preferred requirements, fulfilling accepted levels in all requirements
- Best in evaluated multiple criteria benefit/cost ratio.

2.3.4 Service life planning and optimisation

The objective of service life planning is to ensure, that the service life of the facility is functionally, technically and economically optimised over the design life. The decisive factor that dictates service life varies. It might be either defective performance, or functional, technical or economic obsolescence (Sarja 2002, Guide for service ... 1998). Because of different optimal lengths of service life of different parts of a facility, a modular planning principle can be used (Sarja & Hannus 1995, Sarja & Vesikari 1996, Sarja 2002).

Obsolescence means the inability to satisfy changing functional, technical or economic requirements. Obsolescence can be with regard to an entire traffic system or its structures (Sarja 2002, Sarja & Hannus 1995).

Functional obsolescence is connected to changes in the functions of a building, traffic system or other built facility, when the facility partly or even totally looses their function. Functional obsolescence is often strongly tied to the planning of the infrastructural system.

Technological obsolescence is mostly connected to changes in technical requirements. Typical changing requirements are those of increasing traffic loads. Often this is connected to changes in the entire society, region, technology or technical system. Technological obsolescence can sometimes be avoided or diminished with an estimation of the future technical development in the design.

Economic obsolescence means unacceptably high operation and maintenance costs in comparison to new systems and facilities. This can partly be avoided in design by carefully minimising the operation and maintenance costs through the selection of such materials and structures that need the minimum amount of work and materials in their maintenance and operation. Often this means simple and safely working structures that are not sensitive to defects and their influences.

2.3.5 *Durability design*

Durability design, also titled service life design, means the designing and detailing of structures for a specified design service life. The design service life specifications are produced during the service life planning as a result of life cycle optimisation. Structural service life design can also be called durability design (Sarja 2002, Sarja & Vesikari 1996).

Durability design methods can be classified starting with the most traditional and ending with the most advanced methods as follows:

1. Design based on structural detailing
2. The reference factor method
3. Statistically calculated lifetime safety factor method
4. Statistical durability design

Structural detailing for durability is a dominant practical method applied to all types of materials and structures. The principle is to specify structural design and details as well as materials specifications in such a way, that either deteriorating impacts on structures or the effects of environmental impacts on structures can be eliminated or diminished. The first principle is typically dominating in the design of structures that are sensitive to environmental effects, like wooden structures. The second principle is suited to structures that can be designed to resist even strong environmental impacts, like concrete structures and coated steel or wooden structures.

The methods for durability detailing are presented in current the norms and standards.

The factor method is aimed at estimating the service life of a particular component or assembly under specific conditions. It is based on a reference service life – in essence the expected service life under the conditions that generally apply to that type of component or assembly – and a series of modifying factors that relate to the specific conditions of the case (Guide for service ... 1998). The method uses modifying factors for each of the following:

- A   component quality
- B   design level
- C   work execution level
- D   indoor environment
- E   outdoor environment
- F   in-use conditions
- G   maintenance level

Estimated service life of the component (ESLC) = RSLC $\times$ A $\times$ B $\times$ C $\times$ D $\times$ E $\times$ F $\times$ G where RSLC is the Reference Service Life of the component.

For more detail refer to ISO/CD 15686-1, where the method is described (Guide for service ... 1998).

The lifetime safety factor design procedure is somehow different for structures consisting of different materials, although the basic design procedure is the same for all kinds of materials and structures. The design service life is determined by formula (Sarja & Vesikari 1996):

$$t_d = \gamma_t \cdot t_g \qquad (4)$$

where

t_d is the design service life,
γ_t the lifetime safety factor, and
t_g the target service life.

The lifetime safety factor can be calculated with equation [5]

$$\gamma_t = (\beta \cdot v_D + 1)^{1/n} \qquad (5)$$

where

β is the safety index corresponding to the statistical reliability level
v_D the coefficient of variation of degradation
n the exponent of the degradation function.

The lifetime factor design procedure is as follows:

1. Specification of target service life and design service life
2. Analysis of environmental effects
3. Identification of durability factors and degradation mechanisms
4. Selection of a durability calculation model for each degradation mechanism
5. Calculation of durability parameters using available calculation models
6. Possible updating of calculations of the ordinary mechanical design (e.g. own weight of structures)
7. Transfer of durability parameters into final design

The simplest mathematical model for describing the event "failure" comprises a load variable S and a response variable R (Sarja & Vesikari 1996). In principle the variables S and R can be any quantities and expressed in any units. The only requirement is that they are commensurable. Thus, for example, S can be a weathering effect and can be the capability of the surface to resist the weathering effect without unacceptably large visual damage or loss of the reinforcement concrete cover.

If R and S are independent of time, the event "failure" can be expressed as follows (Sarja & Vesikari 1996):

$$\{failure\} = \{R < S\} \tag{6}$$

The failure probability P_f is now defined as the probability of that "failure":

$$P_f = P\{R < S\} \tag{7}$$

Either the resistance R or the load S or both can be time dependent quantities. Thus the failure probability is also a time dependent quantity. Considering $R(\tau)$ and $S(\tau)$ are instantaneous physical values of the resistance and the load at the moment τ, the failure probability in a lifetime t could be defined as:

$$P_f(t) = P\{R(\tau) < S(\tau)\} \quad \text{for all } \tau \leqslant t \tag{8}$$

The determination of the function $P_f(t)$ according to the formula 6a is mathematically difficult. That is why R and S are considered to be stochastic quantities with time dependent or constant density distributions. By this means the failure probability can usually be defined as:

$$P_f(t) = P\{R(t) < S(t)\} \tag{9}$$

Considering continuous distributions, the failure probability P_f at a certain moment of time can be determined using the convolution integral:

$$P_f = \int_{-\infty}^{\infty} F_R(s) f_S(s) ds \tag{10}$$

where
$F_R(s)$ is the distribution function of R,
$f_S(s)$ the probability density function of S, and
s the common quantity or measure of R and S.

The Integral 10 can be solved by approximative numerical methods.

The statistical method can in principle be used for individual special cases even in praxis, but this will not find any common use. The main use of statistical theory is in the development of deterministic methods. Such a method is the lifetime safety factor method presented above.

2.3.6 *Design for re-use and recycling*
It is important to recognise that the recycling possibilities of the building components, modules and even technical systems shall be reconsidered in connection with design. The higher the hierarchical level of recycling, the higher also the ecological and economical efficiency of recycling (Sarja 2002, Sarja 1997, Schultmann 1999, Müller 1999, Fukuima 1999). Therefore the re-use of entire components, modules or systems has to be preferred, even if there are difficulties in quality requirements and quality control in re-use.

Material recycling is a tool used to save raw materials, but the reduction in environmental burdens and energy consumption is usually small. The components of the environmental profile of the basic materials usually already include the recycling efficiency.

The recycling ability of the structural materials and components depends on the degree and/or the technical level of the desired re-use. Special issues to be treated in the design of structures and materials for re-use and recycling are (Müller 1999):

- separability of the structural components or materials during demolition of a structure (demountable structures)
- structural separation of components, modules or systems with different service lives and different recycling techniques
- reduction in the variety of materials
- separation ability of materials, which cannot be recycled together
- avoidance of insoluble composite substances t.

Several civil engineering structures, like roads and streets, are major consumers of raw materials. In those structures, materials consumption can be reduced with the use of industrial by-products like fly ash and blast furnace slag, and construction and demolition waste materials like crushed concrete and masonry. Detailed quality specifications and an effective quality control are needed for the use of these secondary materials.

2.3.7 *Design for health*
The healthy checking can follow national and international codes, standards and guides. The main issues in the case of buildings are to avoid moisture in structures and on finishing surfaces, and in the case of both building and civil engineering structures, to check that all materials used do not cause emissions or radiation, which is dangerous to the health and comfort of the users. In some regions the radiation from the ground must also be eliminated though the insulation and ventilation of the foundation. Thus, the main tools for health design are: selection of materials, especially finishing materials, eliminating risks of moisture in structures through water proofing, drying under construction and ventilation, and elimination of possible radioactive earth radiation with air proofing and ventilation of ground structures.

2.3.8 *User's manual and maintenance plan*

A building or a civil engineering object, like a car or other piece of equipment, needs a user's manual (Sarja 2002). The manual will be produced gradually during the design process in co-operation with the partners in design, manufacture and construction. Ordinary tasks of the structural designer are:

- Collating a list of maintenance tasks for the structural system
- Collating and applying instructions for the operation, control and maintenance procedures and works
- Checking and co-ordination of operation, control and maintenance instructions for product suppliers and contractors
- Preparing the relevant chapters for the user's guidebook
- Checking relevant parts of the final user's guidebook.

REFERENCES

Sarja, A. 2002. *Integrated Life Cycle Design of Structures*. 224 pp. 42 pp. Spon Press, London. ISBN 0-415-25235-0.

Sarja, Asko. 1997. *A vision of sustainable materials and structural engineering*. In: Selected research studies from Scandinavia. Report TVBM-3078. Lund University, Lund Institute of Technology. Lund, Sweden, pp. 153–163.

Sarja, A., W 105 – Life Time Engineering in Construction. Introduction of the New Working Commission. CIB, *International Council for Research and Innovation in Building and Construction*, Information 1/ 01, pp. 24–25.

Sarja, A. 1997. *Framework and methods of life cycle design of buildings*. Symposium; Recovery, Recycling, Reintegration, R'97, Geneva, Switzerland, February 4–7, 1997. EMPA, Volume VI, pp. 100–105.

Sarja, A. & Vesikari, E. (Editors) 1996. *Durability design of concrete structures*. RILEM Report Series 14. E&FN Spon, Chapman & Hall, 16 pp.

Sarja, A. 1995. *Methods and methodology for the environmental design of structures*. RILEM Workshop on environmental aspects of building materials and structures. Technical Research Centre of Finland, Espoo. 5 pp.

Sarja, A. 1998. *Integrated life cycle design of materials and structures*. CIB World Congress, Gävle, Sweden, June 8–13.

Schultmann, F. 1999. *Material flow based deconstruction and recycling management for buildings*. R'99, Recovery, Recycling, Re-integration Congress, Geneva, Switzerland, February 2–5, 1999. Proceedings. Volume I, pp. 253–258. EMPA.

Müller, C. 1999. *Design for recycling on the example of concrete and masonry*. In: SARJA, Asko (Editor), Guide for integrated life cycle design of materials and structures, pp. 76–85. Manuscript.

Willkomm, W. 1990, *Recyclinggerechtes Konstruieren im Hochbau: Recycling-Baustoffe einsetzen, Weiterverwertung einplanen*. Köln: Verlag TÜV Rheinland.

Fukuima, T. 1999. *Integrated life cycle design of materials*. In: SARJA, Asko (Editor), Guide for integrated life cycle design of materials and structures, pp. 76–85. Manuscript.

Guide for service life design of buildings, 1998. Draft standard ISO/DIS 15686-1. ISO TC 59/SC14. International Standards Organisation.

Sarja, A. & Hannus, M. 1995. *Modular Systematics for the Industrialized Building*. VTT Publications 238, Technical Research Centre of Finland, Espoo, 216 pp.

European Committee for Standardisation (CEN), 1997, *Concrete. Performance, Production and Conformity*, Draft, CEN, Brussels, prEN 202.

Sarja, A. Design of Transportation Structures for Sustainability. *IABSE Congress 2000*, Luzern. International Association of Bridge and Structural Engineering, Zuerich.

System-based Vision for Strategic and Creative Design, Bontempi (ed.)
© *2003 Swets & Zeitlinger, Lisse, ISBN 90 5809 599 1*

Towards lifetime oriented structural engineering

A. Sarja
Technical Research Centre of Finland, VTT Building and Transport

ABSTRACT: Buildings and civil and industrial infrastructures are the longest lasting and most important products of our societies. At the time being the design is mainly focused on the construction phase and the first use, and maintenance and repair are reactive. New and increased demands, e. g. life cycle economy, health and environmental aspects are not enough included. The integrated lifetime engineering methodology concerns the development and use of technical performance parameters to guarantee, that the structures fulfil through the life cycle the requirements arising from human conditions, economy, cultural, social and ecological considerations. This is called the lifetime quality. The lifetime engineering includes: investment planning and decision making, integrated lifetime design, integrated lifetime management, maintenance, repair and rehabilitation (MR&R) planning, reuse, recycling and disposal. In order to realise the lifetime engineering we need new design and management processes, planning and design methods and modelling of long term performance of materials and structures.

1 INTRODUCTION

Sustainable development is aimed to reach a stable social and economic development in harmony with nature and cultural heritage. Against all these aspects: social, economic, ecological and cultural, the construction branch is a major player. As an example, in Europe the construction branch share is: 11% of GNP, 15% of employment, and 40% of raw materials consumption, energy consumption and waste production.

Building and civil engineering structures are the longest lasting products in societies. Typically the real service life of structures lies between 50 years and several hundreds of years. This is the reason why sustainable engineering in the field of buildings and civil infrastructures is especially challenging in comparison to all other areas of technology.

In order to reach these objectives we have to make changes even into paradigm, and especially the frameworks, processes and methods of engineering in all phases of the life cycle: investment planning and decisions, design, construction, use and facility management, demolition, reuse, recycling and wasting. This is the reason for starting to speak on life cycle engineering. The lifetime (also called: "whole life" or "life cycle") principle has been started to introduce into design and management of structures during last years, and this development process is getting increasing interest in practice of structural engineers.

Current goal and trend in all areas of mechanical industry as well as in building and civil engineering is the Lifetime Engineering (also called "Life Cycle Engineering"). The integrated lifetime engineering methodology is aiming at regulating optimisation and guaranteeing the life cycle requirements with technical performance parameters. With the aid of lifetime engineering we can thus control and optimise the human conditions (functionality, safety, health and comfort), the monetary (financial) economy and the economy of the nature (ecology), taking into consideration also local cultural compatibility (Table 1).

Table1. Generic classified requirements of the structures.

1. Human requirements	2. Economic requirements
• functionality in use	• investment economy
• safety	• construction economy
• health	• lifetime economy
• comfort	
3. Cultural requirements	4. Ecological requirements
• building traditions	• raw materials economy
• life style	• energy economy
• business culture	• environmental budens economy
• aesthetics	• waste economy
• architectural styles and trends	• biodiversity
• image	

2 CONTENT OF THE LIFETIME ENGINEERING

2.1 Definition and general content of lifetime engineering

Lifetime Engineering is an innovative idea and a concretisation of this idea for solving the dilemma that currently exists between infrastructures as a very long-term product and short-term approach to design, management and maintenance planning.

Lifetime engineering includes:

Lifetime investment planning and decision making
Integrated lifetime design
Integrated lifetime construction
Integrated lifetime management and maintenance planning
Modernisation, reuse, recycling and disposal.

The integrated lifetime engineering methodology concerns the development and use of technical performance parameters to optimise and guarantee the *lifetime quality* of the structures in relation to the requirements arising from human conditions, economy, cultural and ecological considerations. *The lifetime quality* is the capability of the whole network or an object to fulfil the requirements of users, owners and society over its entire life, which means in the practice the planning period (usually 50 to 100 years).

Integrated lifetime design includes a framework, a description of the design process and its phases, special lifetime design methods with regard to different aspects: human conditions, economy, cultural compatibility and ecology. These aspects will be treated with parameters of technical performance and economy, in harmony with cultural and social requirements, and with relevant calculation models and methods.

Integrated lifetime management and maintenance planning includes continuous condition assessment, predictive modelling of performance, durability and reliability of the facility, maintenance and repair planning and the decision-making procedure regarding alternative maintenance and repair actions.

2.2 Integrated life cycle design

2.2.1 Framework
The objective of the integrated life cycle design is to make concrete the design methods and methodologies for structural design in order to meet the requirements of sustainable development during the entire life cycle of the structures (resources, transports, manufacture, use, recycling and reuse, demolition, wasting). The objective is to design for life cycle quality, which means the ability of the structures to fulfil the requirements of the users, owners and society over the entire design period. The integrated life cycle design includes the mechanical, physical, economical, energy, health and environment aspects. The integrated life cycle design shall manage the multiple requirements in a systematic way.

2.2.2 Design process
The main phases in the model of integrated life cycle design process are: Analysis of the actual requirements, interpretation of the requirements into technical performance specifications of structures, creation of alternative structural solutions, life cycle analysis and preliminary optimisation of the alternatives, selection of the optimal solution between the alternatives and finally the detailed design of the selected structural system and its modules and components.

The conceptual, creative design phase is very decisive in order to utilise the potential benefits of integrated design process effectively. Controlled and rational decision making when optimising multiple requirements with different metrics is possible through the application of systematics of multiple attribute optimisation and decision making. In detailed design phase, life cycle aspects rise needs for total performance over the life cycle, including durability design and design for mechanical and hygro-thermal long term performance.

At the life cycle planning, a modular systematic is preferred. This allows the systematic allocation and optimisation of the target service life as well as life cycle economy and ecology of different parts of the building. A suited modularisation at the highest level of hierarchy is the following: Bearing frame, envelop, foundations, partitions, heating and ventilating services, information, water and sewage system, control, data processing and communication services and waste management system. All of these assemblies are specified during the development or design process on continuously increasing precision starting from general performance specifications and ending into detailed designs.

Ranking between design alternatives ends the sketch design phase, resulting the draft designs. All classes of requirements are systematically taken into account at the ranking. Applying the Multiple Criteria Analysis method, the core properties are mainly calculated quantitatively with numerical equations, but some of added properties are evaluated qualitatively only. The selected alternative can fulfil some of the following criteria:

- Best in all requirements
- Best weighted properties with reasonable cost level
- Best in preferred requirements, fulfilling accepted level in all requirements
- Best in valuated multiple criteria benefit/cost ratio

At the phase of detailed design the durability design is a new approach, which is important for long life structures in a harsh environment. In durability design following methods can be applied:

- durability design with structural detailing rules
- design of the environmental conditions of the structures for durability
- protection of the materials and structures against deterioration
- lifetime safety factor method
- reference factor method.

2.2.4 Indicators of the lifetime quality

The central life cycle quality indicators of a structural system are adaptability in use, changeability during use, reliable safety, technical performance and durability, resistance against obsolescence, healthy and ecological efficiency. In buildings, the compatibility and easy changeability between load-bearing structures, partition structures and building service systems is important. Regarding the life cycle ecology of buildings, the energy efficiency of the building is a dictating factor. Envelope structures are responsible for most of the energy consumption, and therefore the envelope must be durable and have an effective thermal insulation and safe static and hygro-thermal behaviour. The internal walls have a more moderate length of service life length, but they have the requirement of coping with relatively high degrees of change, and must therefore possess good changeability and re-useability. In the production phase it is important to ensure the effective recycling of the production wastes in factories and on site. Finally, the requirement is to recycle the components and materials after demolition. Obsolescence of buildings is either technical or functional, sometimes even aesthetic in nature. Technical and functional obsolescence is usually related to the primary life time quality factors of structures. Aesthetic obsolescence is usually architectural in nature.

Civil engineering structures like harbours, bridges, dams, off shore structures, towers, cooling towers etc. are often very massive and their target service life is long. Their repair works under use are difficult. Therefore their life cycle quality is tied to high durability and easy maintainability during use, saving of materials and selection of environmentally friendly raw materials, minimising and recycling of construction wastes, and finally recycling of the materials and components after demolition. Some parts of the civil engineering structures like waterproof membranes and railings have a short or moderate service life and therefore the aspects of easy re-assembly and recycling are most important. Technical or performance related obsolescence is the dominant reason for demolition of civil engineering structures, which raise the need for careful planning of the whole civil engineering system, e. g. the traffic system, and for selection of relevant and future oriented design criteria.

2.3 Integrated life cycle management and maintenance

Analogously to life cycle design, the life cycle maintenance planning and management system also includes sustainability aspects: life cycle monetary economics, life cycle economics of nature (ecology) and life cycle performance. These indicators can be modelled and optimised applying the same basic methods as presented in the description of the life cycle design methods. The economic and performance models developed in the design stage are the first estimates for the life cycle maintenance planning. In course of use, the results of periodic condition assessments can be used for updating the forecasting models. Thus the models are serving as a basis for repair and renewal plans for each of the next planning periods, until the planning of demolition and recycling. Future long-term maintenance, repair and renewal plans can be optimised. These principles can be concretised in a predictive; life cycle oriented and integrated Life Cycle Management System (LMS).

Integration and Life Cycle principle means the implementation of all planning aspects: LCC (Life Cycle Cost), LCP (Life Cycle Performance), LCE (Life Cycle Ecology), functionality, safety, health and comfort.

As tools for concretising this approach can be applied the following methods of system technology and mathematics: Multi-Attribute Optimisation and Decision Making, Performance systematics and mathematical modelling, Quality Function Deployment (QFD), risk analysis and statistical reliability theory, and modelling of performance and service life of structures. Both numerical calculations and qualitative descriptions (LCE/biodiversity, health, comfort, obsolescence) will be included in the optimisation and decision making

At present state there is no quantitative classification of exposure environment or potential degradation factors in standards. The classification system of environmental degradation loads have to be developed taking into consideration the interaction between environment and structure. The classification has to be mainly quantitative, and it must be compatible with the predictive performance and service life models. The quantitative classification of degradation loads have to be divided regionally, nationally and locally. Quantitative degradation loads will be defined at design and maintenance planning on a structural level (structure as a whole) and on detail level (specific surfaces, joints etc.). A schedule of the "Lifecon" management system is presented in Figure 2 (Guidelines RIL-216–2001).

2.4 Reuse and recycling

Several civil engineering structures, like roads and streets, are major consumers of raw materials. In those structures, materials consumption can be reduced with the use of industrial by-products like fly ash and blast furnace slag, and construction and demolition waste materials like crushed concrete and masonry. Detailed quality specifications and an effective quality control are needed for the use of these secondary materials.

The recycling ability of the structural materials and components depends on the degree and/or the technical level of the desired re-use. It is important to recognise that the recycling possibilities of the building components, modules and even technical systems shall be reconsidered in connection with design. The higher the hierarchical level of recycling, the higher also the ecological and economical efficiency of recycling. Therefore the re-use of entire components, modules or systems has to be preferred, even if there are difficulties in quality requirements and quality control in re-use.

Special issues to be treated in the design of structures and materials for re-use and recycling are:

- separability of the structural components or materials during demolition of a structure, e.g. the use of demountable structural components using suitable connections and joints,
- structural separation of components, modules or systems with different service lives and different recycling techniques,
- reduction in the variety of materials,
- separation ability of materials, which cannot be recycled together,
- avoidance of insoluble composite substances and/or composite substances that are either only slightly soluble or soluble only with a high expenditure or energy input.

Selective dismantling includes the detailed planning of dismantling phases, optimising the work sequences and logistics of the dismantling and selection process. The main goal is to separate the different fractions of materials and different types of components already at the demolition phase in order to avoid multiple actions. The recycling ability of the building materials and structural components depends on the degree and/or the technical level of the desired re-use. In case of reuse of building components the main problem is to guarantee the quality of the reused products

3 CURRENT RESEARCH AND STANDARDISATION WORKS

Lifetime engineering is under very intensive development all over in the world. Most research works are focused on some specific issues, typically degradation modelling of structures, life cycle costing LCC, and environmental impact analysis LCA. Some personal observations of the author on the regionally specific features of these works are as follows:

- In USA most active is the area of investment planning, life cycle economy and multi-attribute decision making. Several ASTM standards on these issues have been published in 1990`s.
- Canadian research is often directed towards ecology "Green building".
- Japanese research and standardisation is especially active in service life planning and design, as well as in lifetime quality issues. Several standards and guidelines have been published and are under work on these issues.
- China and Southeast Asian countries are most active in facility management and condition assessment issues.
- European research and standardisation is active in condition assessment, service life classification, degradation modelling of materials, repair of structures, energy efficiency of buildings and life cycle costing.
- ISO has already published and is working out in worldwide co-operation standards and raft standards on environmental management, service life planning and design and life cycle costing.

The fifth framework program of EU Commission XII DGL has included an objective to collect research projects into larger entities called "Clusters".

Cluster "LIFETIME":"Life time design and management of civil infrastructures and buildings", is working with integration and systemising the development of the lifetime engineering idea. This cluster is consisting of five ongoing projects of the EU Growth program:

LIFECON, which is dealing with life cycle management of infrastructures,

INVESTIMMO, which is concerned with new methods for organising maintenance and refurbishment of processes for residential dwellings,

EUROLIFEFORM, which will focus on the design of new structures, and on life cycle costing (LCC) applied to both new and existing structures,

LICYMIN, which deals with life cycle assessment of mining projects for waste minimisation and long term control of rehabilitated sites and

CONLIFE, dealing with frost resistance modelling of high strength concrete.

The objectives of the Cluster "LIFETIME" are:

- to integrate the knowledge of partners of these three projects for advancing the work and results of all projects
- to co-operate in similar tasks in order to avoid parallel overlapping work and parallel results
- to produce an integrated and generic "European Guide for Life Time Design and Management of Civil Infrastructures and Buildings".

- To integrate the Information Networks through linking between these three projects
- To create and deliver a continuously updated database of produced models for later European exploitation

The Cluster has a common management, which will be carried out by Management Group. The Management Group is consisting of Co-ordinators of the five projects. Each project will produce own results and deliverables on their focus areas. From these results, an editorial group under the guidance and control of Cluster management Group will produce a generic integrated report: "European Guide for Life Time Design and Management of Civil Infrastructures and Buildings".

A large Thematic Network "Lifetime":"Lifetime Engineering of Buildings and Civil Infrastructures" is supporting the dissemination and exploitation of the results of the Cluster "Lifetime". This Network is consisting of 94 partners from 29 countries and is planned to work in the years 2002–2005.

The overall *objective* of the LIFETIME Thematic Network is to contribute to European and worldwide development of a more sustainable built environment. We are willing to generate, promote and support actions to change the current practice towards principles, processes and methods of lifetime engineering. The network is focusing on application of lifetime principles both in buildings and civil infrastructures, as well as in mining and other industrial engineering.

The Network will involve all key stakeholders of buildings and civil infrastructures, including mining; activities concern investment planning, design, facility management and maintenance, reuse and recycling. It will deal with roles and actions of all these stakeholders.

The work of the LIFETIME Network is expected to include:

- benchmarking of current practice, ongoing international and national R&D programs
- state of the art reports, recommendations for future R&D
- bridging gaps between best know-how / research results and their introduction into practice
- raising awareness of all stakeholders in the fields of building and civil infrastructures, informing of best practices and R&D works and results, and development potential
- making proposals and support for development of international and national regulations and standards
- making proposals, support and actions for development of stakeholder education and training
- dissemination, experimentation and exploitation of systematic lifetime principles and methodology in practice of the partners.

4 FUTURE STEPS TOWARDS THE LIFETIME ENGINEERING

4.1 Research of European Union

The sixth Framework Program of EU Commission XII GDL has some sub-programs with a strong focusing on lifetime engineering. As examples of these can be listed the following points of these programs:

- New production processes and devices (1.1.3.iii)
  (a) the development of new processes and flexible and intelligent manufacturing systems incorporating advances in virtual manufacturing technologies, including simulations, interactive decision-aid systems high-precision engineering and innovative robotics;
  (b) systems research needed for sustainable waste management and hazard control in production and manufacturing, including bio-processes, leading to a reduction in consumption of primary resources and less pollution;
  (c) development of new concepts optimising the life cycle of industrial systems, products and services.
- Sustainable energy systems (1.1.6.1)
  (a) in the short and medium term, especially in the urban environment (1.1.6.1.i):
    (i) clean energy, in particular renewable energy sources and their integration in the energy system, including storage, distribution and use;
    (ii) energy savings and energy efficiency, including those to be achieved through the use of renewable raw materials;
    (iii) alternative motor fuels;

4.2 General comments

Central needs for future steps in the development of lifetime engineering are:

- to continue the realisation process integrating the issues of lifetime engineering into general technology of buildings and civil infrastructures in design, construction, management, demolition, recovery, reuse and recycling of buildings and civil infrastructures.
- this requires especially the development of international and national standards and guidelines
- a lot of education and training efforts will be required for development of the capability of all civil engineers to develop, adapt and use this technologies.

5 CONCLUSIONS

Incorporating life cycle principles into the practical design, construction, maintenance and recycling of structures is quite an extensive process. Application of life cycle principles is widening the scope of structural design to the extent that the entire working processes must be re-engineered. The tradition of structural engineers in applying mathematical and physical calculation methods in design will serve as a good basis for applying the additional multiple calculation methods that are needed in lifetime structural engineering.

Concerning materials and structures, new basic knowledge will be needed especially regarding environmental impacts, hygro-thermal behaviour, durability and service life of materials and structures in varying environments. Structural design methods that are capable of life cycle design, multiple analysis decision-making and optimisation will have to be further developed. Recycling design and technology demand further research in design systematics, recycling materials and structural engineering. The knowledge obtained will have to be put into practice through standards and practical guides.

For practical application also IT software tools are needed. These tools shall include all process models, procedures and methods which belong to the lifetime design, maintenance, repair and rehabilitation planning, as well as to the planning of selective demolition, reuse, recycling and wasting. Relevant databases of data will support the use of the lifetime engineering software.

REFERENCES

Sarja, A., 2002, *Integrated Life Cycle Design of Structures*, Spon Press, London 2002, 142 pp.

Sarja, A. 1999, Towards life cycle oriented structural engineering, In: *Construction materials – theory and application*. Ibiidem-Verlag, Stuttgart, pp. 667–676.

Sarja, A. 2000, Design of Transportation Structures for Sustainability, *IABSE Congress 2000*, Luzern. International Association of Bridge and Structural Engineering, Zuerich.

Sarja, A. 1999, Environmental Design Methods in Materials and Structural Engineering, *RILEM Journal: Materials and structures*, Vol. 32, December , pp. 699–707.

Sarja, A. 2000, Integrated life cycle design of concrete structures. *Concrete technology for a sustainable development in the 21st century*. E&FN SPON, London and New York, 2000, pp. 27–40.

Sarja, A. 2000, Durability design of concrete structures – Committee report 130-CSL. *Materials and Structures/ Matériaux et Constructions*, Vol. 33, January–February 2000 pp. 14–20.

Guide for service life design of buildings. Draft standard ISO/DIS 15686-1. ISO TC 59/SC14. International Standards Organisation (1998).

European Committee for Standardisation (CEN), Concrete. Performance, Production and Conformity, Draft, CEN, Brussels, prEN 202 (1997).

Sarja, A. & Vesikari, E. 1996, (Editors and author group chair and secretary). Durability design of concrete structures. *RILEM Report of TC 130-CSL. RILEM Report Series 14*. E&FN Spon, Chapman & Hall, 165 pp.

Sarja, A. (Co-ordinator) 2000, Life Cycle Management of Concrete Infrastructures for improved Sustainability, LIFECON. *Work Description. EU Research Program: Competitive and Sustainable Growth*.

Sarja, A. (Co-ordinator) 2001, Lifetime Engineering of Buildings and Vivil Infrastructure, LIFETIME. Plan of Thematic Network. *EU Research Program: Competitive and Sustainable Growth*.

Guidelines RIL-216–2001 2001, Life cycle engineering of structures. *Association of Finnish Civil Engineers*. Helsinki. 312 p. (in Finnish).

Sarja, A. 2002, Reliability based life cycle design and maintenance planning. *JCSS Workshop on Reliability Based Code Calibration*. ETH, Zürich, March, 2002. 18 pp. Published in Website :http://www.jcss.ethz.ch.

System-based Vision for Strategic and Creative Design, Bontempi (ed.)
© 2003 Swets & Zeitlinger, Lisse, ISBN 90 5809 599 1

An approach to optimal allocation of transportation facilities on expressway

A. O'hashi
Pacific Consultants Co., Ltd, Tokyo, Japan

A. Miyamura
School of Design and Architecture, Nagoya City University, Nagoya, Japan

A. De Stefano
Department of Structural Engineering, Politecnico di Torino, Turin, Italy

ABSTRACT: The present study deals with a mathematical model for the optimal allocation of transportation facilities or interchange-exits on an expressway network in Japan, all of which has been evaluated empirically from the engineering point of view. Consequently, it is important for the present mathematical model to include socio-economical constraints rationally from practical statistics, which includes constrained parameters in terms of demography, industrial output, geography, environment load, and political requirements at times inconsistently to each other. These constraints should be evaluated rationally, and for practical simplification an equivalent density distribution is proposed to be triaged by the maximum entropy criterion that provides each ratio of attribution based upon Frobenius characteristic equation. An effective width along a given route can be assumed, which evaluates these effective parameters into a single equivalent density distribution function in terms of distance. Consequently, the present problem is described as minimization of total sum of weighted distance coordinate to any exit for the optimal allocation. An example optimal allocation is compared to the actual expressway under construction in Japan.

1   INTRODUCTION

The investment of expressway network in Japan has played an important role to boost the economical activity after the end of World War II with the pros and cons. Rapid growth of motorization provided stimuli to construction of the expressway network since 1960 when Meishin expressway was supplied, with length of over 7,000 km in total so far resulting in still the gradual increase of recent traffic congestion and accidents. Although further investment for extensive redeployment of the expressway network is in progress, regional demand for interchange-exits is strong with, in practical, a lack of acceptable decision making ways for the appropriate allocation of such interchange-exits facilities. The allocation of interchange-exits on the expressway in Japan has been determined under various kinds of constraints economically, socially, and sometimes politically on such as demography, industrial output, geography, environment, and political requirements. Since these constraints are interactively correlated and frequently inconsistent to each

other, civil engineers should carry out their decision subject to the past database relevant to these constraints, and frequently they should play a role to fill a gap between them as not an engineering problem to solve but an inconsistent non-engineering problem empirically. Hence, it is indispensable to establish a more rational, scientific model to improve such a decision making process with, herein, a proposal of a mathematical model for the optimal allocation of the given number of interchange-exits on an expressway.

Many mathematical methods are proposed, for example, to deal with optimal allocation of bus-stops in urban area network (Vaughan 1977) closely related to the mathematical modelling of Volonoi allocation. This Volonoi allocation deals with optimal allocation of many facilities on any weighted 2D plane and is applied to the other problems (Okabe 1984, 1988). Some of models focused on engineering design problem belong to the combinatorial optimality that demands basically a generate-and-test procedure with a result in exponential increase of research space. Particularly, when dealt with discrete optimality, due to non-differentiable property it is necessary to develop more problem-oriented strategy for optimality (Polak 1987). For practical engineering problems several effective approaches are discussed elsewhere (Miyamura 1996b). One of effective techniques is the branch-and-bound method when dealt with a monotonically increasing objective function. It can be applied to non-monotonic case by some heuristic or problem-oriented rules to cut futile branches so drastically when compared to the other enumerations (Takada 2001). A characteristic approach to the optimality is realized by the genetic algorithm still with a lack of theoretical guarantee to optimality, which becomes applicable from the engineering point of view (Miyamura 1996a). The multi-objective genetic algorithm becomes successful for a problem of city planning (Balling 1999) with inconsistent constraints thus to be described as a set of maximum, minimum multi-objective functions. A set of Pareto solutions can be obtained.

Herein, the present study consists of two phases. First, the status quo of design constraints in Japan is discussed socio-economically, particularly from the view of the engineering codes and regulations. Second, these constraints are evaluated into the density distribution functions along an expressway in terms of demography, industrial output, and environmental load from practical statistics. Then an equivalent density function is obtained in the maximum entropy criteria, which estimates the most possible ratio of attribution. A practical example by comparison of the present method to actual allocation on the Daini–Tomei expressway between Nagoya and Toyohashi with twelve exchange-exits, is discussed.

2 THE STATUS QUO IN JAPAN

The allocation of exchange-exits is determined under a number of constraints socially, economically, environmentally, and occasionally politically, together with the engineering regulations. This demands consistency between urban and regional development projects, and positive improvement of transportation network due to allocation of transportation facilities or interchange-exits on the expressway. Under the present authorized codes and regulations in Japan it is expected that the following items should be realized.

General direction of allocation requires that they are: at or near an intersection connected to important main routes, suburban area of a city with more than 30,000 in population or influence area from an interchange-exit with 50,000 to 100,000 in population in total, near important harbor, airport, or at or near an intersection with approach road to a place for tourism internationally, allocation with less than 30,000 cars per day by in/out traffic density, allocation between 5 km in

minimum to 30 km in maximum to neighboring interchange-exits.

The number of interchange-exits should be: one for a city with 100,000 in population, less than three for a city with less than 300,000 in population, less than four for a city with less than 500,000 in population, or three for a city more than 500,000 in population.

Standard distance between interchange-exits should be: 5 to 10 km for urban and industrial area, 15 to 25 km for plain area with smaller cities, or 20 to 30 km for local or mountainous area.

Minimum distances of interchange-exit to other facilities should be: 5 km to neighboring interchange-exit, service points, or parking points, or 4 km to bus-stops or tunnels.

As a result these codes and regulations demand the allocation to improve serviceability for users, efficiency of logistics, and facility of maintenance in substance. The allocation could be also evaluated economically by means of cost effectiveness, B/C-factor (B means expected benefit due to interchange-exits, and C, sum of initial and running costs, respectively). Practically, realization of the allocation becomes further restricted by more qualitative constraints on the codes and regulations because of ambiguous prospective demand after completion. Furthermore, such constraints become conclusive to determination at times. The subsequent items are involved positively. First, regarding serviceability towards users, demography distribution on an expressway network is an important factor with its discreteness to be difficult for further mathematical procedure. The same is true to industrial or agricultural output distribution. Second, it is difficult to find an effective theory to synthesize such inconsistent constraints by common scale partly because of difficult estimation of correlation to each other that requires some empirical thumb rules.

3 THE PRESENT CONSTRAINTS

Although there are various kinds of constraints practically, it is preferable to evaluate them in common scale. It is expected that their expenses, which can be calculated by statistics, herein, in Japanese yen per hectare, to interchange-exits should be minimized for the present optimality. Three density distribution factors are evaluated, that is to say demography, industrial output, and environmental load, for a practical example, along the Daini–Tomei expressway between Nagoya and Toyohashi under construction in Figure 1. Subsequently, these three constraints are described in detail.

Demographically, it is assumed that inhabitants within 20 km widths of both sides of the expressway could approach to any nearest exchange-exit, where each time expense becomes 40 yen per hour per person

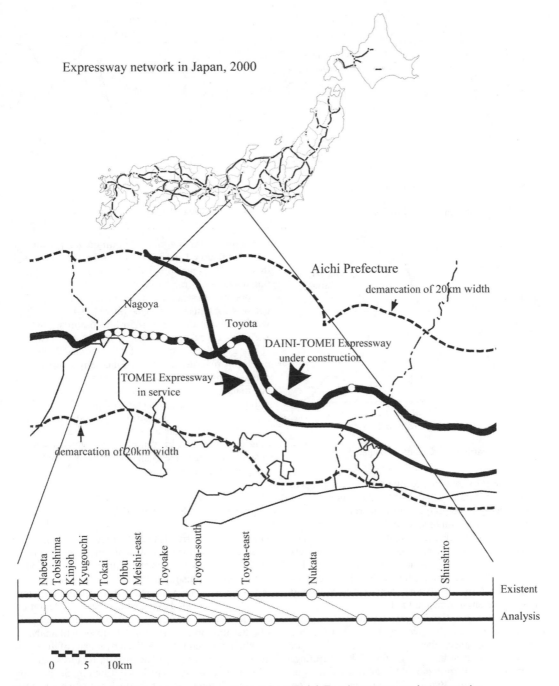

Figure 1. Comparison between actual and theoretical results on Daini–Tomei expressway under construction.

with a result in the total amount of 114,181 million yen per year. Figure 2 shows the resulted distribution aspect from condensation to the normalized route that corresponds to 70 km.

The industrial output is evaluated by the statistics with both effective weight of industrial products of 4 ton and transportation cost of 40 yen per ton-km with a result in the total amount of 35,561 million yen per

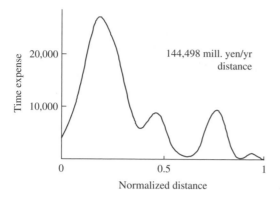

Figure 2. Distribution of time expense by inhabitants.

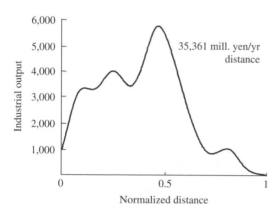

Figure 3. Distribution of industrial output.

year. Figure 3 shows the resulted distribution aspect from condensation by both sides of the route.

The quantitative prediction of environmental load from construction and in-service should be investigated from the time-dependent view of point. Allocation of exchange-exits would provide negative side effects such as air contamination, noise nuisance, and global warming, which deteriorate human health consequently. The Japanese regulations demand the following rules for the environmental load.

For air contamination, the emission of NOX in gram per km per day can be given by $(0.2a_1 + 4.41a_2)Q$. For noise nuisance, the equivalent noise level in $dB(A)$ can be given by $40 + A$. For global warming, the emission of carbon dioxide in gram-c per km per day can be given by $(46a_1 + 137a_2)Q$ where $A = 10\log (a_1 + 4.5a_2) + 10\log(Q/24)$ means the rate of inter-fusion of smaller-sized cars, a_2, of full-sized cars thus $a_1 + a_2 = 1$, and Q, the volume of traffic, respectively. Based upon this evaluation the present expected expenses due to the environmental loads become as follows.

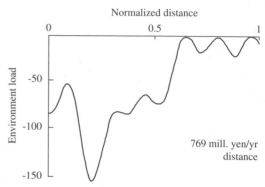

Figure 4. Distribution of environmental load.

For air contamination, 2.92 million yen per ton for densely inhabited district, 0.58 for otherwise inhabited district, 0.2 for non-urban district, and 0.01 for mountainous district, respectively.

For noise nuisance, 2.4 million yen per $dB(A)$ per km per year for densely inhabited district, 0.475 for otherwise inhabited district, 0.1656 for non-urban district, and 0.0072 for mountainous district, respectively.

For global warming effect it is assumed 2,300 yen per ton-c regardless district. Since any interchange-exit should be regulative to less than 30,000 cars per day in maximum, which provides the total sum of 769 million yen per year as shown in Figure 4.

These three density distributions restrictive to the allocation can be evaluated in terms of the corresponding currency unit by yen per year, which makes it possible to synthesize as an equivalent distribution to integrate each attribution.

4   MAXIMUM ENTROPY CRITERIA

The present constraints include three density distributions along the expressway with effective widths of 20 km both sides in terms of constraint factors, namely, demography, industrial output, and environmental load in Japanese yen per year. These three factors are interactively so correlated to each other that any integration criteria is indispensable to provide an equivalent density distribution function for the subsequent analysis, which, herein, could be realized by means of the maximum entropy criteria. The criteria requires that the information entropy, which is defined by Shannon as the sum of product of probability density function, $p(\mathbf{y})$, by its natural logarithm on relevant eventual factors, of these three density distributions should be maximized in regard to ambiguity or non-committalness to be included in various kinds of phenomena naturally or socially, that is, under their constraints. The criteria can be applied to almost all variational principles and its basic role is to characterize equilibrium states. Mathematically, thus, the maximum entropy criteria can be formulated by Lagrangian multiplier, λ_j, as,

$$\min \Pi = \min \left\{ H(\mathbf{y}) - \sum \lambda_j g_j(\mathbf{y}) \right\} \qquad (1)$$

where $H(\mathbf{y})$ means the information entropy of the constraint variables given by,

$$H(\mathbf{y}) = -\iint \cdots \int p(\mathbf{y}) \ln p(\mathbf{y}) \, dy_1 dy_2 \cdots dy_n \qquad (2)$$

where y is a probabilistic variable vector.

$$\mathbf{y} = \{y_1, y_2, \cdots, y_n\}^{lt}$$

Equation 1 is a conditional maximization in terms of variable vector, $\mathbf{y}$. $g_j(\mathbf{y})$ is the j-th function of m constraints. When the ratio pattern of total sums, which could be normalized by the nearest integers, $\{q_1:q_2:\cdots:q_m\}$, as a constraint to maximize the entropy mathematically, the positive root of the subsequent Frobenius equation can provide the resultant ratio of each attribution to occurrence.

$$\sum W^{q_k} = 1 \qquad (3)$$

By the unique positive root, W_0, of Equation 3, each attribution factor, r_k, is given by,

$$r_k = W_0^{q_k} \text{ for } k = 1, \cdots, m \qquad (4)$$

Hence, the equivalent density distribution function is given by,

$$F(s) = \sum_{j=1}^{m} r_j f_j(s) \qquad (5)$$

In the present case, since the sums of output from the demography, industries, and environmental load have the following relationships, $\{q_1:q_2:\cdots:q_n\} = \{114,181:35,561:762\}$ in mill. yen $= \{148:47:1\}$, and Frobenius equation becomes, $W^{-148} + W^{-47} + W^{-1} = 1$. The positive roots become $\{0.00013, 0.05847, 0.94138\}$ for attribution, and the corresponding equivalent density distribution is given by,

$$F(s) = 0.00013f_1(s) + 0.05847f_2(s) + 0.94138f_3(s)$$

Figure 5 shows the present equivalent density distribution along the expressway extended on a straight

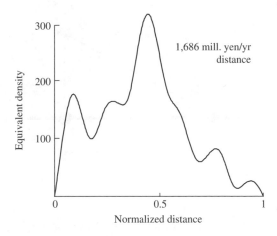

Figure 5. Equivalent density distribution function.

line, which provides prominent appearance of two peaks, the one for high demographic shift to Nagoya, and the other for larger industrial output around Toyota area. Hence, the equivalent density distribution is a function in terms of distance along the route.

5 MATHEMATICAL MODEL

Based upon the Volonoi allocation which requires minimization of relative distance from any point to all of neighboring points for the given number of points on a density distributed plane, the present optimal allocation can be described such that under the given number of interchange-exits to minimize the sum of all relative distance from any point to its two neighboring interchange-exits along an equivalent density distributed route to be cut out by a prescribed length. Hence, for a route in straight line in Figure 6 the total sum of distance along all of component routes of an expressway network can be given by,

$$h(\mathbf{P}) = \sum_{i=1}^{M} \int_{net} \int_{-r_0}^{r_0} \sqrt{r^2 + (s - S_i)^2} f(r,s) dr ds \qquad (6)$$

where s means any point along routes. $h(\mathbf{P})$ by Equation 6 should be minimized in terms of interchange-exits coordinate vector, $\mathbf{S} = \{S_i\}$, that is to say the first derivative should be zero. Consequently,

$$\frac{\partial h(\mathbf{P})}{\partial \mathbf{S}} = \sum_{i=1}^{M} \frac{\partial}{\partial \mathbf{S}} \int_{net} \int_{-r_0}^{r_0} \sqrt{r^2 + (s - S_i)^2} f(r,s) dr ds \qquad (7)$$
$$= \{0\}$$

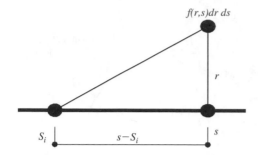

$$f(r,s)dr\,ds$$

$$r$$

$$S_i \downarrow \qquad s-S_i \qquad s$$

Figure 6.   Distance coordinate.

For further simplicity Equation 7 is approximated by linearization as,

$$h(\mathbf{P}) \approx \sum_{i=1}^{M} \int_{\text{act}} \int_{-r_0}^{r_0} 0.75\left(|r|+|s-S_i|\right) f(r,s)dr ds$$

$$=0.75\sum_{i=1}^{M}\int_{s_{ai}}^{s_{bi}}|s-S_i|F_0(s)ds+C$$

$$= 0.75\sum_{i=1}^{M}\left\{ \int_{s_{ai}}^{S_i}(-s+S_i)F_0(s)ds \right.$$

$$\left. + \int_{S_i}^{s_{bi}}(s-S_i)F_0(s)ds\right\}+C \qquad (8)$$

where C represents an integral constant, and

$$F_0(s) = \int_{-r_0}^{r_0} f(r,s)dr \qquad (9)$$

where $\{s_{ai},s_{bi}\}$ is a center coordinate to each neighboring exit, each of which defines the dominant distances at the i-th exit coordinate, hence provided by,

$$s_{ai} = \frac{S_{i-1}+S_i}{2} \text{ and } s_{bi} = \frac{S_i+S_{i+1}}{2} \qquad (10)$$

Equation 10 implies that $\{s_{ai}, s_{bi}\}$ becomes a function of an exit coordinate, S_i, and, herein, assumed as constant in the premise of implementation of converging iteration. Therefore, to minimize $h(\mathbf{P})$ in terms of S_i,

$$\frac{\partial h(\mathbf{P})}{\partial \mathbf{S}} = 0.75\left\{ \int_{s_{ai}}^{S_i}F_0(s)ds + \int_{S_i}^{s_{bi}}F_0(s)ds \right\} \qquad (11)$$

$$= \{0\}$$

Equation 11 becomes,

$$2G(S_i)=G(s_{ai})+G(s_{bi}) \qquad (12)$$

where $G(s) = \int F_0(s)ds$. Equation 12 shows that $G(S_i)$ represents the mean value of $G(s_{ai})$ and $G(s_{bi})$, and

$$S_i=G^{-1}\left(\frac{G(s_{ai})+G(s_{bi})}{2}\right) \qquad (13)$$

Since $G(s)$ in Equation 13 becomes non-linear in general, it is necessary to apply a converging iteration technique such as Newton method. Figure 1 shows a practical example case of Daini–Tomei expressway under construction to be parallel to the Tomei expressway between Tokyo and Nagoya extended to Osaka supplementally almost along the Pacific coastline connected to a large part of active urban districts, thus to become an important artery socially, industrially, and economically after 40 years in service. Maturity of Tomei expressway offers additional or auxiliary interchange-exits, particularly, around so densely populated or industrialized area that it is difficult to make decision of allocation of interchange-exits on Daini–Tomei expressway under the entangled, complicated constraints frequently inconsistent to each other. The result was empirical with a lack of rationality. Figure 1 also shows comparison between the actual allocation under construction and the present analytical result assumed a stretched straight route. The actual allocation requires rather concentrated exits around Nagoya, which is densely inhabited, theoretically in contrast with around Toyota mainly due to high level of industrial output together with influence of environmental constraint. The environmental constraint had not been positively, quantitatively evaluated so far for the actual decision of allocation. This implies importance of environmental constraints towards the decision to prevent aftermath and side effects by such environmental negative load.

6   CONCLUDING REMARKS

Since the decision policy of allocation of interchange-exits should be reflected various kinds of constraints not only technically but economically, socially, and at times politically from the view of either macroscopic or microscopic stances, some empirical decision could be indispensably inevitable though without rationality to some extent. However, as an engineering decision problem the status quo requires any mathematical model at least to rationalize exclusion of inconsistency during the decision process. Still the present model for the optimal allocation could not overcome obstacles from such as the correlation with general road networks or the characteristics of expressway including alignment, bridges or tunnels, geography, but the present mathematical model could be extended more practically to realization of the rational decision of allocation. Hence, the following concluding remarks are obtainable:

a) The Japanese codes and regulations for allocation of interchange-exits on expressway network can provide appropriate, rather conservative direction, but practical constraints are economically, socially, environmentally, and at times politically so complicated that the decision policy including the engineering judgment becomes inclined empirical. Consequently, it is necessary to realize a rational method to synthesize all of these constraints without inconsistency then a mathematical model for optimal allocation.

b) The constraints to confine the optimal allocation consist of enormous amount of information with different degree of accuracy or ambiguity to each other. It is naturally expected that their eventual occurrence should be in maximum independently of their degree of ambiguity. Hence, the maximum entropy criteria can estimate the most possible attribution of the constraints, which can be given by unique positive root of Frobenium equation. Thus, the equivalent dense distribution function along the expressway route can be evaluated.

c) The present optimal allocation is described by a mathematical model to minimize the sum of relative distances to neighboring interchange-exits on an expressway network. A proper numerical method such as GA or Newton method can be applied.

d) Comparison on an actual case under construction shows that discrepancy between theory and practice appears largely because of attribution of the environmental constraint. The present environmental load such as air contamination due to NOX, noise nuisance, and global warming effect should be evaluated from statistics including their effects after completion.

e) The present model can provide a rational solution for the allocation, however there remain still crucial factors. Practically, considering the status quo of highly utilized urban district interactive effects with existing road networks should be evaluated,

particularly, from the engineering point of view. Transitive lapse of the constraints should be evaluated for allocation, at least, to realize reallocation or renovation in the future.

REFERENCES

Balling, R. et al, 1999. City planning using a multi-objective genetic algorithm and interactive Pareto set scanner, Optimization and Control in Civil and Structural Engineering, *Civil-Comp*, pp. 35–39

Miyamura, A. et al, 1996a. Optimal allocation of shear wall at 3D frame by genetic algorithm application, *Proc. of the 3rd Int. Conf. on Computational Structures Technology*, Advances in Optimization for Structural Engineering, Budapest, pp. 73–79

Miyamura, A. et al, 1996b. A combinatorial problem: failure load analysis of rigid frames, *Advances in Engineering Software*, 27, pp. 137–143

Okabe, A. & Miki, F., 1984. A conditional nearest-neighbor spatial association measure for the analysis of conditional locational independence, *Environment and Planning A*, Vol. 16, pp. 107–114

Okabe, A. et al, 1988. The statistical analysis of a distribution of activity points in relation to surface-like elements, *Environment and Planning A*, Vol. 20, pp. 609–620

Polak, E., 1987. On the mathematical foundations of non-differentiable optimization in engineering design, *SIAM Rev.*, Vol. 29, No.1

Suzuki, T. et al, 1991. Sequential location-allocation of public facilities in one- and two-dimensional space: comparison of several policies, *Mathematical Programming*, 52, pp. 125–146

Takada, T. et al, 2001. Optimization of shear wall allocation in 3D frames by branch-and-bound method, *Proc. of the 7th Int. Conf. on Computer Aided Optimum Design of Structures*, Bologna, pp. 107–116

Vaughan, R.J. & Cousins, E.A., 1977. Optimum location of stops on a bus route, *Proc. of the Seventh International Symposium on Transportation and Traffic Theory*, pp. 697–716

Wirasinghe, S.C. & Goneim, N.S., 1981. Spacing of bus-stops for many to many travel demand, *Transportation Science*, Vol. 15, No.3, pp. 210–221

System-based Vision for Strategic and Creative Design, Bontempi (ed.)
© *2003 Swets & Zeitlinger, Lisse, ISBN 90 5809 599 1*

An integrated computational platform for system reliability-based analysis of mixed-type highway bridge networks

F. Akgül & D.M. Frangopol

Dept. of Civil, Environmental and Architectural Engineering, University of Colorado, Boulder, CO, USA

ABSTRACT: Deterioration of aging infrastructure in United States is directing the attention of bridge managers and engineers toward more efficient inspection and evaluation techniques. Recent advances in reliability analysis methods justify new techniques that can be implemented in current practice. This paper discusses a network-level lifetime system reliability analysis method for bridges. A multi-phase research was carried out targeted toward an integrated system reliability-based, fully probabilistic analysis of highway bridge networks. The tasks included identification of random variables for different member types, quantification of random variable data (i.e., determination of statistical descriptors), derivation of standardized code-based limit state equations, investigation and implementation of load (increasing growth in traffic) and resistance (deterioration due to corrosion) models, and development and utilization of a network-level system reliability analysis program. Integration of these tasks was accomplished on a common computational platform. The proposed method for determining reliability profiles is illustrated using the results obtained for existing bridges within a selected highway bridge network.

1 INTRODUCTION

The objective of this research, which led to the development of an integrated computational platform presented herein, was to consider a network of structures, and determine the lifetime reliability profiles of these structures based on system reliability and load and resistance models.

However several issues had to be resolved in order to achieve such an objective. A time-variant reliability analysis program based on first-order system reliability method was needed to calculate the lifetime reliability profiles for the components and the system of a structure. Resistance random functions could be simulated using a simulation algorithm while performing the reliability analysis within the same program unit. Furthermore, live load models and resistance deterioration models could be integrated in the new program. Most importantly, the program was supposed to be able to handle different bridge types existing within a bridge network. To accomplish these tasks, an integrated computational platform is developed containing the desired capabilities.

The program is a network-level, time-variant, system reliability analysis tool capable of incorporating live load and resistance models, and applicable to multiple structure types located within a structure network.

Although the program is designed with a focus on a specific structure type (i.e., bridges), the general concept and the overall algorithmic structure is applicable to other civil engineering structure types with necessary modifications.

In this paper, following the description of an existing bridge network, the phases of the system reliability analysis are explained. The computational platform integrating these phases is summarized, and the lifetime reliability analysis results of a bridge located in the network are presented.

2 THE MIXED-TYPE BRIDGE NETWORK

At the initial phase of this study, an existing bridge network located in Colorado was selected. The network, shown in Figure 1, is located near Denver. Data gathering process was a comprehensive task. The level of detail required for advanced reliability analysis involved precise geometric dimensions of all superstructure bridge members (i.e., slab and girders) and detailed information on material and load properties. The information was gathered from bridge construction plans, bridge records, inventory data, traffic data, inspection histories, correspondence, memorandums, and rating records.

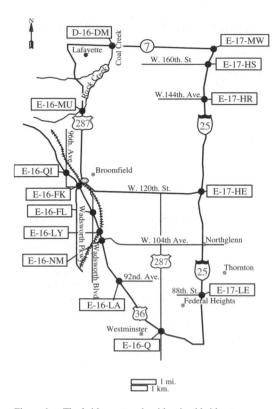

Figure 1. The bridge network with mixed bridge types.

Table 1. Characteristics of mixed-type bridges in the network.

Bridge name	Bridge type	Number of spans	Length (m)	Year built
E-16-MU	Prestressed	1	34.1	1994
E-16-LA	Prestressed	1	77.9	1983
D-16-DM	Prestressed	2	44.5	1990
E-16-QI	Prestressed	2	74.1	1995
E-16-LY	Prestressed	3	74.3	1985
E-16-NM	Prestressed	2	64.6	1991
E-17-MW	Prestressed	2	72.7	1987
E-16-FK	Steel I-Beam	4	69.2	1951
E-16-FL	Steel I-Beam	4	54.0	1951
E-16-Q	Steel I-Beam	5	82.3	1953
E-17-LE	Steel Plate Girder	4	68.6	1972
E-17-HS	Hybrid Conc/ Steel	4	64.5	1963
E-17-HR	Hybrid Conc/ Steel	4	64.0	1962
E-17-HE	Hybrid Conc/ Steel	4	67.7	1962

Characteristics of the bridges in the selected network are listed in Table 1. Bridges E-16-MU and E-16-LA have simply supported girders while the remaining five prestressed bridges have continuous spans. Bridges E-17-HS, E-17-HR, and E-17-HE are four-span hybrid girder bridges having reinforced concrete end spans and steel plate girders at middle spans.

Lifetime reliability indices for superstructure components $\beta_i(t)$ and for the bridge itself $\beta_{sys}(t)$ are determined for each individual bridge in the network. The procedure used for system reliability analysis and the integrated computational platform used for these analyses are described in the following sections.

3 SYSTEM RELIABILITY ANALYSIS

The steps followed for performing the lifetime system reliability analysis of the network bridges can be briefly summarized as follows:

1. Identify the necessary random variables and deterministic parameters for each bridge component type,
2. Determine the mean value and standard deviation for each random variable, and assign values for deterministic parameters,
3. Derive standardized (i.e., parametric) limit state equations based on code requirements and formulas,
4. Define a system failure model for each bridge based on failure of the slab or failure of any two adjacent girders or both,
5. Adopt live load and deterioration models to be used for the lifetime reliability analyses, and
6. Determine the lifetime reliability profiles for the components and systems of bridges in the selected network.

Random variables and deterministic parameters are separately identified for four main categories of bridge types. A detailed discussion of random variables and deterministic parameters for four separate bridge types is presented in Akgül (2002). As a representative example, however, the random variables and deterministic parameters for a steel bridge girder are presented herein in Figure 2, and Tables 2 and 3 list the descriptions of these random variables and deterministic parameters.

Numerous limit state equations are developed for component failure modes of four separate bridge types which are individually derived in Akgül (2002). Since majority of bridge plans and design documents in the United States are in Customary US units, and AASHTO (1996) Specifications in these units was used for this study, the derivations were made in the same format. As a representative example, the limit state equation for flexural failure of a noncomposite compact steel girder section is obtained as:

$$g_{SteelGirder, Flexure_i} = C_{11}F_yS_p\gamma_{mfg} - (C_{12} + C_{15})\lambda_s$$
$$- C_{13}\lambda_c - C_{14}\lambda_a - C_{16} - M_{trk}I_fD_f = 0 \qquad (1)$$

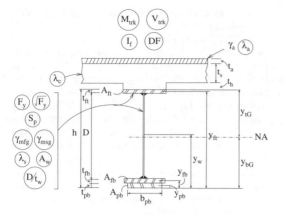

Figure 2. Random variables (circled) and deterministic parameters for a steel girder.

Table 2. Random variables for steel bridge girder.

Random Variable	Description
F_y	Yield strength of steel girder (MPa)
λ_a	Asphalt weight uncertainty factor
λ_c	Concrete weight uncertainty factor
λ_s	Structural steel weight uncertainty factor
A_w	Web area of the steel girder (cm^2)
S_p	Plastic Section Modulus (cm^3)
γ_{mfg}	Modelling uncertainty for flexure in girder
γ_{msg}	Modelling uncertainty for shear in girder
M_{trk}	Moment due to truck load (kN-m)
V_{trk}	Shear due to truck load (kN)
I_f	Impact factor
D_f	Distribution factor
D/t_w	Girder depth to web thickness ratio

Table 3. Deterministic parameters for steel bridge girder.

Deterministic parameter	Description
γ_c	Unit weight of concrete (kN/m^3)
γ_a	Unit weight of asphalt pavement (kN/m^3)
t_a	Thickness of asphalt pavement (cm)
t_s	Thickness of concrete slab (cm)
t_h	Thickness of concrete haunch (cm)
t_{ft}	Thickness of girder top flange (cm)
t_{fb}	Thickness of girder bottom flange (cm)
t_{pb}	Thickness of bottom plate (cm)
h	Height of the girder (cm)
y_{tG}	Distance from girder top to neutral axis (cm)
y_{bG}	Distance from girder bottom to neutral axis (cm)
y_{ft}	Distance from top flange NA to bottom of plate (cm)
y_w	Distance from web NA to bottom of plate (cm)
y_{fb}	Distance from bottom flange NA to bottom of plate (cm)
y_{pb}	Distance from bottom plate NA to bottom of plate (cm)
b_{pb}	Width of bottom plate (cm)
A_{ft}	Area of top flange (cm^2)
A_{fb}	Area of bottom flange (cm^2)
A_{pb}	Area of bottom plate (cm^2)
b_a	Width of asphalt pavement on deck (m)
t_c	Height of the curb (m)
b_c	Width of the curb (m)
s_G	Spacing of the girders (m)
L	Span length of the girder (m)
N_G	Number of girders
β_1	Factor for concrete stress distribution depth

where

$$C_{11} = 1/12 \qquad (2)$$

$$C_{12} = \text{moment due to noncomposite dead load} \qquad (3)$$

$$C_{13} = \frac{\eta_4 w_s}{1000} + \frac{\eta_4 t_c b_c \gamma_c}{72,000 N_G} \qquad (4)$$

$$C_{14} = \frac{\eta_4 t_a b_a \gamma_a}{12,000 N_G} \qquad (5)$$

$$C_{15} = \eta_3 \qquad (6)$$

$$C_{16} = \frac{\eta_4 w_r}{500 N_G} + \frac{\eta_4 w_{pG}}{1,000} \qquad (7)$$

where the random variables and deterministic parameters are listed in Tables 2 and 3, $\eta_4 = \eta_2/w_{DLC}$, and

η_2 and η_3 are the moment values corresponding to uniform composite dead load w_{DLC} and concentrated composite dead load P_{DLC}, respectively, and C_i, $i = 11, ..., 16$, are the constant coefficients.

Other component limit state equations for different bridge types are derived in a similar fashion. Each equation is formulated in terms of random variables and constant coefficients, where constant coefficients are themselves defined by formulas containing deterministic parameters.

Using this approach, it was possible to determine the component reliability indices for a given bridge by simply substituting the values of deterministic parameters into constant coefficient formulas and by defining the statistical distributions of random variables.

Once the values of random variables and constant coefficients are assigned, a system failure model is defined for each bridge. The system failure is based on failure of the slab or failure of any two adjacent girders or both. Other system failure definitions are also possible. However, Estes (1997) determined that the complexity of the system failure model did not significantly affect the system reliability index. The effect of including or excluding the slab in system failure model is demonstrated in the numerical example section.

Based on the random variables, constant coefficients, and the system failure definitions, initial reliability indices for the bridge components and systems in the network are determined. The lifetime reliability index profiles, however, are determined by using a live load model and a deterioration model. Live load model is used to determine the values of load effects in bridge slabs and girders due to increasing truck traffic in time. The model used in this study is based on the live load model reported by Nowak (1993). Deterioration models, however, are used to calculate the decrease in member resistances with time due to deterioration mechanisms. In this study, deterioration is assumed to be caused by corrosion only. Corrosion may be defined as being atmospheric or may be caused by salt applied on bridge decks during winter seasons.

The computer program developed for the purpose of integrating the analysis phases described above (i.e., random variables, deterministic parameters, constant coefficients, limit state equations, time-variant system reliability, and live load and deterioration models) is explained in the following section.

4 COMPUTATIONAL PLATFORM

RELNET (*REL*iability of system *NET*works) is a computer program capable of solving the problem of time-variant reliability with live load and material deterioration models for different bridge types in an actual bridge network. The term *actual* is used here to indicate the level of detail used in structure description of existing bridges during the pre and post-processing phases. The lifetime reliability profiles are computed considering the fact that the time-variant live load increases due to larger number of trucks passing on the bridge and, simultaneously, time-variant resistance of steel reinforcement and structural steel gradually decreases due to corrosion propagation.

RELNET is written in FORTRAN-90 and uses module structures to transfer and store dynamic data and cases for conditional statements. Being the higher level algorithm, RELNET calls two main subroutines: MONTE CARLO and NRELSYS. Program MONTE CARLO is written using some of the subroutines of a previous program MCSC (Enright and Frangopol 1998). The other external subroutine is NRELSYS (Estes et al. 1999). RELNET serves as a shell for NRELSYS. It acts as a higher level main program by calling NRELSYS as a subroutine. It also completely directs the input and output process of NRELSYS and

continually exchanges information with NRELSYS throughout its execution.

RELNET's programs and databases are developed for the broader objective of performing those functions necessary to perform a complete Probabilistic Risk Assessment for civil engineering structures. The core components of RELNET are the Structure Database, Member Type Database, Integrated Reliability Analysis System, Integrated Simulation System, Resistance Formula Database, and the Load Models. The program includes functionalities to allow the user to create a bridge as a system of structural components, to define failure events and system failure models, to solve system reliability analysis over time, and to quantify time-variant random variables and deterministic parameters of the problem.

The program currently contains four separate member types for bridges. These are: the reinforced concrete slab, prestressed concrete girder, reinforced concrete girder, and steel girder (including both steel rolled sections and steel welded plate girders). Integrated reliability analysis system (IRAS) is NRELSYS, however in the future, any other reliability analysis module can be substituted. Integrated simulation system (ISIS) used is the MONTE CARLO program. Ten different resistance formulas (used for the time-variant functions of random variables) belonging to the different member types are also stored in the program. The load model implemented is for bridges used for the calibration of the AASHTO LRFD Bridge Design Specifications (AASHTO 1994).

There exists an internally defined connectivity between member types and the resistance formulas. For each member type, corresponding resistance model functions are sequentially called and simulated. At the end, both descriptors and the outcome probability distributions of the random functions for resistance formulas are determined at each point in time along the lifetime of each bridge. A complete description of RELNET is provided in Akgül and Frangopol (2003).

5 ANALYSIS RESULTS FOR BRIDGE E-16-Q

Detailed discussion of the results for all fourteen bridges in the network are presented in Akgül (2002), whereas, owing to the space limitations, the analysis results of a single bridge in the network are presented herein.

Colorado bridge E-16-Q is located over US Highway 36 on State Highway 287 (Federal Boulevard) between Lowell Boulevard and Pecos Street in Boulder County. E-16-Q is a five span *Concrete* on rolled *I*-beam *Continuous* (CIC) bridge. Figure 3 shows the elevation of the bridge. Total length of the structure is 82.3 m between centerlines of the abutment bearings, and the width is 12.19 m. The end and

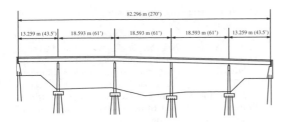

Figure 3. Elevation of Colorado highway bridge E-16-Q (CDOT, 1951).

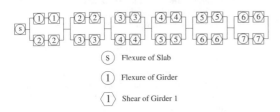

Figure 4. Cross sectional view of the bridge superstructure.

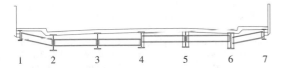

(S) Flexure of Slab

(1) Flexure of Girder

(1⟩ Shear of Girder 1

Figure 5. System failure model for the superstructure.

intermediate spans are 13.26 m and 18.59 m, respectively. The bridge has two lanes and the average daily truck traffic (ADTT) consists of 810 single trucks and 80 combination units, totaling 890 trucks per day.

The deck is 18.3 cm thick reinforced concrete slab covered with 8.9 cm thick asphalt pavement. Reinforced concrete slab is supported by five main and two exterior steel I-beams as shown in Figure 4. Exterior girders are 21WF62 sections in all five spans, while interior girders are made up of 30WF108 sections at end and center spans and 30WF116 sections at intermediate spans. The girders are connected to each other in transverse direction by steel diaphragms. Steel beams are fastened together at their ends by splices. There are a total of four splices along the overall length of the bridge. Each abutment is supported by 34 driven treated timber piles.

Four piers formed by a reinforced concrete cap beam resting on two tapered square columns carry the girders dividing the bridge into five spans. Each pier column rests on a square footing which is supported by 12 timber piles.

A cross sectional view and the system failure model for the superstructure of the bridge are presented in Figures 4 and 5, respectively. The system

Table 4. Limit states and corresponding initial reliability indices.

Limit state	Reliability index β_o
g *Slab, Flexure*	3.96
g *Steel Girder, Flexure, i*	3.76
g *Steel Girder, Shear, j*	6.23
g *Steel Girder, Serviceability, i*	2.36
System with Slab	3.74
System without Slab	3.80

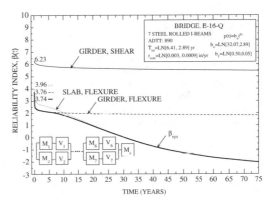

Figure 6. Variation of component and system reliabilities for bridge E-16-Q. System failure model including the slab.

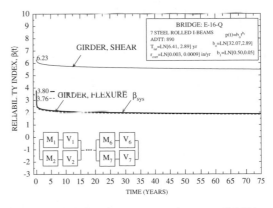

Figure 7. Variation of component and system reliabilities for bridge E-16-Q. System failure model with girders only.

failure model is constructed such that the bridge failure is defined as the failure of the slab or failure of any two adjacent girders (either in flexure or in shear) or both. Table 4 defines the limit states and shows the associated initial reliability indices.

Figures 6 and 7 present the variation of component and system reliabilities for the Bridge E-16-Q. The system reliabilities in Figures 6 and 7 are computed by including and excluding the slab, respectively.

6 CONCLUSIONS

Results of a comprehensive study for evaluating the lifetime performance of different bridge types in an existing network are presented. The most beneficial aspect of the developed computational platform for system reliability-based analysis of mixed-type bridge network is its generality.

ACKNOWLEDGMENTS

The partial financial support of the U.S. National Science Foundation through grants CMS-9912525 and CMS-0217290 is gratefully acknowledged. The support provided by the Colorado Department of Transportation is also gratefully acknowledged. The opinions and conclusions presented in this paper are those of the writers and do not necessarily reflect the views of the sponsoring agencies.

REFERENCES

AASHTO (1994). *AASHTO LRFD Bridge Design Specifications*, American Association of State Highway and Transportation Officials, First Edition, Washington, D.C.

AASHTO (1996). Standard Specifications for Highway Bridges, AASHTO 16th Edition. American Association of State Highway and Transportation Officials, Washington, D.C.

Akgül, F. (2002). *Lifetime System Reliability Prediction of Multiple Structure Types in a Bridge Network*, Ph.D. Thesis, Department of Civil, Environmental and Architectural Engineering, University of Colorado, Boulder, Colorado.

Akgül, F. and Frangopol, D.M. (2003). "Computational Platform for Predicting Lifetime System Reliability Profiles for Different Structure Types in a Network", (submitted for publication).

Akgül, F. and Frangopol, D.M. (2003), Lifetime performance prediction of existing steel bridges: II – Application", (submitted for publication).

CDOT (1951). *Plans for Highway Bridge E-16-Q, Project No: SP 12–382–501*, Plan Archives, Bridge Management System Unit, Colorado Department of Transportation, Denver, Colorado, Drawing pp. 8–12.

Enright, M.P. and Frangopol, D.M. (1998). *MCSC: Monte Carlo Simulation for Corrosion and RELTSYS: Reliability of Time-Variant Systems, Software Documentation,* Report No. 98-2, Structural Engineering and Structural Mechanics Research Series No. CU/SR-98/2, Department of Civil, Environmental, and Architectural Engineering, University of Colorado, Boulder, 252 pp.

Estes, A.C. (1997). *System Reliability Approach to the Lifetime Optimization of Inspection and Repair of Highway Bridges* Ph.D. Thesis, Department of Civil Engineering, University of Colorado, Boulder, Colorado.

Estes, A.C., Imai, K. and Frangopol, D.M. (1999). *NRELSYS: New Reliability of Systems, Software Documentation,* Report No. 99-1, Structural Engineering and Structural Mechanics Research Series No. CU/SR-99/1, Department of Civil, Environmental, and Architectural Engineering, University of Colorado, Boulder, 173 pp.

Nowak, A.S. (1993), "Live load model for highway bridges", *Structural Safety* Vol. 13, pp. 53–66.

25. New didactical strategies and methods in higher technical education

Architects and structures: how much structure is (not) enough for an architect?

D. Simic
Kansas State University, College of Architecture Planning and Design, Manhattan, Kansas, USA

ABSTRACT: Knowledge of the fundamentals of structural engineering for architects is indispensable. However, teaching and learning structures in architectural colleges is an everyday obstacle race. Students and studio professors often expect practical application of the knowledge of structures gained in a short period after the commencement of the course in structures. Moreover, a considerable number of students lack sufficient background knowledge of basic sciences and therefore are reluctant to learn about structural systems. On the other hand, instructors of structures have a very hard time to incorporate all the necessary structural issues in the limited course curricula and prepare students for appropriate application of structural engineering in architectural design. Structural education of architectural students can be vastly improved with the use of computer and related structural engineering software. Computer software allows quick modelling of structural systems with a variety of alternatives in structural design, thus making the course appealing to students.

1 INRODUCTION

Although architects do not have to design structural framework, they must be aware of structural systems and physical properties of materials, to use them effectively in architectural designs. Knowledge of structures, for architects, is essential for productive interaction with structural engineers. A successful collaboration between the two professionals is unlikely if they are unable to comprehend each other. These compelling reasons make courses in structures mandatory in all architectural programs and colleges. However, the most common question in almost all colleges of architecture relates to the level of knowledge in structures for an architect. HOW MUCH STRUCTURE IS (NOT) ENOUGH FOR AN ARCHITECT? or HOW MUCH OF AN EXPERT, IN STRUCTURAL ENGINEERING, SHOULD AN ARCHITECT BE? The primary basis for this question is the time frame required for students to acquire satisfactory knowledge for future. Once, when the time frame is determined, structure courses can be designed and organized in an efficient manner. Unfortunately, there is no simple answer to the previous simple question. The complexity of this issue involves numerous other questions that often lead to confusion. Some of these issues will be analyzed further to present ideas on feasible solutions for this problem.

2 EXPECTATIONS FROM ARCHITECTS AND THEIR KNOWLEDGE OF STRUCTURES

The primary task of architects is to design spaces or buildings with an architectural context. The buildings designed should be built in accordance to the architectural design. Stability of the buildings constructed is a very important aspect that is cornerstone to any architecture. Structural systems are necessary features contributing to stability of buildings. An architect's lack of knowledge in structures and structural systems can be a very crippling impediment towards successful practice of architecture. Ideally, architects must establish a conceptual structural system for building designs very early in design process that is related to and defined in a three-dimensional space. In addition, the architect should be capable to determine approximate sizes of main structural members before consulting a structural engineer. It is not at all an easy task for an architect to achieve without sufficient knowledge in structural materials and structural systems. Rigorous college curriculum in structures complemented by practical experience is the best guarantee for professional success. Active participation by teachers and students can prepare the students for successful future in the profession.

3 GENERAL PROBLEMS AND DIFFICULTIES IN TEACHING AND LEARNING STRUCTURE COURSES IN ARCHITECTURE COLLEGES

Students in architecture colleges are primarily interested in design and art. A majority of them come to college with a strong and wrong opinion that architecture is all about art only. In the early stages of their studies, many students are ignorant about technical disciplines and subject areas that are imperative for quality architecture. The lack of elementary knowledge of mathematics and physics is a typical for a large number of students in architecture and as a result they encounter serious learning difficulties in technical courses.

Structure courses are required in all colleges in architecture. In most colleges they are offered as sequential two to three semester courses. Although rare, The College of Architecture, Planning and Design at Kansas State University require four semesters of coursework in structures, for students enrolled in Bachelor of Architecture program. Although four semesters of structure courses is better than two or three semesters, it is not enough to cover all the important issues related to structures in architecture. It should be noted that two different opinions exist among teachers of structures in architecture colleges on the method of delivery and the duration of the courses.

According to the first group, structure courses should be organized and delivered in a general and encyclopedic manner, where no calculation and structural design is included in the course. Lectures, supported by images of different types of structures and models of structures that students make as part of studio assignment should aid students in gaining knowledge in structures. In this case no more than two semesters is needed for the entire course to be delivered. This opinion is widely received and supported by faculty, whose mathematics and structural engineering background, as well as professional practice experience, is deficient. Some of these teachers even advocate an idea about "the sense of structure" that architects can develop without serious study of structures. This idea is senseless and maybe even dangerous. The sense of structure cannot be developed without complete understanding of principles of structural systems, strong knowledge in all areas of structural engineering, and adequate practice. However, many students favor this method to their disadvantage. By the time when they should use and apply the knowledge acquired in structure courses,

these students are confused and helpless. Some of them will totally ignore structural systems in their design projects if not insisted by studio professor. Students, who try to analyze structural system for design projects that are already past the architectural design stage, will face an unpleasant reality that their design must be radically changed for the purpose of building stability and safety. This is a harsh realization for many students about the importance of knowledge in structures.

The second group, supported by professionals who teach structures in colleges of architecture, believes that structures courses cannot be taught without structural design and calculations, involving intense application of mathematics and physics. This method is arduous and laborious for students and it is not a complete solution.

College algebra and descriptive physics are the only prerequisite courses for students in architecture to enroll in structure courses. Knowledge of trigonometry and vector algebra is also essential, which many students with only the above stated prerequisites do not have. It is clear that this magnitude of subject awareness in mathematics and physics, limits students in their learning and comprehension of structures. It also creates hindrances to teachers in teaching the course because they can seldom fathom the depth of their student's grasp of the subject taught.

Time conflicts between structure courses and design studio courses are also a considerable problem that confuses students and at the same time prompts instructors to hasten coursework all the time. Generally, students take first course in structures in the second year of study. Introduction courses in structures deal mainly with statics and basic principles of statics. Structural systems and applications are not introduced in the introduction courses. At the same time, second year studio design projects are complex and students need to apply proper structural systems and materials but are unable to do so because those materials are covered in the third year. This conflict is indeed a difficult and stressful condition that causes confusion among students and teachers.

4 TEXTBOOKS AND MANUALS

Numerous books and manuals about structures are available to students in architecture. Algebra, trigonometry, and geometry are used in almost all the textbooks and it does not require proficiency in calculus to understand the text or for solving the numerical exercise problems in textbooks. It serves the purpose in Statics of Determinate Systems and the Strength of Materials. However, when considering indeterminate systems and structural design of members and systems, the only possible way is simplified formulae and "rule of thumb" principles. Simplified formulae or shortcuts are developed by mathematical means based on principles of physics and different code requirements. Students are receptive to know the derivation and appropriate use of formulae, as well as the development of formulae rather than just rote repetition of calculations using any particular formulae. However, most of the students lose their initial enthusiasm when mathematics is employed for total explanation.

Charts and tables for preliminary design of structural members are given as appendixes in textbooks or they are printed as separate manuals that can be used as supplements to textbooks. These charts are usually formed on the depth-to-span ratio principle, which is applied to different structural materials and members. Although these charts are simple to use, inexperienced students would be confused while using them. Different books will suggest different cross-sectional sizes for structural members of same materials and spans. Also, the thickness of the cross-sectional sizes that should be selected varies between 50% for the same material and span conditions. A major drawback with the use of these charts is that students acquire the impression that structural systems are determined after cross section of structural members are decided upon.

5 THE USE OF COMPUTER AND APPROPRIATE SOFTWARE AS AN EXCEPTIONAL TOOL IN TEACHING AND LEARNING PROCESS

Drawings and models are essential media that assist in the presentation of ideas in the architectural profession. Computer and appropriate software have replaced the older media to some extent, especially in visual presentation of design projects. Thus, it is not surprising to know that students in architecture start using computers in the design process from a very early stage during college. In many architectural colleges, students are required to have their own computer and to use it in studio design projects from the beginning of the third year in college. An important observation here is that students are excited about the new tool and rarely find its use boring. This fact is a positive proclamation for teachers in structures to introduce computers in structure courses. The benefits are manifold for both, students and teachers. Firstly, students are motivated to learn "routine" subject material to be capable to use computer in the next stage. Secondly, even very complex structural systems can be solved quickly without any strong knowledge in mathematics. A significant advantage of using computers in teaching and learning process is the possibility of creating structural models and modifying these models and their parts at a faster pace.

This allows students to explore different structural applications before arriving at the optimum solution.

Students in architecture are trained and taught to visualize the problem first and then to explore different possibilities to solve it. Present structural engineering software is equipped with robust graphics options, which allow users to see structural members in several different graphical modes. Among others, color rendered structural members, shown with materials from which they are made, is especially helpful for students to visualize their work. Also, applied loads, moment diagrams, and the deflected shape of the structural system under applied loads that can be shown together, helps students visualize and understand relationship between loads and members instantly. It is almost impossible to achieve the same task using traditional methodology of teaching structures. Consequently, it is a wise to accept that computer has become an indispensable pedagogical tool in structure courses in architecture colleges.

6 CONCLUSION

Structure courses are very important courses in the education of an architect. Time frame for the delivery of structure courses is underestimated in many architecture colleges. It is unlikely that this can be corrected by adding more time for structure courses. Most students in architecture come to college without a strong foundation of technical courses because of their lack of knowledge in mathematics and physics. This increases their indifference toward structure courses, which consequently affects structure teachers and their enthusiasm in teaching courses. Anachronistic structure courses with respect to studio courses create an additional problem in this complex situation. Question that is very often asked "HOW MUCH STRUCTURES IS ENOUGH FOR AN ARCHITECT" can be modified to "HOW TO ACHIEVE MORE KNOWLEDGE IN STRUCTURES IN A LIMITED TIME". One of possible answers is "BY INTRODUCING AND USING COMPUTER AND APPROPRIATE SOFTWARE IN STRUCTURE COURSES".

REFERENCES

Allen, E. & Iano, J. 2002. *The Architect's Studio Companion, Rules of Thumb for Preliminary Design*, 3rd ed., John Wiley & Sons, Inc., New York.
Cowan, H.J. 1971. *Architectural Structures: An introduction to Structural Mechanics*, American Elsevier Publishing Company, Inc. New York.
Levy, M. & Salvadori, M. 1994. *Why Buildings Fall Down*, W.W. Norton & Company, New York.

Risa-2D, Rapid Interactive Structural Analysis – 2-Dimensional, Version 5.5, RISA TECHNOLOGIES, Foothill Ranch, CA 92610

Risa-3D, Rapid Interactive Structural Analysis – 3-Dimensional, Version 4.5, RISA TECHNOLOGIES, Foothill Ranch, CA 92610

Salvadori, M. 1990. *Why Buildings Stand Up*, W.W. Norton & Company, New York.

Scheuller, W. 1996. *The Design of Building Structures*, Prentice Hall, Upper Saddle River, N.J.

Shoedek, D.L. 2001. *Structures*, 4th ed., Prentice Hall, Upper Saddle River, N.J.

Spiegel, L. & Limbrunner, G.F. 1999. *Applied Statics and Strength of Materials*, 3rd ed., Prentice Hall, Upper Saddle River, N.J.

Teaching creativity to undergraduate civil engineers

T.M. Lewis
Department of Civil Engineering, The University of the West Indies, St Augustine, Trinidad & Tobago, West Indies

ABSTRACT: There can be little doubt that the engineering industry, as well as an individual engineering firm, needs innovation to improve its performance and to put itself into a position to address the future with confidence. Of course, firms are not innovative, people are. Thus it is important that young engineers be given the tools that can help them be creative, so that they will be innovative in their future careers. Unfortunately the typical undergraduate engineering degree is already so full with technical information that must be learned and analytical tools that must be mastered, that there is no time left to introduce them to the concept of "creativity", or to the tools that can help promote it. This paper describes the way in which creativity has been introduced into a course in civil engineering management taught to final year undergraduates.

1   INTRODUCTION

The technological basis of society today calls for a different type of knowledge than was required two or three decades ago. We can expect the future to be even more different again. The tools, processes and applications of today have changed significantly within our own lifetimes. What we were taught in university thirty or forty years ago was different from what we teach students now. How do we deal with this need for constant vigilance about the way we treat with that changing world? More importantly, how will our students cope with their need to deal with a changing future? How do we prepare them to be able to adapt and to be receptive to new situations? In other words, how do we instill or foster the creativity that our students will need to deal with their future world? (Berglund et al. 1998). The key is creativity because engineers must be flexible and imaginative if they are to deal effectively with the kind of problems that they will face, and that tend to be unique and unstructured.

Technology does not stand still. We do all sorts of things differently now than we used to do them. The economic, technological, social, political and even ecological environments are constantly changing and need a creative response. Yet this crucial creative capability has been largely ignored in engineering education – it has been left to the natural ingenuity of the practicing engineer. It is doubtful whether it is wise to continue in this way.

However, engineering students already tend to have a very full work-load, as Samuel Florman, points out: " … the stultifying influence of engineering schools where the least bit of imagination, social concern or cultural interest is snuffed out under the crushing load of purely technical subjects" (Florman 1976). Is it sensible or feasible to increase the load by adding "creativity" in somewhere? In addition to this, as De Bono (1992) points out, 95% of academics hold the view that creativity is unnecessary, and that all engineering students really need is a good grounding in logic. Many also hold the view that it is, anyway, impossible to teach creativity. It is a happy accident of birth – you can't teach an also-ran to be an Edison (Thompson 1992) in the creativity stakes[1]. These attitudes probably explain why creativity has been neglected for so long.

In response, each of these positions can be repudiated. In the first place it may be more sensible in the longer term to ask whether we can afford not to find room for creativity in our degree programs than to dither about how to fit it in. Secondly, although logic is good and essential for engineers, so is creativity. The very name "engineer" derives from the same root as "ingenious". Ingenuity/creativity is an essential part of the engineer's toolbox.

Thirdly, some reviews of research on the effectiveness of programs designed to teach children to be

---

[1]Thomas Edison holds 1,093 patents but a Dr. Yoshiro NakaMats holds more than 2,300 including patents for such items as the floppy disk, compact disc, compact disc player, and digital watch.

more creative are discouraging – many do not appear to produce a notable increase in children's creative abilities (Mayer 1983). By contrast, however, there is an equivalent body of research[2] which has shown that significant gains in "creative thinking" can be made by effective training (Lazar 2001). Both types of study tend to use the Torrance Tests for measuring creativity (Antonietti 1997), and these tests do have problems with the major assumptions underlying their measures of creativity and the ability of those measures to achieve consistent and reproducible results. These problems raise doubts about both the measures and their conclusions.

However, if it can be agreed that creativity is fundamentally a cognitive process, that is influenced by both heredity and (nature and nurture), then there is no reason why it should be different from other cognitive processes. If we can teach people to be more logical, then there seems no reason why we cannot teach people to be more creative. All children possess the qualities that are displayed by creative personalities, but most either lose or inhibit these qualities as they grow. The pressures of conformity, the need to adopt prescribed patterns of thought and behavior, and the systematized methods of teaching and assessment, all tend to restrain creative instincts. Change some of these pressures and influences and creativity may be able to flourish[3]. Even if it is that "You cannot teach creativity, but you can kill it" (Johnston 1995) – then by removing the factors that inhibit and kill creativity, you should be able to encourage and enhance it.

Given that creativity is an innate capacity in all of us that may flourish given the right conditions, the key then becomes to find those conditions. Many people believe that it is possible to activate that innate creativity using some form of trigger mechanism (Acar 1998) – it is in this belief that various techniques have been developed, like brainstorming, synectics or morphological analysis for example (Biondi 1974), which are designed to prompt our creativity to be expressed.

Engineering students are aware of the need for creativity in their chosen profession, as, when a group[4] was asked to describe the qualities and values they thought professional engineers should possess, and which should be reflected in the curriculum being taught, they listed: responsibility, honesty, an ethical approach, analytical abilities, and last but by no means least, innovation and creativity (Csikszentmihalyi 1996). This also implies their belief that creativity can be improved by appropriate instruction.

In professional life the pressure of time, the risk to human life and and the need to meet the stipulations of codes of practice all tend to push young engineers into a conservative mode, and to be disinclined to take the risk of being more creative. It has also been established that it is easier to be creative when the outcome is not going to be evaluated (Gardner 1993). People produce creative solutions more often when they do something for pleasure, rather than when they are tested on the outcomes or promised rewards. In professional life, this is inconvenient, as the most critical evaluation (and possible reward) is likely to be directed at those activities of design where creativity is most needed.

2 THE MEANING OF CREATIVITY

As the focus here is on creativity and how it may be improved, it relevant to examine briefly what the various disciplines that have an interest in it have made of it. The literature about creativity is enormous and diverse and takes two main perspectives; one being from the perspective of the individual, in terms of cognitive abilities and personality traits – idealized as the lone inventor in his garage (Marton & Saljo 1976), and the other from the perspective of creativity as an interaction between people with different (complimentary) skills within a given context in terms – as exemplified by the creativity of big laboratories and their research teams (Marton & Booth 1996).

The focus within the first option is that "creativity" is basically a set of personal traits – and that it should be possible to identify those traits, test for them, and, ideally, encourage them. Within the second option, the focus is on the social context and the issue becomes less how to deal with the individual than with the context, and the key creative characteristics become those of a supportive environment and efficient networking especially the development of strong interpersonal skills (Tornkvist 1998).

---

[2]In the study, computations were made based on mean pretest scores of seventh grade students using the Torrance Test of Creative Thinking, and their expected progress as predicted by the Torrance Generalized Development Curve. After one year of training, eighth grade students demonstrated a mean raw score gain of 89.49 points; this was 83.30 points more than projected! (See Lazar, 2001).

[3] It is accepted that people who are obviously creative (e.g. Mr NakaMats) also live with the same sorts of pressures and influences, but are affected differently or react differently to them. This suggests other people should be able to do so too.

---

[4]These students were also asked to develop a profile for the ideal engineering student of 2010. The students described their vision as "environmentally, economically and globally aware, professional problem-solvers" who could seek out information for themselves. Creativity was absent from this profile!

The term creativity itself, it has been defined as:

- the ability of human intelligence to produce original ideas and solutions using imagination (Drabkin 1996).
- is not a singular human trait…it is dependent on other traits, for example, academic ability and practical aptitude (Pace 2000).
- a process that enables people to discover new and meaningful ideas, it is a universal human characteristic (Rickards 1990).
- a value word and represents a value judgement – no one ever calls creative something new which he dislikes (De Bono 1967).
- coming up with something novel, something different. And this new idea, in order to be interesting, must be intelligible (Boden 1994). No matter how different it is, we must be able to understand it in terms of what we knew before. Human creativity uses what already exists and changes it in unpredictable ways that bring about a desired enlargement of human experience that goes beyond the usual choices (Navin 1994).
- it has many facets including, in no particular order: judgement, emotions, renewal, rewards, models, intuition, choices, values, motivation, problem-solving approach, creativity stimulators, vision, programming, inhibitors, environment, ethics, relationships, attitude, communication, groups, information, checking, experiment, patents, records, practice (Bailey 1978).

There are certain characteristics that are "typical" of such definitions, such as that it depends upon ideas that are original, and that they satisfy human needs by exploiting resources in a different and more effective manner (Gregory & Monk 1972). It is not a singular human trait in that it depends on other traits, such as intellectual ability, mental preparedness, and practical aptitude (Pace 2000).

Much of creativity depends on new combinations of pre-existing elements, new arrangements of concepts (Taylor 1988), it seems obvious that creativity can be improved by having a number of different "resources" in place, e.g.

- basic knowledge about physical objects and principles,
- basic knowledge of design and problem-solving processes,
- good judgement about what is "reasonable",
- open-mindedness to new ideas and suggestions,
- suitable motivation,
- ability to communicate ideas,
- frustration at problem or recognition of opportunity.

With creativity involving this range of "resources", and the likely variability with which people possess them, there should be little wonder that psychological tests that try to measure a person's "creativity" in terms of a single numerical value seem to be so inconclusive. Even with all of these "resources" in plentiful supply, there is no guarantee that a person will be creative. Very often, creativity can only be recognized after the event. (Elder 1994).

3 CREATIVITY AND DESIGN

Creativity has four characteristics, the primary one being that it is a mental activity, but the others are also important. They are that it is triggered by specific problems or opportunities; it produces in novel solutions; and these solutions often have implications beyond their immediate context. These secondary characteristics support the notion that opportunity favours the prepared mind, through thinking about and being frustrated by problems, or seeing opportunities, creative people respond with novel solutions. It is characteristic of such novel solutions – particularly those that are patented – that they find more use in contexts other than the ones for which they were conceived.

Creative thinking is a form of conceptual thought, and as conceptual thought can be improved by the appropriate instructional techniques, creativity can also be improved by teaching. Students generally need a context for new knowledge or they find it hard to relate what they are learning to the other elements of their programme of studies or to life in general. The natural context for creativity is in a design course, and preferably early on (Cross 1989). The traditional approach to civil engineering design tends to lead the student to believe that there is one correct answer to an engineering problem. The design process is supposed to enable us to find that solution, by a series of iterations that close on the answer. Students, eager to find *the* solution, often short-circuit this process by *satisficing* – either stopping at the first acceptable solution, or optimizing the specific configuration that they have selected (rather than creatively examining alternative approaches to fulfilling the problem's requirements). Equally, students have the preconception that engineering design is an intellectual activity involving deductive and mathematical thinking which is applied systematically to "solve" a problem. Beder (1997) quotes Ferguson (1992) as writing: "It is usually a shock to [engineering] students to discover what a small percentage of decisions made by a designer are made on the basis of the kind of calculation he [or she] has spent so much time learning in school". Because of this attitude, they tend to become frustrated when they are presented with open-ended design problems that require a creative input (Eder 1996).

This is obviously not satisfactory, and to improve matters, the teaching of engineering design should emphasise and encourage the type of creative thinking needed to address open-ended problems that more closely reflect the real world (Ray, 1985) and to encourage students to become conscious of the cognitive processes (Miller 1996) involved in undertaking a design activity (Ramirez 1994).

4 THE ENGINEERING DESIGN PROCESS

The purpose of designing is to take a specification of a set of (functional) needs and transform it into full instructions for manufacturing a product and/or implementing a process. To be effective, this purpose is met by generating a range of alternative solutions and evaluating these with reference to their ability to satisfy the specified needs. This calls for intuitive, iterative and creative thinking in order to generate potential solutions. This process also helps the problem to be better understood.

The process engineering design usually begins with the identification of a problem and the limitations imposed on it that act as constraints. As the design process proceeds, usually by trial-and-error, the definition of the problem and the relevant constraints become more sharply defined, and more tightly focused on the desired functionality of the finished artifact. (Lethbridge & Davies 1995).

Engineering design takes the discoveries and principles of science (Pahl & Beitz 1996), pulls in elements of technology and culture, and addresses the creative issues that arise during the evolution of a product (Dixon 1996) from the initial concept through to the end of its effective life (BSI 1989). Thus, engineering design involves the transformation of ideas and knowledge into the description of an artifact that will satisfy a set of identified needs (Cripps & Smith 1993), and includes detailing the way in which the artifact, whether it is a product or a technical system, can be produced (Hales 1993).

Both the description of the artifact and the process by which it can be produced require the generation, evaluation and use of ideas. Many models of the design process exist (French 1985) (Pugh 1991), most of which follow the same four main phases – clarification of the task, conceptual design, embodiment design and detail design (Pahl & Beitz 1996). These processes are applied to design activities that can be classified as either original, adaptive or variant (Black & Shaw 1991), which are defined as:

- **Original design**: involves the elaboration of a new, original solution for a product or technical system
- **Adaptive design**: involves adapting a known system to a changed task.
- **Variant design**: involves varying the size and/or the arrangement of certain aspects of the chosen

product or system, with the function and approach remaining unchanged.

Each of these design typologies requires creative inputs, with creativity normally needing to be highest in original design and progressively less in adaptive and variant design.

In general, creativity stems from an attempt to solve an unstructured, non-routine problem. You don't need to be creative to decide to switch on the light when it starts to get too dark to read. That is a straightforward, routine problem. If you are trying to decide how best to reduce the traffic congestion between St Augustine and Port of Spain (a real problem in Trinidad), then you are dealing with a much less structured problem and it is very far from being routine. It is a problem that calls for creative thinking and needs a number of alternative possible solutions to be generated that can be evaluated to establish which is "better" than the others.

5 CREATIVITY ON THE TEACHING AGENDA

Most engineering students have an educational background showing specialisation in mathematics and the physical sciences. Such students usually have strong analytical and deductive skills, and can be described as "convergent" rather than "divergent" thinkers (Hudson 1967). Students with a natural instinct for creativity ("divergent" thinkers) will tend to head for the arts (Dyson 1997). So, creative thinking will not tend to be the "strong suit" of most engineering students.

If we accept that they have the innate ability to be creative, then efforts need to be made to help them develop and express that ability by applying the appropriate techniques. This can be by way of promoting "freedom" rather than "order" in their thinking (Ihsen et al. 1998); encouraging them to reformulate problems to see what difference a change of perspective has; requiring them always to generate alternative design solutions to any problem, and helping them access relevant, but not necessarily obvious, background knowledge (Fischer 1994).

The training and application of techniques to enhance creativity should be structured around real, complex problems, which are open-ended, and have many different, but possible solutions. These problems should be addressed by groups, in which each member may address a different aspect of the problem but they have a common interest in its solution. The communications between the group members[5]

---

[5]Piaget wrote that "most researchers tend to neglect the central role played by language and concept in thought" and suggested that poor communication skills will inhibit conceptual thinking skills, thus emphasizing the importance of good communication skills amongst students.

are a necessary part of the creative process (Inhelder & Piaget 1968). The groups should, if possible be inter-disciplinary (though in a University setting this is not often possible) and they should be encouraged to use techniques (Dasgupta 1996) that promote creative thinking like brainstorming, mindmaps, synectics and morphological analysis (Evans & Deeham 1988). The key elements of all these methods are to prepare the mind and to postpone evaluation for as long as possible so as to increase the number of ideas.[6]

Universities often use design or research projects to create a group environment for the students' work.[7] Although the projects do create the kind of work environment that the students will find themselves in when they enter the "real" world, they rarely manage to mobilize the creativity of the students. There are various reasons for this, including, the pressure of time, the fact of evaluation, the conservatism induced by the careful compliance with codes and regulations, the need to get a solution to analyse in detail, and the use of computer software to produce a solution. All of these tend to cause students to produce design project solutions that are "typical" rather than creative.

6 EDUCATING FOR CREATIVITY

If engineers are going to play a significant role in solving the problems which confront the world today and, more so in the future, then creativity must be a crucial weapon in their armoury. Engineering educators, have control over at least some of the factors which influence creativity, and so we have the freedom either to foster or to discourage the creative development of our students.

Many universities do attempt to encourage creativity by using competitions to encourage creative thinking. For example, students are encouraged to design robots to accomplish specific tasks using a restricted set of materials; or to use some unlikely material like ice-cream sticks or spaghetti to construct a bridge, or concrete to construct a boat. These competitions do help students to show their creativity in designing "useful" artifacts under special sets of constraints. Other "creativity" exercises that are used to encourage

students to develop their communicating skills include getting them to write essays giving a different perspective on current topics or designing effective business cards. These activities encourage students to think critically about issues in the news, or about the micro-decisions involved in creating a business card.

Creativity has been introduced into the syllabus for a course in Civil Engineering Management for undergraduates, and Construction Management for postgraduates at the University of the West Indies. Because of the weight of other material that "has" to be covered, it is only accorded two lecture hours in each. In this time, the students cannot learn about theories and theorists, but they are shown how a little flexibility and divergence in their thinking could help solve awkward problems. It is not an element of the course that can be assessed, and so it plays no part in the examination process.

The lecture time is taken up by presenting the class with a number of "problems" and using them to exemplify some aspects of divergent/creative thinking. Many of these "problems" will be familiar to those who have studied creativity. In each case, the "problem" is presented to the class on an overhead slide, and they are asked to think about it and discuss it (quietly) for a few minutes. Then they are asked for possible solutions, which are discussed before *the* solution is presented. These sessions are usually a lot of fun.

Problem 1.   A young engineer, employed by the firm that managed a large high-rise office building, was called to his boss's office. The boss said he had received many complaints about the elevators being too slow, and told the engineer to do what he could to solve the problem. What would you do?

Solution 1.   After a week the young engineer returned to announce that he had a solution. Pleased, the boss sat back to listen, expecting to hear about increased speeds, a new algorithm for servicing the floors, and so on. Instead, the engineer proposed the installation of mirrors on each floor next to the elevator. If people were kept busy checking their appearance, he explained, they would be less likely to notice the wait. Whilst the boss had seen the problem as one of speeding up the elevator service, the young engineer had seen the problem as being the impatience of the people waiting rather than the speed of the elevators.

This is used to show the student that the "obvious" approach to the solution may not be the best one, especially when you start to introduce extra constraints on the speed and operation of the lifts. This is used to introduce the idea of lateral thinking as a way of getting out of some of the traps we build ourselves when looking at problems and looking for solutions. In the situation illustrated above, the obvious way to see the problem is as an engineering issue, related to the efficiency of the lifts. The young engineer saw it

---

[6]An excellent discussion of these various techniques can be obtained in Thompson & Lordan, 1998.

[7]The civil engineering students at The University of the West Indies have to do a final year design project as a group, with individual assignments and design elements and group and individual assessment. They have to manage their interactions and the way they work themselves. Unfortunately the students are rarely creative in the solutions they put forward, usually settling for the obvious, conventional approach without a second thought, i.e. they design buildings as boxes that are easy to analyse.

as a behavioural problem related to human impatience when kept waiting with nothing to do. The "creative" solution was found by not letting the logical consequences of sequential thinking predetermine the shape of the solution environment. As Russell Ackoff wrote "Successful problem solving requires finding the right solution to the right problem. We fail more often because we solve the wrong problem than find the wrong solution to the right problem".

In a similar way, we too often let our emotions colour our perception of problem situations, and this limits the options we consider. This introduction allows the second Problem to be introduced.

Problem 2. You are driving an ambulance with an emergency case on board along a narrow country lane. As you turn a corner you find the road blocked by a flock of sheep. How are you going to get your urgent case to the hospital with the least loss of time?

Solution 2. As you drive towards sheep they just jog along, milling about in front of you. You cannot drive fast or you will run them down. You have to get the ambulance to the other side of the flock somehow. The emotional emphasis is on the need for speed. We tend to see the problem from the perspective of the driver and the vehicle as a unit with an urgent problem, and what we would do from that perspective. To find a "creative" answer we must look at the problem from a different perspective – we must see it as the need to transpose two objects, an ambulance and the sheep. As with any such transposition, you have two options, you can either move the ambulance to the other side of the sheep, or the sheep to the other side of the ambulance. When viewed in this way is becomes obvious that you park the ambulance, walk behind the sheep and drive them past the ambulance then proceed with your patient. The emotional hang up on the need for speed leads us away from the idea of getting out of the vehicle. It should be noted that this problem assumes that people know the behavioural traits of sheep – i.e. that they will move along ahead of you if you move towards them from any direction, but will tend to ignore an object that is stationary.

One of the most limiting habits of engineers is to make assumptions that are not a necessary part of the problem, but are more a habit of experience, and thereby limit potential solutions to the ideas that follow logically from those assumptions. An old familiar example of this trap is illustrated in Problem 3.

Problem 3. The problem is to connect the nine dots by making four straight lines, and without lifting your pen (pencil) from the page.

```
*    *    *
*    *    *
*    *    *
```

Solution 3. The tendency amongst engineers is to see the dots as forming a boundary that cannot be violated, by the lines that are drawn. However, there is no such implied restriction; the lines do not have to stay within the "boundary". The problem can only be solved by breaking this barrier and going outside the area defined by the dots.

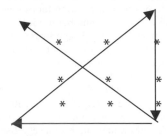

Another, similar (old) example is illustrated by the next problem. Again the lesson here is about recognizing the assumptions you are making and understanding that those assumptions are unnecessary constraints.

Problem 4. Six quarters (or coins of some description) are lined up in two crossing rows, one with three coins, and one with four. The problem is to move one coin so as to make two rows, each having four quarters in it.

Solution 4. The limiting assumption here is that this is a planar problem. This assumption is encouraged by the implied restriction of the plane in which the coins are lying. It is not a limiting criterion. The solution is to pick up the end coin on the right and put it on the top of the coin in the middle vertical line, thus making two lines of four.

The two problems 3 and 4 are both used to show students that they impose restrictions on their thinking that are not necessarily related to the problem they are dealing with, and that to be more creative, they need to examine and make explicit the assumptions that they are dealing with. The next problem is more illustrative, and is used to prompt students into not letting their project work slip during the first few weeks of semester. It is surprising how often the answer "90 kph" is given.

Problem 5. A driver has to average 60 kph for three, ten kilometer long laps of a race course. He

1734

makes a poor start, while getting used to the car and only averages 30 kph for the first lap. He then speeds up so as to average 60 kph for the second lap. What speed should he do on the third lap to achieve the required 60 kph average?

Solution 5. The answer is that it is impossible. He cannot achieve his average because all the time necessary for three laps at the required average has already been used up. He took 20 minutes for the first lap, 10 minutes for the second and so he has already used up all the 30 minutes of time that was available for the three laps. The instinctive response for students is 90 kph until you work it out for them. The lesson is that sometimes we can be trapped by our initial actions, and that these may restrict our future options and abilities. It is a lesson in creativity as well as in the importance of keeping to schedules.

Problem 6. You have two jugs of beer in front of you, one marked "Carib" and the other marked "Stag" (these are local beers, but any beverages of standard specific gravity will do). The first one has 10 ounces of Carib in it, the second one has 10 ounces of Stag in it. Now, suppose you want to try out a mixture of the two, so you pour two ounces of Carib into the "Stag" jug and mix it up thoroughly, then you pour two ounces of this mixture back into the "Carib" jug. Both jugs again have 10 ounces of liquid in them, but which one has the greater amount of "foreign" beer?

Solution 6. This is the "problem" causes the most confusion. The answer the students normally give is that the Stag has more Carib, but the truth, which is obvious in retrospect, is that they are identical. What has come out of the Carib jug has replaced an equal amount of Stag from the Stag jug, and vice versa. The missing Carib and the Stag that has replaced it must be equal. This can easily be shown mathematically for the rigidly convergent, doubting students.

After the two transfers, the Carib jug has

$(10C + 2S) \times 10/12 = 100C/12 + 20S/12$

And the Stag jug has

$8S + (10C + 2S) \times 2/12 = 20C/12 + 100S/12$

In other words, the alien liquid is $20S/12$ in one jug and $20C/12$ in the other. Equal quantities. The point of this "problem" is that it can be solved without needing to do a mathematical "proof". The order in which events occur also tends to cause a bias in our thinking. Creative, divergent thinking lets us see things conceptually, without always having to refer to analysis and mathematics, which can be used as a back-up if necessary.

Problem 7. You are organising a knock-out tennis tournament with 128 players in the first round draw, when, at the last minute six more people want to play.

You now have 134 players in your tournament. How many matches are you now going to have to schedule in order to complete the tournament (there is no "plate" competition nor is there a 3rd place play-off)?

Solution 7. This looks like one of those awkward problems where you have to draw out a table and count up the matches, but in fact it has a simple solution if the logic is reversed. If the problem is restated in a creative way, the solution "falls out". In the new tournament with 134 players, in order to have a single winner, there will have to be 133 losers. There will always be one and only one loser in each match, so there will have to be 133 matches. The same logic, of course, applies for any number of players. This is a neat example of how rephrasing the question can help determine the solution. In civil engineering it is likely to be much more creative to ask "How can the problem of congestion between Port of Spain and St Augustine be relieved?" than by asking "How do we design and build a fly-over at the main intersection?". They address the same problem but from different perspectives. The next problem provides another example of how a solution can be determined by a methodology that appears unlikely.

Problem 8. The Japanese Priest and the Mountain One morning, exactly at sunrise, a Japanese monk began to climb a mountain. A narrow path, a foot or two wide, spiralled around the mountain to a temple at the summit. The monk ascended at varying rates, now pressing ahead, now stopping along the way to rest. He reached the temple shortly before sunset. After several days at the temple, he began his journey back along the same path, starting at sunrise and again walking at variable speeds with many pauses along the way. His average speed of descent was, of course, greater than his average climbing speed. There was one spot along the path, however, that he occupied on both trips at exactly the same time of day. Prove that this was bound to be the case.

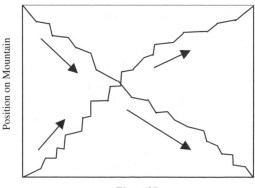

Time of Day

Solution 8. It seems unlikely that you could use a graph to prove that the priest occupies the same spot at the same time of day on both journeys, but it does provide a very good visual image. The diagram here shows a rough image of the priest starting off at the dawn at the bottom of the mountain (bottom left). At the end of the day he is at the top (top right). Then the diagram also shows him at dawn at the top of the mountain (top left) moving down to the bottom of the mountain at the end of the day (bottom right). Where the two paths meet, the priest is at the same point on the mountain at the same time of day. This shows students that a graphical/conceptual method can be used to resolve a problem that appears logical/mathematical. This unusual approach helps to illustrate the use of analogical reasoning, which Jacob Bronowski pointed out as being important in both science and the arts. The ability to see likenesses that most of us miss is the hallmark of a truly creative person. He says "I found the act of creation to lie in the discovery of a hidden likeness. The scientist or the artist takes two facts or experiences which are separate; he finds in them a likeness which had not been seen before; and he creates a unity by showing the likeness."

Problem 9. The final problem I use is one in which the students are told they have been asked to design a clock with no moving features at all on its face. (No hands, no numbers changing, no shadow moving around.) What do you suggest?

Solution 9. This seems to be the easiest "problem" for the students to resolve, as they almost always come up with the "talking clock" solution.

One important aspect of these sessions is that they should be fun. This is not often a strong point of engineering education – it is said that children smile on average 400 times a day while adults smile only about 15 times (Poon Teng Fatt 2000). Both humour and creativity require imagination and the ability to suspend judgement. In addition to these "problems" and "solutions", my final overhead slide shows a list of "instructions" accredited to the Dalai Lama. It invariably surprises the students, and they often ask for extra copies of it. They are good instructions for creativity and life.

Exhibit 10. Instructions for Life in the new millennium from the Dalai Lama:

1. Take into account that great love and great achievements involve great risk.
2. When you lose, don't lose the lesson.
3. Follow the three Rs: Respect for self, respect for others, responsibility for all your actions.
4. Remember that not getting what you want is sometimes a wonderful stroke of luck.
5. Learn the rules so you know how to break them properly.
6. Don't let a little dispute injure a great friendship.
7. When you realize you've made a mistake, take immediate steps to correct it.
8. Spend some time alone every day.
9. Open your arms to change, but don't let go of your values.
10. Remember that silence is sometimes the best answer.
11. Live a good, honorable life. Then when you get older and think back, you'll be able to enjoy it a second time.
12. A loving atmosphere in your home is the foundation for your life.
13. In disagreements with loved ones, deal only with the current situation. Don't bring up the past.
14. Share your knowledge. It's a way to achieve immortality.
15. Be gentle with the earth.
16. Once a year, go someplace you've never been before.
17. Remember that the best relationship is one in which your love for each other exceeds your need for each other.
18. Judge your success by what you had to give up in order to get it.
19. Approach love and cooking with reckless abandon.

REFERENCES

Acar, B. Serpil, 1998. Releasing Creativity In An Interdisciplinary Systems Engineering Course, *European Journal of Engineering Education*, June, Vol. 23 Issue 2, pp 133–41.

Berglund, A., Mats, D., Hedenborg, M. & Tengstrand, A. 1998. Assessment To Increase Students' Creativity: Two Case Studies, *European Journal Of Engineering Education*, March, Vol. 23 Issue 1, pp 45–55.

Antonietti, A. 1997. Unlocking Creativity, *Educational International*, March.

Bailey, R.L. (1978) *Disciplined Creativity for Engineers* (Ann Arbor, Ann Arbor Science Publishers).

Beder, S. 1997. Beyond technicalities: expanding engineering thinking, in: Tornhvist, S. (Ed.) *Teaching Science for Technology. Second International Conference at the Royal Institute of Technology* (Stockholm, Royal Institute of Technology).

Biondi, A.M. 1974. *Have an Affair with your Mind* (Bearly, Buffalo, NY).

Black, I. & Shaw, W.N. 1991. Organizational And Management Aspects Of CAD In Mechanical Design, *Design Studies,* 12, pp 96–101.

Boden, M.A. 1994. Agents And Creativity, *Communications Of The ACM*, 37, pp 117–121.

Bs 7000, 1989. *A Guide To Managing The Design Process* (London, British Standards Institution).

Cripps, R.J. & Smith, F. 1993. An Information System To Support The Design Activity, *Journal of Engineering Design*, 4, pp 3–10.

Cross, N. 1989. *Engineering Design Methods* (New York, John Wiley).

Csikszentmihalyi, M. 1996. *Creativity* (New York, Harper Collins), p 321.

Dasgupta, S. 1996. *Technology And Creativity* (London, Oxford University Press).

De Bono, E. 1967. *Five Day Course in Thinking* (New York: Basic Books, Inc.)

De Bono, E. 1992. *Serious Creativity* (Harper Collins, London).

Dixon, J. 1966. *Design Engineering: Inventiveness, Analysis And Decision Making* (New York, Mcgraw-Hill).

Drabkin, S. 1996. Enhancing Creativity When Solving Contradictory Technical Problems, *Journal Of Professional Issues In Engineering Education And Practice* (New York, ASCE), 122, pp 78–82.

Dyson, J. 1997. The future: a way forward for Britain, *Keynote Address at the Royal Academy of Engineering Visiting Professors in the Principles of Engineering Design, Annual Workshop,* University of Cambridge, Cambridge, UK, 17–18 September 1997.

Eder, W.E. 1994. Developments In Education For Engineering Design: Some Results Of 15 Years Of Workshop Design-Konstruktion Activity In The Context Of Design Research, *Journal of Engineering Design*, Vol. 5, Issue 2.

Eder, W.E. 1996. (Ed.) *Engineering Design And Creativity, Proceedings Of The Workshop Edc*, State Scientific Library, Pilsen, Czech Republic, 16–18 November, Heurista.

Evans, P. & Deeham, G. (1988) *The Keys To Creativity* (London, Grafton).

Ferguson, E.S. (1992) *Engineering and the Mind's Eye* (Cambridge, MA, MIT Press).

Fischer, G. 1994. Turning breakdowns into opportunities for creativity, *Knowledge-based Systems*, 7, pp 221 232.

Florman, S. 1976. *The Existential Pleasures of Engineering* (New York, St Martin Press).

French, M. 1985. *Conceptual Design For Engineers* (*London, Design Council, Springer*).

Gardner, H. 1993. *Creating Minds* (New York, BasicBooks).

Gregory, S.A. & Monk, J.D. 1972. Creativity: definitions and models, in: Gregory, S.A. (Ed.), *Creativity and Innovation in Engineering* (London, Butterworths), pp 52–78.

Hales, C. 1993. *Managing Engineering Design* (London, Longman).

Hudson, L. 1967. *Contrary Imaginations. A Psychological Study of the English Schoolboy* (Methuen, Pelican Books).

Ihsen, S., Isenhardt, I. & Steinhagen De Sanchez, U. 1998. Creativity, Complexity And Self-Similarity: The Vision Of The Fractal University, *European Journal of Engineering Education,* March, Vol. 23 Issue 1, pp 13–23.

Inhelder, I. & Piaget, J. 1968. *The Growth of Logical Thinking*, (Routledge and Keegan Paul, London).

Johnston, C. 1995. *Fostering Deeper Learning* (Economics Department, University Of Melbourne).

Lazar, Ruth S. 2001. *You don't need an umbrella in a brainstorm.* Education.

Lethbridge, D. & Davies, G. 1995. Initiative For The Development Of Creativity In Science And Technology (Crest): An Interim Report On A Partnership Between Schools And Industry, *Technovation,* 15, pp 453–465.

Marton, F. & Booth, S.A. 1996. The Learner's Experience Of Learning, In: Olson, D. & Torrance, N. (Eds), *The Handbook Of Education And Human Development* (Oxford, Blackwell).

Marton, F. & Saljo, R. 1976. On Qualitative Differences Of Learning: 1 Outcomes And Process, *British Journal Of Educational Psychology,* 46, pp 4–11.

Mayer, R.E. 1983. *Thinking, Problem-Solving, Cognition.* (New York: Freeman).

Miller, A. 1996. *Insights Of Genius: Imagery And Creativity In Science And Art* (New York, Copernicus).

Navin, F.P.D. 1994. Engineering Creativity Doctum Ingenium, *Canadian Journal Of Civil Engineering,* 21, pp 499–511.

Pace, Sydney, 2000. *Teaching Mechanical Design Principles On Engineering Foundation Courses*, International Journal of Mechanical Engineering Education, Jan, Vol. 28, Issue 1.

Pahl, G. & Beitz, W. 1996. *Engineering Design: A Systematic Approach, 2nd Ed* (London, Springer).

Poon Teng Fatt, J. 2000. Fostering Creativity In Education, *Education,* Summer, Vol. 120 Issue 4, pp 744–758.

Pugh, S. 1991. *Total Design: Integrated Methods For Successful Product Engineering* (London, Addison Wesley).

Ramirez, M.R. 1994. Meaningful Theory Of Creativity: Design As Knowledge: Implications For Engineering Design, *IEEE Frontiers In Education Conference,* pp 594–597.

Ray, M.S. 1985. *Elements of Engineering Design,* (UK, Prentice-Hall).

Rickards, T. 1990. *Creativity And Problem Solving At Work* (Aldershot, Gower).

Taylor, C.W. 1988. Various approaches to and definitions of creativity, in: Sternberg, R.J. (Ed.), *The Nature of Creativity: Contemporary Psychological Perspectives* (Cambridge, Cambridge University Press), pp 99 124.

Thompson, C. 1992. *What a great idea!* (New York: Harper Perennial).

Thompson, G. & Lordan, M. 1998. A review of creativity principles applied to engineering design, *Proc Inst. Mech. Engrs.* Vol. 213 Part E, pp 17–31.

Tornkvist, S. 1998. Creativity: Can It Be Taught? The Case Of Engineering Education, *European Journal Of Engineering Education,* March, Vol. 23, Issue 1.

Ullman, D.G. 1992. *The Mechanical Design Process* (New York, Mcgraw-Hill).

System-based Vision for Strategic and Creative Design, Bontempi (ed.)
© 2003 Swets & Zeitlinger, Lisse, ISBN 90 5809 599 1

A WWW-based learning framework for construction mediation

T.W. Yiu, S.O. Cheung, K.W. Cheung & C.H. Suen
Construction Dispute Resolution Research Unit (CDRRU), Department of Building and Construction, City University of Hong Kong

ABSTRACT: With increasing demand on teachers' time for teaching, administration and research, time available for interaction with students becomes limited. The use of web-based learning platform has gained remarkable advancement. This paper presents a comprehensive and integrated approach to the study of Construction Mediation, capitalizing on the strength of the World Wide Web. The design and contents of the training programme were based on the Mishra's (2002) online course framework which has been successfully applied in a number of Post-graduate Courses. The framework includes three main elements: Course Contents, Active Participation and Support. The Course Contents include materials which are essential for the study of Construction Mediation. Participation is achieved through on-line discussion forum among teachers and students. Support is effectuated with commonly asked questions and teacher's feedback. In this paper, the overall development of the web-based learning programme is presented. Future developments are also discussed in the paper.

1 INTRODUCTION

Conflict and disputes have been described as an endemic problem in the construction industry (Smith 1992). They may lead to delay of construction programme, loss in money, can be detrimental to business relationships. The use of mediation for resolving construction disputes is now prevalent in Hong Kong. It is largely accepted as part and parcel of the dispute resolution process (Lightburn 2000). This process not only facilities the settlement, but also changing individual's relationships to the problem and maintaining the business firm's public relations (Yuan 1999). In order to avoid the win-lose situation as in the case of litigation, mediation is firmly accepted as the first stage of dispute resolution proceedings. In Hong Kong, the provisions of mediation are incorporated in the Government, MTR and KCRC forms of contract. Thus, it is timely to equip professionals as well as students with mediation skills.

The University plays an important role on preparing students to attain a professional standard on resolving construction disputes. To successfully achieve it, mediation courses require constant monitoring and adjustments to the course design in order to cater for the change of development in mediation. Indeed, with the help of information technology, such goal can be achieved effectively. One of the goals of University is to provide a quality education and help them to develop more advanced and independent ways of learning. With this perspective on teaching, active participation on teaching is absolutely essential. To achieve this, Web-based learning tools are now used to inspire the students to explore the knowledge in effective and virtual way.

2 VIRTUAL LEARNING TOOLS: A WEB-BASED APPROACH

The idea of web-based learning tools is a solution to help lecturers to manage their classroom and maintain a quality teaching environments. The World Wide Web offers a platform for displaying course materials in addition to downloadable files, graphics and multimedia. It also encourage human-computer interaction ranging from navigation to dynamically tailoring the displayed materials to fit student profile (Kekkonen-Moneta 2002). In the view of University, such learning tools can also provide a more accessible and flexible learning for students. Students can be prepared to get into the information age by making use of such learning tools (UNS 2000). Flake (1996) advocated that the use of web-based learning tools and large organization facilitates sharing of ideas and information and help developing positive learning experience. Such learning tools can also enrich the style of presentation, and most importantly, provide a platform for self-taught, self directed learning (Yaverbaum et al. 1997) and can result in higher student performance (Schutte 1997, Zhang 1998, Wegner et al. 1999). With these perceived benefits, the popularity of using Information Technology on education will increase gradually. But, these online-teaching programme must cater for the rapidly changing technological and industrial environment. And, the quality of online learning should not be ignored as well. Due to the lack of basic design consideration, the online courses is simply used as a medium for the delivery instruction. Some of the studies on quality of online courses reveal that some on-line courses tend to be electronic version of the traditional print-based lectures (Dehoney & Reeves 1998). Hence, web-based tools is not just published the lecture materials to the Web, but also make use of the attributes and resources of the World Wide Web to create a meaningful learning environment for students. The application of multi-media and useful hyperlinks is a good example for such purpose. Furthermore, with the amount of information expanding at an ever-increasing rate, education programs must also cater for the rapidly changing technological and industrial environment. If so, the quality of online learning will significantly benefit the students. Indeed, researchers should spend a considerable time to work out a good quality of online learning environment.

3 ARCHITECTURE OF WEB-BASED LEARNING TOOLS

The rationale of the mediation web-based learning tool is based on the theoretical model developed by Villalba and Romiszowski (2000). Three schools of thought have been applied to guide the instructional practice: behaviourism, cognitive psychology and constructivism. While constructivism has been identified as the most suitable for the online learning environment (Hung 2001, Hung & Nichani 2001). Mishra (2002) illustrated the integration of the three schools of thought in graphical form, and is shown in Figure 1.

The design framework as shown in Figure 1 provides ample ideas for developing and devising web-based mediation learning tools. In the previous work of Mishra (2002), such framework has been applied in developing the online learning facilities for a six-month Post-Graduate Certificate in Management of Displacement. However, Mishra (2002) advocated that such framework provides a blend of all the best features of different approaches and expected it would also be applied in online delivery of other subjects.

4 AIM AND OBJECTIVES

The objective of this paper is to create a protocol of a web-based construction mediation learning platform

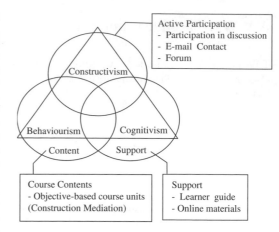

Figure 1. Design framework for web-based learning environments (Mishra 2002)

based on the design framework of Mishra (2002), and some tailor-made features for Construction Mediation training was also incorporated as well.

The development of such protocol was supervised by the Construction Dispute Resolution Research Unit (CDRRU) of the Department of Building and Construction, City University of Hong Kong. A recent development of CDRRU is the online information centre, called Construction Contract Information Service (CCIS), which act as a learning tool for Construction Contract Studies. In this paper, Construction Dispute Resolution Studies (CDRS), which is the further development of CCIS, is introduced. CDRS provides a web-based learning environment for Construction ADR Studies which focuses on Construction Mediation including the Mediation Cases, Various Mediation Rules, Reference Standard Documents, Procedural Matters and Mediation Technique/Tactics. All of such development is based on the mentioned key elements of the online course, Active Participation, Course Content and Support.

4.1 Active participation

In this learning tool, online discussion forum and e-mail facilities are equipped so as to encourage active participation. The attribute of Internet gives students a chance to share what they learn with others. They can also rise questions after lesson and discuss with their classmates or tutors at their will. Students can exchange their different views on some study topics. Hence, they can acquire more information than is often found in textbooks. Such an interactive approach is absolutely essential in the course of learning process and is the centre core of this Construction Mediation virtual learning environment.

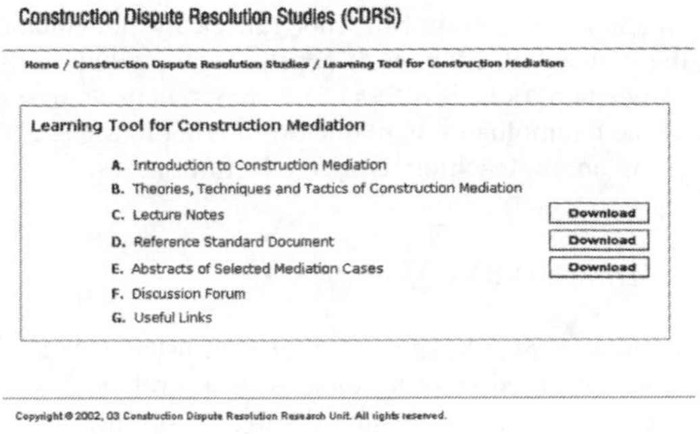

Figure 2.   Screen shot of the main page.

4.2 Support

Having a good support can reinforce the learning processes in an effective way. It is necessary for all stages of learning. Students can strengthen their theoretical concepts via the online mediation cases. And, they can understand the Mediation procedures used in practice. In the present of useful Hyperlinks, students can acquire the updated information regarding the development of Construction Mediation. Such useful links including the connection to the Mediation Group of Hong Kong International Arbitration Center (HKIAC) and some overseas Mediation Councils such as American Arbitration Association (AAA). Electronic Database is also incorporated as a support for students, they can search some contractual or mediation cases by simply inputting the key words of related topics.

4.3 Course contents

All learning processes started from the fundamental. The course contents including general introduction, definitions of some terminology and the general procedures of Construction Mediation. All of these materials were presented in an inter-active way. For example, the application of multi-media, illustrations and flow charts etc. Some standard documentation such as Mediation Rules, Mediation agreement, and Mediator Practice Guides were also incorporated in this sections which provide practical and theoretical supports in terms of preparing, managing, and assessing mediation activities. Hence, by browsing the CDRS learning environment, students are expected to acquire more information then in Lecture and enrich their knowledge in respect of Construction Mediation.

As this web-based learning tool was intended to target Quantity Surveying and Construction Contract students, in order to cater for the need of the students, an elevation has been conducted regarding the content, web-design, presentation method and multimedia prior to the construction of framework model, Figure 2 give the final design. Useful suggestions such as downloadable lecture notes, mediation procedures demonstration and free download of reference document etc. provided an intensive design scheme to this web-based tool.

Apart from the application of Mishra's online course principle, the content and design of this Construction Mediation learning tools was constructed based on the following principles:

1. User-friendly application
2. Secured access
3. Support teaching and learning of Construction Mediation
4. Tailor-made for students
5. Contribution to the industry , and
6. Keep in track with the Construction Industry.

Based on the above principles, four subject areas were defined: 1. Introduction to Construction Mediation; 2. Abstracts of Selected Mediation Cases; 3. Theory, Techniques and Tactics of Construction Mediation; 4. Reference Standard Documents. These subject areas were developed on considering the need of students, teachers and the industry. A brief description of each subject is given below:

4.3.1 Introduction to Construction Mediation

Students can extract the fundamental knowledge of Construction Mediation in this page. It included some general information such as the terminology, definitions and procedure, the nature of mediators etc. All these information was presented in flow charts, illustrations or even multi-media. These interactive presentations can simplify the complicated concept of construction mediation.

4.3.2 Theory, Techniques and Tactics of Construction Mediation

Mediation is a facilitated negotiation. Its techniques/tactics are closely related to some negotiation theories. Several typical negotiation theories such as intervention technique and hypothesis building, were introduced in this section. Furthermore, the application of such techniques/tactics can also demonstrate in the mediation cases which stored in the database of this learning platform.

4.3.3 Abstracts of Selected Mediation Cases

Many mediation theory, technique and tactics are built up due to the unique nature of each case. During the course of the mediation processes, students can establish the concept on the application of some theories regarding mediation. These mediation cases are stored as one of the branch of Hong Kong Construction Cases in CCIS. Students can access these cases by making use of the search engine incorporated in the web-page. Due to the confidentiality of Mediation, it is rather difficult to collect the Mediation Cases, thus, in order to demonstrate the application of some mediation

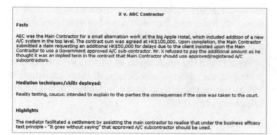

X v. ABC Contractor

Facts

ABC was the Main Contractor for a small alternation work at the big Apple Hotel, which included addition of a new A/C system in the top level. The contract sum was agreed at HK\$100,000. Upon completion, the Main Contractor submitted a claim requesting an additional HK\$50,000 for delays due to the client insisted upon the Main Contractor to use a Government approved A/C sub-contractor. Mr. X refused to pay the additional amount as he thought it was an implied term in the contract that Main Contractor should use approved/registered A/C subcontractors.

Mediation techniques/skills deployed:

Reality testing, caucus: intended to explain to the parties the consequences if the case was taken to the court.

Highlights

The mediator facilitated a settlement by assisting the main contractor to realise that under the business efficacy test principle - "It goes without saying" that approved A/C subcontractor should be used.

Figure 3. Screen shot of abstracts of an Hypothetical Mediation Cases.

theories and procedures in practice, some hypothetical mediation cases were published in this learning tool. A typical example of the abstracts of mediation cases is shown in Figure 3.

4.3.4 *Reference Standard Documents*

This part of the learning tool provides a database of useful information such as Mediation Rules (from HKIC) and (from AAA) and Draft Mediation Agreement, etc. All documents are made in PDF file for ease of download and browse.

5 OUTCOMES/PERCEIVED BENEFITS ON TEACHING AND LEARNING

The benefits of having the Web site can be summarized as follows:

5.1 *Dynamic learning environment*

The WWW-based e-learning platform enables interactive learning environment such as Simulations, models and visualization tools for students. With this learning tool, learning can occur in places where there is none, extends resources where there are few, expends the learning day, and open the learning place.

5.2 *Inspiration of learning to students*

The proposed e-learning platform facilitates the use of interactive multimedia where the students control the navigation of materials that incorporate test, sound, and moving and still images. It can stimulate Students' interest regarding the studies of Construction ADR. Students are positively motivated to learn, and help them to build-up their confidence towards this subject.

5.3 *Effectiveness learning*

The complicated ADR procedures, theories and legal concept are one of the difficulties that students need to overcome in their studies. This learning tool allows the presentation of such complicated information in

an inter-active way. Students can easily understand the underlying concept of construction mediation by using such tools. It offers more comprehensive usage of the technologies available on Internet to assist and complement teaching using the lecturer notes.

6 DISCUSSION

To allow the establishment of the Construction Mediation web-based learning tools, several problems need to be overcome. Firstly, in order to develop such learning tools, programming and graphical design are the two main fundamental skills. Thus, the works such as web-page design, graphic design, management of web server, etc. are conducted by the experienced webmaster. Secondly, due to the confidentiality of mediation, limited sources can be found for the value-added information for this learning tool. Thus, considerable time is needed to develop meditation cases. This may also be one of the obstacles to expand the database in the future. Thirdly, mediation involved many human-behavior, negotiation and psychology, hence, it increases the complexity of the design of the teaching materials. More illustrations are utilized in an attempt to simplify the concept of construction mediation.

7 FUTURE DIRECTION

The information in the learning tools shall be updated regularly in order to cater for the latest development of the Construction Mediation. Hence, it is necessary to update the mediation rules, cases, and useful hyperlinks with relevant sites.

One of the main advantages of Internet is the unlimited extended of information pools. The learning tools can be further developed to other dispute-related issues, such as Contractual Theory, Construction Negotiation and Arbitration.

Furthermore, the learning tools shall seek to benefit the professionals in the construction industry. One of the directions is towards the development of learning kit for candidates taking Professional Competence Assessment of professional institutions in construction. Candidates can utilize the on-line learning platform to learn contractual issues and ADR methods from the case-based construction claims cases. Furthermore, the updated database for construction dispute cases can also visualize the latest trend of dispute sources and ADR used in the industry. These statistics can be used as industry indicators.

It is anticipated that Information Technology is the main stream of development of construction industry. The use of World Wide Web as a media for learning/ teaching will become more and more popular. Many aspects in the building industry should also develop its own learning tools and seeks to link together via Internet so as to form an intensive and informative resource for the professionals.

8 CONCLUSION

A framework for the development of a Web-based Construction Mediation learning platform has been described in this paper. This learning tool is useful for all students in construction study. In addition, it is also useful to professionals in the construction industry. Learning is enhanced through the use of the Internet. Database technology is employed to provide an interactive learning environment. The list of subject areas included in the framework were developed from the Mishra's online course design approach and supported by students and professionals. Four subject areas were considered useful, namely, Introduction to Construction Mediation; Abstracts of Selected Mediation Cases; Theory, Techniques and Tactics of Construction Mediation and Reference Standard Documents. Through this learning platform, students can assess the updated mediation information relevant to their areas of study.

ACKNOWLEDGMENTS

The work described in this study was fully supported by a City University Grant (Project No. 7001469).

REFERENCES

Dehoney, J. & Reeves, T. 1998. Instructional and social dimensions of class web pages. *Journal of Computing in High Education* 10(2): 19–41.

Flake, J. 1996. The world wide web and education. *Computers in the school* 12(1/2): 89–100.

Hung, D. & Nichani, M. 2001. Constructivist and e-learning: balancing between the individual and social levels of cognition. *Educational Technology* 41(2): 40–44.

Hung, D. 2001. Design principles for web-based learning: implications for Vygotskian thought, *Educational Technology* 41(3): 33–41.

Kekkonen-Moeta, S. & B. Moneta. G. 2002. E-learning in Hong Kong: comparing learning outcomes in online multimedia and lecture versions of an introductory computer course. *British Journal of Educational Technology* 33(4): 423–433.

Lightburn, E. 2000. Mediation in international construction disputes. *The International Construction Law Review* 17(1): 202–206.

Mishra, S. 2002. A design framework for online learning environments. *British Journal of Education Technology* 33(4): 493–496.

Schutte, J. 1997. Virtual teaching in higher education: the new intellectual superhighway or just another traffic jam (available at http://www.csun.edu/sociology/virexp.htm).

Smith, M. 1992. Facing up to conflict in construction. *Proceedings of the first international conference on construction conflict: management and resolution.* Manchester, UMIST, 27–34.

The University of Sydney 2000. The University of Sydney News – 7 September 2000: Uni moves towards e-teaching (available at http://www.usyd.edu.au/publications/news).

Villalba, C. & Romiszowski, A. 2001. Current and ideal practice in designing, developing and delivering web-based training. In Khan, B. (ed), *Web-based Training ESP*: 325–342. Englewood Cliff, NJ.

Wegner, S., Holloway, K. & Garton, E. 1999. The effects of internet-based instruction on student learning. *Journal of Asynchronous Learning networks* 3(2) (available at http://www.aln.org/alnweb/journal/vol3_issue2/wegner.htm)

Yaverbaum, G., Kulkarni, M. & Wood, C. 1997. Multimedia project: an exploratory study of student perceptions regarding interest, organization, and clarity. *Journal of educational Multimedia and Hypermedia* 6(2): 139–154.

Yuan, L.L., 1999. The use of meditation for promoting good customer relationship. *Asian Dispute Review* 3: 39–41.

Zhang, P. 1998. A case study of technology used in distance learning. *Journal of Research on Technology in education* 30(4): 398–419.

System-based Vision for Strategic and Creative Design, Bontempi (ed.)
© 2003 Swets & Zeitlinger, Lisse, ISBN 90 5809 599 1

Moving the university to the industry – a successful case study

C. Cosma & Z.J. Herbsman
Department of Civil Engineering, University of Florida, Gainesville, Fl., USA

J.D. Mitrani
Department of Construction Management, Florida International University, Miami, Fl., USA

ABSTRACT: The Army Corps of Engineers (ACE) in the Jacksonville district approached the Department of Civil Engineering at University of Florida (UF) about the possibility to offer a training (Master) program for those employees seeking career development through continuing their education. The paper will discuss the administrative and financial aspects of such a program, from both the university and organization's point of view. The paper will also analyze the conclusions drawn from seven years of conducting this program and will discuss some problems that had to be overcome.

1 INTRODUCTION

A report prepared by the Center for the Study of Human Resources (CSHR) of the Lyndon B. Johnson School of Public Affairs at the University of Texas-Austin for the Texas State Occupational Information Coordinating Committee (SOICC) concluded that the educational requirements for employment appear to be changing. Today higher degree expectations seem to be the norm for the majority of occupations. Employers commonly provide continuous training opportunities on the job to keep their workforce viable and/or maintain licensure. Such training is not associated with promotions; employees themselves fund more academic pursuits that accelerate career advancement. The same report stated that seniority plays virtually no role in promotions; career advancement is associated mainly with ability, experience and education. Continuous education and training, as well as higher education, are crucial components of success in today's labor markets (O'Shea 1999).

Looking at the case of the engineering management profession, a certain particularity can be noticed. From the engineers that graduate with a Bachelor's degree, a few years later, after accumulating practical experience, a significant percentage will express the desire to continue their education towards an advance degree (mainly a Master's degree) in their area of specialization. These engineers have to face many challenges in order to continue their education. One of the main challenges is that they must coordinate the different areas of their lives, which influence each other – their families, full time jobs, spare time, and studies. For this specific segment of prospective students a suitable graduate program has to be made available locally during evenings or weekends. This situation does not show up too often and the most frequently offered option for these engineers is to become full time students at a university for a period of 1–2 years. This option is very expensive and in most cases unfeasible for financial and logistical reasons.

One of the keys to this problem is to "move" the university to the industry by offering an "in-house" graduate degree. This solution will offer the advantage that engineers will be able to keep their full time jobs while pursuing a higher degree at the same time. At their turn the organizations offering convenient access to advanced education, will highly benefit from recruiting top people, as well as motivating existing employees.

The paper will describe such a program that was developed at University of Florida and is offered since 1995 to the Army Corps of Engineers in the Jacksonville district. There are 35 students (army professionals) enrolled in this program and the authors consider it a very successful one.

The program was using three different lecturing tools:

- Professors traveling to organization's location to teach graduate classes there;
- Graduate courses offered on videotape (FEEDS);
- On-line courses (Internet).

The conclusions drawn from almost eight years of conducting this program and some of the problems that had to be overcome will be discussed in this paper. The authors believe that such a program can be developed for almost any midsize to large organization. Steering the experience in outsourcing from this program will enable other universities to develop similar programs in their area. Some administrative and financial aspects of such a program, from both the university and organization's point of view will be also discussed.

2 DELIVERING EDUCATION

In order to provide increased educational opportunities without increased budgets, many educational institutions have developed distance education programs. Distance education (DE) covers a wide range of knowledge acquiring approaches and techniques. The one basic characteristic that these techniques have in common is that the learner and teacher are separated by physical distance. In order to bridge the instructional gap technology in combination with face-to-face communication is used. From a systemic point of view distance education can be defined as being both a system and a process, which connects learners to distributed sources of knowledge.

Acquiring knowledge through learning is a very challenging task for the students, which requires motivation, planning, and the ability to analyze and apply the information being taught. But when knowledge is delivered at a distance, additional challenges result for the DE students. One of the main challenges is the fact that they must coordinate the different areas of their lives: family, jobs, spare time, and studies (Willis 1991). The isolation of the student involved in a distance education program, the lack of motivational factors that usually arise from direct contact or competition with peers is another aspect that has to be considered. Another particularity is that students' needs and difficulties that appear during the process of studying lack the teacher's immediate support in a DE system. Also in a DE system it takes a much longer time to develop an effective and efficient learner-lecturer relationship. Last but not least developing knowledge transfer channels in a DE system relies heavily on a wide range of technological options available to the educator. In his book, "Distance Education – A Practical Guide", Dr. Barry Willis, Associate Dean for Outreach classifies the technological options available to the DE lecturer into four major categories: Voice, Video, Data, and Print. Trying to find the answer to the question "which technology is best", Dr. Willis suggests a systematic approach, consisting in a mix of media, each serving a specific purpose.

The two main categories of distance education delivery systems are: synchronous and asynchronous.

Synchronous system requires the simultaneous participation of all students and instructors. The main advantage of this system is that the interaction among the key players of the system (teachers and students) is done in "real time". Asynchronous instruction does not require the simultaneous participation of students and instructors. Rather, students may choose their own instructional time frame according to their schedules. The principal advantage of asynchronous delivery system is obviously the student choice of location and time and the interaction opportunities for all students, in the case of telecommunications (i.e. e-mail) (Steiner 1997).

Most of the distance education programs share similar characteristics:

- Provide credit for prior learning,
- Offer open entry/exit,
- Offer various administrative options,
- Usually offer open/liberal time for course completion.

Educators have been asking if DE students learn as much as students in the traditional face-to-face instruction. Research indicated that teaching and learning at a distance could be as effective as traditional instruction system when the method and technologies used are appropriated for the instructional tasks, when there is interaction among students, and when there is feedback from teacher to students in a timely manner (Verduin 1991).

3 THE CASE STUDY

3.1 Program assessment

The U.S. Army Corps of Engineers Jacksonville District's 232 person Engineering Division has a large, diversified and complex civil works program. Major engineering features in their $59 million program include flood control, navigation, beach erosion, concrete, rock and earth-fill dams, jetties, levees, dikes, bridges and environmental restoration engineering. The Division utilizes professionals with expertise in hydrology and hydraulics, soil engineering, geology, surveying and mapping, cost engineering, structural, mechanical, electrical, and civil engineering and architecture. Other Corps' districts or private sector Architect Engineering firms operating under the direction of the Division's technical staff accomplish approximately 50% of the Division's work. High technology computer equipment is used to develop complex models for conceptual planning, analysis, visualization and detailed design.

Management recognized that education is a critical variable for this professional engineer based organization. An assessment revealed the following:

- Lack of advanced degrees or course work by many of the current staff.

- Lack of graduate training opportunities in engineering and engineering management in the Jacksonville area.
- Concern over the expense of sending students out of the city for graduate courses.
- Unwillingness of the students to take all or most of their courses by video tape distance learning.
- Potential recruits increasingly expressing the desire to locate where they can continue their education.
- Existing staff's strong desire for graduate training indicating such a program would serve as a retention tool.
- Recognition that many of the people that we deal with from outside the Corps have graduate degrees.

Management had a strong desire for the local courses to become part of a larger continuous learning initiative utilizing multiple forms of training. The University graduate courses offered both locally and by long distance complemented training offered by the Corps and available short courses. These training opportunities were incorporated into a training plan for upgrading each individual's skills. Where possible the University professors would spend periods of time working at the Corps. This had a dual benefit by providing the instructor the opportunity to work on real world problems and also providing the student with an opportunity to interact with the professor in applying higher-level considerations to their regular problem solving.

Management was pleased with the graduate training that people have received as well as the interest in all advanced training that this program has helped foster. Four students received their Masters Degree and several others are moving in that direction. There is a high level of interest in learning and over 60 students explore participating in the after work graduate classes. After much effort by all, the in-house degree program has more than met the expectations.

3.2 *The program*

The Master's degree offered by University of Florida is a regular Civil Engineering Master's degree, which includes 32 credit hours of which 12 credits have to be in the major area of engineering management. The rest of the credits can be from various areas of civil engineering and related management areas. All students have to submit a Master's report before the completion of their studies. As of the end of 1997, twenty students participated in the program. As of now four students were able to complete their degree.

The second cycle of the Army employees pursuing their Master's degree started in 1998 and 10–15 new students joined the program. This experiment has gotten a very positive reputation around Florida, and other organizations like the Department of Transportation, public works organizations (Sarasota County) are trying to participate in similar programs. In Table 1 is

Table 1. Typical Master's program without thesis.

Civil Engineering Management

Technical Management

CGN 6155	Civil Engineering Practice 1	2
CGN 6156	Civil Engineering Practice 2	2
CCE 6037	Civil Engineering Operations 1	2
CCE 5035	Construction planning and Scheduling	2
CGN 5125	Legal aspects of Civil Engineering	3
CGN 5135	Value Engineering	3
CGN 5115	Civil Engineering Feasibility Analysis	3
CGN 6974	Master of Engineering	2

Technical Engineering

CEG 6015	Advanced Soil Mechanics	3
CEG 6125	Soil stabilization	2
TTE 5835	Pavement design	2

Management Related

MAN 3021	Principles of Management	3
MAN 3151	Organizational Behavior	3
MAN 4310	Problems in Personnel Management	4
Total Required Credits		32

presented a typical Master's program of 32 credits. Table 2 presents an example of course outline.

When this program was established, the organizers had to consider several cost factors for the entire system. The cost of technology to be used was one of the main ones considered, cost of knowledge transmission, maintenance costs for equipment, sender-receiver network costs, cost of supporting personnel, cost of adapting the teaching materials, and other miscellaneous expenses needed for the success of the DE system. The most difficult part of this feasibility study was to attach a dollar amount to the benefits offered by the DE system. The organizers considered that facilitating accessibility of mature students to training, assuring the continuity of DE students at their job, and exposing students to the expertise of the most qualified faculty, are major benefits to the learners that justify the development of such a program.

4 LECTURING TOOLS

The lecturing tools used in the Master's program are:

- Distance traveling,
- Video courses (FEEDS) and
- Internet courses.

Their advantages and disadvantages will be discussed further on.

4.1 *Distance traveling*

Some professors traveled to Jacksonville to lecture in-house graduate courses. The distance (73–74 miles driving distance) created some administrative

Table 2. Example of course outline.

UNIVERSITY OF FLORIDA
Department of Civil Engineering

Course Syllabus
FALL 2002

I. Course Number and Title
CCE 5035 Construction Planning and Scheduling
Thursday 2–3 period, 210 Weil Hall
Occasionally on Tuesday 2–3 period, 210 Weil Hall

II. Course Description
Planning, scheduling, organizing, and control of civil
engineering projects with CPM and PERT. Application of
optimization techniques

III. Professor and Office Hours
Dr. Zohar Herbsman
522 Nuclear Science Building (NSB)
   T & R 10:30 a.m. – 12:00 p.m. (Other times may be
arranged by appointment)

IV. General Requirements
Students are expected to complete all class assignments prior
to class. Reading assignments should be completed before
class.
Homework problems and written reports are due at the
beginning of the period on the due date. Work not turned in
by the time the class starts will be counted as late. Late work
is eligible for fi credit maximum. No assignment will be
accepted after one week of its due date.

V. Grading
Tests   60%
Homework   10%
Term project I   15%
Term project II 15%

VI. Required Text
Hinze, Jimmie W., Construction Planning and Scheduling,
Prentice Hall, New Jersey, 1998.

VII. Course Web Page http://www.ce.ufl.edu/~zherb/

problems, but with the use of e-mail, phone, and fax,
most problems have been solved. This method was
the best one regarding interaction between the professor and students. However, this method is very expensive and time consuming. The following figures will
demonstrate the cost dilemma.

Total cost of teaching 2–3 credit hours was $15,000
lecturer's fee included (over the regular salary), travel,
per diem and administrative cost. When the course was
a requirement, 25–30 students participated in the course
and the cost per student was around $500/student per
course, compared to regular fees of $400/student. This
difference was acceptable for the Army.

However, when classes were elective only 10–15
students participated and the cost per student was
very high. The second problem was that most of the
professors consider this method to be very exhausting
(it is a full day of work) and only a few were ready to
continue doing it in the future.

4.2 FEEDS (Florida Engineering Education Delivery Systems)

For many years University of Florida has offered
courses that were video taped in specially equipped
studios in Gainesville. The videocassettes were later
sent by mail to various locations, enabling students to
watch the video at home or at work and participate in
the class. Assigned homework was sent to the lecturer
by mail. For tests special security systems were used
while they were organized in an in-house manner.

Economically this system is very efficient. The cost
for the students was like the regular course tuition.
Because of the extra efforts of the lecturer there has
been university discussion to increase such course
tuition by 25% and to give to the lecturer some monetary incentive.

The major disadvantage of this method is the lack
of any direct interaction professor–students. Also
specialized equipment, facilities and staffing were
required. Both students and professors admitted that this
delivery system is less effective than regular teaching.

Videoconferencing method was also used, where
the lecturer was teaching his regular courses for the
full time students at Gainesville. The course was being
taught in the distance education studio. The transfer
of the lecture to the Army studio in Jacksonville,
Florida was done by using telephone connections.
The Army studio included a few video monitors and
video cameras. When any participant in Jacksonville
had a question or a remark he/she called the studio in
Gainesville establishing an audio-visual communication. There were some delays in communication time
but after some early technical problems, the system was
working very well. The major advantage with this
method is the interaction between the professor and
the students. In the future, this method will become
the major tool for our distance education program.
Economically, it is not expensive when using telephone lines. The cost of the 2 studios is substantial
but it can be distributed over many courses and for
many years. The major disadvantage of this tool is
that the professors have to get used to the new media.
More training and more preparation time are needed.

After testing both techniques, the organizers concluded that it is more effective to videotape the lecture
at the source (university) and to send the tape by fast
mail to the students in Jacksonville. The students got
the videotape the next day. The tapes from the source
were of a much better quality than the recording done
in the Jacksonville studio.

4.3 On-line courses

A few professors started developing courses on
Internet (Fig. 1). This is in an experimental stage and
the idea is that all the course material will be offered
on the Internet.

Figure 1. An Internet Course Offered by University of Florida.

This delivery system has unquestionable advantages if the right type of course is chosen combined with a properly designed home page, that will facilitate and encourage the free thinking, discussion and dynamic participation of the DE student. These are difficult challenges for the Internet course designer. Courses like estimating, scheduling, legal aspects, value engineering, etc. should be developed first. Authors consider that starting with a certain type of course that is easily to fit to this type of media would be the proper approach to take. The major disadvantage of Internet courses is that substantial resources (time, money) are needed in order to develop a good course of 2–3 credits. From the authors' experience it takes at least a year of hard work to develop a course using Internet (Herbsman 1998).

5 CONCLUSIONS AND RECOMMENDATIONS

From the almost eight years of directing the program the major conclusion drawn by the organizers of the DE program was that distance education reach the effectiveness of conventional education as long as a systemic approach is taken when selecting the best delivery system. Organizers, educators, students and supporting staff involved in the DE system have to cooperate in order to find the most appropriate means for knowledge transfer. A focus on meeting the clients' (DE students) needs and requirements is absolutely necessary if the best performance is to be achieved. For the DE deliverer, a benefit/cost analysis properly conducted will offer the answer to many questions.

Face-to-face contact should not be completely abandoned in a DE system. It is essential to have some mechanism that facilitates student–professor direct interaction. Organizing periodical meetings, tests, or any other arrangements would be a solution.

The many practical issues that still have to be addressed when using distance education differ from one university system to another.

The questions that cropped up from University of Florida's experience relate to the following:

- the system to be used to compensate the lecturers involved in DE,
- the location restrictions that might arise, degree recognition,
- tests and homework's security and copyright issues, and
- performance evaluation comparison of DE system versus traditional education system.

Distance education will definitely become a major part of the graduate programs in construction management in the future. The main reason for the increasing use of distance education is the need of the engineering profession for advanced degrees and the cost/benefit of this method.

REFERENCES

Herbsman, Z.J. & Elias, A.M. 1998. Distance Learning – The Solution for Graduate Programs in Engineering Management – A Case Study. *International Conference on Engineering Education ICEE1998*, August 17–20, Rio de Janeiro.

Markowitz, H.Jr. 1990. Distance Education: A Staff Handbook. *University of Illinois at Champaign-Urbana.*

O'Shea, D.P., Betsinger, A.M. & King, C.T. 1999. Successful Career Progression: Exploratory Findings from a Study of Selected Occupations Executive Summary *Ray Marshall Center for the Study of Human Resources,* Lyndon B. Johnson School of Public Affairs, The University of Texas at Austin.

Steiner, V. 1997. What is Distance Education? *The Distance Learning Resource Network (DLRN).* http://www.fwl.org/distance.html

Verduin, J.R. & Clark, T.A. 1991. *Distance Education: The Foundations of Effective Practice.* San Francisco, CA: Jossey-Bass Publisher.

Willis, B. 1991. Distance education at a glance. *Engineering Outreach.* EO Home. University of Idaho. http://www.uidaho.edu/eo/distglan.html

System-based Vision for Strategic and Creative Design, Bontempi (ed.)
© 2003 Swets & Zeitlinger, Lisse, ISBN 90 5809 599 1

InfoBase: a multimedia learning environment to support group work and discourse

R. Stouffs, B. Tunçer & I.S. Sariyildiz

Faculty of Architecture, Delft University of Technology, Delft, the Netherlands

ABSTRACT: We are developing a multimedia learning environment to support group work and discourse. It aims to offer a student the means and tools to organize his or her learning activities in cooperation with others. The functional focus is on information and document management, presentation and publication, communication and discourse, and cooperation and group work. The design of this environment is considered in relationship to an educational process in which the student becomes familiar with the use of ICT for supporting communication and cooperation. In this process, we are targeting four different kinds of uses: as a digital portfolio, as a cooperative database, as a digital logbook, and as a work environment to support discourse and cooperation. We believe that such a system plays a vital role in the application of e-learning to the design context, and that it will facilitate the initiation of networks or working groups that serve as virtual centers of knowledge and experimentation.

1 THE INFOBASE PROJECT

In this paper, we report on an active, university-funded, educational project that aims to employ ICT means to support educational processes in which students are encouraged to learn from one another and work together, at their own initiative (InfoBase 2003, Stouffs et al. 2002). We distinguish four main activities necessary within this project. These are: the design and development of a multimedia learning environment to support group work and discourse; the development of learning processes and activities that use the functionalities of this environment; the embedding of these processes and activities in the educational program; and the evaluation of this embedding towards the future development of the educational process.

We are developing a multimedia learning environment that offers the students the means and tools to create and develop structures for information management, presentation, communication, and cooperation during the learning process. This should allow students, at their own initiative or under the guidance of an instructor, to organize their learning activities and to cooperate amongst themselves and with their instructors. The work processes are hereby at least as important as the information results. Integrating these processes into the didactic approach central to the curriculum review will require both the adaptation of existing learning methods and processes as well as the development of new methods and processes. Our aim is to initiate networks or working groups, within an integration of education and research, that serve as virtual centers of knowledge and experimentation on specific themes, in order to advance both education and research.

2 THE DIDACTIC CONCEPT

We are experimenting with a didactic model whose goal is to be a container for meta-data that can be used as a design backbone. This model is based on four characteristics that are important with respect to social processes: constructive, objective, relational and subjective characteristics (Kooistra & Hopstaken 2002, Stouffs et al. 2003). The space defined by these characteristics can be said to define a domain, a theme, or a culture (Fig. 1). When we apply these characteristics in the context of architecture, we can make architectural analogies about each characteristic. The objective characteristic defines the functionalities or program of a building. The constructive characteristics define the idea or building type. The subjective characteristics define one's reason to go to a building. The relational characteristic defines the energy that is needed to make this building continue to exist. Physical architecture is mostly concerned with the constructive and objective characteristics. However, it is very important for an architecture student to be fully aware of the subjective and relational characteristics.

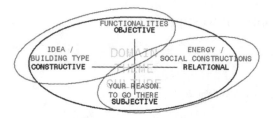

Figure 1. Diagram denoting the space formed by four characteristics of socio-cultural qualities, and their inter pretation for architecture.

These subjective and relational qualities are depended upon the community in which the students operate. These qualities are not only defined within the social processes in the community but also influence these processes. By teaching students how to express and utilize these qualities in their communication with peers, they can learn to become more effective in their use of information from various sources to support such communication. In the InfoBase project, we consider the expression of these qualities through metadata in various forms, such as ratings, semantics structures of keywords, and different types of links between information. We consider a methodology for adding, utilizing and managing metadata embedded in an educational path with increasing responsibility to the student: programmed instruction, networking exercises, and community formation.

The first step is linked to a digital portfolio, in which students present and publish design abstractions to others. The construction of a digital portfolio forms an integral part of the student's learning activities in the first year of the B.Sc. program. The students are encouraged to maintain and update their digital portfolio throughout the remainder of their education. For this purpose, the digital portfolio is conceived as a public portion of a student's electronic workspace. The public is offered various means of providing feedback to the author, with an emphasis on subjective qualities, such that the author can learn to consider the viewpoints of others in the organization and presentation of design information.

The second step involves a cooperative database that serves the management and presentation of design analysis information and construction component and material information on a collective basis. Here, students must consider their place and contribution in a larger process of collective information gathering and presentation. Students are presented with an initial organizational structure that should assist them in presenting their own information as part of a larger collection of information. The emphasis is here on the relational qualities. The construction of a cooperative database serves the design process in the second year (fourth semester) of the B.Sc. program.

In the third step, the aim is to support the formation of networks or working groups through communication and discourse. In the third year (fifth semester) of the B.Sc. program, the process of communication and information sharing supports the design and renovation of an urban site where the different buildings constitute the design subjects of different students or groups. In the project phase (second year) of the M.Sc. program, the process of communication and discourse supports the relating of various design and research projects under a common research theme, in order to advance both education and research.

3   THE INFOBASE ENVIRONMENT

The development of this environment continues from the earlier development of an educational electronic document management system and profits from continuing research into such systems. In the development, we distinguish the basic infrastructure, authoring tools, application-specific interfaces, and its integration with various information sources. These information sources include the university's electronic library, digital image archives developed within the Faculty of Architecture, and course descriptions and other information available within Blackboard®, the digital educational environment selected by the university as a central educational repository and learning environment.

Currently, the InfoBase environment is built around a client-server architecture, using a MySQL database for persistent storage, the PHP scripting language for the implementation of all server-side functionalities, HTML and JavaScript for basic client-side interfacing, XML and related developments for advanced interfaces and visualizations, and Java for additional client-side authoring tools. We intend to redevelop the InfoBase environment based on a commercial learning content managament system.

4   A DESIGN ANALYSIS APPLICATION

In the fourth semester, the students use the InfoBase environment as a cooperative database for design analysis in the context of a design studio. The central design theme of this studio is a "small public building", either a small theater or a museum. It is offered twice a year to about 350 students in total. The students are given a relatively complex functional program and are required to design and work out the materialization of this public building.

The students begin the studio by analyzing selected precedents (historical and contemporary) of the relevant building type with respect to various criteria (composition, program, construction, context, type, etc.) and from structural, formal, and functional points of view. Documentation of these precedents is presented to the students in the form of drawings, pictures, and texts, within the InfoBase environment. The analysis is performed and presented within the same environment. For this purpose, the students are provided within this environment with a number of tools. The result is a common library such that students, in later design activities, can draw upon other students' results for comparisons and relationships between different aspects or buildings.

We are developing tools to create a keyword hierarchy and view it as a semantic map, and to decompose documents and relate these with keywords. The students will be provided with a keyword hierarchy corresponding to a system of architectural types as a structure to hook up their contributions. In general, and depending on their knowledge of the domain, students can collaboratively define or extend this structure. We are also developing tools to draw colored areas, section markers and view markers on plans, sections and elevations, to link these areas and markers to other documents or relate these to keywords, and then to generate web pages from these, as entry pages to analyses. The user interface provides views for individual documents and all their related documents at one glance, and visual overviews of the entire document and keyword structures and their links.

4.1 Prototype

We developed a first prototype for the presentation of architectural analyses on the web in order to illustrate the presented techniques (Tunçer et al. 2001a), using Ottoman Mosques as a case study. The interface allows the user to view both the keyword and document hierarchies and their relationships. These views include both in-world and out-world views (Papanikolaou & Tunçer 1999). An in-world view presents a document (or keyword) together with its immediate neighbors within the hierarchy, and displays all other documents that share a keyword with it (Fig. 2). Such a view allows one to browse the structure, interpret the relationships, and as such lead to interesting out-world views. Out-world views offer an overview of (a part of) the information structure including all its relationships. Such views may be presented as structural maps, providing visual feedback to the users on their traversals and selected views by presenting the location of the currently viewed node within the hierarchy. Such views also give an overview of the scope and depth of the semantic structure guiding the analysis. Figure 3 presents some exemplar out-world views as clickable maps that offer an overview of the entire keyword hierarchy in relationship to the related documents.

While the keywords serve for the most part as binding elements in the structure providing relationships between the documents, when traversing the

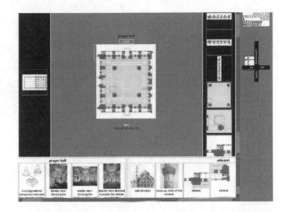

Figure 2. A snapshot of an in-world view from the prototype application. The middle frame contains the currently viewed document. The left frame contains its "parent" document, and the right frame contains its components. The bottom frame contains the document's keywords and all other documents related to these keywords.

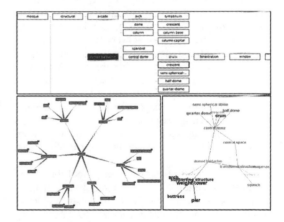

Figure 3. Three snapshots from out-world views of the keyword hierarchy. The focus of this figure is on the graphical representation of the structures, not on the keywords themselves. Top, 2D list view; left, 2D dynamic tree view; right, 3D dynamic network view.

information structure, the content as available in these documents is the most important aspect. As such, while the document's keyword, and its location in the keyword hierarchy, may be presented as properties of the document, the relationships are specified primarily as document-to-document relationships. This not only ensures that links are presented as shortly as possible, facilitating a swift traversal, but also shifts the focus onto the content, rather than the structure that surrounds it. Keywords further serve a role as index to the information structure.

1753

Figure 4. An image decomposition: left, snapshot from the image decomposition tool with two rectangular areas marking two image components; right, an image decomposition hierarchy.

4.2 Tools

Based on the results of this prototype, we are currently developing various tools for the construction and presentation of a body of architectural analyses in the context of a design studio. Specifically, we are developing and implementing a number of tools in order to facilitate the development of keyword structures and the decomposition of images and texts, and to construct image maps that can serve as guides into parts of the information space. A first tool serves to create the keyword structure, and view it as a semantic map (Fig. 3). This tool extends on an existing freeware application for building up and viewing network structures. Another tool assists the user in the decomposition of image abstractions. Image abstractions are decomposed by selecting rectangular areas from the images (Fig. 4), selecting sets of keywords from the keyword hierarchy (Fig. 3), and attaching these to the image components.

The same application also offers a tool for adding hotlinks to images, allowing for the development of image maps that can serve as a content map or index to a collection of related documents. The base image may for example constitute a plan of a building; markers can then be positioned on the image and related to the appropriate documents. From this information, a web page is generated containing the respective image map (Fig. 5). When one moves the mouse pointer over a marker, a preview image of the related document appears. Markers can be clicked to browse to the respective document. Currently we provide for section markers, indicating where on a plan a section is taken, and in which direction (Fig. 5a), and view markers, defining where a picture or an elevation is taken in relation to the plan (Fig. 5b). We also intend to implement markers and hotlinks for annotations, and panoramas, and to distinguish markers for elevations and images. Within the same application, another tool serves to draw areas on plans, sections, and elevations, color them, and link them to the appropriate keywords. This composition will be generated as a new collage. When

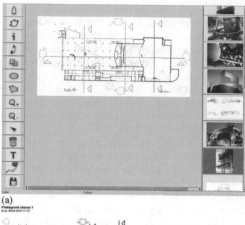

(a)

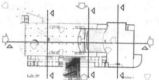

(b)

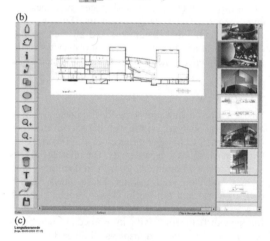

(c)

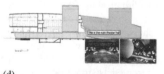

(d)

Figure 5. Toolkit which is used to generate web pages containing image maps that serve as a content map or index to a collection of related documents: (a) toolkit; (b) resulting image map with section, elevation, and view (photo) markers; (c) toolkit; (d) image map with area markers and annotations.

moving the mouse over an area, the related keyword will be shown. Clicking the area will bring up related documents that share the same keyword. These areas will serve in most cases as functional zone markers.

5 THE MEDIATED DISCOURSE EXPERIMENT

In the context of a design studio, students benefit from one another by sharing design information and by providing feedback on each other's design. Such a process can be strengthened, especially during the design conception phase, by facilitating a discourse among students, faculty, and other interested parties. By adopting an electronic environment to support this discourse, both the result and the process can be captured and presented to the participants during and after the discourse. We're currently running an elective course and educational project entitled "Mediated discourse as a form of architectonic intervention," which, in the context of the InfoBase project, serves as a testing ground for the exploration and evaluation for various ideas regarding information, communication and discourse, both in terms of functionality and interaction. The goal of this exploration is to evaluate the strengths of such an environment in supporting a discourse between students, teaching faculty and researchers in the context of a design research laboratory in which students execute individual MSc projects under a common research theme. These ideas are gathered from past experience (Tunçer et al. 2001b), a workshop with students as current and future users, and insights into information characteristics that are important with respect to social processes (see above).

The course objective is the design and development of a multidisciplinary discourse, with a relationship to architecture and the built environment, based in various media, and held between students, faculty, artists, and the public. The theme of the discourse, "cultural transformation," is selected in relation to an urban transformation project in the city of Rotterdam: ParkStad (ParkStad Rotterdam 2003). Specifically, the identity of ParkStad and its relation to the surrounding neighborhoods is under investigation in this discourse. The selection of an actual transformation project anchors the discussion in reality and offers the students both a concrete problem and a public. The selection of the theme "cultural transformation" provides the discourse with a social dimension that is actual and of broad interest.

The elective course is open to sixth semester BSc students and MSc students. Students participate in small groups in two workshops. During the second quarter of the semester, each parallel workshop will be guided both by a faculty member and an artist. The faculty member chooses the medium in which the students develop their contribution. Together, faculty member and artist select a problem statement in relation to the discourse theme to serve as a starting point for the students' investigation.

The InfoBase environment serves in this course as an e-learning environment (BMVK05 Web Environment 2003) in which participants can access information related to the ParkStad project and results from other events accompanying the course, log their own activities, access others' contributions, and provide feedback in various ways. As such, the environment provides immediate access to a current reproduction of the discourse, and serves to broaden the discourse by allowing anybody to participate by relating, rating, and commenting on contributions.

In the "Mediated discourse" web environment, participants are offered the ability to link contributions, either in reaction to a previous contribution or in relationship to others, to rate contributions according to four different and independent aspects, to comment on contributions and their earlier comments, and to associate keywords to contributions. These keywords are organized into a semantic structure in the form of a graph. Using these various metadata, three different views are presented to the user. The first offers the user a list of various guides, where a guide presents a selection of data based on and sorted according specific metadata criteria (Fig. 6). Examples of such criteria are the best or worst rated contributions, contributions having the most links from or to them, and the most or most recently commented contributions. The second view offers the user an overview of the semantic structure of keywords and enables the user to search the data space according to these keywords. The third view offers the user an overview of the information structure composed by the individual contributions and their reaction links (Fig. 7). The user can browse all of these structures.

6 CONCLUSION

Technological advances enable practitioners and students to make the design process more information-intensive, both in their own activities and in collaboration with others. For this purpose, it is important that students familiarize themselves with such technology, including database access, archiving functionalities, portfolio presentations, and a variety of communication tools, and adopt it in ways that meet their needs and requirements. We envision this technology to become commonplace in educational environments, extending the current set of electronic information and communication tools available to students.

Therefore, we aim to develop a flexible environment that provides students with the tools and means to adapt and apply this technology throughout the curriculum, supported by course specific e-learning offerings. For this purpose, these tools should be designed and developed so as to stimulate the student to explore the respective field of study independently of the instructor and to cooperate with others in

Figure 6. Screenshot of the Mediated discourse web environment illustrating the guides view.

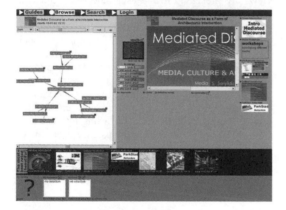

Figure 7. Screenshot of the Mediated discourse web environment illustrating the browse view.

achieving solutions to a problem. As such, the objective of the environment is to support educational processes that motivate students to learn from one another and collaborate with each other.

Applications of this functionality are the submission and management of digital products in the context of digital lab sessions, information and document management in the context of a design studio with the purpose of creating a digital log corresponding to the design process, the creation of a digital portfolio by the students, and the presentation of architectural analyses in the context of an extensible digital library on historic precedents related to the theme of a design studio.

The final aim of the InfoBase development is to support the formation of networks or working groups, within an integration of education and research, leading to the prospective establishment of virtual knowledge centers. While strengthening the scientific character of the education, integrating education and research in this way also enables the students' educational activities and the results thereof to contribute to the research. Constructing such virtual knowledge centers requires both an informational and a social, interactive basis. The development of a multimedia learning environment to support group work and discourse can assist with respect to both aspects.

ACKNOWLEDGEMENTS

The authors would like to thank everybody who contributed, in the past and presently, to the InfoBase project and the Mediated discourse elective course. Among others, we would like to thank Jan Kooistra for his invaluable input with respect to the didactic concept, Henry Kiksen for the implementation of the analysis tool, and Maia Engeli for her role in the conception of the elective course. The authors would also like to acknowledge the role of Ernst Janssen Groesbeek in the development of the ICT curriculum at the Faculty of Architecture.

REFERENCES

BMVK05 Web Environment. 2003. bk-info.bk.tudelft.nl/ INFOBASE20020101/elective/bkmvk05new/.

InfoBase. 2003. www.bk.tudelft.nl/infobase/.

Kooistra, J. & Hopstaken, K. 2002. Ommat: dealing with electronic scientific information. In Peter Brophy, Shelagh Fisher, and Zoë Clarke (eds), *Libraries Without Walls 4: The Delivery of Library Services to Distant Users*: 241–250. London: Facet Publishing.

Papanikolaou, M. & Tunçer, B. 1999. The fake space experience – exploring new spaces. In A. Brown, M. Knight and P. Berridge (eds), *Architectural Computing: from Turing to 2000*: 395–402. UK: eCAADe and The University of Liverpool.

ParkStad Rotterdam. 2003. www.parkstad.rotterdam.nl/.

Stouffs, R., Tunçer, B., Venne, R.F. & Sariyildiz, I.S. 2002. InfoBase: a multimedia learning environment to support group work. In D. Rebolj (ed.), *Construction Information Technology in Education*: 75–82. Rotterdam, The Netherlands: International Council for Research and Innovation in Building and Construction.

Stouffs, R., Engeli, M. & Tunçer, B. 2003. Mediated discourse as a form of architectonic intervention. In *Proceedings of the 4th AVOCAAD Conference 2003*, Brussels, 3–5 April 2003.

Tunçer, B., Stouffs, R. & Sariyildiz, I.S. 2001a. Types and documents: structuring building project information. In *Proceedings of the CIB-W78 International Conference IT in Construction in Afric*a: 20.1–20.13. Pretoria: CSIR.

Tunçer, B., Stouffs, R. & Sariyildiz, I.S. 2001b. Collaborative information structures: educational and research experiences. In *COOP 2000 Workshop Proceedings: Analysing and Modelling Collective Design*: 20–28. Rocquencourt, France: INRIA, 2000.

26. Durability analysis and lifetime assessment

System-based Vision for Strategic and Creative Design, Bontempi (ed.)
© *2003 Swets & Zeitlinger, Lisse, ISBN 90 5809 599 1*

Impact of low sulfate metakaolin on strength and chloride resistance of cement mortar and high strength concrete

J. Suwanpruk, S. Sujjavanich & J. Punyanusornkit

Department of Civil Engineering, Kasetsart University, Bangkok, Thailand

ABSTRACT: The effect of local processed kaolin from the southern part of Thailand, with fineness of about $9,800 \, cm^2/gm$, on strength and resistance to chloride permeability, was investigated. Between 0 to 40% cement replacement in mortar and 0–30% in concrete were studied for use as substitute for imported silica fume in high strength concrete with the target strength of 660–860 kg/cm^2, and as repair material.

Metakaolin increased microstructure denseness. Cement replacement of up to 30% by metakaolin increased the compressive strength of mortar. At 40% the increase dropped slightly, but the strength was still higher than the control mix. The charge in the void system affected permeability development more than it did compressive strength. In the rapid chloride permeability determinations, the charge passed in metakaolin mortar decreased from moderate to very low level as ages or percentage replacement increased, while the control mixes provided high charges passed at all ages. The reduced penetration depth of chloride front indicated the substantial improvement of this material.

For concrete, the strength improvement especially during the first 3 days was observed. The increase in compressive and flexural strength were in the range of 13–18%, and 1–16% respectively. An optimum percentage replacement of 20% was found for strength improvement. Significant microstructure improvement was revealed through the very high level of chloride ingress resistance, compared to the medium level of high strength concrete. The potential as a low cost, locally produced, supplement material for repair material and high strength and durable concrete was high.

1 INTRODUCTION

1.1 *General*

The tropical climate and the high humidity of most part of Thailand is one of various factors causing rapid deterioration of concrete structures. The need for durable and effective material for both new construction and repair purpose is currently of interest as well as the improvement in the codes of practice in the construction industry, to bring about long service life of structures.

Mineral admixtures, especially imported silica fume, have been used in Thailand for about a decade. While the local fly ash which has only been introduced in concrete industry a few years ago, is significantly gaining wider acceptance. However, the high cost of silica fume and the low early strength development of fly ash were recognized as their weak points.

Metakaolin, the product of processed-heat treatment of natural kaolin, is widely reported as a quality and effective pozzolanic material, particularly for the early strength development (1, 2, 3). A few reports investigated the pozzolanic potential of local kaolin from several areas in Thailand, mainly used in ceramic industry, for the use in concrete industry (4, 5, 6). The effective range of burning temperature and grinding time for local kaolin of 750–800°C and 6 hours have been reported. In addition to pozzolanic reaction, the action of microfiller has been reported to partly improve strength development of cement–metakaolin mortar (6, 7).

Low sulfate metakaolin from Ranong, the southern part of Thailand, has been investigated in this paper, as a potential supplement for cementitious material for use as repair material and as a substitute for imported silica fume in high strength concrete. Both strength and improvement, in term of chloride impermeability which mainly influenced corrosion, were studied for mortar and concrete. Information on strength development and chloride permeability of mortar and concrete incorporating local metakaolin under tropical climate was expected. This is essential for further study to develop good quality construction material for high strength and durable concrete structure.

2 TEST PROGRAM

2.1 *Materials and specimen preparations*

The processed metakaolin with fineness of about 9,800 cm^2/gm were used throughout this study. The chemical compositions and physical properties are shown in Table 1. The microstructure, and particle size distribution are shown in Figures 1 and 2.

In this study, two test programs were reported. For repairing aspect, cement mortar incorporating with five levels of cement replacement of 0, 10, 20, 30 and 40% were studied. The cement to sand ratio was 1:2.75 and water to binder ratio was 0.485 with

enough superplasticizer to maintain flowability of 105–115%. Mortar properties in term of compressive and flexural strength as well as resistance to chloride permeability were investigated.

For high performance concrete, properties of metakaolin concrete with the 28-days target strength of 660–860 kg/cm^2 using four replacement level of 0%, 10%, 20% and 30%, were compared with those of conventional concrete. The concrete mix proportions with 2.7% superplastcizer are shown in Table 2.

Aggregates in this study were crushed limestone with maximum size of 3/8″ and coarse river sand. Concrete specimens, cast in 7.5 × 7.5 × 7.5 cm. cubical mold for compressive strength test and 10 × 20 cm. cylindrical mold for Rapid Chloride Permeability test, were moist cured for 24 hours. The 7.5 × 7.5 × 7.5 cm. sample size was chosen to avoid variation between mixes due to the limited capacity of mixer. After demolding, two sets of specimens were water cured at room temperature until test. All specimens for chloride test were continuously water cured at room temperature until the age of testing, 28 and 56 days.

Table 1. Chemical compositions and physical properties of metakaolin, compared to portland cement and ASTM class pozzolan.

Chemical compositions	Portland cement	Metakaolin	Pozzolan type (ASTM C618) N	F	C
SiO$_2$ (%)	21.16	54.64			
Al$_2$O$_3$ (%)	5.09	42.87			
Fe$_2$O$_3$ (%)	3.01	1.01			
SiO$_2$ +Al$_2$O$_3$ + Fe$_2$O$_3 \geq$ (%)	29.26	98.52	70	70	50
CaO (%)	66.22	0.01			
MgO (%)	1.27	NA			
K$_2$O (%)	0.25	1.16			
Na$_2$O (%)	0.04	NA			
SO$_3$ < (%)	2.42	NA	4	5	5
LOI < (%)	0.98	1.10	10	6	6
Moisture content < (%)		0.45	3	3	3
Specific gravity	3.15	2.43	2.10–2.58		
Blaine fineness, cm^2/g	2900–3200	9800	2,400–4,800		

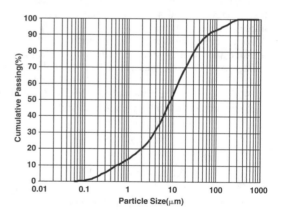

Figure 2. Metakaolin particle size distribution.

Table 2.

Mix No	(kg/m^3) Cement	MK	Water	Limestone	Sand	fc′ 28 days (kg/cm^2)	Slump (cm)
1	500	0	150	975	786	730	17.5
2	450	50	150	975	774	802	14.8
3	400	100	150	975	762	869	10.0
4	350	150	150	975	749	772	5.0
5	500	0	200	975	655	660	27.2
6	450	50	200	975	643	735	24.6
7	400	100	200	975	630	774	21.5
8	350	150	200	975	618	717	15.3

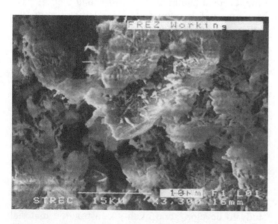

Figure 1. Microstructure of metakaolin.

The cylindrical samples were cut and the 2 in. slices at the mid height were used for testing.

3 RESULT

3.1 Test results and discussion

Because of its particle shape and high surface area, metakaolin remarkably reduced workability and water reducer is needed for the required workability for both mortar and concrete. The dense microstructure and less detected CH than of the control cement paste were observed through the SEM and XRD results of metakaolin-cement paste at 20% replacement at 28 days (not shown in this paper). Compared to the same dosage of superplasticizer in this study, slight segregation at high dosage of mixture with metakaolin did not show significant negative effect in strength development, normally associated with severe segregation. Continued increase in compressive strength was observed. However, the higher rate of slump loss of metakaolin concrete with high w/b (0.35) than that of normal concrete was observed, but reduced slump was observed as the replacement percentage increased (not shown in this paper).

Because of its small particle size, wide range of size distribution as well as chemical composition and amorphology, both microfiller and pozzolanic effects are thought to provide this positive effect, as well as to partly accelerate the cement hydration at the very early age, as previously reported (7). These effect explained the 8–20% increase in compressive strength of cement mortar. The strength development of the mixes with and without metakaolin is shown in Figure 3. Metakaolin did not greatly affect flexural and tensile strength of mortar. It tended to decrease those properties especially at over 30% metakaolin.

The effect of changes of void system was higher in the permeability development than in the compressive strength. The resistance of chloride ingress increased as age and percentage replacement increased as shown in Figure 4. The use of 30–40% metakaolin resulted in the 56 days level of permeability that was in the impermeable class as stated in ASTM C1202, compared to the high level of control mortar at the same age. Also, the observed chloride front from the electrochemical-accelerated test indicated the substantial improvement of this material. The study indicated the potential of using metakaolin of about 20–30% to improve compressive strength and durability of mortar for repair work.

Metakaolin significantly affects strength development of concrete especially in the first 3 days after casting. As in the case of mortar, compressive strength increased significantly. The strength development of metakaolin concrete is shown in Figure 5.

For all ages, all water binder ratios and all replacement percentages, metakaolin concrete showed higher compressive strength than that of the control. The 20% replacement appeared to be optimum, considering the strength improvement. The compressive strength ratio of metakaolin and normal concrete is shown in Figure 6. Concerning permeability, rapid chloride permeability test, according to ASTM C1202 indicated significant improvement of porosity of the matrix. The charge passed through metakaolin concrete, as shown in Figure 7, was in the range of 150–700 coulomb and 100–600 coulomb for 28 and 56 days test ages respectively.

These are classified as very low range, compared to the medium range of the control mixes (1,932–2,478 coulomb) as indicated in Table 3. It was observed that, metakaolin strongly improved chloride permeability of the matrix. Compared to high strength

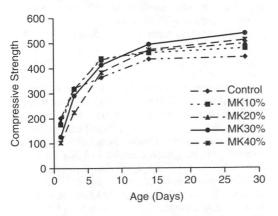

Figure 3. Strength development of mortar with and without metakaolin.

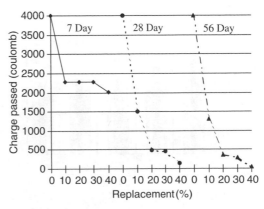

Figure 4. Measured charge passed of mortar with and without metakaolin.

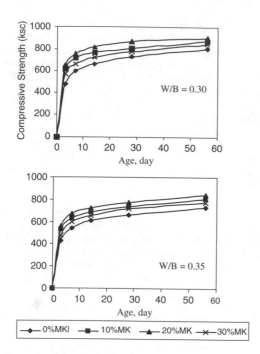

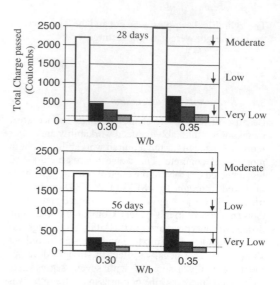

Figure 7. Comparison of W/B and charge pass of metakaolin and conventional concrete.

Figure 5. Strength development of metakaolin and conventional concrete.

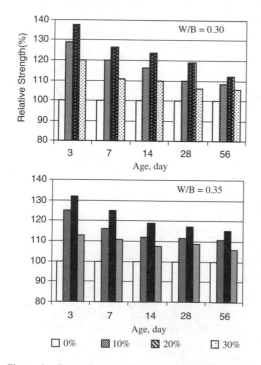

Figure 6. Strength ratio of metakaolin and conventional concrete.

Table 3. Charge passed and chloride permeability level, according to ASTM C120.

Charge passed (coulomb)	Chloride permeability level
>4,000	High
2,000–4,000	Moderate
1,000–2,000	Low
100–1,000	Very low
<100	Impermeability

conventional concrete with the same range of compressive strength, metakaolin efficiently reduced chloride ingress. The observed chloride front from silver nitrate spray technique is shown in Figure 8.

Concerning the effectiveness as a substitute for imported silica fume in high strength concrete, the comparison was made with silica fume concrete with the same strength range at 28day, 620–880 kg/cm^2 (8). Metakaolin concrete with w/b 0.30 and replacement of 20% appeared to have a slower rate of change during the first three days (1.3 vs 1.4). Significant increase in later age strength was observed while the compressive strength gain was almost ceased in silica fume concrete. The effect of metakaolin on the resistance of chloride ingress appeared to be higher than that of silicafume. The chloride resistance level was classified as impermeable compared to moderate and high for metakaolin concrete, silicafume concrete and normal high strength concrete using only superplasticizer.

Control 0%

30% MK

Figure 8. Comparison of chloride front between control mix and mix with 30% Metakaolin.

4 CONCLUSION

The following conclusions can be drawn.

1. Both microfiller and pozzolanic effect benefited both strength and durability aspect of mortar and concrete.
2. Metakaolin greatly affected the resistance of chloride ingress and compressive strength but not flexural and tensile strength of mortar.
3. The study indicated the potential of using about 20–30% metakaolin to improve compressive strength and durability of mortar for repair work.
4. The results indicated improvement of compressive strength of concrete in the range of 5–30%. The effect of metakaolin was high even at the early age, particularly on the compressive strength.
5. The replacement of 20% appeared to be an optimum percentage for strength improvement.
6. Significant microstructure improvement was revealed through the very high level of chloride ingress resistance, compared to the medium level of high strength concrete, according to ASTM C1202.
7. From this study, the potential as a low cost local-supplement material for high strength concrete was confirmed.

REFERENCES

Khatib, J.M. & Wild, S. 1996. Pore Size Distribution of Metakaolin Paste. *Cem. Concr. Res.* 26(10): 1553–1996.

Kostuch, J.A., Walters, G.V. & Jones, T.R. 1993. High performance concrete incorporating metakaolin – a review, *Concrete 2000.* University of Dundee. 1799–1811.

Curcio, F., De Angelis, B.A & Pagliolico, S. 1998. Metakaolin as a Pozzolanic Microfiller for High Performance Mortars. *Cem. Concr. Res.* 28(6): 803–809.

Hensadeekul, T. 1995. Use of Metakaolin from Lampang Province as a Pozzolana. *M.Eng.Thesis,* Asia Inst. Technol.

Suwan, A. 2000, Using Metakaolin for sulfate resistance of cement mortar, *Master thesis,* Technology Mahanakorn.

Sayamipuk, S. 2000. Development of Durable Mortar and Concrete Incorporating Metakaolin form Thailand. *Ph.D. Thesis,* Asia Inst. Technol.

Wild, S., Khatib, J.M. & Jones, A. 1996. Relative Strength, Pozzolanic Activity and Cement Hydration in Superplasticised Metakaolin Concrete. *Cem. Concr. Res.* 26(10): 1537–1544.

Kongkaew, W. 1999. Effect of elevated temperature curing on properties of Silica Fume Concrete, *Master Thesis* for Kasetsart University, Thailand.

System-based Vision for Strategic and Creative Design, Bontempi (ed.)
© 2003 Swets & Zeitlinger, Lisse, ISBN 90 5809 599 1

Influence of environmental variables on moisture transport in cementitious materials

M.K. Rahman, M.H. Baluch, A.H. Al-Gadhib & S. Zafar
King Fahd University of Petroleum & Minerals, Dhahran, Saudi Arabia

ABSTRACT: Concreting in the arid Gulf environment of the Middle East remains a challenge as the insidious forces of corrosion await the formation of cracks to accelerate their destructive process. A primary line of defense against cracking is the appropriate design of concrete mixes to minimize the probability of the occurrence of cracks due to drying shrinkage.

Drying shrinkage in cementitious materials is now known to be influenced by a number of variables, including free shrinkage strain, tensile creep strain, tensile strength and modulus of elasticity. This paper addresses the role of two pertinent parameters that control the magnitude of free shrinkage strain. These parameters are the coefficient of moisture diffusivity D and the convective moisture transfer coefficient h_f (also known as the surface factor). A combined experimental-numerical approach is adopted to measure these parameters for two cement based commercial repair materials. The influence of temperature and wind velocity on diffusivity and convective transfer coefficient is highlighted.

1 INTRODUCTION

Concrete structures in the harsh environment of the Gulf region in the Middle East suffer from deterioration or distress. As the damage propagates, a time comes when the safety and serviceability of the structure gets seriously impaired, hence necessitating repair work to restore safety and serviceability. In the Arabian Gulf region, the potentially aggressive environment has resulted in premature deterioration requiring extensive repairs. It is estimated that repair and maintenance of concrete structures in Saudi Arabia would run into billions in the coming few decades (Al-Gahtani et al. 1995).

Initial cracking in concrete in most cases result from stresses due to restrained shrinkage on drying or the volume changes resulting from ambient and fresh concrete temperature variations. The focus of this work is on the material parameters that influence the moisture transport process in concrete.

Moisture diffusivity is the key physical parameter that is required for computation of moisture transport in cementitious materials. The transport coefficient is largely material specific, i.e. depends exclusively on material porosity, pore structure and moisture content. It is known that in the diffusion of gas through a catalyst, the diffusion paths are tortuous, irregularly shaped channels; accordingly, the flux becomes less than it would be in a uniform pore of the same length and mean radius. The effective coefficient of diffusivity in linear diffusion problems can be expressed in terms of a tortuosity parameter τ, a factor that describes the relationship between the actual path length relative to the nominal length of the porous media (Welly et al. 1984). In lieu of measuring the tortuosity, diffusivity may be established as a regressed function of the water/cement ratio for cementitious materials. However, inasmuch as the diffusion of moisture through concrete is now known to be a non-linear problem, the influence of moisture concentration level on the diffusivity has also to be considered (Meng 1992). External influences like ambient temperature, humidity and wind speed are also believed to have an influence on the diffusivity coefficient and the convective surface transfer coefficient (Sakata et al. 1980). Water transport processes are often accompanied by temperature variation. From the thermodynamic theory, it should be expected that the transport properties increase with the temperature. According to the Hirschfelder equation (Hirschfelder et al. 1954), for diffusion of a gas through a binary gas mixture, the diffusivity varies approximately in the ratio of absolute temperatures to the power 1.5. An application of this to concrete diffusivity has been recently noted by Jooss & Reinhardt (2002).

The recent publication of Jooss & Reinhardt (2002) in which the authors have discussed the influence of temperature on permeability and moisture diffusivity of concrete, diffusivity is measured using DIN 52615 which results in a constant value. It is known that the modelling of stresses and damage associated with the restrained shrinkage of concrete cannot be established using constant values for the coefficient of diffusivity (Rahman et al. 1999, Baluch et al. 2002). The simulation of this problem can only be achieved by treating moisture diffusivity as a function of moisture concentration which renders the boundary value problem non-linear. One feasible approach for this is to calibrate an assumed form for the coefficient of diffusivity in terms of unknown parameters of known functions of moisture concentration level, water/cement ratio and concrete temperature, using data from experiments and numerical results from a finite element driven program (Rahman 1999).

2 GOVERNING EQUATIONS

The governing nonlinear differential equation for moisture diffusion in the domain of a generalized 3-D solid in terms of moisture content in the solid can be written as:

$$\frac{\partial C(x_k,t)}{\partial t} = \frac{\partial}{\partial x_i}\left[D(C)\frac{\partial C(x_k,t)}{\partial x_i}\right] \quad \begin{matrix} k = 1,2,3 \\ i = 1,2,3 \end{matrix} \quad (1)$$

$C = C(x_i,t)$, $i = 1,2,3$ is the moisture content varying in domain and with time

$D(C)$ = isotropic moisture diffusivity coefficient which is function of C.

A simplified treatment considers the transport in the form of a constant flux boundary condition and can be written as:

$$q_i^\Gamma(x_k,t) = h_f(C_e - C_s) \quad (2)$$

q_i^Γ: the rate of moisture transfer per unit area across the boundary Γ in the ith direction

h_f : surface factor or convective transfer coefficient

C_s : moisture content at the surface of the solid

C_e : moisture content/relative humidity of the ambient temperature.

A nonlinear 2-D finite element program DIANA-2D (**DI**ffusion **ANA**lysis) has been developed in this study for numerical simulation of moisture transport in repair materials using nonlinear moisture diffusion equation. This model requires an empirical, moisture dependent diffusivity law. Determination of the diffusivity of a cementitious material directly by experiments is extremely difficult, if not impossible (Xin et al. 1995). Several methods have been reported in

literature for determination of diffusivity of cementitious mortar and concrete. The Boltzmann-Matano method (Sakata 1983) is the most popular approach and has been used in several studies.

A combined experimental-finite element based approach for computation of diffusivity is proposed in this paper. A nonlinear finite element method combined with the least squares fit method was used for the determination of the diffusivity parameters for selected repair materials. In this method, the parameters are obtained by comparing the computed moisture profiles with the experimental drying data as a function of time. Drying tests were conducted on two types of commercially available repair materials. The effect of elevated temperature on the diffusivity of the selected repair materials was also investigated.

3 PROPOSED PROCEDURE FOR COMPUTATION OF DIFFUSIVITY

A combined experimental-numerical approach for computing the diffusivity of repair materials is proposed in this work. A nonlinear finite element method combined with the least squares fit method was used for the computation of the diffusivity parameters. In this method, the parameters are obtained by comparing the computed moisture profiles with the experimental drying data as a function of time.

A functional form of the relationship between the diffusivity and moisture content was assumed (Eq. 4). The selected functional form and preliminary estimates of the parameters were incorporated in the nonlinear finite element based diffusion analysis program DIANA-2D. An iterative technique was then used to find the parameters that best fit the experimental data. This iterative technique involves computation of moisture loss over the domain of the experimental sample with appropriate boundary conditions followed by computation of the mean moisture loss in the sample. A residual, or functional error, is then computed using

$$Error = \sum_{t=0}^{n\,time} \left[W_{ex}(t) - W_{fe}(t)\right]^2 \quad (3)$$

where

$W_{ex}(t)$ – is the mean experimental moisture loss at time t

$W_{fe}(t)$ – is the mean moisture loss computed from finite element run.

The value of the parameters for which the error in Equation 3 is below a certain specified tolerance gives the best fit. The parameters obtained by this procedure are then used to carry out finite element diffusion analysis on samples of different thicknesses to check the validity of the procedure.

4 DRYING TESTS FOR DIFFUSIVITY COMPUTATIONS

Drying tests were conducted on specimens of various thicknesses with unidirectional moisture movement at temperatures of 30°C (RH 60 ± 3%) and 50°C (RH 50 ± 3%). Additional data is presently being generated for the influence of forced convection on the surface factor h_f.

The drying tests were carried out for computing diffusivity of the repair materials. Two commercially available repair materials were selected for this purpose. These selected micro-concretes (FMC1 and FMC2) had earlier been noted to perform the best out of a group of six commercial repair materials (Al-Gadhib et al. 2001). A brief description of each is as follows:

Repair material FMC1
Is a blend of dry products which requires only the site addition of clean water to produce a free-flowing, shrinkage compensated micro-concrete suitable for large volume concrete repairs. The material is based on Portland cement, graded aggregates and additives which impart controlled expansion in both the plastic and hardened states while minimizing water demand. Its self-compacting nature eliminates honeycombing, with the low water requirement ensuring fast strength gain and long-term durability. Its 28-day compressive strength is 60 MPa (8700 psi) and its two-hour water absorption (BS 1881) is 0.0013 ml/m²/sec.

Repair material FMC2
Is a shrinkage compensated, free-flowing precision concrete. It requires only on-site addition of water to provide a free-flowing concrete. It is a pre-packaged blend of Portland cement, carefully graded natural aggregate, specially selected sands, shrinkage control agents and fluidifiers.

The utilization of a low water cement ratio ensures minimum permeability and high durability. The free-flowing nature eliminates honeycombing in areas of congested reinforcement. The self-compacting characteristic ensures complete filling of spalled/damaged areas. The unique formulation provides a dual shrinkage compensating action in both plastic and hardened states. Its 28-day compressive strength is 65 MPa (9400 psi) and water permeability (DIN 1048) is minimal.

Specimen of sizes 100 mm * 100 mm * B mm (B = 20, 40, 50, 60, 75 mm) were used in the drying tests. All specimens were cured in moulds for 24 hours sealed in plastic wraps. They were demoulded and then cured in water for 28 days prior to sealing and drying in the environmental chamber. Appropriate surfaces were sealed in order to ensure uni-dimensional diffusion through thickness B. After removal from

water, specimens were towel dried and kept in plastic foil until sealing.

5 FUNCTIONAL FORM OF DIFFUSIVITY LAW ADOPTED

For the repair materials under investigation, it was observed that the moisture loss-time curve is steep at early ages. This corresponds to the loss of water from large capillary pores. Later the rate of moisture loss becomes extremely slow characterizing a low diffusivity of the material. The functional forms of diffusivity-moisture content relationships used in the literature were explored but were not found to give a satisfactory fit. A trigonometric function of the form given below with three unknown parameters was developed and found to give good results. This functional form was incorporated in the program DIANA-2D.

$$D(C) = b_o \tan (b_1 C^n) \qquad (4)$$

where b_0, b_1 and n are parameters to be evaluated for the best fit.

6 PARAMETERS FOR THE DIFFUSIVITY LAW OF REPAIR MATERIALS

The steps outlined above were then applied to find the parameters b_0, b_1 and n of the diffusivity law (Eq. 4) for the two repair materials. The adopted tangent functional form yielded a good fit for the two repair materials under consideration. Finite element discretization of a diffusivity specimen is shown in Figure 1. A half model of the cross section of the specimen is considered. The iterative finite element runs were first carried out for the specimen of size 100 mm * 100 mm * 40 mm to ascertain the parameters b_0, b_1 and n. The computed parameters were then used to verify the predictability of the model in specimens of other sizes (100 mm * 100 mm * B, where B is the set (20, 50, 60, 75 mm)).

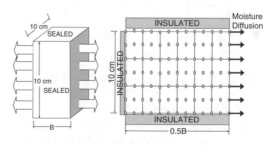

Figure 1. Finite element discretization of diffusivity specimens.

Table 1. Diffusivity parameters.

Material	Chamber temperature (°C)	Environmental humidity C_e	Surface factor h_f (cm/day)	Diffusivity parameters b_0	b_1	n	D_{av} (cm^2/day)
FMC1	30	0.6	0.3	0.05	1.5	26.0	0.010
FMC1	50	0.5	0.6	0.4	0.5	4.0	0.040
FMC2	30	0.6	0.3	0.15	1.1	30.0	0.008
FMC2	50	0.5	0.6	0.3	0.5	4.0	0.030

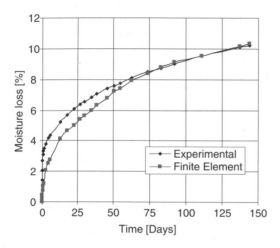

Figure 2. Moisture loss curves for FMC1 at 30°C.

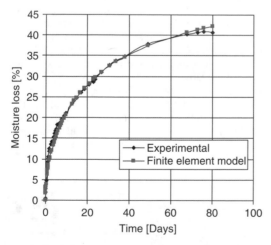

Figure 3. Moisture loss curves for FMC1 at 50°C.

6.1 Diffusivity parameters for FMC1

For the flowing micro-concrete FMC1, the parameters b_0, b_1 and n were established for diffusion at 30°C (RH = 60 ± 3%) by using a finite element model for the 100 * 100 * 40 mm specimen as shown in Figure 1.

The moisture diffusion in this specimen was unidirectional. The iterative technique yielded the value of these coefficients as shown in Table 1. The computed mean moisture loss shows a good fit with the experimental mean moisture loss data as can be seen in Figure 2. The values of parameters b_0, b_1 and n established in this run were then used for other diffusivity specimens of sizes 100 * 100 * B mm, where B included the set (20, 50, 60, 75 mm), which were dried under same environment. Good correlation was observed between experimental and predicted values. The entire procedure was repeated, but with temperature set at 50°C (RH = 50 ± 3%). The correlation between experimental and model prediction for specimen of 100 * 100 * 40 mm is shown in Figure 3. The diffusivity parameters b_0, b_1 and n at this elevated temperature are shown in Table 1. The use of these parameters for specimens of other thicknesses yielded close correlation.

6.2 Diffusivity parameters for FMC2

DIANA-2D was used for the computation of the diffusivity parameters for the repair material FMC2. Results of computations similar to those of FMC1 for this material are presented in Figures 4 and 5. The diffusivity parameters for both temperatures are shown in Table 1.

6.3 Diffusivity law for repair materials

A typical plot of Equation 4 is shown in Figure 6. Whilst the parameter b_0 is merely a measure of the amplitude of the diffusivity, parameters b_1 and n control its shape. It is noted that higher values of the index n result in a rapid decay of diffusivity D versus moisture content, leading to lowered average values of D (D_{av}). Results of Table 1 indicate that to simulate moisture losses at higher temperatures, index n has to be decreased significantly for both materials, resulting in values of D_{av} approximately four-folds greater.

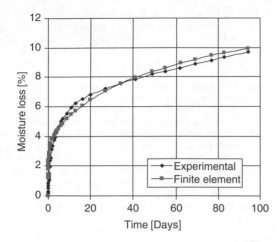

Figure 4. Moisture loss curves for FMC2 at 30°C.

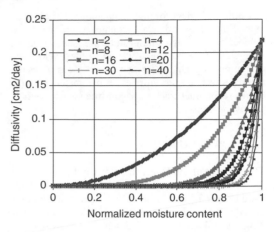

Figure 6. Generic shapes for diffusivity vs. moisture content.

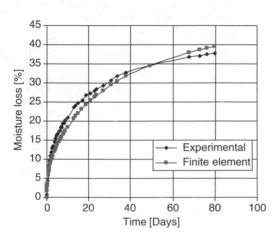

Figure 5. Moisture loss curves for FMC2 at 50°C.

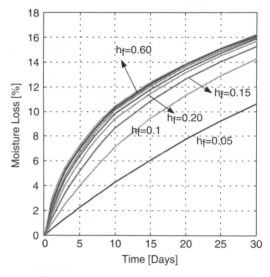

Figure 7. Influence of convective transfer coefficient on moisture loss.

6.4 Effect of wind on convective transfer coefficient h_f

This phase of the work is currently under progress, in which specimens of sizes $100 * 100 * B$ mm ($B = 20$, 40, 50, 60, 75 mm), sealed so as to diffuse only through the direction of the thickness B, are placed in the drying chamber at controlled conditions of temperature, humidity and wind velocity. Initial results indicate that the influence of wind velocity leads to accelerated moisture loss in the early stages which may be predicted by the model by the use of higher values of the convective transfer coefficient (or surface factor) h_f. Figure 7 is a qualitative indicator of the trend of moisture loss curves with increasing values of h_f.

7 CONCLUSIONS

A combined experimental-finite element based approach has been presented for deriving the empirical diffusivity laws of repair materials. A tangent form of diffusivity law with three regression parameters gives an excellent prediction of mean moisture loss in two types of commercially available repair materials.

The diffusivity of cement based repair materials is significantly affected by temperature due to an increase in the energy of the water vapor molecules. The influence of windy conditions leads to increased values of the convective transfer coefficient.

Combined effects of high temperature, low environmental humidity and windy conditions on rate development of shrinkage strain can be predicted using the results of this work. This would allow for development of engineering guidelines that would focus on establishing limits for certain relevant parameters in order to ensure minimal cracking in construction with cementitious materials.

ACKNOWLEDGEMENTS

The support of the Department of Civil Engineering, the Research Institute and the University itself (King Fahd University of Petroleum & Minerals) is acknowledged. This work was completed under a grant from King Abdulaziz City for Science and Technology (KACST) under Project AT-17-36.

REFERENCES

Al-Gadhib, A.H., Baluch, M.H. & Rahman, M.K. 2001. Design guidelines and performance criteria for concrete repair systems. KACST Project AT-17-36, Dept. of Civil Engineering, King Fahd University of Petroleum & Minerals, Dhahran, Saudi Arabia, July.

Al-Gahtani, A.S., Rasheeduzzafar & Al-Musallam, A.A. 1995. Performance of repair materials exposed to fluctuation of temperatures. *J. Materials in Civil Eng.* 7(1).

Baluch, M.H., Rahman, M.K. & Al-Gadhib, A.H. 2002. Risk of cracking and delamination in patch repair. *ASCE J. Materials in Civil Eng.* 14(4): 294–302, Jul./Aug.

Hirschfelder, J.O., Curtiss, C.F. & Bird, R.B. 1954. *Molecular theory of gases and liquids.* New York: John Wiley & Sons.

Jooss, M. & Reinhardt, H.W. 2002. Permeability and diffusivity of concrete as function of temperature. *Cement and Concrete Research* 32: 1497–1504, April.

Meng, B. 1992. Calculation of moisture transport coefficients on the basis of relevant pore structure parameters. *Materials and Structures* 27: 125–134.

Rahman, M.K. 1999. Simulation and assessment of concrete repair system. PhD dissertation, King Fahd Univ. of Petroleum & Minerals, Dhahran, Saudi Arabia.

Rahman, M.K., Baluch, M.H. & Al-Gadhib, A.H. 1999. Modeling of shrinkage and creep stresses in concrete repair. *ACI Materials J.* 96(5): 542–559, Sep./Oct.

Sakata, K. 1983. A study on moisture diffusion in drying and drying shrinkage of concrete. *Cement and Concrete Research* 13(2): 216–224.

Sakata, K., Nishigaki, M. & Kuramoto, O. 1980. Finite element analysis of nonlinear moisture transfer problems in concrete. *Research & Development*, Okayama University, 1(4): 11–33, March.

Welly, J.R., Wicks, C.E. & Wilson, R.E. 1984. *Fundamentals of momentum, heat and mass transfer.* New York: John Wiley & Sons.

Xin, D., Zollinger, D.G. & Allent, G.D. 1995. An approach to determine diffusivity in hardening concrete based on measured humidity profiles. *Advanced Cement Based Materials* 2(4): 138–144.

Stainless clad reinforcing for greater durability of motorways and infrastructure

A.G. Cacace
Stelax Industries Ltd., Wern Works, Briton Ferry, Neath. UK

H.S. Hardy
Stelax Industries Ltd., Texas, USA

ABSTRACT: Stainless steel reinforcing bars, both solid and clad, have now been tested and used in North America and Europe and their use, especially of the SCR (stainless clad bar) to replace galvanized and epoxy coated rebar is growing rapidly in the USA and Europe where over 700 tons of Nuovinox™, a SCR produced in the UK and available at a much lower cost than stainless steel rebar (SS), has been supplied into numerous projects. Corrosion tests on SCR shows an equivalent corrosion resistance to SS and a superior corrosion resistance over traditional reinforcing bar materials and coating alternatives. The US highway authorities are expediting the approval of two international specifications for SCR; ASTM and AASHTO, to expand the use of SCR into applications where mostly epoxy, (ES) and galvanized zinc coated (GS) reinforcing bars are presently being specified.

1 INTRODUCTION

Austenitic stainless steel (SS) rebars have shown very promising corrosion performance in chloride-contaminated concrete (Flint et al. 1988; Cox et al. 1996; Nurnberger 1996; Pedeferri et al. 1997; McDonald et al. 1998; Hewitt et al. 1994) but at a substantially higher cost than conventional plain carbon steel (CS) rebar. Stainless steel clad rebar (SCR) with a carbon steel core offers the potential for performance comparable to that of solid SS rebar but at a much lower cost. (Stelax (UK) Ltd. 1998) Furthermore relatively thick, stainless steel cladding, hot-rolled onto the core contrasts with the thin and more fragile metallic or organic coatings of galvanized (GS) or epoxy coated (ES) rebar and thereby offsets their commonly associated fabrication and handling problems.

Recent market price estimates of typical grades of SS, ES, and CS in Europe show:

	Relative Cost.	US$/Tonne
CS:	1	430
GS: (straight lengths)	2.3	990
GS: (bent bars)	4.5 (!)	1920
ES:	2.7	1150
SCR (lean austenitic)	2.0	880
SCR – 304L	2.7	1150
SCR – 316L	3.0	1320
304 SS	5.5	2350
316 SS	6.8	2920

Nuovinox™, SCR is therefore, between one-third to one-half of the cost of an equivalent SS rebar for 304 and 316 grades.

The concept of SCR is not new and was reported in 1982. (Parkin 1982). The difficulties however, associated with inserting a core of high tensile steel and fusing the metals together added to the cost, which thereby offset the savings resulting from the use of a cheaper core. (European Federation of Corrosion Publications 1996) This earlier SCR was subsequently discontinued in the late 1980's because of having to sell it at too high a price, almost that of solid SS.

Stainless steel clad plate has been commercially available for some time and for which an ASTM specification is available. (ASTM A 264 – 94a 1994) This stainless clad plate however, as with hot-extruded stainless steel clad pipe, is generally more expensive than a 304 and 316 solid stainless steel equivalent, except for SS plate thicker than 25 mm. (Clad Metal Products 1999) These clad products are traditionally used in applications where the performance of the combined properties of the two metals together outweighs the extra costs involved.

Nuovinox™, on the other hand, is made using a solid state technology, which in contrast to other clad products, results in a hot-rolled stainless clad bar product at a substantially lower cost than its solid stainless steel equivalent. The core is made using bushy carbon steel turnings as the raw material. These are first crushed to fragments of less than about 5 mm in size. The fragments are then cleaned and blended by heating them to about 500°C in a rotary kiln.

The cleaned fragments, with small quantities of patented additions are compacted to about 80% density in stainless steel pipes (currently 6 mm wall thickness and 100 mm diameter) to form composite billets. The additions in the core produce a reducing atmosphere inside the billets during reheating for hot rolling. This is essential to ensure that the rolling not only consolidates the core to full density, but also creates sound metallurgical bonding within the core and between the core and the stainless steel cladding.

The Nuovinox™ process is "green" in that it recycles known-source high tensile carbon steel machine shop turnings without melting to form its carbon steel core. Only 17% of the steel-making energy for melting and casting steel into a conventional steel billet is required, to form a Nuovinox™ stainless steel clad or composite "billet" which is then conventionally hot-rolled into a SCR. Almost any stainless steel cladding can be pre-selected to suit an end purpose.

The Nuovinox™ SCR, manufacturing route includes a final, post-fabrication (after all endsealing, bending or cutting) descaling process by acid pickling. Acid descaling has also been found very useful in highlighting any cladding breaks, at the final 100% visual quality control inspection. Even small breaks become highly visible due to the acid-attacked carbon steel core's dark red corrosion products at the breaks. These sites contrast sharply against stainless steel's bright, acid-descaled, appearance and lead to rejection of these bars. The latest ASTM draft specification for SCR (ASTM Designation XXXX, 2003), unlike the specification for ES, specifies rejection of SCR with any cladding breaks whatsoever, regardless of the break size or frequency of occurrence.

This paper will attempt to answer some of the most pertinent questions regarding SCR.

Projects which have used SCR.
Corrosion tests and trials with sound and with accidentally perforated cladding.
Mechanical properties.
The bond between carbon steel/stainless steel.
The bond between SCR and concrete.
Specifications available.

2 PROJECTS WHICH HAVE USED SCR

2.1 Early SCR application

One of the first reported applications reported of SCR was a 9-year field study on a bridge deck constructed in 1983 by the New Jersey DOT (Department of Transport) on I-295 over Arena Drive in Hamilton Township, Trenton, New Jersey, using 22 tons of

Figure 1. Ashland Ave. Bridge, Green Bay, Wisconsin 130t SCR – 316L May 2001.

SCR – 304. None of the bars tested showed any corrosion, except under the plastic end caps. (McDonald 1995)

2.2 Launch of Nuovinox™

Nuovinox™ was launched in the early 1990's as a stainless steel clad flat bar and required some more development before it could be commercially launched as a rebar. It was supplied into its first project in Ontario-Canada, in 1999 as SCR – 316L. Subsequent projects in Alberta-Canada, North Dakota, Minnesota, Wisconsin (See Figure 1), New York State, Hawaii, Ohio, Delaware and Washington State totalled some 700 tons of SCR – 316L in 2000 and 2001. Mostly used for bridge top decks and in abutments and approach slabs, it has also been specified in Florida for marine piers.

A project in Holland used 32 mm diameter SCR – 316L as dowels.

The largest tonnage contract supplied to date however, was for the UK's Southern and Scottish's hydroelectric power station in Nant, Scotland, where 154 tons of 25 mm SCR – 316L was supplied for a SCR reinforced-concrete water tunnel several hundred metres long. See Figure 2.

According to current US Federal Highway Administration opinion, the SCR most closely fits the profile for the ideal non-corroding rebar (Clemena et al. 2001) for a 75–100 year project life. The prestigious ENR describes "The industry's search for a cost effective corrosion resistant reinforcing bar for concrete has taken on proportions akin to the hunt for the Holy Grail." (Engineering News Record 2001)

In Florida, on the Gulf of Mexico, where epoxy-coated rebar usage has been discontinued for many years due to the collapse of a bridge over the Florida

Figure 2. The 280 m water tunnel at the Nant power station in Scotland supplied with 154t of 25 mm SCR in July 2001.

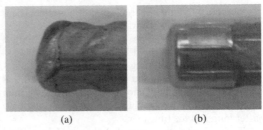

Figure 3. (a) a typical welding overlay (b) a typical SS cap.

Keys, (Sagues et al. 1994) SCR is being regarded as a potential breakthrough: "Nuovinox™ is potentially the most significant development in corrosion control of reinforcing steel that I have seen in my 31 years career" stated Mr. Rod Powers, Florida Department of Transportation's State Corrosion Engineer.

3 CORROSION TESTS AND TRIALS WITH SOUND AND PERFORATED CLADDING

3.1 Earlier corrosion tests with SCR

3.1.1 Test 1

A 20+ year and perhaps the longest term corrosion test done on SCR – 304, in concrete involved outdoor exposure slabs fabricated in 1982 in which SCR – 304, (Clear 1998) was tested against other coated rebars. Concrete slabs were exposed to 3,1 years of daily ponding with 3% NaCl solution after which salting was halted with chloride levels having reached 10 wt% for the top layer of rebars. Exposure to natural weathering occurred since.

Average Times to corrosion-induced concrete cracking (hairline):	Years
Uncoated rebar – Both mats	= 1,3
Galvanised rebar – Top mat only	= 3,0
Galvanised rebar – Both mats	= 3,4
Epoxy coated rebar – Top mat only	= 8,9
Epoxy coated rebar – Both mats	= 8,9
SCR – 304: Both mats	= Over 20

and still passive with Mat/Mat Macrocell Current still 0.

3.1.2 Test 2

A 7 – Year exposure programme of 304 – SCR, CS, GS & ES in chloride – bearing concretes were exposed to the hot and aggressive environment of Eastern Saudi Arabia (Rasheeduzzafar et al. 1992) at 3 levels of chloride mass% of cement; 0.6%; 1.2% & 4.8% which resulted in the following observations:

CS & GS; Severe rust-related damage was observed for all 3 chloride levels. Galvanising merely delayed corrosion.

ES: Although excellent performance was exhibited in 0.6% and 1.2% chloride level concretes there was significant corrosion however in 4.8% chloride level concrete with cracking of the concrete.

SCR – 304 proved to have the best durability with no signs of corrosion after 7 years of embedment for all the levels of chloride.

3.2 Latest corrosion tests with Nuovinox™

The most comprehensive corrosion testing to date which have been running over the past 2–3 years, for SCR – 316L, is contained in a recent paper. The corrosion performance of Nuovinox™, SCR – 316L was investigated on bar which had been terminated with either stainless steel welded ends, or alternatively with "crimped" or "swaged" polyurethane pre-filled stainless steel caps StelCap.™ (Cui et al. 2002 & 2003)

Both at room temperature, and 40°C all SCR specimens were passive and remained free of corrosion for up to one year so far, in extremely corrosive conditions which included both 15% chloride simulated carbonated concrete pore solution and in concrete with up to 8% chloride by weight of cement. The chloride levels used both in solutions and concrete corresponded to nearly chloride-saturated pore water, at the limit of the chloride bearing capacity of concrete. This is a very promising indication of performance for sound SCR – 316L terminated with either SS-welded or Stel-Cap,™ stainless steel (Grade 316L) end caps. See Figure 3. In essence, sound SCR – 316L behaved in the same way as would be expected from (solid) 316 SS.

In the same study SCR specimens with a 1 mm size hole showed active corrosion with both 15% chloride solution and in 8% chloride by weight of cement, for at least one of each of the three specimens, and less active corrosion for the other two.

Nominal corrosion rates however all showed a decrease with time, which suggests that some favourable plugging effect by corrosion product accumulation at the opening was taking place.

Calculations based on a model for SCR with a cladding break, indicated that the resistivity of concrete and size of cladding breaks were critical parameters in establishing the rate of corrosion, and that corrosion of SCR with sub-millimeter breaks in high quality concrete would cause concrete cracking only after long service times.

The calculations suggested that widely spaced cladding breaks of sub-millimeter size would be tolerable in concrete that retains high resistivity.

4 SCR'S MECHANICAL PROPERTIES

The macrocomposite nature of SCR is clearly revealed when bars are sectioned either transversely or longitudinally and ground or polished. The outer layer of stainless steel appears brighter than the carbon steel core. The SS-cladding for a 19 mm diameter bar is fairly uniform in thickness with variations between 0.6 mm and 0.8 mm around the perimeter and along the length. The non-melting process by which the core is produced is also evident, by occasional stringers of inclusions or micropores, in cross section. There are however no concentrations of these features at the interface between the core and cladding. They occur randomly in the core and are always below a minimum level determined by the quality control procedures. The solid state processing route does, however, mean that they cannot be completely eliminated. For reinforcing bars, the important properties are those in the longitudinal direction.

The physical and mechanical properties, are essentially determined by the carbon steel core, while the corrosion properties are those of the stainless steel cladding. In this context, it is noteworthy that the thermal expansion coefficient and magnetic properties are those of normal carbon-manganese steel reinforcing bar.

Mechanical properties consistently satisfy specifications, and the only consequence of the solid state processing on tensile tests is apparent after ductile fracture. Typically the fracture surface of the core has a "woody" appearance, commonly found in all microcomposite materials, such as high technology carbon fibre composites. The high ductility leads to extensive necking of the bars before fracture and the triaxial stresses created in the neck can lead to partial or complete fracture of the bond between core and clad in the local region of the neck. Such triaxial stresses are not met in service; the tensile fracture of the interface does not indicate unsatisfactory bonding.

Nuovinox SCR is currently produced to meet ASTM A 615's tensile and bending requirements, which are the same as for the ASTM draft specification for SCR:

Specification:

Y.S. = 420 MPa U.T.S. = 520 MPa El% = 9%.

SCR:

Y.S. = 479 MPa U.T.S. = 728 MPa El% = 14.8%.

5 THE BOND BETWEEN CARBON STEEL AND STAINLESS STEEL

The SCR has a true, solid-state metallurgical bond between core and cladding, across which atomic transfer of elements occurs. Two boundaries can be observed, with the initial boundary, characterized by a line of well dispersed oxide inclusions, "migrated" into the stainless steel clad, and the second as a new but less distinct demarcation of the 2 different steel structures. In this regard, the "Kirkendall Effect" can be observed, especially if the SCR is reheated. See Figure 4.

Micro-spectral analysis of chemical compositions, confirms furthermore, a Nickel and Chrome diffusion differential across the boundary area from: 0.6% Cr and 0.3% Ni to 3.1% Cr and 0.8% Ni on 10 microns of either side of the boundary. The Ni and Cr compositions then progressively increase further into the SS-clad to reflect ss 316L's standard composition.

In service the bond is subjected only to shear stresses. Fatigue testing over 2 million cycles (Ministry of Transportation Ontario 1998) furthermore, did not result in any debonding.

The SCR has been designed to well exceed the minimum bond shear strength requirements for suitability of purpose. The draft ASTM specification for SCR currently under consideration only requires the

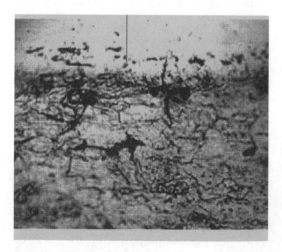

Figure 4. Microphotograph of SCR Interface between carbon steel core and stainless clad, illustrating atom migration across metallurgical bond – Kirkendall Effect.

specification's standard bend test according to bar diameter, without breaks or "wrinkling" of the cladding as a bond strength test.

No claim, however, has been made for tensile fracture toughness of the bond. Localised and extreme thermal differentials such as from abrasive saw cutting and welding may lead to some localized debonding, which is not considered deleterious to the SCR's corrosion resistance.

Shear bond strength tests however were undertaken, on SCR, observing a similar test method described in the specification for stainless clad plate, which resulted in an average shear strength of 290 MPa for 38 mm diameter dowels (no deformations). The ASTM's (ASTM A 264 – 94a, 1994) minimum (optional requirement) bond shear strength specification is only 140 MPa for SS-clad plate.

6 THE BOND BETWEEN SCR AND CONCRETE

Comparative bond strength tests were performed for SCR, SS, and CS based on ASTM C234(14), according to the concept of "Lollipop" failure of rebars in concrete. (Keys et al. 1999)

The failure load being determined when slippage of 2,5 cm of reinforcing bar has occurred and based on the mean of 3 tests:

SCR: 4.7 MPa SS: 4.35 MPa CS: 3.72 MPa.

SCR therefore has at best a better or at least an equivalent concrete bond strength compared to SS or CS.

7 SPECIFICATIONS AVAILABLE FOR SCR

The application was started in 1998 for an ASTM specification. Following several drafts, it appears likely that a full ASTM specification could be granted later this year, based on the current draft. It also appears likely that an AASHTO specification will also be finalized this year.

8 CONCLUSIONS

Nuovinox™ SCR has been described as a third generation reinforcing bar.

Foremost, it is very price effective and highly competitive despite the early stage of its product life cycle, even compared with existing products having inferior corrosion resistances.

It is modern in that, similar to many modern-day composites, it is tailor-made, according to a solid state

process, to suit an end purpose. The type and thickness of SS-cladding can be selected depending on end-product size and corrosion requirements: the right blend of core ingredients can be chosen for Nuovinox™ SCR to conform to a particular rebar specification.

Despite the advances in energy savings made by the modern steel-making industry, the Nuovinox™ non-melting process consumes only a fraction of the energy consumed by traditional steel melting and casting.

Corrosion tests on sound Nuovinox™ SCR with sealed ends via stainless steel end caps, have shown the same corrosion resistance and projected concrete structure life of 75–100 years for SCR – 316L as a 316 SS solid equivalent. Corrosion tests have shown that small and widely spaced cladding breaks can be tolerated in good quality concrete even with high chloride levels.

Core/Clad and Bar/Concrete bond strengths are shown to be more than functionally adequate and fit-for-purpose.

In conclusion, Nuovinox™ SCR is a product that can greatly improve the durability of concrete structures even in the most aggressive environments and at cost levels approaching conventional, coated CS.

REFERENCES

ASTM – Designation: A XXXX/A XXXXM, January, 2003: Rev 3. Standard Specification for Deformed and Plain, Austenitic Stainless Steel-Clad Carbon-Steel Bars for Concrete Reinforcement.

ASTM A 264 – 94a (Re-approved 1999) Standard Specification for Stainless Chromium-Nickel Steel-Clad Plate, Sheet and Strip.

Bertolini, L., Pastore, T. & Pedeferri, P. 1996. Stainless steel behaviour in simulated concrete pore solution. *Brit. Corros. J.* 1996, 3.

Clad Metal Poducts, Inc., Box 11313 Boulder, CO., USA. Ph: 303-581-0621

Clear, K.D.C. 1998, Long Term Outdoor Exposure Slabs with various Rebar Coatings. Fabricated in 1982 – TMB Associates, Inc., Sperryville, Virginia. Report of the 03-12-1998.

Clemena, G. & Virmani, Y.P. 2001. New Generation of Reinforcement for Transportation Infrastructure. *Concrete Construction*. September 2001/Volume 46/Number 9.

Concrete Society. Guidance on the Use of Stainless Steel Reinforcement, *Technical report* No. 51, Page 6.

Cox, R.N. & Oldfield, J.W. 1996. The Long Term Performance of Austenitic Stainless Steel in Chloride Contaminated Concrete. *Proceedings of the Fourth International Symposium on Corrosion of Reinforcement in Concrete Construction*, Eds. Page, C.L., Bamforth, P.B. & Figg, J.W. (Cambridge, UK: Society of Chemical Industry, 1996), p. 662

Cui, F. & Sagues, A.A. 2002. Dept. of Civil and Environmental Engineering, University of South Florida & R.G. Powers – Materials Office, Florida Department of Transportation Corrosion Behaviour of Stainless Steel Clad Rebar. *NACE Proc. 2002*.

Cui, F. & Sagues, A.A. Dept. of Civil and Environmental Engineering, University of South Florida. Corrosion Performance of Stainless Clad Reinforcing Bar in Concrete. *NACE Proc. 2003.*

Darwin, D. 2001. South Dakota Department of Transportation study of Epoxy Coated reinforcing bars, MMFX, conventional black bars. Office of Research: *Project SD2001-05-F* under contract with University of Kansas Center for Research, Inc.

Engineering News Review, July 16, 2001. Page 17 Flint, G.N. & Cox, R.N. *Magazine of Concrete Research* 40, 142 (March 1988): 13.

Hewitt, J. & Tullmin, M. 1994. Corrosion and Stress Corrosion Cracking Performance of Stainless Steel and Other Reinforcing Bar Materials in Concrete, *Proceedings of International Conference: Corrosion and Corrosion Protection of Steel in Concrete*, Ed. Swamy, K.N. (Sheffield, UK: Sheffield Academic Press): 527.

Jernkontoret, Corrosion Tables, Sweden 1979.

Keys & Basheer 1999. A Comparative Study of the Bond Strength and the Corrosion Resistance of Different Types of Steel Reinforcement. Queen's University, Belfast, Northern Ireland 1999.

McDonald, D.B., Sherman, M.R., Pfeiffer D.W. & Virmani Y.P. 1995. Stainless Steel Reinforcing as Corrosion Protection – *Concrete International* – May 1995.

McDonald, D.B., Pfeifer, D.W. & Sherman, M.R. 1998. Corrosion Evaluation of Epoxy-Coated, Metallic-Clad and Solid Metallic Reinforcing Bars in Concrete, U.S. Dept. of Transportation: Federal Highway Administration Report, Publication No. FHWA-RD-98-153.

MEPS – World Steel Market Review, 2002.

Metal Bulletin: Bi-weekly Publication – Ferrous Scrap Prices Ministry of Transportation Ontario – Engineering Materials Office and Integrity Testing Laboratoy ITL – 140198-0006.

MMFX – Product Bulletin, September 2001 Page 10, Estimated Years of Service Life.

Nurnberger, U. 1996. Corrosion Behavior of Welded Stainless Steel Reinforced Steel in Concrete, *Proceedings of the Fourth International Symposium on Corrosion of Reinforcement in Concrete Construction*, Eds. Page, C.L., Bamforth, P.B. & Figg, J.W. (Cambridge, UK: Society of Chemical Industry, 1996), p. 623.

Parkin, G. 1982. Practical Applications of Stainless Steel Reinforcement. Proceedings of Special Steels & Systems for Corrosion Prevention In Reinforced Concrete, *The Concrete Society*, London.

Pedeferri, P., Bertolini, L., Bolzoni, F. &. Pastore, T. Behavior of Stainless Steels in Concrete, *Proceedings of the International Seminar: The State of the Art of the Repair and Rehabilitation of Reinforced Concrete Structures*, Eds. Silva-Araya, W.F., De Rincon, O.T. & O'Neill, L.P. (Reston, Va/Asce, 1997): 192.

Rasheeduzzafar, Dakhil, F.H., Bader, M.A. & Khan, M.M. 1992. Performance of Corrosion Resisting Steels in Chloride-Bearing Concrete. *ACI Materials Journal*, 439–448.

Sagues, A.A. University of South Florida, Powers, R.G. & Kessler, R. Materials Office Florida Department of Transportation. Corrosion 94 – Paper 299 Corrosion Processes and Field Performance of Epoxy-Coated reinforcing Steel in Marine Substructures.

Stainless Steel in Concrete, *European Federation of Corrosion Publications*, No. 18 – 1996.

Stelax (U.K.) Limited, (West Glamorgan), U.K. Nuovinox™, Product Information, 1998.

Treadaway, K.W.J. 1978. Corrosion of steel reinforcements in concrete construction, Materials Preservatiopn Group, Symposium Soc., Chem. Ind. London, 1978.

Treadaway, K.W.J., Cox, R.N. & Brown, B.L. 1989. Durability of corrosion resisting steels in concrete, Proc. Instn. Civ. Engrs., Part 1. 86, 305–331.

System-based Vision for Strategic and Creative Design, Bontempi (ed.)
© 2003 Swets & Zeitlinger, Lisse, ISBN 90 5809 599 1

Design models for deteriorated concrete bridges

G. Bertagnoli, V.I. Carbone, L. Giordano & G. Mancini
Politecnico di Torino, Italy

ABSTRACT: The aim of this work is the evaluation of the loss of bearing capacity due to steel-concrete bond deterioration subsequent to reinforcement corrosion. It has been proved, by international studies, that a slight corrosion, in terms of reinforcement area reduction, involves substantial bond reduction. In this paper a non-linear static analysis of different beams subjected to corrosion is presented; the first is a sample test used to fit parameters, the second is a T shaped bridge girder beam, the third and the fourth are a numerical evaluation of experimental tests. It has come to evidence that a sensible bond reduction leads the beam to an arch-tie mechanism, shifting failure from steel to concrete side with a relevant bearing capacity reduction. It is also shown how difficult is to spot this problem on actual structures, as the serviceability behaviour of damaged beams is almost the same as sound ones.

1 INTRODUCTION

The reinforcement corrosion process in reinforced concrete structures causes, besides the obvious cross section and material ductility reduction, a significant variation in the steel-concrete interaction mechanisms, mainly as regards bond. The reduction of bearing capacity of structures with corroded reinforcement can, then, be larger than expected from the bare variation of steel resistant section, as classical resistant mechanisms can also be subjected to significant modifications.

The following basic aspects in bond variation due to reinforcement corrosion can be drawn from main bibliographical references of the field:

- For modest values of section corrosion, about 1% area ratio, for instance due to natural causes, a small increment in bond can be seen, mainly due to a superficial roughness and confinement increase, subsequent to a light swelling of corroded bars (Kemp et al. 1968), (Maslehuddin et al. 1990);
- For cross section area reductions falling between 1% and 9% strong bond reductions appear, up to the complete loss of it (Al Sulaimani et al. 1990).
- The bond reduction is due to the bar swelling, that causes high stress levels in nearby concrete with consequent effects of delamination, spalling, cover loss. This process has been fairly numerically reproduced by the use of non linear finite elements (Dagher & Kulendran 1998), (Youang & Weyers 1998) (Soh et al. 1999), (Coronelli 1997).

In bibliography relevant contributions to the study of ultimate and serviceability behaviour of reinforced concrete structures subjected to corrosion can also be found. These works, mainly of experimental nature, yielded numerical explanations, but yet in an incomplete way, (Tachibana et al. 1990), (Mangat & Elgarf 1999), (Almusalam et al. 1996), (Yoon et al. 200), (fib State of Art report 2000).

The aim of this work is numerical evaluation of residual strength of r.c. structures subjected to reinforcement corrosion and give a broad outline for the definition of a referring physical model easy to be used.

2 NUMERICAL MODELLING OF CORROSION EFFECTS

The structure can be modelled by finite elements whose behaviour is defined by the following parameters.

2.1 Concrete

The concrete matrix is described using two-dimensional or three-dimensional elements made of isotropic material up to the tensile stress f_{ct}, becoming then ortotropic, after crack opening, with ortotropic directions parallel and orthogonal to the crack, whose direction, once opened, doesn't change with further load variations (Von Grabe & Tworuschka 1997).

In the uniaxial case, this material is described by the Sargin law in compression and a bilinear law in tension according to the following constitutive equations:

$$E_t = E_0 \frac{1 - B\left(\dfrac{\varepsilon}{\varepsilon_c}\right)^2 - 2C\left(\dfrac{\varepsilon}{\varepsilon_c}\right)^3}{\left[1 + A\left(\dfrac{\varepsilon}{\varepsilon_c}\right) + B\left(\dfrac{\varepsilon}{\varepsilon_c}\right)^2 + 2C\left(\dfrac{\varepsilon}{\varepsilon_c}\right)^3\right]^2} \quad (1)$$

where:

$$p = \frac{\varepsilon_u}{\varepsilon_c} \qquad E_0 = 2.15 \cdot 10^4 \left(\frac{f_c}{10}\right)^{1/3}$$

$$E_S = \frac{\sigma_c}{\varepsilon_c}$$

$$A = \frac{\dfrac{E_0}{E_u} + \left(p^3 - 2p^2\right)\dfrac{E_0}{E_S} - \left(2p^3 - 3p^2 + 1\right)}{p\left(p^2 - 2p + 1\right)}$$

$$B = 2\left(\frac{E_0}{E_S} - 3\right) - 2A \quad C = \left(2 - \frac{E_0}{E_S}\right) + A$$

In tension the constitutive law is assumed linear till the tensile failure stress; the post cracking behaviour is on the other end described by a softening linear branch that reduces to zero the stress orthogonal to the cracking plane when, in that direction, a strain eight time the cracking strain is reached.

The failure surface in multiaxial compression is described in agreement with Kostovos proposal (Kotsovos 1979).

2.2 Steel

Reinforcement is treated in two different ways: considering it smeared in the concrete matrix, without the creation of specific finite elements, or punctual, by the introduction in the mesh of fitting truss elements. In both cases, concrete and steel stiffness contributions are added.

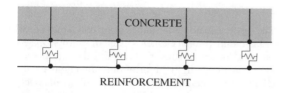

Figure 1.   Steel concrete bond scheme.

2.3 Steel-concrete bond

In order to evaluate the effect of bond reduction on structural response, reinforcement is discretized by introducing truss elements, but steel bar elements are laid on additional nodes, whose position don't coincide with the ones assigned to the continuum concrete matrix. The connection between concrete and reinforcement elements is provided by truss elements, whose first node is on the concrete mesh, while the second one is shared with the reinforcement one.

Connection trusses ensure an elastic-plastic behaviuor in the direction parallel to the reinforcement bar. An infinitely rigid bond in the orthogonal direction (Fig. 1) is also provided. That means to represent the bond, in the line of other numerical proposals in literature (Marti et al. 1998), (Kaufmann & Marti 1998).

3   MODELLING CRITERIA APPLICATION

Modelling criteria described in point 2 are applied to a test structure, which is a simply supported 6 m span beam, with a 150×250 mm rectangular section, reinforced with two continuous $\phi 16$ mm bars and subjected to uniform distributed load. Materials mechanic characteristics are $f_c = 30$ MPa and $f_y = 430$ MPa.

As a first step the beam behaviour without corrosion is analyzed and steel side failure can be appreciated. In Figure 2 the variation of stress in the steel bars is shown in function of distance $\times$ from the beam extremity and the load parameter $\mu = P/P_{ult}$ (where P_{ult} is the collapse load of the sound beam). The upper curve points out that the yielding stress is reached in a broad zone around the midspan of the beam.

On the other side, if we focus on the stresses in the bond truss-elements during the whole load evolution process, can be noticed that the bond strength required, that corresponds to the stress variation in the reinforcement, is amply less than the available one.

A large reduction of the bond limit strength is then necessary to see significant variations in the bearing

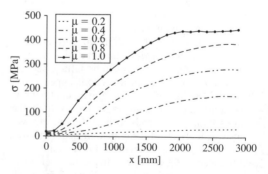

Figure 2.   Stresses in reinforcement in sound beam.

capacity; but, as seen before, even small reductions of steel cross section area due to corrosion, entail large bond spoiling. Therefore we run into a highly hyper-proportional relation between corrosion entity and bearing capacity reduction.

It is now analyzed the behaviour of the same structure assuming a bond reduction of 90%, that is, according to (Al Sulaimani et al. 1990), approximately related to a 7,5% steel cross section reduction, and supposing that the bar slip off can't be reached in the extremity regions (perfect bond in the bearing zone).

It can be noticed that failure is reached in a complete different way: concrete fails under a load that is only 47% of the bearing capacity of the sound beam verified before.

Figure 3, corresponding to Figure 2, shows the stress variation in the reinforcement bars from the bearing to the midspan as a function of load level. Comparing the tensional output of Figure 2 and Figure 3 brings to highlight the following behaviour differences of the beam with corroded reinforcement:

• from the last three load levels, that are provided of the sound curves too, can be noticed that the stress variation is now linear in a significant zone. Its slope is correlated to the maximum bond strength exchangeable per unit length; strength variation in reinforcement is, in fact, a function of this parameter.

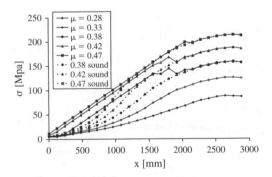

Figure 3.   Stresses in reinforcement in damaged beam.

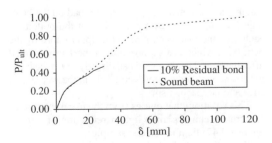

Figure 4.   Midspan deflection of sample beam.

Can be indeed considered, as a limit situation, the complete bond strength absence with working extremity anchorage, that would convert the structure to a tied-arch, with constant strength in the reinforcement and a nil slope line on the diagram in Figure 3.

• in the ultimate condition, this straight line, whose slope is known, turns out to be tangent to the response curve of the sound beam for the same load step. Is then possible to identify the bearing capacity of a damaged beam following the subsequent procedure:

  – residual bond definition, as a function of reinforcement section area reduction, and thus computation of the limit slope of the linear available stress variation law;
  – drawing of the straight line just found with intersection on the ordinate axis corresponding to the limit strength available in the extreme zone;
  – identification of the stress variation curve, tangent to that line, on the sound beam;
  – evaluation of the load level corresponding to that curve, which is the damaged structure ultimate load.

In Figure 4 is shown the relation between applied load (P/P_{ult}) and calculated midspan displacements (δ), both in presence and in absence of corrosion.

It turns clearly out that, for load levels close to damaged beam failure, the deflections of both beams are almost the same, therefore there's no clear evidence in serviceability conditions of the significant safety reduction. Brittle collapse then occurs, preceded by light behaviour discrepancies from the sound element, that take place only beyond 80% of the damaged beam ultimate load.

4   BRIDGE BEAM BEHAVIOUR

Procedure tested in previous paragraphs is now applied to an existing reinforced concrete beam from a 15 m span bridge girder. The beam is reinforced with bent bars following the scheme shown in Figure 5. For this structure a parameter λ has been introduced, defined as characteristic combination of actions multiplier.

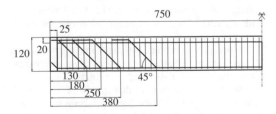

Figure 5.   Structure scheme.

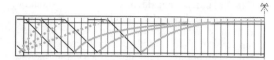

Figure 6.   Resistant mechanism.

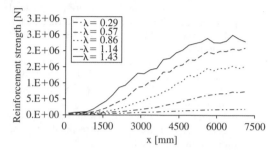

Figure 7.   Strength in bars in sound beam.

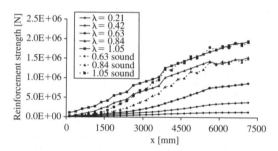

Figure 8.   Strength in bars in damaged beam.

Figure 7 shows the strength in sound reinforcement as a function of distance from beam extremity, for different values of λ parameter; it can be there appreciated that $\lambda_{max} = 1.43$ is reached.

Furthermore, the total strength variation law has a segmental shape, even for low λ values, as the presence of bent bars entails variable reinforcement areas along the span and discontinuous resistance variations.

Is it shown, in Figure 8, the same strength variation in reinforcement bars in presence of a bond reduction of 90% from the sound one. On the same graph are drawn, with a plain line, damaged beam behaviour, and with a dashed line, sound beam behaviour for corresponding λ values.

Before all, can be clearly noticed that the maximum λ value is now 1.05, that means very slight safety margin in the characteristic load combination.

The beam behaviour is then made of a series of straight lines, going from one bent zone to another. Can be therefore deduced that, as a limit situation, the structure will be modified to a sequence of tied-archs,

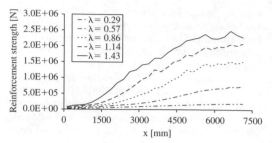

Figure 9.   Strength in bars in sound uniformly reinforced beam.

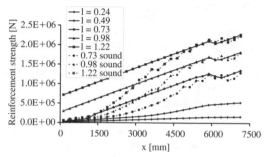

Figure 10.   Strength in bars in corroded uniformly reinforced beam with perfect bond at the extremity.

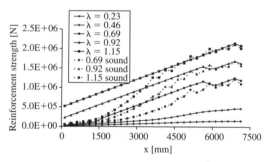

Figure 11.   Strength in bars in corroded uniformly reinforced beam with corroded anchorage length.

suitable to work in series to carry the load on the bearings (Fig. 6). The reason being that, at the bottom of each arch the load is carried to the upper chord by the bent bars, which are supposed to be perfectly bonded after the bent. Then the load can directly reach the bearing, by an inclined compression field in the web with a strut and tie mechanism.

This solution, that comes from the use of bent reinforcement, can explain a lighter reduction of λ (0.73 against 0.47) than in the test sample.

It has been studied, furthermore, the behaviour of an hypothetical beam, just like the previous one, but with uniform reinforcement along the span up to the bearings.

This beam was analyzed in three different situations: sound (Fig. 9), with corroded reinforcement but perfect bond between bars and concrete at the extremity (Fig. 10), corroded and bonded at the ends using the required bond length, that is also subjected to corrosion (Fig. 11).

It comes to evidence that λ parameter keeps its value $\lambda_{max} = 1.43$ on the sound beam, behaving almost as the previous sample, as available bond is much bigger than requested. If reinforcement is corroded but rigidly bonded at the beam edges a $\lambda_{max} = 1.22$ can be evaluated, that drops to $\lambda_{max} = 1.15$ with actual bond in bearing regions.

The opportunity to use a single tied-arch mechanism, and then to fully work all the ties till the bearing, leads to the optimization of strength redistributions in reinforcement along its whole length, even beyond the zones where, in the sound beam, it wouldn't result strictly necessary.

5 EXPERIMENTAL VALIDATION OF PROPOSED METHODOLOGY

The experimental validation of proposed methodology may be achieved using the tests of Molina, Gutierrez and Garcia (Molina et al. 2002) on simple supported beams in which the loss of bond is simulated by means of plastic ducts along some length of reinforcements bars (Fig. 12).

In particular tests A2 and DA3 have been numerically reproduced, corresponding respectively to a sound beam and a bond reduction of 55%.

The comparison results are presented in Table 1, both in terms of ultimate load (Q_{ult}) and deflection at a load level of 50% (δ_1) and 75% (δ_2) of Q_{ult}.

Can be appreciated a very good agreement between experimental and numerical tests in terms of ultimate load (Q_{ult}). Some discrepancies has been found on deflections, being the numerical model more rigid than the experimental one. This difference may be attributed to the rigid-plastic schematization used for bond in our model which can't take account of deformability within the anchorage length, that can play an important role in DA3 test, characterized by 13 debonding regions.

On the other hand the simplicity of rigid plastic approach allows to reach accurate results on the ultimate strength evaluation maintaining the operationally simplified approach of smeared cracking.

6 CONCLUSIONS

Corrosion in reinforcement involves over proportional variations in bearing capacity, with a very high reduction of safety margins and sometimes collapse risks under serviceability conditions. The proposed numerical model is able to follow the actual behaviour of the structure in presence of severe corrosion and gives the grounds for a circumstantial evaluation of residual safety.

It's also possible to use the simplified proposed procedure to gain a reliable evaluation of structural behaviour. Further research will make it possible to appreciate corrosion effect in presence of multiaxial strength situations.

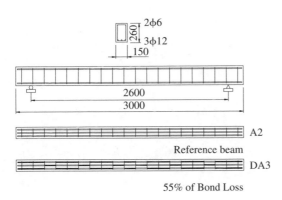

Figure 12.   Plans elevations and section of test beams.

Table 1.   Experimental and numerical tests comparison.

Test	Q_{ult} Exper. [kN]	Q_{ult} Numer. [kN]	δ_1 Exper. [mm]	δ_1 Numer. [mm]	δ_2 Exper. [mm]	δ_2 Numer. [mm]
A2	81.2	85.3	6.2	5.9	10	9.5
DA3	57.0	57.8	5.7	3.6	9.0	6.1

REFERENCES

Al Sulaimani G.J., Kalemullah M., Basumbul I.A. & Rasheeduzzafar 1990. Influence of Corrosion and Cracking on Bond Behaviuor and Strength of Reinforced Concrete Members; *ACI Structural Journal*, March–April 1990, pp. 220–231.
Almusalam A.A., Al Gahtani A.S., Aziz A.R. & Dakhil F.H., Rasheeduzzafar 1996. Effect of Reinforcement Corrosion on Flexural Behaviour of Concrete Slabs; *Journal of Material in Civil Engineering*, August 1996, pp. 123–127.
Coronelli D. 1997. Bond of Corroded Bars in Confined Concrete: Test Results and Mechanical Modelling; *Studi e Ricerche*, Politecnico di Milano-Italia, vol. 18, 1997.
Dagher H.J. & Kulendran S. 1998. Finite Element Modelling of Corrosion Damage in Concrete Structures; *ACI Structural Journal*; November–December 1998, pp. 699–708.

Fib State of Art Report 2000. Bond of Reinforcement in Concrete; Bulletin 10, August 2000, Chapter 4/Bond of Corroded Reinforcements, pp. 188–215.

Kaufmann W. & Marti P. 1998. Structural Concrete: Cracked Membrane Model; *Journal of Structural Engineering, ASCE*, 1998, 124, No. 12, pp. 1467–1475.

Kemp E.L., Brezny F.S. & Unterspan J.A. 1968. Effects of Rust and Scale on the Bond Characteristic of Deformed Reinforcing Bars; *ACI Journal*, September 1968, pp. 743–756.

Kotsovos M.D. 1979. A mathematical Description of the Strength Properties of Concrete under Generalized Stress; *Magazine of Concrete Reserch*, 1979, 31, No. 1, pp. 151–158.

Mangat P.S. & Elgarf M.S. 1999. Flexural Strenght of Concrete Beams with Corroding Reinforcement; *ACI Structural Journal*, January–Febrruary 1999, pp. 149–158.

Marti P., Alvarez M., Kaufmann W. & Sigrist V. 1998. Tension Chord Model for Structural Concrete; *Structural Engineering International*, 1998, 8, No. 4, pp. 287–298.

Maslehuddin M., Allam I.M., Al-Sulaimani G.J., Al-Mana A.I. & Abduljauwad S.N. 1990. Effect of Rusting of Reinforcing Steel on Its Mechanical Properties and Bond with Concrete; *ACI Materials Journal*, September–October 1990, pp. 496–502.

Molina M., Gutierrez P. & Garcia Mª D. 2002. Evaluation of reinforced concrete structures with partial loss of concrete/steel bond; Bond in Concrete – from research to standards. *Proceedings of the third international Symposium held at the Budapest University of Technology and Economics*, Budapest, Hungary, 20–22 November 2002, pp. 182–189.

Soh C.K., Chiew S.P. & Dong Y.X. 1999. Damage Model for Concrete-Steel Interface; *Journal of Engineering Mechanics*, August 1999, pp. 979–993.

Tachibana Y., Maeda K., Kajikawa Y. & Kawamura M. 1990. *Mechanical Behaviour of RC Beams Damaged by Corrosion of Reinforcement*; Ed. Page, Treadaway and Bamforth, Elsevier, 1990, London.

Von Grabe W. & Tworuschka H. 1997. An interface Algorithm for non linear Reliability Analysis of Reinforced Concrete Structures using ADINA; *Computer and Structures*, 1997, 64, No. 5/6.

Yoon S., Wang K., Weiss W.J. & Shah S.P. 2000. Interaction Between Loading Corrosion and Serviceability of Reinforced Concrete; *ACI Material Journal*, November–December 2000, pp. 637–644.

Youang Liu, R.E. & Weyers 1998. Modelling the Time-to-Corrosion Cracking in Chloride Contaminated Reinforced Concrete Structures; *ACI Material Journal*, November–December 1998, pp. 675–681.

System-based Vision for Strategic and Creative Design, Bontempi (ed.)
© 2003 Swets & Zeitlinger, Lisse, ISBN 90 5809 599 1

A case study of a concrete bridge corrosion phenomenon

A. Guettala
Civil Engineering Dept., Biskra University, Algeria

A. Abibsi
Mechanical Engineering Dept., Biskra University, Algeria

ABSTRACT: Certain cement concrete works remain in excellent state after more than one century of exposure to particularly severe climatic conditions. This article describes the diagnosis carried out on the degradation of a reinforced concrete bridge put into service only 15 years back as well as the follow up after repair. The principle causes of the acceleration of its degradation are highlighted. The stages of repair and reinforcement are discussed as well as its state after repair. It's also shown in this study the importance of maintenance and regular inspection of a given cement concrete work or the lack of it.

1 INTRODUCTION

The cement concrete constitutes certainly one of most durable building materials. Certain works remain in excellent state after more than one century of exposure to particularly severe climatic conditions.

However, the exposure of these materials to a severe environment such as: temperature, deicing agents, humidity, industrial gases makes them more vulnerable. Their durability depends mainly on the work design, material selection and their processing. The reinforcement corrosion phenomenon is the principle cause of disorder of the cement concrete structures and it has been the main concern worldwide.

For instance, in the USA alone, a colossal sum of 250 billions dollars will be spent for the next years in order to repair the 650 000 bridges affected by the corrosion phenomenon. The Ministère des Transports du Québec, Info DLC (1998) has proposed a series of measures to combat this corrosion phenomenon, mainly by the use of corrosion inhibitors, galvanized steel, cathodic protection, zinc based paints…

In any investigation, it appears necessary to proceed first in collecting the maximum of information: age, architectural plans, nature of the materials used (cement type, proportioning…) and the nature of the environment. A detailed visual inspection makes it possible to recognize the nature of the disorders, to describe their localization and, if necessary, to follow their evolution.

In general, one of the principle causes of disorders is the corrosion of the reinforcements. The disorders appear in the shape of cracks, bursts. But the corrosion started generally in the zones nearby without causing visible disorders yet. After having identified the origins of corrosion (chlorides, carbonation), and evaluated its extent (penetration depth), the use of non-destructive tests (measurements of wrapping and potential) makes it possible to determine the extent of the corroded zones correctly and to predict its evolution, Larrard & Bouny (2000) and Taché & Vié (1998). The principle parameters which condition the behavior of a work with respect to corrosion are, in the order of their importance: the wrapping, the quality of the concrete and the environment Poineau (1994) and Baron (1992). This article describes the diagnosis carried out on the degradation of a reinforced concrete bridge put into service 15 years ago. The principle causes of the acceleration of its degradation are highlighted. The stages of repair and reinforcement are discussed in later stage. The state of the work after 3 years of repair is also described as well as the importance of regular inspection and maintenance; or the lack of these as shown in this study.

2 BRIDGE DESCRIPTION

The work investigated here is located on the Route Nationale3, (RN3), 25 km from Biskra, (South-east of Algeria) Figure 1, Guettala, Benmenarek & Abibsi (2002). It was completed and brought into service in 1985. It is about a horizontal mixed bridge of an

Figure 1.   General view of the bridge.

Figure 2.   Bandage degradation.

overall length of 340 m, made up of 11 isostatic spans of 30 m each. A 10 m width deck is composed of three metallic girders of 1.50 m height. These girders are linked by two butt-spacers, three intermediate-spacers and surmounted by a reinforced concrete slab. Its profile consists of an 8 m width roadway framed by two pavements of 1 m width. The roadway support is made of two reinforced concrete supports with wing walls and ten intermediate piles consisting of: 1.20 m width and 0.8 m height bandage resting on three circular piers of 1.0 m in diameter. The pile foundations established in the river soil are protected by metal sheeting embankment, Figure 1.

We are primarily interested in the degradation of the pile concrete. This was proportioned with 400 kg/m^3 Portland cement – CPA 325 – with water/cement (W/C) ratio of 0.55; 750 kg/m^3 of round sand 0/5 and 1100 kg/m^3 gravel 5/25 coming from a quarry nearby.

3   DESCRIPTION OF THE DISORDERS OBSERVED ON THE PILES BEFORE REPAIR

Many disorders were noticed on this work mainly cracking of the concrete and corrosion of the reinforcements causing the bursting of the wrapping concrete. These degradations were observed on the bandages and on the piers, particularly, those protected by coffer dams in metal sheeting.

3.1   Bandages

The bandages are mostly affected, some of which badly, see Figure 2. The degradations are of the following forms:

- advanced corrosion of the reinforcements;
- deterioration of the wrapping concrete;
- multi-directional cracking of the concrete.

It is at the bandage ends that degradations are significant. In their median part, they are only slightly degraded. This can be explained simply by the stagnation of water at these ends.

3.2   Piers

We noticed that the intermediate piers are not affected; on the other hand certain bank piers (upstream and downstream) showed degradations. The degradations appear in the following forms:

- Parallel, vertical and equidistant cracking, the position of the cracks generally coincides with the position of steels;
- Bursting by places and beginning of separation of concrete plates;
- Significant corrosion of the reinforcements observed after concrete scouring of cracked wrapping.

3.3   Deck

The deck presents an overall satisfactory state.

4   CAUSES OF DEGRADATION

The construction of the bridge was carried out between 1983 and 1985 by the National Company of the Bridges and Work of Art (SAPTA). In 1994, hidden walls in full breeze blocks were built above the diagrids of the piles and the abutments. This was done for countering possible acts of sabotage.

We have observed that the joints of roadways at the pavements level were not tight and this allowed rainwater to penetrate through the joints and pouring on the bandages where they are retained by the hidden walls before running out on the piers. The badly designed metal sheeting embankments form a basin like, retaining thus water.

We noticed that where there are signs of water flows, the degradations were very significant indicating that the water rain is unusually very aggressive. After investigation, it appeared that the severe aggression of the rain water resulted from salts falling off the trucks making the shuttle between the salt layer Chat Malghigh and Biskra. The rain helps the leaching of the deck and dissolves the salt. The resulted salty water penetrates through the loose inter-spans joints.

Therefore, the principle cause of degradations is the chemical attack by the chloride ions causing the corrosion of the reinforcements and consequently the bursting of the concrete.

The concentrated degradation at the level of the ends of the bandages is probably related to the band-ages roof shape (symmetrical cross fall of 2.5%).

The lack of wrapping in certain places; the bad quality of the concrete used indicated by the state of the facade and the presence of the gravel niches con-tributed enormously to corrosion of steels.

5 CORROSION MECHANISM

The protection of the reinforcements is related to two processes:

- chemical process: the alkalinity produced during the cement hydration;
- physical process: by wrapping, acting like a barrier with respect to the environment.

It is accepted that the mechanism of the corrosion of the reinforcements proceeds in two successive stages:

- Starting phase: this corresponds to the penetration of the aggressive agents (mainly carbon dioxide, air and chlorides) through the layer of wrapping, until the starting of the corrosion of the reinforce-ment; destruction of "a passive film". This period depends mainly on the processes assuring corro-sion elements to the reinforcement. It depends also on the chemical reactions taking place within the concrete. Undoubtedly, the wrapping quality (per-meability, thickness…) have a fundamental role on the aggressive agents penetration, Dhir (1993).
- Growth phase: this corresponds to the formation of expansive iron oxides and to the damaging of the complex reinforcement-concrete. This period is particularly related to the speed of corrosion, Dhir (1993).

In our case, the corrosion observed is in its growth phase state. It is of uniform type (generalized) which occurs when the reinforcement is not protected any more and micro-cells corrosion can be formed over the whole surface.

5.1 Consequences of the reinforcement corrosion

Once the reinforcement has been corroded, it starts expanding, creating thus cracks on the surface and

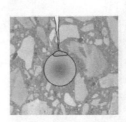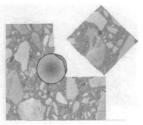

Figure 3.   Cracking and spalling.

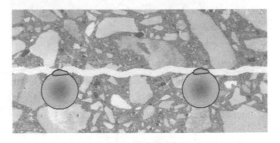

Figure 4.   Separation of the concrete plates.

finally bursting by places and beginning of separation of concrete plates, Figures 3 and 4.

6 REPAIRS

6.1 Elimination of the immediate causes (to be done as an urgent work)

- Demolition of the hidden walls built on the diagrids of the supports;
- Installation of tight roadway joints including in the zones of the pavements;
- Widening of the openings of the gutter-spouts;
- Cutting of the metal sheeting embankments which form a retaining water basin, avoiding therefore the water stagnation.

6.2 Piers

Repair relates to the damaged piers cited previously. The operations to be carried out are:

6.2.1 Surface preparation
- Prick all the disaggregated and removed concrete along the whole wall height (including the buried parts) until reaching the concrete center.
- Expose the reinforcements at least 2 cm depth all around the pier.
- Remove all the corrosion products from the rein-forcement bars using blast sanding. This operation must be done perfectly well.

Figure 5.   Addition of new bars to the reinforcement.

- Rinsing and blast cleaning of all the concrete surfaces;
- The non-exposed bars to the sand blasts should be cleaned with metallic brushes
- Treat the corroded reinforcements using an anti-corrosion product generally in the form of pre-proportioned kit composed of a resin and a hardener, Jonathan (1993).
- Strengthen the affected reinforcements (section diminution) by adding new bars, Figure 5, and making sure of the total covering at pier base, if the existing bars are corroded;

6.2.2 Framing
The framing, preferably, metallic, must be provided with spacers intended to ensure the wrapping of the reinforcements with an increase in the diameter of the piers by 20 cm compared to the initial diameter. This increase in the section will facilitate the casting of the concrete and also reinforce the pier resistance.

6.2.3 Sheaths filling
The filling of the sheaths will be done by a latex concrete. The latex concretes are not completely tight to the chlorides penetration, but they offer a protection quite higher than that obtained with current concretes of type B 30. According to various authors, the lifespan of the latex concrete patching (repair) should be at least fifteen to twenty years, Baron & Ollivier (1997).

The formulation of the latex concretes is identical to that of the cement to take into account the water proportion contained in the latex emulsion and the significant effect of latex plasticization. According to the proportioning and the type of latex employed, the ratio W/C will be between 0.30 and 0.40. Due to the latexes diversity available, one should check the compatibility of the selected product with the cement using tests. The quantity of latex to be used

Table 1.   Typical formulation for a latex concrete for 1 m³.

Cement (kg)	Latex (%)	Gravel (kg)	W/C	Sand (kg)
390	15	985	0.3	805

Table 2.   Concrete characteristics.

Sinking (cm)	Resistance to 28 days (MPa)	Permeability to chloride ions (coulombs)
15.5	45	850

(expressed in dry extract) ranges generally between 10 and 20% of the cement mass.

The improvement of the properties is proportional to the added quantity of latex. With low proportioning, the gain is marginal; there is less of polymer to form film and the reduction of W/C ratio is less significant. However, an excessive proportioning is not economically justified. This causes an exaggerated air drive and in all cases, the improvement of the properties reaches its maximum or even regresses. Tables 1 and 2 give a typical formulation and properties for a latex concrete respectively, Baron & Ollivier (1997).

The interest for the latex concretes lies in the fact that it improves several properties (impermeability, durability, adherence) while preserving an easy application.

6.3 Bandages

As for the bandages repair, the use of cementing matrices reinforced with polypropylene fiber is well recommended.

6.3.1 Surface preparation
- Prick all the disaggregated and removed concrete along the whole bandage length until reaching the concrete centre;
- Rinsing and blast cleaning of all the concrete surfaces;
- Remove all the corrosion products from the reinforcement bars using blast sanding;
- Clean the non-exposed bars, using metallic brushes;
- Treat the corroded reinforcements, using an anti-corrosion product generally in the form of pre-proportioned kit composed of a resin and a hardener, Baron & Ollivier (1997).

6.3.2 Latex based mortar
The durability of mortar depends not only on the mortar fix itself but on the environmental conditions encountered during its service life. In this study, a latex based mortar reinforced with the polypropylene fiber has

been used for the bandages repair. In general, the latex content varies between 10 and 20% in respect to the cement mass. The latex addition gives a good adherence to the support. It gives also the impermeability and the improvement in protection of the reinforcement, thus resistance to chemical attacks.

7 DESCRIPTION OF THE DISORDERS OBSERVED ON THE PILES AFTER REPAIR

The second part of the present work concerns the study of the bridge state after 3 years since its repair. It has been observed practically the same type of degradation for the bandages. However, the piers showed no sign of degradation.

7.1 *Bandages*

Figure 6. Large cracks at the end of the bandage.

The bandages are mostly affected, mainly at their ends. The degradations started by the cracks formation at the end of all the bandages. These cracks can be as large as 10 mm, Figure 6. Some of the bandages have shown cracks in all directions, Figure 7. The formation of these cracks is certainly due to the reinforcement corrosion which arises from the fact that the work repair had not been carried out properly, poor corrosion product removal during repair.

The removal of the corrosion product must be done properly. If this is not the case, it's possible that the corrosion process will continue after repair, especially in the case of chloride induced corrosion as in our case. Special care must be taken to ensure the cleaning efficiency. This has not been done because of probably the difficulties encountered during repair. Besides, the mortar layer used was very thin, only the degraded part concrete has been replaced.

What have caused and accelerated the corrosion phenomenon was the rain water flow through the pavement joints which were not so tight, Figure 8.

We noticed that where there are signs of water flows, the degradations were very significant indicating that the water rain is unusually very aggressive.

Figure 7. Multi-directional cracking of the concrete.

7.2 *Piers*

We noticed that the piers are not affected.

8 INSPECTION AND MAINTENANCE

An inspection program should be drawn for any work in service. And it's more urgent and a must for a work which had been already repaired like in our case. This program should include the following actions:

- Reckon in time the work damages and determine the degradation origins;

Figure 8. A non tight joint.

- Establish a technical file for the work from observations;
- Provide the data for maintenance;
- Take urgent measures imposed.

9 CONCLUSIONS

This article presents the results of a diagnostic and repairs realized on a reinforced concrete bridge as well as the bridge state after repair. The objectives of this study were mainly to determine the causes of degradation of the bandages and the piers of the bridge and also to highlight the importance of the maintenance and inspection of the work. The results of this investigation converge towards a chemical attack by the chloride ions which caused the corrosion of the reinforcements leading inevitably to bursting of the concrete.

For bandages repair, a latex based mortar reinforced with polypropylene fiber has been used whereas a latex concrete has been used for the piers repair.

Three years after repair, practically, the same disorders were noticed on the work mainly cracking of the concrete and corrosion of the reinforcements causing the bursting of the wrapping concrete. These degradations were observed particularly at the bandage ends where there are signs of water flows due to a poor jointing.

These degradations would be avoided if a simple inspection program had been adapted by the concerned Organism Technical Services.

REFERENCES

Baron, J. & Ollivier, J.P. (Eyrolles) 1997. *Les béton bases et données pour leur formulation.* Paris: Eyrolles.

Baron, R. 1992. La durabilité des armatures et du béton d'enrobage. *Presses de l'ENPC*:173–225.

Dhir, R.K. 1993. Concrete: chlorite diffusion rates. *Mag.Con. Res*: 1–9.

Guettala, A., Benmenarek, S. & Abibsi, A. 2002. Diagnosis and Repair of a Reinforced Concrete Bridge: 6th *International Conference on Concrete Technology for Developing Coubtries; Proc.intern.conf.,* Amman, 21–23 October 2002. (3): 819–828.

Info DLC. 1998. Etat de la recherche sur la corrosion des aciers d'armature dans les ouvrages de béton. *Bulletin d'information technique* (11) 3.

Jonathan, G.M.W. 1993. Quelques expériences étrangères en matières de prévention et de diagnostic: spécification pour des grands ouvrages: pont, tunnels et barrages. *Annales de ITBTP*: 3–20.

Larrard, F.D. & Bouny, V.B. 2000. Vieillissement des bétons en milieu naturel. *Bulletin des Laboratoires des Ponts et Chaussées*, Mars-Avril: 51–65.

Poineau, D. 1994. Origine des pathologies, observation, diagnostic dans les ouvrages d'art. *Bulletin de Liaison des Laboratoires des Ponts et Chaussées*: 97–124.

Taché, G. & Vié, D. 1998. Diagnostic des ouvrages en béton armé, facteurs de vieillissement des ouvrages. *Annales du Bâtiment et des Travaux Publics*: 27–37.

System-based Vision for Strategic and Creative Design, Bontempi (ed.)
© 2003 Swets & Zeitlinger, Lisse, ISBN 90 5809 599 1

Estimation of the residual capability of existing buildings subjected to re-conversion of use: a non linear approach

H. Albertini Neto & E. Garavaglia
Department of Structural Engineering, Politecnico di Milano, Milan, Italy

L. Sgambi
Department of Structural and Geotechnical Engineering, Università di Roma "La Sapienza, Rome, Italy

ABSTRACT: The building rehabilitation needs particular attention. It depends on a lot of factors: socio-economic, urban and engineering factors. By structural point of view the rehabilitation of an existing building needs a structural approach different by the approach usually adopted for the project of a new building, in fact the parameters involved in the problem often suffer of uncertainty that cannot be forgotten. Following this way, the possible re-conversion of an existing building is here analysed and to take in account the uncertainties involved in the analysis a non-linear approach has been adopted.

1 INTRODUCTION

The building rehabilitation needs particular attention. The first aspect involved in the building rehabilitation problem is the choice of the re-conversion of use (by industrial use to residential, commercial or public use). In this decision different aspects play an important role: social/cultural aspects, land-planning and urban-planning aspects, structural and technological aspects, etc.; in a set of possible choices, these aspects can lead to prefer a solution to any other. To make a correct choice, a complete analysis of the urban contest and of its current and future requirements is important, but the "best" choice of re-conversion cannot be based only on this analysis: it must, also, involve a correct structural analysis.

The changing of the original use of a building can involve a strong changing of the load history and, as consequence, of the residence and reliability of the whole structural system. It is clear that, a reliability analysis of the existing and future situations cannot be neglected. It can suggest the "best" rehabilitation action between those proposed (the "best" obtained by the analysis of the contest).

In an existing building the reliability analysis is not simple. Usually the structural behaviour of an existing building is not linear and the parameters involved in it change by case to case, therefore they are not univocally defined and, often, present a random behaviour.

By the previous observation emerged that in a project of rehabilitation of existing buildings the structural reliability analysis must be based on a non-deterministic approach.

In this paper the possible re-conversion of use of an existing building is investigate with attention at its residual bearing capability. The evaluation of the residual bearing capability has required the survey of the current deterioration level of the building and the formulation of hypotheses about the current performance and structural capability of the building.

2 STRUCTURAL REHABILITATION OF R.C. EXISTING BUILDING: A NON LINEAR APPROACH

In the analysis of an existing RC building the uncertainties can involve several quantities like the load histories, the geometry of the structure, the quality of connections. For structural systems having non-linear behaviour, a realistic description of the response under all load levels can be obtained only by taking this non-linearity into account.

In this context, though the reliability of the structure as resulting from a general and comprehensive examination of all its possible failure modes, one must pay attention to the following four aspects which define the assessment process:

– Available data.
– Modelling of the uncertainties.
– Non-linear structural analysis.
– Synthesis of the results.

The data needed to operate a structural rehabilitation of R.C. existing buildings are: the structural project tables, the deterioration level reached by the system: position of the damage and its entity, and news about the new destination.

To check the deterioration level reached, important is a robust monitoring action able to give safety answers with few points of measurement.

To model the structure like in origin leads to know, in deterministic way, the performance of the system when it was "new". Starting by this modelling, the uncertainties can be introduced.

2.1 Structural analysis of R.C. building

In most cases the R.C. structure should be analysed by taking material and geometrical non-linearity into account and their performance should generally be described with reference to a specified set of limit states as regards both serviceability and ultimate conditions (Bontempi et al. 1998, Biondini et al. 2001).

2.1.1 Limit states

Splitting cracks and considerable creep effects may occur if the compression stresses in concrete are too high. Besides, excessive stresses either in reinforcing steel can lead to unacceptable crack patterns. Excessive displacements $\mathbf{s}$ may also involve loss of serviceability. They have to be limited within assigned bounds $\mathbf{s}^-$ and $\mathbf{s}^+$. Based on these considerations, the following limitations account for adequate durability at the serviceability stage (*Serviceability Limit States*):

$$\mathbf{1.} \; -\sigma_c \le \alpha_c f_c \quad \mathbf{2.} \; |\sigma_s| \le \alpha_s f_{sy} \quad \mathbf{3.} \; -\mathbf{s}^- \le \mathbf{s} \le \mathbf{s}^+ \qquad (1)$$

where α_c and α_s are suitable reduction factors of the strengths f_c and f_s.

When the strain in concrete ε_c or the reinforcing steel ε_s reach the limit values ε_{cu} and ε_{su} respectively, the collapse of the corresponding cross-section occurs. However, the collapse of a single cross-section doesn't necessarily lead to the collapse of the whole structure, the latter is caused by the loss of equilibrium arising when the reactions r requested for the loads f can no longer be developed. So the following ultimate conditions have to be verified (*Ultimate Limit States*).

$$\mathbf{1.} \; -\varepsilon_c \le -\varepsilon_{cu} \quad \mathbf{2.} \; |\varepsilon_s| \le \varepsilon_{su} \quad \mathbf{3.} \; \mathbf{f} \le \mathbf{r} \qquad (2)$$

Since these limit states refer to internal quantities of the system, a check of the structural performance through a non-linear analysis needs to be carried out at the load level. To this aim, it is useful to assume $\mathbf{f} = \mathbf{g} + \lambda \mathbf{q}$, where $\mathbf{g}$ is a vector of dead loads and $\mathbf{q}$ a vector of live loads whose intensity varies proportionally to a unique multiplier $\lambda \ge 0$. With this position, the safe domains at both serviceability and ultimate states and the previous limit conditions can be synthetically formulated as follows:

$$\lambda \le \lambda_s = \max_{\lambda \in s} \lambda$$

$$S = \left\{ \lambda \mid -\sigma_c \le -\alpha_c f_c, \; |\sigma_s| \le \alpha_s f_{sy}, \; \mathbf{s}^- \le \mathbf{s} \le \mathbf{s}^+ \right\} \qquad (3)$$

$$\lambda \le \lambda_U = \max_{\lambda \in U} \lambda$$

$$U = \left\{ \lambda \mid -\varepsilon_c \le -\varepsilon_{cu}, \; |\varepsilon_s| \le \varepsilon_{su}, \; \mathbf{f} \le \mathbf{r} \right\} \qquad (4)$$

being λ_s and λ_u the limit multipliers which define the failure loads.

2.2 Structural model and non linear analysis

2.2.1 The R.C. beam element

The R.C. structure here presented is modeled using R.C. beam finite element whose formulation, based the Bernoulli-Navier hypothesis, deals with such kinds of nonlinearities (Bontempi et al. 1995, Malerba 1998) (Fig. 1). In particular, both material $\mathbf{K}'_M$ and geometrical $\mathbf{K}'_G$ contributes to the element stiffness matrix $\mathbf{K}'$ and to the nodal force vector $\mathbf{f}'$, equivalent to the applied loads $\mathbf{f}'_0$, are derived by applying the principle of virtual displacements and evaluated by numerical integration over the length l of the beam:

$$\mathbf{K}' = \mathbf{K}'_M + \mathbf{K}'_G \qquad \mathbf{f}' = \int_0^l \mathbf{N}^T \mathbf{f}'_0 dx$$

$$\mathbf{K}'_M = \int_0^l \mathbf{B}^T \mathbf{H} \mathbf{B} dx \qquad \mathbf{K}'_G = \int_0^l N \mathbf{G}^T \mathbf{G} dx \qquad (5)$$

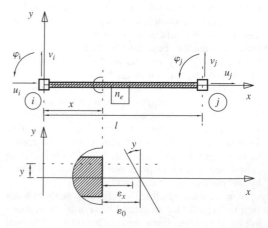

Figure 1. Bernoulli-Navier finite element.

$$\mathbf{N}=\begin{bmatrix} \mathbf{N}_a & 0 \\ \hline 0 & \mathbf{N}_b \end{bmatrix}$$

$$\mathbf{B}=\begin{bmatrix} \partial\mathbf{N}_a/\partial x & 0 \\ \hline 0 & \partial^2\mathbf{N}_b/\partial x^2 \end{bmatrix} \quad \mathbf{G}=\begin{bmatrix} 0 & \dfrac{\partial\mathbf{N}_b}{\partial x} \end{bmatrix} \tag{6}$$

where N is the axial force and $\mathbf{N}$ is a matrix of axial $\mathbf{N}_a$ and bending $\mathbf{N}_b$ displacement functions. In the following, the shape functions of a linear elastic beam element having uniform cross-sectional stiffness $\mathbf{H}$ and loaded only at its ends adopted (Przemieniecki 1968). However, due to material non-linearity, the cross-sectional stiffness distribution along the beam is non uniform even for prismatic members with uniform reinforcement. Thus, the matrix $\mathbf{H}$ has to be computed for each section by integration over the area of composite element or by assembling contributes of concrete and steel. In this way, after the constitutive laws of the materials are specified, the matrix $\mathbf{H}$ of each section can be computed under all load levels. The equilibrium conditions of the beam element are derived from the principle of virtual work. Thus, by assembling the stiffness matrix $\mathbf{K}$ and the vectors of the nodal forces $\mathbf{f}$ with reference to global co-ordinate system, equilibrium can be formally expressed by $\mathbf{Ks} = \mathbf{f}$, where s is the vector of the nodal displacements. It is worth noting that the vectors $\mathbf{f}$ and $\mathbf{s}$ have to considered as total or incremental quantities depending on the nature of the stiffness matrix $\mathbf{K} = \mathbf{K(s)}$, or if a secant or, a tangent formulation is adopted.

2.2.2. *Material properties*

The stress–strain diagram of the concrete is described by the Saenz's law in compression without strength in tension (Fig. 2a). By assuming $E_{c0} = 9500 f_c^{1/3}$ [MPa] the diagram is completely defined by the strain limits ε_{c1}, ε_{cu}, and the compression strength f_c.

The stress–strain diagrams of the reinforcing steel is defined by the elastic modulus $E_s = f_{sy}/\varepsilon_{sy}$, the limit strains ε_{su}, and the yield strengths f_{sy} (Fig. 2b) (Bontempi et al. 1998).

3 APPLICATION

The presented procedure is applied at the rehabilitation analysis of the existing building shown in Figure 3 (Milia et al. 2001, Albertini 2002).

The building was built in Cantù, Milan (Italy), in 1955. Since 1989, year of its abandon, it was the Molteni's furniture farm. Now it is under preservation of work as example of "industrial archeology", therefore it cannot be pulled down. In the current City Urban Plan (CUP) the rehabilitation of this building is considered. Different solutions are proposed in it, each of them have to be evaluated under different

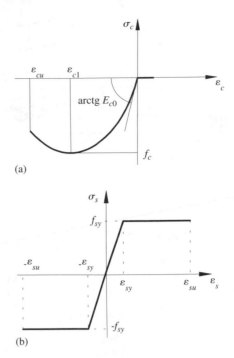

(a)

(b)

Figure 2. Stress–strain diagrams of the materials: (a) concrete; b) reinforcement.

Figure 3. Fossano building (ex Molteni) external view.

points of view, not last the structural compatibility of the choice. One of the solutions proposed in the CUP is the re-conversion of the building in district library. This solution is here analysed.

3.1 *Geometry of the building*

The building is constitutes by four floors (Fig. 3).

The plan shows an "L" form (Fig. 4). In Table 1 the principal dimensions are reported.

The structure is a R.C. frame with R.C. beams-and waffle slabs. Instead, a R.C. cylindrical shell (Fig. 5) constitutes the roof.

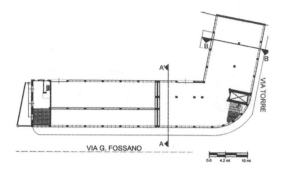

Figure 4. Fossano building: ground level plan (Milia et al. 2001).

Table 1. Building geometry.

Building dimensions		
Transversal width		12.4 m
Length	On Via G. Fossano	47 m
	On Via Torre	19 m
Floor highness	Ground floor	5.0 m
	1st floor	3.5 m
	2nd floor	3.5 m
	3rd floor	3.0 m
Inter-column	On the fronts	4.7 m
span	transversally	varying

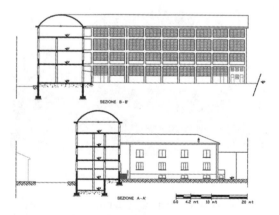

Figure 5. Building sections (Milia et al. 2001).

Though a visual analysis the damage level reached by the structural elements has been recorded (Fig. 6). Of course, more detailed analysis are necessary, but already this simple, not invasive and not expensive analysis is able to give useful information. In our case it has put in evidence that the relevant damage is the carbonation of concrete with the consequent reduction of the resistant section of beams and columns and the reinforcing steel damage; this can seriously compromises the bearing capability of the whole structure or of a part of it. The results of this analysis have been considered in the modelling has uncertainties.

3.2 *The case study: linear and non linear analysis*

The analysis of residual capability of an existing building can be dealt in different ways:

- *Linear analysis* with controls on conformity with structural design code prescriptions.
- *Non linear analysis* with search of ultimate limit state level.
- *Non linear fuzzy analysis* able to take in account the uncertain involved in the problem (Biondini et al. 2002, Sgambi 2003).

All these analyses will require different information; of course, as output, they will give different answers more or less suitable.

3.2.1 *Some code prescription*

Following the Italian structural code prescriptions (D.M. 09/01/1996) concerning the ultimate limit state conditions, the expression $\mathbf{f} = \mathbf{g} + \lambda\mathbf{q}$, could be written in the following terms:

$$\mathbf{f} = \lambda_g\,\mathbf{g} + \lambda_q\,\mathbf{q}, \tag{7}$$

where $\mathbf{g}$ and $\mathbf{q}$ identify the dead load and live load respectively. For the re-conversion of use proposed for the building in study, $\mathbf{q}$ is assumed to be major than $6\,kN/m^2$. Following the code prescription also these values are assumed:

- The materials safe coefficients:
  $\gamma_{cls} = 1.6$ for concrete; $\lambda_{seel} = 1.15$ for reinforced steel (of course, this will modify the constitutive laws of materials described in §2.2).
- The dead load multiplier:
  $\lambda_g = 1.4$

In non-linear analysis, the live load multiplier λ_q has been assumed variable: it increases until the value connected with the collapse of the structural system (or a structural element). Therefore, the ultimate limit load of $\lambda_q\,\mathbf{q}$ represents the maximum live load admitted for the structure.

If the multiplier λ_q obtained by the analysis is:

$$\lambda_q \geq \overline{\lambda}_q \tag{8}$$

where $\overline{\lambda}_q = 1.5$ is the live load multiplier prescribed by Italian structural code, the building can be considered adequate for the re-conversion of use proposed.

3.2.2. *Structural system modelling*

The structural analysis of an existing building start by its structural project tables; for the building investigate

Facade: particular of the structural joint

Columns

Example of deck decay

Volt decay

Figure 6.   Present state of deterioration.

Table 2.   Material properties.

Concrete characteristic strength	$R_{ck} = 25\,Mpa$
Steel type	FeB32 k
Young modulo	$E_c = 25\,MPa$
	$E_s = 210\,MPa$

Table 3.   Structural elements geometry.

	Real size		Resulting size	
Structural element	b(cm)	H(cm)	b(cm)	H(cm)
Boundary roof beams	40	45	40	43.5
Boundary 3rd roof beams	40	50	40	43.5
Boundary 2nd roof beams	40	45	40	36.8
Boundary 1st roof beams	40	45	40	39.6
Boundary ground roof beams	40	30	40	23.2
Beams	40	83	40	63.9
Beams	60	83	60	53.9
Beams	40	60	40	29.0
Beams	40	60	40	49.6
	L(cm)	L(cm)		
Columns (3rd roof)	30	40	Verified for	
Columns (2nd roof)	40	40	different:	
Columns (1st roof)	40	50	$\mu = A_s/A_c$	
Columns (grou. roof)	50	60	A_s = steel area A_c = concrete area	

this step has not been possible, therefore, the structure has been re-projected. The re-project has been made using the measurements made on site and following the Italian structural code in use at the age of constructio. As live load, the original live load has been assumed. Assuming the materials properties reported in Table 2 and supposing simple reinforced beams the results obtained seem to be in accordance with the geometry of elements and structural code prescriptions (Tab. 3) (Fig. 7). The structural system obtained has been assumed as the "real" structural system. On it, through linear and non-linear analyses, the possible

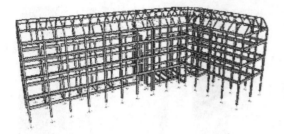

Figure 7. Fossano building: Structural schema.

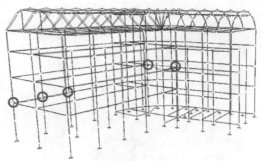

Figure 9. $\lambda_q = 1.8$: formation of other plastic hinges.

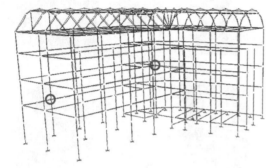

Figure 8. $\lambda_q = 1.3$: formation of the first plastic hinges.

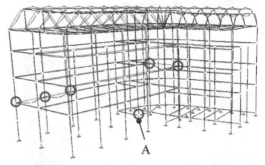

A

Figure 10. $\lambda_q = 2.5$: collapse of column A.

re-conversion has been investigated. The structural computer code used has been SAP2000 based on the finite element theory in the evaluation of strains and displacements.

3.2.3 *Linear analysis*

The first step of analysis has been the evaluation of the strengths level in damaged structure due to the loads connected with the re-conversion of use proposed. This can be made through the evaluation of internal forces of the damaged structure due to actual loads.

This kind of analysis, based on a deterministic linear approach, is not able to give complete response on the real residual strength capability of the damaged building. In the case study proposed, this analysis has been put in evidence that in some damaged elements of the building the structural code prescriptions were not respected, but about the last load level (failure load) no indication are given. Therefore a non-linear analysis is needed.

3.2.4 *Non-linear analysis*

Considering the non-linear materials laws described in §2.2 and the live load limit at collapse defined in §2.1, the non-linear analysis is applied, where the load multiplier λ_q is assumed varying between zero until the value connected with the collapse of the

system. In Figure 8, 9 and 10 the results concerning the different values of λ_q are reported.

The analysis made shows that the first plastic hinges appear with a $\lambda_q = 1.3$ (Fig. 8), but they do not lead at total or partial collapse of the structure. Therefore, the non-linear analysis continues and other plastic hinges appear (Fig. 9). The procedure finish when in a structural element the ultimate strength level is reached: in our case it is happen for a column with $\lambda_q = 2.5$ (Fig. 10).

4 CONCLUSION

In an existing building the re-conversion of use cannot be approached only by an architectonic point of view. In fact, a choice, correct by the architectonic point of view, can be "not suitable" by the structural point of view; it can require the rehabilitation of structural system.

In this paper the re-conversion of an existing building has been analysed by the structural point of view. Since this problem involves a lot of uncertainties, a non-linear approach, able to take into account some of these uncertainties, has been here proposed.

The results obtained show that, for the new destination of use proposed, the building seems in suitable conditions. However, by Figure 6 appear that, on volt and waffle slabs serious works of maintenance are needed. So that the first plastic hinges could appear for a load level greater than that prescribed by the actual structural code, also the reinforcement of the connections shown in Figure 8 is required. The reinforcement of the column A (Fig. 10), cause of collapse, is suggested too. It is true that the collapse seems to happen for a load multiplier greater than the multiplier prescribed by the actual law, however, a failure of a structural element is a dangerous event that it is always better to prevent.

REFERENCES

Albertini Neto, H., 2002. Residual bearing capability in an existing building subjected to re-conversion of use, *M. Sc. Thesi*, Post degree Master *"Aspetti e Tecnologie Strutturali in Architetura"*, Fac. di Architettura Civile, Politecnico di Milano, Milano, Italy (in Italian).

Biondini, F., Bontempi, F. & Garavaglia, E., 2002. Fuzzy optimisation design of concret bridges, *Proc. of Iabmas'02, First Int. Conf. on Bridge Maintenance, Safety and Management,* July 2002, Barcelona, Spain, J.R. Casas, et al. (eds.), CD-ROM, CIMNE Barcelona, Spain, EU.

Bontempi, F., Biondini, F. & Malerba, P.G., 1998. Reliability Analysis of Reinforce Concrete Structures based on a Monte Carlo Simulation, *Stochastic Structural Dynamics*, Spencer, B.F. Jr, Johnson E.A. (eds.), Rotterdam, Balkema, pp. 413–420.

Bontempi, F., Malerba, P.G. & Romano, L., 1995. A Direct Secant Formulation for the Reinforced and Prestressed Concrete Frames Analysis. *Studi e Ricerche*, Scuola di Specializzazione in Costruzioni in Cemento Armato, Politecnico di Milano, 16, pp. 351–386 (in Italian).

Malerba, P.G., 1998. Limit and Non-Linear Analysis of Reinforced Concrete Structures, *CISM*, Udine, Italy (in Italian).

Milia, E. & Pirola, M., 2001. Evoluzione di nuovi modi di lavoro per il professionista digitale: proposta di un polo progettazione/sviluppo per l'industria del design a Cantù. *Degree Thesis*, Fac. di Architettura, Politecnico di Milano, Milano, Italy (in Italian).

Przemieniecki, J.S., 1968. *Theory of Matrix Structural Analysis,* McGraw Hill, New York, NJ, USA.

Sgambi, L., 2003. Fuzzy approach in the three-dimensional non linear analysis of reinforced concrete two-blade slender bridge piers *Second M.I.T. Conference*, Boston, June 17–20, 2003 (to appear).

System-based Vision for Strategic and Creative Design, Bontempi (ed.)
© 2003 Swets & Zeitlinger, Lisse, ISBN 90 5809 599 1

Natural ageing of earth stabilized concrete

A. Guettala
Civil Engineering Dept., Biskra University, Algeria

H. Houari
Civil Engineering Dept., Constantine University, Algeria

A. Abibsi
Mechanical Engineering Dept., Biskra University, Algeria

ABSTRACT: Different stabilizers can be added to earth in order to improve its strenght and durability. In this work, four additions have been made: cement, lime, cement plus lime and cement plus resin and then evaluated by various laboratory tests. A performance of test structures erected at Biskra University was evaluated in real climatic conditions. Even though the artificial simulation is often imperfect, it yields information about the manner by which the climatic factors interact. However, experimentation in natural real climatic conditions imposes itself to validate the results of the so-called accelerated ageing tests.

1 INTRODUCTION

Probably, earth is the first building material used by man. Easton (1996) has suggested that at least 50% of the world's population still live in earth houses However, its main drawback is its deterioration under the action of the climatic conditions. The main deterioration causes are the shrinkage cracking, erosion, the undermining at the basis and mechanical deterioration Mariotti (1983) and Heathcote (1995). Whole villages constructed by earth have been destroyed completely; Nsutam village near Bunso (Ghana) is an example, during the 1970 inundations, Hammond (1973).

At present, there is a growing market for earth walled buildings, with commercial building companies tending to more durable stabilized materials, particular in the area of cement stabilized pressed earth bricks and rammed earth. Various works have been carried out in order to evaluate different stabilizers, Hakimi et al (1996), Gresillon (1976), Rigassi (1998), Guettala et al (1997, 1998) as well as to improve the material properties, Gregory & Kevan (2002), Heathcote (1985) & Keraali (2001).

Houben (1984) recommends the utilization of an earth that doesn't contain an excess of big elements and an exposure of a sufficiently smooth face on which water will have less action in order to get a good resistance with time.

The various accelerated ageing tests are means of comparing different stabilizers performances used under laboratory exposure conditions. These tests are fast but are subject to controversy as one cannot simulate in the laboratory the complex succession of the multiple climatic phenomenon: rain, sun, temperature, humidity, wind … However, little work has been done correlating the performance of samples under conditions similar to that of real buildings, Heathcote (1995), Rubaud et al (1985), Ghoumari (1989), Ogunye (2002) and Venkatarama (2002).

The present work presents a program aiming at the realization of experimental walls carried out in the laboratory of materials at Biskra University, Algeria. This program consists of comparing the performances of different additives: cement, lime, cement + lime and cement + resin.

The project consists of constructing 8 walls on the University roof using different stabilized bricks and then evaluating their performance in real climatic conditions. The bricks have been already evaluated in laboratory conditions in order to establish a correlation between the two different tests.

2 EXPERIMENTAL WORK

All the characterization tests have been carried out according to AFNOR (1984, 2001) and ASTM (1993) standards.

The bricks (20 × 10 × 6 cm) have been prepared out of an earth corrected by 30% sand and compacted

Table 1. Bricks and walls characteristics.

Bricks characteristics	Cement (%)		Lime (%)		Cement (%) + Lime (%)		Cement (%) + Resin (%)	
	5	8	8	12	5 + 3	8 + 4	5 + 50	8 + 50
Compressive strength in dry state, MPa	15.4	18.4	15.9	17.8	17.5	21.5	17.2	19.5
Compressive strength in wet state, MPa	9	12.7	10.1	11.7	12.3	15.6	11.5	14
Water Strength Coefficient	0.58	0.69	0.64	0.66	0.63	0.7	0.67	0.72
Capillary absorption, %	2.35	2.2	3.7	2.9	2.3	2	2.3	2.1
Total absorption, %	8.27	7.35	9.8	9.02	8.1	7.9	5.9	5.3
Weight loss (wet-dry), %	1.4	1.25	2.3	2.1	1.2	1.0	0.9	0.9
Weight loss (freeze-thaw), %	2.35	2.23	3.7	2.9	2.3	2.0	2.3	1.8
Hole depth (mm)	1.0	0.5	2.2	1.0	1.0	0.5	0.25	0.2

using 15 MPa stresses. These parameters have been chosen on the basis of the good performances showed by the bricks during the laboratory tests.

The following materials were used to confection the bricks.

2.1 Soil and sand

A clayey and sandy soil was used, composed mainly of kaolin and illites. A local rolled sand was used to correct the soil granulometry; passed through a 1 mm sieve.

2.2 Stabilizers used

Four different stabilizers were used: cement, lime, cement + lime and cement + resin, Table 1.

2.3 Laboratory tests

The stabilized bricks have undergone different laboratory tests carried out and reported in other publications, Guettala et al (2000, 2002, 2003): Compressive strength in the dry and wet sates, Absorption (Capillary and Total), Wetting-Drying, Freeze-Thaw and Spraying. The results are summarized in Table 1.

2.4 Field test

8 walls have been constructed on the University roof, arranged in a row and sufficiently remote from one another in order to avoid mutual protection, 10 cm wall thickness. They were oriented South and North so that one of their main faces is exposed to the dominant rains. The walls joint were of a cement mortar.

A general check up has been made after the construction in order to detect any defects or damages caused during the construction. Then, two month periodical visits and inspections were programmed for a period of four years.

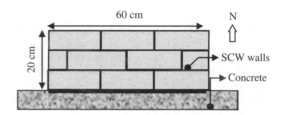

Figure 1. Walls dimensions and orientation.

Figure 2. General view of the confectioned walls.

3 CLIMATIC INFORMATION

As we didn't have any equipment for measuring the meteoroligical parameters on site, we used the local station data. The quantity of rain drop is weak of the order of 150–200 mm per year.

The collected climatic data were spread over four consecutive years 1999, 2000, 2001 and 2002. They reflect a very rigorous, cold and dry semi-arid climate in winter, hot and dry in summer, whose features are as follows:

The relative humidity also knows the important fluctuations. It can vary from 90% in winter (maximum value in January) to a minimum value of 10% in July and August; a value that sometimes can spread over the 4 to 5 hot months.

The main agents of earth wall erosion are principally rain and frost, apparently little present in Biskra.

This doesn't prevent to neglect the deep damages that can be caused by rain dripping and that falls occasionally. During the month of the January 2003, there was a huge quantity of rain fall during 7 days; about 73 mm, a quantity which represents three quarters of that usually falls in one year.

The quantity of precipitation during the years 1999, 2000, 2001 and 2002 were 190, 81, 93 and 95 mm respectively.

There are two types of seasonal dominating winds in the region of Biskra. The cold winds of winter blowing from north-west with an average speed of 35 km/h and the hot and dusty winds of the south-east and the south-west during spring and fall seasons. The winds reach the speed of 80 km/h provoking disasters in the region. The region also knows the dry and hot winds that blow during summer that can reach a maximum value of 47 km/h.

4 EXPERIMENTAL RESULTS

It is important to observe degradation with maximum quantifiable criteria: degradation type, number, shape and dimensions of cracks; depth of the erosion ... All of these will be recorded on cards for each wall and the degradation evolution will be noted. Photographs were also being taken.

4.1 Stabilization using lime

In general, all the walls treated were behaved satisfactorily after 48 months exposure, except those treated with lime which has shown slight erosion degradation. It has been observed for certain bricks of the walls treated with 8% lime a partial crumbling at the north face level of the first and the second row, after four years exposure, as shown clearly in Figure 4. This deterioration provoked erosion reaching a maximal depth of 1 mm and over an area of about 40% of the exposed block surface. On the other south face, no deterioration was observed. Apparently, this is due to the effect of the dominant wind direction.

For the walls treated with 12% lime, no erosion has been recorded but a disappearance of small pieces of the brick of the first row has been noted. At left and right angle levels on the north face, Figure 3.

It has also be noted that during the winter period, the walls have the efflorescence at the base level but in more important manner for the case of the walls treated with lime.

Figure 3.   12% lime treated wall – North Face disappearance of small pieces.

Figure 4.   8% lime stabilized wall. Erosion deterioration after 48 months exposure.

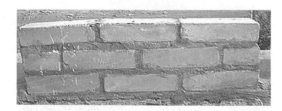

Figure 5.   5% cement stabilized wall, after 48 months exposure.

4.2 Stabilization using cement

For the walls constructed with 5% cement bricks, no deterioration has been observed a part from the light leaching of the joints, Figure 5.

The walls prepared with the 8% cement bricks showed no sign of deterioration.

4.3 Stabilization using cement and lime

Two experimental walls have been prepared using mixed additives, cement and lime. It has been noted that the treatment by 5% cement plus 3% lime has shown a very good behavior.

Exactly, the same remarks can be made about the second treatment concerning the 8% cement plus 4% lime case. No degradation has been observed.

4.4 Stabilization using cement and resin

The resin used for this work has a commercial name of "Medalatex"; supplied by Granitex; private algerian compagny of additives making. Medalatex is an aqueous dispersion of resin of white colour. It's compatible with most of cement as well as lime. The latex addition gives a good adherence to the support. It gives also the impermeability, the durability and the improvement in protection of the reinforcement, thus resistance to chemical attacks.

Again, two walls have been prepared using two different mixtures of cement and resin. It has been noted that a treatment by the 5% cement + 50% resin in compacting water gave excellent results with a slight signs of leaching.

The same observations can be made for the second treatment type; 8% cement + 50% resin in compacting water.

It has been noted an erosion (5 mm) of faces of the wall top whose bricks have been treated with lime. On the other hand, the faces treated by the other products were not affected by the climatic conditions. Usually the test wall top is covered by a hat as it would be in reality. However, this is not the case in this present work as to evaluate the direct effect of the rain falls.

5 VALIDATION OF ACCELERATED AGEING TESTS

According to Gresillon (1976), the wall bricks without protection exposed to rain and wind would be soaked with water. After immersion, spraying, wetting and drying and freeze-thaw, crushing tests are necessary. However, one might wonder on the validity of such tests in the region where the present work has been done:

The immersion of blocks during 24 hours appears very severe. No rain is equivalent to such a regime in the region of Biskra. The spraying is very violent; in this case; we tend to replace the spraying time by the violence of the jet. Is there really compatibility between the spraying time and the pressure of water jet?

The test of wetting-drying appears also very severe because the exposed bricks have not undergone a total immersion. The test of freeze-thaw appears also very severe because the bricks have not been exposed to temperatures below zero.

Among these accelerated ageing tests used to study the performances of the earth treated in mass, one can only keep the spraying test. This latter could establish a correlation effectively with the natural ageing test because the experimental walls have undergone neither a total nor a partial immersion.

The accelerated ageing tests are very severe compared to the natural ageing tets carried out in this present work. This explains the good behavior of the walls after four years exposure to natural conditions of the region of Biskra. Houses in this region constructed in raw earth with the Adobe technique without any outside protection stayed in an acceptable state for more than 60 years of their existence.

6 CONCLUSION

The experimental conditions have been characterized by a weak quantity of rain fall of about 120 mm/year. However, the walls have been exposed to an important and unusual rain during the month of the January 2003 for seven days, of the order of 73 mm.

In general, It has been noted that all treated walls showed no signs of deterioration after 4 years exposure in real climatic conditions. However, a light degradation has been recorder in the case of certain 8% lime bricks. This deterioration provoked erosion reaching a maximal depth of 1 mm and covering up to 40% of the surface of the brick exposed.

It is also worth noting that the laboratory tests conditions seem very severe compared to the natural climatic conditions of the region of Biskra. This explains partly the good behavior of the stabilized walls after four years exposure.

As a result of the present work, the following classiffication of the different products used as earth stabilzers can be made; according to their good durability:

- cement + resin
- cement + lime
- cement
- lime

It is also important to note that the utilization of the resin as a material of protection is not economic. It is practically eight times more expensive than the treatment by cement.

For the type of soil (sandy clayey soil) used for this work; it is recommended to manufacture bricks stabilized with cement (5%) using a compacting stress of the order of 10 MPa as used in the laboratory tests. With these manufacturing conditions, one can assure an acceptable durability for the climatic conditions of the Biskra region.

REFERENCES

Easton, 1996. The rammed earth house. Chelsea Green Pub Co, White River Junction, Vt.

Mariotti, M. 1983. Programme expérimental sur badigeons de protection des murs en béton de terre, *Proceedings symposium Nairobi*, 7–14 nov., Sect. V, Kenia: 402–415.

Heathcote, K.A. 1995. Durability of earthwall buildings. *Building Materials*: 3(9): pp. 185–189.

Hammond, A.A. 1973. Prolongation de la durée de vie des construction en terre sous les tropiques. *CDU 69.03* (213): 167–197.

Hakimi, A., Yamani, N & Ouissi, H. 1996. Résultats d'essais de résistance mécanique sur échantillon de terre comprimée. *Materials and Structures/Matériaux et constructions*: 29: 600–608.

Gresillon, J.M. 1976. Etude sur la stabilisation et la compression des terres pour leur utilisation dans la construction. *Annales de l'Institut Technique du Bâtiment et des travaux Publics Série: Matériaux*. 339, Mai, Paris, France: 21–33.

Rigassi, V. & CRATerre-EAG (Hoehel-Druck). 1995. Blocs de terre comprimée: Manuel de production (1). Germany.

Guettala, A. & Guenfoud, M. 1997. Béton de Terre Stabilisée Propriétés Physico-Mécaniques et Influence des Types d'Argiles, *La technique moderne* 1–2: 21–26.

Guettala, A. & Guenfoud, M. 1998. Influence des Types d'Argiles sur les Propriétés Physico-mécaniques du Béton de Terre Stabilisée au Ciment, *Annales du Bâtiment et des Travaux Publics*: 1: 15–25.

Gregory, M. & Kevan, H. 2002. Earth Building in Australia – Durability Research. *Proceedings of Modern Earth Building.*, Berlin, 19–21 April 2002: 129–139.

Heathcote, K.A. 1985. Durability earthwall buildings. *Construction and Building Materials* Volume 9, Number 3: 185–189.

Keraali, A.G. 2001. Durability of compressed and cementstabilized building blocks. PhD Thesis, University of Warwick, school of Engineering, UK, September.

Houben, H. & Guillaud, H. (CRATerre), 1984. 20 Earth Construction, Primer Brussels, CRATerre/PGC/CRA/ UNCHS/AGCD.

Rubaud, M. & Laurent, J.P. 1985. La durabilité de protection sur terre stabilisée, l'expérience des murets de Dreyfus, *Cahiers du Centre Scientifique et Technique du Bâtiment*: 13.

Ghoumari, F. 1989. Matériau en Terre Crue Compactée: Amélioration de sa Durabilité à l'Eau; Thèse de Doctorat, INSA de Lyon.

Ogunye, F.O. & Boussabaine, H. 2002. Diagnosis of assessment methods for weatherbility of stabilized compressed soil blocks. Construction and building materials. 16: 196–172.

Venkatarama, R.B.V. 2002. Long-term strength and durability of stabilized mud blocks. Vietnam International Conference on non-conventional materials and technology: 422–431.

AFNOR, 1984. Recueil de Norme Françaises. Bâtiment Béton et Constituants du Béton. Paris.

AFNOR XP P 13–901, 2001. Blocs de terre comprimée pour murs et cloisons.

American Society for Testing and Materials 1993. Annual Book of ASTM Standards, Vol. 04.01, Philadelphia.

Guettala, A., Mezghiche, B., Chebili, R. & Houari, H. 2000. *Durability of Blocks of Earth Concrete*. Proceedings of II International Symposium Cement and Concrete Technology, September 6–10 Istanbul, Turkey: 273–281.

Guettala, A., Houari, H., Mezghiche, B. & Chebili, R. 2002. Durability of Lime stabilized Earth Blocks. *Proceedings of international conference University of Dundee*, Scotland UK 9–11 Sep: 145–654.

Guettala, A., Houari, H. & Abibsi, A. 2003. Durability of cement and lime stabilized Earth Blocks. *Al-Azhar Engineering 7th. International Conference University of Cairo*, Egypt April 07–10: 78.

System-based Vision for Strategic and Creative Design, Bontempi (ed.)
© *2003 Swets & Zeitlinger, Lisse, ISBN 90 5809 599 1*

Serviceability assessment of deteriorating reinforced concrete structures

W. Lawanwisut, C.Q. Li, S. Dessa Aguiar & Z. Chen
Department of Civil Engineering, University of Dundee, Dundee, UK

ABSTRACT: This paper attempts to embody the concept of serviceability assessment of deteriorating reinforced concrete structures using the reliability approach. Models of the response of corrosion affected structures for each lifetime are developed based on experimental data produced from a comprehensive testing program. This includes corrosion initiation and corrosion induced concrete cracking. A time-dependent reliability method is employed in the paper to determine the probability of the attainment in each lifetime. The methodology presented in this paper could equip structural engineers and asset managers with confidence in their decision-making with regard to the repairs rehabilitation of deteriorating structures.

1 INTRODUCTION

Reinforcement corrosion in concrete is regarded as the predominant causal factor in the premature deterioration of reinforced concrete (RC) structures, leading ultimately to structural failure (Broomfield 1997, Schiessl 1988). Failure does not necessarily mean structural collapse only, but also includes loss of serviceability, characterised by concrete cracking, spalling, and excessive deflection of structural members. The degree to which performance of reinforced concrete is damaged as a result of reinforcement corrosion is a matter of great concern to those responsible for assessment and maintenance of affected structures. In the UK alone, the annual cost of repairs to concrete structures due to reinforcement corrosion has been estimated at £500,000,000 (Hobbs 1996). Therefore, it is a clear need demand and attention to assess the serviceability of deteriorated structure for assist engineers and owners to plan for cost-effective repair and maintenance strategy.

To meet these need and demand, various theoretical frameworks have been developed to assess the service life performance of corrosion affected RC structures including the employment of advanced reliability theories (e.g., Engelund et al. 1999, Frangopol et al. 1997, Maage et al. 1996, Prezzi et al. 1996). However, lack of rational models for structural response (deterioration) in different lifetimes hampers the application of these advanced theories and in turn hinders the further progress of the framework. As a result, prediction of the service life of a structure remains at a stage of parametric studies. Prediction of the effects of the corrosion process, both initiation and propagation, on structural

behaviour should be based on models derived from realistic and accurate data representative of service conditions (Enright & Frangopol 1998, Maage et al. 1996, Prezzi et al. 1996). Field data tends to be highly variable and laboratory data is rarely produced from tests either under service conditions (i.e., natural salt ingress and simultaneous service loads) or on full size structural members. In order to remedy this situation a comprehensive test program has been undertaken (Li 2001) to produce data that closely represents the real service conditions of RC structures. From this data, complemented by data collected from research literature, rational and practical models of structural deterioration can be now developed.

The intention of this paper is to embody the concept of serviceability assessment of deteriorating RC structures using the reliability approach. In this paper, a performance-based model of each lifetime for corrosion affected concrete structures is proposed. The serviceability assessment is defined as the time from newly built construction until the time to critical limit crack width of the concrete cover. A merit of the proposed model is that lifetimes are derived from the data representing reinforced concrete structures in real service conditions. A time-dependent reliability method is employed in the paper to determine the probability of the attainment of each lifetime. The methodology presented in the paper could equip structural engineers and asset managers with confidence in their decision-making with regard to the maintenance and repairs of corroded concrete structures. Timely maintenance and repairs have the potential to prolong the service life of reinforced concrete structures.

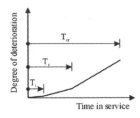

Figure 1. Serviceability assessment profile of corrosion affected structure.

2 SERVICEABILITY ASSESSMENT PROFILE

In serviceability assessment of deteriorating RC structures, it is important to determine each lifetime in service of the structure. Lifetime can be defined based on structural behaviour associated with deterioration. As schematically described in Figure 1, the first lifetime of corrosion affected concrete structures is the time period from completion of the newly built structure to the corrosion initiation in the structure, denoted as $(0, T_i)$. The second lifetime is the time period from corrosion initiation to corrosion induced concrete cracking and denoted as (T_i, T_c). The third lifetime presents the time period from corrosion induced cracking to the critical crack width and denoted as (T_c, T_{cr}). With the formulation in Figure 1, each lifetime can be determined when models of structural response in each time period, i.e., corrosion initiation, concrete cracking, crack width, are available.

3 CORROSION INITIATION

It is well known that chloride ingress in concrete is one of two basic mechanisms that trigger the corrosion of the reinforcing steel in concrete. To predict the time to the initiation of reinforcement corrosion in concrete structures, let C_{Cl} be the chloride content in concrete, which varies with time due to the chloride ingress from the surrounding environment. Also, let δ_{Cl} be a threshold value (concentration) for the chloride content, the attainment of which may lead to the onset of corrosion. Since the corrosion onset is a random phenomenon, even when the chloride content at the surface of reinforcing bars exceeds the threshold value the corrosion of the rebar does not necessarily initiates in the concrete. Therefore, it is well justified that the prediction of corrosion initiation should be formulated in a probabilistic manner:

$$p_i(t) = P[C_{Cl}(t) \geq \delta_{Cl}] \times P[\xi] \tag{1}$$

where P denotes probability of an event; $p_i(t)$ is the probability of corrosion initiation at time t; $P[\xi]$

denotes the probability of corrosion onset given that $C_{Cl}(t) > \delta_{cl}$. Thus, for a given acceptable probability, $p_{i,a}$, whenever:

$$p_i(T_i) \geq p_{i,a} \tag{2}$$

the initiation time of reinforcement corrosion in concrete is determined, i.e., T_i. Clearly $p_{i,a}$ represents the reliability (or confidence) of the prediction. The greater $p_{i,a}$ is, the more reliable the prediction is, and the more confident it is that the corrosion will initiate during the period of $(0, T_i]$. Therefore, Equation 2 can be used to determine the first lifetime of corrosion affected concrete structures.

3.1 Chloride ingress

Although Fick's second law has been widely used to predict the chloride ingress in concrete, an increasing number of published papers have suggested that Fick's second law may not be applicable to practical RC structure, in particular, flexural members, whereby load induced cracks invalidate the mechanism of diffusion. Therefore, in this paper, a model of chloride ingress is developed based on data produced from 255 specimens of comprehensive test program (Li 2001).

Due to the uncertainty of chloride ingress in concrete and its time-variant nature it is well justified to model chloride ingress in concrete as a stochastic process, quantified by the chloride content at the surface of reinforcement bars, $C_{Cl}(t)$. Based on the experiment results (Li 2001), the mean function of chloride content, $\mu_C(t)$, and the coefficient of variation (COV) for chloride content, $V_c(t)$, can be assumed to be in the form of:

$$\mu_C(t) = C_0 \exp(\alpha t) \tag{3a}$$

$$V_C(t) = \beta \cdot t + 0.1433 \tag{3b}$$

where C_0 is the mean value of initial (i.e., $t = 0$) chloride content in concrete at the surface of reinforcement. In Equation 3a, α is a coefficient representing the rate of chloride ingress. It allows for the effects of such factors as concrete properties and external environment. In general, α is difficult to determine. β is a coefficient. In Equations 3a and 3b, t is time (in days). Coefficient α and β can be determined from regression analysis of experimental data from 183 specimens with normal cement and a 0.45 water to cement ratio. The initial chloride content C_0 of the specimens was found to be 0.018% of the concrete weight, $\alpha = 0.01$ and $\beta = 0.0004$.

3.2 Corrosion onset

In the real world of concrete structures, the onset of reinforcement corrosion is a random phenomenon and should be dealt with in a probabilistic manner. The test results on corrosion onset used to derive $P[\xi]$ were obtained via visual inspection after breaking open the test specimens (Li 2001). The range of chloride content from 0.04 to 0.07 is the most sensitive concentration to corrosion onset, which is consistent with the threshold values widely used in practice (Dhir 1999). From regression analysis of the test results $P[\xi]$ can be expressed as:

$$P[\xi] = 62605C_{Cl}^4 - 15003C_{Cl}^3 + 1042.7C_{Cl}^2 - 6.8341C_{Cl};$$
$$\text{where } 0 \le C_{Cl} \le 0.08 \tag{4a}$$

$$P[\xi] = 1 \quad ; \text{where } 0.08 \le C_{Cl} \le 0.1 \tag{4b}$$

It should be noted that the chloride ingress in this work is assumed to follow normal distribution.

4 CORROSION INDUCED CRACKING

After initiation, the reinforcement corrosion propagates in concrete and produces expansive rusts, which increases the pressure (i.e., stress) at the interface of the reinforcement and concrete. Eventually the stress due to the expansion causes concrete to crack. Based on this phenomenon, the probability of corrosion induced concrete cracking $p_c(t)$ at time t can be considered as follows:

$$p_c(t) = P[\sigma_c(t) \ge \sigma_t] \tag{5}$$

where $\sigma_c(t)$ is the stress asserted by the expansive corrosion products and σ_t is the tensile strength of concrete required causing cracking of concrete. At the time that $p_c(t)$ is greater than a minimum acceptable probability, $p_{c,a}$, the time to cracking, T_c, can be determined, i.e.:

$$p_c(T_c) \ge p_{c,a} \tag{6}$$

where $p_{c,a}$ represents the reliability of the prediction. Therefore, Equation 6 can be used to determine the time to surface cracking of corrosion affected RC structures.

4.1 Model of $\sigma_c(t)$

With the propagation of corrosion, its products (i.e., ferrous and ferric oxides) occupy a much greater volume than the original reinforcement, thereby generating

pressures on the surrounding concrete. The pressure builds up to a level that causes internal concrete cracking at the interface of the reinforcement and concrete. The crack eventually extends through the concrete cover. This work considers concrete to be an elastic material and it is modelled as a thick-wall cylinder under internal radial pressure due to rust expansive. The schematic representation of the model is shown in Figure 2a, where D is the diameter of the reinforcement bar, d_0 is thickness of the annular layer of corrosion products (i.e., a pore band) at the interface of the reinforcement and concrete and a and b are the inner and outer radius of the thick-wall concrete cylinder. The inner radius $a = (D + 2d_0)/2$, and the outer radius $b = C + (D + 2d_0)/2$, where C is the concrete cover. As corrosion progresses, the rust products will fill the pore band completely and then generate an expansive pressure on the concrete. Thereafter the inner radius will increase as corrosion increases (i.e., the thickness of rust product) and it can be determined by:

$$r_1(t) = [(D + 2d_0)/2] + d_s(t) \tag{7}$$

where $d_s(t)$ is the thickness of corrosion products needed to generate tensile stress and it can be expressed as (Liu & Weyers 1998):

$$d_s(t) = \frac{W_{rust}}{\pi(D + 2d_0)} \left(\frac{1}{\rho_{rust}} - \frac{\alpha_r}{\rho_{st}} \right) \tag{8}$$

where α_r is a coefficient related to types of rust products; ρ_{rust} is the density of corrosion products; ρ_{st} is

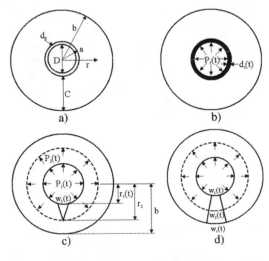

Figure 2. Schematic representation of concrete cracking modelled as a thick-wall cylinder.

the density of the reinforcement; W_{rust} is the mass of corrosion products that generate the pressure. Thereafter the rust growth generates the tensile stress. W_{rust} is related to corrosion rate as measured by corrosion current density I_{corr} in $\mu A/cm^2$ and can be determined by (Liu & Weyers 1998):

$$W_{rust} = \left(2 \int_0^t 0.098(1/\alpha_r)\pi D I_{corr} dt \right)^{1/2} \quad (9)$$

From the test results (Li 2003), I_{corr} can be expressed, for RC flexural members as:

$$I_{corr} = 0.3683 \text{Ln}(t) + 1.1305 \quad (10)$$

where t is the time in years. The trend of I_{corr} over time curve is in good agreement with other tests, e.g. Schiessl (1988) and Andrade et al. (1990). However, the values of I_{corr} should be obtained from the structure to be assessed in practice. Based on the theory of elasticity (Timoshenko & Goodier 1970, Ugural 1986), the expansive stress can be derived as follows:

$$\sigma_c(t) = P_1(t) = \frac{2E_{ef}d_s(t)}{(D + 2d_0)\left(\frac{b^2 + r_1^2(t)}{b^2 - r_1^2(t)} + v_c \right)} \quad (11)$$

where E_{ef} is an effective elastic modulus of concrete; v_c is Poisson's ratio of concrete.

4.2 Model of σ_t

The minimum stress required to induce cracking of concrete cover is apparently related to the tensile strength of concrete and the thickness of the cover. Based on the mechanism of corrosion induced concrete cracking, the volume of the corrosion products is also related to the thickness of the pore band at the interface of the reinforcement and concrete. According to Bažant (1979) and Liu & Weyers (1998), σ_t this can be expressed as:

$$\sigma_t = \frac{2Cf_t}{D + 2d_0} \quad (12)$$

where f_t is the tensile strength of the concrete.

5 CORROSION INDUCED CRACK WIDTH

It has been generally assumed that one of the factors influencing the severity of corrosion is the crack width. It is therefore a basic design criterion prescribed in all building codes. When the corrosion continues, cracks progressively widen with increasing corrosion and delamination of the concrete cover eventually results. Based on this phenomenon, the probability of corrosion induced concrete crack width $p_{cr}(t)$ at time t can be determined as follows:

$$p_{cr}(t) = P[w_c(t) \geq w_{cr}] \quad (13)$$

where $w_c(t)$ is the crack width at the concrete surface by the expansive corrosion products and w_{cr} is the limit crack width on the surface of concrete cover which can be determined based on the building code. At the time that $p_{cr}(t)$ is greater than a maximum acceptable probability, $p_{cr,a}$, the time to limit crack width, T_{cr}, can be determined, i.e.:

$$p_{cr}(T_{cr}) \geq p_{cr,a} \quad (14)$$

Again, $p_{cr,a}$ represents the reliability of the prediction as in the case of time to cracking. Therefore, Equation 14 can be used to determine the crack limit lifetime of corrosion affected RC structures.

5.1 Crack width evolution

As the concrete is considered to be an elastic material, an internal crack will start when the maximum tangential stress exceeds the tensile strength of concrete. The internal cracks develop to the position r_2 where the tangential stresses have reached f_t (see Fig. 2c). From this, it divides the cylinder into an internal cracked ring and an uncracked ring (Tepfers 1979). At this stage, the crack width $w_1(t)$ of the rust front in the internal cracked ring can be determined by:

$$w_1(t) = 2\pi r_1(t)\varepsilon_{\theta_1}(t_1) \quad (15)$$

where t_1 is time corresponding tangential stress at the interface reaches the tensile strength of concrete. The tangential strain ε_{θ_1} at radius $r_1(t)$ can be determined from the theory of elasticity by:

$$\varepsilon_{\theta_1}(t) = [P_1(t) - v_c f_t]/E_{ef} \quad (16)$$

where $P_1(t)$ is the uniform expansive pressure and can be determined by Equation 11. The internal pressure from the corrosion mechanism is still loading and the pressure is now transferred through the inner radius between the internal cracked and the uncracked part of the ring. The transformed pressure on the inner surface of the uncracked ring can be expressed as $P_2(t) = P_1(t)r_1(t)/r_2$ and the maximum stress at the inner surface of the uncracked part of the cylinder of radius r_2 can be expressed as follows:

$$\sigma_\theta(t) = \frac{P_1(t)r_1(t)}{r_2}\left[\frac{b^2 + r_2^2}{b^2 - r_2^2}\right] \qquad (17)$$

Equation 17 can be used when considering the maximum stresses are equal to f_t. By differentiation with respect to r_2, gives $r_2 = 0.486b$.

With the continuing growth of corrosion products, the crack will eventually pierce to the surface of the concrete cover. The concrete cover will be cracked open due to the brittle nature of concrete in tension. This state is shown in Figure 2d. The crack width on the surface of concrete cover can be approximately determined by:

$$w_c(t) = w_1(t) + \frac{w_2(t) - w_1(t)}{r_2 - r_1(t)}(b - r_1(t)) \qquad (18)$$

where

$$w_2(t) = 2\pi r_2 \varepsilon_{\theta_2}(t) \qquad (19a)$$

$$\varepsilon_{\theta_2}(t) = \frac{1}{E_{ef}}\left(\frac{P_1(t)r_1(t)}{r_2} - v_c f_t\right) \qquad (19b)$$

6 LIFETIME ASSESSMENT

6.1 Time to corrosion initiation

A RC flexural member with a typical cross-section shown in Figure 3 is used as example to determine structural response. With the models of both $C_{Cl}(t)$ and $P[\xi]$, it is possible to calculate the probability of corrosion initiation over time using Equation 1 for a given threshold. The typical results are shown in Figure 4. With $p_{i,a} = 0.9$ (ASTM C876 1991) and $\delta_{Cl} = 0.06$ (Li 2001), it can be obtained from Equation 2 that $T_i = 0.49$ years.

6.2 Time to cracking

With the model of $\sigma_c(t)$ and σ_t, the probability of corrosion induced concrete cracking can be determined using Equation 5 and the values of basic variables listed in Table 1. The results are shown in Figure 5. In accordance with a confidence level of 90%, i.e., $p_{c,a} = 0.9$, it can be obtained from Equation 6 that the time to cracking is 1.8 years for a concrete cover of 31 mm with crack appearance delayed for larger concrete cover. These results are consistent with experiments as described by Liu & Weyers (1998).

6.3 Time to limit crack width

With the model of $w_c(t)$ and the criterion of w_{cr} being 0.3 mm (BS 8110), the probability of corrosion induced concrete crack width can be determined using Equation 13 and the results are shown in Figure 6. Again, using a confidence level of 90%, i.e., $p_{cr,a} = 0.9$, it can be obtained from Equation 14 that the time to limit crack width is 7.6 years for a concrete cover of 31 mm. The crack width also delayed with larger concrete cover. The evolution of cracks cause the loss of concrete integrity, which reduces the concrete contribution to the load bearing capacity and affect the external appearance of the structure.

This analysis has assumed that the concrete is modelled as cylinder. Methods for determining concrete crack may be refined by a more accurate modelling. Then experiment results may be used to verify or calibrate the computation, e.g., by the finite element method (FEM). FEM may be used to make a more

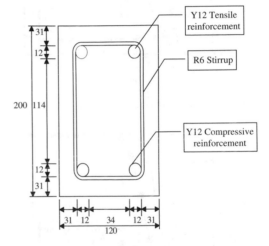

Figure 3. A typical cross-section of RC flexural member.

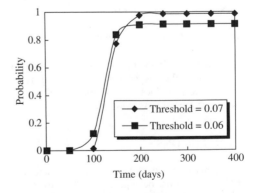

Figure 4. Probability of corrosion initiation.

Table 1. Statistical parameters of basic variables.

Variables	Mean	COV	Distribution
C	31 mm	0.2	Normal
D	12 mm	0.15	Normal
E_c	41.62 GPa	0.12	Normal
f_t	5.725 MPa	0.2	Normal
d_0	12.5 μm	–	–
I_{corr}	Equation 10	–	–
α_r	0.57	–	–
v_c	0.18	–	–
ρ_{rust}	3600 (kg/m^3)	–	–
ρ_{st}	7850 (kg/m^3)	–	–

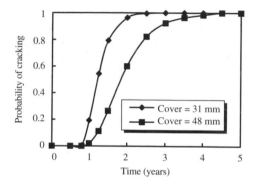

Figure 5. Probability of cracking.

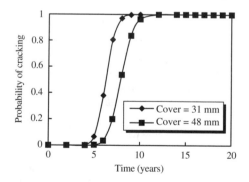

Figure 6. Probability of crack width.

intelligible extrapolation to real cases, e.g., geometry of concrete modelled as rectangular form. This is provided that the soundness of the conceptual basis of the model proposed is ascertained.

7 CONCLUSIONS

The concept of serviceability assessment of deteriorating RC structures has been proposed. A performance-based lifetime formulation for corrosion affected RC structures has been developed. With time-dependent reliability method, the lifetime has been predicted with confidence. It has been found that time to initiation is relatively short for the service life consideration whereas the length of those time to surface cracking and time to limit crack width are much longer. The determination of lifetime depends on the acceptable risk of serviceability failure associated with the different damage levels and minimum performance requirements. The methodology presented in this paper could equip structural engineers and asset managers with confidence in their decision-making with regard to the repairs rehabilitation of deteriorating structures. It is expected more research would be carried out in modelling the structural response of corrosion affected structures.

ACKNOWLEDGEMENTS

Financial support of Engineering and Physical Sciences Research Council (EPSRC), UK with Grant No. GR/R28348 is gratefully acknowledged.

REFERENCES

Andrade, C., Alonso, M.C. & Gonzalez, J.A. 1990. An initial effort to use the corrosion rate measurements for estimating rebar durability. *Corrosion Rates of Steel in Concrete, ASTM STP 1065*, N.S. Berke, V. Chaker and D. Whiting, Eds., American Society for Testing and Materials, Philadephia: 29–37.

ASTM C876 1991. Standard test method for half-cell potentials of uncoated reinforcing steel in concrete. *ASTM*, Philadelphia: 425–430.

Bažant, Z.P., 1979. Physical model for steel corrosion in concrete sea structures – theory. *Journal of Structural Division, ASCE*, Vol. 105, (No. ST6): 1137–1153.

Broomfield, J. 1997. *Corrosion of Steel in Concrete, Understanding, Investigating & Repair*, E & FN Spon, London.

BS 8110, British Standards Structural. 1997. *Use of Concrete – Code of Practice for Design and Construction–Part 1*, UK.

Dhir, R.K. 1999. Concrete durability. *Lecture Notes*, Department of Civil Engineering, University of Dundee, UK.

Engelund, S., Sorensen, J.D. & Sorensen, B. 1999. Evaluation of repair and maintenance strategies for concrete coastal bridges on a probabilistic basis. *ACI Mat. J.*, March–April: 160–166.

Enright, M.P. & Frangopol, D.M. 1998. Service life prediction of deteriorating concrete structures. *J. Struct. Engrg., ASCE*, Vol. 124 (No. 3): 309–317.

Frangopol, D.M., Lin, K.Y. & Estes, A. 1997. Reliability of reinforced concrete girders under corrosion attack. *J. Struct. Engrg., ASCE*, Vol. 123 (No. 3): 286–297.

Hobbs, D.W., 1996. Chloride ingress and chloride–induced corrosion in reinforced concrete members. *Corrosion of Reinforcement in Concrete Construction*, Ed. Page C.L. et al. *Royal Society of Chemistry, Cambridge*: 124–135.

Li, C.Q. 2001. Initiation of chloride induced reinforcement corrosion in concrete structural members–experimentation. *ACI Structural J.* Vol. 98 (No. 4): 501–510.

Li, C.Q. 2003. Life cycle modelling of corrosion affected concrete structures – propagation. *J. Struct. Engrg., A.S.C.E.* (accepted).

Liu, Y. & Weyers, R.E. 1998. Modeling the time-to-corrosion cracking in chloride contaminated reinforced concrete structures. *ACI Mat. Jour.*, Vol. 95 (No. 6): 675–681.

Maage, M., Helland, S., Poulsen, E., Vennesland, O. & Carlsen, J.E. 1996. Service life prediction of existing concrete structures exposed to marine environment. *ACI Material Journal*, Nov.–Dec.: 602–608.

Prezzi, M., Geyskens, P. & Monterio, P.J.M. 1996. Reliability approach to service life prediction of concrete exposed to marine environments. *ACI Mat. Jour.*, Nov.–Dec.: 544–552.

Schiessl, P. 1988. *Corrosion of Steel in Concrete*, Report of the TC60-CSC RILEM, Chapman and Hall, London.

Tepfers, R. 1979. Cracking of concrete cover along anchored deformed reinforcing bars. *Magazine of Concrete Research*, Vol. 31 (No. 106): 3–12.

Timoshenko, S.P. & Goodier, J.N. 1970. *Theory of Elasticity*, McGraw-Hill Book Company, New York.

Ugural, A.C. 1986. *Advanced Strength and Applied Elasticity*, Elsevier Applied Science, London.

Vulnerability assessment of deteriorating reinforced concrete structures

C.Q. Li, W. Lawanwisut & Z. Chen
Department of Civil Engineering, University of Dundee, Dundee, UK

ABSTRACT: Reinforcement corrosion in concrete is the predominant causal factor for the premature deterioration of reinforced concrete structures, causing various degrees of damage to the structure and leading to ultimate structural collapse. The existence of reinforcement corrosion makes reinforced concrete structures increasingly vulnerable over time. This evidently poses a potential risk to the public. The intention of the present paper is to propose a method to assess the vulnerability of reinforced concrete structures deteriorating over time due to reinforcement corrosion. The application of the proposed method to corrosion affected flexural members is demonstrated. It is found that corrosion induced deterioration is the most important single factor that makes reinforced concrete structures vulnerable. The method presented in the paper can equip engineers, operators and asset managers with confidence in developing a risk-informed maintenance strategy in the asset management of corroded concrete infrastructure. Timely maintenance and repairs have the potential to prolong the service life of infrastructure.

1 INTRODUCTION

Experience from past decades has shown that reinforcement corrosion in concrete is the predominant causal factor for the premature deterioration of reinforced concrete (RC) structures, causing various degrees of damage to the structure and leading to ultimate structural collapse (Broomfield 1997, Chaker 1992, Schiessl 1988). The existence of reinforcement corrosion in concrete structures, in particular, the effect of its propagation over time on structural behavior, makes RC structures increasingly vulnerable over time. This will ultimately reduce the expected service life of the structures and hence pose a potential risk to the public. Maintaining the safety and serviceability of corrosion affected RC structures is not only very costly but also interrupts routine life of the public, which has seriously taxed the abilities of engineers, operators and asset managers alike. It is estimated that corrosion related maintenance and repairs for concrete infrastructure cost around $100 billion per annum in the world (Dhir 2000). There is, therefore, a clear need from both the field of scientific research and social demand that a method be developed to assess the vulnerability of RC structures under designed loads and expected future overloading.

Structural reliability theory has been the pivotal tool employed by most researchers to develop various methods of quantitative risk assessment for structures. In this theory the probability of structural failure is used as a measure of structural reliability (both safety and serviceability). It can be expressed as follows (Melchers 1999):

$$p_f(t) = P[G(\mathbf{X},t) \le 0] = \int \cdots \int_{G(X,t) \le 0} f_X(\mathbf{x},t) d\mathbf{x} \qquad (1)$$

where p_f is the probability of structural failure; $G(\ )$ is the limit state function defining the structural failure; $f_X(\ )$ is the joint probability density function of $\mathbf{X}$; the vector of basic random variables and t is time. It is well known that analytical solutions to Equation 1 are limited, numerical solutions are usually viable alternatives, such as various simulation algorithms. In addition to mathematical complexity, lack of statistical information on basic random variables (design variables) hampers its wide application to practical structures. This in turn hinders the further development of structural reliability theory.

Since the full probabilistic information about all design variables of structures is usually not available, it is reasonable to separate those variables whose statistical information can be acquired (by whatever means) from those cannot. With this separation, the theory of conditional probability can be applied, i.e., the resulted probability is conditional on given values of those variables whose probabilistic information is not complete. This concept has recently been employed in risk assessment of nuclear plants (Kennedy & Ravindra 1984), bridges (Shinozuka, et al. 2000a, b), coastal defences (Dawson & Hall 2001), dams (Ellingwood &

Tekie 2001) and wood frame houses (Rosowsky & Ellingwood 2002). In these studies, a term of fragility has been introduced to represent the conditional probability of structural failure (conditional on a given load) and a two-parameter lognormal distribution function has been assumed in determining fragility curves. Also in these studies, the time-variant nature (in structural resistance in particular, such as deterioration) has not been fully explored. With the deterioration of structural resistance, structures become more vulnerable over time under expected future load. This gives rise for further studies on the vulnerability of deteriorating structures.

The intention of this paper is to propose a method to assess the vulnerability of deteriorating RC structures for a given intensity of load. The method is derived within the framework of structural reliability theory and forms an integrated part of full quantitative risk assessment for structures. The application of the proposed method to vulnerability assessment of corrosion affected flexural members is demonstrated, based on models of structural deterioration derived from experimental data. The method presented in the paper can equip engineers, operators and asset managers with confidence in developing a risk-informed maintenance strategy in the asset management of corroded concrete infrastructure.

2 FORMULATION OF STRUCTURAL VULNERABILITY

In assessing structural vulnerability, the basic random variable $\mathbf{X}$ in Equation 1 can be categorized into load effect, denoted as S, and structural resistance, R. With this notation, the structural vulnerability can be defined as the conditional probability of structural failure for a given intensity of loads or hazards (Casciati & Faravelli 1991). Quantitatively, it can be expressed as:

$$V_f(t) = P[G(t) \leq 0 \mid S(t) = s] \cdot P[S(t) = s] \qquad (2)$$

where V_f denotes structural vulnerability. In Equation 2, the limit state function (also known as safety margin), $G(t)$ can be expressed in terms of R and S as (Melchers 1999):

$$G(t) = R(t) - S(t) \qquad (3)$$

The formulation of structural vulnerability in Equation 2 has a number of advantages in comparison with conventional reliability formulation of Equation 1:

(i) It effectively uncouples the correlation between load and structural resistance (Ang & Tang 1984). As is well known, the correlation of random variables is in general the most difficult issue in accurate computation of the probability of structural failure, i.e., Equation 1. This prevents its effective application to practical structures. When both load and resistance are time-dependent variables as shown in Equation 3, it is almost an illusive task to determine the co-variance function of the random variables (Li & Melchers 1993). Also, the "uncorrelated" load and structural resistance makes structural analysis easier, in particular for complex structural systems and with finite element methods.

(ii) A full probabilistic description of both load and structural resistance is in most cases unavailable (Yao 1985). In comparison, the information on the maximum load and its likelihood during the lifetime of a structure may be of more concern to engineers, operators and asset managers (Ellingwood & Tekie 2001). Absent from credited data on load, they might simply inquire about either how vulnerable the structure is were the design load to be exceeded by, say, 20% or for what load intensity it is 95% likely that the structure becomes too vulnerable, i.e., maintenance is imminent.

(iii) The vulnerability assessment is less complex than a comprehensive quantitative risk assessment, but serve the same purpose when statistical information cannot be available. As is known, a full risk assessment for a structure is usually more mathematically and computationally involved to end users (Ellingwood 1994). With the formulation of Equation 2, it is less likely that there is any miscommunication between the assessment team and the end users and decision-makers.

(iv) Finally, vulnerability assessment can show how various postulated events (loading) might affect the likelihood of structural failure in the absence of a complete hazard analysis.

As may be noted there is no essential difference between conventional reliability formulation (Equation 1) and the vulnerability formulation (Equation 2). This can be demonstrated as follows. With the limit state function of Equation 3, Equation 1 can be re-written as:

$$p_f = \int_0^\infty \int_0^s f_{R,S}(r,s) dr ds \qquad (4)$$

where $f_{R,S}(r,s)$ is the joint probability density function of R and S. According to the theory of conditional probability, $f_{R,S}(r,s)$ can be expressed as (Devore 1995):

$$f_{R,S}(r,s) = f_{R|S}(r \mid s) f_S(s) \qquad (5)$$

where $f_{R|S}(r \mid s)$ is conditional probability density function of R given that $S = s$. Thus, for a given load, i.e., $S = s$, Equation 4 becomes:

$$P_{f|S=s} = \int_0^s f_{R|S}(r\,|\,s)dr \int_0^\infty f_S(s)ds \qquad (6)$$

In probability terms, Equation 6 can be expressed as (Ang & Tang 1984):

$$P_{f|S=s} = P[G(t) \leq 0\,|\,S = s] \cdot P(S = s) \qquad (7)$$

which is identical to Equation 2. This indicates that the difference between the vulnerability formulation (Equation 2) and the conventional reliability formulation (Equation 1) is more technical than methodological and/or conceptual. Therefore methods used in reliability computation can apply to vulnerability computation as well. Also as may be noted that vulnerability formulation is complementary to reliability problems formulated in load space. This is beyond the scope of the paper but can be referred to, e.g., Ditlevsen et al. (1988) and Melchers (1992).

In applying Equation 2 to practical structures, limit states defining the structural failure or damage conditions need to be identified. A limit state for a structure is prescribed in design codes and standards and/or specified by users, based on performance requirements for the structure. The commonly used performance criteria can be summarized as follows (Melchers 1999):

(i) Serviceability limit state. Under normal loading, the structure is serviceable at any time;
(ii) Strength limit state. Under code-prescribed maximum loading, the structure is adequate during its service life;
(iii) Extreme limit state. Under improbable extreme loading, the structure does not collapse.

The violation of any of these criteria is considered as a failure and a corresponding limit state function is established. In general, the first two criteria apply to both structural members (locally) and structures as a whole (globally), while the third one is more for structures as a whole (structural systems). As can be seen, a vulnerability assessment requires a thorough understanding of the mechanics of structural response to a variety of loadings, ranging from those encountered in service conditions to those that may occur at levels well above design basis. At high levels of loading, the behavior of the structure is usually highly non-linear in nature which makes it difficult to establish limit state function for the structure. This is another advantage of vulnerability assessment when applied to structural systems (which will not be discussed herein but be referred to Li 1996).

3 APPLICATION TO FLEXURAL MEMBERS

To apply Equation 2 to practical structures, the main effort lies in developing a resistance model that can incorporate structural deterioration. For corrosion affected RC structures, the structural deterioration is mainly induced by reinforcement corrosion in concrete (Broomfield 1997). Since statistical data on resistance deterioration (both strength and serviceability) are available from experiment on full size RC beams (Li 2003), RC flexural members will be used as an example to demonstrate the application of the vulnerability method formulated above.

3.1 Resistance model

As proposed in Li (1995), a general model of structural resistance deterioration can be of the form:

$$R(t) = \varphi(t)R_0 \qquad (8)$$

where $\varphi(t)$ denotes the deterioration function and R_0 is the original structural resistance. One of the advantages of the deterioration model in the form of Equation 8 is that the deterioration function φ is a relative rather than absolute value, i.e. $\varphi(t) = R(t)/R_0 \leq 100\%$. The relative form of the deterioration function can normalize the data produced from experiment or collected on site for structures of different types and original strength. This can maximize the use of available data which is usually scarce.

Given the complexity of the deterioration of RC structures, given the current state of knowledge and understanding of corrosion propagation in concrete and its effect on structural resistance deterioration, the development of deterioration function will be resorted to mathematical regression of experimental data produced from a comprehensive experiment on RC cantilever beams subjected to corrosion and simultaneous service loading. Detailed information of the experiment and the testing methodology has been published in Li (2003) and hence will not be repeated herein. It suffices to note that RC cantilever beams with structurally significant size (Fig. 1) were used as test specimens and a total of thirty specimens were tested, representing different concrete compositions, e.g., water cement ratio and cement type. Based on the analysis of experimental data, the deterioration function for corrosion affected RC flexural members can be modeled by a mean function, $\mu_\varphi(t)$, and a function of coefficient of variation (COV), $V_\varphi(t)$, in the following form:

$$\mu_\varphi(t) = \varphi_0 \exp(-\alpha\,t) \qquad (9a)$$

$$V_\varphi(t) = \beta \cdot t + V_0 \qquad (9b)$$

where φ_0 is initial (i.e., $t = 0$) deterioration function, which is unity according to the definition of deterioration function of Equation 8. α is a parameter

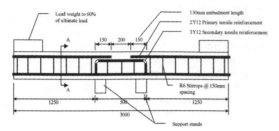

(a) Dimension and reinforcement layout

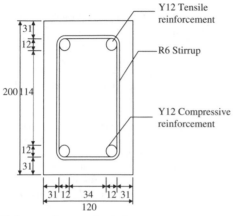

(b) Cross-section at section A-A

Figure 1. Test specimen used in the example.

representing the rate of structural deterioration. It allows for the effects of such factors as corrosion propagation, concrete composition and structural detailing. It may be appreciated that α is difficult to determine in general. In Equation 9b, V_0 is the initial variation of concrete structural properties which can be obtained from research literature (e.g., Mirza et al. 1979). β is a parameter representing the increase of uncertainty during the deterioration process. With Equation 9, mathematical regression can be employed to process the experimental data. For strength deterioration of RC flexural members, the regression analysis of the tested data in Li (2003) results in:

$$\mu_{\varphi,u}(t) = \exp(-0.027t) \tag{10a}$$

$$V_{\varphi,u}(t) = 0.016t \tag{10b}$$

It may be noted that V_0 in Equation 9b has been assumed to be zero for strength deterioration as shown in Equation 10b. This is because it is almost certain that there is no deterioration at the beginning of the service of new built structures. It needs to be noted

that the data represent the mean values of the data obtained from all available test specimens (Li 2003).

Analogy to regression analysis for strength deterioration function, the expressions for the mean and COV of serviceability deterioration function as represented by stiffness (i.e., deflection) can be obtained in the same manner, which are expressed as follows:

$$\mu_{\varphi,s}(t) = \exp(-0.096t) \tag{11a}$$

$$V_{\varphi,s}(t) = 0.014t + 0.1 \tag{11b}$$

It may be noted that V_0 in Equation 9b is not zero for stiffness deterioration as shown in Equation 11b. This is because it is less certain that there is no stiffness deterioration at the beginning of the service of even new built structures, such as creep and shrinkage effects.

3.2 Limit state function

Two limit state functions (performance criteria) will be considered in the vulnerability assessment of RC flexural members exemplified in Figure 1. One is the strength limit state in terms of bending capacity at a critical section, i.e.:

$$G(t) = \phi_u(t)M_o - c_1 S \tag{12}$$

where $\phi_u(t)$ is the strength deterioration function determined by Equation 10; M_o is the original bending capacity determined from relevant design codes; S is the given applied load and c_1 is a coefficient to convert load to bending moment (load effect) from structural analysis. Under this limit state, the 70th, 95th and 99th percentile values of the load, S, are considered to be normal, maximum and extreme loading conditions.

The other is the serviceability limit state represented by deflection at the critical section of the flexural member, i.e.:

$$G(t) = \delta - c_2 \frac{S}{\phi_s(t)EI_o} \tag{13}$$

where δ is the code-prescribed deflection limit; $\phi_s(t)$ is the stiffness deterioration function determined by Equation 11; EI_o is the original flexural stiffness determined from relevant design codes and c_2 is a coefficient to convert load to load effect, i.e., deflection from structural analysis. Under this limit state, only the normal loading condition (i.e., 70th percentile value of S) is of relevance to practical structures.

With the values given in Table 1, the structural vulnerability of the flexural member in Figure 1 can be

Table 1. Values of variables used in vulnerability computation.

Variables	Mean	COV
c_1	1.25 m	–
c_2	0.65 m^3	–
EI_o	3.33×10^6 Nm^2	0.12
M_o	13.86 kNm	0.15
S	4 kN	0.15
δ	6.25 mm (span/200)	–
ϕ_s	Eqn. (11a)	Eqn. (11b)
$\phi_{,u}$	Eqn. (10a)	Eqn. (10b)

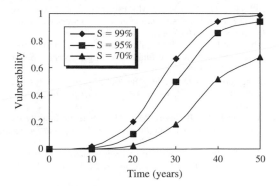

Figure 2. Vulnerability under strength limit state.

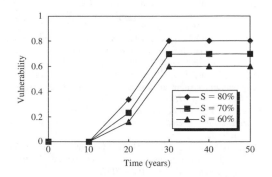

Figure 3. Vulnerability under serviceability limit state.

computed using Equation 2. Typical results are shown in Figures 2 and 3, from which it can be seen that at a given time, the structure is more vulnerable under high levels of load intensity for both strength and serviceability limit states. This makes sense both theoretically and practically. A comparison of structural vulnerability under different limit states (i.e., strength and serviceability) at the same level of load is shown in Figure 4. As can be seen corrosion affected structures are more vulnerable under serviceability requirement than strength requirement. These results are consistent with

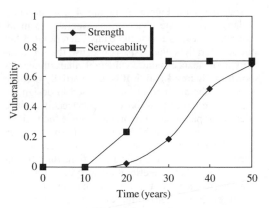

Figure 4. Comparison of vulnerability under different limit states ($S = 70\%$).

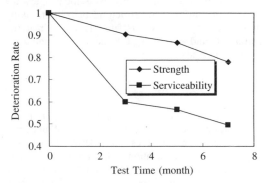

Figure 5. Comparison of structural deterioration under different limit states.

the experimental observation as shown in Figure 5, in which the deterioration rates for strength and serviceability (represented by deflection) are quite different, with serviceability deteriorating much faster than strength. The reason could be that the serviceability (i.e., deflection) is a measure of mechanical properties related to the geometry of a structure and its cross-sections to a larger extent than strength. There are more factors that affect the geometry of RC flexural members. Predominant factors that affect the geometry and are closely related to the corrosion are concrete cracking, delaminating and de-bonding between the concrete and reinforcement. All these effects are prominent once corrosion actively propagates. As a consequence, the deflection increases much faster. The deterioration of strength, on the other hand, is less affected by these "cosmetic" factors which only damage tensile regions of RC flexural members. As is well known, the tensile region is not included in the calculation of cross-sectional strength of RC flexural members.

Together with Figure 5, Figure 4 also provides explanations for site observations whereby so many concrete structures are seen "badly" deteriorated but still structurally sound. This is because the serviceability performance is more vulnerable than strength as shown in Figures 4 and 5. It is also partially because the safety factors (partial factors) used in design for strength are larger than those for serviceability since safety is of paramount importance. Since the maintenance costs are usually high for concrete structures, in particular for strengthening, it is of practical importance to differentiate maintenance actions for different levels of structural deterioration in order to achieve cost-effectiveness in the asset management of RC structures. In this sense Figures 4 and 5 can provide rational guidance for engineers and asset managers in making decisions regarding the maintenance of corrosion affected concrete structures.

The proposed vulnerability assessment opens the way to the study of the structural behaviour, taking into account many aspects such as spatial material variability and system of RC structure with combination of the load effects.

4 CONCLUSIONS

A method for vulnerability assessment of deteriorating structures has been proposed in the paper. With empirical models of structural resistance derived from experimental data, the application of the proposed method to the vulnerability assessment of reinforced concrete flexural members has been demonstrated. Results from vulnerability assessment show that, under normal loading conditions, corrosion affected concrete structures are more vulnerable under serviceability performance requirement than strength. It has been found that the corrosion-induced deterioration is the most important single factor that makes reinforced concrete structures vulnerable. It can be concluded that the proposed method for vulnerability assessment of reinforced concrete structures can serve as a tool for engineers and asset managers in developing a risk-informed management strategy for corroded concrete infrastructure.

ACKNOWLEDGEMENTS

Financial support of Engineering and Physical Sciences Research Council (EPSRC), UK with Grant No. GR/R28348 is gratefully acknowledged.

REFERENCES

Ang, A.H-S. & Tang, W.H. 1984. *Probability Concepts in Engineering Planning and Design: Volume II: Decision, Risk and Reliability.* John Wiley & Sons, New York.

Broomfield, J. 1997. *Corrosion of Steel in Concrete, Understanding, Investigating & Repair.* E & FN Spon, London.

Casciati, F. & Faravelli, L. 1991. *Fragility Analysis of Complex Structural Systems.* RSP, Research Studies Press Ltd., Taunton, Somerset.

Chaker, V. (Ed) 1992. *Corrosion Forms & Control for Infrastructure.* ASTM STP 1137, Philadelphia.

Dawson, R.J. & Hall, J.W. 2001. Improved condition characterisation of coastal defences. *Proc. Coastlines, structures & breakwaters 2001*, ICE, London, 1–12.

Devore, J.L. 1995. *Probability and Statistics for Engineering and the Sciences.* Duxbury Press, Belmont.

Dhir, R.K. 2000. *Concrete Durability.* Lecture Notes, Department of Civil Engineering, University of Dundee, UK.

Ditlevsen, O., Hasofer, A.M., Bjerager, P. & Olesen, R. 1988. Directional simulation in gaussian processes. *Prob. Engineering Mechanics* 3 (4): 207–217.

Ellingwood, B. 1994. Probability-based codified design: past accomplishment and future challenges. *Struct. Safety* 13(3): 159–176.

Ellingwood, B.R. & Tekie, P.B. 2001. Fragility analysis of concrete gravity dams. *Journal of Infrastructures Systems* 7(2): 41–48.

Kennedy, R.P. & Ravindra, M.K. 1984. Seismic fragility for nuclear power plant studies. *Nuclear Engrg. and Design* 79(1): 47–68.

Li, C.Q. 1995. A case study on reliability analysis of deteriorating concrete structures. *Struct. & Bldg., I.C.E.*, 110(4): 269–277.

Li, C.Q. 1996. Computation of time-dependent structural system reliability without using global limit state functions. *Struct. & Bldg., I.C.E.* 116(2): 129–137.

Li, C.Q. 2003. Life cycle modelling of corrosion affected concrete structures – propagation. *J. Struct. Engrg., A.S.C.E.* 129 (6) (to appear).

Li, C.Q. & Melchers, R.E. 1993. Outcrossings from convex polyhedrons for nonstationary gaussian processes. *J. Engrg. Mech., ASCE* 119(11): 2354–2361.

Melchers, R.E. 1992. Load space formulation for time-dependent structural reliability. *Journal of Engineering Mechanics, ASCE* 118(5): 853–870.

Melchers, R.E. 1999. *Structural Reliability Analysis and Prediction, Second Edition.* John Wileys & Sons, Chichester.

Mirza, S.A., Hatzinikolas, M. & MacCgregor, J.G. 1979. Statistical description of strength of concrete. *J. Struct. Engrg., ASCE* 105 (6): 1021–1037.

Rosowsky, D.V. & Ellingwood, B.R. 2002. Performance-based engineering of wood frame housing: fragility analysis methodology. *J. of Structural Engineering, ASCE* 128 (1): 32–38.

Shinozuka, M., Feng, M.Q., Lee, J. & Naganuma, T. 2000a. Statistical analysis of fragility curves. *Journal of Engineering Mechanics, ASCE* 126(12): 1224–1231.

Shinozuka, M., Feng M.Q., Kim, H-K. & Kim, S-H. 2000b. Nonlinear static procedure for fragility curve development. *Journal of Engineering Mechanics, ASCE* 126(12): 1287–1295.

Schiessl, P. 1988. *Corrosion of Steel in Concrete, Report of the TC60-CSC RILEM.* Chapman and Hall, London.

Yao, J.T.P. 1985. *Safety and Reliability of Existing Structures*, Pitman Publishing Inc.

System-based Vision for Strategic and Creative Design, Bontempi (ed.)
© 2003 Swets & Zeitlinger, Lisse, ISBN 90 5809 599 1

Influence of cover on the flexural performance of deteriorated reinforced concrete beams

E.H. Hristova
FaberMaunsell Ltd., Birmingham, UK

F.J. O'Flaherty, P.S. Mangat & P. Lambert
School of Environment and Development, Sheffield Hallam University, Sheffield, UK

ABSTRACT: Reinforcement corrosion is the principal cause of deterioration of reinforced concrete. It has been estimated that around €1.5 bn is spent annually in Europe on concrete repairs to increase the service life of deteriorated structural members. In order to obtain a better understanding of the flexural performance of deteriorated reinforced concrete beams, a series of tests was conducted on 28 laboratory sized specimens. Main steel reinforcement was corroded along the entire span with target corrosion ranging from 0% (control) to 20% in 5% increments. The influence of three different covers to the main steel reinforcement (20, 30 and 50 mm) was investigated and related to the flexural performance.

1 INTRODUCTION

Reinforced concrete has been in use for over 110 years, and in general, the environment provided by concrete protects the reinforcing steel. This is due to the high pH environment present in Portland cement pore solution which passivates the steel (Cornet et al. 1968, Borgard et al. 1990). Corrosion of the steel reinforcement will not occur unless an external agent changes the normal passive state of the steel in this alkaline environment. When this occurs, corrosion becomes a subject of technical and scientific interest as well as of economic interest.

Corrosion phenomena are very complex. When steel reinforcement corrodes, tensile stresses are generated in the concrete as a result of the corrosion products. Since concrete can endure much less tensile stress than compressive stress, tensile cracks are readily nucleated and propagated as a result. The development of corrosion products along the bar surface may affect the failure mode and ultimate strength of flexural members due to two causes: firstly, due to a reduction in the degree of bar confinement caused by an opening of longitudinal cracks along the reinforcement and, secondly, due to significant changes at the steel-concrete interface caused by changes in the surface conditions of the reinforcing steel (Al-Sulaimani et al. 1990). In all corrosion cases, repair is necessary to increase the service life of the member (Mangat & O'Flaherty 1999, 2000) and it

is estimated to cost approximately €1.5 bn in Europe each year (Davies 1996).

2 BACKGROUND

In this investigation, the steel in reinforced concrete beams was subjected to an accelerated corrosion technique in the laboratory using one of the several methods available. The galvanostatic method was used in this study to simulate the field conditions. The method involves passing a direct current through the reinforcement to accelerate corrosion. The galvanostatic corrosion is carried out whilst the beam is unloaded, which is different from the corrosion in actual structures. The corrosion by galvanostatic method is general, whereas actual structures have some specific areas that are more prone to corrosion. Thus in the latter case, there is always the possibility of pitting corrosion whereby the cross-sectional area of the reinforcing bars could be significantly reduced, thus reducing the tensile strength of the reinforcing bars. However, to ensure consistency of results in this investigation, the steel reinforcement was subjected to general corrosion only, which allows easier repeatability compared to pitting corrosion.

According to standard corrosion theory, steel embedded in concrete is largely in a protected state because of the alkalinity of the matrix. The corrosion rate depends on the ratio of the cathodic area to the anodic area. In this investigation, the potential was measured every day to ensure that the steel was corroding. The potential cannot be measured directly, as the available measuring devices can measure only a difference in potential. To overcome this limitation, a Saturated Calomel Electrode (SCE) was added to the system by means of a suitable salt bridge. The potential for the steel reinforcement in this study ranged between $-750\,mV$ and $-500\,mV$ which represents the active state of corrosion process.

3 DESIGN OF THE BEAM SPECIMENS

A total of thirty reinforced concrete beams were tested to examine the influence of reinforcement cover on the flexural behavior. Details of test specimens are given in Figure 1 and Table 1. Beams were 910 mm long with a cross-section of 100 mm $\times$ 150 mm deep. All specimens were detailed for flexural failure; sufficient links were provided to ensure adequate shear capacity at the anticipated maximum load of the corroded beam. Beams 2T8/0/20 (number, type, and diameter of the main steel in mm/target percentage of corrosion/cover

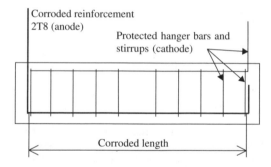

Figure 1. Beam specimens.

Table 1. Variables in test programme.

Main steel	Target corrosion (%)	Cover (mm)	No. of specimens
2T8	0		2
	5	20	3
	10		2
	15		3
2T8	0		2
	5	30	4
	10		2
	15		2
2T8	0		2
	5	50	3
	10		1
	15		2

in mm), 2T8/0/30 and 2T8/0/50 were tested without corrosion to serve as control specimens (Table 1). The number of specimens tested for each target corrosion percentage is also shown in Table 1.

Main reinforcement consisted of high yield (ribbed) bars with a nominal characteristic strength of $460\,N/mm^2$. Shear reinforcement was 6 mm diameter plain round mild steel bars with a yield strength of $250\,N/mm^2$ at 85, 80 or 65 mm spacing for 20, 30 and 50 mm cover respectively. Hanger top bars for all beams consisted of two 6 mm diameter plain round mild steel bars with a yield strength of $250\,N/mm^2$. The steel reinforcement was weighed before casting to enable the actual percentage corrosion to be calculated at a later stage.

Test specimens were cast in the laboratory using a concrete with target cube strength of $40\,N/mm^2$. Mix proportions were 1:1.7:3.8 of Portland cement: fine aggregate: coarse aggregate. Fine and coarse aggregates were oven dried at 100°C for 24 hours. Calcium chloride ($CaCl_2$) was added to the mix (1% by weight of cement) in order to promote corrosion of the reinforcement. The material was placed in steel moulds in three layers, each layer being carefully compacted on a vibrating table. The specimens were then placed in the mist curing room (20°C and 95% $\pm$ 5% Relative Humidity) for 24 hours. The samples were demoulded after 1 day and cured in water at 20°C for a further 27 days (28 days in total). Specimens were then transferred to a tank filled with a saline solution for accelerated corrosion at 28 days age.

4 ACCELERATED CORROSION PROCESS

The beam specimens were immersed in artificial seawater in a plastic tank at the end of the curing period. A 3.5% $CaCl_2$ solution was used as the electrolyte. The direction of the current was arranged so that the main reinforcing steel served as the anode and the hanger bars and the stirrups acted as the cathode.

A constant current density of $1\,mA/cm^2$ was passed through the reinforcement. This current density was adopted on the basis of pilot tests to provide desired levels of corrosion in a reasonable time. The current supplied to each specimen was checked on a regular basis and any drift was corrected.

The relationship between corrosion current density and the weight of metal lost due to corrosion was determined by applying Faraday's law as shown in Equation 1:

$$\Delta\omega = \frac{AIt}{ZF} \qquad (1)$$

where $\Delta\omega$ = weight loss due to corrosion in (g); A = atomic weight of iron (56 g); I = electrical current in (A); t = time in (sec.); Z = valence of iron which is 2; and F = Faraday's constant (96 500 coulombs).

The metal weight loss due to corrosion can also be expressed as:

$$\Delta\omega = a\delta\gamma \qquad (2)$$

where a = rebar surface area before corrosion (cm^2); δ = material loss (cm); and γ = density of material (7.86 g/cm^3).

The corrosion current can be expressed as:

$$I = (i)\,(a) \qquad (3)$$

where i = corrosion current density (amp/cm^2).

Therefore, combining Equations 1, 2 and 3 gives:

$$a\delta\gamma = \frac{A\,I\,t}{Z\,F} = \frac{A\,i\,a\,t}{Z\,F} \qquad (4)$$

Substituting known values into Equation 4 gives:

$$\delta = \frac{A\,i\,t}{\gamma\,Z\,F} = \frac{56*i*365*24*60*60}{7.86*2*96500} =$$
$$= 1165*i \ (\text{cm/year}) \qquad (5)$$

Rewriting Equation 5, where R is defined as the material loss per year (cm/year), gives:

$$R = 1165*i \ (\text{cm/year}) \qquad (6)$$

As an example, for a corrosion rate, i, of 1 (mA/cm^2), R, equals 1.165 (cm/year) (from Equation 6).

If, in a reinforced concrete structure, the period of corrosion after initiation is T years, then:

Metal loss after T years = RT (cm) $\qquad (7)$

Therefore:

% reduction in rebar dia. in T years $= \dfrac{2RT}{D} x100 \qquad (8)$

(Mangat & Elgarf 1999)

The expression [2R(T/D)]%, which represents reduction in rebar diameter due to corrosion in T years, is also defined as the degree of reinforcement corrosion.

Preliminary tests were carried out before commencing the research program to confirm the reliability of the accelerated corrosion technique.

Figure 2 show reinforced concrete specimens undergoing accelerated corrosion using specialised equipment in the laboratory. The first sign of corrosion was

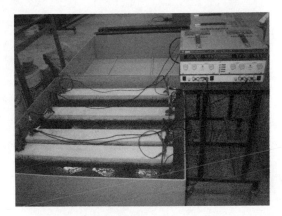

Figure 2. Reinforced concrete beams undergoing accelerated corrosion.

rust staining on the concrete surface, followed by longitudinal cracking in the concrete cover zone.

5 BEAM TESTING

The control specimens (zero percent corrosion) were tested at the age of 28 days but the deteriorated beams were tested at 42, 56 and 63 days for the 5, 10 and 15% target corrosion respectively due to the time taken to reach the desired levels of corrosion. All specimens were tested under four point bending as shown in Figure 3, to determine the ultimate flexural strength. Premature shear failure was prevented by sufficient shear reinforcement. The testing machine loading rate was set at 5 kN/min.

The first load tests were carried out on the control specimens and these behaved as expected and in accordance with the design procedures of BS 8110 (British Standards Institutions). Failure of all 28 beams was in flexure; no shear failure occurred.

Upon completion of the corrosion period and flexural testing, the reinforcing bars were removed from the concrete as shown in Figure 4, cleaned with a wire brush and re-weighed. The percentage loss in weight was subsequently calculated. The resulting degree of corrosion in this investigation, 2RT/D%, ranged between 0% (control) and 18.45% (Table 2).

The corrosion damage was generally spread along the length of the bars. Where serious section loss occurred, it was in the form of localized pitting corrosion rather then general corrosion. This mainly occurred at higher percentages of corrosion.

6 TEST RESULTS AND DISCUSSION

Table 2 shows the results from testing 30 beams in flexure in the laboratory. Each beam is identified by

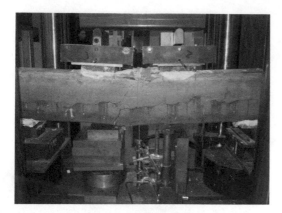

Figure 3. Flexural testing of deteriorated reinforced concrete beams.

Figure 4. Deteriorated main steel (anode) with uncorroded links and hanger bars (cathode).

Table 2. Beam test results.

Beam identification	Actual corrosion (%)	Ultimate load (kN)	Failure mode
2T8/0/20	0	57.40	Flexure
2T8/0/20	0	54.10	Flexure
2T8/5/20	1.0	52.75	Flexure
2T8/5/20	1.4	54.70	Flexure
2T8/5/20	2.3	43.78	Flexure
2T8/10/20	3.4	50.12	Flexure
2T8/10/20	8.9	41.10	Flexure
2T8/15/20	10.0	34.98	Flexure
2T8/15/20	15.5	19.23	Flexure
2T8/15/20	18.5	20.57	Flexure
2T8/0/30	0	50.40	Flexure
2T8/0/30	0	55.20	Flexure
2T8/5/30	0.8	56.80	Flexure
2T8/5/30	0.9	52.10	Flexure
2T8/5/30	1.4	45.03	Flexure
2T8/5/30	4.6	44.60	Flexure
2T8/10/30	8.5	34.70	Flexure
2T8/10/30	9.6	34.20	Flexure
2T8/15/30	15.0	24.57	Flexure
2T8/15/30	17.8	20.90	Flexure
2T8/0/50	0	40.00	Flexure
2T8/0/50	0	42.70	Flexure
2T8/5/50	2.7	42.91	Flexure
2T8/5/50	3.5	39.24	Flexure
2T8/5/50	6.9	34.58	Flexure
2T8/10/50	8.9	26.34	Flexure
2T8/15/50	15.1	17.08	Flexure
2T8/15/50	16.4	10.10	Flexure

Table 3. Comparison of residual strength at different covers and degrees of corrosion.

Degree of corrosion (%)	P_{ult}/P_{con} (%) Cover		
	20 mm	30 mm	50 mm
4	84	86	90
10	63	65	62
16	41	43	33

the amount of main steel, target corrosion and cover (e.g. 2T8/10/20). The actual corrosion (calculated as described in Section 5) is also given along with the ultimate load at failure. Table 2 also shows that the failure of each beam in the three categories (20, 30 and 50 mm cover) was flexural.

It is clear from the ultimate loads given in Table 2 that the strength of the beams decrease with increasing main steel corrosion (compare the control load of beam 2T8/0/20 with that of 2T8/20/20, the ultimate load decreases from 57.40 kN to 20.57 kN). This is also applicable to the other two categories (30 and 50 mm cover, Table 2) which also show significant reductions in ultimate strength due to corrosion.

To gain a better understanding of the influence of cover on the flexural strength of the deteriorated beams, Figures 5–7 show the relationship between P_{ult}/P_{con} and the degree of corrosion to the main steel

reinforcement. P_{ult} is the ultimate load obtained from testing the beams in the laboratory (those exhibiting main steel corrosion) and P_{con} is the average failure load of the control specimens (0% corrosion to the main steel reinforcement). In all cases, the actual percentage of corrosion was used in the analysis of data as opposed to the target corrosion. This led to a better correlation between flexural performance and degree of corrosion as there was some variation between target and actual values (Table 2). Referring to Table 3, comparisons are made between P_{ult}/P_{con} and the degree of

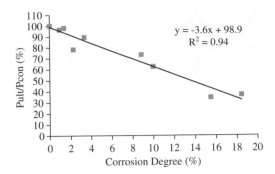

Figure 5. Relationship between P_{ult}/P_{con} and degree of corrosion for beams designed with 20 mm cover.

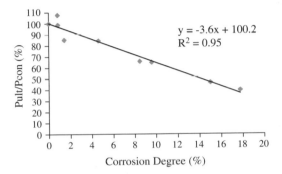

Figure 6. Relationship between P_{ult}/P_{con} and degree of corrosion for beams designed with 30 mm cover.

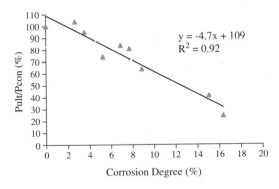

Figure 7. Relationship between P_{ult}/P_{con} and degree of corrosion for beams designed with 50 mm cover.

corrosion at arbitrary values of 4, 10 and 16% corrosion (calculated from the best fit equations in Figures 5–7). It is evident from Table 3 that there is only a small difference in residual strength at 4 and 10% for beam categories 20, 30 and 50 mm cover (84, 86 and 90% at

4% corrosion and 63, 65 and 62% at 10% corrosion respectively). However, at 16% corrosion, the beams with 20 and 30 mm cover exhibit a residual strength of 41 and 43% respectively whereas the beam with 50 mm exhibits a residual strength of only 33%. Therefore, based on these results, beams designed with high reinforcement cover suffer a higher reduction in flexural strength when exposed to significant reinforcement corrosion (up to 15% loss of cross section).

It was also evident during testing that the time taken for the first crack to develop was dependant on cover. For example, the beams with 20 mm cover cracked at 5% corrosion whereas the beams designed with 50 mm cover cracked at a higher degree of corrosion (10%). However, cracking was more severe for beams with 50 mm cover than for beams with 20 mm cover at similar degrees of corrosion, thereby leading to a lower residual strength.

7 CONCLUSIONS

The main conclusions from the results reported in this paper are as follows:

– reinforced concrete beams show a loss in residual strength with increasing corrosion of the main steel reinforcement;
– the cracking in the cover concrete was more severe at 50 mm cover compared to 20 mm cover at similar levels of main steel reinforcement corrosion;
– deteriorated reinforced concrete beams suffer the most reduction in flexural strength when the beams are designed with high reinforcement cover and are subjected to high levels of main steel corrosion.

ACKNOWLEDGEMENTS

The work described in this paper forms a part of a research project entitled "Residual strength of corroded reinforced concrete beams" which was carried out at Sheffield Hallam University. The assistance received from the technical staff in the laboratory is gratefully acknowledged. The authors also acknowledge with thanks the support of FaberMaunsell Ltd and CPI for providing the multi-channel power supply for use in the laboratory.

REFERENCES

Al-Sulaimani, G. et al. 1990. Influence of Corrosion and Cracking on the Bond Behavior and Strength of Reinforced Concrete Members. *ACI Structural Journal* 87 (2): 220–231.

Borgard, B. et al. 1990. Mechanisms of Corrosion of Steel in Concrete. *Corrosion Rates of Steel in Concrete ASTM STR 1065*: 174–188.

British Standard Institution 1997. Structural use of concrete: Code of practice for, design and construction BS 8110-1.

Cornet, I. et al. 1968 *Materials Protection*: 44–47.

Davies, H. 1996. Repair and maintenance of corrosion damaged concrete: Materials strategies, Concrete Repair, Rehabilitation and Protection. London: E&FN Spon.

Mangat, P.S., Elgarf, M.S. 1999. Flexural strength of concrete beams with corroding reinforcement, *ACI Structural Journal* 96 (1): 149–158 Jan–Feb 1999.

Mangat, P.S., O'Flaherty, F.J. 1999. Long-term performance of high stiffness repairs in highway structures, *Magazine of Concrete Research* 51 (5): 325–339.

Mangat, P.S., O'Flaherty, F.J. 2000. Influence of elastic modulus on stress redistribution and cracking in repair patches. *Cement and Concrete Research* 30: 125–136.

Evaluation of atmospheric deterioration of concrete

R. Drochytka & V. Petránek
Brno University of Technology, Faculty of Civil Engineering, Brno, Czech Republic

ABSTRACT: The volume of the present industrial production results in the increased air pollution. This means a negative effect not only on the environment, but even a harmful influence on materials of building structures. The increased deterioration of surroundings causes failures of structures which can significantly decrease the utility value of buildings and their durability. The corrosive effects of outer surroundings on concrete structures are significant, especially if the penetration of acidic gases into concrete is possible. The character of failures of concrete structures exposed to the surrounding of free atmosphere shows that it is necessary to pay attention to the study of the interaction between corrosive gases from the atmosphere and concrete as early as in the preparatory design, and then during the construction of engineering structures. In the text to follow, the attention is particularly paid to a brief summary of the present knowledge concerning the area of atmospheric concrete deterioration and the evaluation of damage caused by the atmospheric deterioration.

1 INTRODUCTION

The defects that originate through the exploitation of concrete structures in the environment of free atmosphere manifest themselves that the knowledge of the durability of concrete and concrete structures are not sufficient for the increased operational claims both in the stage of design and in the stage of construction. Harmful effects of the external corrosive environment on concrete structures especially occur if the concrete was not sufficiently compacted during its placing. The insufficient compaction is reflected in the increased porosity, in an increased permeability of water, water vapour, of air and harmful gases. As far as the reinforcement is concerned, this is insufficiently protected against external harmful effects by a cover layer of concrete which is not thick enough. It is the service life of a reinforced concrete structure that is to a large extent dependent on the corrosion of steel reinforcement. The action of some especially inorganic compounds on reinforced concrete structures results in the deterioration of all components of concrete, cement binder also aggregates and in the corrosion of steel reinforcement. The protective function of the concrete covering layer on the steel reinforcement fails through the corrosion of cement binder and aggregates.

2 CONCRETE DETERIORATION

Since the concrete deterioration is a rather complex problem, our attention will be focused on the deterioration caused by external factors only. These factors, may be classified as mechanical effects (static and dynamic stresses), physical effects (changes of temperature and humidity, frost, high temperatures, fire, electric current, radiation), chemical effects (liquid and gaseous environment) and biological (macroscopic and microscopic) effects.

3 CONCRETE DETERIORATION BY ACIDIC GASES

At the present time, not only carbon dioxide, dust or fly ash are exhausted to the atmosphere, but also other acidic gases, especially SO_2, SO_3, NH_3, H_2S, nitrogen compounds, HF, HCl, chlorine, acetic acid and formic acid, etc. As follows from the above listing, Czech Republic ranks among the countries with the most polluted atmosphere.

The action of other gases of acid nature then CO_2 and SO_2 such as hydrogen sulphide or nitrogen oxide is limited to rather small areas.

The next chapters discuss the chemical effects of gaseous materials, to which realistically the concrete structures can be exposed. These are particularly carbon dioxide CO_2, sulphur dioxide SO_2 and sulphur trioxide respectively.

3.1 *Carbonation*

The concentration of carbon dioxide in the atmosphere is mostly stable and it is about 60 mg CO_2 in 1 m^3 of air.

A great increase of concentration we find mostly near sources of CO_2, for instance of great natural springs, in combustion gases etc.

The normal content of SO_2 in natural milieu is not more than 0,01 mg of SO_2 in 1 m^3 of air. From the hygienic point of view the highest admissible concentration of sulphur dioxide in the Czech Republic is 0,15 mg SO_2 in 1 m^3 of air and intermittently up to 0,5 mg SO_2 in 1 m^3 of air. In larger cities the concentration of SO_2 significantly increases, especially in winter months, because of the low quality brown coal combustion in local heating.

The hydrogen monosulphide in gaseous form is a gas present in agricultural objects and similar places. Nitrogen oxides are mostly products of road transport but we can find these compositions even in air pollution products of the chemical industry

A factor, which can significantly limit the durability of concrete structure, is the influence of atmospheric carbon dioxide. Carbon dioxide causes in contact with cement binder formed by basic hydration products of binders a neutralization reaction normally called carbonization, because the products of this reaction are different carbonates. The course of neutralization reactions we can observe according to the change of pH value. The determination of pH in the inter-grainal liquid is not quite reliable. The cement binder can for dozen years step by step restore by hydration of coarse cement clinker grains the original high pH value of the inter-granular liquid and in this way substitute the hydroxyl ions decrease accompanying the carbonization. In the same time new insoluble forms of $CaCO_3$ are formed which sediment in pores and capillary tubes and fill them successively. This limits the access possibility of new CO_2 or ions forming the sediments. In this way the volume mass of concrete surface layer increases and the microstructure of cement binder changes. In heavy concrete this process manifests in permeability decrease. This is very significant, because the mentioned processes explain why the perfectly dense concrete is resistant against the harmful influence of carbonization and why the steel reinforcement is for dozens of years protected against corrosion by the diffusion of acid gases.

The normal types of concrete, especially of lower and middle strength class have a higher porosity and the newly formed carbonate compositions are not able to decrease this porosity and substantially decrease the permeability of concrete for gases and vapors. The carbonization continues into the depth, attacks here the cement binder and causes a very strong corrosion of the reinforcement.

In the case of concrete saturation by water practically all micro-pores of concrete are completely filled with water and the water film prevents the penetration of gases into the internal parts of concrete in spite of that, that they are soluble in water as it is the case with CO_2.

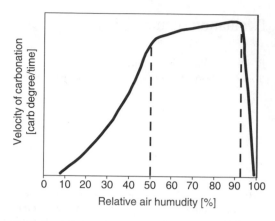

Figure 1. Relation between the carbonization stage °C and the pH value of gravel sand concrete with marking the four carbonization stages.

If the pores are full of water the concrete carbonization is significantly slower. The presence of certain moisture in concrete is the condition for the carbonization to take place, because it is an ionic reaction. The completely dry concrete doesn't react with CO_2 and other gases at all. There exists a certain optimum or range of concrete moisture in which the carbonization of concrete will progress with highest velocity.

Graphically is the dependence of concrete carbonization velocity upon the relative air humidity [φ] as determined by Prof. Matousek [1] presented in Figure 1.

3.1.1 Four stages of carbonization

Following to the long time research and measurements carried out at our institute the carbonization process was divided into four stages.

I. In the first stage of carbonization the Ca $(OH)_2$ or its solution transforms in the inter-granular space to insoluble $CaCO_3$, which fills the pores. In this stage the main physico-mechanical parameters improve.

II. In the second stage of carbonization other hydration products of cement are attacked by carbon dioxide. The product of these reactions is the calcium carbonate $CaCO_3$ together with the amorphous gel of silicic acid. These components remain in a pseudo-morphous state or as very fine-grained crystalline new formations of $CaCO_3$. The properties of concrete, don't change in the second stage of carbonization too much, the mechanical properties fluctuate around the original values.

III. The third period of carbonization is characterized by re-crystallization of earlier formed carbonate formations from the inter-granular solution. During this process numerous and ten times greater crystals of calcite and aragonite are formed.

The physical-mechanical properties of concrete deteriorate during the third period. A significant decrease of pH value of concrete takes place.

IV. The fourth stage is characterized by nearly 100% proof degree of carbonization. During this stage the coarse crystals of aragonite and mainly of calcite pervade the whole structure of cement body. This in an extreme case can cause the loss of its coherence and strength. In this stage the pH value is so low that a significant corrosion of reinforcement takes place

These stages of concrete carbonization are substantial for the correct proposal of concrete structure rehabilitation.

3.2 Concrete sulphation

The higher concentration of sulphur dioxide occurs above all in highly industrialised areas where its concentration increases especially in winter months. The process of concrete corrosion by sulphur dioxide is mostly called sulphation.

The sulphation process of SO_2 and also the synergic effect of CO_2 and SO_2 were examined experimentally. It was proved, that the moisture together with SO_2 concentration determine the quantitative and even the qualitative presence of sulphation products formed in the corroded material in different surroundings.

Generally we can state that with lower concentration of SO_2, the influence of the medium is dominating but with higher concentration of SO_2 the relative humidity is no more of such importance.

More significant decrease of compression strength values take place after a longer exposure (at least 180 days) in all samples with higher relative moisture of surrounding. Here the SO_2 concentration value determines more the decrease of strength.

In the sense of four stages of carbonization it is possible to formulate four or five stages of sulphation.

4 CONCRETE DETERIORATION BY CORROSIVE FLUIDS

Concrete is attacked most frequently by substances dissolved in ground water, i.e. by corrosive carbon dioxide, by sulphates and nitrates. Chlorides come into contact with concrete during the winter maintenance of roads.

Corrosive carbon dioxide CO_2 first reacts with calcium hydroxide together with the creation of $CaCO_3$, which is consequently dissolved to hydrogencarbonate $Ca(HCO_3)_2$, extracted from the cement matrix by water. The concentration of hydroxide ions in the porous solution decreases by the reaction of the corrosive CO_2 and results in the reinforcement corrosion.

Especially the lower parts of structures and also the structures of buildings in the chemical industry come into contact with sulphates because sulphates are very often present in ground water. Sulphates react with components of hydrated cement and cause sulphate corrosion.

The observation of the action of nitrates to concrete is less frequent than the action of other ions. Nitrates are often contained in ground water, in farm buildings and in chemical plants. The activity of nitrifying bacteria occurring in soil and waste water is not negligible. These bacteria oxidise ammonia nitrogen to almost nitrate nitrogen. In industrial areas, nitrogen oxides denoted as Nox neutralise calcium hydroxide contained in the cement matrix.

Chlorides may affect the cement matrix depending on the kind of the ion like nitrates where equimolar amounts of calcium ions, which with chloride ions will form a well soluble calcium chloride, is released. This calcium chloride is gradually washed out from concrete. Chlorides in the porous solution in the form of $CaCl_2$ also react with the aluminate component of the cement binder together with the creation of the so-called Friedl salt composed of $3CaO$, Al_2O_2, $CaCl_2$, $10H_2O$, which causes the deterioration of the cement binder by crystallisation pressure.

5 CONCRETE DETERIORATION DIAGNOSIS

The proper deterioration prognosis of reinforced-concrete structures consists in the following steps: First, it is necessary to carry out a thorough technical and engineering investigation during which concrete sampling for further analyses, and then other non-destructive tests are carried out. The analyses may be divided into the following areas:

- The determination of physical mechanical properties of deteriorated concrete
- Physicochemical determination of properties of deteriorated concrete
- Chemical diagnosis of affected concrete

5.1 Determination of physical-mechanical properties of concrete

1. Compressive strength of concrete
2. Concrete volume weight
3. Visual evaluation of concrete
4. Tensile strength of concrete surface layers
5. Simple tension strength of concrete
6. Modulus of elasticity
7. Concrete surface absorptivity and absorption capacity
8. Carbonation depth of concrete by phenolpthalein test
9. Determination of physical-chemical properties different kinds of concrete

5.1.1 The principle of the method

Chemical and physicochemical measurement is carried out on suitable samples taken from the structure evaluated. Based on this measurement, the following characteristics are determined:

a) Carbonation degree K
b) Degree of modification changes MP
c) Sulphatation degree S
d) Conversion degree (in Al concrete only)
e) Degree of decay (in Al concrete only)
f) The pH value at extract
g) Concrete and cement binder microstructure

Description of praticular characteristics:

a) Carbonation degree K – the ratio of carbonated calcium oxide to the content of total CaO capable of undergoing carbonation.
b) Degree of modification changes MP – is the ratio of coarse-grained and fine-grained crystals of $CaCO_3$ formed during carbonation. The crystals of calcium carbonate may also occur in pseudo-morphoses of hydration products. The degree of modification changes is a non-dimensional number.
c) Suphatation degree S – is the ratio of CaO bonded on gaseous SO_2 and SO_3 in deteriorated concrete to the content of total CaO expressed in per cent.
d) Degree of decay R – used in concrete of aluminate cement or the percentage ratio of gibbsite found by thermal analysis for the gibbsite content which might originate at all
e) Conversion degree Dc – according to the literature written in English or the ratio of gibbsite created to the product of gibbsite and CAH_{10} compound. The variable is applicable in aluminate concrete only.

For testing the physicochemical determination of the atmospheric deterioration of concrete, it is, first of all, necessary to carry out the following measurement:

1. Chemical determination of loss by annealing CaO, SO_3
2. Electrochemical determination of pH extract
3. Derivatographic analysis of concrete
4. Roentgenometric determination of main mineraklogic components (vaterite, calcite, gypsum, $CaSO_3$, $1/2H_2O$, $CaSO_4 \cdot 1/2\ H_2O$, ettringite, in porous concrete tobermorite, in aluminate concrete, C_3AH_6, CAH_{10}, gehlenitehydrate and other minerals.
5. Photographing the concrete microstructure by the scanning electron microscope.

The chemical analysis for the determination and evaluation of carbonation is the basic method. It refers here above all to the determination of the total content of calcium oxide in hydrated cement or in carbonated concrete. It is of great advantage if the present MnO in the concrete sample is also determined because based on its content, we may estimate whether the cement

used contained blast furnace slag, and what was its amount.

The assessment of pH value at the is an important value of the estimated criterion for deterioration.

The use of differential thermal analysis and thermo gravimetric analysis is the most important when physicochemical testing methods are used. The reason is because at these analyses we may distinguish fine-grained and coarse-grained modifications of $CaCO_3$.

In order to identify the present clusters of crystals in carbonated concrete, the roentgenographic analysis the most reliable; in some cases, the use of the infrared spectroscopy is recommended.

Finally, the scanning of the cement binder structure by the electron scanning microscope should be mentioned. With the help of this method, we may follow the microstructure, its changes, origin and the recrystallisation of carbonate new forms.

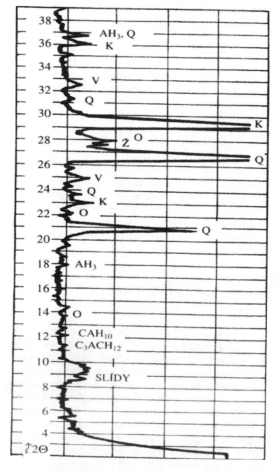

Figure 2. Roentgenogram of carbonated concrete structure, Q quartz, K calcite, V vateritu, Z feldspar.

5.1.2 Evaluation of physicochemical determination of atmospheric deterioration of concrete

The results of particular test determine particular characteristics of the deterioration. It should be noted here that none of the methods itself does not give fully decisive results and that it is always necessary to combine several methods. For illustration, several pictures could be shown. (See Figures 2–3). So that these pictures may be evaluated correctly, it is suitable to explain several calculations which concern particular degrees of deterioration by which the engineering structure is distinguished.

a) *Carbonation degree*

Degree of carbonation K [%] is assessed by its calculation from the content of CaO and SO_3 found by means of the chemical analysis according to the Czech Standard CSN 72 0100. The basic process of silicate analysis, and from the content of CO_2, bonded to coarse-grained and fine-grained $CaCO_3$ found derivatographically. First of all, the content of carbonated CaO is calculated:

$$CaO_{carb} = 1.273\ (CO_{2f} + CO_{2c})\ [\%] \qquad (1)$$

Figure 3.   Typical crystals calcium hydroxide in one year old concrete SEM magnified 6900 times.

where
CO_{2f} – the content of CO_2 bonded to fine-grained $CaCO_3$%
CO_{2c} – the content of CO_2 bonded to coarse-grained $CaCO_3$%

Then the degree of carbonation °K (%) is determined from the relation

$$K = \frac{CaO_{karb}}{CaO - 0,700.SO_3} 100\,[\%] \qquad (2)$$

where

CaO and SO_3 are the contents of the given oxides in the sample in per cent.

b) *Degree of modification changes*

The degree of modifications changes MP [−] is determined by the calculation from the content of CO_2 bonded to fine-grained and coarse-grained $CaCO_3$ which was found at the derivatographic determination as their ratio.

c) *Sulphatation degree*

Sulphatation degree S [%] is determined by the calculation from the chemical assessment of CaO and SO_3 according to the above- mentioned Czech Standard CSN 72 0100. Basic procedures of silicate analysis and from the chemical determination of SO_2 in the sample. Here, the following holds good

$$S = \frac{0,7 \times SO_3 + 0,875 \times SO_2}{CaO} 100\,[\%] \qquad (3)$$

where

SO_3 and CaO are the contents in per cent of the given oxides in the partial sample based on the chemical determination; SO_2 is the content of sulphur dioxide in the partial sample in per cent according to the thermo-chemical assessment. Due to a very small amount, the content of sulphates supplied direct in the production is neglected in the calculation.

When elaborating the results of the measurement, the following operations are carried out:

1. The classification of concrete deterioration in the stage of carbonation, including the assessment of the stage of modification changes, which is to carryout best based on Table 1.
2. The determination of the carbonation depth when evaluating a greater number of principal testing samples from the same core sample or when we determine the dependence of carbonation characteristics (K, MP and pH values at the leach) on the position depth of the partial sample under the surface.

Table 1. Limiting values of carbonation degree of degree of modification changes and pH values of extract in relation to carbonation stages.

Carbonation stage	Carbonation degree /%/	Degree of modification changes /−/	pH value
I	<55	<0.5	>10.8
II	55–73	0.5–0.4	10.8–9.6
III	73–85	0.4–0.8	9.6–8.0
IV	>85	>0	<8.0

3. The determination of the extent of deterioration attack of concrete by sulphur dioxide and the formation of calcium calcium sulphates also when evaluating a greater number of samples from the same core samples always in dependence on the depth of the testing sample under the surface.
4. Finding out the presence of carbonate or sulphate crystallic new forms with the size of the main dimension larger than 1 m, and in porous concrete, through the deformation of tobermorite crystals.
5. In concrete made of aluminate cement, the extent of conversion, carbonation and the decay of the original cement stone.

5.2 Chemical diagnosis of concrete deterioration

The chemical diagnosis of concrete deterioration may include the attack of different salts causing the deterioration as mentioned above. This refers to the determination of the content of

– chlorides
– sulphates
– nitrates

The presence of chlorides in concrete is detected by the precipitation reaction between silver and chloride ions.

6 CONCLUSION

The chemical diagnosis of deterioration is a necessary prerequisite of a correct evaluation of the deterioration attack of reinforced-concrete structures. The reaction of corrosive substances from the surroundings of the structure with the cement binder, aggregates or reinforcement demonstrates its externally by the change of mechanical properties because these demonstrate the chemical bonds formed in the compounds present.

Carrying out necessary tests and the determination of correct diagnosis of the structure material represents a valuable basis for the proposal of its rehabilitation. The economy in the field of diagnosis is leading the designer to propose a more expensive rehabilitation. The costs for the diagnosis will return in the optimalization of the costs for the rehabilitation.

ACKNOWLEDGEMENTS

This work was published with the financial support of the research project CEZ – MSM 26110008. Name of the project is: "Research and Development of New Materials and Securing their Higher Durability in Building Structures" and research project GAR 103/0101314.

REFERENCES

Drochytka, R. & Hela, R. 1996. Defects of cooling concrete in the Czech Republic. *4th International Symposium on Natural Draught Cooling Towers:* Kaiserlautern: pp. 445–449.
Matoušek, M. & Drochytka, R. 1986. *Physico – chemical Determination of Atmospheric Deterioration of Concrete,* Bratislava: TSÚS.
Matoušek, M. & Drochytka, R. 1998 *Atmospheric deterioration of concrete.* Prague: IKAS.

Author index